THIRD CANADIAN EDITION

environment

THE SCIENCE BEHIND THE STORIES

jay withgott   matthew laposata   barbara murck

UNIVERSITY OF TORONTO MISSISSAUGA

PEARSON

TORONTO

Editorial Director: Claudine O'Donnell
Executive Acquisitions Editor: Cathleen Sullivan
Senior Marketing Manager: Kimberly Teska
Program Manager: Darryl Kamo
Project Manager: Jessica Mifsud
Manager of Content Development: Suzanne Schaan
Developmental Editor: Joanne Sutherland
Media Editor: Daniella Balabuk
Media Developer: Kelli Cadet
Production Services: Cenveo® Publisher Services
Permissions Project Manager: Kathryn O'Handley
Photo and Text Permissions Research: Integra-CHI, US
Art Director: Alex Li
Interior and Cover Designer: Anthony Leung
Cover Image: photo © Edward Burtynsky, courtesy Nicholas Metivier Gallery, Toronto / Howard Greenberg Gallery and Bryce Wolkowitz Gallery, New York

Vice-President, Cross Media and Publishing Services: Gary Bennett

Credits and acknowledgments for material borrowed from other sources and reproduced, with permission, in this textbook appear on the appropriate page within the text or on page C-1.

Original edition published by Pearson Education, Inc., Upper Saddle River, New Jersey, USA. Copyright © 2017 Pearson Education, Inc. This edition is authorized for sale only in Canada.

If you purchased this book outside the United States or Canada, you should be aware that it has been imported without the approval of the publisher or the author.

4 16

Library and Archives Canada Cataloguing in Publication

Withgott, Jay, author
        Environment : the science behind the stories / Jay Withgott, Matthew Laposata (Kennesaw State University (KSU)), Barbara Murck (University of Toronto). -- Third Canadian edition.

Includes bibliographical references and index.
ISBN 978-0-321-93146-7 (paperback)

        1. Environmental sciences--Textbooks. I. Laposata, Matthew, author
II. Murck, Barbara W. (Barbara Winifred), 1954-, author III. Title.

GE105.W58 2016   363.7   C2015-904095-7

ISBN 978-0-321-93146-7

Brief Contents

Contents

4 EVOLUTION, BIODIVERSITY, AND POPULATION ECOLOGY   90

5 SPECIES INTERACTIONS AND COMMUNITY ECOLOGY   118

PART TWO
ISSUES, IMPACTS, AND SOLUTIONS   151

6   HUMAN POPULATION   152

7   SOILS AND SOIL RESOURCES   178

9 CONSERVATION OF SPECIES AND HABITATS   242

10 FORESTS AND FOREST MANAGEMENT   282

12 MARINE AND COASTAL SYSTEMS AND FISHERIES 351

13 ATMOSPHERIC SCIENCE AND AIR POLLUTION 383

14 GLOBAL CLIMATE CHANGE   419

15 FOSSIL FUELS: ENERGY USE AND IMPACTS   462

16 ENERGY ALTERNATIVES   499

17 MINERAL RESOURCES AND MINING 542

18 MANAGING OUR WASTE 566

22 STRATEGIES FOR SUSTAINABILITY   700

Preface

Dear Student,

You are coming of age at a unique and momentous time in history. Within your lifetime, our global society must chart a promising course for a sustainable future. The stakes could not be higher.

Today we live long lives enriched with astonishing technologies, in societies more free, just, and equal than ever before. We enjoy wealth on a scale our ancestors could hardly have dreamed of. Yet we have purchased these wonderful things at a price. By exploiting Earth's resources and ecological services, we are depleting our planet's bank account and running up its credit card. We are altering our planet's land, air, water, nutrient cycles, biodiversity, and climate at dizzying speeds. More than ever before, the future of our society rests with how we treat the world around us.

Your future is being shaped by the phenomena you will learn about in your environmental science course. Environmental science gives us a big-picture understanding of the world and our place within it. Environmental science also offers hope and solutions, revealing ways to address the problems we create. Environmental science is not simply some subject you learn in college or university. Rather, it provides you basic literacy in the foremost issues of the twenty-first century, and it relates to everything around you over your entire lifetime.

We have written this book because today's students will shape tomorrow's world. At this unique moment in history, students of your generation are key to achieving a sustainable future for our civilization. The many environmental challenges that face us can seem overwhelming, but you should feel encouraged and motivated. Remember that each dilemma is also an opportunity. For every problem that human carelessness has created, human ingenuity can devise a solution. Now is the time for innovation, creativity, and the fresh perspectives that a new generation can offer. Your own ideas and energy *will* make a difference.

Environmental science helps show us how Earth's systems function and how we influence these systems. It gives us a big-picture understanding of the world and our place within it. Studying environmental science helps us comprehend the problems we create, and it can help us find solutions for those problems. This is not just another course in your university or college program; it relates to everything that is around you, and it will resonate for the rest of your life.

Dear Instructor,

You perform one of our society's most vital jobs by educating today's students—the citizens and leaders of tomorrow—on the fundamentals of the world around them, the nature of science, and the most central issues of our time. We have written this book to assist you in this endeavour because we feel that the crucial role of environmental science in today's world makes it imperative to engage, educate, and inspire a broad audience of students.

In *Environment: The Science behind the Stories*, we strive to implement a diversity of modern teaching approaches and to show how science can inform efforts to bring about a sustainable society. We aim to encourage critical thinking and to maintain a balanced approach as we flesh out the vibrant social debate that accompanies environmental issues. As we assess the challenges facing our civilization and our planet, we focus on providing forward-looking solutions, for we truly feel there are many reasons for optimism.

In crafting this latest Canadian edition, we have incorporated the most current information from this fast-moving field of environmental science, and have streamlined our presentation considerably to promote learning. We have examined every line with care to make sure all content is accurate, clear, and as up to date as possible. We also have introduced a number of changes and new material that we think you will enjoy using in your teaching.

We sincerely hope that our efforts will come close to being worthy of the immense importance of our subject matter. We invite you to let us know how well we have achieved our goals and where you feel we have fallen short. We are committed to continual improvement and value your feedback, as does the team at Pearson Canada. Please feel free to write and send comments or suggestions to Barbara Murck at barbara.murck@utoronto.ca.

—**Jay Withgott,**
Matthew Laposata, and Barbara Murck

The Canadian Edition

When we embarked upon writing a first Canadian edition of *Environment: The Science behind the Stories*, we endeavoured to produce a book that represented a truly Canadian perspective of environmental science, while maintaining the powerful teaching and learning tools of the American edition. We wanted to tell students about the great, sometimes groundbreaking, work done by Canadian environmental scientists. We wanted to celebrate our environmental achievements and history, and familiarize students with the people, locations, and events of that history with examples from coast to coast to coast.

During these years of the Canadian first and second editions, we have enjoyed receiving feedback and suggestions from adopters and reviewers across Canada. You, our users, have told us that you appreciate the truly Canadian focus of the book, as well as the balanced approach and the integration of science with policy. You like the rigour and the breadth and depth of coverage and value our efforts to represent environmental issues from all corners of this vast country and around the world. You also welcome the clarity and liveliness of the writing and the visual program.

We hope you will find that these traditions have continued in the third Canadian edition. We have tackled some of the significant changes that have been happening in Canada, in the physical and biological environment, as well as changes in the political, social, and economic context of our nation that have an impact on the environment. We believe that we have approached these important changes with balance, and a spirit of scientific inquiry.

What's New in the Third Canadian Edition?

For the third Canadian edition we maintain all of the aforementioned strengths, while enhancing the Canadian content, clarifying the overall structure, adding significantly to the cited scientific sources, and updating the book as a whole. We have responded to your suggestions for new examples, additions, and, where needed, conceptual reorganization, and we thank all of the adopters and reviewers who contributed ideas and suggestions.

This has been a deep revision. Some chapters have been streamlined, updated, and strengthened with the addition of new topics; other chapters have been completely restructured. You will now find that the learning objectives in each chapter closely parallel the structure of the chapter headings. In this edition you will also discover scores of new and updated graphs, photographs, and tables, as well as expanded and improved discussions of many topics, new case studies, and scientific focus articles.

Integrated Central Cases

Telling compelling stories about real people and real places is the best way to capture students' interest. Narratives with concrete detail also help teach abstract concepts because they give students a tangible framework with which to incorporate new ideas. We integrate each chapter's "Central Case" into the main text, weaving information throughout the chapter. In this way, the concrete realities of the people and places of the central case study demonstrate the topics we cover. Students and instructors using the book have lauded this approach, and we hope it can continue to bring about a new level of effectiveness in environmental science education. As instructors ourselves, we find the central cases to be extremely effective as a pedagogical tool for opening our lectures and setting the stage for new areas of inquiry.

In the third Canadian edition, 15 of the 22 Central Cases have a specifically Canadian focus. All of the Central Cases have been updated and improved, with additional scientific sources. Four of them are new:

- **Chapter 2:** The Tōhoku Earthquake: Shaking Japan's Trust in Nuclear Power
- **Chapter 13:** "Airpocalypse" in Beijing
- **Chapter 19:** Microplastics: Big Concerns about Tiny Particles
- **Chapter 21:** SARA and the Sage-Grouse

These, along with classic Canadian cases like "Battling over the Last Big Trees at Clayoquot Sound," "Lessons Learned: The Collapse of the Cod Fisheries," and "The Retreat of the Athabasca" tell the important stories—the iconic stories that our students *need* to know if we are to move forward in our search for environmental solutions, rather than repeating the mistakes of the past.

Each chapter now contains questions that are specifically aimed at helping students actively engage with graphs and data. The questions accompany some of the data-driven figures in each chapter, challenging students to practise quantitative skills of interpretation and analysis. To encourage students to test their understanding as they read, answers are provided in Appendix A.

The Science behind the Story

Our goal is not simply to present students with facts, but to engage them in the scientific process of testing and discovery. To do this, we feature "The Science behind the Story" boxes, which expand upon particular studies, guiding readers through details of the research. In this way we show not merely what scientists discovered, but *how* they discovered it. Instructors and students confirm that this feature enhances comprehension of chapter material and deepens understanding of the scientific process itself—a key component of effective citizenship in today's science-driven world.

"Science behind the Story" tells students about the science of the environment in Canada and around the globe, and about the important work of Canadian environmental scientists internationally. This edition showcases many new "Science behind the Story" features, including

- **Chapter 5:** Ecological Recovery at Mount St. Helens
- **Chapter 6:** A Different Population Bomb: The "Household Explosion"
- **Chapter 9:** Counting Species in the World's Most Biodiverse Place
- **Chapter 10:** Assisted Migration: Getting Trees Where They Need to Go in a Changing Climate
- **Chapter 11:** Near-Death Experience at the Experimental Lakes Area
- **Chapter 15:** Keystone XL, Northern Gateway, and the Dilbit Controversy
- **Chapter 16:** Weighing the Impacts of Solar and Wind Development
- **Chapter 17:** Mount Polley Tailings Dam Failure
- **Chapter 18:** Edmonton Showcases Reduction and Recycling
- **Chapter 20:** Ethics and Intergenerationality in Economics: Discounting, Climate Change, and the Stern Review
- **Chapter 21:** The Great Lakes and the International Joint Commission

End-of-Chapter Features

Each chapter concludes with features that facilitate review and develop critical thinking skills. "Reviewing Objectives" summarizes each chapter's main points and relates them to the learning objectives presented at the beginning of each chapter, enabling students to confirm that they have understood the most crucial concepts. "Testing Your Comprehension" questions provide concise study questions targeted at main topics in each chapter, while "Thinking It Through" questions encourage broader creative thinking aimed at finding solutions.

"Interpreting Graphs and Data" uses figures from recent scientific studies to help students build quantitative and analytical skills in reading graphs and tables, and making sense of data.

MasteringEnvironmentalScience

With this edition we are thrilled to offer expanded opportunities through **MasteringEnvironmentalScience**, our powerful yet easy-to-use online learning and assessment platform. We have developed new content and activities specifically to support features in the textbook, thus strengthening the connection between these online and print resources. This approach encourages students to practise their science literacy skills in an interactive environment with a diverse set of automatically graded exercises. Students benefit from self-paced activities that feature immediate wrong-answer feedback, while instructors can gauge student performance with informative diagnostics. By enabling assessment of student learning outside the classroom, **MasteringEnvironmentalScience** helps the instructor maximize the impact of in-classroom time. As a result, both educators and learners benefit from an integrated text and online solution.

- "Pearson eText" gives students access to the text whenever and wherever they have online access to the internet. eText pages look exactly like the printed text, offering powerful new functionality for students and instructors. Users can create notes, highlight text in different colours, create bookmarks, zoom, click hyperlinked words and phrases to view definitions, and view in single-page or two-page view.
- "Process of Science" activities help students navigate the scientific method, guiding them through in-depth explorations of experimental design using "Science behind the Story" features. These activities encourage students to think like a scientist and to practise basic skills in experimental design.
- "Interpreting Graphs and Data": "Data Q" activities pair with the new in-text "Data Analysis Questions" and coach students to further develop skills related to presenting, interpreting, and thinking critically about environmental science data.
- "First Impressions" Pre-Quizzes help instructors determine their students' existing knowledge of environmental issues and core content areas at the outset of the academic term, providing class-specific data that can then be employed for powerful teachable moments throughout the term.
- "Video Field Trips" allow the instructor to kick off class with a short visit to a wind farm, a site tackling invasive species, or a sustainable campus.

- "Interpreting Graphs and Data" exercises and the interactive "GraphIt!" program guide students in exploring how to present and interpret data and how to create graphs.
- "Viewpoints" are paired essays, which are authored by invited experts who present divergent points of view on topical questions.

The Teaching and Learning Package for the Instructor

We have prepared an excellent supplements package to accompany the text.

Instructor Resources The instructor resources are available online via the Instructor Resources section of **MasteringEnvironmentalScience**® and **http://catalogue.pearsoned.ca**. The following supplements are designed to facilitate lecture presentations, encourage class discussions, aid in creating tests, and foster learning:

- The **Instructor's Manual** includes lecture outlines, teaching notes that integrate material from the chapter, discussions of "The Science behind the Story" features, suggestions for supplementary print and online resource material, and solutions to end-of-chapter questions and problems.
- The **Test Item File** is a test bank that contains approximately 1400 questions and includes multiple-choice, short-answer, graphing, and scenario-based items. For all questions, we identify a suggested answer, an associated learning objective, and a difficulty level of easy, moderate, or difficult.

- The **Computerized Testbank (Pearson TestGen™)** presents the testbank in a powerful program that enables instructors to view and edit existing questions, create new questions, and generate quizzes, tests, exams, or homework. TestGen also allows instructors to administer tests on a local area network, have the tests graded electronically, and have the results prepared in electronic or printed reports. The Pearson TestGen is compatible with Windows or Mac systems.
- The **PowerPoint Presentations** are available in Microsoft PowerPoint®. The colourful slides highlight, illuminate, and build on key concepts in the text.
- The **Image Library** is an impressive resource to help instructors create vibrant lecture presentations. Almost every figure and table from the text is provided in electronic format and is organized by chapter for convenience. These images can be imported easily into Microsoft PowerPoint to create new presentations or to add to existing ones.

Pearson's Learning Solutions Managers Learning Solutions Managers work with faculty and campus course designers to ensure that Pearson technology products, assessment tools, and online course materials are tailored to meet your specific needs. This highly qualified team is dedicated to helping schools take full advantage of a wide range of educational resources, by assisting in the integration of a variety of instructional materials and media formats. Your local Pearson sales representative can provide you with more details on this service program.

Acknowledgments

A textbook is the product of *many* more minds and hearts than one might guess from the names on the cover. All three of us have been exceedingly fortunate to be supported and guided, through this and previous editions of the book, by a tremendous publishing team and a small army of experts in environmental science who have generously shared their time and expertise. Although we alone, as authors, bear responsibility for any inaccuracies, the strengths of this book result from the collective labour and dedication of innumerable people.

As the author of the Canadian edition I am extremely grateful to the team at Pearson Canada for its support, advice, and professionalism throughout the development of the Canadian editions. First of all, many thanks to Sherry Zweig, who first got me involved with this project and with Pearson Canada, for her continuing support and friendship. Special thanks go to Joanne Sutherland, Developmental Editor, and Susan Bindernagel, Production Editor and Proofreader, for bearing with me through all of the inevitable delays and frustrations, and never losing faith, and to Cathleen Sullivan and Lisa Rahn, Executive Acquisitions Editors, and Darryl Kamo, Program Manager. Other members of the Pearson Canada team whose contributions to this project (not to mention their patience and professionalism) were crucially important include Söğüt Y. Güleç, Copy Editor; Jessica Hellen and Jessica Mifsud, Project Managers; Kathryn O'Handley, Permissions Project Manager; Kimberley Teska, Marketing Manager; Anthony Leung, Senior Designer; and Vignesh Sadhasivam, photo researcher from Integra. Francine McCarthy, Technical Proofreader, made me shake my fist but also made many valuable contributions to the text. It is a pleasure to work with such a talented and committed group of people.

I also want to extend my deep appreciation to my colleague and friend, Monika Havelka, whose creative ideas, experience teaching with the book, and insights from being the other half of my brain have been invaluable.

And for the third Canadian edition, I would once again like to thank my ever-patient family, who happily put up with a perpetually temporary office on the dining room table. My son Riley King and my daughter Eliza King also contributed editorial and research assistance, at times when they were really needed. I also want to acknowledge the assistance of my colleagues (especially Professors Andrea Olive, Nick Collins, and Kent Moore), Educational Resources Assistant Jennifer Soehner, my friends (especially the Lost Trio Hiking Association), and generations of ENV100 students at the University of Toronto Mississauga.

We dedicate this book to today's students, who will shape tomorrow's world.

—Barbara Murck
(for Jay Withgott and Matthew Laposata)

Reviewers

We have been guided in our efforts by extensive input from colleagues across Canada who have served as reviewers and advisors for the first and second Canadian editions, in addition to the contributions of many reviewers for the U.S. editions. The participation of so many learned and thoughtful experts has improved the book in countless ways and has made this edition much stronger. Sometimes you made me shake my fist at my laptop, but in the end your insightful comments and suggestions led to a much stronger book. If the thoughtfulness and thoroughness of these reviewers are any indication, then the teaching of environmental science in Canada is in excellent hands!

Clem Bamikole, *Georgian College*
Darren Bardati, *Bishop's University*
Sarah Boon, *University of Lethbridge*
John Buskard, *Concordia University*
Steven J. Cooke, *Carleton University*
Mario Corbin, *Champlain College Lennoxville*
Tim Elkin, *Camosun College*
Mariola Maya Janowicz, *Concordia University College of Alberta*
Jeff Lewis, *Vancouver Island University*
Brian O'Neill, *Holland College*
Ben Rubin, *University of Western Ontario*
Yolanda Spithoven, *University of New Brunswick*
Lorraine Vanderzwet, *Mohawk College of Applied Arts and Technology*

About the Authors

Jay H. Withgott has authored *Environment: The Science behind the Stories* as well as its brief version, *Essential Environment*, since their inception. In dedicating himself to these books, he works to keep abreast of a diverse and rapidly changing field and continually seeks to develop new and better ways to help today's students learn environmental science.

As a researcher, Jay has published scientific papers in ecology, evolution, animal behaviour, and conservation biology in journals ranging from *Evolution* to *Proceedings of the National Academy of Sciences*. As an instructor, he has taught university lab courses in ecology and other disciplines. As a science writer, he has authored articles for numerous journals and magazines including *Science, New Scientist, BioScience, Smithsonian,* and *Natural History*. By combining his scientific training with prior experience as a newspaper reporter and editor, he strives to make science accessible and engaging for general audiences. Jay holds degrees from Yale University, the University of Arkansas, and the University of Arizona.

Jay lives with his wife, biologist Susan Masta, in Portland, Oregon.

Matthew Laposata joined the writing team for the fifth U.S. edition of *Environment: The Science behind the Stories*. Matt is a professor of environmental science at Kennesaw State University (KSU). He holds a bachelor's degree in biology education from Indiana University of Pennsylvania, a master's degree in biology from Bowling Green State University, and a doctorate in ecology from The Pennsylvania State University.

Matt is the coordinator of KSU's two-semester general education science sequence titled Science, Society, and the Environment, which enrolls roughly 6000 students per year. He focuses exclusively on introductory environmental science courses and has enjoyed teaching and interacting with thousands of nonscience majors during his career. He is an active scholar in environmental science education and has received grants from state, federal, and private sources to develop and evaluate innovative curricular materials. His scholarly work has received numerous awards, including the Georgia Board of Regents' highest award for the Scholarship of Teaching and Learning.

Matt resides in suburban Atlanta with his wife, Lisa, and children, Lauren, Cameron, and Saffron.

Barbara Murck has authored the Canadian editions of *Environment: The Science behind the Stories* from the beginning. Barb has taught environmental and Earth science at the University of Toronto Mississauga (UTM) for more than 30 years. Her academic background is in geology, with an undergraduate degree from Princeton University and Ph.D. from the University of Toronto.

Barb has worked on a wide variety of environmental management projects in the developing world, from Africa to Asia, mainly as an expert on training and curriculum development. She has published numerous books on topics ranging from physical geology to environmental science to sustainability. She was honoured with the University of Toronto President's Teaching Award in 2010. A current much-loved project is teaching a field course each summer in the Ecuadorian Andes, the Amazon, and the Galápagos Islands. Barb has greatly appreciated having had the opportunity to influence the lives and learning of thousands of UTM students over the years.

Barb lives with her family, including her two kids and the world's best dogs, in a 115-year-old house in Southern Ontario. When not at work, she is likely to be found hiking the Bruce Trail, the oldest and longest marked hiking trail in Canada.

About Our Sustainability Initiatives

This book is carefully crafted to minimize environmental impact. The materials used to manufacture this book originated from sources committed to responsible forestry practices. The printing, binding, cover, and paper come from facilities that minimize waste, energy consumption, and the use of harmful chemicals.

Pearson closes the loop by recycling every out-of-date text returned to our warehouse. We pulp the books, and the pulp is used to produce items such as paper coffee cups and shopping bags. In addition, Pearson has become the first climate-neutral educational publishing company.

The future holds great promise for reducing our impact on Earth's environment, and Pearson is proud to be leading the way. We strive to publish the best books with the most up-to-date and accurate content, and to do so in ways that minimize our environmental impact.

PEARSON

About the Cover ...

Cover image: *Colorado River Delta #2 Near San Felipe, Baja, Mexico, 2011*

"My hope is that these pictures will stimulate a process of thinking about something essential to our survival; something we often take for granted—until it's gone."

—Edward Burtynsky

Canadian photographer Edward Burtynsky has devoted his career to documenting our conflicted relationship with the natural environment, particularly the transformation of landscapes by industrial activity. With his Water project, Burtynsky undertook to understand and document our use and abuse of water; our cover photo is part of the resulting series. It shows the Colorado River delta, once a vast tidal estuary flowing into the Gulf of California. The delta and its unique and biodiverse ecosystems were fundamentally degraded by decades of water diversion from the river. In the late 1990s, however, the delta began a slow revival. Releases of water from reservoirs, return flows of agricultural irrigation water, flood water, and even flows of municipal waste water, although polluted and saline, have restored some parts of the delta ecosystem. Water is beginning to flow, where once there was only a desert.

Like most of Burtynsky's photos, the image on our cover makes us stop and think about the impacts of our activities on the things that we depend upon most fundamentally. His images are complicated and contradictory, like our relationships with the natural environment. We are centrally dependent on environmental goods and services to support our lives and our lifestyles; yet, by our very presence and activities, we degrade and destroy those resources. But human actions are also capable of reversing or mitigating those negative impacts, as we see in this image showing the tentative recovery of the Colorado River delta. Burtynsky's photos thus typically contain an element of despair and even horror amidst the beauty: "We did that... ?!" But the despair is mixed with hope that things can change.

In *Environment: The Science behind the Stories*, Third Canadian Edition, as in the previous editions, we look for scientific evidence of our use of environmental resources, our impacts on those resources, and the effectiveness of our efforts to mitigate the impacts. We try to guide students through the complexities of the discipline, and encourage them to think for themselves about the natural world, what science can and cannot tell us, and the impacts of their own attitudes and lifestyle choices on the environment. We have always tried to distinguish, clearly, between environmental science as a scientific discipline, and environmentalism as an advocacy movement; yet, in the end, they come together. Because we are *of* the world, not just *in* it, we are part of the environment, and everything we do (and even what we think) necessarily has impacts. In our modern world we all have a responsibility to be conscious of the impacts of our choices. Burtynsky's photos address these complexities and contradictions head-on, and that is why we have chosen them to be on the covers of all three Canadian editions.

PART ONE

Foundations of Environmental Science

Combers Beach in Pacific Rim Reserve, British Columbia, is part of Canada's National Parks system.

Rolf Hicker/All Canada Photos

An Introduction to Environmental Science

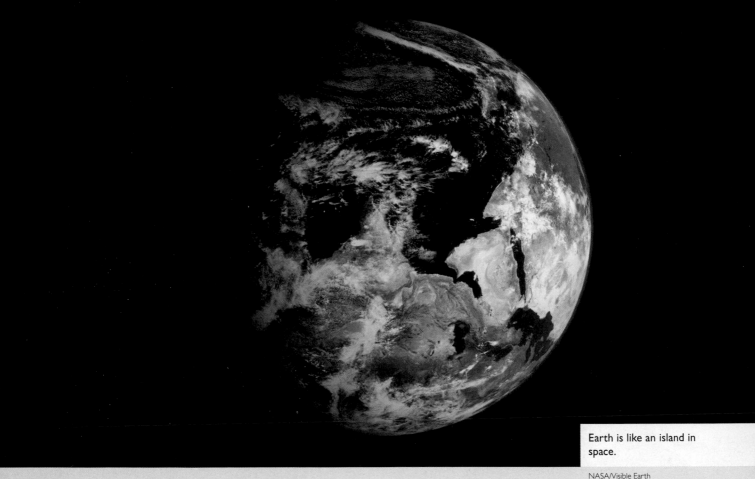

Earth is like an island in space.

Upon successfully completing this chapter, you will be able to

- Define the term *environment*
- Characterize the interdisciplinary nature of environmental science
- Describe several types of natural resources and explain their importance to human life
- Diagnose and illustrate some of the pressures on the global environment
- Articulate the concepts of sustainability and sustainable development

This crescent Earth was one of the very first photos taken of the whole planet.

NASA

CENTRAL CASE
EARTH FROM SPACE: THE POWER OF AN IMAGE

"The two-word definition of sustainability is 'one planet.'"

—MATHIS WACKERNAGEL, ECOLOGICAL ECONOMIST AND CO-DEVELOPER OF THE ECOLOGICAL FOOTPRINT CONCEPT

"We're not the first to discover this, but we'd like to confirm, from the crew of *Apollo 17*, that the world is round."

—EUGENE CERNAN, *APOLLO 17* COMMANDER

Consider the following: Prior to November 9, 1967, no one had *ever seen* a photograph of the whole planet Earth, because no such thing existed.

Those of us who were alive back in 1967 were not completely clueless. We knew that Earth is a planet, surrounded by space. We knew that Earth is round (although visual confirmation of this fact still made a considerable impact on *Apollo 17* astronauts a few years later). Yet a simple photograph of Earth—floating in space, blue and shining and covered by clouds, vegetation, and a whole lot of water—managed to take everyone by surprise and changed both society and history in the process.

Actually, the very first photographs of the whole Earth, taken in 1967, were not the ones that eventually caught the imagination of the general public. The 1967 photographs were taken by automated camera from the unmanned *Apollo 4* spacecraft, the first spacecraft to get far enough away from Earth to photograph the entire planet. Only part of the planet was in sunlight, so the photographs show only a "crescent" Earth (see photo). Not long after, on December 24, 1968, *Apollo 8* astronauts took the first handheld photographs showing Earth rising over the horizon of the Moon (the closing photo of this book). The crew did a live radio broadcast that day, during which astronaut James Lovell commented, "The Earth from here is a grand oasis in the big vastness of space."[1]

It was not until 1972 that the *Apollo 17* mission put astronauts in a position to photograph the entire *illuminated* planet Earth. The result was the famous Blue Marble[2] image, a version of which opens this chapter. The photograph was beautiful, its impact stunning, even unsettling. The original image was oriented with Antarctica at the top of the globe and an "upside-down" Africa in the middle. The unfamiliar perspective caused consternation among those who had never stopped to consider that the convention of orienting maps with north at the top is completely arbitrary.

The Blue Marble photograph is widely credited with kick-starting the modern environmental movement. Just five years elapsed between the first whole-Earth photographs in 1967 and the last ones to be recorded by human hands. Since 1972, no manned space flight has been far enough away for the planet to be photographed in its entirety by astronauts. In that five-year period was the summer of love, and war—the Vietnam War, the Six Day War, the Cold War. The Beatles sang on the first live international satellite television production. Canada celebrated the hundredth year of Confederation. Neil Armstrong became the first person to walk on the Moon. Civil rights activist Martin Luther King, Jr., died; so did J. Robert Oppenheimer, the "father of the atomic bomb." The first handheld calculator was sold (for almost $400).

Society changed dramatically during those five years, and it was a period of dawning awareness and public involvement in environmental issues. The first major oil spill happened in 1967 when the *Torrey Canyon* ran aground near England with 120 000 metric tons of crude oil on board. The first hints of trouble began to surface (literally) from hazardous chemicals stored underground at Love Canal, New York. Within a few years the site would be infamous, leading to the first declaration of an environmental state of emergency in the United States and making a grassroots hero of local activist Lois Gibbs. Books on environmental topics began to appear on bestseller lists, including *Limits to Growth,*[3] *The Population Bomb,*[4] *Small Is Beautiful,*[5] and their predecessor, *Silent Spring.*[6] The 1970s opened with the signing of the first federal environmental legislation, the United States' *Environmental Protection Act* (1970). The first Earth Day was held (1970). Greenpeace was founded (1971). The *United Nations Environment Programme* was established (1972).[7]

British astronomer Sir Frederick Hoyle is reputed to have said in 1948, "Once a photograph of the Earth, taken from outside, is available—once the sheer isolation of the Earth becomes known—a new idea as powerful as any in history will be let loose." To what extent were the milestones of environmental history descended from the first glimpses of our planet from space, with all of its fragility and limitations? We will never know, but the Blue Marble is still considered to be one of the most influential photographs in history—possibly the most widely distributed image of all time—and it remains an iconic symbol of the modern environmental movement.

Our Island, Earth

Viewed from space, our home planet appears suspended against a vast inky-black backdrop. Although few of us will ever witness that sight directly, photographs taken from space convey a sense that Earth is like an island—small, isolated, and finite. Yet this island supports all of the vastness and complexity of life as we know it.

The environment is more than just our surroundings

A photograph from space reveals a great deal, but it does not adequately convey the complexity of the environment. Our **environment** is more than water, land, and air; it is the sum total of our surroundings. It includes all of Earth's **biotic** components, or living things, and **abiotic** components, the nonliving things with which we interact. The abiotic constituents include the continents, oceans, clouds, rivers, and icecaps that you can see in a photo of Earth from space. The biotic constituents are the animals, plants, forests, soils, microbes, and people that occupy the landscape. In a more inclusive sense, the environment also encompasses the built environment, including the urban centres, living spaces, and physical infrastructure that humans have created. In its *most* inclusive sense, the environment includes the complex webs of scientific, ethical, political, economic, and social relationships and institutions that shape our daily lives.

People often use the term *environment* in a narrower sense, though, referring to a "natural" world that stands apart from human society. This connotation is unfortunate, because it masks the important fact that humans exist within the environment and are an integral part of the interactions that characterize and shape it. As just one of many species, we share with the others a fundamental dependence on a healthy, functioning planet. The limitations of language make it all too easy to speak of "people and nature," or "society and the environment," or even "environment versus economy," as though they were separate, not interconnected, or in conflict. Fundamentally, we exist as part of the natural world, and our interactions with the other parts matter a great deal.

"Environment" has legal, social, economic, and scientific aspects

Why is it important that we give such careful consideration to the meaning of the term *environment*? Back in 1971, when the federal government passed Canada's first environmental legislation, the environmental awareness of most North Americans was limited. If they thought about it at all, most people would have equated *environment* with *wilderness*. This oversimplification changed as public consciousness of environmental issues grew. Wilderness preservation is still an important concern, but our understanding of the environment, our impacts on it, and its role in our health and daily lives has broadened dramatically since then.

Today our definition of the term *environment* must be sufficiently comprehensive to include its legal, social, economic, and scientific aspects. Business management, politics, ethics, international relations, economics, social equity, engineering, law enforcement, and chemical, physical, geological, and biological sciences—all of these play a role in managing and protecting both people and the natural environment. Consequently, the mandate of **Environment Canada**, the department of the federal government that is most directly responsible for the protection of the environment, is also very comprehensive. Among other things, the role of Environment Canada includes preserving and enhancing the quality of the natural environment, protecting and conserving renewable resources and water resources, enforcing Canada's sovereignty over our boundary waters, and forecasting weather conditions and warnings.[8]

To accomplish all of this, our environmental leaders and policymakers need to know what they are talking about; this is the main reason that Environment Canada was established as a science-based organization. As a community, we must constantly improve and refine our basic scientific understanding of water, air, land and soils, wildlife, weather and climate, and the dynamic interactions among all the components of which ecosystems are composed. This is where *environmental science*—the central focus of this book—comes in.

Environmental science explores interactions between people and the natural world

Environmental science is the study of how the natural world works, how our environment affects us, and how we affect our environment. Appreciating how we interact with the physical and biological environment is crucial for a well-informed view of our place in the world, and for a mature awareness that we are one species among many on a planet full of life. As our population, our technological powers, and our consumption of resources increase, so does our ability to alter our planet and damage the very systems that keep us alive. We need to understand these impacts more thoroughly and manage them more effectively. Environmental science emerged in the latter half of the twentieth century in response to this need.

Understanding the functioning of the natural environment and our role in it, our interactions with it, and our impacts on it is the essential first step toward finding solutions to our most pressing environmental problems. Part 1 of this book, *Foundations of Environmental Science*, takes that first step by providing an introduction to the materials and processes that characterize the biotic and abiotic components of the environment, and the basic concepts and principles of science as applied to the study of the environment.

It can be daunting to reflect on the number and magnitude of environmental dilemmas that confront us today. Many environmental scientists are trying to apply their knowledge to develop practical solutions to the environmental challenges we face. We examine these challenges and issues in Part 2, *Issues, Impacts, and Solutions*, starting with a look at the human population itself and how it has grown and changed over time.

Fortunately, with problems also come opportunities for devising creative solutions. Right now, global conditions are changing more quickly than ever. Right now, through science, we as a civilization are gaining knowledge more rapidly than ever. And right now, the window of opportunity for acting to solve problems is still open. With such bountiful challenges and opportunities, this particular moment in history is an exciting time to be studying environmental science.

The Nature of Environmental Science

Environmental scientists strive to understand how Earth's natural systems function, how humans are influenced by those systems, and how we are influencing those systems. In addition, many environmental scientists are motivated by a desire to develop solutions to environmental problems. The solutions themselves (such as new technologies, policy decisions, or resource management strategies) are applications of environmental science. The study of such applications and their consequences is also part of environmental science.

Science is a systematic process for learning about the world

Environmental science is part of the broader human endeavour of **science**, a systematic process for learning about the world and testing our understanding of it. The term *science* also refers to the accumulated body of knowledge that arises from this dynamic process of observation, testing, and discovery, which we will explore in greater detail in the chapter *Matter, Energy, and the Physical Environment*.

Knowledge gained from science can be applied to societal problems. Among the most important applications of science are its use in developing new technologies, and informing policy and management decisions (**FIGURE 1.1**). These pragmatic applications in themselves are not science, but they must be informed by science in order to be effective. Many scientists are motivated simply by a desire to know how the world works, and others are motivated by the potential for developing useful applications and solutions to problems.

Why does science matter? The late American astronomer Carl Sagan wrote the following in his 1995 treatise *The Demon-Haunted World: Science as a Candle in the Dark:*

> We've arranged a global civilization in which the most crucial elements—transportation, communications, and all other industries; agriculture, medicine, education, entertainment, protecting the environment; and even the key democratic institution of voting—profoundly depend on science and technology.[9]

Sagan and many others have argued that science is essential if we hope to develop solutions to the problems—environmental and otherwise—that we face today. We can go a step further and suggest that the *democratization* of science—making the science of our world accessible and understandable to as many people as possible—is also essential if we are to make informed decisions about the management of this planet. That is one reason why it is important for you to learn as much as possible about the science of the environment, and to pass along some of that knowledge to others.

Scientific ideas and methods change and evolve as new information is discovered, ideas are tested, and knowledge grows. Understanding how science works is especially relevant in environmental science, a young field that is changing rapidly as we gather vast amounts of new information, as human impacts on the planet multiply, and as lessons from the consequences of our actions become apparent. Because so much remains unstudied and undone, and because so many issues we cannot foresee are likely to arise in the future, environmental science will remain an exciting frontier for you to explore as a student and as an informed citizen throughout your life.

Environmental science is an interdisciplinary pursuit

Like science in general, environmental science informs practical applications and can be motivated by them. Studying natural systems and addressing environmental problems are complex endeavours that require expertise from many disciplines. Environmental science is thus an **interdisciplinary** field of study—one that employs concepts and techniques from numerous disciplines and brings research results from these disciplines together into a broad synthesis (**FIGURE 1.2**). Traditional disciplines (such as biology, geology, and chemistry) are valuable because their scholars delve deeply into topics, uncovering new

FIGURE 1.1
Scientific knowledge can be applied in policy and management decisions and in technology. Prescribed burning, shown here, is a management practice that is used to restore healthy forests, and is informed by scientific research into forest ecology.

Courtesy of Jay Withgott

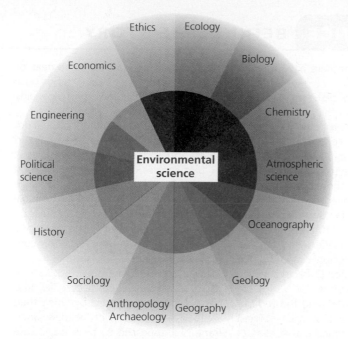

FIGURE 1.2

Environmental science is an interdisciplinary pursuit, involving input from many different established fields of study across the natural and social sciences.

knowledge and developing expertise in particular areas. Interdisciplinary fields are valuable because their practitioners take specialized knowledge from different disciplines, consolidate it, synthesize it, and apply it in a broad context to serve the multifaceted interests of society.

Environmental science is especially broad because it encompasses not only the *natural sciences* (disciplines that study the natural world) but also the *social sciences* (disciplines that study human interactions and institutions). The natural sciences provide us with the means to gain accurate information about the physical environment and to interpret it reasonably. Addressing environmental problems, however, also involves weighing values and understanding human behaviour, and this requires the social sciences. Most environmental science programs in universities focus predominantly on the natural sciences as they pertain to environmental issues. Programs that heavily incorporate the social sciences often prefer the term *environmental studies* or *environmental management* to describe their academic umbrella. Whichever approach we take, these fields reflect many diverse perspectives and sources of knowledge.

Just as an interdisciplinary approach to studying issues can help us better understand them, an integrated approach to addressing problems can help us produce effective and lasting solutions. For example, consider how the Canadian mining industry approaches the problem of acid mine drainage, which can occur when sulphur is present at a mine site. Sulphur is a common constituent of coal and metal ores, both of which are important to the Canadian economy. If sulphur-bearing waste rock at a mine site interacts with rain or surface water, sulphuric acid is formed; if not contained, the acid can enter local streams, where it is devastating to affected ecosystems. To solve a problem involving *acid drainage*, a mining company could consult a biologist or an ecologist regarding the impacts of the acid on local plants and animals. A hydrologist would be helpful, to understand the flow of water at the site. A mining engineer could help decide how best to contain and isolate the waste rock piles. The company could consult with a chemist about the nature and behaviour of the acidic solution, and how it interacts with rocks and soils. Someone skilled at management would be helpful, to act as a liaison between the scientists and the mine management team. Canadian mining companies routinely make use of teams like this in their efforts to control acid mine drainage.

Environmental science is not the same as environmentalism

Although many environmental scientists are interested in solving problems, it is incorrect to confuse environmental science with environmentalism or environmental activism. They are *not* the same. Environmental science is the pursuit of scientific knowledge about the workings of the environment and our interactions with it. **Environmentalism** is a social movement dedicated to protecting the natural world—and, by extension, humans—from undesirable changes brought about by human choices (**FIGURE 1.3**).

Sean Kilpatrick/The Canadian Press

FIGURE 1.3

Environmental scientists play roles very different from those of the environmental activists shown here. Some scientists do become activists to promote what they feel are workable solutions to environmental problems; most try to keep their advocacy separate from their scientific work. This photograph shows Greenpeace activists protesting on Parliament Hill in Ottawa. Greenpeace was founded in Vancouver in 1971.

THE SCIENCE BEHIND THE STORY

These immense *moai* statues are iconic symbols of Rapa Nui, or Easter Island.
Viktorus/Shutterstock

The Lesson of Rapa Nui

Rapa Nui (or Easter Island) is a small island in the Pacific and one of the most remote inhabited places on the globe. When European explorers reached the island in 1722, they found a barren landscape populated by fewer than 2000 people living a marginal existence in caves. The island featured gigantic statues of carved stone, called *moai*, evidence that a sophisticated civilization had once lived there.

How could people without wheels or rope, on an island without trees, have moved statues 10 m high, weighing 80 metric tons? The answer lies in the fact that the island did not always lack trees, and its people were not always without rope. The island was once lushly forested, supporting a prosperous society of 6000 to 30 000 people. What hap-

pened? Many researchers have tried to solve the mystery.

A key discovery, based on many lines of evidence, was that the island was once forested. British scientist John Flenley and colleagues[10] drilled cores deep into lake sediments and examined ancient pollen grains preserved there. Finding a great deal of palm pollen, they inferred that the island had been covered with tall palm trees. Archaeologists also found ancient palm nut casings buried in soil near carbon-lined channels made by palm roots, and researchers deciphering script on stone tablets discerned characters etched in the form of palm trees.

By studying pollen and the remains of wood from charcoal, archaeologist Catherine Orliac[11] found that at least 21 other plant species—now gone—had also been common. Clearly, the island had supported a diverse forest. Forest plants would have provided fuelwood, building material for houses and canoes, fruit to eat, fibre for clothing—and, researchers guessed, logs and fibrous rope to help move the enormous *moai* statues.

Pollen analyses showed that trees declined and were replaced by ferns and grasses. Charcoal in the soil showed that forests had been burned, perhaps for slash-and-burn farming. Researchers concluded that the islanders, desperate for forest resources and cropland, had deforested the island. With the forest gone, the soil eroded away, confirmed

by data that showed a great deal of sediment accumulating on lake bottoms. Erosion would have lowered crop yields, perhaps leading to starvation.

Wild animals also disappeared. Archaeologist David Steadman analyzed 6500 bones and found that at least 31 bird species originally provided food for the islanders.[12] Today only one native bird species is left. Remains from charcoal fires show that early islanders also feasted on fish, sharks, porpoises, turtles, octopus, and shellfish—but in later years they consumed little seafood.

As resources declined, some researchers concluded, people fell into clan warfare, suggested by unearthed weapons and skulls with head wounds. Rapa Nui appeared to be a tragic case of ecological suicide: A once-flourishing civilization depleted its resources and destroyed itself. In this interpretation, Rapa Nui seemed to offer a clear lesson: We, on planet Earth, had better learn to use our limited resources wisely.

Canadian economists Scott Taylor and James Brander took a different approach.[13] They developed a computer model of the interplay between renewable resources and population. The model is based on standard ecological predator–prey models, with people in the role of predator, and resources as their prey. This scenario generates "feast-and-famine" cycles of rising and falling population and resource stocks. The researchers

---

Environmental scientists study many of the same processes, locations, and issues that environmentalists care about, but as scientists they attempt to maintain an objective approach in their work. Ideally, science informs and responds to political and social influences, without being overly influenced by them. Remaining as free as possible from personal, political, or ideological bias—and open to whatever conclusions the data demand—is a hallmark of the effective scientist.

Both environmental scientists and individuals in non-scientific professions can make important contributions to the understanding, protection, management, and sustainable use of the natural environment. These people work in a wide range of positions, from policymaker to activist, artist, journalist, business person, hunter, or animal rescuer. Many of them are scientists *and* writers, or scientists *and* filmmakers, or gardeners or politicians or

musicians or managers—and, yes, many of them are also environmentalists.

Consider, for example, David Suzuki, who has been called "Canada's environmental conscience." We know him best as a journalist, writer, TV broadcaster, and environmental activist, but Suzuki originally trained and started his career as a scientist—a professor of genetics. His background in science informs his advocacy; however, he has consciously given up "doing" science on an everyday basis, choosing instead to focus on more political questions. David Suzuki's career demonstrates that environmental science and environmentalism are different, but not entirely or necessarily separate.

Science is a human endeavour; it can never be entirely free of political or social influence. We want our leaders to incorporate scientific understanding into their social decisions, but there is no foolproof way to ensure that

Archaeological investigations, like this one by the Easter Island Statue Project, aim to understand and preserve the cultural legacy of the islanders.

The rats ate palm nuts—perhaps so many that the trees could not regenerate. With no young trees growing, the palm went extinct. Despite the forest loss, Hunt and Lipo argue that islanders were able to persist and thrive, adapting to Rapa Nui's poor soil and windy weather by developing rock gardens to protect crops. Tools that previous researchers viewed as weapons were actually farm implements, they concluded; lethal injuries were rare, and no evidence of battle or defensive fortresses was uncovered.

The evidence led Hunt and Lipo to propose that the islanders had acted as responsible stewards of the island's resources. The eventual collapse of this civilization, they argue, came with the arrival of Europeans, who unwittingly brought contagious diseases to which the islanders had never been exposed. Before that, Hunt and Lipo say, Rapa Nui's people maintained a peaceful and resilient society for 500 years.

This interpretation represents a shift in how we view Rapa Nui. Were the early inhabitants good environmental stewards who were overcome by insurmountable outside forces? Or did they overuse and deplete the island's resource base, initiating the decline and collapse of the civilization? Debate between the two camps remains heated, and research continues as scientists look for new ways to test the differing hypotheses. In the long term, data from additional studies should lead us closer and closer to the truth.

speculated that such cycles may account for the decline and eventual collapse of civilizations like that of Rapa Nui, as a result of rapid population growth and consequent resource degradation.

Anthropologist Terry Hunt and archaeologist Carl Lipo drew entirely different conclusions from their scientific findings.[14] When they began their research on Rapa Nui in 2001, they expected simply to help fill gaps in a well-understood history. But science is a process of discovery, and sometimes evidence leads research-ers far from where they anticipated. Hunt and Lipo ended up convinced that nearly everything about the traditional "ecocide" interpretation was wrong.

Their findings suggested that deforestation occurred rapidly, shortly after the arrival of the first colonists. How could so few people have destroyed so much forest so fast? Their answer: rats. When Polynesians settled new islands, they brought crop plants, domestic animals such as chickens, and rats. Rats multiply quickly, and they soon overran Rapa Nui.

science is not misused to serve political ends. By becoming aware of the complex relationships among science, society, and politics, we can work to ensure that an appropriate balance is maintained. Environmental science is *distinct* from politics, law, commerce, philosophy, religion, art, and activism, but is it *exclusive* of these human undertakings? You will have to judge for yourself, but you can count on this book to help you make a more informed judgment of what you read, hear, and experience in your encounters with the natural environment.

Environmental science can help us avoid mistakes made in the past

Today we are confronted with news and predictions of environmental catastrophes on a regular basis, but it can be difficult to assess the reliability of such reports. It is even harder to evaluate the causes and effects of environmental change. Perhaps most difficult is to devise effective solutions to environmental problems. Studying environmental science will outfit you with the tools to evaluate information on environmental change, and think critically and creatively about possible actions to take in response. These tools can help us do better in the future, and avoid some of the mistakes that have been made in the past in our interactions with the environment.

There is historical evidence that civilizations may crumble when pressures from population and consumption overwhelm resource availability (see "The Science behind the Story: The Lesson of Rapa Nui"). Many great civilizations have fallen after depleting resources or damaging their environment. The Greek and Roman empires show evidence of this, as do the civilizations of the Maya, the Anasazi, and other New World peoples. Plato wrote of the deforestation and environmental degradation accompanying ancient Greek cities, and today further evidence is accumulating from research by archaeologists, historians, and paleoecologists who study past societies and landscapes. The arid deserts of today's Near Eastern and Middle Eastern countries were far more lushly vegetated when the great ancient civilizations thrived there.

Researchers have now learned enough about ancient civilizations and their demise that scientist and author Jared Diamond—in his 2005 book, *Collapse*—could hypothesize why civilizations succeed and persist, or fail and collapse.[15] Diamond identified five critical factors that determine the survival of civilizations: climate change, hostile neighbours, trade partners, environmental problems, and, finally, the society's response to environmental problems. It is interesting to note that only one of these factors—the response to environmental problems—is wholly controllable, and it is this factor that has been the crucial determinant of survival. Success and persistence, it turns out, depend largely on how societies interact with their environments.

Earth's Natural Resources

Planet Earth is very large, but its material resources—like those of any island—are finite. Specifically, there are limits to the availability of many **natural resources**, the substances and energy sources provided by the environment that are of economic value, and that we need for survival and for the functioning of our modern society.

Resources range from inexhaustible to nonrenewable

We can view natural resources as lying on a continuum from the most to the least renewable (**FIGURE 1.4**). Natural resources that are replenishable over short periods are known as **renewable natural resources**. Some resources, such as sunlight, wind, and wave energy, are continuously replenished and essentially inexhaustible. Others are replenished more slowly, and may be depleted if we use them at a rate that exceeds the rate at which they are renewed or replenished. For example, living animals and plants that we harvest from the wild are typically renewable if we do not harvest them too quickly, but may vanish if we do.

Resource management is strategic decision-making and planning aimed at balancing the use of a resource with its protection and preservation. The basic premise of renewable resource management—for both living and nonliving resources—is to balance the rate of use with the rate of renewal or regeneration. If the **stock**, the harvestable portion of a resource, is harvested or withdrawn at a faster rate than it can be replenished, then the stock will eventually be depleted. If trees are cut faster than new trees can be seeded and grow to maturity, or if fish are caught faster than new ones can be born and grow to harvestable age, the stocks of these resources will run out. For this reason, renewable resources are sometimes called *stock-and-flow resources*, highlighting the importance of this balance in their management.

Some nonliving renewable resources, including ground water and soil, can be utilized according to principles similar to those that govern living resources; they will continue to be available as long as they are not extracted more quickly than they are replenished. However, the rates of replenishment of such resources are limited by the rates of physical processes, such as the infiltration of ground water to replenish an aquifer, or the weathering of rock to produce soil. Because these rates can be quite slow—up to 10 000 years for soil formation in cold climates like those of northern Canada, for example—it may take a very long time for these resources, once damaged or depleted, to be replenished.

FIGURE 1.4
Natural resources lie on a continuum. Inexhaustible resources such as sunlight are perpetually renewed. Nonrenewable resources such as oil and coal are not renewed on a humanly accessible time scale. Resources such as fish, timber, ground water, and soil are renewable, and can continue to be available if we are careful not to deplete or damage them.

Renewable natural resources

- Sunlight
- Wind energy
- Wave energy
- Geothermal energy

- Agricultural crops
- Fresh water
- Forest products
- Soils

Nonrenewable natural resources

- Crude oil
- Natural gas
- Coal
- Copper, aluminum, and other metals

In contrast to inexhaustible and renewable resources, **nonrenewable natural resources** such as fossil fuels and mineral deposits are finite, lying at the other end of the continuum in **FIGURE 1.4**. They are depletable because they are formed much more slowly than we use them. For example, it can take 100 million years for natural geological processes to form an ore deposit or a petroleum deposit. Once we use them up, nonrenewable resources are no longer available because they will not be replenished on a humanly accessible time scale. Simply by withdrawing from the stock, we deplete the resource.

Our civilization depends on numerous nonrenewable mineral and energy resources: Oil, natural gas, and coal are extracted to drive the machinery of modern society. Iron is mined and processed to make steel. Copper is used in pipes, electrical wires, and a variety of other applications. Aluminum is extracted from bauxite ore and used in packaging and other end products. Lead is used in batteries, to shield medical patients from radiation, and in many other ways. Zinc, tungsten, phosphate, uranium, gold, silver—the list goes on and on.

Although we rely on these nonrenewable resources, we do not manage their extraction in the same way that we manage renewable natural resources. Fossil fuels and minerals are *mined*, rather than *harvested*. The oil and mining industries benefit by extracting as much material as possible, as fast as possible, and then, once extraction becomes too inefficient to be profitable, moving on to new sites. From a consumer's perspective, therefore, the management of a nonrenewable resource demands conservation, reuse, and recycling to extend the useful lifetime of the resource as long as possible.

Some resources are truly and permanently nonrenewable, nonreplenishable, and nonreplaceable: Once an atom has been split to release its nuclear energy, it will never return to its original state. Once a piece of coal has been burned to generate heat or electricity, its energy content can never again be harvested. Once a species has become extinct, it will never return to life.

We need to manage the resources we take from the natural world carefully and effectively. Resource managers are guided in their decision-making by research in the natural sciences, but their decisions are also influenced by political, economic, and social factors. A key question in managing resources is whether to focus narrowly on the resource of interest, or to look more broadly at the environmental system of which the resource is a part. Taking a broader view can often help avoid damaging the system and can thereby help sustain the availability of the resource in the long term.

Preserving natural resources is an important consideration for the future, but it also speaks to the past and to our shared history as Canadians. Our economy, our identity, and even our national symbols have always been closely linked to the abundant physical resources of our environment. In recent decades, however, the consumption of natural resources has increased greatly—in Canada and throughout the world—driven by rising affluence and the growth of the largest global human population in history.

The environment provides goods and services but has intrinsic value

The natural environment supports the life and well-being of humans—and all other organisms—by the provision of both goods and services. **Goods** are tangible material things that can be extracted from the environment. This includes, for example, food, water, mineral and energy resources, and materials for shelter. In a human context, we usually think of goods as having real or potential economic value, although not all goods are tradable commodities (like crops, lumber, oil, and minerals) that are easily valued in monetary terms. Other organisms don't have money-based economies, of course, but they, too, benefit from life-supporting materials provided by the environment.

The environment also provides **services**—functions and processes that are useful or even vital in the support of living organisms. In human economic terms, services are the "intangible" equivalent of goods, so it is theoretically possible (though not always easy or straightforward) to assign dollar values to them. Environmental services that are valuable to people include things like nutrient cycling by soils in support of crop growth; detoxification of contaminated ground water by wetlands; and climate regulation by the ocean and atmosphere. Typically it is much more difficult to place an economic or monetary value on environmental services than it is for environmental goods, many of which can at least be visualized, if not actually traded, as marketable commodities.

Something else to keep in mind—and we will explore this idea in more detail in the chapter entitled *Environmental Ethics and Economics: Values and Choices*—is that there are many ways to "value" something, aside from assigning it a monetary or market value. For example, a beautiful view might offer *aesthetic* value, or a particular location might have *spiritual* value for certain people. However, the idea of applying a value to environmental goods and services is fundamentally flawed, because it implies that nothing in the world has value unless it is valuable to humans. In contrast, many people argue that other species, the environment, the biosphere, and even the planet as a whole have *intrinsic* or *existence* value—they are important and even possess rights simply because they exist, rather than by virtue of any utility that they provide for humans.

Earth's carrying capacity is limited

When we think about Earth's limited resources, it is useful to consider the idea of carrying capacity. **Carrying capacity** is a measure of the ability of a system to support life. Environmental scientists quantify carrying capacity in terms of the number of individuals of a particular species that can be sustained by the biological productivity of a given area of land without incurring permanent damage. When the carrying capacity of the land (or water) system is exceeded—that is, when there are simply too many individuals for the system to support—one of two things will typically happen: Either the population of that species will decline or collapse, or the system itself will be altered, damaged, or depleted.

Ecologist Garrett Hardin of the University of California, Santa Barbara, illustrated this process while disputing the economic theory that the unregulated exercise of individual self-interest serves the public good. According to Hardin's best-known essay, "The Tragedy of the Commons," published in the journal *Science* in 1968,[16] resources that are open to unregulated exploitation inevitably become overused and, as a result, are damaged or depleted.

Hardin based his argument on the simple scenario of a public pasture, or "commons," open to unregulated grazing. He argued that each person who puts animals to graze on the commons will be motivated by selfish interests to increase the number of his or her animals in the pasture. Because no single person owns the pasture, no one has any incentive to limit the number of grazing animals or to expend money or effort to care for the pasture. This is the **tragedy of the commons**: Each individual withdraws whatever benefits are available from the common property as quickly as possible until the resource becomes overused and depleted. Ultimately, the carrying capacity of the pasture will be exceeded, and its food production capacity will collapse.

In some situations, private ownership may address this problem. In China, for example, private land ownership—illegal for many decades under the Communist system—has recently become possible in some rural areas. These limited experiments with private ownership have shown that landowners tend to be better environmental stewards than are short-term tenants, primarily because they are willing to make long-term investments in land management. In other cases, people who share a common resource may voluntarily organize and cooperate in enforcing its responsible use. In other cases it may require government regulation of the use of resources held in common by the public, from forests to air to fresh water. Each approach has its own strengths and weaknesses.

Human Activities and the Environment

Inevitably, our activities modify the environment, whether intentionally or not. Many of these activities have enriched our lives, bringing us longer lifespans, better health, and greater material wealth, comfort, mobility, and leisure time. However, many have damaged the very natural systems that sustain us. Such impacts as air and water pollution, soil erosion, and species extinction may compromise the well-being of living organisms, pose risks to human life, and threaten our ability to build a society that will survive and thrive in the long term.

People differ in their perception of environmental problems

Just as people differ in the value they place on different aspects of the environment, the perception of what constitutes an environmental problem differs from one person or group of people to another, and from one situation to another. A person's age, gender, class, race, nationality, employment, and educational background can all affect whether he or she considers a given environmental impact to be a "problem."

For instance, DDT is an insecticide that was widely used throughout the first half of the twentieth century. In the 1960s it became apparent that DDT was having persistent toxic effects on humans and other animals, leading to a widespread ban. People today are much more likely to view the spraying of DDT as a problem than people did in the 1940s, because more is known today about its health risks (**FIGURE 1.5**). However, a person living today in a malaria-infested village in Africa or India may still welcome the use of DDT because it kills mosquitoes that transmit malaria, a more immediate health threat. Thus, an African and a North American who have each knowledgeably assessed the pros and cons of pesticides may, because of differences in their circumstances, differ in their judgment of the severity of DDT as an environmental problem.

People also differ in their awareness of environmental problems. For example, in many cultures women are responsible for collecting water and fuelwood. As a result, they are often the first to perceive environmental degradation affecting these resources. In most urbanized societies, information about environmental health risks tends to reach wealthy people more readily than poor people. Thus, who you are, where you live, what you do, your income, your gender, and your socioeconomic status can have a huge effect on how you perceive your environment, how you react to change, and what impact those changes may have on how you live your life.

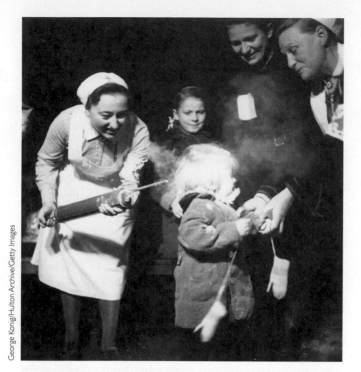

George Konig/Hulton Archive/Getty Images

FIGURE 1.5
How a person or society views an environmental problem can differ with time and circumstance. In 1945, health hazards from the pesticide DDT were not yet known, so children were doused with the chemical to treat head lice. Today, knowing of its toxicity to people and wildlife, many nations (including Canada) have banned DDT. However, in some countries where malaria is a threat, DDT is still used to eradicate mosquitoes that transmit the disease.

Population growth has driven our environmental impacts

For nearly all of human history, only a few million people populated Earth at any one time. Our numbers have nearly quadrupled in the past 100 years, reaching the milestone of 7 billion in 2011[17] and surpassing 7.3 billion in 2015.[18] We add more than 80 million people to the planet each year—more than 200 000 per day. Today the rate of population growth is slowing, but our absolute numbers continue to increase and shape our interactions with one another and with our environment. The steep and sudden rise in human population has amplified nearly all of our environmental impacts (**FIGURE 1.6**).

Four significant periods of societal change appear to have triggered remarkable increases in population size, along with greatly increased environmental impacts. The first happened as many as 2.5 million years ago during the *Paleolithic Period* (or *Old Stone Age*), when early humans began to shape and use stones as tools. Fire, too, was used by early humans to modify their environment, perhaps as long ago as 1.7 million years. The second was the transition from a nomadic, hunter-gatherer lifestyle to a settled, agricultural way of life. This change began to occur around 10 000 to 12 000 years ago, and it is formally known as the *Neolithic* or *Agricultural Revolution*.

The third major societal change, the *Industrial Revolution*, began in the mid-1700s and entailed a shift from rural life, animal-powered agriculture, and artisanal manufacturing to an urban society powered by fossil fuels. Life improved in many ways as a result of the Industrial Revolution. It also marked the beginning of industrial-scale pollution and many other environmental and social problems. Air quality declined dramatically as a result of the new reliance on coal. Land and water quality declined with the emergence of densely populated city centres. Workplace health and safety suffered as factories were hastily erected and expanded.

NASA Earth Observatory/NOAA NGDC

FIGURE 1.6   As a result of the rapid increase in human population in the past 100 years, people now occupy much of Earth's habitable surface. Some areas are more densely populated than others, as shown by the lights of human settlements visible from outer space in this famous NASA image of Earth at night.

The modern environmental movement had its roots in efforts by concerned citizens during the Industrial Revolution to ensure a cleaner, safer environment.

Today we are in the midst of a fourth transition, which some have labelled the modern *Medical–Technological Revolution*. Advances in medicine and sanitation, the explosion of communication technologies, and the shift to modern agricultural practices have allowed more people to live longer, healthier lives. However, we are now facing new environmental challenges. For example, new approaches to food production could bring an end to hunger but have the potential for significant environmental and health impacts.

Each major societal transition introduced technological advancements that made life easier and resources more available, effectively increasing the capacity of the environment to support human life, and allowing the population to increase dramatically. The modern Medical–Technological Revolution is ongoing, and the ultimate impacts on population and the environment are as yet unknown.

Consumption and technology also make an impact

Population growth is unquestionably at the root of many environmental problems. However, patterns and habits of resource consumption and the impacts of new technologies are also responsible for environmental impacts. The Industrial Revolution enhanced the material affluence of many of the world's people, raising standards of living. It led to an increase in population, just as new technologies and increased levels of consumption placed additional pressures on the environment. We can expect that the same will be true of the Medical–Technological Revolution.

One approach to quantifying these impacts is called the **IPAT model**, which represents our total impact (I) on the environment as the product of population (P), affluence (A), and technology (T):

$$I = P \times A \times T$$

Thus, environmental impact is a function not only of population but also of affluence (i.e., consumption) and technology. An increase in the number of people has impacts on the environment, but we must also concern ourselves with the increased consumption of natural resources and manufactured goods as a consequence of affluence and purchasing power.

New technologies can also have deep impacts on the environment, sometimes in ways that we cannot foresee. Consider, for example, the "scrubbing microbeads" in many body and face wash products. These tiny beads of plastic, which appeared in personal care products only a few years ago, were originally thought to be inert and harmless. It turns out, however, that they are toxic if consumed by fish or other organisms, and have a tendency to accumulate other toxins from the water. Microplastics degrade exceedingly slowly and they have collected very rapidly in natural water bodies; in 2013 a group of marine pollution researchers documented more than 466 000 microbeads per km^2 in some areas of the Great Lakes.[19]

Every new technology has the potential to change our impacts, sometimes positively and sometimes negatively. Sometimes we face a double-edged sword. For example, catalytic converters in automobiles contain platinum; their introduction led to greatly improved air quality, but caused an increase in platinum-mining activity and a reduction in the rate of recycling of platinum. The development of modern hormonal methods of birth control helped curb global population growth; now, however, there is concern about the negative effects of residual hormones on organisms in the aquatic environment. All technologies have impacts that can be translated and magnified, for better or worse, through the complex cycles of the natural environment.

Ecological footprints help us quantify our impacts

We can quantify our impacts and our resource consumption relative to the planet's carrying capacity using a concept developed in the 1990s by Mathis Wackernagel and William Rees at the University of British Columbia. The **ecological footprint (EF)** expresses the environmental impact of an individual or a population in terms of the area of land and water required to provide the raw materials that person or population consumes, and to absorb or recycle their wastes, including direct and indirect impacts.

Ecological footprint is essentially the *inverse* of carrying capacity—it is a measure of the land (and water) required to support an individual, rather than the number of individuals that can be supported by an area of land (or water). The capacity of a terrestrial or aquatic system to be biologically productive and to absorb waste, especially carbon dioxide, is also a useful measure called **biocapacity**. When a population exceeds or overshoots the biocapacity of a system, the system will be at risk of permanent damage; this can happen if biocapacity is damaged, or if the ecological footprint of the population increases, or both.

Researchers calculate that our species is now using about 47% more resources than are presently available (**FIGURE 1.7**).[20] That is, we are exceeding the biocapacity of the planet and depleting renewable resources faster than they are being replenished, and 1.47 "Earths" would be required in order to sustain this level of resource use. This

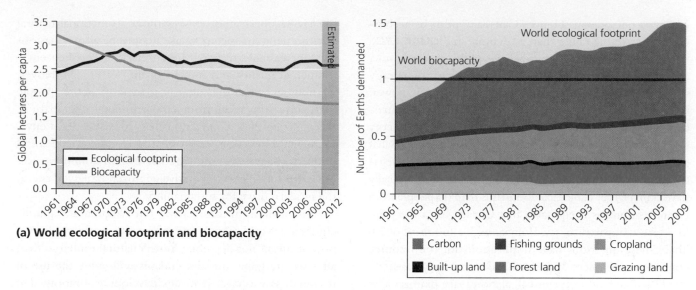

(a) World ecological footprint and biocapacity

(b) Components of the ecological footprint

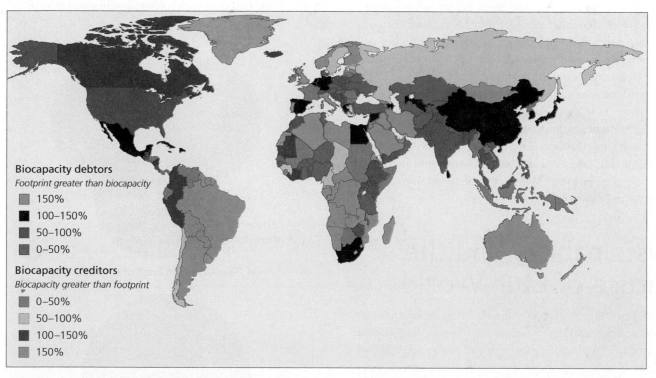

Biocapacity debtors
Footprint greater than biocapacity
- 150%
- 100–150%
- 50–100%
- 0–50%

Biocapacity creditors
Biocapacity greater than footprint
- 0–50%
- 50–100%
- 100–150%
- 150%

(c) Biocapacity "debt"

FIGURE 1.7 The world ecological footprint now exceeds the biocapacity of the planet **(a)**. It would require 1.47 Earths to meet the current demand for resources imposed by the human population **(b)**. Some nations, where demand for resources exceeds availability, are biocapacity "debtors"; others are biocapacity "creditors" **(c)**.
Source: Global Footprint Network, 2012. "World Ecological Footprint and Biocapacity"; "Components of the Ecological Footprint"; "Biocapacity 'debt'"; *National Footprint Accounts, 2012 Edition.* http://www.footprintnetwork.org/en/index.php/GFN/page/footprint_data_and_results, pp. 5-6. © 2015 Global Footprint Network. www.footprintnetwork.org.

DATA Q On graph **(b)**, why is biocapacity shown as a straight line, when graph **(a)** clearly shows that it has changed over time?

is like drawing the principal out of a bank account, rather than living off the interest.

The ecological footprint of Canada is approximately 7 gha (global hectares) per person.[21] This is far larger than the average world ecological footprint of 2.7 gha per person, which already exceeds the global biocapacity of 1.8 gha. However, Canada is a large country with abundant natural resources, and an available biocapacity of about

14.9 gha per person. Hence, Canada appears as a "bio-capacity creditor" rather than a "debtor" in **FIGURE 1.7C**.

Footprint calculations can have political, economic, and even ethical aspects. For example, various energy sources—fossil fuels, nuclear energy, hydroelectric power, wind, or solar energy—have very different environmental impacts. How should the impacts of different energy sources be accounted for in the calculations? Under what circumstances should one impact be weighted "more negatively" than another? People from wealthy nations typically have much larger ecological footprints than do people from poorer nations; is it equitable that Canada has a footprint of 7 gha, compared to, say, Bolivia, with a footprint of 2.6 gha? These questions have implications for the resource consumption choices that we all make on a daily basis.

The results of footprint calculations vary dramatically; you will even find variations among examples mentioned in this book. This is because the calculations are complicated and depend heavily on how certain components are defined. For example, different approaches use different methodologies to account for the surface area of the ocean, which—though supportive of life—clearly does not have the same significance as land area does for humans, as "living space" or even "biologically productive space." Different methodologies can lead to very different results in EF calculations. The work of researchers to standardize ecological footprint calculations should make the calculations more robust and their application to questions of environmental impact more effective.

Sustainability and the Future of Our World

Throughout this book you will see examples of environmental scientists asking questions, developing hypotheses, conducting experiments, gathering and analyzing data, and drawing conclusions about the causes and consequences of environmental change. Environmental scientists are studying some of the most centrally important issues of our time. The primary challenge we face is how to live within our planet's means, such that Earth and its resources can sustain us, and the rest of Earth's biota, for the foreseeable future. This is the challenge of **sustainability**, a guiding principle of modern environmental management.

Sustainability meets environmental, social, and economic goals

Environmental protection is often cast as being in opposition to the economic and social needs of human society, but environmental scientists have long recognized that our civilization cannot exist without a functional natural environment. In recent years, people of all persuasions have increasingly realized the connection between environmental quality and human quality of life. Moreover, we now recognize that often it is society's poorer people who suffer the most from environmental degradation. Recognizing these connections has led advocates of environmental protection, economic development, and social justice to begin working together toward common goals (**FIGURE 1.8**). This cooperative approach has given rise to the modern drive for sustainability.

Sustainability means leaving our children and grandchildren a world as rich and full as the world we live in now. It means not depleting Earth's natural capital so that, after we are gone, our descendants will enjoy the use of resources as we have. It means developing solutions that will work in the long term. Sustainability requires maintaining fully functioning ecological systems, because we

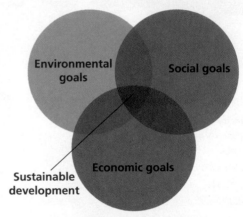

(a) "Triple bottom line" of sustainability

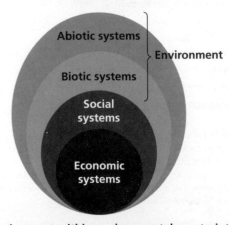

(b) Development within environmental constraints

FIGURE 1.8
This familiar diagram envisions three separate but interacting sets of goals: environmental, economic, and social **(a)**. Sustainable development requires all three sets of goals to be maximized. An alternate model envisions economic and social systems functioning, as they must, within the limitations of the natural world **(b)**. This is referred to as a "storng sustainability" model.

cannot sustain human civilization without sustaining the natural systems that nourish it.

Economists employ the term *development* to describe the use of natural resources for economic and social advancement (as opposed to simple subsistence or survival). Construction of homes, schools, hospitals, power plants, factories, and transportation networks are all examples of development. The United Nations has defined **sustainable development** as development that " ... meets the needs of the present without sacrificing the ability of future generations to meet their needs."[22] This definition came from the United Nations–sponsored *Brundtland Commission* (named after its chair, Norwegian prime minister Gro Harlem Brundtland), which popularized the term sustainable development in an influential 1987 report entitled *Our Common Future.*

Prior to the Brundtland Report, people aware of human impact on the environment might have thought *sustainable development* to be an oxymoron—a phrase that contradicts itself. Although development involves making purposeful changes intended to improve the quality of human life, environmental advocates have long pointed out that development often so degrades the natural environment that it threatens the very improvements for human life that were intended. Conversely, many people remain under the impression that protecting the environment is incompatible with serving people's economic needs.

Fortunately, sustainable development efforts by governments, businesses, industries, organizations, and individuals everywhere—from students on campus to international representatives at the United Nations—are beginning to alter these perceptions. These efforts are generating sustainable solutions that meet environmental, economic, and social goals simultaneously, satisfying the so-called triple bottom line (see **FIGURE 1.8**).

Sustainability and the triple bottom line demand that our current human population limit its environmental impact while also promoting economic well-being and social equity. These aims require us to make an ethical commitment to our fellow citizens and to future generations. They also require that we apply knowledge from the sciences to help us devise ways to limit our impact and maintain the functioning environmental systems on which all life depends.

We face many environmental challenges

The dramatic growth in human population and consumption is due in part to our successful efforts to expand and intensify the production of food. Since the

FIGURE 1.9
Indoor and outdoor air pollution contribute to millions of premature deaths each year. Environmental scientists and policymakers are working to reduce these problems in a variety of ways.

origins of agriculture and the Industrial Revolution, new technologies have enabled us to grow increasingly more food per unit of land. These advances in agriculture must be counted as one of humanity's great achievements, but they have come at some cost. We have converted nearly half the planet's land surface for agriculture; our extensive use of chemical fertilizers and pesticides negatively affects organisms and alters natural systems; and erosion, overgrazing, climate change, and poorly managed irrigation are degrading the productive farmland needed to feed the 80 million people added to the population each year.

Meanwhile, pollution from farms, industries, households, and individual actions dirties our land, water, and air (**FIGURE 1.9**). Outdoor air pollution, indoor air pollution, and water pollution contribute to the deaths of millions of people each year. Environmental toxicologists are chronicling the impacts on people and wildlife of the many synthetic chemicals and other pollutants we emit into the environment.

Perhaps our most pressing challenge is to address the looming spectre of global climate change, which has the potential to influence all other aspects of life on this planet. Scientists have firmly concluded that human activity is altering the composition of the atmosphere and that these changes are affecting Earth's climate.[23] Since the start of the Industrial Revolution, atmospheric carbon dioxide concentrations have risen to a level not present for millions of years. This increase has resulted from our reliance on burning fossil fuels to power our civilization. Carbon dioxide and several other gases absorb heat and warm Earth's surface. The anthropogenic enhancement of this process is likely responsible for glacial melting, sea-level rise, impacts on wildlife and crops, and increased episodes of destructive weather. Atmospheric carbon dioxide also contributes to the acidification of ocean water.

The combined impact of human actions such as climate change, overharvesting, pollution, the introduction of non-native species, and particularly habitat alteration has driven many aquatic and terrestrial species out of large parts of their ranges and toward the brink of extinction. Today Earth's biological diversity, or **biodiversity**, the cumulative number and diversity of living things, is declining dramatically. Many biologists say we are already in a mass extinction event comparable to only five others documented in all of Earth's history. Biologist Edward O. Wilson has warned that the loss of biodiversity is our most serious and threatening environmental dilemma, because it is not the kind of problem that responsible human action can remedy. The extinction of species is irreversible; once a species has become extinct, it is lost forever.

A comprehensive scientific assessment of the present condition of the world's ecological systems and their ability to continue supporting our civilization was completed in 2005, when more than 2000 of the world's leading environmental scientists from nearly 100 nations completed the *Millennium Ecosystem Assessment*. Four of the main findings of this exhaustive project are summarized in **TABLE 1.1**. The Assessment makes clear that our degradation of the world's environmental systems is having negative impacts on all of us, but that with care and diligence we can still turn many of these trends around.

TABLE 1.1   Main Findings of the Millennium Ecosystem Assessment

- Over the past 50 years humans have changed ecosystems more rapidly and extensively than in any comparable period of time in human history, largely to meet rapidly growing demands for food, fresh water, timber, fibre, and fuel. This has resulted in a substantial and largely irreversible loss in the diversity of life on Earth.

- The changes made to ecosystems have contributed to substantial net gains in human well-being and economic development, but these gains have been achieved at growing costs. These costs include the degradation of ecosystems and the services they provide for us, and the exacerbation of poverty for some groups of people.

- This degradation could grow significantly worse during the first half of this century.

- The challenge of reversing the degradation of ecosystems while meeting increasing demands for their services can be partially overcome, but doing so will involve significantly changing many policies, institutions, and practices.

Source: From Millennium Ecosystem Assessment, *Synthesis Report*, 2005. Copyright © 2005 by World Resources Institute. Reprinted by permission..

Solutions to environmental problems must be global and sustainable

We cannot live without exerting *any* impact on Earth's systems. In trying to solve environmental problems, we face many trade-offs; the challenge is to develop solutions that increase the quality of life for people everywhere in the world, while minimizing harm to the environment that supports us.

The nature of virtually all environmental issues is being changed by the set of ongoing phenomena commonly dubbed *globalization*. Our increased global interconnectedness in trade, politics, and the movement of people and other species poses many challenging problems. However, it also sets the stage for novel and effective solutions on a global scale. Many workable solutions are at hand, and we can achieve many more solutions with further effort.

Energy choices will be particularly important in helping us, as a world community, to develop in a more sustainable manner. Our reliance on fossil fuels to power our civilization has intensified virtually every negative impact we have on the environment, from habitat alteration to air pollution to climate change. Fossil fuels have also brought us the material affluence we enjoy. We have been able to power the machinery of the Industrial Revolution, produce the chemicals that boosted agricultural yields, run the vehicles and transportation networks of our mobile society, and manufacture and distribute our countless consumer products. It is little exaggeration to say that the lives we live today are a result of the availability of fossil fuels.

However, in extracting fossil fuels, we are splurging on a one-time bonanza. Scientists calculate that we have depleted half the world's oil supplies and that we are in for a rude awakening very soon, once the supply begins to decline while the demand continues to rise. Coal and natural gas are also nonrenewable and available in limited supply, although we are likely farther from reaching the limits of those resources. The search is now on for alternative sources of energy that will allow us to maintain an acceptable standard of living while minimizing the environmental impacts of energy use (**FIGURE 1.10**). How we handle the imminent crisis of fossil fuel depletion and the search for replacements will largely determine the nature of our lives in the twenty-first century.

Recycling and production efficiencies in industry are helping to minimize resource use and solve our waste problem. In response to agricultural problems, scientists and others have developed and promoted soil conservation, high-efficiency irrigation, and organic agriculture. Technological advances have greatly reduced the

FIGURE 1.10
Although fossil fuels have powered our civilization since the Industrial Revolution, many renewable energy sources exist, such as solar energy, which can be collected with panels like these. Such alternative energy sources could be further developed for sustainable use now and in the future.

pollution emitted by industry and automobiles, especially in wealthier countries. Advances in conservation biology are enabling scientists and policymakers to work together to try to protect habitat, slow extinction, and safeguard endangered species (**FIGURE 1.11**).

These are a few of the many solutions we will explore in the course of this book, and we will examine some of the new structures, programs, processes, and technologies that are emerging in support of these solutions. Canadian scientists have been at the forefront of many of these technological advances and have made fundamental contributions to global environmental management theories and to our current understanding of the human–environment relationship.

FIGURE 1.11
Efforts to save endangered species and reduce biodiversity loss include many approaches, but all require that adequate areas of appropriate habitat be preserved in the wild. The habitat of polar bears, for example, is increasingly threatened by global climate change.

Are things getting better or worse, and how can we tell?

Despite the myriad challenges we face, some people maintain that the general conditions of human life and the environment are getting better, not worse. Some believe that with modern technology we will always find ways to make Earth's natural resources meet all our needs indefinitely and that human ingenuity will see us through any difficulty. In contrast, others predict doom and disaster for the world because of our impacts on it.

As usual, the truth lies somewhere between these two extremes. Some aspects of human development and environmental quality have improved noticeably over the past few decades; nevertheless, as a world community we face daunting challenges. New technologies and new scientific approaches have made it easier than ever before to monitor, understand, and evaluate environmental change (see "The Science Behind the Story: Mission to Planet Earth").

Scientists always aim to find quantitative answers. In 2008, there was a gathering of a group of scientists led by Johan Rockström from the Stockholm Resilience Centre, and their goal was to find out exactly how humanity is doing in managing its impacts on the environment. The group identified nine key systems that are crucially important to the Earth system as a whole:

1. stratospheric ozone layer
2. biodiversity
3. toxic chemicals dispersion
4. climate change
5. ocean acidification
6. freshwater consumption and the global hydrological cycle
7. land system change
8. nitrogen and phosphorus inputs to the biosphere and oceans
9. atmospheric aerosol loading

These nine systems were chosen because of their core importance in maintaining Earth and its life-supporting functions, for their global influence, and for the potential for their rapid or irreversible change. They are illustrated in **FIGURE 1.12**.

The scientists' premise was that keeping human impacts within the "safe" operating boundaries of these nine systems would ensure the continued functioning of Earth's life-supporting systems, but exceeding these boundaries could generate rapid or irreversible environmental damage. They hoped that the identification of a safe operating space could eventually help with science-based policy decisions about the environment. The original proposal and discussion

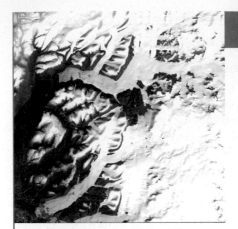

Canada's Ellesmere Island National Park Reserve, seen in this satellite image, is the most northerly park on Earth.
NASA/GSFC/MITI/ERSDAC/JAROS, and U.S./Japan ASTER Science Team

THE SCIENCE BEHIND THE STORY

Mission to Planet Earth: Monitoring Environmental Change

Science begins with observation. One of the most significant legacies of the program to explore outer space is the ability to observe Earth from afar. The development of this ability changed our perspective on this planet permanently, leading to a new sense of respect, care, and concern for the environment that continues to grow as a social priority.

Since the last handheld photograph of the whole planet was taken in 1972 by *Apollo 17* astronauts, scientists and technologists have dramatically improved their ability to capture and interpret a wide variety of images of this planet. Handheld photography continues to provide an important part of the data returned by manned space missions; you can see many of these photographs archived by NASA at http://eol.jsc.nasa.gov/. Today, satellites and the sophisticated instrumentation they carry provide us with the opportunity to observe, study, monitor change, and gather an unprecedented amount of information about the planet through technologies and processes that are collectively referred to as **remote sensing**.

By the early 1980s, NASA scientists and administrators had faced the reality that they were unlikely to obtain the necessary funding to return to the Moon any time in the near future. Instead, they began to turn their attention to another nearby planetary object: Earth. They called the new scientific approach **Earth system science** and defined as its goal "to obtain a scientific understanding of the entire Earth system on a global scale by describing how its component parts and their interactions have evolved, how they function, and how they may be expected to continue to evolve on all timescales."[24] The observation and interpretation of environmental change was fundamental to this new scientific approach, right from the beginning.

To achieve the goals of Earth system science, scientists began to use technologies that had been developed for other purposes—space exploration, communications, even warfare—to observe Earth and its component parts over time, and they called the endeavour *Mission to Planet Earth*. The scientific results of this mission have been and continue to be spectacular. Satellites and remote detection and measuring technologies have provided a massive amount of information about the environment, how it changes over time, and how human activity affects it—information with a depth and breadth that would have been unimaginable back when those first whole-planet photographs were taken. A great deal of what environmental scientists now know about this planet as a coherent system, how its various parts interact, and how it is changing and evolving has been based on information derived from the *Mission to Planet Earth*.

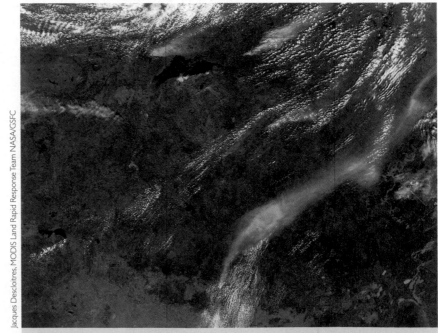

Jacques Descloitres, MODIS Land Rapid Response Team NASA/GSFC

This satellite image shows Lake Athabasca, the dark, irregular patch straddling the border between Alberta (west) and Saskatchewan (east). Numerous active summer forest fires are indicated by red dots. A large smoke plume stretches across Saskatchewan and into Manitoba, to the east.

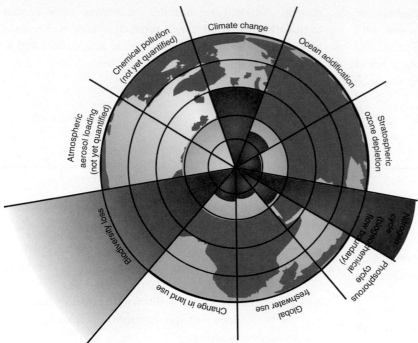

FIGURE 1.12 The inner circle represents the safe operating space for the nine key systems. The red wedges indicate the best estimate of the current situation. Three boundaries have already been crossed: climate change, the nitrogen cycle, and biodiversity loss.
Source: Azote Images/Stockholm Resilience Centre.

of the planetary boundaries were published in *Nature* and other scholarly journals.[25, 26]

In subsequent studies, scientists have attempted to quantify the boundaries more precisely; to determine how close humanity has come to surpassing the boundaries; and to find out more about the risks of exceeding the boundaries, and the potential for rapid or irreversible change. This work is ongoing, but it appears that at least three of the "safe" boundaries have already been crossed: climate change, the nitrogen cycle, and biodiversity loss. Consequently, some scientists propose that we have now entered a new geological epoch, characterized by human impacts on environmental systems; they have named it the *Anthropocene Epoch*.

Is this possibility that we have exceeded three important boundaries irretrievably bad news? Or is it comparatively good news that we're not doing so badly with the other six? It can be very confusing to interpret and evaluate media reports and scientific data about the environment. For reasons that may or may not be valid, "experts" often present differing or even contradictory viewpoints and interpretations. To ensure that you are thinking comprehensively about Earth systems and that you have a long-term perspective, these three questions are worth asking each time you are confronted with seemingly conflicting statements:

1. Do the impacts being debated pertain only to humans, or have other organisms and natural systems been adequately considered?
2. Are the debaters thinking in the short term or the long term?
3. Are they considering all costs and benefits relevant for the question at hand, including environmental costs, or only some of them?

As you proceed through this book and encounter many contentious issues, consider how a person's perception of them may be influenced by these three factors. The cases and examples in the book will help you gain an understanding of how science allows us to understand and document environmental changes and impacts. You also will learn to become an informed and critical consumer of news, and to detect when the reporting of environmental information is incorrect, strongly biased, or not based on a foundation of scientific rigour.

Conclusion

Identifying a problem is the first step in devising a solution to it. Many of the trends detailed in this book may cause us worry, but others give us reason to hope. One often-heard criticism of environmental science courses and books is that they emphasize the negative. Recognizing the validity of this criticism, in this book we attempt to balance the discussion of environmental problems with a corresponding focus on potential solutions.

Solving environmental problems can move us toward health, longevity, peace, and prosperity. Finding effective ways to live peacefully, healthfully, and sustainably on our diverse and complex planet will require a thorough scientific understanding of both natural and social systems. Science in general, and environmental science in particular, can aid us in our efforts to develop balanced and workable solutions to the many environmental dilemmas we face today.

"Will we develop in a sustainable way?" may well be the single most important question in the world today.

Environmental science holds a crucial key to addressing it: By helping us understand our intricate relationship with the environment, it informs our attempts to solve and prevent environmental problems, and to create a better world for ourselves and our children. Because so much remains unstudied and undone, and because it is so central to our modern world, environmental science will remain an exciting frontier for you to explore as a student and as an informed citizen throughout your life.

REVIEWING OBJECTIVES

You should now be able to

Define the term *environment*

- Our environment consists of everything around us, including living and nonliving things. Humans are a part of the environment, not separate from it.
- How we define "environment" is important because it has legal, economic, social, and scientific implications.
- Environmental science is the study of how the natural world works, how the environment affects us, and how we affect the environment.

Characterize the interdisciplinary nature of environmental science

- Science is the set of approaches that we use to study and understand the world, and the body of knowledge that results from that study.
- Environmental science is a scientific endeavour that relies on the approaches and insights of numerous disciplines from the natural sciences and social sciences.
- Environmentalism and environmental science may inform each other, but they are not the same thing.
- Basing our environmental management decisions on science and an understanding of the natural world should help us find robust solutions to environmental problems, and avoid repeating mistakes of the past.

Describe several types of natural resources and explain their importance to human life

- Natural resources are essential to human life and civilization. Some are inexhaustible or perpetually renewable; others are nonrenewable; and still others are renewable if we are careful not to exploit them at too fast a rate.
- The environment is valuable to humans because it provides us with goods and services, but it also has its own intrinsic or existence value.
- Earth's carrying capacity—the number of individuals that can be sustained by a given area of productive land—is large, but ultimately finite.

Diagnose and illustrate some of the pressures on the global environment

- People's interpretations of the severity of environmental problems differ according to time, awareness, and circumstance.
- The rapidly increasing human population has been the main driver of our environmental impacts and resource extraction.
- Patterns of consumption, increasing affluence, and new technologies also influence and may exacerbate human impacts on the environment.
- The ecological footprint, a measure of the productive area or biocapacity required to support an individual or population at a certain level of consumption, can help us quantify our environmental impacts.

Articulate the concepts of sustainability and sustainable development

- Sustainability means living within the planet's means, so that Earth's resources can sustain us—and other species—for the foreseeable future. Sustainable development means pursuing environmental, economic, and social goals in a coordinated way.
- Human activities, including agriculture, industry, and the use of fossil fuels for energy, while central to our modern life, are causing diverse environmental impacts such as resource depletion, air and water pollution, habitat destruction, climate change, and the diminishment of biodiversity.
- Globalization has exacerbated some environmental problems, but it also has opened many opportunities for innovative solutions. Making good energy choices will be particularly important for our future on this planet.
- It can be confusing to interpret the wide array of information, both positive and negative, about the state of the environment. Thinking about core planetary boundaries may help scientists quantify our impacts. It is especially important to take a long-term approach and account for costs and benefits from a broad, ecosystem-based perspective.

TESTING YOUR COMPREHENSION

1. What is *environmental science?* Name several disciplines that contribute to environmental science.

2. What do renewable resources and nonrenewable resources have in common? How are they different? Identify two renewable and two nonrenewable resources.

3. What is *the tragedy of the commons?* Explain how the concept might apply to an unregulated industry that is a source of water pollution, or to fishing in open ocean waters.

4. How and why did the Agricultural Revolution and the Industrial Revolution affect human population size and the environment?

5. What is an *ecological footprint,* and what are some of the challenges in making this type of calculation?

6. What is the relationship between *carrying capacity* and *ecological footprint?*

7. What is *biocapacity?* How can a country that has a large per capita ecological footprint, like Canada, be a biocapacity "creditor" rather than a biocapacity "debtor"?

8. Why are our energy choices going to be particularly important in setting a more sustainable course for society?

9. Give examples of three major environmental problems in the world today, along with their immediate and root causes.

10. How would you define the term *sustainable development?*

THINKING IT THROUGH

1. Many resources are renewable if we use them in moderation, but can become nonrenewable if we overexploit them. Order the following resources on a continuum of renewability (see **FIGURE 1.4**), from most renewable to least renewable: soils, timber, fresh water, food crops, and biodiversity. What factors influenced your decision? For each of these resources, what might constitute overexploitation, and what might constitute sustainable use?

2. What environmental problem do *you* feel most acutely? Do you think there are people in the world who do not view your issue as an environmental problem? Who might they be, and why might they take a different view?

3. If the human population were to stabilize tomorrow at the current population, would that solve our environmental problems? Which types of problems might be alleviated, and which might continue to become worse?

4. If you were an environmental scientist and were asked to provide an assessment of a complex environmental situation, what kinds of experts would you call upon? Think about specialists and team members who might be needed to contribute their expertise to the following situations: construction of a new hydroelectric dam in Canada's far north; proposed draining of a wetland to build a new subdivision; proposal to permit moose hunting in a national park; management of a large oil spill just offshore from a pristine beach.

INTERPRETING GRAPHS AND DATA

Environmental scientists study phenomena that range in size from individual molecules to the entire Earth and that occur over time periods lasting from fractions of a second to billions of years. To simultaneously and meaningfully represent data covering so many orders of magnitude, scientists have devised a variety of mathematical and graphical techniques, such as exponential notation and logarithmic scales. On the next page are two graphical representations *of the same data,* representing the growth of a hypothetical population from an initial size of 10 individuals at a rate of increase of approximately 2.3% per generation. The graph in part **(a)** uses a conventional linear scale for the population size; the graph in part **(b)** uses a logarithmic scale.

1. Using the graph in part **(a)**, what would you say was the population size after 200 generations? After 400? After 600? After 800? How would you answer the same questions by using the graph in part **(b)**? What impression does the graph in part **(a)** give about population change for the first 600 generations? What impression does the graph in part **(b)** give?

2. Compare graphs **(a)** and **(b)** to the graph of human population growth in **(c)**. Does this graph use a linear scale, or a logarithmic scale? What does the human population appear to be doing between 10 000 B.P. and 2000 B.P.? How would you describe the growth pattern after that time?

3. The size of a population that is growing by a constant rate of increase will plot as a straight line on a logarithmically scaled graph like the one in part **(b)**, but if the annual rate of increase changes, the line will curve. Do you think the data for the human population over the past 12 000 years would plot as a straight line on a logarithmically scaled graph? If not, when and why do you think the line would bend?

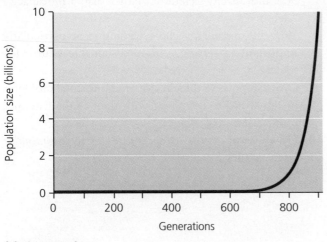

(a) Linear scale

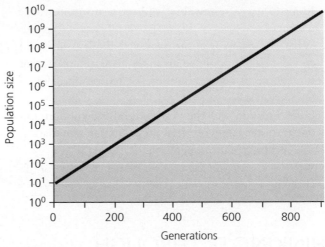

(b) Logarithmic scale

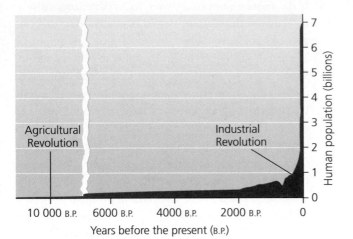

(c) World population growth

Hypothetical population growth curves **(a)** and **(b)** both assume an initial population size of 10 and a constant rate of increase of approximately 2.3% per generation. Graph (c) shows human population growth from about 12 000 years ago to the present.

Source: Based on data compiled from U.S. Census Bureau, UN Population Division, and other sources.

MasteringEnvironmentalScience®

STUDENTS
Go to **MasteringEnvironmentalScience** for assignments, the eText, and the Study Area with practice tests, videos, current events, and activities.

INSTRUCTORS
Go to **MasteringEnvironmentalScience** for automatically graded activities, current events, videos, and reading questions that you can assign to your students, plus Instructor Resources.

2 Matter, Energy, and the Physical Environment

Rescue workers in northeastern Japan clear bodies and rubble after the 2011 Tōhoku earthquake and tsunami.

Paula Bronstein/Getty Images

Upon successfully completing this chapter, you will be able to

- Describe how scientists use the scientific method to test ideas and investigate the nature of the environment
- Summarize the basic properties of matter, the foundation for all materials on Earth
- Differentiate among various types of energy, the fundamental properties of energy, and the role of energy in environmental systems

- Explain how plate tectonics and the rock cycle shape the landscape around us and Earth beneath our feet
- Summarize the characteristics of early Earth and the main hypotheses for the origin of life

Tsunami overtops a seawall following the Tōhoku earthquake in 2011.

Ho New/Reuters

Sea of Japan (East Sea)

JAPAN

Fukushima Daiichi

North Pacific Ocean

CENTRAL CASE
THE TŌHOKU EARTHQUAKE: SHAKING JAPAN'S TRUST IN NUCLEAR POWER

"This used to be one of the best places for a business. I'm amazed at how little is left."

—TAKAHIRO CHIBA, SURVEYING THE DEVASTATED DOWNTOWN AREA OF ISHINOMAKI, JAPAN, WHERE HIS FAMILY'S SUSHI RESTAURANT WAS LOCATED

"Fukushima should not just contain lessons for Japan, but for all 31 countries with nuclear power."

—TATSUJIRO SUZUKI, VICE-CHAIR, JAPAN ATOMIC ENERGY COMMISSION

At 2:46 p.m. on March 11, 2011, the land along the northeastern coast of the Japanese island of Honshu began to shake violently—and continued to shake for six minutes. The tremors were caused when a large section of seafloor along a fault line 125 km offshore suddenly lurched, releasing huge amounts of energy through the crust and generating an earthquake of magnitude 9.0 on the Richter scale.[1] Almost 16 000 people were killed,

and property damage was extensive. But no one knew, at the time, that this quake would initiate a series of events that would affect not only Japan, but also the future of nuclear power in Japan and around the world.

The Tōhoku earthquake, as it was later named, was not the first major earthquake to strike Japan. In 1923, an earthquake devastated the cities of Tokyo and Yokohama, resulting in approximately 140 000 deaths.[2] Losses of life and property from the Tōhoku quake were far less extensive than the losses from the Tokyo and Yokohama event, thanks to new stringent building codes that enable buildings to resist crumbling and toppling over during earthquakes. But even when the earth stopped shaking, the residents of northeastern Japan knew that further danger might still await them—from a tsunami.

A **tsunami** ("harbour wave" in English) is a powerful surge of sea water generated when an offshore earthquake displaces large volumes of rocks and sediment on the ocean bottom, suddenly pushing the overlying ocean water upward. This upward movement of water creates waves that speed outward from the earthquake site in all directions. These waves are hardly noticeable at sea, but can rear up to staggering heights when they enter the shallow waters near shore and can sweep inland with great force. The fear of a tsunami was well founded, as strong ocean surges followed the 1923 Tokyo–Yokohama earthquake, pushing walls of debris in front of them and drowning victims still trapped in the wreckage from the earthquake.

The Japanese had built seawalls to protect against tsunamis, but the Tōhoku quake caused the island of Honshu to sink, lowering the height of the seawalls by up to 2 m in some locations. Waves reaching up to 14 m in height then overwhelmed these defences (see photo). The raging water swept up to 9.6 km inland, scoured buildings from their foundations, and inundated towns, villages, and productive agricultural land. As the water's energy faded, it receded, carrying structural debris, vehicles, livestock, and human bodies out to sea.

When the tsunami overtopped the 5.7-m seawall protecting the Fukushima Daiichi nuclear power plant, it flooded the diesel-powered emergency generators responsible for circulating water to cool the plant's nuclear reactors. With the local electrical grid knocked out by the earthquake and the backup generators offline, the nuclear fuel in the cores of the three active reactors at the plant began to overheat. The water that normally kept the nuclear fuel submerged within the reactor cores boiled off, exposing the nuclear material to the air and further elevating temperatures inside the cores. As the overheated nuclear fuel underwent a meltdown, chemical reactions within the reactors generated hydrogen gas. This set off explosions in each of the three reactor buildings, releasing radioactive material into the air.

As events worsened over several tense days, Japanese authorities became desperate to cool the reactors, contain the radioactive emissions, and prevent a full-blown catastrophe that could render large portions of their nation uninhabitable. They sent in teams of engineers and soldiers who risked their lives amid the radiation and flooded the reactor cores with sea water pumped in from the ocean.

The 1–2–3 punch of the earthquake–tsunami–nuclear accident left many thousands of people dead or missing and caused $300 billion in material damage. Around 340 000 people were displaced from their homes, including 100 000 people from towns near the Fukushima Daiichi plant where radioactive fallout contaminated the soil to unsafe levels. An area of 20-km radius around the Fukushima Daiichi plant has been permanently evacuated, and the full extent of nuclear contamination is still being determined. Some of the greatest concerns centre on contaminated food and water, so domestically produced crops and seafood will require testing for radiation for years to come. Full recovery from these events is expected to take decades.

One of the longest-lasting legacies of these events may be the impact on the future of nuclear power in Japan and around the world. The Japanese government had championed a view that nuclear power was perfectly safe, but the events at Fukushima have severely shaken public support for nuclear power in Japan. More than a year after the events, half of the 70 000 residents of Minamisoma, a town on the edge of the nuclear evacuation zone, were still refusing to return because of fears of radiation.[3] In the summer of 2012, over 100 000 people marched in Tokyo to protest the restarting of nuclear reactors that had been shut down after the Tōhoku quake, and public opinion surveys found 70% of Japanese wished for their nation to rely less on nuclear power.

In North America and Europe, the events at Fukushima Daiichi caused a public already skeptical of the safety of nuclear power to become further wary of its use. But with the challenges facing us in climate change, many energy analysts, scientists, and policymakers caution that we should not abandon a carbon-free source of energy like nuclear power, but rather refocus efforts on maximizing its safety. Whatever the eventual outcome of these analyses, the events of March 11, 2011, will not soon be forgotten—in Japan or elsewhere.

How Scientists Investigate the Environment

In this chapter we examine some of the basic principles of matter and energy, and consider how matter and energy, together, shape the physical environment of this planet. But first, let's take some time to think about the processes by which we gather scientific information and seek to verify our understanding of this planet and how it works.

Scientists develop and refine their ideas about how the world works by designing tests to determine whether the ideas are supported by evidence. If a particular statement or explanation is testable and resists repeated attempts to disprove it, scientists are likely to accept it as a useful explanation. Scientific inquiry is thus an incremental approach to the truth.

Scientists test ideas by critically examining evidence

To test their ideas, scientists generally follow a process called the **scientific method**, a technique for testing ideas by making observations and by gathering evidence. There is nothing mysterious about the scientific method; it is merely a formalized version of the procedure any of us might naturally take, using common sense, to answer a question.

The scientific method is a theme with variations, however, and scientists pursue their work in many different ways. Because science is an active, creative, imaginative process, an innovative scientist may find good reason to depart from the traditional scientific method when a particular situation demands it.

Scientists from different fields also approach their work differently because they deal with dissimilar types of information. A natural scientist, such as a chemist, will conduct research quite differently from a social scientist, such as a sociologist. Because environmental science includes both natural and social sciences, in our discussion here we use the term *science* in its broad sense, to include both.

Scientists of all persuasions broadly agree on the fundamental elements of scientific inquiry. The scientific method relies on the following assumptions:

- The universe functions in accordance with fixed natural laws that do not change from time to time or from place to place.
- All events arise from some cause and, in turn, lead to other events.

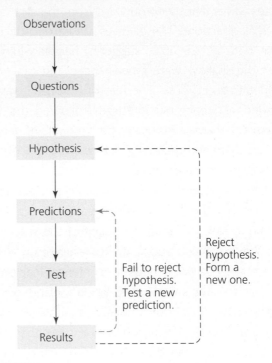

Scientific method

Observations → Questions → Hypothesis → Predictions → Test → Results

Fail to reject hypothesis. Test a new prediction.

Reject hypothesis. Form a new one.

FIGURE 2.1
The scientific method is an observation-based approach that scientists use to learn how the world works. This simplified diagram shows the basic steps of the scientific method, but cannot convey the dynamic and creative nature of science. Researchers from different disciplines pursue their work in ways that legitimately vary from this model.

- We can use our senses and reasoning abilities to detect and describe natural laws that underlie the cause-and-effect relationships we observe in nature.

As practised by individual researchers or research teams, the scientific method (**FIGURE 2.1**) typically consists of the steps outlined below.

Make observations   Advances in science typically begin with the observation of a phenomenon that the scientist wants to explain. Observations set the scientific method in motion and function throughout the process.

Ask questions   Scientists are naturally curious about the world and love to ask questions. Why are certain plants or animals less common today than they once were? Are storms becoming more severe, or flooding more frequent? What causes excessive growth of algae in local ponds? Do the impacts of pesticides on fish or frogs mean that people could be affected in the same ways? All of these are questions environmental scientists have asked and attempted to answer.

Develop a hypothesis   Scientists attempt to answer their questions by devising explanations that can be tested. A **hypothesis** is an educated guess to explain a phenomenon or answer a scientific question. For example, a scientist investigating why algae are growing excessively in local ponds might observe chemical fertilizers being applied on farm fields nearby. The scientist might then state a hypothesis as follows: "Agricultural fertilizers running into ponds cause the algae in the ponds to increase." Sometimes this takes the form of a *null hypothesis*, a statement that the scientist expects no relationship between variables, such as between fertilizer and algal growth in a pond.

Make predictions

The scientist uses the hypothesis to generate **predictions**, which are specific statements that can be directly and unequivocally tested. In our algae example, a prediction might be "If agricultural fertilizers are added to a pond, the quantity of algae in the pond will increase." A null hypothesis also can lead to predictions; for example, the scientist might predict that adding agricultural fertilizer to a pond will cause no change in the amount of algae growing in the pond.

Test the predictions

Predictions are tested by gathering evidence, and the strongest form of evidence comes from experimentation. An **experiment** is an activity designed to test a prediction by manipulating **variables**, conditions that can change. Depending on the results, the experiment may either support or not support the hypothesis on which the prediction was based. Experiments are especially useful because they can establish *causal relationships*, by showing that changes in one variable cause predictable changes in another variable.

For example, a scientist could test the hypothesis linking algal growth to the use of fertilizer by selecting two identical ponds and adding fertilizer to one while leaving the other in its natural state. In this example, fertilizer input is the *independent variable*, a variable the scientist manipulates. The quantity of algae that results is the *dependent variable*, one that depends on the fertilizer input. If the two ponds are identical except for a single independent variable (fertilizer input), then any differences that arise between the ponds can be attributed to that variable. Such an experiment is known as a *controlled experiment* because the scientist controls for the effects of all variables except the one being tested—the dependent variable.

Whenever possible, it is important to reproduce the experiment; that is, to stage multiple tests of the same comparison of control and treatment. Our scientist could perform a *replicated experiment* on, say, 10 pairs of ponds, adding fertilizer to one of the ponds in each pair. It is also important to compare the results of the experiment to an identical circumstance that has not been manipulated; this is called the **control**. In our example, the pond that is left unfertilized serves as the control—the unmanipulated point of comparison for the manipulated or treated pond.

Experiments are not the only means of testing a hypothesis. Sometimes a hypothesis can be convincingly addressed through *correlation*—that is, searching for relationships and patterns among variables. Suppose our scientist surveys 50 ponds, 20 of which are fed by fertilizer runoff from nearby farm fields and 30 of which are not. Let's also say he or she finds seven times as much algal growth in the fertilized ponds as in the unfertilized ponds. The scientist would conclude that algal growth is correlated with fertilizer input—that is, that one tends to increase along with the other—thus supporting the prediction and the hypothesis.

Although generally not as strong as evidence of causal relationships from experimentation, a correlation study is sometimes the best approach; sometimes it is the only feasible approach. For example, in studying the effects of global climate change, we could hardly run an experiment that involved adding carbon dioxide to 10 treatment planets and comparing the result to 10 control planets.

Analyze and interpret results

Scientists record **data**, or pieces of information, from their studies. They particularly value *quantitative* data, information expressed numerically, because numbers provide precision and are easy to compare. The scientist running the fertilization experiment, for instance, might quantify the area of water surface covered by algae in each pond, or measure the dry weight of algae in a certain volume of water taken from each pond. Even with the precision that numbers provide, however, a scientist's results may not be clearcut. Experimental data may differ from control data only slightly, or different replicates may yield different results. The scientist must therefore analyze the data by using statistical tests. With these mathematical methods, scientists can determine objectively and precisely the strength and reliability of the patterns they find. If the results are unreliable or cannot be replicated, it may be necessary to attempt a different kind of test.

Some research, especially in the social sciences, involves information that is *qualitative*, or not easily expressible in terms of numbers. Research involving historical texts, personal interviews, surveys, detailed examination of case studies, or descriptive observations of behaviour can include qualitative data on which statistical analyses may not be possible. Such studies are still scientific in the broad sense, because the data can be interpreted systematically by using other accepted methods of analysis.

If experiments fail to support a hypothesis, the scientist will reject the hypothesis and may develop a new one to replace it. If experiments support the hypothesis, this still does not *prove* it is correct. The scientist may then choose to generate new predictions to test the hypothesis in a different way. Thus, the scientific method loops back on itself, often giving rise to repeated rounds of hypothesis revision, prediction, and testing (see **FIGURE 2.1**).

If repeated tests fail to disprove a particular hypothesis, and evidence in its favour is accumulating, the researcher may conclude that the idea is well supported. Ideally, a scientist would want to test all possible explanations for the question of interest. For instance, our scientist might propose an additional hypothesis that algae increase in fertilized ponds because numbers of fish or invertebrate animals that eat algae decrease. It is possible, of course, that both hypotheses could be correct and that each may explain some portion of the initial observation that local ponds were experiencing algal blooms.

There are different ways to test hypotheses

An experiment in which the researcher actively chooses and manipulates the independent variable is known as a *manipulative experiment* (**FIGURE 2.2A**). Physics and chemistry typically involve manipulative experiments, but many other fields deal with entities less easily manipulated than are physical forces and chemical reagents. This is true of historical sciences, such as cosmology, which deals with the history of the universe, and paleontology, which explores the history of past life. It is difficult to experimentally manipulate a star thousands of light years away, or a mastodon that lived 15 000 years ago.

Disciplines that do not quite fit the "physics model" of science sometimes rely on *natural experiments* rather than manipulative ones (**FIGURE 2.2B**). For instance, an evolutionary biologist might want to test whether animal species isolated on oceanic islands tend to evolve large body sizes over time. The biologist cannot run a manipulative experiment by placing animals on islands and waiting long enough for evolution to do its work; however, this is exactly what nature has already done. The biologist might test the idea by comparing pairs of closely related species in which one of each pair lives on an island and the other on a continental mainland, or one is a modern species and the other an ancient, fossilized relative. The experiment has been conducted naturally, and it is up to the scientist to interpret the results.

In many disciplines, both manipulative and natural experimentation are used. For example, an ecologist wanting to measure the importance of a certain insect in pollinating the flowers of a given crop plant might fit some flowers with a device to keep the insects out while leaving other flowers accessible, and later measure the fruit output of each group. Other ecological questions that involve large spatial scales or long time scales may instead require natural experiments.

The social sciences generally involve less experimentation than the natural sciences, depending more on careful observation and statistical interpretation of patterns in data. For example, a sociologist studying how people from different cultures conceive of the notion of wilderness might conduct a survey and analyze responses to the

FIGURE 2.2
A researcher wanting to test how temperature affects the growth of a crop might run a manipulative experiment in which the crop is grown in two identical greenhouses: one kept at 20°C and the other kept at 25°C **(a)**. Alternatively, the researcher might run a natural experiment in which he or she compares the growth of the crop in two fields at different latitudes: a cool northerly location and a warm southerly one **(b)**. The researcher would most likely collect data on a number of northern and southern fields and correlate temperature and crop growth using statistical methods.

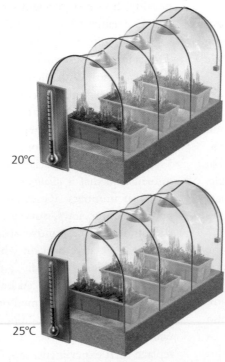
20°C

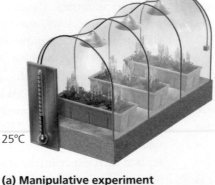
25°C

(a) Manipulative experiment

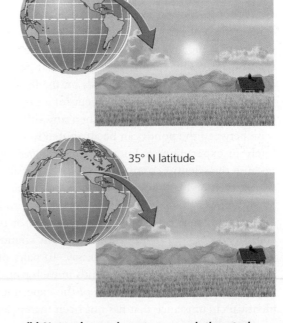
50° N latitude

35° N latitude

(b) Natural experiment, or correlative study

questions, looking for similarities and differences among respondents. Such analyses may be either quantitative or qualitative, depending on the nature of the data and the researchers' particular questions and approaches.

Manipulative experiments provide strong causal information, but natural experiments and correlative studies preserve real-world complexity that manipulative experiments may sacrifice. Because large-scale manipulations are difficult, some of the most important questions in environmental science tend to be addressed with correlative data. The large scale and complexity of many questions in environmental science also mean that few studies, manipulative or correlative, come up with neat and absolute results. As such, scientists are not always able to give policymakers and society definitive answers to questions. Even when science is able to provide answers, deciding upon the optimal social response to a problem can still be very difficult.

The scientific process does not stop with the scientific method

Scientific work takes place within the context of a community of peers, and to have any impact, a researcher's work must be published and made accessible to this community. Thus, the scientific method is embedded within a larger process that takes place at the level of the scientific community as a whole (**FIGURE 2.3**).

Peer review When a researcher's work is done and the results have been analyzed, he or she writes up the findings and submits them to a journal for publication. Other scientists who specialize in the topic examine the manuscript, provide comments and criticism (generally anonymously), and judge whether the work merits publication. This procedure, known as **peer review**, is an essential part of the scientific process.

Peer review is a valuable guard against faulty science contaminating the literature on which all scientists rely. However, because scientists are human and may have their own personal biases and agendas, politics can sometimes creep into the review process. Fortunately, just as the vast majority of individual scientists strive to be accurate and objective in their research, the scientific community does its best to ensure fair review of all work before it can be published in scholarly journals.

In addition to publishing, scientists also frequently present their work at professional conferences, where they interact with colleagues and receive comments informally on their research. Feedback from colleagues can help improve the quality of a scientist's work before it is submitted for publication.

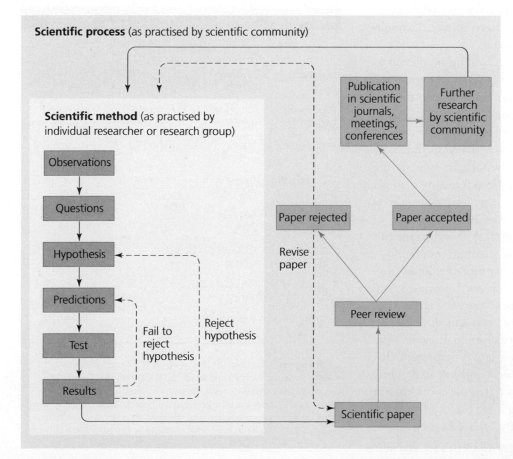

FIGURE 2.3
The scientific method (inner box) followed by individual researchers or teams exists within the context of the overall process of science (outer box) in the scientific community. This process includes peer review and publication of research, acquisition of funding, and the development of theories through the cumulative work and consensus of many researchers.

Grants and funding   Research scientists spend large portions of their time writing grant applications, requesting money to fund their research from private foundations or government agencies such as the Natural Sciences and Engineering Research Council or the Social Sciences and Humanities Research Council in Canada. Grant applications undergo peer review just as scientific papers do, and competition for funding is intense.

Scientists' reliance on funding sources can lead to potential conflicts of interest. A scientist who obtains data showing his or her funding source in an unfavourable light may be reluctant to publish the results for fear of losing funding—or, worse yet, may be tempted to doctor the results. This situation can arise, for instance, when an industry funds research to test its products for safety or environmental impact. Most scientists do not succumb to these pressures, but it can happen. This is why as a student or an informed citizen, when critically assessing a scientific study, you should always try to find out where the researchers obtained their funding.

Repeatability, consensus, and theories   Sound science is based on doubt rather than certainty, and on repeatability rather than one-time occurrences. Even when a hypothesis appears to explain observed phenomena, scientists are inherently wary of accepting it. The careful scientist will test a hypothesis repeatedly in various ways before submitting the findings for publication. Following publication, other scientists usually will attempt to reproduce the results in their own experiments and analyses.

If a hypothesis survives repeated testing by numerous research teams and continues to predict experimental outcomes and observations accurately, it may potentially be incorporated into a theory. A **theory** is a widely accepted, well-tested explanation of one or more cause-and-effect relationships, which has been extensively validated by extensive research. Whereas a hypothesis is a simple explanatory statement that may be refuted by a single experiment, a theory consolidates many related hypotheses that have been tested and supported by a large body of experimental and observational data. A theory represents a very strong consensus within the scientific community.

Note that scientific use of the word *theory* differs from popular usage of the word. In everyday language, when we say something is "just a theory," we are suggesting it is a speculative idea without much substance. Scientists, however, mean just the opposite when they use the term; to them, a theory is a conceptual framework that effectively explains a phenomenon and has undergone extensive and rigorous testing, such that the confidence in it and the consensus surrounding it are extremely strong.

For example, Darwin's *theory of evolution by natural selection* has been supported and elaborated upon by many thousands of studies over 150 years of intensive research. Such research has shown repeatedly and in great detail, through both manipulative and natural experimentation, how plants and animals change over generations, or evolve, to express characteristics that best promote survival and reproduction. Because of its strong support and explanatory power, the evolutionary theory is the central unifying principle of modern biology. Another example is the *theory of plate tectonics*; it took many decades of scientific research before a strong consensus emerged in support of this unifying theory of geology that explains so much about how our physical planet operates.

Matter

The powerful geological processes that shape our planet provide the physical basis for life itself, but they also can lead to tragic events like those in northeastern Japan in 2011. Environmental scientists regularly study these types of physical and geological processes to understand how our planet works. Because all large-scale processes are made up of small-scale components, however, environmental science—the broadest of scientific fields—must

TABLE 2.1   Earth's Most Abundant Chemical Elements, by Mass

Earth's crust	Oceans	Air	Organisms
Oxygen (O), 49.5%	Oxygen (O), 85.8%	Nitrogen (N), 78.1%	Oxygen (O), 65.0%
Silicon (Si), 25.7%	Hydrogen (H), 10.8%	Oxygen (O), 21.0%	Carbon (C), 18.5%
Aluminum (Al), 7.4%	Chlorine (Cl), 1.9%	Argon (Ar), 0.9%	Hydrogen (H), 9.5%
Iron (Fe), 4.7%	Sodium (Na), 1.1%	Other, <0.1%	Nitrogen (N), 3.3%
Calcium (Ca), 3.6%	Magnesium (Mg), 0.1%		Calcium (Ca), 1.5%
Sodium (Na), 2.8%	Sulphur (S), 0.1%		Phosphorus (P), 1.0%
Potassium (K), 2.6%	Calcium (Ca), <0.1%		Potassium (K), 0.4%
Magnesium (Mg), 2.1%	Potassium (K), <0.1%		Sulphur (S), 0.3%
Other, 1.6%	Bromine (Br), <0.1%		Other, 0.5%

DATA Q As seen in this table, the crust is about 49.5% oxygen by mass (or weight-fraction), and about 25.7% silicon by mass. However, oxygen doesn't weigh very much compared to silicon, which has a much higher density (check this by referring to the Periodic Table of the Elements). What would be your prediction as to the most abundant element in the crust *by volume*? Hint: The ionic radius of oxygen is 126 pm and the ionic radius of silicon is 54 pm (1 pm or picometre = 1×10^{-12} m).

Potential pathways for oil to reach bottom sediments

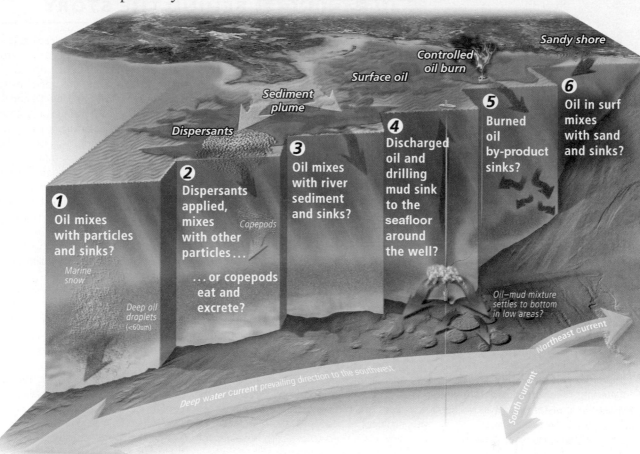

FIGURE 2.4 Matter never just disappears, as much as we might sometimes wish it. Oil from the BP Deepwater Horizon oil spill in the Gulf of Mexico (2010) moved around and changed its form along all of the pathways shown here. All of the oil ended up somewhere—in rock and sediment at the bottom of the gulf, mixed with sediment on the shore, dispersed in the water, ingested by marine organisms, or volatilized into the air
Source: U.S. Department of Commerce/National Oceanic and Atmospheric Administration/Illustration by Kate Sweeney and Mary Baker.

also study small-scale phenomena. At the smallest scale, an understanding of matter itself helps us fully appreciate the physical, geological, chemical, and biological processes of our world.

Matter cannot be created or destroyed

All of the material in the universe that has mass and occupies space, whether solid, liquid, or gaseous, is termed **matter**. Matter may be transformed from one type of substance into others, but it cannot be created or destroyed. This principle is referred to as the law of **conservation of matter**.

In environmental science, this principle helps us understand that the amount of matter stays constant as it is recycled in nutrient cycles and ecosystems. It also makes clear that we cannot simply wish away matter (such as waste and pollution) that we want to eliminate. Any piece

of garbage or drop of spilled oil (**FIGURE 2.4**) or billow of smokestack pollution or canister of nuclear waste that we dispose of will not simply disappear. Instead, we will need to take responsible steps to mitigate its impacts.

Atoms, isotopes, and ions are chemical building blocks

All matter is composed of elements. An **element** is a fundamental type of matter, a chemical substance with a given set of properties, which cannot be broken down into substances that have other properties. Chemists recognize 92 elements occurring in nature, as well as more than 20 others that have been artificially created. Elements that are especially abundant in living organisms include carbon, nitrogen, hydrogen, and oxygen (**TABLE 2.1**). Each element has its own abbreviation, or chemical symbol; for example, the symbol for carbon is C, and the symbol for nitrogen in N. The *Periodic Table of the Elements* summarizes

THE SCIENCE **BEHIND THE STORY**

Scientists use isotopes to investigate a wide variety of Earth materials and processes. Here a scientist collects water samples for isotopic studies that will provide information about the chemical transformation and cycling of nitrogen.

United States Geological Survey (USGS)

How Isotopes Reveal Secrets of Earth and Life

Isotopes are one of the most powerful instruments in the environmental scientist's toolkit. They enable scientists to date ancient materials, reconstruct past climates, and study the lifestyles of prehistoric humans. They also allow researchers to work out photosynthetic pathways, measure animals' diets and health, and trace flows of materials through organisms and ecosystems, both in the past and in the present.

Researchers studying the past often use radiocarbon dating. Carbon's most abundant isotope is ^{12}C, but ^{13}C and ^{14}C also occur in nature. Carbon-14 is a **radioactive isotope**, which means that it decays spontaneously. It occurs in organisms at the same low concentration that it occurs in the atmosphere. Once an organism dies, no new ^{14}C is incorporated into its structure, and the radioactive decay process gradually reduces its store of ^{14}C, converting these atoms to ^{14}N (nitrogen-14).

The decay of ^{14}C is slow and steady, acting like a clock. Scientists can date ancient organic materials by measuring the percent of carbon that is ^{14}C and matching this value against the clocklike progression of decay. In this way, archaeologists and palaeontologists have dated prehistoric human remains; charcoal, grain, and shells found at ancient campfires; and bones and frozen tissues of recently extinct animals, such as mammoths. The most recent ice age has been dated from ^{14}C analysis of trees overrun by glacial ice sheets. Other radioactive isotopes that decay more slowly are used to establish timelines for much older rocks and minerals.

Researchers interested in present-day processes often use **stable isotope** analysis. Unlike radioactive isotopes, stable isotopes occur in nature in constant ratios, and they do not decay radioactively. For instance, nitrogen occurs as 99.63% nitrogen-14 and 0.37% nitrogen-15. These ratios, or *isotopic signatures*, of various environmental materials and processes are diagnostic—almost like fingerprints. For example, organisms tend to retain ^{15}N in their tissues but readily excrete ^{14}N. As a result, animals higher in the food chain show isotopic signatures biased toward ^{15}N, as do animals that are starving.

Keith Hobson, an ecologist with the Prairie and Northern Wildlife Research Centre, Canadian Wildlife Service of Environment Canada, and the University of Saskatchewan, along with scientific colleagues, has used nitrogen or other stable isotopic signatures to analyze the diets, eating habits, and food-chain positions of seabirds and marine mammals, track migration patterns, determine the sources of harvested wildlife, and trace artificial contaminants in food chains.[4]

When animals eat plants, they incorporate the plants' isotopic signatures into their own tissues, and this signature passes up the food chain. Grasses have higher ratios of ^{13}C to ^{12}C than oak trees do, for

information about the properties of the elements in a comprehensive and elegant way.

Elements are composed of **atoms**, the smallest components that maintain the chemical properties of that element (**FIGURE 2.5A**). Every atom has a nucleus consisting of *protons* (positively charged particles) and *neutrons* (particles with no electric charge). The atoms of each element have a defined number of protons, referred to as the element's *atomic number*. Elemental carbon, for instance, has six protons in its nucleus; thus, its atomic number is 6. An atom's nucleus is surrounded by negatively charged particles known as *electrons*, which balance the positive charge of the protons.

All atoms of a given element contain the same number of protons, but not necessarily the same number of neutrons. Atoms of the same element with differing numbers of neutrons are called **isotopes** (**FIGURE 2.5B**). The atomic number of carbon is 6, so all isotopes of carbon have six protons (if they had a different number of protons, they wouldn't be carbon). But different numbers of neutrons are

possible, leading to different mass numbers for the various isotopes of carbon. Isotopes are denoted by their mass number, followed by their elemental symbol. For example, ^{14}C (carbon-14) is an isotope of carbon that has 8 neutrons and 6 protons in the nucleus, rather than the more common 6 neutrons and 6 protons of ^{12}C (carbon-12).

Because they differ slightly in mass, isotopes also differ slightly in their behaviour. This fact has turned out to be very useful for researchers. Scientists have been able to use isotopes to study a number of phenomena that help illuminate the history of Earth's physical environment. Researchers also have used them to study the flow of nutrients within and among organisms and the movement of organisms from one geographical location to another (see "The Science behind the Story: How Isotopes Reveal Secrets of Earth and Life").

Some isotopes are **radioactive** and decay spontaneously, changing their chemical identity as they shed subatomic particles and emit high-energy radiation. These radioisotopes decay into lighter radioisotopes until they instance, whereas cacti have intermediate ratios. As a result, carbon isotope studies can tell ecologists what an animal has been eating. Similarly, archaeologists have used isotopic signatures in human bone to determine when ancient people switched from a hunter-gatherer diet to an agricultural one.

Researchers like Hobson can also use isotopes to track the movements of birds and other animals that migrate thousands of kilometres. This is possible because isotopic signatures in rainfall vary systematically across large geographical regions. This signature gets passed from rain water to plants and from plants to animals, leaving a fingerprint of geographical origin in an animal's tissues. For example, Hobson and colleagues used a combination of isotopic data from hydrogen and carbon to pinpoint the geographical origins of monarch butterflies that had migrated to Mexico, providing important information for their conservation (see **FIGURE 1**).

Stable isotope analysis is used to study many different environmental materials and processes, in the present and the past. There remains much more we can learn from the use of these subtle chemical clues.

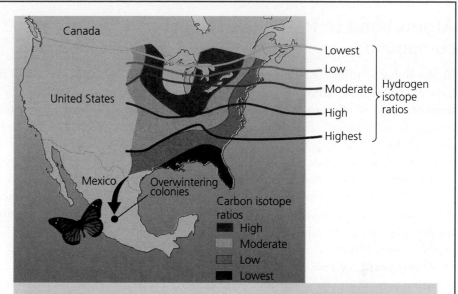

FIGURE 1

Plants in different geographical areas have different carbon and hydrogen isotopic ratios, which caterpillars incorporate into their tissues when they eat the plants. When the caterpillars metamorphose into butterflies and migrate, they carry these isotopic signatures with them. This map of isotopic ratios across eastern North America, from measurements of monarch butterflies, shows decreasing ratios of ^{13}C to ^{12}C (that is, a decrease in the heavier isotope of carbon) from north to south (coloured bands). They also measured increasing ratios of ^{2}H to ^{1}H (that is, an increase in the heavier isotope of hydrogen) from north to south (grey lines). By measuring carbon and hydrogen isotopic ratios in monarchs wintering in Mexico and matching the numbers against this map, researchers were able to pinpoint the geographical origin of many of the butterflies *Source*: Based on Wassenaar, L. I., and K. A. Hobson (1998). *Proceedings of the National Academy of Sciences of the USA, 95*, pp. 15436–15439.

eventually become stable isotopes. Each radioisotope decays at a rate determined by its **half-life**, the amount of time it takes for half of the atoms to decay. Each radioisotope has its own characteristic half-life, and the half-lives of various radioisotopes range from fractions of a second to billions of years. For example, the naturally occurring radioisotope uranium-235 (^{235}U) is the principal source of energy for most commercial nuclear reactors. With a half-life of about 700 million years, ^{235}U decays into a series of daughter isotopes, eventually forming lead-207 (^{207}Pb).

In addition to having stable or radioactive isotopic variants, involving differences in the nucleus, atoms also may gain or lose electrons from their outer shells. In so doing, they become **ions**, electrically charged atoms or combinations of atoms (**FIGURE 2.5C**). Ions are denoted by their elemental symbol followed by their ionic charge. For instance, Ca^{2+} is a calcium atom that has lost two electrons and so has a positive charge of 2 (mussels and clams use this common ion to form shells). Ions that form when an atom loses electrons, and therefore carry a positive charge, are

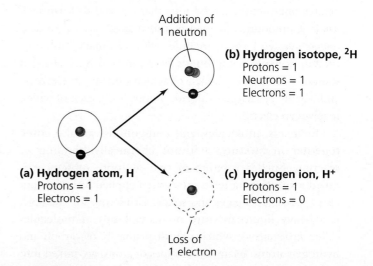

FIGURE 2.5

Hydrogen **(a)** is an element; it has 1 proton + 0 neutron (mass number = 1). Deuterium **(b)**, an isotope of hydrogen, with 1 proton + 1 neutron; it thus has greater mass than a hydrogen atom. Because deuterium has 1 proton, though, it is still a form of hydrogen. If a hydrogen atom loses its 1 electron **(c)**, it gains a positive charge and becomes a hydrogen ion, H^+.

called *cations*. Ions that form when an atom gains electrons, and therefore carry a negative charge, are called *anions*.

Atoms bond to form molecules and compounds

Atoms link chemically to form **molecules**, combinations of two or more atoms. Molecules may contain one element or several. Common molecules that contain only a single element include oxygen (O_2) and nitrogen (N_2), both of which are abundant in air. A molecule composed of atoms of two or more different elements is called a **compound**. Water is a compound; it is composed of two hydrogen atoms bonded to one oxygen atom, denoted by the chemical formula H_2O. Another common compound is carbon dioxide, consisting of one carbon atom bonded to two oxygen atoms; its chemical formula is CO_2.

Atoms **bond**, or combine chemically, because of an attraction for one another's electrons. The strength of this attraction varies among elements, so atoms may be held together in different ways according to whether and how they share or transfer electrons.

When atoms in a molecule share electrons, they generate a *covalent bond*. For instance, two atoms of hydrogen bond to form hydrogen gas, H_2, by sharing electrons equally. Atoms in some covalent bonds share electrons unequally, with one atom exerting a greater pull. Such is the case with water, in which oxygen attracts electrons more strongly than does hydrogen.

If the strength of attraction between the atoms in a compound is sufficiently unequal, an electron may actually be transferred from one atom to another. Such a transfer creates oppositely charged ions that are said to form *ionic bonds*. Compounds formed in this manner are called *ionic compounds*, and these include *salts*. Ordinary table salt (NaCl) contains ionic bonds between positively charged sodium cations (Na^+), each of which donates an electron, and negatively charged chloride anions (Cl^-), each of which receives an electron.

Elements, molecules, and compounds can also come together in mixtures without chemically bonding or reacting. Such a mixture of two or more substances is called a **solution**, a term most often applied to liquids but also applicable to gases and solids. Crude oil, for example, is a heavy liquid mixture of many kinds of molecules called *hydrocarbons*, which consist primarily of carbon and hydrogen atoms. Many other types of atoms are mixed into the solution or bonded to the carbon and hydrogen. Air in the atmosphere is also a solution, composed of molecules of nitrogen (N_2), oxygen (O_2), water (H_2O), carbon dioxide (CO_2), methane (CH_4), ozone (O_3), and many others. Blood, ocean water, mud, plant sap, most minerals, and metal alloys such as brass are all solutions.

The chemical structure of the water molecule facilitates life

Water dominates Earth's surface, covering over 70% of the globe, and its abundance is a primary reason why Earth is hospitable to life. Scientists think life originated in water and stayed there for 3 billion years before moving onto land. Today every land-dwelling creature remains tied to water for its existence.

The water molecule's amazing capacity to support life results from its unique chemical properties. As mentioned previously, the oxygen atom in a water molecule attracts electrons more strongly than do the two hydrogen atoms, resulting in a polar molecule in which the oxygen end has a partial negative charge and the hydrogen end has a partial positive charge. Because of this, water molecules can adhere to one another in a special type of interaction called a *hydrogen bond*, in which the oxygen atom of one water molecule is weakly attracted to the hydrogen atoms of another (**FIGURE 2.6**).

Its chemical structure gives water several properties that are important for supporting life and stabilizing Earth's climate (**FIGURE 2.7**):

■ Water remains liquid over a wide range of temperatures. At Earth's surface, water exists in liquid form from 0°C all the way to 100°C. This means that water-based biological processes can occur in a very wide range of environmental conditions.

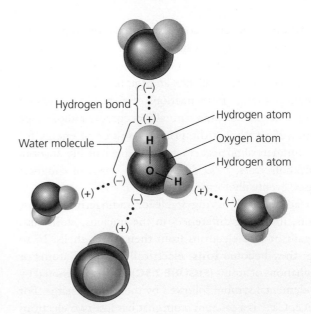

FIGURE 2.6
Water is a unique compound that has several properties crucial for life. Shown here, hydrogen bonds give water cohesion by enabling water molecules to adhere loosely to one another.

Ice

Liquid water

(a) Why ice floats on water

FIGURE 2.7

(a) Ice floats on water because solid ice is less dense than liquid water. This is an unusual property of H_2O—it is far more common for the solid form of a material to be denser than the liquid form. **(b)** Water can dissolve many chemicals, especially polar and ionic compounds. Sea water holds sodium and chloride ions, among others, in solution.

Sodium ion surrounded by negatively charged regions of water molecules

Na⁺

Water molecule

Cl⁻

Chloride ion surrounded by positively charged regions of water molecules

Salt (sodium chloride, NaCl)

(b) Water as a solvent; how water dissolves salt

■ Water exhibits strong cohesion. (Think of how water holds together in drops and how the drops on a surface join together when you touch them to one another.) This facilitates the transport of chemicals, such as nutrients and waste, in plants and animals and in the physical environment.

■ Water has a high heat capacity, which means that it can absorb a large amount of heat with only small changes in its temperature. This quality helps stabilize systems against change, whether those systems are organisms, ponds, lakes, or climate systems.

■ Water molecules in ice are farther apart than in liquid water (**FIGURE 2.7A**), so ice is less dense than liquid water—the reverse of most other compounds, which become denser as they freeze. In ice, each molecule is connected to neighbouring molecules by stable

hydrogen bonds that form a spacious crystal lattice. In liquid water, the molecules are closer together and less well organized. This explains why ice floats on water. Floating ice is important because it has an insulating effect that can prevent water bodies from freezing solid in winter.

■ Fresh water reaches its maximum density at 4°C; water cooled to this temperature increases in density, but when cooled below this temperature it becomes gradually less dense. This means that surface water cooled below 4°C by winter air will tend to remain at the surface, promoting the formation of ice and further insulating deeper waters from the frigid air.

■ Water molecules bond well with other polar molecules because the positive end of one molecule bonds readily to the negative end of another. As a result, water can hold in solution, or dissolve, many other molecules, including chemicals necessary for life (**FIGURE 2.7B**). Most biologically important solutions involve water, and we sometimes informally refer to water as the "universal solvent," meaning that it is able to dissolve many things.

■ Another important characteristic of water is its transparency to light; without this feature there would be no photosynthesis, which is an ultimate source of energy for almost all organisms living in water.

Hydrogen ions control acidity

In any *aqueous solution* (a solution in which water is present as a solvent), a small number of water molecules *dissociate*, or split apart, each forming a hydrogen ion (H^+) and a hydroxide ion (OH^-). The product of hydrogen and hydroxide ion concentrations is always the same; as the concentration of one increases, the concentration of the other decreases. Pure water contains equal numbers of these ions, and we refer to it as *neutral*. Most aqueous solutions, however, contain different concentrations of these two ions. Solutions in which the H^+ concentration is greater than the OH^- concentration are called *acidic*; the stronger the acid, the more readily dissociation occurs and

H^+ ions are released. Solutions in which the OH^- concentration is greater than the H^+ concentration are called *basic*.

The **pH** scale (**FIGURE 2.8**) quantifies the acidity or basicity of solutions. The scale runs from 0 to 14; pure water (neutral) has a hydrogen ion activity of 10^{-7} and it is said to have a pH of 7. Solutions with pH less than 7 are acidic; those with pH greater than 7 are basic. The pH scale is logarithmic, so each step on the scale represents a tenfold difference in hydrogen ion concentration. **FIGURE 2.8** shows the pH for a number of common substances. For example, normal rain water is slightly acidic, but industrial air pollution has intensified the acidity of precipitation. As a result, the pH of rain in parts of south-central Canada and the northeastern and midwestern United States now frequently dips to 4 or lower.

Matter is composed of organic and inorganic compounds

Beyond their need for water, living things also depend on organic compounds, which they create and of which they are composed. **Organic compounds** consist of carbon atoms (and generally hydrogen atoms) joined by covalent bonds, often with other elements, such as nitrogen, oxygen, sulphur, and phosphorus. Carbon's unusual ability to build elaborate molecules has resulted in millions of different organic compounds, many of which are highly complex.

Chemists differentiate organic compounds from **inorganic compounds**, which also are fundamentally important in the support of life. (Water is one example of a crucial inorganic compound.) Some inorganic compounds may contain carbon, but they are not organic because they lack the carbon–carbon bonds that are characteristic of organic compounds.

It is important to remember that in scientific terminology, *organic* doesn't mean "natural" or "environmentally friendly" or "pesticide-free," as we have come to use it in everyday language. The term *organic* doesn't even imply that a compound is or was once alive—it simply refers to the presence of carbon–carbon bonding in the chemical compound.

Most biological materials are composed of organic compounds called **hydrocarbons**, which consist primarily of carbon and hydrogen (although other elements may enter the compounds, often as impurities). The simplest hydrocarbon is **methane (CH_4)**, the key component of natural gas; it has one carbon atom bonded to four hydrogen atoms (**FIGURE 2.9A**). Adding another carbon atom and two more hydrogen atoms gives us ethane (C_2H_6), the next-simplest hydrocarbon (**FIGURE 2.9B**). The smallest (and therefore lightest-weight) hydrocarbons exist in a gaseous state at normal temperatures and

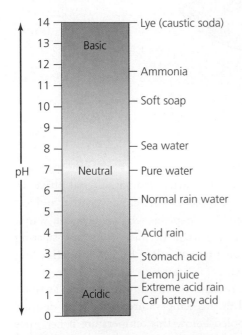

FIGURE 2.8

The pH scale measures how acidic or basic a solution is. The pH of pure water is set as 7, the midpoint of the scale. Acidic solutions have higher hydrogen ion (H^+) concentrations and lower pH, whereas basic solutions have lower hydrogen ion (H^+) concentrations and higher pH.

DATA Q Noting that the pH scale is logarithmic, what concentration of hydrogen ions would be present in a solution with pH of 6 *relative to* a solution with pH of 7? What about pH of 5 compared to pH of 7?

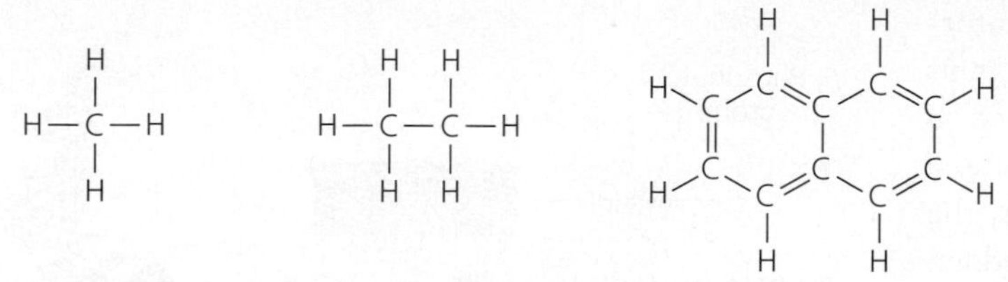

(a) Methane, CH₄   **(b) Ethane, C₂H₆**   **(c) Naphthalene, C₁₀H₈**

FIGURE 2.9
The simplest hydrocarbon is methane **(a)**. Many hydrocarbons consist of linear chains of carbon atoms with hydrogen atoms attached; the shortest of these is ethane **(b)**. Volatile hydrocarbons with multiple rings, such as naphthalene **(c)**, are called polycyclic aromatic hydrocarbons (PAHs).

pressures. Larger (and therefore heavier) hydrocarbons are liquids, and those with more than 20 carbon atoms are normally solids.

Some hydrocarbons pose health hazards to wildlife and people. For example, the polycyclic aromatic hydrocarbons, or PAHs (**FIGURE 2.9C**), are volatile molecules with a structure of multiple carbon rings. These toxic compounds can evaporate from spilled oil and gasoline, and they can mix with water. The eggs of fish and young offspring of other aquatic creatures, such as beluga whales, are often most at risk (see "Central Case: The Plight of the St. Lawrence Belugas," in the chapter on environmental systems and ecosystems). PAHs also occur in particulate form in various combustion products, including cigarette smoke, wood smoke, and charred meat.

Macromolecules are building blocks of life

Organic compounds sometimes combine to form long chains of repeated molecules. Some of these chains, called *polymers*, play key roles as building blocks of life. Three types of polymers are essential to life: proteins,

nucleic acids, and carbohydrates. Lipids, not considered to be polymers, are also fundamental to life. These four important types of molecules are referred to as *macromolecules* because of their large size.

Proteins consist of long chains of organic molecules called *amino acids* (**FIGURE 2.10A**). Organisms combine up to 20 different types of amino acids into long chains to build proteins (**FIGURE 2.10B**), with parts of the chain exposed and others hidden inside complex folds (**FIGURE 2.10C**). The folding pattern affects the protein's function because the position of each chemical group helps determine how it interacts with other molecules. Some proteins help produce tissues, such as skin, hair, muscles, and tendons, and provide structural support for the organism. Some help store energy, and others transport substances. Some act in the immune system to defend the organism against foreign attackers. Still others act as hormones, molecules that serve as chemical messengers within an organism. Finally, proteins can serve as enzymes, molecules that catalyze, or promote, certain chemical reactions, such as digestion.

Nucleic acids direct the production of proteins. Two nucleic acids—**deoxyribonucleic acid (DNA)** and

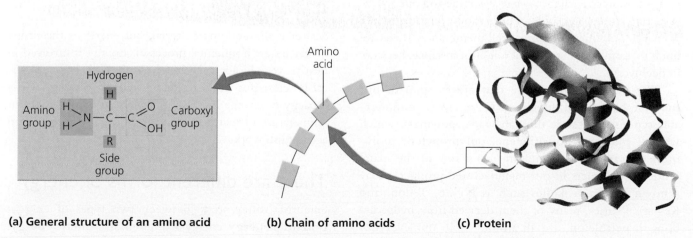

(a) General structure of an amino acid   **(b) Chain of amino acids**   **(c) Protein**

FIGURE 2.10 Proteins are polymers that are vital for life. They are made up of long chains of amino acids **(a, b)** that form complex shapes **(c)** that help determine their functions.

ribonucleic acid (RNA)—carry the hereditary information for organisms and are responsible for passing traits from parents to offspring. Nucleic acids are composed of *nucleotides,* structural units made of a sugar molecule, a phosphate group, and a nitrogenous base. In DNA, the nucleotides link together in a double strand, like a ladder, that is twisted into a spiral (**FIGURE 2.11**). Hereditary information is encoded in the sequencing of the nucleotides that form the rungs of the ladder. Regions of DNA coding for particular proteins that perform particular functions (that is, segments of the ladder) are called **genes**. RNA molecules use this genetic information to direct the order in which amino acids assemble to build proteins, in turn influencing the structure, growth, and maintenance of the organism.

Carbohydrates, a third type of biologically important macromolecule, consist of atoms of carbon, hydrogen, and oxygen. Glucose ($C_6H_{12}O_6$), a simple sugar, is one of the most common and important carbohydrates, providing energy that fuels plant and animal cells. Glucose also serves as a building block for complex carbohydrates, such as starch. Plants use starch to store energy, and animals eat plants to acquire starch. Both plants and animals use complex carbohydrates to build structure. Insects and crustaceans form hard shells from the carbohydrate chitin. *Cellulose,* the most abundant organic compound on Earth, is a complex carbohydrate found in the cell walls of leaves, bark, stems, and roots of plants, as well as in some bacteria, fungi, and algae.

A fourth important type of macromolecule is the chemically diverse group of compounds called **lipids**, classified together because they do not dissolve in water. These include *fats* and *oils,* which are convenient forms of energy storage, especially for mobile animals. Their hydrocarbon structures resemble gasoline, a similarity echoed in their function: to effectively store energy and release it when burned. *Phospholipids* are similar to fats but have one water-repellent (*hydrophobic*) side and one water-attracting (*hydrophilic*) side, which allows them to be the primary component of animal cell membranes. *Waxes* are lipids that often play structural roles (for instance, beeswax in beehives).

The polymers in nature that are so vital to our survival have inspired chemists to create innumerable types of synthetic (human-made) polymers, which we call **plastics**. Polyethylene, polypropylene, polyurethane, and polystyrene are just a few of the many synthetic polymers in our manufactured products today (known by brand names such as Nylon, Teflon, and Kevlar). Plastics, many of them derived from hydrocarbons in petroleum, are all around us in our everyday lives, from furniture to food containers to fibre optics to fleece jackets.

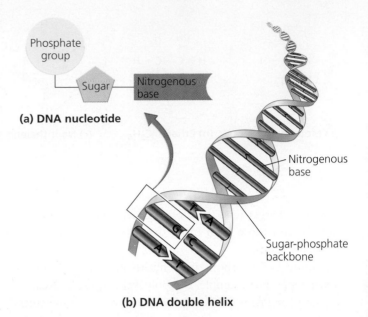

(a) DNA nucleotide

(b) DNA double helix

FIGURE 2.11
Nucleic acids are macromolecules that encode genetic information in the sequence of nucleotides, small molecules that pair together like rungs of a ladder. DNA includes four types of nucleotides **(a)**: adenine (A), guanine (G), cytosine (C), and thymine (T). The ladderlike pairs form a twisted shape called a double helix **(b)**.

We value synthetic polymers because they are versatile and they resist chemical breakdown. Although plastics make our lives easier, the waste and pollution they create when we discard them is long-lasting as well. In future chapters we will see how pollutants that resist breakdown can cause problems for wildlife and human health, for water quality, for marine animals, and for waste management. Fortunately, chemists, policymakers, and citizens are finding more ways to design and use less-polluting substances and to recycle materials effectively.

Energy

Matter is the building material, but energy is the driver of Earth's environmental processes. Energy is involved in nearly every biological, chemical, physical, and geological process. But what, exactly, is energy? **Energy** is the capacity to change the position, physical composition, or temperature of matter—in other words, a force that can accomplish work.

There are different forms of energy

Scientists differentiate between two types of energy: **potential energy**, energy of position; and **kinetic energy**, energy of motion. Consider river water held behind a dam. By preventing water from moving downstream, the dam

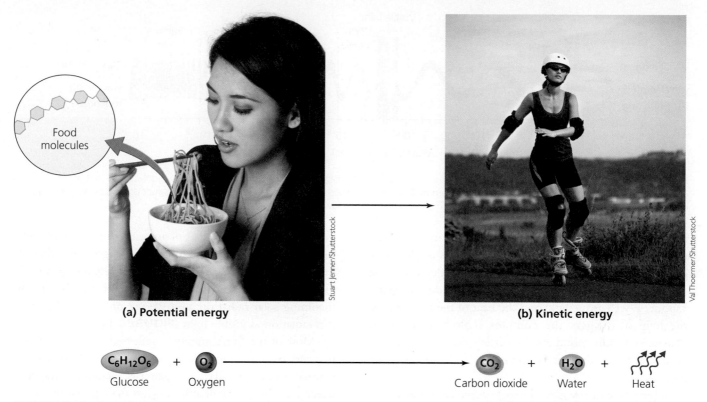

(a) Potential energy

(b) Kinetic energy

FIGURE 2.12 Energy is released when potential energy is converted to kinetic energy. Potential energy stored in sugars, such as glucose, in the food we eat **(a)**, combined with oxygen, becomes kinetic energy when we exercise **(b)**, releasing carbon dioxide, water, and heat as by-products.

causes the water to accumulate potential energy. When the dam gates are opened, the potential energy is converted to kinetic energy in the form of water's motion as it rushes downstream.

Such energy transfers also take place at the atomic level every time a chemical bond is broken or formed. Chemical energy is potential energy held in the bonds between atoms. Converting a molecule with high-energy bonds (such as the carbon–carbon bonds of sugars) into molecules with lower-energy bonds (such as the bonds in water or carbon dioxide) releases energy by changing the potential energy into kinetic energy, which produces motion, action, or heat. Just as automobile engines split the hydrocarbons of gasoline to release chemical energy and generate movement, our bodies split glucose molecules in food for the same purpose (**FIGURE 2.12**).

Besides occurring as chemical energy, potential energy can occur as nuclear binding energy (the energy that holds atomic nuclei together and is released when an atom is split), and as stored mechanical energy (the energy in a compressed spring, or a tree bending in the wind, for example). Kinetic energy also takes a variety of forms. These include thermal energy, light energy, electrical energy, and sound energy, all of which involve movement of electrons, atoms, molecules, or objects.

Energy is always conserved but can change in quality

Although energy can change from one form to another, it cannot be created or destroyed. The total energy in the universe remains constant and thus is said to be conserved. Scientists call this principle the *first law of thermodynamics*. The potential energy of the water behind a dam will equal the kinetic energy of its eventual movement down the riverbed. Similarly, burning converts the potential energy in a log of firewood to an equal amount of energy produced as heat and light. We obtain energy from the food we eat, which we expend in exercise, put toward the body's maintenance, or store as fat; we do not somehow create additional energy or end up with less than the food gives us. Any individual system can temporarily increase or decrease in energy, but the total amount in the universe remains constant.

Although the overall energy is conserved in any energy transfer, the *second law of thermodynamics* states that the nature or quality of the energy will change from a more-ordered state to a less-ordered state if no force counteracts this tendency. That is, systems tend to move toward increasing disorder (termed *entropy*). For instance, after death every organism undergoes decomposition and loses

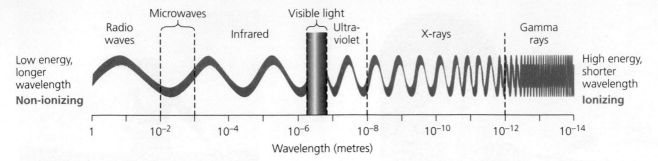

FIGURE 2.13 The Sun radiates energy from many portions of the electromagnetic spectrum. Visible light makes up only a small part of this energy. Some radiation that reaches our planet is reflected back; some is absorbed by air, land, and water; and a small amount powers photosynthesis. High-energy radiation (right) can strip electrons from atoms, causing ionization; lower-energy radiation does not do this.

its structure. A log of firewood—the highly organized and structurally complex product of many years of slow tree growth—transforms in the campfire to a residue of carbon ash, smoke, and gases such as carbon dioxide and water vapour, as well as the light and the heat of the flame. With the help of oxygen, the complex biological polymers that make up the wood are converted into a disorganized assortment of rudimentary molecules and into heat and light energy.

The nature of any given energy source helps determine how easily humans can harness it. Sources such as fossil fuels and the fuels used in nuclear power plants contain concentrated energy; it is relatively easy to withdraw large amounts of energy efficiently from these high-quality sources. In contrast, sunlight and the heat stored in ocean water are more diffuse energy sources. Each day the world's oceans absorb heat energy from the Sun equivalent to that of 250 billion barrels (about 40 trillion litres) of oil—more than 3000 times as much as our global society uses in a year. But because this energy is spread out across such vast spaces, it is difficult to harness effectively.

In every transfer of energy, some energy is lost—not destroyed, but converted into a less-usable form. The inefficiency of some of the most common energy conversions that power our society can be surprising. When we burn gasoline in an automobile engine, only about 16% of the energy released is used to power the automobile's movement; the rest is converted to heat and escapes without being used. Incandescent light bulbs are even less efficient; only 5% of their energy is converted to the light that we use them for, while the rest is lost as heat. Viewed in this context, the 15% efficiency of some current solar technologies and 7% to 15% efficiency of geothermal power plants look pretty good.

Light energy from the Sun powers most living systems

The energy that powers Earth's environmental systems comes primarily from the Sun. The Sun releases energy as radiation across large portions of the **electromagnetic spectrum**, the range of wavelengths of radiation from shortest (gamma radiation) to longest (radio waves). However, our atmosphere filters out much of the incoming solar radiation, and we see only a small part of this radiation as visible light (**FIGURE 2.13**).

Most of the Sun's energy is reflected or absorbed and re-emitted by the atmosphere, land, or water. Solar energy drives our weather and climate patterns, including winds and ocean currents. A small amount (less than 1% of the total) powers plant growth, and a still smaller amount flows from plants into the organisms that eat them and the organisms that decompose dead organic matter. A minuscule amount of energy, relatively speaking, is eventually deposited below ground in the form of the chemical bonds in fossil fuels.

The Sun's light energy is used directly by some organisms to produce their own food. Organisms that produce their own food energy, called **autotrophs** or **primary producers**, include green plants, algae, and cyanobacteria (a type of bacteria named for their characteristic blue-green, or cyan, colour). *Photoautotrophs* turn light energy from the Sun into chemical energy via the process of **photosynthesis** (**FIGURE 2.14**). In photosynthesis, sunlight powers a series of chemical reactions that convert carbon dioxide and water into sugars, transforming low-quality energy from the Sun into high-quality energy the organism can use.

Photosynthesis is a complex process, but the overall reaction can be summarized with the following equation:

$$\text{Photosynthesis: } 6CO_2 + 12H_2O + \text{solar energy} \rightarrow C_6H_{12}O_6 \text{ (glucose)} + 6O_2 + 6H_2O$$

The numbers that precede each molecular formula indicate how many molecules of each type are involved in the reaction. Note that the sums of the numbers on each side of the equation for each element are equal: There are 6 C, 24 H, and 24 O on each side. This illustrates how chemical equations are balanced, with each atom recycled and matter conserved. No atoms are lost; they are simply rearranged into different molecules. Note also that water appears on both sides of the equation. The reason is that for

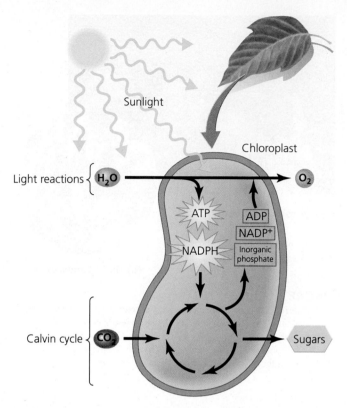

FIGURE 2.14

In photosynthesis, autotrophs such as plants, algae, and cyanobacteria use sunlight to convert carbon dioxide and water into sugars and oxygen. This diagram summarizes the complex sets of chemical reactions involved. Water is converted to oxygen in the presence of sunlight, creating high-energy molecules (ATP and NADPH). These help drive reactions in the Calvin cycle, in which carbon dioxide is used to produce sugars.

every 12 water molecules that are dissociated in the process, 6 water molecules are newly created. We can streamline the equation by showing only the net loss of 6 water molecules:

Photosynthesis: $6CO_2 + 6H_2O$ + solar energy $\rightarrow$
$$C_6H_{12}O_6 \text{ (glucose)} + 6O_2$$

Thus in photosynthesis, water, carbon dioxide, and light energy from the Sun are transformed to produce sugar (glucose) and oxygen. To accomplish this, green plants draw up water from the ground through their roots, absorb carbon dioxide from the air through their leaves, and harness sunlight. With these ingredients, they create sugars for their growth and maintenance and release oxygen as a by-product.

Animals, in turn, depend on oxygen and sugars from photosynthesis (**FIGURE 2.15**). Animals survive by being **consumers**, or **heterotrophs**, organisms that gain their energy by feeding on other organisms. They eat plants (thus becoming *primary consumers*), or they eat animals that have eaten plants (thus becoming *secondary consumers*), and they take in oxygen. Animals, plants, and all organisms other than bacteria appeared on Earth's surface only after the atmosphere had been supplied with oxygen by cyanobacteria, the earliest photoautotrophs.

Utilizing the chemical energy created during photosynthesis requires a process called **cellular respiration**. Photosynthesis and respiration belong to a wide range of life-supporting and growth-supporting biochemical processes that happen within organisms, known collectively as *metabolism*. To release the chemical energy of glucose, cells use the reactivity of oxygen to convert glucose back into its original starting materials; in other words, respiration can be thought of as a "reverse photosynthesis"—it oxidizes glucose to produce carbon dioxide and water. The net equation for cellular respiration is thus the exact opposite of that for photosynthesis:

Respiration: $C_6H_{12}O_6$ (glucose) $+ 6O_2 \rightarrow$
$$6CO_2 + 6H_2O + \text{energy}$$

The energy gained per glucose molecule in respiration is only two-thirds of the energy input per glucose molecule in photosynthesis—a prime example of the second law of thermodynamics. Cellular respiration is a continuous process that occurs in all living things; it is

FIGURE 2.15

When an animal such as this deer eats the leaves of a plant, it consumes the sugars and starches the plant produced through photosynthesis and gains energy from those compounds.

essential to life. It occurs both in the autotrophs that create sugars and in the heterotrophs that eat them.

Geothermal energy also powers Earth's systems

Although the Sun is Earth's primary power source, it is not the only one. A minor additional source is the gravitational pull of the Moon, which causes ocean tides. A more significant additional energy source is **geothermal energy**, or *terrestrial energy*—heat that emanates from Earth's interior.

Geothermal energy is powered primarily by radioactivity. When we think of radioactivity, nuclear power plants and atomic weapons may come to mind, but it is a natural phenomenon that involves the release of radiation by radioisotopes as their nuclei spontaneously decay. Radiation from naturally occurring radioisotopes deep inside Earth heats the inside of the planet, and this heat gradually makes its way to the surface. This internal heat energy drives geological processes, heats magma that erupts from volcanoes, and warms ground water and thermal hot springs (**FIGURE 2.16**). Geothermal heat from within the planet is now being harnessed for commercial power, especially in locations where it is particularly concentrated at the surface.

Long before humans came along, geothermal energy was powering other biological communities. In certain locations on the ocean's floor, jets of geothermally heated water—essentially underwater geysers—gush into the icy-cold depths. One of the amazing scientific discoveries of recent decades was the realization that these heated-water, or *hydrothermal*, vents can host entire communities of organisms that thrive in the extreme high-

(a) Hydrothermal vent

(b) Giant tubeworms

FIGURE 2.17

Hydrothermal vents on the ocean floor **(a)** send spouts of hot, mineral-rich water into the cold blackness of the deep sea. Specialized biological communities thrive in these unusual conditions. Odd creatures such as these giant tubeworms, *Riftia pachyptila* **(b)**, survive thanks to bacteria that produce food from hydrogen sulphide by chemosynthesis.

temperature, high-pressure conditions. Gigantic clams, immense tubeworms, and odd mussels, shrimps, crabs, and fish all flourish in the seemingly hostile environment near scalding water that shoots out of tall chimneys of encrusted minerals (**FIGURE 2.17**).

These locations are so deep underwater that they completely lack sunlight, so their communities cannot fuel themselves through photosynthesis. Instead, bacteria in deep-sea vents use the chemical-bond energy of hydrogen sulphide (H_2S) to transform inorganic carbon into organic carbon compounds in a process called **chemosynthesis**. There are many types of chemosynthesis, but note how this particular reaction closely resembles the photosynthesis reaction:

Chemosynthesis: $CO_2 + 6H_2O + 3H_2S \rightarrow$
$$C_6H_{12}O_6 \text{ (glucose)} + 3H_2SO_4$$

The two processes use different energy sources, but each combines water and carbon dioxide to produce a sugar and a by-product, and each produces potential energy that is later released during cellular respiration. Energy from chemosynthesis passes through the deep-sea-vent animal community as heterotrophs such as clams, mussels, and shrimp gain nutrition from *chemoautotrophic* bacteria that use the oxidation of chemicals like hydrogen sulphide as a source of energy to generate organic matter. Seafloor hydrothermal vent communities have excited scientists not only because they were novel and unexpected, but also because they may hold clues to the early Earth, and might even help us answer the question of how life itself originated.

FIGURE 2.16

These thermal pools at Rabbitkettle Hot Springs in Nahanni National Park Reserve, Northwest Territories, are heated year-round by geothermal energy from deep below the ground.

Geological Systems: The Physical Basis for the Environment

If we want to understand how our planet functions, a good way to start is to examine the rock, soil, and sediment beneath our feet. The physical processes that take place at and below Earth's surface shape the landscape around us and lay the foundation for most environmental systems and for life. Understanding the physical nature of our planet also benefits society, for without the study of Earth's rocks and the processes that shape them, we would have no energy from geothermal sources or from fossil fuels. We are constantly drawing on resources, materials, and processes from beneath the surface of our planet and putting them to use in our everyday lives.

Earth consists of layers

Most geological processes take place near Earth's surface, but our planet consists of multiple layers (**FIGURE 2.18**). At the planet's centre is a dense **core** consisting mostly of iron, solid in the inner core and molten in the outer core. Earth's core is intensely hot—perhaps 6000°C, according to recent studies.[5] As discussed previously, this is partly because of the presence of radioactive materials that are releasing energy as they decay, but it also results from the intense pressure of the materials at this great depth in the planet.

Surrounding the core is a thick layer of rock called the **mantle**. A portion of the upper mantle, called the *asthenosphere*, contains softer rock, which is close to its melting temperature and actually molten in some areas. The harder rock above the asthenosphere is what we know as the **lithosphere**. The lithosphere includes the uppermost mantle and the **crust**, the thin, brittle, low-density layer of rock that covers Earth's surface, and on which we live.

Our planet is dynamic, and this dynamism motivates *geology*, the study of Earth's physical features, processes, and history. A human lifetime is just a blink of an eye in the long course of geological time, and Earth as we currently experience it is merely a snapshot in our changing planet's long history. We can begin to grasp this long-term dynamism as we consider two processes of fundamental importance: plate tectonics and the rock cycle.

Plate tectonics shapes the geography of oceans and continents

The intense heat from inside the planet rises from core to mantle to crust and eventually dissipates at the surface.

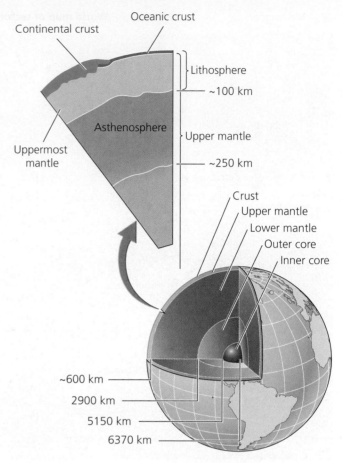

FIGURE 2.18
Earth's three primary layers—core, mantle, and crust—are themselves internally layered. The inner core of solid iron is surrounded by an outer core of molten iron. The rocky mantle has several layers (not shown here), including the relatively soft asthenosphere in the upper part. The lithosphere consists of the uppermost mantle (above the asthenosphere) and the crust. The crust is of two major types: dense, thin oceanic crust; and less-dense, thicker continental crust.

This heat drives great movements in the mantle, pushing the mantle rock upward (as it warms) and downward (as it cools), like a gigantic conveyor belt system. This process, called **convection**, is seen wherever materials move around as a result of heating and cooling; for example, we will encounter convection currents again in both the atmosphere and the ocean. As the rock of the mantle moves, it drags large blocks of brittle lithosphere, known as *plates*, along its surface. This movement of lithospheric plates is **plate tectonics**, a process of extraordinary importance to our planet.

Our planet's surface consists of about 15 major tectonic plates, which fit together like pieces of a jigsaw puzzle (**FIGURE 2.19**). Imagine peeling an orange and then placing the pieces of peel back onto the fruit; the ragged pieces of peel are like the lithospheric plates that form Earth's surface. However, the plates are even thinner relative to the planet's size—more analogous to the thickness of the skin of an apple, or the glass of a lightbulb.

World map of tectonic plates

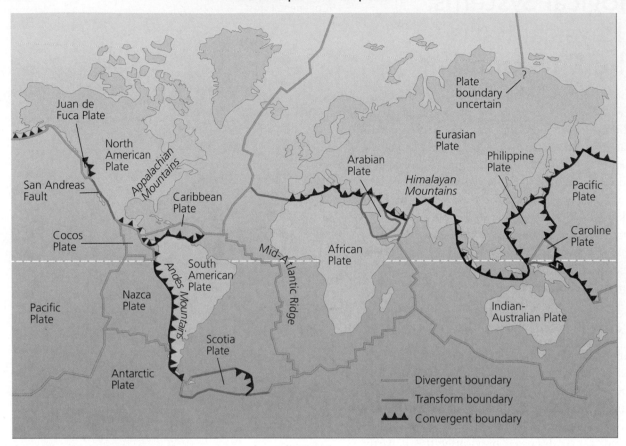

FIGURE 2.19 Earth's lithosphere has broken into approximately 15 major plates (shown here) and a number of minor plates that all move very slowly relative to one another as part of the process of plate tectonics. Along divergent boundaries, plates move apart from each other; along transform boundaries, they move past one another laterally; and along convergent boundaries, plates move toward one another.

The lithospheric plates move at rates of roughly 2 to 10 cm per year, driven by convection currents in the mantle. This slow movement has influenced Earth's climate and life's evolution throughout our planet's history as the continents combined, separated, and recombined in various configurations. By studying ancient rock formations throughout the world, geologists have determined that at least twice in Earth's history nearly all landmasses were joined together in a "supercontinent," before splitting and moving apart from one another once again.

There are three main types of plate boundaries

The processes that occur at the boundaries between plates have major consequences. We can categorize these boundaries into three types: divergent, transform, and convergent (**FIGURE 2.20**).

At **divergent plate boundaries**, tectonic plates move apart from one another, or diverge. In these locations, **magma** (rock heated to a molten, liquid state) rises upward to the surface, forming new crust as it

cools and solidifies (**FIGURE 2.20A**). An example is the Mid-Atlantic Ridge, part of a 74 000–km-long system of divergent plate boundaries. Iceland is a volcanic island built when magma from underwater volcanoes along the Mid-Atlantic Ridge was extruded above the ocean surface and cooled. Iceland possesses numerous active volcanoes, including the Eyjafjallajökull and Grimsvötn volcanoes, both of which disrupted air travel extensively by erupting large clouds of ash in 2010 and 2011, respectively. In Iceland, geothermally heated water is piped throughout towns and cities, and nine out of ten people heat their homes with geothermal energy.

A **fault** is a fracture in Earth's crust, along which the blocks of rock on either side are displaced relative to one another. Where two plates meet along a strike-slip fault, they slip and grind alongside one another horizontally in opposite directions. Plate boundaries that are marked by strike-slip faults are called **transform plate boundaries** (**FIGURE 2.20B**). The horizontal movement of the plates creates friction that typically generates earthquakes. The Pacific Plate and the North American Plate are sliding past one another along the San Andreas Fault,

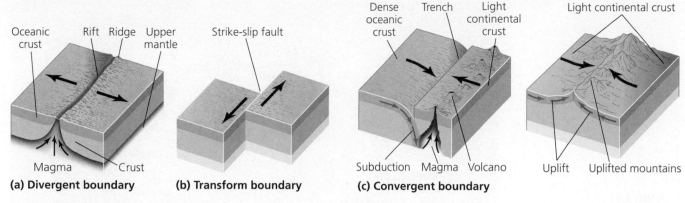

FIGURE 2.20   There are three basic types of plate boundaries: divergent, transform, and convergent. At a divergent boundary, such as a mid-ocean ridge **(a)**, the two plates move gradually away from the boundary, and magma rises to the surface through cracks along the boundary. At a transform plate boundary **(b)**, two plates slide alongside each other, creating friction that leads to earthquakes. Where plates collide at a convergent plate boundary involving oceanic crust **(c)**, one plate may be subducted beneath another, leading to volcanism. If only continental plates are involved in the convergence, high mountain ranges are formed.

which runs roughly north–south along the west coast of California. Southern California is slowly creeping toward the northwest relative to the rest of North America.

Convergent plate boundaries occur where two plates come together, that is, converge or collide (**FIGURE 2.20C**). Along such plate boundaries, some of the world's most powerful earthquakes take place, including the Tohōku earthquake discussed in the Central Case.

Oceanic crust consists mainly of the rock type *basalt*, which is denser than the *granite* that composes much of the continental crust. Whenever oceanic crust is involved in a convergent plate boundary, a process called **subduction** occurs, in which the dense oceanic plate descends into the mantle. The subducted plate is heated and pressurized as it sinks. Water vapour escapes from the subducting plate, helping to melt the overlying rock by lowering its melting temperature. The molten rock rises, and this magma may erupt through the surface in the form of volcanoes.

When a plate of oceanic lithosphere subducts beneath another plate of oceanic lithosphere, the resulting volcanism may form lines of volcanic islands, called a *volcanic arc*; examples include Japan and the Aleutian Islands. Subduction also creates deep trenches, such as the Mariana Trench in the western Pacific, our planet's deepest abyss—at 10 971 m deep, it is significantly deeper than Mount Everest is high.

When oceanic lithosphere subducts beneath less dense continental lithosphere, it leads to the formation of volcanic mountain ranges that parallel the coastline (**FIGURE 2.20C**, left). The Cascades, where Mount Saint Helens erupted violently in 1980 and renewed its activity in 2004, are fuelled by magma from the subduction of the Juan da Fuca Plate under the North American Plate along the coast of British Columbia, Washington, and Oregon. Another example is South America's Andes Mountains, where the Nazca Plate is subducting beneath the South American Plate.

When two plates of continental lithosphere converge, the continental crust on both sides resists subduction and instead crushes together, bending, buckling, and deforming layers of rock from both plates in a **continental collision** (**FIGURE 2.20C**, right). Portions of the accumulating masses of buckled crust are forced upward as they are pressed together, and mountain ranges result. The Himalayas, the world's highest mountains, result from the collision of the Indian-Australian Plate with the Eurasian Plate, which began 40 to 50 million years ago; these mountains are still rising today as the plates continue to converge. The Appalachian Mountains of eastern North America, once the world's highest mountains themselves, resulted from several ancient collisions with the edge of what is today Africa.

Tectonic processes build Earth's landforms

The processes of plate tectonics create the landforms around us by building mountains; shaping the geography of oceans, islands, and continents; and giving rise to earthquakes and volcanoes. They also help determine the locations of geothermal energy resources and the formation of ore deposits of various types.

The topography created by tectonic processes, in turn, shapes climate systems by altering patterns of rainfall, wind, ocean currents, heating, and cooling—all of which affect rates of weathering and erosion and the ability of plants and animals to inhabit different regions. Thus, the locations of plant and animal communities are fundamentally influenced by plate tectonics. Tectonics has even affected the history of life's evolution; for example, the convergence of landmasses into supercontinents is thought to have contributed to widespread extinctions by reducing the area of species-rich coastal regions and by creating arid continental interiors with extreme temperature swings.

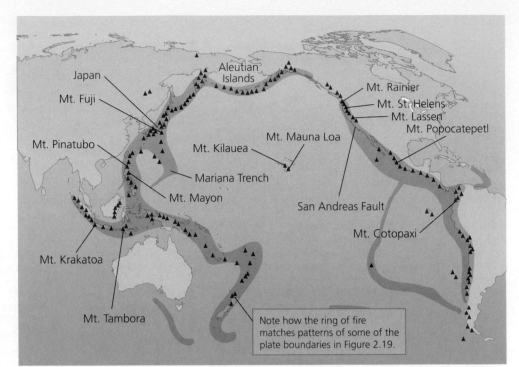

FIGURE 2.21
Many of our planet's earthquakes and volcanic eruptions occur along the circum-Pacific "ring of fire," the system of subduction zones and other plate boundaries that encircles the Pacific Ocean. The red symbols indicate major active volcanoes, and the grey shading indicates earthquake-prone areas. Compare the distribution of these hazards with the plate boundaries shown in **FIGURE 2.19**.

Only in the last several decades have scientists learned about plate tectonics—this environmental system of such fundamental importance was unknown just half a century ago. Our civilization was sending people to the Moon by the time we were coming to understand the movement of land under our very feet. One of the early pioneers in developing the theory of plate tectonics in the 1950s and 1960s was J. Tuzo Wilson, a Canadian geophysicist.[6] The tectonic cycle of diverging oceanic lithosphere, creation of new oceanic crust, horizontal plate motion, and eventual subduction of the lithosphere back into the mantle is called a *Wilson cycle* in his honour. Geologists and geophysicists continue to refine our understanding about how the complex processes of the tectonic cycle shape our environment.

Plate movements also cause geological hazards

The movements of tectonic plates give rise to the creative forces that shape our planet, yet they can also pose hazards to us. Earthquakes and volcanic eruptions are examples of natural geological processes that are vital to the natural functioning of the environment, but risky when they occur in populated regions. We can see how such hazards relate to tectonic processes by examining a map of the circum-Pacific belt, or "ring of fire" (**FIGURE 2.21**). Compare the route of the circum-Pacific belt to some of the tectonic plate boundaries shown in **FIGURE 2.19**, and note how closely they match. Nine out of 10 earthquakes and over

half the world's volcanoes occur along this 40 000-km arc of subduction zones and fault systems.

Along plate boundaries, and in other places where faults occur, built-up pressure is relieved in fits and starts. Each release of energy causes what we experience as an **earthquake**. Most earthquakes are barely perceptible, even by sensitive instrumentation, but occasionally they are powerful enough to do tremendous damage to human life and property, like the devastating Tōhoku earthquake discussed in our Central Case. Earthquakes are a manifestation of the tremendous power of plate tectonic motion; they also provide a window into Earth's interior. By mapping the passage through Earth's interior of the energy released during earthquakes, geophysicists have built up a detailed picture of the planet's internal layering.

Submarine earthquakes like the 2011 Tōhoku quake may displace huge volumes of ocean water, causing tsunamis. Submarine volcanic eruptions and large coastal landslides also can trigger tsunamis. The world's attention was drawn to this hazard on December 26, 2004, when a massive tsunami was triggered by an earthquake off Sumatra. It devastated the coastlines of countries all around the Indian Ocean, including Indonesia, Thailand, Sri Lanka, India, and several African nations. Roughly 228 000 people were killed, 1–2 million were displaced, and whole communities were destroyed. The 2011 Tōhoku tsunami was triggered by a megathrust earthquake in a subduction zone off the coast of Japan; it destroyed coastal towns and damaged nuclear power plants, leading to one of the worst nuclear crises in history.

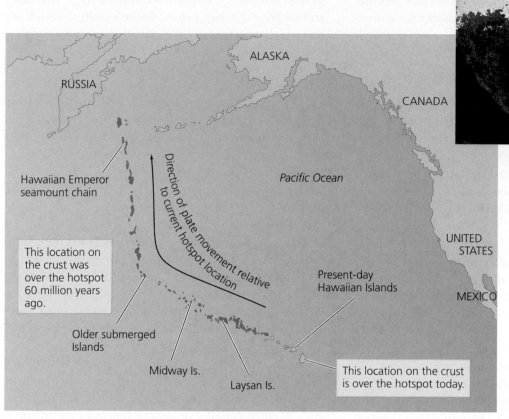

(a) **Current and now-submerged islands in the Hawaiian chain**

(b) **Kilauea erupting**

FIGURE 2.22
The Hawaiian Islands **(a)** were formed by repeated eruptions from a plume of magma rising from the mantle. The island of Hawaii itself is the youngest; it is still volcanically active because it sits on top of the magma plume, or "hotspot." The other islands are older and have begun to erode. To the northwest stretches a long series of former islands, now submerged, that were carried away from the hotspot by the movement of the Pacific Plate. Active volcano Kilauea **(b)**, on Hawaii's southeast coast, is currently located above the edge of the hotspot.

Where molten rock (lava), hot gas, or ash erupts through Earth's surface, a **volcano** is formed, often creating a mountain over time as the cooled volcanic materials accumulate. As we have seen, lava can extrude along mid-ocean ridges or over subduction zones as one tectonic plate dives beneath another. Lava may also be emitted at *hotspots*, localized areas where plumes of molten rock rise from the mantle and erupt through the crust. As a tectonic plate moves across a hotspot, repeated eruptions from this source may create a linear series of volcanoes. The Hawaiian Islands provide an example of this process (**FIGURE 2.22**). At some volcanoes, particularly those at hotspots and along divergent plate boundaries, lava flows slowly downhill; an example is Kilauea in Hawaii, which has been erupting almost continuously since 1983. Other volcanoes, particularly those in subduction zones, may erupt violently, releasing large amounts of ash and cinder in a sudden explosion, such as during the massive 1980 eruption of Mount Saint Helens in the Cascades Range (**FIGURE 2.23**).

Landslides occur when large amounts of rock or soil flow, fall, or slide downhill. Landslides are a severe and often sudden manifestation of the more general phenomenon of **mass wasting**, the downslope movement of soil and rock due to gravity. Mass wasting occurs everywhere, naturally, but major landslides are most common in areas of rugged terrain; for example, the Himalayan Mountains, rising as a

result of the collision of the Indian–Australian and Eurasian tectonic plates, experience frequent large landslides and rockfalls. Heavy rains may contribute to the hazard by saturating soils; earthquake and volcanic eruptions also trigger massive mudslides and volcanic avalanches. Mass wasting

FIGURE 2.23
The explosive eruption of Mt. St. Helens in Washington state in 1980 was a natural consequence of the subduction of a small tectonic plate (the Juan da Fuca Plate) underneath the North American Plate.

also can be brought about by human land use practices that expose or loosen soil or steepen slopes, making them more prone to collapse.

The rock cycle modifies Earth's physical environment

Just as plate tectonics shows geology's dynamism at a global scale, the rock cycle shows it at a more local scale. We tend to think of rock as pretty solid stuff. However, rocks and the minerals that compose them are heated, melted, cooled, broken down, and reassembled in a series of very slow steps, which, altogether, make up the **rock cycle** (**FIGURE 2.24**).

Rock is a naturally occurring, solid aggregate; most rock is composed primarily of mineral grains and fragments. A **mineral** is a naturally occurring solid element or inorganic compound that has a crystal structure, a specific chemical composition, and distinct physical properties. Minerals are thus the building blocks of rocks. The type of rock in a given region affects soil composition and characteristics, and thereby influences the region's plant community. Understanding the rock cycle enables us to appreciate the formation and conservation of soils, mineral resources, fossil fuels, groundwater sources, geothermal energy sources, and other natural resources, and to predict many natural hazards—all of which we discuss in later chapters.

Igneous rock   All rock can melt. At high enough temperatures, rock will enter the molten, liquid state called *magma*. If magma is ejected to the surface during a volcanic eruption, it may flow or spatter across Earth's surface in the form of **lava**. Rock that forms when magma or lava cools is called **igneous rock** (from the Latin *ignis*, meaning "fire").

Igneous rock comes in two main classes, because magma can solidify in different ways. When magma cools slowly and solidifies while it is below Earth's surface, it forms *intrusive* or *plutonic* igneous rock. This process created the Coast Range Mountains of British Columbia (**FIGURE 2.25A**). Granite is the best-known type of intrusive rock. A slow cooling process allows minerals of different types to aggregate into large crystals, giving granite its multicoloured, coarse-grained appearance.

In contrast, when molten rock is ejected or extruded from a volcano, it cools quickly, so minerals have little time to grow into coarse crystals. This class of igneous rock is called *extrusive* or *volcanic* igneous rock. Its most

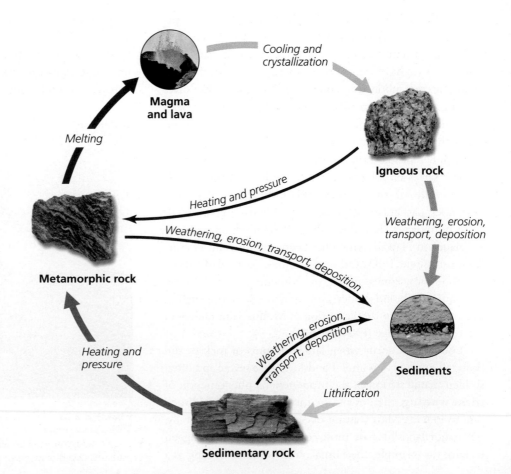

FIGURE 2.24
In the rock cycle, igneous rock is formed when rock melts and the resulting magma or lava then cools. Sedimentary rock is formed at or near the surface, either by the deposition of minerals in aquatic environments, or when rock is weathered and eroded, and the resulting sediment is compressed and cemented to form new rock. Metamorphic rock is formed when rock is subjected to intense heat and pressure underground. Refer back to this schematic diagram of the rock cycle as we summarize the main rock types and processes.

common representative is basalt, the principal rock type of the Hawaiian Islands (**FIGURE 2.25B**) and the main rock type of oceanic crust.

Sedimentary rock All exposed rock is worn away with time. The relentless forces of wind, water, freezing, thawing, and chemical dissolution cause **weathering**, stripping off one tiny grain (or large chunk) after another. Through **erosion**, weathered particles of rock are blown by wind, washed away by water, dissolved and transported by flowing ground water or surface water, or carried by slowly flowing glacial ice. The transported particles are called **sediment** (from the Latin *sedimentum*, "settling or sinking") when they come to rest, either by settling out downhill, downstream, or downwind from their source, or by precipitating from a solution.

Sediment layers accumulate over time, causing the weight and pressure of overlying layers to increase. **Sedimentary rock** is formed when sediments are physically pressed together (*compaction*) and dissolved minerals seep through sediments and bind the sediment particles

(a) Intrusive igneous rock—granite

(b) Extrusive igneous rock—basalt

(c) Sedimentary rock—limestone

(d) Sedimentary rock—sandstone

(e) Metamorphic rock—gneiss

FIGURE 2.25
The Coast Range Batholith of British Columbia **(a)**, the core of a great mountain range, is made of granite, an intrusive igneous rock. The volcanic islands of Hawaii **(b)** are built of basalt, an extrusive igneous rock. These classic cliffs in Ha Long Bay, Vietnam **(c)**, are made of the sedimentary rock limestone. These "hoodoos" in Dinosaur Provincial Park, Alberta **(d)**, are erosional features made of layered sandstone, a sedimentary rock. The ancient, folded rock shown here **(e)** is a metamorphic rock called gneiss; it began as sedimentary rock but was then compressed, contorted, heated, and recrystallized during several intense periods of mountain-building.

together (*cementation*). The formation of rock through these processes of compaction and cementation is termed *lithification*. Examples of sedimentary rock include limestone (**FIGURE 2.25C**), formed as dissolved calcite precipitates from water or as the remains of marine organisms settle to the bottom; sandstone (**FIGURE 2.25D**), made of cemented sand particles; and shale, with still smaller mud particles. Related processes lead to the fossilization of organisms and, in just the right circumstances, the fossil fuels we use for energy. Because sedimentary layers accumulate in chronological order, geologists and palaeontologists can assign relative ages to fossils they find in sedimentary rock, and thereby they make inferences about Earth's history.

Metamorphic rock   Geological forces may bend, uplift, compress, or stretch rock. When any type of rock is subjected to great heat or pressure, it may alter its form, becoming **metamorphic rock** (from the Greek for "changed form"). The forces that metamorphose rock generally occur deep underground, at temperatures lower than the rock's melting point but high enough to change its appearance, chemical composition, and physical properties. Metamorphic rock (**FIGURE 2.25E**) includes gneiss, formed when shale is subjected to very high heat and pressure; slate, formed when shale is subjected to somewhat lower heat and pressure; and marble, formed when limestone is heated and pressurized.

The oldest rocks ever found on this planet, so far, are metamorphic rocks. In fact, the oldest known rock is from northwestern Canada—it is part of a formation called the Acasta Gneiss, and it has been dated at 4.0 billion years.[7] Individual grains of the mineral zircon 4.4 billion years old have been discovered in Western Australia, also in gneiss. These extremely old Earth materials give scientists an important glimpse into what early Earth may have been like.

Early Earth and the Origin of Life

How Earth was formed, how the physical environment came to be as it is today, and how life originated are among the most centrally important questions in modern science. In searching for answers, scientists have learned a great deal about the history of this planet and what early Earth was like. That scientific interest has extended to other planets, such as Mars, which may prove to have once harboured life. We study the early geological and chemical environments of both planets to learn more about what this part of the solar system was like billions of years ago, when life first took hold on this planet.

Early Earth was a very different place

Earth formed more than 4.5 billion years ago in the same way as the other planets of our solar system: Dispersed bits of material whirling through space around our Sun were drawn by gravity into one another, coalescing into a series of planetary spheres. For several hundred million years after the planets formed, there remained enough stray material in the solar system that Earth and the other young planets were regularly bombarded by large chunks of debris in the form of asteroids, meteorites, and comets. The largest impacts were probably so explosive that they vaporized the newly formed oceans.

This extraterrestrial rain of debris, combined with extreme volcanic activity, intense ultraviolet radiation, and a powerful greenhouse effect, made early Earth a pretty hostile place. Any life that got underway during this "bombardment stage" might easily have been killed off. Only after most debris was cleared from the solar system was life able to gain a foothold.

Earth's early atmosphere also was very different from our atmosphere today. It was chemically reducing, and "free" oxygen (oxygen that was not part of a compound) was largely absent until photosynthesizing microbes started producing it. Whereas today's atmosphere is dominated by nitrogen and oxygen (see **TABLE 2.1**), Earth's early atmosphere is thought to have contained large amounts of hydrogen, ammonia (NH_3), methane, carbon dioxide, carbon monoxide (CO), and water vapour.

Figuring out how Earth's atmosphere evolved into its current state is an interesting and challenging area of research. We know where some of the constituents that were so abundant early in Earth's history have gone; for example, much of the carbon dioxide from the early atmosphere is now bound up in thick sequences of carbonate rock—limestone. This long-term storage of carbon, called **sequestration**, is a good thing for life on this planet; if the carbon dioxide were released from carbonate rocks, we would have an atmosphere of 95% carbon dioxide—similar to that of Venus—and that would not be conducive to life as we know it.

Several hypotheses have been proposed to explain life's origin

Most scientists interested in life's origin think that it must have begun when inorganic chemicals linked themselves into small molecules and formed organic compounds.[8] Some of these compounds gained the ability to replicate, or reproduce themselves, whereas others found ways to group together into proto-cells. There is much debate and ongoing research on the details of this process, especially concerning the location of the first chemical reactions and the energy source(s) that powered them.

The heterotrophic hypothesis

A hypothesis that has been traditionally favoured is that life evolved from a "primordial soup" of simple inorganic chemicals—carbon dioxide, oxygen, and nitrogen—dissolved in the ocean's surface waters or tidal shallows. Scientists since the 1930s have suggested how simple amino acids might have formed under these conditions and how more complex organic compounds could have followed, including simple ribonucleic acids that could replicate themselves. This hypothesis is termed *heterotrophic* because it proposes that the first life forms used organic compounds from their environment as an energy source.

In 1953, biochemists Stanley Miller and Harold Urey passed electricity through a mixture of water vapour, hydrogen, ammonia, and methane, which they thought to be representative of the early atmosphere. The experiments produced organic compounds, including amino acids. Subsequent experiments confirmed the basic validity of these experiments, but it is important to remember that amino acids, on their own, are not life. Scientists since then also have modified their ideas about early atmospheric conditions and chemistry.

The panspermia hypothesis

Another hypothesis proposes that microbes from elsewhere in the solar system travelled on meteorites that crashed to Earth, "seeding" our planet with life. Scientists had long rejected this idea, believing that even if amino acids or bacteria were to exist in space, the searing temperatures that comets and meteors attain as they enter our atmosphere should destroy them before they reach the surface. However, the Murchison meteorite, which fell in Australia in 1969, was found to contain many amino acids.

Since then, experiments simulating impact conditions have shown that organic compounds and some bacteria can withstand a surprising amount of abuse. Recent astrobiology research suggests that comets have brought large amounts of water, and possibly organic compounds, to Earth throughout its history. As a result of such findings, long-distance travel of microbes through space and into our atmosphere seems more plausible than previously thought.

The chemoautotrophic hypothesis

In the 1970s and 1980s, several scientists proposed that life originated at deep-sea hydrothermal vents, like those in **FIGURE 2.17**. In this scenario, the first organisms were chemoautotrophs, creating their own food from hydrogen sulphide, abundant in the deep-sea vent environment. A related hypothesis suggests that life originated in the hot, moist environment of thermal pools and hot springs, like those in **FIGURE 2.16**—an environment that is currently favoured by specifically adapted types of bacteria. Current research on *extremophiles*—organisms that are adjusted to conditions of extreme heat, cold, pressure, acidity, or salinity—is helping to further our understanding of the earliest life forms on Earth and the environmental conditions in which they survived.

Genetic analysis of the relationships of present-day organisms suggests that some of the ancient ancestors of today's life forms did live in extremely hot, wet environments. The extreme heat of hydrothermal vents could act to speed up chemical reactions that link atoms together into long molecules, a necessary early step in life's formation. Scientists have shown experimentally that it is possible to form amino acids and begin a chain of steps that might potentially lead to the formation of life under high-temperature, high-pressure conditions similar to those of hydrothermal vents.

Conclusion

The chemical basis of matter, the nature and transformation of energy, and the geological processes that have shaped our planet (and continue to do so) provide the physical foundations for our present-day environment, and support the existence of life. Matter, energy, and the physical environment are tied to every significant process in environmental science.

An understanding of the characteristics and interactions of matter can provide tools for finding solutions to environmental problems, whether one wants to analyze agricultural practices, manage water resources, conduct toxicological studies, or find ways to mitigate global climate change. Likewise, an understanding of energy is of fundamental scientific importance, as well as considerable practical relevance. Physical processes of geology such as plate tectonics and the rock cycle create some hazards, but they are centrally important. They shape Earth's terrain, influence oceanic and atmospheric processes, and form the foundation for living systems.

Life has flourished on Earth for billions of years, stemming from an origin that scientists are eagerly attempting to understand. Deciphering how life originated requires an understanding of the geological processes that have shaped our physical environment over the course of this planet's history. Knowledge in all of these areas also enhances our understanding of how present-day organisms interact with one another, how they relate to their nonliving environment, and how environmental systems function.

REVIEWING OBJECTIVES

You should now be able to

Describe how scientists use the scientific method to test ideas and investigate the nature of the environment

- The scientific method consists of a series of steps, including making observations, formulating questions, stating a hypothesis, generating predictions, testing predictions, and analyzing the results obtained from the tests.
- The scientific method has many variations, and there are many different ways to test questions scientifically.
- Scientific research occurs within a larger process that includes peer review of work, journal publication, and interaction with colleagues.

Summarize the basic properties of matter that are the building blocks for all materials on Earth

- Matter may be transformed from one type of substance into others, but it cannot be created or destroyed.
- An atom is the smallest component of an element that retains all of the properties of the element. Changes at the atomic level can result in alternative forms of elements, such as ions and isotopes.
- Atoms combine, by bonding together, to form molecules.
- The characteristics of the water molecule, including its high heat capacity, strong cohesion, and other chemical properties, help facilitate life.
- Hydrogen ions perform a very important function by controlling the acidity of aqueous solutions.
- Living things depend on organic compounds, which are carbon based, as well as inorganic compounds.
- Proteins, nucleic acids, carbohydrates, and lipids are key building blocks of life.

Differentiate among various types of energy, the fundamental properties of energy, and the role of energy in environmental systems

- The total amount of energy in the universe is conserved; it cannot be created or destroyed.
- Energy can be either potential or kinetic. Energy can occur in a variety of forms, including chemical, nuclear-binding, mechanical, light, sound, thermal, and electrical energy.
- Earth's systems are powered by mainly radiation from the Sun. In photosynthesis, autotrophs use solar energy, carbon dioxide, and water to produce the sugars they need, as well as oxygen. In respiration, the process is reversed.
- Geothermal heating from the planet's interior and tidal interactions among Earth, the Sun, and the Moon are also important sources of energy.

Explain how plate tectonics and the rock cycle shape the landscape around us and Earth beneath our feet

- Earth consists of distinct layers that differ in composition, temperature, density, and other characteristics. The three main layers are the core, mantle, and crust.
- Plate tectonics is a set of fundamental processes that shape Earth's surface through the movement of lithospheric plates, driven by convection in the mantle.
- Tectonic plates meet at three types of boundaries: divergent, transform, and convergent.
- The ocean basins, high mountains, and deep valleys, and even the atmosphere, climate system, soils, and distribution of plants and animals—all are fundamentally influenced by plate tectonics.
- Geological hazards, including earthquakes, tsunamis, volcanic eruptions, and landslides, are closely related to the processes of plate tectonics.
- Matter is cycled and rocks are transformed from one type to another through the processes of the rock cycle.

Summarize the characteristics of early Earth and the main hypotheses for the origin of life

- Early Earth was a very different place and was not very hospitable to life.
- The heterotrophic hypothesis proposes that life arose from chemical reactions in surface waters. The panspermia hypothesis proposes that substances needed for life's origin arrived from space. The chemoautotrophic hypothesis proposes that life arose from chemical reactions near deep-sea hydrothermal vents.

TESTING YOUR COMPREHENSION

1. What is the typical sequence of steps in the scientific method? Explain the difference between a manipulative experiment and a natural experiment.

2. How does an ion differ from an isotope? Differentiate among atoms, molecules, and compounds, and give an example of each.

3. Name four ways in which the chemical nature of the water molecule facilitates life.

4. What are three important types of biological polymer, and what are their functions?

5. Describe the two major forms of energy, and give examples of each.

6. What are the three major sources of energy that power Earth's environmental systems?

7. Name the primary layers that make up our planet. Which portions does the lithosphere include?

8. What are the three major types of plate boundaries? Compare and contrast the types of processes that typically happen at each of them.

9. Name the three main types of rock, and describe how each type may be converted to the others via the rock cycle.

10. Compare and contrast three competing hypotheses for the origin of life.

THINKING IT THROUGH

1. What has to happen to a researcher's results before they are publishable? Why is this process important? Do you think the scrutiny of published scientific work is effective in preventing undue influence by funding agencies, such as corporations, on the results of scientific research?

2. Think about the ways we harness and use energy sources in our society—both renewable sources such as geothermal energy and nonrenewable sources such as coal, oil, and natural gas. What implications does the first law of thermodynamics have for our energy usage? How is the second law of thermodynamics relevant to our use of energy?

3. Describe how plate tectonics accounts for the formation of (a) mountains, (b) volcanoes, and (c) earthquakes. Why do you think it took so long for scientists to discover an environmental system of such fundamental importance as plate tectonics?

4. Which lines of evidence in the debate over the origin of life strike you as the most convincing, and why? Which strike you as the least convincing, and why? Can you think of any further scientific research that could be done to address the question of how life originated?

INTERPRETING GRAPHS AND DATA

After the nuclear accident of 2011, a group of scientists led by Dr. Ken Buesseler tested the radioactivity of air, water, plankton, and other free-swimming organisms in the ocean off Japan. They found water radiation levels of more than 100 000 becquerels per cubic metre (Bq/m^3; a becquerel is a unit of measurement for radioactivity) in early April, up from a pre-accident level of about 1.5 becquerels per cubic metre. The post-accident ocean radiation levels were about 100 times greater than those found in the Black Sea after the Chernobyl nuclear meltdown of 1986 (see graph on the next page), even though the release of radioisotopes from Chernobyl was roughly five times larger than that from Fukushima.

Radioactivity levels in the water did not decline to pre-accident levels, but remained high for many months after the accident, suggesting that radioactive material might still have been leaking from the plant and/or radioactively contaminated ground water flowing into the ocean. Radioactivity levels in locally captured fish also had not declined one and a half years after the accident. Radioactive materials with long half-lives could accumulate in offshore sediments and be ingested by species that live on the ocean bottom or filter water for food. Sediments in areas where radiation was concentrated are likely to remain contaminated for several decades, posing a long-term threat to humans consuming seafood from these areas.

The work of Buesseler and his team shed light on the movements of radioisotopes released into the ocean by the Fukushima accident,[9] but there are still many questions, such as how radioactive contaminants will move through aquatic food webs. One of the most problematic radioactive contaminants generated by nuclear fission processes in reactors is cesium-137. It is highly water soluble, so it moves around quite easily in the natural environment. Continued monitoring, likely for many decades, will be necessary to determine threats to human health and ecosystems from cesium-137 and other radioactive contaminants.

1. What was the trend in ocean radioactivity levels from cesium-137 from 1960 to 2010?

2. Did releases from the Chernobyl accident significantly alter this trend?

3. What long-term trends would you expect to see following the releases from Fukushima in 2011?

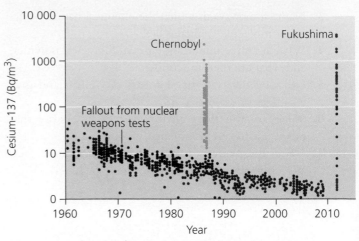

Ocean concentrations of radioactive cesium-137 show evidence of the nuclear accidents at Chernobyl and Fukushima. Note the y axis is logarithmic, such that each unit is 10 times greater than the previous unit.
Source: From Ken Buesseler. Copyright © by Woods Hole Oceanographic Institution. Reprinted by permission.

MasteringEnvironmentalScience®

STUDENTS
Go to **MasteringEnvironmentalScience** for assignments, the eText, and the Study Area with practice tests, videos, current events, and activities.

INSTRUCTORS
Go to **MasteringEnvironmentalScience** for automatically graded activities, current events, videos, and reading questions that you can assign to your students, plus Instructor Resources.

3 Earth Systems and Ecosystems

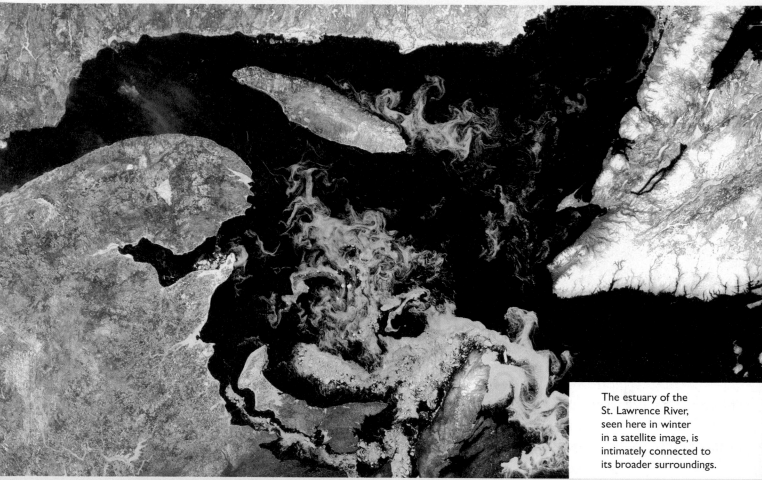

The estuary of the St. Lawrence River, seen here in winter in a satellite image, is intimately connected to its broader surroundings.

Jeff Schmaltz/MODIS Rapid Response Team/ NASA

Upon successfully completing this chapter, you will be able to

- Describe the fundamental properties of systems, and the importance of linkages and flows of matter and energy among environmental systems
- Outline the characteristics of Earth's major subsystems
- Discuss how living and nonliving entities interact and how energy and matter move around in ecosystems

- Recognize the importance and complexity of the differing spatial and temporal scales of Earth processes
- Summarize the main global biogeochemical cycles and their human impacts, especially the global water, carbon, nitrogen, and phosphorus cycles

Researchers attach a satellite-linked transmitter to a beluga whale. Satellite tracking is being used to study beluga whale migrations in and out of the St. Lawrence estuary.

Hudson Bay

CANADA

St. Lawrence Estuary

CENTRAL CASE

THE PLIGHT OF THE ST. LAWRENCE BELUGAS

"Nature does not show us any isolated 'building blocks,' but rather appears as a complicated web of relations between the various parts of the whole."

—FRITJOF CAPRA, THEORETICAL PHYSICIST

"The concept of a subtly interconnected world, of a whispering pond in and through which we are intimately linked to each other and to the universe ... is part of humanity's response to the challenges that we now face in common."

—ERVIN LASZLO, SYSTEMS THEORIST

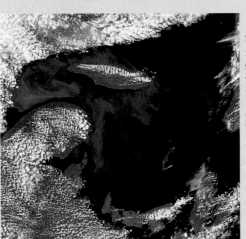

The lighter-coloured water shows massive algal blooms in this satellite image of the St. Lawrence estuary, where it joins the Gulf of St. Lawrence and the Atlantic Ocean.

The St. Lawrence is one of the great river systems of Canada. From its origin in Lake Ontario (see map and satellite image) it flows approximately 1200 km to the Gulf of St. Lawrence, the world's largest estuary. The cold Labrador Current brings an abundant food source of plankton and fish from the Atlantic Ocean into the estuary, supporting a small population of beluga whales.

The playful, sociable beluga or white whale, *Delphinapterus leucas*, prefers cold saltwater estuaries as its habitat.

Health problems have plagued the St. Lawrence belugas for decades, causing their population to decrease to about 1000[1]—much less than the many thousands that occupied the estuary at the beginning of the twentieth century. The population was first challenged by hunting and commercial whaling, which was banned in the St. Lawrence in 1979.[2] Now the belugas are threatened by habitat degradation, food shortages, entanglement in nets, and interactions with shipping vessels; they also seem to be dying of cancer.[3]

Daniel Martineau and a team of veterinarians from the Université de Montréal carried out autopsies on more than 100 dead whales from the St. Lawrence. They observed cancers, mainly gastrointestinal, in 27% of the autopsied belugas, young and old.[4] Cancer occurrence and mortality in the St. Lawrence belugas is comparable to that of humans and domestic animals such as dogs, but much higher than in dolphins and other wild mammals. In addition to the cancers, the St. Lawrence belugas have very low reproductive rates and other health issues, including cysts and bacterial infections.

Toxicological studies by Martineau and colleagues showed that the dead whales had been exposed to organochloride pollutants, notably *polycyclic aromatic hydrocarbons*, or *PAHs*.[3,4] PAHs come primarily from the burning of fossil fuels and other combustion sources. They do not break down easily and have become one of the most widespread contaminants in aquatic and marine environments. Deposited from the air, they are carried to waterways by runoff, accumulating in the bottom sediments of rivers and near shorelines. Belugas feed on organisms that live in these sediments. The concentration of contaminants increases in belugas and other animals that eat higher up the food chain. PAHs are also *lipophilic*, or "fat-loving" compounds, so they combine easily with fats and build up over time, or **bioaccumulate**, in the blubber of the belugas.

Where do these pollutants come from? The answer may be found by looking at the Great Lakes and St. Lawrence River as a single connected system, of which the beluga whales are a small but integral part. Any changes upstream in the system will have impacts downstream. This includes pollutants from as far away as the Golden Horseshoe industrial zone in Ontario that eventually make their way to the estuary and into the food chain of the beluga whales.

It is not clear to what extent the various health problems plaguing the belugas are directly attributable to exposure to PAHs. Other organochloride compounds and toxic heavy metals from industrial, urban, and airborne sources are also concentrated in the waters of the estuary. Agricultural development along the St. Lawrence River contributes pesticides; some of these contaminants find their way into the thousands of tonnes of fish harvested from St. Lawrence commercial fisheries each year.

Other factors may have additional indirect effects on the health of belugas and other organisms in the estuary. For example, excess organic matter and plant nutrients from fertilizer runoff and animal waste have contributed to a sharp drop in oxygen concentration in the deepest waters of the estuary, where dissolved oxygen levels have declined by half since the 1930s. The lack of dissolved oxygen, or **hypoxia**, is fairly common in estuarine and coastal waters around the world.

Nutrients are elements and compounds, both organic and inorganic, that organisms consume and require for survival. Nutrient over-enrichment in water bodies can lead to algal overgrowths, called *blooms* (as seen in the satellite image of the St. Lawrence that opens this chapter). The subsequent decay of organic matter consumes the available oxygen, leading to ecosystem degradation. The overall process of nutrient over-enrichment in a water body (either fresh or salt), with consequent hypoxia, is called **eutrophication**; it can result from both natural and human (anthropogenic) influences. A natural contributor to hypoxia in the Gulf of the St. Lawrence is the influx of warm, oxygen-depleted water from the Gulf Stream, which displaces the cold, oxygen-rich water of the Labrador Current.

All of these influences on the habitat, food, and health of the belugas come from sources that are external to the St. Lawrence Estuary but are intimately connected to it by a wide variety of environmental and anthropogenic processes. The St. Lawrence beluga population is classified as "threatened" by COSEWIC, the Committee on the Status of Endangered Wildlife in Canada.[5] The health of the whales is an indicator of the overall health of the estuary, and a reminder of the interconnectedness of the Great Lakes–St. Lawrence ecosystem. Recognizing this, scientists are striving to understand the ecosystem as a whole and to develop strategies for protecting the estuary and the organisms that inhabit it.

Environmental Systems

Our environment consists of complex networks of interlinked systems. These systems include the webs of relationships among living species and the interactions of living species with the nonliving entities around them. Earth's environmental systems also include **cycles**—flows of key chemical elements and compounds that move substances from one place to another within the system, facilitate environmental processes, regulate climate, and support life. We depend on these systems and cycles for our very survival.

Taking a "systems approach" to investigating problems such as the decline of the St. Lawrence beluga population is helpful in environmental science because so many of these issues are multifaceted and complex. Before we turn our attention to specific Earth systems and cycles, let's take a look at the general properties of systems.

Systems are networks of relationships

A **system** is a network of relationships among parts, elements, or components that interact with and influence one another through the exchange of energy, matter, or information. Systems receive inputs, process these inputs, and produce outputs. Systems that receive inputs of both energy and matter and produce outputs of both are called **open systems**. Systems that receive inputs and produce outputs of energy, but not matter, are called **closed systems**. In a closed system, matter cycles among the various parts of the system but does not leave or enter the system. It is scientifically more straightforward to deal with closed systems, but in nature no system is truly, perfectly closed.

As discussed in the chapter *Matter, Energy, and the Physical Environment*, energy moves among Earth's various environmental subsystems in a variety of forms, including solar radiation, heat released by geothermal activity, and tidal energy. Energy is converted for use by organisms through photosynthesis, respiration, and other processes of metabolism, and by human activities such as fossil fuel combustion. Inputs and flows of matter occur when chemicals or physical material moves among subsystems, such as when a gaseous chemical is dispersed through the air, when seeds are carried long distances by the wind, when migratory animals deposit waste far from where they consumed food, when water flows in a river channel to the ocean, or when plants convert carbon from the air into living tissue by photosynthesis.

Essentially all environmental systems are open systems, meaning that both matter and energy can pass into and out of them. As an open system, for example, the estuary of the St. Lawrence receives inputs of fresh water, sediments, nutrients, and pollutants from the St. Lawrence and other rivers, as well as salt water from the Atlantic Ocean and rain from the atmosphere. Large animals, such as the belugas, harvest some of the system's material output, and so do human fishers, withdrawing both matter and potential energy in the form of fish and plankton. This output subsequently becomes input to the human economic system and to the digestive systems of the people and whales that consume seafood from the St. Lawrence.

What about Earth as a whole? Technically, Earth, too, is an open system. We receive inputs of solar energy and we return energy, mainly in the form of heat, to outer space. We receive inputs of matter in the form of incoming meteorites, and we lose occasional molecules to space from the outer part of the atmosphere. However, the inputs and outputs of matter are so incredibly tiny compared to the mass of Earth as a whole that we can say that Earth is effectively a closed system: Energy passes into and out of the system, driving environmental processes and residing for differing periods in different parts of the system, but the matter within the system is effectively finite and limited.

Feedbacks are common in environmental systems

Sometimes a system's output can serve as input to that same system. This circular process is described as a **feedback loop**. Feedback loops are of two types, negative and positive.

In a **negative feedback loop**, output that results from a system moving in one direction acts as input that moves the system in the other direction. Input and output essentially neutralize one another's effects, stabilizing the system. A thermostat, for instance, stabilizes a room's temperature by turning the furnace on when the room gets cold and shutting it off when the room gets hot. Similarly, negative feedback regulates our body temperature. If we get too hot, our sweat glands pump out moisture that evaporates to cool us down, or we may move from sun to shade. If we get too cold, we shiver, creating heat, or we move into the sunlight or put on more clothing. Another example of negative feedback would be a predator–prey system in which the populations of predator and prey rise and fall in response to one another (**FIGURE 3.1A**). Most systems in nature involve negative feedback loops. Negative feedback loops enhance stability, and in the long run, only those systems that are stable will persist.

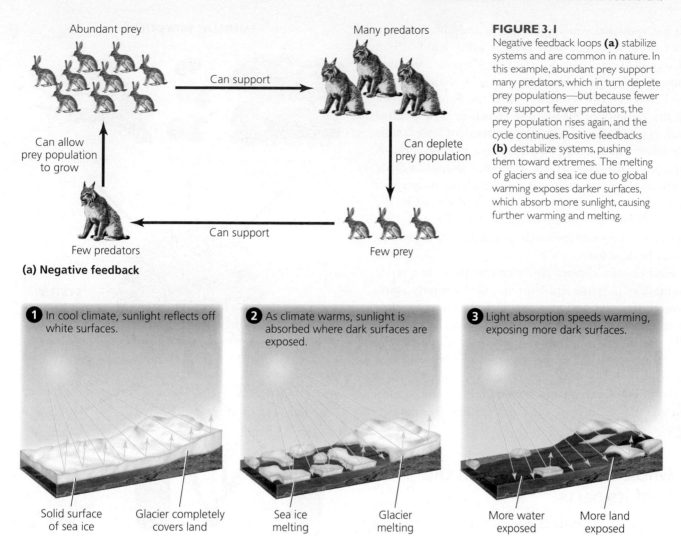

Abundant prey

Can support

Many predators

Can allow prey population to grow

Can deplete prey population

Few predators

Can support

Few prey

(a) Negative feedback

FIGURE 3.1
Negative feedback loops **(a)** stabilize systems and are common in nature. In this example, abundant prey support many predators, which in turn deplete prey populations—but because fewer prey support fewer predators, the prey population rises again, and the cycle continues. Positive feedbacks **(b)** destabilize systems, pushing them toward extremes. The melting of glaciers and sea ice due to global warming exposes darker surfaces, which absorb more sunlight, causing further warming and melting.

❶ In cool climate, sunlight reflects off white surfaces.

❷ As climate warms, sunlight is absorbed where dark surfaces are exposed.

❸ Light absorption speeds warming, exposing more dark surfaces.

Solid surface of sea ice

Glacier completely covers land

Sea ice melting

Glacier melting

More water exposed

More land exposed

(b) Positive feedback

Positive feedback loops have the opposite effect. Rather than stabilizing a system, they drive it further toward one extreme or another. Erosion, the removal of soil by water or wind, can lead to a positive feedback. If erosion removes vegetation, the exposed soil will be more susceptile to further erosion. If deeper soils are exposed and allowed to dry out, erosion may become progressively more severe. Another example is climatic warming leading to the melting of ice, which exposes underlying darker surfaces. Darker surfaces absorb more sunlight, causing further warming and leading to additional melting (**FIGURE 3.1B**). Positive feedbacks can alter a system substantially. Positive feedback loops are rare in nature, but they are common in natural systems that have been altered by human impact.

The inputs and outputs of complex natural systems usually occur simultaneously, keeping the system constantly active. When processes within a system move in opposing directions at equivalent rates so that their effects balance out, the process is said to be in a state of **dynamic equilibrium**. The term *dynamic* is used to indicate that even though the system is in balance or at equilibrium, it is an ever-changing, ever-adjusting balance, not a static or unchanging condition.

Homeostasis is a state of balance

Processes in dynamic equilibrium can contribute to **homeostasis**, the tendency of a system to maintain constant or stable internal conditions. **Resistance** is a property of homeostatic systems; it refers to the strength of the system's tendency to remain constant—that is, to resist disturbance. **Resilience**, another characteristic of homeostatic systems, is a measure of how readily the system will return to its original state once it has been disturbed.

To illustrate the concepts of resistance and resilience, let's say that someone gives you a sharp push. If you are resistant, you might sway a little bit, but you won't stumble or fall over. On the other hand, if you did stumble or fall but got right back up to a standing position with no trouble, then you were demonstrating the characteristic of resilience—after the disturbance you had the capacity to recover and return to your original standing position. In nature, we can think of the example of a forest that is being subjected to a pest invasion. A forest with the property of resistance would be little altered by the pest invasion. A forest with the property of resilience, on the other hand, would be affected by the pest invasion—tree health would likely suffer, and some trees might die—but the forest would quickly recover its former healthy state.

Homeostatic systems are often said to be in a stable condition of dynamic equilibrium, called a **steady state**; however, the state itself may change over time, even while the system maintains its ability to stabilize conditions internally. For instance, organisms grow, mature, and change, yet at each stage of life the organism can be said to be in a stable state. Similarly, Earth has experienced changes in the composition of the atmosphere over geological time; yet life has adapted, and Earth remains, by most definitions, a homeostatic system.

A whole may be more than the sum of its parts

It is difficult to understand systems fully by focusing on their individual components because systems can show **emergent properties**, characteristics that are not evident in the individual components on their own. Stating that systems possess emergent properties is a lot like saying, "The whole is more than the sum of its parts."

For example, if you were to reduce a tree to its component parts (leaves, branches, trunk, bark, roots, fruit, and so on), you would not be able to predict the whole tree's emergent properties, which include the role the tree plays as habitat for birds, insects, parasitic vines, and other organisms (**FIGURE 3.2**). You could analyze the tree's chloroplasts (photosynthetic cell organelles), diagram its branch structure, and evaluate its fruit's nutritional content, but you would still be unable to understand the tree as habitat, as a provider of shade and shelter, as a part of a forest landscape, or as a reservoir for carbon.

Complex systems have multiple subsystems

Systems seldom have well-defined boundaries, so deciding where one system ends and another begins can be difficult.

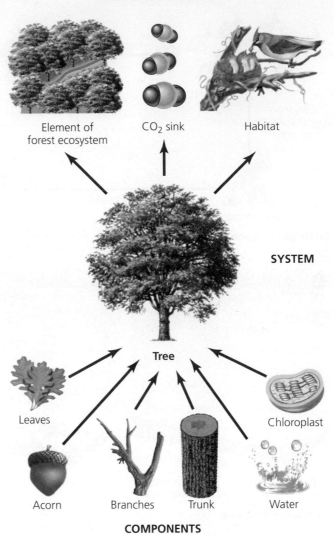

EMERGENT PROPERTIES

Element of forest ecosystem

CO_2 sink

Habitat

SYSTEM

Tree

Leaves

Chloroplast

Acorn

Branches

Trunk

Water

COMPONENTS

FIGURE 3.2
A system's emergent properties are not evident when we break the system into its component parts. A tree serves as wildlife habitat and plays roles in forest ecology and global climate regulation, but you would not know that from considering the tree as a collection of leaves, branches, and chloroplasts.

No matter how we attempt to isolate or define a system, we soon see that it has many connections to systems larger and smaller than itself. Systems may exchange energy, matter, and information with other systems, and they may contain or be contained within other systems. Where we draw boundaries depends on the spatial or temporal scale we wish to consider.

Consider a desktop computer system. It is certainly a network of parts that interact and exchange energy and information, but what are its boundaries? Is the system just what arrives in a packing crate and sits on top of your desk? Or does it include the network you connect it to? What about the energy grid you plug it into, with

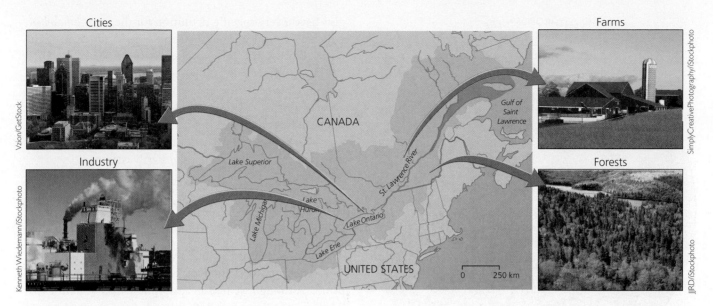

FIGURE 3.3 The watershed of the Great Lakes and St. Lawrence is a geographically large and complex system. Runoff from the land into the river carries water, sediment, and pollutants from a variety of sources downstream to the Gulf of the St. Lawrence, where pollution has given rise to a hypoxic zone and other environmental problems. Farms, cities, and industry are all contributors; so are natural sources, such as forests and soils.

its distant power plants and transmission lines? What about the internet? Browsing the web, you are drawing in digitized text, light, and sound from around the world. And what about the smaller systems that together compose the computer, such as the motherboard, the keyboard, and the mouse? All of these systems, large and small, contribute and are linked to the complex system that is your desktop computer.

The Great Lakes, the St. Lawrence River, and the Atlantic Ocean are examples of environmental systems that interact with one another. On a map, the river appears as a branched network of water channels (**FIGURE 3.3**). But where are this system's boundaries? You might argue that the river consists primarily of water, originates in Ontario, and ends in the Atlantic Ocean. But what about the rivers that feed it, and the farms, cities, and forests that line its banks? Major rivers such as the Ottawa and Saguenay flow into the St. Lawrence. Hundreds of smaller tributaries drain vast expanses of farmland, woodland, fields, cities, towns, and industrial areas before their water joins the St. Lawrence. These waterways carry with them millions of tonnes of sediment, hundreds of species of plants and animals, and numerous pollutants. The St. Lawrence River system is also intimately interconnected with the entire Great Lakes system—together they constitute an integrated hydrological system.

For an environmental scientist interested in runoff and the flow of water, sediment, or pollutants, it would make sense to view the Great Lakes–St. Lawrence River watershed as one great system. One must consider the entire area of land a river drains to comprehend and solve problems of river pollution. In contrast, for a scientist interested in the estuary's hypoxic zone, it might make the best sense to view only the river + the Gulf + the coastal waters of the Atlantic as the system of interest, because their interaction is central to the problem. For a scientist interested in the question of beluga whale contamination, another useful system might be the whale + the fish and plankton eaten by the whale + the bottom sediments and local water in which the contaminants have accumulated. For some scientists, like Daniel Martineau, the body of a single beluga whale can be the system of interest for an investigation of how pollutants affect the internal organs and health of the animal. In environmental science, the delineation of a system necessarily depends on the questions one is addressing, and on the temporal and spatial scale of interest.

Earth's Major Subsystems

There are many ways to define and delineate natural systems. Your choice as a scientist will depend on the particular issues in which you are interested, and the temporal and spatial scales on which you wish to study these processes. Categorizing environmental systems can help make Earth's dazzling complexity comprehensible to the human brain and more accessible to study and problem-solving.

FIGURE 3.4

For purposes of study and monitoring, scientists divide the complex Earth system into smaller, interacting subsystems. Both materials and energy flow among the four major subsystems: geosphere, atmosphere, hydrosphere, and biosphere. The anthroposphere—the human realm—is shown as encompassing the rest, because the impacts of human activity are now observable in all other parts of the Earth system.

Photo Credit Top: Paha I/Dreamstime/GetStock
Photo Credit Centre: LP7/E+/Getty Images
Photo Credit Left: Icefield/Dreamstime/GetStock
Photo Credit Right: Itsajoop/Dreamstime/GetStock

For this reason, scientists divide Earth's functional components into several broad subsystems (**FIGURE 3.4**). This is somewhat counterintuitive; as environmental scientists, aren't we interested in emphasizing the interactions among the parts of the Earth system? Yes, but splitting the Earth system into smaller subsystems can make it easier to think about how each part functions on its own, and how flows of material and energy connect the various parts of the larger system together.

The geosphere is the ground beneath our feet

The **geosphere** is the rock and sediment of the solid Earth. Sometimes it is useful to think of the whole planet as the geosphere, but more commonly the term is used to refer only to the outermost solid layer. The geosphere is sometimes called the lithosphere; however, that usage is not quite correct. As you may recall from the chapter on matter, energy, and the physical environment, geolo-

gists have a very specific definition for the term *lithosphere*, which technically refers to the topmost portion of the mantle (that is, the asthenosphere) and the crust, together.

The geosphere is composed of solid rock, but it also includes broken-up rock—the product of weathering and erosion in the rock cycle—and soil. It provides the physical and chemical foundation for life on this planet, and it is the source for the mineral nutrients and other materials that cycle through the Earth system.

The atmosphere is our planet's gaseous envelope

The **atmosphere** is a thin envelope of gases (along with some water droplets and dust particles) that surrounds our planet. Although we had an atmosphere from the time that Earth was first formed, billions of years ago, its chemical composition and other properties were very different from those of the present-day atmosphere, and this has had many implications for life on this planet.

A number of other planets in our solar system (and even some moons) also have atmospheres, so we are not unique in that regard. However, the composition of our atmosphere—which consists mainly of nitrogen and oxygen—is unique, as far as we know. We use the term **air** to refer to the specific collection of gases that makes up Earth's atmosphere.

Other elements in our atmosphere are present in much lower concentrations than nitrogen and oxygen, but their roles can be just as crucial for life. For example, atmospheric carbon dioxide is essential in the functioning of the natural greenhouse effect and the process of photosynthesis—both essential for life as we know it. The atmosphere also provides a medium through which energy is conveniently transferred, in the form of both heat and light, among the various physical subsystems of Earth.

Surface and near-surface waters compose the hydrosphere

The **hydrosphere** encompasses all of the water—salt or fresh, liquid, ice, or vapour—that resides in surface water bodies (ocean, lakes, rivers, wetlands, and even soil moisture), glaciers, and ground water. The perennially frozen parts of the hydrosphere comprise a subsystem that has its own name: It is called the **cryosphere**, and it includes all of the glaciers, snow, land ice, and sea ice.

Water in the atmosphere technically does not belong to the hydrosphere, but the two systems are closely linked through the water cycle. We can think of this connection from a systems perspective: When raindrops form in the atmosphere and fall to the surface, matter in the form of liquid water is transferred from the atmosphere into the hydrosphere. Both are open systems, and they are closely interconnected in terms of both materials and energy. Water also resides in living organisms, though for vanishingly short periods of time on a geological scale; this also is not part of the hydrosphere, although the very close connections are evident.

There is also water deep within our planet, in tiny pore spaces in rock and tightly bound up in mineral structures. Technically, this water also is not part of the hydrosphere, which conceptually is focused mainly on surface and near-surface waters. Some of the water deep inside the planet is ocean water that has been carried from the surface back down into the mantle by the tectonic process of subduction. Some of it is water that has been inside the planet for billions of years and has never participated in the near-surface processes of the hydrosphere; we refer to this as *juvenile water* or *primordial water*.

The biosphere is the living sphere

The **biosphere** consists of all the planet's living organisms, and recently deceased and decaying organic matter. The existence of a biosphere is obviously one of the unique features of our planet; as far as we know, no other planets host life forms (although this may change as scientists discover more and more potentially habitable planets).

We have seen that energy and matter flow within and among the various physical subsystems of Earth. Energy and matter also flow into, through, and out of the biosphere, connecting the biosphere with all of the physical Earth systems. These flows and interconnections are the basis for the life-sustaining processes of our environment.

For example, picture a robin plucking an earthworm from the ground after a rain. You are witnessing an organism (the robin, part of the biosphere) consuming another organism (the earthworm) by removing it from part of the geosphere (the soil) that the earthworm had been modifying. This is possible because rain (from the atmosphere) recently wet the ground (moving from the atmosphere into the hydrosphere). The robin might then fly through the air (the atmosphere) to a tree (an organism that converts matter and sunlight from the atmosphere into food energy, via photosynthesis). As the bird flies and breathes, it converts potential energy (gained from eating the worm) into kinetic energy through the process of respiration, in which oxygen from the atmosphere is combined with glucose from the organism, releasing water to the hydrosphere and carbon dioxide and heat to the atmosphere. Eventually, the bird will defecate, returning organic matter to the geosphere.

The anthroposphere is the human realm

The **anthroposphere** (sometimes called the *technosphere*) encompasses the parts of the Earth system that are modified by humans or constructed for human use, including the built environment in which we live, work, and study. The relatively recent explosion in human population also has intensified our impacts. Today we can detect human impacts within all of the other parts of the Earth system, as suggested in **FIGURE 3.4**. In many respects, environmental science is the study of the interactions of the anthroposphere with the other subsystems of Earth.

There is even a movement among some scientists to designate the sliver of geological time in which we are currently living as the *Anthropocene Epoch*. Some scientists feel that human impacts, primarily in the past few hundred years since the Industrial Revolution, are strong enough and rapid enough to warrant naming a new geological epoch after ourselves (**FIGURE 3.5**).[6]

For example, humans have caused a sharp increase in erosion rates worldwide. By clearing forests and raising crops, we have sent immense amounts of soil downwind and downstream from continents into the ocean. This rapid deposition of sediment in the ocean will be noticeable in the rock record far into the future as today's sediments become compacted into tomorrow's sedimentary rock layers.

We also have rapidly altered the composition of the atmosphere by emitting greenhouse gases as a result of deforestation, agriculture, and especially our combustion of coal, oil, and natural gas. These releases have brought atmospheric carbon dioxide and methane to their highest levels in at least 800 000 years. Consequently, Earth's mean temperature has risen 0.7°C in the past century and is predicted to rise by 1.8–4.0°C in the current century. Rising temperatures are melting polar ice; the influx of melt water into the ocean, combined with the fact that warmed water expands in volume, means that the sea level is rising. The ocean also is becoming more acidic as a result of taking in excess carbon dioxide from the atmosphere.

All of these changes—along with the pollution and habitat disturbances we are inflicting on Earth's biotic communities—are causing extinctions of animals and plants. If we step back to view our impacts in deep geological time, it becomes clear that we are setting in motion a mass extinction event. All of this is happening in the blink of an eye, geologically speaking, so it will appear very sudden to a geologist of the far future. Perhaps it is fitting to recognize this unprecedented time of rapid change in Earth's history by designating it the Anthropocene Epoch.

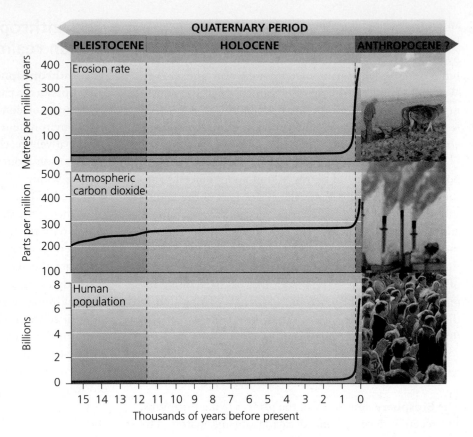

FIGURE 3.5
Global soil erosion rates (top) and atmospheric carbon dioxide concentrations (middle) have increased sharply in just the last few hundred years, along with human population (bottom). These patterns have persuaded some scientists that we should recognize a new epoch in Earth history and call it the Anthropocene
Source: Adapted from Zalasiewicz, J., et al. (2008). Are We Now Living in the Anthropocene? *GSA Today*, 18(2), pp. 4–8, Figure 1.

Ecosystems

The boundaries of all these subsystems overlap, and they interact in innumerable ways. Materials and energy flow through and among them, and from them, into and out of the biosphere. The study of such dynamic, cyclical interactions among the **biotic**, or living, components and the **abiotic**, or nonliving, components of the environment is a key part of the study of life on this planet, as scientists increasingly approach Earth systems and biological systems holistically.

Ecosystems are systems of interacting biotic and abiotic components

The systems concept can be applied to interactions that involve living systems, in much the same way as we have applied it to interactions among the physical parts of the Earth system. An **ecosystem** consists of all organisms and nonliving entities that occur and interact in a particular area at the same time. The ecosystem concept is related to the idea of a biological community (a group of interacting organisms of various types, living together in a specific habitat), which we will address in subsequent chapters.

The idea of ecosystems originated in the early twentieth century, championed by British ecologist Arthur Tansley, who saw that biological entities are tightly intertwined with chemical and physical entities. Tansley and others felt

that there was so much interaction and feedback between organisms and their abiotic environments that it made most sense to view living and nonliving elements together. The scientific discipline that deals with the abundance and distribution of organisms, the interactions among them, and their interactions with the abiotic environment is **ecology**.

Since then, **ecosystem ecology** has come to refer to the study of energy and material flows among living and nonliving components of systems. Ecosystem ecologists analyze both the *structure* of ecosystems—that is, the individual components (organisms) and their relationships—and the *functional processes* of ecosystems. Ecosystems are open systems that receive inputs of energy, process and transform the energy while cycling matter internally, and produce a variety of outputs (such as heat, water flow, and animal waste products) that can move into other ecosystems. Both matter and energy move from one level to another within the ecosystem, mainly through the feeding relationships of the constituent organisms.

Energy flows *through* ecosystems. Energy most typically arrives as radiation from the Sun, powers the system, is transformed (primarily through photosynthesis and respiration or decay), and exits in the form of heat (**FIGURE 3.6A**). Matter, in contrast, is generally recycled *within* ecosystems (**FIGURE 3.6B**). Matter is recycled because when organisms die and decay, their nutrients remain in the system. In contrast, most energy that organisms take in is later lost through respiration and the

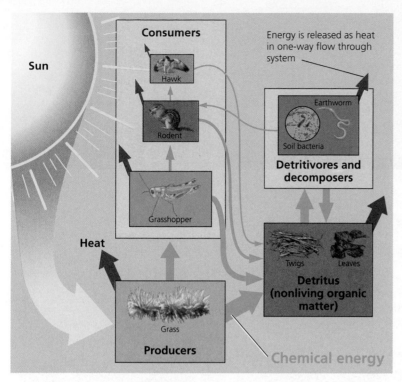

(a) Energy flows through an ecosystem

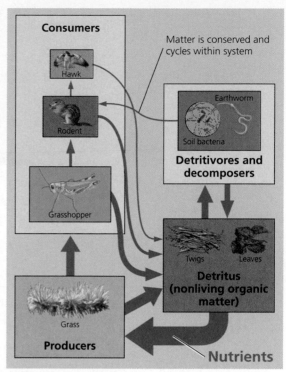

(b) Matter cycles within an ecosystem

FIGURE 3.6 Energy enters, flows through, and exits an ecosystem. In **(a)**, light energy from the Sun (yellow arrow) drives photosynthesis in producers, which begins the transfer of chemical energy (green arrows) among organisms. Energy exits through respiration and other chemical conversion processes, in the form of heat (red arrows). Matter cycles within an ecosystem. In **(b)**, blue arrows show the movement of nutrients among organisms by way of their feeding relationships. In both diagrams, box sizes represent relative magnitudes of energy or matter content, and arrow widths represent relative magnitudes of energy or matter transfer. Some abiotic components (such as water, air, and inorganic soil content) of ecosystems are omitted from these schematic diagrams.

 Based on the figure, which transfer of chemical energy is the largest in ecosystems? Which is the largest transfer of nutrients in ecosystems?

excess energy released to the surroundings as heat. The two flows are intimately related; the flow of energy through the ecosystem drives the constant recycling of matter.

Energy is converted to biomass through primary productivity

Energy flow in most ecosystems begins with radiation from the Sun. You have learned how organisms such as green plants and phytoplankton use photosynthesis to capture the Sun's energy and produce food. The result of this process is the production of **biomass**, organic material of which living organisms are formed.

The conversion of solar energy to the energy of chemical bonds in sugars by autotrophs is termed **gross primary production (GPP)**. Autotrophs use a portion of this production to power their own metabolism by respiration. The energy that remains after respiration, which is used to generate biomass, is the **net primary production (NPP)**. Thus, net primary production equals gross primary production minus respiration (or NPP = GPP – respiration by autotrophs). Net primary production can be

measured by the organic matter stored by plants, algae, and other autotrophs after they have metabolized enough for their own maintenance.

Another way to think of net primary production is that it represents the energy or biomass that is available for consumption by heterotrophs. For example, heterotrophs that eat plants use the energy they gain from the plants for their own metabolism, growth, and reproduction (or they might eat animals that have eaten plants and gain energy in that way). The total biomass that heterotrophs generate by consuming autotrophs is termed **secondary production**.

Ecosystems vary in the rate at which plants convert energy to biomass. The rate at which production occurs is termed **productivity**, and ecosystems whose plants convert solar energy to biomass rapidly are said to have high net primary productivity. (Note that because productivity is a *rate*, it is generally described in terms of an amount of production in a given area, *per unit of time*.)

Freshwater wetlands, tropical forests, coral reefs, and algal beds tend to have the highest net primary productivities, whereas deserts, tundra, and open ocean tend to have the lowest (**FIGURE 3.7A**). Variation in net primary

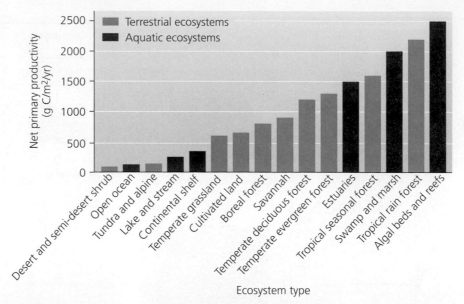

(a) Net primary productivity (NPP) for major ecosystem types

Source: Data from Whittaker, R. H. (1975). *Communities and Ecosystems*, 2nd ed. New York, MacMillan.

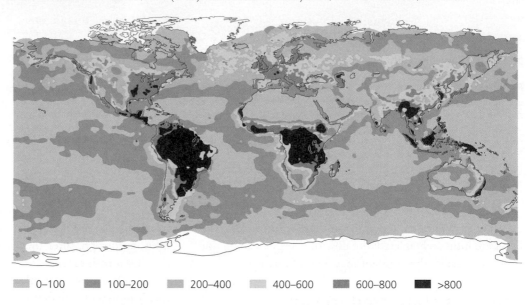

0–100 100–200 200–400 400–600 600–800 >800

(b) Global map of NPP (g C/m²/yr)

Source: From satellite data presented by Field, C. B., et al. (1998). Primary Production of the Biosphere: Integrating Terrestrial and Oceanic Components. *Science. 281*, pp. 237–240.

FIGURE 3.7 (a) Freshwater wetlands, tropical forests, coral reefs, and algal beds show high net primary productivities (NPP) on average. Deserts, tundra, and the open ocean are lower. **(b)** On land, NPP varies geographically with temperature and precipitation. In the ocean, NPP is highest around the margins of continents, where nutrients (of both natural and human origin) run off from land, and where deep, cold, nutrient-rich waters commonly well up to the surface. This map **(c)** shows human appropriation of net primary production as a percentage of the local NPP. The map provides insight into the percent of plant resource used by people in an area, compared to the amount of plant resource that is actually available locally.

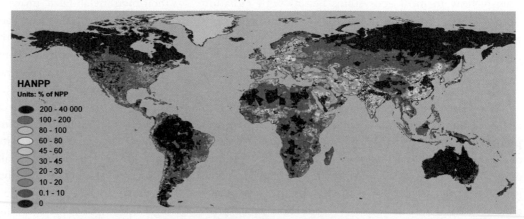

(c) Human appropriation of NPP

Source: From NASA. http://www.nasa.gov/vision/earth/environment/0624_hanpp.html.

productivity among ecosystems results in geographical patterns of variation across the globe (**FIGURE 3.7B**). In terrestrial ecosystems, net primary productivity tends to increase with temperature and precipitation. In aquatic ecosystems, net primary productivity tends to rise with light and the availability of nutrients.

Marc Imhoff and colleagues from NASA's Goddard Space Flight Center found that humans annually utilize approximately 20% of the terrestrial net primary productivity in harvesting food, fuel, fibre, and wood.[7] The study did not consider ocean primary productivity. The percentage of terrestrial NPP appropriated for human use (**FIGURE 3.7C**) depends on the type and intensity of use in a region, such as food or biomass fuel harvesting, compared to the amount of locally generated NPP (that is, plant growth). For example, people in sparsely populated but highly productive areas, like the Amazon, consume only a very small percentage of the local NPP. However, large urban areas consume many times more NPP than the local area produces.

Nutrient availability limits primary productivity

As discussed previously, nutrients are the elements and compounds that organisms consume and require for survival. Organisms need several dozen naturally occurring chemical elements to survive. Elements and compounds required in relatively large amounts are called *macronutrients*; these include nitrogen, carbon, and phosphorus. Nutrients needed in small amounts are called *micronutrients*.

Nutrients stimulate primary production by autotrophs. The availability of nitrogen or phosphorus frequently acts as a **limiting factor**; in other words, if one of these nutrients is present in less-than-ideal abundance, it will place a limitation on plant or algal growth. When these nutrients are added to a system, on the other hand, primary production is stimulated, with producers typically showing the greatest response to whichever nutrient had been in shortest supply. Phosphorus tends to be the main limiting factor in freshwater systems; nitrogen in marine systems. Thus marine hypoxic zones result primarily from excess nitrogen, whereas freshwater ponds and lakes tend to suffer eutrophication and hypoxia when they contain too much phosphorus.

Canadian ecologist David Schindler and colleagues demonstrated the effects of nutrient variability on freshwater systems in the 1970s by experimentally manipulating entire lakes.[8] In one experiment the researchers divided a 16-ha lake in Ontario in half with a plastic barrier. To one half they added carbon, nitrate, and phosphate; to the other they

added only carbon and nitrate. Soon after the experiment began, they saw a dramatic increase in algae in the half of the lake that received phosphate, whereas the other half hosted algal levels typical for lakes in the region (**FIGURE 3.8**). This difference held until shortly after they stopped fertilizing seven years later, when algae decreased to normal levels in the half that had previously received phosphate. Such experiments showed clearly that phosphorus addition can markedly increase primary productivity in lakes.

Similar experiments in coastal ocean waters show nitrogen to be the more important limiting factor for primary productivity. In experiments in the 1980s and 1990s, Swedish ecologist Edna Granéli and colleagues took samples of ocean water from the Baltic Sea and added phosphate, nitrate, or nothing.[9] Chlorophyll and phytoplankton increased greatly in the flasks with nitrate, whereas those with phosphate did not differ from the controls. Experiments by other researchers have shown similar results. For open ocean waters far from shore, research indicates that iron is a highly effective nutrient.

Because nutrients run off from land into the Baltic Sea, Gulf of the St. Lawrence, Gulf of Mexico, and other coastal waters around the world, primary productivity in the ocean tends to be greatest in nearshore waters, and lowest in open ocean areas far from land (see **FIGURE 3.7B**). Satellite imaging technology that reveals phytoplankton densities has given scientists an improved view of productivity at regional and global scales; this, in turn, has helped them track blooms of algae that may contribute to harmful coastal hypoxic zones.

FIGURE 3.8
A portion of this lake in the Experimental Lakes Area in Ontario was experimentally treated with the addition of phosphate. The treated portion experienced an immediate, dramatic, and prolonged algal bloom, visible in the opaque water in the topmost part of this photo.

TABLE 3.1   Ecosystem Services

Ecosystems provide many services that benefit us. Among other things, they

- regulate oxygen, carbon dioxide, stratospheric ozone, and other atmospheric gases
- regulate temperature and precipitation with ocean currents, cloud formation, and so on
- protect against storms, floods, and droughts, mainly with vegetation
- store and regulate water supplies in surface water and ground water
- prevent soil erosion
- form soil by weathering rock and accumulating organic material
- cycle carbon, nitrogen, phosphorus, sulphur, and other nutrients
- filter waste, remove toxins, recover nutrients, and control pollution
- pollinate plant crops and wild plants so they reproduce
- control crop pests with predators and parasites
- provide habitat for organisms to breed, feed, rest, migrate, and winter
- produce fish, game, crops, nuts, and fruits that people eat
- supply lumber, fuel, metals, fodder, and fibre
- furnish medicines, pets, ornamental plants, and genes for resistance to pathogens and crop pests
- provide recreation such as ecotourism, fishing, hiking, birding, hunting, and kayaking
- provide aesthetic, artistic, educational, spiritual, and scientific amenities

FIGURE 3.9 Ecosystems provide countless valuable services that support both natural and human systems; some of these services are illustrated here.

Ecosystems provide vital services

When scientists try to understand how ecosystems function, it is not simply out of curiosity about the world. They also know that human society depends on healthy, functioning ecosystems. When Earth's ecosystems function normally, they provide goods and services that we could not survive without. We rely not just on natural resources (which can be thought of as goods from nature), but also on the ecosystem services that our planet's systems provide (**TABLE 3.1**).

Ecological processes form the soil that nourishes our crops, purify the water we drink and the air that we breathe, store and stabilize supplies of water that we use, pollinate the food plants we eat, and receive and break down (some of) the waste we dump and the pollution we emit. The negative feedback cycles that are typical of ecosystems regulate and stabilize the climate and help dampen the impacts of the disturbances we create in natural systems. On top of all these services that are vital for our very existence, ecosystems also provide services

that enhance the quality of our lives, ranging from recreational opportunities to scenery for aesthetic enjoyment to inspiration and spiritual renewal. Ecosystem goods and ecosystem services (**FIGURE 3.9**) support our lives and society in profound and innumerable ways.

Environmental Systems in Space and Time

Environmental systems come in all shapes and sizes. A single atom or cell is a system; so is a giant redwood forest. A pond is a system; so are the Great Lakes, the ocean, and the hydrosphere. Earth as a whole is a system; so is the solar system, and even the universe.

In addition to this enormous range in the spatial scales of natural systems, the time or *temporal* scales of natural processes also differ immensely (**FIGURE 3.10**). Some processes happen nearly instantaneously; others operate

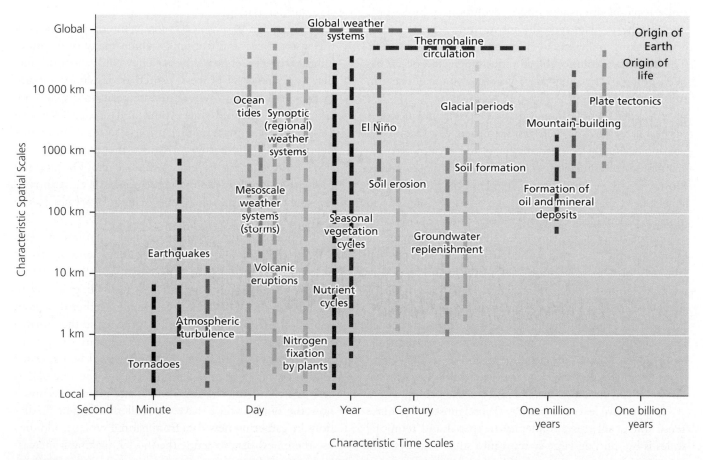

FIGURE 3.10 Natural systems and Earth processes operate on an extremely wide range of temporal and spatial scales. Some examples are illustrated here, with their (roughly) characteristic time and space scales.

DATA Q Do you notice a trend among the processes shown on this figure, from lower left to upper right? What do you think is the significance of this trend?

DATA Q Note that the characteristic time scale for soil formation is a bit longer (that is, slower) than the time scale for soil erosion. Can you think of any problems that might arise from this difference?

over millions or even billions of years. All this variation in temporal and spatial scales adds immense complexity to the scientific study of natural systems and processes. As environmental scientists, we have to adjust our approaches to study and experimentation so that we can handle this great variety.

Temporal and spatial scales of natural processes differ dramatically

Some processes within organisms, such as chemical transfers involved in photosynthesis or nitrogen fixation, happen within a fraction of a second. An earthquake may last just 10 seconds, although the tectonic processes leading to the earthquake may have been operating over centuries or millennia. Turbulence in the atmosphere can start in seconds, building to a tornado in minutes; a weather system like El Niño builds over a period of months.

In contrast to these relatively rapid processes, the evolution of species and the emergence of new species are very slow (exactly how rapidly evolution can occur is still a topic of discussion among evolutionary biologists). A lithospheric plate typically moves only 4 to 10 cm in a year, taking millennia to move any significant distance. The chemical evolution of our atmosphere and ocean has been going on for billions of years, as you will learn in subsequent chapters.

An added complication is that local events and processes may have regional or even global impacts. For example, a volcano is a highly localized landform, and a typical volcanic eruption directly impacts an area of perhaps 10-km radius. However, if the eruption cloud is so large that it reaches the stratosphere, then the volcanic materials can be transported globally, affecting the climate for a year or more.

Furthermore, if they are repeated, local, short-timescale events and processes add up to **cumulative effects** that can be global in scale and long-lasting in duration. For example, photosynthesis in a single leaf—a process that is highly localized and almost instantaneous (though seasonally limited)—is repeated over and over again in all vegetated areas of the world. The cumulative result is global change in the chemistry of the atmosphere on a seasonal basis. Through the history of life on our planet, photosynthesis has fundamentally modified the chemistry of the atmosphere.

As you have learned, one way that scientists try to make sense out of all this complexity in spatial and temporal scales is by splitting large systems into smaller subsystems. It is much more feasible to study the characteristics and processes of a smaller subsystem, and then to consider the nature of its interactions with other systems.

Therefore, scientists choose to focus on systems and processes that operate at spatial scales relevant to the types of questions they are asking. If you were interested in finding out about the role of earthworms in forest ecosystems, you might choose to study a small patch of forest soil; you probably wouldn't choose to study a system the size of North America. You might look at processes on a scale of seconds to months or perhaps years (maybe even longer, if you were interested in the cumulative effects of earthworms in soil formation); however, temporal scales encompassing millions or billions of years would likely be much less useful to your study.

Models help scientists understand complex systems

Another way that ecologists and other environmental scientists seek to make sense of complex systems is by working with models. In science, a **model** is a simplified representation of a complex natural process, designed to help us understand how the process occurs and to make predictions. Modelling is the practice of constructing and testing models that aim to explain and predict how natural systems function.

Models can be physical, such as building a model of the water cycle in a fish tank, which many of us did as children in elementary school. They can be artistic renderings of physical processes, such as drawing a picture of the water cycle showing the interrelationships among lakes, rivers, ground water, rain, and so on. They can be more conceptual, such as using a series of interconnected boxes and arrows to show the movement and storage of water in the global water cycle. Or they can be mathematical and computerized, with equations to represent and link each of the driving processes in the cycle.

Mathematical models of complex natural systems such as ecosystems or the global climate system can be extremely complicated, with many levels of detail. However, the general approach of modelling is easy to understand (**FIGURE 3.11**). Researchers gather data from nature on relationships that interest them and then form a hypothesis about what those relationships are. In a computerized model, these relationships would be expressed mathematically, using equations. Ideally, the researchers will be able to use the model to make predictions about how the system will behave. Modellers test their predictions by gathering new data from natural systems, and they use these new data to refine the model, making it increasingly accurate.

The process illustrated in **FIGURE 3.11** resembles the scientific method in general, in that it moves from observation through hypothesis, prediction, testing, and refinement of the hypothesis. This is because models are

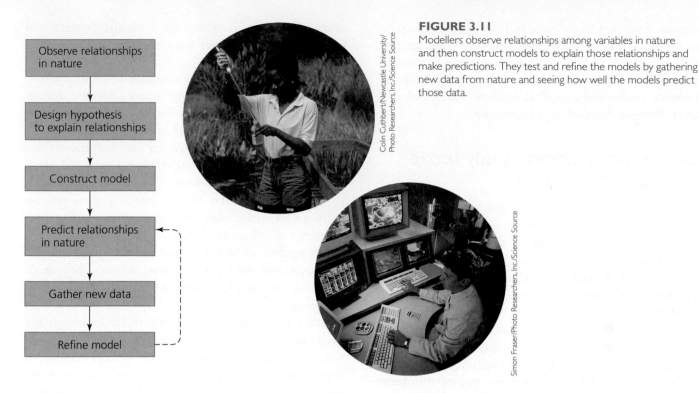

FIGURE 3.11
Modellers observe relationships among variables in nature and then construct models to explain those relationships and make predictions. They test and refine the models by gathering new data from nature and seeing how well the models predict those data.

Flow chart (left):
- Observe relationships in nature
- Design hypothesis to explain relationships
- Construct model
- Predict relationships in nature
- Gather new data
- Refine model

Image credits: Colin Cuthbert/Newcastle University/Photo Researchers, Inc./Science Source; Simon Fraser/Photo Researchers, Inc./Science Source

essentially, themselves, hypotheses about how systems function. Accordingly, the use of models is a key part of environmental research today.

For example, Lael Parrott from the Complex Systems Laboratory at Université de Montréal and her colleagues developed a multidimensional, mathematical model of the St. Lawrence Estuary, with the goal of assessing the effects of maritime traffic on marine mammals, including whales.[10] The model utilizes inputs from a variety of data sets, including spatial and temporal data about marine mammal movement patterns, and real-time tracking data from different types of vessels, including cargo ships and commercial whale-watching boats. Models of this type allow scientists and managers to investigate and test various scenarios and hypotheses without actually having to experiment on the animals in the estuary. In this particular example, the model is integrated into a decision-making support system that allows environmental managers to test scenarios and then use the results to support better decision-making in the management of maritime traffic and marine mammal interactions in the estuary.

Ecosystems are studied on a variety of spatial scales

Like all systems, we can conceptualize ecosystems at very different scales. An ecosystem can be as small as an ephemeral puddle of water where brine shrimp and tadpoles feed on algae and detritus. Or an ecosystem might be as large as a bay, a lake, or a forest. For some purposes, scientists even view the entire biosphere as a single all-encompassing ecosystem.

The term *ecosystem* is most often used, however, to refer to systems of moderate geographical extent that are somewhat self-contained. For example, the salt marshes that line the outer part of the St. Lawrence Estuary where its waters mix with those of the Atlantic Ocean may be classified as ecosystems. The individual salt marshes, in combination with the river, form part of the larger estuarine ecosystem.

Adjacent ecosystems often share components and interact extensively. For instance, a pond ecosystem is very different from a forest ecosystem that surrounds it, but salamanders that develop in the pond live their adult lives under logs on the forest floor until returning to the pond to breed. Rain water that nourishes forest plants may eventually make its way to the pond, carrying with it nutrients from the forest's leaf litter. Likewise, coastal dunes, the ocean, and a lagoon or salt marsh all may interact, as do forests and prairies where they converge.

The transitional zones where two ecosystems meet and interact are called **ecotones**. In an ecotone, the components of the adjacent ecosystems mix. The salt marshes of the St. Lawrence Estuary are ecotones, transitional between the oceanic, saltwater environment and the terrestrial, freshwater environment. In salt marshes, the water may be fresh part of the time and salty part of the time, or it may be *brackish* (saltier than fresh water, but less salty than ocean water). Like all systems, ecotones occur on a wide range of spatial scales, from a sharp, local boundary between a forest and a pond, to much broader transitional zones between regional-scale ecosystems.

Landscape ecologists study broad geographical patterns

Because components of different ecosystems may intermix, ecologists often find it useful to view these systems on a larger geographical scale that encompasses multiple ecosystems. For instance, if you are studying large mammals such as black bears, which move seasonally from mountains to valleys or between mountain ranges, you need to consider the overall landscape that includes all these areas. If you study fish such as salmon, which move between marine and freshwater ecosystems, or belugas, which have been known to follow migrating salmon quite far into estuaries and freshwater rivers, you need to know how these systems interact.

In such a broad-scale approach, called **landscape ecology**, scientists study how landscape structure affects the abundance, distribution, and interaction of organisms. A landscape-level approach is also useful for scientists, citizens, planners, and policymakers in planning for sustainable regional development. For example, the Government of Canada's strategy document for the recovery of the St. Lawrence beluga whale population[11] considers the estuarine habitat of the belugas, and takes into account all the possible inputs from the surrounding areas that might affect the welfare of the whales.

For a landscape ecologist, a landscape is made up of a spatial array of *patches*, which may be ecosystems or areas of habitat for a particular organism. Patches are spread spatially over a landscape in a *mosaic*. This metaphor reflects how natural systems often are arrayed across landscapes in complex patterns, like an intricate work of art. Thus, a forest ecologist might refer to a mosaic of forested patches within an agricultural landscape; a butterfly biologist might speak of a mosaic of grassland patches that acts as habitat for a particular species of butterfly; or a marine mammal ecologist might examine patches of aquatic habitat within a broader aquatic landscape (**FIGURE 3.12A**).

One can view landscapes at different spatial scales. **FIGURE 3.12B** illustrates a landscape consisting of four ecosystem types, with ecotones along their borders indicated by thick red lines. At this scale, we perceive a mosaic consisting of four patches plus a river. However, the inset shows a magnified view of an ecotone. At this finer resolution, we see that the ecotone consists of patches of forest and grassland in a complex arrange-

ment. The scale at which an ecologist focuses will depend on the particular questions or organisms he or she is studying.

Every organism has specific habitat needs, so when its habitat is distributed in patches across a broader mosaic, individuals may need to expend energy and risk predation travelling from one to another. If the patches are distant enough, the organism's population may become divided into subpopulations, each occupying a different patch in the mosaic. Such a network of subpopulations, most of whose members stay within their respective patches but some of whom move among patches or mate with members of other patches, is called a *metapopulation*. When patches are still more isolated from one another, individuals may not be able to travel between them at all. In such a case, smaller subpopulations may be at risk of extinction.

Because of this extinction risk, metapopulations and landscape ecology are of great interest to *conservation biologists*, who study the loss, protection, and restoration of species and their habitats. Of particular concern is the fragmentation of habitat into small and isolated patches—something that often results from human impact. Establishing corridors to link patches is one approach that conservation biologists pursue as they attempt to maintain biodiversity in the face of human impact. We will return to these issues when we consider conservation biology and habitat fragmentation in subsequent chapters.

Remote sensing and GIS are important tools

Remote sensing technologies—that is, technologies that collect information about a target object from a distance—are improving our ability to take a landscape perspective on complex ecosystems. Satellites orbiting Earth are sending us more and better data than ever before on what the surface of our planet looks like (the chapter-opening photo is an example). By helping us monitor our planet from above, satellite imagery has become a vital tool in modern environmental science. (See also "The Science behind the Story: Mission to Planet Earth: Monitoring Environmental Change," in Chapter 1.)

A common tool that makes use of both remotely sensed and ground-based data is the **geographic information system (GIS)**. A GIS consists of computer software that takes multiple types of spatially referenced data (for instance, on geology, hydrology, vegetation, animal species, and human development) and combines them on a common set of geographical coordinates. The idea is to create a complete picture of a landscape and to analyze how elements of the different data sets are arrayed spatially and how they may be correlated.

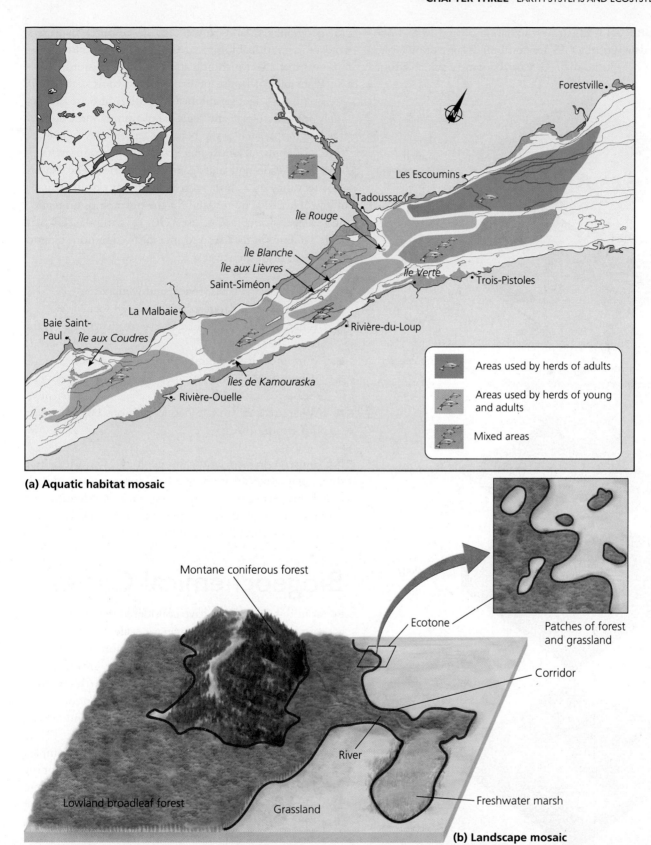

(a) Aquatic habitat mosaic

Legend:
- Areas used by herds of adults
- Areas used by herds of young and adults
- Mixed areas

Montane coniferous forest

Ecotone

Patches of forest and grassland

Corridor

River

Freshwater marsh

Lowland broadleaf forest

Grassland

(b) Landscape mosaic

FIGURE 3.12 Landscape ecology deals with broad spatial patterns and interactions among ecosystems. Within the St. Lawrence Estuary ecosystem, a variety of habitat patches **(a)** host different herds of beluga whales. Each patch offers slightly different water salinity, temperature, and depth conditions within the broader mosaic of the estuary. The landscape in **(b)** consists of a mosaic of five ecosystem types (three terrestrial types, a marsh, and a river). Thick red lines indicate ecotones. Zooming in on the forest–grassland ecotone (magnified), we see that it also consists of variegated patches on a smaller scale. Source for part (a): Recovery strategy for the beluga whale (*Delphinapterus leucas*) St. Lawrence Estuary population in Canada, p. 10, Figure 6: Summer distribution of St. Lawrence belugas by herd composition, adapted from Michaud (1993). http://www.sararegistry.gc.ca/virtual_sara/files/plans/ rs_st_laur_beluga_0312_e.pdf.

FIGURE 3.13 illustrates in a simplified way how different data sets of a GIS are combined, layer upon layer, to form a composite map. GIS has become a valuable tool for geographers, landscape ecologists, resource managers, and conservation biologists. GIS technology also brings insights that affect planning and land-use decisions.

Principles of landscape ecology and tools such as GIS are useful for scientific inquiries, but they also are increasingly applied in local and regional planning. Some conservation groups, such as the Nature Conservancy of Canada (NCC), apply a landscape ecology approach in their land acquisition and management strategies. The Nature Conservancy is a land trust—an organization, usually nonprofit, that acquires land for the purpose of protection and conservation. NCC uses GIS to compile and map large volumes of data, and to investigate questions like the following, from their website:[12]

- Where are the lakes, rivers, and wetlands?
- Where have endangered species been seen?
- Where are the different ecosystems that make up the landscape?
- What areas have the highest diversity of species?
- Where is the most suitable habitat for a particular species?
- What areas are in the best condition, but are under the greatest threat?

NCC uses GIS and landscape ecology as tools to help them prioritize areas in need of conservation, decide which properties to acquire, keep track of scientific and other data about those properties, and track both the human and natural features of their properties.

Biogeochemical Cycles

So far in this chapter we have considered the general characteristics of systems; the planetary-scale subsystems that make up the Earth system; the interaction of biotic and abiotic components in ecosystems; and the great diversity in temporal and spatial scales among Earth's natural systems and processes. Now we turn our attention to the movement of materials and energy through and among these systems.

Materials move through the environment in complex and fascinating ways. Many of these flows are crucial for the support of life on this planet. Energy enters an ecosystem from the Sun, flows from one organism to another, and is dissipated to the atmosphere as heat, but physical matter is circulated and recirculated through natural systems, over and over again. The cycling of nutrients is one of the most important and fundamental of the ecosystem services that support life on this planet. Through the processes that take place within and among ecosystems, the chemical elements and compounds that we need—carbon, nitrogen,

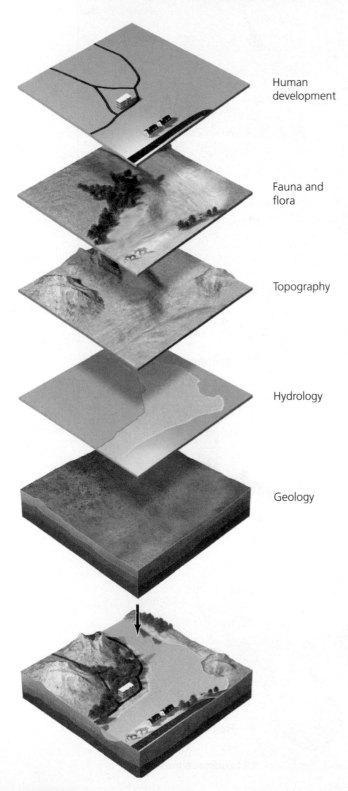

Human development

Fauna and flora

Topography

Hydrology

Geology

FIGURE 3.13
Geographic information systems allow scientists to layer different types of data on natural landscape features and human land uses, from both remote sensing and ground-based observations, and to produce maps integrating this information. GIS can be used to explore correlations among these data sets.

phosphorus, water, and many more—move through the environment in complex, global biogeochemical cycles.

Nutrients and other materials move in biogeochemical cycles

Materials move through ecosystems in **nutrient cycles**, also called **biogeochemical cycles**. They are called *biogeochemical* because the processes involved are biological, geological, and chemical (as well as physical). In global biogeochemical cycles, materials travel through the atmosphere, hydrosphere, and geosphere, moving from one organism to another in dynamic equilibrium. A carbon atom in your fingernail today might have resided in the muscle of a cow a year ago, in a blade of grass before that, in the atmosphere before that, in soil organic matter before that, and in a dinosaur's body 100 million years before that. After we die, the nutrients in our bodies will spread widely through the physical environment, eventually being incorporated by an untold number of organisms in the biosphere, far into the future.

Nutrients and other materials (including metals and other toxins, as you will see in subsequent chapters) move from one *pool*, or **reservoir**, to another, remaining for varying amounts of time—the **residence time**—in each reservoir. The dinosaur, the soil, the atmosphere, the grass, the cow, and you are each reservoirs for carbon atoms. The average residence time for an atom of carbon in your body is longer than the average residence time for an atom of carbon in a blade of grass, which has a short lifespan and will soon die, releasing its carbon back to the surrounding environment.

Virtually *all* materials, not just nutrients, move from reservoir to reservoir through biogeochemical cycles. Take mercury, for example. Mercury is an element (Hg). It occurs in inorganic form in the mineral cinnabar (mercury sulphide, HgS). Once it has been weathered from its host rock in the geosphere, mercury moves among reservoirs in the atmosphere, hydrosphere, and biosphere in a wide variety of forms. Some are inorganic compounds, and some are organic; many forms of mercury are toxic to humans. Some mercury compounds bioaccumulate in organisms, like PAHs that bioaccumulate in whales. When the organism dies, the mercury compound is released to the surrounding environment, moving from one reservoir to another and changing form as it moves through the biogeochemical cycles of the natural environment.

The movement of materials among reservoirs is termed a **flux** (**FIGURE 3.14**). Fluxes are rates, so they are stated in terms of a mass or volume of material moving among reservoirs *per unit of time*. The flux of a material between reservoirs can change over time. Human activity also has greatly influenced the fluxes of certain materials. For example, we have increased the flux of nitrogen from the atmosphere to terrestrial reservoirs, and we have shifted the flux of carbon in the opposite direction.

Reservoirs that release more nutrients (or any other material of interest) than they accept are called **sources**; reservoirs that accept more nutrients than they release are called **sinks**. Carbon sinks are of particular importance today, as we struggle to lower the rate at which carbon is released into the atmosphere, which has significant impacts on the global climate system.

The time it would take for all of the atoms (or particles) of a particular material to be flushed through a particular reservoir is called the **turnover time**. Turnover time is a function of the balance between fluxes *into* the reservoir (from a source) and fluxes *out of* the reservoir (to a sink). If we stop all new sources of material incoming into the reservoir, then the turnover time is the amount of time it would take for the material to be completely flushed through and out of the system.

Turnover time also depends on processes that influence the residence time of the material, including any processes that might hold or bind the material within the reservoir or cause it to be flushed through more quickly. For example, let us say that we are interested in mercury in the water of a particular lake. The turnover time for mercury

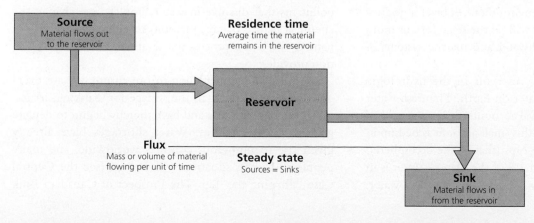

Source
Material flows out to the reservoir

Residence time
Average time the material remains in the reservoir

Reservoir

Flux
Mass or volume of material flowing per unit of time

Steady state
Sources = Sinks

Sink
Material flows in from the reservoir

FIGURE 3.14 The properties of reservoirs and how cyclical fluxes move materials into and out of reservoirs are fundamentally important concepts in environmental science today.

in the lake will depend on how quickly fresh water runs into the lake, and how quickly the mercury-laden water leaves the lake. It will also depend on other sources of mercury inputs, such as mercury deposited onto the lake surface from the air, or taken up from mercury-bearing rocks in the lake bed. Now, what if some of the mercury in the water sinks to the bottom and binds chemically to bottom sediments? This would greatly increase both the residence time (how long the mercury stays in the lake) and the turnover time (how quickly the mercury will be flushed through the lake).

These are extremely important concepts in environmental science today. For example, we are concerned about the presence in the atmosphere of substances that damage the stratospheric ozone layer. Many of these substances have very long residence times in the atmosphere. Even if we stop producing them altogether—which we have almost accomplished as a result of the Montreal Protocol on Ozone-Depleting Substances—the turnover time for these materials to be cleared out of the atmosphere by natural processes will be measured in decades. As we discuss various biogeochemical cycles, think about how human actions can influence fluxes and generate feedback loops in these cycles.

Let's start with a look at the most fundamental of the biogeochemical cycles: the water cycle.

The hydrologic cycle influences all other cycles

Water is so integral to life that we take it for granted. The essential medium for all manner of biochemical reactions, water plays key roles in nearly every environmental system, including each of the nutrient cycles we have just discussed. Water carries nutrients and sediments from the continents to the ocean via rivers, streams, and surface runoff, and it distributes sediments onward in ocean currents. Increasingly, water also distributes artificial pollutants.

The *water cycle*, or **hydrologic cycle** (**FIGURE 3.15**), summarizes how water—in liquid, gaseous, and solid forms—flows through our environment. A brief introduction to the hydrologic cycle will set the stage for our more in-depth discussion of freshwater and marine systems in subsequent chapters.

The ocean is the main reservoir in the hydrologic cycle, holding 97% of all water on Earth. The fresh water we depend on for our survival accounts for less than 3% of the total, and two-thirds of this small amount is tied up in glaciers, snowfields, and ice caps (that is, the cryosphere). Thus, considerably less than 1% of the planet's water is in a form that we can readily use—shallow ground water,

surface fresh water, and rain from atmospheric water vapour.

Water moves from the ocean, lakes, ponds, rivers, and moist soil into the atmosphere by **evaporation**, the conversion of a liquid to gaseous form. Warm temperatures and strong winds speed rates of evaporation. A greater degree of exposure has the same effect; an area logged of its forest or converted to agriculture or residential use will lose water more readily than a comparable area that remains vegetated. Water also enters the atmosphere by **transpiration**, the release of water vapour by plants through their leaves. Transpiration and evaporation act as natural processes of distillation, effectively creating pure water by filtering out minerals carried in solution.

Water returns from the atmosphere to Earth's surface as **precipitation** when water vapour condenses and falls as rain or snow. Precipitation may be taken up by plants and used by animals, but much of it flows as **runoff** into streams, rivers, lakes, ponds, and the ocean. Amounts of precipitation vary greatly from region to region globally, helping give rise to the great variety of ecosystems.

Some precipitation and surface water soaks down through soil and rock to recharge underground reservoirs called *aquifers*. Aquifers are porous bodies of rock and soil that hold **ground water**, water found underground beneath layers of soil. Aquifers may hold ground water for long periods of time, so the water may be quite ancient. In some cases ground water can take hundreds or even thousands of years to recharge fully after being depleted. Where ground water intersects the surface, the exposed water becomes surface runoff or evaporates into the atmosphere.

Human activity affects every aspect of the water cycle. By damming rivers to create reservoirs, we increase evaporation and, in some cases, infiltration of surface water into aquifers. By altering Earth's surface and its vegetation, we increase surface runoff and erosion. By spreading water on agricultural fields, we can deplete rivers, lakes, and streams and can increase evaporation. By removing forests and other vegetation, we reduce transpiration and may lower water tables. By emitting into the atmosphere pollutants that dissolve in water droplets, we change the chemical nature of precipitation, in effect sabotaging the natural distillation process that evaporation and transpiration provide.

Perhaps most threatening to our future, we are overdrawing ground water to the surface for drinking, irrigation, and industrial use and have thereby begun to deplete groundwater resources. Water shortages have already given rise to numerous conflicts worldwide, and many people think this situation will worsen (see the Central Case "Turning the Tap: The Prospect of Canadian Bulk

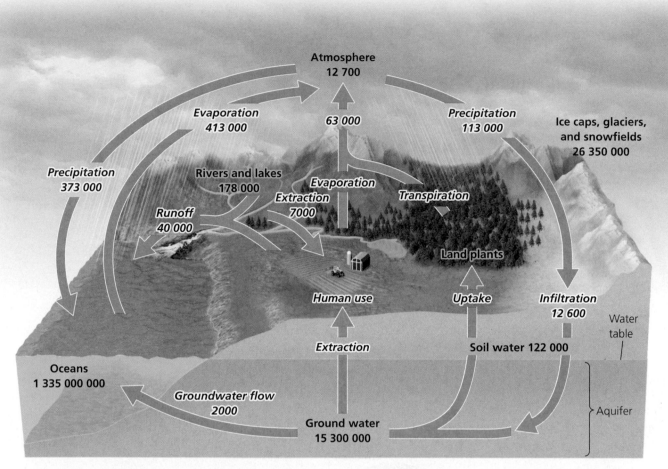

FIGURE 3.15 The hydrologic cycle summarizes the many routes that water molecules take as they move through the environment. Grey arrows represent fluxes. The hydrologic cycle is a system itself, but it also plays key roles in other biogeochemical cycles. Reservoir names are in black type, and the black numbers represent reservoir sizes expressed in units of cubic kilometres (km³). Transfer processes (red) give rise to fluxes, expressed in km³ per year.
Source: Data from Schlesinger, W.H. (2013). *Biogeochemistry: An Analysis of Global Change,* 3rd edition. Academic Press, London.

 Examine the inputs and outputs of water in the ocean, via precipitation and evaporation. Do they balance? Why (or why not)?

Water Exports" in the chapter *Freshwater Systems and Water Resources*).

The carbon cycle provides the foundation for living organisms

As the definitive component of organic molecules, **carbon (C)** is an ingredient in carbohydrates, fats, and proteins, and in the bones, cartilage, and shells of all living things. From fossil fuels to DNA, from plastics to pharmaceuticals, carbon atoms are everywhere. The **carbon cycle** describes the routes that carbon atoms take through the environment (**FIGURE 3.16**).

Producers, including terrestrial and aquatic plants, algae, and cyanobacteria, pull carbon dioxide out of the atmosphere and surface water to use in photosynthesis.

Photosynthesis breaks the bonds in carbon dioxide (CO_2) and water (H_2O) to produce oxygen (O_2) and carbohydrates (e.g., glucose, $C_6H_{12}O_6$). Autotrophs use some of the carbohydrates to fuel their own respiration, releasing some of the carbon back into the atmosphere and ocean as CO_2. When producers are eaten by primary consumers, which in turn are eaten by secondary consumers, more carbohydrates are broken down, producing carbon dioxide and water. The same process occurs when decomposers consume waste and dead organic matter. Respiration from all these organisms releases carbon back into the atmosphere and ocean.

All organisms use carbon for structural growth, so a portion of the carbon taken in becomes incorporated into their tissues. Plants are a major reservoir for carbon because they take in so much carbon dioxide for photosynthesis,

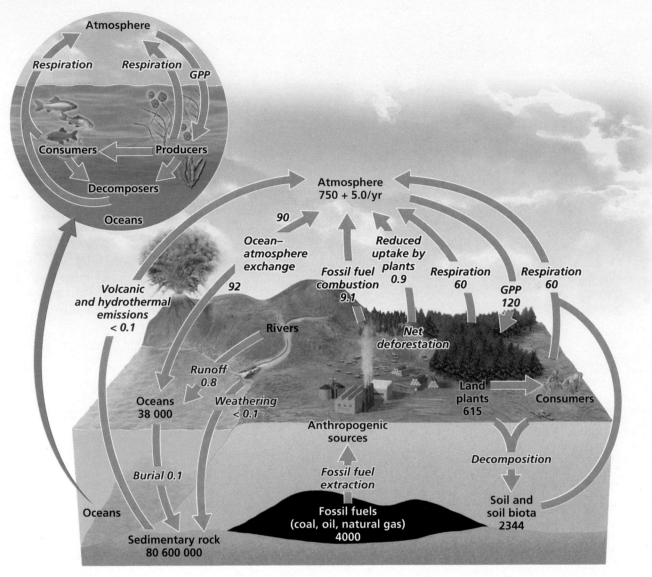

FIGURE 3.16 The carbon cycle summarizes the many routes that carbon atoms take as they move through the environment. Grey arrows represent fluxes among carbon reservoirs. Reservoir names are in black, and the black numbers represent reservoir sizes expressed in petagrams (units of 10^{15} g) of C. Transfer processes give rise to fluxes, in red, expressed in petagrams (billion metric tons) of C per year. Fluxes of carbon to the atmosphere from human sources (fossil fuels and deforestation) have increased in recent decades.
Source: Data from Schlesinger, W.H. (2013). *Biogeochemistry: An Analysis of Global Change,* 3rd edition. Academic Press, London.

and they are so abundant. Because CO_2 is a greenhouse gas of primary concern, much research on global climate change is directed toward measuring the amount of CO_2 held by plants. Scientists are working toward understanding exactly how and to what extent this portion of the carbon cycle influences Earth's climate.

As organisms die, their remains may settle in sediments in ocean basins or coastal wetlands. As layers of sediment accumulate, older layers are buried more deeply, experiencing high pressure over long periods of time. These conditions can lead to the conversion of soft tissues into fossil fuels—coal, oil, and natural gas—and shells and skeletons into sedimentary rock, such as limestone. Sedimentary rock makes up the largest single reservoir in

the carbon cycle. A given carbon atom spends a relatively short time in the atmosphere, but carbon trapped in sedimentary rock may reside there for hundreds of millions of years. Carbon trapped in sediments and fossil fuel deposits may eventually be released into the ocean or atmosphere by geological processes such as tectonic uplift and volcanic eruptions, or by the burning of fossil fuels.

After sedimentary rock, the ocean is the second-largest reservoir in the carbon cycle. Ocean water absorbs carbon compounds from the atmosphere, from terrestrial runoff, from undersea volcanoes, and from the waste and detritus of marine organisms. Carbon atoms absorbed by the ocean in the form of carbon dioxide, carbonate ions (CO_3^{2-}), and bicarbonate ions (HCO_3^-) combine with calcium ions

(Ca^{2+}) to form calcium carbonate ($CaCO_3$), an essential ingredient in the skeletons and shells of microscopic marine organisms. As these organisms die, their calcium carbonate shells sink to the ocean floor and begin to form sedimentary rock. The rate at which ocean water absorbs and releases carbon depends on many factors, including temperature and the number of marine organisms converting CO_2 into carbohydrates and carbonates.

By mining fossil fuel deposits, we are removing carbon from an underground reservoir in the geosphere, where it has a residence time of millions of years. When we combust fossil fuels in our automobiles, homes, and industries, we release carbon dioxide and greatly increase the flux of carbon from the geosphere to the atmosphere. Since the mid-eighteenth century, fossil fuel combustion has added over 250 billion metric tons of carbon to the atmosphere. Cutting down forests and burning fields also remove carbon from the vegetation reservoir and release it to the air. Meanwhile, the movement of CO_2 *from* the atmosphere *back* to the hydrosphere, geosphere, and biosphere has not kept pace.

As a result, scientists estimate that today's atmospheric carbon dioxide reservoir is the largest that Earth has experienced in at least the past 800 000 years, and perhaps in the past 20 million years. The anthropogenic flux of carbon out of the fossil fuel reservoir and into the atmosphere is a driving force behind global climate change. Some of the excess CO_2 in the atmosphere is now being absorbed by ocean water. This is causing ocean water to become more acidic, leading to problems that threaten many marine organisms. We will consider these issues in much greater detail in the chapter on global climate change.

Our understanding of the carbon cycle is incomplete. For example, scientists have long been baffled by the so-called "missing" carbon sink. Of the carbon dioxide we emit by fossil fuel combustion and deforestation, scientists have measured how much goes into the atmosphere and ocean, but there remain roughly 2.3 to 2.6 billion metric tons per year unaccounted for. Many researchers think this is taken up by plants or soils of the northern temperate and boreal forests. But they'd like to know for sure—because if forests are acting as a major sink for carbon, it would be a good idea to keep it that way. If forests that today are sinks were to turn into sources and begin releasing the "missing" carbon, climate change could accelerate drastically.

The nitrogen cycle involves specialized bacteria

Nitrogen (N) makes up 78% of our atmosphere by mass and is the sixth most abundant element on Earth. It is an essential ingredient in the proteins that build our bodies, and an essential nutrient for plant growth. Thus the **nitrogen cycle** (**FIGURE 3.17**) is of vital importance to all organisms.

Despite its abundance in air, nitrogen gas (N_2) is chemically inert and cannot cycle out of the atmosphere and into living organisms without assistance from lightning, highly specialized bacteria, or human intervention. For this reason, nitrogen in elemental form is scarce in the geosphere and hydrosphere and in organisms. However, once nitrogen undergoes the right kind of chemical change, it becomes biologically active and available to the organisms that need it. Its scarcity makes biologically active nitrogen a limiting factor for plant growth.

To become biologically available, inert nitrogen gas (N_2) must be "fixed," or combined with hydrogen in nature to form ammonia (NH_3), whose water-soluble ions of ammonium (NH_4^+) can be taken up by plants. **Nitrogen fixation** in nature is accomplished in two ways: by the intense energy of lightning strikes, or (much more importantly) by the action of specialized bacteria. One such group is aquatic cyanobacteria, either free-living or living inside corals; another group of specialized bacteria fixes nitrogen on land, in the top layer of soil. These bacteria live in a mutually beneficial relationship with many types of land plants, including clover, soybeans, and other legumes, providing them with nutrients by converting nitrogen to a usable form. Farmers have long nourished their soils by planting crops that host nitrogen-fixing bacteria among their roots.

Other types of specialized soil bacteria then perform a process known as **nitrification**, in which ammonium ions are first converted into nitrite ions (NO_2^-), and then into nitrate ions (NO_3^-). Plants can take up these ions, which can also become available after atmospheric deposition on soils or in water or after application of nitrate-based fertilizer. Animals obtain the nitrogen they need by consuming plants or other animals, and decomposers obtain nitrogen from dead and decaying plant and animal matter and from animal urine and feces. Once the decomposers process the nitrogen-rich compounds they take in, they release ammonium ions, making these available to nitrifying bacteria to convert again to nitrates and nitrites.

The final step in the nitrogen cycle occurs when denitrifying bacteria convert nitrates in soil or water back into gaseous nitrogen, via the multistep process of **denitrification**. This completes the cycle by releasing nitrogen back into the atmosphere as a gas. We will revisit these processes and the role of soil in the nitrogen cycle in our chapter on soil resources. The nitrogen biogeochemical cycle is an interesting example of a cycle with processes that operate on a local—even microscopic—scale. But when you consider together all of the billions and billions of microscopic transformations of nitrogen that are happening in soils at any given time, it adds up to a global cycle of enormous importance.

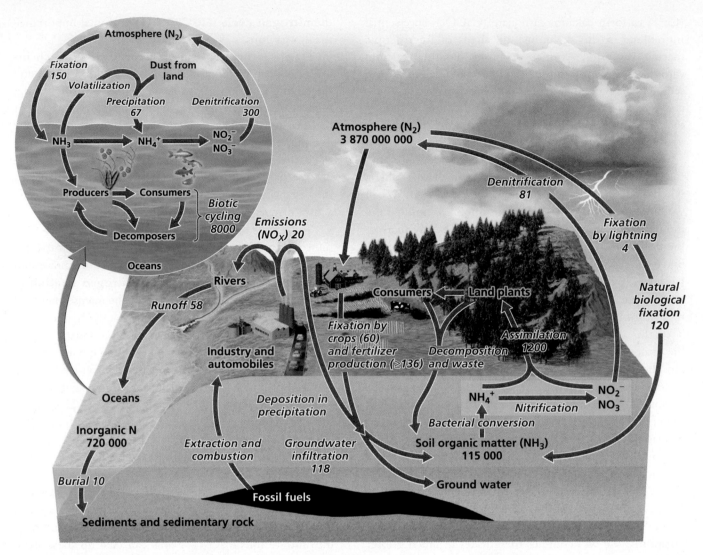

FIGURE 3.17   The nitrogen cycle summarizes the many routes that nitrogen atoms take as they move through the environment. Grey arrows represent fluxes among reservoirs for nitrogen. Reservoir names are in black, and the black numbers are reservoir sizes expressed in teragrams (units of 10^{12} g) of N. Transfer processes give rise to fluxes, in red, expressed in teragrams of N per year.
Source: Data from Schlesinger, W.H. (2013). *Biogeochemistry: An Analysis of Global Change*, 3rd edition. Academic Press, London.

Human activities have had deep impacts on the nitrogen cycle. One of the greatest advances in human history was the discovery of a mechanism for fixing nitrogen on an industrial scale. This **Haber-Bosch process**, a chemical process for synthesizing ammonium, thereby fixing nitrogen on an industrial scale, enabled people to overcome the limits on agricultural productivity long imposed by nitrogen scarcity in nature, but also greatly increased our impacts on the nitrogen cycle. Its widespread application has made modern agriculture possible, but it has also led to dramatic alterations of the nitrogen cycle. Today our species is fixing more nitrogen artificially than is fixed naturally; we have doubled the natural rate of nitrogen fixation on Earth.

By fixing atmospheric nitrogen with fertilizers, we increase the flux of nitrogen from the atmosphere to Earth's surface. We also enhance this flux by cultivating legume crops whose roots host nitrogen-fixing bacteria.

Moreover, we reduce nitrogen's return to the air when we destroy wetlands that filter nutrients; wetland plants host denitrifying bacteria that convert nitrates to nitrogen gas, so wetlands can mop up a great deal of nitrogen pollution.

When farming practices speed runoff and allow soil erosion, nitrogen flows from farms into terrestrial and aquatic ecosystems. Fertilizer-laden runoff increases the nitrogen available to aquatic plants and algae, boosting their growth (**FIGURE 3.18**). Algal populations soon outstrip the availability of other required nutrients and begin to die and decompose. This large-scale decomposition uses up oxygen in the water, placing what is called a *biochemical oxygen demand* (or *BOD*) on the water body. The reduction of dissolved oxygen in the water can lead to hypoxia, robbing aquatic organisms of oxygen and leading to shellfish die-offs and other significant impacts on ecosystems. Excess nitrogen from agricultural fertilizers in the Great Lakes and St. Lawrence system have had a negative

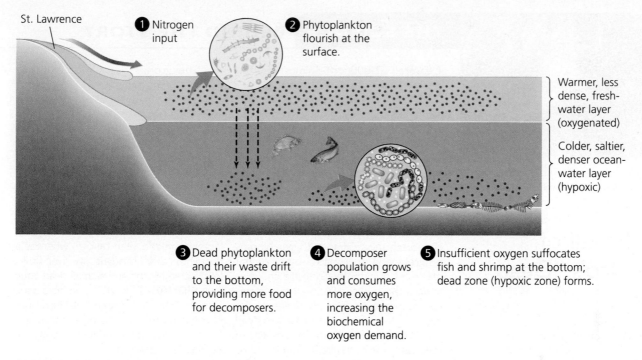

St. Lawrence

① Nitrogen input

② Phytoplankton flourish at the surface.

Warmer, less dense, fresh-water layer (oxygenated)

Colder, saltier, denser ocean-water layer (hypoxic)

③ Dead phytoplankton and their waste drift to the bottom, providing more food for decomposers.

④ Decomposer population grows and consumes more oxygen, increasing the biochemical oxygen demand.

⑤ Insufficient oxygen suffocates fish and shrimp at the bottom; dead zone (hypoxic zone) forms.

FIGURE 3.18 Excess nitrogen can cause eutrophication in marine and freshwater systems. Coupled with stratification (layering) of the water, eutrophication can severely deplete dissolved oxygen. Surface water is warmer and fresher, and thus less dense; deeper water is colder and saltier, and thus denser. Nitrogen from river water (**1**) boosts the growth of phytoplankton in surface layers (**2**), which eventually die and are decomposed at the bottom by bacteria (**3**). Phytoplankton in surface water absorb light, preventing photosynthesis from occurring in deeper layers (thus inhibiting oxygen production in deeper waters). Stability of the surface layer also prevents deeper water from absorbing atmospheric oxygen to replace oxygen consumed by decomposers (**4**). Oxygen depletion in lower layers suffocates or drives away bottom-dwelling marine life (**5**), and gives rise to hypoxic or "dead" zones.

impact on both water quality and the health of marine organisms in downstream areas, as discussed in the Central Case. Similar impacts in the Mississippi River watershed have become painfully evident to shrimpers and scientists with an interest in the Gulf of Mexico (see "The Science behind the Story: The Gulf of Mexico's 'Dead Zone'").

Human activity also affects fluxes in other parts of the nitrogen cycle. When we burn forests and fields, we force nitrogen out of soils and vegetation and into the atmosphere. When we burn fossil fuels, we increase the rate at which nitric oxide (NO) enters the atmosphere and reacts to form nitrogen dioxide (NO_2), a precursor to acid precipitation. We introduce nitrous oxide (N_2O) by allowing anaerobic bacteria to break down the tremendous volume of animal waste produced in agricultural feedlots. We have accelerated the introduction of nitrogen-rich compounds into terrestrial and aquatic systems by destroying wetlands and by cultivating crops that host nitrogen-fixing bacteria in their roots. As these examples show, human activities have affected the nitrogen cycle in diverse and often far-reaching ways.

The phosphorus cycle circulates a key plant nutrient

The element **phosphorus (P)** is a key component of cell membranes and of several molecules vital for life,

including DNA, RNA, ATP, and ADP. Although phosphorus is indispensable for life, the amount of phosphorus in organisms is dwarfed by the vast amounts in rocks, soil, sediments, and the ocean. Unlike the carbon and nitrogen cycles, the **phosphorus cycle** (**FIGURE 3.19**) has no appreciable atmospheric component, aside from tiny amounts of windblown dust and sea spray.

The vast majority of Earth's phosphorus is contained in rocks and mobilized by weathering, which releases phosphate ions (PO_4^{3-}) into water. Phosphate dissolved in lakes or in the ocean precipitates into solid form, settles to the bottom, and re-enters the geosphere in the form of sediment. Because most phosphorus is bound up in rock, environmental concentrations of phosphorus available to organisms tend to be very low. This relative rarity explains why phosphorus is frequently a limiting factor for the growth of plants and algae, and why an artificial influx of phosphorus can produce immediate and dramatic effects.

Plants can take up phosphorus through their roots only when phosphate is dissolved in water. Primary consumers acquire phosphorus from water and plants and pass it on to secondary consumers. Consumers release phosphorus through the excretion of waste. Decomposers break down phosphorous-rich organisms and their wastes and, in so doing, return phosphorus to the soil.

Fishers in the Gulf of Mexico have been dealt numerous challenging blows in recent years.

Emily Michot/MCT/Landov

The Gulf of Mexico's "Dead Zone"

Fishers in Louisiana ply the rich waters of the Gulf of Mexico to send shrimp, fish, and shellfish to dinner tables around the world. But fishing in the Gulf has become difficult; catches of shrimp and other common species are half of what they were in the 1980s. The fisheries began declining years before oil gushed from the BP *Deepwater Horizon* drilling platform and fouled the region in 2010, and before Hurricanes Katrina and Rita pummelled the region in 2005 and left boats, docks, and marinas in ruins. Those disasters worsened a long decline that was already underway.[13]

The reason for the decline? Billions of marine organisms have been suffocating in a hypoxic "dead zone," a region of water in the Gulf that is so depleted of oxygen that organisms are killed or driven away. Determining the causes of such dead zones involves understanding environmental systems and the complex behaviour they exhibit.

Aquatic animals obtain oxygen through their gills and, like us, will asphyxiate if deprived of oxygen. Fully oxygenated water contains up to 10 parts per million (ppm) of oxygen. When concentrations drop below 2 ppm, creatures will leave or die. The Gulf's hypoxic dead zone appears each spring and grows through the summer and fall, beginning in Louisiana waters offshore from the mouths of the Mississippi and Atchafalaya rivers. The zone reached its largest size in 2002 when

it covered 22 000 km² —almost half the size of Nova Scotia.

What is starving these waters of oxygen? Scientists studying the dead zone are pointing a finger at human activities hundreds of kilometres away.

Rain and runoff carry excess nitrogen and phosphorus from fertilized agricultural fields and into streams and rivers, eventually flushing these nutrients down the Mississippi and Atchafalaya rivers, which drain into the Gulf. Once the excess nutrients reach the Gulf, they trigger blooms of plankton in the surface waters. As the masses of plankton begin to die and drift toward the bottom, they nourish bacteria, which also become abundant. As the bacteria decompose the masses of dead plankton, they use up dissolved oxygen in the bottom waters. Urban runoff, industrial discharges, fossil fuel emissions, and municipal sewage outflow add additional nitrogen and phosphorous pollution to the rivers as they head toward the Gulf.

In 1985, Dr. Nancy Rabalais and other researchers from the Louisiana Universities Marine Consortium (LUMCON) began tracking oxygen levels at dozens of sites in the Gulf and nearshore locations.[14] Sensors lowered into the water were used to measure oxygen levels, sending continuous readings back to a shipboard computer. Further data come from fixed, submerged oxygen meters that continuously measure dissolved oxygen and store the data. They collected hundreds of water samples and measured nitrogen, salt, bacteria, and phytoplankton. They monitored more than 70 sites in the Gulf, and donned scuba gear to observe the condition of shrimp, fish, and other sea life.

The long-term data collection allowed the researchers to build a "map" of the dead zone, tracking its location and its consequences. In 1991, Rabalais made that map public, earning immediate headlines. That year, her group mapped the size of the dead zone at more than 10 000 km². Bottom-dwelling shrimp were stretching out of their burrows, straining for oxygen. Many fish had fled. The bottom waters, infused with sulphur from bacterial decomposition, smelled of rotten eggs.

The group's years of monitoring enabled them to predict the dead zone's emergence. As rivers rose each spring, and as fertilizers were applied on Midwestern farms, oxygen would start to disappear in the northern Gulf. The hypoxia would last through the summer or fall, until seasonal storms mixed oxygen into hypoxic areas. The monitoring also linked the dead zone's size to the volume of river flow and its nutrient load. The 1993 flooding of the Mississippi created a zone much larger than the year before, whereas a drought in 2000 brought low river flows, low nutrient loads, and a small dead zone (see **FIGURE 1**). In 2005, the dead zone was predicted to be large, but Hurricanes Katrina and Rita stirred oxygenated surface water into the depths, decreasing the dead zone that year.

In response to these findings, U.S. government regulators proposed that farmers in Midwestern states be required to cut down on fertilizer use. Farmers' advocates protested that farmers were being singled out while urban pollution sources were ignored. Meanwhile, scientists have documented coastal dead zones in 400 other areas throughout the world, from Chesapeake Bay to the St. Lawrence Estuary to the Black Sea.

Midwestern farming advocates and some scientists, such as Derek Winstanley, chief of the Illinois State Water Survey, challenged the link to farm practices. They argued that the Mississippi naturally carries high loads of nitrogen from the rich prairie soil, and that Rabalais's team had not ruled out upwelling in the Gulf as a source of nutrients. But sediment analyses showed that Mississippi River mud contained many fewer nitrates early in the twentieth century. In 2000, a federal assessment team of dozens of scientists laid the blame for the dead zone on nutrient runoff from fertilizers and other sources.

Then in 2004, U.S. Environmental Protection Agency water quality scientist Howard Marshall suggested that the best way to alleviate the dead zone would be to reduce phosphorous pollution from industry and sewage treatment. His reasoning was that phytoplankton need both

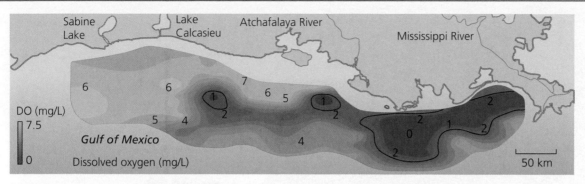

(a) Dissolved oxygen at bottom, July 2009

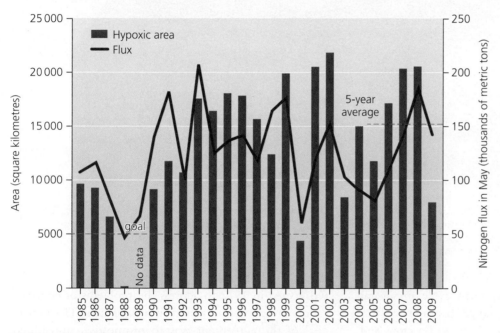

(b) Area of hypoxic zone in the northern Gulf of Mexico

FIGURE 1

The map **(a)** shows dissolved oxygen concentrations in bottom waters of the Gulf of Mexico off the Louisiana coast from the July 2009 survey. Areas in red indicate the lowest oxygen levels. Regions considered hypoxic (< 2 mg/L) are encircled with a black line. The size of the Gulf's hypoxic zone varies **(b)** as a result of several factors. Floods increase its size by bringing additional runoff (as in 1993), whereas tropical storms decrease its size by mixing oxygen-rich water into the dead zone (as in 2003 and 2005). Between 2005 and 2009, the hypoxic zone averaged 15 670 km² in size. *Source:* Data from Nancy Rabalais, LUMCON.

nitrogen and phosphorus, but there is now so much nitrogen in the Gulf that phosphorus has become the principal limiting factor on phytoplankton growth.

Research has supported this contention, and most scientists now agree that nitrogen and phosphorus should be managed jointly. Large-scale restoration of wetlands along the river and at the river's delta would best effectively filter pollutants before they reach the Gulf. The research is guiding a plan to reduce farm runoff, clean up the Mississippi, restore wetlands, and ultimately, hopefully, shrink the Gulf's dead zone. It has also led to a better understanding of hypoxic zones around the world.

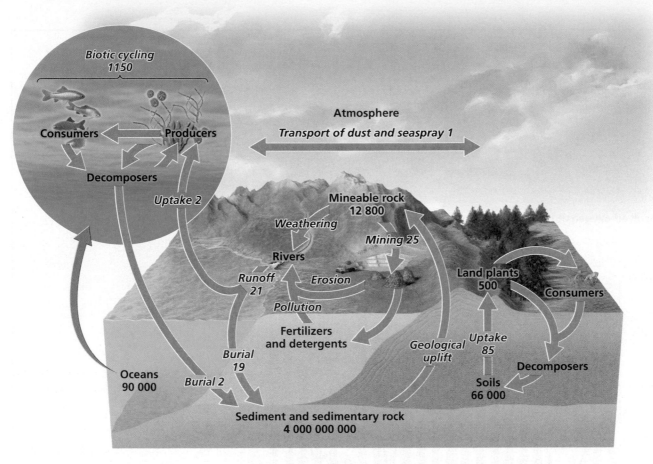

FIGURE 3.19   The phosphorus cycle summarizes the many routes that phosphorus atoms take as they move through the environment. Grey arrows represent fluxes among reservoirs for phosphorus. Reservoir names are in black, and the black numbers represent reservoir sizes expressed in teragrams (units of 10^{12} g) of P. Transfer processes give rise to fluxes, in red, expressed in teragrams of P per year.
Source: Data from Schlesinger, W.H. (2013). *Biogeochemistry: An Analysis of Global Change,* 3rd edition. Academic Press, London.

Humans influence the phosphorus cycle in several ways. We mine rocks containing phosphorus to extract this nutrient for the inorganic fertilizers we use on crops and lawns. Our wastewater discharge also tends to be rich in phosphates. Phosphates that run off into waterways can boost algal growth and cause eutrophication, leading to murkier waters and altering the structure and function of aquatic ecosystems. Phosphates are also present in detergents, so one way each of us can reduce phosphorous input into the environment is to purchase phosphate-free detergents.

In the 1970s Lake Erie began to exhibit signs of hypoxia. As in the St. Lawrence Estuary and Gulf of Mexico examples that we have touched on, the problem was traced to human inputs of plant nutrients; for Lake Erie, this meant mainly phosphate from fertilizers, detergents, and municipal sewage. The sources were land-based, but runoff carried the phosphate-bearing materials into the lake from surrounding farms and towns in Canada and the United States.

The International Joint Commission (IJC), set up under the International Boundary Waters Treaty of 1909, adopted the Great Lakes Water Quality Agreement

(GLWQA) in 1972 to address the problems of eutrophication and phosphate runoff into Lake Erie. The broader goal of the agreement was to promote an integrated, cooperative, scientific, and ecosystem-based approach to the management of the international waters of the Great Lakes. Initially, IJC activities under the GLWQA focused on reducing phosphate from detergents and municipal sewage, as these were the best understood pollutants from a scientific perspective. Since then, the agreement has grown in both strength and scope, and now deals with all threats to water quality and habitat in the Great Lakes system.

In the 1970s, Lake Erie responded quickly to controls placed on phosphate runoff, and recovered from the worst of its eutrophication and hypoxia. However, runoff of phosphate and other pollutants into the Great Lakes continues to be problematic, and all of the lakes still face periodic episodes of oxygen depletion (even Lake Superior, although its great depth and the cold water temperature help it to remain oxygen-saturated). In the 1990s, Lake Erie again began to exhibit signs of eutrophication and hypoxia, linked mainly to phosphate loading in agricultural runoff. Donald Scavia and colleagues investigated recent trends in oxygen levels, documenting severe hypoxia in the central basin of Lake Erie.[15] They estimate that returning the central basin to pre-1990s condition will require reducing phosphate loading in runoff by at least 46%.

Conclusion

Thinking in terms of systems is important in understanding Earth's dynamics, so that we may learn how to avoid disrupting its processes. Addressing problems such as those in the St. Lawrence Estuary and the Gulf of Mexico requires us to study the environment from a systems perspective, and to integrate scientific findings with the policy and decision-making processes.

Earth hosts many interacting systems. The way we perceive them and define them depends on the questions we want to investigate. Life interacts with its nonliving environment in ecosystems, systems through which energy flows and matter is recycled. Understanding the biogeochemical cycles that describe the movement of nutrients within and among ecosystems is crucial, because human activities are causing significant changes in the ways those cycles function.

Unperturbed ecosystems use renewable solar energy, recycle nutrients, and are stabilized by negative feedback loops. The environmental systems we see on Earth today are those that have survived the test of time. Our industrialized civilization is young in comparison. Natural systems provide us a blueprint to mimic as we move toward greater sustainability in modern society.

REVIEWING OBJECTIVES

You should now be able to

Describe the fundamental properties of systems, and the importance of linkages and flows of matter and energy among environmental systems

- Systems are networks of interacting components. Most environmental systems are open systems.
- Positive feedback loops are self-reinforcing; negative feedback loops are self-limiting.
- Natural systems are often in a state of dynamic equilibrium, which can contribute to homeostasis. Resistance and resilience are properties of homeostatic systems.
- Natural systems often have emergent properties.
- Earth's natural systems are complex, so environmental scientists often take a holistic approach to studying environmental systems. How we delineate and define systems depends on the questions we are interested in investigating.

Outline the characteristics of Earth's major subsystems

- The geosphere includes the solid Earth.
- The atmosphere is the envelope of gases that surrounds our planet.
- The ocean, glaciers and ice caps, ground water, and surface water bodies are the main components of the hydrosphere. The frozen parts of the hydrosphere are referred to as the cryosphere.
- Living and recently deceased and decaying organisms compose the biosphere.
- The anthroposphere consists of human activities and impacts, and the built environment.

Discuss how living and nonliving entities interact and how energy and matter move around in ecosystems

- Ecosystems consist of all organisms and nonliving entities, that is, biotic and abiotic components that occur and interact in a particular area at the same time.
- Energy flows in one direction through ecosystems, whereas matter is recycled. Energy is converted to biomass through primary productivity, and ecosystems vary in their productivity.
- Nutrient availability limits primary productivity. Input of nutrients can boost productivity, but an excess of nutrients can alter ecosystems in ways that cause severe ecological and economic consequences.
- Ecosystems provide "goods" that we know as natural resources and a wide variety of services that we depend on for everyday living.

Recognize the importance and complexity of the differing spatial and temporal scales of Earth processes

- Earth systems and processes differ dramatically in their characteristic spatial and temporal scales.
- Studying subsystems and developing models can give scientists the ability to study highly complex systems, or systems with unwieldy temporal or spatial scales.
- Landscape ecologists take a broad perspective on how landscape structure influences organisms. Landscapes consist of ecosystem patches spatially arrayed in a mosaic.
- Remote sensing technology and GIS are assisting the use of landscape ecology in conservation and regional planning.

Summarize the main global biogeochemical cycles and their human impacts, especially the global water, carbon, nitrogen, and phosphorus cycles

■ Water moves throughout the global environment as a result of processes in the hydrologic cycle. Depletion of groundwater resources is one example of a human impact on the hydrologic cycle.

■ The carbon flux between organisms and the atmosphere occurs via photosynthesis and respiration. Most carbon is contained in sedimentary rock, but substantial amounts also occur in the ocean and in soil. Human activity has moved carbon from long-term fossil fuel reservoirs into the atmosphere.

■ Nitrogen is a vital nutrient for plant growth. Most nitrogen is in the atmosphere and must be "fixed" by specialized bacteria or lightning before plants can use it. At least as much nitrogen is now fixed by human industrial processes as by natural processes.

■ Phosphorus is most abundant in sedimentary rock, with substantial amounts in soil and the ocean. Phosphorus, a key nutrient for plant growth, has no appreciable atmospheric pool. Runoff from agricultural land has contributed phosphates to surface water bodies, resulting in widespread eutrophication and hypoxia.

TESTING YOUR COMPREHENSION

1. Which type of feedback loop is most common in nature, and which more commonly results from human action? How might the emergence of a positive feedback loop affect a system in homeostasis?
2. Name and briefly describe Earth's major subsystems.
3. What is an ecosystem?
4. Describe the typical movement of energy through an ecosystem. Describe the typical movement of matter through an ecosystem.
5. What is the difference between net primary productivity and gross primary productivity?
6. Why are patches in a landscape mosaic important to scientists with an interest in conservation?

7. What role does each of the following play in the carbon cycle?
   ■ cars
   ■ photosynthesis
   ■ the ocean
   ■ Earth's crust
8. Contrast the function performed by nitrogen-fixing bacteria with that performed by denitrifying bacteria.
9. How has human activity altered the carbon cycle? The phosphorus cycle? The nitrogen cycle? To what environmental problems have these changes given rise?
10. What is the difference between evaporation and transpiration? Give examples of how the hydrologic cycle interacts with the carbon, phosphorus, and nitrogen cycles.

THINKING IT THROUGH

1. Once vegetation is cleared from a riverbank, water begins to erode the bank away. This erosion may dislodge more vegetation. Would you expect this to result in a feedback process? If so, which type—negative or positive? Explain your answer. How might we halt or reverse this process?
2. For an ecologist interested in studying populations of the organisms listed below, why might it be helpful to take a landscape ecology perspective?
   ■ a forest-breeding warbler that suffers poor nesting success in small, fragmented forest patches
   ■ a bighorn sheep that must move seasonally between mountains and lowlands
   ■ a toad that lives in upland areas but travels cross-country to breed in localized pools each spring

3. A simple change in the flux rate between just two reservoirs in a single nutrient cycle can potentially have major consequences for ecosystems and, indeed, for the globe. Explain how this can be, using one example from the carbon cycle and one example from the nitrogen cycle.
4. Imagine that you are a fisher in the St. Lawrence Estuary, and your income is decreasing because nutrient pollution is causing algal blooms that affect your catch. One day your MP comes to town, and you have a one-minute audience with her. What steps would you urge her to take to alleviate the nutrient pollution in the St. Lawrence, and maintain the quality of the fishery?

Now imagine that you are a farmer in rural Quebec who has learned that the government is insisting that you use 30% less fertilizer on your crops each year. In good growing years you could do without that fertilizer, and you'd be glad not to have to pay for it. But in bad growing years, you need the fertilizer to ensure a harvest so that you can continue making a living. You must apply the fertilizer each spring before you know whether it will be a good or bad year. What will you tell your MP when he comes to town?

INTERPRETING GRAPHS AND DATA

The use of PCBs (polychlorinated biphenyls) has been closely regulated in Canada since 1970. In contrast, PBDEs (polybrominated diphenyl ethers) have only recently been regulated in a few jurisdictions in North America. Environment Canada is still working with industry to regulate the production and release of PBDEs in Canada. The graph documents trends in the concentrations of two major pollutant groups (PCB and PBDE) in the fatty tissues of beluga whales from the St. Lawrence. (The unit "ng/g lw" refers to "nanograms per gram of lipid weight," which is a unit of measure for the concentration of a material in fat.) The pie chart shows the causes of death of beluga whales in the St. Lawrence from 1983 to 2002, based on studies of 148 whale carcasses.

1. Are the data shown on the graph consistent with the regulatory histories of these two groups of pollutants? Why (or why not)? Are there other factors that should be considered in interpreting the graph?
2. Are there any causes of death shown on the pie chart that you think might be directly attributable to the whales' exposure to toxic pollutants? Are there any that might be indirectly related to exposure to toxins? Are there any causes of death illustrated here that definitely do not seem to be related to exposure to toxins? (Note that the term *neoplasia* refers to the abnormal and uncontrolled growth of cells, resulting in the formation of tumours.)

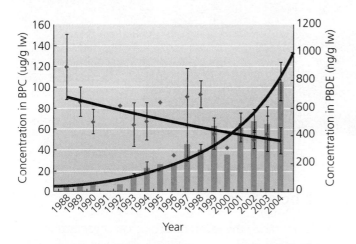
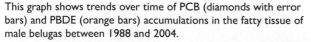

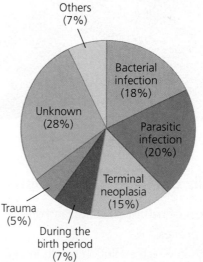

This chart shows the principal causes of death of stranded belugas in the Estuary and Gulf of St. Lawrence from 1983 to 2002.

This graph shows trends over time of PCB (diamonds with error bars) and PBDE (orange bars) accumulations in the fatty tissue of male belugas between 1988 and 2004.

Source: Adapted from Fisheries and Oceans Canada, *Beluga Whale Population of the Estuary,* 2nd ed. Maurice Lamontagne Institute. This does not constitute an endorsement by Fisheries and Oceans Canada of this product.

MasteringEnvironmentalScience®

STUDENTS
Go to **MasteringEnvironmentalScience** for assignments, the eText, and the Study Area with practice tests, videos, current events, and activities.

INSTRUCTORS
Go to **MasteringEnvironmentalScience** for automatically graded activities, current events, videos, and reading questions that you can assign to your students, plus Instructor Resources.

4 Evolution, Biodiversity, and Population Ecology

The Monteverde cloud forest in Costa Rica is a hotspot of biodiversity.

olof van der steen/iStockphoto

Upon successfully completing this chapter, you will be able to

- Explain the process of natural selection and cite evidence for this process
- Describe the ways in which evolution results in biodiversity and what the fossil record has taught us about evolution
- Discuss reasons for species extinction and mass extinction events

- Summarize the levels of ecological organization
- Define logistic growth, carrying capacity, limiting factors, and other fundamental concepts of population ecology

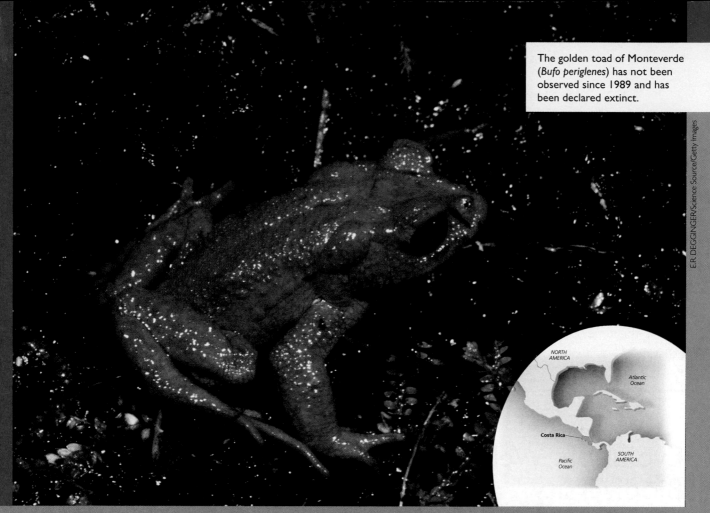

The golden toad of Monteverde (*Bufo periglenes*) has not been observed since 1989 and has been declared extinct.

CENTRAL CASE
STRIKING GOLD IN A COSTA RICAN CLOUD FOREST

"What a terrible feeling to realize that within my own lifetime, a species of such unusual beauty, one that I had discovered, should disappear from our planet."

—DR. JAY M. SAVAGE, DESCRIBING THE GOLDEN TOAD IN 1998

"To keep every cog and wheel is the first precaution of intelligent tinkering."

—ALDO LEOPOLD, 1949

During a 1963 visit to Central America, biologist Jay Savage heard rumours of a previously undocumented toad living in Costa Rica's mountainous Monteverde region. The elusive amphibian, according to local residents, was best known for its colour: a brilliant golden yellow-orange. Savage was told the toad was hard to find because it appeared only during the early part of the region's rainy season.

Monteverde means "green mountain" in Spanish, and the name couldn't be more appropriate. The settlement of Monteverde sits beneath the verdant slopes of the Cordillera de Tilarán, mountains that receive over 400 cm of annual rainfall. The lush forests above Monteverde, which begin at an altitude of around 1600 m, are known as *lower montane rain forests*. They are also called "cloud forests" because much of the moisture they receive arrives in the form of low clouds that blow inland from the Caribbean Sea.

Monteverde's cloud forest was not fully explored at the time of Savage's first visit, and researchers who had been there described the area as pristine, with a rich bounty of ferns, liverworts, mosses, clinging vines, orchids, and other organisms that thrive in cool, misty environments. Savage knew that such conditions create ideal habitat for many toads and other amphibians.

In May of 1964, Savage organized an expedition into the muddy mountains above Monteverde to try to document the existence of the previously unknown toad species in its natural habitat. Late in the afternoon of May 14, he and his colleagues found what they were looking for. Approaching the mountain's crest, they spotted bright orange patches on the forest's black floor. In one area that was only 5 m in diameter, they counted 200 golden toads. The discovery received international attention, making a celebrity of the tiny toad—which Savage named *Bufo periglenes*[1] (literally, "brilliant toad")—and making a travel destination of its mountain home.

At the time, no one knew that the Monteverde ecosystem was about to be transformed—by warming and drying of the cloud forest as a result of climate change, or by fungal diseases that caused amphibian numbers to crash, or perhaps both. No one could have guessed that this newly discovered species of toad would be extinct within fewer than 25 years. By 1988 the annual count of golden toads had dropped to one individual; the count was the same in 1989, and not a single golden toad has been observed anywhere since then, in spite of extensive searches. The extinction of *Bufo periglenes* (now known as *Incilius periglenes*) was declared final in 2007 on the International Union for Conservation of Nature (IUCN) *Red List of Threatened Species* by Dr. Jay Savage, who first found the new species in 1964.[2]

Evolution: Wellspring of Biodiversity

Although the golden toad was new to science, and countless species still await discovery, scientists understand in much detail how the world became populated with the remarkable diversity of organisms we see today. We know that the process of biological evolution has brought us from a stark planet inhabited solely by microbes to a lush world of perhaps 1.8 million (and likely millions more) species (**FIGURE 4.1**).

Evolution in the broad sense means "change over time"; in this book, for example, we discuss the chemical evolution of the atmosphere and hydrosphere over the course of Earth history. However, scientists most commonly use the term **evolution** to refer specifically to biological evolution, which consists of genetic changes in organisms across generations. This genetic change often leads to modifications in the appearance, functioning, or behaviour of organisms through time.

The theory of evolution is one of the best-supported and most illuminating concepts in all of science. Recall

(a) Resplendent quetzal

(b) Heliconia flower

(c) Harlequin frog

(d) Scutellerid bug

FIGURE 4.1   Much of our planet's rich biological diversity resides in tropical rain forests. Monteverde's cloud-forest community includes organisms such as the **(a)** resplendent quetzal (*Pharomachrus mocinno*); **(b)** heliconia flower (*Heliconia wagneriana*); **(c)** harlequin frog (*Atelopus varius*); and **(d)** scutellerid bug (*Pachycoris torridus*).

that in order to be designated as a theory, a scientific hypothesis must be widely accepted and extensively validated, surviving repeated testing by numerous research teams, and successfully predicting experimental outcomes and observations accurately. Many, many scientists have contributed to our understanding of evolution and have helped refine it and move it from the status of a hypothesis to an accepted theory.

From a scientific standpoint, evolutionary theory is indispensable; it is the foundation of modern biology. Perceiving how organisms adapt to their environments and change over time is crucial for understanding the history of life. Understanding evolution is also vital for a full appreciation of environmental science. Evolutionary processes are relevant to many aspects of environmental science, including pesticide resistance, agriculture, medicine, and environmental health.

Natural selection shapes organisms and diversity

Biological evolution results from random genetic changes, and it may proceed randomly or may be influenced by natural selection. **Natural selection** is the process by which traits that enhance survival and reproduction are passed on more frequently to future generations than those that do not, altering the genetic makeup of entire populations of organisms through time. In 1858, Charles Darwin and Alfred Russel Wallace each independently proposed the concept of natural selection as the *mechanism* for evolution, and as a way to explain the great variety of living things. Darwin and Wallace were English naturalists who had studied plants and animals in such exotic locales as the Galápagos Islands and the Malay Archipelago.

Both Darwin and Wallace recognized that organisms face a constant struggle to gain sufficient resources to survive and reproduce. They observed that organisms produce more offspring than can possibly survive, and they realized that some offspring may be more likely than others to survive and reproduce. Furthermore, they recognized that whichever characteristics give certain individuals an advantage in surviving and reproducing might be inherited by their offspring. These characteristics, they reasoned, would tend to become more prevalent in the population in future generations.

Natural selection is a simple concept that offers an astonishingly powerful explanation for patterns apparent in nature. The idea of natural selection follows logically from a few straightforward premises (**TABLE 4.1**). One is that individuals of the same species vary in their characteristics. Although not known in Darwin and Wallace's time, we

TABLE 4.1   The Logic of Natural Selection

- Organisms produce lots of offspring, not all of which survive.
- Individuals vary in their characteristics; some characteristics are accidental, but many characteristics are inherited by offspring from parents.
- The natural environment can be challenging; resources are limited, and environmental change is common.
- Some individuals will be better adapted to succeed in a challenging environment than others; as a result, they tend to survive longer and to have better reproductive success than poorly adapted individuals.
- By producing more offspring and/or offspring of higher quality, better-adapted individuals transmit more genes to future generations than poorly adapted individuals.
- Future generations will contain more genes (and thus more of the inherited characteristics) of the better-adapted individuals, and as a result the population as a whole will slowly shift its general characteristics.
- Over time, the genes (and thus the inherited characteristics) of the better-adapted individuals will come to dominate the population.

now know that variation in characteristics is due to differences in genes, the environments within which genes are expressed, and the interactions between genes and environment. (Recall that a *gene* is a sequence of nucleotides, part of the "twisted-ladder" or "double helix" structure of DNA. This structure stores or encodes, in its sequencing, all of the information that will be used to direct the building and maintenance of the proteins that comprise an organism's cells, and to pass along characteristics from parents to offspring.)

In other words, individuals vary, even when they belong to the same species. As a result of this variation, some individuals within a species will happen to be better suited to their particular environment than others and thus will be able to survive longer and/or have greater reproductive success. It is important to note, however, that the same genes that make an individual particularly well suited to one environment might be disadvantageous in a different environment.

Many characteristics are passed from parent to offspring through the genes. A parent that is long-lived, robust, and produces many offspring will pass on genes to more offspring than a weaker, shorter-lived individual that produces only a few offspring. In the next generation, therefore, the genes of better-adapted individuals will be more prevalent than those of less well-adapted individuals. From one generation to another through time, species will evolve to possess characteristics that lead to better and better success in a given environment. A trait that promotes success is called an **adaptive trait**, or an **adaptation**. A trait that reduces success is said to be *maladaptive*.

Natural selection acts on genetic variation

For an organism to pass a trait along to future generations—that is, for the trait to be *heritable*—genes in the organism's DNA must code for the trait. Accidental alterations that arise during DNA replication give rise to genetic variations among individuals. In an organism's lifetime, its DNA will be copied millions of times by millions of cells. During all of this copying and recopying, sometimes a mistake is made. Accidental changes in DNA, called **mutations**, can range in magnitude from the addition, deletion, or substitution of single nucleotides to the insertion or deletion of large sections of DNA. If a mutation occurs in a sperm or egg cell, it may be passed on to the next generation. Most mutations have little effect, but some can be deadly, whereas others can be beneficial. Mutations that are not lethal provide the genetic variation on which natural selection acts.

Sexual reproduction also generates variation. In sexual organisms, genetic material from two individuals is mixed, or recombined, so that a portion of each parent's genome is included in the genome of the offspring. This process of *recombination* produces novel combinations of genes, generating variation among individuals.

Natural selection can act on genetic variation and alter the characteristics of organisms through time in three fundamental ways, illustrated in **FIGURE 4.2**. Selection that drives a feature in one direction rather than another—for example,

toward larger or smaller, faster or slower, longer or shorter—is called **directional selection**. In contrast, **stabilizing selection** produces intermediate traits, in essence preserving the status quo. Under **disruptive selection**, traits diverge from their starting condition in two or more directions.

Environmental conditions influence adaptation

Environmental conditions determine what pressures natural selection will exert, and these selective pressures affect which members of a population will survive and reproduce. Over many generations, this results in the evolution of traits that enable success within the environment in question. Individuals and populations that live in even slightly different environments experience different selective pressures; eventually, they will tend to diverge in their traits as differing pressures drive their adaptations (**FIGURE 4.3A**). This process is called **divergent evolution**.

On the other hand, sometimes completely unrelated species may develop similar traits as a result of adapting to selective pressures from similar environments. This process is called **convergent evolution** (**FIGURE 4.3B**).

Environments change over time; organisms also sometimes move to new places, where they can encounter new conditions. In either case, a trait that promotes success at one time or location may not do so at another. For example, the population of golden toads that had adapted to the moist conditions of Monteverde's cloud forest did

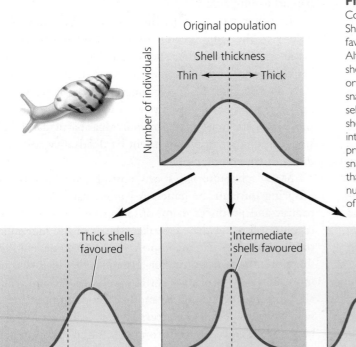

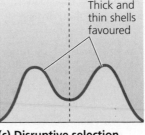

(a) Directional selection   **(b) Stabilizing selection**   **(c) Disruptive selection**

FIGURE 4.2 Natural selection can act in three different ways. Consider a population of snails with shells of different thicknesses (top). Shells protect snails against predators, so snails with thick shells may be favoured over those with thin shells, through directional selection **(a)**. Alternatively, suppose that a shell that is too thin breaks easily, but a shell that is too thick wastes resources that are better used for feeding or reproduction; in such a case, stabilizing selection **(b)** could favour snails with shells that are neither too thick nor too thin. In disruptive selection **(c)**, extreme traits are favoured. For example, perhaps thin-shelled snails are so resource-efficient that they can out-reproduce intermediate-shelled snails, whereas thick-shelled snails are so well protected from predators that they out-reproduce intermediate-shelled snails. In such a case, each of the "extreme" strategies is more effective than a compromise, and natural selection increases the relative numbers of thin- and thick-shelled snails, while reducing the number of intermediate-shelled snails.

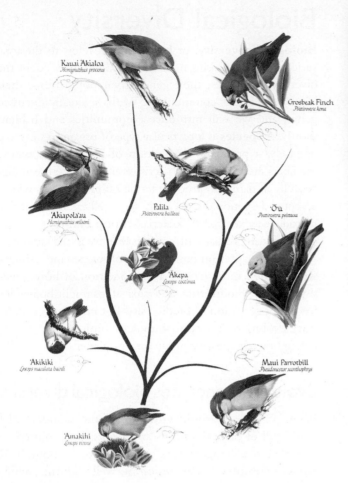

(a) Divergent evolution of Hawaiian honeycreepers

(b) Convergent evolution of cacti (Arizona; top) and euphorbias (Canary Islands; bottom)

FIGURE 4.3   Natural selection can cause closely related species to diverge in appearance and characteristics as they adapt to selective pressures from different environments. In Hawaiian honeycreepers **(a)**, a single ancestral species adapted to different food sources and habitats, evolving into a number of species with specialized bill shapes and plumage colours. Natural selection also can cause distantly related species to develop similar appearances **(b)** if the pressures of their environments are similar, as in the convergent evolution of cacti (family *Cactaceae*) in the Americas and spurges (family *Euphorbia*) in Africa. Plants in each family independently adapted to arid environments through the evolution of tough succulent stems to hold water, thorns to keep thirsty animals away, and photosynthetic stems without leaves to reduce surface area and water loss.

not persist after Monteverde's climate started to become drier 25 years ago. Varying environmental conditions in time and space make adaptation a moving target.

In all these ways, variable genes and variable environments interact as species engage in a perpetual process of adapting to the changing conditions around them. During this process, natural selection does not simply weed out unfit individuals. It also helps diversify traits that in the long term may lead to the formation of new species and whole new types of organisms.

Evidence of natural selection is all around us

The results of natural selection are all around us, visible in every adaptation of every organism. In addition, countless lab experiments (mostly with fast-reproducing organisms, such as bacteria and fruit flies) have demonstrated rapid evolution of traits. The evidence for selection that may be most familiar to us is that which Darwin himself cited prominently in his work 150 years ago: our breeding of domestic animals. In our dogs, our cats, and our livestock, we have conducted our own version of selection. We have chosen animals with traits we like and bred them together, while not breeding those with variants we do not like. Through such *selective breeding*, we have been able to exaggerate particular traits we prefer.

Consider the great diversity of dog breeds, all of which are variations on a single subspecies, *Canis lupus familiaris*. From Great Dane to Chihuahua, they can interbreed freely and produce viable offspring, yet breeders maintain the striking differences between them by allowing only

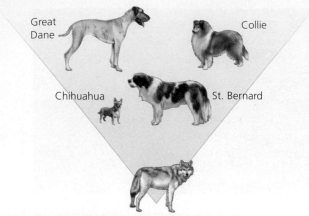

(a) Ancestral wolf and derived dog breeds

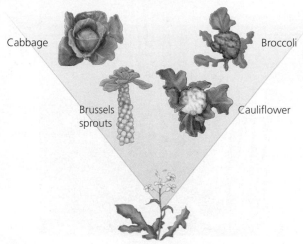

(b) Ancestral *Brassica oleracea* and derived crops

FIGURE 4.4 Selection imposed by people (selective breeding, or artificial selection) has resulted in numerous breeds of dogs **(a)**. Starting with the grey wolf (*Canis lupus*) as the ancestral wild species, and by breeding and selecting for the traits we prefer, we have produced breeds as different as Great Danes and Chihuahuas. By this same process we have created the immense variety of crop plants **(b)** that we depend on for food. Cabbage, Brussels sprouts, broccoli, and cauliflower were all generated from a single ancestral species, *Brassica oleracea*.

like individuals to breed with like (**FIGURE 4.4A**). The process of selection conducted under human direction is termed **artificial selection**.

Artificial selection has also given us the many crop plants we depend on for food, all of which were domesticated from wild ancestors and carefully bred over years, centuries, or millennia. Through selective breeding, we have created corn with larger, sweeter kernels; wheat and rice with larger and more numerous grains; and apples, pears, and oranges with better taste. We have diversified single types into many—for instance, breeding variants of the plant *Brassica oleracea* to create broccoli, cauliflower, cabbage, and Brussels sprouts (**FIGURE 4.4B**). Our entire modern agricultural system is based on artificial selection.

Biological Diversity

Biological diversity, or **biodiversity**, refers to the total variety of all organisms in an area, taking into account the diversity of species, their genes, their populations, their habitats, and their communities. We have already discussed genes, and we will introduce communities and habitats shortly. A **species** is a particular type of organism or, more precisely, a population or group of populations whose members share certain characteristics and can interbreed successfully in nature to produce fertile offspring. (This widely accepted definition doesn't apply very well to organisms that reproduce asexually, like bacteria.) A **population** is a group of individuals of a particular species that live in the same area.

In the chapter on conservation of species and habitats, we will look more closely at the question of how scientists measure biodiversity, and what steps might be needed for the preservation of biodiversity. For now, let's consider how evolution has contributed to biological diversity throughout the history of life on this planet.

Evolution generates biological diversity

When Charles Darwin wrote about the wonders of a world full of diverse animals and plants, he conjured up the vision of a "tangled bank" of vegetation harbouring all kinds of creatures. Such a vision fits well with the arching vines, dripping leaves, and mossy slopes of the tropical cloud forest of Monteverde. Indeed, tropical forests and cloud forests worldwide teem with life and harbour immense biological diversity.

Scientists have described about 1.8 million species, but many more remain undiscovered or unnamed. Estimates for the total number of species in the world range up to 100 million, with many of them thought to occur in tropical forests. In this light, the discovery of a new toad species in Costa Rica in 1964 doesn't seem too surprising. Although Costa Rica covers a tiny fraction (0.01%) of Earth's surface area, it is home to 5–6% of all species known to scientists. Of perhaps 500 000 species scientists estimate may exist in the country, only a small fraction of them have been inventoried and described.

Cloud forests such as Costa Rica's are by no means the only places rich in biodiversity. Step outside anywhere on Earth, even in a major city, and you will find numerous species within easy reach. They may not always be large and conspicuous like polar bears or blue whales or elephants, but they will be there. Plants poke up from cracks in asphalt in every city in the world, and even Antarctic ice harbours microbes. In a handful of backyard soil there may exist an entire miniature world of life, including several insect species, several types of mites, a millipede or two, many nematode worms, a few plant seeds, countless fungi, and millions upon millions of bacteria.

Speciation produces new types of organisms

How did Earth come to have so many species? Whether there are 1.8 million or 100 million, such large numbers require scientific explanation. The process by which new species are generated is termed **speciation**. Speciation can occur in a number of ways, but most biologists consider the main mode of species formation to be *allopatric speciation,* the emergence of a new species as a result of the physical separation of populations over some geographical distance.

To understand allopatric speciation, begin by picturing a population of organisms. Individuals within the population possess many similarities that unify them as a species because they are able to reproduce with one another and share genetic information. However, if the population is broken up into two or more populations that become isolated from one another, individuals from one population cannot reproduce with individuals from the others.

When a mutation or natural genetic variation occurs in the DNA of an organism in one of these isolated populations, it cannot spread to the other populations. Over time, each population will independently accumulate its own set of variations. Eventually, the populations may diverge, or grow different enough, that their members can no longer mate with one another. Individuals from the two differing populations may no longer recognize one another as being the same species because they have diverged so much in appearance or behaviour or they may simply become genetically incapable of producing viable offspring.

Once this has happened, there is no going back; the two populations cannot interbreed, and they have embarked on their own independent evolutionary trajectories as separate species (**FIGURE 4.5**). The populations will continue diverging in their characteristics as chance variations accumulate that confer traits causing the populations to become different in random ways. If environmental conditions happen to be different for the two populations, then natural selection may accelerate the divergence. Through the speciation process, single species can generate multiple species, each of which can in turn generate more.

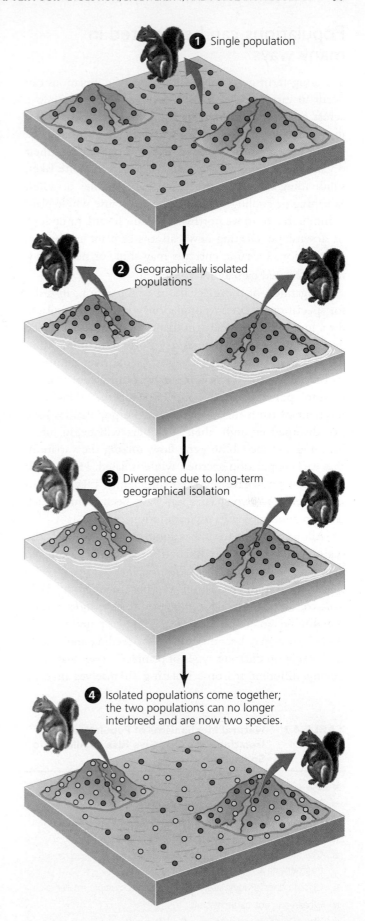

FIGURE 4.5
In allopatric speciation, part of a population becomes isolated from the rest. In this example, two mountaintops **(1)** are turned into islands by rising sea level, isolating populations of squirrels **(2)**. Each isolated population accumulates its own set of genetic changes over time, until individuals become genetically distinct and unable to breed successfully with individuals from the other population **(3)**. The two populations now represent separate species and will remain so, even if the geographical barrier is removed and the new species intermix **(4)**.

Populations can be isolated in many ways

The long-term geographical isolation of populations can occur in various ways (**TABLE 4.2**). Ice sheets may move across continents during glaciations and split populations in two. Major rivers may change course and do the same. Mountain ranges may rise and divide regions and their organisms. Drying climate may partially evaporate lakes, subdividing them into multiple smaller bodies of water. Warming or cooling temperatures may cause whole plant communities to move northward or southward, or upslope or downslope, creating new patterns of plant and animal distribution. Oceanic currents may shift, or new islands may be created by seafloor volcanism.

Regardless of the mechanism of separation, in order for speciation to occur, populations must remain isolated for a long time, generally thousands of generations. If the geological or climatic process that isolated the populations reverses itself—the glacier recedes, or the river returns to its old course, or warm temperatures turn cool again—then the populations can come back together. This is the moment of truth for speciation. If the populations have not diverged enough, their members will begin interbreeding and reestablish gene flow, mixing the variations that each population accrued while isolated. However, if the populations have diverged sufficiently, they will not interbreed, and two species will have been formed, each fated to continue on its own evolutionary path.

Allopatric speciation has long been considered the main mode of species formation, but speciation can happen in other ways. *Sympatric speciation* can occur when species form from populations that become reproductively isolated, occupying a new ecological role, or *niche*, within the same geographical area. For example, populations of some insects may become isolated by feeding and mating exclusively on different types of plants. Or they may mate during different seasons, isolating themselves in time

rather than space. In some plants, speciation apparently has occurred as a result of hybridization between species. In others, it seems to have resulted from natural variations or from mutations that changed the numbers of chromosomes, creating plants that could not mate with plants with the original number of chromosomes. Biologists still actively debate the relative importance of each of these modes of speciation.

We can infer the history of life's diversification by comparing organisms

Innumerable speciation events have generated complex patterns of diversity at levels above the species level. Such patterns are studied by evolutionary biologists who examine how groups of organisms arose and how they evolved the characteristics they show. For instance, how did we end up with plants as different as mosses, palm trees, daisies, and redwoods? Why do fish swim, snakes slither, and sparrows sing? How and why did the ability to fly evolve independently in birds, bats, and insects? To address such questions, one needs to know how the major groups diverged from one another, and this pattern ultimately results from the history of individual speciation events.

Scientists represent this history of divergence using branching diagrams called **phylogenetic trees** (**FIGURE 4.6A**), which show relationships among species, among major groups of species, among populations within a species, or even among individuals. Scientists construct these trees by analyzing patterns of similarity among the genes or external characteristics of present-day organisms and inferring which groups share similarities because they are related.

By mapping traits onto a phylogenetic tree according to which organisms possess them, one can trace how the traits themselves may have evolved. For instance, the tree of life shows that birds, bats, and insects are distantly related, with many other flightless groups between them. It makes far more sense to conclude that the three groups evolved flight independently than to conclude that the many flightless groups all lost an ancestral ability to fly. Because phylogenetic trees help biologists make such inferences about so many traits, they have become one of the modern biologist's most powerful tools.

A major advance was made in recent years as scientists acknowledged an entire new domain of life, the archaea, single-celled organisms that are genetically different from bacteria. Today most biologists view the tree of life as a three-pronged edifice consisting of *Bacteria*, *Archaea*, and *Eukaryota* (**FIGURE 4.6B**).

TABLE 4.2　Natural Mechanisms of Population Isolation That Can Give Rise to Allopatric Speciation

- Glacial ice sheets advance.
- Mountain chains are uplifted.
- Major rivers change course.
- Sea level rises, creating islands (see **FIGURE 4.5**).
- Climate warms, pushing vegetation up mountain slopes and fragmenting it.
- Climate dries, dividing large single lakes into multiple smaller lakes.
- Ocean current patterns shift.
- Islands are formed in the sea by volcanism.

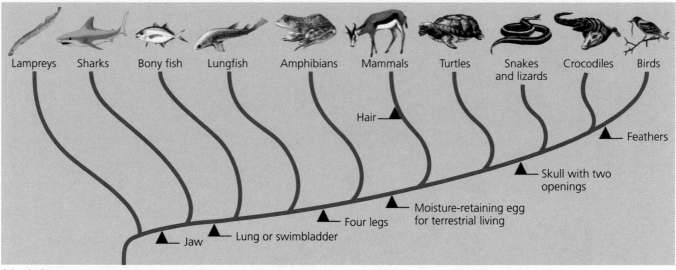

(a) Phylogenetic tree with major traits

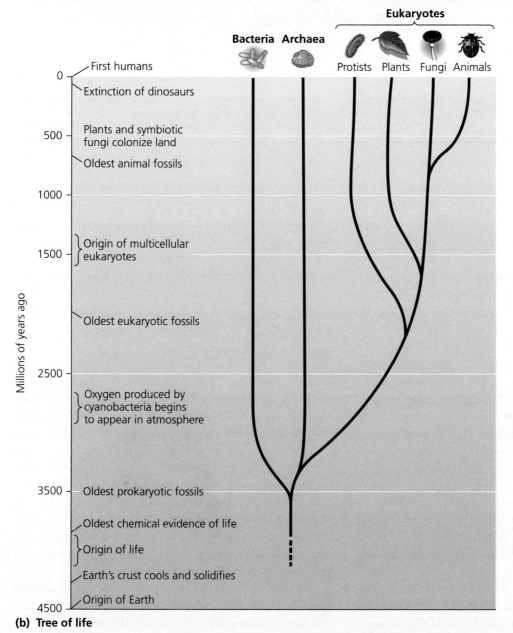

(b) Tree of life

FIGURE 4.6
Phylogenetic trees and similar types of diagrams show the history of life's divergence, and illustrate relationships among groups of organisms. Each branch from the main trunk represents a speciation event. By mapping traits onto phylogenetic trees, biologists can study how traits have evolved over time. In **(a)**, several major traits are mapped; triangular arrows indicating the point at which they originated. For instance, all vertebrates "above" the point at which jaws are indicated have jaws, whereas lampreys diverged before jaws originated and thus lack them. In **(b)**, each fork denotes the divergence of major groups of organisms, each group of which in this greatly simplified diagram includes many thousands of species. All are classified as Bacteria, Archaea, or Eukaryota, the last of which includes all plants, fungi, and animals.

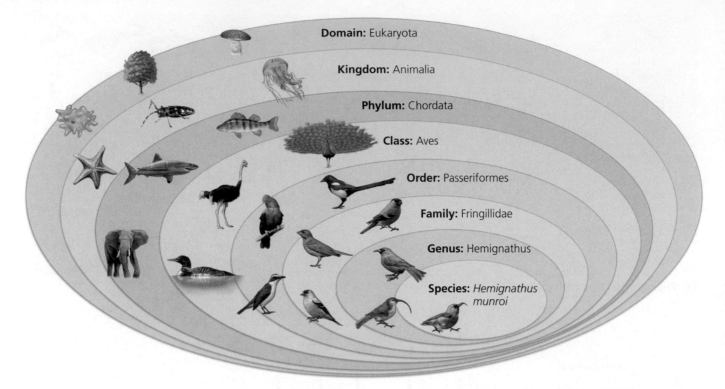

FIGURE 4.7 Scientists classify organisms using a hierarchical system that reflects evolutionary relationships. Species similar in appearance, behaviour, and genetics (because they share recent common ancestry) are placed in the same genus. Organisms of similar genera are placed in the same family. Families are placed within orders, orders within classes, classes within phyla, phyla within kingdoms, and kingdoms within one of the three domains: Bacteria, Archaea, or Eukaryota (see **FIGURE 4.6B**). Illustrated here is the taxonomic classification of the Hawaiian honeycreeper *Hemignathus munroi*, a member of the finch family, Fringillidae.

Knowing how organisms are related to one another also helps scientists classify them and name them, so that we can make sense of the life around us and communicate effectively. Biological scientists use an organism's physical characteristics and genetic makeup to determine its species. These scientists then group species by their similarity into a hierarchy or *taxonomy* of categories meant to reflect evolutionary relationships. Related species are grouped together into *genera* (singular, *genus*), related genera are grouped into families, and so on (**FIGURE 4.7**). Each species is given a two-part Latin or Latinized scientific name denoting its genus and species.

For example, the Hawaiian honeycreeper 'akiapōlā'au, *Hemignathus munroi*, is similar to other honeycreepers in the genus *Hemignathus* (honeycreepers from the genus *Hemignathus* are illustrated at the top left of **FIGURE 4.3A**). All of the species in this genus are closely related in evolutionary terms, as indicated by the genus name they share. They are more distantly related to honeycreepers in other genera, but all honeycreepers are classified together in the family Fringillidae, the finches.

This system of naming and classification was devised by Swedish botanist Carl Linnaeus (1707–1778) long before Darwin's work on evolution. Today biologists use

evolutionary information from phylogenetic trees to help classify organisms under the Linnaean system's rules. We will revisit taxonomic classification in a later chapter, when we consider the conservation of species and their habitats.

The fossil record teaches us about life's long history

Scientists also decipher life's history by studying the preserved remains of organisms that lived long ago: **fossils**. As organisms die, some are buried by sediment. The hard parts of their bodies, most commonly bones, shells, and teeth, but sometimes even the soft or delicate parts, such as feathers, eggshells, and skin, may be preserved as the sediment is compressed into rock. Minerals replace the organic material, leaving behind a fossil, a preservation or imprint in stone of the dead organism (**FIGURE 4.8**). Trails, footprints, burrows, and even fossilized feces can tell scientists a lot about the behaviours, physiology, and habitats of long-dead organisms.

In countless locations throughout the world, geological processes across millions of years have buried sedimentary rock layers and later brought them to the surface, revealing

Dorling Kindersley, Ltd.

FIGURE 4.8
The fossil record reveals the history of life on Earth. The abundance of trilobite fossils, like the one shown here, suggest that these animals were abundant in the oceans from roughly 540 million to 250 million years ago, after which they died out in the Permian-Triassic Extinction, the greatest mass extinction in Earth history.

assemblages of fossilized plants and animals from different time periods. By determining the ages of rock layers that contain fossils, palaeontologists (scientists who study the history of Earth's life) can learn when particular organisms lived. The cumulative body of fossils worldwide, spanning most of geological time, is known as the **fossil record**.

The fossil record shows that

- Life has existed on Earth for at least 3.5 billion years.
- Modern organisms developed, or evolved, from earlier ancestral organisms, some of which are no longer extant.
- The number of species existing at any one time has generally increased through time.
- The species living today are a tiny fraction of all species that ever lived; the vast majority are long extinct.
- There have been several episodes of mass extinction, or simultaneous loss of great numbers of species.

Across life's 3.5 billion years on Earth, complex structures have evolved from simple ones, and large sizes from small ones. However, simplicity and small size have also evolved when favoured by natural selection, and it is easy to argue that Earth still belongs to the bacteria and other microbes, some of them little changed over eons. Even fans of microbes, however, must marvel at some of the exquisite adaptations of animals, plants, and fungi: the heart that beats so reliably for an animal's entire lifetime that we take it for granted; the complex organ system to which the heart belongs; the stunning plumage of a peacock in full display; the ability of plants on the planet to lift water and nutrients from the soil, gather light from the Sun, and turn it into food; the staggering diversity of beetles and other insects; the human brain and its ability to reason. All these and more have resulted as the process of evolution has generated new species and whole new branches on the tree of life.

Extinction

Although speciation generates Earth's biodiversity, it is only one part of the equation. The appearance of species and the disappearance of species—together—determine Earth's biodiversity. The disappearance of a species from one part of its normal range is called **extirpation**; the complete disappearance of a species from Earth is called **extinction**.

Extinction is a natural process, profoundly influenced by humans

From studying the fossil record, palaeontologists calculate that the average time a species spends on Earth is 1 to 10 million years. The number of species in existence at any one time is equal to the number added through speciation minus the number removed by extinction. As we know from the fossil record of the history of life, the vast majority of species that once lived on this planet are now gone.

In general, extinction occurs when environmental conditions change rapidly or severely enough that a species cannot adapt genetically to the change; the slow process of natural selection simply does not have enough time to work. All manner of events can cause extinction—climate change; the arrival of a new, competitive species; severe weather events; lack of availability of food; and more. In general, small populations are vulnerable to extinction because fluctuations in their size could, by chance, bring the population size to zero. Species narrowly dependent on some particular resource or way of life are also vulnerable, because environmental changes that make that resource or way of life unavailable can doom them.

Species that are **endemic** to a region, meaning that they occur nowhere else on the planet, also face elevated risks of extinction because all their members belong to a single, sometimes small, population. For example, *Bufo periglenes*, the golden toad, was endemic to one small area of

FIGURE 4.9 Until 10 000 years ago, the North American continent teemed with a variety of large mammals, including mammoths, camels, giant ground sloths, lions, sabre-toothed cats, and various types of horses, antelope, bears, and others. Nearly all of this megafauna went extinct suddenly, about the time that humans first arrived on the continent. Similar extinctions coincident with the arrival of humans occurred in other areas, suggesting that overhunting or other human impacts may have been at least partly responsible.

a cloud forest in Monteverde, Costa Rica. The Hawaiian honeycreeper *Hemignathus munroi* (not extinct, but rare and endangered[3]) is endemic to the island of Hawaii.

Extinction is a natural process, which has happened repeatedly throughout the long history of life on Earth. However, human activities can profoundly affect the rate at which extinctions occur (**FIGURE 4.9**). The apparent extinction of the golden toad made headlines worldwide, but unfortunately it was not such an unusual occurrence.

Species extinction brought about by human impact may well be the single biggest environmental problem we face, because the loss of a species is irreversible. The biological diversity that makes Earth a unique planet is being lost at an astounding rate. This loss affects humans directly because other organisms provide us with life's necessities—food, fibre, medicine, and ecosystem services.

Earth has seen several episodes of mass extinction

Most extinction occurs gradually, one species at a time. The rate at which this type of extinction occurs is referred to as the *background extinction rate*. However, Earth has seen at least five events of staggering proportions that killed off massive numbers of species at once. These episodes, called **mass extinctions**, have occurred at widely spaced intervals in Earth history and have wiped out up to 95% of our planet's species each time (**TABLE 4.3**).

These events represent times when the biological diversity on Earth dropped dramatically in a geologically short period of time. The causes of all five major mass extinctions and other less dramatic events are still more or less speculative. In some cases, rapid global cooling and widespread glaciations appear to have played a role. Massive asteroid impacts, widespread volcanism, ocean acidification, and other causes have been investigated. In at least two major extinction events (the Triassic-Jurassic and Cretaceous-Paleogene), a synchronous decline in the rate of speciation may also have contributed to the drop in biodiversity. As we have seen, biological diversity results from a balance between speciations and extinctions.

The best-known of the "big five" mass extinctions occurred 66 million years ago, and essentially brought an end to the age of dinosaurs (although birds remain as the modern descendants of dinosaurs). Evidence suggests that the impact of a gigantic asteroid caused this event, the Cretaceous-Paleogene (popularized as the Cretaceous-Tertiary or K-T) event.

The K-T event, as massive as it was, was moderate compared to the mass extinction at the end of the Permian Period 252 million years ago. Palaeontologists estimate that 95% of marine species and 70% of terrestrial species—almost 60% of all genera—perished during this event. Precisely what caused the Permian-Triassic extinction is unknown. The evidence for extraterrestrial impact is much weaker than it is for the K-T event, and other ideas abound. The hypothesis with the most support so far is that massive volcanism threw into the atmosphere a global blanket of volcanic ash and sulphur aerosols, smothering

TABLE 4.3   Mass Extinctions: The "Big Five" and the Current Extinction

Event	Date (millions of years ago)	Cause	Types of life most affected	Relative proportion of loss[4]
Ordovician-Silurian	≈450 mya	Possibly plate tectonics, global cooling, glaciation	Marine organisms; terrestrial record is unknown	>40% decrease in genera
Late Devonian*	374 mya	Unknown, possibly multiple environmental changes contributing	Marine organisms; terrestrial record is unknown	>20% decrease in genera
Permian-Triassic	252 mya	Possibly widespread volcanism	Marine organisms; terrestrial record is less known	Almost 60% decrease in genera; 80–95% of species**
Triassic-Jurassic*	201 mya	Possibly climate change or ocean acidification	Marine organisms; terrestrial record is less known	>30% decrease in genera
Cretaceous-Paleogene (or Cretaceous-Tertiary***)	66 mya	Probably asteroid impact	Marine and terrestrial organisms, including dinosaurs	Almost 35% decrease in genera; >50% of species**
Current or Anthropocene	Beginning 0.01 mya	Human impact, through habitat destruction and other means	Large animals, specialized organisms; island organisms; organisms hunted or harvested by humans	Ongoing

*The loss of biodiversity in these events may have been at least partly caused by a marked decrease in origination of new species, in combination with a high rate of extinction.[4]

**Species losses often seem more dramatic, but for a number of reasons the loss of genera (the ecological classification level above species) is considered to be more representative.

***The term "Tertiary" is no longer in common use by geologists.

the planet, reducing sunlight, altering the chemistry of the ocean, and inducing severe climate changes.

Human activities have initiated another mass extinction

Many biologists have concluded that Earth is currently entering its sixth mass extinction event—and that we are the cause. Changes to Earth's natural systems set in motion by human population growth, development, and resource depletion have driven many species extinct and are threatening countless more. The alteration and outright destruction of natural habitats, the hunting and harvesting of species, and the introduction of invasive species from one place to another where they can harm native species—these processes and many more have combined to threaten Earth's biodiversity.

When we look around us, it may not appear as though a human version of an asteroid impact is taking place, but we must not judge such things on a human timescale. On the geological timescale, extinction over 100 years or over 10 000 years appears every bit as instantaneous as extinction over a few days.

Many scientists say that biodiversity loss is our biggest environmental problem today. But why should we be concerned about loss of biodiversity and extinction of species? The Vancouver Island marmot, a small mammal whose

natural range is entirely within Canada, is on the brink of extinction as a result of habitat alteration. What would be lost if this animal were to become extinct? (Had you ever even heard of it?) Other endangered species in Canada include the northern abalone, the whooping crane, and the Atlantic halibut—each with its own appeal, certainly, but marmots are cute and furry; does that make them more compelling as examples of the value or importance of biodiversity?

Another of Canada's endangered species is the blue whale (*Balaenoptera musculus*), the largest animal that has *ever lived*. Is the blue whale more worthy of preservation than the Vancouver Island marmot, or the whooping crane, or the northern abalone? How should we define the criteria for what is "worthy" of preservation? Should we be concerned about the value of one species, or is the true importance that these species are part of a larger whole? We will return to questions like these in subsequent chapters.

Levels of Ecological Organization

The extinction of species, their origination through speciation, and other evolutionary mechanisms and patterns are of substantial importance in ecology. It's often said that ecology provides the stage on which the play of evolution unfolds; certainly they are tightly intertwined in many ways.

(a) Hierarchy of matter within organisms

	Organism	An individual living thing
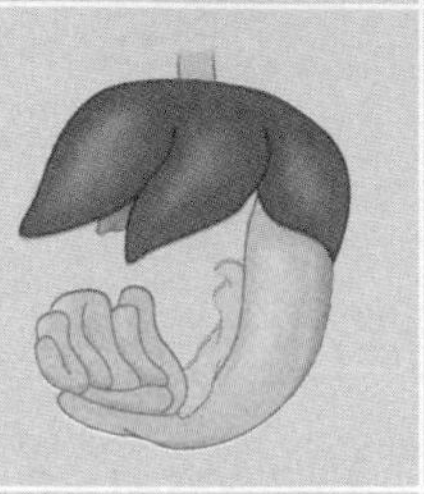	Organ system	An integrated system of organs whose action is coordinated for a particular function
	Organ	A structure in an organism composed of several types of tissues and specialized for some particular function
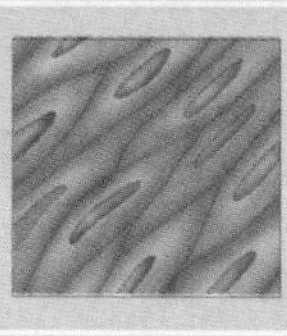	Tissue	A group of cells with common structure and function
	Cell	The smallest unit of living matter able to function independently, enclosed in a semi-permeable membrane
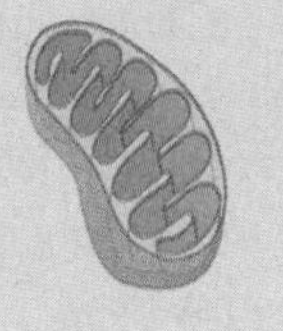	Organelle	A structure inside a eukaryotic cell that performs a particular function
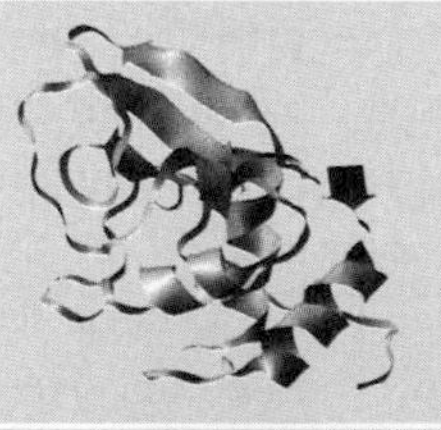	Macro-molecule	A large organic molecule (includes proteins, nucleic acids, carbohydrates, and lipids)
	Molecule	A combination of two or more atoms chemically bonded together
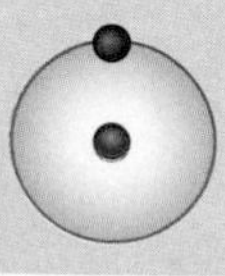	Atom	The smallest component of an element that maintains the element's chemical properties

We study ecology at several levels

Life occurs in a hierarchy of levels. Atoms and molecules (for our purposes) represent the lowest levels in this hierarchy (**FIGURE 4.10A**). Organisms use natural polymers and macromolecules to build *organelles*, which are grouped together in **cells**, one of the basic functional units of life. All living things are composed of cells, and organisms range in their complexity from single-celled bacteria to plants and animals that contain millions of cells. Aggregations of cells of particular types form tissues, and tissues form organs, and organs function together as organ systems. All of these, together, make up an individual living organism.

Ecologists study relationships on the higher levels of this hierarchy (**FIGURE 4.10B**), namely on the organismal, population, community, and ecosystem levels. At the organismal level, the science of ecology describes relationships between organisms and their physical environments. It helps us understand, for example, what aspects of the golden toad's environment were important to it, and why.

We have already defined a population as a group of individuals of a particular species that live in the same area.

(b) Levels of ecological organization

	Biosphere	The sum total of living things on Earth and the areas they inhabit
	Ecosystem	A functional system consisting of a community, its nonliving environment, and the interactions between them
	Community	A set of populations of different species living together in a particular area
	Population	A group of individuals of a species that live in a particular area
	Organism	An individual living thing

FIGURE 4.10 Life exists in a hierarchy of levels. Within an individual organism **(a)**, matter is organized in a hierarchy of levels, from atoms through cells through organ systems. Ecology **(b)** includes the study of the organismal, population, community, and ecosystem levels and, increasingly, the level of the biosphere.

Population ecology investigates the quantitative dynamics of how individuals of the same species interact with one another, and how their populations change over time. It helps us understand why populations of some species (such as the golden toad) decline, while populations of others (such as ourselves) increase.

Communities are different from populations, in that they include multiple interacting species that live in the same area. A population of golden toads, a population of resplendent quetzals, and populations of ferns and mosses, together with all of the other interacting plant, animal, fungal, and microbial populations in the Monteverde cloud forest, would comprise a community. Community ecology focuses on interactions among species, from one-to-one interactions to complex interrelationships involving entire communities of organisms. In the case of Monteverde, it allows scientists to study how the golden toad and many other species of its cloud-forest community interacted.

Ecosystems, as you have learned, encompass both communities and populations, and the abiotic materials with which their members interact. Monteverde's cloud-forest ecosystem consists of the community plus the air, water, soil, nutrients, and energy that support the community's organisms. In ecosystem ecology we focus on patterns of energy and nutrient flow by studying living and nonliving components of systems in conjunction.

As improved technologies such as remote sensing allow scientists to learn more about the complex operations of natural systems on a global scale, ecologists are increasingly expanding their horizons beyond individual and regional ecosystems (or *biomes*), to the biosphere as a whole. We will consider these levels of study in subsequent chapters.

Each organism has habitat needs

Individual organisms relate to the surrounding environment in ways that tend to maximize survival and reproduction. One key relationship involves the specific environment in which an organism lives—its **habitat**. The habitat of a species consists of both living and nonliving elements— rock, soil, leaf litter, and moisture, as well as the other organisms around it. The golden toad lived in a habitat of cloud forest—more specifically, on the moist forest floor, using seasonal pools for breeding and burrows for shelter.

Plants known as *epiphytes* use other plants as their habitat; they grow on trees for physical support, obtaining water from the air and nutrients from organic debris that collects among their leaves. Epiphytes thrive in cloud forests because they require a habitat with high humidity. The Monteverde cloud forest in Costa Rica (like most cloud forests) hosts more than 330 species of epiphytes, mostly ferns, orchids, and bromeliads. By collecting pools of rain water and pockets of leaf litter, epiphytes create

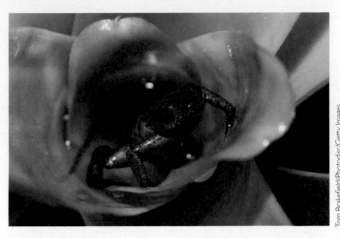

Tom Brakefield/Photodisc/Getty Images

FIGURE 4.11

The strawberry poison dart frog (*Dendrobates pumilio*) places her tadpoles in pools of rain water that collect in the leaves of epiphytic plants such as bromeliads. The pools supply breeding habitat for mosquitoes, whose larvae are eaten by the tadpoles. Epiphytes provide habitat for many small animals, and trees, in turn, provide habitat for epiphytes.

habitat for many other organisms, including many invertebrates and even frogs that lay their eggs in the rainwater pools (**FIGURE 4.11**). The abundance of habitats and niches in just this one small system is an indicator of the rich complexity of tropical cloud forests.

Habitats vary with the body size and needs of the species. A tiny soil mite may use less than a square metre of soil in its lifetime. A vulture, elephant, or whale, in contrast, may traverse many kilometres of air, land, or water in just a day. Species may also have different habitat needs at different times of year; many migratory birds use distinct breeding habitats, wintering habitats, and migratory habitats.

The criteria by which organisms favour some habitats over others vary greatly. To a soil mite, the important characteristics of a habitat might be the chemistry, moisture, and compactness of the soil and the percentage and type of organic matter. A vulture may ignore not only soil but also topography and vegetation, focusing solely on the abundance of dead animals in the area that it scavenges for food. To a whale, water temperature and salinity, light level, and abundance of marine microorganisms might be the critical characteristics. Every species assesses habitats differently because every species has different needs.

Each organism thrives in certain habitats and not in others, leading to nonrandom patterns of habitat use. *Motile* organisms (those that are able to move about freely) actively select habitats from among the range of options they encounter, a process called *habitat selection*. In the case of plants and *sessile* animals (those that are not freely mobile and whose progeny disperse passively), patterns of habitat use result from success in some habitats and failure in others.

Habitat use is an important consideration in environmental science because the availability and quality of habitat are crucial to an organism's well-being. Indeed, because habitats provide everything an organism needs, including nutrition, shelter, breeding sites, and mates, the organism's very survival depends on the availability of suitable habitats. Often this creates conflict with people who want to alter or develop a habitat for their own purposes.

Niche and specialization are key concepts in ecology

Another way in which an organism relates to its environment is through its niche. A species' **niche** reflects its use of resources and its functional role in a community. This includes its habitat use, its consumption of certain foods, its role in the flow of energy and matter, and its interactions with other organisms. Niche is a multidimensional concept, a kind of summary of everything an organism does. Eugene Odum, who pioneered the science of ecology, once wrote that "habitat is the organism's address, and the niche is its profession."[5] We will examine the niche concept more fully in the chapter on species interactions and community ecology.

Organisms vary in the breadth of their niches. Species with narrow breadth, and thus very specific requirements, are said to be **specialists**. Those with broad tolerances, able to use a wide array of habitats or resources, are **generalists**. The koala, which eats almost nothing but eucalyptus leaves, is a specialist; so is the panda, which eats only bamboo shoots. Species that survive well in urban environments, like raccoons, coyotes, rats, and cockroaches, tend to be generalists.

In a study of eight insect-eating bird species from Costa Rica, ornithologist T. Scott Sillett found that all eight species spent considerable time and effort (at least 30% of their foraging time) feeding on insects from epiphytes. Four of the bird species were found to be generalists; they were simply drawn to the epiphytes because insect prey were abundant there. The other four species were specialists, spending at least 75% of their foraging efforts feeding only on insects found in very specific types of ephiphytes.[6]

Specialist and generalist strategies each have advantages and disadvantages. Specialists can be successful over evolutionary time by being extremely good at the things they do, but they are vulnerable when conditions change and threaten the habitat or resource on which they have specialized. Generalists succeed by being able to live in many different places and weather variable conditions, but they may not thrive in any one situation to the degree that a specialist does. An organism's habitat, niche, and degree of specialization each reflect the adaptations of the species and are products of natural selection.

Population Ecology

The central focus of population ecology is to study the dynamics of populations—usually of non-human species, although many of the approaches used by population ecologists are adapted from or applicable to the study of the human population. Population ecologists consider all of the factors that might lead a population to grow or to decrease in size. This includes internal factors, such as the age, sex, and health of individual members of the population, as well as interactions with other populations and other species, and with the surrounding abiotic environment.

Populations show characteristics that help predict their dynamics

All populations—from humans to golden toads to Hawaiian honeycreepers—have characteristics that help population ecologists predict the future dynamics of that population. Attributes such as density, distribution, sex ratio, age structure, and birth and death rates all help the ecologist understand how a population may grow or decline. The ability to predict growth or decline is especially useful in monitoring and managing threatened and endangered species. It is also vital for understanding the dynamics of the human population, one of the prime challenges for our society today.

Population size   Expressed as the number of individual organisms present at a given time, **population size** may increase, decrease, undergo cyclical change, or remain the same over time. A dramatic, precipitous population decline may precede extinction. For example, as late as 1987, scientists documented a golden toad population at Monteverde in excess of 1500 individuals; by 1989, scientists were able to find only a single toad. After many intensive searches with no sightings, the species was finally declared extinct.

The passenger pigeon (*Ectopistes migratorius*), also now extinct, illustrates the extremes of population decline (**FIGURE 4.12**). This was once the most abundant bird in North America; flocks of passenger pigeons literally darkened the skies, nesting in gigantic colonies in the forests of the U.S. Midwest and southern Canada. According to essayist Jerry Sullivan

John James Audubon rode the 55 miles from Henderson, Kentucky, to Louisville one day in autumn 1813, and through the whole long day, he rode under a sky darkened from horizon to horizon by a cloud of passenger pigeons. He estimated that more than a billion birds had passed over him. In 1866, a cloud of birds passed into southern Ontario. It was a mile wide, 300 miles long, and took 14 hours to pass a single point.[7]

(a) Passenger pigeon, *Ectopistes migratorius*

(b) Nineteenth-century lithograph of pigeon hunting in Iowa

FIGURE 4.12 The passenger pigeon **(a)** was once North America's most numerous bird. Its flocks literally darkened the skies when millions of birds passed overhead **(b)**. However, hunting and clearing of forests drove the species to extinction within a few decades.

Once people began cutting the forests, the birds' great concentration made them easy targets for hunters, who shipped them to market by the wagonload. By the end of the nineteenth century, the passenger pigeon population had declined to such a low number that they could not form the large colonies they needed for protection from predators and to breed effectively. In 1914, the last passenger pigeon (named "Martha") died in the Cincinnati Zoo, bringing the continent's most numerous bird species to extinction within just a few decades.

Population density The once-enormous flocks and breeding colonies of passenger pigeons demonstrated high population density, another attribute that ecologists assess to better understand populations. **Population density** describes the number of individuals in a population per unit of area. For instance, the 1500 golden toads counted in 1987 within 4 km^2 indicated a density of 375 toads per km^2. In general, larger organisms tend to have lower population densities because they require more resources to survive.

High population density can make it easier for organisms to group together, protect themselves from predators, and find mates, but it can also lead to conflict in the form of competition if space, food, or mates are in limited supply. A large mass of overcrowded organisms may attract predators that feed on them, and close contact among individuals can increase the transmission of infectious disease. For these reasons, organisms sometimes leave an area when densities become too high. In contrast, at low population densities, organisms benefit from more space and resources but may find it harder to locate mates and companions.

High population densities in small remnants of habitat may have doomed Monteverde's harlequin frog (see **FIGURE 4.1C**), an amphibian that disappeared from the cloud forest around the same time as the golden toad (see "The Science behind the Story: Climate Change, Disease, and the Amphibians of Monteverde"). The harlequin frog (*Atelopus varius*) is a specialist, favouring "splash zones," areas alongside rivers and streams that receive spray from waterfalls and rapids. As Monteverde's climate grew warmer and drier in the 1980s and 1990s, water flow decreased, and many streams dried up. Splash zones grew smaller and fewer, and harlequin frogs were forced to cluster together in what remained of the habitat. Researchers J. Alan Pounds and Martha Crump recorded frog population densities up to 4.4 times higher than normal. Overcrowding likely made the frogs more vulnerable to disease transmission, predator attack, and assault from parasitic flies. The researchers concluded that these factors contributed to the harlequin frog's disappearance from Monteverde.[8]

A "remnant" population of harlequin frogs was found in 2003 on a private reserve in Costa Rica, so there is still hope that the species may survive. The frog was rediscovered by University of Delaware student Justin Yeager, who was doing field research during his study abroad trip in Costa Rica that summer. The harlequin frog remains critically endangered, however. From more than 100 populations known in Costa Rica before 1988, the species has declined to just one, although some populations may still be surviving in Panama.[9]

Population distribution The spatial arrangement of organisms within an area also influences

Dr. J. Alan Pounds (left) and Dr. Luis Coloma look for harlequin frogs.

J. Alan Pounds, Monteverde Cloud Forest Reserve

THE SCIENCE BEHIND THE STORY

Climate Change, Disease, and the Amphibians of Monteverde

Soon after the golden toad's disappearance, scientists began to investigate the potential role of global climate change in driving cloud-forest species toward extinction. The period from July 1986 to June 1987 was the driest on record at Monteverde, with unusually high temperatures and record-low stream flows. These conditions caused the golden toad's breeding pools to dry up in the spring of 1987, likely killing nearly all of the eggs and tadpoles in the pools.

By reviewing weather data, scientists found that the number of dry days and dry periods each winter in the Monteverde region had increased between 1973 and 1998. Because amphibians breathe and absorb moisture through their skin, they are susceptible to dry conditions. Based on these facts, herpetologists J. Alan Pounds and Martha Crump in 1994[10] hypothesized that hot, dry conditions were to blame for high adult mortality and breeding problems among golden toads and other amphibians.

Pounds and others reviewed the scientific literature on atmospheric and ocean science to analyze the impacts on

(a) Cool ocean conditions

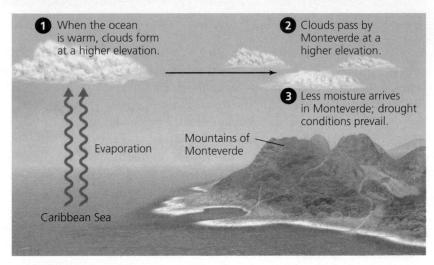

(b) Warm ocean conditions

FIGURE 1   Monteverde's cloud forest gets its name and life-giving moisture from clouds that sweep inland from the oceans. When ocean temperatures are cool **(a)**, the clouds keep Monteverde moist. Warmer ocean conditions **(b)** resulting from global climate change cause clouds to form at higher elevations and pass over the mountains, drying the cloud forest.

population dynamics. This is called the **population distribution** or **population dispersion**. Ecologists define three main distribution types: random, uniform, and clumped (**FIGURE 4.13**).

In *random distribution,* individuals are located haphazardly in space in no particular pattern. This type of distribution can occur when the resources an organism needs are found throughout an area, and other organisms or the characteristics of the abiotic environment do not strongly influence where members of a population settle. In nature, however,

the distribution of organisms is virtually never random; even the apparently randomly distributed trees in the forest shown in **FIGURE 4.13A** have grown in specific locations that are beneficial for access to the resources they need.

A *uniform distribution* is one in which individuals are evenly spaced. This can occur when individuals hold territories or otherwise compete for space. For instance, in a desert where there is little water, each plant may need a certain amount of space for its roots to gather adequate moisture. As a result, each individual plant may be equidistant from others.

Monteverde's local climate of warming patterns in the ocean regions around Costa Rica. Warmer oceans, the researchers found, caused clouds to pass over at higher elevations, where they were no longer in contact with the trees. Once the cloud forest's moisture supply was pushed upward, out of reach of the mountaintops, the forest began to dry out (see **FIGURE 1**).

In a 1999 paper in the journal *Nature*, Pounds and two colleagues[11] reported that climate modification was causing local changes at the species, population, and community levels. They argued that higher clouds and decreasing moisture in the forest could explain the disappearance of the golden toad and harlequin frog and also the concurrent population crashes and subsequent disappearance of 20 other species of frogs and toads from the Monteverde region. As the forests dried out, drought-tolerant species of birds and reptiles shifted upslope, and moisture-dependent species were stranded at the mountaintops by a rising tide of aridity.

Pounds and his colleagues expanded the story further in *Nature* in 2006.[12] Although clouds had risen higher in the sky, the extra moisture evaporating from warming oceans was increasing cloud cover overall, blocking sunlight during the day and trapping heat at night. As a result, at Monteverde and other tropical locations, daytime and nighttime temperatures were becoming more similar. Such conditions are optimal for chytrid fungi, pathogens that can lethally infect amphibians. In recent years the chytrid fungus *Butrachochytrium dendrobatidis* is thought to have contributed to the extinction of 67 of the world's 113 species of harlequin frogs. At Monteverde and elsewhere, Pounds's team concluded with "very high confidence (>99%)" that

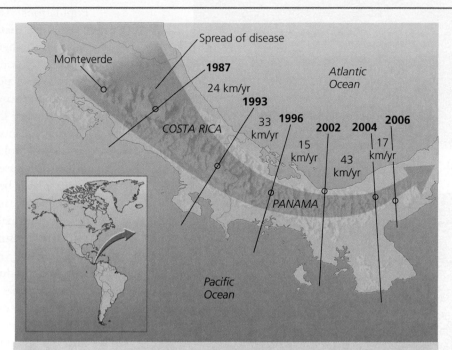

FIGURE 2   Dr. Karen Lips's research team used known dates of decline (years in figure) in harlequin frog populations at sites in Costa Rica and Panama to infer the spread of a wave of infection (arrow) by chytrid fungus across the region. By their analysis, chytrid reached Monteverde before 1987 and was well into Panama by 2000
Source: From Karen R. Lips; Jay Diffendorfer; Joseph R. Mendelson III; Michael W. Sears. (2008). Riding the Wave: Reconciling the Roles of Disease and Climate Change in Amphibian Declines. *PLoS Biology 6,* pp. 441–454. Published by PLoS (Public Library of Science).

climate change is promoting disease epidemics that are driving extinct many of the world's amphibians.

Other researchers agree that chytrid fungus is a major threat but are more cautious about the connection to climate change. In 2008, one team led by biologist Karen Lips of Southern Illinois University at Carbondale mapped amphibian declines and inferred the rapid spread of non-native and invasive chytrid fungus in waves across Central America and South America in

recent years (see **FIGURE 2**).[13,14] The analysis by Lips and colleagues suggested that the dramatic amphibian die-offs at Monteverde and elsewhere are directly due to the arrival of this devastating pathogen, perhaps promoted by climate change.

Clearly more research is needed into the effects of both disease and climate change on amphibian populations. Biologists today are racing to find the answers, hoping to save many of the world's amphibian species from extinction.

In *clumped distribution*, the pattern most common in nature, organisms arrange themselves according to the availability of the resources they need to survive. Many desert plants grow in patches around isolated springs or along *arroyos*, dry river valleys, that flood with water after rainstorms. During their mating season, golden toads were found clumped at seasonal breeding pools. Humans, too, exhibit clumped distribution; people frequently aggregate together in urban centres. Clumped distributions often arise from habitat selection.

Distributions depend on the scale at which one measures them. At very broad scales, all organisms show clumped or patchy distributions because some parts of the total area they inhabit are bound to be more hospitable than others.

Sex ratio and age structure   For organisms that reproduce sexually and have distinct male and female individuals, the sex ratio of a population can help determine whether it will increase or decrease in size over time. A population's **sex ratio** is its proportion of males to

(a) "Random"

(b) Uniform

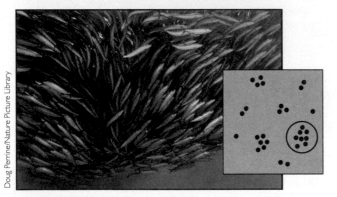

(c) Clumped

FIGURE 4.13

Individuals in a population can be spatially distributed over a landscape in three main ways. In random distribution **(a)**, organisms are dispersed randomly throughout the environment. In nature, the distribution of organisms is virtually never random, though; the trees in this forest look randomly distributed, but they have grown in specific locations that are beneficial for access to the resources they need. In uniform distribution **(b)**, individuals are spaced evenly, at equal distances from one another. Territoriality can result in such a pattern. In clumped distribution **(c)**, individuals occur in patches, concentrated more heavily in some areas than in others. Habitat selection or flocking to avoid predators can result in such a pattern.

females. In monogamous species (in which each sex takes a single mate), a 50/50 sex ratio maximizes population growth, whereas an unbalanced ratio leaves many individuals of one sex without mates. Most species are not

monogamous, though, so sex ratios vary from species to species.

Populations also consist of individuals of different ages. The **age distribution** or **age structure** describes the relative numbers of individuals of each age within a population. By combining this information with data on the reproductive potential of individuals in different age classes, a population ecologist can predict how the population may grow or shrink.

For many plants and animals that continue growing in size as they age, older individuals reproduce more; a tree that is large because it is old can produce more seeds, and a fish that is large because it is old may produce more eggs. In some animals, such as birds, the experience they gain with age often makes older individuals better breeders. Humans are unusual because we often survive long past our reproductive years. A population made up largely of older (post-reproductive) individuals will tend to decline over time, whereas one with many young individuals (of reproductive or pre-reproductive age) will tend to increase. We will use diagrams to explore these ideas further when we turn our attention to human population growth.

Birth and death rates   All the preceding factors can influence the rates at which individuals within a population are born and die. A convenient way to express birth and death rates is to measure the number of births and deaths per 1000 individuals for a given time period, termed the *crude birth rate* or *crude death rate*.

Just as individuals of different ages have different abilities to reproduce, individuals of different ages have different probabilities of dying. For instance, people are more likely to die at old ages than at young ages; if you were to follow 1000 10-year-olds and 1000 80-year-olds for a year, you would find that at year's end more 80-year-olds had died than 10-year-olds. However, this pattern does not hold for all organisms. Amphibians, like the golden toad, produce large numbers of young, which suffer high death rates. For a toad, death is less likely (and survival more likely) at an older age than at a very young age.

To show how the likelihood of death can vary with age, population ecologists use graphs called **survivorship curves (FIGURE 4.14)**. There are three fundamental types of survivorship curves. Humans, with higher death rates at older ages, show a *type I* survivorship curve; the same is true of other large mammals that have low numbers of offspring. Toads, with higher death rates at younger ages, show a *type III* survivorship curve in which there is a rapid die-off among young offspring; the same is true of many invertebrates. A *type II* survivorship curve is intermediate and indicates equal rates of death at all ages. Many birds and some small mammals can be characterized with type II curves.

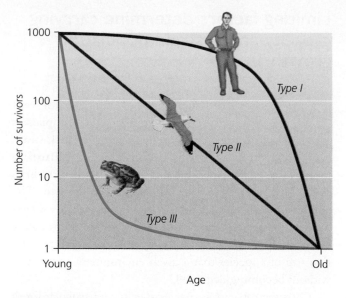

FIGURE 4.14

In a type I survivorship curve, survival rates are high when individuals are young and decrease sharply when individuals are old. In a type II survivorship curve, survival rates are equivalent regardless of an individual's age. In a type III survivorship curve, most mortality takes place at young ages, and survival rates are greater at older ages.

 On the basis of this graph, which organism has the highest rate of survival at a young age: a toad, a bird, or a human being?

Populations may grow, shrink, or remain stable

Now that we have outlined some key attributes of populations, we are ready to take a quantitative view of population change by examining some simple mathematical concepts used by both population ecologists and demographers (scientists who study human populations). Population growth, or decline, is determined by four factors:

1. births within the population, or **natality**
2. deaths within the population, or **mortality**
3. **immigration**, the arrival of individuals from outside the population
4. **emigration**, the departure of individuals from the population

Births and immigration add individuals to a population, whereas deaths and emigration remove individuals. If we are not interested in the effects of migration, we can measure the **natural rate of population growth** by subtracting the crude death rate from the crude birth rate:

(Crude birth rate − Crude death rate) = Natural rate of population growth

The natural rate of population growth reflects the degree to which a population is growing or shrinking as a result of its own internal factors.

To obtain an overall **population growth rate**, the total rate of change in a population's size per unit of time, we must also take into account the effects of migration. Thus, we include terms for immigration and emigration (each expressed per 1000 individuals per year) in the formula for the population growth:

(Crude birth rate − Crude death rate) + (Immigration rate − Emigration rate) = Population growth rate

The resulting number tells us the net change in a population's size per 1000 individuals per year. For example, a population with a crude birth rate of 18 per 1000/yr, a crude death rate of 10 per 1000/yr, an immigration rate of 5 per 1000/yr, and an emigration rate of 7 per 1000/yr would have a population growth rate of 6 per 1000/yr:

$$(18/1000 - 10/1000) + (5/1000 - 7/1000) = 6/1000$$

Thus, a population of 1000 in one year will reach 1006 in the next. If the population is 1 000 000, it will reach 1 006 000 the next year. Such population increases are often expressed as percentages, which we can calculate using the following formula:

Population growth rate × 100%

Therefore, a growth rate of 6/1000 would be expressed as

$$6/1000 \times 100\% = 0.6\%$$

By measuring population growth in terms of percentages, scientists can compare increases and decreases in species that have far different population sizes. They can also project changes that will occur in the population over longer periods.

Unregulated populations increase by exponential growth

When a population, or anything else, increases by a fixed percentage each year, it is said to undergo *geometric growth*, or **exponential growth**. A savings account is a familiar frame of reference for describing exponential growth. If at the time of your birth your parents had invested $1000 in a savings account earning 5% interest compounded each year, with no additional investments you would have $1629 by age 10, $2653 by age 20, and over $30 000 when you turn 70. If you could wait just 10 years more, that figure would rise to nearly $50 000. Only $629 was added during your first decade, but approximately $19 000 was added during the decade between ages 70 and 80.

The reason for the very rapid growth in later years is that a fixed percentage of a small number produces a small increase, but that same percentage of a large number

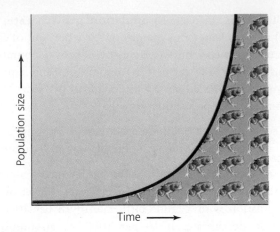

FIGURE 4.15
No species can maintain exponential growth indefinitely, but some may grow exponentially for a time when colonizing an unoccupied environment or exploiting an unused resource.

produces a large increase. Thus, as populations (or savings accounts) become larger, each incremental increase likewise gets larger. Such acceleration is a fundamental characteristic of exponential growth.

In contrast, if your parents added a fixed amount to your savings account every year, your savings would still grow, but by *arithmetic growth*, or **linear growth**. If both accounts (linear growth, with interest at a fixed *amount* per year, and exponential growth, with interest at a fixed *percentage* per year) were allowed to proceed unchecked, the exponential-growth account would necessarily outstrip the linear-growth account. This will be the case even if the balance in the linear-growth account is higher for the first few years.

We can visualize such changes in population size using population growth curves. The J-shaped curve in **FIGURE 4.15** shows an example of an exponential population increase. Populations of organisms can increase exponentially until they meet constraints. Each organism reproduces by a certain amount, and as populations get larger, more individuals reproduce by that amount. If there were no external limits on growth, ecologists theoretically would expect exponential growth to occur.

Exponential growth usually occurs in nature when a population is small, competition is minimal, and environmental conditions are ideal for the organism in question. Most often, these conditions occur when organisms are introduced to a new environment. Mould growing on a piece of bread or fruit, and bacteria colonizing a recently dead animal are cases in point. But species of any size may show exponential growth under the right conditions. We sometimes see this in urban areas where animals such as squirrels, raccoons, Canada geese, coyotes, and even moose can become nuisances when their populations expand quickly as a result of easy access to food supplies.

Limiting factors determine carrying capacity and restrain population growth

Exponential growth rarely lasts long. If even a single species in Earth's history had increased exponentially for very many generations, it would have blanketed the planet's surface, and nothing else could have survived. Instead, every population eventually is constrained by **limiting factors**—physical, chemical, and biological characteristics of the environment that restrain population growth. The combination of these factors exerts *environmental resistance* on the population. Environmental resistance ultimately determines the **carrying capacity**, the maximum population size of a species that a given environment can sustain without becoming degraded.

Carrying capacity is species-specific; a particular forest has a carrying capacity for wolves, for example, and a different carrying capacity for rabbits. They are closely interconnected, of course, because the two species interact with each other, with other species in the community, and with similar factors in the abiotic environment. Environments and communities are complex and ever-changing, so the carrying capacity of a particular environment for a given species can increase or decrease over time. If a fire destroys the forest, for example, the carrying capacities for many forest species will decline, at least temporarily. In contrast, species that are well adapted to fire will benefit from the change. The jack pine (*Pinus banksiana*) is specifically fire-adapted; its cones open to

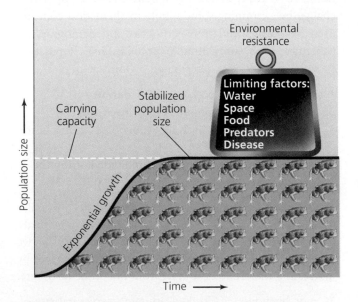

FIGURE 4.16
The logistic growth curve shows how population size may increase rapidly at first, then grow more slowly, and finally stabilize at a carrying capacity. Carrying capacity is determined both by the biotic potential of the organism and by various external limiting factors, which collectively exert *environmental resistance* that constrains population growth.

release seeds only in conditions of extreme heat. Jack pines not only benefit from forest fires, they depend on them for successful reproduction.

Ecologists use an S-shaped (or *sigmoidal*) curve like the one shown in **FIGURE 4.16** to show how an initial exponential increase is slowed and finally brought to a standstill by limiting factors. Called a **logistic growth curve**, it rises sharply at first but then begins to level off as the effects of limiting factors become stronger. Eventually the force of these factors stabilizes the population size at its carrying capacity.

The logistic curve is a simplified model; real populations can behave quite differently. Some may cycle indefinitely above and below the carrying capacity. Some may show cycles that become less extreme and approach the carrying capacity. Others may overshoot the carrying capacity and then crash, fated either for extinction or recovery (**FIGURE 4.17**).

Space is one of the factors that limits the number of individuals a given environment can support; if there is no physical room for additional individuals, they are unlikely to survive. Other limiting factors for animals in the terrestrial environment include the availability of food, water, mates, shelter, and suitable breeding sites; temperature extremes; prevalence of disease; and abundance of predators. Plants are often limited by amounts of sunlight and moisture, and the availability of nutrients that can be extracted from the soil, as well as disease and attack by plant-eating animals. In

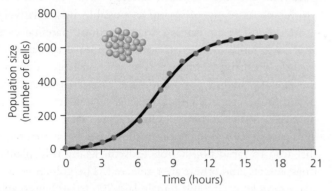

(a) Yeast cells, *Saccharomyces cerevisiae*

Source: Data from Pearl, R. (1927). The Growth of Populations. *Quarterly Review of Biology 2*, pp. 532–548.

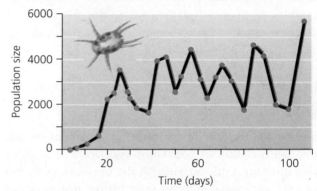

(b) Mite, *Eotetranychus sexmaculatus*

Source: Data from Utida, S. (1967). Damped Oscillation of Population Density at Equilibrium, *Researches on Population Ecology 9*, pp. 1–9.

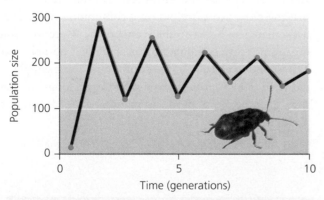

(c) Stored-product beetle, *Callosobruchus maculatus*

Source: Data from Huffaker, C. B. (1958). Experimental Studies on Predation: Dispersion Factors and Predator-Prey Oscillations, *Hilgardia 27*, pp. 343–383.

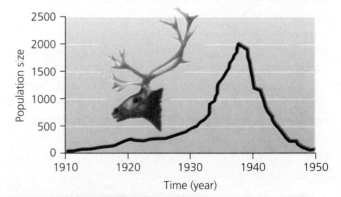

(d) St. Paul reindeer, *Rangifer tarandus*

Source: Data from Scheffer, V.C. (1951). Rise and Fall of a Reindeer Herd, *Scientific Monthly 73*, pp. 356–362.

FIGURE 4.17   Population growth in nature often departs from the stereotypical logistic growth curve, and it can do so in several fundamental ways. Yeast cells from a lab experiment show logistic growth **(a)** that closely matches the theoretical model. Some organisms, like the mite **(b)**, show cycles in which population fluctuates indefinitely above and below the carrying capacity. Population oscillations can also lessen in intensity, eventually stabilizing at carrying capacity, as in a lab experiment **(c)** with the stored-product beetle. Populations that rise too fast and deplete resources may crash just as suddenly **(d)**, like the population of reindeer introduced to the Bering Sea island of St. Paul.

DATA Q   The average lifespan of a reindeer (*R. tarandus*) in the wild might be 30 to 40 years. How many generations did it take for the population of reindeer introduced to the island of St. Paul (illustrated in **(d)**) to soar and then crash completely? The mite *E. sexmaculatus*, on the other hand, lives anywhere from about 20 to 40 days. On average, what was the duration, in generations, of each cycle of rapid population growth and decline in the study illustrated in **(b)**?

aquatic systems, limiting factors include salinity, sunlight, temperature, dissolved oxygen, and water chemistry, including the availability of nutrients. To determine limiting factors, an ecologist might conduct experiments in which they increase or decrease a hypothesized limiting factor to observe its effects on population size.

Population density can enhance or diminish the impact of other limiting factors on that population. Recall that high population density can help organisms find mates, but also increases competition and the risk of disease and some types of predation. Such factors are said to be *density-dependent*, because their influence waxes and wanes according to population density. *Density-independent* factors are limiting factors whose influence is not affected by population density. Temperature extremes and catastrophic events such as floods, fires, and landslides are examples of density-independent factors because they affect populations without being influenced by population density.

Reproductive strategies vary from species to species

Limiting factors from an organism's environment provide only half the story of population regulation. The other half comes from the attributes of the organism itself. For example, organisms differ in their **biotic potential**, or maximum capacity to produce offspring under ideal environmental conditions. A fish with a short gestation period that lays thousands of eggs at a time has high biotic potential, whereas a whale with a long gestation period that gives birth to a single calf at a time has low biotic potential. The interaction between an organism's biotic potential and the environmental resistance to its population growth helps determine the fate of its population.

Giraffes, elephants, humans, and other large animals with low biotic potential produce a relatively small number of offspring and take a long time to gestate and raise each of their young. Species that take this approach to reproduction compensate by devoting large amounts of energy and resources to caring for and protecting the relatively few offspring they produce during their lifetimes (**FIGURE 4.18**). Such species are said to be **K-selected** (or are called *K-strategists*). K-selected species are so named because their populations tend to stabilize over time at or near their carrying capacity, and *K* is a commonly used abbreviation for carrying capacity. Because their populations stay close to carrying capacity, natural selection in these species favours individuals that invest in producing offspring of high quality that can be good competitors.

In contrast, species that are **r-selected** focus on quantity, not quality. Species that are considered to be r-selected (or are called *r-strategists*) have high biotic potential and devote their energy and resources to producing as many offspring as possible in a relatively short time. Their offspring do not require parental care after birth, and r-strategists commonly leave the survival of their offspring to chance. The abbreviation *r* denotes the *rate* at which a population increases in the absence of limiting factors. Population sizes of r-selected species fluctuate greatly, so they are often below carrying capacity. This is why natural selection in these species favours traits that lead to rapid population growth. Many fish, plants, frogs, insects, and others are r-selected. The golden toad is one example; each adult female laid 200 to 400 eggs, and its tadpoles spent five weeks unsupervised in the breeding pools metamorphosing into adults.

TABLE 4.4 summarizes stereotypical traits of r-selected and K-selected species. However, it is important to note that these are two extremes on a continuum; most species fall somewhere between these endpoints. Moreover, some organisms show combinations of traits that do not clearly correspond to a place on the continuum. For example, the redwood tree (*Sequoia sempervirens*) is large, long-lived, and slow-growing, yet it produces many small seeds and offers no parental care.

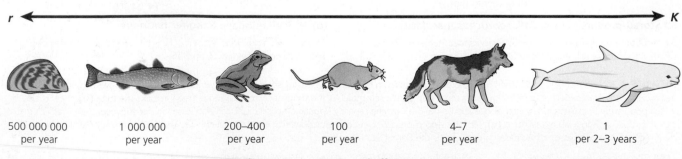

r					*K*
500 000 000 per year	1 000 000 per year	200–400 per year	100 per year	4–7 per year	1 per 2–3 years

Rough estimates—number of offspring per year

FIGURE 4.18 Species that are said to be K-selected (right) typically are large; they produce few offspring per year, but invest energetically in the care and raising of the offspring. Species that are r-selected (left) produce many offspring per year, but fewer of the offspring survive.

TABLE 4.4   Typical Characteristics of r-Selected and K-Selected Species

r-selected species	K-selected species
Small size	Large size
Fast development	Slow development
Short-lived	Long-lived
Reproduction early in life	Reproduction later in life
Many, small offspring	Few, large offspring
Fast population growth rate	Slow population growth rate
No parental care	Parental care
Weak competitive ability	Strong competitive ability
Variable population size, often well below carrying capacity	Constant population size, close to carrying capacity
Variable and unpredictable mortality	More constant and predictable mortality

Conclusion

Changes in ecosystems, communities, and populations have been taking place naturally as long as life has existed. Studies of both living and long-extinct organisms have helped illuminate fundamental concepts of evolution and population ecology that are integral to environmental science. The evolutionary processes of natural selection, speciation, and extinction have been the driving forces behind the development of Earth's biodiversity.

Unlike the golden toad, harlequin frog, and other species that have struggled to adapt to changing environments, our own species has proved capable of overcoming limiting factors. We have learned to alter environment so as to reduce environmental resistance and raise the carrying capacity for our species. When our ancestors began to build shelters and use fire for heating and cooking, they reduced the environmental resistance of areas with cold climates and were able to expand into new territory. People have managed so far to increase the planet's carrying capacity, but we have done so by appropriating immense proportions of the planet's resources. In the process, we have reduced the carrying capacities for countless other organisms and have called into question our own long-term survival.

Today human development, resource extraction, and population pressure are speeding the rate of environmental change, altering the types of change, and threatening specific organisms and biodiversity as a whole. The ways we modify our environment cannot be understood in a scientific vacuum, however. Understanding how ecological processes work at the population level is crucial to protecting biodiversity threatened by the mass extinction event that many biologists maintain is already underway. The factors that threaten biodiversity have complex social, economic, and political roots, and environmental scientists appreciate that we must understand these aspects if we are to develop solutions.

Fortunately, there are things people can do to forestall population declines of species threatened with extinction. Millions of people around the world are already taking action to safeguard the biodiversity and ecological and evolutionary processes that make Earth such a unique place. We will look at many of these efforts in future chapters.

REVIEWING OBJECTIVES

You should now be able to

Explain the process of natural selection and cite evidence for this process

- Natural selection is the mechanism for evolution. Because individuals vary in their traits, and many traits are inherited, some individuals will prove better at surviving and reproducing. Their genes will be passed on and become more prominent in future generations.
- Mutations and recombination provide the genetic variation for natural selection.
- Environmental conditions exert pressures that can drive natural selection.
- Natural selection is all around us, in all of the adaptations displayed by organisms. We have produced our pets, farm animals, and crop plants through artificial selection.

Describe the ways in which evolution results in biodiversity and what the fossil record has taught us about evolution

- Natural selection can act as a diversifying force as organisms adapt to their environments in myriad ways.
- Speciation is the process whereby new species emerge.
- Populations can become isolated by geographical separation and by other means, such as reproductive isolation.
- The branching patterns of phylogenetic trees reflect the historical pattern by which lineages of organisms have diverged.
- By preserving some of the long history of life on this planet, the fossil record has shown us that later organisms evolved from earlier ancestral organisms.

Discuss reasons for species extinction and mass extinction events

- Extinction is a natural process. It may occur when species that are highly specialized or that have small populations encounter rapid environmental change.
- Life on Earth has experienced five known major episodes of mass extinction, which are attributable to factors such as asteroid impact, volcanism, and climatic change.
- Today, human impact may be initiating a sixth great extinction.

Summarize the levels of ecological organization

- Life is organized hierarchically, starting with the atoms, molecules, and cells that make up individual organisms.
- Ecologists classify organisms in a hierarchical structure. They study phenomena on the organismal, population, community, and ecosystem levels—and, increasingly, on the biosphere level.
- Habitat, niche, and specialization are important ecological concepts.

Define logistic growth, carrying capacity, limiting factors, and other fundamental concepts of population ecology

- Populations are characterized by population size, population density, population distribution, sex ratio, age structure, and birth and death rates.
- Immigration and emigration, as well as birth and death rates, determine how a population will grow or decline.
- Populations undergo exponential growth until limiting factors generate environmental resistance that constrains population growth.
- Logistic growth describes the effects of density dependence; exponential growth slows as population size increases, and population size levels off at a carrying capacity.
- K-selection and r-selection describe theoretical extremes in how organisms achieve growth and reproduction.

TESTING YOUR COMPREHENSION

1. Explain the premises and logic that support the concept of natural selection.
2. How does allopatric speciation occur?
3. Name two examples of evidence for natural selection.
4. Name three organisms that have gone extinct, and give a probable reason for each extinction.
5. What is the difference between a species and a population? Between a population and a community?
6. Contrast the concepts of habitat and niche.
7. List and describe each of the five major population characteristics discussed in this chapter. Explain how each shapes population dynamics.
8. Could any species undergo exponential growth forever? Explain your answer.
9. Describe how limiting factors relate to carrying capacity.
10. Explain the difference between K-selected species and r-selected species. Can you think of examples of each that were not mentioned in the chapter?

THINKING IT THROUGH

1. In what ways has artificial selection changed people's quality of life? Give examples. Can you imagine a way in which artificial selection could be used to improve our quality of life further? Can you imagine a way it could be used to lessen our environmental impact?
2. What types of species are most vulnerable to extinction, and what kinds of factors threaten them? Can you think of any species in your region that are threatened with extinction today? What reasons lie behind their endangerment?
3. If you were given the task of counting all of the species alive on Earth, how would you go about it? If you were asked to measure the biodiversity of one area and contrast it with that of another area, what indicators of biodiversity would you measure? (We will return to this question in the chapter on conservation of species and habitats.)
4. Let us say that you are a population ecologist studying animals in a national park. The government is asking for advice on how to focus its limited conservation funds. How would you rate the following three species, from most vulnerable (and thus most in need of attention) to least vulnerable? Give reasons for your choices.
   - a bird with an even sex ratio that is a habitat generalist
   - a salamander that is endemic to the park and lives only in high-elevation forest
   - a fish that specializes on a few types of invertebrate prey and has a large population size

INTERPRETING GRAPHS AND DATA

Amphibians are sensitive biological indicators of climate change because their reproduction and survival are so closely tied to water. One way in which drier conditions may affect amphibians is by reducing the depth of the pools of water in which their eggs develop. Shallower pools offer less protection from UV-B (ultraviolet) radiation, which some scientists maintain may kill embryos directly or make them more susceptible to disease.

Herpetologist Joseph Kiesecker and colleagues conducted a field study of the relationships among water depth, UV-B radiation, and survivorship of western toad (*Bufo boreas*) embryos in the Pacific Northwest.[15] In manipulative experiments, the researchers placed toad embryos in mesh enclosures at three different depths of water. The researchers placed protective filters that blocked all UV-B radiation over some of these embryos, while leaving other embryos unprotected without the filters. Some of the study's results are presented in the accompanying graph.

1. If the UV-B radiation at the surface has an intensity of 0.27 watts/m², approximately what is its intensity at depths of 10 cm, 50 cm, and 100 cm?

2. Approximately how much did survival rates at the 10-cm depth differ between the protected and unprotected treatments? Why do you think survival rates differed significantly at the 10-cm depth but not at the other depths?

3. What do you think would be the effect of drier-than-average years on the western toad population if the average depth of pools available for toad spawning dropped? How do the data in the graph address your hypothesis? Do they support cause-and-effect relationships among water depth, UV-B exposure, disease, and toad mortality?

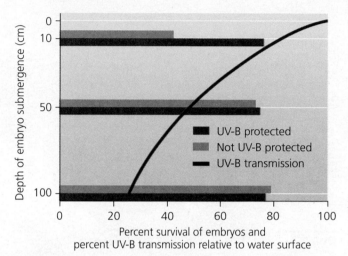

The graph shows embryo survivorship in western toads (*Bufo boreas*) at different water depths and UV-B light intensities. Red bars indicate embryos protected under a filter that blocked UV-B light; orange bars indicate unprotected embryos. The blue line indicates the amount of UV-B light reaching different depths in the water column, expressed as a percentage of the UV-B radiation at the water surface.
Source: Data from Kiesecker, J.M., et al. (2001). Complex Causes of Amphibian Population Declines. *Nature 410*, pp. 681–684.

MasteringEnvironmentalScience®

STUDENTS
Go to **MasteringEnvironmentalScience** for assignments, the eText, and the Study Area with practice tests, videos, current events, and activities.

INSTRUCTORS
Go to **MasteringEnvironmentalScience** for automatically graded activities, current events, videos, and reading questions that you can assign to your students, plus Instructor Resources.

5 Species Interactions and Community Ecology

This marsh–wetland community is one of the few that remain along the shores of Lake Ontario.

Jim West/Alamy

Upon successfully completing this chapter, you will be able to

- Compare and contrast the major types of species interactions
- Characterize feeding relationships and energy flow in ecological communities, using them to construct trophic pyramids and food webs

- Describe how communities respond to disturbances, referencing the concepts of resistance, resilience, succession, and restoration
- Describe and illustrate the terrestrial biomes of the world

This is a typical aggregation of zebra mussels, like those that have invaded the Great Lakes.

CANADA

UNITED STATES

Great Lakes

CENTRAL CASE
BLACK AND WHITE AND SPREAD ALL OVER: ZEBRA MUSSELS INVADE THE GREAT LAKES

"The zebra mussel is helping us understand what makes a good invader."

—ANTHONY RICCIARDI, MCGILL UNIVERSITY

"The zebra mussel has altered aquatic ecosystems beyond recognition."

—MICHAEL BARDWAJ, *CANADIAN GEOGRAPHIC*[1]

As if the Great Lakes had not been through enough, the last thing they needed was the zebra mussel. The pollution-fouled waters of Lake Erie and the other Great Lakes had become gradually cleaner in the years following the establishment of the International Joint Commission and the signing of the Great Lakes Water Quality Agreement between Canada and the United States in 1972. As these international efforts brought industrial discharges under control, people once again began to use the lakes for recreation, and populations of fish rebounded.

Then the zebra mussel arrived. Black-and-white-striped shellfish the size of a dime (see photo), zebra mussels (*Dreissena polymorpha*) attach to hard surfaces and feed on algae by filtering water through their gills. This mollusc is native to the Caspian Sea, Black Sea, and Azov Sea in western Asia and eastern Europe. It made its North American debut in 1988 when it was discovered in Canadian waters at Lake St. Clair, which connects Lake Erie with Lake Huron. Evidently, ships arriving from Europe had discharged ballast water containing the mussels or their larvae into the Great Lakes.

Within two years of their discovery in Lake St. Clair, zebra mussels had reached all five of the Great Lakes.[2] The next year, they entered New York's Hudson River to the east and the Illinois River at Chicago to the west. From the Illinois River and its canals, they soon reached the

Mississippi River, giving them access to a vast watershed covering 40% of the United States. By 2014, zebra mussel colonies and sightings had been confirmed in Ontario, Quebec, Manitoba, and 30 U.S. states and waterways.

How could a mussel spread so quickly? The zebra mussel's larval stage is well adapted for long-distance dispersal. Its tiny larvae drift freely for several weeks, travelling as far as the currents take them. Adults that attach themselves to boats and ships may be transported from one place to another, even to isolated lakes and ponds well away from major rivers. They can survive out of the water for several days and are known to have been transported overland to many locations. In North America the mussels encountered none of the predators, competitors, and parasites that had evolved to limit their population growth in the Old World.

Why the fuss? Zebra mussels are best known for clogging up water intake pipes at factories, power plants, municipal water supplies, and wastewater treatment facilities. At one power plant, workers counted 700 000 mussels per square metre of pipe surface. Great densities of these organisms can damage boat engines, degrade docks, foul fishing gear, and sink buoys that ships use for navigation. Through such impacts, it is estimated that zebra mussels cost hundreds of millions of dollars each year. The total cost to Great Lakes economies is estimated to have reached $5 billion, with ongoing annual costs of $20 000 to $350 000 per industrial facility.[3] These figures include only costs to industry, not to individuals or cottagers, who also suffer costs such as clogged water pipes, ruined motorboats, and fouled beaches.

Zebra mussels also have severe impacts on the ecological systems they invade. They eat **phytoplankton**, microscopic algae that drift in open water. Because each mussel filters a litre or more of water every day, they consume so much phytoplankton that they can deplete populations. Phytoplankton are the foundation of the Great Lakes food web, so its depletion is bad news for **zooplankton**, the tiny aquatic animals that eat phytoplankton—and for the fish that eat both. Water bodies with zebra mussels have fewer zooplankton and open-water fish than water bodies without them, researchers are finding.

However, zebra mussels also provide benefits to some bottom-feeding invertebrates and fish. By filtering algae and organic matter from open water and depositing nutrients in their feces, they shift the community's nutrient balance to the bottom and benefit the species that feed there. Once they have cleared the water, sunlight penetrates deeper, spurring the growth of large-leafed underwater plants and algae. Such changes have ripple effects throughout the community that scientists are only beginning to understand.

In the past several years, scientists have noticed a surprising twist: One invader is being displaced by another. The quagga mussel (*Dreissena buensis*), a close relative of the zebra mussel, is spreading through the Great Lakes, replacing the zebra mussel in many locations.[4] What consequences this may have for ecological communities, scientists are only beginning to understand.

Species Interactions

By interacting with many species in a variety of ways, new arrivals like the zebra mussels have set in motion an array of changes in the ecological communities they have invaded. Interactions among species are the threads in the fabric of communities, holding them together and determining their nature and their stability. Ecologists have organized species interactions into several fundamental categories. Most prominent are competition, predation, parasitism, herbivory, and mutualism.

TABLE 5.1 summarizes the positive (+) and negative (−) impacts of each type of interaction for each participant.

TABLE 5.1   Effects of Species Interactions on Their Participants

Type of interaction	Effect on species 1	Effect on species 2
Mutualism	+	+
Commensalism	+	0
Predation, parasitism, herbivory	+	
Amensalism	−	0
Competition	−	−

"+" denotes a positive effect; "−" denotes a negative effect; "0" denotes no effect.

An interaction with no impact is shown by a "0" in the table. In competitive interactions, each participant has a negative effect on other participants, because each takes resources the others could have used. This is reflected in the two minus signs shown in **TABLE 5.1**. In other types of interactions, some participants benefit while others are harmed (the +/− interactions in the table). We can think of interactions in which one member exploits another for its own gain as *exploitative interactions*. Such interactions include predation, parasitism, herbivory, and related concepts, outlined below.

Competition can occur when resources are limited

When multiple organisms seek the same limited resource, their relationship is said to be one of **competition**. Competing organisms may fight with one another directly and physically, but there are other ways to compete. Competition is often more subtle and indirect, involving the consequences of one organism's ability to match or outdo others in procuring resources. The resources for which organisms compete can include just about anything an organism might need to survive, including food, water, space, shelter, mates, sunlight, and more.

Amensalism is a variation of a competitive relationship, in which one organism is harmed and the other is apparently unaffected (that is, not benefited). Amensalism has been difficult to pin down, because it is hard to prove that the organism doing the harm is not, in fact, outcompeting a competitor for a resource. For instance, some plants release poisonous chemicals that harm nearby plants (a phenomenon called *allelopathy*), and some experts have suggested that this is an example of amensalism. However, allelopathy can also be viewed as one plant investing in chemicals to outcompete others for space.

Competitive interactions can take place among members of the same species (*intraspecific competition*) or among members of two or more different species (*interspecific competition*). We have already discussed intraspecific competition, without naming it as such. Recall that density dependence limits the growth of a population; individuals of the same species compete with one another for limited resources, such that competition is more acute when there are more individuals per unit area (denser populations). Thus, intraspecific competition is really a population-level phenomenon.

In contrast, interspecific competition—competition between organisms of different species—can have substantial effects on the composition of communities. If one species is a very effective competitor, it may exclude another species from resource use entirely. This outcome, called **competitive exclusion**, occurred in Lake St. Clair and western Lake Erie as zebra mussels outcompeted a native mussel species.

Alternatively, if neither competing species fully excludes the other, the species may live side by side at a certain ratio of population sizes. This result, called *species coexistence*, may produce a stable point of equilibrium, in which the population size of each remains fairly constant through time.

Coexisting species that use the same resources tend to adjust to their competitors to minimize competition with them. Individuals can do this by changing their behaviour so as to use only a portion of the total array of resources they are capable of using. In such cases, individuals are not fulfilling their entire *niche*, or ecological role (discussed in Chapter 4). The full niche of a species is called its **fundamental niche** (**FIGURE 5.1A**). An individual that plays only part of its role because of competition or other species interactions is said to be displaying a **realized niche** (**FIGURE 5.1B**), the portion of its fundamental niche that is actually filled, or realized.

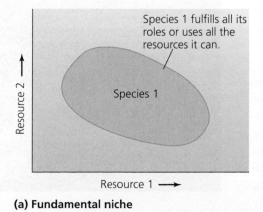

(a) Fundamental niche

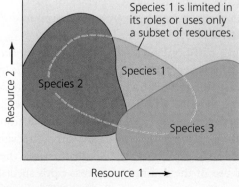

(b) Realized niche

FIGURE 5.1 An organism facing competition may be forced to play a lesser ecological role or use fewer resources than it would in the absence of its competitor. With no competitors, an organism can exploit its full fundamental niche **(a)**. When competitors restrict what an organism can do or what resources it can use, the organism is limited to a realized niche **(b)**, which covers only a subset of its fundamental niche.

White-breasted nuthatch climbs down trunk looking for insects.

Yellow-bellied sapsucker drills rows of holes and consumes sap and insects stuck in sap.

Pileated woodpecker digs deeply into wood to find large insects.

Brown creeper climbs up trunk looking for tiny insects.

FIGURE 5.2
When species compete, they tend to partition resources, each specializing on a slightly different resource or way of attaining a shared resource. A number of types of birds—including the woodpecker, creeper, sapsucker, and nuthatch shown here—feed on insects from tree trunks, but use different portions of the trunk, seeking different foods in different ways.

Species make similar adjustments over evolutionary time. They adapt to competition by evolving to use slightly different resources or to use their shared resources in different ways. If two bird species eat the same type of seeds, one might come to specialize on larger seeds and the other to specialize on smaller seeds. Or one bird might become more active in the morning and the other more active in the evening, thus avoiding direct interference. This process is called **resource partitioning**, because the species divide, or partition, the resource they use in common by specializing in different ways (**FIGURE 5.2**).

Resource partitioning can lead to *character displacement*, which occurs when competing species evolve physical characteristics that reflect their reliance on the portion of the resource they use. By becoming more different from one another, two species reduce their competition. For example, through natural selection, birds that specialize on larger seeds may evolve larger bills that enable them to make best use of the resource, whereas birds specializing on smaller seeds may evolve smaller bills. This is precisely what extensive research has revealed about the finches from the Galápagos Islands that were first described by Charles Darwin.

Predators kill and consume prey

Every living thing needs to procure food and, for most animals, that means eating other living organisms. **Predation** is the process by which individuals of one species—the **predator**—hunt, capture, kill, and consume individuals of another species, the **prey** (**FIGURE 5.3**). Along with competition, predation has traditionally been viewed as one of the primary organizing forces in community ecology. Interactions between predators and prey structure the food webs that we will examine shortly, and they influence community composition by helping determine the relative abundance of predators and prey.

Zebra mussel predation on phytoplankton reduced phytoplankton populations by up to 90%, according to studies in the Great Lakes and Hudson River.[5] Zebra mussels also consume the smaller types of zooplankton. This predation, combined with the competition mentioned above, has caused zooplankton population sizes and biomass to decline by up to 70% in Lake Erie and the Hudson River since zebra mussels arrived. Meanwhile, the mussels do not readily digest some cyanobacteria, so concentrations of these cyanobacteria rise in lakes with zebra mussels. Most predators are also prey, however, and zebra mussels have become a food source for a number of North American species since their introduction. These include diving ducks, muskrats, crayfish, flounder, sturgeon, eels, and several types of fish with grinding teeth, such as carp and freshwater drum.

Predation can sometimes drive population dynamics by causing cycles in population sizes. An increase in the population size of prey creates more food for predators, which may survive and reproduce more effectively as

Александр Просвиров/123RF

FIGURE 5.3
Predator–prey interactions have ecological and evolutionary consequences for both prey and predator. Here, a fire-bellied snake (*Liophis epinephalus*) devours a frog in the Monteverde cloud forest.

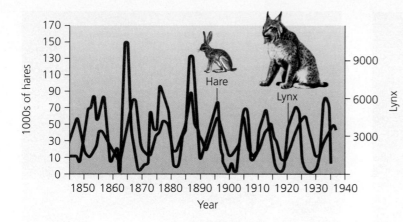

FIGURE 5.4
Predator–prey systems sometimes show paired cycles, in which increases and decreases in one organism apparently drive increases and decreases in the other. Such cycles are predicted by theory and are seen in lab experiments, but they are difficult to document conclusively in natural systems.
Source: Data from Maclulich, D.A. (1937). *Fluctuation in the Numbers of Varying Hare (Lepus americanus).* Univ. Toronto Stud. Biol. Ser. No. 43. Toronto, University of Toronto Press.

a result. As the predator population rises, additional predation drives down the population of prey. Fewer prey in turn cause some predators to starve, so that the predator population declines. This allows the prey population to begin rising again, starting the cycle anew. Most natural systems involve so many factors that such cycles do not last long, but in some cases we see extended cycles (**FIGURE 5.4**).

Predation also has evolutionary ramifications. Individual predators that are more adept at capturing prey will likely live longer, healthier lives and be better able to provide for their offspring than will less adept individuals. Thus, natural selection on individuals within a predator species leads to the evolution of adaptations that make them better hunters. Prey are faced with an even stronger selective pressure—the risk of immediate death. For this reason, predation pressure has caused organisms to evolve an elaborate array of defences against being eaten (**FIGURE 5.5**).

Parasites exploit living hosts

Organisms can exploit other organisms without killing them. **Parasitism** is a relationship in which one organism, the **parasite**, depends on another, the **host**, for nourishment, support, or some other benefit while simultaneously doing the host harm. Unlike predation, parasitism usually does not result in an organism's immediate death, although it sometimes contributes to the host's eventual death.

Many parasites live in close contact with their hosts. These parasites include disease pathogens, such as the protists that cause malaria and dysentery, as well as animals, such as tapeworms, that live in the digestive tracts of their hosts. Other parasites live on the exterior of their hosts, such as ticks that attach themselves to their hosts' skin, and the sea lamprey (*Petromyzon marinus*), another invader of the Great Lakes (**FIGURE 5.6A**). Sea lampreys are tube-shaped vertebrates that grasp the bodies of fish by using a suction-cup mouth and a rasping tongue, sucking blood

(a) Cryptic coloration **(b) Warning coloration** **(c) Mimicry**

FIGURE 5.5 Natural selection to avoid predation has resulted in many adaptations. Some prey hide from predators by *crypsis*, or camouflage, such as this leopard gecko (*Eublepharis macularius*) on tree bark **(a)**. Other prey are brightly coloured to warn predators that they are toxic or distasteful, such as this monarch butterfly (*Danaus plexippus*) **(b)**. Still others fool predators with mimicry. Some, like walking sticks imitating twigs, mimic for crypsis. Others mimic toxic, distasteful, or dangerous organisms, like the viceroy butterfly, a non-toxic species that looks virtually identical to a monarch butterfly. When disturbed, this caterpillar of a spicebush swallowtail butterfly (*Papilio troilus*) **(c)** swells and curves its tail end and shows eye-spots, to look like a snake.

(a) Sucker mouth of a sea lamprey (*Petromyzon marinus*)

FIGURE 5.6 Parasites harm their host organism. With its suction-like mouth and rasping tongue, the sea lamprey (*Petromyzon marinus*) **(a)** attaches itself to a fish and sucks its blood, sometimes killing the fish. In **(b)**, an ant (*Pachycondlyla*) is infected by a fungus (*F. cordyceps*) that will eventually kill it. Fruiting bodies of the fungus are sprouting from the ant. The fungus will soon alter the ant's behaviour, causing it to climb to the top of a nearby plant so its spores will attain the broadest possible distribution.

(b) *Fungus cordyceps* infecting an insect

from the fish for days or weeks. Sea lampreys invaded the Great Lakes from the Atlantic Ocean after people dug canals to connect the lakes for shipping, and the lampreys soon devastated economically important fisheries of chub, lake herring, whitefish, and lake trout. Since the 1950s, Great Lakes fisheries managers have reduced lamprey populations by applying chemicals that selectively kill lamprey larvae.

Other types of parasites are free-living and come into contact with their hosts only infrequently. For example, the cuckoos of Eurasia and the cowbirds of the Americas parasitize other birds by laying eggs in their nests and letting the host bird raise the parasite's young.

Some parasites cause little harm, but others may kill their hosts. Many insects parasitize other insects, often killing them in the process, and are called *parasitoids*. Various species of parasitoid wasps lay eggs on caterpillars. When the eggs hatch, the wasp larvae burrow into the caterpillar's tissues and slowly consume them. The wasp larvae metamorphose into adults and fly from the body of the dying caterpillar.

Just as predators and prey evolve in response to one another, so do parasites and hosts, in a process termed **coevolution**. Hosts and parasites can become locked in a duel of escalating adaptations, a situation sometimes referred to as an "evolutionary arms race." Like rival nations racing to stay ahead of one another in military technology, host and parasite may repeatedly evolve new responses to the other's latest advance. In the long run,

though, it may not be in a parasite's best interests to become too harmful to its host. Instead, a parasite might leave more offspring in the next generation—and thus be favoured by natural selection—if it allows its host to live a longer time, or even to thrive.

Herbivores exploit plants

One of the most common types of exploitation is **herbivory**, which occurs when animals feed on the tissues of plants. Insects that feed on plants are the most widespread type of herbivore; just about every plant in the world is attacked by some type of insect (**FIGURE 5.7**). In most cases, herbivory does not kill a plant outright, but may affect its growth and reproduction.

Like animal prey, plants have evolved a wide array of defences against the animals that feed on them. Many plants produce chemicals that are toxic or distasteful to herbivores. Others arm themselves with thorns, spines, or irritating hairs. In response, herbivores may evolve ways to overcome these defences, and the plant and the animal may embark on an evolutionary arms race.

Some plants go a step further and recruit certain animals as allies to assist in their defence. Many such plants encourage ants to take up residence by providing thorns or swelled stems for the ants to nest in or nectar-bearing structures for the ants to feed from. These ants protect the plant in return by attacking other insects that land or crawl on it. Other plants respond to herbivory by releasing volatile chemicals when

FIGURE 5.7
Herbivory is a common way to make a living. The world holds many thousands, perhaps millions, of species of plant-eating insects, such as this caterpillar, part of the larval growth stage of a monarch butterfly (*Danaus plexippus*)

mycorrhizae. In these symbioses, the plant provides energy and protection to the fungus, while the fungus assists the plant in absorbing nutrients from the soil through nitrogen fixation (discussed in Chapter 3). Fungi in such relationships may also detoxify harmful substances and provide protection from pathogens. In the ocean, coral polyps, the tiny animals that build coral reefs, share beneficial arrangements with algae known as *zooxanthellae*. The coral provide housing and nutrients for the algae in exchange for a steady supply of food—up to 90% of their nutritional requirements.

You, too, are part of a symbiotic association. Your digestive tract is filled with microbes that help you digest food—microbes for which you are providing a place to live. Indeed, we may owe our very existence to symbiotic mutualisms. It is now widely accepted that the eukaryotic cell originated after certain prokaryotic cells engulfed other prokaryotic cells and established mutualistic symbioses. Scientists have inferred that some of the engulfed cells eventually evolved into cell organelles.

Not all mutualists live in close proximity. One of the most important mutualisms in environmental science involves free-living organisms that may encounter each other only once in their lifetimes. This is **pollination** (**FIGURE 5.8**), an interaction of key significance to agriculture and our food supply. Bees, birds, bats, and other animals transfer pollen (carrying the male sex cells) from one flower to the ova (female sex cells) of another, fertilizing the egg, which subsequently grows into a fruit. The pollinating animals visit flowers for their nectar, a reward the plant uses to entice them. The pollinators receive food,

they are bitten or pierced. The airborne chemicals attract predatory insects that may attack the herbivore. Such cooperative strategies as trading defence for food are examples of our next type of species interaction, mutualism.

Mutualists help one another

Mutualism is a relationship in which two or more species benefit from interaction with one another. Generally each partner provides some resource or service that the other needs.

Many mutualistic relationships—like many parasitic relationships—occur between organisms that live in close physical contact. (Indeed, biologists hypothesize that many mutualistic associations evolved from parasitic ones.) Such physically close association is called **symbiosis**. For example, thousands of terrestrial plant species depend on mutualisms with fungi. The plant roots and some fungi form symbiotic associations called

FIGURE 5.8
In mutualism, organisms of different species benefit one another. One important mutualistic interaction is pollination. Here a juvenile ruby-throated hummingbird (*Archilochus colubris*) visits flowers to gather nectar. In the process it transfers pollen between flowers, helping the plant reproduce. Pollination is of key importance to agriculture.

and the plants are pollinated and reproduce. Various types of bees alone pollinate 73% of our crops, one expert has estimated—from soybeans to potatoes to tomatoes to beans to cabbage to oranges.

Commensalism is a species interaction in which one species benefits while the other is unaffected. One association commonly cited as an example of commensalism occurs when the conditions created by one plant happen to make it easier for another plant to establish and grow. For instance, palo verde trees in the Sonoran Desert create shade and leaf litter that allow the soil beneath them to hold moisture longer, creating an area that is cooler and moister than the surrounding sun-baked ground. Young plants find it easier to germinate and grow in these conditions, so seedling cacti and other desert plants generally grow up directly beneath "nurse" trees such as the palo verde. This phenomenon, called *facilitation*, influences the structure and composition of communities and how they change through time.

Ecological Communities

We have defined a *community* as a group of populations of organisms that live in the same place at the same time. The members of a community interact with one another in the ways described above, and the direct interactions among species often have indirect effects that ripple outward to affect other community members. The strength of interactions also varies, and together species' interactions determine the species composition, structure, and function of communities. The concept of *ecological community* is slightly different from the concept of *ecosystem*, which includes interactions with the abiotic environment, as well as interactions among organisms. *Community ecologists* are interested in which species coexist, how they relate to one another, how communities change through time, and why these patterns exist.

Energy passes among trophic levels

The interactions among members of an ecological community are many and varied, but some of the most important involve who eats whom. The energy that drives such interactions in most systems comes ultimately from the sun via photosynthesis. As organisms feed on one another, this energy moves through the community, from one rank in the feeding hierarchy, or **trophic level**, to another (**FIGURE 5.9**). The word *trophic* comes from a Greek root that means "nourishment" or "food." In the context of ecosystems, it also carries a connotation of "energy" because in ecosystems, food is energy and energy is food.

Producers   **Producers** or **autotrophs** ("self-feeders") compose the first trophic level. Terrestrial green plants, cyanobacteria, and algae capture solar energy and use photosynthesis to produce sugars. The chemosynthetic bacteria of hot springs and deep-sea hydrothermal vents use geothermal energy in a similar way to produce food.

Consumers   Organisms that derive their food energy from other organisms are **heterotrophs**, and they are also called **consumers**. Consumers that eat producers are known as **primary consumers**; they compose the second trophic level. Grazing animals, such as deer and grasshoppers, are primary consumers. The third trophic level consists of **secondary consumers**, which prey on primary consumers. For example, wolves that prey on deer are secondary consumers, as are rodents and birds that prey on grasshoppers. Predators that feed at even higher trophic levels are known as *tertiary consumers*. Examples of tertiary consumers include hawks and owls that eat rodents that have eaten grasshoppers.

Note that most primary consumers are **herbivores** because they consume plants, whereas secondary and tertiary consumers are **carnivores** because they eat animals. Animals that eat both plant and animal food are referred to as **omnivores**, from root words that mean "all-consuming."

Detritivores and decomposers   Some organisms consume nonliving organic matter. **Detritivores**, such as millipedes and soil insects, scavenge the waste products or the dead bodies of other community members. **Decomposers**, such as fungi and bacteria, break down leaf litter and other nonliving matter further into simpler constituents that can then be taken up and used by plants. These organisms play an essential role as the community's recyclers, making nutrients from organic matter available for reuse by living members of the community. Often, detritivores and decomposers are *saprotrophs*—organisms that eat dead organisms.

In Great Lakes communities, phytoplankton are the main producers, floating freely and photosynthesizing using sunlight that penetrates the upper layer of the water. Zooplankton are primary consumers, feeding on the phytoplankton. Phytoplankton-eating fish are primary consumers, and zooplankton-eating fish are secondary consumers. At higher trophic levels are tertiary consumers, such as larger fish and birds that feed on plankton-eating fish. Zebra mussels, by eating both phytoplankton and zooplankton, function on multiple trophic levels. When any of these organisms dies and sinks to the bottom, detritivores scavenge its tissues, and microbial decomposers recycle its nutrients.

Aquatic examples   **Terrestrial examples**

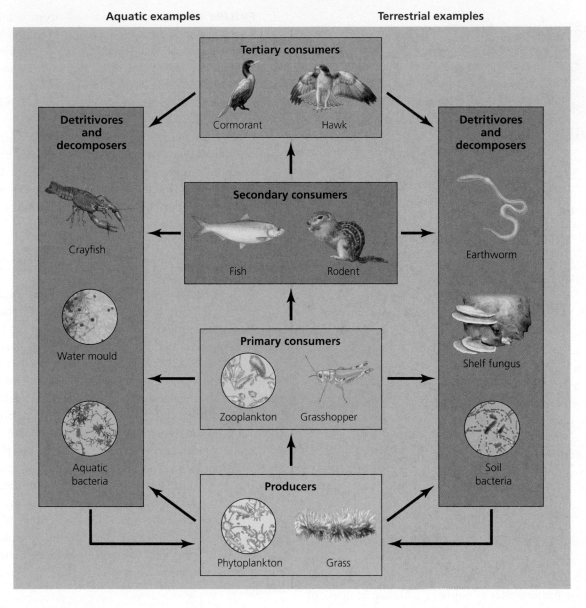

FIGURE 5.9 Ecologists organize species hierarchically by their feeding rank, or trophic level. The diagram shows aquatic (left) and terrestrial (right) examples. Arrows indicate directions of energy flow. Producers produce food by photosynthesis; primary consumers (herbivores) feed on producers; secondary consumers eat primary consumers; and tertiary consumers eat secondary consumers. Detritivores and decomposers feed on nonliving organic matter and the remains of dead organisms from all trophic levels, and "close the loop" by returning nutrients to the soil or the water column for use by producers.

Energy, biomass, and numbers decrease at higher trophic levels

At each trophic level, most of the energy that organisms use is lost through respiration. Only a small amount of the energy is transferred to the next trophic level through predation, herbivory, or parasitism. As energy is transferred from species on one trophic level to species on other trophic levels, it is said to pass up a **food chain**.

The first trophic level (producers) contains a large amount of energy, but the second (primary consumers) contains less energy—only that amount gained from consuming producers. The third trophic level (secondary consumers) contains still less energy, and higher trophic levels (tertiary consumers) contain the least. A general rule of thumb is that each trophic level contains roughly 10% of the energy of the trophic level below it in the food chain, although the actual proportion can vary greatly. This energy inefficiency—the loss of energy with each step up the chain—is one of the fundamental characteristics of food chains.

This pattern, which can be visualized as a **trophic pyramid**, generally also holds for the numbers of organisms

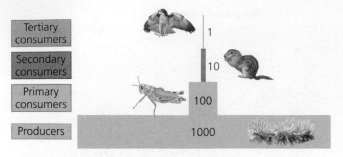

FIGURE 5.10

A trophic pyramid is a rule of thumb for the way ecological communities are structured. Organisms at lower trophic levels generally exist in far greater numbers, with greater energy content and greater biomass, than organisms at higher trophic levels. The example shown here is generalized; the actual shape of any given pyramid may be different.

 Using the ratios shown in this graph, let's suppose that a system has 3000 grasshoppers. How many rodents would be expected?

at each trophic level (**FIGURE 5.10**). Typically, fewer organisms exist at higher trophic levels than at lower trophic levels. A grasshopper eats many plants in its lifetime, a rodent eats many grasshoppers, and a hawk eats many rodents. Thus, for every hawk in a community there must be many rodents, still more grasshoppers, and an immense number of plants.

The same pyramid-like relationship also often holds true for biomass. Even though rodents are larger than grasshoppers, and hawks larger than rodents, the sheer number of prey relative to the predators means that prey biomass will likely be greater overall. Unlike the pyramid of energy, however, the pyramids of numbers and biomass can occasionally be inverted. For example, the number of trees is much smaller than the number of insects feeding on them, and in the ocean, a small biomass of producers may support a higher biomass of zooplankton.

Food webs show feeding relationships and energy flow

Plant, grasshopper, rodent, and hawk make up a food chain (as in **FIGURE 5.10**), a linear series of feeding relationships. Thinking in terms of food chains is conceptually useful, but in reality ecological systems are far more complex than simple linear chains. A more accurate representation of the feeding relationships in a community is a **food web**, a visual map of feeding relationships and energy flow, showing the many paths by which energy passes among organisms as they consume one another. Whereas food chains provide a way to focus on the movement of energy and distribution of biomass in the system, food webs allow us to look more broadly at ecological and feeding relationships.

FIGURE 5.11 shows a food web from a temperate deciduous forest of eastern North America. Like virtually all diagrams of ecological systems, it is greatly simplified, leaving out the vast majority of species and interactions that occur. Note that even within this simplified diagram, we can pick out a number of different food chains involving different sets of species.

A Great Lakes food web would involve phytoplankton and cyanobacteria that photosynthesize near the water's surface; the zooplankton that eat them; the fish that eat all these; the larger fish that eat the smaller fish; and the lampreys that parasitize the fish. It would include a number of native mussels and clams and, since 1988, the zebra and quagga mussels that are crowding them out. It would include diving ducks that used to feed on native bivalves and now are preying on mussels.

This food web would also show that an array of bottom-dwelling invertebrates feed from the refuse of the exotic mussels. These waste products promote bacterial growth and disease pathogens that harm native bivalves, but they also provide nutrients that nourish crayfish and many smaller *benthic* (bottom-dwelling) invertebrate animals. Finally, the food web would include underwater plants and macroscopic algae, whose growth is promoted by the non-native mussels. The mussels clarify the water by filtering out plankton, and sunlight penetrates deeper into the water column, spurring photosynthesis and plant growth. Thus, zebra and quagga mussels alter this food web fundamentally, by shifting productivity from the open-water regions to the benthic and *littoral* (nearshore) regions. In so doing, the mussels affect fish indirectly, helping benthic and littoral fishes and making life harder for open-water fishes.

Keystone organisms play especially important roles in communities

"Some animals are more equal than others," George Orwell wrote in his 1945 book *Animal Farm*. Although Orwell was making wry sociopolitical commentary, his remark hints at a truth in ecology. In communities, ecologists have found, some species exert greater influence than do others. A species that has a particularly strong or far-reaching impact is often called a **keystone species**. A keystone is the wedge-shaped stone at the top of an arch that is vital for holding the structure together; remove the keystone, and the arch will collapse. In an ecological community, removal of a keystone species will have substantial ripple effects and will alter a large portion of the food web.

FIGURE 5.11 Food webs are conceptual representations of feeding relationships in a community. This food web for a North American temperate deciduous forest includes organisms on several trophic levels. Arrows indicate the direction of energy flow as a result of predation, parasitism, or herbivory. For example, an arrow from the grass to the cottontail rabbit indicates that cottontails consume grasses. The arrow from the cottontail to the tick indicates that parasitic ticks derive nourishment from cottontails. Communities include so many species and are so complex that most food web diagrams are oversimplifications.

Some keystone species have been removed from their natural communities with unintended consequences, in what are essentially uncontrolled large-scale experiments. A well-known example is the elimination of wolves (intentionally, for the most part) from many parts of North America. Wolves are voracious predators; when they are eliminated from an area, the populations of large herbivores, such as elk and moose, can grow out of control. This can have far-reaching impacts on vegetation and, consequently, on all other animals in the area.

Ecologists also have verified the keystone species concept by careful observation and controlled experiments. For example, classic work by marine biologist Robert Paine established that the predatory starfish *Pisaster ochraceus* has great influence on the community composition of intertidal organisms on the Pacific coast of North America.[6] When *Pisaster* is present in this community, species diversity is high, with several types of barnacles, mussels, and algae. When *Pisaster* is removed, the mussels it preys on become numerous and displace other species, suppressing species diversity. More recent work off the Atlantic coast published in 2007 suggests that the reduction of shark populations by commercial fishing has allowed populations of certain skates and rays to increase, which has depressed numbers of bay scallops and other bivalves they eat.

Sea otters provide another example. Regions with abundant otters off the coast of British Columbia and Alaska also host dense "forests" of kelp, a tall brown alga (seaweed) that anchors to the seafloor. Kelp forests are important because they provide complex physical structures in which diverse communities of fish and invertebrates find shelter and food. In regions without sea otters, scientists have found kelp forests to be less abundant, or absent. One hypothesis to explain this relationship is that in the absence of otters, sea urchins become so numerous that they eat every last bit of kelp, creating empty seafloors called "urchin barrens" that are relatively devoid of life. Ecologists have determined that otters promote the presence of the kelp forests simply by keeping urchin numbers in check through predation. This research, by scientists like James Estes of the University of California at Santa Cruz and Jane Watson of Vancouver Island University, established sea otters as a prime example of a keystone species.[7]

To understand the significance of one or two large species in an ecosystem, and the delicacy of the balance, we can also consider the case of moose and wolves in the boreal forest of Cape Breton Highlands National Park (**FIGURE 5.12**). Moose were rare by 1900 and had completely disappeared from the area by 1924, because of

excessive hunting and habitat destruction. Parks Canada reintroduced moose to the park during 1947 and 1948, by importing and releasing 18 animals from Elk Island National Park. The reintroduction was highly successful, and moose are currently plentiful in the park, possibly too plentiful.[8] In the absence of the moose's natural predator, the wolf, which disappeared from the area as early as the mid-1800s, there are few natural controls on the moose population within the park. Moose are selective eaters; their preferred winter food is the balsam fir. By browsing heavily on certain types of food but not others, they can alter the forest landscape dramatically, leading to significant changes in the ecosystem.

The success of the reintroduced moose population in Cape Breton Highlands National Park may now be leading to changes in the composition of the boreal forest there. In addition to dramatic impacts on vegetation, moose may have contributed to the decline of native caribou in some areas by preventing regeneration of the mature forests, limiting the caribou's food, and opening the landscape to predators.

Animals at high trophic levels, such as wolves, starfish, and sea otters, are most often seen as keystone species. Other species attain keystone status as "ecosystem engineers" by physically modifying the environment shared by community members. For example, beavers build dams and turn streams into ponds, flooding vast expanses of dry land and turning it to swamp. Prairie dogs dig burrows that aerate the soil and serve as homes for other animals. Bees are crucial, even for human food security, because they are pollinators, moving pollen from male to female plants to facilitate the plants' sexual reproduction, as well as ensuring plant genetic diversity.

Less conspicuous organisms and those toward the bottoms of food chains can potentially be viewed as keystone species, too. Remove the fungi that decompose dead matter or the insects that control plant growth or the phytoplankton that are the base of the marine food chain and a community may change very rapidly indeed. Because there are usually more species at lower trophic levels, however, it is less likely that any one of them alone might have wide influence; if one species is removed, other species that remain may be able to perform many of its functions.

Identifying keystone species is no simple task, and there is no cut-and-dried definition of the term to help us. Community dynamics are complex, species interactions differ in their strength, and the strength of species interactions can vary through time and space. It is worthwhile to consider the important role of keystone species in ecosystems for conservation efforts in the future.

wildnerdpix/Fotolia

FIGURE 5.12

Moose (*Alces alces*) are a keystone species in Cape Breton Highlands National Park. They dramatically alter the composition of the plant community by selectively browsing on certain plants. The reintroduction of moose into the park in the 1940s led to dramatic changes in the ecosystem, especially given the absence of their main predator, wolves.

Disturbance, Resilience, and Succession

The removal of a keystone species is just one of many types of disturbance that can modify the composition, structure, or function of an ecological community. Over time, any given community may experience natural (or anthropogenic) disturbances ranging from gradual, such as climate change, to sudden, such as fires, hurricanes, floods, volcanic eruptions, or avalanches.

Communities respond to disturbance in different ways

Communities are dynamic systems and may respond to disturbance in several ways. Sometimes a community is so altered by a disturbance that it never returns to its original state, and instead settles into a new equilibrium condition.

For example, the Atlantic cod is a top-level or "apex" predator in the North Atlantic, part of a set of large benthic (bottom-water) predators. In the late 1980s and early 1990s, overfishing led to the collapse of the cod population, as discussed in the Central Case in Chapter 12. Subsequently, researchers noted that pelagic organisms (those that inhabit the upper layers of the open ocean) such as the northern shrimp, northern snow crab, and small pelagic fishes increased dramatically in number as a result of the absence of their main predators.[9] This signalled a shift from benthic to pelagic—a fundamental change in the characteristics of the ecological community. This type of situation, in which the disappearance of a top predator leads to dramatic changes at lower trophic levels, is referred to as a *trophic cascade*. Some researchers have speculated that, even if the cod population were to recover in the absence of commercial fishing (which it has not yet done), the community would likely never return to its former condition.

In other circumstances, a community may resist change and remain stable despite major disturbances. In that case, the community or ecosystem is said to show **resistance** to the disturbance. For example, a healthy forested area may resist change for many seasons, even though drought settles in.

Alternatively, a community may show **resilience**, meaning that it does change in response to the disturbance, but returns to its original state fairly quickly. For example, if a fire sweeps through a forested area, there will clearly be significant changes to the community. Many trees will die, animals will be killed or will flee. Soil will be exposed. However, many forests are adapted to fire and are resilient; even after undergoing such a major disturbance, they may make a substantial recovery over a relatively short time.

Succession follows severe disturbance

If a disturbance is severe enough to eliminate all or most of the species in a community, the affected site will undergo a somewhat predictable series of changes that ecologists call **succession**. In the traditional view of this process, ecologists described two types of succession.

Primary succession follows a disturbance so severe that no vegetation or soil life remains from the community that occupied the site. Primary succession starts with a clean slate; the biotic community is built essentially from scratch. In contrast, **secondary succession** begins when a disturbance dramatically alters an existing community but does not destroy all living things or all organic matter in the soil. In secondary succession, vestiges of the previous community remain; the surviving organisms have a head start over other organisms and these building blocks help shape the process of reconstruction.

At terrestrial sites, primary succession takes place after a bare expanse of rock, sand, or sediment becomes newly exposed to the atmosphere. This can occur when glaciers retreat, lakes dry up, or volcanic deposits spread across the landscape (see "The Science behind the Story: Ecological Recovery at Mount St. Helens"). Species that arrive first and colonize the new ground are referred to as **pioneer species** (**FIGURE 5.13** on page 134). Pioneer species are well adapted for colonization, having such traits as spores or seeds that can travel long distances.

The pioneers best suited to colonizing bare rock are mutualistic aggregates of fungi and algae known as **lichens**. Lichens succeed because their algal component provides food and energy via photosynthesis, while the fungal component takes a firm hold on rock and captures the moisture that both organisms need to survive. As lichens grow, they secrete acids that break down the rock surface. The resulting waste material forms the beginnings of soil, and once soil begins to form, small plants, insects, and worms find the rocky outcrops more hospitable. As new organisms arrive, they provide more nutrients and habitat for future arrivals. As time passes, larger plants establish themselves, the amount of vegetation increases, and species diversity rises.

Secondary succession on land begins when a fire, a hurricane, logging, or farming removes much of the biotic community. Consider a farmed field in eastern North America that has been abandoned (**FIGURE 5.14** on page 134). In the first few years after farming ends, the site will be colonized by pioneer species of grasses, herbs, and

THE SCIENCE **BEHIND THE STORY**

Mount St. Helens in Washington erupted explosively in 1980, shown here.

Jim Valance, Cascades Volcano Observatory, U.S. Geological Survey

Ecological Recovery at Mount St. Helens

Step outside, look around, and you'll find secondary succession occurring somewhere nearby. The nearest weedy lot or overgrown field will show plants colonizing a disturbed area and building a new community from the foundations of the old. But finding primary succession is not as easy. It's unusual to come across a place where all life has been extinguished and a brand-new community is being built from scratch as new organisms arrive from far away.

The eruption of the volcano Mount St. Helens in Washington state offered ecologists a rare opportunity to study how communities recover from catastrophic disturbance. On May 18, 1980, the volcano erupted with sudden and spectacular violence. The massive explosion obliterated an entire forested landscape as a scalding mix of gas, steam, ash, and rock was hurled outward for miles. A pyroclastic flow sped downslope, along with the largest landslide in recorded history. Rock and ash rained down, and mudslides and volcanic mudflows, called *lahars*, raced down river valleys, devastating everything in their paths (**FIGURE 1A**). Altogether, 4.1 km³ of material was ejected from the mountain, severely altering 1650 km².

In the aftermath of the blast, ecologists moved in to take advantage of the natural experiment of a lifetime. The eruption provided an extraordinary chance to study primary succession on a fresh volcanic surface. Which organisms would arrive first? What kind of community would emerge? How long would it take? Researchers set up study plots to examine how populations, communities, and ecosystems would respond.

Today, over 35 years later, the barren grey moonscape that resulted from the blast is a vibrant green (**FIGURE 1B**), carpeted with colourful flowers each summer. What ecologists have learned has modified our view of primary succession and informed the entire study of disturbance ecology.

Given the ferocity and scale of the eruption, most scientists initially presumed that life had been wiped out completely over a large area. They expected that pioneer species would colonize the area gradually, spreading slowly from the outside margins inward, and that over many years a community would be rebuilt in a systematic and predictable way. Instead, researchers discovered that some plants and animals had survived the blast. Some were protected by deep snowbanks. Others were sheltered on steep slopes facing away from the blast. Still others were dormant underground when the eruption occurred. These survivors, it turned out, would play key roles in rebuilding the community.

(a) 1980

(b) 2012

FIGURE 1   The photos show Mount St. Helens **(a)** after the eruption in 1980, and **(b)** in 2012. The crater at the top of the volcano shows where rock and lava were violently ejected during the eruption, essentially blowing the top off of the mountain.

forbs that were already in the vicinity and that disperse effectively. As time passes, shrubs and fast-growing trees, such as aspens, rise from the field. Pine trees subsequently rise above the aspens and shrubs, forming a pine-dominated forest. This pine forest develops an understorey of hardwood trees, because pine seedlings do not grow well under mature pines, whereas some hardwood seedlings do. Eventually the hardwoods outgrow the pines, creating a hardwood forest.

Processes of succession occur in many diverse ecological systems, from ponds to rocky intertidal areas to the carcasses of animals. A lake or pond that originates as

Many of the ecologists drawn to Mount St. Helens studied plants. Virginia Dale of Oak Ridge National Laboratory in Tennessee and colleagues examined the debris avalanche, a landslide of rock and ash as deep as a 15-storey building. The region appeared barren, yet plants of 20 species had survived, growing from bits of root, stem, or seed carried down in the avalanche. However, most plant regrowth occurred from seeds blown in from afar. Dale's team used sticky traps to sample these seeds as they chronicled the area's recovery.[10]

Plant regrowth was very slow for several years and then accelerated. After 20 years, 150 species of plants covered 65% of the ground. One important pioneer species was the red alder. This tree germinates on debris, grows quickly, and produces many seeds at a young age. As a result, it has become the dominant tree species on the debris avalanche. Because it fixes nitrogen, the red alder enhances soil fertility and thereby helps other plants grow. Researchers predict that red alder will remain dominant for years or decades, and that conifers such as Douglas fir (which today are moving in and beginning to seed) will eventually outgrow them and establish a conifer-dominated forest.

Patterns of plant growth varied in different areas. Roger del Moral of the University of Washington and colleagues compared ecological responses in barren pumice, mixed ash and rock, mudflows, and the "blowdown zone" where trees were toppled like matchsticks.[11] Numbers of species and percent of plant cover increased in different ways on each surface (**FIGURE 2**), affected by a diversity of factors. Windblown seeds accounted for most regrowth, but plants that happened to survive in sheltered "refugia" within the impact zone helped repopulate areas nearby.

Chance played a large role in determining which organisms survived and how vegetation recovered, del Moral and others found. Had the eruption occurred in late summer instead of spring, there

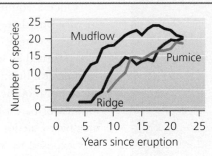

(a) Species richness

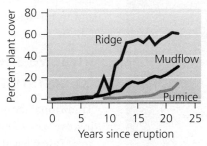

(b) Percent plant cover

FIGURE 2 Plants recovered differently at mudflow, ridge, and pumice sites at Mount St. Helens. In the 25 years after the eruption, **(a)** species richness of plants and **(b)** percentage of ground covered by plants both increased

Source: Data from del Moral, R., et al. (2005). Proximity, Microsites, and Biotic Interactions during Early Succession, pp. 93–109 in V. Dale et al., eds., *Ecological Responses to the 1980 Eruption of Mount St. Helens.* Springer, New York.

DATA Q Which substrate gained the greatest species richness, and what was its highest number of species? On which substrate was the increase in percent plant cover the slowest?

would have been no snow, and many of the plants that survived would have died. Had the eruption occurred at night instead of in the morning, nocturnal animals would have been hit harder.

Animals played major roles in the recovery right from the beginning.

Researchers showed that insects and spiders arrived in the impact zone in great numbers before plants did.[12] Insects fly, while spiders disperse by "ballooning" on silken threads. Trapping and monitoring at Mount St. Helens in the months following the eruption showed that insects and spiders landed in the impact zone by the billions. Researchers estimated that over 1500 species arrived in the first few years, surviving by scavenging or by preying on other arthropods. Most individuals soon died, but the nutrients from their bodies enriched the soil, helping the community to develop.

Once plants took hold, animals began exerting influence through herbivory. Caterpillars fed on plants, occasionally extinguishing small populations as they began to establish. As elk from surrounding forests moved into the region, Dale and her team fenced off "exclosure" plots to compare how plants grew within the ungrazed exclosures versus in plots grazed by elk outside the exclosures. Both types of plots saw increases in plant cover and similar amounts of species diversity, because elk herbivory can spur plant growth that compensates for what they eat. Non-native species did best in the grazed plots, whereas species important for forest recovery did best in the ungrazed exclosures. Overall, elk herbivory did not stall plant regrowth.

All told, research at Mount St. Helens has shown that succession is not a simple and predictable process. Instead, communities recover from disturbance in ways that are dynamic, complex, and highly dependent on chance factors affecting which species survive to repopulate the new landscape.

The results from Mount St. Helens also show life's resilience. Even when the vast majority of organisms perish in a natural disaster, a few may survive, and their descendants may eventually build a new community. Ecological change at Mount St. Helens is still in its early stages and will continue for many decades more and ecologists will continue to study and learn from this tremendous natural experiment.

nothing but water on a lifeless substrate begins to undergo succession as it is colonized by algae, microbes, plants, and zooplankton. As these organisms grow, reproduce, and die, the water body slowly fills with organic matter. The lake or pond acquires further organic matter and sediments from the water it receives from rivers, streams, surface runoff,

and even from the atmosphere. Eventually, the water body fills in, becoming a bog or even a terrestrial ecosystem.

In this traditional view of succession that we have described, the transitions between stages of succession eventually lead to a **climax community**, which remains in place, with little modification, until some disturbance

Charles D.Winters/Photo Researchers, Inc./Science Source

FIGURE 5.13
As this glacier retreats, small pioneer plants (foreground) begin the process of primary succession.

restarts succession. Early ecologists felt that each region had its own characteristic climax community, determined by the region's climate.

Today, ecologists recognize that succession is far more variable and less predictable than originally thought. The trajectory of succession can vary greatly according to chance factors, such as which particular species happen to gain an early foothold. The stages of succession blur into one another and vary from place to place, and some stages may sometimes be skipped completely. In addition, climax communities are not predetermined solely by climate but may vary with other conditions from one time or place to another.

Once a community is disturbed and succession is set in motion, there is no guarantee that the community will ever return to that original state. Many communities disturbed by human impact have not returned to their former conditions. An example that may be familiar to those from southern Ontario is the Cheltenham Badlands, an area of bare rock and soil in the Niagara Escarpment area near Caledon, Ontario. This formerly forested region was exposed to intensive agriculture and poor soil management in the early 1900s that led to accelerated erosion and exposure of bare rock. The region has never recovered and to this day the bare rock remains unforested.

Invasive species pose new threats to community stability

Traditional concepts of communities and successions involve sets of organisms understood to be native to an area. But what if a new organism arrives from elsewhere? And what if this non-native **exotic** (or *alien*) organism spreads widely and becomes dominant? Such **invasive species** can alter a community substantially and are one of the central ecological forces in today's world.

Most often, invasive species are non-native species that people have introduced, intentionally or by accident, from elsewhere in the world. Any non-native organism introduced into an ecosystem will require adjustments, but species become invasive **pests** when the negative impacts outweigh the benefits, especially when the limiting factors that might regulate their population growth are absent. Thus, the main characteristics of *problematic* invasive species include the ability to spread rapidly, and unimpeded, in the new environment, and the ability to have a negative impact on native species, communities, and ecosystems into which it has been introduced.

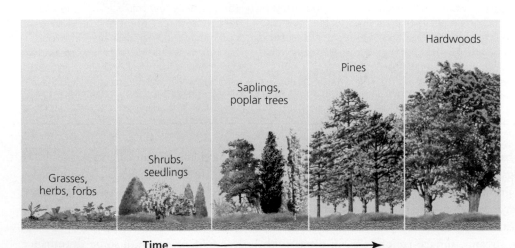

Hardwoods

Pines

Saplings, poplar trees

Shrubs, seedlings

Grasses, herbs, forbs

Time

FIGURE 5.14
Secondary succession occurs after a disturbance, such as fire, landslides, or farming, removes much of the vegetation from an area. Shown here is a typical series of changes in a plant community of eastern North America following the abandonment of a farmed field.

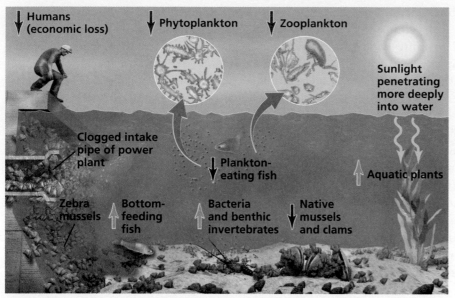

(a) Impacts of zebra mussels on members of a Great Lakes nearshore community

FIGURE 5.15

The zebra mussel is a prime example of a biological invader that has modified an ecological community. By filtering phytoplankton and small zooplankton from open water, it generates a number of impacts, both negative (red downward arrows) and positive (green upward arrows) **(a)**. This map **(b)** shows the range of zebra and quagga mussels in North America as of 2014. In less than three decades the zebra mussel has been reported in Ontario, Quebec, Manitoba, and 30 U.S. states and waterways. They have mainly been dispersed by boats, but in some cases by overland transport *Source:* Adapted from U.S. Geological Survey. Map showing 2014 distribution of zebra mussel sightings, Benson, A. J. (2011). Zebra Mussel Sightings Distribution. Retrieved from http://nas.er.usgs.gov/taxgroup/mollusks/zebramussel/zebramusseldistribution.aspx.

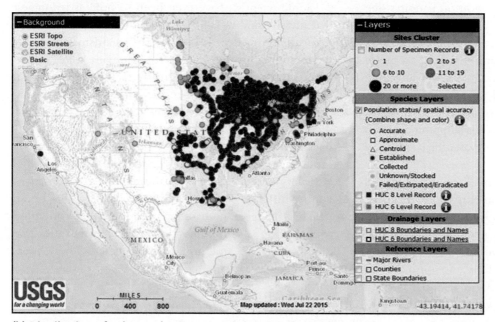

(b) Distribution of zebra and quagga mussels in North America, 2014

Plants and animals brought to one area from another may leave their main predators, parasites, and competitors behind, freeing them from natural constraints on population growth (**FIGURE 5.15**). If there happen to be few organisms in the new environment that can act as predators, parasites, or competitors, the introduced species may do very well. As it proliferates, it may exert diverse influences on its fellow community members. An example is the chestnut blight, an Asian fungus that killed nearly every mature American chestnut (*Castanea dentata*), the dominant tree species of many forests of eastern North America, in the quarter-century preceding 1930. Asian trees had evolved defences against the fungus over long millennia of coevolution, but the American chestnut had not.

In other cases, a species may be considered a pest even in regions where it is native. An example is the Asian long-horned beetle (*Anoplophora glabripennis*), a voracious wood-eating insect that is native to China but is also a pest, with few natural predators. The Asian long-horned beetle feeds on many different species of temperate hardwood trees, including maple, birch, horse chestnut, poplar, willow, elm, ash, and black locust. The beetle, introduced to North America from China via packaging materials, has the potential to cause widespread destruction. To date, the only approach to controlling the spread of the beetle is to cut down and burn any infested trees.[13]

So far we have considered examples of the impacts of a non-native insect, a fungus, and a mollusc, but virtually any type of organism can become an invasive pest, given circumstances that facilitate its spread. Introduced grasses and shrubs, such as the Scotch broom (*Cytisus scoparius*), have had dramatic impacts on the Garry oak (*Quercus garryana*)–based ecosystem of southeastern Vancouver Island and the Gulf Islands, a disappearing ecosystem that supports many rare plant species.[14] Fish introduced into streams purposely for sport or accidentally via shipping compete with and exclude native fish.

A particularly troublesome example of a native invasive species is the sea lamprey (*Petromyzon marinus*), an eel-like fish described in our section on parasites (see **FIGURE 5.6**). The sea lamprey is native to the Atlantic Ocean and was probably introduced to the Great Lakes as early as the 1830s by ocean-going vessels. It's a pretty unpleasant creature to begin with, as it lives by attaching itself to the flanks of other fish and feeding parasitically on their blood. But lampreys have had tremendous success as they have spread—mainly via shipping channels and canals—into the Great Lakes, devastating some local fish populations.

Hundreds of native island-dwelling animals and plants worldwide have been driven extinct by the goats, pigs, and rats intentionally or accidentally introduced by human colonists. The cane toad (*Bufo marinus*), introduced to Australia to control insects in sugar cane fields (which it never did very successfully), is poisonous to just about anything that tries to eat it and has been extremely damaging to a wide variety of indigenous animal populations.

The impacts of invasive species on native species and ecological communities are severe and growing year by year with the increasing mobility of humans and the globalization of our society. Global trade helped spread zebra mussels, which were unintentionally transported in the ballast water of cargo ships. To maintain stability at sea, ships take water into their hulls as they begin their voyage and discharge that water at their destination. Decades of unregulated exchange of ballast water have ferried hundreds of species across the oceans.

In North America, zebra mussels—and the media attention they generated—helped put invasive species on the map as a major environmental and economic problem. Scientific research into introduced species has proliferated, and many ecologists view invasive species as the second-greatest threat to species and natural systems, behind only habitat destruction.

Funding has now become more widely available for the control and eradication of invasive species, and control mechanisms are widely researched and shared across jurisdictional boundaries. Managers at all levels of government have been trying a variety of techniques to control the spread of zebra and quagga mussels—removing them manually, applying toxic chemicals, drying them out, depriving them of oxygen, introducing predators and diseases, and stressing them with heat, sound, electricity, carbon dioxide, and ultraviolet light. However, most of these are localized and short-term fixes that are not capable of making a dent in the huge populations at large in the environment. In case after case, managers are finding that controlling and eradicating invasive species are so difficult and expensive that preventive measures (such as ballast water regulations) represent a much better investment.

Some scientists have even been so bold as to suggest that invasive and non-native species should not be automatically considered as "bad." Organisms certainly were moving around long before humans were around (although their dispersal is greatly facilitated by human activity). The amount of money spent preventing and battling against invasive species is so large; some have suggested that the money could be better spent protecting native species in other ways. One proposal is that we should adjust to the fact that exotic and invasive species are everywhere; rather than aiming to restore "native" communities in an unaltered state, which may be an unachievable goal, perhaps we need to turn our attention to managing so-called *novel* or *hybrid* communities that combine native and non-native species. These ideas remain controversial.

Some altered communities can be restored to a former condition

Invasive species are adding to the tremendous transformations that humans have already forced on natural landscapes through habitat alteration, deforestation, hunting of keystone species, pollution, and other impacts. With so much of Earth's landscape altered by human impact, it is impossible to find areas that are truly pristine. This realization has given rise to the conservation effort known as **ecological restoration**. The practice of ecological restoration is informed by the science of **restoration ecology**. A related practice is the rehabilitation of a former mine site or industrial site, in which case the term *reclamation* is more appropriate; we will explore land reclamation in greater detail when we turn our attention to mining and related activities in Chapter 17.

FIGURE 5.16

Garry oak (*Quercus garryana*) ecosystems in south Vancouver Island and the Gulf Islands of British Columbia are being invaded by aggressive exotic species, including scotch broom and spurge-laurel. This photo shows part of the ecological restoration of a Garry oak ecosystem at Fort Rodd Hill National Historic Site in Victoria, BC, implemented by Parks Canada.

Restoration ecologists wrestle with some of the same issues as environmental managers who deal with invasive species. They research the historical conditions of ecological communities as they existed before our industrialized civilization altered them. They then try to devise ways to restore some of these areas to an earlier condition, often to a natural "presettlement" condition. But what is the condition that is being restored? Is it pre-modern? Or pre-industrial? Or perhaps pre-Aboriginal? These are challenging questions; often it is more accurate to think of an ecosystem or ecological community as being "rehabilitated," and returned to a healthy, functional state rather than being restored to a particular preexisting state.

We can see one example of this in activities underway at Chatterton Hill Park in Saanich, British Columbia, which are aimed at the ecological restoration of Garry oak–associated ecosystems (**FIGURE 5.16**). As mentioned above, these delicate, complex ecosystems, which occur almost exclusively in south Vancouver Island and the Gulf Islands, are being invaded by aggressive exotic species that suppress the growth of native plants. In 2006, local restoration ecologists compared a Garry oak restoration site with a similar nearby site where no restoration activities had taken place; they concluded that biodiversity at the restoration site was increasing. At the comparison site, the tree layer was similar, but the understorey vegetation was still dominated by grasses and invasive exotic plants such as the Scotch broom.[15]

But the question remains: To which earlier state is the ecosystem being restored? Traditionally, Aboriginal people maintained the Garry oak meadows with frequent burnings. However, with European settlement and the resulting control and suppression of fires,

Douglas firs have been increasingly successful at invading the meadows. Some Garry oak restoration projects have undertaken to remove Douglas firs physically or using controlled fires, but when the Douglas firs are removed, the resulting disturbance of the soil can facilitate the establishment of other invasive species.[16] This case demonstrates that ecological restoration is neither easy nor simple.

Perhaps the world's largest ecological rehabilitation project is the ongoing effort to restore parts of the Florida Everglades. The Everglades, a unique 7500 km^2 ecosystem of interconnected marshes and seasonally flooded grasslands, has been drying out for decades because the water that feeds it has been heavily managed for flood control and overdrawn for irrigation and development. The water management system inadvertently caused extensive degradation of the environment, resulting in the loss of more than half of the Everglades and the elimination of whole classes of ecosystems. Populations of wading birds have dropped by 90–95%, and economically important fisheries have suffered greatly. Extensive engineering of river channels also led to a loss of aesthetic appeal, problems with stagnancy and pollution of water in the channels, and a host of other problems. The 30-year, $7.8 billion restoration project intends to restore water by undoing damming and diversions of 1600 km of canals, 1150 km of levees, and 200 water control structures.

Ecosystem restoration and rehabilitation are almost always expensive and only sometimes, or partially, successful. Regardless, the more our population grows and development spreads, the more ecological restoration will become a vital conservation strategy for the future.

Earth's Biomes

Across the world, each portion of each continent has different sets of species, leading to endless variety in community composition. However, communities in far-flung places often share strong similarities in their structure and function. This allows us to classify communities into broad types. A **biome** is a major regional complex of similar communities—a large ecological unit recognized primarily by its dominant plant type and vegetation structure. The world contains a number of biomes, each covering large geographical areas (**FIGURE 5.17**).

A term that is often used interchangeably with biome, but probably should not be, is ecoregion. An **ecoregion** is defined by the World Wildlife Fund as a large area of land or water that contains a geographically distinct assemblage of natural communities that share a large majority

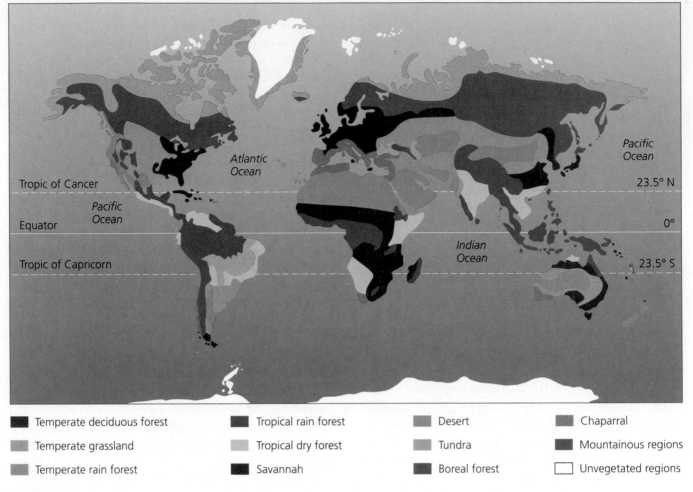

■ Temperate deciduous forest	■ Tropical rain forest	■ Desert	■ Chaparral
■ Temperate grassland	■ Tropical dry forest	■ Tundra	■ Mountainous regions
■ Temperate rain forest	■ Savannah	■ Boreal forest	□ Unvegetated regions

FIGURE 5.17 Biomes are distributed around the world according to temperature, precipitation, atmospheric and oceanic circulation patterns, and other factors.

of their species and ecological dynamics, share similar environmental conditions, and interact ecologically in ways that are critical for their long-term persistence.[17] A particular ecoregion—such as the short grasslands of the Canadian Prairies of southern Alberta and Saskatchewan, for example—is thus a representative of a biome (the "temperate grasslands" biome) that is broader in scope and occurs in numerous localities around the world.

The difference between an ecoregion and a biome might be easiest to grasp for a highly distinctive environment, such as a desert. The desert biome has certain climatic and ecological characteristics in common (see the section on the desert biome that follows), notably the lack of precipitation. There are many representatives of the desert biome around the world, including the Mojave Desert, the Gobi Desert, and the Sahara Desert, among many others. Each of these individual desert environments constitutes or is part of an ecoregion, with its own particular characteristics and flora and fauna, but it is still consistent with the broad characteristics that define the desert biome.

Climate influences the locations of biomes

Which biome covers any particular portion of the planet depends on a variety of abiotic factors, including temperature, precipitation, atmospheric circulation, and soil characteristics. Among these factors, temperature and precipitation exert the greatest influence (**FIGURE 5.18**), because, as you will recall from our discussion about productivity, these two factors have the dominant effect on net primary productivity of terrestrial ecosystems. Because biome type is largely a function of climate, and because average monthly temperature and precipitation are among the best indicators of an area's climate, scientists often use climate diagrams, or **climatographs**, to depict such information.

Global climate patterns cause biomes to occur in large patches in different parts of the world. For instance, temperate deciduous forest occurs in eastern North America, north central Europe, and eastern

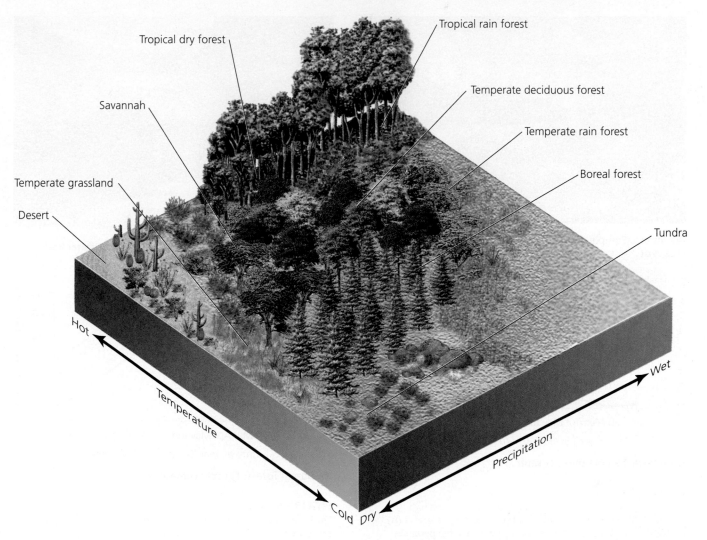

FIGURE 5.18 As precipitation increases, vegetation generally becomes taller and more luxuriant. As temperature increases, plant communities change. Together, temperature and precipitation are the main factors determining which biome occurs in a given area. For instance, deserts occur in dry regions; tropical rain forests occur in warm, wet regions; and tundra occurs in the cold, dry regions.

China. Note how patches representing the same biome tend to occur at similar latitudes. This is due to the north–south gradient in temperature and to atmospheric circulation patterns.

We can divide the world into 10 major terrestrial biomes

Each biome encompasses a variety of communities that share similarities. For example, the eastern United States and the southernmost part of eastern Canada support part of the temperate deciduous forest biome. From New Hampshire to the Great Lakes to eastern Texas, precipitation and temperature are similar enough that most of the region's natural plant cover consists of broad-leafed trees that lose their leaves in winter. Within this region, however, exist many different types of temperate deciduous forest, such as oak–hickory, beech–maple, and

pine–oak forests, each sufficiently different to be designated a separate community.

Let us look briefly at the characteristics that define each of the major terrestrial biomes.

Tundra The **tundra** (**FIGURE 5.19**) is a dry biome—nearly as dry as a desert—but located at very high latitudes along the northern edges of Russia, Canada, and Scandinavia. Extremely cold winters with little daylight and moderately cool summers with lengthy days characterize this landscape of lichens and low, scrubby vegetation without trees. The great seasonal variation in temperature and day length results from this biome's high-latitude location, and the low angle of sunlight striking the surface.

Because of the cold climate, underground soil remains more or less permanently frozen and is called *permafrost*. During the long, cold winters, the surface soils freeze as well; then, when the weather warms, they melt and

(a) Tundra: Sample location: Cumberland Sound, Baffin Island, Nunavut

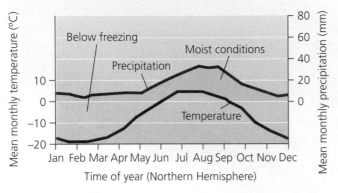

(b) Tundra: Sample climatograph

FIGURE 5.19
Tundra is a cold, dry biome found near the poles and atop high mountains at lower latitudes **(a)**. Scientists use climate diagrams **(b)** to illustrate an area's average monthly precipitation and temperature.[18] Typically in these diagrams, the *x*-axis marks months of the year (beginning in January for regions in the Northern Hemisphere and in July for regions in the Southern Hemisphere). Paired *y*-axes denote average monthly temperature and average monthly precipitation. The twin curves plotted on a climate diagram indicate trends in precipitation (blue) and in temperature (red) from month to month.

(a) Boreal forest: Sample location: Jasper National Park, Alberta

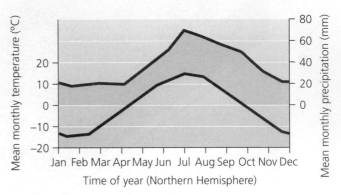

(b) Boreal forest: Sample climatograph

FIGURE 5.20
The huge boreal forest biome is defined by long, cold winters, relatively cool summers, and moderate precipitation.

produce seasonal accumulations of surface water that make ideal habitat for mosquitoes and other biting insects. The swarms of insects benefit bird species that migrate long distances to breed during the brief but productive summer. Caribou (*Rangifer tarandus*) also migrate to the tundra to breed, and then leave for the winter. Only a few large animals, such as polar bears (*Ursus maritimus*) and musk oxen (*Ovibos moschatus*), can survive year-round in this extreme climate. Tundra also occurs as **alpine tundra** at the tops of high mountains in temperate and tropical regions.

Because of the extreme climate in Canada's North, Alaska, Russia, and Scandinavia, much of the tundra biome remains intact and relatively unaltered by direct human occupation and interference. For example, the World Wildlife Fund reports the tundra ecoregions of

North America to be 95–98% intact. However, much of the tundra is unprotected by national or provincial or territorial legislation, rendering it potentially susceptible to alteration through human activities. Furthermore, the indirect effects of human modification of the global environment, especially the climate, are increasingly evident in the tundra. One problem that is of external origin in tundra regions is atmospheric fallout, which results in the deposition of heavy metals, and pesticide pollution. In many areas of the tundra, seasonal ice connects the many islands with the mainland; with climatic warming and the associated melting of sea ice, these habitats and the animals that depend on them will be in increasing peril.

Boreal forest   The northern coniferous forest, or **boreal forest**, also called **taiga** (**FIGURE 5.20**), stretches in a broad band across much of Canada, Alaska, Russia, and Scandinavia. In Canada the boreal forest encompasses nearly 6 million km². A few species of **coniferous** or evergreen trees, which create seed cones, have needle-like leaves, and typically remain green year-round, dominate large stretches of the boreal forest, interspersed with bogs and lakes. The black spruce (*Picea mariana*) is a common evergreen species.

The boreal forest's uniformity over huge areas reflects the climate common to this latitudinal band of the globe: These forests develop in cooler, drier regions than do temperate forests, and they experience long, cold winters and short summers. Soils are typically nutrient-poor and somewhat acidic. As a result of the strong seasonal variation in day length, temperature, and precipitation, many organisms compress a year's worth of feeding, breeding, and rearing of young into a few warm, wet months. Year-round residents of boreal forests include mammals, such as moose (*Alces alces*), wolves (*Canis lupus*), bears (such as the black bear, *Ursus americanus*), lynx (*Felis lynx*), and many burrowing rodents (such as the bank vole, *Clethrionomys glareolus*). This biome also hosts many insect-eating birds that migrate from the tropics to breed during the brief, intensely productive summer season.

The boreal forest, one-third of which resides in Canada, is the largest continuous forest ecosystem. It hosts more wetland area than any other ecosystem on Earth, providing invaluable habitat for many species. The enormous global importance of the boreal forest highlights Canada's role as a steward for much of the world's forested area. We will revisit this role in the chapter on forests and forest management.

Temperate deciduous forest

The **temperate deciduous forest** (**FIGURE 5.21**) is characterized by broad-leafed trees that are **deciduous**, meaning that they lose their leaves each fall and remain dormant during winter, when hard freezes would endanger leaves. These mid-latitude forests occur in much of Europe and eastern China as well as in eastern North America—all areas in which precipitation is spread relatively evenly throughout the year. Although soils of the temperate deciduous forest are relatively fertile, the biome generally consists of far fewer tree species than are found in tropical rain forests. Oaks, beeches, and maples are a few of the most abundant types of trees in these forests. A sampling of typical animals of the temperate deciduous forest of eastern North America is shown in **FIGURE 5.11**.

Much of the temperate deciduous or broad-leafed forest in North America has been greatly altered since European settlement; for example, it has been estimated that only about 5% of New England–Acadian mixed broad-leafed forest in Canada remains intact, with about 50% of the habitat in this region described as "heavily altered" by human activity.[19] However, forest cover has been making a comeback in the New England–Acadia

(a) **Temperate deciduous forest: Sample location: Simon Lake, Naughton, Ontario**

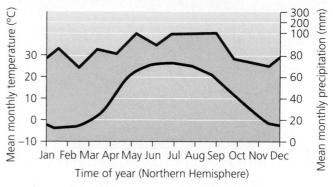

(b) **Temperate deciduous forest: Sample climatograph**

FIGURE 5.21
Temperate deciduous forests experience relatively stable seasonal precipitation but more variation in seasonal temperatures. When the precipitation curve falls well above the temperature curve, as shown here, the region experiences relatively moist conditions, as indicated by the green shading.

region in recent decades, thanks to changing land use and conservation efforts.

Temperate grassland

Moving westward from the Great Lakes, we find **temperate grasslands** (**FIGURE 5.22**). This is because temperature differences between winter and summer become more extreme, and rainfall diminishes. The limited amount of precipitation in the Prairies and the Great Plains region can support grasses (Poaceae family) more easily than trees. Also known as *steppes* or *prairies*, temperate grasslands were once widespread throughout parts of North and South America and much of central Asia.

Today people have converted most of the world's grasslands for agriculture, greatly reducing the abundance of native plants and animals. Both the tallgrass prairies that characterize the midwestern United States and the shortgrass prairies of southern Alberta and Saskatchewan are described by the World Wildlife Fund as having been "virtually converted," mainly for wheat production and

(a) Temperate grassland: Sample location: Moose Jaw, Saskatchewan

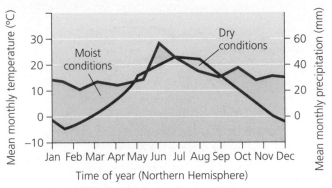

(b) Temperate grassland: Sample climatograph

FIGURE 5.22
Temperate grasslands experience temperature variations throughout the year, and too little precipitation for many trees to grow. This climatograph indicates both "moist" (green) and "dry" (yellow) climate conditions. When the temperature curve is above the precipitation curve, as is the case in May and mid-June through September, the climate conditions are "dry."

(a) Temperate rain forest: Sample location: Queen Charlotte Islands, British Columbia

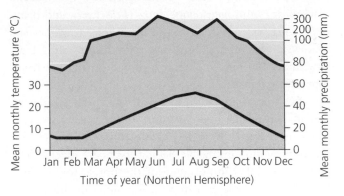

(b) Temperate rain forest: Sample climatograph

FIGURE 5.23
Temperate rain forests receive a great deal of precipitation year-round, and feature moist, mossy interiors.

grazing, with only small undisturbed patches remaining (less than 2% remaining "intact" in Canada). However, restoration ecology efforts in this biome have seen some success, with efforts to reestablish populations of native species, such as the black-footed ferret (*Mustela nigripes*) and bison (*Bison bison*). Other vertebrate animals characteristic of North American grasslands include prairie dogs, pronghorn antelope (*Antilocapra americana*), and ground-nesting birds, such as meadowlarks.

Temperate rain forest   Moving still further west in North America, the topography becomes more varied, and biome types are intermixed. The coastal Pacific region, with heavy coastal mountain rainfall, features **temperate rain forest** (**FIGURE 5.23**), a forest type known for its potential to produce large volumes of commercially important forest products, such as lumber and paper. Coniferous trees such as cedars, spruces, hemlocks,

and Douglas fir (*Pseudotsuga menziesii*) grow very tall in the temperate rain and damp. In the Queen Charlotte Islands, for example, moisture-loving animals, such as the bright yellow banana slug (*Ariolimax columbianus*) are common, and old-growth conifer stands provide habitat for the endangered spotted owl (*Strix occidentalis*). The soils of temperate rain forests are usually quite fertile but are susceptible to landslides and erosion if forests are cleared.

Temperate rain forests have been the focus of controversy in Pacific coastal regions, where overharvesting has driven some species toward extinction. Clear-cut logging and road building into forested areas remain the greatest threats to temperate rain forest habitat in these areas (see "Central Case: Battling over the Last Big Trees at Clayoquot Sound," in the chapter on forests and forest management).

Tropical rain forest   In tropical regions we see the same pattern found in temperate regions: Areas of high rainfall grow rain forests, areas of intermediate rainfall host dry or deciduous forests, and areas of lower rainfall become dominated by grasses. However, tropical biomes differ from their temperate counterparts in other ways. They are closer to the equator and therefore warmer on average year-round, yielding an *aseasonal* environment in which there is no major difference from summer to winter. They also hold far greater biodiversity.

The **tropical rain forest** biome (**FIGURE 5.24**) is found in Central America, South America, Southeast Asia, West Africa, and other tropical regions, and is characterized by year-round rain and uniformly warm temperatures. Tropical rain forests have dark, damp interiors, lush vegetation, and highly diverse biotic communities, with greater numbers of species of insects, birds, amphibians, and various other animals than any other biome.

These forests are not dominated by a few species of trees, as are forests closer to the poles, but instead consist of very high numbers of tree species intermixed. Any given tree may be draped with vines, enveloped by strangler figs, and loaded with *epiphytes* (orchids and other plants that grow in trees), such that trees occasionally collapse under the weight of all the life they support.

Despite this profusion of life, tropical rain forests have very poor, acidic soils that are low in organic matter. Nearly all nutrients present in this biome are contained in the trees, vines, and plants—not in the soil. An unfortunate consequence is that once tropical rain forests are cleared, the nutrient-poor soil can support agriculture for only a short time. As a result, farmed areas are abandoned quickly, and the soil and forest vegetation recover very slowly.

Tropical dry forest   Tropical areas that are warm year-round but where rainfall is lower overall and highly seasonal give rise to **tropical dry forest**, or tropical deciduous forest (**FIGURE 5.25**), a biome widespread in India, Africa, South America, and northern Australia. Wet and dry seasons each span about half a year in tropical dry forest. Rains during the wet season can be extremely heavy and, coupled with erosion-prone soils, can lead to severe soil loss when forest clearing occurs over large areas. Across the globe, much tropical dry forest has been converted to agriculture. Clearing for farming or ranching is made easier by

(a) Tropical rain forest: Sample location: Bogor, Java, Indonesia

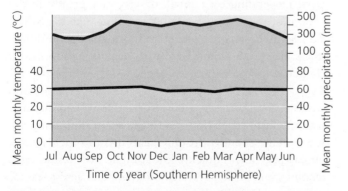

(b) Tropical rain forest: Sample climatograph

FIGURE 5.24
Tropical rain forests, famed for their biodiversity, grow under constant, warm temperatures and a great deal of rain. This is a largely aseasonal biome; conditions in the winter months are pretty much the same as in the summer months.

(a) Tropical dry forest: Sample location: Darwin, Australia

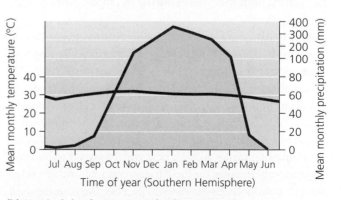

(b) Tropical dry forest: Sample climatograph

FIGURE 5.25
Tropical dry forests experience significant seasonal variations in precipitation and relatively stable warm temperatures.

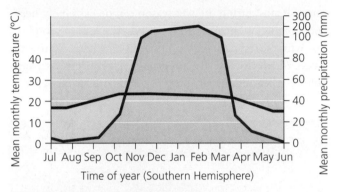

(a) Savannah: Sample location: Harare, Zimbabwe

(b) Savannah: Sample climatograph

FIGURE 5.26
Savannahs are grasslands with sparse clusters of trees. They experience slight seasonal variation in temperature but significant variation in rainfall.

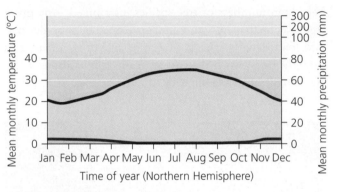

(a) Desert: Sample location: Cairo, Egypt

(b) Desert: Sample climatograph

FIGURE 5.27
Deserts are dry year-round, but they are not always hot. Precipitation can arrive in intense, widely spaced storm events, and temperatures can vary dramatically within a 24-hour period.

the fact that vegetation heights are much lower and canopies less dense than in tropical rain forest. Organisms that inhabit tropical dry forests have adapted to seasonal fluctuations in precipitation and temperature. For instance, plants are deciduous and often leaf out and grow profusely with the rains, then drop their leaves during the driest times of year.

Savannah Drier tropical regions give rise to **savannah** (**FIGURE 5.26**), tropical grassland interspersed with clusters of acacias (genus *Acacia*) or other trees. The savannah biome is found today across stretches of Africa (the ancestral home of our species), South America, Australia, India, and other dry tropical regions. Precipitation in savannahs usually arrives during distinct rainy seasons and concentrates grazing animals near widely spaced water holes. Common herbivores on the African savannah include zebras (most commonly *Equus quagga*), gazelles (such as the well-known Thomson's gazelle, *Eudorcas thomsoni*), and giraffes (*Giraffa camelopardalis*). The predators of these grazers include lions (*Panthera leo*), hyenas (such as the spotted hyena, *Corcuta corcuta*), and other highly mobile carnivores.

Desert Where rainfall is very sparse, **desert** (**FIGURE 5.27**) forms. This is the driest biome on Earth; most deserts receive well under 25 cm of precipitation per year, much of it during isolated storms months or years apart. Depending on rainfall, deserts vary greatly in the amount of vegetation they support. Some, like the Sahara and Namib deserts of Africa, are mostly bare sand dunes; others, like the Sonoran Desert of Arizona and northwest Mexico, are quite heavily vegetated. Deserts are not always hot; the high desert of the western United States is one example. Because deserts have low humidity and relatively little vegetation to insulate them from temperature extremes, sunlight readily heats them in the daytime, but daytime heat is quickly lost at night. As a result, temperatures vary widely from day to night and across seasons of the year. Desert soils can often be quite saline and are sometimes known as *lithosols*, or stone soils, for their high mineral and low organic-matter content.

Desert animals and plants have evolved many adaptations to deal with the harsh climatic conditions. Most reptiles and mammals, such as rattlesnakes (family Viperidae) and kangaroo mice (such as *Microdipodops pallidus*), are active in the cool of night, and many Australian desert birds are nomadic, wandering long distances to find areas of recent rainfall and plant growth. Many desert plants have thick leathery leaves to reduce water loss, or green trunks so that the plant can photosynthesize without leaves, which would lose water. The spines of cacti and many other desert plants guard those plants from being eaten by herbivores desperate for the precious water they hold. These are examples of convergent evolution of plants and animals, adapting separately to the dry conditions that are characteristic of the desert biome (as shown in **FIGURE 4.3**).

Mediterranean   In contrast to the boreal forest's broad, continuous distribution, *chaparral* or **Mediterranean** woodland (**FIGURE 5.28**) is limited to fairly small patches widely flung around the globe. Scrub woodland consists mostly of evergreen shrubs and is densely thicketed. This biome is also highly seasonal, with mild, wet winters and warm, dry summers. This type of climate is induced by oceanic influences; in addition to

(a) Mediterranean or Chaparral: Sample location: Baja California Peninsula, Mexico

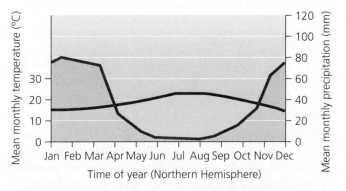

(b) Mediterranean or Chaparral: Sample climatograph

FIGURE 5.28
The Mediterranean or chaparral biome is a highly seasonal biome dominated by shrubs, influenced by marine weather, and dependent on fire.

ringing the Mediterranean Sea, chaparral occurs along the coasts of California, Chile, and southern Australia. In Europe it is called *maquis*; in Chile, *matorral*; and in Australia, *mallee*. Mediterranean-type communities experience frequent fires, and their plant species are adapted to resist fire or even to depend on it for germination of their seeds.

Altitude creates patterns analogous to latitude

As any hiker or skier knows, climbing in elevation causes a much more rapid change in climate than moving the same distance toward the poles. Vegetative communities change along mountain slopes in correspondence with this small-scale climate variation (**FIGURE 5.29**). These changes with altitude define the **alpine** biomes. *Altitudinal zonation*, as it is referred to, is independent of latitude—variation of temperature, precipitation, and ecological communities from low to high altitude can occur anywhere.

For example, a hiker ascending one of southern Arizona's higher mountains would begin in the Sonoran Desert or desert grassland and proceed through oak woodland, pine forest, and finally spruce–fir forest—the equivalent of passing through several biomes, without any change in latitude. A hiker scaling one of the great peaks of the Andes in Ecuador, near the equator, could begin in tropical rain forest and end amid glaciers in alpine tundra.

Characteristics that are typical of the alpine biome on a high-mountain peak, compared with lowlands or foothills that surround it, include lower temperatures, lower atmospheric pressures (and less oxygen), higher exposure to ultraviolet radiation, and higher precipitation. The foothills of the Canadian Rockies, for example, are characterized by meadows, grasslands, and riparian (riverside) woodlands, which grade upward through boreal forests and into alpine tundra and glaciers on the mountain peaks; precipitation on the peaks can be as much as twice that of the foothill areas.

The zonation of ecological communities on mountain slopes is also greatly influenced by secondary climatic effects related to topographic relief, such as rainshadow effects and exposure to or shelter from the prevailing winds and sunlight. For example, when moisture-laden air ascends a steep mountain, it releases precipitation as it cools; this explains the wet temperate rain forest on the ocean side of the mountain slopes in British Columbia, for example. By the time the air flows over the top of the mountain and down the other side, it can be very dry, creating a *rainshadow desert*; this explains environments like California's Death Valley, one of the driest locations on Earth.

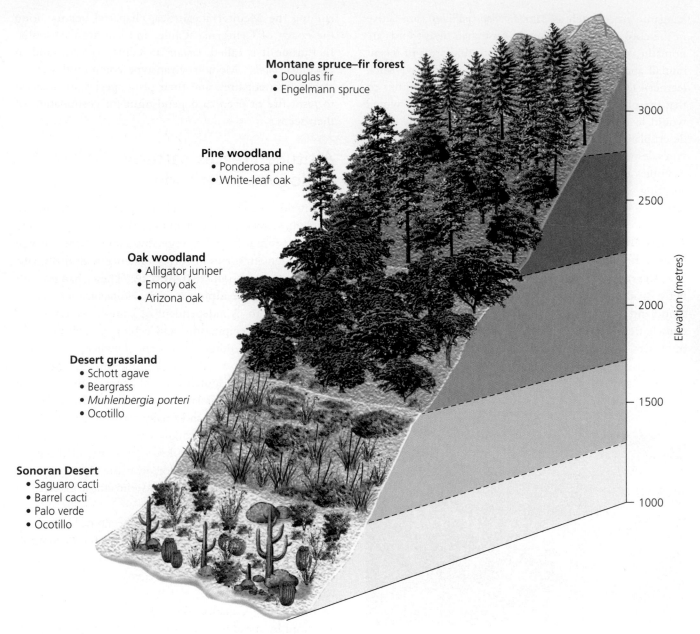

Montane spruce–fir forest
- Douglas fir
- Engelmann spruce

Pine woodland
- Ponderosa pine
- White-leaf oak

Oak woodland
- Alligator juniper
- Emory oak
- Arizona oak

Desert grassland
- Schott agave
- Beargrass
- *Muhlenbergia porteri*
- Ocotillo

Sonoran Desert
- Saguaro cacti
- Barrel cacti
- Palo verde
- Ocotillo

Elevation (metres)

3000

2500

2000

1500

1000

FIGURE 5.29   As altitude increases, vegetation changes in ways similar to how it changes as one moves toward the poles, taking a hiker through the local equivalent of several biomes.

Aquatic and coastal systems also show biome-like patterns

In our discussion of biomes, we have focused exclusively on terrestrial systems, because the biome concept, as traditionally developed and applied, has been limited to terrestrial systems. Areas equivalent to biomes also exist in both freshwater aquatic environments and in the oceans. However, the forces that determine the productivity of aquatic systems have very different geographical patterns than those that dominate the productivity of terrestrial systems.

In the oceanic environment, we can consider the thin strips along the world's coastlines to represent one aquatic system, the continental shelves another, and the open ocean, deep sea, coral reefs, and kelp forests as still other distinct sets of communities. There are also many coastal systems that straddle the line between terrestrial and aquatic, such as salt marshes, rocky intertidal communities, mangrove forests, and estuaries. And, of course, there are freshwater systems, such as those of the Great Lakes.

Unlike terrestrial biomes, aquatic systems are shaped not by air temperature and precipitation but by such factors as water temperature, salinity, dissolved oxygen and nutrients, wave action, currents, depth, and type of substrate (e.g., sandy, muddy, or rocky bottom). Light levels can play a role, too. Surface water tends to be less salty (although salinity can fluctuate as a result of evaporation from the surface), with higher light levels and warmer temperatures, in comparison with deep-water environments. Coastal waters tend to be both warmer and fresher (because of the influx of fresh water from rivers), but may be more turbid (because of sediment load) compared to open-ocean environments. Marine communities are also more clearly delineated by their animal life than by their plant life.

In freshwater aquatic systems we can distinguish between large and small lakes, large and small rivers, estuaries, wetlands, headwaters, and others. We will examine freshwater, marine, and coastal systems in greater detail in subsequent chapters.

Conclusion

The natural world is so complex that we can visualize it in many ways and at various scales. Dividing the world's communities into major types, or biomes, is informative at the broadest geographical scales. Understanding how communities function at more local scales requires understanding how species interact with one another. Species interactions, such as predation, parasitism, competition, and mutualism, give rise to effects that are both weak and strong, direct and indirect. Feeding relationships can be represented by the concepts of trophic levels and food webs, and particularly influential species are sometimes called keystone species. Increasingly, humans are altering communities, in part by introducing non-native species that may turn invasive. But increasingly, through ecological restoration, we are also attempting to undo the changes we have caused.

REVIEWING OBJECTIVES

You should now be able to

Compare and contrast the major types of species interactions

- Competition results when individuals or species vie for limited resources. It can occur within or among species and can result in coexistence or exclusion. It also can lead to realized niches, resource partitioning, and character displacement.
- In predation, one species kills and consumes another. It is the basis of food webs and can influence population dynamics and community composition.
- In parasitism, one species derives benefit by harming (but usually not killing) another.
- Herbivory is an exploitative interaction whereby an animal feeds on a plant.
- In mutualism, species benefit from one another. Some mutualists are symbiotic, whereas other mutualists are free-living.

Characterize feeding relationships and energy flow in ecological communities, using them to construct trophic pyramids and food webs

- Energy is transferred in food chains among trophic levels in ecosystems.
- Lower trophic levels generally contain more energy, biomass, and numbers of individuals than higher trophic levels.

- Food webs illustrate feeding relationships and energy flow among species in a community.
- Keystone species have impacts on communities that are far out of proportion to their abundance. Top predators are frequently keystone species, but other organisms may be thought of as keystones in some circumstances.

Describe how communities respond to disturbances, referencing the concepts of resistance, resilience, succession, and restoration

- Disturbances like fires, pest invasions, and climate change affect ecological communities in different ways. A community may display resistance to change, or may be resilient—succumbing to the disturbance, but recovering quickly.
- Succession is a stereotypical pattern of change within a community through time. Primary succession begins with an area devoid of life. Secondary succession begins with an area that has been severely disturbed.
- Invasive species, such as the zebra mussel, have altered the composition, structure, and function of communities. Humans are the cause of most modern species invasions, but we can respond with prevention and control measures.
- Ecological restoration aims to restore communities to a more "natural" state, variously defined as before human or industrial interference.

Describe and illustrate the terrestrial biomes of the world

- Biomes represent major classes of communities spanning large geographical areas. Their distribution and characteristics are largely determined by temperature and precipitation.
- There are ten major terrestrial biomes: tundra, boreal forest, temperate deciduous forest, temperate grassland, temperate rain forest, tropical rain forest, tropical dry forest, savannah, desert, and Mediterranean.
- With increasing altitude, ecosystems change in ways that parallel the types of changes that occur with latitude.
- The biome concept by tradition refers to terrestrial systems. Aquatic systems can be classified in similar ways, determined by different factors.

TESTING YOUR COMPREHENSION

1. How does competition lead to a realized niche? How does it promote resource partitioning?
2. Contrast the several types of exploitation. How do predation, parasitism, and herbivory differ?
3. Give examples of symbiotic and nonsymbiotic mutualisms. Describe at least one way in which mutualisms affect your daily life.
4. Explain how trophic levels, food chains, and food webs are related.
5. Name several ways in which a species could be considered a keystone species.
6. Explain and contrast primary and secondary terrestrial succession.
7. What are the general characteristics of *invasive* species? Name five changes to Great Lakes communities that have occurred since the invasion of the zebra mussel.
8. What is restoration ecology?
9. What factors most strongly influence the type of biome that forms in a particular place on land? What factors determine the type of aquatic system that may form in a given location?
10. Draw generalized climate diagrams (climatographs) for a boreal forest, a temperate rain forest, and a desert. Label all parts of the diagrams, and describe all the types of information an ecologist could glean from such diagrams.

THINKING IT THROUGH

1. Imagine that you spot two species of birds feeding side by side, eating seeds from the same plant, and you begin to wonder whether competition is at work. Describe how you might design a scientific research project or experiment to address this question. What observations would you try to make at the outset? Would you try to manipulate the system to test your hypothesis that the two birds are competing? If so, how?
2. From year to year, biomes are stable entities, and our map of world biomes appears to be a permanent record of patterns across the planet. But are the locations and identities of biomes permanent, or could they change over time? Provide reasons and explanations for your answers.
3. Can you devise possible responses to the zebra mussel invasion in the Great Lakes? What strategies would you consider if you were put in charge of the effort to control this species' spread and reduce its impacts? Name some advantages of each of your ideas, and identify some obstacles it might face in being implemented.
4. Consider this real-life example of invasive-species management and restoration ecology, and then answer the questions that follow, in your new role as an environmental land manager:

The Garry oak meadows on Trinity Western University's Crow's Nest Ecological Research Area are experiencing encroachment by the Douglas fir (*Pseudotsuga menziesii*), which threatens the health and survival of the meadow. Traditionally, Aboriginal people maintained the Garry oak meadows with frequent burnings. However, with European settlement and the resulting control and suppression of fires, Douglas firs have been increasingly successful at invading the meadows. Some Garry oak restoration projects have undertaken to remove Douglas firs physically or using controlled fires, but when the Douglas firs are removed, the resulting disturbance of the soil can facilitate the establishment of other invasive species.[16]

(a) How would you design a comprehensive scientific study to determine the best ways of removing Douglas firs without enhancing conditions for encroachment by other invasive species?

(b) Removing Douglas firs from the Garry oak meadows would not return the meadows to their "natural," pristine, or pre-human state; instead, it would represent a return to an earlier phase of human (Aboriginal) ecosystem management. Is this the best approach? Do you think the Douglas firs should be allowed to advance naturally, without interference by fire or other removal techniques? Or is it worthwhile to preserve the meadows—which are disappearing rapidly, and host a variety of unusual and threatened species—by returning to an earlier land management approach?

(c) Can you think of some general guidelines that might be used to make ecosystem management decisions in other cases of this type?

INTERPRETING GRAPHS AND DATA

The grey wolf (*Canis lupus*) is a keystone species in Yellowstone National Park's ecosystem. Wolf packs hunt elk, gorge themselves on the kill, and leave the carcass as carrion for scavenger species, such as ravens, magpies, eagles, coyotes, and bears. As the global climate has warmed, winters in Yellowstone have become shorter over the past 55 years. Fewer elk weaken and die in milder weather, and so less carrion is available to scavengers during warmer, shorter winters. Biologists Christopher Wilmers and Wayne Getz studied the links among climate change, wolves, elk, and scavenger populations in Yellowstone. They used empirical field data on wolf predation rates and elk carrion availability recorded over 55 years to develop a model that estimated carrion availability with and without wolves for each winter month. Some of their findings are presented in the graph.

1. How much less carrion is available in April than in November when wolves are present? When wolves are not present?

2. Wolves were hunted nearly to extinction in the 1930s and were reintroduced to Yellowstone only in 1995. How, would you suspect, has their reintroduction affected scavenger populations since then? Why?

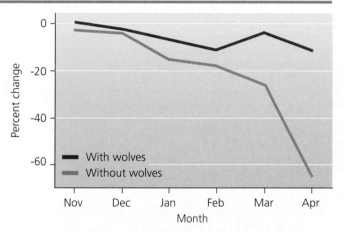

Changes in amount of winter carrion available to scavengers in Yellowstone National Park, with and without wolves, according to the model. Differences in March and April are statistically significant.

Source: Data from Wilmers, C.C., and W.M. Getz (2005). Gray Wolves as Climate-Change Buffers in Yellowstone. *PLoS Biology 3*(4), p. e92. Published by PLoS (Public Library of Science).

3. Predict what effect continued shorter, warmer winters would have on scavenger populations. Why? Are the predicted effects of wolf reintroduction and of climate change compounded, or do they tend to cancel one another out?

MasteringEnvironmentalScience®

STUDENTS
Go to **MasteringEnvironmentalScience** for assignments, the eText, and the Study Area with practice tests, videos, current events, and activities.

INSTRUCTORS
Go to **MasteringEnvironmentalScience** for automatically graded activities, current events, videos, and reading questions that you can assign to your students, plus Instructor Resources.

PART TWO

Issues, Impacts, and Solutions

Twenty-year-old Meenakshi Diwan carries out maintenance in the solar village of Tinginapu, India.
Abbie Trayler Smith/Panos

6 Human Population

This crowded street is in Guangzhou, one of China's largest cities.

Redlink/Corbis

Upon successfully completing this chapter, you will be able to

- Assess the scope and historical patterns of human population growth, and evaluate how human population, affluence, and technology affect the environment
- Explain and apply the fundamental concepts of demography and the demographic transition

- Describe how population growth is connected to environmental, social, and economic factors, and how sustainable development goals are seeking to address disparities

A billboard in Chengdu promotes China's one-child policy.

ASIA

China

CENTRAL CASE
CHINA'S ONE-CHILD POLICY

"As you improve health in a society, population growth goes down. You know ... before I learned about it, I thought it was paradoxical."

—BILL GATES (2003)

"Either we limit our population growth or the natural world will do it for us."

—SIR DAVID ATTENBOROUGH (2013)

"People often focus on the downsides of population growth but neglect the upsides. These upsides may even outweigh the downsides, making a larger population a good thing overall."

—IAN GOLDIN (2011)

The People's Republic of China is the world's most populous nation, home to one-fifth (1.4 billion) of the 7.3 billion[1] people living on Earth as of 2015.

The first significant increases in China's population in the past 2000 years of the nation's history resulted from enhanced agricultural production and a powerful government during the Qing (Manchu) Dynasty in the 1800s. Population growth began to outstrip food supplies by the mid-1850s, and quality of life for the average Chinese peasant began to decline. From the mid-1800s (an era of increased European intervention in China) until 1949, China's population grew very slowly, at about 0.3% per year. This slow population growth was due, in part, to food shortages, war, and political instability, which caused high mortality and a decline in birth rates. Population growth rates rose again following the establishment of the People's Republic and have declined once again since the establishment of the one-child policy.

When Mao Zedong founded the country's current regime in 1949, roughly 540 million people lived in a mostly rural, war-torn, impoverished nation. Mao believed population growth was desirable, and under his leadership

China grew and changed. By 1970, China's population had grown to approximately 800 million people. At that time, the average Chinese woman gave birth to 5.8 children in her lifetime.

Unfortunately, the country's burgeoning population and its industrial and agricultural development were eroding the nation's soils, depleting its water, levelling its forests, and polluting its air. Chinese leaders realized that the nation might not be able to feed its people if their numbers grew much larger. They saw that continued population growth could exhaust resources and threaten the stability and economic progress of Chinese society. The government decided to institute a population control program that prohibited most Chinese couples from having more than one child.

The program began with education and outreach efforts encouraging people to marry later and have fewer children. Along with these efforts, the Chinese government increased the accessibility of contraceptives and abortion. By 1975, China's annual population growth rate had dropped from a peak of around 2.7% to 1.5% (see **FIGURE 1**). To further decrease the birth rate, in 1979 the government took the more drastic step of instituting a system of rewards and punishments to enforce a one-child limit. One-child families received better access to schools, medical care, housing, and government jobs, and mothers with only one child were given longer maternity leaves. Families with more than one child, meanwhile, were subjected to social scorn and ridicule, employment discrimination, and monetary fines. In some cases, the fines exceeded half the offending couple's annual income.

Beginning in 1984, the one-child policy was loosened, strengthened, and then loosened again as government leaders sought to maximize population control while minimizing public opposition. Through the 1990s and into the 2000s, the policy applied mostly to urban couples, whereas many rural farmers and ethnic minorities were exempt, especially if their first child was a girl. In late 2015 the policy was finally terminated, making it possible, for the first time in more than 35 years, for any couple in China to have two children.

In enforcing these policies, China has, in effect, been conducting one of the largest and most controversial social

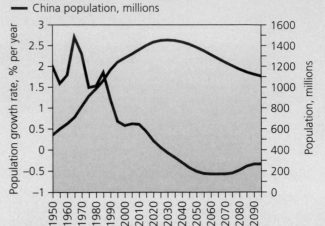

FIGURE 1

Population growth rates in China declined rapidly after the one-child policy was enforced in 1980.

Why did the population continue to increase after 1980, even though the growth rate of the population was declining? Why do the projections show population continuing to decline after 2070, even though the growth rates are increasing?

Source: Data from United Nations Department of Economic and Social Affairs, Population Division, Population Estimates and Projections Section, http://esa.un.org/unpd/wpp/unpp/panel_population.htm. Accessed 16 March 2015.

experiments in history. In purely quantitative terms, the experiment has been a major success; the nation's growth rate is now down to less than 0.5%, making it easier for the country to deal with its many social, economic, and environmental challenges.

However, China's population control policies have also produced unintended social consequences, such as female infanticide, an unbalanced sex ratio, and a black-market trade in teenaged girls. It is expected to lead to further problems in the future, including an aging population and shrinking workforce. Moreover, the policies have elicited intense criticism from those who oppose government intrusion into personal reproductive choices.

As other nations become more crowded, might their governments also feel forced to turn to drastic policies that restrict individual freedoms? In this chapter, we examine human population dynamics worldwide, consider their causes, and assess their consequences for the environment and our society.

The Human Population at over 7 Billion

While China works to slow its population growth and speed its economic growth, populations continue to rise in most nations of the world. Most of this growth is occurring in poor nations that are ill-equipped to handle it. India (**FIGURE 6.1**) is currently on course to surpass China as the world's most populous nation. Although the *rate* of global growth is slowing, we are still increasing in absolute numbers, and today more than 7.3 billion of us inhabit the planet.

Just how much is 7.3 billion? We often have trouble conceptualizing huge numbers like a billion; 1 billion is 1000 times greater than 1 million. If you were to count once each second without ever sleeping, it would take more than 30 years to reach 1 billion. To travel 7.3 billion kilometres, you would have to fly to the Moon and back more than 9600 times.

Population is still growing, but more slowly

The human population has grown rapidly in the past several decades. The world's population has doubled since 1970 and is still growing by roughly 80 million people annually (approximately 2.5 people are added—that is, births minus deaths—*every second*).[2] This is the equivalent of adding more than twice the population of Canada to the world each year. It took until after 1800—most of human history— for our population to reach 1 billion. Yet we reached 2 billion by 1927, and 3 billion in just over 30 more years, in 1960. Our population added its next billion in just 14 years (1974), its next billion in a mere 13 years (1987), and

the most recent billion in another 12 years (**FIGURE 6.2**). Think about when you were born and how many people have been added to the planet since that time. No previous generations have ever lived amid so many other people.

What accounts for such unprecedented growth of population in the most recent two centuries of existence of the human species? Exponential growth—

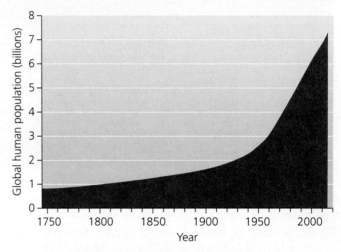

(a) Human population growth, 1700–present

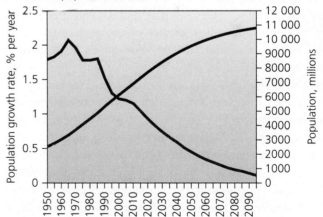

(b) Population growth and growth rate, 1950–2100

FIGURE 6.2
Global population has grown exponentially, rising from less than 1 billion in 1800 to approximately 7.3 billion today **(a)**. Nearly all growth has occurred in the last 200 years. The rate of growth of the world population peaked in the mid-1960s and has been declining since then **(b)**. This graph (and the graph of population and growth rate in China in the Case Study) is based on a "medium" growth-rate scenario from the United Nations Population Division.
Source: Data from U.S. Bureau of the Census and United Nations Department of Economic and Social Affairs, Population Division, Population Estimates and Projections Section, http://esa.un.org/unpd/wpp/unpp/panel_population.htm. Accessed 16 March 2015.

FIGURE 6.1
India is on course to surpass China soon as the world's most populous nation. In this photo, Indian women wait in line for immunizations for their babies.

About how many people have been added to the world's population since you were born? Since your parents were born? On the basis of graph **(b)**, in what year will the growth rate of the population drop to about 1.0%/yr? How many people will be added to the world's population that year?

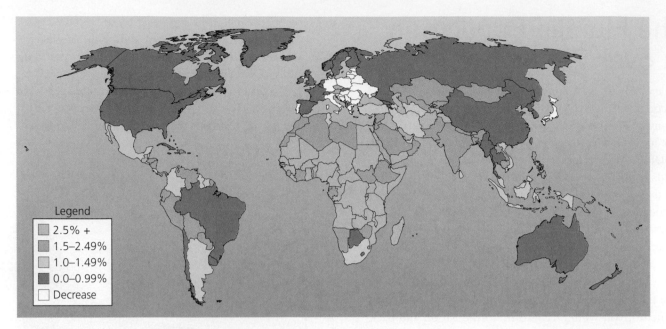

FIGURE 6.3 Population growth rates vary greatly from place to place. Population is growing fastest in poorer nations of the tropics and subtropics but is now beginning to decrease in some industrialized nations, even falling below zero in some countries. Shown are natural rates (that is, population growth rates as a function of births and deaths, not including immigration and emigration), of population change as of 2014.

Source: Carl Haub and Toshiko Kaneda, 2014 World Population Interactive Map, Washington, DC: Population Reference Bureau, 2014.

the increase in a quantity by a fixed percentage per unit time—accelerates the absolute increase of population size over time, just as compound interest accrues in a savings account. The reason is that a given percentage of a large number is a greater quantity than the same percentage of a small number. Thus, even if the growth *rate* remains steady, population *size* will increase by greater increments with each successive generation.

In fact, the world's population growth rate has not remained steady. During much of the twentieth century the growth rate actually rose from year to year. It peaked at 2.1% per year during the 1960s and has declined since then, reaching about 1.2% at the turn of the twenty-first century. Although 1.2% may sound small, exponential growth endows small numbers with large consequences. For instance, a hypothetical population starting with one man and one woman, growing at 1.2% per year, gives rise to a population of 2939 after 40 generations and 112 695 after 60 generations. In today's world, rates of annual growth vary greatly from region to region. **FIGURE 6.3** maps this variation.

At a 2% annual growth rate, a population (or your bank account) doubles in just 35 years. For low rates of increase, we can estimate doubling times with a handy rule of thumb. Just take the number 70, and divide it by the annual percentage growth rate: 70 ÷ 2 = 35. Had China not instituted its one-child policy—that is, had its growth rate remained unchecked at 2.8%—it would have taken only 26 years to double in size (70 ÷ 2.7 = 26). Had population growth continued at this rate, China's population would have exceeded 2 billion people by 2003. That milestone was not

reached, and likely will never be reached, as China's population is expected to stabilize at around 1.45 billion by 2030.

Perspectives on human population have changed over time

At the outset of the Industrial Revolution in England of the 1700s, population growth was regarded as a good thing. For parents, a high birth rate meant more children to support them in old age. For society, it meant a greater pool of labour for factory work. British economist Thomas Malthus (1766–1834) had a different opinion. Malthus claimed that unless population growth was limited by laws or other social controls, the number of people would outgrow the available food supply until starvation, war, or disease arose and reduced the population (**FIGURE 6.4**). Malthus's most influential work, *An Essay on the Principle of Population*, published in 1798, argued that a growing population would eventually be checked by either limits on births or increases in deaths. If limits on births (such as abstinence and contraception) were not implemented soon enough, Malthus wrote, deaths would increase.

Malthus's thinking was specifically shaped by the rapid urbanization and industrialization he witnessed during the early years of the Industrial Revolution. More recently, biologist Paul Ehrlich of Stanford University has been called a "neo-Malthusian" because—like Malthus—he warned that population growth would have disastrous effects on the environment and human welfare. In his 1968 book, *The Population Bomb*, Ehrlich predicted that

(a) Eighteenth-century London, England

(b) Thomas Malthus

FIGURE 6.4
The England of Thomas Malthus's era (1766–1834), shown in this engraving **(a)**, favoured population growth as the society industrialized. Malthus **(b)** argued that the pressure of population growth on the availability of resources could lead to disaster.

the rapidly increasing human population would unleash widespread famine and conflict that would consume civilization by the end of the twentieth century. Ehrlich and other neo-Malthusians have argued that population is growing much faster than our ability to produce and distribute food and that population control is the only way to prevent massive starvation, environmental degradation, and civil strife.

Is population growth really a "problem" today?

The world's ongoing population growth has been made possible by technological innovations, improved sanitation, better medical care, increased agricultural output, and other factors that have led to a decline in death rates, particularly a drop in rates of infant mortality. Birth rates have not declined as much, so births have outpaced deaths for many years now. Thus, the population "problem" actually arises from a very good thing—our ability to keep more people alive longer.

The mainstream view in Malthus's day held that population increase was a good thing. Today there are still many people who argue that population growth poses no problems. Even though human population has nearly quadrupled in the past 100 years—the fastest it has ever grown—Malthusian and neo-Malthusian predictions have not materialized on a devastating scale.

This is due, in part, to enormous increases in crop yields and advances in agricultural technology associated with the Green Revolution. Increasing material prosperity and control of reproductive health has also helped bring down birth rates—something that Malthus and Ehrlich did not foresee.

Some "cornucopian" thinkers believe that resource depletion caused by population increase is not a problem, and that new resources can be found or new technologies developed to replace depleted resources. Libertarian writer Sheldon Richman argues that there is no population problem, that overpopulation does not exist, and that the concept of carrying capacity (introduced in Chapter 4, *Evolution, Biodiversity, and Population Ecology*) does not apply to people. He expressed this view as follows:

> *There is no population problem . . . Human beings create resources. We find potential stuff and human intelligence turns it into resources.*[3]

The premise that the human population is not subject to the limitations of carrying capacity leads to some pretty ridiculous conclusions. For example, a simple calculation shows that if the human population continued to grow at the rate of 1.2% per year, in 2500 years the people of Earth would weigh as much as Earth itself (assuming an average weight of 60 kg per person). This is an impossible scenario; environmental limitations *do* exist.

In contrast to the idea that humankind will always be able to save itself with technology, most environmental scientists recognize that not all resources can be replaced or reinvented once they have been depleted. For example, once a species has gone extinct, we cannot replicate its exact function in ecosystems, or know what medicines or other practical applications we might have obtained from it, or regain the educational and aesthetic value of observing and interacting with it. Another irreplaceable resource is land; we cannot expand Earth like a balloon to increase its surface area.

Even if resource substitution could enable population growth to continue indefinitely, could we maintain the *quality* of life that we desire for our descendants and ourselves? Surely some of today's resources are easier or cheaper to use, and less environmentally destructive to harvest or mine, than the resources that might replace them. Unless resource availability keeps pace with population growth, the average person in the future will have less space in which to live, less food to eat, and less material wealth than the average person does today. Thus population increases are indeed a problem if they create stress on resources, social systems, or the natural environment, such that our quality of life declines.

In today's world, population growth is much more strongly correlated with poverty than with wealth. In spite of this, many governments have found it difficult to let go of the notion that population growth increases a nation's economic, political, or military strength. Many national governments still offer financial and social incentives to encourage their citizens to produce more children. Governments of countries currently experiencing population declines (such as many in Europe) feel especially uneasy. According to the Population Reference Bureau, more than three of every five European national governments take the view that their birth rates are now too low, and none states that its rate is too high.

Researchers in the Malthusian tradition are still making important contributions to our understanding of population and the environment today. Research has shown that environmental degradation and scarcity, particularly in situations of overcrowding in sensitive environments, can lead to migrantism, refugeeism, and even armed conflict. For example, political scientist Thomas Homer-Dixon of the University of Waterloo investigated the linkages between agricultural land scarcity and ethnic tensions in initiating the genocide in Rwanda in the mid-1990s, which cost the lives of 800 000 people. In other active conflicts we can see that global population growth and resulting environmental scarcity have played a central role in causing famine, disease, and social and political conflict around the world.

The dire predictions of Malthus, Ehrlich, and other neo-Malthusians were not accurate; however, the idea that a "technological fix" will always emerge in time to rescue humans from problems brought on by population growth is equally unsatisfactory. Our current understanding of the limitations of the resources of this planet suggests that the real answer lies somewhere between these extremes.

Population is one of several factors that affect the environment

The extent to which population increase can be considered a problem involves more than just numbers of people. One widely used formula gives us a handy way to think about factors that affect the environment. Nicknamed the **IPAT model**, it is a variation of a formula proposed in 1974 by Paul Ehrlich; Barry Commoner, a professor of biology; and John Holdren, a professor of environmental policy. The IPAT model represents how our total impact (I) on the environment results from the interaction among population (P), affluence (A), and technology (T):

$$I = P \times A \times T$$

Increased population intensifies impact on the environment as more individuals take up space, use natural resources, and generate waste. Increased affluence magnifies environmental impact through the greater per capita resource consumption that generally has accompanied enhanced wealth. Changes in technology may either decrease or increase human impact on the environment. Technology that enhances our abilities to exploit minerals, fossil fuels, old-growth forests, or ocean fisheries generally increases impact, but technology to reduce smokestack emissions, harness renewable energy, or improve manufacturing efficiency can decrease impact.

We might also add a sensitivity factor (S) to the equation to denote how sensitive a given environment is to human pressures:

$$I = P \times A \times T \times S$$

For instance, the arid lands of western China are more sensitive to human disturbance than the moist regions of southeastern China. Plants grow more slowly in the arid west, making deforestation and soil degradation more likely. Thus, adding an additional person to western China should have more environmental impact than adding one to southeastern China (all other conditions being equal).

Various population researchers have refined the IPAT equation by adding terms for the effects of social institutions, such as education, laws, and their enforcement; stable and cohesive societies; and ethical standards that promote environmental well-being. Factors like these affect how population, affluence, and technology translate into environmental impact. Dr. Jack Liu and colleagues from the University of Michigan have discovered that even the way the population is distributed—specifically, the sizes of households—has a major impact on levels of resource use (see "The Science behind the Story: A Different Population Bomb: The 'Household Explosion'").

Impact can be thought of in various ways, but it generally boils down to either pollution or resource consumption. Pollution became a problem in the modern world once our population grew large enough that we produced great quantities of waste. The depletion of resources by larger and hungrier populations has been a focus of scientists and philosophers since before Malthus's time. History offers examples of cases in which resource depletion helped bring an end to civilizations, from the Mayans to the Mesopotamians. Some environmental scientists have predicted similar problems for our global society in the near future if we do not manage to embark on a path toward sustainability (**FIGURE 6.5**).

As discussed, the neo-Malthusians have not seen their direst predictions come true. The main reason is that we have developed technology—the T in the IPAT equation—time and again, to alleviate our strain on resources and allow us to further expand our population. For instance, we have employed technological advances to increase global agricultural production faster than our population has risen.

Dr. Jianguo "Jack" Liu, Director of the Michigan State University Center for Systems Integration and Sustainability, examines historical patterns in the way we live and the environmental impacts of those patterns.
Dr. Jianguo "Jack" Liu, Michigan State University

A Different Population Bomb: The "Household Explosion"

There is another type of population time bomb that continues to tick, even as we start to put the brakes on population growth. As a global population, our progress in curbing population growth rates has not resulted in reduced consumption of natural resources. Although fertility rates are declining globally and family sizes are shrinking, our increasing tendency to live in smaller and smaller households is putting higher demands on our natural resources.

Professor Jianguo Liu is the Rachel Carson Distinguished Professor of Sustainability at Michigan State University. Part of the focus of his research is on the connections between households and resource consumption. "Long-term dynamics in human population size as well as their causes and impacts have been well documented," says Liu, "but little attention has been paid to long-term trends in the numbers of households, even though households are basic consumption units."

Liu and his team collected data spanning centuries (1600–2000 CE) and from 213 nations (and colonies, territories, and protectorates) to examine trends in household sizes and numbers. They published their research in 2014 in the journal *Population and Environment*. Through analysis of the data, they found that household size has been in decline in some countries for centuries (see **FIGURE 1**). In more economically developed countries, household size peaked at 5.0 individuals per household around the late 1800s, and then began a decline to about 2.5 individuals per household today. In developing nations, the decline did not begin until the late 1980s, but the trend is apparent in almost every country studied by Liu and his team.

Many sociological factors can lead to a proliferation of households: increasing affluence, aging populations, increasing divorce rates, changing preferences for privacy, and so on. One result of the shrinking size of households, along with a growing or stable population, has meant that the rate of growth in the number of households has outpaced population growth since 1985.

The consequences of adding households to the planet, even as population growth itself starts to slow, are immense. Houses require lumber and other building materials. Smaller households tend to be less efficient, so per-capita demands on land, water, and energy go up. More houses require other types of infrastructure such as roads, sewers, and commercial development—all of which significantly increase our ecological footprint. In fact, the research carried out by Liu's group suggests that the number of households is a more useful number than population size itself in predicting important environmental variables such as CO_2 emissions, fuelwood consumption, and per capita automobile use.

The IPAT model could be adapted to include the effect of household dynamics on the environment. Liu and his team proposed the following modification of the IPAT model:

$$I = PHoG$$

where

Impact (I) = population × personal goods (P) + households × household goods (HoG)

Current demographic trends suggest that as industrialization and urbanization proceed, 800 million new homes could be added to the planet *without* population growth. Each of these homes would demand more household goods and more services, and would have lower efficiency of resource use (if the trend toward smaller numbers of inhabitants continues). This "household population bomb" may be one that we cannot afford to ignore.

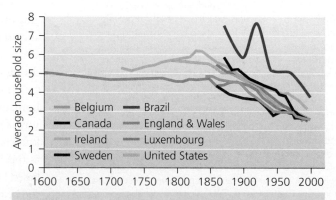

FIGURE 1 This graph shows changes in average household size over time in various countries.
Source: Population and environment by American Psychological Association. Reproduced with permission of Springer-Verlag Dordrecht in the format Book via Copyright Clearance Center.

On the basis of this graph, what was the approximate household size in Canada around 1860? What was it in 2000? By what percentage did the average household size change over that period?

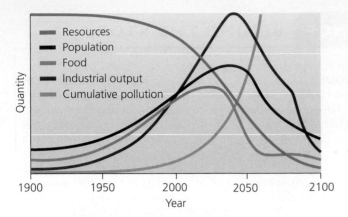

(a) Projection based on status quo policies

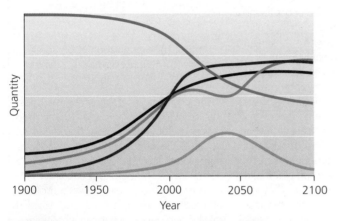

(b) Projection based on policies for sustainability

FIGURE 6.5
Environmental scientist Donella Meadows and colleagues used computer simulations to generate projections of trends in human population, resource availability, food production, industrial output, and pollution. The projections are based on data from the past century and current scientific understanding of the environment's biophysical limits. In **(a)** are projections for a world in which society proceeds "without any major deviation from the policies pursued during most of the twentieth century." Population and production increase until declining nonrenewable resources make further growth impossible, causing population and production to decline rather suddenly. In a scenario with policies aimed at sustainability **(b)**, population levelled off at 8 billion, production and resource availability levelled off at medium-high levels, and pollution declined to low levels.
Source: Reprinted from *Limits to Growth: The 30-Year Update,* copyright 2004 by Donella Meadows, Jorgen Randers, and Dennis Meadows, used with permission from Chelsea Green Publishing (www.chelseagreen.com)

Modern-day China shows how the elements of the IPAT formula can combine to cause tremendous environmental impacts in very little time. While millions of Chinese are increasing their material wealth and consumption of resources, the country is battling unprecedented environmental challenges. Intensive agriculture has expanded westward out of the country's historic moist rice-growing areas, causing farmland to erode and literally blow away. China has overpumped many of its aquifers and has drawn so much water for irrigation from the Huang He (Yellow River) that the once-mighty waterway now dries up in many stretches. The country faces new urban pollution threats from rapidly increasing numbers of automobiles. As the world's developing countries try to attain the level of material prosperity that industrialized nations enjoy, China is a window on some of the problems they may encounter.

The environment has a carrying capacity for humans

It is a fallacy to think of people as being somehow outside of nature, or unaffected by its limitations. Humans exist within their environment as one species out of many. The principles of population ecology that apply to toads, frogs, and passenger pigeons apply to humans as well. Environmental factors set limits on our population growth, and the environment has a carrying capacity for our species, as it does for every other. We happen to be a particularly successful organism, however—one that has repeatedly increased the carrying capacity of the environment by developing technology to overcome natural limits on population growth.

Four significant periods of societal change appear to have fundamentally altered the human relationship with the environment and increased the carrying capacity, triggering remarkable increases in population size (**FIGURE 6.6**).

The first transition happened in the *Paleolithic period* (or *Old Stone Age*), when early humans gained control of fire (perhaps as much as 1.7 million years ago) and began to shape and use stones (as much as 2.5 million years ago) as tools with which to modify their environment. We can speculate that this transition made life so much easier and the environment so much more manageable for our ancestors that their population grew substantially, although we have little direct evidence about world population dating from that period.

The second major change was the transition from a nomadic hunter-gatherer lifestyle to a settled agricultural way of life. This change began around 12 000 years ago and is known as the **Agricultural Revolution**, in what is known as the *Neolithic* (or *New Stone Age*) period. This agriculture-based lifestyle was much more intensive and manipulative in the production of resources from the land. As people began to grow their own crops, raise domestic animals, and live settled, sedentary lives in villages, they found it easier to meet their nutritional needs. As a result, they began to live longer and to produce more children who survived to adulthood. The Agricultural Revolution initiated a permanent change in the way humans relate to the natural environment. Agriculture also created surplus food that could be passed on to people not directly involved in food acquisition; in other words, agriculture made cities, trades, science, armies, and other aspects of modern culture possible.

The third major societal change, the **Industrial Revolution**, began in the mid-1700s. It entailed a shift

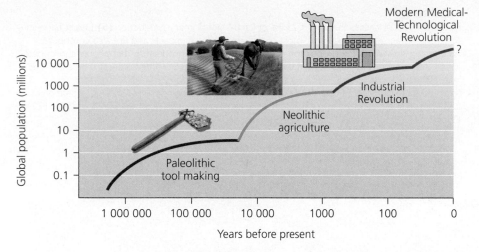

FIGURE 6.6
Tool making, agriculture, and industrialization each allowed our species to increase the global carrying capacity. The logarithmic scale of the axes makes it easier to visualize this pattern. We are currently in the midst of a fourth transition, involving the globalization of modern medical-technological advances; its impacts on population and the environment are as yet unknown.
Source: Data from Goudie, A. (2000). *The Human Impact.*

from rural life, animal-powered agriculture, and manufacturing by craftsmen, to an urban society powered by fossil fuels. The Industrial Revolution introduced improvements in sanitation and medical technology. Another very important aspect was the impact on agriculture and animal husbandry. Agricultural production was greatly enhanced during the Industrial Revolution by the introduction of fossil-fuel-powered equipment, steam engines, and synthetic fertilizers, along with advances in plant and animal breeding.

We are currently in the midst of a fourth major transition, involving the globalization of modern medical and technological advancements. The **Medical-Technological Revolution** is marked by developments in medicine, sanitation, and pharmaceuticals; the explosion of communication technologies; and the shift to modern agricultural practices known as the *Green Revolution* that have collectively allowed more people to live longer, healthier lives. This transition is still in progress, and the long-term implications for the human population, individual health, and the environment are unknown. Perhaps this will also be a period during which human society makes the transition to more sustainable, renewable energy sources and away from dependence on fossil fuels.

Environmental scientists who have tried to quantify the human carrying capacity of this planet have come up with wildly differing estimates. Estimates range from 1 billion to 2 billion people living prosperously in a healthy environment to 33 billion living in extreme poverty in a degraded world of intensive cultivation without natural areas. As our population climbs to 8 billion and probably beyond, we may yet continue to find ways to increase carrying capacity. Given our knowledge of population ecology, however, we have no reason to presume that human numbers can go on growing indefinitely. Indeed, populations that exceed their carrying capacity can crash.

Humans place heavy demands on the planet's primary production

Burgeoning numbers of people are making heavy demands on Earth's natural resources and ecosystem services. How can we quantify and map the environmental impacts our expanding population is exerting? One way is to ask: Of all the biomass that Earth's plants can produce, what proportion do human beings use (for food, clothing, shelter, etc.) or otherwise prevent from growing?

Helmut Haberl and colleagues from the Institute of Social Ecology in Austria have attempted to measure our consumption of net primary production (NPP), the net amount of energy stored in plant and algal biomass as a result of photosynthesis (discussed in Chapter 3, *Earth Systems and Ecosystems*). Human overuse of NPP diminishes resources available for other species; alters habitats, communities, and ecosystems; and threatens our future ability to derive ecosystem services.

The researchers first mapped how vegetation varies with climate across the globe, producing a detailed world map of "potential NPP"—vegetation that would exist if there were no human influence. Then they gathered data on crop harvests, timber harvests, grazing pressure, and other human uses of vegetation from global databases maintained by the United Nations Food and Agriculture Organization (FAO) and other sources. They also gathered data on how people affect vegetation indirectly, such as through fires, erosion and soil degradation, and other changes due to land use. To calculate the proportion of NPP that people appropriate, the researchers divided the amounts used up in these impacts by the total "potential" amount.

The researchers concluded that people are harvesting 12.5% of global NPP. Land use reduces it 9.6% further, and fires another 1.7% (**FIGURE 6.7A**). This makes us responsible for using up fully 23.8% of the planet's NPP—a staggeringly large amount for just a single species! Half of this

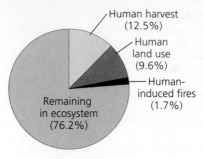

(a) Human use of the planet's net primary production

FIGURE 6.7

Humanity uses or causes the loss of 23.8% of Earth's net primary production **(a)**. Direct harvesting (of crops, timber, etc.) accounts for most of this, and land-use impacts and fire also contribute. As the map shows **(b)**, the proportion of Earth's net primary production that people appropriate varies from region to region. Regions that are densely populated or intensively farmed exert the heaviest impact.

Source: Data from Haberl, H., et al. (2007). Quantifying and Mapping the Human Appropriation of Net Primary Production in Earth's Terrestrial Ecosystems. *Proc. Natl. Acad. Sci. 104*, pp. 12942–12947. Copyright (2007) National Academy of Sciences, U.S.A.

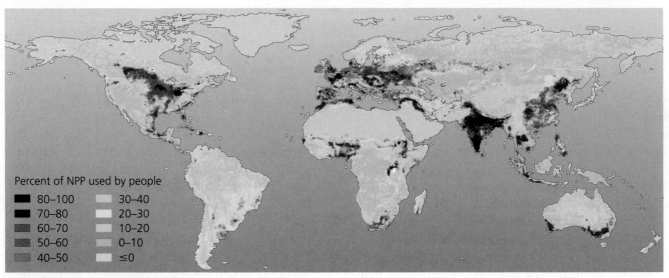

(b) Appropriation of NPP, by region

use occurred on cropland, where 83.5% of NPP was used. In urban areas, 73.0% of NPP was consumed; on grazing land, 19.4%; and in forests, 6.6%.

A global map showing human use of NPP (**FIGURE 6.7B**) reveals that densely populated and heavily farmed regions such as India, eastern China, and Europe have the greatest proportional use of NPP. The influence of population is clear. For instance, although people in southern Asia consume very little per capita, dense populations here result in a 63% use of NPP. In contrast, in sparsely inhabited regions of the world (such as the boreal forest, Arctic tundra, Himalayas, and Sahara Desert), humans consume only a tiny proportion of NPP. In North America, NPP use is heaviest in the East, Midwest, and Great Plains. In general, the map shows heavy appropriation of NPP in areas where population is dense relative to the area's vegetative production.

The map does not fully show the effects of resource consumption due to affluence. Wealthy societies commonly import food, fibre, energy, and products from other places, and this consumption can drive environmental degradation in poorer regions. For instance, North Americans and Europeans import timber logged from the Amazon basin, as well as soybeans and beef grown in areas where

Amazonian forest was cleared. Through global trade, we redistribute the products we gain from the planet's NPP. As a result, the environmental impacts of our consumption are often felt far from where we consume products.

Demography

The study of statistical change in human populations is the focus of the social science of **demography**. The field of demography developed along with and partly preceded the academic discipline of population ecology, and the two fields of study have influenced and borrowed from each other.

Demography is the study of human population

Demographic data help us understand how differences in population characteristics and related phenomena (for instance, decisions about reproduction) affect human communities and their environments. Demographers study population size, density, distribution, age structure, sex ratio, and rates of birth, death, immigration, and emigration of humans, just as population ecologists study these characteristics in other organisms. Each of these factors helps

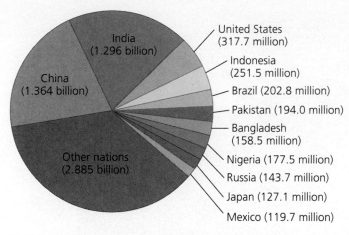

FIGURE 6.8
Almost one in five people in the world is Chinese, and more than one of every six people lives in India. Three of every five people live in nations with populations above 100 million.
Source: Data are for mid-2014, based on Population Reference Bureau: 2014 World Population Data Sheet.

demographers understand population dynamics and the potential environmental impacts of population changes.

Population size   The global human population of over 7 billion today comprises more than 200 nations, with populations ranging from China's almost 1.4 billion and India's almost 1.3 billion (**FIGURE 6.8**) to a number of island nations with populations below 100 000. The size that our global population will eventually reach remains to be seen (**FIGURE 6.9**). However, **population size** alone—

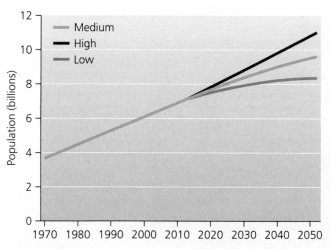

FIGURE 6.9
The United Nations predicts trajectories of world population growth, presenting its estimates in several scenarios based on different assumptions of fertility rates. In this version, the *middle* line represents the current population trajectory, projected to reach 8.1 billion in 2025 and 9.6 billion in 2050. Small variations in fertility have a big impact on future trajectories; just 0.5 additional children per woman would result in a population of 11 billion in 2050 (*high* scenario, the top line on graph). However, UN demographers expect fertility rates to continue falling. In the *low* scenario, if women each have 0.5 children less than in the current trajectory, the world will have 8 billion people in 2050, stabilizing at that level by about the year 2030.
Source: From *Concise Report on World Population in 2014*. Copyright © 2014 United Nations. Reprinted with permission of the United Nations.

the absolute number of individuals—does not tell the whole story. Rather, a population's environmental impact depends on its density, distribution, and composition, as well as on affluence, technology, level of consumption, and other factors outlined earlier.

Population density and distribution   People are distributed very unevenly over the globe. In ecological terms, our distribution is clumped at all spatial scales. At the global scale, **population density**—the number of people per unit of land area—is particularly high in regions with temperate, subtropical, and tropical climates, such as China, Europe, Mexico, southern Africa, and India (**FIGURE 6.10A**). Population density is low in regions with extreme-climate biomes, such as desert, deep rain forest, and tundra. Dense along seacoasts and rivers, human population is less dense at locations far from water. At intermediate scales, we cluster together in cities and suburbs and are spread more sparsely across rural areas, especially as the world becomes more and more urbanized (**FIGURE 6.10B,C**). At even smaller scales, we cluster in neighbourhoods and in individual households.

This uneven distribution means that certain areas bear far more environmental impact than others. High population densities and urbanization entail significant land conversion and impacts on both terrestrial and aquatic ecosystems, packaging and transport of goods, intensive fossil fuel consumption, and hotspots of pollution. However, the concentration of people in cities also increases efficiency and economies of scale, and relieves pressure on ecosystems in less-populated areas by releasing some of them from some human development.

Urban centres where population, cultural activities, and political power are concentrated have been part of human culture for several thousand years. However, what is new in the urban setting of today is the sheer scale of today's metropolitan areas. Today, 36 cities are **mega-cities**—home to more than 10 million residents—and the number increases every year. The greater metropolitan area of the world's most populous city, Tokyo, Japan, hosts almost 38 million people. North America's largest cities are Mexico City and New York City with populations of about 22 million and 20 million, respectively. In comparison, the Metropolitan Toronto area just passed the 6 million mark in 2015. Using the broadest possible definition of *metropolitan area*, which bundles everything from Hamilton to Oshawa into one large Toronto-centred economic area called the Golden Horseshoe, the total population is still under 9 million.

Most of the fastest-growing cities today are in the developing world. Some of this growth and rural–urban migration are occurring because industrialization is decreasing the need for farm labour and promoting commerce and jobs in cities. Sadly, another reason is that

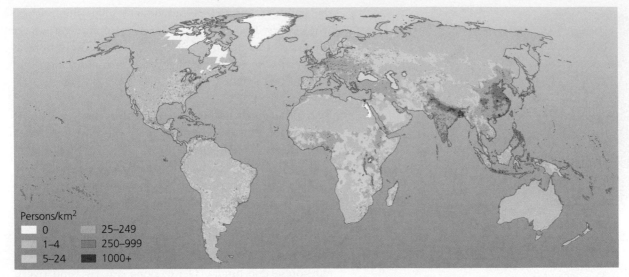

Persons/km²

0	25–249
1–4	250–999
5–24	1000+

(a) World population density

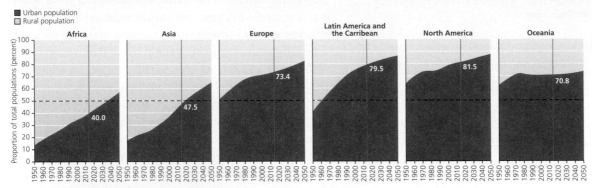

■ Urban population
□ Rural population

Africa   40.0
Asia   47.5
Europe   73.4
Latin America and the Carribean   79.5
North America   81.5
Oceania   70.8

(b) Rural and urban populations by region

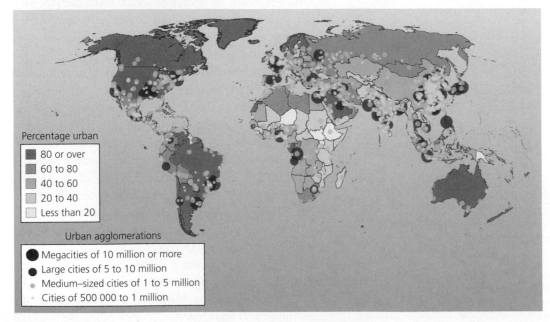

Percentage urban

■	80 or over
■	60 to 80
■	40 to 60
■	20 to 40
□	Less than 20

Urban agglomerations

- Megacities of 10 million or more
- Large cities of 5 to 10 million
- Medium–sized cities of 1 to 5 million
- Cities of 500 000 to 1 million

(c) Urban agglomerations

FIGURE 6.10   Human population density varies tremendously from one region to another **(a)**. Arctic and desert regions have the lowest population densities, whereas areas of India, Bangladesh, and eastern China have the densest populations. Populations are also distributed among rural and urban areas. The graphs **(b)** show rural and urban populations in various regions, as a proportion of total population. The map **(c)** shows the percentage of urban residents in each country, and the location of urban agglomerations with at least 500 000 inhabitants, as of 2014.

Source: From Department of Social and Economic Affairs. *World Urbanization Prospects: The 2014 Revision Highlights.* (b) Figure 3; (c) Map 1. Copyright © 2014 United Nations. Reprinted with the permission of the United Nations.

wars, conflict, and environmental degradation are driving millions of people out of the countryside and into cities. Cities like Mumbai (India), Lagos (Nigeria), and Cairo (Egypt) are growing in population even more quickly than North American cities did prior to the 1970s. Many cities in the developing world are growing at rates of 3 to 5% per year and even higher. These rates have been matched by some North American cities during their fastest growth, but only for short periods. For example, Calgary—which has more than once been North America's fastest-growing city in recent years—grew by about 3.5% in 2014. Recall how exponential growth works; on a sustained basis, even these seemingly small growth rates imply doubling times of just 14 to 23 years for many cities in the developing world.

These large, highly dense metropolitan populations have significant environmental impacts, but many areas with low population density also are highly sensitive to anthropogenic impacts and resulting environmental degradation. Some locations are sparsely populated, in the first place, because they are sensitive ecosystems that cannot support many people (a high S value in our revised IPAT model). Deserts, for instance, are easily affected by development that commandeers a substantial share of available water. Grasslands can be turned to deserts if they are farmed too intensively, as has happened across vast stretches of the Sahel region bordering Africa's Sahara Desert, in the Middle East, and in parts of China. The Arctic tundra is another environment that is highly sensitive to environmental change and human impacts. For example, a disturbance of vegetation—something as simple as a set of car tracks—can cause deep melting of permafrost and collapse of soil, the scars of which may last for years or even decades.

Age structure

Data on the age structure or age distribution of human populations are especially valuable to demographers trying to predict future dynamics of populations. As you learned in the context of non-human populations, large proportions of individuals in young age groups portend a great deal of reproduction and, thus, rapid population growth. Age structure diagrams, commonly called **population pyramids**, are visual tools scientists use to illustrate age structure (**FIGURE 6.11**). The width of each horizontal bar represents the number of people in each age class. A pyramid with a wide base denotes a large proportion of people who have not yet reached reproductive age—and this indicates a population soon capable of rapid growth. In this respect, a wide base of a population pyramid is like an oversized engine on a rocket—the bigger the booster, the faster the increase.

Examine the age structure diagrams, or age pyramids, for Canada and Afghanistan (**FIGURE 6.12**). Not surprisingly, Afghanistan has the greater population growth rate. In fact, its annual growth rate is more than twice that of Canada.

By causing a dramatic reduction in the number of children born since 1980 through the one-child policy, China virtually guaranteed that its population age structure would change (**FIGURE 6.13**). In 1995 the median age in China was 27; by 2030 it will be 39. In 1997 there were 125 children under age five for every 100 people aged 65 or older in China, but by 2030 there will be only 32. The number of people older than 65 will rise from 100 million in 2005 to 236 million in 2030.

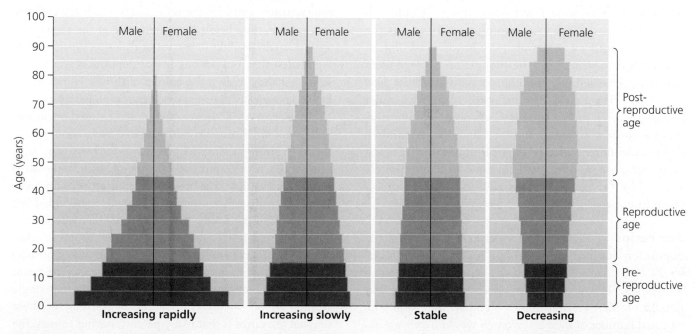

FIGURE 6.11   Age structure diagrams show numbers of individuals of different age classes in a population. A diagram like that on the left is weighted toward young age classes, indicating a population that will grow quickly. A diagram like that on the right is weighted toward old age classes, indicating a population that will decline. Populations with balanced age structures, like the one shown in the third diagram, will remain stable.

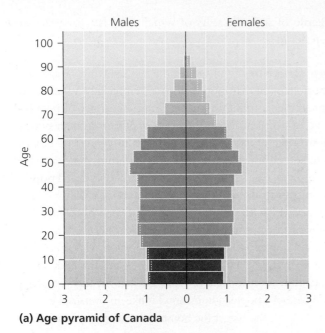

(a) Age pyramid of Canada

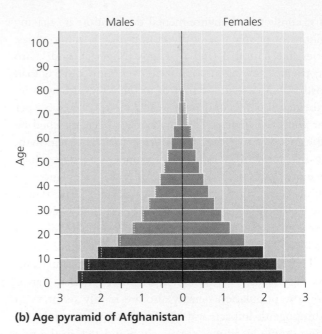

(b) Age pyramid of Afghanistan

FIGURE 6.12 Canada's age structure **(a)** shows relatively even numbers of individuals in various age classes. The tiny dashed lines show the differences in sex ratios (relative number of males and females) in each age cohort. Afghanistan **(b)** shows an age distribution heavily weighted toward young people. Afghanistan's population growth rate is more than twice as high as that of Canada. The post–World War II "baby boom" is visible as a "bump" in the age pyramid for Canada, between the ages of 45 and 65. In future years the nation will experience an aging population, as baby boomers grow older. *Source:* From Department of Economic and Social Affairs, Population Division (2013) World Population Prospects: The 2012 Revision, Volume II, Demographic Profiles (ST/ESA/SER.A/345). Copyright © 2013 United Nations. Reprinted with the permission of the United Nations.

This pattern of aging in the population is occurring in many countries, including Canada (see **FIGURE 6.12A**). Older populations will present new challenges for many nations, as increasing numbers of older people require the care and financial assistance of relatively fewer working-age citizens.

This dramatic shift in age structure will challenge China's economy, health care systems, families, and military forces because fewer working-age people will be available to support social programs that assist the increasing number of older people. However, the shift in age structure also reduces the proportion of dependent children. The reduced number of young adults may mean a decrease in the crime rate. Moreover, older people are often productive members of society, contributing volunteer activities and services to their children and grandchildren. Clearly, in terms of both benefits and drawbacks, life in China will continue to be profoundly affected by the particular approach its government has taken to population control.

Sex ratios   The ratio of males to females also affects population dynamics. Imagine two islands, one populated by 99 men and 1 woman and the other by 50 men and 50 women. Where would we be likely to see the greatest population increase over time? Of course, the island with an equal number of men and women would have a greater number of potential mothers and thus a greater potential for population growth.

The naturally occurring sex ratio in human populations at birth features a slight preponderance of males; for every 100 female infants born, 105 to 106 male infants are born. (You can see the difference in males and females in each age cohort if you look closely at the dashed lines in **FIGURES 6.12** and **6.13**.) This may be an evolutionary adaptation to the fact that males are slightly more prone to death during any given year of life. It usually ensures that the ratio of men to women is approximately equal by the time people reach reproductive age; women then begin to predominate as the population ages, generally leading to a ratio of males to females that is slightly less than one-to-one in the population as a whole. Thus, a slightly uneven sex ratio at birth may be beneficial. However, a greatly distorted ratio can lead to problems.

In recent years, demographers have noted an unsettling trend in China, also observed in other nations where there is a strong traditional preference for boys: The ratio of newborn boys to girls has become skewed. In the 2010 census, for example, 118 boys were reported born for every 100 girls, with some provinces reporting even higher sex ratios. The overall ratio of males per 100 females for the Chinese population as a whole is currently around 107, but that will change as the younger, male-dominated portion of the population ages, possibly ballooning to the point where young single men will outnumber young single women by 50-60%.[4] The leading hypothesis for these unusual sex ratios is that many parents, having learned the sex of their fetuses by ultrasound, are selectively aborting female fetuses; this is called *prenatal sex selection.*

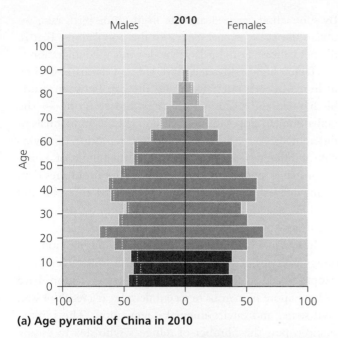

(a) **Age pyramid of China in 2010**

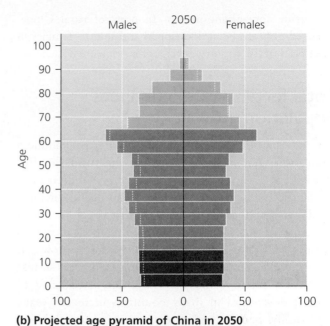

(b) **Projected age pyramid of China in 2050**

(c) **Young female factory workers in China**

(d) **Elderly Chinese**

FIGURE 6.13   As China's population ages, older people will outnumber the young. Age pyramids show the predicted greying of the Chinese population between 2010 **(a)** and 2050 **(b)**. Again, the tiny dashed lines show the differences in sex ratios in each age cohort. Today's children may, as working-age adults **(c)**, face pressures to support greater numbers of older citizens **(d)** than has any previous generation.
Source: Graphs (a) and (b) from Department of Economic and Social Affairs, Population Division (2013) World Population Prospects: The 2012 Revision, Volume II, Demographic Profiles (ST/ESA/SER.A/345). Copyright © 2013 United Nations. Reprinted with the permission of the United Nations.

DATA Q   Can you point out on graph **(a)** the beginning of the one-child policy and on graph **(b)** the cohort that represents those children in 2050?

Traditionally, Chinese culture has valued sons because they can carry on the family name, assist with farm labour in rural areas, and care for aging parents. Daughters, in contrast, will most likely marry and leave their parents, as the culture dictates. As a result, they will not provide the same benefits to their parents as will sons. Sociologists hold that this cultural gender preference, combined with the government's one-child policy, has led some couples to abort female fetuses or to abandon female infants. The Chinese government reinforced this gender discrimination when in 1984 it exempted rural peasants from the one-child policy if their first child was a girl, but not if the first child was a boy.

China is, of course, not the only nation in the world to experience the phenomenon of a skewed sex ratio. According to the World Bank, the United Arab Emirates and Qatar both have population sex ratios of more than 200, meaning that two boys survive to adulthood for every one girl, and Oman, Bahrain, and Kuwait are all more than 120.[5]

The unbalanced sex ratio in China and elsewhere may have the effect of further lowering population growth rates, as today's under-15 children grow up. However, it has already proven tragic for some of the "missing girls." It is also beginning to have the undesirable social consequence of leaving many Chinese men single. This, in turn, has resulted in a grim new phenomenon: In parts of rural China, teenaged girls are being kidnapped and sold to families in other parts of the country as brides for single men.

Total fertility rate influences population growth

One key statistic demographers calculate to examine a population's potential for growth is the **total fertility rate (TFR)**, or the average number of children born per female member of a population during her lifetime. **Replacement fertility** is the TFR that keeps the size of a population stable. For humans, replacement fertility is equal to a TFR of 2.1. When the TFR drops below 2.1, population size, in the absence of immigration, will shrink.

Various factors influence TFR and have acted to drive it downward in many countries in recent years. Historically, people tended to conceive many children, which helped ensure that at least some would survive. Lower infant mortality rates have made this less necessary. Increasing urbanization has also driven TFR down; whereas rural families need children to contribute to farm labour, in urban areas children are usually excluded from the labour market, are required to go to school, and impose economic costs on their families. If a government provides some form of social security, as most do these days, parents need fewer children to support them in their old age when they can no longer work. Finally, with greater education and changing roles in society, women tend to shift into the labour force, putting less emphasis on child rearing.

All these factors have come together in Europe, where TFR has dropped from 2.6 to 1.5 in the past half-century. Every European nation now has a fertility rate below the replacement level, and populations are declining in 28 of 44 European nations. Worldwide by 2015, 72 countries had fallen below the replacement fertility rate of 2.1. These countries made up roughly 45% of the world's population and included China (with a TFR of 1.6).

Population change results from birth, death, immigration, and emigration

Rates of birth, death, immigration, and emigration help determine whether a human population grows, shrinks, or remains stable. The formula for measuring population growth that we used for non-human populations also pertains to humans: Birth and immigration add individuals to a population, whereas death and emigration remove individuals.

It is convenient to express birth and death rates as the number of births and deaths per 1000 individuals for a given period—the **crude birth rate** (also called *nativity* or *natality*) and **crude death rate** (more commonly called *mortality*).

By subtracting crude death rate from crude birth rate, we obtain the *rate of natural increase* of the population—that is, the net increase from births alone, leaving aside migration.

Technological advances have led to a dramatic decline in human death rates, widening the gap between crude birth rates and crude death rates and resulting in the global human population expansion. Just as individuals of different ages have different abilities to reproduce, individuals of different ages show different probabilities of dying. For instance, people are more likely to die at old ages than young ages; if you were to follow 1000 10-year-olds and 1000 80-year-olds for a year, you would find that at year's end more 80-year-olds had died than 10-year-olds.

In today's ever-more-crowded world, immigration and emigration are playing increasingly large roles. Refugees, people forced to flee their home country or region, have become more numerous in recent decades as a result of war, civil strife, and environmental degradation. The United Nations puts the number of refugees who flee to escape poor environmental conditions in the millions per year. It is also widely acknowledged that environmental degradation and resource shortages often contribute to other causes of refugeeism, including internal and international conflicts.

The movement of refugees also causes significant environmental problems in receiving regions, as desperate victims try to eke out an existence with no livelihood and no cultural or economic attachment to the land or incentive to conserve its resources. The millions who fled Rwanda following the genocide there in the mid-1990s, for example, inadvertently destroyed large areas of forest while trying to obtain fuelwood, food, and shelter to stay alive once they reached the Democratic Republic of Congo (**FIGURE 6.14**).

All of the demographic factors that we have discussed so far come together to determine the growth rate of a

FIGURE 6.14
The United Nations High Commissioner for Refugees (UNHCR) estimates that more than 4 million people have fled from conflicts in Syria, and another 7.6 million are displaced within Syria, as of 2015.

Khalil Mazraawi/AFP/Getty Images

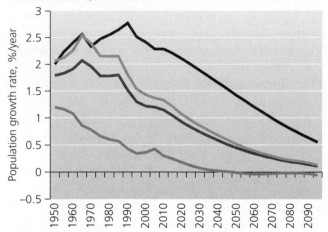

— World
— More developed nations
— Less developed nations
— Least developed nations

FIGURE 6.15
The annual growth rate of the global human population (blue line) peaked in the 1960s and has declined since then. Growth rates of more economically developed, industrialized nations (orange line) have fallen since the 1950s, while those of developing nations (green line) have fallen since the global peak in the late 1960s. For the world's least-developed nations (red line), growth rates began to fall in the 1990s. Recall from **FIGURE 6.2** that, although growth *rates* are declining, the absolute size of the global population is still growing, because smaller percentages of ever-larger numbers will still add to the absolute population.

population. A positive rate of growth indicates that the population size is growing; a negative rate of growth indicates that the population size is shrinking. Since 1970, growth rates in many countries have been declining, even without population control policies, and the global growth rate also has declined (**FIGURE 6.15**). This decline has come about, in part, from a steep drop in birth rates.

Note, however, that this is the *rate of growth* that is slowing, while the *absolute size* of the population continues to increase.

Some nations have experienced a demographic transition

Many nations that have lowered their birth rates and TFRs have been going through a suite of interrelated changes. In countries with good sanitation, good health care, and reliable food supplies, more people than ever before are living long lives. As a result, over the past 50 years the life expectancy for the average person has increased from 46 to 68 years as the global crude death rate has dropped from 20 deaths per 1000 people to 8 deaths per 1000 people. Strictly speaking, **life expectancy** is the average number of years that an individual in a particular age group is likely to continue to live, but often people use this term to refer to the average number of years a person can expect to live from birth. Much of the increase in life expectancy is due to reduced rates of infant mortality. Societies going through these changes are mostly the ones that have undergone urbanization and industrialization and have been able to generate personal wealth for their citizens.

To make sense of these trends, demographers developed a concept called the **demographic transition** (**FIGURE 6.16**). This is a model of economic and cultural change proposed in the 1940s and 1950s by demographer Frank Notestein, and elaborated on by others, to explain the declining death rates and birth rates that have occurred in Western nations as they became industrialized. Notestein observed that nations tend to move from a stable pre-industrial state of high birth and high death rates to a stable post-industrial state of low birth and low

FIGURE 6.16
The demographic transition is an idealized process that has taken some populations from a pre-industrial state of high birth rates and high death rates to a post-industrial state of low birth rates and low death rates. In this diagram, the wide green area between the two curves illustrates the gap between birth and death rates that causes rapid population growth during the middle portion of this process. *Source:* Data from Kent, M. M., and K. A. Crews. (1990). *World Population: Fundamentals of Growth.* Washington, DC: Population Reference Bureau.

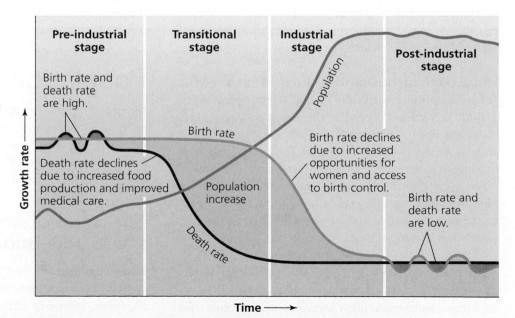

death rates. Industrialization, he proposed, caused these rates to fall naturally by first decreasing mortality and then lessening the need for large families. Parents would thereafter choose to invest in quality of life rather than quantity of children. Because death rates fall before birth rates fall, a period of net population growth results.

Thus, under the demographic transition model, population growth is seen as a temporary phenomenon that occurs as societies move from one stage of development to another. Those who study international development now recognize that not all societies follow the same development pathway as the older industrialized societies did. However, the demographic transition model remains a useful way to understand the interrelationships among birth rate, death rate, and population size, and the various social and economic factors that influence them. The United Nations also uses the demographic transition model to help with projecting future population; possible population trajectories are produced by estimating how quickly (or slowly) nations can be expected to move through the demographic transition in various fertility scenarios.

The pre-industrial stage
The first stage of the demographic transition model is the *pre-industrial stage*, characterized by conditions that have defined most of human history. In pre-industrial societies, both death rates and birth rates are high. Death rates are high because disease is widespread, medical care rudimentary, and food supplies unreliable and difficult to obtain. Birth rates are high because people must compensate for high mortality rates in infants and young children by having several children. In this stage, children are valuable as additional workers who can help meet a family's basic needs. Populations within the pre-industrial stage are not likely to experience much growth, which is why the human population was relatively stable from Neolithic times until the Industrial Revolution.

Industrialization and falling death rates
Industrialization initiates the second stage of the demographic transition, known as the *transitional stage*. This transition from the pre-industrial stage to the industrial stage is generally characterized by declining death rates because of increased food production and improved medical care. Birth rates in the transitional stage remain high, however, because people have not yet grown used to the new economic and social conditions. As a result, population growth surges.

The industrial stage and falling birth rates
The third stage in the demographic transition is the *industrial stage*. Industrialization increases opportunities for employment outside the home, particularly for women. Children become less valuable, in economic terms, because they do not help meet family food needs as they did in the pre-industrial stage. If couples are aware of this, and if they have access to birth control, they may choose to have fewer children. Birth rates fall, closing the gap with death rates and reducing the rate of population growth.

The post-industrial stage
In the final stage, the *post-industrial stage*, both birth and death rates have fallen to low levels. Population sizes stabilize or decline slightly. The society enjoys the fruits of industrialization without the threat of runaway population growth.

The demographic transition has occurred in many European countries, the United States, Canada, Japan, and several other developed nations over the past 200 to 300 years. China has passed through the demographic transition quickly; partly by industrialization, and partly by force of government policy. However, the demographic transition model may not apply to all developing nations as they industrialize. Some social scientists point out that population dynamics may be different for developing nations that adopt the Western world's industrial model rather than devising their own. Some demographers assert that the transition will fail in cultures that place greater value on childbirth or grant women fewer freedoms.

Moreover, natural scientists warn that there are not enough resources in the world to enable all countries to attain the standard of living that developed countries now enjoy. It has been estimated that for people of all nations to have the quality of life that Canadians do, we would need the natural resources of at least two to three more Earths. Whether developing nations, which include the vast majority of the planet's people, will pass through the demographic transition as developed nations have is one of the most important and far-reaching questions for the future of our civilization and Earth's environment.

Population and Society

Demographic transition theory links the statistical study of human populations with various societal factors that influence, and are influenced by, population dynamics. Let us now examine a few of these major societal factors more closely.

The status of women greatly affects population growth rates

Many demographers had long believed that fertility rates were influenced largely by degrees of wealth or poverty. However, affluence alone doesn't determine the total

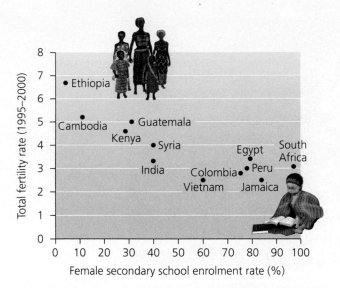

FIGURE 6.17

Increasing female literacy is strongly associated with reduced birth rates in many nations.

Source: Data from McDonald, M., and D. Nierenberg (2003). *Linking Population, Women, and Biodiversity. State of the World 2003.* Washington, DC: Worldwatch Institute.

DATA Q Is the correlation between birth rate and female literacy a *positive* correlation (as variable 1 increases, so does variable 2), or a *negative* correlation (as variable 1 increase, variable 2 decreases)? Does this correlation demonstrate *causation*?

fertility rate. Recent research is highlighting factors pertaining to the social empowerment of women. Drops in TFR have been most noticeable in countries where women have gained access to contraceptives and education, particularly family planning education (see **FIGURE 6.17**).

In 2014, about 65% of women worldwide (married or in a civil union, aged 15–49) reported using some form of contraception to plan or prevent pregnancy. Interestingly (and unlike the situation just a few years ago), China did not have the highest rate of contraceptive use of any nation; that was the Czech Republic, at 95%, followed by Norway, Malta, Belgium, and then China. Canada's level of contraceptive use in 2014 was 80%. At the other end of the spectrum, 39 nations had rates of 10% or lower; most of these are in Africa, and most are among the poorest and least-developed nations in the world.[6]

These data clearly demonstrate that in societies where women have little power, substantial numbers of pregnancies are unintended. Demographer Leontine Alkema and colleagues estimate that as of 2010, the unmet need for worldwide contraception was over 12%, down from 15% in 1990; but this still amounts to almost 150 million women in marriages or civil unions with an unmet need for contraception.[7] Studies show that when women are free to decide whether and when to have children, fertility rates have fallen and the resulting children are better cared for, healthier, and better educated. Still, in many societies,

by tradition, men restrict women's decision-making abilities, including decisions about how many children they will bear.

The gap between the power held by men and by women is just as obvious at the highest levels of government. Worldwide, as of 2015, only 22% of elected government officials in national legislatures are women. Canada (in forty-eighth place, at about 25%) is slightly above the world average and on par with most of Europe, but lags behind the Scandinavian countries (e.g., Sweden at 44%) and many developing nations (e.g., Rwanda at 64% and Bolivia at 53%) in the proportion of women in positions of power in government.[8]

Many nations have policies to control population growth

Data show that funding and policies that encourage family planning have been effective in lowering population growth rates in all types of nations, even those that are least industrialized. No nation has pursued a population control program as extreme as China's, but other rapidly growing nations have implemented population-restrictive programs.

When policymakers in India introduced the idea of forced sterilization as a means of population control in the 1970s, the resulting outcry brought down the government. Since then, India's efforts have been more modest and far less coercive, focusing on family planning and reproductive health care. A number of Indian states also run programs of incentives and disincentives promoting a "two-child norm," and current debate centres on whether this is a just and effective approach. Regardless, unless India strengthens its efforts to slow population growth, it seems set to overtake China and become the world's most populous nation by about the year 2030.

The government of Thailand relies on an education-based approach to family planning that has reduced birth rates and slowed population growth. In the 1960s, Thailand's growth rate was 2.3%, but today (as of 2014) it stands at 0.32%. This decline was achieved without a one-child policy. It has resulted, in large part, from government-sponsored programs devoted to family-planning education and increased availability of contraceptives. Brazil, Mexico, Iran, Cuba, and many other developing countries have instituted active programs to reduce their population growth. These programs entail setting targets and providing incentives, education, contraception, and reproductive health care.

In contrast, some nations have instituted programs to encourage greater fertility and increase birth rates (**FIGURE 6.18**). This highlights the different attitudes toward population and population growth and their role in the economies and societies of nations. In general, the

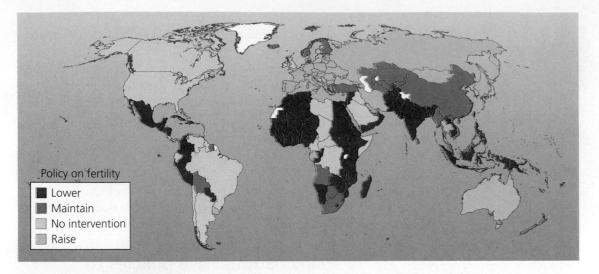

FIGURE 6.18   This map shows countries that have instituted programs or policies at the national level to either promote or reduce population growth.

Source: From Department of Economic and Social Affairs, Population Division (2013) World Fertility Policies 2013. www.unpopulation.org. Copyright © 2013 United Nations. Reprinted with the permission of the United Nations.

United Nations Population Division has turned away from older notions of command-and-control population policy intended to lower population to pre-set targets. Instead, it urges governments to offer better education and health care, and to address social needs that bear indirectly on population (such as alleviating poverty and disease, and improving the access of women to education and reproductive health care).

Poverty is correlated with rapid population growth

In general, poorer societies tend to show higher population growth rates than do wealthier societies. This pattern is consistent with demographic transition theory. **FIGURE 6.19** shows that poorer nations (lower per capita GDP) tend to have higher fertility rates, as well

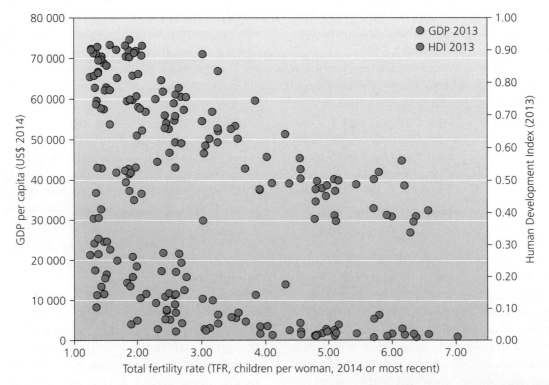

FIGURE 6.19   Total fertility rate has typically been higher in poorer countries, and lower in countries with higher income and higher Human Development Index, a composite indicator of life expectancy, education and income used by the United Nations Development Programme to rank countries on overall human well-being. This plot shows number of women per child plotted against GDP per capita and HDI for countries with populations over 5 million.
Source: Based on United Nations Development Programme Human Development Reports. 2014 Human Development Statistical Tables, http://hdr.undp.org/en/data.

as lower results on the United Nations Development Programme's Human Development Index. They also tend to have higher overall growth rates, higher birth and infant mortality rates, and lower rates of contraceptive use.

Such trends as these have affected the distribution of people on the planet. In 1950, 7.7% of all people lived in the least developed nations. By 2010, more than 12% of the world's population was living in the poorest countries of the world. Fully 99% of the next billion people to be added to the global population will be born in poor, less-developed regions of the world, the highest proportion of which is in Africa. The United Nations estimates that there is a 90% probability that 1 billion people will be added to the current population of 1.1 billion in Africa by 2045. By comparison, Asia—even with population growth driven by the already-large populations of China and India—is only anticipated to increase by a half-billion by then.[9]

It is unfortunate from a social standpoint that so many people will be added to countries that are poorly equipped to provide for them. It is also unfortunate from an environmental standpoint, because poverty often results in environmental degradation. People dependent on agriculture in an area of poor farmland, for instance, may need to farm just to stay alive, even if doing so degrades the soil and is not sustainable. This is one reason that Africa's once-productive Sahel region, like many regions of western China, is turning to desert (**FIGURE 6.20**). Poverty also drives the hunting of many large mammals in Africa's forests, including the great apes that are now disappearing as local settlers and miners kill them for "bush meat."

FIGURE 6.20
In the semi-arid Sahel region of Africa, where population is increasing beyond the land's ability to handle it, dependence on grazing agriculture has led to environmental degradation.

Demographic change is linked to environmental, social, and economic factors

Poverty can lead people into environmentally destructive behaviour, but wealth can produce even more severe and far-reaching environmental impacts. The affluence that characterizes such societies as Canada, the United States, Japan, or the Netherlands is built on massive and unprecedented levels of resource consumption. Much of this chapter has dealt with numbers of people rather than with the amount of resources each member of the population consumes or the amount of waste each member produces. The environmental impact of human activities, however, depends not only on the number of people involved but also on the way those people live; recall the A for "affluence" in the IPAT equation. Affluence and consumption are spread unevenly across the world, and affluent societies generally consume resources from other localities, as well as from their own.

Earlier in the book, we introduced the concept of the *ecological footprint*, the cumulative amount of Earth's surface area required to provide the raw materials a person or population consumes and to dispose of or recycle the waste that they produce. Individuals from affluent societies leave a considerably larger per capita ecological footprint. In this sense, the addition of 1 Canadian to the world has about as much environmental impact as the addition of 6 Chinese, or 12 Indians or Ethiopians, or 40 Somalians. This fact should remind us that the "population problem" does not lie entirely with the developing world.

Indeed, just as population is rising, so is consumption, and some environmental scientists have calculated that we are already living beyond the planet's means to support us sustainably. One recent analysis concluded that humanity's global ecological footprint surpassed Earth's capacity to support us in the early 1970s, and that our species is now living as much as 50% beyond its means.[10] The rising consumption that is accompanying the rapid industrialization of China, India, and other populous nations makes it all the more urgent for us to find a path to global sustainability.

The contrast between affluent and poorer societies in today's world (**FIGURE 6.21**) is the cause of social as well as environmental stress. According to the World Bank, about a billion people are currently living on less than $1.25 per day, the internationally defined limit for extreme poverty.[11] The richest one-fifth of the world's people earns more than 80 times the income of the poorest one-fifth (42% of world income), and controls almost 90% of the world's resources, including energy, food, and water resources. As the gap between rich and poor grows wider and as the sheer numbers of those living in poverty continue to

(a) A family living in North America

(b) A family living in India

FIGURE 6.21

A typical North American family **(a)** may own a large house, keep numerous material possessions, and have enough money to afford luxuries, such as vacation travel. A typical family in a developing nation such as India **(b)** may live in a smaller dwelling with few material possessions and little money for luxuries.

increase, it seems reasonable to predict increasing tensions between the "haves" and the "have-nots."

The rising material wealth and falling fertility rates of many industrialized nations today are slowing population growth in accordance with the demographic transition model. Some other nations, however, are not following this script. This is especially the case in countries where the HIV/AIDS epidemic has taken hold. African nations are being hit hardest. Of the 35 million people in the world infected with HIV/AIDS as of 2014, two-thirds live in the nations of sub-Saharan Africa. One in every 20 adults in sub-Saharan Africa is infected with HIV, and for southern African nations, the figure is more than one in five.

The AIDS pandemic is having the greatest impact on human population of any disease since the Black Death killed roughly one of three people in fourteenth-century

Europe, and since smallpox was brought by Europeans to the New World. AIDS took roughly 3000 lives in Africa every day in 2013, but this is down from just a few years earlier.[12] The pandemic is unleashing a variety of demographic changes, including generations of children in Africa who are being raised by their grandparents. Infant mortality in sub-Saharan Africa has risen to 15 times the rate in the developed world. The high numbers of infant deaths and premature deaths of young adults have caused life expectancy in parts of Africa to fall to levels not experienced since the 1950s.

Everywhere in sub-Saharan Africa, AIDS is undermining the ability of developing countries to make the transition to modern technologies, because it is removing many of the youngest and most productive members of society. The loss of productive household members causes families and communities to break down as income and food production decline. Valuable environmental and farming knowledge is being lost as an entire generation of Africans is decimated.

These problems are hitting many countries at a time when their governments are already experiencing what has been called *demographic fatigue*. Demographically fatigued governments face overwhelming challenges related to population growth, including educating and finding jobs for their swelling ranks of young people. With the added stress of problems like HIV/AIDS and Ebola, these governments face so many demands that they are stretched beyond their capabilities to address problems. As a result, the problems grow worse, and citizens lose faith in their governments' abilities to help them.

No region of the world has completely conquered AIDS, but the turnaround point may be in sight. Improved public health efforts (including sex education, contraceptives, and policies to address intravenous drug abuse) across the world have slowed HIV transmission rates, and improved medical treatments are lengthening the lives of people who are infected. The World Health Organization and UNAIDS have set a "90-90-90" target for 2020, in which 90% of people living with HIV would know their HIV status, 90% of those who know their status would be able to access HIV treatment, and 90% of people on treatment would achieve viral suppression.

The Sustainable Development Goals address many of these issues

In 2000, world leaders came together to adopt the *Millennium Declaration*, which set out a framework of basic goals for humanity over the next decade and a half. The *Millennium Development Goals* (MDG) set a target date of 2015 to achieve many of the fundamental goals for sustainable development that aid organizations have worked so hard to achieve over the past few decades. The original eight MDGs were to[13]

- eradicate extreme poverty and hunger
- achieve universal primary education
- promote gender equality and empower women
- reduce child mortality
- improve maternal health
- combat HIV/AIDS, malaria, and other diseases
- ensure environmental sustainability
- develop a global partnership for development

Perhaps tellingly, population control was *not* one of the Millennium Development Goals. However, the interconnections we have discussed in this chapter should make it clear that in order to achieve the *other* goals, both population growth and resource consumption levels will need to be addressed.

In the MDG target year of 2015, the global community again convened to set new goals for 2030. The new goals for the Post-2015 Development Agenda are being called *Sustainable Development Goals* (SDGs). These have been greatly expanded from the original eight MDGs, and much learning has been incorporated in the design of the new goals for 2030. We will examine the Sustainable Development Goals more closely in Chapter 22, *Strategies for Sustainability*.

If humanity's overarching goal is to generate a high standard of living and quality of life for all the world's people, then developing nations must find ways to reduce their population growth. However, those of us living in the industrialized world must also be willing to reduce our consumption; otherwise, the goal of achieving environmental sustainability will elude us. Earth does not hold enough resources to sustain all 7 billion of us at the current North American standard of living, nor can we go out and find extra planets. We must make the best of the one place that supports us all.

Conclusion

Today's human population is larger than at any time in the past. Our growing population, as well as our growing consumption, affects the environment and our ability to meet the needs of all the world's people. Approximately 90% of children born today are likely to live their lives in conditions far less healthy and prosperous than most of us in the industrialized world are accustomed to.

However, there are at least two major reasons to be encouraged. First, although global population is still rising, the *rate* of growth has decreased nearly everywhere. Most developed nations have passed through the demographic transition, showing that it is possible to stabilize population while still lowering death rates and creating more-prosperous societies. Second, women are slowly being treated more equitably, receiving better education, obtaining more economic independence, and gaining more ability to control their reproductive decisions. Aside from the clear ethical progress these developments entail, they are helping to slow population growth; where aggressive programs to control population growth have failed in many countries, educating girls and giving reproductive rights to women are succeeding.

Human population cannot continue to rise forever. The question, however, is how it will stop rising: through the gentle and benign process of the demographic transition; through restrictive governmental intervention, such as China's one-child policy; or through Malthusian checks of social conflict caused by overcrowding and competition for scarce resources. Moreover, sustainability demands a further challenge—that we stabilize our population size in time to avoid destroying the natural systems that support our economies and societies.

We are indeed a special species. We are the only one to come to such dominance as to fundamentally change so much of Earth's landscape and even its climate system. We are also the only species with the intelligence needed to turn around an increase in our own numbers before we destroy the very systems on which we depend.

REVIEWING OBJECTIVES

You should now be able to

Assess the scope and historical patterns of human population growth, and evaluate how human population, affluence, and technology affect the environment

- The current global population of 7.3 billion people adds about 80 million people per year (2.5 people every second). The world's population growth rate peaked at 2.1% in the 1960s and now stands at just over 1%. Growth rates vary among regions of the world.
- Attitudes toward the "population problem" have changed over time. The Malthusian perspective holds that population is a problem to the extent that it depletes resources, intensifies pollution, stresses social systems, or degrades ecosystems, such that the natural environment or our quality of life declines.

- The IPAT model summarizes how environmental impact (I) results from interactions among population size (P), affluence (A), and technology (T). Rising population and rising affluence (leading to greater consumption) each increase environmental impact. Technological advances have frequently exacerbated environmental degradation, but they can also help mitigate our impact.

- Four major societal transitions have fundamentally altered the way the human population interacts with the environment. They are the Paleolithic use of tools and control of fire; Neolithic development of agriculture; Industrial Revolution switch to fossil fuels and mechanization; and the modern Medical-Technological and Green Revolutions.

- Human overuse of primary productivity diminishes resources available for other species; alters habitats, communities, and ecosystems; and threatens our future ability to derive ecosystem services.

Explain and apply the fundamental concepts of demography and the demographic transition

- Demography applies principles of population ecology to the statistical study of human populations. Demographers study size, density, distribution, age structure, and sex ratios of populations, as well as rates of birth, death, immigration, and emigration.

- Total fertility rate (TFR) contributes greatly to change in a population's size and rate of growth.

- Birth rate, death rate, immigration, and emigration also contribute to increases or decreases in the size of a population.

- The demographic transition model explains why population growth has slowed in industrialized nations. The demographic transition may or may not be applicable to all nations, and different countries pass through the transition at different rates.

Describe how population growth is connected to environmental, social, and economic factors, and how sustainable development goals are seeking to address disparities

- When women are empowered and achieve equality with men, fertility rates fall, and children tend to be better cared for, healthier, and better educated.

- Many nations have put in place family planning and other programs and policies to control population.

- Poorer societies tend to have higher population growth rates than do wealthier societies.

- The high consumption rates of affluent societies may make their ecological impact greater than that of poorer nations with larger populations. Demographic change is linked to many other social and economic factors, including epidemics like HIV/AIDS, which has greatly influence population dynamics in places like sub-Saharan Africa.

- The UN Sustainable Development Goals do not explicitly mention population, but they do address many environmental, social, and economic factors that are related to demographic change.

TESTING YOUR COMPREHENSION

1. What is the approximate current human global population? How many people are being added to the population each day? How many have been added since you were born?

2. Why has the human population continued to grow in spite of environmental limitations?

3. Contrast the views of environmental scientists with those of libertarian writer Sheldon Richman over whether population growth is a problem. Why does Richman think the concept of carrying capacity does not apply to human populations?

4. Explain the IPAT model. How can technology either increase or decrease environmental impact? Provide at least two examples.

5. What characteristics and measures do demographers use to study human populations? Which of these help determine the impact of human population on the environment?

6. What is the total fertility rate (TFR)? Can you explain why the replacement fertility for humans is approximately 2.1? Why would it not be exactly 2.0? How is Europe's TFR affecting its natural rate of population change?

7. Why have fertility rates fallen in many countries, and where are they currently the highest?

8. In the demographic transition model, why is the pre-industrial stage characterized by high birth and death rates, and the industrial stage by falling birth and death rates?

9. How does the demographic transition model explain the increase in population growth rates in recent centuries? How does it explain the decrease in population growth rates in recent decades?

10. Why do poorer societies have higher population growth rates than wealthier societies? How does poverty affect the environment? How does affluence affect the environment?

THINKING IT THROUGH

1. Apply the IPAT model to the example of China provided in the chapter. How do population, affluence, technology, and ecological sensitivity affect China's environment? Now consider your own country, region, province, or territory. How do population, affluence, technology, and ecological sensitivity affect your environment? How can we regulate the relationship between population and its effects on the environment?

2. Do you think that all of today's developing nations will complete the demographic transition and come to enjoy a permanent state of low birth and death rates? Why or why not? What steps might we as a global society take to help ensure that they do? Now think about developed nations, including Canada. Do you think these nations will continue to lower their birth and death rates in a state of prosperity? What factors might affect whether they do so?

3. India's prime minister has put you in charge of that nation's population policy. India currently has a population growth rate of 1.46% per year, a TFR of about 2.5, a 55% rate of contraceptive use, and a population that is 68% rural. Their ranking in the Human Development Index is .586 ("medium"). What policy steps would you recommend, and why?

4. Now imagine that you have been tapped to design population policy for Germany. Germany is losing population, with an annual growth rate of −0.07% and has a TFR of 1.37, a 66% rate of contraceptive use, and a population that is 75% urban. Their ranking in the Human Development Index is .911 ("very high"). What policy steps would you recommend, and why?

INTERPRETING GRAPHS AND DATA

At right are graphed data representing the economic condition of the world's population data for 2006. The *y* axis indicates the per capita income for each country or region expressed as standardized purchasing power (termed *gross national income in purchasing power parity*, or *GNI PPP*). The *x* axis indicates the cumulative percentage of the world population whose per capita GNI PPP is equal to or greater than that country's or region's per capita GNI PPP. The horizontal dotted line indicates the global average per capita GNI PPP.

1. What percentage of the world population is at or below the global average per capita GNI PPP? What percentage is at or below one-half of the global average per capita GNI PPP? What percentage is at or above a level that is four times the global average?

2. Given a global average per capita GNI PPP of about $9000 and a world population of 7 billion people (at the time of this data set, 2006), what was the total global GNI PPP? What would the global GNI PPP be if everyone lived at the level of affluence of Canada?

3. How do you personally think we can resolve the conflict between the desirable goal of raising the standard of living of the billions of desperately poor people in the world, and the likelihood that increasing their affluence (A in the IPAT equation) will have a negative impact on the environment?

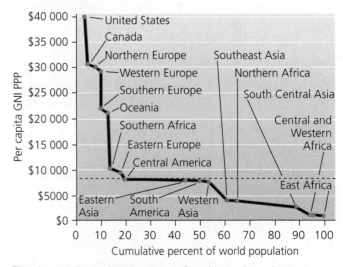

This diagram shows the percentage of world population living at various income levels, as of 2006.
Source: Data from Population Reference Bureau (2006). World Population Data Sheet 2006.

7 Soils and Soil Resources

This is part of the Mer Bleue provincial wetland near Ottawa, Ontario.

Cameron Proctor

Upon successfully completing this chapter, you will be able to

- Describe important properties of soil and soil-forming processes
- Characterize the role of soils in biogeochemical cycling and in supporting plant growth

- Identify the causes and predict the consequences of soil erosion and soil degradation
- Outline the history and explain the basic principles and methods of soil conservation

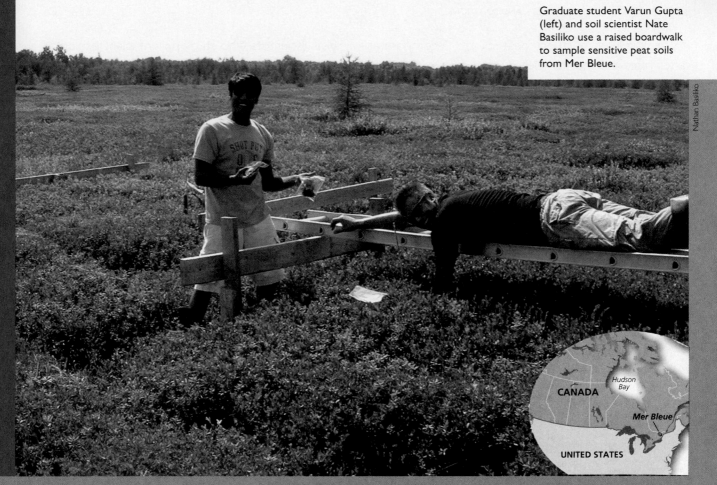

Graduate student Varun Gupta (left) and soil scientist Nate Basiliko use a raised boardwalk to sample sensitive peat soils from Mer Bleue.

Nathan Basiliko

CANADA

Hudson Bay

Mer Bleue

UNITED STATES

CENTRAL CASE
MER BLEUE: A BOG OF INTERNATIONAL SIGNIFICANCE

"This soil of ours, this precious heritage, what an unobtrusive existence it leads! To the rich soil let us give the credit due. The soil is the reservoir of life."

—J.A. TOOGOOD, CANADIAN SOIL SCIENTIST

"The nation that destroys its soil destroys itself."

—FORMER U.S. PRESIDENT FRANKLIN D. ROOSEVELT

The Mer Bleue Conservation Area is a 35-km² provincially protected wetland—technically, a type of wetland called a *bog*—situated just east of Ottawa, Ontario. It is located in an ancient, now-abandoned channel of the Ottawa River and hosts a number of plant species that are specially adapted to moist, boggy, acidic conditions, including *Sphagnum* moss, bog rosemary, blueberry, cottongrass, cattails, and tamarack. The area, classified as an open bog, has been recognized under the Ramsar Convention as a wetland of international importance.[1]

The Mer Bleue wetland provides an example of a specific type of soil—peat—that has been accumulating in many northern areas since the end of the last ice age. Canada has some of the most extensive peatlands in the world, covering 14% of our land area, and the peat deposits in some parts of the Mer Bleue Bog, which formed over the past 8000 years, are up to 6 m thick.

Northern peatlands are extremely important storage reservoirs for carbon and are thought to hold about one-third of all the carbon stored in soils. Through decomposition of organic matter, peat produces soil gases such as CO_2 and CH_4, which function as greenhouse gases in the atmosphere. Understanding the potential reaction of these very sensitive soils to climate change, particularly changes in water content and temperature, is of great interest to scientists. For example, if warming leads northern peat soils to decompose at faster rates and thus to release more soil gases, it could have a major impact on the concentration of these carbon-based gases in the atmosphere. This could set up a positive feedback loop and have a reinforcing influence on greenhouse warming.

The storage of carbon in peat depends on the balance between net primary production and decomposition. Plants store or sequester carbon as a result of photosynthesis, then contribute the stored carbon to the peat soil, where it accumulates in the form of plant litter. Temperature and light levels are of obvious importance to this balance, because the process of photosynthesis is involved. Moisture is another very important factor. When water levels are high, CH_4, the by-product of anaerobic (reduced) decomposition, is produced in large quantities; when conditions are drier, respiration tends to be aerobic (oxidized), producing less CH_4 but more CO_2.

The Peatland Carbon Study (PCARS) was initiated by a group of Canadian scientists in 1997; it is now continuing through a project called FluxNet-Canada. Researchers involved in the project, linked by an interest in ecosystem structure and function, have included soil scientists, microclimatologists, hydrologists, palynolo-gists (scientists who study pollen), plant ecologists, and graduate students from fields as varied as geochemistry, botany, physical geography, and microbial ecology.

The work is part of a larger network of research scientists who are interested in measuring and modelling the influence of climatic and seasonal changes on the carbon balance and other biogeochemical cycling in peatlands. Scientific activities at the Mer Bleue site include an instrument tower equipped for meteorological measurements, including energy balance, water vapour, and carbon dioxide and methane fluxes, combined with field investigations on plant growth and decomposition, hydrology, and plots for experimentation with factors such as drainage and excess nutrients. The experimental plots shown in the opening photo are being used to study the impacts of changes in nitrogen deposition on biogeochemical cycling.

On the basis of these ongoing measurements, now one of the longest-standing continuous sets of measurements of a northern peatland, scientists are developing a series of comprehensive ecosystem models for peatlands.[2] By studying this typical Canadian ecosystem, these scientists are contributing to our understanding of how soils may behave in a global context in response to climate change.

Soil as a System

We generally overlook the complexity of soil. In everyday language we tend to equate *soil* with *dirt*. **Soil**, however, is not merely loose material derived from rock; it is a complex plant-supporting system that consists of weathered rock, organic matter, water, gases, nutrients, and microorganisms (**FIGURE 7.1**). Soil is also fundamental to the support of life on this planet and the provision of food for the growing human population. As a resource it is renewable if managed carefully, but it is easily degraded and is currently at risk in many locations around the world.

Soil is a complex, dynamic mixture

Soil—very roughly—consists of half solid material (mostly mineral matter with varying proportions of organic matter), and the remainder is pore space filled by air, water, and other soil gases. The mineral particles in the soil are mostly inherited from the **parent material**, the base geological material in a given location from which the soil is formed. The parent material thus determines the starting composition of the soil.

The organic matter in soil includes living and dead microorganisms as well as decaying material derived from plants and animals. A single teaspoonful of soil can contain 100 million bacteria, 500 000 fungi, 100 000 algae, and 50 000 *protozoans* (simple eukaryotic microorganisms). Soil also provides habitat for earthworms, insects, mites, millipedes, centipedes, nematodes, sowbugs, and other invertebrates, as well as burrowing mammals, amphibians, and reptiles.

Water partially fills the open spaces, or *pore spaces*, between the mineral grains and particles of organic matter in soils. This is never "pure" water; it contains a variety of dissolved constituents, both minerals and organics, and it is variously referred to as the *soil solution*, *soil water*, or *soil moisture*. These solutions are very important for the

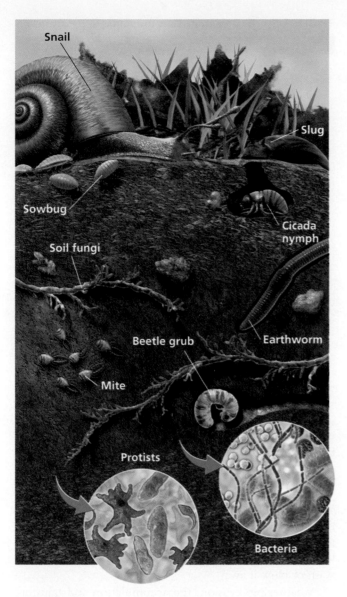

FIGURE 7.1
Soil is a complex mixture of organic and inorganic components and is full of living organisms whose actions help keep it fertile. Entire ecosystems exist in soil. Most soil organisms, from bacteria to fungi to insects to earthworms, decompose organic matter. Many, such as earthworms, also help aerate the soil.

Soil can have as much influence on a region's ecosystems as do the climate, latitude, and elevation. In fact, because soil is composed of living and nonliving components that interact with each other and with their surroundings in complex ways, soil itself meets the definition of an ecosystem. Together the mineral, organic, aqueous, and gaseous components of soil constitute a dynamic, ever-changing system that links the solid geosphere to the atmosphere, hydrosphere, and biosphere.

Soil formation is slow and complex

The formation of soil begins when the parent material is exposed to the effects of the atmosphere, hydrosphere, and biosphere. Parent material can be lava or volcanic ash; rock or sediment deposited by glaciers; wind-blown dunes; sediments deposited by rivers, in lakes, or in the ocean; or, perhaps most commonly, any type of **bedrock**, the continuous mass of solid rock that makes up Earth's crust.

The processes most responsible for soil formation are weathering, erosion, and the deposition and decomposition of organic matter. **Weathering** describes the physical, chemical, and biological processes that break down rocks and minerals, turning large particles into smaller particles and sometimes altering their composition (**FIGURE 7.2**). These small, loose particles of mineral matter—collectively called *regolith*—are the precursors of soils.

Physical or mechanical weathering breaks rocks down without triggering a chemical change in the parent material. Temperature, wind, rain, and ice are the main agents of physical weathering. Daily and seasonal temperature

support of plant growth because they dissolve and mobilize soil constituents that plants require as nutrients.

Similarly, the "air" that partially fills soil pore spaces is not exactly the same as the air that we normally breathe. Like the atmosphere, *soil gas* contains oxygen and nitrogen, as well as carbon dioxide, methane, and other gases that reflect its chemical equilibrium with the liquid and solid constituents of the soil, including soil organisms. Soils also contain gases that are released from the underlying rock, such as radon, and gases that infiltrate from above, such as volatile constituents derived from spilled gas and oil.

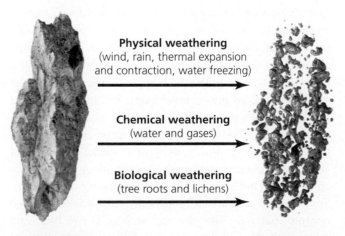

Parent material
(rock)

Physical weathering
(wind, rain, thermal expansion and contraction, water freezing)

Chemical weathering
(water and gases)

Biological weathering
(tree roots and lichens)

Smaller particles
of parent material

FIGURE 7.2
The weathering of parent material is the first step in soil formation. Rock is broken down into finer particles and modified by physical, chemical, and biological processes.

variation aids their action by causing the thermal expansion and contraction of parent material; areas with extreme temperature fluctuations experience rapid rates of physical weathering. Flowing water, wind, and glacial ice move rock particles that scrape and abrade other rock surfaces, causing physical weathering. Water freezing and expanding in cracks in rock is another common cause of physical weathering.

Chemical weathering results when water or other substances chemically interact with parent material. If you have ever visited an old cemetery and noticed that some of the headstones seem worn and smooth, this is likely the influence of chemical weathering, slowly dissolving and carrying away the mineral constituents of the headstones. Notice that the headstones made of limestone are the ones that most clearly demonstrate chemical weathering; limestone dissolves easily in normal, slightly acidic precipitation, whereas granite (the other common material used for headstones) is resistant to chemical weathering. Conditions where precipitation or ground water is unusually acidic promote chemical weathering, as do warm, wet conditions.

Biological weathering occurs when living things break down parent material by physical or chemical means. For instance, lichens initiate primary terrestrial succession by producing acid, which chemically weathers rock. A tree may accelerate weathering through the physical action of its roots as they grow into fissures in rock. It may also accelerate weathering chemically through the decomposition of its leaves and branches or with chemicals it releases from its roots.

Biological activity further contributes to soil formation through the deposition, decomposition, and accumulation of organic matter. As plants, animals, and microbes die or deposit waste, this material mixes with minerals from the underlying weathered parent rock. The deciduous trees of temperate forests, for example, drop their leaves each fall, making leaf litter available to the detritivores and decomposers that break it down and incorporate its nutrients into the soil. In decomposition, complex organic molecules are broken down into simpler ones, including those that plants can take up through their roots.

Partial decomposition of organic matter creates **humus**, a dark, spongy, crumbly mass of material made up of complex organic compounds. Soils with a high humus content hold moisture well and are productive for plant life. Soils that are dominated by partially decayed, compressed organic material—like the soil at Mer Bleue Bog—are called **peat**. Peat is characteristic of (though not exclusive to) northern climates because cool temperatures and saturation by surface water slow the decay process, allowing great thicknesses of organic material to accumulate.

Weathering produces fine particles and is the first step in soil formation. Another process often involved is

TABLE 7.1   **Five Factors That Influence Soil Formation**

Factor	Effects
Climate	Soil forms faster in warm, wet climates. Heat speeds chemical reactions and accelerates weathering, decomposition, and biological growth. Moisture is required for many biological processes and can speed weathering.
Organisms	Earthworms and other burrowing animals mix and aerate soil, add organic matter, and facilitate microbial decomposition. Plants add organic matter and affect a soil's composition and structure.
Topography	Hills and valleys affect exposure to sun, wind, and water, and they influence where and how soil and water move. Steeper slopes result in more runoff and erosion and in less leaching, slower accumulation of organic matter, and less differentiation of soil layers.
Parent material	Chemical and physical attributes of the parent material influence properties of the resulting soil.
Time	Soil formation takes decades, centuries, or millennia. The four factors above change over time, so the soil we see today may be the result of multiple sets of factors.

erosion, the movement of particles from one location to another. When soil or regolith is transported by wind, water, or ice and then deposited somewhere else, it is generally referred to as **sediment**. The transport process itself can promote physical weathering as the transported particles collide and scrape against one another. Erosion is particularly prevalent when soil is denuded of vegetation, leaving the surface exposed to water and wind that may wash or blow it away.

Weathering, erosion, the accumulation and transformation of organic matter, and the other processes that contribute to soil formation are all influenced by environmental factors. Soil scientists cite five primary factors that influence the formation of soil (**TABLE 7.1**).

A soil profile consists of layers known as horizons

Once weathering has produced an abundance of small mineral particles, wind, water, and organisms begin to move and sort them, and water moves through, picking up and transporting soluble materials. Eventually, distinct layers develop. Each layer of soil is known as a **horizon**, and the cross-section as a whole, from surface to bedrock, is known as a **soil profile**.

Minerals are carried downward through the developing soil profile as a result of **leaching**, a process in which materials suspended or dissolved in liquid are transported through the subsurface. Soil that undergoes leaching is a

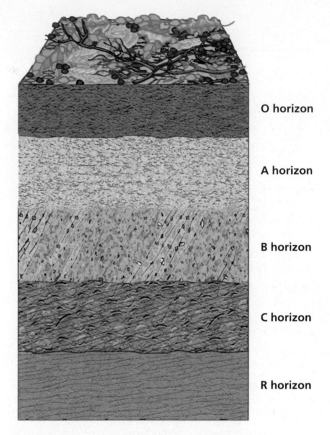

FIGURE 7.3
Mature soil consists of layers, or horizons, with different compositions and characteristics. The number and depth of horizons vary from place to place. The O horizon is mainly organic matter. The A horizon (or topsoil), is the uppermost mineral horizon, with some organic material. Material leached from the A horizon is deposited in the B horizon. The C horizon consists largely of weathered material that is still identifiable as parent rock, overlying an R horizon of pure parent material.

O horizon

A horizon

B horizon

C horizon

R horizon

bit like coffee grinds in a drip filter. When it rains, water infiltrates the soil (just as it infiltrates coffee grounds), dissolves some of its components, and carries them downward into the deeper horizons. At depth in the soil profile, some of this material may be deposited in zones of *accumulation*. Minerals that are commonly leached are those that are the most *soluble* (easily dissolvable), including various iron, aluminum, and silicate minerals. In some soils, minerals may be leached so rapidly that plants are deprived of nutrients. Minerals that leach rapidly from soils may be carried into ground water, where they can affect water quality.

Soil scientists subdivide horizons according to their characteristics and the processes that take place within them. For our purposes we will discuss five major horizons, known as the O, A, B, C, and R horizons (**FIGURE 7.3**). Soils from different locations vary, and few soil profiles contain all of these horizons, but any given soil contains at least some of them. Generally, the degree of weath-

ering and the concentration of organic matter decrease downward in the soil profile.

Peat deposits are classified as **O horizons**. Surface deposits of leaves, branches, mosses, animal waste, collectively termed **litter**, form another type of organic horizon common in upland forests. These surface deposits (including *L*, *F*, and *H* horizons) are classified and subdivided on the basis of the degree of decomposition of the organic material.

Just below the litter layer lies the **A horizon**, the uppermost mineral horizon, which consists of inorganic mineral components with organic matter and humus from above mixed in. The A horizon is often referred to as **topsoil**, that portion of the soil that is most nutritive for plants and therefore most vital to ecosystems and agriculture. Topsoil takes its loose texture and dark colour from its humus content. The A horizon is home to most of the countless organisms that give life to soil.

Minerals and organic matter that are leached from the topsoil move down into the **B horizon**, or *subsoil*, where they accumulate. If leached minerals are deposited in the B horizon, they can lead to the development of hard, mineral-rich layers that are variously called *hardpan*, *claypan*, *duripan*, or *caliche*, depending on their specific composition and structure. These hard layers can cause problems for plant growth because they interfere with drainage and prevent plant roots from penetrating to lower, nutrient-rich layers of the soil.

The **C horizon**, if present, is a transition zone located below the B horizon. It consists of broken-up parent material only slightly altered by the processes of soil formation, and contains rock particles that are larger and less weathered than the layers above. The C horizon may sit directly above an **R horizon** of unaltered parent material (*R* for *rock*). Some soils also are characterized by the presence of a distinct layer of water, called a *W horizon*, and some arctic soils contain a perennially frozen layer called **permafrost**.

Soils vary in colour, texture, structure, and pH

The horizons presented above depict an idealized, "typical" soil, but soils display very great variety. Canadian soil scientists classify soils into 10 major groups, based largely on the processes through which the soils are formed (**TABLE 7.2**). Within these 10 *soil orders*, there are dozens of "great groups," hundreds of "subgroups," and thousands of soils belonging to lower categories, all arranged in a hierarchical or taxonomic system. Scientists classify soils into these various categories using properties such as colour, texture, structure, and pH.

TABLE 7.2 **The Canadian System of Soil Classification**

Categories, or "Taxa," in the Canadian System of Soil Classification

Taxa	Principles used	No. of classes
Order	Dominant soil-forming process	10
Great group	Strength of soil-forming	31
Subgroup	Kind and arrangement of horizons	231
Family	Parent material characteristics	About 10 000
Series	Detailed features of the soil	About 100 000

Orders in the Canadian System of Soil Classification

Regosolic	Weakly developed soils, lacking a B horizon and typically forming in recent deposits
Brunisolic	Poorly developed soils (i.e., lacking in horizon development, sometimes only lightly weathered, but slightly more developed than regosols, see above) that typically form under boreal forests
Chernozemic	Well-drained to imperfectly drained soils with surface horizons darkened by the accumulation of organic matter from the decomposition of grasses; typical of the Interior Plains of Western Canada
Luvisolic	Soils with light-coloured eluvial (E, or leached) horizons and B horizons in which clay has accumulated; characteristic of well-drained to imperfectly drained sites; in sandy loam to clay, base-saturated parent materials under forest vegetation in subhumid to humid and mild to very cold climates; from the southern extremity of Ontario to the zone of permafrost and from Newfoundland to British Columbia
Vertisolic	Unstable soils with high clay contents, characterized by shrinking–swelling or wetting–drying cycles that either disrupt or inhibit the formation of soil horizons
Solonetzic	Soils with a B horizon that is very hard when dry and swells to a sticky mass of very low permeability when wet; occur on saline parent materials in some areas of the Interior Plains in association with Chernozemic soils, mostly associated with a vegetative cover of grasses
Podzolic	Soils with a B horizon in which the dominant accumulation product is amorphous material composed mainly of humified organic matter combined in varying degrees with Al and Fe; typically form in coarse soils under forest or heath vegetation in cool to very cold climates
Gleysolic	Soils that are mottled (i.e., patchy) in colour, as a result of intermittent or continuous saturation with water and reducing (i.e., nonoxygenated) conditions; saturation may result from either a high groundwater table or temporary accumulation of water, or both
Organic	Soils that are composed largely of organic materials, including soils commonly known as peat, muck, or bog and fen soils; commonly saturated with water for prolonged periods
Cryosolic	Soils that form in either mineral or organic materials that have permafrost within 1 m of the surface; occupy much of the northern third of Canada

Source: Based on Soil Classification Working Group (1998). *The Canadian System of Soil Classification,* 3rd ed. Agriculture and Agri-Food Canada Publication 1646, 187 pp. http://sis.agr.gc.ca/cansis/taxa/cssc3/index.html. Accessed 27 November 2014.

Soil colour The colour of soil (**FIGURE 7.4**) can indicate its chemical composition and sometimes its organic content. For example, the famously red colour of soils on Prince Edward Island is a result of the high iron content of the soil. Black or dark brown soils are usually rich in organic matter, whereas a pale grey to white colour often indicates a chalky composition, calcium carbonate deposits, salt deposits, or leaching. Colour variation occurs among soil horizons in any given location, and also among soils from different geographic locations. Long before modern analytical tests of soil content were developed, the colour of topsoil provided farmers and ranchers with information about a region's potential to support crops and provide forage for livestock.

Soil texture **Soil texture** is determined by the size of particles and is the basis on which soils are assigned to one of three general categories (**FIGURE 7.5**). **Clay** consists of particles less than 0.002 mm in diameter, **silt** of particles 0.002 to 0.05 mm, and **sand** of particles 0.05 to 2 mm. Sand grains, as any beachgoer knows, are large enough to see individually and do not adhere to one another. Clay particles, in contrast, readily adhere to one another and give clay a sticky feeling when moist. Soil with a relatively even mixture of the three particle sizes is known as **loam**.

For a farmer, soil texture determines a soil's *workability*, the relative ease or difficulty of cultivation. Texture also greatly influences the soil's **porosity**, a measure of the volume of spaces between particles in the soil, as well as its

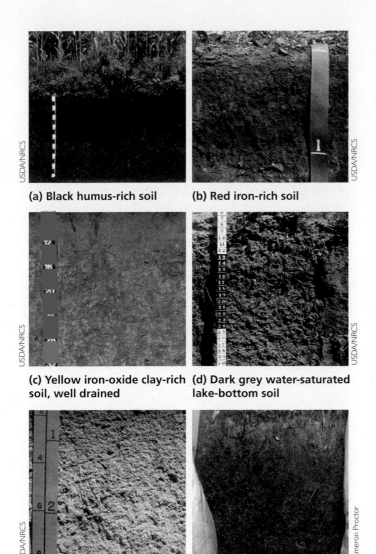

(a) Black humus-rich soil

(b) Red iron-rich soil

(c) Yellow iron-oxide clay-rich soil, well drained

(d) Dark grey water-saturated lake-bottom soil

(e) White calcium-carbonate-rich soil

(f) Dark brown peat soil

FIGURE 7.4
The colour of soil can vary dramatically, influenced by composition and other factors. Soils high in organic matter and humus tend to be dark brown or black **(a)**. Iron-rich soil **(b)** is typically bright red. Clay-enriched soils with iron oxides **(c)** can range from greenish-grey to buff (yellow)-coloured, as seen here. Waterlogged soils like this fine, clay-rich soil deposited on a glacial lake bottom **(d)** tend to be dark in colour. White horizons **(e)** typically indicate the presence of calcium carbonate. The dark brown peat soil of Mer Bleue Bog **(f)** is wet and loaded with partially decomposed organic matter.

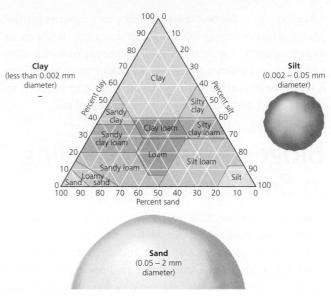

FIGURE 7.5
Scientists classify soil texture according to the relative proportions of sand, silt, and clay. After measuring the percentage of each type of particle size in a soil sample, a scientist can trace the appropriate white lines extending inward from each side of the triangular graph to determine the soil texture.

 DATA Q What type of soil contains 20% clay, 60% silt, and 20% sand?

and shaped like flat plates, so they can pack together very efficiently, with many but very tiny spaces in between. Conversely, soils with large, round particles tend to have larger spaces that are more interconnected, allowing water to pass through too quickly for plants to access it. Thus, crops planted in sandy soils require frequent irrigation. Silty soils with medium-sized pores, or loamy soils with mixtures of pore sizes, are generally best for plant growth and crop agriculture.

Soil structure   **Soil structure** is a measure of the organization or "clumpiness" of soil. Some degree of structure encourages soil productivity, and organic matter, biological activity, and clay help promote this structure. However, soil clumps that are too large can discourage plant roots from establishing if the soil particles are compacted too tightly together. Repeated tilling can compact soil and make it less able to absorb water. When farmers repeat-edly till the same field at the same depth, they may end up forming a crusty hardpan or *ploughpan* layer that resists the infiltration of water and the penetration of roots.

Soil pH   In addition to the compositional aspects of soils that we have already discussed, the degree of acidity or alkalinity of a soil also influences the soil's ability to support plant growth. Plants can die in soils that are too acidic (low pH) or too alkaline (high pH), but even a moderate

permeability, a measure of the interconnectedness of the spaces and the ease with which fluids can move through. In general, the finer the texture of a sediment or soil, the smaller the pores. The smaller the pores, the harder it is for water and air to travel through the soil, slowing infiltration and reducing the amount of oxygen available to soil biota.

It is possible for a material to have a fairly high porosity but low permeability. This is typical of soils with high clay content. Clay mineral particles are extremely fine variation can influence the availability of nutrients for the plants' roots. During leaching, for instance, acids from organic matter may remove some nutrients from the sites of exchange between plant roots and soil particles. Water leaches these nutrients, carrying them to deeper levels in the soil, and making them less available for plants.

Biogeochemical Cycling in Soil

You have learned that soil is a complex mixture of organic and inorganic, living and nonliving, solid, liquid, and gaseous components. Soil and regolith physically blanket Earth's surface, and act as an important interface for exchanges of material through biological, geological, chemical, atmospheric, and hydrologic processes.

Soils support plant growth through ion exchange

Materials that move through soils and plants gain many of their nutrients through a set of processes called **ion exchange** (**FIGURE 7.6**), in which positively charged particles (*cations*) and negatively charged particles (*anions*) are exchanged between the soil and the soil solution. Nutrients are held on *exchange sites*, along the positively or negatively charged edges of soil particles; in clay and humus particles, negatively charged sites prevail. Soil particle surfaces that are negatively charged hold on to positively charged nutrients, such as calcium (Ca^{2+}), magnesium (Mg^{2+}), potassium (K^+), and ammonium (NH_4^+). Cations and anions move from the exchange sites into the soil solution; from there they are taken up by

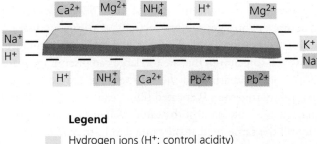

Legend

▢ Hydrogen ions (H^+; control acidity)

▢ Contaminants such as heavy metals (e.g., Pb^{2+})

▢ Plant nutrients (e.g., NH_4^+)

FIGURE 7.6
Negatively charged soil solids (mainly humus or clay, shown here in brown) in soils attract positively charged nutrients (cations), such as Ca^{2+}, NH_4^+, Mg^{2+}, and K^+, making them available when needed for plant growth. Heavy metals such as lead (Pb^{2+}) can be attracted to negatively charged soil particles, making it challenging to clean up contaminated soils.

roots and other soil organisms. Nutrients are resupplied to the exchange sites and the soil solution by weathering of minerals and the decomposition of organic matter.

Cation exchange capacity is a measure of a soil's ability to hold cations, preventing them from leaching away, thus making them available to plants. It is a useful indicator of *fertility*, the soil's ability to support plant growth. Soils with fine texture (lots of clay particles) and soils rich in organic matter have the greatest cation exchange capacity. As soil pH becomes lower (more acidic), cation exchange capacity diminishes, nutrients leach away, and soil instead may supply plants with harmful aluminum ions. This is one way in which acid precipitation can harm soils and plant communities.

Many pollutants are also positively charged, notably heavy metals such as cadmium, lead, arsenic, and mercury. These cations are attracted to the negatively charged clay and humus particles in soil, which can make it difficult to remediate soil that has been contaminated with heavy metals. It also can mean that heavy metal contaminants are held in soil instead of being released into aquatic ecosystems, which can be beneficial.

Soil is a crucial part of the nitrogen cycle

Soil is the locus for a crucial set of processes that are part of the global biogeochemical cycle of nitrogen (**FIGURE 7.7A**). Nitrogen gas (N_2) is chemically inert and must undergo a series of chemical changes in order to become biologically available to the organisms that need it. In its biologically active form, nitrogen is a powerful plant nutrient; when scarce, it is a limiting factor for plant growth.

Nitrogen fixation To become biologically available, inert nitrogen gas from the atmosphere must be "fixed"—that is, combined with hydrogen to form ions of ammonium (NH_4^+). This transformation, called **nitrogen fixation**, can be accomplished when air in the top layer of soil comes in contact with various specialized nitrogen-fixing bacteria. Some of these bacteria are free-living in the soil. Others live in a symbiotic, mutualistic relationship with plants, particularly *leguminous* plants, such as beans, peas, and clover, which host nitrogen-fixing bacteria in nodules attached to their roots (**FIGURE 7.7B**). Farmers routinely promote this process by planting leguminous crops to boost soil fertility.

Other natural processes that lead to nitrogen fixation are combustion (such as in forest fires) and lightning strikes. A significant amount of nitrogen is also fixed by the industrial production of nitrate fertilizers, using the *Haber-Bosch process*. However, the most important nitrogen-fixing process is still biologically mediated nitrogen fixation by

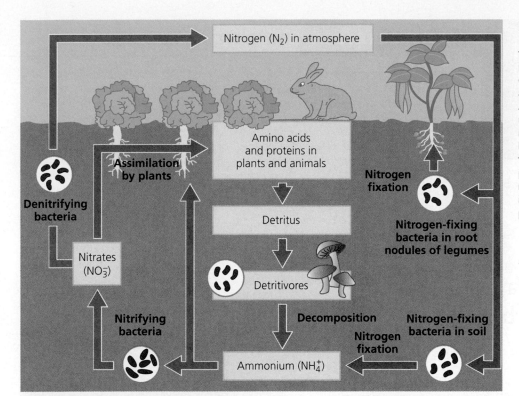

(a) Nitrogen cycle in soil

FIGURE 7.7
Many of the important processes in the global nitrogen cycle take place in soil **(a)**. Specialized bacteria in soils convert atmospheric nitrogen into ammonium through nitrogen fixation. Ammonium is transformed into nitrate, useable as a plant nutrient, by nitrification. Excess nitrate is converted back to the atmospheric gas N_2 through denitrification. All of these processes are biologically mediated by microorganisms that live in soils and in nodules among the roots of certain plants **(b)**, especially legumes.

(b) Nitrogen fixation

soil- and root-dwelling bacteria. These microorganisms thus play a crucial role in supporting life.

Nitrification Once atmospheric N_2 has been converted into ammonium, other specialized bacteria perform a process known as **nitrification**. In this process, ammonium ions are first converted into nitrite ions (NO_2^-), then (by yet another group of specialized microorganisms) into nitrate ions (NO_3^-). Plants can take up nitrate ions as nutrients, through their roots. The microorganisms that perform the task of nitrification are

chemoautotrophs, because they are producing their own food energy (*autotroph*) not through photosynthesis but through chemical reactions.

Animals, in turn, obtain the nitrogen they need by consuming plants or other animals. Decomposers obtain nitrogen from dead and decaying plant and animal matter, and from animal urine and feces (as shown in **FIGURE 7.7**). Once decomposers process the nitrogen-rich compounds they take in, they release ammonium ions, again making these available to nitrifying bacteria to convert into nitrates and nitrites for plant use.

THE SCIENCE ⟩ BEHIND THE STORY

Prof. Johannes Lehmann holds a bowl of wood chips (right), biomass that can be turned into biochar (left).
Cornell University Photography

Dark Earth: A New (Old) Way to Sequester Carbon

About 15 years ago, while doing field-work in the central Amazon of Brazil, Dr. Johannes Lehmann encountered patches of unusually dark, rich, fertile soil, locally called *terra preta*, or "black earth" in Portuguese. To Dr. Lehmann, a professor in the Department of Crop and Soil Sciences at Cornell University, these dark patches of soil stood out amongst the agriculturally degraded, nutrient-poor soils of the Amazon.

Further research on *terra preta* by Dr. Lehmann and other researchers has revealed that the black soil is a remnant of pre-Columbian Indigenous residents of the area, from as long ago as 450 BCE. It is probable that the charcoal-like *terra preta* was not made intentionally, but rather accumulated over time as a black, carbon-rich residue of cooking fires and pottery kilns. Indigenous farmers are thought to have used this residue as an addition to soils to increase their fertility. Hundreds and even thousands of years later, the black earth is still in place, an indication of its surprising stability.

Dr. Lehmann, who is particularly interested in nutrient cycling and carbon storage in soils, recognized in *terra preta* the potential for a carbon reservoir that would be significantly longer lasting than trees or even natural soils. He began to research methods for producing black earth. The most promising technology for this has turned out to be *pyrolysis*, which is a relatively straightforward, low-temperature heating process, whereby biomass (either plant or animal matter) is reduced to a black, carbon-rich, charcoal-like residue. This residue has been named *biochar*.

A photomicrograph of biochar (see photo) reveals that it is riddled with tiny, micron-sized pore spaces. The complexity of the spaces and surfaces partly explain the efficiency of biochar at holding onto plant nutrients and thus enhancing the fertility of soils. Biochar also appears to be very effective at absorbing and holding some pollutants, notably heavy metals, potentially contributing to improvements in water quality.

But the most exciting prospect of all, according to Professor Lehmann and other biochar reasearchers, is the possibility of using biochar to enhance the ability of soils to act as long-term reservoirs for carbon. Soils are already the most important terrestrial reservoir for carbon—even larger than forests and grasslands (see "Interpreting Graphs and Data" at the end of this chapter). If adding biochar could greatly increase the longevity and stability of the soil reservoir, thus sequestering carbon from the atmosphere, it could help in the fight against rising atmospheric carbon and global warming.

The success of biochar as a method of carbon sequestration depends on the technologies that are used. They

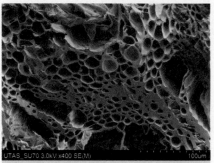

Biochar (seen here in a photomicrograph, magnified thousands of times) has many complex, micron-sized pore spaces and surfaces. This partly accounts for some of biochar's properties, such as the capacity to improve water and soil quality by absorbing pollutants and holding on to fertilizers and plant nutrients.

Denitrification   The next step in the nitrogen cycle is **denitrification**, which occurs when bacteria convert nitrates in soil or water into gaseous forms of nitrogen (either N_2, nitrous oxide, or nitric oxide). This, too, is a multistep process that is carried out by several varieties of microorganisms in anaerobic (nonoxygenated) conditions in soil. Denitrification thereby completes the cycle and balances the nitrogen-fixation process by releasing nitrogen back to the atmosphere.

Soil is an important terrestrial reservoir for carbon

In studies of the carbon cycle, much attention is justifiably paid to the role of forests and the atmospheric and oceanic carbon reservoirs. However, soils also play a crucial role in the global carbon cycle.

Soil represents the largest terrestrial reservoir for carbon—about four times as large as the entire reservoir in terrestrial biota, including forests (**FIGURE 7.8**). The rock reservoir (below a depth of 1 metre or so) is considerably larger; however, carbon sequestered in rock is generally not active in the short-term carbon cycle unless it has been remobilized as a result of human activities, mainly through the burning of fossil fuels and manufacture of cement from carbonate rocks. (See "The Science behind the Story: Dark Earth: A New [Old] Way to Sequester Carbon" for further discussion about carbon sequestration in soils.)

The main carbon fluxes in which soil is involved are driven by photosynthesis and the production of require energy inputs, which contribute to carbon dioxide emissions. The key is to balance the carbon emitted to the atmosphere as a result of the pyrolysis process with the carbon sequestered through the production of biochar. Dr. Lehmann's calculations show that with low-temperature pyrolysis of biomass it should be possible to halve the amount of carbon being released to the atmosphere, as compared to natural decomposition of the biomass.

Using pyrolysis to produce biochar also has an economic cost, which represents a potential barrier to its adoption as an approach for mitigating carbon emissions. One way to improve its economic competitiveness is to promote the use of biochar as an agricultural amendment for the improvement of soil fertility (see **FIGURE 1**). Another is to capture the heat released during the pyrolysis process and use it as a form of bioenergy. One of the most intriguing aspects of biochar is that the technologies are simple and adaptable even in remote settings. It is possible that someday soon farmers all over the world will be enhancing their soils with biochar while capturing and sequestering carbon at the same time.

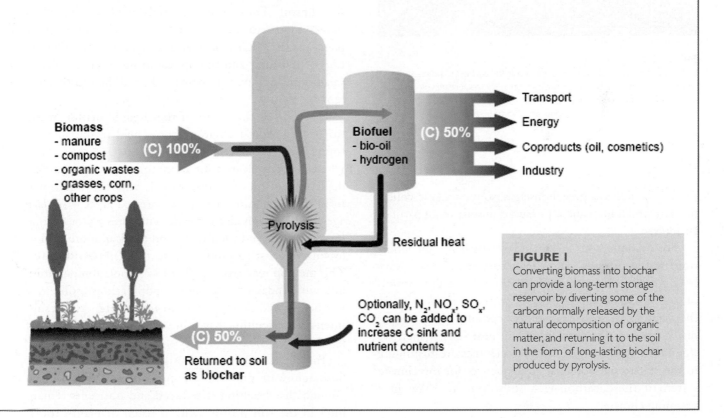

FIGURE 1
Converting biomass into biochar can provide a long-term storage reservoir by diverting some of the carbon normally released by the natural decomposition of organic matter, and returning it to the soil in the form of long-lasting biochar produced by pyrolysis.

organic matter, followed by respiration and decay of organic matter. Carbon fluxes from the atmosphere to soils therefore generally pass through photosynthetic plants, which contribute litter and other organic matter to soils, either directly or via consumption by animals. The subsequent alteration of litter, animal waste, and soil organic matter results in the production of humus and the accumulation of carbon, especially in organic-rich soils such as peat.

In addition to producing humus, the decay of soil organic matter produces soil gas that contains carbon species. These decay processes are complex, and their specific chemistry depends on the decomposing microorganisms that are involved and on the physical-chemical conditions within the soil. For example, the decay of organic matter in aerobic (oxygenated) conditions typically produces soil gas with a higher concentration of carbon dioxide (CO_2), whereas decay in anaerobic conditions tends to be slower and produces soil gas with a higher concentration of methane (CH_4). Wet soils are more likely to be anaerobic than dry soils, so the soil moisture conditions are also important to consider.

Carbon-bearing soil gases are released to the atmosphere (as shown in **FIGURE 7.8**) along with other soil gases, including the N_2 and nitrous oxides produced by denitrification, and radon produced by the radioactive decay of uranium in underlying bedrock. Both carbon dioxide and methane are greenhouse gases, so these fluxes are of significance in understanding the global climate system. In fact, methane is many times more effective than

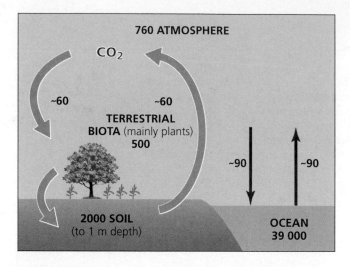

760 ATMOSPHERE

CO_2

~60   ~60

TERRESTRIAL BIOTA (mainly plants)
500

~90   ~90

2000 SOIL
(to 1 m depth)

OCEAN 39 000

FIGURE 7.8
Soil is the largest active terrestrial reservoir for carbon. (The rock reservoir is considerably larger, but carbon in the rock reservoir is not active in the short-term global carbon cycle unless it is remobilized by human activity.) Reservoir contents are shown in gigatonnes, and fluxes are shown in gigatonnes of carbon per year.

carbon dioxide as a greenhouse gas, so the specific conditions in which methane is produced in soils are of particular interest.

For example, soils in wetlands—which are (by definition) water-saturated for at least part of the year—vary both spatially and temporally in the relative proportions of carbon dioxide and methane gas they produce and in their ability to store and release carbon. Many perennially frozen soils (permafrost) are peat soils, which are significant reservoirs of carbon; scientists are beginning to investigate the potential consequences for the climate system if these carbon-rich soils were to thaw in a warming climate.

In short, most people recognize the importance of soils in supporting plant growth, particularly in the agricultural context (although this was not always the case—as you will see, in North America it took the disaster of the Dust Bowl in the 1930s for governments to take seriously the challenge of soil conservation). Given the extensive environmental impacts of modern human activities, it is increasingly important that we also strive to understand the crucial role of soils in the global biogeochemical cycling of carbon, nitrogen, phosphorus, sulphur, and other elements.

Regional differences affect soil fertility

The characteristics of soil and soil profiles and the health and fertility of the soil vary from place to place. One striking example is the difference between soils of tropical rain forests and those of temperate grasslands. Although rain forest ecosystems have high primary productivity, most of their nutrients are tied up in plant tissues and are not in the soil (**FIGURE 7.9**). The soil of Amazonian rain forest is much less productive than the soil of the grasslands in Saskatchewan.

To understand how this can be, consider the main differences between the two regions: temperature and rainfall. The enormous amount of rain that falls in the Amazon readily leaches minerals and nutrients out of the topsoil. Those not captured by plants are taken quickly down to the water table, out of reach of most plants' roots. High temperatures speed the decomposition of leaf litter and the uptake of nutrients by plants, so amounts of humus remain small, and the topsoil layer remains thin.

This means that most of the organic matter in the tropical rain forest environment is held in the aboveground plant biomass (as shown in **FIGURE 7.9**). Thus when forest is cleared for farming, cultivation quickly depletes the soil's fertility. This is why the traditional form of agriculture in tropical forested areas is *swidden* agriculture, in which the farmer cultivates a plot for one to a few years and then moves on to clear another plot, leaving the first to grow back to forest (**FIGURE 7.10**). This method may work well at low population densities, but with today's high human populations, soils may not be allowed enough time to regenerate. As a result, intensive agriculture has ruined the soils and forests of many tropical areas.

In temperate grassland areas such as the Saskatchewan prairies, in contrast, rainfall is low enough that leaching is reduced and nutrients remain high in the soil profile, within reach of plants' roots. Plants take up nutrients and then return them directly to the topsoil when they die; this cycle maintains the soil's fertility. In addition, cool temperatures slow the rate of decomposition of organic matter. Most of the organic matter in this system is stored below ground, in the soil (as shown in **FIGURE 7.9**), and the thick, rich topsoil of temperate grasslands can be farmed repeatedly with minimal loss of fertility if proper farming techniques are used.

However, even in the rich soils of temperate grasslands, growing and harvesting crops without returning adequate organic matter to the soil gradually depletes soil organic matter. Leaving soil exposed to the elements also increases the rate of erosion of topsoil. It is such consequences that farmers in many locations around the world have sought to forestall through the use of agricultural approaches to soil conservation.

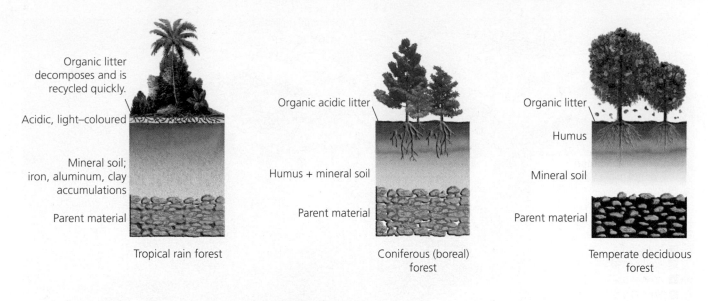

Tropical rain forest

- Organic litter decomposes and is recycled quickly.
- Acidic, light–coloured
- Mineral soil; iron, aluminum, clay accumulations
- Parent material

Coniferous (boreal) forest

- Organic acidic litter
- Humus + mineral soil
- Parent material

Temperate deciduous forest

- Organic litter
- Humus
- Mineral soil
- Parent material

Grassland

- Dark, alkaline soil very rich in humus
- Clay and calcium accumulations
- Parent material

Desert

- Thin humus–mineral mixture
- Dry mineral soil; accumulations of clay, carbonate, and soluble salts
- Parent material

FIGURE 7.9 Soil horizons and productivity vary regionally. In tropical forests, most of the organic content is held above the ground in plant biomass, rather than in soils, because decomposition proceeds rapidly in the warm, moist climate, and decomposed organic matter is carried away by runoff. In temperate and boreal forests, and especially in grasslands, a much greater percentage of the organic matter is stored in soils. In desert biomes there is little organic matter because of the relatively low biomass available.

Ron Gilling/Photolibrary/Getty Images

FIGURE 7.10

A traditional form of farming is swidden agriculture, seen here in Suriname. Forest is cleared, often by burning. The plot is farmed for a few years; the farmer then moves on to clear another plot, leaving the first to regenerate. Frequent movement is necessary because tropical soils are nutrient-poor. Burning the cut vegetation temporarily adds nutrients to the soil, which is why this practice is called "slash-and-burn" agriculture. At low population densities, this form of farming had little impact on forests, but at today's high population densities, it is a leading cause of deforestation.

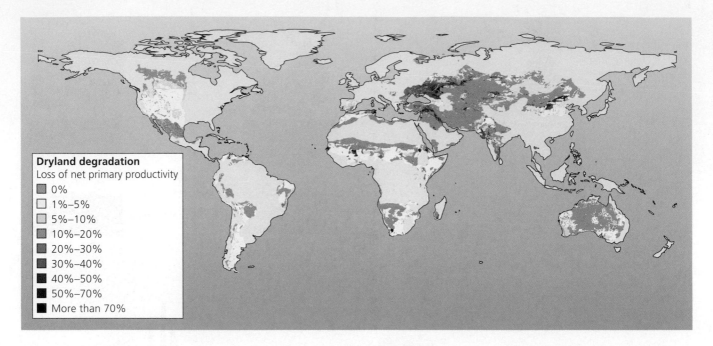

Dryland degradation
Loss of net primary productivity
- 0%
- 1%–5%
- 5%–10%
- 10%–20%
- 20%–30%
- 30%–40%
- 40%–50%
- 50%–70%
- More than 70%

FIGURE 7.11   Soils are becoming degraded in many areas worldwide. Degradation is increasing in many agricultural regions, especially in dryland areas in Africa and Asia.
Source: From *Global Environmental Outlook GEO-5*. http://www.unep.org/geo/geo5.asp. Accessed 27 November 2014. Copyright 2014 by United Nations Environment Programme. Reprinted by permission.

Soil Erosion and Degradation

Healthy soil is vital for agriculture, for the growth of forests, and for the biogeochemical functioning of Earth's natural systems, including the climate system. Productive soil is a renewable resource, but not on a human time scale. If we abuse it through careless or uninformed practices, we can greatly reduce its productivity. Like other renewable resources, if soil is degraded or depleted at a rate that is faster than the rate at which it can be renewed, it effectively becomes nonrenewable.

Throughout the world, especially in drier regions, soils have become eroded and degraded (**FIGURE 7.11**). **Soil degradation**, that is, damage to or loss of soil, worldwide has resulted from erosion caused by roughly equal parts forest removal, poorly managed cropland agriculture, and overgrazing of livestock, with a much smaller (but still significant) contribution from industrial contamination (**FIGURE 7.12A**). Additional causes of soil degradation include mining, construction, acid rain, and other sources of chemical contamination; we will consider all of these in subsequent chapters.

Scientists' studies of soil and the practical experience of farmers have shown that the most desirable soil for agriculture is a loamy mixture with a pH close to neutral, which

FIGURE 7.12
(a) Most of the world's soil degradation results from cropland agriculture, overgrazing by livestock, and deforestation.
(b) Various factors limit the agricultural productivity of soil.
Source: (a) Data based on information from United Nations Environment Programme (UNEP). (b) Data based on information from the United Nations Food and Agriculture Organization.

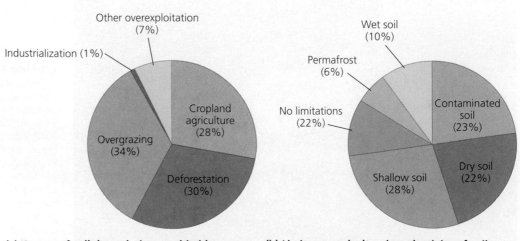

Other overexploitation (7%)
Industrialization (1%)
Overgrazing (34%)
Cropland agriculture (28%)
Deforestation (30%)

(a) Causes of soil degradation worldwide

Wet soil (10%)
Permafrost (6%)
No limitations (22%)
Contaminated soil (23%)
Shallow soil (28%)
Dry soil (22%)

(b) Limits on agricultural productivity of soils

is workable and capable of holding nutrients. Many soils deviate from this ideal and prevent land from being arable or limit the productivity of arable land (**FIGURE 7.12B**). Increasingly, limits to productivity are being set by human impact that has degraded many once-excellent soils.

Soil erosion can degrade ecosystems and agriculture

Erosion, as we have noted, is the removal of weathered material from one place and its transport to another by the action of wind, water, or glacial ice. **Deposition** of sediment is the other end of this process—the arrival of the eroded material at its new location.

Erosion and deposition are both natural processes, integral to the normal geological cycles of our planet. In the long run, they help create new soil. For example, flowing water can deposit eroded sediment in river valleys and deltas, producing rich and productive soils. This is why floodplains are excellent for farming and, in fact, hosted the dawn of agriculture in human history; it is also why flood control measures can decrease the productivity of agriculture in the long run.

Accelerated erosion is associated with human activity. It is a problem locally for ecosystems and agriculture, because it takes place much more quickly than soil is formed. Furthermore, erosion removes topsoil, the most valuable soil layer for living things. People have increased the vulnerability of fertile lands to erosion through three widespread practices:

1. overcultivating fields through poor planning or excessive ploughing, particularly when land is left bare of vegetative cover
2. overgrazing rangelands with more livestock than the land can support, resulting in reduced vegetative cover
3. clearing forested areas, especially on steep slopes or with large clear-cuts

Erosion can be gradual and hard to detect, but still devastating in the long run. For example, an erosion rate of 12 tonnes/ha/yr removes only a dime's thickness of soil per year. That doesn't seem like much until you realize that it would have taken anywhere between 75 and 1500 years to produce a layer of soil the thickness of a penny. In many parts of the world, scientists, farmers, and extension agents are measuring erosion rates in hopes of identifying areas in danger of serious degradation before they become too badly damaged.

Erosion happens by several mechanisms

Several types of erosion can occur, including wind erosion and four principal kinds of water erosion (**FIGURE 7.13**).

Glaciers can also be an agent of erosion; we will consider glacial processes in a later chapter.

Research indicates that rill erosion has the greatest potential to move topsoil, followed by sheet erosion and splash erosion, respectively. All types of water erosion—particularly gully erosion—are more likely to occur where slopes are steeper. In general, steeper slopes, greater precipitation intensities, and sparser vegetative cover all lead to greater water erosion.

The *Universal Soil Loss Equation (USLE)* was developed as a tool for estimating erosion losses by water from cultivated fields and to show how different soil and management factors influence soil erosion (**TABLE 7.3**). A revised version of the USLE is used in Canada to estimate losses from soil erosion, mainly for the purpose of soil conservation planning. It utilizes the same variables as the original format, but with refinements for quantifying their values in situations that are more common in Canada (for example, if frost is present, some of the factors will be applied differently).[3]

Wind, like water, is a moving fluid. Wind often flows very quickly over the surface, but it does not have the same ability to pick up and transport soil particles that water has. Nevertheless, wind can be a highly effective agent of erosion. Wind erosion, also called *aeolian erosion*, operates mainly by *deflation*, in which all loose, fine-grained material is picked up from the surface and carried

TABLE 7.3 Universal Soil Loss Equation

A = R K LS C P		
A	Predicted soil loss due to water erosion	Results in tonnes/ha/yr
R	Erosivity factor	Quantifies the erosive energy of rainfall and runoff; takes into account both total amount of rainfall and its intensity
K	Soil erodibility factor	Represents the ease with which a soil is eroded; based on the cohesiveness of a soil and its resistance to detachment and transport
LS	Slope length and steepness factor	Steeper slopes lead to higher flow velocities; longer plots accumulate runoff from larger areas and result in higher flow velocities
C	Vegetative cover and management factor	Considers the type and density of vegetative cover as well as all related management practices such as tillage, fertilization, and irrigation
P	Erosion control practices factor	Influence of conservation practices such as contour planting, strip cropping, grassed waterways, and terracing relative to the erosion potential of simple up–down slope cultivation

Source: Wischmeier, W.J. and D.S. Smith (1978). *Predicting Rainfall Erosion Loss—A Guide to Conservation Planning. Agricultural Handbook* no. 537, Washington, DC, USDA.

Greg Dale/Getty Images

(a) Splash erosion

David Cole/Alamy

(b) Sheet erosion

U.S. Department of Agriculture

(c) Rill erosion

U.S. Department of Agriculture

(d) Gully erosion

FIGURE 7.13
Splash erosion **(a)** occurs as raindrops strike the ground with enough force to dislodge small amounts of soil. *Sheet erosion* **(b)** results when thin layers of water traverse broad expanses of sloping land. *Rill erosion* **(c)** leaves small pathways along the surface where water has carried topsoil away. *Gully erosion* **(d)** cuts deep into soil, leaving large gullies that can expand as erosion proceeds.

away by wind. Another important mechanism of wind erosion is *abrasion*, whereby wind-transported particles become "projectiles," striking other rocks at the surface and causing them to break up.

The *Wind Erosion Prediction Equation* shows that wind erosion is a function of five factors and their interactions (**TABLE 7.4**).

Grasslands, forests, and any other healthy plant cover protect soil from both wind and water erosion. Vegetation breaks the wind and slows water flow, and plant roots hold soil in place and take up water. Removing plant cover will nearly always accelerate erosion.

Accelerated soil erosion is widespread

In today's world, humans are the primary cause of erosion, and we have accelerated it to unnaturally high rates. Geologist Bruce Wilkinson analyzed prehistoric erosion rates from the geological record and compared these with modern rates.[4] He concluded that humans are over 10 times more influential at moving soil than are all other natural processes on the surface of the planet combined.

The global status of soil erosion and degradation is extremely difficult to pin down, and suffers from a lack of

TABLE 7.4   Wind Erosion Prediction Equation

E = f (I C K L V)		
E	Predicted soil loss due to wind erosion	Results in tonnes/ha/yr
I	Soil erodibility factor	Represents the ease with which a soil is eroded by the abrasive action of wind-carried particles; dependent upon a number of factors including soil texture and aggregation
C	Local wind erosion climate factor	Quantified as the wind energy available to erode soil; takes into account moisture, wind speed, and wind direction. Wind erosion is most common in arid and semi-arid regions
K	Roughness factor	Describes the surface roughness of the soil; greater roughness indicates greater resistance to erosion
L	Length of field factor	A measure of the unsheltered length of the field; a longer field will have higher wind velocities and thus greater erosion potential
V	Vegetative cover factor	Accounts for cover type and density, including cover from crop residues; erosion risk is greatest when there is least cover

Source: Wischmeier, W.J. and D.S. Smith (1978). Predicting Rainfall Erosion Loss—A Guide to Conservation Planning. *Agricultural Handbook* no. 537, Washington, DC, USDA.

data on top of its inherent complexity as an environmental issue.[5] It is widely reported that more than 19 billion ha of the world's croplands suffer from erosion and other forms of soil degradation resulting from human activities. In the last half of the twentieth century, China lost as much arable farmland as exists in Denmark, France, Germany, and the Netherlands combined. In Kazakhstan, industrial cropland agriculture has caused tens of millions of hectares to be degraded by wind erosion. For Africa, soil degradation over the next 40 years could reduce crop yields by half. Couple these declines in soil quality and crop yields with the rapid population growth occurring in many of these areas, and we begin to see why some observers describe the future of agriculture as a crisis situation.

Erosion of agricultural soil has been a significant concern in Canada for the past 25 years or more, but improvements are occurring. Soil researchers from Agriculture and Agri-Food Canada in the late 1990s determined the on-farm cost of agricultural land degradation in Canada to be almost $670 million per year.[6] The risk of soil erosion on Canadian cropland steadily declined in subsequent years, and by 2006, 80% of cropland area was assessed to be in the very low-risk category. This reflected improvements in all areas of soil erosion risk, including wind and water erosion.

A combination of reduced tillage, less intensive crop production, and removal of *marginal land* (that is, land ill-suited to agriculture) from production is cited as

contributing to lower erosion rates in Canada in recent years.[7] Overall, changes in farming techniques since the recognition of a soil erosion crisis in the early 1980s have led to slow but steady improvements in the health and stability of Canada's soils.

Soils can also be degraded by chemical contamination

Accelerated erosion is not the only cause of soil degradation. Over-application of fertilizers can chemically damage soils. Even organic fertilizers such as manure, when applied in amounts needed to supply sufficient nitrogen for a crop, may introduce excess phosphorus that can run off into waterways. Inorganic (mineral-based, industrially produced) fertilizers are generally more susceptible than are organic fertilizers to leaching and runoff, and they are somewhat more likely to cause soil degradation and off-site impacts.

Nitrogen and phosphorous runoff from farms and other sources can lead to phytoplankton (algal) blooms, creating oxygen-depleted "dead zones" in river mouths, lakes, and marine coastal zones throughout the world. Moreover, nitrates readily leach through soil and contaminate ground water, and components of some nitrogen fertilizers can even volatilize (evaporate) into the air. Nitrate and phosphate buildup in soil and water systems also pose human health risks, including cancer and methemoglobinemia, or "blue-baby" syndrome, which reduces the oxygen-carrying capacity of the blood. Many negative side effects of fertilizer use can be reduced by proper timing and appropriate application methods, and by regular testing of the soil to determine whether, what type, and how much fertilizer is needed, thus matching the fertilizer delivery to the amount that can be utilized by crops.

Pesticides applied to agricultural fields are another important source of contamination of soil and adjacent waterways. These chemicals, designed to attack specific weeds, insect pests, and crop fungi, also affect nontarget species, including humans. Most insecticides, for example, are *neurotoxins*, so they affect the central nervous systems of organisms. Excess pesticide that is not absorbed by plants or animals can leave the system via runoff, infiltration into the soil, or volatilization. The amount, type, and timing of application are all crucially important in preventing pesticide-related contamination problems. For example, a pesticide that is water soluble and is applied just before a rainstorm will infiltrate quickly into the soil and be carried away by surface runoff, increasing the risk of contamination of soil and adjacent water bodies.

A growing problem worldwide is the contamination of soil as a result of industrial activity, particularly from inappropriate disposal or poorly designed storage of hazardous industrial wastes. An example occurred in Elmira, Ontario, in 1988 when a carcinogenic (cancer-causing) chemical, nitrosodimethylamine (NDMA), was found in the town's drinking water supply. The source of the chemical was traced to the Uniroyal Chemical Company, a producer of agrochemicals. The company had been disposing of hazardous chemical wastes in an aboveground storage pond (which met the legal requirements of Ontario's Ministry of Environment). The bottom of the pond was lined with clay, a common practice because of its capacity to block liquids and absorb contaminants. But when the bottom layers became saturated, the liquid waste began to seep through the clay layer and into the underlying soil. It percolated into the ground water, eventually contaminating several of the town's wells and nearby streams. Water for drinking and cooking had to be trucked in, and local residents were even advised not to bathe in water from the municipal wells.

The example of Elmira illustrates that soil and water contamination are so closely connected that they cannot be considered separately. We will return to these issues in greater detail when we look at freshwater contamination, waste management, and environmental health concerns.

Desertification damages formerly productive lands

Much of the world's population lives and farms in arid environments, where desertification is a concern. The terms *desertification* and *degradation* are often confused, and the United Nations is careful to distinguish them.[8]

Land degradation refers to a change in soil health that results in the reduction or loss of biological or economic productivity of land; the term "degradation" usually carries the implication that it is caused or exacerbated by human activities. **Desertification**, in principle, is a natural process by which formerly productive land becomes a desert as a result of climatic change or prolonged drought. *Accelerated desertification* is a type of land degradation that occurs in arid and semi-arid areas, brought on by human activities. This term generally is used to describe a loss of more than 10% productivity as a consequence of erosion, soil compaction, forest removal, overgrazing, drought, salinization, climate change, depletion of water sources, and other factors.

Severe desertification can result in the expansion of desert areas or creation of new ones in areas that once supported fertile land (**FIGURE 7.14A**). This process has occurred in many areas of the Middle East that have been inhabited, farmed, and grazed for long periods of time. To appreciate the cumulative impact of centuries of traditional agriculture, we need only look at the present desertified state of that portion of the Middle East where agriculture

originated, nicknamed the "Fertile Crescent." These arid lands—in present-day Iraq, Syria, Turkey, Lebanon, and Israel—are not so fertile anymore.

Arid and semi-arid lands are prone to desertification because their precipitation is too meagre to meet the greater demands in productivity from a growing human population. According to the United Nations Environment Programme (UNEP), 40% of Earth's land surface can be classified as drylands, that is, arid areas that are particularly subject to degradation. Although soil degradation and desertification are global problems, almost half of the world's most degraded land lies within the poorest areas.[9] Declines of soil quality in these areas have endangered the food supply or well-being of billions of people around the world.

In the affected lands, most degradation results from wind and water erosion. In recent years, gigantic dust storms from denuded land in China have blown across the Pacific Ocean to North America, and dust storms from Africa's Sahara Desert have blown across the Atlantic Ocean to the Caribbean Sea. Similarly massive dust storms occurred in the Canadian Prairies and the Great Plains of the United States during the 1930s, when desertification shook North American agriculture and society to their very roots (**FIGURE 7.14B**).

It is estimated that desertification affects fully one-third of the planet's land area, impinges on the lives of 1.5 billion people, and costs $490 billion per year overall.[10] China alone loses billions of dollars annually from desertification. In its western reaches, desert areas are expanding and joining one another because of overgrazing from over 400 million goats, sheep, and cattle. In the Sistan Basin, on the border of Iran and Afghanistan, an oasis that supported a million livestock recently turned barren in just five years, and windblown sand buried over 100 villages. In Kenya, overgrazing and deforestation fuelled by rapid population growth has left 80% of its land vulnerable to desertification. In a positive feedback cycle, the soil degradation forces ranchers to crowd onto more marginal land and farmers to reduce fallow periods, both of which further exacerbate soil degradation.

Although better soil management has slowed erosion in many parts of the world, including in Canada, the situation for drylands in particular may be exacerbated by the effects of climate change. The United Nations estimates that desertification, worsened by climate change, could result in the displacement of 50 million people within the next 10 years.[11] Overall, land degradation affects approximately 1.9 billion ha of land worldwide, and 12 million ha are lost each year as a consequence of drought and desertification (equivalent to 23 ha per minute). Had it continued to be agriculturally productive, this land could have yielded as much as 20 million metric tons of grain. Thus soil

Tor Eigeland/Alamy

(a) Present-day desertification, Mauritania

Library and Archives Canada PA-139645/CP Photo

(b) 1930s Dust Bowl, Canada

FIGURE 7.14
(a) Desertification occurs when formerly productive land turns into desert, as shown here in this photo from Mauritania, Africa. **(b)** Canada is not exempt from the impacts of wind erosion, as shown in this photo from the Dust Bowl of the 1930s.

erosion, land degradation, and accelerated desertification are not just problems of the physical environment; they affect the economic and social well-being of people, and the food security of the world population.

The Dust Bowl was a monumental event in North America

North America is not immune to threats from accelerated erosion and desertification. Prior to large-scale cultivation of the Prairies and the Great Plains, native prairie grasses of this temperate grassland region held erosion-prone soils in place. The American bison played a significant role as a keystone species in this ecosystem. In the late nineteenth and early twentieth centuries, however, many homesteading settlers arrived with hopes of making a living there as farmers. Between 1879 and 1929, cultivated area in the region soared, driven primarily by rapid increases in the price of wheat. Bison were hunted almost to extinction. Farmers in the region grew abundant wheat, and ranchers

grazed many thousands of cattle, sometimes expanding onto unsuitable land. Both types of agriculture contributed to erosion by removing native grasses and breaking down soil structure.

At the end of 1929 the stock market crashed, sending the price of wheat lower than the price of seed; the Great Depression began, and with it an inexorable cycle of poverty and land degradation that would last most of the decade. Starting in the early 1930s, a prolonged period of drought in the region exacerbated the ongoing human impacts on the soil from overly intensive agricultural practices. Strong winds began to carry away millions of tonnes of topsoil, and often newly planted seed as well. Dust storms travelled up to 2000 km, blackening skies and coating the skins of farm workers. Some areas lost as much as 10 cm of topsoil in a few short years.

The affected region in the Prairies and Great Plains became known as the *Dust Bowl*, a term now also used for the historical event itself. The "black blizzards" of the Dust Bowl destroyed livelihoods and caused many people to suffer a type of chronic lung irritation and degradation known as dust pneumonia, similar to the silicosis that afflicts coal miners exposed to high concentrations of coal dust. Large numbers of farmers were forced off their land; those who stayed faced infestations of grasshoppers so thick that they clogged car radiators and made the roads slippery. Chickens and turkeys ate the grasshoppers, which caused their meat and eggs to develop a bad taste. There were no pesticides (chemical pesticides had not yet been invented) and no way to control the grasshoppers.

By 1937, Dust Bowl conditions had reached their peak. Since the price of wheat was so low, farmers began to plant alternative crops, such as oats, rye, flax, peas, and alfalfa. They adapted to the dry weather with reduced tilling, crop rotations (to allow the soil a chance to replenish itself), and fertilizer applications. By the end of the decade, the slow recovery from the Dust Bowl had begun.[12]

Protecting Soils

As environmental problems go, soil erosion has two good things going for it: Most of the approaches that can be used to protect soils are technologically simple and reasonably inexpensive. Most often, soil can be protected by modifying practices in farming and forestry. In both Canada and the United States, the beginnings of better soil management date back to the experience of drought during the Dust Bowl years and subsequent serious droughts. These efforts continue today.

The Soil Conservation Council emerged from the experience of drought

In 1935, the Prairie Farm Rehabilitation Administration (PFRA) was set up; interestingly, however, it was not until the early 1980s that the issue of soil erosion and the degradation of agricultural soils really took centre stage in Canada on a nationwide basis. This occurred partly as a result of another serious drought in the late 1970s, followed by the publication in 1984 of a book by the federal government titled *Soil at Risk: Canada's Eroding Future*.[13] In that publication, the Standing Committee on Agriculture, Fisheries, and Forestry (also known as the Sparrow Commission) concluded, "Canada risks permanently losing a large portion of its agricultural capability if a major commitment to conserving the soil is not made immediately by all levels of government and by all Canadians." The following year saw the first National Soil Conservation Week and the establishment of the National Soil Conservation Program.

The Soil Conservation Council of Canada was established in 1987, with the goals of improving understanding and awareness about the causes of soil degradation in Canada; facilitating communication among soil conservation groups, government, and industry; and encouraging sustainable use of soil and related resources.[14]

Today a number of federal government agencies (as well as provincial and territorial agencies) provide services to assist farmers with all aspects of soil management and conservation in Canada. The PFRA still exists, but it is now one component of a broader organization, the Agri-Environment Services Branch (AESB) of Agriculture and Agri-Food Canada, which was set up specifically to address agricultural and soil-related environmental issues. The organization was an integration of three previously existing federal agencies: the PFRA, the National Land and Water Information Service (NLWIS), and the Agri-Environmental Policy Bureau. The AESB promotes "an integrated approach to sustainable agriculture in Canada, which recognizes that environmentally responsible agriculture and competitive agriculture are part of an integrated system."[15]

Internationally, the United Nations promotes soil conservation and sustainable agriculture through a variety of programs of its Food and Agriculture Organization (FAO). The FAO's Farmer-Centred Agricultural Resource Management Program (FAR) is a project that supports innovative approaches to resource management and sustainable agriculture in China, Thailand, Vietnam, Indonesia, Sri Lanka, Nepal, the Philippines, and India. The program studies agricultural success stories and tries to help other farmers duplicate the successful efforts. Rather than following a top-down, government-controlled approach, the FAR program relies on the creativity of local communities to educate and encourage farmers throughout Asia to conserve soils and secure their food supply.

Erosion-control practices protect and restore plant cover

Farming and forestry methods to control erosion make use of the general principle that maximizing vegetative cover will protect soils, and this principle has been applied widely. It is common to stabilize eroding banks along creeks and roadsides by planting vegetation to anchor the soil or in rows to protect open fields from the wind. In areas with severe and widespread erosion, some nations have planted vast plantations of fast-growing trees. For example, China has embarked on the world's largest tree-planting program to slow its soil loss (**FIGURE 7.15**). Although such reforestation efforts do help slow erosion, they do not at the same time produce ecologically functioning forests, because tree species are selected only for their fast growth and are planted in monocultures.

Several farming techniques can reduce the impacts of conventional cultivation on soils. Such measures have

China Photos/Getty Images

FIGURE 7.15
Vast swaths of countryside in western and northern China have been planted with row upon row of fast-growing trees, each surrounded by its own tiny water catchment basin. These reforestation efforts can slow erosion but do not create ecologically functional forests because the plantations are too biologically simple.

been widely shared and applied in many places around the world, and some have been practised by traditional farmers for centuries.

Crop rotation The practice of alternating the kind of crop grown in a particular field from one season or year to the next is called *crop rotation* (**FIGURE 7.16A**). Rotating crops can return nutrients to the soil, break cycles of disease associated with continuous cropping, and minimize erosion that can come from letting fields lie exposed. Some farmers in Canada have returned to an earlier farming approach in which they plant alternating swaths of land each year, but leave the field stubble from the previous year to protect fields from exposure to wind and water erosion. Many farmers rotate between wheat or corn and soybeans from one year to the next. Soybeans are legumes, with nitrogen-fixing bacteria among their roots, so they can revitalize soil that the previous crop had partially depleted of nutrients. Crop rotation also reduces insect pests; if an insect is adapted to feed and lay eggs on one particular crop, planting a different crop will leave its offspring with nothing to eat.

Intercropping and agroforestry Farmers may also gain protection against erosion by *intercropping*, planting different types of crops in alternating bands or other spatially mixed arrangements (**FIGURE 7.16B**). Intercropping helps slow erosion by providing more complete ground cover than does a single crop. Like crop rotation, intercropping offers the additional benefits of reducing vulnerability to insect and disease incidence, and, when a nitrogen-fixing legume is one of the crops, of replenishing the soil. Cover crops can be physically mixed with primary food crops,

which include maize, soybeans, wheat, onions, cassava, grapes, tomatoes, tobacco, and orchard fruit.

When crops are interplanted with trees, called **agroforestry**, even more benefits can be realized. Agroforestry systems are generally more biologically productive than farm systems in which food crops are grown alone. One reason is that trees draw nutrients and water from deep in the soil through their root systems, cycling them into the shallower layers of soil. They also contribute organic material to the topsoil, in the form of tree litter, such as fallen branches and leaves. Trees can also provide partial shade for crops, although light levels must be managed by appropriate pruning and spacing of the trees. In many agricultural regions, especially in the developing world where small-scale agriculture is still practised, agroforestry provides a low-cost way to close the nutrient cycle and rehabilitate soils while providing other sustainably harvested forest products such as fruits, nuts, and timber.

Contour farming and terracing Water running down a hillside can easily carry soil away, particularly if there is too little vegetative cover to hold the soil in place. Thus, sloped land is especially vulnerable to erosion. Several methods have been developed for farming on slopes. *Contour farming* (**FIGURE 7.16C**) consists of ploughing furrows sideways across a hillside, perpendicular to its slope, to help prevent formation of rills and gullies. The technique is so named because the furrows follow the natural contours of the land. In contour farming, the downhill side of each furrow acts as a small dam that slows runoff and catches soil before it is carried away. Contour farming is most effective on gradually sloping land with crops that grow well in rows.

On extremely steep terrain, *terracing* (**FIGURE 7.16D**) is the most effective method for preventing erosion. Terraces are level platforms, sometimes with raised edges, that are cut into steep hillsides to contain water from irrigation and precipitation. Terracing transforms slopes into series of steps like a staircase, enabling farmers to cultivate hilly land without losing huge amounts of soil to water erosion. Terracing is common in ruggedly mountainous regions, such as the foothills of the Himalayas and the Andes, and has been used for centuries by farming communities in such areas. Terracing is labour-intensive to establish but in the long term is likely the only sustainable way to farm in mountainous terrain.

Shelterbelts and buffer zones   A widespread technique to reduce soil erosion, especially from wind. is to establish **shelterbelts** (**FIGURE 7.16E**). Shelterbelts act as *windbreaks*—rows of trees or tall shrubs that are planted along the edges of fields to slow the wind. Shelterbelts have been widely planted across the Prairies and the Great Plains, where fast-growing species such as poplars are often used. Statistics Canada reports more than 61 000 shelterbelt installations (about 30% of all farms) in Canada, the majority in Ontario, Saskatchewan, and Alberta.[16]

Shelterbelts are often combined with intercropping in a practice known as *alley cropping*. In this approach, fields planted in rows of mixed crops are surrounded by or interspersed with rows of trees that provide fruit, wood, or protection from wind. Agroforestry and alley cropping are widely used in India, Africa, China, and Brazil, where coffee growers commonly combine farming and forestry.

Another related approach identified by Agriculture and Agri-Food Canada as a "best practice" for soil management is the establishment of a *buffer zone* between planted fields and the edge of adjacent waterways. Buffer zones usually consist of a strip around the edge or *riparian zone* of the water body, which is planted with trees, shrubs, or grasses (depending on the local situation). They help stabilize eroding shorelines, and create a physical barrier between agricultural activities and sensitive aquatic areas. Almost 43 000 farms in Canada (about 21% of all farms) have planted buffer zones between their cropped fields and waterways.[16]

Reduced tillage   Repeated cycles of ploughing and planting over many decades diminish the productivity of the soil and render it more susceptible to erosion. Overturning the soil loosens it and makes it more likely to move downslope under the influence of gravity; the result is called *tillage erosion*, and it is one of the three main erosion types monitored by Agriculture and Agri-Foods Canada (the others are wind and water erosion).

To plant using the *reduced-tillage* or *no-till* method (**FIGURE 7.16F**), a tractor pulls a drill that cuts long, shallow furrows through the litter of dead weeds and crop residue and the upper levels of the A horizon. The device drops seeds into the furrow and closes the furrow over the seeds. Often a localized dose of fertilizer is added to the soil along with the seeds. By increasing organic matter and soil biota while reducing erosion, no-till and reduced-tillage farming can build soil up, aerate it, restore it, and improve it. Proponents of no-till farming claim that the practice offers a number of benefits, including higher crop yields, reduced erosion, reduced costs of land preparation, and a reduced carbon footprint.

The no-till or reduced-tillage practice (also called *conservation tillage*) has been widely adopted in Canada, but not with unmitigated success. The approach can save money, prevent soil loss, and decrease carbon loss from soils in some situations. However, some soil textures and some crops—notably corn—are not amenable to reduced tillage, which tends to keep the soil colder and moister for longer than conventional tillage methods (because litter mulch is applied).[17] Critics of no-till and reduced-tillage farming also note that these techniques often require substantial use of chemical herbicides (because weeds are controlled chemically rather than physically removed from fields) and synthetic fertilizer (because other plants take up a significant portion of the soil's nutrients).

Better irrigation technologies can prevent soil salinization

Poorly designed and improperly implemented irrigation also can result in soil degradation. If the climate is very dry, or too much water evaporates or runs off before it can be absorbed into the soil, crops may require irrigation. The soil's ability to hold water and make it available to plant roots also influences the amount of irrigation required. Some crops, such as rice and cotton, require lots of water, whereas others, such as beans and wheat, require relatively little.

By irrigating croplands, people have managed to turn previously dry and unproductive regions into fertile farmland. Currently about 70% of all fresh water withdrawn by people is used for irrigation.[18] Irrigated land area has doubled in size over the past 50 years, according to the UN FAO.[19] The rate of increase of irrigated area has declined more recently as a result of limited availability of both fresh water and new arable land.

If some water is good for plants and soil, it might seem that more must be better. But this is not necessarily the case; there is indeed such a thing as too much water.

(a) Crop rotation

(b) Intercropping

(c) Contour farming

(d) Terracing

(e) Shelterbelts

(f) No-till farming

FIGURE 7.16 Farmers have adopted many strategies to conserve soil. Rotating crops such as soybeans and corn **(a)** helps restore soil nutrients and reduce impacts of crop pests. Intercropping **(b)** can reduce soil loss while maintaining soil fertility. Contour farming **(c)** reduces erosion on hillsides, whereas terracing **(d)** is most useful in steep mountainous areas. Shelterbelts **(e)** protect against wind erosion. In **(f)**, corn grows up from amid the remnants of a "cover crop" used in reduced-tillage agriculture.

Overirrigation in poorly drained areas can cause or exacerbate certain soil problems. Soils too saturated with water may become waterlogged. When **waterlogging** occurs, the water table is raised to the point that water bathes plant roots, depriving them of access to gases and essentially suf-

focating them. If it lasts long enough, waterlogging can damage or kill plants.

An even more frequent problem is **salinization** (or *salination*), the buildup of salts in surface soil layers. In dryland areas where precipitation is minimal and evaporation rates

FIGURE 7.17
Salinization can result from overirrigation, especially in arid and semi-arid regions. It is a widespread problem affecting agricultural soils, as shown here in Alberta. The white crust on the surface is a layer of salts.

are high, water evaporating from the A horizon may pull saline water up from lower horizons by capillary action. As this water rises through the soil, it carries dissolved salts; when the water evaporates at the surface, those salts precipitate and are left at the surface. Eventually, high salinity levels can make the soil inhospitable to plants. Irrigation in arid areas generally hastens salinization, because it provides repeated doses of water. Moreover, because irrigation water is often ground water that contains some dissolved salt in the first place, it can be a new source of salt into the soil. In areas of farmland where overirrigation has occurred, soils may turn white with encrusted salt (**FIGURE 7.17**).

The best way to prevent soil salinization is to avoid planting crops that require a great deal of water in areas that are prone to the problem. A second way is to irrigate with water that is as low as possible in salt content. A third way is to irrigate efficiently, supplying no more water than the crop requires, thus minimizing the amount of water that evaporates and hence the amount of salt that accumulates in the topsoil. This can be accomplished through the use of highly efficient drip systems that deliver measured amounts of water directly to the plant roots just when the plant needs it.

The remedies for correcting soil salinization once it has occurred are much more expensive and difficult to implement than the techniques for preventing it in the first place. One approach is to stop irrigating and wait for rain to infiltrate and leach the salts from the soil. However, this solution is often unfeasible because salinization is mainly a problem in dryland areas, where precipitation is inadequate to flush salts from the soil. A better option may be to plant salt-tolerant plants, such as barley, that can be used as food or pasture. A third option is to bring in large quantities of less-saline water with which to flush the soil; however, using too much water may cause waterlogging. The most effective, but most expensive, option is to install

artificial drainage to lower the level of the saline water table below a depth where it can be drawn up to the soil surface. Rain or irrigation can then flush the remaining salts away from the rooting zone.

To protect soil it is important to avoid overgrazing

When sheep, goats, cattle, or other livestock graze on the open range, they feed primarily on grasses; sometimes the grasslands are planted and maintained, but often they are natural grasslands. As long as livestock populations do not exceed a range's carrying capacity, and do not consume grasses faster than they can be replaced or grow, grazing may be sustainable. However, when too many animals eat too much of the plant cover, impeding plant regrowth and preventing the replacement of biomass, the result is *overgrazing*.

Rangeland scientists have shown that overgrazing causes a number of impacts, some of which give rise to positive feedback cycles that exacerbate damage to soils, natural communities, and the land's productivity for grazing (**FIGURE 7.18**). When livestock remove too much of an area's plant cover, more soil surface is exposed and made vulnerable to erosion. Soil erosion makes it difficult for vegetation to regrow, perpetuating the lack of cover and giving rise to more erosion. Moreover, non-native weedy plants may invade denuded soils. These invasive plants are usually less palatable to livestock and can outcompete native vegetation in the new, modified environment, further decreasing native plant cover.

Overgrazing can also compact soils and alter their structure. Soil **compaction**, in which pore space in the soil is reduced, makes it harder for water to infiltrate, for soils to be aerated, for plants' roots to expand, and for roots to conduct cellular respiration. All of these effects further decrease the growth and survival of native plants. Soil compaction also can be caused by over-tilling, use of heavy agricultural machinery, clear-cut logging, and rapid withdrawal of ground water.

As a cause of soil degradation, overgrazing is equal to poorly managed cropland agriculture, and it is an even greater cause of desertification. Humans worldwide keep almost 4 billion cattle, sheep, and goats.[19] Rangeland classified as degraded now adds up to 680 million ha, although some estimates put the number as high as 2.4 billion ha, fully 70% of the world's rangeland area. Rangeland degradation is estimated to cost $23.3 billion per year. Grazing exceeds the sustainable supply of grass in India by 30% and in parts of China by up to 50%. To relieve pressure on rangelands, both nations are now beginning to feed crop residues to livestock.

Range managers do their best to assess the carrying capacity of rangelands and inform livestock owners of these limits, so that herds are rotated from site to site as

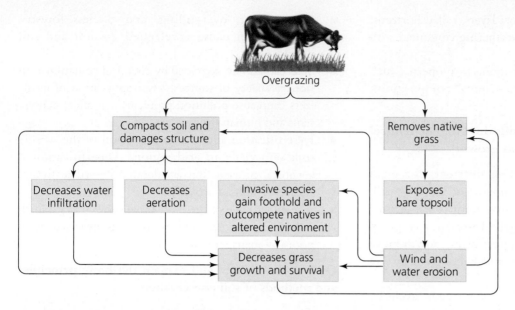

FIGURE 7.18 When grazing by livestock exceeds the carrying capacity of rangelands and their soil, overgrazing can set in motion a series of consequences and positive feedback loops that degrade soils and grassland ecosystems.

DATA Q What is the immediate cause of exposure of bare topsoil? What is the immediate consequence of exposing bare topsoil? How many immediate consequences of wind and water erosion are shown on this diagram?

needed to conserve grass cover and soil integrity. Managers also can establish and enforce limits on grazing on publicly owned land when necessary. Today increasing numbers of ranchers are working cooperatively with government agencies, environmental scientists, and even environmental advocates to find ways to ranch more sustainably and safeguard the health of the soil.

Conclusion

Soil is a complex system that functions as the interface between the geosphere, hydrosphere, and atmosphere. The importance of soil as a resource is often underestimated, but the preservation of arable soil is crucial for the maintenance of global food security.

Many of the policies enacted and the practices developed to combat soil degradation in Canada and worldwide have been quite successful, particularly in reducing the erosion of topsoil. However, soil is still being degraded at a rate that threatens the sustainability of the resource. Despite all we have learned about soil degradation and conservation, many challenges remain.

The role of soil as a reservoir in biogeochemical cycling is also of increasing interest to scientists. As long ago as 1945, Russian scientist Vladimir Vernadsky, considered to be one of the founders of biogeochemistry, declared that humans had become "a mighty and ever-growing geological force." This has never been truer than it is today, as we race to document, explain, and mitigate the global and potentially permanent environmental impacts of some human actions. Understanding the characteristics and behaviour of soils in this context will be a crucial part of this task.

REVIEWING OBJECTIVES

You should now be able to

Describe important properties of soil and soil-forming processes

■ Soil is a complex system that consists of mineral fragments with varying proportions of organic matter, with the rest of the pore space taken up by soil water

and gases. Biotic communities in soil include microorganisms and larger organisms such as earthworms, burrowing mammals, amphibians, and reptiles.

■ Soil formation begins with the breakdown of parent rock by physical (mechanical), chemical, or biological weathering. Climate, organisms, relief, parent material, and time are factors that influence soil formation.

- Soil profiles consist of distinct layers, called horizons, which form as a result of weathering combined with leaching.
- Soil can be categorized according to properties such as colour (influenced by chemical composition), texture, structure, and pH.

Characterize the role of soils in biogeochemical cycling and in supporting plant growth

- Soil is crucial for providing nutrients for plant growth and, thus, for the support of life on Earth. Soil properties affect (and may limit) the potential for plant growth and agriculture in any given location. Materials that support plant growth move from soil to soil solutions and back again by ion exchange.
- Soils play a crucial role in the nitrogen cycle by hosting free-living and symbiotic microorganisms that mediate nitrogen fixation, nitrification, and denitrification.
- Soils represent the largest terrestrial reservoir for carbon in the active carbon cycle—larger than all terrestrial vegetation and the atmosphere combined.
- There are significant regional differences in the composition, health, and fertility of soils. The differences between soils in tropical and temperate biomes illustrate this.

Identify the causes and predict the consequences of soil erosion and soil degradation

- As the human population grows, pressures from agriculture and other activities are degrading Earth's soil, and we are losing topsoil from productive cropland at an unsustainable rate.
- The main mechanisms of soil loss are splash, sheet, rill, and gully erosion by water, and deflation and abrasion by wind.

- Overgrazing, over-tilling, and careless forestry practices can cause accelerated erosion and soil degradation.
- Soils can also be degraded by chemical contamination from a variety of sources. Over-application of fertilizers can cause pollution problems that affect ecosystems and human health.
- Desertification affects a large portion of the world's soils, especially in arid regions. Desertification is a natural process, but accelerated desertification is caused by human activities.
- The experience of the Dust Bowl showed that North America is not immune to problems associated with poor soil management.

Outline the history and explain the basic principles and methods of soil conservation

- The experience of the Dust Bowl encouraged scientists and farmers to develop ways of better protecting and conserving topsoil. Around the world, including in Canada, governments are devising innovative policies and programs to deal with soil degradation.
- Techniques such as crop rotation, contour farming, intercropping, terracing, shelterbelts, buffer zones, and reduced tillage are relatively simple ways to help protect soils from erosion.
- Overirrigation can cause salinization and waterlogging, which lower crop yields and are difficult to mitigate. Carefully designed irrigation can prevent salinization of soils.
- Overgrazing is a major cause of soil compaction and degradation worldwide. It is important to preserve grasses on rangelands in order to protect soils from the effects of overgrazing.

TESTING YOUR COMPREHENSION

1. What is soil?
2. Describe the three types of weathering that may contribute to the process of soil formation.
3. What processes most influence the formation of soil? What is leaching, and what is its role in soil formation?
4. Name the five primary factors thought to influence soil formation, and describe one effect of each.
5. How are soil horizons created? What is the general pattern of distribution of organic matter in a typical soil profile?

6. Why is erosion generally considered a destructive process? Name three human activities that can promote soil erosion.
7. Describe the principal types of soil erosion by water and by wind.
8. How does terracing effectively turn very steep and mountainous areas into arable land?
9. How can fertilizers and irrigation contribute to soil degradation?
10. Describe the effects on soil of overgrazing.

THINKING IT THROUGH

1. How and why might actual soils differ from the idealized five-horizon soil profile presented in the chapter? How might departures from the idealized profile indicate the impact of human activities? Provide at least three examples.

2. Some pollutants, such as heavy metals, are positively charged and adhere to clay and other negatively charged soil particles. This can make it difficult to clean up contaminated soils and can reduce nutrient availability for plants. Explain why this is so, making reference to the process of cation exchange in soils.

3. You are a land manager with your provincial government and you have just been put in charge of 200 000 ha of public lands that have been severely degraded by overgrazing. Soil is eroding, creating large gullies. Shrubs have encroached on grassland areas because fire was suppressed. Environmentalists want an end to ranching on the land, and they want to bring back wolves and other endemic species. Ranchers want continued grazing and are strongly opposed to the reintroduction of native species, especially wolves. However, the ranchers are concerned about the land's condition and are willing to entertain new ideas. What steps would you take to assess the land's condition and begin restoring its soil and vegetation? Would you allow grazing, and if so, would you set limits on it? Would you try to reintroduce wolves to the area?

4. You are the head of an international granting agency that assists farmers with soil conservation and sustainable agriculture. You have $10 million to disburse. Your agency's staff has decided that the funding should go to (1) farmers in an arid area of Africa prone to salinization, (2) farmers in an area of Indonesia where swidden agriculture is practised, (3) farmers in southern Brazil practising no-till agriculture, and (4) farmers in a dryland area of Mongolia undergoing desertification. What types of projects would you recommend funding in each of these areas, how would you apportion your funding among them, and why?

INTERPRETING GRAPHS AND DATA

Dr. Henry Janzen is a soil scientist at the Lethbridge Research Centre in Alberta and an adjunct professor at the University of Manitoba. He is an expert on carbon cycling and the emission of greenhouse gases from soils, and he carries out research on the effects of different management approaches on carbon storage in agricultural soils. In a 2004 paper in the journal *Agriculture, Ecosystems and Environment*, Dr. Janzen summarized many of the main scientific questions concerning carbon storage in terrestrial reservoirs.

One concern outlined in Dr. Janzen's study was the impact of land-use changes on carbon storage in terrestrial reservoirs. The figure from the paper summarizes major land-use changes since 1700.

1. On the basis of this graph, which of the biomes or managed ecosystems (forest/woodland, steppe/savannah/grassland/shrubland, tundra/desert, or pasture/cropland) *increased* the most, in terms of percentage increase, from 1700 to 1990? Which one of them *decreased* the most, in terms of percentage decrease? What happened to the area covered by tundra/desert over the time period represented on the graph?

2. The table on the next page presents some information about carbon storage in various terrestrial reservoirs, based on data from Janzen's study. It also provides the current total area occupied by each of the biome types.

For each biome type and pasture/cropland, calculate and fill in the last column of the table, which shows the amount of carbon stored per hectare (in units of t/ha, or tonnes per hectare). One calculation (for the

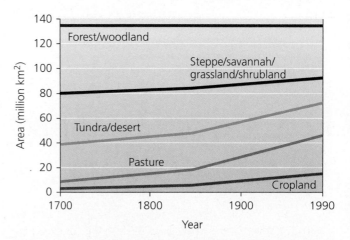

This graph shows the change in terrestrial area for three groups of major biome types and the increase in area of pasture and croplands since 1700.

Source: Republished with permission of Elsevier Science, Inc., from Figure 2 from Janzen, H.H. (2004). Carbon Cycling in Earth Systems—A Soil Science Perspective. *Agriculture, Ecosystems and Environment, 104,* pp. 399–417; permission conveyed through Copyright Clearance Center, Inc.

steppe/savannah grouping) has been completed as an example.

3. From the calculations you made and the information that is now in the final column of your table, which biome type stores the *most* carbon per hectare? Which stores the *least* carbon per hectare? How does the pasture/cropland carbon storage per hectare compare to that in the natural ecosystems?

4. Based on the data provided in the table, which of the natural ecosystems stores the highest proportion of its carbon in *plants?* Which of the natural ecosystems stores the highest proportion of its carbon in the *soil?* How does the proportion of carbon stored in plants vs. soils in pasture/cropland compare to the natural systems? Do these results surprise you? Why or why not?

5. In terms of carbon storage in terrestrial reservoirs, what do you think would be the overall result of the shift, shown in the graph, to increasing pasture/cropland at the expense of forests and grasslands?

Biome type	Global Carbon Reservoir (in Gt, or tonnes × 10^9)			Current total area (ha × 10^9)	C stored per hectare (t/ha)
	In Plants	**In Soils**	**Total**		
Forest/woodland	359	787	1146	4.17	
Steppe/savannah/grassland/shrubland	75	559	634	3.50	181.1
Tundra/desert (incl. ice-covered)	14	312	326	5.50	
Pasture/cropland	3	128	131	1.60	

This table shows the approximate area of three sets of biome types and pasture/cropland, in billions of hectares (ha × 10^9), and the size of the carbon stock in each of them, in gigatonnes (Gt, or tonnes × 10^9).

Source: Based on information in Table 1 from Janzen, H.H. (2004). Carbon Cycling in Earth Systems—A Soil Science Perspective. *Agriculture, Ecosystems and Environment, 104,* pp. 399–417.

MasteringEnvironmentalScience®

STUDENTS
Go to **MasteringEnvironmentalScience** for assignments, the eText, and the Study Area with practice tests, videos, current events, and activities.

INSTRUCTORS
Go to **MasteringEnvironmentalScience** for automatically graded activities, current events, videos, and reading questions that you can assign to your students, plus Instructor Resources.

8 Agriculture, Food, and Biotechnology

An agricultural engineer checks tomato plants in a greenhouse.

sezer66/Shutterstock

Upon successfully completing this chapter, you will be able to

- Outline the historical development of agriculture, the transition to industrialized agriculture, and the challenge of feeding a growing human population
- Identify the main approaches and summarize the environmental impacts of the Green Revolution
- Summarize the challenges, strategies, and impacts of pest management and the critical importance of pollination

- Describe the science and evaluate the controversies associated with genetically modified food
- State the importance of crop diversity and outline some approaches to preservation
- Assess the positive and negative aspects of feedlots and aquaculture, and the energy trade-offs in raising animals for food
- Summarize the main goals and methods of sustainable agriculture

A farmer works in a maize field in Oaxaca, Mexico.

CENTRAL CASE
GM MAIZE AND ROUNDUP-READY CANOLA

"If you desire peace, cultivate justice, but at the same time cultivate the fields to produce more bread; otherwise there will be no peace."

—NORMAN BORLAUG, NOBEL PRIZE WINNER AND "FATHER" OF THE GREEN REVOLUTION

"Worrying about starving future generations won't feed them. Food biotechnology will. . . . At Monsanto, we now believe food biotechnology is a better way forward."

—MONSANTO COMPANY ADVERTISEMENT

"I never put those plants on my land. The question is, where do Monsanto's rights end and mine begin?"

—PERCY SCHMEISER

Corn is a staple grain of the world's food supply. We can trace its ancestry back roughly 5500 years, when people in the highland valleys of what is now the state of Oaxaca in southern Mexico first domesticated that region's wild maize plants. The corn we eat today arose from some of the many varieties that evolved from the early selective crop breeding conducted by the people of this region.

Today Oaxaca remains a world centre of biodiversity for maize, with many native varieties growing in the rich soil (see photo). Preserving such varieties of crops in their ancestral homelands is important for securing the future of our food supply, scientists maintain, because these varieties serve as reservoirs of genetic diversity—reservoirs we may need to draw on to sustain or advance our agriculture.

In 2001, Mexican government scientists conducting routine genetic tests of Oaxacan farmers' maize announced that they had turned up DNA that matched *transgenes*, that is, genes transferred from genetically modified (GM) corn, even though Mexico had banned its cultivation since 1998. Corn is one of many crops that scientists have genetically engineered to express desirable traits such as large size, fast growth, and resistance to insect pests.

Activists opposed to GM food trumpeted the disturbing news and urged a ban on imports of transgenic crops from producer countries such as the United States into developing nations. The agrobiotech industry defended the safety of its crops and questioned the validity of the research. Further research by Mexican government scientists confirmed the presence of transgenes in Mexican maize. Those studies were controversial and still have not been definitively verified, although the findings were accepted by a special commission of experts convened under the North American Free Trade Agreement (NAFTA). The commission concluded that corn imported from the United States was the source for the transgenes, which spread by wind pollination and interbreeding with native crop plants once in Mexico.

Meanwhile, back in Canada, Monsanto, a producer of GM seeds, was engaged in a high-publicity battle with 74-year-old Saskatchewan farmer Percy Schmeiser (see photo) over Schmeiser's canola crop. Canola is an edible oil derived from rapeseed. It is widely grown throughout the Canadian prairies, and 80% of the rapeseed grown in Canada is genetically modified.

Schmeiser maintained that pollen from Monsanto's Roundup Ready® canola had blown from a neighbouring farm onto his land and pollinated his non-GM canola. Schmeiser had not purchased the patented seed and said that he did not want the crossbreeding. Monsanto investigators charged him with violating a Canadian law that makes it illegal for farmers to reuse or grow patented seed without a contract. The courts sided with Monsanto, ordering the farmer to pay the corporation $238 000. Schmeiser appealed to the Supreme Court of Canada, which ruled that Monsanto's patent had indeed been violated but acknowledged that the farmer had not benefited from the GM seeds and had not intended their use. The Court exempted him from paying any fines or fees to the company.

In spite of the Supreme Court loss, Schmeiser received wide public support. He and his wife, Louise, were given the 2007 Right Livelihood Award in recognition of their struggle. A government committee sought a revision in the patent law, and the National Farmers

Saskatchewan farmer Percy Schmeiser was accused by Monsanto Canada of planting its patented Roundup Ready® canola without a contract with the company. Schmeiser said his non-GM plants were contaminated with Monsanto's transgenes from neighbouring farms. He became a hero to small farmers and anti–GM food activists worldwide.

John Schmeiser

Union of Canada called for a moratorium on GM food. Schmeiser says that the most difficult part of the entire saga was the loss, through contamination by transgenes, of the local variety of rapeseed that he had planted throughout his 60-year farming career.[1]

The larger question of the legality of holding patents on living organisms remains unresolved. In 2002, the Supreme Court of Canada refused to allow the patenting of a "higher organism" (a plant or an animal, as opposed to a gene or a single-celled organism; a genetically modified mouse was the organism at the centre of that decision). Monsanto continues to demand that farmers heed patent laws. In lawsuits similar to the Schmeiser case, one of the outcomes was that farmers convicted of having planted Monsanto products without purchasing them were banned from ever being allowed to purchase or use any Monsanto-patented technology in the future. In an interesting turn of events for *Monsanto v. Schmeiser*, the farmer contacted the company in 2005 to report that he had found Roundup Ready® canola in his field, and to request that it be cleaned out. An out-of-court settlement was reached in March of 2008, in which the company agreed to pay the remediation cost of $660.

The Race to Feed the World

Although human population growth has slowed, we can still expect that our numbers may swell to 9 billion or more by the middle of this century. For every two people living in the year 2000, there will likely be three in 2050. Feeding 50% more mouths will require significant advances in farming technology and food distribution processes. Much of the planet's **arable** land (land that is suitable for the annual planting of crops) has already been brought into production, so expanding the cultivated area likely won't be sufficient to meet the need. Soil, water, and ecosystems are already under stress; increasing agricultural production while protecting these resources will require that sustainability be a guiding principle.

Agriculture is the practice of raising crops and livestock for human use and consumption. We obtain most of our food, fibre, and (increasingly) biofuels from **cropland**, land used to raise plants for human use, and **rangeland**, land used for grazing livestock. As the human population has increased, so have the amounts of land and resources we devote to agriculture, which currently occupies more than 38% of Earth's land surface (cropland and rangeland combined).[2]

Agriculture is not something that people have "always done." The development of agriculture was a major achievement—perhaps the most important technological leap forward by humans in our history as a species. In the past 60 years or so, agricultural technology has undergone massive, rapid changes. These changes have allowed food production to keep pace with population growth, more or less, but they have led to some unanticipated environmental impacts. In this chapter we look at the development of agricultural technologies over time and in recent decades, the challenge of offering food security to the world's population, and the environmental implications of both.

Agriculture first appeared around 10 000 years ago

During most of our species' 200 000-year existence[3] we were hunter-gatherers, dependent on wild plants and animals. Between 12,000 and 10 000 years ago, as the climate warmed following a period of glaciation, people in some cultures began to **cultivate**, or raise and breed plants (*cultivars*) from seeds, and to **domesticate** or tame animals as a source of both food and labour.

Agriculture most likely began as hunter-gatherers brought back to their encampments wild fruits, grains, and nuts. Some of these foods fell to the ground, were thrown away, or were eaten and survived passage through the digestive system. The plants that grew from these seeds near human encampments likely produced fruits that were on average larger and tastier than those in the wild because they sprang from seeds of fruits people selected for their size and flavour. As these plants bred with others nearby that shared their characteristics, they gave rise to subsequent generations of plants with large and flavourful fruits.

Eventually, people realized that they could guide this process and began intentionally planting seeds from the plants whose produce was most desirable. This is *artificial selection*, or *selective breeding*, at work. The practice of selective breeding continues to the present day and has produced the many hundreds of crops we enjoy, all of which are artificially selected versions of wild plants. People followed the same process of selective breeding with animals, creating domesticated livestock from wild species—by accident at first, then by intention.

Evidence from archaeology and paleoecology suggests that agriculture was invented independently by different cultures in at least four and possibly 10 or more areas of the world at around the same time (**FIGURE 8.1**). The earliest widely accepted archaeological evidence for plant domestication is from the "Fertile Crescent" region of the Middle East about 10 500 years ago, and the earliest evidence for animal domestication, also from that region, is just 500 years later.

Once our ancestors learned to cultivate crops and domesticate animals, they began to settle in more permanent camps and villages near water sources. Agriculture and a *sedentary* (that is, settled) lifestyle likely reinforced one another. The need to harvest crops kept people sedentary, and once they were sedentary it was necessary to plant more crops to support the population. Population increase resulted from these developments and further promoted them, in a positive feedback cycle.

Agriculture—even small-scale agriculture—is a form of resource **intensification**, a way to increase the productivity from a given unit of land. A hunter-gatherer lifestyle requires a very large land area to support a given population; switching to a sedentary lifestyle based on agriculture increased the carrying capacity of the land, and allowed for larger groups to be supported on much smaller areas of land.

The development of agriculture thus permitted—or possibly spurred—a sudden dramatic increase in population. The ability to grow excess farm produce enabled some people to leave farming and to live off the food that others produced. This led to the development of professional specializations, commerce, technology, densely populated urban centres, social stratification, and politically powerful elites. For better or worse, the

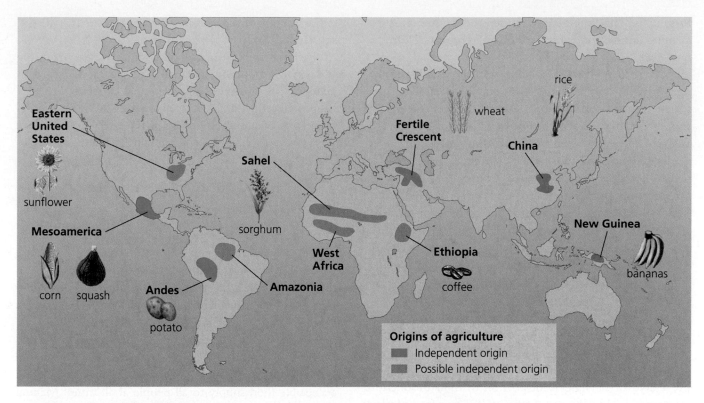

FIGURE 8.1 Agriculture originated independently in multiple locations as different cultures domesticated plants and animals from wild species. Green areas on the map are locations where people are thought to have independently invented agriculture; blue represents regions where people either invented agriculture independently or obtained the idea from cultures of other regions. A few of the many crop plants domesticated in each region are shown. *Source:* Based on data from syntheses in Diamond, J. (1997). *Guns, Germs, and Steel.* New York: W.W. Norton; and Goudie, A. (2000). *The Human Impact*, 5th ed. Cambridge, MA: MIT Press.

advent of agriculture eventually brought us the civilization we have today.

For thousands of years, crops have been cultivated, harvested, stored, and distributed by human and animal muscle power, using hand tools and simple machines

(**FIGURE 8.2**). This small-scale, biologically powered farming is known as **traditional agriculture**, and it still goes on today. Traditional agriculture sometimes uses draft animals and can employ significant quantities of irrigation water and fertilizer, but stops short of using

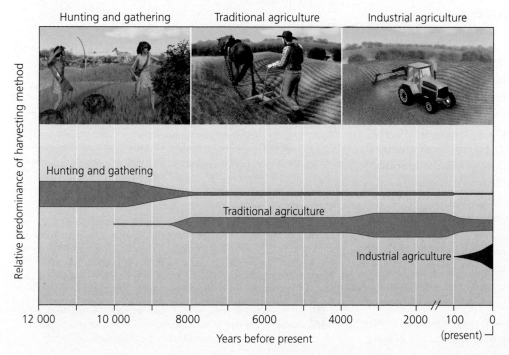

FIGURE 8.2
Hunting and gathering characterized the human lifestyle until the onset of agriculture and sedentary living in farms and villages, beginning around 10 000 years ago. Societies practising traditional agriculture gradually replaced hunter-gatherer cultures. Within the past century and a half, industrialized agriculture has spread, replacing much traditional agriculture.

fossil fuels. This type of agriculture produces food for the farming family, and extra food or cash crops to sell in the market.

In the oldest form of traditional agriculture, **subsistence agriculture**, farming families produce only enough food for themselves and do not make use of modern irrigation, farm machinery, or teams of labouring animals. Because this type of farming is commonly *rain-fed*, that is, dependent upon rain rather than irrigation for water, it does not place stress on water resources—an important consideration in the context of a changing climate. Today there are still many subsistence and traditional farmers, especially in the developing world. They are often called *smallholder farms* in reference to the fact that most are very small in area. There are about 500 million smallholder farms in the world today, supporting about 2 billion people with jobs and food.[4]

It is very difficult to pin down how much of the world's cropland is occupied by subsistence and traditional farms, and how much agricultural production can be attributed to them. In addition to their small size and great diversity, a significant portion of their production is consumed locally and never makes it into a market setting. These factors make it difficult to quantify the contributions of subsistence farmers to world agricultural output. A commonly cited statistic is that smallholder farms are responsible for about 70–80% of agricultural production in the developing world, especially in terms of food supply (as opposed to export and non-food crops), but it is difficult to substantiate this number.[5]

Industrialized agriculture is more recent

The Industrial Revolution introduced large-scale mechanization and fossil fuels to agricultural fields, enabling farmers to replace horses and oxen with faster and more powerful means of cultivating, harvesting, transporting, and processing crops. Other advances facilitated irrigation and fertilization, while the (later) invention of chemical pesticides reduced competition from weeds and herbivory by insects and other crop pests. Modern farming is also called *high-input farming*, because of its reliance on irrigation, fossil-fuel-powered machinery, and chemical fertilizers and pesticides.

To be efficient and productive, modern **industrialized agriculture** requires that vast fields be planted with single types of crops. The uniform planting of a single crop type over a large expanse of land, termed **monoculture**, is distinct from the *polyculture* approach of much traditional agriculture, such as Native

American farming systems that mixed maize, beans, squash, and peppers in the same fields.

Farming took a great technological leap during the Industrial Revolution, but many of the most important changes in modern, industrialized agriculture have happened just within the last few decades. In the rest of this chapter we will examine many of these changes and their implications for both the environment and food security.

We are producing more food per person

Over the past half century, our ability to produce food has grown even faster than population (**FIGURE 8.3**). However, largely because of political obstacles and inefficiencies in distribution, today, according to the United Nations, more than 800 million people in the world do not have enough to eat and are chronically undernourished.[6] Every five seconds, somewhere in the world, a child starves to death.

Agricultural scientists and policymakers pursue the goal of **food security**—the guarantee of an adequate and acceptable food supply to all people at all times. Making a food supply sustainable depends on maintaining water resources and crop biodiversity, and on the long-term ability of the world's soils to support crops and livestock. However, food supply is about more than agricultural productivity—it also requires having the policies, food

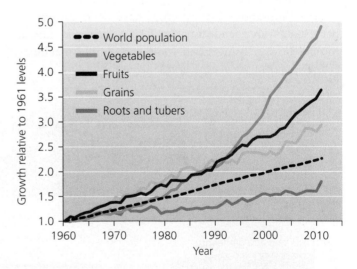

FIGURE 8.3
Global production of most foods has risen more quickly than world population. This means that we have produced more food per person each year. Trend lines show cumulative increases relative to 1961 levels. For example, a value of 2.0 means twice the 1961 amount. Food is measured by weight.
Source: Data from U.N. Food and Agriculture Organization (FAO).

DATA Q According to this graph, in which year did the global population reach a value double that of the population in 1960? In which year did the production of grains double, compared to the 1960 level?

TABLE 8.1	The Six As of Food Security
Availability	Sufficient food is available at all times.
Affordability	Food is economically affordable by all people.
Adequacy	Food is safe and of high enough quality nutritionally to keep people healthy and productive.
Acceptability	The available food is culturally acceptable and does not violate peoples' dignity, religious requirements, or human rights.
Accessibility	People can access non-emergency food supplies without physical blockages.
Agency	Government structures and policies are in place to ensure food security for everyone.

storage and distribution systems, and infrastructure in place to ensure that people have both economic and physical access to non-emergency food sources (**TABLE 8.1**).

In the 1960s there were predictions of widespread starvation and catastrophic failure of agricultural systems, amidst concerns that the human population could not continue to grow without outstripping its food supply (a "neo-Malthusian" perspective, as discussed in Chapter 6). Population has continued to increase well past the predictions; tragically, many people are still chronically hungry today. However, in percentage terms hunger has been reduced by half—from 26% of the population in 1970 to 11.3% or 805 million today, down from 900 million people just in the past decade.[7] Progress is slow, but there is progress.

We face undernourishment, overnutrition, and malnutrition

Although many people lack access to adequate food, others are affluent enough to consume more than is healthy. People who are *undernourished*, receiving less than 90% of their daily caloric needs, mostly live in the developing world. Meanwhile, in the developed world, many people suffer from *overnutrition*, taking in too many calories each day. In Canada, where food is available in abundance and people tend to lead sedentary lives with little exercise, 48% of adults exceed healthy weight standards, and 25% of adults and almost 9% of children are obese (according to standard definitions).[8]

For most people who are undernourished, the dominant reasons are economic. Over a billion of the world's people (14.5%) live on less than $1.25 per day (the current international standard for "absolute poverty"), according to the World Bank.[9] Hunger is a problem even in Canada, where over 830 000 people used the services of a food bank during a typical month in 2013, according to Hunger Awareness Week and Food Banks Canada; and almost 40% of them were children.[10] This measure underrepresents the true situation, particularly for rural areas that lack access to food banks. Food insecurity—the inability to procure

sufficient food when needed—is a factor in 13% of Canadian households, affecting 1.7 million households and over 4 million people, of whom 1.1 million are children.[11]

In a wealthy, food-producing nation such as Canada, the cause of hunger is more than just lack of available food. Many factors coalesce to cause hunger insecurity around the world, and these factors have as much to do with poverty and the weaknesses of our food delivery systems as with the abundance and availability of food. In Canada, the highest prevalence of food insecurity is in Nunavut and the Northwest Territories, with more than 45% and more than 20% of households (respectively) affected. In Nunavut, more than 62% of children live in households that experience food insecurity.[12] These statistics reflect the particular challenges of ensuring an adequate food supply for the people of Canada's North.

Just as the *quantity* of food a person eats is important for health, so is the *quality* of food. **Malnutrition**, a shortage of nutrients the body needs, including a complete complement of vitamins and minerals, can occur in both undernourished and overnourished individuals. Malnutrition can lead to disease (**FIGURE 8.4**). Vitamin A

Tony Karumba/AFP/Getty Images

FIGURE 8.4
Millions of children, including this child in Somalia, suffer from forms of malnutrition such as marasmus.

deficiency, which can lead to blindness, is a common type of malnutrition in parts of the developing world. A diet without enough protein or essential amino acids can result in *kwashiorkor*, which causes bloating of the abdomen, deterioration and discoloration of hair, mental disability, anemia, immune suppression, developmental delays, and reduced growth in children. Protein deficiency together with a lack of calories can lead to *marasmus*, which causes wasting or shrivelling among millions of children in the developing world.

Impacts of the Green Revolution

Beginning around 1950, agricultural advancement allowed new technologies, crop varieties, and farming practices to be implemented in North America and exported to the developing world. Although many people saw growth in production and increased agricultural efficiency as the key to ending starvation, hunger still exists in the world. We now realize that boosting agricultural production is only part of the solution to hunger. Nevertheless, technological advances allowed farmers to dramatically increase yields per hectare of cropland and helped millions of people avoid starvation.

The Green Revolution led to dramatic increases in agricultural productivity

The desire for greater quantity and quality of food for the growing human population led directly to the so-called **Green Revolution**, with enormous increases in agricultural productivity during the mid- to late twentieth century. We have increased food production and per capita food consumption worldwide by devoting more energy (especially fossil fuel energy) to agriculture; by planting and harvesting more frequently; by greatly increasing the use of irrigation, fertilizers, and pesticides; by planting monocultures; by increasing the amount of cultivated land; and by developing (through crossbreeding and genetic engineering) more productive crop and livestock varieties.

Prior to the Green Revolution, the best way to increase agricultural productivity was to plant more land with crops or to increase the size of a herd. This is a form of **extensification**—increasing resource productivity by bringing more land into production (as opposed to *intensification*, discussed previously, in which new technologies permit greater resource productivity from each unit of land). In the mid-twentieth century, agricultural scientists

FIGURE 8.5
Norman Borlaug holds examples of the wheat variety he bred, which helped launch the Green Revolution. The high-yielding disease-resistant wheat helped increase agricultural productivity in many developing countries. Borlaug, quoted at the beginning of this chapter, won the Nobel Peace Prize for his work toward ending world hunger.

consciously worked to develop technologies for increasing crop output per unit area of cultivated land. In the end, the Green Revolution was characterized by both extensification *and* intensification of agricultural production.

The transfer of agricultural technology to the developing world that marked the Green Revolution began in the 1940s and 1950s, when U.S. agronomist Norman Borlaug introduced Mexico's farmers to a specially bred type of wheat (**FIGURE 8.5**). This strain of wheat produced large seed heads, was short in stature to resist wind, was resistant to diseases, and produced high yields. Within two decades of planting and harvesting this specially bred crop, Mexico tripled its wheat production and began exporting wheat.

The stunning success of this program inspired others. Next were India and Pakistan, and soon many developing countries were increasing their crop yields using selectively bred strains of wheat, rice, corn, and other crops from developed nations. Some varieties yielded three or four times as much per hectare as did their predecessors.

Along with the new grains, developing nations adopted the methods of modern, industrialized agriculture. They began applying large amounts of synthetic fertilizers and chemical pesticides to their fields, irrigating crops with generous amounts of water, and using heavy equipment powered by fossil fuels. This high-input agriculture allowed farmers to harvest dramatically more corn, wheat, rice, and soybeans from each hectare of land. From 1900 to 2000, humans expanded the world's total cultivated area by 33% but increased energy inputs into agriculture by 800%.

The Green Revolution has had both positive and negative impacts

The developments associated with the Green Revolution had mixed impacts on the environment. On the positive side, the intensified use of already-cultivated land reduced pressures to convert additional natural lands for new cultivation. Between 1961 and 2002, food production rose 150% and population rose 100%, while land area converted for agriculture increased only about 10%. The Green Revolution can be said to have prevented some degree of deforestation and habitat conversion in many countries while those countries were experiencing their fastest population growth rates. In this sense, the Green Revolution was beneficial for natural ecosystems.

Despite its successes, the Green Revolution has exacted a high price. The intensive cultivation of farmland has created new environmental problems and exacerbated some old ones. Many of these problems pertain to the integrity of soil and water supplies, which are the very foundation of our food supply. The intensive use of water, fossil fuels, and chemical fertilizers and pesticides had negative environmental impacts in the form of pollution, salinization, and desertification. The social and economic impacts of the Green Revolution on small-scale farmers in the developing world—many of whom are disadvantaged by lack of income, lack of education, or both, preventing them from accessing or benefiting from these technologies—are particularly controversial.

Fertilizer impacts   One hallmark of the Green Revolution was greatly increased use of industrial fertilizers. Plants remove nutrients from soil as they grow, and water also carries away nutrients. If agricultural soils come to contain too few nutrients, crop yields decline. Therefore, a great deal of effort has aimed to enhance productivity in nutrient-limited soils by adding **fertilizer**, any of various substances that contain essential nutrients.

There are two main types of fertilizers (**FIGURE 8.6**). **Inorganic** (or industrial) **fertilizers** are mined or synthetically manufactured mineral supplements, mainly various combinations of nitrogen, phosphorus, and potassium. **Organic fertilizers** consist of natural materials, including animal manure; crop residues; fresh vegetation (or "green manure"); and *compost*, a mixture produced when decomposers break down organic matter, including food and crop waste, in a controlled environment. Organic fertilizers offer some benefits that inorganic fertilizers cannot, but they are not a panacea.

Applying substantial amounts of fertilizer to croplands can have impacts far beyond the boundaries of the fields (**FIGURE 8.7**). Nitrogen and phosphorous runoff from farms and other sources can lead to phytoplankton blooms, creating oxygen-depleted "dead zones." Such eutrophication

(a) Inorganic fertilizer

Paul Debois/AGE Fotostock

(b) Organic fertilizer

FIGURE 8.6
Two main types of fertilizer exist. Inorganic fertilizer **(a)** consists of synthetically manufactured granules. Organic fertilizer **(b)** includes substances such as compost crawling with earthworms.

Rachel Husband/Alamy

occurs at many river mouths, lakes, and ponds throughout the world. Moreover, nitrates readily leach through soil and contaminate ground water, and components of some nitrogen fertilizers can even volatilize (evaporate) into the air.

Through these processes, excess nitrate and phosphate spread through ecosystems and pose human health risks, including cancer and methemoglobinaemia, or "blue-baby" syndrome. Health Canada and the U.S. Environmental Protection Agency both have determined that nitrate concentrations below 10 mg/L in drinking water are safe,[13] yet many sources around the world exceed even the looser standard of 50 mg/L set by the World Health Organization. Careful timing and regular

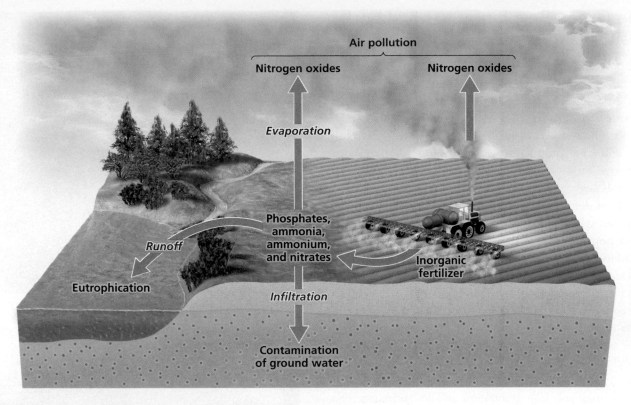

FIGURE 8.7 Over-application of fertilizers can have effects beyond the farm field, because nutrients that are not taken up by plants may end up accumulating in other reservoirs. Anthropogenic inputs of nitrogen have greatly modified the nitrogen cycle and now account for one-half the total nitrogen flux on Earth.

soil testing can help minimize applications of fertilizers that exceed what plants need and can absorb, thus limiting the amount of nitrate that enters water bodies.

Irrigation impacts   Another key feature of the Green Revolution was an enormous increase in the amount of irrigated cropland, with the result that agriculture is the main reason for extraction and use of fresh water worldwide. Unfortunately, irrigation efficiency worldwide is quite low; only 43% of the water applied actually gets used by plants. *Drip irrigation* systems (**FIGURE 8.8**) that target water directly to plants are one solution to the problem. These systems allow more control over where water is aimed; they waste far less water, and typically increase yields. Once considered expensive to install, they are becoming less costly.

Poorly designed irrigation also can lead to waterlogging and salinization of soils, especially in hot, dry regions. As described in Chapter 7, *Soils and Soil Resources,* when croplands are overwatered, soils can become soggy and saturated. As this water evaporates, it moves up through the soil profile, carrying dissolved mineral salts toward the surface. At the surface, only the fresh water evaporates, leaving a concentrated layer of salts encrusting the surface. Salinized agricultural soils are problematic in many parts of the world, and once soil has been degraded in this way it is almost impossible to restore it.

Salinization now inhibits agricultural production on one-fifth of all irrigated cropland globally, costing more than $11 billion annually. The Food and Agriculture Organization of the United Nations estimated that the total area of salinized soil worldwide is 397 million ha. Of 230 million ha of irrigated land, 45 million ha (or 19.5%) of soil are affected by salt, and of 1500 million ha of dryland agriculture, 32 million ha (or 2.1%) of soil are affected by salinization due to human activities.[14]

Monoculture impacts   One key aspect of Green Revolution techniques has had particularly negative consequences for biodiversity and mixed consequences for crop yields. Monoculture, the planting of large expanses of single crop types (**FIGURE 8.9**), has made planting and harvesting more efficient and has thereby increased output. However, when all plants in a field are genetically similar, all will be equally susceptible to viral diseases, fungal pathogens, or insect pests that can spread quickly from plant to plant. For this reason, monocultures require large amounts of pesticides and bring significant risks of catastrophic failure.

Monocultures have also contributed to a narrowing of the human diet. Globally, 90% of the food we consume now comes from only 15 crop species and 8 livestock species—a drastic reduction in diversity from earlier times. Amazingly, of the 50 000 known crop species, 60% of

(a) Conventional irrigation

(b) Drip irrigation

FIGURE 8.8
Plants take up only about half of the water applied through conventional irrigation methods, which lose a great deal of water to evaporation **(a)**. In more efficient drip irrigation approaches such as this **(b)**, hoses are arranged so that water drips from holes in the hoses directly onto the plants that need the water.

FIGURE 8.9
Most agricultural production in industrialized countries comes from monocultures—large stands of single types of crop plant, such as this wheat field. Clustering crop types in uniform fields on large scales greatly improves the efficiency of planting and harvesting, but it also decreases biodiversity and makes crops more susceptible to outbreaks of pests that specialize on particular crops.

the calories consumed globally come from just three of them: rice, wheat, and corn (maize).[15] One reason that farmers and scientists were so concerned about transgenic contamination of Oaxaca's native maize is that Oaxacan maize varieties are a valuable source of genetic variation in a world where so much variation is being lost to monocultural practices. The nutritional dangers of such dietary restriction have been alleviated by the fact that expanded global trade has provided many people access to a wider diversity of foods from different locations around the world. However, this effect has benefited wealthy people far more than poor people.

Pesticide impacts
Throughout the history of agriculture, the insects, fungi, viruses, rodents, and weeds that eat or compete with our crops have taken advantage of the clustering of food plants into agricultural fields. These organisms, in making a living for themselves, cut

crop yields and make it harder for farmers to make a living. As just one example of thousands, various species of moth caterpillars known as armyworms decrease yields of everything from beets to sorghum to millet to canola to pasture grasses. Pests and weeds have always caused problems in traditional agriculture; they pose an even greater threat in a monoculture situation, where a pest adapted to specialize on that particular crop can easily move from one individual plant to many others of the same type.

To prevent pest outbreaks and to limit competition with weeds, people have developed thousands of artificial chemicals to kill insects (*insecticides*), plants (*herbicides*), and fungi (*fungicides*). Poisons that target pest organisms are collectively termed **pesticides**. Enormous increases in the use of a wide variety of pesticides were another key feature of the Green Revolution. In the past 50 years, population has more than doubled, but cultivated land has only increased by 10%; the growing use of pesticides has been part of the response to the need to increase agricultural production with higher yields. Billions of kilograms of pesticides are applied to cropland globally each year, averaging about 2.4 kg per ha, at an annual cost of about $45 billion.[16]

Public concern over the potential effects of pesticides dates back to the publication of *Silent Spring* by Rachel Carson in 1962. Essentially a chemistry treatise on the effects of pesticides on wildlife, the book became a bestseller and has been credited with launching the modern environmental movement. The book focused on neurotoxicity (effects on the nervous system) of DDT and related chemicals in songbirds. More recently, researchers have found increasing evidence of the cumulative effects of pesticides on both humans and other wildlife of all types.[17] In addition to neurotoxicity, common effects include altered growth or metabolism, and endocrine disruption (that is, hormonal effects). An additional concern has been the apparent decrease in effectiveness of pesticides as some pests develop resistance.

Pests and Pollinators

What humans term a **pest** is any organism that damages crops that are valuable to us, and a **weed** is any plant that competes with crops. These are subjective categories that we define entirely by our own economic interests. There is nothing inherently malevolent in the behaviour of a pest or a weed; these organisms are simply trying to survive and reproduce. From the viewpoint of an insect that happens to be adapted to feed on corn, grapes, or apples, a grain field, vineyard, or orchard represents an endless buffet.

Thousands of chemical pesticides have been developed

Currently more than $2 billion is expended each year on pesticides in Canada (out of total farm input expenditures

of about $37.5 billion).[18] Most pesticides used in Canada are for agricultural purposes (91% of sales), while the remaining nonagricultural pest management products—primarily for domestic use, but also used in the forestry and industrial sectors, for the control of mosquito-borne diseases like malaria, and for the management of golf courses and other landscapes—represent 9% of total sales. Eighty-five percent of the total pesticides sold in Canada are herbicides, followed by fungicides, insecticides, and other specialty pest-management chemicals such as rodenticides and algicides.

Thousands of pesticides are registered for use in Canada. Many of the active ingredients in these pesticides have not been evaluated for health or environmental impacts for many years—over 150 were approved for use in Canada prior to 1960. The *Pest Control Products Act*, which came into effect as federal legislation in Canada in 2006, requires products to be reevaluated 15 years after they are initially approved for use, among other provisions designed to improve the safety and minimize the environmental impacts of pesticide use. The *Pest Management Regulatory Agency* of Health Canada is currently revisiting older registrations to bring them up to date.[19] **TABLE 8.2** shows the main categories of chemical pesticides used in Canada.

Pesticides are, by definition, designed to be toxic to organisms, and the toxic effects may not be limited to

TABLE 8.2   Categories of Pesticides Used in Canada

Class of Chemical Pesticide	First Used	Examples	Types	Current Status	Effects
Organochlorines	1942	Aldrin; chlordane; dieldrin; endrin; heptachlor; lindane; methoxychlor; toxaphene; HCB; PCP; DDT	Mostly insecticides	Some registered in Canada; others (such as DDT) discontinued in Canada but still used in developing nations	Persistent; bioaccumulative; affect ability to reproduce, develop, and withstand environmental stress
Organophosphates	Very early 1940s	Schradan; parathion; malathion	Insecticides	Schradan discontinued in 1964, resulting in a move toward less toxic groups (malathion, parathion)	Nonpersistent; systemic; not very selective; toxic to humans
Carbamates	First appeared in 1930; large-scale use in 1950s	Carbaryl; methomyl; propoxur; aldicarb	Fungicides, insecticides	Aldicarb discontinued in 1964; the others are registered in Canada	Nonpersistent; not very selective; toxic to birds and fish
Phenoxy	1946	2,4-D 2,4,5-T	Herbicides	2,4-D is widely used; 2,4,5-T banned in Canada	Selective effects on humans and mammals are not well known; some potential to cause cancer in laboratory animals
Pyrethroids	1980	Fenpropanthrin; deltamethrin; cypermethrin	Insecticides	Fenpropanthrin is not registered in Canada, unlike the two other pesticides	Nonpersistent, target-specific: more selective than organophosphates or carbamates; not acutely toxic to birds or mammals, but particularly toxic to aquatic species

Source: Government of Canada (2000). *Pesticides: Making the Right Choice for Health and the Environment.* Report of the Standing Committee on Environment and Sustainable Development. http://cmte.parl.gc.ca/cmte/CommitteePublication.aspx?COM=173&Lang=1&SourceId=36396. Accessed 28 November 2014. Copyright ©2014 by Parliament of Canada. Reprinted by permission.

the target organisms. Consequently, the application of synthetic pesticides can have health consequences for both humans and other nontarget organisms (see the "Effects" column of **TABLE 8.2**). There is evidence to suggest that pesticide use on farms may have peaked in North America (although not in most other parts of the world); this may be coincident with the uptake of crop varieties that have been engineered to be pest-resistant, along with rising concerns about the ecological and human health impacts of pesticide use.

Pests can evolve resistance to pesticides

Despite the toxicity of these chemicals, their usefulness in controlling target species tends to decline with time as pests evolve resistance to them. Recall from our discussion of natural selection that organisms within populations vary in their traits. Because most insects and microbes occur in huge numbers, it is likely that a small fraction of individuals may have genes that confer some degree of immunity to a given pesticide. Even if a pesticide application kills 99.99% of the insects in a field, 1 in 10 000 survives. If an insect survives by being genetically resistant to a pesticide, and if it mates with other resistant individuals of the same species, the resistant population may grow. As a result, pesticide applications will diminish in effectiveness over time (**FIGURE 8.10**).

In many cases, industrial chemists are caught up in an "evolutionary arms race" with the pests they battle, racing to increase or retarget the toxicity of their chemicals while the armies of pests evolve ever-stronger resistance to their efforts. The number of species known to have evolved resistance to pesticides has grown over the decades, and some have evolved resistance to multiple pesticides. Resistant pests can take a significant economic toll on crops.

In Canada, the United States, and Mexico, resistance to herbicides is of great concern; both the number of herbicide-resistant weeds and the area of land covered by such weeds are increasing. Agriculture and Agri-Foods research scientist Hugh Beckie and colleague Linda Hall of the University of Alberta reviewed a chain of events that has led to widespread herbicide resistance.[20] For example, rapid adoption by farmers of crops genetically modified to be tolerant of the herbicide glyphosate (trademarked by Monsanto as Roundup) resulted in a surge in the use of glyphosate to control weeds in fields with the Roundup-Ready GM crops. This led to an "unrelenting rise" in glyphosate-resistant and multiple-herbicide-resistant weeds, which, in turn, necessitated further increases in herbicide use to control resistant weeds. They concluded that the only solution is to set targets for herbicide-use reductions, supported by financial incentives and penalties.

In 2013, the Government of Canada introduced new regulations on resistance management, stating that, "The management of pesticide resistance development is an important part of sustainable pest-management and this, in conjunction with alternative pest-management strategies and integrated pest-management (IPM) programmes, can make significant contributions to reducing risks to humans and the environment from pesticide use."[21] The new legislation is also aimed at promoting a uniform approach to the management of pest resistance across North America.

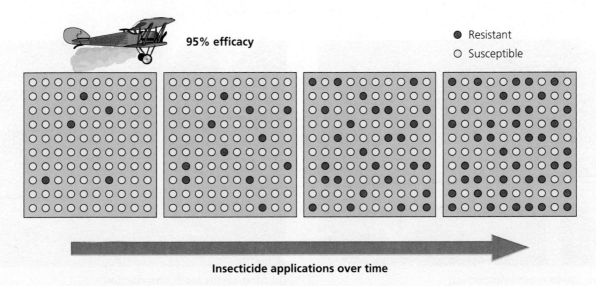

95% efficacy

● Resistant
○ Susceptible

Insecticide applications over time

FIGURE 8.10   Through natural selection, pest species can evolve resistance to the poisons we apply to kill them. This simplified diagram shows that when a pesticide is applied to an outbreak of insect pests, it may kill all individuals except those few with an innate immunity to the poison. The surviving individuals may found a population with genes for resistance to the poison. Future applications of the pesticide may then be ineffective, forcing us to develop a more potent poison or an alternative means of pest control.

Biological control pits one organism against another

Because of pesticide resistance and the health risks of some synthetic chemicals, agricultural scientists are increasingly battling pests and weeds with organisms that eat or infect them. This strategy is called **biological control**, or *biocontrol* for short. For example, parasitoid wasps are natural enemies of many caterpillars. These wasps lay eggs on a caterpillar, and the larvae that hatch from the eggs feed on the caterpillar, eventually killing it. Parasitoid wasps have been used as biocontrol agents in many situations. Some such efforts have succeeded at pest control and have led to steep reductions in chemical pesticide use.

One classic case of successful biological control is the introduction of the cactus moth, *Cactoblastis cactorum*, from Argentina to Australia in the 1920s to control invasive prickly pear cactus that was overrunning rangeland (**FIGURE 8.11**). Within just a few years, the moth managed to free millions of hectares of rangeland from the cactus.

A widespread modern biocontrol tool has been the use of *Bacillus thuringiensis* (Bt), a naturally occurring soil bacterium that produces a protein that kills many caterpillars and the larvae of some flies and beetles. Farmers have used the natural pesticidal activity of this bacterium to their advantage by spraying its spores on their crops. If used correctly, Bt can protect crops from pest-related losses. Crops such as corn have now been genetically modified so that they contain Bt, giving them an internal defence against pests. This allows farmers to decrease the amount of insecticide sprayed on their fields.

In most cases, biological control involves introducing an animal or microbe into a foreign ecosystem. Such relocation helps ensure that the target pest has not already evolved ways to deal with the biocontrol agent, but it also introduces risks. In some cases, biocontrol can produce unintended consequences if the biocontrol agent becomes invasive and begins to affect nontarget organisms. Following the cactus moth's success in Australia, for example, it was introduced in other countries to control prickly pear; however, it is now feared that the moth larvae could decimate native and economically important species of cacti.

Scientists debate the relative benefits and risks of biocontrol measures. If biocontrol works as planned, it can be a permanent solution that requires no further maintenance and is environmentally benign. However, if the agent has nontarget effects, the harm done may become permanent. Removing the agent from the system once it is established is far more difficult than simply stopping a chemical pesticide application. The potential impacts of releasing a biocontrol agent into the natural environment are basically the same as for any alien or non-native species.

Because of concerns about unintended impacts, researchers now study biocontrol proposals carefully before putting them into action, and government regulators must approve these efforts. Canada has been a world leader in this regard. However, there will never be a sure-fire way of knowing in advance whether a given biocontrol program will work as planned.

(a) Before cactus moth introduction

(b) After cactus moth introduction

Department of Natural Resources, Queensland, Australia

FIGURE 8.11 In one of the classic cases of biocontrol, larvae of the cactus moth, *Cactoblastis cactorum*, were used to clear an invasive, non-native prickly pear cactus from millions of hectares of rangeland in Queensland, Australia. These photos from the 1920s show an Australian ranch before **(a)** and after **(b)** the introduction of the moth.

IPM combines biocontrol and chemical methods

Since both chemical and biocontrol approaches have drawbacks, agricultural scientists and farmers developed a more sophisticated strategy, combining the best attributes of each approach. In **integrated pest management (IPM)**, numerous techniques are integrated to achieve long-term suppression of pests, including biocontrol, use of chemicals, close monitoring of populations, habitat alteration, crop rotation, use of genetically modified crops, alternative tillage methods, and mechanical pest removal. IPM is broadly enough defined that it encompasses a wide variety of strategies.

IPM has now become popular in many parts of the world. Indonesia (**FIGURE 8.12**) subsidized pesticide use heavily for years, but its scientists came to understand that pesticides were actually making pest problems worse. They were killing the natural enemies of the brown planthopper, which began to devastate rice fields as its populations exploded. Concluding that pesticide subsidies were costing money, causing pollution, and decreasing yields, the Indonesian government in 1986 banned the importation of 57 pesticides, slashed pesticide subsidies, and encouraged IPM. Within four years, pesticide production fell to below half its 1986 level, imports fell to one-third, and subsidies were phased out (saving $179 million annually), but rice yields actually rose. Unfortunately, despite being widely perceived as successful, Indonesia's national IPM policy was terminated in 1999, in the wake of the financial crisis in Asia.[22]

Farms are not the only places where IPM is practised. For example, the Integrated Pest Management Council of Canada runs an accreditation program for organizations that practise IPM to maintain golf courses, sports fields, and public parks and landscapes.[23]

We are critically dependent on insects to pollinate crops

Managing insect pests is such a major issue in agriculture that many people fall into a habit of thinking of all insects as somehow bad or threatening. But in fact, most insects are harmless to agriculture, and some are absolutely essential. The insects that pollinate agricultural crops are one of the most vital, yet least understood and least appreciated, factors in cropland agriculture.

Pollination is the process by which male sex cells of a plant (contained in *pollen*) fertilize female sex cells of a plant. Many plants achieve pollination by wind distribution. Millions of minuscule pollen grains are blown long distances, and by chance a small number land on the female parts of other plants of their species. The many kinds of plants that sport showy flowers, however, typically are pollinated by animals, such as hummingbirds, bats, and insects, primarily bees (**FIGURE 8.13**). Flowers are, in fact, evolutionary adaptations that function to attract pollinators. The sugary nectar and protein-rich pollen in flowers serve as rewards to lure these sexual intermediaries, and the sweet smells and bright colours of flowers are signals that advertise these rewards.

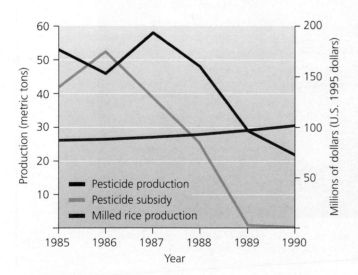

FIGURE 8.12
Once the Indonesian government threw its weight behind integrated pest management (IPM) in 1986, pesticide production and pesticide imports were reduced drastically, pesticide subsidies were phased out, and yields of rice increased. The national IPM strategy was cut in 1999, despite being widely seen as successful.

DATA Q Use the graph to carefully read the rice production for 1986 (when IPM was adopted) and 1989 (when pesticide subsidies were ended). By what percentage did rice production increase over that period of time?

FIGURE 8.13
Many agricultural crops depend on insects or other animals to pollinate them. Our food supply, therefore, depends partly on conservation of these vital organisms. These apple blossoms are being visited by a European bee. Plants use flowers with colours and sweet smells to advertise nectar and pollen, enticements that attract pollinators.

THE SCIENCE BEHIND THE STORY

This leafcutter bee is pollinating an alfalfa flower.
Peggy Greb/United States Department of Agriculture

The Alfalfa and the Leafcutter

In the first half of the twentieth century, land clearing for agriculture destroyed many nesting sites of native leafcutter bees in the Canadian Prairies. As a result, Canadian alfalfa seed production decreased dramatically, virtually collapsing by mid-century, in spite of a large increase in alfalfa planted area. Alfalfa is an important forage crop for livestock, and it is used to control moisture and nutrient levels in agricultural fields. Honeybees (*Hymenoptera apis*) are ineffective alfalfa pollinators because they can steal nectar without "tripping" the alfalfa flower, a process that uncovers the plant's stigma and is required for the pollination to be successful. With the loss of the most effective native pollinators, by 1950 Canada was importing alfalfa seed to meet 95% of its domestic needs.

In response to this crisis, the European alfalfa leafcutter bee (*H. Megachile rotun-*

data) was introduced to Canada in 1961 (see photo). Scientists, beekeepers, and seed growers worked together to develop a management system for the new bees, which eventually resulted in a six-fold increase in alfalfa yields. By the end of the twentieth century, Canada was meeting or exceeding its demand for alfalfa seed, thanks to the alfalfa leafcutter bee. The bees are now being used by blueberry producers in eastern Canada and to pollinate buckwheat and hybrid canola in the Prairies.

The importation of leafcutter bees also led to a new kind of beekeeper who sells bee larvae to other growers for pollination. The leafcutter bee is gentler than the honeybee and typically will sting only if it is squeezed. Leafcutters are solitary (rather than gregarious, like other kinds of bees) and less likely to wander than honeybees. These characteristics make leafcutters easier to manage and handle than other bees. Management of leafcutters, which do not build colonies or store honey, involves building large nesting arrays of layered, grooved materials. These arrays mimic the sites in which the bees would naturally build their individual nests, using leaves to fill cracks and grooves in soft, decaying wood (see photo below). The

nests must be carefully managed; the density and close proximity of the bees make them susceptible to the development of problems such as fungal infections.[24]

By controlling temperature, the emergence of the adult bees can be synchronized with the alfalfa bloom. The bees are kept in "cold storage" during the winter, remaining dormant at 4°C until about three weeks before the expected crop bloom. At that point, the temperature is turned up to 29°C to trigger the development of adult leafcutter bees. This management system has made Canada the leading producer of alfalfa leafcutters, currently producing 4 billion bees for pollination of domestic and international crops each year. Scientists continue to investigate the threshold and optimal temperature conditions for overwintering survival of leafcutter bees.[25]

Interestingly (but not surprisingly, given their name) leafcutters can also become a pest species. To line their nests, the bees carefully incise small, round pieces of leaves and carry them back to the nesting site, which can damage the leaf. The damage is usually minor, unless there is an unusually large population of feral (escaped domesticated) or native leafcutter bees in a small area.[26]

Photo by Margriet Dogterom

In nature, leafcutter bees use pieces of leaves to line their nests, creating cylinder-shaped nests in decaying wood.

Although our staple grain crops are grasses and are wind-pollinated, many other crops depend on insects for pollination. Approximately 1500 crop species depend on insect pollination,[27] which translates into about 3 to 8% of global crop production.[28] This environmental service has economic, not just ecological, value; recent estimates of

the global value of insect pollination range up to hundreds of billions of dollars.

Many pollinating insects are at risk from the same pressures that threaten other species. Nick Hanley and colleagues reviewed the current main threats to pollinating insects:[29]

- landscape change in agricultural settings, which includes loss of wild food sources for bees
- growing use of certain pesticides; particularly concerning is a group of insecticides called neonicotinoids, linked in some studies to honeybee colony collapse disorder
- introduction of invasive plant species
- pathogens and parasites
- climate change, affecting the ranges of the pollinator species

These are not new phenomena; for example, landscape change and habitat destruction in the 1930s caused the failure of native leafcutter bee populations in Manitoba; this, in turn, led to the collapse of alfalfa seed production in the Canadian Prairies (see "The Science behind the Story: The Alfalfa and the Leafcutter"). Increases in the rate and extent of environmental change now put these ecological relationships at even greater risk. The relationship between flowering plants and insect pollinators is sometimes so specific (the precise shape of an insect's appendages can determine its suitability for pollinating a particular plant species) that even minor changes can have devastating impacts.

Conservation of pollinators is crucial

Preserving the biodiversity of native pollinators is especially important today. North American farmers regularly hire beekeepers to bring colonies of this introduced honeybee to their fields when it is time to pollinate crops (**FIGURE 8.14**). The domesticated workhorse of pollination, the European honeybee (genus *Apis*), is being devastated by parasites and, more recently, by *colony collapse disorder* (or *CCD*), the causes of which are poorly understood.

Some possible mechanisms that have been proposed to explain CCD include impacts of pesticides, especially *neonicotinoids*, a class of neurotoxic pesticide similar to nicotine; infections by mites and other pathogens; malnutrition; immunodeficiency; and loss of habitat. Jeffrey Pettis of the Bee Research Institute and his colleagues studied the occurrence of fungicides on plants that are pollinated by honeybees, and detected 35 different pesticides in the sampled pollen. Even though fungicides are generally considered to be safe for bees, the researchers found an increased probability of infection by a gut parasite in bees that consumed pollen with high fungicide loads.[30]

Research also indicates that honeybees are sometimes less effective pollinators than many native species[31] but often outcompete them, keeping the wild species away from the plants. This may diminish the overall effectiveness of bee pollination.

Farmers and homeowners can help maintain populations of pollinating insects by reducing or eliminating pesticide use. All insect pollinators, including honeybees, are vulnerable to insecticides that are applied to crops, lawns, and gardens. Some insecticides are designed to specifically target certain types of insects, but many are not. Without full and detailed information on the effects of pesticides, farmers and homeowners trying to control the "bad" bugs that threaten the plants they value all too often kill the "good" insects as well.

Homeowners, even in the middle of a city, can encourage populations of pollinating insects by planting gardens of flowering plants and by providing nesting sites for bees. By allowing noncrop flowering plants (such as clover) to grow around the edges of their fields, farmers can maintain a diverse community of insects—some of which will pollinate their crops.[7]

FIGURE 8.14
European honeybees are widely used to pollinate crop plants, and beekeepers transport hives of bees to crops when it is time for flowers to be pollinated. However, honeybees have recently suffered devastating epidemics of parasitism and colony collapse disorder, making it increasingly challenging to conserve native species of pollinators.

Genetically Modified Food

The Green Revolution enabled us to feed a greater number and proportion of the world's people, but relentless population growth demands still more. A new set of potential solutions began to arise in the 1980s and 1990s, as advances in genetics enabled scientists to directly alter the genes of organisms, including crop plants and livestock. The genetic modification of organisms that provide us food holds promise for increasing nutrition and the efficiency of agriculture while lessening the impacts of agriculture on the planet's environmental systems.

However, genetic modification of food organisms raises concerns and may pose risks that are not yet thoroughly understood. This has given rise to protest around the globe from consumer advocates, small farmers, opponents of big business, and environmental activists. Because genetic modification of food organisms has generated so much emotion and controversy, it is vital at the outset to clear up the terminology and clarify exactly what is involved in the process.

Genetic modification of organisms depends on recombinant DNA

The genetic modification of crops and livestock is one type of **genetic engineering**, any process whereby scientists directly manipulate an organism's genetic material in the lab by adding, deleting, or changing segments of its DNA. To genetically engineer organisms, scientists extract genes from the DNA of one organism and transfer them into the DNA of another to create a **genetically modified organism (GMO)**. The technique uses *recombinant DNA* technology, referring to DNA that has been patched together from the DNA of multiple organisms. In this process, scientists break up DNA from multiple organisms and then splice segments together, trying to place genes that produce certain proteins and code for certain desirable traits (such as rapid growth, disease or pest resistance, or higher nutritional content) into the genetic information, or *genome*, of organisms lacking those traits.

Recombinant DNA technology was developed in the 1970s by scientists studying the bacterium *Escherichia coli*. As shown in **FIGURE 8.15**, scientists first isolate *plasmids*, small, circular DNA molecules, from a bacterial culture. DNA containing a gene of interest is removed from the cells of another organism. Scientists insert the gene of interest into the plasmid to form recombinant DNA. This recombinant DNA enters new bacteria, which then reproduce, generating many copies of the desired gene. These copies are then introduced into the cells of the organism that is to be genetically modified.

An organism that contains DNA from another species is called a **transgenic** organism, and the genes that have moved between them are called *transgenes*. The creation of transgenic organisms is one type of **biotechnology**, the application of biological science to create products derived from organisms.

Recombinant DNA and other types of biotechnology have helped us develop medicines, clean up pollution, understand the causes of cancer and other diseases, dissolve blood clots after strokes and heart attacks, and make better beer and cheese. **FIGURE 8.16** details several notable developments in GM foods. These examples and the stories behind them illustrate both the promises and the pitfalls of food biotechnology.

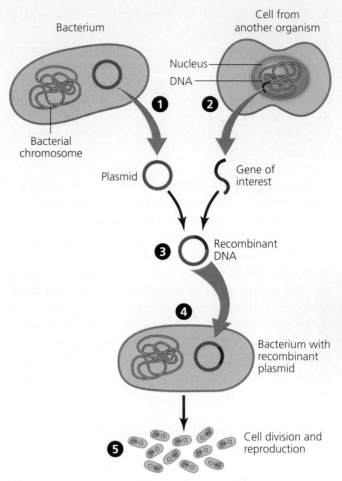

FIGURE 8.15

To create recombinant DNA, a gene is excised from the DNA of an organism and inserted into a stretch of bacterial DNA called a plasmid. The plasmid is then introduced into cells of the organism to be modified. If all goes as planned, the new gene will be expressed in the GM organism as a desirable trait, such as rapid growth or high nutritional content in a food crop.

Genetic engineering is like, and unlike, traditional breeding

The genetic alteration of plants and animals by humans is nothing new; we have been influencing the genetic makeup of our livestock and crop plants for thousands of years. Our ancestors altered the gene pools of domesticated plants and animals through selective breeding by preferentially mating individuals with favoured traits, so that offspring would inherit those traits. Early farmers selected plants and animals that grew faster, were more resistant to disease and drought, and produced large amounts of fruit, grain, or meat.

Proponents of GM crops often stress this continuity with our past and say there is little reason to expect that today's GM food will be any less safe than traditionally bred food. However, as biotech's critics are quick to point out, the techniques used to create GM organisms differ from traditional selective breeding in several important ways.

For one, selective breeding generally mixes genes of individuals of the same species, whereas with recombinant

Several notable examples of genetically modified food technology

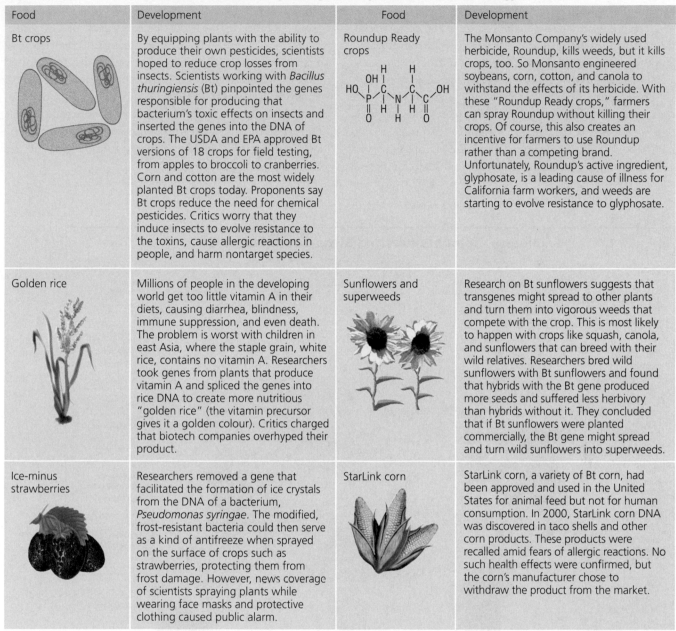

Food	Development	Food	Development
Bt crops	By equipping plants with the ability to produce their own pesticides, scientists hoped to reduce crop losses from insects. Scientists working with *Bacillus thuringiensis* (Bt) pinpointed the genes responsible for producing that bacterium's toxic effects on insects and inserted the genes into the DNA of crops. The USDA and EPA approved Bt versions of 18 crops for field testing, from apples to broccoli to cranberries. Corn and cotton are the most widely planted Bt crops today. Proponents say Bt crops reduce the need for chemical pesticides. Critics worry that they induce insects to evolve resistance to the toxins, cause allergic reactions in people, and harm nontarget species.	Roundup Ready crops	The Monsanto Company's widely used herbicide, Roundup, kills weeds, but it kills crops, too. So Monsanto engineered soybeans, corn, cotton, and canola to withstand the effects of its herbicide. With these "Roundup Ready crops," farmers can spray Roundup without killing their crops. Of course, this also creates an incentive for farmers to use Roundup rather than a competing brand. Unfortunately, Roundup's active ingredient, glyphosate, is a leading cause of illness for California farm workers, and weeds are starting to evolve resistance to glyphosate.
Golden rice	Millions of people in the developing world get too little vitamin A in their diets, causing diarrhea, blindness, immune suppression, and even death. The problem is worst with children in east Asia, where the staple grain, white rice, contains no vitamin A. Researchers took genes from plants that produce vitamin A and spliced the genes into rice DNA to create more nutritious "golden rice" (the vitamin precursor gives it a golden colour). Critics charged that biotech companies overhyped their product.	Sunflowers and superweeds	Research on Bt sunflowers suggests that transgenes might spread to other plants and turn them into vigorous weeds that compete with the crop. This is most likely to happen with crops like squash, canola, and sunflowers that can breed with their wild relatives. Researchers bred wild sunflowers with Bt sunflowers and found that hybrids with the Bt gene produced more seeds and suffered less herbivory than hybrids without it. They concluded that if Bt sunflowers were planted commercially, the Bt gene might spread and turn wild sunflowers into superweeds.
Ice-minus strawberries	Researchers removed a gene that facilitated the formation of ice crystals from the DNA of a bacterium, *Pseudomonas syringae*. The modified, frost-resistant bacteria could then serve as a kind of antifreeze when sprayed on the surface of crops such as strawberries, protecting them from frost damage. However, news coverage of scientists spraying plants while wearing face masks and protective clothing caused public alarm.	StarLink corn	StarLink corn, a variety of Bt corn, had been approved and used in the United States for animal feed but not for human consumption. In 2000, StarLink corn DNA was discovered in taco shells and other corn products. These products were recalled amid fears of allergic reactions. No such health effects were confirmed, but the corn's manufacturer chose to withdraw the product from the market.

FIGURE 8.16 As genetically modified foods were developed, a number of products ran into public opposition or trouble in the marketplace. A selection of such cases serves to illustrate some of the issues that proponents and opponents of genetically modified foods have been debating.

DNA technology, scientists mix genes of different species, as different as spiders and goats. For another, selective breeding deals with whole organisms living in the field, whereas genetic engineering involves lab experiments dealing with genetic material apart from the organism. Whereas traditional breeding selects from among combinations of genes that come together on their own, genetic engineering creates novel combinations directly. Thus, traditional breeding changes organisms through the process of selection, whereas genetic engineering is more akin to the process of mutation.

In just a few decades, GM foods have gone from science fiction to big business. As recombinant DNA technology

was first developed in the 1970s, scientists debated among themselves whether the new methods were safe. They collectively regulated and monitored their own research until most scientists were satisfied that reassembling genes in bacteria did not create dangerous superbacteria. Once the scientific community declared itself confident that the technique was safe in the 1980s, industry leaped at the chance to develop hundreds of applications, from improved medicines to designer plants and animals.

Most GM crops today are engineered to resist herbicides, so that farmers can apply herbicides to kill weeds without having to worry about killing their crops. Other crops are engineered to resist insect attacks. Some are modified for both types of resistance. Crop resistance to herbicides and pests enables large-scale commercial farmers to grow crops more efficiently and economically, and to intensify yields from a fixed amount of hectares. As a result, sales of GM seeds to farmers have risen quickly.

Today three-fourths of the world's soybean plants are transgenic, as is one of every four corn plants, one of every five canola plants, and half of all cotton plants. Globally in 2013, 18 million farmers grew GM crops on over 175 million ha of farmland, more than a 100-fold increase over the 1996 hectare coverage and about 12% of all crops planted globally.[32] Of the 19 "mega"-countries growing more than 50 000 hectares each of GM crops in 2014, 13 are developing nations. The top five overall in hectares (in decreasing order) are the United States, Brazil, Argentina, India, and Canada, at almost 12 million ha (**FIGURE 8.17**). Four crops (soybean, cotton, maize, and canola), and two controlled traits (herbicide tolerance and insect resistance) account for by far the majority of the area devoted to the production of GM crops worldwide.

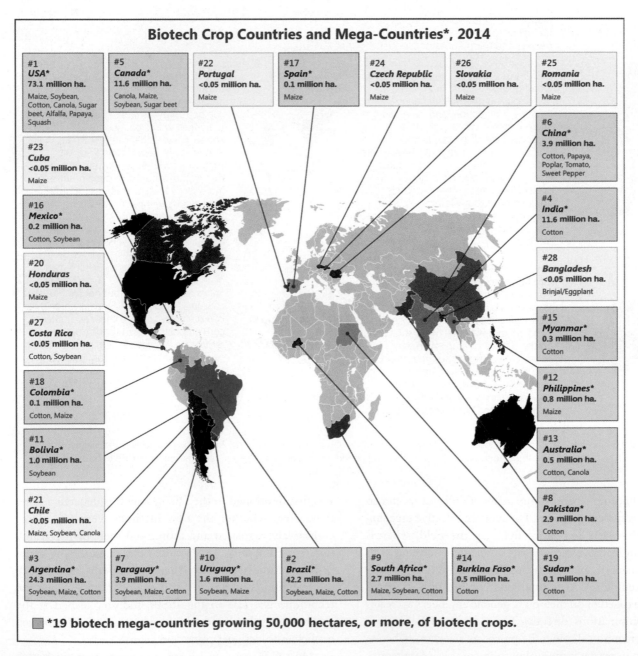

FIGURE 8.17 Of the world's agricultural nations, the United States devotes the most land area to GM crops. Soybeans constitute the majority of GM crops, followed by maize (corn), cotton, and canola.

Source: James, C. (2014). Global Status of Commercialized Biotech/GM Crops: 2014. ISAAA Brief 49. ISAAA, Ithaca, NY. http://www.isaaa.org.

What are the impacts of GM crops?

As GM crops have been adopted more widely, as research proceeded, and as biotech business expanded, many citizens, scientists, and policymakers became concerned. Some feared the new foods might be dangerous for people to eat—what if there were unexpected health consequences, such as unanticipated allergic reactions to transgenes in GM foods? But all of us in North America—virtually without exception—have been exposed to genetically modified food ingredients for decades, and there are essentially no fully documented, reliable cases of negative human health impacts. Many scientific studies have documented no noticeable effects on human health from the ingestion of genetically modified food ingredients.[33]

However, some activists have argued that because it would be impossible to test for all possible human health impacts, we can never be certain that GMOs will not negatively impact human health. In a 2013 study, Sándor Spisák and colleagues argued that the "uptake and fate of foreign DNA ingested with the daily food intake in the gastrointestinal tract of mammals is not a completely understood topic."[34] They also investigated the possibility of foreign genetic material entering the bloodstream after ingestion. Such uncertainty may argue for continued caution.

Other observers have been concerned that transgenes might escape, pollute ecosystems, and damage nontarget organisms. Still others worry that pests could evolve resistance to genetically engineered supercrops and become "superpests," or that transgenes would be transferred from crops to other plants and turn them into "superweeds." Some worried that transgenes might ruin the integrity of native ancestral races of crops.

Because the technology is relatively new and its large-scale introduction into the environment is newer still, there remains a lot scientists don't know about how transgenic crops behave in the field. Millions of North Americans eat GM foods every day without any obvious signs of harm, and evidence for negative ecological effects is limited so far. However, it is still too early to dismiss all concerns without further scientific research. There are numerous mechanisms whereby transgenes can "escape" from the confines of the organism into which they have been implanted and move out into native populations. Therefore, critics argue that we should adopt the *precautionary principle*, the idea that one should not proceed until the ramifications of an action are well understood.

Debate over GM foods involves more than science

Far more than science is involved in the debate over GM foods. Ethical issues have played a large role. For many people, the idea of "tinkering" with the food supply seems dangerous or morally wrong. Even though our agricultural produce is the highly artificial product of thousands of years of selective breeding, people tend to think of food as natural. Furthermore, because every person relies on food for survival and cannot choose *not* to eat, the genetic modification of dietary staples such as corn, wheat, and rice essentially forces people to consume GM products or to go to special effort to avoid them.

The perceived lack of control over one's own food has driven widespread concern about domination of the global food supply by a few large businesses. Gigantic agrobiotech companies, among them Monsanto, Syngenta, Bayer CropScience, Dow, DuPont, and BASF, develop GM technologies. Many activists say these multinational corporations threaten the independence and well-being of the small farmer and raise concerns about the global food supply being dominated by a few large corporations. This perceived loss of democratic control is a driving force in the opposition to GM foods, especially in Europe and the developing world. Critics of biotechnology also voice concern that much of the research into the safety of GM organisms is funded, overseen, or conducted by the corporations that stand to profit if their transgenic crops are approved for human consumption, animal feed, or ingredients in other products.

So far, GM crops have not lived up to their promise of feeding the world's hungry—perhaps because they haven't been allowed to. Nearly all commercially available GM crops have been engineered to express either pesticidal properties (e.g., Bt crops) or herbicide tolerance. Often, these GM crops are tolerant to herbicides that the same company manufactures and profits from (e.g., Monsanto's Roundup Ready crops). Crops with traits such as increased nutritional content, drought tolerance, and salinity tolerance, which might benefit poor small-scale farmers in developing countries, have not been widely commercialized, perhaps because corporations have less economic incentive to do so. Crops such as "golden rice"—engineered with a high content of vitamin A and proposed as the solution to vitamin A deficiencies throughout the developing world—have met with much social and environmental opposition, and limited success.

The development of GM crops has been largely driven by market considerations of companies selling proprietary products. When the U.S.-based Monsanto Company began developing GM products in the mid-1980s, it foresaw public anxiety and worked hard to inform, reassure, and work with environmental and consumer advocates, whom the company feared would otherwise oppose the technology. Monsanto even lobbied the U.S. government to regulate the industry so the public would feel safer about it. These efforts were undermined, however, when the company's first GM product, a growth hormone to spur milk production in cows, alarmed consumers concerned about children's health. Then, when the company went through a leadership change, its new head changed tactics and pushed new products aggressively without first reaching out to opponents. Opposition built, and the company lost the public's trust, especially in Europe and in the developing world.

David-and-Goliath battles that pitted giant Monsanto against lone farmers such as Canadian Percy Schmeiser have not helped to repair the company's public image. Since 1997, Monsanto has pursued 145 such lawsuits in the United States alone.[35] Monsanto says it is merely demanding that farmers heed the patent law. Given such developments, the future of GM foods seems likely to hinge on legal, social, economic, and political factors as well as scientific ones.

European consumers have been particularly vocal in expressing their unease about the possible risks of GM technologies. Opposition in nations of the European Union resulted in a *de facto* moratorium on GM foods from 1998 to 2003, blocking the importation of hundreds of millions of dollars in agricultural products. This prompted the United States to bring a case before the World Trade Organization in 2003, complaining that Europe's resistance was hindering free trade. Europeans now widely demand that GM foods be labelled as such and criticize the United States and Canada for not joining more than 160 other nations in signing the Cartagena Protocol on Biosafety (part of the United Nations Convention on Biodiversity), a treaty that lays out guidelines for open information about exported crops. Canada has been a party to the Convention on Biodiversity since 1992 but has never ratified the Cartagena Protocol.[36]

Transnational spats will surely affect the future direction of agriculture, but consumers and the governments of the world's developing nations could exert the most influence in the end. Decisions by the governments of India and Brazil to approve GM crops (following long and divisive debates) are already adding greatly to the world's transgenic agriculture, and China is aggressively expanding its use of transgenic crops.

A counterexample is Zambia, one of several African nations that refused U.S. food aid meant to relieve starvation during a drought in late 2002. The governments of these nations worried that their farmers would plant some of the GM corn seed that was meant to be eaten and that GM corn would thereby establish itself in their countries. They viewed this as undesirable because African economies depend on exporting food to Europe, which has put severe restrictions on GM food. In the end, Zambia's neighbours accepted the grain after it had been milled (so none could be planted), but Zambia held out.

FIGURE 8.18
Debate over GM foods reached a dramatic climax in Zambia in 2002, when the government refused U.S. shipments of GM corn that were intended to relieve starvation due to drought. Here a Zambian mother and child wait in a line for food assistance.

Citing health and environmental risks, uncertain science, and the precautionary principle, the Zambian government declined the aid, despite the fact that 2 to 3 million of its people were at risk of starvation (**FIGURE 8.18**). Intense debate followed within the country and around the world, and eventually the United Nations delivered non-GM grain to the country. Zambia has largely stuck to its non-GMO policy, and continues to produce and even export non-GMO maize.

The Zambian experience demonstrates some of the ethical, economic, and political dilemmas modern nations face. The corporate manufacturers of GM crops naturally aim to maximize their profits, but they also aim to develop products that can boost yields, increase food security, and reduce hunger. Although industry, activists, policymakers, and scientists all agree that hunger and malnutrition are problems and that agriculture should be made environmentally safer, they often disagree about the solutions to these dilemmas and the risks that each proposed solution presents.

Preserving Crop Diversity

One concern many people harbour about GM crops is that transgenes might move, by pollination, into local native races of crop plants. There is certainly now abundant evidence that this has already happened in some localities. The monocultures of modern industrial agriculture essentially place all our eggs in one basket, such that any single catastrophic cause could potentially wipe out entire crops. Preserving crop diversity—domesticated varieties and the wild relatives of crop plants—gives us the genetic diversity that may include ready-made solutions to unforeseen problems.

Crop diversity provides insurance against failure

Preserving the integrity of native variants provides insurance against widespread commercial crop failure. The regions where crops first were domesticated generally remain important repositories of crop biodiversity. Although modern industrial agriculture relies on a small number of plant types, its foundation lies in the diverse varieties that still exist in places like Oaxaca. These varieties contain genes that, through conventional crossbreeding or genetic engineering, might confer resistance to disease, pests, inbreeding, and other pressures that challenge modern agriculture.

Because accidental interbreeding can decrease the diversity of local variants, many scientists argue that we need to protect areas like Oaxaca. For this reason, the Mexican government helped create the Sierra de Manantlán Biosphere Reserve around an area harbouring the localized plant thought to be the direct ancestor of maize. For this reason, too, it imposed a national moratorium in 1998 on the planting of transgenic corn (although that ban was lifted in 2005).

We have lost a great deal of genetic diversity in our crop plants already. The number of wheat varieties in China is estimated to have dropped from 10 000 in 1949 to 1000 by the 1970s, and Mexico's famed maize varieties now number only 30% of what was grown in the 1930s. In the United States, many fruits and vegetables have decreased in diversity by 90% in less than a century. Note, however, that the number of varieties that exist is not, on its own, indicative of the robustness of biodiversity. For example, in recent years the number of wheat varieties in China has actually increased, but the genetic diversity among those varieties has narrowed.[37]

A primary cause of the loss of crop diversity is that market forces discourage diversity in the appearance of fruits and vegetables. Commercial food transporters and processors prefer items to be similar in size and shape, for convenience. Consumers, for their part, have shown preferences for uniform, standardized food products over the years. Now that local organic agriculture is growing in affluent societies, however, consumer preferences for diversity are increasing.

Seed banks are living museums for seeds

Protecting areas with high crop diversity is one way to preserve genetic assets for our agricultural systems.

(a) Traditional food plants of the desert southwest

(b) Pollination by hand

FIGURE 8.19
Seed banks preserve genetic diversity of traditional crop plants. Native Seeds/SEARCH of Arizona preserves seeds of food plants important to traditional diets of Native Americans of Arizona, New Mexico, and northwestern Mexico. Beans, chiles, squashes, gourds, maize, cotton, and lentils are all in its collections, as well as less-known plants such as amaranth, lemon basil, and devil's claw **(a)**. Traditional foods such as mesquite flour, prickly pear pads, chia seeds, tepary beans, and cholla cactus buds help fight diabetes, which has become more common in Native Americans since they adopted a Western diet. At the farm where seeds are grown, care is taken to pollinate varieties by hand **(b)** to protect their genetic distinctiveness.

Another is to collect and store seeds from crop varieties and periodically plant and harvest them to maintain a diversity of cultivars. This is the work of **seed banks** and **gene banks**, institutions that preserve seed types as a kind of living museum of genetic diversity (**FIGURE 8.19**). In total, these facilities hold roughly 6 million seed samples, keeping them in cold, dry conditions to encourage long-term viability. The $300 million in global funding for these facilities is not adequate for proper storage and for the labour of growing out the seed periodically to renew the stocks. Therefore, it is questionable how many of these 6 million seeds are actually preserved.

The Royal Botanic Garden's Millennium Seed Bank in Britain holds over 2 billion seeds from more than 34 000 wild plant species, and aims to bank seed from 25% of the world's plants by 2020.[38] In Arctic Norway, construction has begun on a "doomsday vault" seed bank, intended to hold seeds from around the world as a safeguard against global agricultural calamity. Other major efforts include large seed banks such as the U.S. National Seed Storage Laboratory at Colorado State University, Seed Savers Exchange in Iowa, Plant Gene Resources of Canada, and the Wheat and Maize Improvement Centre (CIMMYT) in Mexico.

Raising Animals for Food

Food from cropland agriculture makes up a large portion of the human diet, but most people also eat animal products. People don't *need* to eat meat or other animal products to live full, active, healthy lives, but for most people it is difficult to obtain a balanced diet without incorporating animal products. Many of us do eat animal products, and this choice has environmental, social, agricultural, and economic impacts.

Consumption of animal products is growing

As wealth and global commerce have increased, so has our consumption of meat, milk, eggs, fish, and other animal products (**FIGURE 8.20**). The world population of domesticated animals and animals raised in captivity for food rose from 7.3 billion animals in 1961 to 27.5 billion animals in 2011. Most of these animals are chickens, although the most-eaten meat per unit of weight is pork. Global meat production has increased to almost 312 million metric tons per year as of 2014, per capita meat consumption is now at approximately 42.8 kg per capita per year, and the FAO expects this to continue to increase to 2050.[39]

Like other domesticated species, livestock and other farm animals can be at risk of biodiversity loss and even extinction. The FAO's Global Databank for Animal Genetic Resources for Food and Agriculture contains information on almost 8000 livestock breeds, of which about 20% are classified as "at risk." During the first six years of the twenty-first century, for example, 62 livestock breeds became extinct, which amounts to a loss of almost one breed per month.[40] This represents a tremendous loss of diversity in livestock animals.

High demand has led to feedlot agriculture

In traditional agriculture, livestock were kept by farming families near their homes or were grazed on open grasslands by nomadic herders or sedentary ranchers. These traditions have survived, but the advent of industrial agriculture, responding to the pressure of global population growth, has added a new method. **Feedlots**, also known as "factory farms" or *concentrated animal feeding operations (CAFOs)*, are essentially huge warehouses or pens designed to deliver energy-rich food to animals living at extremely high densities (**FIGURE 8.21**). Today over half of the world's pork and most of the poultry come from feedlots.

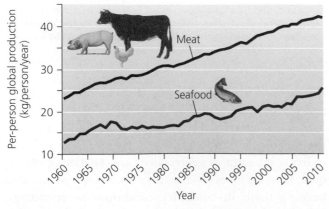

FIGURE 8.20
Per capita consumption of meat from farm animals has increased steadily worldwide over the past few decades, as has per capita consumption of seafood (marine and fresh water, harvested and farmed).
Source: Data from Food and Agriculture Organization of the United Nations (FAO).

FIGURE 8.21
These chickens at a factory farm are housed in crowded conditions and have been "debeaked," the tips of their beaks cut off, to prevent them from pecking one another. The hens cannot leave the cages and spend their lives eating, defecating, and laying eggs, which roll down slanted floors to collection trays. The largest chicken farms house hundreds of thousands of individuals.

Feedlot operations allow for greater production of food and are necessary to keep up with current levels of meat consumption in Canada and the United States. Feedlots have one overarching benefit for environmental quality: Taking cattle, sheep, goats, and other livestock off the land and concentrating them in feedlots reduces the impact they would otherwise exert on large portions of the landscape. Overgrazing can degrade soils and vegetation, and hundreds of millions of hectares of land are considered overgrazed. Animals that are densely concentrated in feedlots do not contribute to overgrazing and soil degradation.

However, feedlots can have significant environmental impacts. Wastes from feedlots emit strong odours and can pollute surface water and ground water with excess nitrogen and phosphorus. Livestock produce prodigious amounts of feces and urine; one dairy cow can produce about 20 400 kg of waste in a single year. Poor waste containment practices at some feedlots have been linked to outbreaks of disease, including virulent strains of *Pfiesteria*, a microbe that poisons fish. In 2000 in Walkerton, Ontario, a deadly strain of *Escherichia coli* (*E. coli*) bacteria, thought to have originated from the contamination of municipal water wells by runoff from factory farms, caused the deaths of seven people and serious illness in hundreds of others. Livestock waste can release other bacteria and viruses that can sicken people, including *Salmonella*, *E. coli*, *Giardia*, *Microsporidia*, and other pathogens that cause diarrhea, botulism, and parasitic infections.

The crowded and dirty conditions under which animals are often kept at factory farms necessitate the use of antibiotics to control disease. These chemicals can be transferred up the food chain, and their overuse can cause microbes to evolve resistance to them, making the drugs less effective. Hormones are administered to livestock as well, and feed is spiked with heavy metals that spur growth. Livestock excrete most of these chemicals, which end up in waste water and may be transferred up the food chain in downstream ecosystems. Some of the chemicals that remain in livestock meat are transferred to us when we eat the meat. Crowded conditions also can exacerbate outbreaks of diseases such as *avian influenza* ("bird flu") and *bovine spongiform encephalitis* (*BSE*, or "mad cow" disease), which are now known to be transferable to humans in serious and even deadly forms.

Feedlot impacts can be minimized when properly managed, and both the federal and provincial governments regulate feedlots in Canada. Feedlot manure can be applied to farm fields, reducing the need for chemical fertilizers. Manure in liquid form can be injected into the ground where plants need it, and farmers can conduct tests to determine amounts that are appropriate to apply. Most importantly, proper waste containment and monitoring can prevent the release of harmful substances into the surrounding environment.

Our food choices are also energy choices

What we choose to eat has significant ramifications for how we use energy, water, and the land that supports agriculture. Whenever energy moves from one level to the next in a trophic pyramid, as much as 90% of the energy is lost. For example, if we feed grain to a cow and then eat beef from the cow, we lose a great deal of the grain's energy to the cow's digestion and metabolism. Energy is used up when the cow converts the grain to tissue as it grows and as the cow uses its muscle mass on a daily basis to maintain itself. For this reason, eating meat is far less energy-efficient than relying on a vegetarian diet. The lower in the food chain we take our food sources, the greater the proportion of the Sun's energy we put to use as food.

Some animals convert grain feed into milk, eggs, or meat more efficiently than others (**FIGURE 8.22**). Scientists have calculated relative energy conversion efficiencies for

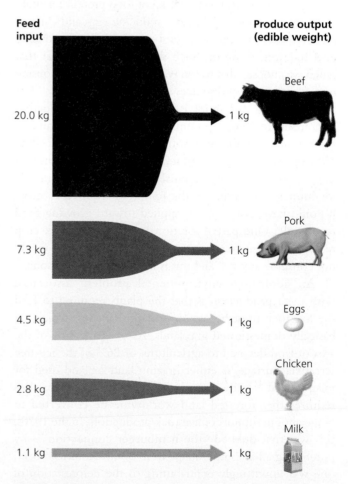

FIGURE 8.22
Different animal food products require different amounts of feed. Chickens must be fed 2.8 kg of feed for each 1 kg of resulting chicken meat, for instance, whereas 20 kg of feed must be provided to cattle to produce 1 kg of beef.
Source: Data from Smil, V. (2001). *Feeding the World: A Challenge for the Twenty-First Century.* Cambridge, MA: MIT Press.

TABLE 8.3 Land and Water Required to Produce 1 kg of Protein

Protein Source	Land (per kg protein)	Water (per kg protein)*
Pork	90 m²	175 kg
Eggs	22 m²	15 kg
Chicken	14 m²	50 kg
Milk	23.5 m²	250 kg
Beef	245 m²	750 kg

*These numbers increase significantly if the water required to grow the feed for the animals is included in the calculation.

Source: Smil, V. (2001). *Feeding the World: A Challenge for the Twenty-First Century.* Cambridge, MA: MIT Press.

FIGURE 8.23
People practise many types of aquaculture. Here, fish farmers in China tend their animals.

different types of animals. Such energy efficiencies have ramifications for land use—land and water are required to raise food for the animals, and some animals require more than others. **TABLE 8.3** shows the area of land and weight of water required to produce 1 kg of food protein for milk, eggs, chicken, pork, and beef. Producing eggs and chicken meat requires the least space and water, whereas producing beef requires the most. Such differences make clear that when we choose what to eat, we are also indirectly choosing how to make use of resources such as land and water.

In 1900 we fed about 10% of global grain production to animals. In 1950 this number had reached 20%, and by the beginning of the twenty-first century we were feeding 45% of global grain production to animals. Although much of the grain fed to animals is not of a quality suitable for human consumption, the resources required to grow it could have instead been applied toward growing food for people. One partial solution is to feed livestock crop residues, plant matter such as stems and stalks that we would not consume anyway, and this is increasingly being done.

An additional environmental problem associated with meat production is that the plants required to feed the livestock must be grown on large ranches, which are basically domesticated grasslands. As of 2014, 80% of the area of land devoted to agriculture, or 26% of the ice-free terrestrial surface, is either grazing land or land used for raising feed.[41] The growth in meat consumption thus requires that forested land worldwide be converted to rangelands in support of livestock production. In the 1970s, this was first dubbed "the hamburger connection"—by purchasing a hamburger made from South American beef, one was unwittingly contributing to the deforestation of tropical rain forests and their conversion into rangelands.

We also raise fish on "farms"

In addition to plants grown on croplands and animals raised on rangelands and in feedlots, we rely on aquatic organisms for food. Wild fish populations are plummeting throughout the world's oceans as increased demand and new technologies have led us to overharvest most marine fisheries. This means that raising fish and shellfish on "fish farms" may be the only way to meet the growing demand for these foods.

Raising aquatic organisms in controlled environments is called **aquaculture**. Many aquatic species are grown in open water in large, floating net-pens. Others are raised in land-based ponds or holding tanks. People pursue freshwater, brackish, and marine aquaculture, growing a variety of plants, animals, and algae (seaweed). Aquaculture is the fastest-growing type of food production; in the past 20 years global output has increased sevenfold, doubling in just the past decade. Aquaculture today provides a third of the world's fish for human consumption, is most common in Asia, and involves over 220 species ranging from shrimp to clams to seaweeds, as well as non-food products such as pearls (**FIGURE 8.23**). Some, such as carp, are grown for local consumption, whereas others, such as salmon and shrimp, are exported to affluent countries.

Aquaculture has benefits and drawbacks

When conducted on a small scale by families or villages, aquaculture helps ensure people a reliable protein source. Small-scale aquaculture can be sustainable, and it is compatible with other activities. For instance, uneaten fish scraps make excellent fertilizers for crops. Aquaculture on larger scales can help improve a region's or nation's food security by increasing overall amounts of fish available.

Aquaculture on any scale has the benefit of reducing fishing pressure on overharvested and declining wild stocks, as well as providing employment for fishers who can no longer fish from depleted natural stocks. Reducing fishing pressure also reduces *bycatch*, the unintended catch of nontarget organisms that results from commercial fishing.

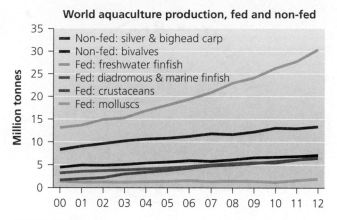

World aquaculture production, fed and non-fed

— Non-fed: silver & bighead carp
— Non-fed: bivalves
— Fed: freshwater finfish
— Fed: diadromous & marine finfish
— Fed: crustaceans
— Fed: molluscs

FIGURE 8.24
The farmed production of fish that require feeding, often with meal made from wild-caught fish, is increasing much faster than non-fed farmed fish. This is especially true of freshwater finned fish, such as perch. The use of fishmeal made from wild-caught fish puts additional pressure on wild stocks. *Diadromous* fish spend part of their lifecycle in fresh water, and part in the ocean.
Source: Food and Agriculture Organization of the United Nations, 2014, State of the World's Fisheries and Aquaculture, www.fao.org/3/a-i3720e. pdf. Reproduced with permission.

Furthermore, aquaculture relies far less on fossil fuels than do fishing vessels and provides a safer work environment. Fish farming can also be remarkably energy-efficient, producing as much as 10 times more fish per unit area than is harvested from oceanic waters on the continental shelf and up to 1000 times as much as is harvested from the open ocean.

Along with its benefits, however, aquaculture has some serious disadvantages. Counterintuitively, aquaculture can increase the pressure on wild fish stocks if the feed for the cultured fish is meal made from fish caught in the wild, which is often the case. The proportion of fish that require feeding is increasing, compared to non-fed farmed fish, exacerbating this problem (**FIGURE 8.24**).

Dense concentrations of farmed animals can increase the incidence of disease, which reduces food security, necessitates antibiotic treatment, and results in additional expense. A virus outbreak wiped out half a billion dollars in shrimp in Ecuador in 1999, for instance. If farmed aquatic organisms escape into ecosystems where they are not native, they can spread disease to native stocks or may outcompete native organisms for food or habitat. The opposite has also occurred—recent research suggests that wild Pacific salmon swimming near aquaculture pens have passed on parasites, which then are able to spread rapidly as a result of the high densities of organisms in the pens.

The possibility of competition also arises when farmed animals have been genetically modified (**FIGURE 8.25**).

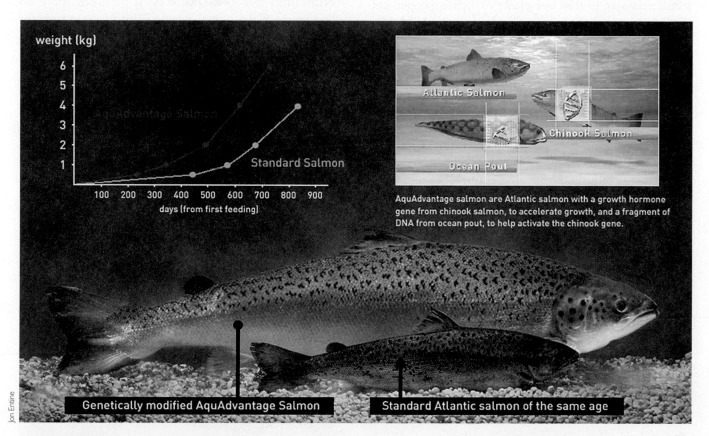

AquAdvantage salmon are Atlantic salmon with a growth hormone gene from chinook salmon, to accelerate growth, and a fragment of DNA from ocean pout, to help activate the chinook gene.

Genetically modified AquAdvantage Salmon

Standard Atlantic salmon of the same age

Jon Entine

FIGURE 8.25
Genetic modification of fish has resulted in transgenic fish that can grow considerably faster and larger than wild fish of the same age and species. In November of 2015, the fast-growing AquAdvantage salmon, shown here, became the first genetically modified animal to be approved by the U.S. Food and Drug Administration for human consumption.

For example, genetic engineering of Pacific salmon has produced transgenic fish that weigh up to 11 times more than nontransgenic ones. Transgenic Atlantic salmon raised in Scotland have been engineered to grow to 5 to 50 times the normal size for their species. GM fish such as these could outcompete their non-GM wild cousins. They may also interbreed with native and hatchery-raised fish and weaken already troubled stocks. Researchers have concluded that, under certain circumstances, escaped transgenic salmon may increase the extinction risk that native populations of their species face, in part because the larger male fish have better odds of mating successfully.

Sustainable Agriculture

Post–Green Revolution industrialized agriculture has allowed food production to keep pace with the growing population, but it also has caused negative environmental impacts. These range from the degradation of soils to reliance on fossil fuels to problems arising from pesticide use, genetic modification, and intensive feedlot and aquaculture operations. Although developments in intensive commercial agriculture have alleviated some environmental pressures, they have exacerbated others. Industrial agriculture in some form seems necessary to feed our planet's 7 billion people, but many feel we might be better off in the long run by practising less-intensive methods of raising animals and crops.

Farmers and researchers have made great advances toward sustainable agriculture in recent years. **Sustainable agriculture** is agriculture that does not deplete soils faster than they form. It is farming and ranching that does not reduce the amount of healthy soil, clean water, and genetic diversity essential to long-term crop and livestock production. It is, simply, agriculture that can be practised in the same way far into the future. For example, *no-till agriculture*, in which the depth and frequency of ploughing and tilling are kept to a minimum to protect soil moisture and prevent compaction, appears to fit the notion of sustainable agriculture and it can be implemented on the scale of modern agriculture.

Sustainable agriculture is closely related to *low-input agriculture*, agriculture that uses smaller amounts of pesticides, fertilizers, growth hormones, water, and fossil fuel energy than are currently used in industrial agriculture. Food-growing practices that use no synthetic fertilizers, insecticides, fungicides, or herbicides—but instead rely on biological approaches such as composting and biocontrol—are termed **organic agriculture**.

Organic agriculture is on the increase

Citizens, government officials, farmers, and agricultural industry representatives have debated the meaning of the

FIGURE 8.26
The organic certification logo designates food products that meet the Canadian standards for organic production.

Copyright © by Canadian Food Inspection Agency. Reprinted by permission

word *organic* for many years. Experimental organic gardens began to appear in North America in the 1940s, but in Canada the federal Organic Products Regulations only came into effect in December 2006, as a new part of the *Canadian Agricultural Products Act*. This law establishes national standards for organic products and facilitates the labelling, quality, and sale of organic food. The organic certification logo (**FIGURE 8.26**) is permitted only on food products that meet specific Canadian standards for organic production, such as using natural fertilizers and raising animals in conditions that mimic nature as much as possible. Products must also contain at least 95% organic ingredients.

Long viewed as a small niche market, the market for organic foods is on the increase. Although it still accounts for only a small percentage of food expenditures in Canada, with a market share less than 2% (**TABLE 8.4**), sales of organic products have tripled since 2006, making

TABLE 8.4 **Estimated Value of Total Canadian Organic Sales in 2012**

	Sales value ($ millions)	Market share
Organic food and beverage (excluding alcohol)	2,823.08	1.70%
Organic alcohol	135.0	0.67%
Organic supplements	34.4	1.25%
Organic fibres and clothing	24.2	0.15%
Organic personal care	41.1	0.45%
Organic flowers	3.0	0.10%
Organic exports from Canada	458.0	
Total Canadian Organic Market	458.0	

Source: Shauna MacKinnon: *Canada Organic Trade Association in The National Organic Market Growth, Trends & Opportunities - 2013.*

organic foods the fastest-growing agri-product sector.[42] Fruits and vegetables are the leaders in the organic sector, followed by dairy products.

Production is increasing along with demand. Although organic agriculture takes up less than 1% of cultivated land worldwide (37 million ha in 2011), this area is rapidly expanding. In North America, the amount of land used in organic agriculture has recently increased 15–20% each year. Today there are about 5000 organic producers in Canada, and certified organic farms have increased by more than 66% since 2001, in spite of an overall decrease in the number of farms.[43]

Several motivating forces have fuelled these trends. Many consumers favour organic products because of better taste, the desire to buy locally, and concern that consuming produce grown with pesticides may pose risks to their health. Consumers also buy organic produce out of a desire to improve environmental quality by reducing chemical pollution and soil degradation. Other consumers do not buy organic produce because it usually is more expensive, and often looks less uniform and aesthetically appealing in the supermarket aisle compared to the standard produce of high-input agriculture.

Overall, enough consumers are willing to pay more for organic meat, fruit, and vegetables that businesses are making such foods more widely available. In addition to food products, many textile makers (among them The Gap, Levi's, and Patagonia) are increasing their use of organic cotton. (We tend to think of cotton as a "natural" fibre, but it actually has intensive impacts on agricultural land. It takes about a half-kilogram of pesticides and fertilizers and as many as 2700 litres of water to grow, by conventional agricultural means, the cotton required to manufacture one T-shirt.[44]) Roots, a company founded in Canada, was one of the first major clothing manufacturers to begin experimenting with large-scale use of organic cotton, in 1989.[45]

Government initiatives also have spurred the growth of organic farming. For example, several million hectares of land have undergone conversion from conventional to organic farming in Europe since the European Union adopted a policy to support farmers financially during the first years of conversion. Such support is helpful because conversion often means a temporary loss in income for farmers. More and more studies, however, suggest that reduced inputs and higher market prices can, in the long run, make organic farming more profitable for the farmer than conventional methods. In the end, consumer choice will determine the future of organic agriculture. Falling prices and wider availability suggest that organic agriculture will continue to increase. In addition, sustainable agriculture, whether organic or not, will sooner or later need to become the rule rather than the exception.

Organic approaches reduce inputs and pollution

Organic agriculture succeeds in part because it alleviates many problems introduced by high-input agriculture, even while passing up many of the benefits. For instance, although in many cases more insect pests attack organic crops because of the lack of chemical pesticides, biocontrol methods can often keep these pests in check. Moreover, the lack of synthetic chemicals maintains soil quality and encourages helpful pollinating insects.

This makes common sense, but has it been demonstrated scientifically? In Switzerland, where one in every nine hectares of agricultural land is managed organically (the fifth-highest rate in the world), Swiss researchers have established experimental farms where they have carried out long-term studies of tracking crop production and soil quality. At the Swiss research site in Therwil, near the city of Basel, wheat, potatoes, and other crops are grown in plots using different treatments:

- conventional farming using chemical pesticides, herbicides, and inorganic fertilizers
- conventional farming that also uses organic fertilizer (cattle manure)
- organic farming using only manure, mechanical weeding, and plant extracts to control pests
- organic farming that also adds natural boosts, such as herbal extracts in compost

The researchers record crop yields at harvest each year. They analyze the soil regularly, measuring nutrient content, pH, structure, and other variables. They also measure the biological diversity and activity of microbes and invertebrates in the soil. Such indicators of soil quality help researchers assess the potential for long-term productivity.

Paul Mäder and colleagues reported results from 21 years of data in the journal *Science*[46] (**FIGURE 8.27**). Over this time, the organic fields yielded 80% of what the conventional fields produced. Organic crops of winter wheat yielded 90% of the conventional yield. Organic potato crops averaged 58–66% of conventional yields because of nutrient deficiency and disease. Although the organic plots produced 20% less, they did so while receiving 35–50% less fertilizer than the conventional fields and 97% fewer pesticides. The team also found that soil in the organic plots had better structure, better supplies of some nutrients, and much more microbial activity and invertebrate biodiversity.

In Pennsylvania, scientists at the Rodale Institute compared organic and conventional fields of corn and soybeans in a large-scale experiment running since 1981. In 2011 they released results from 30 years' worth of data.[47]

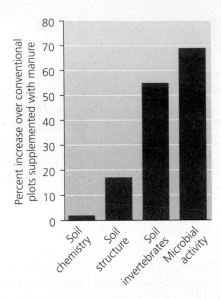

FIGURE 8.27

Organic fields investigated in longitudinal studies in Switzerland developed better soil quality than conventional fields supplemented with manure. Values for soil chemistry (6 variables), structure (3 variables), invertebrates (5 variables), and microbial activity (6 variables) were compared. Organic fields outperformed conventional fields without manure (not shown) still more.

Source: Data from Mäder, P., et al. (2002). Soil Fertility and Biodiversity in Organic Farming. *Science, 296,* pp. 1694–1697.

Averaged across the 30 years, yields of organically grown crops equalled yields of conventionally grown crops, but the organic crops required 30% less energy input. For this reason, raising them released 35% fewer greenhouse gas emissions. Because of lower energy inputs and higher crop prices for organic produce, farming organically created three times the profit for the farmer.

Shorter-term experiments elsewhere have shown similar results. Researchers comparing organic and conventional farms have found that organic farming produces soils with more microbial life, earthworm activity, water-holding capacity, topsoil depth, and naturally occurring nutrients. Given such differences in soil quality, researchers expect that organic fields should perform better and better relative to conventional fields as time goes by—in other words, that they are more sustainable.

As more long-term data are published and new studies begin, we are learning more and more about the benefits of organic agriculture for soil quality—and about how we might improve conventional methods to maximize crop yields while protecting the long-term sustainability of agriculture.

Locally supported agriculture is growing

Increasing numbers of farmers and consumers are also supporting local small-scale agriculture. Farmers' markets

FIGURE 8.28

Farmers' markets, like this one in Toronto, have become more widespread as consumers have rediscovered the benefits of buying fresh, locally grown produce. There has been farmers' market activity at this particular site, the St. Lawrence Market, since 1803.

(**FIGURE 8.28**) are becoming more numerous as consumers rediscover the joys of fresh, locally grown produce. The average food product sold in North American supermarkets travels about 2300 km between the farm and the shelf, and supermarket produce is often chemically treated to preserve freshness and colour. At farmers' markets, consumers can buy fresh produce in season from local farmers and often have a wide choice of organic items and unique local varieties.

The local food or "locavore" movement was kick-started by Alisa Smith and James MacKinnon. After experimenting with a meal of food from the immediate surroundings of their cabin in northern British Columbia, they embarked upon a venture to learn more about the food they ate—specifically, where it comes from and how far it has to travel to get to the dinner table. They pledged to spend one year eating only foods that were grown and produced within 100 miles of their Vancouver apartment. The result was a book, *The 100-Mile Diet: A Year of Local Eating,* and a new concept in eating.

They were surprised by the benefits of the 100-mile diet ("It rolls off the tongue easier than 160-km diet"), and by the enthusiastic response they received from people around the world. Some of the benefits—on top of the environmental benefits of cutting back on the long-distance transportation of food products—included fresher taste, more fruits and vegetables in a generally healthier diet, seasonal awareness of food, and support of the local economy.[48]

However, there were many challenges. It took a while to figure out how to find some local products and how to tell the provenance of ingredients (which usually don't specify their origin, even if they are listed on food labels). Among the items they could not find in local production were sugar, rice, lemons, ketchup, olive oil, peanut butter, and orange juice. One of the only exceptions, or "cheats," they allowed themselves during that first year was the occasional beer (it may be brewed locally, but the ingredients come from elsewhere).

The 100-mile diet is just one entry—though perhaps the best-known—in the growing movement of locavores and "slow" food enthusiasts. MacKinnon and Smith firmly believe that local eating will continue to grow in popularity, especially with rising prices for grains and other basic food products and the rising cost of transporting them. In fact, they believe that local eating will fundamentally transform our approach to food in the coming decades.

Some consumers are partnering with local farmers in a phenomenon called **community-supported agriculture (CSA)**. In this practice, consumers pay farmers in advance for a share of their yield, usually in the form of weekly deliveries of produce. Consumers get fresh seasonal produce, while farmers get a guaranteed income stream up front to invest in their crops—an alternative to taking out loans and being at the mercy of the weather. The CSA network in Quebec, alone, counts over 100 farm members supplying 20 000 people with local organic meats and produce.[49] Sometimes CSA farms market their produce at "pop-up" markets and night markets in urban settings—temporary or seasonal small markets that are highly flexible and adaptable to farmers' needs.

Organic agriculture can even succeed in cities

One surprising place that organic agriculture is making inroads is within cities. Urban farming has long been a way of life in countries of the developing world, where ensuring one's own food security can be a necessity. But many urban areas in more economically developed countries, including Canada, now offer community-run gardens in which residents can grow small plots of fruits and vegetables.

Community garden projects now typically extend far beyond just the organic produce grown in urban plots. Many organizations, such as Toronto's The Stop or Ottawa's Just Food, offer educational workshops, integrated job training, night markets and pop-up neighbourhood markets in underserved communities, skills-building and certification courses for new farmers, and much more. Often these organizations have close relationships with local food banks, some even offering cooking lessons, nutritional information, shared meals, and even catering services staffed by local people dealing with unemployment. This represents an integrated field-to-table approach to food, with a strong social justice component.

An example of the integrated approach is the Mississauga Sustainable Urban Agriculture project (MSURA), run by a not-for-profit organization called EcoSource. MSURA was started in 2010 in response to discussions with local food banks, schools, and university students about the

A community garden coordinated by EcoSource, an Ontario-based environmental organization that inspires the community to become more environmentally responsible through creative public education. www.ecosource.ca

FIGURE 8.29

Organic gardening is taking place in some surprising places in urban areas, even in Canada. This community garden is part of the Mississauga Sustainable Urban Agriculture project of Ecosource, an environmental not-for-profit organization.

need for more access to organic food and agricultural education in the community (**FIGURE 8.29**). MSURA is one of many such projects across Canada's cities that seek new ways to address the needs of low-income residents, helping local food bank clients access fresh, healthy food and useful nutritional information on a more regular basis. Community volunteers and students gather regularly at the Iceland Community Teaching Garden, the main garden site for the project, to help with planting, cultivating, and harvesting food for the Eden Community Food Bank.

Some farmers are even growing commercially marketable crops in urban settings. The largest commercial urban farming operation in Canada is the Fresh City farm located at Downsview Park in Toronto. Fresh City also has a hydroponic farming site in downtown Toronto, called Aquaponic, which at 185 m² is Canada's largest installation of its type.[50]

In urban areas around the world, people are even farming on rooftops. Green roofs have been around for a long time; their benefits are well known, including decreasing urban runoff, minimizing energy expenditures for heating and cooling of the building, and mitigating the urban heat island effect by naturalizing the reflective properties of rooftop surfaces. Translating a roof into an urban garden can involve technologies as simple as planter boxes, or as complex as integrated irrigation systems and specialized geofabric liners. Rooftop gardens with dense planting typically require structural reinforcement.

Lufa Farms in Montreal is one example of an organization that is committed to carrying out large-scale agriculture on urban rooftops. They established their first rooftop garden in Ahuntsic, Quebec, in 2010. They now harvest 70 metric tons of produce from the site and another 120 metric tons from a rooftop farm in Laval each year.

Conclusion

Many of the intensive commercial agricultural practices we have discussed have substantial negative environmental impacts. At the same time, it is important to realize that many aspects of industrialized agriculture have had positive environmental effects by relieving pressures on land and resources.

Whether Earth's natural systems would be under more pressure from 9 billion people practising traditional agriculture or from 9 billion people living under the industrialized agriculture model is a very complicated question. Additional unanswered questions remain, such as whether genetic modification of food organisms will ultimately prove to be beneficial and whether the environmental and nutritional benefits of locally grown foods will outweigh the efficiencies of modern agriculture.

What is certain is that if our planet is to support 9 billion people by mid-century without further degradation of the soil, water, pollinators, and other ecosystem services that support our food production, we must find ways to shift to more sustainable agricultural practices. Approaches such as biological pest control, organic agriculture, pollinator conservation, preservation of native crop diversity, sustainable aquaculture, and likely some degree of careful and responsible genetic modification of food may all be parts of the game plan we will need to set in motion. What remains to be seen is the extent to which individuals, governments, and corporations will be able to put their own interests and agendas in perspective to work together toward a sustainable future.

REVIEWING OBJECTIVES

You should now be able to

Outline the historical development of agriculture, the transition to industrialized agriculture, and the challenge of feeding a growing human population

- Agriculture emerged as a human technology about 10 000 years ago. The transition to a sedentary lifestyle increased the carrying capacity of land and led to a significant increase in population.
- The Industrial Revolution brought mechanization to farming, along with advances in farm equipment and artificial selection. There are still many traditional and subsistence farmers, but much of the world's cropland is now devoted to modern, industrialized farming.
- Food production has outpaced the growth of our population, yet there are still over 800 million hungry people in the world.
- Ensuring food security for all people requires a combination of increased agricultural productivity, a decrease in poverty, and better food distribution methods. Problems include undernourishment, malnutrition, and obesity.

Identify the main approaches and summarize the environmental impacts of the Green Revolution

- The Green Revolution greatly increased agricultural productivity per unit area of land. The stated goal was to feed the world's people. Agricultural scientists used selective breeding to develop strains of crops that grew quickly, were more nutritious, or were resistant to disease or drought.
- Greatly expanded use of fossil fuels, chemical fertilizers, and irrigation—hallmarks of the Green Revolution—caused unintended environmental consequences, including pollution and soil degradation. Monoculture has contributed to reductions in crop diversity.

Summarize the challenges, strategies, and impacts of pest management and the critical importance of pollination

- There are many thousands of synthetic chemicals, developed to control "pests" and "weeds."
- Pests tend to evolve resistance to chemical pesticides, forcing chemists to design ever more toxic poisons.
- Natural enemies of pests can be employed against them in the practice of biological control.
- Integrated pest management includes a combination of techniques, and attempts to minimize use of synthetic chemicals.
- Insects and other organisms are essential for ensuring the reproduction of many of our crop plants. Conservation of native pollinating insects is particularly important to our food supply.

Describe the science and evaluate the controversies associated with genetically modified food

- Genetic modification depends on the technology of recombinant DNA. Genes containing desirable traits are moved from one type of organism into another.

- Modification through genetic engineering is both like and unlike traditional selective breeding. Selective breeding is not new, but modern genetic modification can introduce genes from completely unrelated organisms, creating novel combinations.
- GM crops may have ecological impacts, including the spread of transgenes and indirect impacts on biodiversity. More research is needed to determine how widespread or severe these impacts may be. However, little evidence exists so far for human health impacts from GM foods, but anxiety over health impacts inspires wide opposition to GM foods.
- Many people have ethical qualms about altering the food we eat through genetic engineering, and opponents view multinational biotechnology corporations as a threat to the independence of small farmers.

State the importance of crop diversity and outline some approaches to preservation

- Protecting regions of diversity of native crop varieties, such as Oaxaca, can provide insurance against failure of major commercial crops.
- Seed banks preserve rare and local varieties of seed, acting as storehouses for genetic diversity.

Assess the positive and negative aspects of feedlots and aquaculture, and the energy trade-offs in raising animals for food

- Increased consumption of animal products has driven the development of high-density feedlots.

- Feedlots create tremendous amounts of waste and other environmental impacts, but they also relieve pressure on lands that could otherwise be overgrazed.
- Our food choices are also energy choices. For example, it requires far more energy, land, and water to raise beef cattle for food than to feed the same number of people with grain.
- Aquaculture provides economic benefits and food security, can relieve pressures on wild fish stocks, and can be sustainable.
- Aquaculture also creates pollution, habitat loss, and other environmental impacts.

Summarize the main goals and methods of sustainable agriculture

- Organic agriculture is a small part of the market but is growing rapidly.
- Organic agriculture generally has lower output but lower inputs of fossil fuels, pesticides, and inorganic fertilizers, and fewer environmental impacts than industrial agriculture.
- Locally supported agriculture, as shown by farmers' markets and community-supported agriculture, is also growing.
- Urban farming is not new in the developing world, but now even cities in more economically developed countries are seeing the emergence of community gardens, rooftop gardens, and integrated food services in the urban environment.

TESTING YOUR COMPREHENSION

1. What kinds of techniques have people employed to increase agricultural food production? How did agricultural scientist Norman Borlaug help inaugurate the Green Revolution?
2. Explain how pesticide resistance occurs.
3. Explain the concept of biocontrol. List several components of a system of integrated pest management (IPM).
4. About how many and what types of cultivated plants are known to rely on insects for pollination? Why is it important to preserve the biodiversity of native pollinators?
5. What is recombinant DNA? How is a transgenic organism created? How is genetic engineering different from traditional agricultural breeding? How is it similar?

6. Describe several reasons why many people support the development of genetically modified organisms, and name several uses of such organisms that have been developed so far.
7. Describe the scientific concerns of those opposed to genetically modified crops. Describe some of the other concerns.
8. Name several positive and negative environmental effects of feedlot operations. Why is beef an inefficient food from the perspective of energy consumption?
9. What are some of the economic benefits and negative environmental impacts of aquaculture?
10. What are the objectives of sustainable agriculture, and what factors are causing organic agriculture to expand?

THINKING IT THROUGH

1. What factors make for an effective biological control strategy of pest management? What risks are involved in biocontrol? If you had to decide whether to use biocontrol against a particular pest, what questions would you want to have answered before you decide?

2. Those who view GM foods as solutions to world hunger and pesticide overuse often want to speed their development and approval. Others adhere to the precautionary principle and want extensive testing for health and environmental safety. How much caution do you think is warranted before a new GM crop is introduced?

3. Can we call the Green Revolution a success? Has it solved problems, or delayed our resolution of problems, or just created new ones? How sustainable are Green Revolution approaches? Norman Borlaug hoped that the Green Revolution would give us "breathing room" in which to deal with what he called "the Population Monster." Have we dealt effectively with population during the breathing room that the Green Revolution has given us?

4. Imagine that it is your job to make the regulatory decision as to whether to allow the planting of a new genetically modified strain of cabbage that produces its own pesticide and has twice the vitamin content of regular cabbage. What questions would you ask of scientists before deciding whether to approve the new crop? What scientific data would you want to see, and how much would be enough? Would you also consult nonscientists or take ethical, economic, and social factors into consideration?

INTERPRETING GRAPHS AND DATA

Researchers from the Grey Bruce Centre for Agroecology undertook to investigate whether there is a quantifiable difference in greenhouse gas emissions between produce purchased from small-scale organic growers and from larger, more conventional providers like Loblaw.[51] This is not as simple a calculation as it might sound. Greenhouse gas emissions from agriculture can come from a variety of sources: direct emissions related to agricultural production; the production and use of agricultural inputs (pesticides, fertilizers, and energy); conversion of forested land or wetlands into agricultural land; and post-harvest emissions, including refrigeration, storage, packaging, transport, food processing, retail activities, cooking, and waste disposal. Many actors are involved in each of these activities, and agricultural production and marketing processes vary widely.

The overall finding of the study was that the transportation, packaging, storage, delivery, and consumer pickup of a food box from the small, local distributor accounts for greenhouse gas emission of 1.057 kg of emissions in CO_2 equivalent, whereas if the same box of produce were purchased at a regular supermarket, the emissions per box would be 4.246 kg CO_2 equivalent. The study did not consider food production itself in this calculation.

1. In graph **(a)**, what is the largest source of greenhouse gas emissions calculated for the conventional providers? What percentage of the total is represented

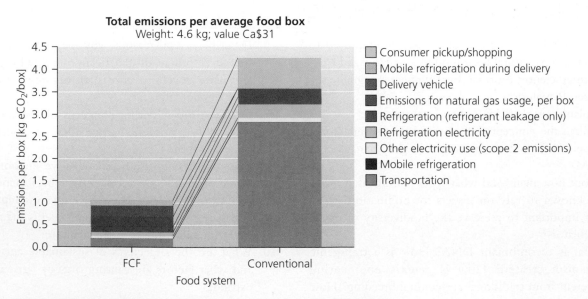

(a) Greenhouse gas emissions: local organic vs. conventional supplier

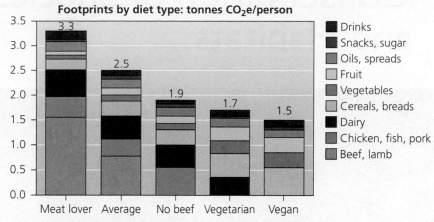

Footprints by diet type: tonnes CO₂e/person

Legend:
- Drinks
- Snacks, sugar
- Oils, spreads
- Fruit
- Vegetables
- Cereals, breads
- Dairy
- Chicken, fish, pork
- Beef, lamb

Meat lover — 3.3
Average — 2.5
No beef — 1.9
Vegetarian — 1.7
Vegan — 1.5

Note: All estimates based on average food production emissions for the United States. Footprints include emissions from supply chain losses, consumer waste, and consumption. Each of the four example diets is based on 2600 kcal of food consumed per day, which in the United States equates to around 3900 kcal of supplied food.

(b) Carbon footprints of standard diets

The bar graph **(a)** compares total greenhouse gas emissions, for acquisition of a box of produce from a small-scale organic grower (Fresh City Farms) and a large conventional producer. The graph includes transportation, packaging, storage, delivery, and consumer pickup of a food box from the local provider, versus purchase of the same produce from a conventional grocery store. The carbon footprint of various diets is shown in **(b)**, calculated per person per year, based on data from the U.S. Department of Agriculture's Economic Research Service.
Source: (a) From Thorsten, Arnold, Brenda Hsueh, Kristine Hammel, and Laurie Thomson. Fresh City: Impacts of local food. TECHNICAL REPORT: Assessment of Greenhouse Gas Emissions by Food Box Distribution. Copyright © by Grey Bruce Centre for Agroecology. Reprinted by permission. (b) Reprinted by permission from Lindsay Wilson.

by this one source? Why do you think this one source of emissions is so high?

2. Can you think of any sources of greenhouse gas emissions that are involved in food production and consumption, but not represented in **(a)**? Hint: Consider processes before and during food production, and during disposal of farm wastes. Would you predict noticeable difference between small-scale, local, organic producers and large conventional providers for any of these sources?

3. What proportion of the greenhouse gas emissions associated with the "average" diet **(b)** is accounted for by meat and poultry? How many tonnes of CO_2 equivalent is this per person per year?

9 Conservation of Species and Habitats

The critically endangered Vancouver Island marmot (*Marmota vancouverensis*) is endemic, that is, it is found only in the highlands of Vancouver Island. Both natural and anthropogenic environmental changes have caused declines in the marmot's habitat.

Jared Hobbs/All Canada Photos/Alamy

Upon successfully completing this chapter, you will be able to

- Describe the scope of biodiversity on Earth and how we measure it
- Evaluate the primary causes of biodiversity loss and species extinction
- Specify the benefits of biodiversity

- Summarize the challenges of conserving habitat and the main approaches to conservation
- Outline reasons for setting aside parks, reserves, and other protected areas

The polar bear (*Ursus maritimus*) is native to the Arctic Circle and depends on sea ice for hunting its main prey, seals. Photographs of polar bears perched on disappearing Arctic ice, like this one, have become iconic.

Arctic Ocean

Greenland (DENMARK)

Arctic Circle

Hudson Bay

CANADA

CENTRAL CASE
SAVING THE POLAR BEAR: WHAT WILL IT TAKE?

"There will be no polar ice by 2060. Somewhere along that path, the polar bear drops out."

—LARRY SCHWEIGER, PRESIDENT AND CHIEF EXECUTIVE OFFICER OF THE NATIONAL WILDLIFE FEDERATION

"Polar bears are constrained in that the very existence of their habitat is changing and there is limited scope for a northward shift in distribution. . . . The future persistence of polar bears is tenuous."

—ANDREW DEROCHER, PROFESSOR OF BIOLOGICAL SCIENCE, UNIVERSITY OF ALBERTA (WITH CO-AUTHORS)[1]

"Scientists come here for just a few days and don't know everything. . . . There are more bears now. [We] hear about global warming, but [we] see more polar bears."

—COMMENT AT A PUBLIC CONSULTATION IN ARCTIC BAY, NUNAVUT[2]

"There aren't just a few more bears. There are a hell of a lot more bears. Scientific knowledge has demonstrated that Inuit knowledge was right. . . . Right now, the bears are so abundant there's a public safety issue."

—MITCHELL TAYLOR, POLAR BEAR BIOLOGIST WITH THE GOVERNMENT OF NUNAVUT, CLYDE BAY

The twenty-first century has been interesting for polar bears and for the people who study them, love them, hunt them, and live in uneasy proximity to them in places like Canada's North. Things became especially complicated in 2008 when, partly in response to a lawsuit by environmental groups, the U.S. Fish and Wildlife Service (FWS) added the polar bear to the Endangered Species List as a "threatened" species, that is, a species that will likely become endangered in the foreseeable future in all or part of its natural range.[3]

In that same year, 2008, the government of Nunavut announced that it would proceed to allow a larger-than-normal cull of polar bears by hunters.[4] The Committee on the Status of Endangered Species in Canada (COSEWIC) also reconfirmed in 2008 that the polar bear would not be listed as "threatened," but would retain its (less critical) status as a species of "special concern," which remains its status in Canada as of 2015. The **Red List of Threatened Species**, maintained by IUCN, an international nongovernmental organization, recognizes the polar bear as "vulnerable" (of more concern than "threatened," but less than "endangered").[5]

The official reason given for "threatened" status in the United States was the recognition that polar bears—unlike other bears, which tend to be adaptable generalists—have a highly specialized diet that requires sea ice for their hunting, and sea ice is disappearing, even more rapidly than the Intergovernmental Panel on Climate Change had first predicted (see **FIGURE 1**). Climate models predict that this trend will continue, ending with a largely ice-free Arctic Ocean well before the end of this century; some researchers predict an ice-free summer Arctic by 2020 or even earlier.[6] In identifying polar bears as "threatened," the U.S. Fish and Wildlife Service also recognized the ingestion of or exposure to toxic contaminants, mostly transported from lower latitudes, as a threat to their survival.[7]

The polar bear (*Ursus maritimus*), native to the Arctic Circle and the world's largest land carnivore, is uniquely adapted to life on the ice. The individual hairs of its outer fur layer are hollow, providing both waterproofing and insulation against the cold. They have a thick layer of blubber that contributes to their ability to retain heat. The fur, which is actually transparent, appears white against their underlying black skin, providing camouflage. On the pads of their paws are small bumps, which, along with their thick, strong claws, provide traction on the ice. Polar bears are born on land but spend most of their lives hunting bearded seals and other marine prey from sea ice platforms. What will happen to this highly specialized animal if its habitat of sea ice disappears?

Also native to the Arctic in Canada, Greenland, Russia, and Alaska are the Inuit, a group of culturally similar Indigenous peoples and one of the three Aboriginal groups officially recognized by Canadian law. The Inuit traditionally hunt polar bears, as well as seals, walrus, narwhals, and whales (see painting). They also eat seabirds, fish, and land mammals, such as caribou and Arctic hares. Arctic plants such sorrel, willow, blueberry, soapberry, wintergreen, and lichens complete the traditional diet, which varies seasonally. Inuit hunters seek to maintain an ecological balance between hunting on the sea and on the land. They believe that if this balance is not maintained, and if they do not respect the animals they are hunting, the resource may disappear. They also carefully use all parts of the organisms they harvest.

The right of the Inuit to carry out subsistence hunting in support of their traditional way of life is protected by federal law in Canada as well as in Alaska. However, other laws, both federal and provincial, are in place to protect

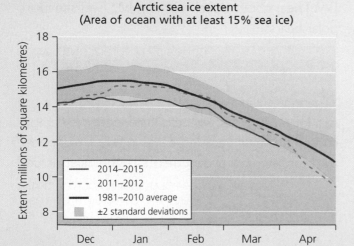

Arctic sea ice extent
(Area of ocean with at least 15% sea ice)

Legend:
- 2014–2015
- 2011–2012
- 1981–2010 average
- ±2 standard deviations

Y-axis: Extent (millions of square kilometres), from 8 to 18
X-axis: Dec, Jan, Feb, Mar, Apr

FIGURE 1
Sea ice, which is monitored by satellite, has been declining rapidly for the past couple of decades. As of the end of March, the annual maximum sea ice extent for 2015 was on track to be both the earliest and the lowest seasonal ice maximum ever recorded, well below the 20-year average.

Source: National Snow and Ice Data Center. http://nsidc.org/data/seaice_index/images/daily_images/N_stddev_timeseries.png.

Polar bear is a traditional staple in the winter diet of Inuit people in Canada's North.

www.kimhunter.ca

wildlife and other aspects of the environment, and this is where it gets really complicated. There are 19 polar bear subpopulations in the world, 13 of which have been identified for management purposes in Canada. Most of these populations lack long-term monitoring data because of the difficulties in tracking and counting polar bears.[8]

The total global population of polar bears is currently estimated to be around 20 000 to 25 000, of which about two-thirds live in Canada.[9] This estimate has remained the same for at least the past decade, although scientists freely admit that the estimates are inconclusive and that large patches of data are missing from some subpopulations. Estimates are mostly based on studies in which bears are captured, marked, and released. The number of marked bears recaptured from year to year provides a basis for the estimates. Recently, Seth Stapleton from the University of Minnesota and colleagues have experimented with the use of remote sensing, a less intrusive and potentially more accurate method for counting and tracking the bears.[10]

Canada cooperates with other national governments, state and provincial governments, international organizations, and Aboriginal groups in setting harvest quotas and managing polar bear populations. Some agreements are internal, some are international, and some are bilateral. One international group that contributes to the scientific discussion is the Polar Bear Specialist Group, which includes members from Canada, Greenland, Denmark, Norway, Russia, and the United States.

In Canada, traditional hunters must deal with a confusing bureaucratic landscape to find out what their annual hunting quota will be. For example, early in 2011 the Amarok Hunters and Trappers Association in Iqaluit announced that its members would be allowed a significant increase, from the usual quota of 23 bears to 41, on the basis of accumulated hunting credits. The Qikiqtaaluk Wildlife Board subsequently announced that the hunters would not be allowed to use all of the credits, and approved an increase of only 10 polar bears to the annual hunt. This came in the context of a decision by the Government of Nunavut Environment Minister and Nunavut Wildlife Management Board to *reduce* the total allowable harvest by 10 bears per year.[11]

The complexity is hard to appreciate. There is a strong consensus among wildlife scientists that the polar bear is threatened by the disappearance of its habitat.

The prospect of an ice-free Arctic Ocean fundamentally calls into question the ability of the polar bear to survive as a species. With warming, sea ice is also changing in its physical character, making hunting more precarious. If sea ice continues to deteriorate, polar bear populations will be forced to migrate and possibly even merge. As the bears migrate, they are encountering other types of bears, such as brown bears and grizzlies, with which they are interbreeding; this, too, may affect the survival chances of the species.[12]

Meanwhile, polar bears are appearing much more frequently in human settlements, necessitating the killing of nuisance bears for public safety. Most scientists interpret the appearance of the bears in settlements as a sign of their increasing stress and desperation for food, and some have reported increases in the number of nutritionally stressed bears.[13] However, many Inuit and other northern residents interpret their abundance in towns as a sign that the number of bears is increasing. This is indicative of a fundamental mistrust that many Inuit hold for scientists, whom they feel do not spend sufficient time on the land to understand the real status of polar bears.

Traditional Inuit hunting practices also are being affected by the loss of sea ice; some hunters have changed their patterns, or the timing of the hunt, in response to these changes. This will eventually affect bear populations but also impacts the survival of traditional Inuit cultures, which depend fundamentally on traditional wildlife harvest. Many Inuit feel that global warming is not a problem of their making and that finding a solution to the problem should not come at the risk of their cultural survival.

As of 2015 the legal status of the polar bear remains unchanged: They are considered "vulnerable" by the IUCN Red List and "threatened" in the United States, but of lesser concern in Canada. Environment Canada enumerates the current threats to polar bear survival, including climate change and its impacts on sea ice; overharvesting; contaminants; mining and oil drilling activities; shipping in the Arctic; and human–bear conflicts, which are anticipated to increase in the future. They also outline the challenges to polar bear conservation, including the broad and complex nature of the threats, lack of information, difficulties in preserving their habitat, and jurisdictional complexities in setting harvest quotas.[14]

What will it take to save the polar bear, or even to come to a more thorough understanding of the potential impacts of climate change and loss of sea-ice habitat on their populations? This case illustrates some of the complexity inherent in challenges related to the conservation of species, which almost never present simple solutions. Conservation entails not only ecological concerns, but political, legal, economic, social, and ethical ones that cross jurisdictional boundaries.

Our Planet of Life

Changes in biological populations and communities have been taking place naturally as long as life has existed. But today, as never before in the history of this planet, population pressure, human development, and resource extraction are speeding the rate of change and altering the types of changes being imposed on Earth's flora and fauna. Some of our actions are diminishing Earth's diversity of life, the very quality that makes our planet so special. Fortunately, there are steps that we can take to safeguard the diversity of species and habitats, and the ecological and evolutionary processes that make Earth such a unique place.

The ways in which we modify our environment, the impacts of these modifications on species, and the steps we take to mitigate our impacts cannot be understood in a scientific vacuum. Actions that threaten species and habitats have complex social, economic, and political roots, as you have seen in the case of the polar bear. Environmental scientists appreciate that we must understand all of these aspects if we are to develop viable solutions that will contribute to the conservation of species and habitats.

In the chapter *Evolution, Biodiversity, and Population Ecology,* we introduced the concept of **biological diversity**, or **biodiversity**, and defined it as the sum total of all organisms in an area, taking into account the diversity of species, their genes, their populations, and their communities. We also looked at many of the processes involved in speciation, extinction, and population change. In this chapter we will refine our understanding of biological diversity, and examine current biodiversity trends and their relevance to our lives. We will explore some solutions to biodiversity loss, and consider some modern approaches to the conservation and preservation of species and habitats.

Some groups hold more species than others

As you will recall from previous discussions, a *species* is a distinct type of organism, a set of individuals that uniquely share certain characteristics and can breed with one another to produce fertile offspring. Biologists use differing criteria to delineate species boundaries; some emphasize characteristics shared because of common ancestry, whereas others emphasize ability to interbreed. In practice, however, scientists broadly agree on species identities.

Taxonomists, scientists who classify species, use an organism's physical appearance and genetic makeup to determine its species. Taxonomists then group species into a hierarchy of categories meant to reflect evolutionary relationships. Related species are grouped together into *genera* (singular: *genus*). Related genera are grouped into *families*, which are grouped into *orders*, and so on (**FIGURE 9.1**).

Every species is given a two-part Latin-based scientific name denoting its genus and species. The polar bear, *Ursus maritimus*, differs from other species of bear, such as the brown bear (*Ursus arctos*, of which the grizzly is a subspecies), the American black bear (*Ursus americanus*), and the Asian black bear (*Ursus thibetanus*). These four species are closely related in evolutionary terms, as indicated by the genus name they share, *Ursus*. They are more distantly related to species such as the giant panda (*Ailuropoda melanoleuca*) and the spectacled bear (*Tremarctos ornatus*), even though they are all classified together in the family Ursidae.

Below the species level is the *subspecies*, populations of a species that occur in different geographic areas and differ from one another in some characteristics. Subspecies are formed by the same processes that drive speciation but result when divergence does not proceed far enough to create separate species. Scientists denote subspecies with the addition of a third scientific name. For example, the grizzly, classified as *Ursus arctos horribilis*, is a subspecies of the brown bear (*Ursus arctos*, which has a large number of subspecies). Grizzlies have recently been documented as breeding successfully with polar bears, producing offspring colloquially (and somewhat unattractively) referred to as "pizzlies."

Species are not evenly distributed among taxonomic groups. In terms of number of species, insects show a staggering predominance over all other forms of life

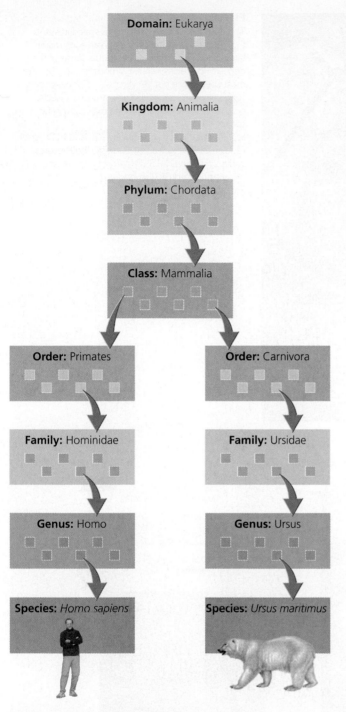

FIGURE 9.1

Taxonomists classify organisms using a hierarchical system meant to reflect evolutionary relationships. Species that are similar in appearance, behaviour, and genetics are placed in the same genus. Organisms of similar genera are placed within the same family. Families are placed within orders, orders within classes, classes within phyla, phyla within kingdoms, and kingdoms within domains. Humans (*Homo sapiens*) and polar bears (*Ursus maritimus*) are both in the class Mammalia. However, the differences between our two species are great enough that we belong to different orders and families.

(**FIGURES 9.2** and **9.3**). Within insects, about 40% are beetles. Beetles outnumber all non-insect animals and all plants. No wonder the twentieth-century British biologist J.B.S. Haldane famously quipped that God must have had "an inordinate fondness for beetles."

Biodiversity encompasses several levels

Biodiversity is a concept as multifaceted as life itself, and definitions of the term are plentiful. Different biologists use different working definitions according to their own aims, interests, and values. Nonetheless, there is broad agreement that the concept applies across several major levels in the organization of life (**FIGURE 9.4**). We will look, in turn, at the concepts of genetic diversity, species diversity, and ecosystem and habitat diversity.

Genetic diversity   Scientists designate a subspecies when they recognize substantial genetic differences among individuals from different populations of a species. However, all species and populations consist of individuals that vary genetically from one another to some degree, and this diversity is an important component of biodiversity. **Genetic diversity** encompasses the varieties in DNA present among individuals within species, subspecies, and populations.

Genetic diversity provides the raw material for adaptation to changes in local conditions. In the long term, populations with more genetic diversity may stand better chances of persisting because their variation better enables them to cope with environmental change, such as changes in the climate, the availability of prey, or the quality of habitat. Populations with little genetic diversity are vulnerable to environmental change for which they are not genetically prepared. Populations with depressed genetic diversity may also be more vulnerable to disease and may suffer *inbreeding depression*, which occurs when genetically similar parents mate and produce weak or defective offspring.

Scientists have sounded warnings over low genetic diversity in many species that have dropped to very low population sizes in the past, but the full consequences of reduced diversity in these species remain to be seen. For example, the northern elephant seal (*Mirounga angustirostris*) was hunted almost to extinction by the end of the nineteenth century. They were decreased to one breeding population, and the global number may have fallen to as low as 20 individuals. Since the early twentieth century they have been protected by law, and through various conservation efforts their numbers have recovered to over 170 000; the IUCN Red List now designates the animal as of "least concern."[15]

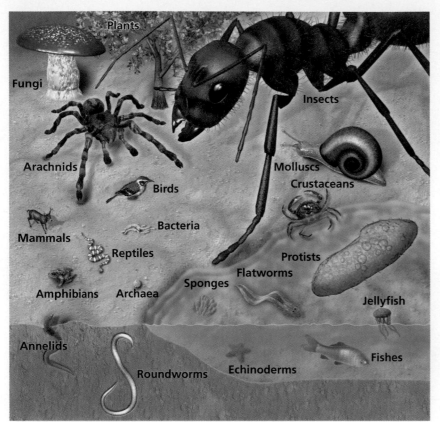

FIGURE 9.2
This illustration shows organisms scaled in size to the number of species known from each major taxonomic group. This gives a visual sense of the disparity in species richness among groups. Because most species are not yet discovered or described, some groups (such as bacteria, archaea, insects, nematodes, protists, and fungi) may contain far more species than we now know of.
Source: Data from Groombridge, B., and M.D. Jenkins (2002). *Global Biodiversity: Earth's Living Resources in the 21st Century.* UNEP World Conservation Monitoring Centre. Cambridge, UK: Hoechst Foundation.

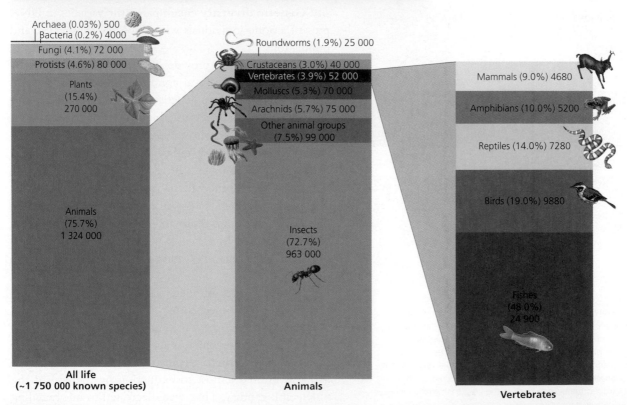

FIGURE 9.3 In the left portion of the figure, we see that three-quarters of known species are animals. The central portion subdivides animals, revealing that nearly three-quarters of animals are insects and that vertebrates comprise only 3.9% of animals. Among vertebrates (right), nearly half are fishes, and mammals comprise only 9%. Overall, most species are not yet discovered or described, so some groups may contain far more species than we now know of.
Source: Data from Groombridge, B., and M.D. Jenkins (2002). *Global Biodiversity: Earth's Living Resources in the 21st Century.* UNEP World Conservation Monitoring Centre. Cambridge, UK: Hoechst Foundation.

DATA Q What percentage of the world's total species do mammals comprise?

Ecosystem diversity

Species diversity

Genetic diversity

FIGURE 9.4
The concept of biodiversity encompasses several levels in the hierarchy of life. *Genetic diversity* refers to variety of genes among individuals within a given population or species. *Species diversity* refers to the number or variety of species in a population or region. *Ecosystem diversity* and related concepts refer to variety at levels above the species level, such as ecosystems, communities, and habitats.

A dramatic decrease in population can cause a **genetic bottleneck**, in which a limited variety of genetic material is available to be passed along by the small number of surviving individuals to their descendants. Even if the population number rebounds, as in the case of the northern elephant

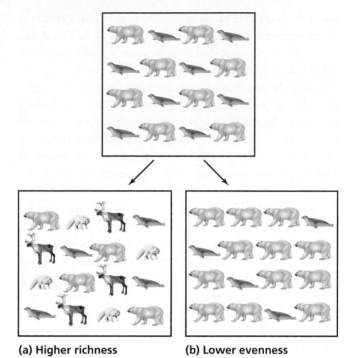

(a) Higher richness　　　　**(b) Lower evenness**

FIGURE 9.5
Compared with the boxed area at the top, which has an even number of individuals of each of the two species, **(a)** has greater species richness because it contains four species instead of just two. In contrast, **(b)** has the same species richness as the top box but reduced evenness, because one species shows a greater relative abundance than the other.

seal, the basic genetic diversity of the population will be limited. This limits the ability of the organism to adapt, and can cause it to be vulnerable to environmental changes.

Species diversity　　Another important type of diversity is **species diversity**, which can be expressed in several different ways to quantify the number and variety of species in the world or in a particular region. One component of species diversity is **species richness**, the number of species in a particular area. Another component is **evenness** or **relative abundance**, the extent to which the population numbers of individuals of each different species are equal or skewed. These are illustrated in **FIGURE 9.5**.

Speciation generates new species, adding to global species richness. *Extinction*, the disappearance of a species, decreases species richness. *Immigration* (migration of a species into an area), *emigration* (migration of a species away from an area), and *extirpation* (the local extinction of a species) may increase or decrease species richness locally. However, only speciation and extinction change it in a global sense.

Ecosystem and habitat diversity　　Biodiversity also encompasses levels above the species level. **Ecosystem diversity** refers to the number and variety of ecosystems in a given area based on variations in climate, topography, soil type, and other physical factors. Ecosystem diversity is directly related to the community types and habitat availability within the specified area, so some scientists prefer to call it *habitat diversity*. If the area is large, scientists may also consider the geographic arrangement of habitats, communities, or ecosystems at the landscape level, including the sizes, shapes, and interconnectedness of patches of these entities. Ecosystem diversity has a direct influence on species richness, because a wide range and variety of habitats provide opportunities for species to specialize.

Generally, habitats that are structurally diverse allow for more ecological niches and support greater species richness and evenness. In any given geographic area, species diversity tends to increase with diversity of habitats because each habitat supports a somewhat different community of organisms. Thus, *ecotones*, where different types of habitat intermix, often have high biodiversity. Because human disturbance (such as clearing plots of forest) can sometimes increase habitat diversity (which ecologists call *habitat heterogeneity*), species diversity may be higher in disturbed areas. However, this is true only at local scales. At larger scales, human disturbance decreases diversity because species that rely on large unbroken expanses of single habitat will disappear.

Measuring biodiversity is not easy

Coming up with precise quantitative measurements to express a region's biodiversity is difficult. This is partly why scientists often express biodiversity in terms of its most easily measured component, species diversity, especially species richness. Species richness is a good gauge for overall biodiversity, but we still are profoundly ignorant of the number of species that exist worldwide. So far, scientists have identified and described about 1.75 million[16] species of plants, animals, and microorganisms. Estimates for the actual total number of species range from 3 million to 10 million, and even as high as 100 million.[17]

Our knowledge of species numbers is incomplete for several reasons. First, some areas of Earth remain little explored. We have barely sampled the ocean depths, hydrothermal vents, or the tree canopies and soils of tropical forests. Second, many species are tiny and easily overlooked. These inconspicuous organisms include bacteria, nematodes (roundworms), fungi, protists, and soil-dwelling arthropods. Third, many organisms are so difficult to identify that ones thought to be identical sometimes turn out, once biologists look more closely, to be multiple species. This is frequently the case with microbes, fungi, and small insects, but also sometimes with organisms as large as birds, trees, and whales.

Smithsonian Institution entomologist Terry Erwin pioneered one method of estimating species numbers.

In 1982, Erwin's crews fogged rain forest trees in Central America with clouds of insecticide and then collected insects, spiders, and other arthropods as they died and fell from the treetops. Using this method, Erwin concluded that 163 beetle species specialized on the tree species *Luehea seemannii*. If this were typical, he calculated, then the world's 50 000 tropical tree species would hold 8 150 000 beetle species and—since beetles represent 40% of all arthropods—20 million arthropod species. If canopies hold two-thirds of all arthropods, then arthropod species in tropical forests alone would number 30 million. Many assumptions were involved in this calculation, and several follow-up studies have revised Erwin's estimate downward,[18] but it remains one of the most robust methods for estimating numbers of species.

Measuring and quantifying biodiversity also require weighting the relative importance of different types of diversity. For example, which would you consider to be more biologically diverse: (1) a region that has high species diversity (lots of different species), in which one species is overwhelmingly dominant and the others are present in vanishingly small numbers; or (2) a region with lower species diversity (a smaller number of species), but in which each species is well represented with numerous individuals? Scientists have devised a number of indexes in their attempts to quantify these differences in meaningful ways, but none of them conveys a complete picture of biodiversity.

By any measure, one of the most biodiverse places on Earth is Yasuní Biosphere Reserve in Ecuador, located in the easternmost part of the Amazon rain forest. Here, researchers have painstakingly compiled and analyzed data on species identifications, concluding that this area hosts a unique and extraordinary collection of species that may be unrivalled on the planet (see "The Science behind the Story: Counting Species in the World's Most Biodiverse Place").

Biodiversity is unevenly distributed on the planet

Living things are distributed across our planet unevenly, and scientists have long sought to explain the distributional patterns they see. For example, species richness generally increases as one approaches the equator (**FIGURE 9.6**). This pattern of variation with latitude, called the *latitudinal gradient*, is one of the most obvious patterns in ecology, but it also has been one of the most difficult ones for scientists to explain.

Hypotheses abound for the cause of the latitudinal gradient in species richness. It seems likely that plant productivity and climate stability play key roles (**FIGURE 9.6B**). Greater amounts of solar energy, heat, and humidity at tropical latitudes lead to more plant growth, making areas nearer the equator more biologi-

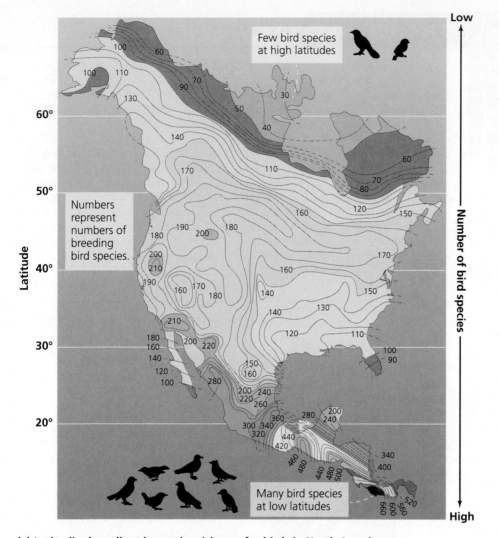

Low

Few bird species at high latitudes

Numbers represent numbers of breeding bird species.

Many bird species at low latitudes

High

(a) Latitudinal gradient in species richness for birds in North America

Temperate and polar latitudes: Variable climate favours fewer species, and species that are widespread generalists.

Tropical latitudes: Greater solar energy, heat, and humidity promote more plant growth to support more organisms. Stable climate favours specialist species. Together these encourage greater diversity of species.

(b) The niche hypothesis

FIGURE 9.6 For many types of organisms, the number of species per unit area tends to increase toward the equator. For example **(a)**, in any one spot in arctic Canada and Alaska, 30 to 100 bird species can be counted; in areas of Costa Rica and Panama, the number is over 600. One hypothesis **(b)** is that the variable climates of polar and temperate latitudes favour organisms that can survive a wide range of conditions. Such generalist species have expansive niches; they can do many things well enough to survive, and they spread over large areas. The stable climates of equatorial regions favour specialist species, which have restricted niches but do certain things very well. Together these factors promote greater species richness in the tropics. *Source:* Based on Variation in Species Density in North American Birds. *Systematic Biology, 18,* pp. 63–84. (Originally published as *Systematic Zoology.*) By permission of Oxford University Press.

cally productive and able to support larger numbers of animals. In addition, the relatively stable climates and *aseasonality* of equatorial regions help ensure that single species won't dominate ecosystems, but instead that numerous species can coexist. Variable environmental conditions favour *generalists*—species that can deal with a wide range of circumstances but that do no single thing extremely well. In contrast, stable conditions favour *specialists*—organisms with very specialized niches that do particular things very well.

Another reason that polar and temperate regions may be relatively lacking in species is that glaciation events repeatedly forced organisms out of these regions and

toward more tropical latitudes. In other words, species in the tropics have simply had a longer period of geological and climatic stability in which to evolve.

The latitudinal gradient influences the species diversity of Earth's biomes overall. Tropical dry forests and rain forests tend to support far more species than tundra and boreal forests, for instance. Tropical biomes typically show more evenness, as well, whereas in high-latitude biomes with low species richness, particular species may greatly outnumber others. For example, the Canadian boreal forest is characterized by immense expanses dominated by black spruce, whereas Panamanian tropical forest contains hundreds of tree species, no one of which greatly out-

Researcher and Station Manager Diego Mosquera checks a trail camera at Tiputini Biodiversity Station in Ecuadorian Amazon.
Diego Mosquera / Tiputini Biodiversity Station/USFQ

Counting Species in the World's Most Biodiverse Place

Counting species is an enormous challenge. Various international organizations, including the IUCN Red List, the Smithsonian Institution Tropical Research Institute, and the Tropical Ecology Assessment and Monitoring Network[19] attempt to keep current lists of new species identifications that are published in the scientific literature each year. Other groups focus on keeping track of newly identified species in particular groups; for example, AmphibiaWeb[20] maintains a rigorous, up-to-date compilation of all discoveries of amphibian species.

Scientists make discoveries of new species and keep track of changes in the status and distribution of known species in a very wide variety of ways, depending on the environment and the type of organism. For example, as described previously, tree-fogging is one approach to discovering new insect species. Insecticide is sprayed into a tree where new species are suspected, and the insects that fall are captured in a tarp, described, and counted, and any new species are identified and sampled. With many types of organisms the process is even more laborious, involving fieldwork in tough conditions, and many hours of watching and waiting for a previously unidentified species to appear.

THE SCIENCE | BEHIND THE STORY

Of all the places in the world where many species are known to exist in a small area, the one that has earned the nickname of "most biodiverse place on Earth" is Yasuní National Park and Biosphere Reserve in Ecuador. The Park (covering an area of 10 227 km²) and Biosphere (27 000 km²) are located in the western part of the Amazon rain forest. The horseshoe-shaped protected area (see **FIGURE 1**) wraps itself around the adjacent Waorani Ethnic Reserve of more than 6000 km², which was established to protect the traditional way of life of the Waorani people, indigenous to the area. Unfortunately, this important protected area is also the focus of intense oil exploitation, which puts much unique habitat at risk.

The first comprehensive study of biodiversity in the Yasuní area was carried out by Margot Bass and colleagues, and published in the open-source journal *PLoS One* in 2010.[21] The study focused mainly on amphibian, bird, mammal, and plant occurrences. The researchers found that Yasuní is uniquely situated in a location where *species richness centres*

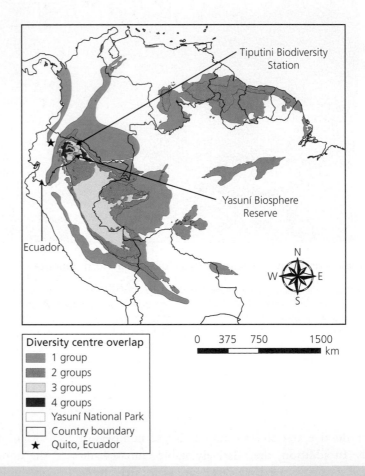

FIGURE 1

Species richness centres—the highest-diversity locations—for each of four focus taxonomic groups (amphibians, birds, mammals, and vascular plants) are shown on this map. The "4 groups" area, shown in red, indicates where richness centres for all four taxonomic groups overlap. Yasuní Biosphere Reserve is the small, horseshoe-shaped area shown on the map in western Ecuador, uniquely situated in an area of overlapping diversity centres. *Source:* From Margot S. Bass, Matt Finer, Clinton N. Jenkins, Holger Kreft, Diego F. Cisneros-Heredia, Shawn F. McCracken (2010). Global Conservation Significance of Ecuador's Yasuní National Park. Published by PLoS (Public Library of Science).

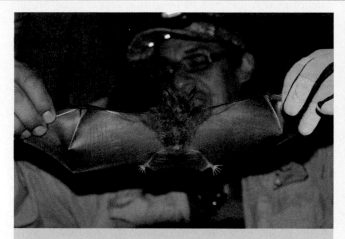

Bat researcher Jaime Guerra uses extremely fine *mist nets* to capture bats in midflight. Of the approximately 100 species of bat thought to be present in Yasuní, Dr. Guerra has identified 82 of them—many previously unknown to the area, and some new to science. The bats are measured and then released unharmed.

Q. Fang, University of Toronto

This photo of a jaguar (*Panthera onca*) was taken by trail cameras triggered by motion sensors. Jaguars are highly elusive and difficult to spot by normal means. Yasuní is thought to be a hotspot for jaguars and other wild cats.[23]

Diego Mosquera/Tiputini Biodiversity Station/USFQ

(the highest-biodiversity areas) for all four taxonomic groups meet and overlap. They noted world records for species richness diversity in the 4-group area for bats (see photo), amphibians, reptiles, and trees, as well as the presence of both endemic and endangered species. The protected area of Yasuní covers just 14% of this particularly important species richness area; unfortunately, however, 79% of the area coincides with the current area of active oil exploitation.

In a remote location on the northern edge of Yasuní is the Tiputini Biodiversity Station (TBS), a research facility run by the Universidad San Francisco de Quito, with Station Manager Diego Mosquera (see photo on previous page). TBS is located in pristine tropical rain forest on the banks of the Tiputini River. The station is only open to researchers and students; no tourists are able to visit. This allows the station to maintain its relatively untouched environment.

Since 2004, biologist Mosquera has been maintaining a trail camera program with 20 or so motion-triggered cameras installed at any given time on the station's trail system, and off-trail at salt licks in the forest. Salt licks (*saladeros* in Spanish) are frequented by animals of all types looking for mineral-rich water with which to supplement their diets.[22] Since 2004, the camera-trap project has processed more than 100 000 photos. The project has now moved into motion-triggered videos, as well, with 10 000 videos from another 15 or so cameras.

Through the camera-trap project, Mosquera has captured on film and identified many species that are difficult or impossible to see in person, as well as some unusual animal behaviours. The project's photo archives contain pictures of 63 identifiable species: 36 mammals (including 2 species of rare wild dogs) and 27 different bird species, as well as several amphibians and reptiles. The videos have captured even more species (75), including some birds and mammals that pass by too quickly to trigger the motion-sensors of the still cameras.

At the salt licks, the cameras capture interesting interactions, such as shared usage of the resource by species that would normally be wary of each other, such as spider monkeys with howler monkeys, or sloths with porcupines. Perhaps most notable of all the species photographed by the camera traps are the jaguars (see photo), which are elusive and very difficult to spot in person. Nine jaguar individuals are regularly observed as resident individuals at TBS, with a number of other individuals passing through and caught on camera.

Other researchers have also made good use of data from the trail cameras. For example, primate researcher Nelson Galvis and colleagues used camera traps to study population changes and life history of spider monkeys (*Ateles belzebuth*) at TBS.[24] Ten species of wild monkeys are seen on a regular basis at TBS. Tracking them and studying changes in groups normally requires many hours of fieldwork in difficult conditions, often combined with sedating, tagging, and releasing the animals so that they can be tracked using radio collars and GIS. The camera traps allow the researchers to observe the animals close up, unobtrusively, and at much less cost in terms of human hours and discomfort.

Yasuní Biosphere Reserve is of enormous global importance for the conservation of species because of its unique location, large size, extraordinary level of species richness, and protected status. The opening of oil concessions in the Yasuní and Tiputini areas by the government of Ecuador in 2013 brought into clear focus the risks of oil exploitation in this particularly rich and uniquely biodiverse place.

numbers the others. Understanding such patterns of bio-diversity is vital for landscape ecology, regional planning, and forest management, as well as for the conservation of species and habitats.

Biodiversity Loss and Species Extinction

Extinction occurs when the last member of a species dies and the species ceases to exist. Once lost to extinction, a species can never return. As mentioned earlier, the disappearance of a particular population from a given area, but not the entire species globally, is termed **extirpation**. For example, the Siberian tiger, once native across much of Asia, has been extirpated from most of its historic range, but it is not yet extinct—it still exists in one small mountainous location in the far east of Russia. The sage-grouse, grey whale, black-footed ferret, and grizzly bear have been extirpated from some parts of their natural ranges in Canada, but they are not extinct. The passenger pigeon, Dawson's caribou, and Labrador duck, among others, used to inhabit parts of Canada but are now altogether extinct.

Although a species that has been extirpated from one place may still exist in others, extirpation is an erosive process that can, over time, lead to extinction. A species that is in imminent danger of extirpation or extinction is referred to as **endangered**. As discussed in the Central Case, one that is likely to become endangered in the near future, if limiting factors are not reversed, is called **threatened**. These categories—threatened, endangered, extirpated, and extinct—are the main classifications used by the Canadian *Species at Risk Act* (also known as **SARA**), and, with some variations in definition, by most organizations that keep track of that status of species. Any species that is agreed to have fallen within one of these categories is considered to be *at risk*, and is listed on the SARA Public Registry, by a process that is described in greater detail later in this chapter.

Extinction and extirpation occur naturally

If organisms did not naturally go extinct, we would be up to our ears in dinosaurs, trilobites, ammonites, and the millions of other types of creatures that vanished from Earth long before humans appeared. Palaeontologists estimate that roughly 99% of all species that have ever lived are now extinct, leaving only about 1% as the wealth of species on our planet today.

Most extinctions prior to the appearance of humans occurred one by one for independent reasons, at a rate that palaeontologists refer to as the *background rate of extinction*. The fossil record indicates that for mammals and marine animals, one species out of 1000 would typically become extinct every 1000 to 10 000 years. This translates to a background rate of about one extinction per 1 to 10 million species per year.

In the past 440 million years, our planet has experienced five distinct episodes of *mass extinction* (**FIGURE 9.7**), each of which has eliminated at least half of existing species. There is evidence for further mass extinctions in the Cambrian period and earlier, more than half a billion years ago. The best-known episode of mass extinction occurred at the end of the Cretaceous period, 65 million years ago, when the global side effects of a large asteroid impact drove the dinosaurs and many other groups to extinction. The most severe episode occurred at the end of the Permian period,

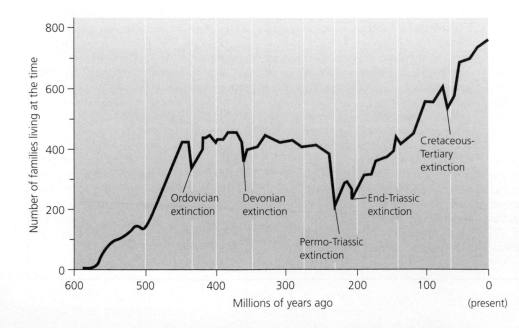

FIGURE 9.7

The fossil record shows evidence of at least five mass extinctions during the past half-billion years of Earth history. At the end of the Ordovician, Devonian, Permian, Triassic, and Cretaceous periods, 50–95% of the world's species became extinct. Each time, biodiversity later rebounded to equal or higher levels, but the rebound required millions of years in each case. *Source:* Data from Raup, D.M., and J.J. Sepkoski (1982). Mass Extinctions in the Marine Fossil Record, *Science, 215,* pp. 1501–1503.

248 million years ago, when close to 54% of all families, 90% of all species, and 95% of marine species went extinct. The cause of this extinction is still being researched.

Some species are more vulnerable to extinction than others

In general, extinction occurs when environmental conditions change rapidly or severely enough that a species cannot adapt genetically to the change; natural selection simply does not have enough time to work. All manner of environmental events can lead to extinction, from climate change to the rise and fall of sea level, to the arrival of new harmful species, to severe weather events such as extended droughts. In general, small populations and species narrowly specialized on a particular resource or way of life are most vulnerable to extinction from environmental change. Thus, **vulnerable** is yet another SARA category, referring to species that are of particular concern because of characteristics that make them particularly sensitive to human activities or natural events.

The golden toad was a prime example of a vulnerable species that did become extinct (see "Central Case: Striking Gold in a Costa Rican Cloud Forest" in Chapter 4). The golden toad was **endemic** to the Monteverde cloud forest in Costa Rica, meaning that it occurred nowhere else on the planet. Endemic species face relatively high risks of extinction because all of their members belong to a single, sometimes small, population. At the time of its discovery in 1964, the golden toad was known in an area of only 4 km^2. It required very specific conditions to breed successfully. The golden toads gathered to breed in springtime in small, root-bound pools. Monteverde provided an ideal living environment for the golden toad, but the minuscule extent of that environment meant that any stresses that deprived the toad of the resources it needed might doom the entire world population of the species. Less than 25 years after its discovery, the golden toad was extinct, probably because of several factors including climate change, habitat alteration, and disease.

Today many amphibian species around the globe are at a particularly high level of vulnerability and risk of extinction. In some cases this is because of specialized habitat and breeding requirements, and the loss of the required habitats. In others, causes were exposure to contaminants, diseases, or the influence of invasive species. The Global Amphibian Assessment was undertaken in 2004 as a comprehensive global assessment of the (now) more than 7000 known species of amphibians, including frogs, toads, salamanders, and caecilians (a less well-known order of amphibians that outwardly resemble worms or snakes).[25] That assessment has been updated periodically since 2004, through the IUCN's Red List process, with more than 650 scientists having contributed data so far.

The Global Amphibian Assessment and subsequent updates found that at least 42% of amphibian species are in decline, and less than 1% is documented as increasing in population. Of Canada's 40+ amphibian species,[26] 31 (about 70%) are designated by COSEWIC as being "at risk" in some regard; only 14 species are considered "not at risk."[27]

Vulnerability resulting from restricted or specialized habitat, lifestyle, or resource requirements is not limited to amphibians. Consider the Vancouver Island marmot (*Marmota vancouverensis*), one of the rarest mammals in North America, and one of only five species of mammals whose natural range is entirely within Canada (i.e., endemic to Canada). Marmots require small patches of south- to southwest-facing, steeply sloping, boulder-filled meadow, between 800 and 1500 m in altitude—pretty specific!

The marmot's range has been drastically reduced over the past few decades, primarily by habitat alteration resulting from logging. As of 2015 only about 30 mature individuals of this species were thought to be alive in the wild, according to the SARA Registry, with only a handful of colonies in a few locations in southern Vancouver Island (**FIGURE 9.8**). This number has increased over the past few years as the result of conservation efforts, including captive breeding programs at the Toronto Zoo, Calgary Zoo, and Mountainview Conservation and Breeding Centre in Langley, British Columbia. However, the Vancouver Island marmot was declared as endangered in 2008 and continues to be listed as endangered on the SARA Registry, with loss of habitat, inbreeding, and climate change listed as reasons for the threat.[28]

Humans may have started a sixth mass extinction

Biodiversity at all levels is currently being lost as a result of human impacts, most irretrievably in the extinction of species. If the current trend continues, the modern era may see the extinction of more than half of all species. Although similar in scale to previous mass extinctions, today's ongoing mass extinction is different in three primary respects. First, humans are causing it. Second, humans will suffer as a result of it. And third, it may be happening even more rapidly than some of the previous "Big Five" mass extinctions.

There have been many instances of human-induced species extinction over the past few hundred years. Sailors documented the extinction of the dodo on the Indian Ocean island of Mauritius in the seventeenth century, and we still have a few of the dodo's body parts in museums. Among North American birds in the past two centuries, we have driven into extinction the Carolina parakeet, great auk, Labrador duck, and passenger pigeon, and probably the Bachman's warbler, Eskimo curlew, and

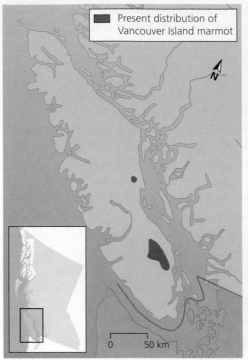

Present distribution of Vancouver Island marmot

0   50 km

Jared Hobbs/All Canada Photos/Alamy

FIGURE 9.8
An extremely restricted natural range with very specific habitat requirements can leave some species highly vulnerable, especially if environmental changes occur. The Vancouver Island marmot's range, shown in purple on the map **(a)**, has been reduced to only a few locations with a handful of colonies, primarily as a result of logging and climate change. This marmot **(b)** is one of the rarest mammals in North America, and one of only five species of mammals that are endemic to Canada. The drastic reduction in the number of mature adults has raised concerns about inbreeding as another threat to existing colonies.

(a) Present range of Vancouver Island marmot   (b) *Marmota vancouverensis*

ivory-billed woodpecker. Several more species, including the whooping crane, California condor, and Kirtland's warbler, are teetering on the brink of extinction.

However, species extinctions caused by humans precede written history. Indeed, people may have been hunting and outcompeting species into extinction for thousands of years. Archaeological evidence shows that, in case after case, a wave of extinctions followed close on the heels of human arrival on islands and continents (**FIGURE 9.9**). After Polynesians reached Hawaii, half of its birds went extinct. Birds, mammals, and reptiles vanished following human arrival on many other oceanic islands, including large islands such as New Zealand and Madagascar. Dozens of species of large vertebrates died off in Australia after Aborigines arrived roughly 50 000 years ago. North America lost 33 genera of large mammals (the so-called Pleistocene *megafauna*, which included animals like the giant beaver, giant sloth, mammoth, mastodon, and sabre-toothed tiger) after people arrived on the continent about 10 000 years ago.

Today, species loss is accelerating as our population growth and resource consumption put increasing strain on habitats and wildlife. Scientists working with the Millennium Ecosystem Assessment calculated that the current global extinction rate is 100 to 1000 times greater than the background rate. They noted a decrease in genetic diversity as well as declining population sizes and numbers of species, accompanied by greater demands on ecosystem services in the past few decades. Moreover,

they projected that the rate of species extinctions would increase tenfold or more in future decades.[29]

As mentioned previously, the World Conservation Union (known as IUCN, International Union for Conservation of Nature), an international nongovernmental organization, regularly updates its Red List of species that are facing high risks of extinction.[30] The 2014 version of the Red List reports that 1 730 725 species have been described altogether, and 76 199 of those have been evaluated by the Red List process. Of these, 22 413 are threatened, including at least 22% (1199) of known mammal species, 13.1% (1373) of bird species, and 26.8% (1957) of amphibian species.[31] For all of these figures, the *actual* numbers of threatened species, like the actual number of total species in the world, are doubtless greater than the *known* numbers.

Anthony Barnosky of the University of California at Berkeley and colleagues carried out a detailed statistical comparison of the current rate of species extinction with the rates of extinction during the "Big Five" mass extinctions. They published their findings in the journal *Nature*.[32] It was an extremely complex undertaking; data limitations and complexities make the analysis very difficult. For example, as the authors pointed out, some 49% of bivalves went extinct at the end of the Cretaceous, but only 1% of the bivalve species alive today have even been assessed, making meaningful comparison difficult. Furthermore, some assessments of threatened and extinct species are inflated because endangered species tend to be

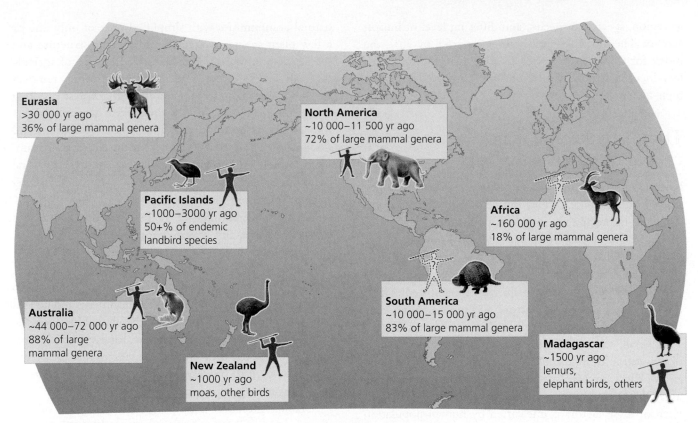

FIGURE 9.9 This map shows for each region the time of human arrival and the extent of the recent extinction wave, with representative extinct megafauna. The human hunter icons are sized according to the degree of evidence that human hunting was a cause of extinctions; larger icons indicate more certainty that humans (as opposed to climate change or other forces) were the cause. Data for South America and Africa are too sparse to be conclusive; future archaeological and palaeontological research could alter these interpretations.
Source: Adapted from Barnosky, A.D., et al. (2004). Assessing the Causes of Late Pleistocene Extinctions on the Continents. *Science, 306,* pp. 70–75; and Wilson, E.O. (1992). *The Diversity of Life.* Cambridge, MA: Belknap Press.

assessed first, whereas others are probably grossly underestimated because so few living species have been assessed. The authors concluded from their study that extinctions in the past 500 years do not yet constitute a mass extinction, but they also concluded that rates of extinction today are much higher than in similar periods during the "Big Five," and that conservation efforts are urgently needed.[33]

Extinction is only part of the story of biodiversity loss. The larger part of the story is the decline in population sizes of many organisms. Declines in numbers are accompanied by shrinkage of species' geographical ranges. Thus, many species today are less numerous and occupy less area than they once did. Tigers numbered well over 100 000 worldwide in the nineteenth century but number only about 5000 today. The Vancouver marmot population dipped to 21 individuals in 2003, in a greatly reduced geographical range, making a small comeback only as a result of captive breeding. The northern elephant seal population may have been as low as 20 at the end of the nineteenth century, in just one breeding population.

To measure and quantify this degradation, scientists at the World Wildlife Fund and the United Nations Environment Programme (UNEP) developed a metric

called the *Living Planet Index.* The index summarizes trends in representative populations of more than 10 000 species of vertebrates (mammals, birds, amphibians, fish, and reptiles) that are well-enough monitored to provide reliable data. Between 1970 and 2014, the Living Planet Index fell by roughly 52%.[34] This means that the monitored populations overall are less than half the size that they were in 1970 (**TABLE 9.1**), although the decline varies from region

TABLE 9.1   Living Planet Index

		Number of Species Monitored	Change from 1970 to 2014
Global	Global	3038	−52%
	Temperate	1606	−36%
	Tropical	1638	−56%
Systems	Terrestrial	1562	−39%
	Freshwater	757	−76%
	Marine	910	−39%

Source: Data from WWF, GFN, WFN, and ZSL (2014). *Living Planet Report 2014: Species and Spaces, People and Places.*

to region, species to species, and differing level of human activity. The index for freshwater species fell by 35%, the index for tropical species fell by 60%, and the index for low-income regions fell by 58% (as compared to 5% in high-income regions).[35]

There are several major causes of biodiversity loss

Reasons for the decline of any given species are often multi-faceted and complex, so they can be difficult to determine. The current precipitous decline in populations of amphibians throughout the world provides an example. Frogs, toads, and salamanders worldwide are decreasing drastically in abundance. Many have already gone extinct, and scientists are struggling to explain why. Recent studies have implicated a wide array of factors, and most scientists now suspect that such factors may be interacting synergistically.

Overall, scientists have identified four primary causes of population decline and species extinction: habitat alteration or loss, invasive species, pollution, and overharvesting. Global climate change is increasing in importance. Each of these factors is intensified by human population growth and by our increase in per capita consumption of resources. The Living Planet Index quantifies the principle causes of population decline in their monitored species as shown in **FIGURE 9.10**.

Habitat alteration   Nearly every human activity alters the habitat of organisms. Farming replaces diverse natural communities with simplified ones of only one or a few plant species. Grazing modifies the structure and species composition of grasslands. Either type of agriculture can lead to desertification. Clearing forests removes the food, shelter, and other resources that forest-dwelling organisms need to survive. Hydroelectric dams turn rivers into reservoirs upstream and affect water conditions and floodplain communities both upstream and downstream. Urbanization and suburban sprawl supplant diverse natural communities with simplified human-made ones, driving many species from their homes.

Because organisms are adapted to the habitats in which they live, any major change is likely to render it less suitable for them. Of course, human-induced habitat change benefits some species. Animals such as starlings, house sparrows, pigeons, raccoons, and grey squirrels do very well in urban and suburban environments and benefit from our modification of natural habitats. However, the species that benefit are relatively few; for every species that gains, more lose. Furthermore, the species that do well in our midst tend to be cosmopolitan generalists that are in little danger of disappearing any time soon.

Habitat alteration is by far the greatest cause of biodiversity loss today (as shown in **FIGURE 9.10**). As just one example of thousands, the prairies native to central North America have been almost entirely converted to agriculture in Canada but especially south of the border in the United States, where the area of prairie habitat has been reduced by more than 99%. As a result, both prairie grasses and grassland bird populations have declined dramatically. Many grassland species have been extirpated from large areas, and the two species of prairie chickens still persisting in pockets of the Great Plains could soon go extinct.

In Canada, one prairie species of particular concern is the greater sage-grouse (*Centrocercus urophasianus*), native to southern Alberta and Saskatchewan. The greater sage-grouse also was native to British Columbia, but is now extirpated. The sage-grouse lives in prairie grassland areas, but these have been widely converted for agriculture or ranching, or degraded as a result of oil exploitation. With a 90% reduction in range between 1988 and 2015, the lack of suitable breeding locations led to a dramatic decrease in population.[36] In an unprecedented move, the federal government stepped in with an emergency order to protect the species and its habitat. We will consider the special case of the greater sage-grouse in the Central Case for Chapter 21, *Environmental Policy: Decision-Making and Problem-Solving*.

Habitat destruction has occurred widely in every biome on Earth. Over half of temperate forests, grass-lands, and shrublands already had been converted by the year 1950 (mostly for agriculture and rangeland). Today

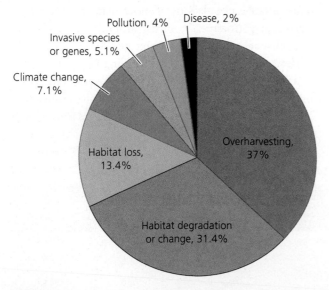

FIGURE 9.10
The pie chart shows the principal reasons for decline of populations of species monitored for the Living Planet Index as of 2014. Habitat degradation or change, combined with habitat loss, is the main cause of population decline.
Source: Data from WWF, GFN, WFN, and ZSL (2014). *Living Planet Report 2014: Species and Spaces, People and Places.*

habitat is being lost most rapidly in tropical rain forests, tropical dry forests, and savannahs. And the Arctic sea ice habitat required by polar bears is disappearing faster than ever in the history of scientific monitoring of its extent. All of these can be expected to worsen with climate change, so protecting habitat has become more urgent than ever.

Invasive species   Our introduction of non-native species to new environments, where some may become invasive (**TABLE 9.2**), has also pushed native species toward extinction. Some introductions have been accidental. Examples include aquatic organisms such as the zebra mussel, transported among continents in the ballast water of ships; animals that have escaped from the pet or food trade; and weed seeds that cling to us as we travel from place to place. Many other introductions have been intentional. People have brought with them food crops, domesticated animals, and other organisms as they colonized new places, generally unaware of the ecological consequences that could result.

Species native to islands are especially vulnerable to disruption from introduced species because native species have been in isolation for so long with relatively few parasites, predators, and competitors. As a result, they have not evolved the defences necessary to resist invaders that are better adapted to these pressures.

Most organisms introduced to new areas perish, but those that survive may do very well, especially if they find themselves without the predators and parasites that attacked them back home or without the competitors that had limited their access to resources. Once released from the limiting factors of predation, parasitism, and competition, an introduced species may increase rapidly, spread, and displace native species. Invasive species cause billions of dollars in economic damage each year.

Pollution   Pollution can harm organisms in many ways. Air pollution can degrade forest ecosystems; water pollution can adversely affect fish and amphibians; agricultural runoff (including fertilizers, pesticides, and sediments) can harm many terrestrial and aquatic species. Heavy metals, PCBs, endocrine-disrupting compounds, and various other toxic chemicals can poison both people and wildlife, and the effects of oil and chemical spills on wildlife are dramatic and well known. Oil exploitation and associated spills and contamination are of concern in parts of the Amazon, and particularly in the biodiverse region of Yasuní Biosphere Reserve.

We saw an example of this in "Central Case: The Plight of the St. Lawrence Belugas" in Chapter 3, in which we examined the effects of contaminants on the health of beluga whales in the St. Lawrence Estuary. Agricultural,

industrial, and urban pollution from as far away as the Greater Toronto area end up being transported into the estuary via the St. Lawrence River, ultimately ending up in the habitat of the beluga populations. The rate of early deaths of beluga whales from cancer has led scientists to draw linkages to this systemic problem.

Exposure to contaminants, mostly transported by atmospheric and oceanic processes from low latitudes, is cited as one of the potential threats to the survival of the polar bear. Canada's National Polar Bear Conservation Strategy lists organic pollutants (such as organochlorines) and inorganic contaminants (such as mercury) as substances that affect polar bears at both the individual and population levels. The strategy also recognizes additional threats from marine oil spills in the Arctic, as well as the potential for contaminant pathways to be changed or amplified by climate change.[37]

Overharvesting   For most species, a high intensity of hunting or harvesting by humans will not *in itself* pose a threat of extinction, but for some species it can. The polar bear may be one such species. Large in size, few in number, long-lived, and raising few young in its lifetime—a classic *K-strategist* species—the polar bear is the type of animal to be vulnerable to population reduction by hunting. Inuit traditional hunters have always carefully balanced and monitored their harvest, informed by their deep ecological knowledge of the environment. The traditional and sport hunt for polar bears is now closely monitored and limited by governments in Canada, the United States, Russia, and other Arctic nations, in conjunction with Aboriginal groups. Whether the balance of sustainable harvesting will be tipped by the impacts of climate change on the polar bear's habitat remains to be seen.

Over the past century, hunting has led to steep declines in the populations of many other K-selected animals. The Atlantic grey whale has gone extinct, and several other whales remain threatened or endangered. Illegal harvesting, poaching, and the sale of contraband wildlife products on the black market contribute to the problem. For example, three of eight tiger species are now extinct; remaining species (such as the Siberian tiger) have been extirpated from much of their natural range. Tiger body parts fetch a high price on the black market (**FIGURE 9.11**). Gorillas and other primates that are killed for their meat may be facing extinction soon. Thousands of sharks are killed each year for their fins, which are used in soup.

It is very difficult to stop trade in endangered animal parts. The *Convention on International Trade in Endangered Species of Wild Fauna and Flora (CITES)*, to which Canada is a signatory, has banned the international trade in elephant ivory since 1990. Some had feared that the ban would simply drive up the price of ivory on the black market,

TABLE 9.2 Invasive Species

Invasive species are species that thrive in areas where they are introduced, outcompeting, preying on, or otherwise harming native species. Of the many thousands of invasive species, this chart shows just a few.

European gypsy moth
(*Lymantria dispar*)

Introduced to Massachusetts in the hope it could produce silk. The moth failed to do so, and instead spread across the eastern United States and then into southern Canada, where its outbreaks defoliate trees over large regions every few years.

Asian long-horned beetle
(*Anoplophora glabripennis*)

Since the 1990s, has repeatedly arrived in North America in imported lumber. These insects burrow into wood and can kill the majority of trees in an area. Chicago, Seattle, Toronto, New York, and other cities have cleared thousands of trees to eradicate these invaders.

European starling
(*Sturnus vulgaris*)

Introduced in the 1800s by Shakespeare devotees intent on bringing every bird mentioned in Shakespeare's plays to America. Outcompeting native birds for nest holes, within 75 years starlings became one of North America's most abundant birds.

Emerald ash borer
(*Agrilus planipennis*)

Discovered in Michigan in 2002, this wood-boring insect reached 12 U.S. states, Ontario, and Quebec by 2010, killing millions of ash trees in the upper Midwest. Billions of dollars will be spent in trying to control its spread.

Cheatgrass
(*Bromus tectorum*)

Cheatgrass is native to Europe, southwestern Asia, and northern Africa. It has now spread across the prairies of the western United States and Canada, and many other places. It crowds out other plants, uses up the soil's nitrogen, and burns readily. Fire kills many native plants, but not cheatgrass, which grows back stronger without competition.

Sudden oak death
(*Phytophthora ramorum*)

Not all invasive species are macroorganisms. This fungal disease has killed millions of oak trees in California and Oregon since the 1990s. The pathogen (a water mould) was likely introduced via infected nursery plants. It has now become established in western Canada and along coastal British Columbia.

Brown tree snake
(*Boiga irregularis*)

Nearly every native forest bird on the South Pacific island of Guam has disappeared, eaten by these snakes that arrived from Asia as stowaways on ships and planes after World War II. Guam's birds had not evolved with snakes, and had no defences against them.

Nile perch
(*Lates niloticus*)

A large fish from the Nile River, introduced into Lake Victoria in Kenya, Uganda, and Tanzania in the 1950s, it proceeded to eat its way through hundreds of species of native cichlid fish, driving a number of them to extinction. People value the perch as food, but it has radically altered the lake's ecology.

Kudzu
(*Pueraria montana*)

A Japanese vine that can grow 30 m in a single season, kudzu was introduced in the 1930s for landscaping and to help control erosion. Kudzu took over forests, fields, and roadsides throughout the southeastern United States. It was first detected in Canada in Southern Ontario in 2009.

Polynesian rat
(*Rattus exulans*)

One of several rat species that have followed human migrations across the world. Polynesians transported this rat to islands across the Pacific, including Easter Island. On each island it caused ecological havoc, and has driven extinct birds, plants, and mammals.

FIGURE 9.11
Body parts from tigers have long been used as medicines or aphrodisiacs in some traditional Asian cultures. Hunters and poachers have illegally killed countless tigers through the years to satisfy market demand for these items. Here a street vendor in northern China displays tiger body parts for sale.

Konrad Wothe/Getty Images

leading to even more poaching and killing of elephants for their ivory. However, after decades of vigorous enforcement, it is now widely accepted that the ban has been successful.

Climate change The preceding four types of human impacts affect biodiversity in discrete places and times. In contrast, our manipulation of Earth's climate system is beginning to have global impacts on both habitat and biodiversity. As we will explore in subsequent chapters, our emissions of carbon dioxide and other "greenhouse gases" that trap heat in the atmosphere are causing average temperatures to warm worldwide, modifying global weather patterns and increasing the frequency of extreme weather events. Scientists foresee that these effects, together termed *global climate change*, will accelerate and become more severe in the years ahead until we find ways to reduce our emissions from fossil fuels.

Climate change is beginning to exert effects on plants and animals. Extreme weather events such as droughts put increased stress on populations, and warming temperatures are forcing species to move toward the poles and to higher altitudes. Some species will be able to adapt, but others will not. Mountaintop organisms cannot move further upslope to escape warming temperatures, so they will likely perish. Trees may not be able to move poleward fast enough. Polar bears may find their sea ice habitat disappearing altogether, with no available replacement. Animals and plants may find themselves among different communities of prey, predators, and parasites to which they are not adapted. The most significant impacts will likely be seen in the Arctic, in the marine environment, and perhaps in tropical rain forests like the Amazon if the predicted impacts on precipitation and the hydrological cycle come to pass.

All five of these primary causes of population decline are intensified by human population growth and rising per capita consumption. More people and more consumption mean more habitat alteration, more invasive species, more pollution, more overharvesting, and more climate change. Growth in population and growth in consumption are the ultimate or root causes behind the proximate threats to biodiversity.

Benefits of Biodiversity

Just as we now have a solid scientific understanding of the causes of biodiversity loss, we are also coming to appreciate its consequences as we begin to erode the many benefits that biodiversity brings us. The loss of one species may or may not affect us as individuals in any discernible way, but it is important to consider biodiversity from a holistic perspective. A comparison has been made (probably first by population biologist Paul Ehrlich) to the rivets in an airplane wing. The loss of one rivet, or two, or three, will not cause the plane to crash. But at some point the structure will be compromised, and the loss of just one more rivet will cause it to fail. If individual species are like the rivets in the airplane wing, then we might well ask how many more we can afford to lose before the structure is compromised.

This suggests the question, "Why does biodiversity matter?" There are many ways to answer this question, but we can begin by considering the ways that biodiversity benefits people. Scientists have offered a number of tangible, pragmatic reasons for preserving biodiversity, showing how biodiversity directly or indirectly supports human society. In addition, many people feel that organisms have an intrinsic right to exist and that ethical and aesthetic dimensions to biodiversity preservation cannot be ignored. (We will consider these questions in greater detail in subsequent chapters.)

Biodiversity provides ecosystem services

Contrary to popular opinion, some things in life can indeed be free, as long as we choose to protect the living systems that provide them. Intact forests provide clean air and buffer hydrologic systems against flooding and drought. Native crop varieties provide insurance against disease and drought. Abundant wildlife can attract tourists and boost the economies of developing nations. Intact ecosystems provide these and other valuable processes, known as *ecosystem services*, for all of us, free of charge.

Maintaining these ecosystem services is one clear benefit of protecting biodiversity. According to the United Nations Environment Programme (UNEP), biodiversity

- provides food, fuel, and fibre
- provides shelter and building materials
- purifies air and water
- detoxifies and decomposes wastes
- stabilizes and moderates Earth's climate
- moderates floods, droughts, wind, and temperature extremes
- generates and renews soil fertility and cycles nutrients
- pollinates plants, including many crops
- controls pests and diseases
- maintains genetic resources as inputs to crop varieties, livestock breeds, and medicines
- provides cultural and aesthetic benefits
- gives us the means to adapt to change

Organisms and ecosystems support a vast number of vital processes that humans could not replicate or would need to pay for if nature did not provide them. In 2012, Rudolf deGroot and colleagues undertook an exhaustive analysis of over 320 publications and 1350 estimates of the monetary value of ecosystem services[38] from 10 major biomes: open oceans, coral reefs, coastal systems, coastal wetlands, inland wetlands, lakes, tropical forests, temperate forests, woodlands, and grasslands.[39] They combined the information from all of these studies into the Ecosystem Service Value Database. Their conclusion was that the total value of ecosystem services is considerable, ranging from $490/ha/yr for ecosystem services provided by the "average" hectare of open ocean, to $350 000/ha/yr for the "average" hectare of coral reef. The researchers also concluded that assigning a monetary value might not be the best way to evaluate ecosystem services, since most of these services are non-monetary, fall outside of financial markets, and are not straightforward to evaluate.

Biodiversity helps maintain ecosystem integrity

Functioning ecosystems are vital, but does biodiversity really help them maintain their function? Ecologists have found that the answer appears to be yes. Research has demonstrated that high levels of biodiversity tend to increase the *stability* of communities and ecosystems. Research has also found that high biodiversity tends to increase the *resilience* of ecological systems—their ability to weather disturbance, bounce back from stresses, and adapt to change. Most of this research has dealt with species diversity, but new work is finding similar effects for genetic diversity. Thus, a decrease in biodiversity could diminish a natural system's ability to function and to provide services to our society.

What about the extinction of individual species, however? Skeptics have asked whether the loss of a few endangered species will really make much difference in an ecosystem's ability to function. Ecological research suggests that the answer to this question depends on which species are removed. Removing a species that can be functionally replaced by others may make little difference.

Recall, however, our previous discussion of *keystone species*. Like the keystone that holds together an arch, a keystone species is one whose removal results in significant changes in an ecological system. If a keystone species is extirpated or driven extinct, other species may disappear or experience significant population changes as a result.

Top predators such as polar bears and jaguars are often keystone species. A single top predator may prey on many other carnivores, each of which may prey on many herbivores, each of which may consume many plants. Thus the removal of a single individual at the top of a food chain can have impacts that multiply as they cascade down the food chain. Moreover, top predators such as tigers, wolves, and grizzly bears are among the species most vulnerable to human impact. Large animals are frequently hunted, and also need large areas of habitat, making them susceptible to habitat loss and fragmentation. Top predators are also vulnerable to the buildup of toxic pollutants in their tissues through the process of biomagnification, as seen in the example of beluga whales in the St. Lawrence estuary and polar bears in the Arctic.

Ecosystems are complex, though, and it is difficult to predict which particular species may be important. The influence of tiny "ecosystem engineers," such as ants and earthworms, can be every bit as far-reaching as those of keystone species. Thus, many people prefer to apply the precautionary principle in the spirit of ecologist Aldo Leopold, who advised, "To keep every cog and wheel is the first precaution of intelligent tinkering."[40]

Biodiversity enhances food security

Biodiversity benefits agriculture as well. Genetic diversity within crop species and their ancestors is enormously valuable. California's barley crops annually receive $160 million in disease resistance benefits from Ethiopian strains of barley. During the 1970s a researcher discovered a maize species in Mexico known as *Zea diploperennis*. This maize is highly resistant to disease, and it is a perennial, meaning it will grow back year after year without being replanted. Yet this valuable plant had almost been lost; at the time of its discovery, its entire range was limited to a 10-ha plot of land in the mountains of the Mexican state of Jalisco.

Other potentially important food crops await utilization (**TABLE 9.3**). The babassu palm *(Orbignya phalerata)* of the Amazon produces more vegetable oil than any other plant. The serendipity berry *(Dioscoreophyllum cumminsii)* produces a sweetener that is 3000 times sweeter than table sugar. Several species of salt-tolerant grasses and trees are so hardy that farmers can irrigate them with salt water. These same plants also produce animal feed, a substitute for conventional vegetable oil, and other economically important products. Such species could be immeasurably beneficial to areas undergoing soil salinization due to poorly managed irrigation.

Biodiversity provides drugs and medicines

People have made medicines from plants for centuries, and many of today's widely used drugs were discovered by studying chemical compounds present in wild plants, animals, and microbes (**TABLES 9.4** and **9.5**). Each year pharmaceutical products owing their origin to wild species generate up to $150 billion in sales and save thousands of lives.

It can truly be argued that every species that goes extinct represents one lost opportunity to find a cure for cancer or AIDS. The rosy periwinkle *(Catharanthus roseus)* produces compounds that treat Hodgkin's disease and a particularly deadly form of leukemia. Had this native plant of Madagascar become extinct prior to its discovery by medical researchers, two deadly diseases would have claimed far more victims than they have to date.

TABLE 9.3 **Potential New Food Sources**

By protecting biodiversity, we can enhance food security. The wild species shown here are a tiny fraction of the many plants and animals that could someday supplement our food supply.

Amaranths
(three species of *Amaranthus*)

Grain and leafy vegetable; livestock feed; rapid growth, drought resistant; quinoa is an examle of a species in the same family

Capybara
(*Hydrochoeris hydrochaeris*)

World's largest rodent; meat esteemed in South America; easily ranched in open habitats near water

Buriti palm
(*Mauritia flexuosa*)

"Tree of life" to Amerindians; vitamin-rich fruit; pith as source for bread; palm heart from shoots

Vicuna
(*Lama vicugna*)

Threatened species related to llama; source of meat, fur, and hides; can be profitably ranched

Maca
(*Lepidium meyenii*)

Cold-resistant root vegetable resembling radish, with distinctive flavor; near extinction

Chachalacas
(*Ortalis,* many species)

Tropical birds; adaptable to human habitations; fast-growing

TABLE 9.4 **Natural Plant Sources of Pharmaceuticals**

By protecting biodiversity, we can enhance our ability to treat illness. Shown here are just a few of the plants that have so far been found to provide chemical compounds of medical benefit.

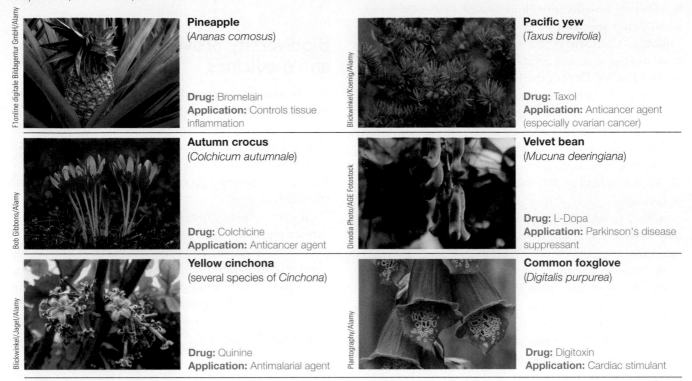

Pineapple
(*Ananas comosus*)

Drug: Bromelain
Application: Controls tissue inflammation

Pacific yew
(*Taxus brevifolia*)

Drug: Taxol
Application: Anticancer agent
(especially ovarian cancer)

Autumn crocus
(*Colchicum autumnale*)

Drug: Colchicine
Application: Anticancer agent

Velvet bean
(*Mucuna deeringiana*)

Drug: L-Dopa
Application: Parkinson's disease suppressant

Yellow cinchona
(several species of *Cinchona*)

Drug: Quinine
Application: Antimalarial agent

Common foxglove
(*Digitalis purpurea*)

Drug: Digitoxin
Application: Cardiac stimulant

In Australia, where the government has placed high priority on research into products from rare and endangered species, a rare species of cork, *Duboisia leichhardtii*, now provides hyoscine, a compound that physicians use to treat cancer, stomach disorders, and motion sickness. Another Australian plant, *Tylophora*, provides a drug that treats lymphoid leukemia. Researchers are now exploring the potential of the compound prostaglandin E2 in treating gastric ulcers. This compound was first discovered in two frog species unique to the rain forest of Queensland, Australia. Scientists believe that both species are now extinct.

A compound that forms the basis for the anticancer drug Taxol is derived from the bark of the Pacific yew (genus *taxus*), native to British Columbia. At first, overharvesting threatened not only the very slow-growing yew but also another endangered species, the spotted owl, which relies on the yew as part of its natural habitat. Today the basic ingredient for Taxol is still extracted from the bark of the Pacific yew, but the tree is cultivated specifically for this purpose.

Biodiversity boosts economies through recreation and tourism

Besides providing products and contributing to our food security and health, biodiversity can represent a direct source of income through tourism, particularly for developing countries in the tropics that have impressive species diversity. Many people like to travel to experience protected natural areas, and in so doing they create economic opportunity for residents living near those natural areas. Visitors spend money at local businesses, hire local people as guides, and support the parks that employ local residents.

Costa Rica is an example of a country that has benefited in this way, as one of the first countries to wholeheartedly embrace the ecotourism movement. Following a period of rapid population growth, Costa Rica had become an international example for the loss of forested land to agriculture. By 1991, Costa Rica was losing its forests faster than any other country in the world—nearly 140 ha per day. As a result, populations of innumerable species were declining, and some were becoming endangered (**FIGURE 9.12**). Few people foresaw the need to conserve biological resources until it became clear that they were being rapidly lost.

In 1970, the Costa Rican government and international representatives came together to create the country's first national parks and protected areas. In 1972 the efforts of local residents, along with contributions from international conservation organizations, provided the beginnings of what is today the Monteverde Cloud

TABLE 9.5 **Some Animals at Risk That Offer Potential Medical Uses**

Animals at risk may offer the potential for medical uses that could greatly benefit human health.

SPECIES AT RISK	POTENTIAL MEDICAL USES
Amphibians 30% of all species are threatened with extinction. 	• Antibiotics, alkaloids for painkillers, chemicals for treating heart disease and high blood pressure • Natural adhesives for treating tissue damage • Ability to regenerate organs and tissues could suggest how we might, too. • "Antifreeze" compounds that allow frogs to survive freezing might help us preserve organs for transplants.
Sharks Overfishing has reduced populations of most species. Some risk extinction. 	• Squalamine from sharks' livers could lead to novel antibiotics, appetite-suppressants, drugs to shrink tumours, and drugs to fight vision loss. • Study of salt glands is helping address kidney diseases.
Horseshoe crabs Overfishing is sharply diminishing populations. 	• A number of antibiotics are being developed. • The compound T140 may treat AIDS, arthritis, and several cancers. • Cells from blood can help detect cerebral meningitis in people.
Bears Nine species are at risk of extinction. 	• An acid from bears' gallbladders already treats gallstones and liver disease, and prevents bile buildup during pregnancy. • While hibernating, bears build bone mass. If we learn how, we could treat osteoporosis and hip fractures, which lead to 740,000 deaths per year. • Hibernating bears excrete no waste for months. Learning how could help treat renal disease.
Cone snails Most live in coral reefs, which are threatened ecosystems. 	Compounds from these snails include one that may prevent death of brain cells from head injuries or strokes, and a painkiller 1000 times more potent than morphine. So far just a few hundred of the 70,000–140,000 compounds these snails produce have been studied.

(a) Green sea turtle

(b) Red-backed squirrel monkey

FIGURE 9.12

Costa Rica is home to a number of species classified as globally threatened or endangered. The green sea turtle (*Chelonia mydas*) **(a)**, found in many locations in the Pacific Ocean, has undergone steep population declines. The red-backed squirrel monkey (*Saimiri oerstedii*) **(b)** is endemic to a tiny area in Costa Rica and is vulnerable to forest loss because of its small geographic range.

Forest Biological Reserve, a privately managed reserve that was established to protect the forest and its biodiversity, including the golden toad.

Costa Rica and its citizens have reaped the benefits of their conservation efforts—not only ecological benefits, but also economic ones. Because of its parks and its reputation for conservation, tourists from around the world now visit Costa Rica for **ecotourism**—travel whose main purpose is to experience relatively pristine, undisturbed natural areas (**FIGURE 9.13**). The ecotourism industry draws more than 1 million visitors to Costa Rica

FIGURE 9.13
Costa Rica has protected a wide array of its diverse natural areas. This protection has stimulated the nation's economy through ecotourism. Here, visitors experience a walkway through the forest canopy in one of the nation's parks.

each year, provides thousands of jobs, and is a major contributor to the country's economy. Today the Costa Rican economy is fuelled in large part by commerce and ecotourism, whose contributions outweigh those of industry and agriculture combined. In recent years, Costa Rica has been losing its prime position in the field of ecotourism, as many other countries have rushed to join the ranks of ecotourism destinations.[41]

It remains to be seen how effectively ecotourism can help preserve natural systems in Costa Rica and other countries in the long term. As forests outside of parks disappear, the parks are beginning to suffer from illegal hunting and timber extraction. Conservationists say the parks are poorly protected and underfunded, and that the money generated by ecotourism often does not benefit local people. There are also concerns that ecotourism, like any form of tourism, leads to environmental damage just by the sheer numbers of people that it draws. Ecotourism will likely need to generate still more money to preserve habitat, protect endangered species, and restore altered communities to their former condition.

In the meantime, ecotourism has become a vital source of income for Costa Rica, with its rain forests; Australia, with its Great Barrier Reef; Belize, with its reefs, caves, and rain forests; and Kenya and Tanzania, with their savannah wildlife. Ecuador has invested heavily in ecotourism in the Andes, where Indigenous people are starting small ecotourism lodges; in the Amazon, where rain forest "adventure" tours are booming; and of course in the Galápagos Islands, almost all of which are protected as national park. Canada, too, benefits from ecotourism; our national and provincial parks draw millions of visitors annually.

Ecotourism is one way to provide financial incentives to preserve natural areas and reduce impacts on the landscape and on native species. As ecotourism increases in popularity, however, critics warn that too many visitors to natural areas can degrade the outdoor experience and disturb wildlife; anyone who has been to a park on a crowded summer weekend can attest to this. As ecotourism continues to increase, so will the debate over its costs and benefits for local communities and for biodiversity.

People value connections with nature

Not all of the benefits of biodiversity to humans can be expressed in the hard numbers of economics or the day-to-day practicalities of food and medicine. Some scientists and philosophers argue that there is a deeper importance to biodiversity. E.O. Wilson (**FIGURE 9.14**) has described a phenomenon he calls **biophilia**, "the connections that human beings subconsciously seek with the rest of life."[42] Wilson and others have cited as evidence of biophilia our affinity for parks and wildlife, our keeping of pets, the high value of real estate with a view of natural landscapes, and our interest—despite being far removed from a hunter-gatherer lifestyle—in hiking, bird-watching, fishing, hunting, backpacking, and similar outdoor pursuits.

Writer Richard Louv adds that as today's children are increasingly deprived of outdoor experiences and direct contact with wild organisms, they suffer what he calls "nature-deficit disorder."[43] Although it is not a medical condition, this alienation from biodiversity and the natural environment, Louv argues, may damage childhood development and lie behind many of

FIGURE 9.14
Edward O. Wilson is the world's most recognized authority on biodiversity and its conservation and has inspired many people who study our planet's life. An expert on ants, Wilson has written over 20 books and has won two Pulitzer prizes.

the emotional and physical problems young people in developed nations face today.

If Wilson, Louv, and others are right, then biophilia may not only affect ecotourism and real estate prices, but also may influence our ethics. We humans are part of nature, and like any other animal we need to use resources and consume other organisms to survive. In that sense, there is nothing immoral about our doing so. However, we have reasoning ability and are able to control our actions and make conscious decisions. Our ethical sense has developed from this intelligence and ability to choose. As our society's sphere of ethical consideration has widened over time, more people have come to believe that other organisms have intrinsic value and an inherent right to exist.

Despite our ethical convictions, however, and despite biodiversity's many benefits—from the pragmatic and economic to the philosophical and spiritual—the future of biodiversity is far from secure. Even our protected areas and parks are not big enough or protected well enough to ensure that biodiversity is fully safeguarded within their borders. The search for solutions to today's biodiversity crisis is an exciting and active one, and scientists are playing a leading role in developing innovative approaches to maintaining the diversity of life on Earth.

Approaches to Conservation

In his 1994 autobiography, *Naturalist*, E.O. Wilson wrote, "In one lifetime exploding human populations have reduced wildernesses to threatened nature reserves. Ecosystems and species are vanishing at the fastest rate in 65 million years. Troubled by what we have wrought, we have begun to turn in our role from local conqueror to global steward."[44] Today, more and more scientists and citizens perceive a need to do something to stem the loss of biodiversity.

In theory, there is a distinction between the concepts of conservation and preservation. *Preservation* implies the maintenance of a natural area or species in a pristine or unaltered state, or as close to it as possible. *Conservation* implies that natural habitat and species should be cared for and maintained for multiple purposes, and that they have not only their own intrinsic value, but also multiple values (potential or actual) for people. You will learn more about these distinctions, and the modern history of these concepts in environmental and resource management, in the chapter on environmental ethics and economics. For now, let's look at some of the approaches that environmental scientists and managers have developed to help us conserve both habitats and species.

Conservation biology addresses habitat degradation and species loss

The loss of biodiversity, the urge to act as responsible stewards of natural systems, and the desire to use science as a tool in that endeavour helped spark the rise of conservation biology. **Conservation biology** is a scientific discipline devoted to understanding the factors, forces, and processes that influence the loss, protection, and restoration of biological diversity. It arose as scientists became increasingly alarmed at the degradation of the natural systems they had spent their lives studying.

Conservation biologists choose questions and pursue research with the aim of developing solutions to the problems of habitat degradation and species loss (**FIGURE 9.15**). Conservation biology is thus an applied and goal-oriented science, with implicit values and ethical standards. This perceived element of advocacy sparked some criticism of conservation biology in its early years. However, as scientists have come to recognize the scope of human impact on the planet, more of them have directed their work to address environmental problems. Today conservation biology is a thriving pursuit that is central to environmental science and to achieving a sustainable society.

Conservation biologists integrate an understanding of evolution and extinction with ecology and the dynamic nature of environmental systems. They use field data, lab data, theory, and experiments to study the impacts of humans on other organisms. They also attempt to design, test, and implement ways to mitigate human impact.

These researchers address the challenges facing biological diversity at all levels, from genetic diversity to species diversity to ecosystem diversity. At the genetic level, *conservation geneticists* study genetic attributes of organisms, to infer the status of their populations. If two populations of

FIGURE 9.15
Conservation biologists integrate lab and field research to develop solutions to biodiversity loss. Here, a conservation biologist checks on a jabiru stork nest in the Pantanal region of Brazil.

Joel Sartore/National Geographic Stock

a species are found to be genetically distinct enough to be considered subspecies, they may have different ecological needs and may require different types of management. Conservation geneticists also ask how small a population can become and how much genetic variation it can lose before running into problems such as inbreeding depression. By determining a minimum viable population size for a given population, conservation geneticists and population biologists provide wildlife managers with an indication of how important it may be to increase the population.

Many conservation research efforts also revolve around habitats, communities, ecosystems, and landscapes. Organisms are sometimes distributed across a landscape as a network of subpopulations. Because small and isolated subpopulations are most vulnerable to extirpation, conservation biologists pay special attention to them. By examining how organisms disperse from one habitat patch to another, and how their genes flow among subpopulations, conservation biologists try to learn how likely a population is to persist or succumb in the face of habitat change or other threats.

Island biogeography can help us understand habitat fragmentation

Safeguarding habitat for species and conserving communities and ecosystems require thinking and working at the landscape level. One key conceptual tool for doing so is the *equilibrium model of island biogeography*. This model, introduced by conservation biologist E.O. Wilson and

ecologist Robert MacArthur in 1963, originally explained how species came to be distributed among oceanic islands.[45] Since then, researchers have also applied it to "habitat islands"—patches of one habitat type isolated within "seas" of others.

The island biogeography model explains how the number of species on an island results from an equilibrium balance between the number added by immigration and the number lost through extirpation. It predicts an island's species richness based on its size and distance from the mainland:

- The farther an island is located from a continent, the fewer species tend to find and colonize it; this is sometimes called the *distance effect*. Thus, remote islands host fewer species because of lower immigration rates (**FIGURE 9.16A**).
- Large islands have higher immigration rates because they present fatter targets for wandering or dispersing organisms to encounter (**FIGURE 9.16B**). The location of the island relative to ocean and air currents also can affect the chance that organisms will find and colonize them.
- Large islands have lower extinction rates because more space allows for larger populations, which are less vulnerable to dropping to zero by chance (**FIGURE 9.16C**).

Together, these latter two trends give large islands more species at equilibrium than small islands—a phenomenon called the *area effect*. Large islands also tend to contain more species because they generally possess more

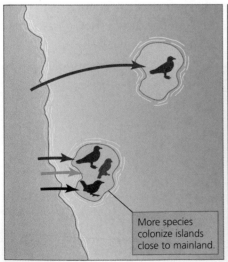
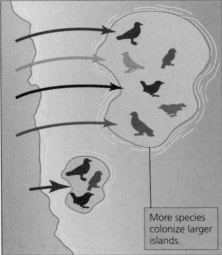
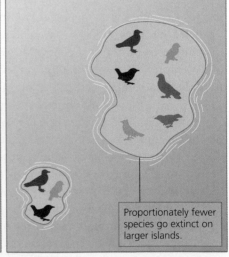

More species colonize islands close to mainland.

More species colonize larger islands.

Proportionately fewer species go extinct on larger islands.

(a) Distance effect   **(b) Target-size effect**   **(c) Area effect**

FIGURE 9.16   Islands located close to a continent receive more immigrants than islands that are distant **(a)**, so that near islands end up with more species. Large islands present fatter targets for dispersing organisms to encounter **(b)**, so that more species immigrate to large islands than to small islands; their location relative to ocean currents and winds is important, too. Large islands also offer more potential habitat and experience lower extinction rates **(c)** because their larger area can support larger populations.

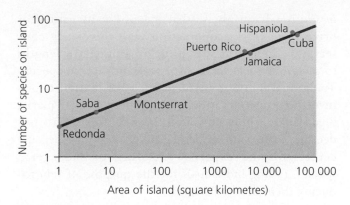

FIGURE 9.17

The larger the island, the greater the number of species—a prediction borne out by data from around the world. By plotting the number of amphibians and reptile species on Caribbean islands as a function of the areas of these islands, the species-area curve shows that species richness increases with area. The increase is not linear, but logarithmic; note the scales of the axes.

Source: Data from MacArthur, R. H., and E. O. Wilson (1967). *The Theory of Island Biogeography.* Princeton University Press.

habitats than smaller islands, providing suitable environments for a wider variety of arriving species. Very roughly, the number of species on an island is expected to double as island size increases tenfold. This effect can be illustrated through *species-area curves*, which quantify the number of species per area in a particular habitat (**FIGURE 9.17**).

The patterns established by island biogeography and the species-area relationship hold up for terrestrial habitat "islands," as well, such as forests cut into smaller areas by logging and road building (**FIGURE 9.18**), a process known as **habitat fragmentation**. Expanding agriculture, spreading cities, highways, logging, and many other impacts have chopped up large contiguous expanses of habitat into small, disconnected ones. Fragmentation of forests and other habitats constitutes one of the prime threats to biodiversity. In response, conservation biologists have designed landscape-level strategies to try to optimize the arrangement of areas to be preserved.

A significant problem with habitat fragmentation is that it creates a greater proportion of *edge* habitat relative to *core* habitat. Edge habitat can be quite different in character from habitat in the core of a forest or other type of ecosystem, particularly in terms of properties such as light levels, density of vegetation, and moisture. Organisms that are adapted to core habitat may suffer or may not survive when fragmentation leads to the replacement of core by edge environments.

Habitat fragmentation has the greatest impact on large species and, especially, migratory species. Bears, mountain lions, elephants, caribou, and other animals that need large ranges in which to roam may disappear.

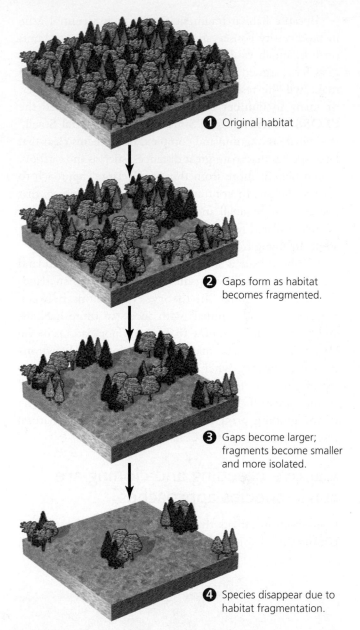

FIGURE 9.18

Forest clearing, farming, road building, and other types of human land use and development can fragment natural areas, leaving small "islands" of habitat. Habitat fragmentation begins when gaps are created within a natural habitat. As development proceeds, these gaps expand, join together, and eventually dominate the landscape, stranding islands of habitat in their midst. In fragmented habitat, fewer populations can persist, and numbers of species decrease with time.

Bird species that thrive in the interior of forests may fail to reproduce when forced near the edge of a fragment of habitat. Their nests are often attacked by predators and parasites that favour open habitats surrounding the fragment or that travel along habitat edges. Avian ecologists judge forest fragmentation to be a main reason why populations of many songbirds of eastern North America are declining; an additional significant impact is predation by feral (escaped) house cats.

Because habitat fragmentation is such a central issue in biodiversity conservation, and because there are limits on how much land can be set aside, conservation biologists have argued heatedly about whether it is better to make wildlife reserves large in size and few in number, or many in number but small in size. Nicknamed the **SLOSS dilemma**, for "**S**ingle **L**arge **o**r **S**everal **S**mall," this debate is ongoing and complex, but it seems clear that large species that roam great distances, such as the caribou, would benefit more from the "single large" approach to reserve design. In contrast, creatures such as insects that live as larvae in small areas may do just fine in a number of small isolated reserves, especially if they can disperse as adults by flying from one reserve to another.

A related issue is whether **corridors** of protected land are important for allowing animals to travel between islands of protected habitat. In theory, connections between fragments provide animals with access to more habitats, and help enable gene flow to maintain populations in the long term. Many land management agencies and environmental groups try, when possible, to join new reserves to existing reserves for these reasons. The establishment and maintenance of greenbelts and natural corridors is one of the guiding principles in environmental and natural resource management in Canada today.

Captive breeding and cloning are single-species approaches

Conservation efforts that aim to save species by maintaining their habitat are examples of *in situ* conservation; *in situ* is a Latin phrase that means "in its natural or original place." In conjunction with this, many conservation biologists are going to impressive lengths to save individual threatened and endangered species. Traditional *ex situ* (out-of-place) conservation efforts involve the preservation of species in zoos, aquaria, seed banks, arboretums, and the like. In the past few decades, many zoos and botanical gardens have become centres for the **captive breeding** of endangered species, raising individuals for the purpose of reintroducing them into the wild.

One example of a captive breeding program is aimed at saving the California condor, North America's largest bird (**FIGURE 9.19**). Condors were persecuted in the early twentieth century, collided with electrical wires, and succumbed to lead poisoning from scavenging carcasses of animals killed with lead shot. By 1982, only 22 condors remained, and biologists decided to take all the birds into captivity in hopes of boosting their numbers and then releasing them. The ongoing program is succeeding, according to various state and federal parks organizations in the United States. Over 100 of the 250 birds raised in captivity have been released into the wild at sites in California, Utah, and Arizona, where a few pairs have begun nesting.

Other reintroduction programs have been more controversial. Reintroducing wolves to Yellowstone National Park—an effort that involved imported wolves from across the border in Canada—has proven popular with the public. However, reintroducing wolves to sites in Arizona and New Mexico met stiff resistance

Corbis

FIGURE 9.19

To save the California condor *(Gymnogyps californianus)* from extinction, biologists have raised chicks in captivity with the help of hand puppets designed to look and feel like the heads of adult condors. Using the puppets, biologists feed the growing chicks in an enclosure and shield them from all contact with humans, so that when the chick is grown it does not feel an attachment to people.

from ranchers who fear the wolves will attack their livestock. The program is making slow headway, and several of the wolves have been shot. China is considering a reintroduction program for the Siberian tiger, which has been extirpated from China, part of its natural range. The Chinese government has attempted to raise Siberian tigers in captivity for release into the forests in the far northeastern part of the country. So far they have not had success with breeding attempts. Critics have noted that the forests are so fragmented that efforts would be better focused on improving habitat first.

Some reintroduction programs require international cooperation. For example, the only naturally occurring wild population of whooping cranes nests and breeds each spring in Wood Buffalo Park, straddling the border between Alberta and the Northwest Territories. The flock then migrates and winters over in the Aransas National Wildlife Refuge on the Gulf coast of Texas. The population of almost 2000 cranes in the late 1800s dipped to an all-time low of just 15 in 1941, after being decimated by hunting. In the past few years the population has slowly increased to about 180 birds as a result of careful reintroductions, led by the International Crane Recovery Team consisting of scientists from both Canada and the United States.[46]

Another example of (tentatively) successful captive breeding and reintroduction is that of the swift fox (*Vulpes velox*), a small, slender, tan-coloured fox whose natural habitat is the prairies of Manitoba, Saskatchewan, and Alberta, as well as the western grasslands of the United States (**FIGURE 9.20**). It is mainly nocturnal and is named for its speed (*velox*); it races through the prairie grasslands at up to 60 km per hour. In the first part of the twentieth century, the swift fox disappeared entirely from the wild in Canada, mostly as a result of the loss of habitat. The last swift fox was captured in Canada in 1928; it was considered extirpated by 1938 and officially declared as such by COSEWIC in 1978.[47]

In 1971, Miles and Beryl Smeeton of the Wildlife Reserve of Western Canada imported two pairs of swift foxes from the United States, with the intention of starting a captive breeding and release program. Over the next decade or so the organization (later renamed the Cochrane Ecological Institute) developed partnerships, first with the University of Calgary, then with the Canadian Wildlife Service, and finally with the government of Saskatchewan and Fish and Wildlife Service of Alberta. In 1983 a reintroduction program was started to replace swift foxes back into parts of their native territory in Canada.[48] Many of the foxes that were reintroduced came from wild populations in the United States; some were bred in captivity in Canada.

FIGURE 9.20
The swift fox has been successfully reintroduced into parts of its natural range in Canada, where it had been declared extirpated.

Robertharding/Alamy

In the mid- to late 1980s, 155 reintroduced foxes were radio-collared and tracked by program-related researchers, who determined that relocation was considerably more successful than captive breeding and relocation; the survival rate after the first year for the relocated foxes was 85%, as compared with 25% for the captive-bred foxes.[49] But with fewer and fewer foxes available for relocation, program scientists began to experiment with different release approaches for the captive-bred foxes. In a "soft" release, young foxes are kept in an outdoor pen over the winter to acclimatize, then released. In a "hard" release, the young foxes are released straight into the wild. Researchers determined that the hard release is as successful as the soft release in terms of the survival rates for the animals, and more cost-effective.[50]

According to Joel Nicholson, a biologist with Alberta Fish and Wildlife, "the swift fox reintroduction program has been one of the most successful *canid* reintroductions in the world."[51] A 2005–2006 count of the swift fox population in Alberta, Saskatchewan, and Montana counted 1162 foxes, almost all born in the wild. In 2013–2014, the Swift Fox Recovery Team from Wildlife Preservation Canada started fieldwork to update the 2006 census.[52] The swift fox is no longer considered extirpated or endangered, but it is still designated as threatened under SARA in Canada.

The newest idea for saving species from extinction is to create more individuals by cloning them. In this technique, DNA from an endangered species is inserted into a cultured egg without a nucleus, and the egg is implanted into a closely related species that can act as a surrogate mother. So far two Eurasian mammals have been cloned in this way. With future genetic technology, some scientists even talk of recreating extinct species from DNA recovered from preserved body parts. However, even if cloning can succeed from a technical standpoint, most biologists agree that such efforts are not an adequate response to biodiversity loss. Without ample habitat and protection in the wild, having cloned animals in a zoo does little good.

Some species act as "umbrellas" to protect communities

Sometimes individual species can be used as tools for the broader conservation of communities and ecosystems. Species-specific legislation can provide legal justification and resources for species conservation, but no such laws exist for communities, habitats, or ecosystems. Large species that roam great distances, such as tigers, bears, and elephants, require large areas of habitat. Meeting the habitat needs of these so-called *umbrella species* automatically helps meet those of thousands of less charismatic animals, plants, and fungi that would never elicit as much public interest.

Environmental advocacy organizations have found that using large, charismatic vertebrates as spearheads for biodiversity conservation has been an effective strategy. This approach of promoting particularly visible or charismatic *flagship species* is evident in the long-time symbol of the World Wide Fund for Nature (World Wildlife Fund in North America, or WWF), the giant panda. The panda is a large endangered animal requiring sizeable stands of undisturbed bamboo forest. Its lovable appearance has made it a favourite with the public—and an effective tool for soliciting funding for conservation efforts that protect far more than just the panda.

At the same time, many conservation organizations today are moving beyond the single-species approach. The Nature Conservancy, for instance, is a *land trust* that focuses on whole communities and landscapes by acquiring large tracts of land for conservation. The most ambitious effort may be the Wildlands Network, which proposes to restore huge amounts of North America's land to its presettlement state in an interconnected network of habitats.

Canada protects species at home and internationally

The United States was the first nation to enact federal-level legislation to protect species with the *Environmental Protection Act (EPA)* in 1970; Canada enacted its long-awaited endangered species law, the *Species at Risk Act (SARA)* in 2002. Both acts are considered to have strengths and weaknesses; SARA is generally seen as placing more emphasis on science and Aboriginal traditional knowledge,

and as being more consultative in process, compared to the EPA. We will consider this legislation in greater depth in Chapter 21, *Environmental Policy*.

Designating a species or changing a species designation under SARA is not a quick process. For one thing, it does depend on having sufficient scientific information; species are listed under SARA on the basis of scientific assessment and Aboriginal traditional knowledge. Recovery strategies also are based on scientific understanding of species and habitats. The action plan (the step after the recovery strategy) is the stage at which political and economic questions can be put forward. As such, there is a legal mandate for science and Aboriginal traditional knowledge to be respected and used through the COSEWIC process. Species are also assessed and given status by the *National General Status Working Group (NGSWG)*. These assessments inform the COSEWIC process; however, the group was established to fulfill an entirely different purpose related to Canada's role in international agreements to protect biodiversity.

Today many nations have laws protecting species, although they are not always well enforced. At the international level, the United Nations has facilitated several treaties to protect biodiversity. The *Convention on International Trade in Endangered Species of Wild Fauna and Flora (CITES)* protects endangered species by banning the international transport of their body parts. When nations enforce it, CITES can protect tigers, elephants, and other rare species whose body parts are traded internationally.

In 1992, leaders of many nations, including Canada, agreed to the *Convention on Biological Diversity*. This treaty is the cornerstone of international efforts to protect biodiversity. The National General Status Working Group, mentioned above, was established in partial fulfillment of Canada's obligations under the Convention. The Convention embodies three goals: to conserve biodiversity, to use biodiversity in a sustainable manner, and to ensure the fair distribution of biodiversity's benefits. The Convention aims to help

- provide incentives for biodiversity conservation
- manage access to and use of genetic resources
- transfer technology, including biotechnology
- promote scientific cooperation
- assess the effects of human actions on biodiversity
- promote biodiversity education and awareness
- provide funding for critical activities
- encourage every nation to report regularly on their biodiversity conservation efforts

The treaty's many accomplishments so far include ensuring that Ugandan people share in the economic

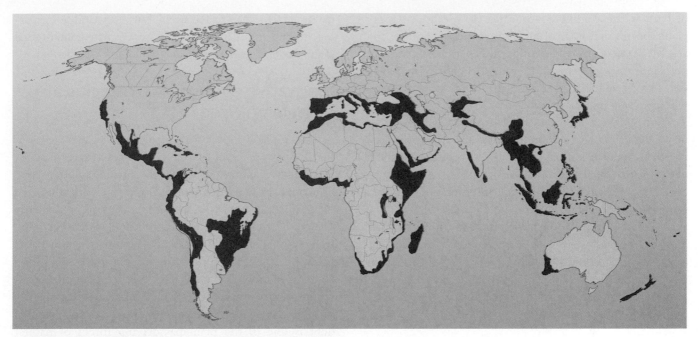

(a) Biodiversity hotspots

(b) Golden lion tamarind

FIGURE 9.21
Shown in red on the map **(a)** are the hotspot areas identified by Conservation International. The golden lion tamarin (*Leontopithecus rosalia*) **(b)**, endemic to Brazil's rain forest, is one of the world's most endangered primates. Captive breeding programs exist, but the tamarin's natural habitat is fast disappearing.
Source: Information from Conservation International.

U.S. government has lost some reputation as a leader in conservation efforts.

Hotspots highlight vulnerable areas of high biodiversity

One international approach oriented around geographic regions, rather than single species, has been the effort to map **biodiversity hotspots**. The concept of biodiversity hotspots was introduced in 1988 by British ecologist Norman Myers as a way to prioritize regions that are most important globally for biodiversity conservation. A hotspot is an area that supports an especially great number of species that are endemic, that is, found nowhere else in the world (**FIGURE 9.21**). To qualify as a hotspot, a location must harbour at least 1500 endemic plant species, or 0.5% of the world total. In addition, a hotspot must have already lost 70% of its habitat as a result of human impact and be in danger of losing more—in other words, it must be at risk.

The nonprofit group Conservation International maintains a list of 35 biodiversity hotspots. Because of habitat loss, the ecosystems of these areas together cover only 2.3% of Earth's land surface. This small amount

benefits of wildlife preserves, increasing global markets for "shade-grown" coffee and other crops grown without removing forests, and replacing pesticide-intensive farming practices with sustainable ones in some rice-producing Asian nations. As of 2015, the Convention on Biological Diversity had 196 parties (signatory nations that are legally bound to the articles of the treaty). Among those choosing *not* to be legally bound by the treaty is the United States. This decision is one example of why the of land is the exclusive home for 50% of the world's plant species and 43% of all bird, mammal, reptile, and amphibian species as endemics.[53] The hotspot concept gives incentive to focus on these areas of endemism, where the greatest number of unique species can be protected with the least amount of effort.

The World Wide Fund for Nature (WWF), another major player in conservation internationally, has organized its conservation efforts around the concept of the **ecoregion**, a large area of land or water with a geographically distinct assemblage of natural communities that share similar environmental conditions and ecological dynamics, and interact ecologically in ways that are critical for their long-term persistence.[54] The organization has used scientific criteria to identify a "Global 200" list of ecoregions that are priorities for conservation (there are actually 238 of them on the list), including ecoregions from terrestrial, freshwater, and marine settings.

Community-based conservation is increasingly popular

Taking a global perspective and prioritizing optimal locations to set aside as parks and reserves makes good sense. However, setting aside land for preservation affects the people that live in and near these areas. In past decades, many conservationists from developed nations, in their zeal to preserve ecosystems in other nations, too often neglected the needs of people in the areas they wanted to protect. Many developing nations came to view this international environmentalism as a kind of neocolonialism.

Today this has largely changed, and many conservation biologists actively engage local people in efforts to protect land and wildlife in their own backyards, in an approach sometimes called *community-based conservation*. Setting aside land for preservation deprives local people of access to natural resources, but it can also guarantee that these resources will not be used up or sold to foreign corporations and can instead be sustainably managed. Moreover, parks and reserves draw ecotourism, which can support local economies.

In the small Central American country of Belize, conservation biologist Robert Horwich and his group Community Conservation have helped start a number of community-based conservation projects.[55] The Community Baboon Sanctuary consists of tracts of riparian forest that farmers have agreed to leave intact to serve as homes and travelling corridors for the black howler monkey, a centrepiece of ecotourism. The fact that the reserve uses the local nickname for the monkey signals respect for residents, and today a local women's cooperative is running the project. A museum was built, and residents receive income for guiding and housing visiting researchers and tourists. Community-based conservation has not always been so successful, but in a world of increasing human population, locally based management that meets people's needs sustainably will likely be essential.

Innovative economic strategies are being employed

As conservation moves from single-species approaches to the hotspot approach to community-based conservation, innovative economic strategies are also being attempted. One strategy that has increased over the past couple of decades is the *debt-for-nature swap*, conceived in the early 1980s by Thomas Lovejoy and pioneered by Conservation International almost 30 years ago with partners in Costa Rica. In debt-for-nature swaps, an environmental group or a corporation takes on a portion of the debt of a developing country, usually in exchange for some form of environmental protection or conservation. Because national debt is one of the principal driving forces of habitat destruction in developing nations (because of the need to extract natural resources and generate foreign capital), debt-for-nature swaps hold the promise of protecting the environment while addressing one of its fundamental threats. So far, in spite of millions of dollars devoted to debt-for-nature swaps worldwide, results have been mixed.

A newer strategy that Conservation International has pioneered is the *conservation concession*. Nations often sell concessions to foreign multinational corporations, allowing them to extract resources from the nation's land. A nation can, for instance, earn money by selling to an international logging company the right to log its forests. Conservation International has stepped in and paid nations for concessions that are more favourable to conservation than to resource extraction. The nation gets the money *and* keeps its natural resources intact. The South American country of Surinam, which still has extensive areas of pristine rain forest, entered into such an agreement and has slowed logging while pulling in $15 million. It remains to be seen how large a role such strategies will play in the future protection of biodiversity.

Parks and Reserves

As resources dwindle; as forests, grasslands, and soils are degraded; as species disappear; and as the landscape fills with more people, the arguments for conservation of resources—for their conservation, as well as their sustainable use—have grown stronger. Also growing stronger is the argument for the preservation of land—setting aside tracts of relatively undisturbed land and habitat intended to remain forever undeveloped.

Why do we create parks and reserves?

Historian Alfred Runte cited four traditional reasons that parks and protected areas have been established:[56]

1. Enormous, beautiful, or unusual features such as the Rocky Mountains and Clayoquot Sound inspire people to protect them—an impulse termed *monumentalism* (**FIGURE 9.22**).
2. Protected areas offer recreational value to tourists, hikers, fishers, hunters, and others.
3. Protected areas offer *utilitarian* benefits and ecosystem services. For example, undeveloped watersheds provide cities with clean drinking water and a buffer against floods.
4. Parks make use of sites lacking economically valuable material resources or that are hard to develop; land that holds little monetary value is easy to set aside.

In Canada, the protection of exploitable resources and the benefits to human health have also been historical reasons for the establishment of parks. To these traditional reasons, another has been added in recent years: the preservation of biodiversity and ecosystems. A park or reserve is widely viewed as an island of habitat that can, scientists hope, maintain species that might otherwise disappear.

Yellowstone National Park in the United States was the very first national park, established in 1872, followed soon after by Yosemite National Park. Canada's first national park was established in Banff, Alberta, in 1885. Today (as of 2015) there are 45 national parks in the Parks Canada system, in every province and territory, with several proposed new parks. National parks cover a total of more than 30 million ha, or about 3% of the total land area of Canada.[57] More than 20 million people visit Canada's national parks each year.[58] Provincial parks in Canada number in the hundreds, and cover more area than national parks. Canada's parks system includes other types of protected areas as well, such as marine conservation areas and cultural, historic, and natural heritage sites.

The Canadian Wildlife Service, part of Environment Canada, contributes to the management and scientific understanding of wildlife and habitat management in Canada. Many sites in the parks system also serve as **wildlife refuges**, which are protected or semi-protected havens for the conservation of wildlife and habitat, as well as, in some cases, being available for hunting, fishing, wildlife observation, photography, environmental education, and other public uses.

Some wildlife advocates find it objectionable that hunting is allowed in many parks and refuges. However, in fact, hunters have long been in the forefront of the conservation movement and have traditionally supplied the bulk of funding for land acquisition and habitat management for refuges. Ducks Unlimited Canada is an example of a nonprofit, nongovernmental organization founded by hunters, but with the specific goal of conserving wetlands. Canada's Aboriginal peoples also retain some of their traditional hunting rights in federal and provincial parklands.

Not everyone supports land set-asides for wilderness and wildlife protection. The restriction of activities in some wilderness areas has generated opposition among those who seek to encourage resource extraction and development, as well as hunting and increased motor vehicle access, on protected Crown lands. The drive to extract more resources, secure local control of lands, and expand recreational access to public lands is epitomized by the *wise-use movement*, a loose confederation of individuals and groups that coalesced in the 1980s and 1990s in response to the increasing success of environmental advocacy.

"Wise-use" advocates are dedicated to protecting private property rights; opposing government regulation; transferring federal lands to state, local, or private hands; and promoting motorized recreation on public lands. The wise-use movement, which has been described as "anti-conservation," includes many farmers, ranchers, trappers, and mineral prospectors, as well as groups representing the industries that extract timber, mineral, and fossil fuel resources. Debate between mainstream environmental groups and wise-use spokespeople has been vitriolic, although more so in the United States than in Canada. Each side claims to represent the will of the people and paints the other as oppressive. Wise-use advocates have played key roles in ongoing debates over policy issues, such as whether recreational activities that disturb wildlife should be permitted.

ImagineGolf/iStockphoto/Getty Images

FIGURE 9.22
The awe-inspiring beauty of some regions of Canada was one reason for the establishment of national parks. Scenic vistas such as this one in Banff National Park have inspired millions of people to visit them.

Nongovernmental organizations (NGOs), including private nonprofit groups, are also involved in land preservation. **Land trusts** are local or regional organizations that purchase land with the aim of preserving it in its natural condition. The Nature Conservancy is the world's largest land trust, but smaller ones are springing up throughout North America. Probably the earliest private land trust in Canada was the Hamilton Naturalists Club in Ontario, which began to acquire land for conservation purposes in 1919.[59] A few of the many examples of land trusts that are currently operating in Canada include the Bruce Trail Conservancy, Georgian Bay Land Trust, Edmonton and Area Land Trust, and Wildlife Preservation Canada.

Protected areas are increasing internationally

Many nations have established national park systems and are benefiting from ecotourism as a result—from Costa Rica to Ecuador to Thailand to Tanzania. The total worldwide area in protected parks and reserves has increased dramatically. In 2015 the world's protected areas—at least one in every nation—covered 32 million km^2, or 15.4% of the planet's land area, and 3.4% of the ocean area.[60]

However, parks in developing countries do not always receive the funding, legal support, or enforcement support they need to manage resources, provide for recreation, and protect wildlife from poaching and timber from logging. Thus many of the world's protected areas are merely *paper parks*—protected on paper but not in reality. The Global Conservation Fund is an international NGO that works with governments around the world, helping with financial resources to boost their capacity to enforce laws and adequately protect and manage their designated parks and reserves.[61]

For example, the parks system in Costa Rica—now a significant source of national income and pride—initially received little real support from the government. According to Costa Rican conservationists, in their early years the parks were granted only five guards, one vehicle, and no funding. Government support for protected areas in Costa Rica improved, to the point where now about 26% of the nation's area is contained in national parks, wildlife and conservation reserves, and marine reserves. (In comparison, in Canada the total of all protected terrestrial areas, both land and aquatic, is 10.4%, and marine areas only 0.9%.[62]) The recent opening of some protected areas in Costa Rica to resource extraction activities, as well as continued illegal fishing and hunting, and the ongoing pressures of tourism highlight some of the challenges associated with habitat protection efforts.

Some types of protected areas fall under national sovereignty but are designated or partly managed internationally by the United Nations. The *UNESCO World Heritage Sites* are an example; over 1000 sites across 161 countries (including a number of cross-boundary sites) are listed for their natural or cultural value.[63] Canada has 17 listed sites (9 for natural heritage and 8 for cultural heritage), including Gros Morne National Park and Nahanni National Park. A cross-boundary example is the Mountain Gorilla Reserve shared by three African countries. This reserve integrates national parklands of Rwanda, Uganda, and the Democratic Republic of Congo. The Waterton Glacier International Peace Park on the Canada–U.S. border is another example. Transboundary parks like these can be quite large, and face numerous management challenges.

Biosphere reserves are tracts of land with exceptional biodiversity that couple preservation with sustainable development to benefit local people. They are designated by UNESCO (the United Nations Educational, Scientific, and Cultural Organization), following application by local stakeholders. UNESCO describes these as locations that "reconcile the conservation of biodiversity with its sustainable use."[64] The unique approach of the biosphere reserve is the focus on integrating human use with protection of species, habitats, and ecosystems.

Each biosphere reserve consists of (1) a *core* area that is isolated from the surroundings, and serves to preserve habitat and biodiversity, (2) a *buffer zone* that allows local activities and limited development that do not hinder the core area's function, and (3) an outer *transitional zone* in which agriculture, human settlement, and other land uses can be pursued in a sustainable way (**FIGURE 9.23**). The design of biosphere reserves seeks to reconcile human needs and sustainable use with the interests of wildlife and the goal of habitat preservation.

Canada has 16 designated biosphere reserves, the most recent of which was Bras d'Or Lake in Nova Scotia (**FIGURE 9.24**).[65] Clayoquot Sound in British Columbia's

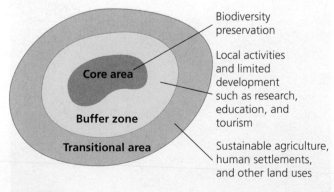

FIGURE 9.23
Biosphere reserves couple preservation with sustainable development to benefit local residents. Each reserve includes a core area that preserves biodiversity, a buffer zone that allows limited development, and a transition zone that permits various uses.

FIGURE 9.24
Bras d'Or Lake, an inland sea (part fresh water, and part salt water) on Cape Breton Island, Nova Scotia, is Canada's latest UNESCO Biosphere Reserve.

temperate rain forest was Canada's 12th biosphere reserve in 2000, and the area has focused on building cooperation among environmentalists, timber companies, native people, and local residents and businesses (see "Central Case: Battling over the Last Big Trees at Clayoquot Sound" in Chapter 10). The core area consists of provincial parks and Pacific Rim National Park Reserve. Environmentalists hoped the designation would help promote stronger land preservation efforts. Local residents supported it because outside money was being offered for local development efforts. The timber industry did not stand in the way once it was clear that harvesting operations would not be affected. The designation has brought Clayoquot Sound international attention, but it has not created new protected areas, has not brought international funding, and has not significantly altered land use policies.

In Chapter 12, *Marine and Coastal Systems and Fisheries*, we consider *marine reserves*, also known as *marine protected areas (MPAs)*. Marine reserves commonly allow some uses, including boating and fishing, with restrictions. In Canada, eight marine protected areas have been established so far, including the Endeavour Hydrothermal Vents (the first MPA in Canada), with eight more under consideration, including the St. Lawrence Estuary.[66] MPAs are designated in Canada under the federal *Oceans Act*.

Parks and green spaces are also key to liveable communities

City dwellers often desire some sense of escape from the noise, commotion, and stress of urban life. Natural lands, public parks, and open space provide greenery, scenic beauty, freedom of movement, and places for recreation. These lands also keep ecological processes functioning by regulating climate, producing oxygen, filtering air water pollutants, and providing habitat for wildlife. The animals and plants of urban parks and natural lands also serve to satisfy our natural affinity for contact with other organisms.

Protecting natural lands and establishing public parks become more important as our societies become more urbanized because many urban dwellers come to feel increasingly isolated and disconnected from nature. In the wake of urbanization and sprawl, people of every industrialized society in the world today have, to some degree, chosen to conserve land in public parks. In the late 1800s, urban public parks began to be established, using aesthetic ideals borrowed from European parks, gardens, and royal hunting grounds. The lawns, shaded groves, curved pathways, and pastoral vistas we see today in many city parks and cemeteries originated with these ideals.

At times during the historical development of urban parks, the aesthetic interests of the educated elite, the recreational interests of the broader citizenry, and the ecological interests of urban wildlands have come into conflict—a friction that survives today in debates over recreation in city parks. For example, Riverwood Conservancy is a 60-ha urban park located in the middle of Mississauga, Ontario (**FIGURE 9.25**). Park planners have had to balance many competing interests in designing the park, including the interests of naturalists, gardeners, artists, mountain bikers, fishers, historians (interested in the heritage buildings on site), schoolchildren, seniors, residents, students and teachers, and the park's wildlife and natural vegetation.

Greenways or corridors, strips of land that connect parks or neighbourhoods, are often located along rivers, streams, or canals, and they may provide access to networks of walking trails. (Compare this use of "corridor" with the

FIGURE 9.25
Riverwood Conservancy is a 60-ha, multiuse urban park in Mississauga, Ontario. Two-thirds of the property is designated to remain in its natural state (or be restored as such); the remainder is designated for a variety of uses by artists, hikers, fishers, bicyclists, school groups, and others.

concept of habitat and migration corridors for wildlife.) They can protect water quality, boost property values, and serve as corridors for the movement of birds and wildlife. For example, the Rails-to-Trails Conservancy has spearheaded the conversion of abandoned railroad rights-of-way into trails for walking, jogging, and biking. To date, nearly 24 000 km of 1200 rail lines and other passageways have been mapped, connected, and converted into trails across North America.[67]

Besides creating new types of urban spaces, many cities are working to enhance the "naturalness" of their parks through ecological restoration, the practice of restoring native communities. In Mississauga's Riverwood Conservancy, for example, volunteer teams gather periodically to remove garlic mustard, an invasive plant that smothers native plants on the forest floor. In Vancouver's Hastings Park, restoration activities have included tearing down a number of old park buildings and establishing a 4-ha natural area with a pond. On Vancouver Island, extensive restoration activities have been undertaken to reestablish degraded Garry oak ecosystems. The devotion of volunteers to the preservation of natural spaces, even in urban neighbourhoods, speaks to our human connection to nature, to wild spaces, and to the biosphere.

Conclusion

The erosion of biological diversity on our planet is threatening to result in a modern mass extinction event equivalent to the mass extinctions of the geological past. Human-induced habitat alteration, invasive species, pollution, and overharvesting of biotic resources, now amplified by global climatic change, are the primary causes of biodiversity loss.

The loss of biodiversity matters. Human society could not function without biodiversity's pragmatic benefits. As a result, conservation biologists and environmental scientists are rising to the challenge of conducting science aimed at saving endangered species, preserving their habitats, restoring their populations, and keeping natural ecosystems intact. The innovative strategies of these scientists and of environmental and natural resource managers around the world hold promise to slow the degradation of habitat and loss of biodiversity that threatens life on Earth.

REVIEWING OBJECTIVES

You should now be able to

Describe the scope of biodiversity on Earth, and how we measure it

- Some taxonomic groups (such as insects) hold far more diversity than others.
- Biodiversity is important not only at the species level, but also at the genetic level and at the levels of ecosystems, communities, and habitats.
- Roughly 1.75 million species have been described so far, but scientists agree that the world holds millions more. Estimates are based on extrapolations from scientific assessments.
- Diversity is unevenly spread across different habitats and areas of the world.

Evaluate the primary causes of biodiversity loss and species extinction

- Extinction occurs naturally. Species have gone extinct at a background rate of roughly one species per 1 to 10 million species each year. Earth's life has experienced five mass extinction events in the past 440 million years.
- Species that are highly specialized in their habitat or food needs, or are K-selected species that are slow to reproduce, are often more vulnerable to extinction than other species.
- Human impact is now causing the beginnings of a sixth mass extinction.
- Habitat loss and degradation is the main cause of current biodiversity loss. Invasive species, pollution, and overharvesting are also important causes. Climate change threatens to become a major cause of environmental change and biodiversity loss.

Specify the benefits of biodiversity

- Biodiversity is vital for functioning ecosystems and the services they provide us.
- Biodiversity contributes to the resiliency and stability of ecosystems.
- Wild species are potential sources of food.
- Many modern medicines are derived from wild species, and many others hold great potential for medical uses.
- Biodiversity contributes to economies by providing opportunities for recreation and tourism, particularly ecotourism.
- Many people feel that humans have a psychological need to connect with the natural world.

Summarize the challenges of conserving habitat and the main approaches to conservation

- Conservation biology is an applied science that studies biodiversity loss and seeks ways to protect and restore biodiversity at all its levels.

- The fragmentation of habitat is a growing problem, partly because it increases the proportion of edge to core habitat. The establishment of natural corridors can help. The equilibrium model of island biogeography and the area-size relationship explain how size and distance influence the number of species occurring on islands, and in fragmented terrestrial landscapes.

- *Ex situ* conservation efforts, in addition to the preservation of species in zoos, seed banks, and aquaria, include captive breeding and reintroduction programs. *In situ* conservation efforts involve the preservation of habitat so that species can continue to exist in their natural state. Increasingly, landscape-level conservation is being pursued in its own right.

- Most conservation efforts and laws so far have focused on threatened and endangered species. Species that are charismatic and well known are often used as tools to conserve habitats and ecosystems.

- Canada's main vehicle for designating threatened species is the *Species at Risk Act (SARA)*. Canada also participates in international conservation efforts,

including treaties such as CITES and the Convention on Biological Diversity.

- The identification of biodiversity hotspots has drawn attention to areas of the world that are highly and uniquely diverse, as well as vulnerable.

- Community-based conservation is increasingly practised, to the mutual benefit of ecosystems and local people.

- Some economic strategies for conservation include debt-for-nature swaps and conservation concessions.

Outline reasons for setting aside parks, reserves, and other protected areas

- Public demand for conservation, preservation, and recreation has led to the creation of parks, reserves, and wilderness areas in North America and across the world.

- Biosphere Reserves and World Heritage Sites are two types of internationally designated protected lands. It is challenging to monitor protected sites and to enforce the laws intended to protect them. Biosphere reserves aim to integrate protection of natural areas with sustainable human use.

- Parks and green spaces, including corridors that connect them, are increasingly important as key components of liveable communities.

TESTING YOUR COMPREHENSION

1. What is biodiversity? List and describe three levels of biodiversity.
2. What are the five primary causes of biodiversity loss? Can you give a specific example of each?
3. List and describe five invasive species and the adverse effects they have had.
4. Define the term *ecosystem services*. Give five examples of ecosystem services that humans would have a hard time replacing if their natural sources were eliminated.
5. What is the relationship between biodiversity and food security? Between biodiversity and pharmaceuticals? Give three examples of potential benefits

of biodiversity conservation for food security and medicine.
6. Describe four reasons why people suggest biodiversity conservation is important.
7. What is the difference between an umbrella species and a keystone species? Could one species be both an umbrella species and a keystone species?
8. Explain the island biogeography model. In what way is this model relevant to the management of habitat fragmentation?
9. What is a biodiversity hotspot?
10. Describe community-based conservation.

THINKING IT THROUGH

1. Biologist E.O. Wilson has said that, "Except in pockets of ignorance and malice, there is no longer an ideological war between conservationists and developers. Both share the perception that health and prosperity decline in a deteriorating environment." Do you agree or disagree? How do people in your community view biodiversity?

2. According to most scientists, the polar bear population in Canada's North is declining, but some Inuit hunters believe that the population is stable or even increasing. Offer two reasons why it might be extremely difficult to settle this question.
3. What would you say are some advantages of focusing on conserving single species, versus trying to conserve

broader communities, ecosystems, or landscapes? What might be some of the disadvantages? Which do you think is the better approach, or should we use both?

4. Bioprospectors for pharmaceutical companies visit biodiversity-rich countries, searching for organisms that might provide new drugs, foods, or other valuable products. They have been criticized for harvesting indigenous species to create commercial products that do not benefit the country of origin. To make sure it would not lose the benefits of its own biodiversity, Costa Rica reached an agreement with the Merck pharmaceutical company in 1991. The nonprofit National Biodiversity Institute of Costa Rica (INBio)

allowed Merck to evaluate a limited number of Costa Rica's species for their commercial potential in return for $1.1 million, a small royalty rate on any products developed, and training for Costa Rican scientists.

Do you think both sides won in this agreement? What if Merck discovers a compound that could be turned into a billion-dollar drug? Does this provide a good model for other countries, and for other companies? What if (as appears to have happened since this agreement was inked) pharmaceutical companies give up bioprospecting in favour of developing synthetic drugs; where would this leave countries like Costa Rica?

INTERPRETING GRAPHS AND DATA

Here is the 2014 summary of species identified in Canada as being of concern. (Many additional species have been examined and assessed, but either not recommended or not accepted for designation as a species at risk.) Use the data from this table to answer the questions below.

Summary of COSEWIC Assessment Results as of November 2014: Status According to Taxonomic Group

Status	Mammals	Birds	Reptiles	Amphibians	Fishes	Arthropods*	Molluscs	Vascular plants	Mosses	Lichens	Total
Extinct	3	3	0	0	7	0	1	0	1	0	**15**
Extirpated	2	2	5	2	2	4	2	3	1	0	**23**
Endangered	26	29	19	10	56	38	20	101	8	5	**312**
Threatened	14	29	10	6	40	7	4	50	3	4	**167**
Special concern	32	25	10	9	54	10	8	44	5	7	**204**
Total	**77**	**88**	**44**	**27**	**159**	**59**	**35**	**198**	**18**	**16**	**721**

*COSEWIC states that only a very small proportion of the many thousands of species in this taxonomic group in Canada has been assessed.

Source: Data from Summary Table of Wildlife Species Assessed by COSEWIC, May 2015. http://www.cosewic.gc.ca/rpts/Full_list_species.htm.

1. How many species altogether, among those that have been assessed by COSEWIC, are either extinct or extirpated from Canada?
2. What proportion of the total number of designated species is considered to be endangered?
3. Which taxonomic group has the highest proportion of endangered designations, relative to the total assessments for that group?

The table on the next page is an excerpt from the IUCN Red List, summarizing the status of major taxonomic groups of animals as of 2015. Use the data

in this table to answer the remaining questions. For the purpose of this exercise, although there are differences, consider the following to be roughly equivalent:

- IUCN's "critically endangered" and "endangered" categories equivalent to COSEWIC's "endangered" designation
- IUCN's "threatened" category equivalent to COSEWIC's "vulnerable" designation
- IUCN's "near threatened" category equivalent to COSEWIC's "special concern" designation

IUCN Red List Status Category Summary by Major Taxonomic Group (Animals)*

Status	Mammals	Birds	Reptiles	Amphibians	Total**
Extinct	77	140	21	34	**733**
Extirpated (extinct in the wild)	2	5	1	2	**32**
Critically endangered	213	213	174	518	**2510**
Endangered	477	419	356	789	**3706**
Vulnerable	509	741	397	650	**5602**
Near threatened	319	959	309	398	**3535**

*This table only includes a small fraction of the data from the original table; columns and rows will not add up because of missing data.

**These data refer to the totals for all species assessed by IUCN Red List—not just mammals, birds, reptiles, and amphibians.

Source: Based on IUCN Red List Summary Statistics. http://www.iucnredlist.org/about/summary-statistics#Tables_1_2. Accessed 22 March 2015.

4. How many species altogether, among those that have been assessed by IUCN, are either extinct or extirpated in the wild? What does it mean to say that a species is "extirpated in the wild"?

5. How many species in each of the major taxonomic groups shown here are either extinct or extirpated in the wild?

6. For each group, what proportion of the total number of designated species (right-hand column) is either extinct or extirpated in the wild?

7. For each group, what proportion of the total number of designated species is either critically endangered or endangered?

8. Which taxonomic group has the highest proportion of critically endangered and endangered designations, relative to the total assessments for that group?

9. Summarize how these results compare to the proportions you found in the data for species designated as endangered in Canada.

MasteringEnvironmentalScience®

STUDENTS

Go to **MasteringEnvironmentalScience** for assignments, the eText, and the Study Area with practice tests, videos, current events, and activities.

INSTRUCTORS

Go to **MasteringEnvironmentalScience** for automatically graded activities, current events, videos, and reading questions that you can assign to your students, plus Instructor Resources.

10 Forests and Forest Management

The old-growth temperate rain forest of Clayoquot Sound, BC, has long been the focus of conflicts between logging companies and environmentalists.

Lynn Amaral/Shutterstock

Upon successfully completing this chapter, you will be able to

- Describe the basic functional processes of trees, the principal forest biomes in Canada, and the role of forests in biogeochemical cycling
- Summarize the economic contributions of forests and the main approaches to forest product harvesting
- Trace the history and scale of forest loss and identify the current drivers of deforestation
- Explain the fundamentals of forest management, and identify current forest management approaches in Canada and internationally

Anti-logging protestors rallied at Clayoquot Sound in 1993.

CANADA

Hudson Bay

Clayoquot Sound

CENTRAL CASE
BATTLING OVER THE LAST BIG TREES AT CLAYOQUOT SOUND

"Forests and trees sustain and protect us in invaluable ways. They provide the clean air that we breathe and the water that we drink. They host and safeguard the planet's biodiversity and act as our natural defence against climate change. Life on Earth is made possible and sustainable thanks to forests and trees."

—UN FAO, INTERNATIONAL DAY OF FORESTS 2015[1]

"Our forests form an important part of our roots as a nation and a big part of our future."

—NATURAL RESOURCES CANADA, THE STATE OF CANADA'S FORESTS, 2014

It was the largest act of civil disobedience in Canadian history, and a pivotal event in Canadian environmental history. It played out along a seacoast of majestic beauty, at the foot of some of the world's biggest trees. At Clayoquot Sound on the western coast of Vancouver Island, British Columbia, protestors blocked logging trucks, preventing them from entering stands of ancient temperate rain forest. The activists chanted slogans, sang songs, and chained themselves to trees.

Loggers complained that the protestors were keeping them from doing their jobs and making a living. By the time the protest was over, 850 of the 12 000 protestors were arrested, and this remote, mist-enshrouded land of cedars and hemlocks became ground zero in the debate over how we manage forests.

That was in 1993, and the activists were opposing **clear-cutting**, the logging practice that removes all trees from an area. Most of Canada's old-growth temperate rain forest had already been cut, and the forests of Clayoquot Sound were among the largest undisturbed stands of temperate rain forest left on the planet.

Timber from **old-growth forests**—complex, primary forests in which the trees are at least 150 years old—had long powered British Columbia's economy. Historically, one in five jobs in BC depended on its $13 billion timber industry, and many small towns would have gone under without it. By 1993, however, the timber industry was cutting thousands of jobs a year because of mechanization, and the looming depletion of old growth threatened to slow the industry. Meanwhile, Greenpeace was convincing overseas customers to boycott products made from trees clear-cut by multinational timber company MacMillan-Bloedel. Soon British Columbia's premier found himself trying to persuade European nations not to boycott his province's main export.

In 1995, the provincial government called for an end to clear-cutting at Clayoquot Sound, after its appointed scientific panel of experts submitted a new forestry plan for the region. The plan recommended reducing harvests, retaining 15–70% of old-growth trees in each stand, decreasing the logging road network, designating forest reserve areas, and managing *riparian* (water's-edge) zones. Two years later, the provincial government reversed many of these regulations, and a new premier pronounced forest activists "enemies of British Columbia."

The antagonists struck a deal; wilderness advocates and MacMillan-Bloedel agreed to log old growth in limited areas, using environmentally friendly practices. In 1998, First Nations people of the region formed a timber company, Iisaak Forest Resources, in agreement with MacMillan-Bloedel's successor, Weyerhaeuser, and began logging at Clayoquot Sound in a more environmentally sensitive manner (see photo).

In the Nuu-chah-nulth language of First Nations people from the Clayoquot Sound area, *iisaak* (pronounced *E-sock*) means "respect," which became a guiding principle for forestry in Clayoquot Sound. The *variable retention harvesting* they applied—logging selectively with the goal of retaining a certain percentage and particular characteristics of the forest ecosystem—is more expensive than normal clear-cutting. Iisaak Forest Resources hoped to recoup some of the extra cost by achieving a premium price for the cut timber and through ecotourism and the sustainable exploitation of other forest resources.

Leaving most of the trees standing accomplished what forest advocates had predicted: People from all over the world are now visiting Clayoquot Sound for its natural beauty and are kayaking and whale-watching

Logging continues in some parts of Clayoquot Sound.

Dewitt Jones/Corbis

in its waters. Ecotourism (along with fishing and aquaculture) has surpassed logging as a driver of local economies. From the perspective of ecotourism, the trees appear to be worth more standing than cut down. The United Nations designated the site as an international biosphere reserve in 2000, encouraging land protection and sustainable development. The designation cited the dangers of fragmentation of forest and alpine ecosystems, and the desire to preserve coastal rainforest biodiversity by providing a refuge for the natural dispersion and re-establishment of species.[2]

Tensions continue today, however, and logging has never completely stopped at Clayoquot Sound—even in areas near park and biosphere reserve boundaries—to the dismay of environmental activists.[3] Local forest advocates worry that the provincial government's Working Forest Policy will increase logging, and the town of Tofino petitioned the province to exempt Clayoquot Sound's forests from the policy.

Ultimately, Iisaak Forest Resources found it difficult to make money doing sustainable forestry and entered into an agreement with the environmental organization Ecotrust Canada in 2006. Today logging is being done more sustainably, and at a profit, under this arrangement. The provincial government is considering new forestry plans that would shift logging out of old-growth forests and into younger forests that were already logged in the past. As long as our demand for lumber, paper, and forest products keeps increasing, pressures will keep building on the remaining forests on Vancouver Island and around the world.

The Forest and the Trees

Forest covers roughly 31% of Earth's land surface (**FIGURE 10.1**), about 4 billion hectares.[4] Forests provide habitat for countless organisms and help maintain soil, air, and water quality. They play key roles in our planet's biogeochemical cycles, serving as one of the most important reservoirs in the carbon cycle. Forests have also long provided humanity with wood for fuel, construction, paper production, and more.

Trees have several basic requirements

Trees are the fundamental biological component of forests, although, as you will learn, forests have many other crucial components, both biotic and abiotic, and not all forests are dominated by trees.

Trees, like all other plants, are autotrophs. They create their own food energy by photosynthesis, through which they extract carbon from atmospheric carbon dioxide and recombine it with water to make carbohydrates (sugars, such as glucose; **FIGURE 10.2**). For photosynthesis to occur and the tree to survive, there are several fundamental requirements:

- an amenable temperature (the specific temperature range varies by species)

- air (with which the tree exchanges carbon, hydrogen, oxygen, and nitrogen)
- light (the energy source for photosynthesis)
- soil (the source for mineral nutrients)
- water (needed for a variety of reasons)

The nutrients required for plant growth are supplied through the tree's roots from the soil, using soil water as the transfer medium. Nutrients that are required by living organisms only in small amounts are called **micronutrients**; for trees, these include iron (Fe), manganese (Mn), zinc (Zn), copper (Cu), boron (B), chlorine (Cl), and molybdenum (Mo). Nutrients that are required in relatively large amounts are called **macronutrients**; for trees, these include nitrogen (N), phosphorus (P), potassium (K), magnesium (Mg), calcium (Ca), and sulphur (S), in addition to the carbon (C) that the tree acquires from the atmosphere.

Recall (from Chapter 7, *Soils and Soil Resources*) that the process of *nitrogen fixation* by soil-dwelling bacteria converts atmospheric nitrogen into a form that is usable by plants, including trees. Lightning is another natural process that leads to the fixation of atmospheric nitrogen in the soil. Thus trees acquire much of their nitrogen from the atmosphere, by way of the soil. Nitrogen also can come from the decomposition of organisms, fecal matter from animals, and artificial fertilizers. The rest of the macronutrients come mainly from the dissolution of mineral grains in soil water.

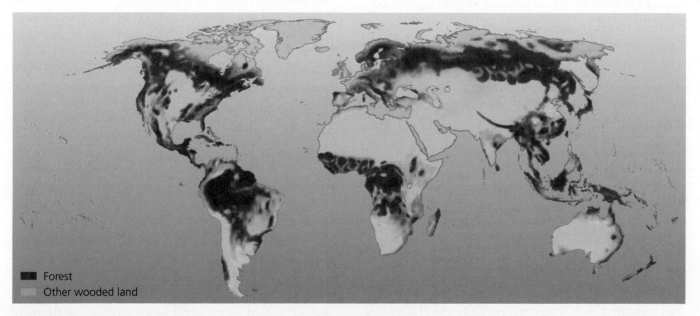

■ Forest
☐ Other wooded land

FIGURE 10.1 About 31% of Earth's land surface is covered by forest. Most of this consists of the boreal forests of the north and the tropical forests of South America and Africa. Other lands (including tundra, shrubland, and savannah) can be classified as "wooded land," implying a more open forest type that supports trees, but at sparser densities.
Source: Data from UN Food and Agriculture Organization (FAO) (2010). *Global Forest Resources Assessment 2010*.

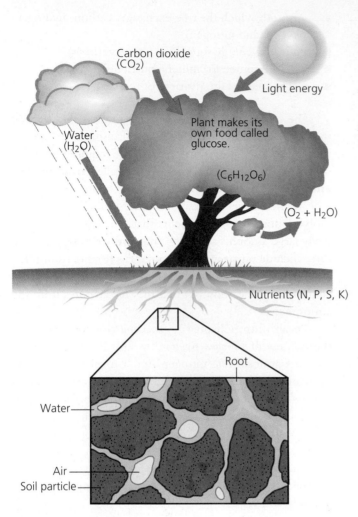

FIGURE 10.2

Trees acquire carbon from the air via photosynthesis. They utilize light energy to carry out photosynthesis for the production of carbohydrates, such as glucose, from atmospheric carbon dioxide. Macronutrients such as phosphorus, sulphur, and potassium are provided by the mineral components of the soil, through the tree's roots by way of soil water. Nitrogen comes primarily from the atmosphere via nitrogen fixation.

Trees, like all plants, need water for several reasons. First, water is used in photosynthesis, for which the general simplified reaction is

$$6CO_2 + 6H_2O = C_6H_{12}O_6 + 6O_2$$

Second, water acts as a solvent, dissolving mineral constituents from particles in the soil. As shown in **FIGURE 10.2**, these are taken up by the tree's roots, to be used as nutrients. Third, in addition to transporting nutrients, water transports chemicals from one part of the tree to wherever they are needed to carry out the metabolic processes that keep the tree alive. Fourth, water provides support for cells. All plant cells require internal water pressure (called *turgidity*) in their cells, or they will wilt and eventually die. Turgidity is the main way that nonwoody plants stay upright; trees have additional structural support in the form of woody trunks that help them remain upright.

Trees (and other plants) require water to pass from their roots through their trunks and branches and evaporate from their leaf surfaces (**FIGURE 10.3**); this is called **transpiration**. Transpiration cools the plant, as well as assisting in the movement of nutrients. It also helps small openings in the leaves, called *stomata*, to open, allowing for the intake of carbon dioxide during photosynthesis. Transpiration occurs when solar energy causes water to evaporate from leaf surfaces, resulting in negative internal water pressure. Water is drawn up through the roots and trunk through narrow, strawlike tubes called *xylem*. Soil water moves into the roots by *osmosis* and is drawn up through the xylem by *adhesion*, upward movement that results from the surface tension between water (a polar molecule, you will recall from an earlier chapter) and the capillary-like walls of the xylem. *Cohesion*, the attraction between water molecules, then serves to draw additional water into the xylem.

This helps explain why it is useful to plant trees in close proximity to crops, called **agroforestry**. Crop plants have much shallower root systems than trees, in general; a tree can draw water and minerals from depth, making them available to nearby crops. At the same time, some trees can be harvested for wood, fruits, nuts, honey, and other products, and domesticated animals can graze on fallen leaf litter.

Trees also function as an important link between the biogeochemical cycles of the atmosphere, hydrosphere, biosphere, and geosphere. Transpiration is an important step in the hydrologic cycle, moderating the movement of water from the atmosphere to the ground and from the ground back to the atmosphere. When trees draw mineral nutrients and water from depth through their root systems, they deliver them to near-surface soil layers, where they become available for other plants (**FIGURE 10.4**). This is sometimes referred to as *water and nutrient pumping*. Trees also deliver organic material back to the topsoil, in the form of **litter**, which consists of fallen branches and leaves.

There are three major groups of forest biomes

A **forest**, strictly speaking, is a land area with significant tree cover, in which the **canopy** (the upper level of leaves and branches defined by the majority of the treetops) is largely **closed**, or continuous. A **woodland** is a wooded (treed) area in which the canopy is more **open**, or discontinuous; that is, there are some openings between the trees that allow light to penetrate to the ground, or **floor**, of the forest (**FIGURE 10.5**).

There are three major types of forest biomes, corresponding roughly to the high, middle, and

How water moves through a tree

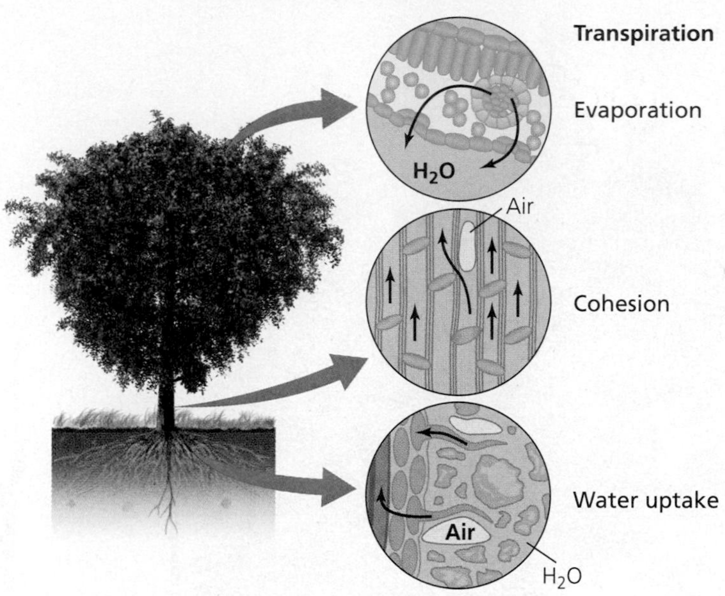

Transpiration

Evaporation

H_2O

Air

Cohesion

Water uptake

Air

H_2O

FIGURE 10.3 The driving force for transpiration is evaporation from leaf surfaces. Water is drawn into root hairs and upward through the xylem by adhesion and cohesion.

equatorial latitudes. However, there are *many* local variations, as well as altitudinal variations; we will look more closely at the forests of Canada later in this chapter.

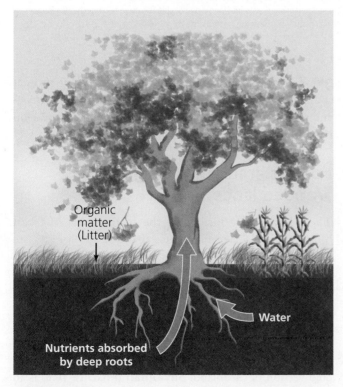

FIGURE 10.4
Trees act like pumps, drawing nutrients and water from depth through their root systems. In this way, they can provide nutrients and water to plants with shallower root systems. They also act as a link in biogeochemical cycles and the hydrologic cycle.

Boreal forest   The **boreal forest** is a high-latitude forest type (mainly in the Northern Hemisphere) that is characterized by cold, relatively dry climates with short growing seasons. The boreal forest biome stretches across much of Canada, Russia, and Scandinavia. Further to the north, the boreal forest grades through the **taiga**, the northernmost part of the boreal forest, into the more open northern **tundra** biome. Boreal forests are characterized by evergreen, **coniferous** trees—trees whose "leaves" take the form of needles and that produce seed pods in the form of cones. Evergreen, waxy-coated needles and cones are energy- and water-saving adaptations that help trees cope with short growing seasons and low precipitation.

Temperate forest   The **temperate forest**, the second major forest biome type, occurs in midlatitude areas of seasonal climate, which typically experience a distinct winter season and summer growing season. Temperate forests occur throughout eastern North America, northeastern Asia, and western and central Europe. Temperate forests cover much less area globally than boreal forests, in part because people have already cleared so many temperate forests. Trees in temperate forests must be adapted to a seasonal climate and wide ranges in temperature and precipitation. Temperate forests are often characterized by **deciduous** trees—trees whose leaves turn colour (*senesce*) and drop off in the fall, in preparation for a period of winter dormancy.

Tropical forest   The third major forest biome type is the **tropical forest**. Tropical rain forests, which

Emergent trees

Canopy

Treefall gap

Snag

Subcanopy

Understorey

Shrub layer

Forest floor

Soil

Ground-cover   Leaf litter   Moss and epiphytes   Roots   Fallen log

FIGURE 10.5   In this generalized cross-section of a mature forest, the crowns of the largest trees form the canopy, and smaller trees beneath them form the shaded subcanopy and understorey. Emergent trees protrude above the main canopy. Shrubs and groundcover grow just above the forest floor, which may be covered in litter. Vines, mosses, lichens, and epiphytes cover portions of trees and the forest floor. Snags (standing dead trees) provide food and nesting and roosting sites for birds and other animals. Fallen logs nourish the soil and young plants and provide habitat for invertebrates as the logs decompose. Gaps caused by fallen trees let light through the canopy and create small openings, allowing new plants to grow within the mature forest.

host extremely diverse flora and fauna, occur in the wet, tropical climates of equatorial South and Central America, equatorial Africa, and Indonesia and Southeast Asia. Trees in rain forests are often evergreen, but not for the energy-saving reasons that boreal trees are evergreen; in tropical forests, trees remain green because they have year-round growing conditions.

For **rain forests** we tend to think first of the tropical environment, but in fact their central characteristic is not high temperature, but high rainfall. Thus, rain forest variations include the evergreen rain forests of the cool, wet Pacific Northwest, including those of Clayoquot Sound. Similarly, there are tropical forests in which the climate is warm but not wet year-round, alternating instead between a rainy season and a dry season. As the dry season gets longer, the character of the forest changes and other wooded biome types emerge, such as the more open tropical dry forest and savannah biomes.

An additional unique type of forest biome comprises coastal forests, such as *mangroves*. Trees in mangrove forests are adapted to conditions of constant standing water and fluctuating water salinity. They provide valuable habitat

for shallow-water marine organisms and they protect shorelines from the battering of storm waves. Mangroves and other coastal forests are discussed in the chapter on marine and coastal systems and fisheries.

Forests grade into open wooded lands

These broad descriptions of the major types of forest biomes should make it obvious that not all "forests" are completely dominated by trees with a continuous canopy cover. At the drier end of the climatic spectrum (in cold, high-latitude regions as well as hot, low-latitude regions), the canopies of wooded areas tend to be quite open. **Shrublands** are wooded areas that are covered by smaller, bushier trees, or shrubs, often interspersed with occasional taller trees. Tundra is a high-latitude (and high-altitude), cold version of shrubland. **Savannah** (**FIGURE 10.6**) is an open woodland area with scattered trees and lots of grass, typically grading into **grasslands,** large open stretches of land dominated by grasses. Again, all of these basic biome types have many local variations.

FIGURE 10.6
Savannah is an open-canopy woodland, characterized by a relatively dry climate, lots of grasses, and dispersed trees.

Savannahs and shrublands are not forests, strictly speaking, although any of them can be partially "wooded." It is common to group these biome types, together with grasslands (areas where the vegetation is dominated by grasses), under the category of **drylands**, emphasizing their central defining characteristic of low precipitation. This is a broad category that includes some areas with a relatively long dry season alternating with a rainy season; some semi-arid regions that experience low precipitation year-round; and—at the extreme end of the dryland spectrum—the arid *desert* biome.

Because drylands are characterized by low overall precipitation, they tend to be extremely sensitive to environmental change and are easily damaged if land use practices become overly intensive. Therefore, *desertification* and *land degradation* are major environmental issues in dryland management. Much of the world's dry woodland and grassland has been converted for the purpose of agriculture or rangeland. We will discuss this in greater detail below.

Canada is a steward for much of the world's forest

Canada's current 348 million ha of forested and other wooded land (310 million ha of which is "true" forest, rather than other types of woodland)[5] represent about 9% of the world's forest cover, 35% of Canada's total land area, 25% of the world's natural (rather than planted) forest, 30% of the world's boreal forest, and 20% of the world's temperate rain forest. Canada's wooded lands include some of the world's largest intact forest ecosystems. Canada has the highest amount of forested land per capita in the world; in contrast, 64 countries in the world have either no forest at all, or less than 10% of forested land

overall.[6] About one-third of the Canada's forested land is in British Columbia, and another third or so is in Quebec and Ontario.[7]

According to the Food and Agriculture Organization of the United Nations (FAO), which monitors the status of the world's forests, more than 50% of Canada's primary forest (that is, long-standing natural forest) remains more or less intact.[8] In comparison, of the original 4 million km[2] (400 million ha) of forested land in the United States, the majority was deforested by the late nineteenth century. (In the early 1900s, forest cover in the United States began to stabilize, however, and in the past few decades, the United States has seen a small annual increase in forested land.)

Together with the other four most forest-rich nations of the world (the Russian Federation, Brazil, the United States, and China[9]), Canada clearly has an obligation to the rest of the world to manage our forests as effectively and sustainably as possible. The Boreal Forest Conservation Framework puts our responsibilities with regard to the boreal forest in perspective, stating, in part, that its goal is to conserve the cultural, sustainable, economic, and natural values of the entire Canadian boreal region by promoting ecosystem-based resource management practices and state-of-the-art stewardship practices.[10]

The members of the Boreal Leadership Council[11] are described as "historically unlikely partners," and include representatives from industry and finance; nongovernmental organizations, nonprofits, and conservation and environmental groups; and Aboriginal organizations and their governing bodies. This type of collaborative, cross-disciplinary, cross-sectoral management process has been typical of Canada's historical approach to the management of complex and sometimes thorny environmental issues.

Canada's forests are varied

Canada's forest biomes include many regional variations, as shown on the map in **FIGURE 10.7**.

Forests of the north   The boreal forest, the largest forested region of Canada, stretches through all of the provinces and territories except Nova Scotia and Prince Edward Island (**FIGURE 10.8A**). White spruces, tamarack, and jack and lodgepole pines are the main coniferous species in the boreal forest, and white birch, aspen, and balsam poplars are the main deciduous trees. In the north the boreal forest merges with the *tundra*, an open woodland biome characterized by scrubby vegetation. Tree growth is limited in the tundra because the growing season is short and temperatures are cool. The term *taiga* is a synonym for boreal forest; in Canada and Alaska taiga

FIGURE 10.7
Canada's forested area is dominated by the great northern boreal forest, as seen on this map, with temperate forest variations in the east and west, grasslands in the central southern regions, and tundra in the far north. *Source:* NRCAN Forests in Canada: Forest Regions. http://www.nrcan.gc.ca/forests/canada/classification/13179.

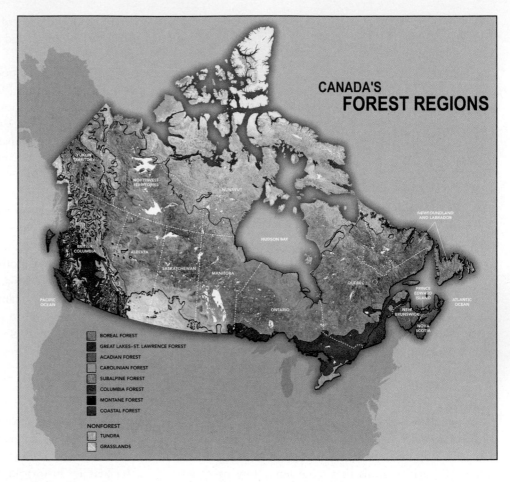

is sometimes considered to be a transitional zone between the northernmost boreal forest and the northern tundra.

Forests of the west   In the west is the *subalpine forest region* of the mountains of British Columbia and western Alberta, with characteristic Engelmann spruce, alpine fir, and lodgepole pines, and the *montane forest region* in British Columbia's central plateau, with Rocky Mountain Douglas fir, lodgepole pine, trembling aspen, and ponderosa pine (**FIGURE 10.8B**). The *coast forest region*, found at Clayoquot Sound and elsewhere on Vancouver Island, is the temperate rain forest, characterized by western red cedar, western hemlock, Sitka spruce, yellow cypress, and deciduous big-leaf maple, red alder, cottonwoods, Garry oak, and arbutus. The *Columbia forest region* of the Kootenay, Thompson, and Fraser River valleys includes species like western white pine, Engelmann spruce, western larch, and grand fir.

Forests of the east   In the east, the *deciduous forest region* north of Lake Erie and Lake Ontario is the smallest forest region in Canada, characterized by deciduous species such as sugar maple, beech, hickory, elm, and oak and by conifers such as the eastern white pine and eastern hemlock (**FIGURE 10.8C**). In this region there are also pockets of "Carolinian" species such as the tulip

tree and black gum, which are more common further to the south in the eastern United States. The existence of Carolinian forest in southern Ontario is mainly due to the moderating influence of the Great Lakes on the climate of the area.

The highly diverse *Great Lakes–St. Lawrence forest region* extends from northwestern New Brunswick, through the St. Lawrence, Lac St. Jean, and Saguenay river valleys, over southern and central Ontario, and into Manitoba. Typical conifers include the eastern white and red pine, eastern hemlock, and white cedar. The characteristic deciduous species is yellow birch, but maple and oak are also common. Finally, the *Acadian forest region* of Nova Scotia, New Brunswick, and Prince Edward Island is typified by spruce and balsam fir, with common deciduous sugar maple, yellow birch, and beech.[12]

Forests are ecologically valuable

Because of their structural complexity and their ability to provide many niches for organisms, forests compose some of the richest ecosystems for biodiversity (**FIGURE 10.9**). Trees furnish food and shelter for an immense diversity of vertebrate and invertebrate animals. Countless insects, birds, mammals, and other organisms subsist on the leaves, fruits, and seeds that trees produce.

(a) Northern boreal forest

(b) Western montane forest

(c) Eastern deciduous forest

FIGURE 10.8

Canada's forest biomes include many regional variations, some of which are seen here.

growth, give a forest structural complexity and provide habitat for still more organisms. Moreover, the leaves, stems, and roots of forest plants are colonized by an extensive array of fungi and microbes, in both parasitic and mutualistic relationships. Much of a forest's diversity resides on the forest floor, where the soil is generally nourished by fallen leaves and branches, called *litter*. There, myriad soil organisms help decompose plant material and cycle nutrients.

An additional habitat consideration in forests is the difference between the forest **core**, in the middle of a large forested area, and the forest **edge**. Edge habitat—even if it is still forested—can be quite different in character from habitat in the forest core. This is particularly so for light

Some animals are adapted for living in the dense treetop canopy, where beetles, caterpillars, and other leaf-eating insects abound, providing food for birds such as tanagers and warblers, while arboreal mammals from squirrels to sloths to monkeys consume fruit and leaves. Other animals specialize on the *subcanopies* of trees, and still others utilize the bark, branches, and trunks. Cavities in trunks provide nest and shelter sites for a wide variety of vertebrates. Dead and dying trees are valuable for many species; these *snags* are decayed by insects that, in turn, are eaten by woodpeckers and other animals.

Meanwhile, the shrubs and groundcover plants of the **understorey**, the forest floor and the lowest levels of

FIGURE 10.9

Forests are ecologically valuable. They are especially important as habitat because of their structural complexity, as shown here in the old-growth forest of Cathedral Grove on Vancouver Island.

levels, density of vegetation, and moisture. Research has demonstrated that wildlife adapted to forest core habitats declines when forced to occupy edge habitats in parts of the forest that are immediately adjacent to roads or cleared areas. Fragmentation of wooded areas greatly increases the ratio of edge to core habitat, even if the total wooded area is not greatly reduced.

In general, forests with a greater diversity of plants, such as tropical rain forests, host a greater diversity of organisms overall. And in general, fully mature forests, such as the undisturbed old-growth forests remaining at Clayoquot Sound, contain more biodiversity than younger forests. Older forests offer more structural diversity and thus more microhabitats and resources to support more species.

Trees provide ecosystem services of value to people

In addition to hosting a significant proportion of the world's biodiversity, trees and forests provide all manner of vital ecosystem services that are of value to people.

FIGURE 10.10
Forests provide us with a variety of ecosystem services, as well as resources that we can harvest.

By regulating the hydrologic cycle, trees and other forest plants slow runoff, lessen flooding, and purify water as they take it in from the soil and release it to the atmosphere. Tree branches and leaves physically block and soften the fall of rain, which further protects the soil from degradation. Forests also store carbon, release oxygen, and act as a moderating influence on climate.

Forest vegetation also stabilizes soil and prevents erosion. The principal direct cause of soil erosion and degradation is the removal of vegetation. This is especially true in tropical rain forests where, counterintuitively, soils are not particularly fertile because most of the biomass of the system resides in the trees and other forest plants. Once the trees have been removed, the soil is exposed to wind and water and can quickly erode. By protecting soil and performing other ecological functions, forests are indispensable for human survival and well-being.

Additionally, people draw many direct socioeconomic benefits from forests—everything from food, shelter, and fuel, to employment and income from the harvesting of forest products, to spiritual fulfillment (**FIGURE 10.10**). In the next section we will consider the many products provided to us by forests.

Harvesting Forest Products

In addition to the immense value of their ecological services, forests provide people with economically valuable harvestable products. It has been estimated that over 1.6 billion people worldwide depend on forests directly for their livelihood in one way or another, but all of us use forest products in our daily lives.

Forest products are economically valuable

For millennia, wood from forests has fuelled our fires, keeping people warm and well fed. Today forest products contribute to shelter for 1.3 billion people, and 2.4 billion cook with woodfuels.[13] Wood also built the ships that carried people and cultures from one continent to another. It allowed us to produce paper, the medium of the first information revolution. In recent decades, industrial harvesting has allowed the extraction of more timber than ever before, supplying all these needs of a rapidly growing human population and its expanding economy. The exploitation of forest resources has been instrumental in helping our society achieve the standard of living we enjoy today.

Most commercial logging today takes place in Canada, Russia, and other nations that hold large expanses of boreal forest and in tropical countries with large amounts of rain forest, such as Brazil and Indonesia. Timber harvested from coniferous trees (such as those that dominate the boreal forest) is called **softwood**, whereas timber that comes from deciduous trees is called **hardwood**. (The terms are not related to the actual hardness of the wood.) The softwood lumber industry is extremely important to Canada's economy.

Forests also supply non-wood products in abundance. Some of these *NTFPs* (for *non-timber forest products*) include medicinal and herbal products, such as ginseng, echinacea, and St. John's wort; decorative products, such as Christmas trees, wreaths, and other greenery; and many edible products, including fruits, honey, maple syrup, edible mushrooms including truffles, and a large variety of nuts. Many Aboriginal and Indigenous peoples make their livelihoods by harvesting non-timber forest products. The *seringeiro* rubbertappers of the Brazilian Amazon are one example of a group of people whose lifestyle is adapted to the sustainable extraction of forest resources.

Nations maintain and use forests for all these economic and ecological reasons. The UN FAO estimates that, globally, about 30% of forests are designated primarily for timber production and harvesting of other forest products.[14] Others are designated for a variety of functions, including conservation of biodiversity, protection of soil and water quality, and social services such as recreation, tourism, education, and conservation of culturally important sites (**FIGURE 10.11**).

Timber is harvested by several methods

When they harvest trees, timber companies use any of several methods. Prior to the 1990s, timber harvests in Canada were mostly conducted using the clear-cutting method, in which all trees in an area are cut, leaving only stumps. Clear-cutting is generally the most cost-efficient method in the short term, but it has the greatest impacts on forest ecosystems (**FIGURE 10.12**). Clear-cut logging continues in Canada (and in many other parts of the world); however, after the late 1990s both the amount of clear-cutting and the sizes of cleared plots began to decrease.

In the best-case scenarios, clear-cutting of small areas may mimic natural disturbance events such as fires, tornadoes, or windstorms that knock down trees across large areas. In the worst-case scenarios, entire communities of organisms are destroyed or displaced, soil erodes, and the penetration of sunlight to ground level changes microclimatic conditions such that new types of plants replace those that had composed the native forest. Essentially, clear-cutting sets in motion an artificially driven process of succession in which the resulting climax community may turn out to be quite different from the original climax community.

Clear-cutting occurred widely across North America at a time when public awareness of environmental problems was blossoming. The combination produced public outrage toward the timber industry and public forest managers. Eventually the industry developed and integrated other harvesting methods (**FIGURE 10.13**). A set of approaches dubbed **new forestry** called for timber cuts that came closer to mimicking natural disturbances.

World forest functions

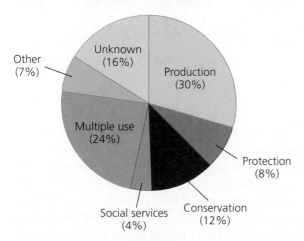

FIGURE 10.11
Worldwide, nations utilize over one-third of forests for production of timber and other forest products. Smaller areas are designated for conservation of biodiversity, protection of soil and water quality, and "social services" such as recreation, tourism, education, and conservation of culturally important sites.
Source: UN Food and Agriculture Organization (FAO) (2010). *Global Forest Resources Assessment 2010*, Key Findings, p. 10. http://www.fao.org/docrep/013/i1757e/i1757e.pdf. Reproduced with permission.

FIGURE 10.12
Clear-cutting is the most cost-efficient method for timber companies, but it can have severe ecological consequences, including soil erosion and species turnover. Although certain species do use clear-cut areas as they regrow, most people find these areas aesthetically unappealing, and public reaction to clear-cutting has driven changes in forestry methods.

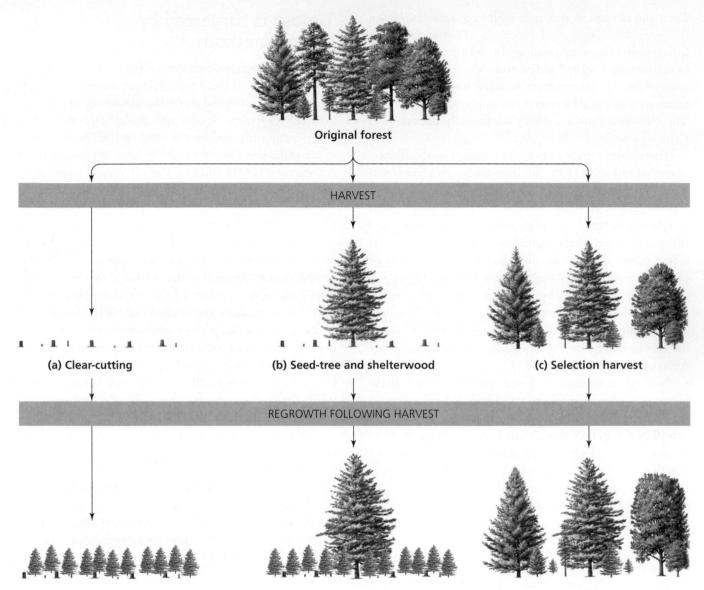

FIGURE 10.13   In clear-cutting **(a)**, all trees are cut, extracting a great deal of timber inexpensively but leaving a vastly altered landscape. In seed-tree and shelterwood systems **(b)**, small numbers of large trees are left to help reseed the area or provide shelter for growing seedlings. In selection systems **(c)**, a minority of trees is removed, while most are left standing. The latter methods involve less environmental impact than clear-cutting, but all timber harvesting causes significant changes to the structure and function of natural forest communities.

For instance, "sloppy clear-cuts" that leave a variety of trees standing were intended to mimic the changes a forest might experience if hit by a severe windstorm.

Clear-cutting (**FIGURE 10.13A**) is still widely practised, but other methods involve cutting some trees and leaving some standing. In the *seed-tree* approach (**FIGURE 10.13B**), small numbers of mature and vigorous seed-producing trees are left standing so that they can reseed the logged area. In the *shelterwood* approach (also **FIGURE 10.13B**), small numbers of mature trees are left in place to provide shelter for seedlings as they grow. These three methods all lead to even-aged stands of trees.

Selection systems, in contrast, allow uneven-aged stand management. In selection systems (**FIGURE 10.13C**),

such as the variable retention harvest system practised in Clayoquot Sound, only some trees in a forest are cut at any one time. The stand's overall rotation time may be the same as in an even-aged approach because multiple harvests are made, but the stand remains mostly intact between harvests. Selection systems include single-tree selection, in which widely spaced trees are cut one at a time, and group selection, in which small patches of trees are cut.

It was a form of selection harvesting that Iisaak Forest Services and other logging organizations pursued at Clayoquot Sound after old-growth advocates applied pressure and the scientific panel published its guidelines. Not wanting to bring a complete end to logging when so many local people depended on the industry for work, activists and scientists instead promoted a more environmentally friendly method of timber removal. However, selection systems are by no means ecologically harmless. Moving trucks and machinery over an extensive network of roads and trails to access individual trees compacts the soil and disturbs the forest floor, not to mention opening roadways and fragmenting core habitat. Selection methods are unpopular with timber companies, too, because they are expensive.

The bottom line, from an ecological perspective, is that all methods of logging result in habitat disturbance, which invariably affects the plants and animals inhabiting an area. All methods change forest structure and composition, increasing edge and diminishing core habitat. Most methods cause increased soil erosion, leading to siltation of waterways, which can degrade habitat and affect drinking water quality. Most methods also speed runoff, sometimes leading to flooding. In extreme cases, as when steep hillsides are clear-cut, landslides can result.

Plantation forestry has increased

The North American timber industry is largely centred on production from managed forests, or *plantations*, of fast-growing tree species that are single-species monocultures. Forest plantations make up about 7% of forested land globally (**FIGURE 10.14A**).[15] Modern logging in Canada is offset by **reforestation**, the planting of trees after logging, and **afforestation**, the planting of trees where forested cover has not existed for some time (over 50 years), so the total forested area does not decrease (and, in fact, may increase) from year to year.

Because all trees in a managed stand are planted at the same time, the stands are **even-aged** (**FIGURE 10.14B**). Stands are cut after a certain number of years (called the *rotation time*), and the land is replanted with seedlings. However, some harvesting methods aim to maintain **uneven-aged** stands, where a mix of ages (and often a mix of tree species) makes the stand more similar to a natural forest.

0   100

Primary forests	Other naturally regenerated forests	Planted forests
36%	57%	7%

(a) Primary and planted forests

(b) Tree plantations

FIGURE 10.14
(a) Approximately 7% of global forested land today is forest plantation, and this is increasing. **(b)** Even-aged tree stand management is practised on tree plantations where all trees are of equal age, as seen in the stand in the foreground that is regrowing after clear-cutting. In uneven-aged tree stand management, harvests are designed to maintain a mix of tree ages, as seen in the more mature forest in the background. The increased structural and species diversity of uneven-aged stands provides superior habitat for most wild species and makes these stands more akin to ecologically functional forests.

It is important to acknowledge that planting new trees will not replace complex old-growth forests that may have taken hundreds of years to develop. Even when regrowth outpaces removal, the character of forests may still change. In North America and worldwide, primary forest continues to be lost and to be replaced by younger second-growth forest. Most ecologists and foresters view these plantations more as crop agriculture than as ecologically functional forests. Because there are few tree species and little variation in tree age, plantations do not offer many forest organisms the habitat they need.

The principle of **maximum sustainable yield (MSY)**, a basic principle of renewable resource management, is based on harvesting only as much wood as can be regenerated within a year. This principle (also applied to the harvesting of fish and other renewable resources) argues for cutting trees shortly after they have gone through their fastest stage of growth, and trees often grow most quickly at intermediate ages. Thus, trees may be cut long before they have grown as large as they would in the absence of harvesting. Although this practice may maximize timber production and harvesting efficiency over time, it can cause drastic changes in the ecology of a forest by eliminating habitat for species that depend on mature trees.

The FAO has found that the forestry policies of most nations emphasize the expansion of timber production capacity, rather than improvements in production efficiency.[16] FAO's concern in this context is to promote sustainable forest management by encouraging efficiency and waste minimization in the harvesting of forest products. One approach, used in Quebec since 2013, has been to allow for auctioning of up to 30% wood harvested from public forests (rather than retaining the wood as part of individual companies' timber allocations). The concept is that by allowing the establishment of a free market for wood (rather than a market that is artificially constrained by timber quotas for private companies), it will encourage the maximization of harvesting efficiency, and minimization of waste.

Land Conversion and Deforestation

The harvesting of timber and other forest products is not new; it has occurred throughout human history. We all depend in some way on wood, and people have cleared forests for millennia to exploit forest resources. Historically, as agriculture emerged and some cultures began to adopt a sedentary or settled lifestyle, the clearing of forested land for settlement and farming would have been one of the very first significant human-generated environmental impacts. Forest clearing has even been used as an approach in warfare (both modern and ancient) and to flush out game for hunting purposes. Land conversion, combined with some bad practices in agriculture, ranching, and forestry, has led to **deforestation**, the loss of forested area worldwide.

The growth of Canada was fuelled by land clearing and logging

Historically, logging for timber and the clearing of land for settlement and farming propelled the growth of both Canada and the United States throughout the phenomenal expansion westward across the North American continent over the past 400 years. The vast deciduous forests of the eastern United States were largely stripped of their trees by the mid-nineteenth century, making way for countless small farms. Timber from these forests built the cities of eastern North America.

As the farming economy shifted to an industrial one, wood was used to stoke the furnaces of industry. Once most of the mature trees were removed from the eastern hardwood forests, timber companies moved to the south and west, eventually harvesting some of the continent's biggest trees in the Rocky Mountains and the Pacific Coast ranges (**FIGURE 10.15**).

By the early twentieth century, little **primary forest**—long-standing natural forest, uncut by people—was left in the United States. Today, the largest oaks and maples found in eastern North America, and even most redwoods of the California coast, are **second-growth** trees: trees that have sprouted and grown to partial maturity after old-growth timber has been cut. The size of the gargantuan trees they replaced can be seen in the enormous stumps that remain in the more recently logged areas of the Pacific Coast. The scarcity of old-growth trees on the North American continent today explains the concern that scientists have for old-growth ecosystems and the passion with which environmental advocates have fought to preserve ancient forests in areas such as Clayoquot Sound.

Because trees are renewable and grow back, deforestation must be considered in terms of the net loss to forested area—that is, the amount lost in excess of new growth. This is a function of the loss of trees as a result of both natural causes (such as fire or insect infestations) and human activities (such as timber harvesting or land clearing). In either case, the areas in which trees fail to grow back (**FIGURE 10.16A**) constitute forested-area loss, that is, deforestation. The forested area gained and lost each year in Canada is tracked by a number of organizations. One purpose for collecting this information is so

Library and Archives Canada

FIGURE 10.15
Huge trees were harvested in many locations in Canada in the first part of the nineteenth century, including the pines shown here in Madawaska River Valley near Ottawa. Early timber harvesting practices in North America caused significant environmental impacts and removed virtually all the virgin timber from one region after another.

that we can report to the global community on the greenhouse gas (GHG) emissions for which we are responsible as a result of the conversion of forest (a carbon reservoir) for other purposes. In the UN's Framework Convention on Climate Change, this category of GHG emissions is referred to as *Land Use, Land-Use Change and Forestry*, also known as *LULUCF*.

In spite of vigorous historical logging, much of Canada is still covered by primary forest. The principal cause of deforestation in Canada today is not logging, which is largely balanced by replanting and reseeding, but land clearing for urban development (**FIGURE 10.16B, C**). Deforestation continues in Canada, but according to Natural Resources Canada (NRCAN) it affects less than 0.02% (approximately 46 000 kha in annual net loss) of Canada's forested area annually.[17] Fire, drought, and, increasingly, outbreaks of invasive pest species are also significant causes of deforestation. Flooding of reservoirs for large hydroelectric dams is an episodic cause; if a large dam is closed, reservoir flooding can lead to significant forest loss, but this doesn't happen every year. Many scientists expect that climate change will be an important driver of deforestation in the near future.

Agriculture is the major cause of forest and grassland conversion globally

Agriculture now covers more of the planet's surface than does forest. Thirty-nine percent of Earth's terrestrial surface is devoted to agriculture—more than the area of

North America and Africa combined. Of this land, about 26% supports rangeland (mostly for beef production), and 13% consists of cropland.[18] Agriculture is the most widespread type of human land use and the principal driver of land conversion today, causing tremendous impacts on land and ecosystems. Although agricultural methods such as organic farming and no-till farming can be sustainable, the majority of the world's cropland hosts either intensive traditional agriculture or monocultural industrial agriculture, involving heavy use of fertilizers, pesticides, and irrigation.

In Canada, clearing for agriculture was once the primary cause of forest loss, but this has slowed considerably in the past few decades. In many parts of the developing world, however, forests continue to be cleared at a rapid rate, both for modern industrial agriculture and for traditional agriculture.

In traditional *swidden* farming, an area of forest is cleared (often by *slash-and-burn*), and crops are planted. After one or two seasons of planting, when the soil has been depleted of nutrients, the farmer moves on to clear another patch of forest, leaving the first clearing in a *fallow* or resting state, giving it time to replenish itself. This can be a sustainable practice if the clearings are given sufficient time—as much as seven years may be needed—in which to replenish the nutrient content of the soil. However, social and economic pressures in the developing world, including population pressure, have led to shorter fallow times, with the result that the cleared forest soils are stripped of their nutrients and erode away, rather than regenerating. After that, the soil will no longer support either crops or forests.

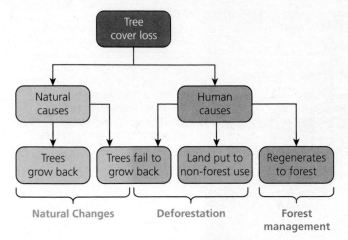

(a) Deforestation: Net loss of forested land

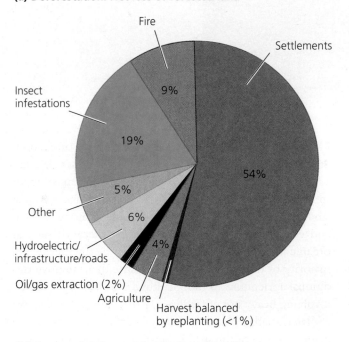

(b) Causes of deforestation in Canada today

(c) Conversion of forested land for agriculture

FIGURE 10.16

Deforestation is net loss of forested area **(a)**, that is, the amount of forested area lost as a result of natural causes and human activities, minus the amount that regenerates. Much of the land cleared by natural processes (fire and some insect infestations) will eventually regenerate. Of the approximately 46 000 kha of forested land lost each year in Canada, the principal cause now **(b)** is urban development. Agriculture, as seen here **(c)** in Quyon, Quebec, was traditionally the main cause of conversion of forested land in Canada on an annual basis, but lately this has declined significantly.

Source: Diagram in (a) is based on "How Remote Sensing Innovations Are Revolutionizing Forest Monitoring and Management in Canada," *The State of Canada's Forests: Annual Report 2014,* Natural Resources Canada, 2014. Reproduced with the permission of the Minister of Natural Resources Canada, 2015.

The total amount of deforested area in Canada in 2013 (the year represented by most of the data in **FIGURE 10.16B**) was approximately 46 000 kha (in kilohectares, equivalent to 46 000 000 ha). According to this figure, how much of this (in land area) was caused by fire + insect infestations? How many hectares of forested land were converted for urban settlements? Of the causes of forest conversion represented in **FIGURE 10.16B**, for which ones is the forest most likely to regenerate? For which causes is forest regeneration least likely?

Most cattle in North America today are raised in high-density feedlots, but they have traditionally been raised by grazing on open **rangelands**—grasslands or wooded areas converted for the purpose of supporting livestock (**FIGURE 10.17**). Grazing can be sustainable, but overgrazing damages soils, waterways, and vegetative communities. Range managers are responsible for regulating ranching on public lands, and they advise ranchers on sustainable grazing practices.

Cropland agriculture uses less than half the land taken up by livestock grazing, which covers a quarter of the world's land surface. In addition to leading to forest loss, poorly managed grazing can have adverse impacts on soil and grassland ecosystems. In Central and South America, the conversion of forested land and grasslands to rangelands occurred with phenomenal rapidity during

the decades following the 1940s. The dramatic loss of forested land led a few countries, notably Costa Rica, to institute some tough restrictions on the clearing of forested land.

Pest infestations have become increasingly problematic

An increasing cause of forest loss, in Canada and elsewhere, is pest infestations. Insect pests are a natural aspect of life in forests; infestations come and go. However, as the climate changes, forest ecosystems are changing, too, and this is having an impact on pests. Some examples of insects that have been problematic in Canadian forests include the emerald ash borer, Asian long-horned beetle, spruce budworm, and mountain pine beetle.

FIGURE 10.17
Grazing of livestock, shown here in Brazil, covers a quarter of Earth's land surface. Conversion of forested land into pasture for ranching has been one of the principal causes of deforestation globally.

The mountain pine beetle (**FIGURE 10.18A**) has been one of the highest-profile forestry problems in British Columbia since the early 1990s. According to NRCAN, the pine beetle has killed 50% of the total commercial volume of lodgepole pine in the province.[19] It has now been spotted in Alberta, and has spread from its usual host into jack pine, the main species of the boreal forest. Warmer winters and hotter summers have both promoted the survival and spread of the beetle. In spite of control efforts, the area deforested by the mountain pine beetle continues to increase.

Like the mountain pine beetle, the western spruce budworm (**FIGURE 10.18B**) is native to Canada. It is a parasite that feeds on new shoots, mainly of Douglas firs in coastal British Columbia and Vancouver Island. Climate has a complicated effect on the spruce budworm. An increase in overall winter temperatures on southern Vancouver Island has nearly eliminated spruce budworm outbreaks in the area. Budworm development is triggered by temperature, whereas development of Douglas firs is triggered more by light and day length. As a result, the budworm larvae are emerging from their shelters earlier than a century ago, whereas development of Douglas

fir buds has remained constant. The limited food supply for the larvae means that the western spruce budworm is no longer the problem that it once was on southern Vancouver Island. However, British Columbia's interior forests are still experiencing large-scale outbreaks.[20]

The Asian long-horned beetle is an alien invasive, which originated in China and probably came into Canada in wooden packing crates and pallets.[21] It has no natural enemies in North America, and began attacking maple, poplar, birch, willow, and elm trees in the Toronto area, where it was first noticed in 2003. The Canadian Forestry Service, Canadian Food Inspection Agency, and Great Lakes Research Centre worked collaboratively to eradicate the pest by establishing a quarantine zone of 150 km^2 around the infected area, and inspecting all wood products that went in or out of the area. About 13 000 trees in the quarantined zone were destroyed. The control program appears to have succeeded; no infested trees have been found since 2007, and steps have been taken to prevent reintroduction of the insect into Canada.[22]

Of current concern in the temperate deciduous forests of eastern Canada is the emerald ash borer. This is also an exotic invasive with no natural enemies, and it was first

(a) Mountain pine beetle

(b) Spruce budworm

FIGURE 10.18
Mountain pine beetles **(a)** have killed millions of cubic metres of timber in British Columbia. Warmer summers and milder winters promote the activity of these invasive beetles. In **(b)**, a spruce budworm larva sits on a Douglas fir branch.

seen in Canada in 2002 in Windsor, Ontario. It has now spread to ash-dominated forests as far north as Sault Ste. Marie, and has been detected in several places in Quebec. According to NRCAN, 99% of the trees in a stand will be dead within six years of the insect's arrival. Scientists are trying to learn more about the spread of this pest, both by natural processes and by human activities, such as the movement of firewood, and to develop methods for controlling its dissemination.[23]

Scientists studying invasive insects and their impacts say there are two primary reasons for the current extraordinary outbreaks. One is that past forest management has resulted in even-aged forests across large regions, and many trees in these forests are now at a prime age for beetle infestation. Plantation forests dominated by single species that the beetles prefer are most at risk. The second reason is climate change. Milder winters allow beetles to overwinter further north, and warmer summers speed up their consumption and reproduction. Meanwhile, droughts in recent years have stressed and weakened trees, making them vulnerable to attack. On top of all this, beetle outbreaks create a positive feedback loop; by killing trees, they reduce the amount of carbon dioxide pulled from the air, and thereby intensify climate change.

Deforestation is most rapid in developing nations, for a number of reasons

Deforestation from land conversion, resource extraction, urbanization, and pest infestations has altered the landscapes and ecosystems of much of our planet. Forest resources can be harvested sustainably, but unfortunately this doesn't always happen. Impacts are greatest in tropical areas because of the potentially massive loss of biodiversity, and in dryland regions because of the vulnerability of these lands to desertification. Deforestation can cause soil degradation, habitat loss, and even species extinction. In addition, deforestation adds carbon dioxide (CO_2) to the atmosphere: CO_2 is released when plant matter is burned or decomposed, and thereafter less vegetation remains to grow and act as a sink for atmospheric CO_2. Deforestation is thereby a contributor to further global climate change.

Globally, about 13 million ha are deforested each year—equivalent to the area of Nova Scotia and New Brunswick combined. This has been the average yearly rate of deforestation from 2000 to 2014, down from an average of about 16 million ha per year in the 1990s.[24] In Canada, as in some other economically developed nations, reforestation and afforestation have almost offset losses to forested area in the past decade or so.

Today forests are being felled at the fastest rates in the tropical rain forests of developing nations in Latin America

and West Africa (**FIGURE 10.19**). These nations are in the position Canada faced a century or two ago: having a vast frontier that they can develop for human use. Developing countries are striving to expand areas of settlement for their burgeoning populations, and to boost their economies by extracting natural resources and selling them abroad. Moreover, many people still cut trees for fuelwood for their daily cooking and heating needs. In Asia, forests are increasing only because of the planting of vast plantations in China; otherwise primary tropic rain forest is being decimated. In contrast, areas of Europe and eastern North America are holding steady, or even slowly gaining forest cover as they recover from severe deforestation of past decades and centuries. Overall, however, the world is losing its forests.

Developing nations are often so desperate for economic development, and for foreign capital with which to maintain the interest payments on enormous national debt loads, that they impose few or no restrictions on logging. Where there are restrictions, it is very challenging to uphold them and prevent illegal logging. Often their timber is extracted by foreign multinational corporations, which have paid fees to the developing nation's government for a *concession*, or right to extract the resource. In such cases, the foreign corporation has little or no incentive to manage forest resources sustainably. Many of the short-term economic benefits are reaped not by local residents but by the corporations that log the timber and export it elsewhere. Local people may or may not receive temporary employment from the corporation, but once the timber is harvested they no longer have the forest and the ecosystem services it once provided.

In Sarawak, the Malaysian portion of the island of Borneo, foreign corporations that were granted logging concessions have deforested several million hectares of tropical rain forest since 1963 (**FIGURE 10.20A**). The clearing of this forest—one of the world's richest, hosting such organisms as orangutans and the world's largest flower, *Rafflesia arnoldii* (**FIGURE 10.20B**)—has had direct impacts on the 22 tribes of people who live as hunter-gatherers in Sarawak's rain forest. The Malaysian government did not consult the tribes about the logging, which decreased the wild game on which these people depended. Oil palm agriculture was established afterward, leading to pesticide and fertilizer runoff that killed fish in local streams. The tribes protested peacefully and finally began blockading logging roads. The government, which at first jailed them, now is negotiating, but it insists on converting the tribes to a farming way of life.

Throughout Southeast Asia and Indonesia today, vast swaths of tropical rain forest are being cut to establish plantations of oil palms (**FIGURE 10.21**). Oil palm fruit produces palm oil, which we use in snack foods, soaps, and cosmetics, and now as a biofuel. In Indonesia, the world's largest palm oil producer, oil palm plantations have

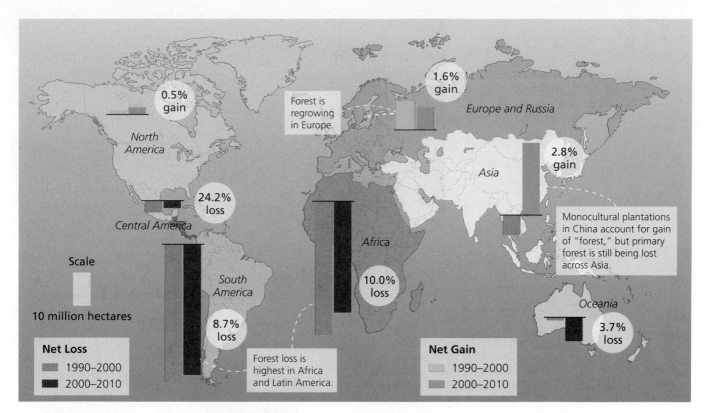

FIGURE 10.19 South America and Africa are experiencing rapid deforestation as they develop, extract resources, and clear new agricultural land for growing populations. In Europe, forested area is slowly increasing as some formerly farmed areas are allowed to grow back into forest. North and Central America reflect a balance of regrowth and loss. In Asia, primary natural forests are being lost, but the extensive planting of tree plantations (here counted as forests) in China has increased forest cover for Asia since 2000.
Source: Data from UN Food and Agriculture Organization, *Global Forest Resources Assessment, 2010.*

displaced over 6 million ha of rain forest. Clearing for plantations encourages further development and eases access for people to enter the forest and conduct logging illegally.

The palm oil boom represents a conundrum for environmental advocates. Many people eager to fight climate change had urged the development of biofuels to replace fossil fuels. Yet grown at the large scale that our society is demanding, monocultural plantations of biofuel crops such as oil palms are causing severe environmental impacts by displacing natural forests.

The World Resources Institute (WRI) is working with palm oil companies in Indonesia that own concessions

(a) Logging in Borneo

(b) *Rafflesia arnoldii* in bloom

FIGURE 10.20 Logging, both illegal and legal **(a)**, and associated deforestation are rampant in Borneo, Malaysia. The principal habitat of *Rafflesia arnoldii*, the world's largest flower **(b)**, and that of many other important and threatened species, is in Borneo.

FIGURE 10.21
Oil palm plantations are replacing primary forests across Southeast Asia and Indonesia. Forest clearing for plantations promotes further development, illegal logging, and forest degradation. Since 1950, the immense Indonesian island of Borneo (maps) has lost most of its forest.
Source: Data from Radday, M., WWF-Germany, 2007. Designed by Hugo Ahlenius, UNEP/GRID-Arendal. Extent of Deforestation in Borneo 1950–2001, and Projection towards 2020. http://maps.grida.no/go/graphic/extent-of-deforestation-in-borneo-1950–2005-and-projection-towards-2020.

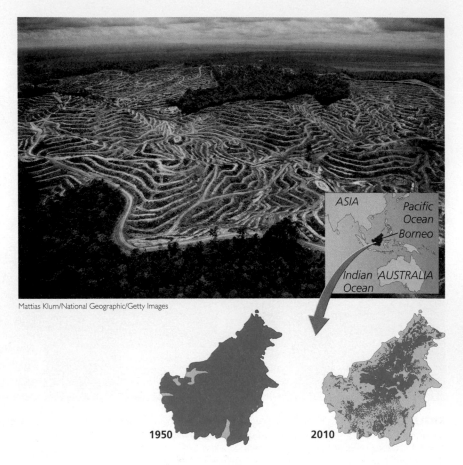

Mattias Klum/National Geographic/Getty Images

1950   2010

to clear primary rain forest, and is steering them instead toward land that is already logged and degraded. WRI will then protect the forests that were slated for conversion or else allow certified sustainable forestry activities to take place in them. WRI hopes this will encourage oil palm plantations to become sustainable, while preserving primary rain forest. Because primary forest stores more carbon than oil palm plantations, these land swaps can reduce Indonesia's greenhouse gas emissions and qualify for credits in the form of carbon offsets.

Forest Management Principles

Professionals who manage forests through the practice of **forestry** (or *silviculture*) must balance the central importance of forests as ecosystems with civilization's demand for wood products. Sustainable forest management, like the management of other renewable natural resources, is based on maintaining equilibrium between *stocks* and *flows*. In principle, the removal or **harvesting** of material from the resource by logging should not occur at a rate that exceeds the capability of the resource to replenish or regenerate itself.

Public forests in Canada are managed for many purposes

Nearly 94% of Canada's forest is publicly owned; the majority of this is under provincial jurisdiction. Only about 6% of forested land in Canada is privately owned, about 1.5% by logging companies, with the remainder under federal or territorial control.[25] An increasing amount of land in Canada, including forested land, is under Aboriginal jurisdiction as land claims are settled.[26]

In Canada, timber is extracted by private timber companies from both privately owned and publicly held forests. In fact, much of the resource extraction industry in Canada—both logging and mining—is carried out on Crown lands (mainly provincial). The provinces and territories and the federal government have different requirements governing the acquisition of rights to carry out resource extraction on Crown lands. Under the Constitution, provinces and territories own and regulate the natural resources within their boundaries.

The federal role in forestry is based on its responsibility for the national economy, trade, science and technology, the environment, and federal Crown lands and parks, as well as Aboriginal and treaty rights, which are protected by the federal *Constitution Act of 1982*.

The Canadian Forest Service, part of Natural Resources Canada, was established in 1899, and given the responsibility to preserve timber on Crown lands and to develop policies to encourage tree culture. Since then, the Canadian Forest Service has been involved in the scientific study and monitoring of Canada's forests and in managing the extraction of timber and non-timber forest products from national forests. The National Forest Strategies began to be officially developed by the Council of Forest Ministers in 1988. The current National Forest Strategy[27] makes reference to the central role of forests in Canada's society and economy, as well as our obligation to the world to maintain and preserve the vast forests of which we are the stewards.

For the past half-century, forest management throughout North America has been guided by the policy of **multiple use**, meaning that forests were to be managed for recreation, wildlife habitat, mineral extraction, and various other uses. In recent decades, increased awareness of the problems associated with resource extraction and logging has prompted many citizens to protest the way public forests are managed in Canada. They have urged that provincial forests be managed for wildlife and ecosystem integrity, rather than principally for timber, oil, or minerals. The Canadian Council of Forest Ministers tried to balance these often competing goals in its vision document, which recognizes sustainable forestry, utilization of Aboriginal knowledge for ecosystem management, and managing for climate change as central goals.[28]

Many forest managers practise adaptive and ecosystem-based management

In addition to applying the principle of maximum sustainable yield, many renewable resource managers today use **ecosystem-based management**, which attempts to manage the harvesting of resources in ways that minimize impacts on the ecosystems and ecological processes of the forest. Ecosystem-based management aims to preserve forest health, structure, functions, composition, and biodiversity. In Canada, this has been partly achieved by the establishment, over time, of a system of provincially, federally, and internationally protected areas (discussed later in this chapter). An additional goal of ecosystem-based forest management is to maintain forests as viable reservoirs, or sinks, for atmospheric carbon. This is crucial for managing carbon emissions in the global effort to control climate change.

By carefully managing ecologically important areas such as riparian corridors, considering patterns at the landscape level, and affording protection to some forested areas, ecosystem-based management plans aim to preserve the functional integrity of the forest ecosystem, even while allowing for some resource-extraction activities. Although ecosystem-based management has gained a great deal of support in recent years, it is challenging for managers to determine how best to implement this type of management. Ecosystems are complex, and our understanding of how they operate is limited. Thus, ecosystem-based management has often come to mean different things to different people.

Another guiding principle in forest management recently has been **adaptive management**, which involves systematically testing different management approaches, and learning from approaches that work or don't work. It entails monitoring the results and continually adjusting practices as needed, based on what has been learned. Adaptive management is a partnership of science and management, because scientific hypotheses about how best to manage resources are explicitly tested and the results are used to design and modify management approaches.

Adaptive management can be time-consuming and complicated, however. It has posed a challenge for many managers because those who adopt new approaches must often overcome resistance to change from proponents of established practices. In British Columbia, the Ministry of Forests and Range has embraced adaptive management, promoting the approach through collaborative projects that range from testing alternative forestry practices, to monitoring whole watersheds, to evaluating the effectiveness of various land and resource management strategies.[29]

Fire management has stirred controversy

Ecological research shows that many ecosystems depend on fire, particularly in the boreal forest. Certain plants such as jack pines have seeds that germinate only in response to fire. This makes evolutionary sense; when a seedling grows after a fire, there are no big trees to take the light from it, and it can benefit from nutrients contained in the ash of the burnt forest. Researchers studying tree rings have documented that many ecosystems historically experienced frequent fires. Burn marks in tree rings reveal past fires, giving scientists an accurate history of fire events extending back hundreds or even thousands of years.

Many wooded dryland ecosystems also depend on fire. For example, researchers have found that North America's grasslands and open pine woodlands burn regularly in the natural system. Ecosystems dependent on fire are adversely affected by its suppression; pine woodlands become cluttered with hardwood understorey that ordinarily would be cleared away by fire, for instance, and animal diversity and abundance decline.

(a) Ground fire

(b) Crown fire

(c) Slope fire

FIGURE 10.22
Fires need fuel, oxygen, and a heat source. The buildup of dry litter and organic matter in soil layers can lead to ground fires **(a)**, in which the litter layer on top of and within the soil itself burns. Fires that move into the top part of the canopy **(b)** are called crown fires. Fires can move very quickly up slopes **(c)**, fanned by wind and fuelled by heat from below.

Fires depend on the triad of oxygen, heat, and fuel to progress. In the forest context, branches, fallen logs, sticks, and leaf litter accumulate on the forest floor, producing kindling and fuel for future fires. Climate and weather are crucial components. An overall dry climate or unusually dry weather can cause litter and upper soil layers to be dry, and organic matter to be easily ignited. This leads to *ground fires* (**FIGURE 10.22A**), in which the litter layer itself burns, as opposed to *crown fires*, in which the upper tree canopy is ignited (**FIGURE 10.22B**). Storms also bring winds that can fan fires.

Topography is another important component of forest fires. A fire burning up the slope of a canyon (**FIGURE 10.22C**) can be particularly challenging to contain. Winds blowing up the hillslope can fan the fire, whereas the heat from below ignites dry fuel on upslope areas. The damaging fires in Kelowna, British Columbia, in 2003 were fast-moving fires that swept upslope from the valley bottoms.

Lightning is responsible for igniting the majority of naturally induced forest fires. Other causes of forest fires are volcanic eruptions and, of course, human carelessness. Devastation of large forested areas by invasive insects can also set the stage for large areas to burn. The management of fires is one of the most controversial aspects of forest management today. For over a century, land management agencies throughout North America quickly suppressed fire whenever and wherever it broke out.

In the long term, however, fire suppression leads to a buildup of dead wood, which can fuel catastrophic fires that truly damage forests, destroy human property, and threaten human lives. Fuel buildup helped cause the 1988 fires in Yellowstone National Park, the 2003 fires in

FIGURE 10.23
Forest fires are natural phenomena to which many plants and ecosystems are adapted. However, climate change combined with the suppression of fire over the past century has led to a buildup of dry litter and woody debris, which serve as fuel to increase the severity and frequency of fires. As a result, catastrophic wildfires have become more common in recent years. Shown here, a devastating forest fire ripped through Slave Lake, Alberta, in May 2011, destroying much of the town.

southern California, the 2003 fires in British Columbia, and thousands of other wildfires across the continent each year (**FIGURE 10.23**). Fire suppression and fuel buildup have made catastrophic fires significantly greater problems than they were in the past. Now, global climate change is bringing drier weather to much of the Canadian Prairies, further worsening the wildfire risk. At the same time, increasing residential development on the edges of forested land is placing more homes in fire-prone situations.

To reduce fuel load and improve the health and safety of forests, forest management agencies have in recent years been burning areas of forest under carefully controlled conditions. These **prescribed burns**, or **controlled burns**, have worked effectively, but they have been implemented on only a relatively small amount of land.

Another significant and controversial aspect of fire management concerns what happens after a fire, which may include the physical removal of small trees, underbrush, and dead trees by timber companies. The removal of dead trees, or snags, following a natural disturbance is called **salvage logging**. From an economic standpoint, salvage logging may make good sense. Proponents of salvage logging argue that forests regenerate best after a fire if they are logged and replanted with seedlings. Moreover, salvage logging may reduce future fire risk by removing woody debris that could serve as fuel. However, snags have immense ecological value; the insects that decay them provide food for wildlife, and many birds, mammals, and reptiles depend on snags for nesting and roosting sites. Removing timber from recently burned land can also cause severe erosion, collapse of stream banks, and soil damage.

Faced with all of these competing pressures, fire managers must balance the damaging aspects of fires—the loss of harvestable wood and threats to communities—with the important natural role of fires in forest ecosystems. In the past few decades, fires have burned approximately 2.3 million ha of forest per year in Canada.[30] The year 2014 represented a departure; the number of fires was approximately the same, but the area burned was almost twice as large as the average (4.2 million ha). Recent research has suggested that increases in wildfire activity are seen when summers are hotter; if climate change leads to hotter summers, perhaps this indicates a trend toward increasing fire activity.[31]

Climate change poses new forest management challenges

Some have suggested that the huge burned area of the 2014 fire season in the boreal forest may reflect a "new normal" for forests in a world of changing climate zones. It remains to be seen whether changes in fire extent, frequency, and behaviour will become established as a trend. What is clear, though, is that changes in temperature and precipi-

tation will have major impacts on forests worldwide, and we may already be seeing some of those impacts. This has implications for the future management of forests.

The preservation of forests is important, not just because of their biodiversity, ecosystem services, and commercially valuable products, but also because of their role in the climate system. Of all the ecosystem services that forests provide, the storage of carbon is of greatest importance in the context of global climate change. Because trees absorb carbon dioxide from the air during photosynthesis and then store carbon in their tissues, forests serve as a major reservoir for carbon. Scientists estimate that the world's forests store over 280 billion metric tons of carbon in living tissue, which is more than the atmosphere contains. When plant matter is burned or when plants die and decompose, carbon dioxide is released—and thereafter less vegetation remains to soak it up. Carbon dioxide is the primary greenhouse gas contributing to global climate change. Therefore, when we cut forests, we worsen climate change. The more forests we preserve or restore, the more carbon we keep out of the atmosphere, and the better we can address climate change.

Climate change benefits some forest species while harming others. As climate change interacts with pests, diseases, and management strategies, many of our forest ecosystems could be altered in profound ways. Already many dense, moist forests devastated by beetles have been replaced by drier woodlands, shrublands, or grasslands. Further changes to our forests could create novel types of ecosystems not seen today.

As temperatures and precipitation rates change, especially in the North, the natural ranges of trees are also shifting. For the past two decades, scientists have been considering the limitations on the migration and seed dispersal mechanisms of trees. What will happen to forests as tree ranges shift toward the north and toward higher altitudes? Will trees be able to migrate quickly enough to keep up with the changes, and reestablish themselves successfully in new locations? Is there something that forest managers could, or should, be doing to help this process, and prevent the loss of forest biodiversity? One possible answer to this is *assisted migration*—literally helping trees and other organisms to migrate to more appropriate habitats by purposely introducing them into new locations. The pros and cons of this management approach are discussed in "The Science behind the Story: Assisted Migration: Getting Trees Where They Need to Go in a Changing Climate."

Carbon offsets are central to emerging international plans to curb deforestation and climate change. Forest loss accounts for as much as 20% of the world's greenhouse gas emissions—nearly as much as all the world's vehicles emit.[32] (Forest loss is part of the LULUCF portion of GHG calculations, mentioned above.) For this reason, international negotiators have outlined a program called *Reducing*

Assisted migration, as shown here in a trial with hardwood tree species led by Natural Resources Canada and the Ontario Ministry of Natural Resources, is controversial but may one day help ecosystems adapt to climate change.

"Photo showing black walnut trees in the middle of their third growing season", Natural Resources Canada, Forests website: http://www.nrcan.gc.ca/forests/climate-change/adaptation/13121. Reproduced with the permission of the Minister of Natural Resources Canada, 2015.

Assisted Migration: Getting Trees Where They Need to Go in a Changing Climate

As I did my stand upon the hill, I look'd toward Birnam, and anon, methought, The wood began to move. —*Shakespeare, Macbeth, Act 5, Scene 5*

As climate change progresses, one expected effect is that the natural ranges of plant and animal species will shift in response to changes in temperature and precipitation. This will likely have a significant impact on biodiversity in forest ecosystems. It may lead to species extirpations and extinctions, as many species will not be able to migrate quickly enough to keep up with the rate of change in their ranges.

What can be done about this? One possible response, which has been investigated by a number of researchers, is to

help species adjust by essentially picking them up and moving them to new ranges that suit their needs. This is called *assisted migration*. Researcher Nina Hewitt and her colleagues at York University's Institute for Research and Innovation in Sustainability define it as "the intentional translocation or movement of species outside of their historic ranges in order to mitigate actual or anticipated biodiversity losses caused by anthropogenic climatic change."[33] Many equivalent terms have been used in the scientific literature to describe the same idea, including facilitated migration, assisted colonization, managed relocation, and assisted range expansion.

The idea of assisted migration may have been first proposed in 1985 by conservation biologist Robert Peters and ecologist Joan Darling in a paper (the source of the Shakespeare quotation at the beginning of this piece) in which they consider the impacts of global warming— at the time, a relatively new concern—on species in wildlife reserves.[34] They proposed that the impacts of climate change would affect all species but would have the greatest impact on those species restricted to nature reserves of limited range, and with limited opportunities for genetic interaction with other populations (see **FIGURE 1**).

Climate-induced shifts in plant and animal ranges could lead to extinctions in two different ways. The first is physiological; if an organism simply can't survive in the new temperature and moisture conditions, and can't migrate quickly enough to keep up with the changes, it will die out. The second is related to interactions among different species; for example, if climate change causes a prey species to move or become extirpated or extinct, organisms that depend on that species as a food source also will encounter difficulties.[35]

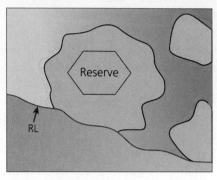

(a) Range limit before climate change

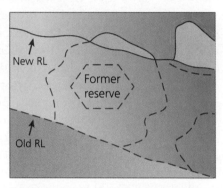

(b) Range limit after climate change

FIGURE 1
In their work on assisted migration, Peters and Darling described how shifts in species ranges resulting from climate change would most severely affect species confined to geographically limited reserves, as shown here. *Source:* After Peters, R.L., and J.D.S. Darling (1985). The Greenhouse Effect and Nature Reserves: Global Warming Would Diminish Biological Diversity by Causing Extinctions among Reserve Species. *Bioscience, 35*(11) pp. 707–717.

Nina Hewitt's research has to do with forest species and their migration and dispersal mechanisms and rates, especially among fragmented patches of forest.[36] Trees and other forest plants disperse by

Emissions from Deforestation and Forest Degradation (REDD), whereby wealthy industrialized nations provide financial incentives to developing nations to conserve forests. Under this plan, coordinated by the United Nations, poor nations gain income while rich nations receive carbon credits to offset their emissions in an international cap-and-trade system. Leaders of rich nations have proposed to transfer

$100 billion per year to poor nations by 2020, and much of this could end up going toward REDD.

The global economic slowdown and protracted debates concerning international climate agreements have slowed progress on REDD initiatives, but they are beginning to gain momentum. The program is now in the "REDD+" phase, which takes into account not only

Trees disperse their seeds by a number of different mechanisms. The seeds of conifers, like this pine tree, are released from the cone and dispersed by wind, gravity, water, or animals. The potential for long-distance seed dispersal is of great interest in assessing the natural ability of trees and forests to keep up with the pace of climate change.

a wide variety of mechanisms (see photo), including seed dispersal by wind, gravity, water, and animals (for example, by attaching to the animal's fur, or by ingestion and later dispersal in feces). Hewitt and her colleagues have been particularly interested in investigating the potential for long-range seed dispersal in trees, which will be crucial if forests are to keep up with the pace of shifts in temperature and precipitation as climate change proceeds.

Animals are sometimes more mobile than plants, but even the migration rates of animals vary dramatically. Consider the difference in migration rate between a mussel and a bird; a worm and a butterfly; or a mole and a caribou. But even larger migratory animals will have trouble keeping up with shifting ranges if patches of appropriate habitat are small and fragmented, especially in human-dominated landscapes. Assisted migration is seen by its supporters as being most appropriate in cases where species are particularly vulnerable to the impacts of climate change; have low dispersal rates; or have habitats that are

geographically small, fragmented, or isolated, especially in Arctic or high alpine areas (where there is literally nowhere else to go).[37] In these cases, assisted migration may be the only option for preserving vulnerable species.

We are already seeing the impacts of warming in the expansion and northward shifts of range in both native and alien invasive species in the forest ecosystems of North America.[38] Examples include the mountain pine beetle (native) and the gypsy moth (alien), imported for its possible silk potential and then accidentally released. Invasives have socioeconomic impacts, in addition to their obvious ecological impacts. These include the loss of commercially valuable resources, and human impacts such as Lyme disease, carried by deer ticks, whose range is expanding northward. Range shifts are expected to increase with climate change, with the potential for positive feedbacks if wholesale changes to forest ecosystems lead to further climatic changes.

Helping species and ecosystems by physically transplanting them to a more amenable environment seems like an attractive option. However, there are significant risks and benefits that must be considered before assisted migration is translated into policy. The benefits are clear: rescuing species from extinction; protecting biodiversity; preserving ecosystems and ecosystem services; and maintaining stocks of commercially valuable resources, including timber.[39]

The risks of assisted migration are complex, including the significant potential for a translocated species to become invasive in its new ecosystem. Other potential risks include the diversion of funds away from other biodiversity protection measures, such as ecosystem restoration, but above all the lack of scientific understanding is of particular concern.[40]

People don't have a very successful history with the intentional movement of species; many of the most harmful invasive species have been introduced intentionally. Assisted migration is also seen by some researchers as running counter to the most basic principle of modern conservation: to protect species and ecosystems *in situ* (in place).

In spite of the risks, experiments in assisted migration have already taken place, and sometimes not under closely supervised scientific circumstances. For example, the "Torreya Guardians,"[41] a group of a "self-organized" group of naturalists, botanists, ecologists, and environmental activists, have undertaken to protect an endangered conifer tree, *Torreya taxifolia*, which is native to northern Florida, by purposely introducing it into areas much farther to the north. The group argues that it is simply helping the tree deal with climate-induced changes in its historically natural habitat by "rewilding" it in appropriate habitat in the mountains of North Carolina. Natural Resources Canada and the Ontario Ministry of Natural Resources have carried out a trial of assisted migration, in which six hardwood tree species were transported from four southern locations to new locations in Southern Ontario, as much as 1400 km farther north. The results are being monitored in an effort to understand the ecological implications of long-range assisted migration.[42]

Like many aspects of climate change, assisted migration is controversial and complex. There is, as yet, no strong consensus in the scientific literature as to whether it presents a desirable or even feasible option for the mitigation of climate change impacts on species and ecosystems.[43] It will take even more discussion and analysis before the idea may one day be implemented as policy and translated into action.

initiatives to reduce emissions and invest in low-carbon technologies, but also encompasses funding for programs aimed at conservation, sustainable management of forests, and enhancement of forest carbon stocks.[44] Since 2007, more than 100 REDD+ projects have been launched, many of them in Indonesia. These have involved a range of activities, from consultations with Indigenous people on

community forest management, to afforestation and reforestation projects, to establishing a carbon credit market.

The small South American nation of Guyana—poor financially, but rich in forests—has taken a leading role in international discussions of REDD. Guyana's president commissioned a consulting firm to calculate the amount of money his nation would make if it cut down its forests for agriculture. The figure came to $580 million per year over 25 years. In a free market under purely financial considerations, this is the amount that wealthy nations "should" pay Guyana to forego cutting its forest. Guyana subsequently forged a deal with Norway in which Norway is paying Guyana for conserving its forest—up to $250 million in total by 2015, depending on Guyana's progress in reducing forest loss.

REDD+ is clearly not going to solve all problems related to forests and climate change. However, it has the potential to encourage financial transfers between developed and developing nations that are directed toward beneficial and sustainable forest management and carbon reservoir solutions.

Canada is no longer a party to the UNFCCC (Kyoto Protocol), but is involved in the REDD+ negotiations nonetheless. Canada's position has been to support programs that will promote better monitoring of forests and changes in forest ecosystems; more focus on human activities that are controllable; and the creation of international incentives for sustainable forest management, to reduce emissions and increase carbon sinks.[45] The Canadian Forest Service is preparing for the forest management impacts of climate change by carrying out research to[46]

- study the influence of Canada's forests on the global carbon balance
- assess the past, present, and future impacts of climate change on Canada's forests
- identify options for using Canada's forests to mitigate climate change
- identify options for helping Canada's forest sector adapt to climate change, including the possibility of assisted migration

At this point, the only sure things are that forests play an important role in climate; that climate change will have an influence on forests; and that we need forests, both as a world community and as Canadians, and need to do what we can to preserve the health and sustainability of forest ecosystems.

Sustainable forestry is gaining ground

In the last several years, a consumer movement has grown that is making it possible for consumers to make informed choices to promote the sustainability of forest products. Any company can claim that its timber harvesting practices are sustainable, but how is the purchaser of wood products to know whether they really are? Several organizations now examine the practices of timber companies and offer **sustainable forestry certification** to products produced using methods they consider sustainable (**FIGURE 10.24**).

FIGURE 10.24

A Brazilian woodcutter places identifying marks on timber harvested from a forest certified for sustainable management in the Brazilian Amazon. A consumer movement centred on independent certification of sustainable wood products is allowing consumer choice to promote sustainable forestry practices.

Reuters/Corbis

Organizations such as the International Organization for Standardization (ISO), the Sustainable Forestry Initiative (SFI) program, and the Forest Stewardship Council (FSC) have established standards for the certification of forest management practices. Consumers can look for the logos of these organizations on forest products they purchase. The FSC is widely perceived to have the strictest certification standards. FSC-certified timber harvesting operations are required to protect rare species and sensitive habitats, safeguard water sources, control erosion, minimize pesticide use, and maintain the diversity of the forest and its ability to regenerate after harvesting. FSC certification is the best way for consumers of forest products to know that they are supporting sustainable practices that protect forests. In 2001, Iisaak, the Aboriginal-run timber company established in Clayoquot Sound, became the first tree farm licence holder in British Columbia to receive FSC certification.

Consumer demand for sustainable wood has been great enough that The Home Depot and other major retail businesses have begun selling sustainable wood. The decisions of such retailers are influencing the logging practices of many timber companies. In British Columbia, 70% of the province's annual harvest now is certified or meets ISO requirements. Overall in Canada over 41% of the forested area is certified for sustainable management and harvesting practices by third-party certification.[47] Sustainable forestry is more costly for the timber industry, but if certification standards can be kept adequately strong, then consumer choice in the marketplace can be a powerful driver for good forestry practices for the future.

Conclusion

Forests and other terrestrial biomes provide crucial ecosystem services, supporting a vast diversity of species and providing goods that have economic value to humans as well. Managing natural resources sustainably is particularly important for resources like trees, which otherwise can be carelessly exploited, degraded, or overharvested. Canada and many other nations have established agencies to oversee and manage forests on publicly held land, and the natural resources that are extracted from them. Early emphasis on resource extraction in forest management in North America has now evolved into policies on maximum sustainable yield, multiple use, ecosystem-based management, and adaptive management.

Forests today are managed not only for timber production, but also for recreation, wildlife habitat, and ecosystem integrity, and with their role in the climate system firmly in mind. These are positive developments because the conservation of forests is essential if we wish our society to be sustainable and to thrive in the future. Forest management faces the traditional challenges of resource extraction, pest infestations, and urban development, as well as new challenges, like climate change. Scientists, policymakers, and the general public are all becoming more aware of the value of forests and the need to protect them.

REVIEWING OBJECTIVES

You should now be able to

Describe the basic functional processes of trees, the principal forest biomes in Canada, and the role of forests in biogeochemical cycling

- Trees have the same basic requirements as other plants: sunlight, water, nutrients, air, and an amenable temperature. Trees function as a link between the biogeochemical cycles of the atmosphere, hydrosphere, biosphere, and geosphere.

- There are three main sets of forest biomes: northern or boreal forests, temperate forests, and tropical forests.

- Forests can have closed or open canopies. Forests dominated by trees grade into more open woodlands, savannahs, and grasslands.

- As host to 9% of the world's forests and 30% of the boreal forest, Canada plays an important role as steward for much of the world's forested land.

- Canada's north is dominated by coniferous boreal forest. The west is characterized by subalpine, montane, and coastal forest types, and the east by temperate deciduous forests. The central Prairies are characterized by more open woodland and grassland ecosystems.

- Forests provide habitat and support biodiversity, in addition to playing important roles in climate and biogeochemical cycles.

- Forests contribute ecosystem services that are of great value to people, including protection of soils, moderation of the climate system and the hydrologic cycle, carbon storage, and oxygen cycling.

Summarize the economic contributions of forests and the main approaches to forest product harvesting

- Forests provide economically important timber, as well as a wide variety of non-timber forest products such as fruit, nuts, honey, rubber, materials for shelter, and many other benefits.

- Harvesting methods for timber include clear-cutting and other even-aged techniques, as well as selection strategies that maintain uneven-aged stands that more closely resemble natural forest.

- Harvesting can be sustainable as long as the principle of maximum sustainable harvest is maintained, so the stock does not become depleted. Tree plantations and managed forests can replace forested areas, but cannot duplicate the habitat complexity and biodiversity of an old-growth forest.

Trace the history and scale of forest loss and identify the current drivers of deforestation

- Forests have been cleared since the beginnings of human civilization, for a wide variety of reasons. Developed nations, including Canada, deforested much land during the process of settlement, farming, and industrialization.

- Agriculture has contributed greatly to deforestation and the conversion of grasslands, and has had enormous impacts on landscapes and ecosystems worldwide.

- Today deforestation is taking place most rapidly in developing nations, driven by factors such as logging (both legal and illegal) and pest infestations, and by underlying pressures that are largely economic or political.

Explain the fundamentals of forest management, and identify current forest management approaches in Canada and internationally

- Forest managers increasingly focus not only on extraction of forest products, but also on sustaining the ecological systems that make resources available.

- Forest managers are beginning to implement ecosystem-based management and adaptive management.
- Fire policy has been politically controversial, but scientists agree that we need to address the impacts of a century of fire suppression.
- Climate change is bringing increased pressures on forests and forest species, and new challenges for forest

managers. Assisted migration is a possible approach if forest species are threatened by changes in their natural range as a result of climate change.

- Certification of sustainable forest products allows consumer choice in the marketplace to influence forest management.

TESTING YOUR COMPREHENSION

1. Where do trees acquire the nitrogen that they need? What about the carbon?
2. How does water move through trees, and why is water important for tree survival?
3. What is a forest, and how does a forest differ from a woodland or savannah?
4. How do minerals differ from timber when it comes to resource management?
5. Compare and contrast maximum sustainable yield, adaptive management, and ecosystem-based management. Why would pursuing maximum sustainable yield sometimes conflict with what is ecologically desirable?

6. Name several major causes of deforestation. Where is deforestation most severe today?
7. Compare and contrast the major methods of timber harvesting.
8. Describe several ecological effects of logging. How has the Canadian Forest Service responded to public concern over the ecological effects of logging?
9. Are forest fires a bad thing? Explain your answer.
10. What are some of the organizations—both governmental and nongovernmental—that are important in forest management today, both in Canada and internationally? Name three and briefly describe their roles.

THINKING IT THROUGH

1. Do you think assisted migration represents an appropriate approach for forest managers in addressing the impacts of climate change on forest ecosystems? What are some of the pros and cons, and benefits and risks involved?
2. People in developed countries are fond of warning people in developing countries to stop destroying the rain forest. People in developing countries often respond that this is hypocritical because the developed nations became wealthy by deforesting their land and exploiting its resources in the past. What would you say to the president of a developing nation that might be seeking to clear much of its forest to raise money, or to make room for farming?
3. Your town is proposing a development that would cut through a large forested area. The size of the forested area would not be substantially reduced, but it would be cut into several smaller pieces by road. Explain to

the town council the difference between core and edge habitat and why they should reconsider the development.

4. You have just become the supervisor of a national forest. Timber companies are requesting to cut as many trees as you will let them, and environmentalists want no logging at all. Ten percent of your forest is old-growth primary forest, and the remaining 90% is secondary forest. Your forest managers are split among preferring maximum sustainable yield, ecosystem-based management, and adaptive management. What management approach(es) will you take? Will you allow logging of all, none, or some old-growth trees? Will you allow logging of secondary forest? If so, what harvesting strategies will you encourage? What would you ask your scientists before deciding on policies on fire management and salvage logging?

INTERPRETING GRAPHS AND DATA

Fire is a natural part of the boreal forest. Many forest species are keenly adapted to fire, and even dependent upon it for germination and other aspects of their survival and reproduction. Natural wildfires burn about 20 to 30 million ha (20 000 to 30 000 km^2) of forested in Canada each year, most of it in the boreal forest and taiga. This

annual average for the past two decades is approximately double the annual average for the preceding two decades.[48] In 2014, fires in Canada's boreal forest burned more than 40 million ha—well over the annual average.

The boreal forest plays an important role in the global carbon cycle, which, in turn, is critical in the regulation of the climate system. Canada's boreal forest acts both as a sink for atmospheric carbon, and a source; most years, these functions approximately balance. It is a sink for carbon if new tree growth takes in more carbon than is released. It acts as a source if fires, insect damage, and removal of trees for other purposes release more carbon than is taken in by new growth (see graph).[49]

What if climate change causes more and larger wildfires in the boreal forest? Researchers are finding a correlation between hotter summers and both the number of wildfires and area burned.[50] If climate change continues to bring hotter summers, perhaps we can anticipate an increasing trend of more and larger fires affecting the boreal forest. Fire is a significant cause of carbon emissions from the boreal forest (though not the main cause, as shown in the graph). Fire releases carbon from forests by combustion, rapidly converting biomass into (mainly) carbon dioxide. Fire also affects the carbon storage capacity of forests, by removing trees that could otherwise continue to grow and incorporate carbon into their biomass. If more and larger areas are burned as summers become hotter, it could shift the carbon balance in boreal forests.

Use the two diagrams shown here to answer the following questions.

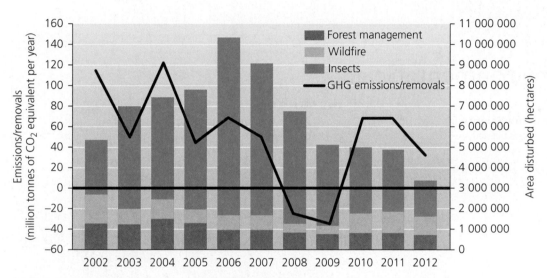

(a) Emissions and removals of carbon by Canadian forests

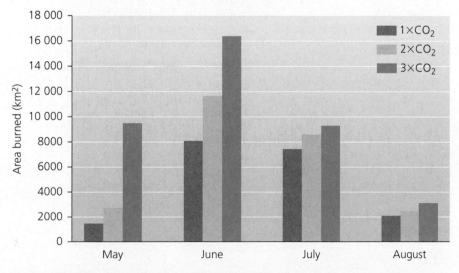

(b) Burned area in Canadian boreal forests in three future climate scenarios

Canada's forests act as both a source and a sink for carbon **(a)**. The main causes of carbon emission and the balance of greenhouse gas emissions and removals are shown here. A positive number indicates net emission of carbon dioxide in Canada's managed forests for that year, while a negative number indicates net removal. Model results **(b)** suggest that the monthly area burned by fire in the boreal forest will increase if atmospheric carbon dioxide continues to increase, under three future climate scenarios (1×, 2×, and 3× present levels of CO_2).

Source: Data for (a) from "Forest Carbon," Natural Resources Canada. http://www.nrcan.gc.ca/forests/climate-change/13085. Reproduced with the permission of the Minister of Natural Resources Canada, 2015. Data for (b) from Future Emissions from Canadian Boreal Forest Fires. *Canadian Journal of Forest Research, 39*, pp. 383-395. Reprinted by permission. © Canadian Science Publishing or its licensors.

1. What is the main cause of carbon emission from boreal forests from year to year?

2. In graph **(b)**, in which month does the model show the largest proportional increase in moving from a climate scenario with present-day carbon dioxide levels ($1 \times CO_2$) to one in which carbon dioxide doubles ($2 \times CO_2$)? What about a climate scenario in which carbon dioxide triples (from $1 \times CO_2$ to $3 \times CO_2$)?

3. From question 2, why do you think there is a shift in the month with the greatest burned area between these two scenarios?

4. In graph **(a)**, only two years are shown in which the net carbon balance was such that more carbon was removed than emitted (2008 and 2009). What does that suggest about tree growth in the forests in those years? What would be the effect on this balance if fire-burned areas were to increase?

5. In 2013 and 2014, over 4 million ha of Canadian forested land were disturbed by fire. Assuming that the areas disturbed by other causes stayed the same as in 2012, sketch in the bars for 2013 and 2014 on the graph. What would you predict for the carbon balance in those years? Do you think it was above zero (indicating net emission of carbon), or below zero (indicating net removal of carbon from the atmosphere)?

MasteringEnvironmentalScience®

STUDENTS

Go to **MasteringEnvironmentalScience** for assignments, the eText, and the Study Area with practice tests, videos, current events, and activities.

INSTRUCTORS

Go to **MasteringEnvironmentalScience** for automatically graded activities, current events, videos, and reading questions that you can assign to your students, plus Instructor Resources.

11 Freshwater Systems and Water Resources

Hundreds of waterfalls cascade from the Niagara Escarpment, a UNESCO World Biosphere Reserve. This one is Tew's Falls near Hamilton, Ontario.

Upon successfully completing this chapter, you will be able to

- Outline the major reservoirs and pathways of the hydrologic cycle, and the major human uses of fresh water
- Describe the major types of freshwater ecosystems
- Discuss how we divert and withdraw water for human use, and how these practices alter freshwater systems

- Assess problems of water supply and propose solutions to address depletion of freshwater sources
- Summarize the types and sources of water pollution
- Propose solutions to prevent or remediate water pollution

About 9% of Canada's surface is covered by water, more than any other nation. The Great Lakes, seen here in a satellite image, straddle the boundary between Canada and the United States, and contain 21% (one-fifth) of the world's total available fresh water.

NASA

CENTRAL CASE

TURNING THE TAP: THE PROSPECT OF CANADIAN BULK WATER EXPORTS

"Water promises to be to the twenty-first century what oil was to the twentieth century: the precious commodity that determines the wealth of nations."

—*FORTUNE MAGAZINE*, MAY 2000

"The wars of the twenty-first century will be fought over water."

—WORLD WATER COMMISSION CHAIR ISMAIL SERAGELDIN

"The time to make these decisions is now, not after our water is gone."

—DAVID SCHINDLER, KILLAM MEMORIAL PROFESSOR OF BIOLOGICAL SCIENCES (EMERITUS), UNIVERSITY OF ALBERTA

There are few topics more controversial for Canadians than the subject of fresh water. Some argue that water is our legacy, our natural capital, and that its abundance defines us as Canadians. They fear that we will place our sovereignty at risk if we allow large-scale diversions of fresh water, or bulk water exports, from Canadian water bodies. Once bulk water exports are allowed to begin flowing to the thirsty southwestern United States, they maintain, they will be impossible to stop.

For others, access to fresh water is a fundamental human right. They argue that those who possess it in abundance have the moral duty to provide water to those who lack it. As the stewards of 25% of the world's wetlands, 7% of the world's renewable flowing water, and 21% of the world's surface fresh water (mainly in the Great Lakes), Canada truly has abundant resources to manage.

For some, water is a valuable, marketable commodity, which Canadians possess in surplus, and for which prices will continue to increase in the coming decades. Canada could be in an enviable position of economic and strategic strength in a world water market. Some even argue that the very thirsty states of the American Southwest will come looking for our water before long anyway, and that perhaps we would be wise to sell it to them first.

Others maintain that we should not even consider exporting our water to serve those who have mistreated, mismanaged, and depleted their own water supplies. Transporting large quantities of water from Canadian water bodies would cause massive changes, perhaps even permanent damage, to our natural ecosystems. And to what end, if those who would import our water have such a poor track record in appropriately managing this precious resource?

For example, desert areas of the southwestern United States are home to the enormous and rapidly increasing populations of cities like Los Angeles, Phoenix, and Las Vegas. The Imperial Valley of California—a natural desert (the sand dune scenes from *Star Wars: Return of the Jedi* were filmed there)—is one of the most fertile agricultural areas in the world, turning out water-intensive fruits like strawberries all winter. This is possible only because vast quantities of water are transported to the area from the Colorado River, via the 132-km All-American Canal. These deserts could never, under natural circumstances, sustainably support such large populations and water-dependent human activities.

As an environmental issue and a political issue it is confusing, complicated, and controversial, but it is crucially important. Canada possesses some of the most enviable water resources in the world, but we also are some of the most wasteful users of water; per capita daily use of water in Canada (about 343 L/day) is surpassed only by the United States.

Water is already exported to the United States from Canada. The 65 bottlers of water in Canada produced approximately 2.3 billion litres of bottled water in 2009; more than a third of it was exported (worth $4 billion), mostly to the United States.[1] Significant *interbasin transfers*—the transportation of water from one drainage basin to another—already occur between Canada and the United States. Most of this—about 97% of the volume—is for electrical power production.[2]

The Government of Canada has not yet approved the wholesale bulk export or massive diversion of water to the United States. The North American Free Trade Agreement (NAFTA) identifies water as a marketable and tradable commodity, which effectively means that Canada is prohibited from restricting water for use exclusively within its national boundaries. Interestingly, Simon Reisman, Canada's chief trade negotiator for NAFTA, was a director of the GRAND Canal Company, a private-sector proponent of bulk water diversion.[3] The "GRAND" (Great Recycling and Northern Development) Canal scheme would involve damming James Bay and diverting the 20 rivers that flow into it toward the south. Another large-scale bulk water export plan, NAWAPA (North American Water and Power Alliance), would divert the Yukon, Peace, and Liard rivers through an 800-km-long canal running along the Rocky Mountain Trench and into the United States.

Canada's fresh water is currently protected against bulk exports by a watershed-based approach in which each province and territory individually controls bulk water exports under the *International Boundary Waters Treaty Act*. Because Canada's position on bulk water exports has been relatively firm, some American companies are becoming more interested in the significant water resources in Alaska as an alternative.

This discussion is ongoing, and it will continue in the coming decades. You should think about it, and reach an informed decision: What is *your* position on bulk water exports? As of 2015, with the State of California suffering from an unprecedented drought that has necessitated severe limitations on water use, the question is taking on even greater urgency.

Earth's Fresh Water

"Water, water, everywhere, nor any drop to drink." The well-known line from Coleridge's poem *The Rime of the Ancient Mariner* describes the situation on our planet quite well. Water may seem abundant to us, but water that we can drink is actually quite rare and limited (**FIGURE 11.1**). Roughly 97.5% of Earth's water resides in the oceans and is too salty to drink or use to water crops. Only 2.5% is considered **fresh water**, water that is relatively pure, with few dissolved salts. Because most of the fresh water is tied up in glaciers, ice caps, and underground aquifers, just over 1 part in 10 000 of Earth's water is easily accessible for human use.

Water moves through the hydrologic cycle

Water is constantly moving among the reservoirs shown in **FIGURE 11.1** via the hydrologic cycle. Recall from our chapter on environmental systems that Earth's hydrosphere encompasses all of the water—salt or fresh, liquid, ice, or vapour—stored in surfacewater bodies and glaciers, the near subsurface, and underground. Water in the atmosphere technically does not belong to the hydrosphere, but the two systems are closely linked, and the atmosphere plays a central role in the hydrologic cycle. As water moves through the hydrologic cycle, it stores and redistributes heat, erodes mountain ranges, builds river deltas, maintains ecosystems, supports living organisms, shapes civilizations, and gives rise to political conflicts.

The reservoirs in the hydrologic cycle differ in the amounts of water they store and transmit, and how long they typically hold water in storage. In general, the largest reservoirs have the longest storage times—or *residence times*—for water. For example, the average residence time for a molecule of water in the ocean, the largest reservoir in the hydrologic cycle, is more than 3000 years. Residence times for water stored in glaciers and ground water, the next largest reservoirs, are hundreds to thousands of years. In contrast, water tends to reside in the atmosphere (a large "container," but a small reservoir in terms of its water content) for only a matter of days. In biological organisms, the average residence time for water is even shorter—hours or less.

Water is distributed unevenly among the reservoirs of the hydrologic cycle. Different regions, even different areas within the same country, can possess vastly different

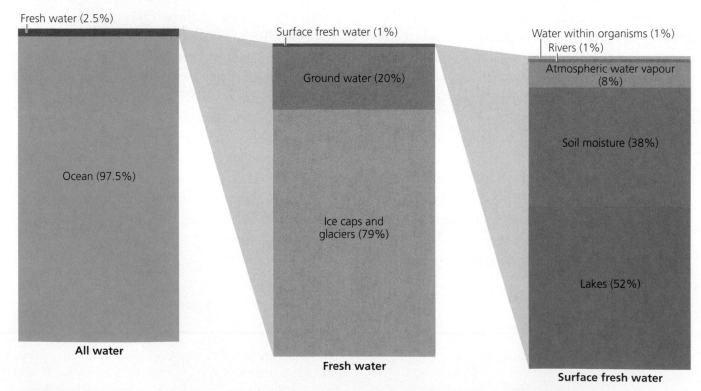

FIGURE 11.1   The world ocean is by far the largest reservoir in the hydrosphere, and it is a saltwater reservoir. Only 2.5% of Earth's water is fresh water. Of that 2.5%, most is tied up in glaciers and ice caps. Of the 1% that is surface water, most is in lakes and soil moisture.
Source: Data from United Nations Environment Programme (UNEP) and World Resources Institute.

DATA Q Ground water represents 20% of the fresh water in Earth's hydrosphere—the second most important freshwater reservoir (after ice caps and glaciers). What proportion of the total water in the hydrosphere is ground water?

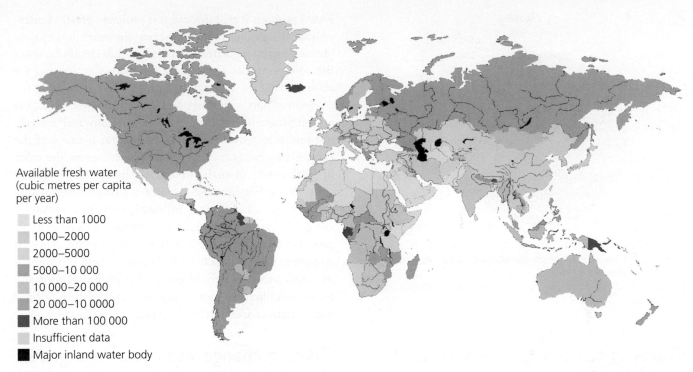

Available fresh water
(cubic metres per capita
per year)

Less than 1000
1000–2000
2000–5000
5000–10 000
10 000–20 000
20 000–10 0000
More than 100 000
Insufficient data
Major inland water body

FIGURE 11.2 Nations vary tremendously in the amount of fresh water per capita available to their citizens. Iceland, Papua New Guinea, Gabon, and Guyana (dark blue in this map) each have more than 100 times as much water per person as many Middle Eastern and North African countries. *Source:* Data from UN Environment Programme and World Resources Institute, as presented by Harrison, P., and F. Pearce (2000). *AAAS Atlas of Population and the Environment.* Copyright © by American Association for the Advancement of Science (AAAS). Reprinted by permission.

amounts of ground water, surface water, and precipitation. For example, the region of the southwestern United States is very dry, but the country as a whole has water resources roughly equivalent to those of Canada; however, the main concentration of water resources in the United States is in Alaska. In terms of global extremes, precipitation ranges from a high of about 1200 cm per year at Mount Waialeale on the Hawaiian island of Kauai to virtually zero in Chile's Atacama Desert. Some polar areas also receive very little precipitation; thus, they qualify as deserts, even though they are cold rather than hot.

People also are not distributed across the globe in accordance with water availability. Many areas with high population density are water poor, leading to inequalities in per capita water resources among and within nations (**FIGURE 11.2**). For example, Canada has twenty times as much water available for each of its citizens as China does. The Amazon River carries 15% of the world's runoff, but its watershed holds less than half a percent of the world's human population. Asia possesses the most water of any continent but has the least water available per person, whereas Australia, with the least amount of water, boasts the most water available per person. Because of this mismatched distribution of water and population, it has always been a challenge to transport fresh water from its source to where people need it.

Fresh water is distributed unevenly in time as well as geographically. India's monsoon season brings concentrated storms in which half of a region's annual rain may fall in a few hours. Northwest China receives three-fifths of its annual precipitation during three months when crops do not need it. The uneven distribution of water across time is one reason that people have built irrigation systems and erected dams to manage floods and store water, so that it can be distributed when needed.

Water supplies our households, agriculture, and industry

We all use water at home for drinking, cooking, and cleaning (**FIGURE 11.3**). Farmers and ranchers use water to irrigate crops and water livestock. Most manufacturing and industrial processes require water. Globally, we spend about 70% of our annual freshwater allotment on agriculture. Industry accounts for roughly 20%, and residential and municipal uses for only 10%. The proportions of each of these three types of use—residential/municipal, agricultural, and industrial—vary dramatically among nations (**FIGURE 11.4**). Nations with arid climates tend to use more fresh water for agriculture, and heavily industrialized nations use a great deal for industry.

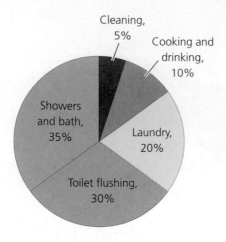

FIGURE 11.3
According to Environment Canada, this is how Canadians apportion their use of water in the home.
Source: Data from Environment Canada. Water Use: Withdrawal Uses (as of 2006).

FAO's position is that this catch is probably greatly underestimated, because inland fisheries are much less closely monitored and regulated than marine fisheries. Freshwater fish also dominate fish farming globally, accounting for about 56% of the total world aquaculture production.

Increasingly, scientists and environmental managers are emphasizing the importance of freshwater systems not just as a resource for human use, but because of the ecosystem services they provide. This recognizes the value of fresh water in maintaining the integrity of aquatic ecosystems; but healthy, resilient ecosystems ultimately support human life and interests, too. For example, the FAO lists a number of ecosystem services that are provided by inland capture fisheries, if properly managed and maintained (**TABLE 11.1**). Some of these services are of direct value to people (such as food provision), and some are indirect benefits, such as the regulation of food web dynamics and nutrient cycling.

Water *in situ* is important for both people and ecosystems

We withdraw water from reservoirs in the hydrosphere for our use in agriculture, industry, and households. This water is critically important as a resource for the support of human life and activities; however, it is increasingly recognized that water holds immense value *in situ* ("in place") in undisturbed freshwater systems.

From a utilitarian perspective (focusing on human uses and values), water *in situ* in rivers and lakes provides opportunities for transportation, boating and recreation, cultural and spiritual uses, and of course freshwater fishing. The Food and Agriculture Organization (FAO) of the United Nations reports that freshwater fishing (known as *inland capture fisheries*) accounted for about 11.2 million tonnes in 2010.[4] This is only about 7% of the total world capture fishery production, but a dramatic increase over a period of just a few years, mostly in Asian countries. The

Climate change will cause changes in the hydrologic cycle

We have seen that water availability and human need are poorly matched, both geographically and temporally. As if this were not enough, climate change will introduce further stresses over the next few decades by altering precipitation patterns, melting glaciers and permafrost, intensifying droughts and floods, decreasing river flows, lowering groundwater levels, and raising sea levels (potentially leading to saltwater incursions into coastal freshwater bodies).

Environment Canada reports that climate change is expected to affect fresh water and the hydrologic cycle in Canada in several ways. Snowmelt and spring runoff can be expected to occur earlier than they do at present. Rain belts will shift, and the continental interior will experience drier summers. In combination with higher summertime evapotranspiration, starting earlier and lasting longer than it does currently, this will increase the threat of drought.[5]

FIGURE 11.4
Freshwater consumption differs by nation. Industry consumes the most water in Canada, agriculture uses the most in India, and most water in Lithuania goes toward domestic use. This reflects differences in the economies of these nations, among other things.
Source: Data are based on information from UN Food and Agriculture Organization (FAO, 2000) and from Environment Canada (2005), *The Management of Water: Water Use.*

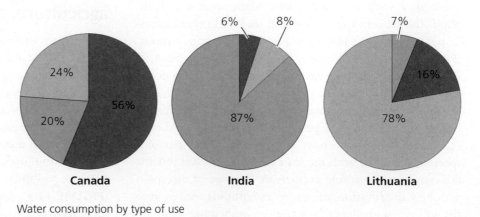

Water consumption by type of use

 Industry ▨ Agriculture ▨ Domestic

TABLE 11.1 Ecosystem Services Provided by Freshwater Fisheries

Ecosystem Service Type	Specific Service Provided by Inland Capture Fisheries
Regulation	Regulation of food web dynamics Nutrient transport and cycling Control of pest organisms
Support	Maintenance of genetic, species, and ecosystem biodiversity Resilience and resistance—life support by the freshwater environment and its response to pressures, including maintenance of ecosystem balance
Provisioning	Food provision—extraction of aquatic organisms for human consumption and nutrition Livelihood provision—contribution to employment and income, including recreational and ornamental fisheries Aquaculture seed provision—inputs to aquaculture for grow-out
Cultural and scientific	Cultural heritage and identity—value associated with freshwater fisheries themselves Recreational fisheries—the noncommercial perspective Cognitive values—education and research resulting from the fisheries Catch composition and species as bio-indicators of health of ecosystem

Source: Food and Agriculture Organization of the United Nations (2012). *The State of World Fisheries and Aquaculture 2012, Part I: World Review of Fisheries and Aquaculture*, FAO Fisheries and Aquaculture Department, Rome. http://www.fao.org/docrep/016/i2727e/i2727e00.htm. Accessed 3 April 2014. Reproduced with permission.

The last major drought in the Canadian Prairies occurred in 2001–2002; it lasted for two years, and was unusually large in its spatial extent. If climate model predictions of warmer temperatures and decreasing precipitation in interior continental regions come to pass, it may mean that droughts in the Prairies will become more common, more severe, and more extended; this could have significant impacts on agricultural productivity. Climatic warming could include warmer river temperatures, damaging aquatic ecosystems and freshwater fish populations. Water temperatures in the Great Lakes also are expected to increase, and water levels to decrease.[6]

There are numerous potential secondary problems related to the impacts of climate change on water systems and resources. For example, if water levels drop, it may become necessary to dredge channels more often to accommodate shipping. This would stir up and re-suspend sediments, increasing the turbidity of the water and potentially reactivating toxic chemicals that had settled into bottom sediments. An anticipated problem in Atlantic Canada, because of rising sea levels, is saltwater intrusion into coastal groundwater aquifers, as well as saltwater disturbance of coastal estuaries.[7]

Freshwater Systems

We have considered water in the hydrosphere from a global perspective and its importance in maintaining both human life and healthy ecosystems. Now we turn our attention to the characteristics of the reservoirs in which fresh water resides in the hydrosphere, and the processes by which it is transferred among these reservoirs in the hydrologic cycle. Let's first examine the parts of the hydrologic cycle that are most conspicuous to us—surfacewater bodies and the ecological systems they support.

Rivers and streams wind through landscapes

Water from rain, snowmelt, and springs runs downhill and converges where the land dips lowest, forming streams, creeks, or brooks. These water courses merge into rivers, whose waters eventually reach the ocean (or sometimes end in a landlocked water body). A smaller river flowing into a larger one is a **tributary**, and the area of land drained by a river and all its tributaries is that river's **drainage basin** or **watershed**.

Rivers shape the landscape through which they run. The force of water rounding a river's bend gradually eats away at the outer shore, eroding soil from the bank. Meanwhile, sediment is deposited along the inside of the bend, where water currents are weaker. In this way, over time, river bends become exaggerated in shape (**FIGURE 11.5**). Eventually, a bend may become such an extreme loop (called an *oxbow*) that water erodes a shortcut from one end of the loop to the other, pursuing a direct course. The bend is cut off and remains as an isolated U-shaped water body called an *oxbow lake*.

Over thousands or millions of years, a river may shift from one course to another, back and forth over a large area, carving out a flat valley. The flat areas nearest to the river's course are flooded periodically, and are said to be within the river's **floodplain**. Frequent deposition of silt from flooding makes floodplain soils especially fertile. As a result, agriculture thrives in floodplains, and **riparian** (riverside) forests are productive and species-rich.

Floodplains demonstrate that flowing water is only one part of riverine and riparian systems. Erosion, transport, and deposition of sediment by the water in these systems are also vitally important. Another thing to remember about river systems is that any changes that occur in *upstream* areas, near the source of the stream, will inevitably have impacts in *downstream* areas, farther along the river's course. These impacts can include changes in water quality, clarity,

FIGURE 11.5
Rivers and streams flow downhill, shaping landscapes, as shown by the oxbow curve of this meandering river in Alberta.

Blackfox Images/Alamy

level, temperature, or velocity of flow, as well as changes in the amount of sediment that is carried downstream.

Rivers and streams host diverse ecological communities. Algae and detritus support many types of invertebrates, from water beetles to crayfish. Insects such as dragonflies, mayflies, and mosquitoes develop as larvae in streams and rivers before maturing into adults that take to the air. Fish consume aquatic insects; and birds, such as kingfishers, herons, and ospreys, dine on fish. Many amphibians spend their larval stages in streams, and some live their entire lives in streams. Salmon migrate from oceans up rivers and streams to spawn.

Wetlands are diverse and complex

Systems that combine elements of fresh water and dry land can be enormously rich and productive. Often grouped together under the term **wetlands**, such areas are characterized by the presence of water, but not necessarily "standing" water (as in a lake), and not necessarily year-round. Wetlands are inundated with water for enough of the year that the soil is water-saturated, and the plants and animals they host are adapted to an aquatic environment. The water in upland wetlands is typically fresh water, but coastal wetlands can hold brackish or even salt water.

The environments in which wetlands occur are varied. They include

- palustrine (inland)
- fluvial, riparian, or riverine (related to a river)
- lacustrine (related to a lake)
- estuarine (related to an estuary)
- coastal or marine (related to an oceanic coastal zone)

Wetland ecosystems thus are very diverse. In the *Canadian Wetlands Classification System*,[8] wetlands are grouped into five *classes* based on their broad characteristics and the environment in which they occur. The five classes are marsh, swamp, bog, fen, and shallow open water. Within each of these five classes, wetlands are further divided according to their *form*, based on more specific characteristics of the water and mineral soils they contain. And finally each class/form combination can be further described by *type*, according to the specific vegetation communities that they host. In Canada, especially in the north, an additional variant is the amount of time that the wetland is frozen during each year.

Let's briefly look at the five major classes of wetlands.

In **marshes** (**FIGURE 11.6**), shallow water allows grasses and seasonal herbaceous plants to grow above the water's surface. Marshes usually have water levels that fluctuate either daily, seasonally, or annually. Cattails and bulrushes are plants typical of North American marshes. There are several types of freshwater marshes (such as *riverine marshes, lacustrine marshes,* and *wet meadows*), but marshes also can contain brackish water (*tidal* and *estuarine marshes*) or even salt water (*salt marsh*), depending on their location and environment.

Swamps also contain shallow, standing water that is rich in vegetation, but they occur in forested or wooded areas. The cypress swamps of the southeastern United States are an example, like the Everglades, where huge cypress trees grow in standing water. Swamps can be created when beavers build dams across streams with limbs from trees they have cut, flooding wooded areas upstream. In the Amazon, flooded forest swamps are referred to as *varzea* or *igapo* wetlands, depending on the chemistry,

FIGURE 11.6
Shallow, ephemeral, or seasonal water bodies with aquatic vegetation are called wetlands and marshes, like this one in Botswana, Africa.

Blickwinkel/Alamy

nutrient content, and source of the water. Great rivers like the Amazon and the Mississippi often have forested swamps along their banks.

Bogs are **peatlands** in which the water table is at or just below the bog surface. Recall that *peat* is a type of carbon-rich soil, characterized by accumulated plant matter typically including (though not exclusively) *Sphagnum* mosses. An example of a large bog with a thick accumulation of carbon-rich soil is the Mer Bleue Bog outside of Ottawa, Ontario, profiled in the Central Case in Chapter 7, *Soils and Soil Resources*. Bog waters are replenished entirely by precipitation, so they tend to be mineral-poor and slightly acidic (characteristics inherited from their plant matter and from the precipitation that feeds them). These characteristics distinguish bogs from **fens**, which are fed by ground water. Fens, like bogs, are peatlands, but they are mineral-rich because of their association with ground water, which typically contains a lot of dissolved minerals.

Shallow-water wetlands are also known as ponds, sloughs, oxbows, or vernal pools (among many other names). This class of wetland contains open, standing, or flowing water that is typically less than 2 m in depth. They are transitional between permanent, deep bodies of water (that is, lakes) and the other types of wetlands that are water-saturated or have standing water that fluctuates. They can occur in any of the environments mentioned above, and they are typically *ephemeral*, that is, short-lived or seasonal.

Wetlands overall are extremely valuable as habitats for wildlife. Because they vary so much both temporally and geographically, they offer a diverse assortment of habitats and niches for organisms. They also provide important ecosystem services by slowing runoff, reducing flooding, recharging aquifers, and filtering pollutants. They also provide locations

for recreation and tourism. Coastal wetlands protect shorelines by buffering the impacts of storm waves. Most importantly, wetlands are vital to the functioning of both global and local water cycles, contributing to water security and overall human well-being.[9] Despite these vital services, people have drained and filled wetlands extensively, mostly for agriculture. For example, southeastern Canada has lost well over half of all wetlands since European colonization, with up to 90% loss in some areas.

Lakes contain open, standing water

Lakes are bodies of open, standing water, usually fairly large. There is no universally accepted definition for a lake (as compared to a pond, for example), but usually the term applies to water bodies with a surface area greater than at least 2 ha. The physical conditions and types of life within any given lake vary with depth and the distance from shore. As a result, scientists have described several zones typical of lakes and ponds (**FIGURE 11.7**).

Canadian scientists have been world leaders in research on lakes and lake-related processes. One of the most important facilities for scientific research on lakes is the Experimental Lakes Area (ELA), a unique research laboratory located in a pristine forest of northwestern Ontario. At the ELA, scientists have carried out longitudinal studies (that is, scientific experiments that last many years) on lakes and freshwater fish and their responses to disturbances and contamination. These studies are important because they deal with the real lakes, fish, and forests—not models in a laboratory or on a computer. Such research is necessary in order to understand how lakes respond to and recover from the impacts of human activity over time (see

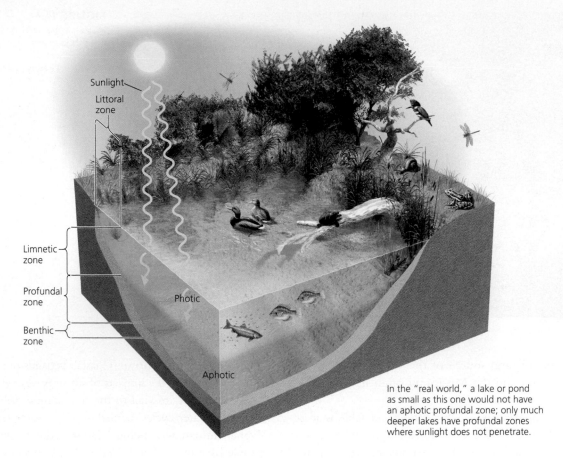

In the "real world," a lake or pond as small as this one would not have an aphotic profundal zone; only much deeper lakes have profundal zones where sunlight does not penetrate.

FIGURE 11.7 In lakes, emergent plants grow in shallow water around the shoreline in the littoral zone. The limnetic zone is the open, sunlit water where photosynthesis takes place. Sunlight does not reach the deeper profundal (aphotic) zone, which is only present in the deepest lakes. The benthic zone, at the bottom, is muddy, rich in detritus and nutrients, and low in oxygen.

DATA Q Light of different wavelengths penetrates to different depths through clear water, as shown below. The aphotic zone is variously defined, but it always refers to the depth at which very little or no light penetrates. For the sake of this exercise, let's define the aphotic zone as the zone where only the deepest indigo or violet light is able to penetrate. Based on this diagram, and assuming that the water is clear (because turbidity greatly affects light penetration), which of the Great Lakes would you expect of have an aphotic or profundial zone? Which would you expect to have a benthic zone?

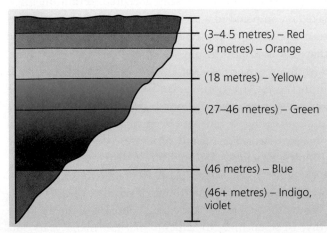

Depth of light penetration in clear water, by colour

Lake	Maximum Depth (m)	Average Depth (m)
Erie	64	19
Huron	228	59
Michigan	282	85
Ontario	245	86
Superior	407	147

"The Science behind the Story: Near-Death Experience at the Experimental Lakes Area").

The region fringing the edge of a lake is the **littoral zone**. Here the water is shallow enough that aquatic plants can be emergent, that is, they grow from the mud but reach above the water's surface. The nutrients and productive plant growth of the littoral zone make it rich in invertebrates—such as insect larvae, snails, and crayfish—on which fish, birds, turtles, and amphibians feed. The **benthic zone** extends along the bottom of the entire water body, from shore to the deepest point. Many invertebrates live in the mud on the bottom, feeding on detritus or preying on one another.

THE SCIENCE BEHIND THE STORY

A researcher adds silver nano-particles to a lake enclosure at the Experimental Lakes Area.

Photo by M. Paterson, IISD-ELA

Near-Death Experience at the Experimental Lakes Area[10]

The Experimental Lakes Area (ELA) is a unique, world-class research facility located in a pristine forested area in northwestern Ontario (see photo). The facility, established in 1968, was jointly operated by Fisheries and Oceans Canada and Environment Canada until funding was cut by the federal government in 2012.

At the ELA, scientists have the ability to carry out ecosystem-scale experiments on 58 lakes and surrounding drainage basins (see photo). The studies are unusual because they can last from years to decades.

This researcher is carrying out work at the Experimental Lakes area.

Department of Fisheries and Oceans Canada

The pristine setting allows researchers to investigate problems and processes in the natural environment, without the direct influence of human activities. Test lakes are manipulated (see **TABLE 1**) and compared to natural lakes, which have been monitored at the ELA for up to 47 years. The ability to compare experimental lakes to natural lakes greatly increases the sensitivity of the studies, and the scientific robustness and practical applicability of the results.

The research provides important scientific understanding of how ecosystems operate in the real world. Many of the experimental studies at the ELA focus on impacts to lake trout, whitefish, and northern pike, which are important in Canada for commercial, recreational, and subsistence fisheries. The ability to manipulate entire ecosystems has produced unexpected results that could not have been found in laboratory experiments. These results have provided valuable input into the development of more effective environmental management and regulatory approaches.

For example, early work at the ELA showed the importance of controlling phosphorus to reduce algal blooms in lakes. These results influenced the decisions of government agencies to remove phosphorus alone from waste water, rather than removing both phosphorus and nitrogen—a much more expensive process. Long-term research on acidification at the ELA has demonstrated the sensitivity of the lake food chain to acid rain, and consequent negative effects on fish populations. The results of this work influenced the regulation of sulphur dioxide emissions from smokestacks in Canada and the United States. These examples, and others included in the table on the next page, highlight the importance of understanding the types of impacts on aquatic systems that can only be effectively understood through long-term, whole-ecosystem studies.

Many scientific publications have resulted from ELA research projects, by some of Canada's preeminent environmental researchers, such as Professor David Schindler of the University of Alberta.[11] The research has had measurable effects on policies and regulations, leading to improved fish habitat, cleaner water, and healthier ecosystems. ELA research also has saved billions of dollars in unnecessary and costly

regulations, such as the needless control of nitrogen in wastewater treatment to control algal blooms in lakes.

In 2012 the Government of Canada announced that it would no longer fund the Experimental Lakes Area. The Polar Environment Atmospheric Research Lab (PEARL), another world-class Canadian research facility, and the groundbreaking National Round Table on the Environment and the Economy were also put on the chopping block. The cuts led to an outcry from environmental scientists; at a rally in July 2012, 2000 scientists in lab coats converged on Parliament Hill to decry the "death of evidence." The rationale cited for the cuts was the need for savings, even though the annual cost to maintain ELA was only $2 million out of a federal science-and-technology budget of about $11 billion.[12]

In 2013 the Province of Ontario came forward with an offer of partial financial support for the facility. At the last possible moment, on April 1, 2014 (when the federal government had already begun to dismantle research facilities on-site), an agreement was reached between the governments of Ontario and Canada, and the International Institute for Sustainable Development (IISD). Through this agreement, IISD became the official operator of the facility.[13]

No other facility like the ELA exists, and it is unlikely that it could ever be recreated anywhere else. Thankfully, because of support from Ontario and IISD, the ELA has survived its near-death experience. However, according to Professor Schindler and Diane Orihel, Director of the Coalition to Save ELA, "It may not be a smooth ride for the ELA in the years to come. The first hurdle will be to rebuild the project's world-class, but now defunct science team. Then, money to fund scientific experiments must be found … [But] the most worrisome issue for the future of Canada's freshwater ecosystems is not at the blossoming IISD, but in the empty offices at Fisheries and Oceans Canada's Freshwater Institute. They were once occupied by highly skilled scientists who conducted the powerful ecosystem science required to support the department's mandate to protect and manage Canada's water resources. The brilliant minds who guided the federal government in making smart decisions on fisheries management and protection of aquatic ecosystem protection are now gone."[14]

(Continued)

TABLE I   Some Examples of Past, Present, and Ongoing Research at the Experimental Lakes Area

Title	Dates	Lead Organizations and Funding Partners	Brief Description
Climate Impacts on Lakes and Their Watersheds	1968–present	Fisheries and Oceans Canada; University of Alberta	Changes in climate will dramatically affect lakes in terms of water quality and quantity and the ability of lakes to support recreational and commercial fisheries. Cold-water fish species, such as lake trout, will be sentinels for climate change, as cold-water habitat is reduced. This study has resulted in one of the longest and most complete data sets in existence, covering all aspects of the response of lakes to changes in climate.
Impacts of Nanosilver on Lakes	2011–present	Trent University; Fisheries and Oceans Canada	Nanosilver is the most widely used engineered nanomaterial. Silver nanoparticles can enter the aquatic environment through wastewater discharges, where they can be toxic to aquatic bacteria and algae. The impacts of nanosilver on lake ecosystems and food webs are not known. This study aims to investigate these impacts, and provide scientific context for policies concerning nanomaterials.
Effects of Acid Rain on Lakes	1976–2004	Fisheries and Oceans Canada; University of Alberta; Oak Ridge National Laboratory, USA; University of Manitoba	In the 1970s and 1980s, concern developed about the loss of fish populations in hundreds of thousands of lakes in eastern North America and northern Europe, probably the result of lake acidification caused by emissions of sulphur dioxide and nitrogen from coal-fired power plants. To test this, scientists added sulfuric acid to lakes at ELA to mimic the effects of acid rain. They demonstrated that even moderate increases in acidification resulted in dramatic impacts to aquatic food webs and caused the collapse of fish populations.
Effects of Endocrine-Disrupting Compounds on Fish	2000–2003	Fisheries and Oceans Canada	In the 1990s, concerns developed about the possible effects of artificial estrogen (from birth control pills) and other common chemicals that mimic reproductive hormones in both humans and wildlife. Populations of fish and other wildlife downstream of sewage treatment plants often show signs of "feminization" of male organisms. At ELA, artificial estrogen was added to a small lake for several years at levels found in effluents from sewage treatment plants. Male fish became feminized and were unable to properly develop sperm; eggs produced by female fish did not develop correctly. These effects led to the collapse of the fish population. Once additions of the estrogen stopped, fish reproduction returned to normal and the population rebounded.
Ecological Impacts of Transgenic Fish	2008–present	Fisheries and Oceans Canada and Centre for Aquatic Biotechnology Regulatory Research	Growing demand for fish protein to feed humans has resulted in dramatic increases in commercial aquaculture. Selective breeding has allowed fish species reared specifically for aquaculture to achieve much higher growth rates than they would naturally. Genetically modified (transgenic) fish have the potential for even greater growth rates. This study addresses the fate and ecological impacts of transgenic fish in the natural environment, which are poorly understood.

TABLE I *Continued*

Title	Dates	Lead Organizations and Funding Partners	Brief Description
Linking Atmospheric Mercury Deposition and Mercury in Fish	2001–present	Fisheries and Oceans Canada; Electric Power Research Institute (EPRI), USA; Natural Science and Engineering Research Council (NSERC); Environment Canada; U.S. Environmental Protection Agency; U.S. Geological Survey; U.S. Department of Energy; State of Wisconsin; Southern Company (USA); Manitoba Hydro; University of Manitoba; Alberta Heritage Fund; Reed Harris Environmental Ltd.; University of Alberta; University of Wisconsin; Carleton University; University of Maryland; Smithsonian Environmental Research Center; University of Toronto; Trent University; Oak Ridge National Laboratory; University of Connecticut; Tetra Tech Inc. (USA); University of Montreal	Fish from remote lakes around the world are contaminated with mercury, a neurotoxin and endocrine disruptor that can be transported long distances and deposited by precipitation. Regulations to force power companies to add mercury scrubbers to smokestacks carry a potential cost of billions of dollars. Because of the large costs, power companies have sought hard evidence that reductions of mercury deposition from the atmosphere would actually reduce mercury in fish. The experiments in this project have demonstrated clearly that controlling atmospheric emissions of mercury will result in a positive response (decline) of mercury concentrations in fish.
Understanding Lake Eutrophication—the Role of Nitrogen	1990–present	University of Alberta; Fisheries and Oceans Canada; York University; Environment Canada; University of Waterloo; Environment Canada; Natural Science and Engineering Research Council (NSERC); University of Guelph	Since 1968, researchers at ELA have continuously added phosphorus to a small lake in conjunction with different amounts of nitrogen. Since 1990, no nitrogen has been added to the lake. Despite the absence of artificial nitrogen inputs, algal blooms have not diminished. These studies demonstrate that efforts to control blue-green algae should focus on the control of phosphorus. Because treatment for nitrogen is more expensive than treatment for phosphorous, the results of this research will potentially save municipalities and governments billions of dollars.
Impacts of Hydro Reservoir Development on Greenhouse Gases	1999–2003	University of Manitoba; Fisheries and Oceans Canada; University of Alberta	Studies at ELA demonstrated that hydroelectric reservoirs produce greenhouse gases, caused by the decomposition of flooded soils and vegetation. Flooded wetlands produced much more greenhouse gases in the long term than did flooded upland areas. Avoiding the flooding of wetland areas could minimize this problem. With the strong support of the results of the ELA experiments, reservoirs are now included in global accounts of human greenhouse gas emissions.
Ecosystem Recovery from Lake Acidification	1992–2008	Fisheries and Oceans Canada; University of Alberta; Ducks Unlimited; Environment Canada; Saskatchewan Watershed Authority; University of Western Ontario	Little was known about the ability of Canada's boreal forest lakes to recover from chronic acidification. To better understand the recovery potential of these lakes, researchers studied a lake that had been acidified experimentally for years and was allowed to return to natural pH. The study lake did recover once the stress of acidification was reduced, but even full pH recovery did not result in complete ecosystem restoration within the timeframe of the study. This demonstrates that it is necessary to lower expectations of the rate of recovery of aquatic habitat after acidification.

In the open portion of a lake or pond, away from shore, sunlight penetrates the shallow waters of the **limnetic zone**. Because light in the *photic zone* enables photosynthesis and plant growth, the limnetic zone supports phytoplankton, which in turn support zooplankton, both of which are eaten by fish. Within the limnetic zone, sunlight intensity (and therefore water temperature) decreases with depth. The water's turbidity affects the depth of this zone; water that is clear allows sunlight to penetrate deeply, whereas turbid water does not.

Below the limnetic zone is the **profundal zone**, the volume of open water that is in the *aphotic zone*, that is, below the depth to which sunlight is able to penetrate. This zone, only found in the deepest lakes (and in the ocean), lacks plant life, is lower in dissolved oxygen and supports fewer animals. Aquatic animals rely on dissolved oxygen, and its concentration depends on the amount released by photosynthesis and the amount removed by animal and microbial respiration, among other factors. Note that the profundal zone is, by definition, aphotic, whereas the benthic zone may be photic or aphotic. "Benthic" simply refers to "the bottom" of the water body, which may or may not be shallow enough for light to penetrate.

Lakes and ponds change naturally over time, as streams and runoff bring them sediment and nutrients. **Oligotrophic** lakes and ponds, which have low-nutrient and high-oxygen conditions, may slowly give way to the high-nutrient, low-oxygen conditions of **eutrophic** water bodies (see **FIGURE 11.23** on p. 339). Eventually, water bodies may fill in completely by the process of aquatic succession. As lakes or ponds change over time, species of fishes, plants, and invertebrates that are adapted to oligotrophic conditions may give way to those that thrive under the eutrophic conditions of a marsh.

Some lakes are so large that they differ substantially in their characteristics from small lakes. These large lakes are sometimes known as inland freshwater seas; the Great Lakes (sometimes referred to as the Laurentian Great Lakes) are a prime example. Because they hold so much water, most of their biota is adapted to open water. For example, major fish species of the Great Lakes include lake sturgeon, lake whitefish, northern pike, alewife, bass, walleye, and perch. Lake Baikal in Asia is the world's deepest lake, at 1637 m deep, and the Caspian Sea is the world's largest enclosed body of water, at 371 000 km².

Ground water plays key roles in the hydrologic cycle

Any precipitation reaching the land surface that does not evaporate, flow into waterways, or get taken up by organisms infiltrates the surface. Most percolates downward through the soil to become **ground water** (**FIGURE 11.8**). Ground water makes up one-fifth of

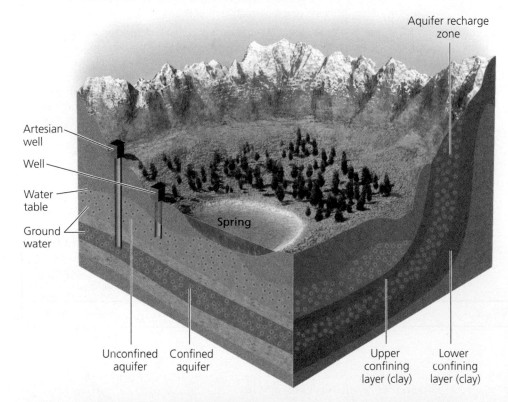

FIGURE 11.8
Ground water may occur in unconfined aquifers above impermeable layers, or in confined aquifers under pressure between impermeable layers. Ground water rises naturally to the surface through springs and wetlands, and through the wells we dig.

Earth's freshwater supply and plays a key role in meeting human water needs.

Ground water is contained within **aquifers**: porous formations of rock, sand, or gravel that hold water. An aquifer's upper layer, or *zone of aeration*, consists of rock or sediment in which the pore spaces are only partly filled with water. In the lower layer, or *zone of saturation*, the spaces are completely filled with water. The boundary between these two zones is the **water table**. Below the level of the water table, all pore spaces and fractures in the rock or sediment are completely filled with water; above the water table, typically some water is present in the pore spaces, but the ground is not completely saturated. Picture a sponge resting partly submerged in a tray of water; the lower part of the sponge is completely saturated, whereas the upper portion may be moist but also contains air in its pores.

There are two broad categories of aquifer. A **confined aquifer**, or *artesian aquifer*, exists when a water-bearing porous layer of rock, sand, or gravel is trapped between overlying and underlying layers of less permeable substrate (often clay). In such a situation, the water is under great pressure. In contrast, an **unconfined aquifer** has no impermeable layers that confine it, so its water is under less pressure and can be readily recharged by surface water.

Any area where water infiltrates Earth's surface, percolates downward, and reaches an aquifer below is known as a **recharge zone**. To prevent ground water from becoming contaminated, it is important to consider what types of development activities are permitted in recharge zones.

Ground water flows downhill and from areas of high pressure to areas of low pressure, emerging to join surfacewater bodies at **discharge zones**. Just as surface water becomes ground water by infiltration and percolation, ground water becomes surface water by emerging through springs, wetlands, and human-drilled wells, often keeping streams flowing when surface conditions are otherwise dry. Because ground water has to flow through the rock or sediment of the aquifer, a typical rate of flow might be only about 1 m per day; this means that ground water may remain in an aquifer for a long time. The average age of ground water has been estimated at 1400 years, and some is many tens of thousands or even millions of years old.

The world's largest known aquifer is the Ogallala Aquifer, which underlies the mostly semi-arid Great Plains of the United States. Water from this massive aquifer has enabled American farmers to create a bountiful grain-producing region. The volumes, the specific characteristics, and even the exact areal extents of the major aquifers in Canada are not yet thoroughly known; they are being studied under the Natural Resources Canada (NRCAN) Groundwater Mapping Program (**FIGURE 11.9**).[15]

Among the most significant aquifers in Canada—both in terms of extent or volume of water and in terms of the areas and populations they service—are the Paskapoo

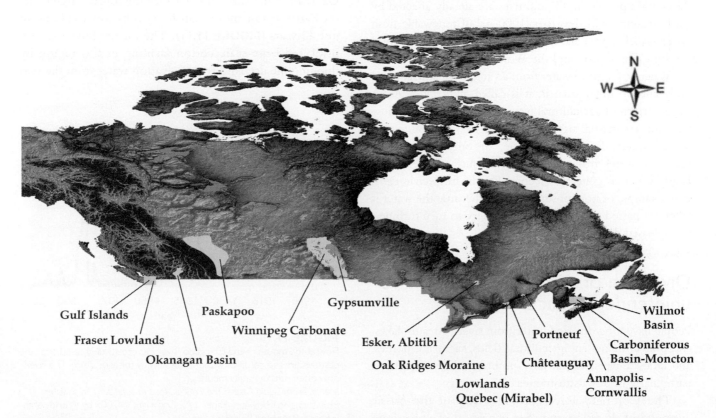

FIGURE 11.9   Some of the significant aquifers that are being studied in the Groundwater Mapping Program include the Paskapoo Formation, Oak Ridges Moraine, and Annapolis–Cornwallis Valley aquifers. Reproduced with the permission of Natural Resources Canada 2015.

Formation, which underlies more than 10 000 km² of southwestern Alberta; the Oak Ridges Moraine, a series of glacial deposits that covers 1900 km² and provides much of the Greater Toronto area with water; and the Annapolis–Cornwallis Valley Aquifers in Nova Scotia, with a surface area of 2400 km² in a valley running parallel to the Bay of Fundy.[16] Ground water is of particular importance on Prince Edward Island, which draws nearly all of its water from aquifers.

Diversion and Consumption of Fresh Water

In our attempts to harness fresh water for human purposes, we have achieved impressive engineering accomplishments. In so doing, we have altered many environmental systems, some irreversibly. It is estimated that 60% of the world's largest 227 rivers (and 77% of those in North America and Europe) have been strongly or moderately affected by artificial dams, canals, and **diversions**, that is, the rerouting of water from its natural river channel or drainage basin by means of built structures.

We also are using too much water. Data indicate that at present our consumption of fresh water in much of the world is unsustainable, and we are depleting many sources of surface water and ground water. One-third of the world's people in 44 countries are already affected by water scarcity, with less than 1000 m³ of renewable fresh water available per person per year, according to the World Health Organization and the World Bank.[17]

When we remove water from an aquifer or surfacewater body and do not return it, it is called *consumptive use*. A large portion of agricultural irrigation and of many industrial and residential uses is consumptive. *Nonconsumptive use* of water does not remove, or only temporarily removes, water from the aquifer or surfacewater body. Using water to generate electricity at hydroelectric dams is an example of nonconsumptive use, because the water is taken in, passed through dam machinery to turn turbines, and released downstream.

Diversion of surface water can have unintended impacts

People have long diverted water from rivers, streams, lakes, and ponds for use in agricultural fields, factories, homes, and cities. Diversion can have unintended consequences, particularly in downstream areas.

The Colorado River provides one of the classic examples of diversion and over-allocation of water from

a major river. Upstream, some Colorado River water is piped through a mountain tunnel and down the Rockies' eastern slope to supply the city of Denver. More is removed for Las Vegas and other cities and for farmland as the water proceeds downstream. When it reaches the California–Arizona state line, large amounts are diverted into the Colorado River Aqueduct, which brings water to millions of people in Los Angeles and San Diego. Arizona also draws water, transporting it in the large canals of the Central Arizona Project. Farther south, water is diverted into the Coachella and All-American Canals, to support agriculture in California's Imperial Valley. To make this desert bloom, Imperial Valley farmers soak the soil with subsidized water for which they pay one penny per 795 L.

The water that is left in the Colorado River after all the diversions is just a trickle making its way to the Gulf of California and Mexico. On some days, water does not reach the gulf at all. This reduction in flow (**FIGURE 11.10**) has drastically altered the ecology of the lower river and the once-rich delta, changing plant communities, wiping out populations of fish and invertebrates, and devastating fisheries. It also led to a tense international incident between the United States and Mexico in the 1970s, necessitating high-level international talks and agreements concerning water reallocation, which improved the situation somewhat.

Nowhere are the effects of surfacewater depletion so evident as at the Aral Sea, on the border of present-day Uzbekistan and Kazakhstan. Once the fourth-largest lake on Earth, it lost more than four-fifths of its volume in just 45 years (**FIGURE 11.11**). The former Soviet Union instituted large-scale cotton farming in this region by flooding the dry land with irrigation water from the two

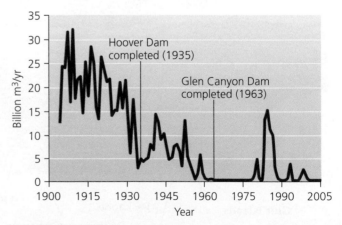

FIGURE 11.10

Flow at the mouth of the Colorado River has greatly decreased over the past century as a result of withdrawals, mostly for agriculture. The river now often runs dry at its mouth.
Source: From *Liquid Assets: The Critical Need to Safeguard Freshwater Ecosystems.* Worldwatch Paper 170. Copyright © 2005 by Worldwatch Institute. Reprinted by permission.

NASA

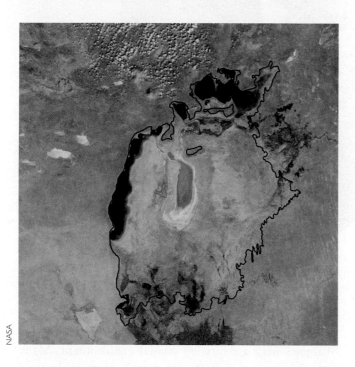

NASA

FIGURE 11.11

Ships lie stranded in the sand **(a)** because the waters of the Aral Sea have receded so far and so quickly as a result of over-withdrawal for irrigation. The Aral Sea was once the world's fourth-largest lake. This satellite image **(b)** from 2013 shows the outline of the Aral Sea in 1960 (black line) for comparison. Today restoration efforts are beginning to reverse the decline in the northern portion of the sea, and waters there are slowly rising.

rivers leading into the Aral Sea. For a few decades this boosted Soviet cotton production, but it caused the Aral Sea to shrink, and the irrigated soil became waterlogged and salinized. Today 60 000 fishing jobs are gone, winds blow pesticide-laden dust up from the dry lake bed, and the regional economy is suffering. However, all may not be lost: Scientists, engineers, and local people have

focused their efforts on saving the northern portion of the Aral Sea, and there are indications that they may have finally begun to reverse the decline in these damaged ecosystems.

We have erected thousands of dams, dikes, and levees

Artificial, engineered modifications of river channels are collectively referred to as **channelization**. Channelization, which includes straightening, deepening, widening, or concrete-lining of a channel, is often associated with other water-control structures like dams, floodwalls, embankments, and other engineered interventions.

A **dam** is an obstruction placed in a river or stream channel to block the flow of water, so that water can be stored in a reservoir. We build dams to prevent floods, provide drinking water, facilitate irrigation, and generate electricity (**FIGURE 11.12**; **TABLE 11.2**). (Power generation with hydroelectric dams is discussed in greater detail in the chapter on energy alternatives.)

Worldwide, there are more than 45 000 large dams (greater than 15 m high) across rivers in more than 140 nations, and tens of thousands of smaller dams. In Canada there are 933 large dams and thousands of small dams.[18] Only a few major rivers in the world remain undammed and free-flowing, in the northern tundra and taiga of Canada, Alaska, and Russia, and remote regions of Latin America and Africa.

Our largest dams are some of humanity's greatest engineering feats. The Gardiner Dam in Saskatchewan (**FIGURE 11.13**) is the largest in Canada in terms of water-holding capacity; the Mica Dam in British Columbia is the tallest. The two behemoths in the United States are the Hoover and Glen Canyon dams. The Hoover Dam, probably the most recognizable icon of the great dam-building period in the United States, holds 35.2 km^3 of water in a reservoir 177 km long and 152 m deep. The Glen Canyon's reservoir is almost as large; together they store four times as much water as flows in the Colorado river in an entire year.

The complex mix of benefits and costs that dams produce is exemplified by the Three Gorges Dam on China's Yangtze River. The dam, 186 m high and 2 km wide, was completed in 2006 to facilitate transportation, flood control, and power generation. Its reservoir is as long as Lake Superior, holding more than 38 trillion litres of water. However, the dam cost $25 billion to build. The filling of its reservoir flooded 22 cities and displaced 1.24 million people, requiring the largest resettlement project in China's history (**FIGURE 11.14A**). The major earthquake in southern China in 2008 raised fears of potential damage; the consequences of a collapse of a dam this large would be devastating.

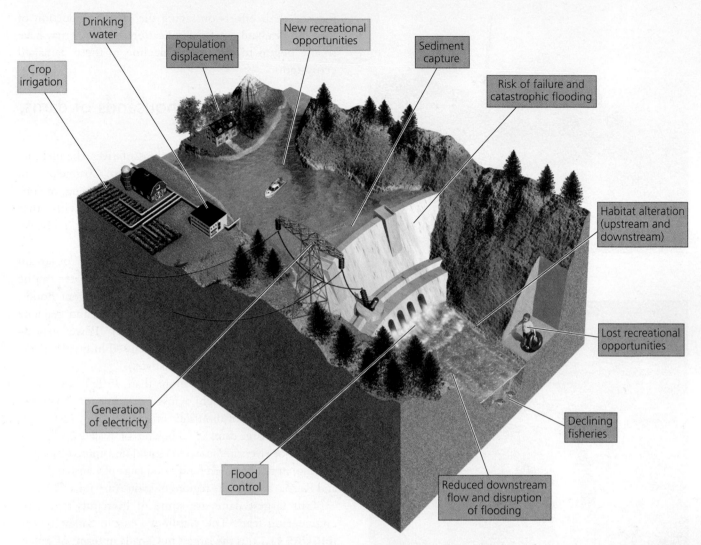

FIGURE 11.12 Damming rivers has diverse consequences for people and the environment. The generation of clean and renewable electricity is one of several major benefits (green boxes); habitat alteration is one of several negative impacts (red boxes).

The filling of the reservoir behind the Three Gorges Dam also submerged archaeological sites, farmlands, and wildlife habitat. The reservoir slows the river's flow, and sediment is settling behind the dam. Because the river downstream is deprived of sediment, the tidal marshes at the Yangtze's mouth are eroding, leaving the city of Shanghai with a degraded coastal environment. Many scientists worry that the Yangtze's many pollutants will also be trapped in the reservoir, making the water undrinkable.

Worldwide, the construction of new dams has slowed. Some dams are even being removed to restore riparian ecosystems, reestablish economically valuable fisheries, and revive recreational activities like fly-fishing and whitewater rafting. Another common reason for the decommissioning of dams is that many aging dams are in need of costly repairs or have outlived their economic usefulness.

Almost 500 dams have been removed in the United States in the past few decades, and more will follow. In Canada only a handful of dams have been decommissioned, but the concept of dam removal for river restoration and ecological recovery is beginning to take hold. Dam removals also provide opportunities for

FIGURE 11.13

At 64 m height and 5000 m length, the Gardiner Dam in Saskatchewan is one of the largest hydroelectric dams in the world. (The largest dam in the world is not a water dam, but the Syncrude tar sand tailings dam in Alberta.)

TABLE 11.2 Major Benefits and Costs of Dams

Benefits

- Power generation: Hydroelectric dams provide inexpensive, clean electricity.

- Emissions reduction: Hydroelectric power produces no greenhouse gases in its operation (although some are produced during construction and maintenance of infrastructure). By replacing fossil fuel combustion as an electricity source, hydropower reduces air pollution and climate change, and their health and environmental consequences.

- Crop irrigation: Reservoirs can release irrigation water when farmers most need it and can buffer regions against drought.

- Drinking water: Many reservoirs store plentiful, reliable, and clean water for municipal drinking water supplies, provided that watershed lands draining into the reservoir are not developed or polluted.

- Flood control: Dams can prevent floods by storing seasonal surges, such as those following snowmelt or heavy rain.

- Shipping: By replacing rocky river beds with deep, placid pools, dams enable ships to transport goods over longer distances.

- New recreational opportunities: People can fish from boats and use personal watercraft on reservoirs in regions where such recreation was not possible before.

Costs

- Habitat alteration: Reservoirs flood riparian habitats and displace or kill riparian species. Dams modify rivers downstream. Shallow warm water downstream from a dam is periodically flushed with cold reservoir water, stressing or killing many fish.

- Fisheries declines: Salmon and other fish that migrate up rivers to spawn encounter dams as a barrier. Although "fish ladders" at many dams allow passage, most fish do not make it.

- Population displacement: Reservoirs generally flood fertile farmland and have flooded many human settlements. An estimated 40–80 million people globally have been displaced by dam projects over the past half century.

- Sediment capture: Sediment settles behind dams. Downstream floodplains and estuaries are no longer nourished, and reservoirs fill with silt.

- Disruption of flooding: Floods create productive farmland by depositing rich sediment. Without flooding, topsoil is lost, and farmland deteriorates.

- Risk of failure: There is always risk that a dam could fail, causing massive property damage, ecological damage, and loss of life.

- Lost recreational opportunities: Tubing, whitewater rafting, fly-fishing, and kayaking opportunities are lost.

scientific study of the response of the river and aquatic communities to changes in flow rate, water temperature, sedimentation, and other factors that accompany decommissioning.

Lest you have the impression that large channelization projects with extensive social and environmental impacts only happen in developing countries, consider the case of the St. Lawrence Seaway (**FIGURE 11.14B**). Although

(a) Displaced person in Sichuan Province, China

(b) Archival photo of the construction of the St. Lawrence Seaway

FIGURE 11.14 Construction of China's Three Gorges Dam caused the displacement of more than a million people. Archaeological treasures were lost forever and whole towns levelled, as shown here in Sichuan Province **(a)**. The construction of the St. Lawrence Seaway **(b)** from Lake Ontario to Montreal, completed in 1959, displaced 6500 people and flooded 14 000 hectares of land, 7 villages, and 225 farms.

it took more than 50 years of discussion, once underway, the St. Lawrence Seaway took only 5 years to complete; it opened for deep-draft navigation in 1959. The seaway is a series of canals that connects the Great Lakes to the Atlantic Ocean, following the route of the St. Lawrence River. Its construction required the displacement of thousands of people from farms and homes along the canal route. Whole villages were flooded, as well as historically important battlefields from the War of 1812. The Lost Villages Historical Society website (**http://lostvillages.ca/**) documents the construction of the St. Lawrence Seaway and the farms and villages that were affected.

In addition to transportation, power generation, and irrigation, another very common purpose for controlling the movement of water is flood control. People have always been attracted to riverbanks for their water supply and for the flat topography and fertile soil of floodplains. Flooding is a normal, natural process caused by snowmelt or heavy rain, and flood waters spread nutrient-rich sediment over large areas, benefiting both natural systems and human agriculture. In the short term, however, floods can do tremendous damage to the farms, homes, and property of people who choose to live in floodplains.

To protect against floods, it is common to construct *dikes* (or floodwalls), floodways, levees (long raised mounds of earth), and reservoirs to store rising flood waters. The flood diversion wall that protects the city of Winnipeg from the Red River is a major example of a flood protection structure (**FIGURE 11.15**); another is the massive

levee system that failed in New Orleans after Hurricane Katrina. Although such structures are intended to prevent flooding, they can sometimes force water to accumulate in channels, building up enormous energy and leading to occasional catastrophic overflow events.

Draining of wetlands is a form of diversion

Throughout recent history, governments have encouraged the draining of wetlands to promote settlement and farming. Many of today's crops grow on the sites of former wetlands that people have drained and filled in. The Prairie Potholes region (**FIGURE 11.16**) of the northern United States and south-central Canada is a highly productive agricultural region, and host of about 4.5 million hectares of wetlands. The vast grasslands and small wetlands of the Potholes region have been widely converted to agricultural production, such that only about half of the wetlands that were present in the late 1700s (prior to European settlement) remain today.[19]

Wetlands have historically been seen as "swamps"—insect-infested, smelly, and useless for any kind of industrial or agricultural development. This attitude began to change dramatically in North America with the rise of the modern environmental movement in the 1970s, and has continued to change as a result of research efforts to determine the economic value of services provided by wetlands.

In 1971, an international agreement was reached in Ramsar, Iran, concerning the documentation and protection of wetlands around the world. It is known as the *Ramsar Convention*, or more accurately, the *Convention on Wetlands*

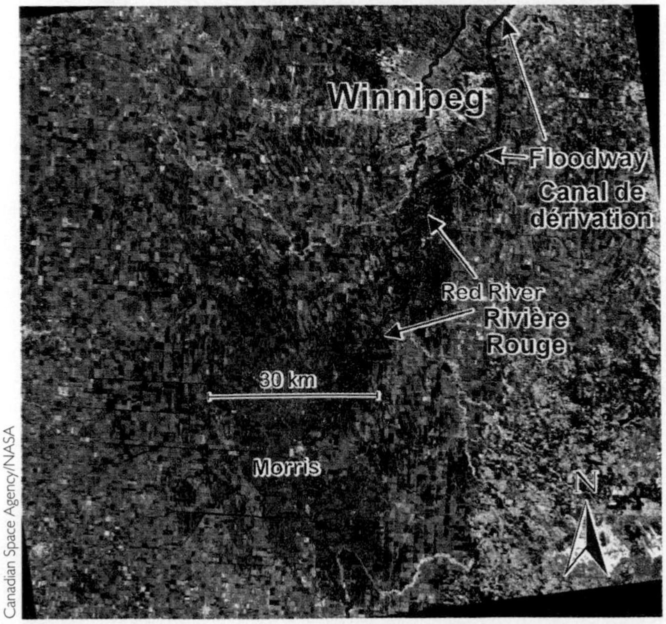

FIGURE 11.15
A flood diversion wall protected the city of Winnipeg during the 1997 Red River flood, but exposed other areas to flooding. In this satellite image the floodway can be seen in the upper right-hand corner, keeping the water from entering the core of Winnipeg.

FIGURE 11.16
Many of North America's wetlands have been drained, and the land converted to agricultural use. Thousands of small water-filled depressions in the Prairie Potholes region have served as nesting sites for waterfowl. Here, farmlands encroach on prairie potholes in North Dakota.

of International Importance, Especially as Waterfowl Habitat. The Ramsar Convention, to which Canada is a signatory, demonstrates the global concern regarding wetland loss and degradation. The mission of the treaty is the "conservation and wise use of all wetlands through local, regional, and national actions and international cooperation."[20] Canada has lost many wetlands but still has the largest wetland area of any country in the world, with 37 sites designated as Ramsar sites of international importance.

Many people now have a different view of wetlands. Rather than seeing them as worthless swamps, science has made it clear that wetlands are valuable ecosystems. This scientific knowledge, along with a preservation ethic, has induced policymakers to develop regulations to safeguard remaining wetlands. Yet, because of loopholes, differing laws, development pressures, and even debate over the legal definition of wetlands, many of these vital ecosystems are still being lost.

Inefficient irrigation wastes water

As we have seen, the agriculture sector is by far the largest consumptive user of water, including both surface water and ground water. The Green Revolution required significant increases in irrigation, and 70% more water is withdrawn for irrigation today than in 1960. During this period, the amount of land under irrigation has doubled.

Expansion of irrigated agriculture has kept pace with population growth; irrigated area per capita has remained stable for at least four decades at around 460 m^2.

Irrigation can more than double crop yields by allowing farmers to apply water when and where it is needed. The world's 274 million hectares of irrigated cropland make up only 18% of world farmland but yield fully 40% of world agricultural produce, including 60% of the global grain crop. Still, most irrigation remains highly inefficient. Only about 45% of the fresh water we use for irrigation actually is taken up by crops. Inefficient "flood and furrow" irrigation, in which fields are liberally flooded with water that may evaporate from standing pools, accounts for 90% of irrigation worldwide. Overirrigation leads to waterlogging and salinization, which affect one-fifth of farmland today and reduce world farming income by $11 billion.

Many national governments subsidize irrigation to promote agricultural self-sufficiency. Unfortunately, inefficient irrigation methods in arid regions are using up huge amounts of ground water for little gain. Worldwide, roughly 15–35% of water withdrawals for irrigation are thought to be unsustainable. In areas where agriculture is demanding more fresh water than can be sustainably supplied, *water mining* is taking place, that is, withdrawing water faster than it can be replenished (**FIGURE 11.17**). In these areas, aquifers are being depleted or surface water is being piped in from other regions.

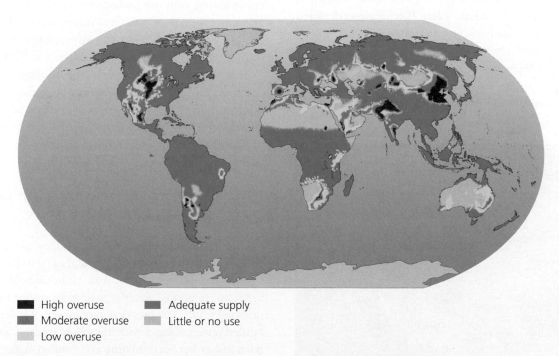

High overuse
Moderate overuse
Low overuse
Adequate supply
Little or no use

FIGURE 11.17 The map shows regions where overall use of fresh water exceeds the available supply, requiring groundwater depletion or diversion of water from elsewhere. The map understates the problem because it does not reflect seasonal shortages.
Source: Data from UNESCO (2006). *Water: A Shared Responsibility. World Water Development Report 2.* Paris and New York: UNESCO and Berghahn Books.

Ground water is easily depleted

Ground water is more easily depleted than surface water because most aquifers recharge very, very slowly. Comparing an aquifer to a bank account, if we are making more withdrawals than deposits, the balance will shrink. Today we are extracting 160 km³ more water each year than is finding its way back into the ground. This is a major problem because one-third of Earth's human population—including 26% of the population of Canada—relies on ground water for its needs. Since ground water is hidden from view under the surface, specific information about remaining reserves is often lacking. As aquifers are depleted, water tables drop. Ground water becomes more difficult and expensive to extract, and eventually it may run out.

When ground water is overpumped in coastal areas, salt water can intrude into aquifers, making water undrinkable. This has occurred widely in the Middle East and in localities as varied as Florida, Turkey, and Bangkok. As aquifers lose water, their substrate can become weaker and less capable of supporting overlying strata, and the land surface above may subside. Mexico City's downtown has sunk over 10 m since the time of Spanish arrival; streets are buckled, old buildings lean at angles, and underground pipes break so often that 30% of the system's water is lost to leaks.

Sometimes land subsides suddenly in the form of **sinkholes**, areas where the ground gives way with little warning, occasionally even swallowing homes, roads, or cars (**FIGURE 11.18**). Once the ground subsides, soil can undergo *compaction*, becoming compressed and losing the porosity that enabled it to hold water. Recharging a depleted aquifer may thereafter become much more difficult.

Falling water tables also do vast ecological harm. Permanent wetlands exist where water tables are high enough to reach the surface, so when water tables drop, surface wetland systems dry up. In Jordan, the Azraq Oasis covered

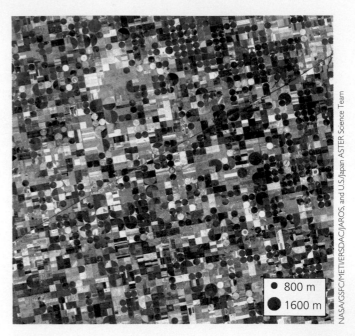

FIGURE 11.19
The Ogallala Aquifer underlies much of the agricultural land of the American Great Plains. In this satellite image, water from the aquifer feeds centre-pivot irrigation systems in Kansas, resulting in circular fields of corn, wheat, and sorghum. The fields are 800 and 1600 metres in diameter.

● 800 m
● 1600 m

NASA/GSFC/METI/ERSDAC/JAROS, and U.S./Japan ASTER Science Team

7500 ha and enabled migratory birds and other animals to find water in the desert. The water table beneath this oasis dropped 2.5–7 m during the 1980s because of increased well use by the city of Amman. As a result, the oasis dried up altogether during the 1990s. Today international donors are collaborating with the Jordanian government to try to find alternative sources of water and restore this oasis.

The great Ogallala Aquifer, which supports agriculture throughout most of the American Great Plains, has had its water-saturated volume depleted by about 9% over the past few decades. Some areas have experienced declines of almost 85 m in the level of the water table, with an average overall decline of almost 4 m.[21] These changes are primarily the result of significant increases in mechanized pumping for irrigation (**FIGURE 11.19**).

Our thirst for bottled water seems unquenchable

It seems to fit our busy lifestyle. Canadians' per capita use of bottled water is surpassed only by that of Americans. Statistics Canada reports that almost 3 in 10 households in Canada today utilize bottled water as their main source for domestic drinking water. The proportion of households dependent on bottled water increases with household income, although the relationship between income and bottled water consumption is complex. Interestingly, in a 2008 study by Statistics Canada, university-educated households were shown to be *less* likely to consume bottled water.[22]

AP/The Canadian Press

FIGURE 11.18
When too much ground water is withdrawn, especially in areas underlain by soluble rocks such as limestone, the land above it may collapse in a sinkhole.

Some 820 million litres of water were bottled for Canadian consumption in 2000; by 2003 the amount had increased to almost 1.5 billion litres. This means that the average per capita consumption of 17.9 L of bottled water per year in 1995 had risen to 27.6 L by 2000, and then jumped to almost 50 L by 2003![23] From an environmental perspective, it takes a considerable amount of energy to produce bottled water, and there are concerns about the sustainability of groundwater withdrawal and lack of protection of the source water.

An interesting thing about bottled water is that much of it is just ordinary tap water—in fact, it has been estimated that as much as a quarter of all bottled water comes straight from a municipal water tap, sometimes with additional filtering or other treatment. Advertising and labelling legislation prohibits bottling companies from misrepresenting what is in the bottle, of course, so if you read the label carefully you should be able to determine whether the water has come from a "natural" source, such as ground water or a spring, or from a municipal supply.

But is bottled water from a natural source necessarily more healthful, or safer? Although most people think so, in fact there are fewer checks on bottled water and the bottling process than on municipal water supplies, which are rigorously monitored and regulated. Canada's *Food and Drugs Act* does not require a manufacturer to obtain a licence in order to bottle water (although the water itself is expected to meet the requirements established for food products).

In 2008, a study of 10 major brands of bottled water detected 38 chemical pollutants, including traces of heavy metals, radioactive isotopes, caffeine and pharmaceuticals from wastewater pollution, nitrate and ammonia from fertilizer, and various industrial compounds such as solvents and plasticizers (**FIGURE 11.20**). Each brand in the study contained eight contaminants on average.[24]

Concerns also have been raised about chemicals that might leach into the water from plastic bottles. In 2009, German researchers Martin Wagner and Jörg Oehlmann tested bottled water for the presence of hormone-disrupting chemicals that mimic estrogen. Endocrine disruptors such as bisphenol A, phthalates, and other compounds found in plastics can exert a wide array of health impacts, even at very low doses (see Chapter 19, *Environmental Health and Hazards*). The researchers analyzed nine brands packaged in glass bottles, nine brands packaged in plastic bottles, and two brands packaged in "Tetra Pak" paperboard boxes with an inner plastic coating. Estrogenic contamination was detected in 60% of the samples. Both Tetra Pak brands and seven of nine plastic brands contained hormone-mimicking substances that apparently leached from the packaging. So did three of the glass-bottled brands, presumably from contamination at the bottling plant.[25]

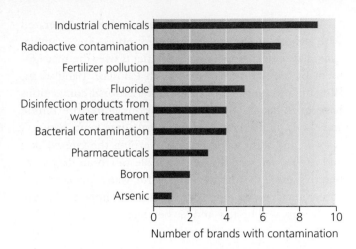

FIGURE 11.20
Of 10 leading brands of bottled water tested, most contained industrial chemicals, radioactive isotopes, and fertilizer pollution, as well as other contaminants.
Source: Data from Naidenko, O., et al. (2008). Bottled Water Quality Investigation: 10 Major Brands, 38 Pollutants. Environmental Working Group, Washington, D.C.

The next time you reach for a bottle of water, consider what's in the bottle, where it came from, how it was extracted and monitored, what contaminants it might to contain, and exactly what you are paying for. Then decide whether you think it is worth generating an empty plastic bottle—keeping in mind that about three out of four of those bottles are typically thrown away, rather than recycled. Many colleges and universities in Canada, after reflecting on these impacts, have decided to ban the sale of bottled water on their campuses.

Dealing with Depletion of Fresh Water

Population growth, expansion of irrigated agriculture, and industrial development doubled our annual freshwater use in the last 50 years. We now use an amount equal to 10% of total global runoff. The hydrologic cycle makes fresh water a renewable resource, but if our usage exceeds what a lake, a river, or an aquifer can provide, we must reduce our use, find another water source, or be prepared to run out of water.

Scarcity of vital resources like water can lead to conflicts. Many predict that water's role in regional conflicts will increase as human population continues to grow in water-poor areas, and as climate change alters regional patterns of precipitation. A total of 261 major rivers, whose watersheds cover 45% of the world's land area, are *transboundary* waterways—that is, they cross or flow along national borders, and disagreements are common. Water is already a key element in the hostilities among Israel, the Palestinian people, and neighbouring nations.

On the positive side, many nations have cooperated with neighbours to resolve water disputes. For example, India has struck cooperative agreements over the management of transboundary rivers with Pakistan, Bangladesh, Bhutan, and Nepal. In Europe, international conventions have been signed by multiple nations along the Rhine and the Danube rivers. International agreements between Canada and the United States governing the Great Lakes and other water bodies that straddle our boundary have been examples of largely successful water management agreements. Such progress gives reason to hope that future water wars will be few and far between.

Solutions can address supply or demand

To address depletion of fresh water, we can aim either to increase supply or reduce demand by stressing conservation and efficiency measures. Lowering demand is more difficult politically in the short term but may be necessary in the long term. In the developing world, international aid agencies are increasingly funding demand-based solutions because they offer better economic returns and cause less ecological and social damage than supply-based solutions.

To increase supply in a given area, people transport water through pipes and aqueducts from areas where it is more plentiful or accessible. In many instances, water-poor regions have forcibly appropriated water from other communities. For instance, Los Angeles grew by using water imported from less-inhabited regions of California. This caused desertification in the environments of those areas, creating dust bowls and destroying rural economies. In 1941, Los Angeles needed water and diverted streams that fed Mono Lake, more than 565 km away in northern California. The lake level fell 14 m over 40 years, salt concentrations doubled, and aquatic communities suffered. Other desert cities in the American Southwest—such as Las Vegas, Phoenix, and Denver—are expected to double in population in the coming decades. Las Vegas is planning a 450-km pipeline to import ground water from sparsely populated eastern Nevada, where local residents and wildlife advocates oppose the diversion plan.

Desalination "makes" more fresh water

Another supply-side strategy is to develop technologies to "make" more fresh water. The best-known technological approach to generate fresh water is **desaliniza-**

FIGURE 11.21

Jubail Desalinization Plant in Saudi Arabia is the largest facility in the world that turns salt water into fresh water. Piped-in sea water undergoes treatments to remove algae and suspended particles. It is desalinated by reverse osmosis filtration, and the resulting water is stored in a reservoir. The plant, which cost about US$3.8 billion to build, will supply more than 900 000 m³ of water per day, in addition to generating electricity.

tion, or *desalination*, the removal of salt from sea water or other water of marginal quality. One method of desalination mimics the hydrologic cycle by hastening evaporation from allotments of ocean water with heat, and then condensing the vapour—essentially distilling fresh water. Another method involves forcing water through membranes to filter out salts; the most common process of this type is called *reverse osmosis*.

More than 7500 desalination facilities are operating worldwide, most in the arid Middle East and some in small island nations that lack ground water. The largest plant, in Saudi Arabia, is designed to produce 900 million litres of fresh water every day (**FIGURE 11.21**). However, desalination is expensive, requires large inputs of fossil fuels, and generates concentrated salty waste.

We can reduce agricultural, residential, and industrial water use

Because most water is used for agriculture, it makes sense to look first to agriculture for ways to decrease demand. Farmers can improve efficiency by lining irrigation canals to prevent leaks, levelling fields to minimize runoff, and adopting efficient irrigation methods.

Low-pressure spray irrigation directs water downward toward plants, and drip irrigation systems target individual plants and introduce water directly onto the soil. Both methods reduce water lost to evaporation and surface runoff. Low-pressure precision sprinklers in use in a number of arid localities have efficiencies of 80–95% and have resulted in water savings of 25–37%. Experts have estimated that drip irrigation, which has efficiencies as high as 90%, could cut water use in half while raising yields by 20–90% and giving developing-world farmers $3 billion in extra annual income.

Choosing crops to match the land and climate in which they are being farmed can save huge amounts of water. Currently, crops that require a great deal of water, such as cotton, rice, and alfalfa, are often planted in arid areas with government-subsidized irrigation. As a result of the subsidies, the true cost of water is not part of the costs of growing the crop. Eliminating subsidies and growing crops in climates with adequate rainfall could greatly reduce water use. Finally, selective breeding and genetic modification can result in crop varieties that require less water.

Personal choices are also important. We can each help reduce agricultural water use by decreasing the amount of meat we eat because producing meat requires far greater water inputs than producing grain or vegetables. In households, we can reduce water use by installing low-flow faucets, showerheads, washing machines, and toilets. Automatic dishwashers, a European study showed, can use less water than does washing dishes by hand.[26] If your home has a lawn, it is best to water it at night, when water loss from evaporation is minimal. Better yet, you can replace a water-intensive lawn with native plants adapted to the region's natural precipitation patterns. An example is *xeriscaping*—landscaping with plants that are adapted to a dry environment.

Many manufacturers are shifting to processes that use less water and in doing so are reducing their economic costs. In 2011 Labatt Breweries of Canada received a Water Efficiency Award from the Ontario Water Works Association, the sector association for drinking-water professionals, for halving the amount of water it uses to make beer. Beer-making is surprisingly water-intensive; before initiating its water conservation program, Labatt was using the equivalent of seven bottles of water for every one bottle of beer produced. One of the big water savings realized by Labatt came when they started to use the final rinse water from one cycle of bottle-washing for the initial rinse in the next cycle. Through this one change, the company was able to save 86 million litres of water per year.

Many efficiencies can be achieved by municipalities, too. Some cities are recycling municipal waste water for irrigation and industrial uses, or capturing excess surface runoff during their rainy seasons and pumping it into aquifers. Finding and patching leaks in pipes have saved some cities large amounts of water—and money. A program to repair and retrofit pipes enabled Massachusetts to reduce water demand by 31%, and allowed them to avoid an unpopular $500 million river diversion scheme.

Economic approaches to water conservation are being debated

Economists who want to use market-based strategies to achieve sustainable water use have suggested ending government subsidies of inefficient practices and letting water become a commodity whose price reflects the true costs of its extraction. Others worry that making water a fully priced commodity would make it less available to the world's poor and increase the gap between rich and poor. Because industrial use of water can be 70 times as profitable as agricultural use, market forces alone might favour uses that would benefit wealthy and industrialized people, companies, and nations at the expense of the poor and less industrialized.

Similar concerns surround another potential solution, the privatization of water supplies. In the past two decades, many public water systems were partially or wholly privatized, with their construction, maintenance, management, or ownership being transferred to private companies. This was done in the hope of increasing the systems' efficiency, but many firms have little incentive to allow equitable access to water for rich and poor alike. Already in some developing countries, rural residents without access to public water supplies, who are forced to buy water from private vendors, end up paying on average 12 times more than those connected to public supplies.

Other experiences indicate that decentralization of control over water, from the national level to the local level, may help conserve water. In Mexico, the effectiveness of irrigation systems improved dramatically once they were transferred from public ownership to the control of 386 local water user associations.

One new economic approach focuses on the *virtual water* trade. This is an extension of the concept of *product lifecycle analysis*, in which all of the environmental inputs into a product, from beginning to end, are taken into consideration. The idea of virtual water is that the water required to produce crops or manufacture products is counted as an integral aspect of production. If a water-intensive product or crop is exported, the water involved in production (the portion that isn't recycled) has also effectively been exported. For example, Israel, an arid nation, discourages the exportation of oranges, a water-intensive crop. In this way, Israel keeps the water involved in the production of oranges within its national boundaries. Similarly, if a water-scarce nation imports a water-intensive crop or product, they are effectively importing the water used in its production. The virtual water trade is more conceptual than practical; implementation (such as in the pricing of commodities) is complicated because of legal, economic, political, and environmental aspects that cut across national boundaries.

Regardless of how demand is addressed, the ongoing shift from supply-side to demand-side solutions is beginning to pay dividends. The new focus on demand (through government mandates and public education) has decreased public water consumption in many municipalities, and industries are becoming more water-efficient.

Freshwater Pollution: Types and Sources

The quantity and distribution of fresh water poses one set of environmental and social challenges; safeguarding the *quality* of water involves another collection of environmental and human health dilemmas. To be safe for consumption by human beings and other organisms, water must be relatively free of disease-causing organisms and toxic substances.

Although developed nations have made admirable advances in cleaning up water pollution over the past few decades, the World Commission on Water for the 21st Century concluded that more than half the world's major rivers are "seriously depleted and polluted, degrading and poisoning the surrounding ecosystems, threatening the health and livelihood of people who depend on them." The largely invisible pollution of ground water, meanwhile, has been termed a "covert crisis."

Water pollution takes many forms

The term **pollution** describes the release into the environment of matter or energy that causes undesirable impacts on the health and well-being of humans or other organisms. Pollution can be physical, chemical, or biological and can affect water, air, or soil.

Water pollution comes in many forms and can cause diverse impacts on aquatic ecosystems and human health. Some water pollution is emitted from **point sources**—

discrete locations, such as a factory outflow or sewer pipe. Water pollution also can arise from **nonpoint sources**, multiple cumulative inputs over larger areas, such as farms, city streets, and residential neighbourhoods (**FIGURE 11.22**). Many common activities give rise to nonpoint-source water pollution, such as applying fertilizers and pesticides to lawns, applying salt to roads in winter, and changing automobile oil. To minimize nonpoint-source pollution of drinking water, governments can limit development on watershed land in riparian areas and surrounding reservoirs.

We can categorize water pollution into several types, including nutrient pollution, biological pollution by disease-causing organisms, toxic chemical pollution, physical pollution by sediment, and thermal pollution.

Nutrient pollution   You learned in previous chapters how nutrient pollution from fertilizers and other sources can lead to **eutrophication** and **hypoxia** in coastal marine areas. Eutrophication proceeds in a similar fashion in freshwater systems, where phosphorus is usually a limiting factor, so its presence spurs growth. When excess phosphorus enters surface waters, it fertilizes algae and aquatic plants, boosting their growth rates and populations. Although such growth provides oxygen and food for other organisms, algae can cover the water's surface, depriving deeper-water plants of sunlight; this harmful situation is called an *algal bloom*. As the algae die off, they provide food for decomposing bacteria. Decomposition requires oxygen, so the increased bacterial activity drives down levels of dissolved oxygen. These levels can drop too low to support fish and shellfish, leading to dramatic changes and the development of hypoxic ("dead") zones in aquatic ecosystems.

Eutrophication (**FIGURE 11.23**) is a natural process, and part of the process whereby lakes can change into wetlands and eventually into dry land. However, excess nutrient inputs from agricultural runoff, golf courses, lawns, and sewage can dramatically increase the rate at which it occurs. We can reduce nutrient pollution by treating waste water, reducing fertilizer applications, planting vegetation to increase nutrient uptake, and purchasing phosphate-free detergents.

Probably the most famous (or infamous) case of eutrophication occurred in the 1970s, when Lake Erie turned a sickly green colour because of extensive algal blooms, and was declared to be "dead." The lake recovered in the 1980s as a result of a binational agreement on lake management between Canada and the United States, which involved stricter nutrient management and runoff regulations. However, the problem resurfaced in the mid-1990s and since then the lake experienced some of the worst algal blooms in recorded history.

Pathogens and waterborne diseases   Disease-causing organisms (pathogenic viruses, protists, and bacteria)

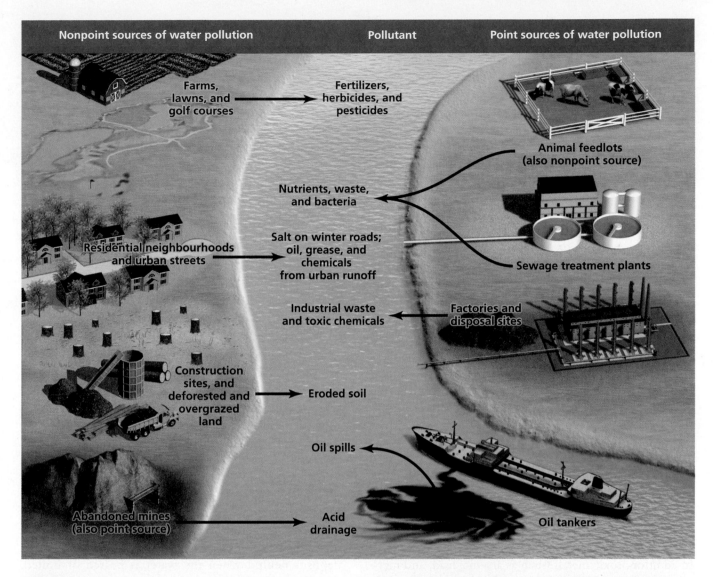

Nonpoint sources of water pollution

Pollutant

Point sources of water pollution

Farms, lawns, and golf courses → Fertilizers, herbicides, and pesticides

Animal feedlots (also nonpoint source)

Nutrients, waste, and bacteria

Sewage treatment plants

Residential neighbourhoods and urban streets → Salt on winter roads; oil, grease, and chemicals from urban runoff

Industrial waste and toxic chemicals ← Factories and disposal sites

Construction sites, and deforested and overgrazed land → Eroded soil

Oil spills

Abandoned mines (also point source) → Acid drainage

Oil tankers

FIGURE 11.22 Point-source pollution comes from discrete facilities or locations, usually from single outflow pipes. Nonpoint-source pollution (such as runoff from streets, residential neighbourhoods, lawns, and farms) originates from numerous sources spread over large areas.

(a) Oligotrophic water body

(b) Eutrophic water body

FIGURE 11.23 An oligotrophic water body **(a)** with clear water and low nutrient content may eventually become a eutrophic water body **(b)** with abundant algae and high nutrient content. Pollution of freshwater bodies by excess nutrients accelerates the process of eutrophication.

can enter drinking water supplies when these are contaminated with human waste from inadequately treated sewage or with animal waste from feedlots. Specialists monitoring water quality can tell when water has been contaminated by waste when they detect fecal coliform bacteria, which live in the intestinal tracts of people and other vertebrates. These bacteria serve as indicators of fecal contamination, which may mean that the water holds other pathogens that can cause serious ailments, such as giardiasis, typhoid, or hepatitis A.

Worldwide, biological pollution by pathogens causes more human health problems than any other type of water pollution. According to the World Health Organization, despite advances in many parts of the world, major problems still exist. On the positive side, 6.1 billion people (89% of the population) now have access to safe water as a result of some form of improvement in their water supply—an increase from 4.1 billion (79% of the population) in 1990.

Treating sewage is one approach to reducing the risks that waterborne pathogens pose and maintaining a safe separation between sewage and drinking water supplies. Another is using chemical or other means to disinfect drinking water. Others include government enforcement of regulations on agricultural runoff and monitoring the quality of municipal and rural water supplies.

Toxic chemicals   Our waterways have become polluted with toxic organic substances of anthropogenic origins, including pesticides, petroleum products, and other synthetic chemicals. Many of these can poison animals and plants, alter aquatic ecosystems, and cause a wide array of human health problems, including cancer. In addition, toxic metals (such as arsenic, lead, and mercury) and acids (from acid precipitation and acid drainage from mine sites) cause negative impacts on human health and ecosystems. Legislating and enforcing more stringent regulations of industry can help reduce releases of toxic inorganic chemicals. Better yet, we can modify our industrial processes and our purchasing decisions to rely less on these substances.

Suspended matter   Although floods build fertile farmland, sediment and other suspended matter that are transported by rivers can also impair aquatic ecosystems. Mining, clear-cutting, land clearing for housing development, and careless cultivation of farm fields all expose soil to wind and water erosion. Some water bodies, such as China's Huang He (Yellow River), are naturally sediment-rich, but many others are not. When a clear-water river receives a heavy influx of eroded sediment, aquatic habitat can change dramatically, and fish adapted to clear-water environments may not be able to adjust. We can reduce sediment pollution by better managing farms and forests and avoiding large-scale disturbance of vegetation.

Thermal pollution   Water's ability to hold dissolved oxygen decreases as temperature rises, so some aquatic organisms may not survive when human activities raise water temperatures. When we withdraw water from a river and use it to cool an industrial facility, we transfer heat energy from the facility back into the river where the water is returned. People also raise surfacewater temperatures by removing streamside vegetation that shades water.

Too little heat can also cause problems. Water at the bottoms of reservoirs is colder than water at the surface. When dam operators release water from the depths of a reservoir, downstream water temperatures drop suddenly. These low water temperatures may favour cold-loving invasive species over endangered native species.

Scientists use several indicators of water quality

Most forms of water pollution are not visible to the human eye, so scientists and technicians measure certain physical, chemical, and biological properties of water to characterize **water quality**. Biological properties include the presence of fecal coliform bacteria and other disease-causing organisms, as already discussed. Algae and aquatic invertebrates are also commonly used as biological indicators of water quality.

Chemical properties include nutrient concentration, pH, taste, odour, and hardness. "Hard" water contains naturally high concentrations of calcium and magnesium ions, which prevent soap from lathering and leave chalky deposits behind when the water is heated or boiled. Another important chemical characteristic is dissolved oxygen content. Dissolved oxygen is a powerful indicator of ecosystem health, because surface waters that are low in dissolved oxygen are less capable of supporting aquatic life.

Among physical characteristics, *turbidity* refers to the amount of suspended particles in a water sample. If scientists can measure only one parameter, they will often choose turbidity because it tends to correlate with many others, and is therefore a good indicator of overall water quality. Fast-moving rivers that cut through arid or easily eroded landscapes, like the Yellow River in China, carry a heavy load of sediment and are turbid and muddy-looking as a result. The colour of water can reveal specific substances present in a sample; for example, some forest streams are the colour of tea because of chemicals called tannins that occur naturally in decomposing leaf litter.

Finally, temperature can be used as a simple indicator of water quality. High temperatures can interfere with some biological processes. Warmer water also holds less dissolved oxygen.

Groundwater pollution is difficult to detect

Most efforts at pollution control have focused on surface water. Yet increasingly, groundwater sources once assumed to be pristine have been contaminated by pollution from industrial and agricultural practices. Groundwater pollution is largely hidden from view and is extremely difficult to monitor (**FIGURE 11.24A**); it can be out-of-sight, out-of-mind for decades until widespread contamination of drinking supplies is discovered.

Rivers flush their pollutants fairly quickly, but ground water generally retains contaminants until they decompose. In the case of persistent pollutants, that can be many years or decades. The long-lived pesticide DDT, for instance, is still found widely in aquifers in North America, even though it was banned more than 40 years ago. Chemicals break down much more slowly in aquifers than in surface water or soils. Ground water generally contains less dissolved oxygen, microbes, minerals, and organic matter, so decomposition is slower. For example, concentrations of the herbicide alachlor decline by half after 20 days in soil, but in ground water this can take 4 years.

There are many sources of groundwater pollutants, including natural sources. Some chemicals that are toxic at high concentrations, including aluminum, fluoride, nitrates, and sulphates, occur naturally in ground water. After all, water that resides in aquifers is surrounded by rocks—natural sources for many chemical compounds—for hundreds to tens of thousands of years. During that time of close contact many compounds are leached from the rocks into the ground water (including calcium and magnesium that lead to water "hardness").

Some of these chemicals, even ones that are naturally occurring, are harmful. In the 1980s, patients in West Bengal, India, and parts of Bangladesh began to exhibit symptoms of arsenic poisoning. After exhaustive testing and research, it was discovered that as many as 2.5 million wells in Bangladesh are contaminated with arsenic. The arsenic is of natural origin, leached from rocks and soils; however, many scientists believe that human activity, probably related to agricultural runoff and irrigation, accelerated the process of arsenic leaching into ground water.

Although there are natural sources for some contaminants, there is no escaping the fact that groundwater pollution from human activity is widespread. Industrial, agricultural, and urban wastes—from heavy metals to petroleum products to industrial solvents to pesticides—can leach through soil and seep into aquifers. Pathogens and other pollutants can enter ground water through improperly designed wells and from the pumping of liquid hazardous waste below ground.

Leakage from underground septic tanks, tanks of industrial chemicals, and tanks of oil and gas also pollutes ground water (**FIGURE 11.24B**). According to Environment Canada, without adequate corrosion protection, more than half of underground gasoline storage tanks can be expected to begin leaking by the time they are 15 years old, and just 1 L of gasoline can contaminate up to 1 million L of ground water.[27] It is crucial to intercept carcinogenic or otherwise toxic pollutants, such as chlorinated solvents and gasoline, before they reach aquifers. Once an aquifer is contaminated with a toxin, it is extremely difficult to remediate.

Agriculture also contributes to groundwater pollution. Nitrate from fertilizers has leached into ground water in agricultural areas throughout Canada and in 49 of the 50 U.S. states. Nitrate in drinking water has been linked to cancers, miscarriages, and "blue baby syndrome"

(a) Monitoring ground water

(b) Leaking underground storage tanks

FIGURE 11.24

Groundwater supplies must be closely monitored **(a)**, especially in areas where the potential for contamination is high. It is much easier to prevent contamination of ground water than to remediate after contamination has occurred. A common source of groundwater contamination is leaky underground gasoline storage tanks **(b)**.

THE SCIENCE BEHIND THE STORY

PART ONE

REPORT OF THE WALKERTON INQUIRY

The Events of May 2000 and Related Issues

The Honourable Dennis R. O'Connor

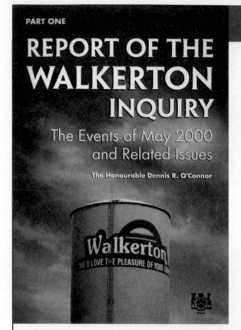

The Report on the Walkerton Inquiry, 2002, by Justice Dennis O'Connor, investigated both the events surrounding the Walkerton tragedy and broader issues in Ontario's approach to water quality management.
Frank Gunn/The Canadian Press

When Water Turns Deadly: The Walkerton Tragedy[28]

Until May 2000, there was little to distinguish Walkerton from dozens of small towns in southern Ontario.[29]

That's how Justice Dennis O'Connor started his 2002 report entitled *Report of the Walkerton Inquiry: The Events of May 2000 and Related Issues.* He continued with a matter-of-fact summary of events:

In May 2000, Walkerton's drinking water system became contaminated with deadly bacteria, primarily Escherichia coli O157:H7. Seven people died, and more than 2,300 became ill. The community was devastated. The losses were enormous. There were widespread feelings of frustration, anger, and insecurity.[30]

In response to these events, the Ontario government ordered the public inquiry, which found that contaminants entered Walkerton's municipal water wells around May 12, via any of a number of pathways. The transport of contaminants and their eventual entry into three wells were probably facilitated by heavy rains. The main source for the contaminants was animal wastes from livestock on a nearby farm. The farmer had followed appropriate procedures, the inquiry concluded, and was not at fault.

The contaminants were *Escherichia coli* and, less importantly, *Campylobacter jejuni.* Both are short, curved, rod-shaped bacteria (see photo) that occur commonly in the feces of birds, bats, and other warm-blooded animals. Symptoms of infection in humans include bloody diarrhea and stomach pain. Some strains of *E. coli,* including the O157:H7 strain found at Walkerton, can be deadly. *C. jejuni* also has been linked to the development of a debilitating and sometimes fatal condition, Guillain-Barré syndrome. These are seriously pathogenic organisms.

Events began to unfold in Walkerton about six days after the initial contamination. Twenty children missed school on May 18; two were admitted to hospital. Multiple cases of intestinal illness prompted the medical officer of health, Dr. Murray McQuigge, to inquire about the safety of the water supply. Three times he was assured of its safety by Stan Koebel, manager of the Walkerton Public Utilities Commission (PUC). On May 21, Dr. McQuigge issued a boil-water advisory, in spite of the PUC's assurances.

The inquiry later concluded that Ontario's procedures for water safety testing were faulty and inadequate, and that PUC operators were insufficiently trained, had not done the daily testing they were supposed to do, and had falsified data entries to give the impression that testing had taken place.

The inquiry concluded that chlorine residual testing and turbidity monitoring would have allowed for the detection and resolution of the problem. Chlorine residual testing is a simple approach that indicates whether sufficient chlorine has been added to the water to inactivate all the bacteria. If no "leftover" or residual chlorine remains in the water, then all of the added chlorine was used up and it is possible that some bacteria remain active. *Turbidity* refers to lack of clarity in the water; sometimes this is caused by suspended sediment, but a concentration of microorganisms can also cause turbidity. If chlorine residual testing and turbidity monitoring had been done on a daily basis, the problems would have been detected earlier and the outbreak could have been contained.

If a problem had been detected, further testing would have been triggered to identify the cause. This would have included technologies that detect an enzyme called beta-glucuronidase, which is produced by almost all forms of *E. coli* and very few other bacteria.

The sample is placed in a medium that changes colour or fluoresces if it is exposed to the enzyme. Sometimes it is necessary to culture the bacteria for a few days to have a sufficient concentration in the sample.

Most tests reveal the presence of *E. coli,* but neither the exact amount nor the strain. Professor Ulrich Krull and colleagues at the University of Toronto Mississauga have developed field detection technologies based on DNA. These *biosensors* contain short sequences of single-stranded DNA or *ssDNA,* which link or *hybridize* with complementary sequences of ssDNA in the sample, if they are present.[31] An indicator, such as a fluorescing gel, shows that hybridization has occurred. Biosensors are useful because they instantly reveal not only the presence of the bacteria, but also the specific strain.

In his report, Justice O'Connor brought forward some pressing concerns about Ontario's approach to drinking water. Health Canada sets guidelines for drinking water quality; the provinces and territories have the responsibility to meet these guidelines and ensure drinking water safety. In some provinces this responsibility falls to the health ministry, in others the environment ministry. In 2002 Ontario passed the *Safe Drinking Water Act* to address the concerns raised by Justice O'Connor through the Walkerton Inquiry.

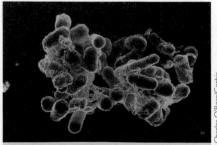

The contaminants found in the municipal water supply in Walkerton in 2000 included the bacteria *C. jejuni* and the deadly O157:H7 strain of *E. coli,* shown here.

Charles O'Rear/Corbis

(methemoglobinemia), which reduces the oxygen-carrying capacity of the blood. Agriculture can also contribute pathogens, primarily originating from animal wastes. In 2000, the groundwater supply of Walkerton, Ontario, became contaminated with the bacterium *Escherichia coli*, or *E. coli*. Two thousand people became ill, and seven died (see "The Science behind the Story: When Water Turns Deadly: The Walkerton Tragedy").

Mitigating and Remediating Water Pollution

As numerous as our freshwater pollution problems may seem, it is important to remember that some of them were even worse a few decades ago, when Lake Erie was declared officially "dead" and hypoxic, covered with sickly green algal mats.

Legislative and regulatory efforts can help reduce pollution

Citizen activism and government response during the 1960s and 1970s resulted in fundamental changes in environmental practices and legislation that made it illegal to discharge pollution from a point source without a permit, set standards for industrial waste water and for contaminant levels in surface waters, and funded construction of sewage treatment plants. In Canada most such legislation is enacted and enforced at the provincial level, although the federal government sets environmental guidelines through the *Canadian Environmental Protection Act* and other federal legislation, and regulates any interprovincial transfers of hazardous materials.

For many years, the most powerful law that served to protect water quality in Canada was the federal-level *Fisheries Act*. The law originally stated that, with a few exemptions, "No person shall carry on any work or undertaking that results in the harmful alteration, disruption or destruction of fish habitat." This progressive, ecosystem-based wording served to protect fish habitat—and thus to protect water quality and aquatic ecosystems generally in Canada.

However, in 2012 the Government of Canada amended the wording of the act to read, in part, that "No person shall carry on any work, undertaking or activity that results in serious harm to fish that are part of a commercial, recreational or Aboriginal fishery, or to fish that support such a fishery." Note that the act now refers to "serious harm" (rather than "harm"), that habitat is no

longer explicitly protected, and that the protected fish must be part of a fishery.

Many scientists were (and are) deeply troubled by these changes, which expose many of Canada's formerly protected aquatic systems to potential damage from harmful activities. A letter to Prime Minister Harper from more than 600 concerned scientists, including many of Canada's most senior ecologists and aquatic scientists, expressed this concern by stating, in part, "We believe that the weakening of habitat protections in Section 35 of the Fisheries Act will negatively impact water quality and fisheries across the country, and could undermine Canada's attempt to maintain international credibility in the environment . . . We should be strengthening, not weakening the habitat protection provisions of the Fisheries Act (and other environmental laws, including the Species at Risk Act and the Migratory Birds Convention Act), in order to protect our dwindling fisheries and species at risk."[32]

The Great Lakes are something of a success story in transnational water management. Much of the work of managing the lakes has been carried out through the International Joint Commission (IJC), the *Great Lakes Water Quality Agreement*, and the *International Boundary Waters Treaty Act*. In the 1970s the Great Lakes, which hold 18% of the world's surface fresh water, were badly polluted with waste water, fertilizers, and toxic chemicals. Coordinated efforts of the Canadian and U.S. governments have paid off. Releases of toxic chemicals are down, phosphorous runoff has decreased, and bird populations are rebounding.

The Great Lakes' troubles are by no means over—sediment pollution is still heavy, PCBs and mercury still settle on the lakes from the air, and fish are not always safe to eat. Recently, Lake Erie has begun to experience episodes of eutrophication rivalling those that "killed" the lake in the 1970s. However, historically we have seen that conditions can improve when citizens push their governments to take action.

Other developed nations have also reduced water pollution. In Japan, Singapore, China, and South Korea, legislation, regulation, enforcement, and investment in wastewater treatment have brought striking water quality improvements. However, nonpoint-source pollution, eutrophication, and acid precipitation remain major challenges globally.

Safe drinking water is a Millennium Development Goal

Technological advances have improved our ability to control pollution and bring safer drinking water to people in all parts of the world. One of the Millennium Development Goals (MDG #7) was to halve, by 2015, the number of people who lack access to safe drinking water. That goal was achieved ahead of the target date. The World Health Organization and UNICEF estimate that approximately 6.1 billion people, or 89% of the world population, had access to improved drinking water by 2010—five years ahead of schedule, and just over the target of 88% set in 2000. However, 11% of the world's people (almost 800 million people) still do not have a safe drinking water source, and some 2.5 billion people lack access to basic sanitation facilities (another of the MDGs).[33]

The treatment of drinking water is a mainstream practice in developed nations today. Before being sent to your tap, water brought in from a reservoir, lake, or aquifer is treated with chemicals to remove particulate matter; passed through filters of sand, gravel, and charcoal; and probably disinfected with small amounts of a chemical agent, such as chlorine.

Health Canada publishes standards for drinking water contaminants, which local governments and water suppliers are obligated to meet. Categories for standards include microbiological parameters (viruses, bacteria, protozoa, turbidity), chemical and physical parameters (including both health and aesthetic guidelines), and radiological parameters. More than 80 characteristics are considered in these guidelines; some have numerical standards associated with them, while others do not. The guidelines are set by the Federal–Provincial–Territorial Committee on Drinking Water, which includes members from Health Canada, Environment Canada, and the Council of Environment Ministers, as well as the Canadian Advisory Council on Plumbing.

It is better to prevent pollution than to remediate the impacts after it occurs

In many cases, solutions to pollution will need to involve prevention, not simply **"end-of-pipe"** treatment and cleanup. With groundwater contamination, in particular, preventing pollution in the first place is by far the best strategy when one considers the other options for dealing with the problem: Filtering ground water before distributing it can be extremely expensive; pumping water out of an aquifer, treating it, and then injecting it back in, repeatedly, takes an impracticably long time; and restricting pollutants on lands above selected aquifers would simply shift pollution elsewhere.

There are many things ordinary people can do to help minimize freshwater pollution. One is to exercise the power of consumer choice in the marketplace by purchasing phos-

phorus-free detergents and other "environmentally friendly" products. (The use of phosphate in both laundry and dishwasher detergents has been limited in Canada since 2010, but phosphate-containing products are still manufactured elsewhere.) Another is to become involved in protecting local waterways. Locally based "riverwatch" groups or watershed associations enlist volunteers to collect data and help provincial and federal agencies safeguard the health of rivers and other water bodies. Such programs are proliferating as citizens and policymakers increasingly demand clean water.

Municipal waste water can be treated and improved

Waste water refers to water that has been used by people in some way. It includes water carrying sewage; water from showers, sinks, washing machines, and dishwashers; water used in manufacturing or cleaning processes by businesses and industries; and stormwater runoff.

Natural systems can process moderate amounts of waste water, but the large and concentrated amounts generated by our densely populated areas can harm ecosystems and pose threats to human health. Thus, attempts are now widely made to treat waste water before releasing it into the environment. In rural areas, **septic systems** are the most popular method of wastewater disposal. In a septic system, waste water runs from the house to an underground septic tank, inside which solids and oils separate from water. The clarified water proceeds downhill to a drain field of perforated pipes laid horizontally in gravel-filled trenches underground. Microbes decompose the waste water these pipes emit. Periodically, solid waste needs to be pumped from the septic tank and taken to a landfill.

In more densely populated areas, municipal sewer systems carry waste water from homes and businesses to centralized treatment locations. There, pollutants in waste water are removed by physical, chemical, and biological means (**FIGURE 11.25**). It is technologically possible (though often not economically feasible or socially acceptable) to return *any* water—no matter how soiled it was originally—to the status of drinking water.

At a treatment facility, *primary treatment*, the physical removal of contaminants in settling tanks or clarifiers, generally removes about 60% of suspended solids from waste water. Waste water then proceeds to *secondary treatment*, in which water is stirred and aerated so that aerobic bacteria degrade organic pollutants. Roughly 90% of suspended solids may be removed after secondary treatment. Finally, the clarified water is treated with chlorine, and sometimes ultraviolet light, to kill bacteria.

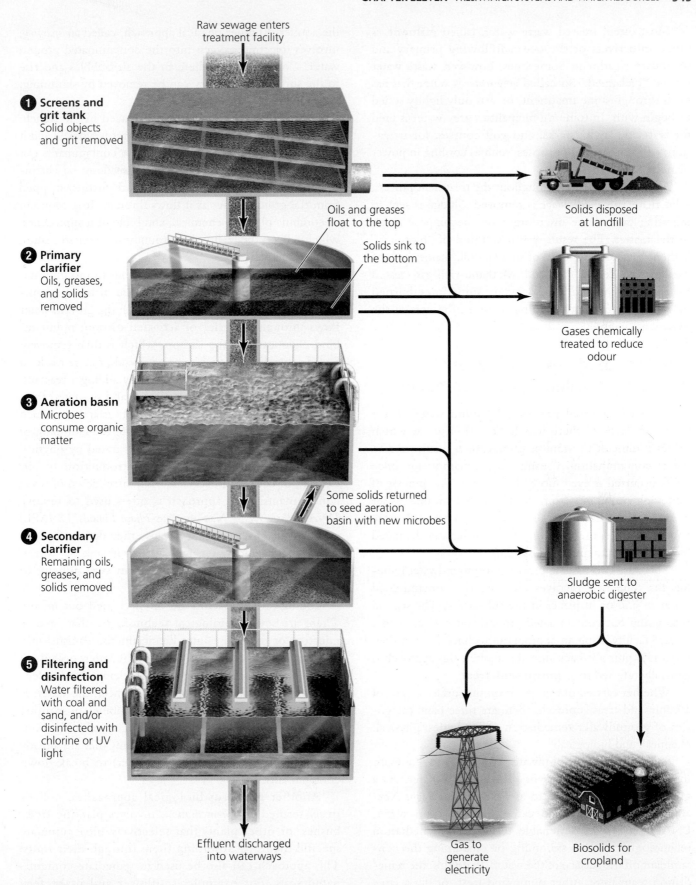

Raw sewage enters treatment facility

1 Screens and grit tank
Solid objects and grit removed

Oils and greases float to the top

Solids sink to the bottom

Solids disposed at landfill

2 Primary clarifier
Oils, greases, and solids removed

Gases chemically treated to reduce odour

3 Aeration basin
Microbes consume organic matter

Some solids returned to seed aeration basin with new microbes

4 Secondary clarifier
Remaining oils, greases, and solids removed

Sludge sent to anaerobic digester

5 Filtering and disinfection
Water filtered with coal and sand, and/or disinfected with chlorine or UV light

Effluent discharged into waterways

Gas to generate electricity

Biosolids for cropland

FIGURE 11.25 In a modern wastewater treatment facility, waste passes through screens and into grit tanks to remove large debris **(1)**, then into the primary clarifier **(2)** in which solids settle to the bottom and oils and greases float to the top for removal. Clarified water proceeds to aeration basins **(3)** that oxygenate the water to encourage decomposition, then into secondary clarifiers **(4)** for removal of further solids and oils. Next, the water may be purified by chemical treatment with chlorine, carbon filtering, and/or exposure to ultraviolet light **(5)**, which reduces the oxygen-carrying capacity of the blood. The treated water may be piped into natural water bodies, used for urban irrigation, flowed through an artificial wetland, or used to recharge ground water.

Most often, treated waste water, called **effluent**, is piped into rivers or the ocean following primary and secondary treatment. Sometimes, however, waste water can be "reclaimed." So-called *grey water* is water that has been through some treatment, or was only lightly soiled to begin with. In some municipalities grey water is used for watering lawns, parks, and golf courses; for irrigation; or for industrial purposes, such as cooling in power plants.

As water is purified throughout the treatment process, solid material called *sludge* is removed. Sludge is sent to digesting vats, where microorganisms decompose much of the matter. The result, a wet solution of "biosolids," is then dried and disposed of in a landfill, incinerated, or used as fertilizer on cropland. Methane-rich gas created by the decomposition process is sometimes burned to generate electricity, helping to offset the cost of the treatment facility.

Contaminated ground water is especially challenging to remediate

As we have discussed previously, ground water occurs in the subsurface where it is hidden from view, which makes it difficult to monitor groundwater quality and to detect contamination. Cleaning up contamination once it has occurred is even more challenging. The process of removing pollutants from contaminated ground water is called **remediation** (and the term can be used more generally to denote cleaning or restoring any degraded habitat, environment, or ecosystem).

There are two basic approaches to groundwater remediation. The first involves cleaning the contaminated water *in situ*, or in place, in the subsurface. The second is to pump out contaminated ground water, clean it in a surface facility using an appropriate technology, and then (typically) pump it back into the aquifer; this approach is generally referred to as **pump-and-treat**.

Whether carried out *in situ* or on the surface as part of a pump-and-treat approach, there are three basic categories of groundwater remediation technologies: physical, chemical, and biological.

Some common groundwater contaminants—notably, many of the constituents of oil—float on water. As a group, these are referred to as *LNAPLs*, or *Light Non-Aqueous-Phase Liquids*. Because they float on water, LNAPLs are often amenable to physical remediation technologies, such as skimming or vacuuming the contaminant off the surface of the water. To do this, the remediation team must either pump-and-treat, or dig a large well or trench to access the ground water, which would otherwise be contained in small cracks and pore spaces in

the aquifer. Another physical approach, called *air sparging*, involves injecting oxygen into the contaminated ground water. Contaminants adhere to the air bubbles and rise to the surface, where they can be removed by skimming, vacuuming, or soil gas extraction.

Sometimes a physical barrier is placed in the trench. The downslope flow of the ground water is blocked and it accumulates behind the barrier, where contaminants can be skimmed off. Barriers also can be made to be chemically reactive, so that contaminants are further stripped from the ground water as it flows through. It is common to combine physical, chemical, and biological approaches, which reflects the reality that groundwater contamination is often complex.

There are many chemical approaches to groundwater remediation, depending on the nature of the contaminants. One is carbon filtering, in which the ground water flows through a barrier of activated carbon; pollutants bind chemically to the carbon, which is then removed. Some contaminants, such as heavy metals, can be made to precipitate out of the ground water by adding a reactant; the solid precipitate is then collected and separated from the rest of the ground water. Chemical surfactants, which act sort of like detergents, can facilitate the clumping of contaminants, which can then be separated by physical means. Oxidation, through the introduction of air, oxygen, or ozone, can accelerate the breakdown of some contaminants. This approach is often used to remove *DNAPLs*, or *Dense Non-Aqueous-Phase Liquids*. DNAPLs are particularly challenging because they do not float on top of the ground water, but rather will sink or (worse) mix with ground water, making them very difficult to separate out.

Biological approaches are often carried out *in situ*. There are several biological technologies that involve injecting or otherwise introducing microorganisms into the contaminated ground water. The microorganisms are selected for their capacity to accelerate the biodegradation of specific contaminants present in the water. A related approach, called *biosparging*, combines air sparging with nutrients to enhance the capacity of indigenous microorganisms (that is, microorganisms that are already naturally present in the ground water) to break down contaminants.

Another group of biological approaches, collectively termed *phytoremediation*, involves planting trees, bushes, or other plants that selectively bioaccumulate specific toxins by absorbing them through their roots. This approach can also be used to remediate contaminated soils. For example, sunflower and brake fern selectively accumulate arsenic, storing it in their leaves; willow selectively accumulates heavy metals, such as cadmium, in roots and shoots; Alpine pennycress accumulates zinc at levels that are toxic to many other plants; and Indian mustard and ragweed remove and bioaccumulate lead.

Artificial wetlands can aid in remediation

Natural wetlands already perform the ecosystem service of water purification. Wastewater treatment engineers are now manipulating wetlands and even constructing wetlands to use them as tools to cleanse waste water. The practice of treating waste water and other types of polluted runoff with so-called **constructed** (or **artificial** or **engineered**) **wetlands** is growing quickly.

Generally in this approach, waste water or discharge that was not highly soiled or has gone through primary treatment is pumped into the wetland. There, physical processes (such as settling of solids), chemical processes (such as oxidation, precipitation of dissolved substances, and acid buffering), and biological processes act to reduce the contamination. Biological processes such as decomposition are facilitated by microbes that live amid the root structures of aquatic plants. Water cleansed in the wetland can be released into waterways or allowed to percolate underground.

Constructed and "enhanced natural" wetlands have been used in Canada and elsewhere to remove contaminants from agricultural and livestock runoff, and from mine discharge. For example, the Campbell gold mine in Ontario utilizes a constructed wetland to remove ammonia, copper, and other contaminants from the mine's discharge. Ammonia (NH_3), a reduced nitrogen compound carried in the discharge, is oxidized in the wetland to form nitrate. Nitrate acts as a plant fertilizer and any remainder is converted by denitrification into harmful nitrogen gas. The result from the Campbell Mine is a completely nontoxic discharge.

Constructed wetlands also serve as havens for wildlife and areas for human recreation. Restored and artificial wetlands in Ontario, Alberta, Nova Scotia, and many other locations in Canada are serving as wetland habitats for birds and wildlife, while helping to recharge depleted aquifers.

Conclusion

Citizen action, government legislation and regulation, new technologies, economic incentives, and public education are all enabling us to confront what will surely be one of the great environmental challenges of the new century: ensuring adequate quantity and quality of fresh water for ourselves and for the planet's ecosystems.

Accessible fresh water is only a minuscule percentage of the hydrosphere, but we take it for granted. With our expanding population and increasing water usage, we are approaching conditions of widespread scarcity. Water depletion and water pollution are already taking a toll on the health, economies, and societies of the developing world, and they are beginning to do so in arid areas of the developed world. However, there is reason to hope that we may yet attain sustainability in our water usage. Potential solutions are numerous, and the issue is too important to ignore.

REVIEWING OBJECTIVES

You should now be able to

Outline the major reservoirs and pathways of the hydrologic cycle, and the major human uses of fresh water

- Only about 1% of all water on Earth is readily available for our use. Water availability varies both spatially and temporally, and is not well matched to the distribution of population.
- We use water for agriculture, industry, and residential purposes. The relative importance of these uses varies among societies, but globally 70% is used for agriculture.

- A functioning hydrologic cycle is vital, not only for human interests but for the maintenance of ecosystem integrity. Water *in situ* performs many crucially important services.
- Climate change will cause changes in the hydrologic cycle and may lead to water shortages in some regions.

Describe the major types of freshwater ecosystems

- Rivers and streams carry water and sediment toward the ocean and modify landscapes by erosion and deposition.

- The five main wetland classes are marshes, swamps, bogs, fens, and shallow-water wetlands. These occur in a wide variety of physical environments and provide important ecosystem services and diverse habitat.

- Lakes are open, standing bodies of surface water. Lake water can be categorized into the littoral, limnetic, and benthic zones, and (only in the largest, deepest lakes) the profundal zone.

- Ground water resides in the pore spaces and cracks of subsurface aquifers. It is replenished in recharge zones by precipitation and infiltration, and it flows out at discharge zones to join surfacewater bodies.

Discuss how we divert and withdraw water for human use, and how these practices alter freshwater systems

- Diversion or withdrawal of water from river channels, other surfacewater bodies, or aquifers can generate unintended environmental impacts; these are typically worse in downstream areas.

- Most of the world's large rivers are dammed. Dams bring a diverse set of benefits and costs—not just environmental, but social and economic as well. Other common types of channel interventions include dikes, floodways, and levees for flood control.

- Wetlands have been aggressively drained and in-filled for agriculture, coastal development, and other activities. The international agreement to protect wetlands is the Ramsar Convention.

- Agriculture is the biggest user of water, and inefficient irrigation systems are a big reason for this.

- Ground water is hidden from view and therefore difficult to monitor. A very slow rate of recharge means that some aquifers are easily depleted.

- Bottled water can contribute to groundwater depletion and generates a lot of plastic waste. Studies have shown that bottled water is not necessarily better for our health, and in some cases may even be worse than tap water.

Assess problems of water supply and propose solutions to address depletion of freshwater sources

- To address depletion of fresh water, we can either increase supply, such as by piping in water from areas where it is more abundant, or we can reduce demand, such as by stressing conservation and efficiency.

- The only way to "make" more fresh water is by desalination, but this can be an expensive solution.

- High-efficiency irrigation can reduce agricultural demand by directing just the right amount of water to the plant's roots. Personal choices such as eating less meat, using xeriscaping, and installing water-efficient toilets can reduce household water use. Industries are finding that water-efficient manufacturing approaches save money.

- Most market-based strategies for encouraging water conservation involve some form of full-cost pricing. The virtual water concept accounts for water use as an integral part of the manufacturing process.

Summarize the types and sources of water pollution

- Water pollution comes from point and nonpoint sources. Common pollutants are excessive nutrients, microbial pathogens, toxic chemicals, sediment, and thermal pollution.

- Scientists who monitor water quality use a variety of biological, chemical, and physical indicators, such as presence of fecal coliform bacteria, pH and hardness, and turbidity.

- Ground water is subject to many of the same types and sources of pollution as surface water, but groundwater pollution can be more persistent and much more difficult to detect.

Propose solutions to prevent or remediate water pollution

- Legislation and regulation have improved water quality in developed nations in recent decades. Many scientists believe that recent changes to Canada's *Fisheries Act* have weakened the protection of aquatic systems and habitat.

- Many people—almost a billion, worldwide—still lack access to improved drinking water and sanitation, even though these are specified in the Millennium Development Goals.

- Preventing water pollution before it happens is better than remediating it afterward.

- Septic systems are used to treat waste water in rural areas. At municipal treatment facilities, waste water is treated in a series of physical, biological, and chemical steps. The effluent can be released or used for urban irrigation or groundwater recharge.

- Contaminated ground water is especially challenging to remediate. There are many different approaches, which make use of chemical, physical, and biological technologies, either *in situ* or through pump-and-treat.

- Artificial wetlands can enhance wastewater treatment while restoring habitat for wildlife.

TESTING YOUR COMPREHENSION

1. Define *ground water*. What role does ground water play in the hydrologic cycle?
2. Why are sources of fresh water unreliable for some people and plentiful for others?
3. Describe three benefits of damming rivers and three costs. What particular environmental, health, and social concerns has China's Three Gorges Dam and its reservoir raised?
4. Why do the Colorado, Rio Grande, Nile, and Yellow rivers now slow to a trickle or run dry before reaching their deltas?
5. Why are water tables dropping around the world? What are some environmental costs of falling water tables?
6. Name three major types of water pollutants, and provide an example of each. List three properties of water that scientists use to determine water quality.
7. Why do many scientists consider groundwater pollution a greater problem than surfacewater pollution?
8. What are some anthropogenic (human) sources of groundwater pollution?
9. Describe how drinking water is treated. How does a septic system work?
10. Describe and explain the major steps in the process of wastewater treatment. How can artificial wetlands aid such treatment?

THINKING IT THROUGH

1. Discuss possible strategies for improving the distribution of water throughout the world; consider both supply and transport issues. Have our methods of distributing and storing water changed very much throughout history? How is the scale of the effort affecting the availability of water supplies?
2. Describe some ways in which we can reduce household, agricultural, and industrial water use.
3. Let's say that you have been put in charge of water policy for your region. The aquifer has been over-pumped to support agricultural production, and many wells have already run dry. Meanwhile, the region's largest city is growing so fast that more water is needed for its burgeoning population. What policies would you consider to restore your region's water supply? Would you take steps to increase supply, to decrease demand, or both? Explain why you would choose such policies.

   Having solved the water depletion problem, your next task is to deal with worsening pollution of the ground water that provides the region's drinking water. What steps would you consider to safeguard the quality of the groundwater supply, and why?
5. In May 2011, during record-high water levels on the Assiniboine River, the government of Manitoba deliberately breached a dike near Portage-la-Prairie to release pressure in the Portage Reservoir and avert a much larger, uncontrolled flood. The controlled release put 150 homes and a considerable area of farmland at risk. The strategy worked; no homes were significantly harmed. A similar situation occurred during the Red River flood of 1997, during which the Red River Floodway diverted flood water to protect the city of Winnipeg, at the cost of flooding in smaller communities such as St. Agathe. Both events caused significant stress for those in the path of diverted flood waters.

   If you were a decision-maker in Manitoba, would you have reached the same decision? Is it worthwhile to protect certain areas, even if it means risking others? What might you do to compensate those who were put at risk?

INTERPRETING GRAPHS AND DATA

Close to 75% of the fresh water used by people is used in agriculture, and about 1 of every 14 people lives where water is scarce. By the year 2050, scientists project that two-thirds of the world's population will live in water-scarce areas, including most of Africa, the Middle East, India, and China. How much water is required to feed 7 billion people a basic dietary requirement of 2700 calories (11.3 kJ) per day? The answer depends on the efficiency with which we use water in agricultural production and on the type of diet we consume.

Review the provided graph to answer the following questions:

1. How many litres of water are needed to produce 2300 calories (9.6 kJ) of food from vegetable sources? How many litres of water are needed to produce 400 calories (1.7 kJ) of food from animal sources? How many litres of water are needed daily to provide this diet? Annually?

2. How many litres of water would be saved daily, compared to the diet in the graph, if the 2700 calories were provided entirely by vegetables? Annually?

3. Reflect on one of the quotes at the beginning of this chapter: "Water promises to be to the twenty-first century what oil was to the twentieth century: the precious commodity that determines the wealth of nations." How do you think the demographic pressure on the water supply could affect world trade, particularly trade of agricultural products? Do you think it could affect prospects for peace and stability in and among nations? How so?

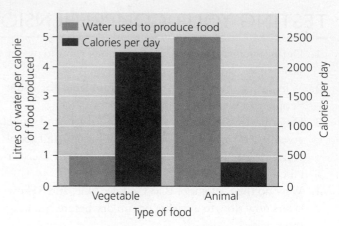

Amount of water needed to produce vegetable and animal food (orange), and global average calories per day consumed of vegetable and animal food (red) (1000 calories = 0.004184 kJ).
Source: Data from Wallace, J.S. (2000). Increasing Agricultural Water Use Efficiency to Meet Future Food Production. *Agriculture, Ecosystems and Environment, 82,* pp. 105–119.

MasteringEnvironmentalScience®

STUDENTS
Go to **MasteringEnvironmentalScience** for assignments, the eText, and the Study Area with practice tests, videos, current events, and activities.

INSTRUCTORS
Go to **MasteringEnvironmentalScience** for automatically graded activities, current events, videos, and reading questions that you can assign to your students, plus Instructor Resources.

12 Marine and Coastal Systems and Fisheries

Marine protected areas (MPAs) can help provide safe havens for marine life.

National Oceanic and Atmospheric Administration

Upon successfully completing this chapter, you will be able to

- Identify physical, geographical, chemical, and biological aspects of the ocean and various marine environments
- Describe major types of marine ecosystems
- Outline human uses of marine resources and their impacts on marine environments

- Review the current state of ocean fisheries and reasons for their decline
- Evaluate marine protected areas and reserves as innovative solutions for marine conservation

Cod fishing has been a way of life for generations in Newfoundland and Labrador. Monster cod like these are no longer there to be harvested.

CENTRAL CASE
LESSONS LEARNED: THE COLLAPSE OF THE COD FISHERIES

"It is almost universally recognized that the future of sustainable fisheries lies with much less fishing effort, lower exploitation rates, larger fish stocks, dramatic reduction in bycatch, increased concern about ecosystem impacts of exploitation, elimination of destructive fishing practices, and much more spatial management of fisheries, including a significant portion of marine ecosystems protected from exploitation. I believe this vision is broadly shared within the fisheries management and the ecological communities."

—RAY HILBORN, IN 2007

Europeans exploring the coasts of North America 500 years ago found such an abundance of Atlantic cod that they caught them by dipping baskets over the railings of their ships. The race to harvest this resource helped spur the colonization of the New World. Starting in the early 1500s, schooners captured countless millions of cod, and the fish became a dietary staple on both sides of the Atlantic.

For many decades, cod fishing was the economic engine for hundreds of coastal communities in eastern Canada and New England. In many Canadian coastal villages, cod fishing was a way of life for generations (see photo). So it came as a shock when the cod all but disappeared, and governments had to step in and close the fisheries.

Atlantic cod (*Gadus morhua*) is a type of *groundfish*, a fish that lives or feeds along the bottom. People have long coveted groundfish such as halibut, pollock, haddock, and flounder. Adult cod eat smaller fish and invertebrates, commonly grow to 60–70 cm long, and can live 20 years. A mature female cod can produce several million eggs. The cod inhabit cool ocean waters on both sides of the North Atlantic, in discrete population groupings called *stocks*. One major stock inhabits the Grand Banks off the Newfoundland coast, and another lives on Georges Bank off the coast of Massachusetts (**FIGURE 12.1A**).

The Grand Banks cod stock provided ample fish for centuries. With advancing technology, however, ships became larger and more effective at finding fish. By the

1960s, massive industrial trawlers from Europe were vacuuming up unprecedented numbers of groundfish. In 1977, Canada exercised its legal right to the waters 200 nautical miles from shore, the **Exclusive Economic Zone (EEZ)** established by the UN Convention on the Law of the Sea. Foreign fleets were expelled from most of the Grand Banks, and Canada's fishing industry was revved up like never before.

Catches began to dwindle in the 1980s (**FIGURE 12.1B**).[1] Too many fish had been taken, and bottom-trawling had destroyed much underwater habitat. Environmental

variability also played a role. By 1992 the situation was dire: Scientists reported that mature cod were at just 10% of their long-term abundance. On July 2, Fisheries Minister John Crosbie announced a two-year ban on commercial cod fishing off Newfoundland and Labrador, where the $700 million fishery supplied income to 16% of the province's workforce. To compensate fishers, the government offered 10 weekly payments of $225, along with training for new job skills and incentives for early retirement. Over the next two years, 40000 fishers and processing-plant workers lost their jobs.

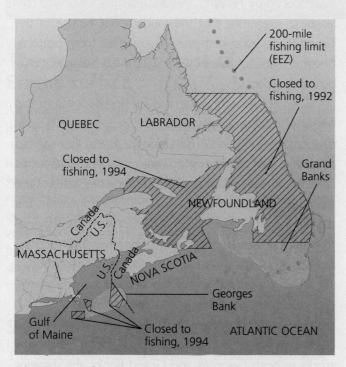

(a) Atlantic cod stocks, Grand Banks and Georges Bank

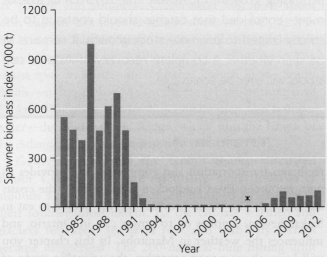

(c) Offshore spawning stock biomass, Grand Banks Atlantic cod, since the early 1980s

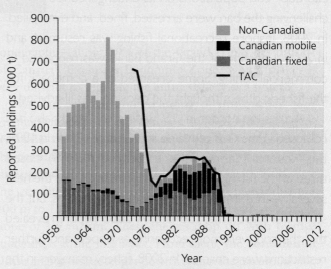

(b) Total allowable and actual catch, Grand Banks Atlantic cod, 1959–2012

(d) Atlantic cod (*Gadus morhua*)

FIGURE 12.1

Ten stocks of Atlantic cod inhabit areas of the northwestern Atlantic Ocean **(a)**, including Grand Banks and Georges Bank. Some areas were closed to fishing when cod populations collapsed after being overfished. Graph **(b)** shows the recent history of the Atlantic cod fishery in Canada's EEZ on the Grand Banks, in Total Allowable Catch (TAC) and actual catch (Reported Landings, thousands of tonnes) in 1959–2012. Graph **(c)** shows the estimated offshore spawning stock biomass of Atlantic cod stocks on the Grand Banks since the early 1980s. The asterisk indicates partial estimates from incomplete survey coverage. *Sources:* Adapted from **(b)** Figure 2 on p. 3 and **(c)** Figure 7 on p. 12 of Fisheries and Oceans Canada, *Stock Assessment of Northern (2J3KL) Cod in 2013.* This does not constitute an endorsement by Fisheries and Oceans Canada of this product.

In **FIGURE 12.1B**, what was the TAC in 1973 (the first year for which a TAC is shown on this graph)? What was the actual reported catch that year? If you had been a fishery manager in 1973, what conclusions would you have drawn from this information?

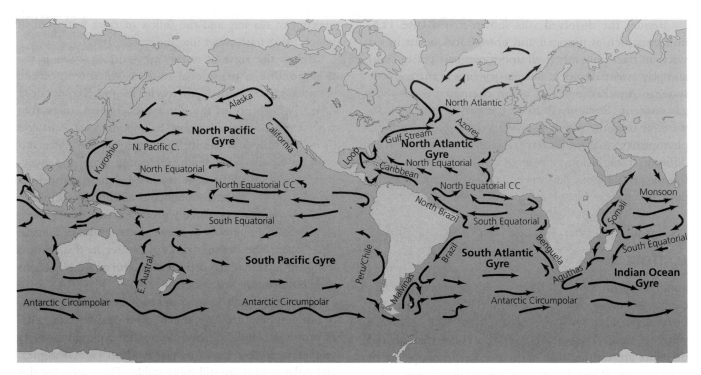

FIGURE 12.7 The upper waters of the ocean flow in currents, which are long-lasting and predictable global patterns of water movement. Warm- and cold-water currents interact with the planet's climate system, and people have used them for centuries to navigate the ocean. *Source:* Based on Garrison, T. (2005). *Oceanography*, 5th ed. Belmont, CA: Brooks/Cole.

DATA Q If you released a buoy into the Pacific Ocean from the southeastern coast of Japan, and it travelled on the surface ocean currents shown here, would it likely reach Australia or North America first? On which currents would it be carried?

Atlantic coast and flows past the eastern edges of Georges Bank and the Grand Banks at a rate of 160 km per day (nearly 2 m/s). Averaging 70 km across, the Gulf Stream continues across the North Atlantic, bringing warm water to Europe and moderating that continent's climate, which otherwise would be much colder.

Some of the ocean's surface currents take the form of enormous gyres. A **gyre** is an oceanic current that flows in a circular motion as a result of the *Coriolis force* (an artifact of Earth's rotation, which we will discuss in greater detail in the chapter on atmospheric science and air pollution). There are five great gyres in the world ocean: the North

FIGURE 12.8 Debris from the 2011 tsunami was carried from Japan across the Pacific Ocean to the western shores of Vancouver Island. Currents carried this 20-m dock to the northwest coast of the United States, where it washed ashore in Oregon more than a year after the tsunami. The algae and invertebrates attached to the dock were removed by wildlife officials to prevent them from becoming invasive species.

Robin Loznak/ZUMA Press/Corbis

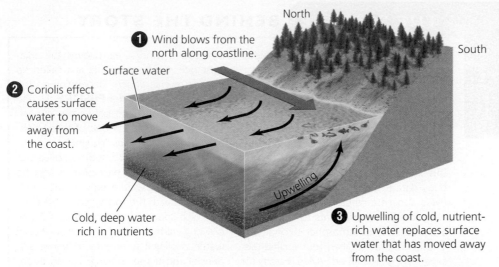

North

South

1 Wind blows from the north along coastline.

Surface water

2 Coriolis effect causes surface water to move away from the coast.

Cold, deep water rich in nutrients

Upwelling

3 Upwelling of cold, nutrient-rich water replaces surface water that has moved away from the coast.

FIGURE 12.9 Upwelling is the movement of bottom waters upward. This often brings nutrients up to the surface, creating rich areas for marine life. As shown here, winds blow from the north along the western coast of Vancouver Island (**1**). The Coriolis effect diverts the wind, which pushes surface water away from the coast (**2**). Water wells up from the bottom (**3**), replacing the water that moves away from shore.

Atlantic, South Atlantic, North Pacific, South Pacific, and Indian Ocean gyres (as shown in **FIGURE 12.7**).

Ocean water also flows vertically. The vertical flow of cold, deep water toward the surface, called an **upwelling**, occurs where horizontal currents diverge, or flow away from one another. Bottom water is rich in nutrients, so upwellings are often sites of high primary productivity and lucrative fisheries. Upwellings also occur where strong winds blow away from or parallel to coastlines (**FIGURE 12.9**). An example is seen along the western coast of Vancouver Island, where north winds and the Coriolis effect move surface waters away from the shore, raising nutrient-rich water from below and creating a biologically rich region. The cold water also chills the air along the coast, giving Vancouver Island its famous cool, rainy summers.

In areas where surface currents converge, or come together, surface water sinks and **downwelling** occurs. Downwelling transports water rich in dissolved gases, providing an influx of oxygen for deep-water life. Vertical currents also occur in the deep zone, where differences in water density can lead to rising and falling convection currents, similar to those seen in molten rock and in air.

Linking the ocean's horizontal and vertical currents, like an enormous conveyor belt, is the global **thermohaline circulation** (**FIGURE 12.10**). The thermohaline circulation connects warm surface water flows to cold deeper water flows, with far-reaching effects on global climate. The term *thermohaline* comes from root words that mean "heat" and "salinity." In this worldwide circulatory system, warmer, fresher water moves along the surface and water in the deep zones, which is colder, saltier, and denser, circulates far beneath the surface.

In the Atlantic Ocean, warm surface water flows northward from the equator in the Gulf Stream, carrying heat to high latitudes and keeping Europe warmer than it would otherwise be. As the surface water of this thermohaline conveyor belt system releases heat energy and cools, it becomes denser and sinks, creating **North Atlantic Deep Water (NADW)**. This downwelling of cold, dense water near Greenland and Norway keeps the northern part of the Atlantic basin connected to the global thermohaline circulation system; without it, the climate in areas bordering the North Atlantic would be very different (see "The Science behind the Story: Tip Jets and NADW off the Coast of Greenland"). A similar phenomenon happens in the Southern Ocean, where extremely cold, salty water sinks to form **Antarctic Bottom Water (AABW)**.

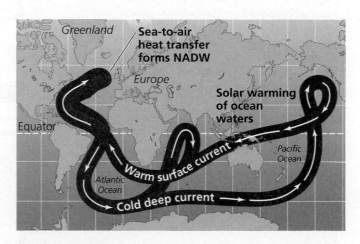

Greenland

Sea-to-air heat transfer forms NADW

Europe

Solar warming of ocean waters

Equator

Atlantic Ocean

Warm surface current

Pacific Ocean

Cold deep current

FIGURE 12.10
The global thermohaline circulation connects warm surface currents with cold deep currents, greatly influencing climate.

Dr. Kent Moore of the University of Toronto in a research plane over the North Atlantic.
Courtesy of Professor G. W. K. Moore

THE SCIENCE BEHIND THE STORY

Tip Jets and NADW off the Coast of Greenland

The formation of North Atlantic Deep Water (NADW), in which extremely cold water sinks in the northern part of the Atlantic to join the deep oceanic circulation, is an integral part of the global thermohaline circulation. It is also crucial for keeping the northern Atlantic basin connected to the rest of the global circulation system. If NADW formation were suppressed, the warm Gulf Stream current—responsible for the relatively mild climate of northeastern North America and Western Europe—could be cut off from entering the North Atlantic, with devastating climatic consequences for this region and globally.

Given its importance, it is surprising to learn that ocean water sinks to form NADW only under extremely specific conditions, and these conditions occur in just a handful of geographical locations in the North Atlantic. What's more, as Professor Kent Moore from the University of Toronto and his colleagues have discovered, the sinking of cold ocean water is also dependent upon highly specialized conditions in the atmosphere off the coast of Greenland.

To investigate the conditions under which NADW forms, Dr. Moore deploys scientific buoys in the North Atlantic, which measure the temperature and other characteristics of ocean water (yes, pretty much like the buoys in the first part of the movie *The Day After Tomorrow*).

With data from these buoys, he is able to pinpoint the exact locations where ocean water becomes cold and dense enough to sink, becoming NADW.

Dr. Moore, an atmospheric physicist, and colleagues have also investigated the role of the atmosphere in oceanic circulation in this part of the Atlantic. They do this by comparing satellite imagery and buoy data to information they collect by flying aboard a research plane equipped with scientific instruments for measuring a wide variety of atmospheric conditions. The aircraft, a modified four-engine passenger jet, is nicknamed FAAM (Facility for Airborne Atmospheric Measurements).

The scientists have discovered that the topography of Greenland plays a fundamental role in oceanic circulation in the North Atlantic, and specifically determines the locations where NADW circulation can be triggered.[4,5] High-speed wind events, with wind speeds greater than 25 m/sec, are common near the southern end of Greenland, which acts as a massive topographic barrier. The topographic barrier funnels and directs the winds in one direction or another, forming very narrow high-speed wind streams over the adjacent ocean. These events are called *tip jets* (or *reverse tip jets*, depending on the direction of flow—see **FIGURE 1**), and Professor Moore describes the experience of flying through one of these high-speed wind streams as "rough."

What Dr. Moore and his colleagues have discovered, in particular, is that tip jets and related wind events are intimately associated with the formation of NADW. The high-speed winds have the effect of drawing warmth away from surface ocean water, cooling it sufficiently to increase its density and trigger sinking. The locations where this occurs are extremely localized because of the narrow channelling of winds that must pass either around or over the topographic barriers formed by the coast of Greenland.

Given the importance of NADW and the thermohaline circulation system in regulating climate, particularly around the North Atlantic, Professor Moore agrees that there is much fundamental research that needs to be carried out to improve our understanding of these atmospheric and oceanic processes and connections.

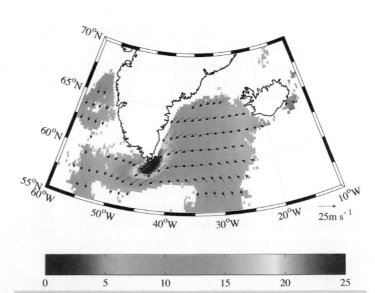

FIGURE 1
Extremely focused, high-speed wind events called tip jets, like this one off Cape Farewell on the southern coast of Greenland, are crucial in the formation of North Atlantic Deep Water because they draw heat away from surface water. In this figure, based on satellite imagery from February 2007, wind speeds are shown in metres per second.
Source: Reprinted by permission from Prof. G. W. K. Moore.

Without or with a weaker thermohaline circulation, Earth would be a very different place. The climate of France would be similar to that of Newfoundland; tropical regions would become even hotter, as their heat would not be moved north by the Gulf Stream. The deep ocean could become lifeless, lacking the oxygen-rich waters provided by the thermohaline circulation. Biological productivity in the Pacific and Indian Oceans would suffer, as fewer nutrients from depth would make it to the surface layer where photosynthesis happens. The removal of greenhouse gases from the atmosphere to the deep ocean would cease, thus accelerating global climate change.

El Niño demonstrates the atmosphere–ocean connection

Horizontal ocean currents can have far-reaching impacts on climate. An important example is the La Niña–El Niño cycle, which demonstrates the linkages between the oceanic and atmospheric systems. Under normal conditions, prevailing winds blow from east to west along the equator, from a region of high pressure in the eastern Pacific to one of low pressure in the western Pacific, forming a large-scale convective loop, or atmospheric circulation pattern (**FIGURE 12.11A**). The winds push surface waters westward, causing warm water to "pile up" in the western Pacific. As a result, water near Indonesia can be 50 cm higher and 8°C warmer than water near South America. The westward-moving surface waters allow cold water to rise up from the deep in a nutrient-rich upwelling along the coast of Peru and Ecuador.

El Niño conditions are triggered when the normal air pressure scenario reverses itself by rising in the western Pacific and falling in the eastern Pacific. This change in air pressure causes the equatorial trade winds to weaken. Without these winds, the warm surface water is not pushed toward the western Pacific (**FIGURE 12.11B**). Instead, it collects along the Pacific coast of South, Central, and North America, suppressing the upwelling of cold water and shutting down the delivery of nutrients that support marine life and fisheries. Coastal industries, such as Peru's anchovy fisheries, are devastated by El Niño events; the 1982–1983 event caused more than $8 billion in economic losses worldwide. El Niño alters weather patterns around the world, creating rainstorms and floods in areas that are generally dry (such as southern California), and causing drought and fire in regions that are typically moist (such as Indonesia). In Canada, El Niño tends to produce weather that is drier and warmer than normal, with more frequent droughts.

La Niña events are the opposite of El Niño—similar to the "normal" conditions in the equatorial Pacific, but more pronounced. Under these conditions, cold surface

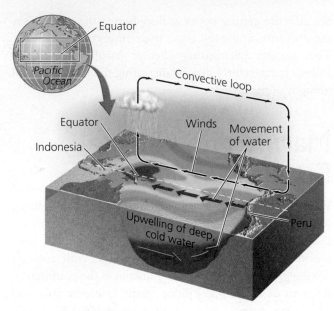

(a) Normal conditions

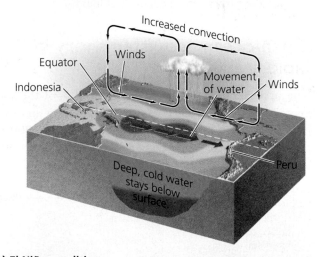

(b) El Niño conditions

FIGURE 12.11
Under normal or La Niña conditions **(a)**, prevailing winds push warm surface waters toward the western Pacific. Red and orange colours denote warmer water; blue and green colours denote colder water. Under El Niño conditions **(b)**, the winds weaken and the warm water flows back across the Pacific toward South America. This suppresses the cold upwelling along the American coast and alters precipitation patterns regionally and globally.
Source: Adapted from National Oceanic and Atmospheric Administration, Tropical Atmospheric Ocean Project.

waters extend far westward in the equatorial Pacific, and weather patterns are affected in opposite ways. La Niña–influenced weather tends to be abnormally cool and wet all the way from British Columbia to southern Quebec.

These cycles, called the **El Niño–Southern Oscillation (ENSO)**, are periodic but irregular, occurring roughly every two to eight years. Southern Oscillation refers specifically to the see-sawing variation in atmospheric pressure between the eastern and western equatorial Pacific.

Scientists are getting better at deciphering the triggers for these events, and predicting their impacts on weather. They are also investigating whether globally warming air and sea temperatures may be increasing the frequency and strength of these cycles.

Marine and Coastal Ecosystems

With their variation in topography, temperature, salinity, nutrients, and sunlight, marine and coastal environments feature a variety of ecosystems. Most marine and coastal ecosystems are powered by solar energy, with sunlight driving photosynthesis by phytoplankton in the photic zone. Regions of ocean water differ greatly in biodiversity, and some zones support more life than others (**FIGURE 12.12**); yet even the darkest ocean depths host life.

Open-ocean ecosystems vary in biological diversity

The uppermost 10 m of water absorb 80% of the solar energy that reaches the water's surface. Nearly all of the ocean's primary productivity occurs in the top 200 m, called the **photic zone**, where light easily penetrates.

Generally, the warm, shallow waters of continental shelves are most biologically productive and support the greatest species diversity. Habitats and ecosystems occurring in the water column between the ocean's surface and floor are termed **pelagic**, whereas those that occur on the ocean floor are called **benthic**.

Biological diversity in pelagic regions of the open ocean is highly variable in its distribution. Primary productivity and animal life near the surface are concentrated in regions of nutrient-rich upwelling. Marine animals that actively swim are referred to as **nekton**. They are contrasted with **plankton**, microscopic organisms that float rather than swim, including *phytoplankton* (typically algae rather than plants, even though the prefix *phyto-* means "plant") and *zooplankton* (animals).

Phytoplankton are particularly important, because they constitute the base of the marine food chain in the pelagic zone. These photosynthetic algae and cyanobacteria feed zooplankton, which in turn become food for nektonic fish, jellyfish, whales, and other free-swimming animals (**FIGURE 12.13**). Predators at higher trophic levels include larger fish, sea turtles, and sharks. Many fish-eating birds, such as puffins, petrels, and shearwaters, feed at the surface of the open ocean, returning periodically to nesting sites on islands and coastlines.

In recent years biologists have been learning more about animals of the very deep ocean, although tantalizing

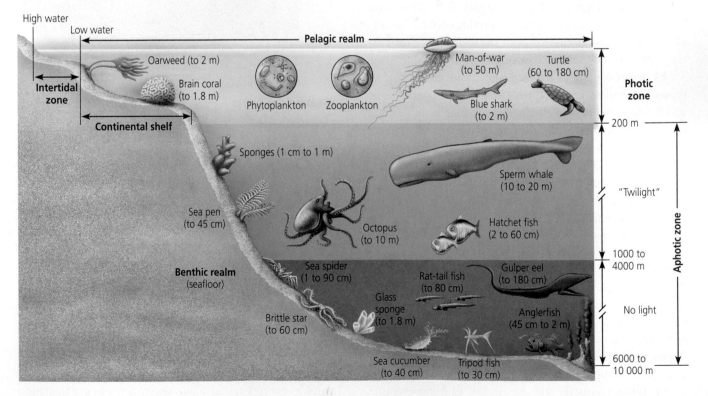

FIGURE 12.12 The various zones of the ocean differ dramatically in the amount and type of life that they support. The organisms shown here are not to scale.

Source: Campbell, Neil A.; Reece, Jane B.; Taylor, Martha R.; Simon, Eric J.; Dickey, Jean L. *Biology: Concepts and Connections*, 6th ed., © 2009, p. 688. Reprinted and electronically reproduced by permission of Pearson Education, Inc., New York, NY.

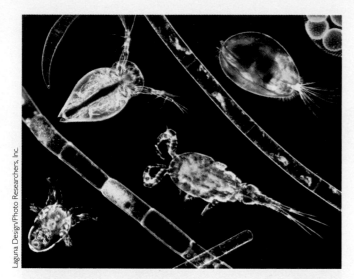

FIGURE 12.13
The surface zone of ocean water contains billions of phytoplankton—tiny photosynthetic algae, protists, and bacteria at the base of the marine food chain—as well as zooplankton, small animals and protists that eat phytoplankton and comprise the next trophic level.

FIGURE 12.14
Life is scarce in the dark depths of the deep ocean, but the creatures that do live there are well adapted. This anglerfish (*Lophius piscatorius*) lures prey toward its mouth with a bioluminescent (glowing) organ that protrudes from the front of its head.

questions remain. In deep-water ecosystems, animals have adapted to deal with the extreme water pressures that characterize the very deep ocean. They also live in the dark without food from photosynthetic organisms, below the depth of light penetration in the **aphotic zone** (*aphotic* means "no light"). Some of these often bizarre-looking creatures scavenge carcasses or organic detritus that falls from above. Others are predators, and still others attain food from symbiotic mutualistic bacteria. Some species carry bacteria that produce light chemically by *bioluminescence* (**FIGURE 12.14**).

Some unusual deep-water ecosystems form around hydrothermal vents, where heated water spurts from the seafloor, often carrying minerals that precipitate to form large rocky structures. Tubeworms, shrimp, and other creatures in these recently discovered systems use symbiotic bacteria to derive their energy by chemosynthesis from chemicals in the heated water, rather than from sunlight. They manage to thrive within amazingly narrow zones between scalding-hot and icy cold water.

Shallow-water systems are highly productive

Masses of giant brown algae, or **kelp**, grow from the floor of continental shelves, reaching upward toward the sunlit surface. Some kelp reaches 60 m in height and can grow 45 cm in a single day. Dense stands of kelp form underwater "forests" along many temperate coasts (**FIGURE 12.15**). Kelp forests supply shelter and food for invertebrates and

FIGURE 12.15
Tall "forests" of kelp grow from the floor of the continental shelf. Seen here is *Macrocystis pyrifera* in California's Channel Islands National Marine Sanctuary. Numerous fish, sea urchins, and other organisms eat kelp and find refuge among its fronds.

fish, which in turn provide food for higher trophic level predators, such as seals and sharks. Sea otters, a keystone species, control sea urchin populations; when otters disappear, urchins overgraze the kelp, destroying the forests and creating "urchin barrens." Kelp forests also absorb wave energy and protect shorelines from erosion. People eat some types of kelp, and it provides compounds known as alginates, which serve as thickeners in a wide range of consumer products, including cosmetics, paints, paper, and soaps.

In shallow subtropical and tropical waters, coral reefs occur. A **coral reef** is a mass of calcium carbonate composed of the skeletons of tiny colonial marine organisms. A coral reef may occur as an extension of a shoreline, along a barrier island paralleling a shoreline, or as an *atoll*, a ring around a submerged island.

Corals themselves are tiny invertebrate animals related to sea anemones and jellyfish. They remain attached to rock or existing reef and capture passing food with stinging tentacles. Corals also derive nourishment from microscopic symbiotic algae, known as *zooxanthellae*, which inhabit their bodies and produce food through photosynthesis. Most corals are colonial, and the colourful surface of a coral reef consists of millions of densely packed individuals. As the corals die, their skeletons remain part of the reef while new corals grow atop them, increasing the reef's size.

Like kelp forests, coral reefs protect shorelines by absorbing wave energy. They also host tremendous biodiversity (**FIGURE 12.16A**). Reefs provide complex physical structures (and thus many varied habitats) in shallow nearshore waters, which are regions of high primary productivity. Besides the staggering diversity of anemones, sponges, hydroids, tubeworms, and other *sessile* (stationary) invertebrates, innumerable molluscs, crustaceans, flatworms, sea stars, and urchins patrol the reefs, while thousands of fish species find food and shelter in reef nooks and crannies.

Coral reefs are now experiencing worldwide declines. Many have undergone **coral bleaching**, a process that occurs when the coloured symbiotic zooxanthellae leave the coral, depriving it of nutrition. Corals lacking zooxanthellae lose colour and frequently die, leaving behind ghostly white patches in the reef (**FIGURE 12.16B**). Coral bleaching is not entirely understood. For example, it is not known exactly why the zooxanthellae leave during coral bleaching—or even *if* they leave; it has been hypothesized, alternatively, that they may lose their pigmentation (hence the "bleaching"), possibly as a result of *photodamage* caused by exposure to ultraviolet radiation.[6]

Coral bleaching may result from stress caused by increased sea surface temperatures associated with global climate change, from changes in light levels in some shallow-water areas, from an influx of pollutants, or from some

(a) Coral reef community

(b) Bleached coral

FIGURE 12.16
Coral reefs provide food and shelter for a tremendous diversity of fish and other creatures **(a)**. However, reefs face multiple environmental stresses from human impacts, and many corals have died as a result of bleaching **(b)**, in which corals lose their zooxanthellae. Bleaching is evident in the whitened portion of this coral.

combination of these and other unknown factors, both natural and anthropogenic. In addition, coral reefs sustain significant damage when divers stun fish with cyanide or by throwing explosives over the side of the boat, a common fishing and fish collection practice (for the pet trade) in the waters of Indonesia and the Philippines. Tourism also has been a significant burden on coral reefs globally; each scuba diver who breaks off a small piece of a reef contributes to its demise. Even with ecotourism, which advocates taking only photographs, divers' feet and hands on the reefs can take a toll.

A more recently emerging problem, as global climate change proceeds, is that the ocean is becoming more acidic (**FIGURE 12.17**). Carbon dioxide from the atmosphere is exchanged freely with surface ocean water. As the carbon dioxide concentration in the atmosphere increases, the excess carbon dioxide reacts with sea water to form

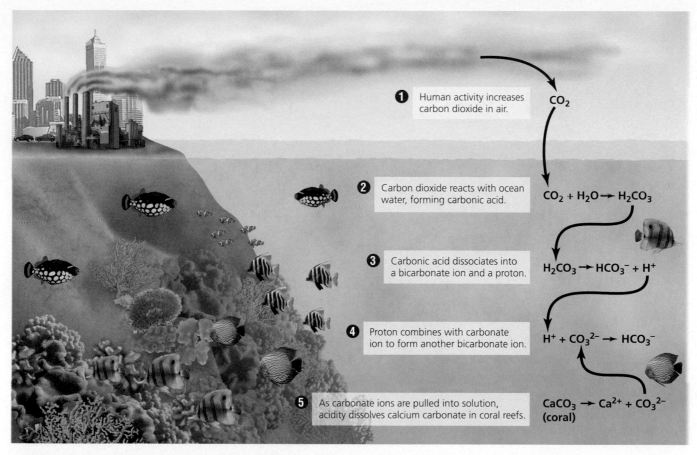

1 Human activity increases carbon dioxide in air.

CO_2

2 Carbon dioxide reacts with ocean water, forming carbonic acid.

$CO_2 + H_2O \rightarrow H_2CO_3$

3 Carbonic acid dissociates into a bicarbonate ion and a proton.

$H_2CO_3 \rightarrow HCO_3^- + H^+$

4 Proton combines with carbonate ion to form another bicarbonate ion.

$H^+ + CO_3^{2-} \rightarrow HCO_3^-$

5 As carbonate ions are pulled into solution, acidity dissolves calcium carbonate in coral reefs.

$CaCO_3 \rightarrow Ca^{2+} + CO_3^{2-}$
(coral)

FIGURE 12.17 Excess carbon dioxide from the atmosphere changes the ocean's chemistry. Carbon dioxide reacts with sea water to form carbonic acid, diminishing the number of carbonate ions available for corals and other organisms to use to build their shells and skeletons. *Source:* Republished with permission of American Association for the Advancement of Science (AAAS), from Coral Reefs under Rapid Climate Change and Ocean Acidification. *Science, 318*, pp. 1737–1742; permission conveyed through Copyright Clearance Center, Inc.

carbonic acid (H_2CO_3), and **ocean acidification** occurs. Even though the mass of the ocean is huge, the acidity of the entire surface layer has changed from a pH of 8.2 some 50 years ago to a present-day pH of 8.08—a 30% increase in acidity, measured by hydrogen ion concentration.[7] Acidification threatens to deprive corals of the carbonate ions they need to produce their structural parts. Other marine organisms that build shells and skeletons from calcium carbonate are also at risk.

Intertidal zones undergo constant change

Where the ocean meets the land, **intertidal** or **littoral** ecosystems (**FIGURE 12.18**) lie between the uppermost reach of the high tide and the lowest limit of the low tide. **Tides** are the periodic rising and falling of the ocean's height at a given location, caused by the gravitational pull of the Moon and Sun. The difference in water level between high and low tide in a location is called the *tidal range*. High and low tides occur roughly six to twelve hours apart, depending on the coastal location and characteristics. Consequently, intertidal organisms

spend part of each day submerged in water, part of the day exposed to the air and sunlight, and part of the day being lashed by waves. Subject to tremendous extremes in temperature, moisture, light exposure, salinity, and tidal range, these creatures must also protect themselves from marine predators at high tide and terrestrial predators— even opportunistic predators like seagulls and raccoons— at low tide.

The intertidal environment is a tough place to make a living, but it is home to a remarkable diversity of organisms. Rocky shorelines can be full of life among the crevices, which provide shelter and pools of water (*tidepools*) during low tides. Sessile animals, such as anemones, mussels, and barnacles, live attached to rocks, filter-feeding on plankton in the water that washes over them. Urchins, sea slugs, chitons, and limpets eat intertidal algae scraped from the rocks. Sea stars (starfish) creep slowly along, preying on the filter-feeders and herbivores at high tide. Crabs clamber around the rocks, scavenging detritus.

The rocky intertidal zone is so diverse because environmental conditions, such as temperature, salinity, and moisture, change dramatically from the high to the low reaches. This environmental variation gives rise to horizontal

Supratidal zone (splash zone)

Level of high tide

Intertidal zone

Level of low tide

Subtidal zone

Tidal zones

FIGURE 12.18 The intertidal zone is the swath of shoreline between the lowest and highest reaches of the tides. This is an ecosystem rich in biodiversity. Areas high on the shoreline are exposed to the air more frequently and for longer periods, so organisms that can best tolerate exposure specialize in the upper intertidal zone. The lower intertidal zone is exposed less frequently and for shorter periods, so organisms less able to tolerate exposure thrive there.

bands formed by dominant organisms as they array themselves according to their habitat needs. Sandy intertidal areas, such as those of Cape Cod, host less biodiversity, yet plenty of organisms burrow into the sand at low tide to await the return of high tide, when they emerge to feed.

Coastal ecosystems protect shorelines

Along many of the world's coasts at temperate latitudes, **salt marshes** occur where the tides wash over gently sloping sandy or silty substrates. Rising and falling tides flow into and out of channels called *tidal creeks* and at highest tide spill over onto elevated marsh flats, called *benches* (**FIGURE 12.19**).

Salt marshes boast very high primary productivity and provide critical habitat for shorebirds, waterfowl, and the adults and young of many commercially important fish and shellfish species. Salt marshes also filter out pollution (hence the use of constructed or artificial wetlands for wastewater management). Coastal marshes stabilize shorelines against **storm surges**, temporary increases in sea level that accompany the low atmospheric pressures of

intense storms, and storm waves, the very wind-driven large waves that crash onto the shore during a major storm.

However, people like to live along coasts, and coastal sites are highly desirable for commerce. As a result, vast expanses of salt marshes have been altered to make way for

U.S. Fish and Wildlife Service

FIGURE 12.19
Salt marshes, like the one shown here, occur in intertidal zones where the substrate is muddy, allowing salt-adapted grasses to grow. Tidal waters generally flow through marshes in tidal creeks, sometimes partially submerging the grasses.

Elizabeth A. Sellers/USGS

FIGURE 12.20
Mangrove forests are important ecosystems along tropical and subtropical coastlines throughout the world. Mangrove trees, such as these red mangroves (*Rhizophora mangle*) at Lizard Island, Australia, have specialized adaptations for growing in salt water and provide diverse habitats.

coastal development. When salt marshes are destroyed, we lose the ecosystem services they provide. When Hurricane Katrina struck the Gulf Coast of Louisiana, for instance, the flooding was made worse because vast areas of salt marshes had vanished because of development, subsidence from oil and gas drilling, and dams that had held back marsh-building sediment.

In tropical and subtropical latitudes, mangrove forests occur along gently sloping sandy and silty coasts. The **mangrove** is a tree with a unique type of root system that curves upward (like a snorkel) to obtain oxygen lacking in the mud in which the tree grows, and downward (like stilts) to support the tree in changing water levels (**FIGURE 12.20**). Fish, shellfish, crabs, snakes, and other organisms thrive among the root networks, and birds feed and nest in the dense foliage of these coastal forests. Besides serving as nurseries for fish and shellfish that people harvest, mangroves also provide materials that people use for food, medicine, tools, and construction.

Mangrove forests in tropical areas have been destroyed as people have developed coastal areas for residential, commercial, and recreational uses. Shrimp farming, in particular, has driven the conversion and destruction of large areas of mangroves. Half the world's mangrove forests have been eliminated, and their area continues to decline by 2–8% per year. When mangroves are removed, coastal areas lose the ability to slow runoff, filter pollutants, and retain soil. As a result, offshore systems, such as coral reefs, are more readily degraded.

Mangrove forests protect coastal communities against storm surges and tsunamis (seismic sea waves). This was demonstrated when the 2004 Indian Ocean tsunami devastated areas where mangroves had been removed, but

caused less damage where mangroves were intact. The loss of coastal mangroves may also have played a role in the scale of devastation from Hurricane Nargus in Burma (Myanmar). Despite these important ecosystem services provided by mangroves, only about 1% of the world's remaining mangroves are protected against development.

Fresh water meets salt water in estuaries

Many salt marshes and mangrove forests occur in or near **estuaries**, water bodies where rivers flow into the ocean, mixing fresh water with salt water. In an estuary, water from the ocean is diluted by water coming in via river channels from land-based runoff, so the resulting water is *brackish*—saltier than fresh water, but not as salty as sea water. Estuaries are biologically productive ecosystems that experience fluctuations in salinity as tides and freshwater runoff vary daily and seasonally. For shorebirds and for many commercially important shellfish species, estuaries provide critical habitat. For *anadromous* fishes (fishes, such as salmon, that spawn in fresh water and mature in salt water), estuaries provide a transitional zone where young fish make the passage from fresh water to salt water.

Estuaries around the world have been affected by urban and coastal development, water pollution, habitat alteration, and overfishing. The Gaspé Peninsula of Quebec, where the St. Lawrence River flows into the Gulf of St. Lawrence and from there into the Atlantic Ocean, is one area in Canada where estuaries and salt marshes are important ecosystems. Florida Bay, where fresh water from the Everglades system mixes with salt water, is a major example in the United States. This estuary has suffered pollution and a reduction in freshwater flow caused by irrigation and fertilizer use by sugarcane farmers, as well as housing development, septic tank leakage, and other human impacts.

Marine Resources: Human Use and Impacts

People have a long history of interacting with the ocean. Our ancestors travelled across ocean waters, clustered their settlements along coastlines, and were fascinated, as we are, by the beauty, power, and vastness of the seas. We have also left our mark upon them by exploiting the ocean for resources and polluting the waters with our waste. Coastal ecosystems, even more than the open ocean, have borne the brunt of human impacts, because two-thirds of Earth's people live within about 150 km of the ocean.

The ocean provides transportation routes

We have used the ocean for transportation for thousands of years, and the ocean continues to provide affordable means of moving people and products over vast distances. Even today, with other means of transportation available, countries that are land-locked and have no access to the ocean are economically disadvantaged.

Ocean shipping has had substantial impacts on the environment as well. Heavily used shipping lanes are by far the most polluted areas of the open ocean. The thousands of ships plying the world ocean today carry everything from cod to cargo containers to crude oil. Ships transport ballast water, which, when discharged at ports of destination, may transplant aquatic organisms picked up at ports of departure. Some of these species, such as the zebra mussel, establish themselves and become invasive.

We extract energy and minerals

Today we mine the ocean for commercially valuable energy. Worldwide, about 30% of our crude oil and nearly half of our natural gas come from seafloor deposits. Most offshore oil and gas is concentrated in petroleum-rich regions, such as the North Sea and the Gulf of Mexico, but energy companies extract smaller amounts of oil and gas from diverse locations, among them the Grand Banks and adjacent Canadian waters.

Ocean sediments also contain a novel potential source of fossil fuel energy. **Methane hydrate** is an ice-like solid consisting of molecules of methane (CH_4, the main component of natural gas) embedded in a crystal lattice of water molecules. Methane hydrates are stable at temperature and pressure conditions found in many sediments on the Arctic seafloor and the continental shelves. It is estimated that the world's deposits of methane hydrates may hold twice as much carbon as all known deposits of oil, coal, and natural gas combined.

Could methane hydrates be developed as an energy source to power our civilization through the twenty-first century and beyond? Perhaps, but a great deal of research remains before scientists and engineers can be sure how to extract these energy sources safely. Destabilizing a methane hydrate deposit could lead to a catastrophic release of gas. This could cause a massive landslide and tsunami, and would release huge amounts of methane, a potent greenhouse gas, into the atmosphere, exacerbating global climate change.

Fortunately, the ocean also holds potential for providing renewable energy sources that do not emit greenhouse gases. Engineers have developed ways of harnessing energy from waves, tides, and the heat of ocean water. These promising energy sources await further research, development, and investment.

We can extract minerals from the ocean floor, as well, although the logistical difficulty has kept seafloor mining only marginally economical thus far. Using large, vacuum cleaner–like hydraulic dredges, miners collect sand and gravel from beneath the sea. Also extracted are sulphur from salt deposits in the Gulf of Mexico and phosphorite from offshore areas near the California coast and elsewhere. Other valuable minerals found on or beneath the seafloor include calcium carbonate (used in making cement) and silica (used as fire-resistant insulation and in manufacturing glass), as well as rich deposits of manganese, copper, zinc, silver, and gold ore. Many minerals are concentrated in manganese nodules, small ball-shaped accretions that are scattered across parts of the ocean floor. More than 1.5 trillion tonnes of manganese nodules may exist in the Pacific Ocean alone, and their reserves of metal may exceed all terrestrial reserves.

Seafloor mineral development has been one of the most controversial factors in the very long history of attempts to convene a comprehensive, binding Law of the Sea, an international legal discussion that actually began as early as the 1600s. The United States still abstains from participating in some aspects of the Law of the Sea, primarily because of proposed restrictions on seafloor mining. The existing Law of the Sea, or *United Nations Convention on the Law of the Sea (UNCLOS)*, is based on a series of international conferences that took place between 1973 and 1982. This version of UNCLOS established the 200-nautical-mile Exclusive Economic Zones of nations (replacing the previous 12-mile zone), which greatly enhanced the ability of nations to control and manage their own coastal zones.

Marine pollution threatens resources and marine life

People have long used the ocean as a sink for waste and pollution. Even into the mid-twentieth century, it was common for coastal cities to dump trash and untreated sewage along their shores. Halifax only began to treat municipal sewage outflow into Halifax Harbour—approximately 181 million litres per day[8]—in the early 2000s. A surprising number of Canadian towns continue to discharge raw sewage into coastal and inland waters, even today. Fort Bragg, a bustling town on the northern Californian coast, boasts of its Glass Beach, an area where beachcombers collect sea glass, the colourful surf-polished glass sometimes found on beaches after storms. But Glass Beach is in fact the site of the former town dump, and besides well-polished glass, the perceptive visitor may also spot old batteries, rusting car frames, and other trash protruding from the bluffs above the beach.

Oil, plastic, industrial chemicals, and excess nutrients all eventually make their way from land into the ocean. Raw sewage and trash from cruise ships and abandoned fishing gear from fishing boats add to the input. The scope of trash in the sea can be gauged by the amount picked up each September by volunteers who trek beaches in the Ocean Conservancy's International Coastal Cleanup. In this nonprofit organization's twenty-fifth annual cleanup, almost 500 000 people from more than 100 nations picked up 16 million kilograms of trash.

Plastic bags and bottles, cigarette butts, discarded fishing nets, gloves, fishing line, buckets, floats, abandoned cargo, and much else that people transport on the sea or deposit into it can harm marine organisms. Because most plastic is not biodegradable, it can drift for decades before washing up on beaches. Marine mammals, seabirds, fish, and sea turtles may mistake floating plastic debris for food and can die as a result of ingesting material they cannot digest or expel. Fishing nets and crab and lobster traps that are lost or intentionally discarded can continue snaring animals for decades (**FIGURE 12.21**).

The so-called "Great Pacific Garbage Patch" is actually one of several large masses of anthropogenic debris, roughly 20 million km², floating in the pelagic zone, trapped in the circular currents of the North Pacific gyre (**FIGURE 12.22A**). The existence and extent of ocean "garbage patches" have been documented since 1979 by the U.S. National Oceanic and Atmospheric Administration. Much of the debris that accumulates in the gyres consists of trash items, but much of it is very small plastic particles—often so fine that they are invisible to the human eye (**FIGURE 12.22B**). Plastics are designed to last for a very long time, and plastic debris is increasing in volume. The impacts of microplastic particles on

FIGURE 12.21
This grey seal became entangled in a discarded fishing net. Each year, many thousands of marine mammals, birds, and turtles are killed by plastic debris, abandoned fishing nets, and other trash that people have dumped in the ocean.

animals that ingest them are not well known, so plastic in marine debris is a growing concern.

Of 115 marine mammal species, 49 are known to have eaten or become entangled in marine debris, and 111 of 312 species of seabirds are known to ingest plastic. All five species of sea turtle in the Gulf of Mexico have died from consuming or contacting marine debris. Marine debris affects people, as well. Surveys of fishers have shown that more than half have encountered equipment damage and other problems from plastic debris.

Marine oil pollution comes from many sources

Major oil spills, such as the 1989 *Exxon Valdez* spill in Prince William Sound, Alaska, and the massive BP Deepwater Horizon spill of 2010 in the Gulf of Mexico, make headlines and cause serious environmental problems. However, the majority of oil pollution in the ocean comes not from large spills in a few particular locations but from the accumulation of innumerable, widely spread small sources (called *nonpoint sources*), including leakage from small boats and runoff from human activities on land. Moreover, the amount of petroleum spilled into the ocean in a typical year is equalled by the amount that seeps into the water from naturally occurring seafloor deposits (**FIGURE 12.23**). (This was not the case in 2010, when the massive spill from the BP Deepwater Horizon rig exceeded all other human sources of oil spilled that year.)

Minimizing the amount of oil we release into coastal waters is important, because petroleum pollution is detrimental to marine life and to our economies. Petroleum

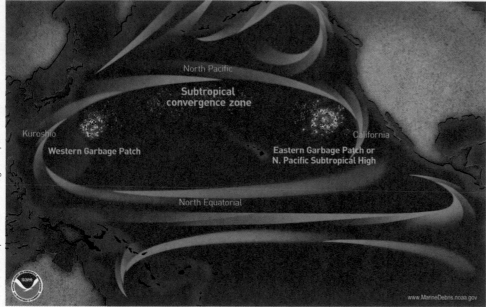

From Great Pacific Garbage Patch, National Oceanic and Atmospheric Administration. Retrieved from http://marinedebris.noaa.gov/info/patch.html#8.

(a) Anthropogenic debris in the ocean

National Oceanic and Atmospheric Administration

(b) Plastic marine debris

FIGURE 12.22
Anthropogenic debris in the ocean is transported by surface currents, and becomes trapped within the major gyres in the ocean, sometimes remaining for many years **(a)**. A large proportion of marine debris is plastics **(b)**. Plastic particles wear down, becoming smaller and smaller. Eventually, as "microplastics," they are ingested by marine organisms at all levels in the oceanic food chain.

can physically coat and kill marine organisms and can poison them when ingested. In response to headline-grabbing oil spills, governments worldwide have begun to implement more stringent safety standards for tankers, such as requiring industry to pay for tugboat escorts in sensitive and hazardous coastal waters, double hulls to preclude punctures, and the development of prevention and response plans for major spills. The oil industry has resisted many such safeguards. Today, the ship that oiled Prince William Sound in the *Exxon Valdez* spill is still sailing, renamed the *Sea River Mediterranean*, and still featuring only a single hull.

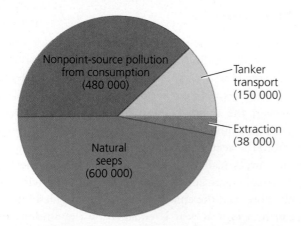

FIGURE 12.23
Of the 1.3 million metric tons of petroleum entering the ocean in a typical year, nearly half is from natural seeps. Nonpoint source pollution from petroleum consumption by people accounts for 38% of total input, spills during petroleum transport account for 12%, and leakage during petroleum extraction accounts for just 3%. In 2010 the BP Deepwater Horizon spill released about 800 000 metric tons of oil into the Gulf of Mexico.
Source: Data from National Research Council (2003). *Oil in the Sea III. Inputs, Fates, and Effects.* Washington, DC: National Academies Press.

Pollutants can contaminate seafood

Marine pollution can make some fish and shellfish unsafe for people to eat. One prime concern today is mercury contamination. Mercury is a toxic heavy metal that is emitted in coal combustion and from other sources. After settling onto land and water, mercury bioaccumulates in animals' tissues and biomagnifies as it makes its way up the food chain. As a result, fish and shellfish at high trophic levels (that is, fish that are large, long-lived, and high on the food chain, such as tuna, swordfish, and shark) can contain substantial levels of mercury. Eating seafood high in mercury is particularly dangerous for young children and for pregnant or nursing mothers, because the fetus, baby, or child can suffer neurological damage as a result.

Pollution from fertilizer runoff or other nutrient inputs can spur unusually rapid growth of phytoplankton, causing eutrophication and hypoxia. The number of known marine hypoxic zones is increasing globally, with about 500 documented so far.[9] In North America, the Gulf of Mexico and Chesapeake Bay may be the most severely affected. Excessive nutrient concentrations sometimes give rise to population explosions among several species of marine algae that produce powerful toxins that attack the nervous systems of vertebrates. Blooms of toxic algae occur periodically on both the east and west coasts of Canada.

Some algal species produce reddish pigments that discolour surface waters, and blooms of these species are nicknamed **red tides** (**FIGURE 12.24**). Harmful algal blooms can cause illness and death among zooplankton, birds, fish, marine mammals, and people as their toxins are passed up the food chain. They also cause economic loss

FIGURE 12.24
In a harmful algal bloom, certain types of algae multiply to great densities in surface waters, producing toxins that can bioaccumulate and harm organisms. Red tides are a type of harmful algal bloom in which marine microorganisms produce a pigment that turns the water red.

for communities dependent on fishing or beach tourism. Reducing nutrient runoff into coastal waters can lessen the frequency of these outbreaks. We can minimize the health impacts by monitoring blooms to prevent human consumption of affected organisms.

Emptying the Ocean

The ocean and its biological resources have provided for human needs for thousands of years, but today we are placing unprecedented pressure on marine resources. According to the UN Food and Agriculture Organization (FAO), half the world's marine fish populations are fully exploited, meaning that we cannot harvest them more intensively without depleting them. An additional 28% of marine fish populations are overexploited and already being driven toward extinction. Thus only one-fifth of the world's marine fish populations can yield more than they are already yielding without being driven into decline.

We have long overfished

People have always harvested fish, shellfish, turtles, seals, and other animals from the ocean. Although much of this harvesting was sustainable, scientists are learning that people began depleting some marine species centuries ago. Overfishing accelerated during the colonial period of European expansion and intensified further in the twentieth century.

Ancient overharvesting likely affected ecosystems in ways we only partially understand today. Several large animals, including the Caribbean monk seal, Steller's sea cow, and the Atlantic grey whale, were hunted to extinction prior to the twentieth century—before scientists were able to study them or the ecological roles they played. Overharvesting of the vast oyster beds of the Chesapeake Bay led to the collapse of its oyster fishery in the late nineteenth century; eutrophication and hypoxia resulted, because there are no oysters to filter algae and bacteria from the water. In the Caribbean, green sea turtles ate sea grass and likely kept it cropped low; with today's turtle population a fraction of what it was formerly, sea grass grows thickly, dies, and rots, giving rise to disease. The best-known case of historical overharvesting is the near-extinction of many species of whales, which resulted from commercial whaling that began centuries ago. Since limitations were imposed in 1986 some species of whale have been recovering, but others have not.

As discussed in our Central Case, groundfish in the Northwest Atlantic historically were so abundant that the people who harvested them never imagined they could be depleted. Yet careful historical analysis of fishing records has revealed that even in the nineteenth century fishers repeatedly experienced locally dwindling catches. Each time, they needed to introduce a new approach or technology to extend their reach and restore their catch rate.

Even so, fish catches continued to increase decade after decade, largely because of significant improvements in fish-finding and harvesting technologies. Eventually, after decades of increases, the total global fisheries catch levelled off after about 1988 (**FIGURE 12.25**), despite continued increases in fishing effort. Fishery collapses, such as those off Newfoundland and Labrador and New England, have taken a severe economic toll on communities that depend on fishing, and they are ecologically devastating. Recent scientific studies have predicted additional catastrophic stock collapses. A controversial 2006 study by Boris Worm of Dalhousie University and others, published in the journal *Science*, predicted that if trends in biodiversity losses in the ocean were to continue, populations of *all* ocean species that we fish for today would collapse by the year 2048.[10]

As our population grows, we will become even more dependent on ocean bounty. This makes it vital, say many scientists and fisheries managers, to turn immediately to more sustainable fishing practices. Integrated fisheries management approaches that bring a variety of stakeholders together in decision-making, focusing on the biological limitations of ecosystems, and making use of lessons learned from successful recoveries will make this transition possible.[11]

FIGURE 12.25
After increasing for many decades, the total global fisheries catch has stalled. Many scientists fear that a global decline is imminent if conservation measures are not taken soon. The figure shows trends with and without data on China's substantial fishing industry, which had been withheld from international scrutiny for many years.
Source: Data from the Food and Agriculture Organization of the United Nations, 2012. *The State of World Year Fisheries and Aquaculture 2012.*

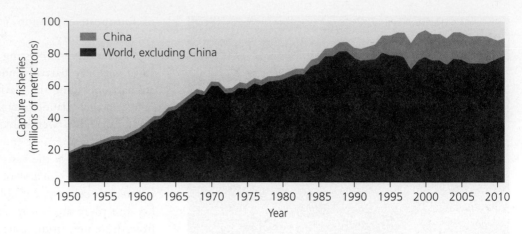

Modern commercial fishing is highly efficient

Modern commercial fishing fleets are highly industrialized and efficient. So-called *factory fishing* vessels are huge ships that use large amounts of fossil fuels and employ powerful fish-finding technologies to capture fish in volumes never dreamed of by nineteenth-century mariners (**FIGURE 12.26A**). These vessels can travel enormous distances, and even process and freeze their catches on board while at sea. The global reach of today's industrialized fleets makes their impacts much more rapid and intensive than in past centuries.

Smaller vessels (**FIGURE 12.26B**) are still economically important in commercial fishing. For example, the salmon fishery in Bristol Bay, Alaska, which produces about half of the entire global harvest of sockeye salmon, consists mainly of driftnetters that range from 25 to 40 m in length. These smaller boats periodically drop their catch at floating

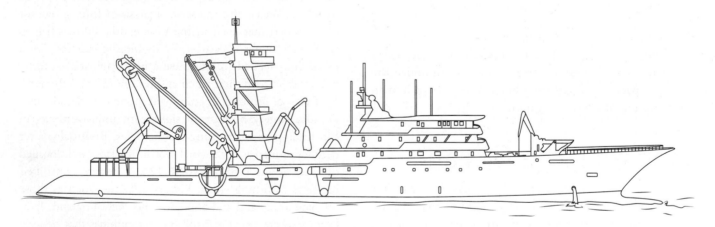

(a) Tuna seiner

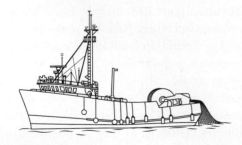

(b) Salmon driftnetter

(c) Dory

FIGURE 12.26
Commercial fishing vessels come in a very wide range of types and sizes. The three different examples shown here are drawn approximately to scale. They are **(a)** an enormous tuna seiner (up to 115 m in length); **(b)** a salmon driftnetter (ranging from about 25 to 40 m in length); and **(c)** a dory used for small-scale commercial and recreational fishing (traditional boats can range from a couple of metres to 15 m or more in length).

processing facilities called *tenders* so that they can remain in the active fishing area while the fishery is open.

Still smaller vessels (**FIGURE 12.26C**) differ widely in size and type in different parts of the world. These can be used for commercial fishing (and small-scale vessels can still use high-tech fish-finding equipment), for recreational fishing, or for subsistence fishing. Small-scale fishing, sometimes called *artisanal fishing*, is extremely important in the global economy. Although it requires much lower fossil fuel and dollar investments, artisanal fishing employs vastly greater numbers of fishers and produces harvests that rival those of the large factory vessels, overall.

The modern fishing industry uses a number of methods to capture fish at sea. Some vessels set out long *driftnets* that span large expanses of water (**FIGURE 12.27A**). These strings of nets are arrayed strategically to drift with currents so as to capture passing fish and are held vertical by floats at the top and weights at the bottom. Driftnetting usually targets species that traverse the open water in immense schools (flocks), such as herring, sardines, and mackerel. Specialized forms of driftnetting are used for sharks, shrimp, and other animals.

Longline fishing (**FIGURE 12.27B**) involves setting out extremely long lines with up to several thousand baited hooks spaced along their lengths. Tuna and swordfish are among the species targeted by longline fishing.

Trawling entails dragging immense cone-shaped nets through the water, with weights at the bottom and floats at the top to keep the nets open. Trawling in open water captures pelagic fish, whereas *bottom-trawling* (**FIGURE 12.27C**) involves dragging weighted nets across the floor of the continental shelf to catch groundfish and other benthic organisms, such as scallops.

Some fishing practices kill nontarget animals and damage ecosystems

Unfortunately, some fishing practices catch more than just the species they target. **By-catch** refers to the accidental capture of nontarget animals, and it accounts for the deaths of many thousands of fish, sharks, marine mammals, and birds each year. For example, the FAO reports that in South and Central American shrimp fisheries in the 1990s, for each 1 kg of shrimp caught, 10–33 kg of other animals were caught; 30% of the by-catch was used, while the rest was discarded as trash.[12]

Driftnetting captures substantial numbers of dolphins, seals, and sea turtles, as well as countless nontarget fish. Most of these animals end up drowning (mammals and turtles need to surface to breathe) or dying from air exposure on deck (fish breathe through gills in the water). Many nations have banned or restricted driftnetting

(a) Driftnetting

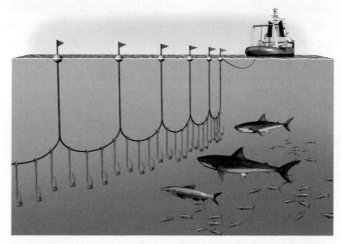

(b) Longlining

(c) Bottom-trawling

FIGURE 12.27
Commercial fishing vessels, both large and small, use several main methods of capture. In driftnetting **(a)**, huge nets are set out to drift through the open water to capture schools of fish. In longlining **(b)**, lines with numerous baited hooks are set out in open water. In bottom-trawling **(c)**, weighted nets are dragged along the floor of the continental shelf. All methods result in the capture of nontarget animals. These illustrations are schematic and do not capture the immense scale that these technologies can attain; for instance, industrial trawling nets can be large enough to engulf multiple jumbo jets.

because of excessive by-catch. The widespread deaths of dolphins in tuna driftnets motivated consumer efforts to label tuna as "dolphin-safe" if its capture uses methods that avoid dolphin by-catch. Such measures helped reduce dolphin deaths from an estimated 133 000 per year in 1986 to about 2000 per year since 1998, although statistics about dolphin by-catch are difficult to access.

Similar by-catch problems exist with longline fishing, which kills turtles, sharks, and albatrosses, magnificent seabirds with wingspans up to 3.6 m. Several methods are being developed to limit by-catch from longline fishing, but an estimated 300 000 seabirds of various species die each year when they become caught on hooks while trying to ingest bait.

Bottom-trawling can destroy entire communities and ecosystems. The weighted nets crush organisms in their path and leave long swaths of damaged sea bottom. Trawling is especially destructive to structurally complex areas, such as reefs, that provide shelter and habitat for many animals. In recent years, underwater photography has begun to reveal the extent of structural and ecological disturbance done by trawling (**FIGURE 12.28**). Trawling is often likened to clear-cutting and strip-mining, and in heavily fished areas, the bottom may be damaged more than once. At Georges Bank, it is estimated that the average expanse of bottom has been trawled three times.

We can see the effects of large-scale industrialized fishing in the catch records of groundfish from the Northwest Atlantic, discussed in our Central Case. Although cod had been harvested since the 1500s on the Grand Banks, catches more than doubled once immense industrial trawlers from Europe, Japan, and the United States appeared in the 1960s (**FIGURE 12.29**). These record-high catches lasted only a decade; the industrialized approach removed so many fish that the stock has not recovered. A similar pattern was observed on Georges Bank.

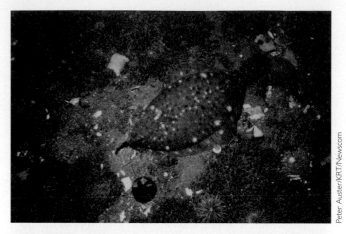

(a) Before trawling

(b) After trawling

FIGURE 12.28
Bottom-trawling causes severe structural damage to reefs and benthic habitats, and it can decimate underwater communities and ecosystems. A photo of an untrawled location **(a)** on the seafloor of Indonesia shows bottom-dwelling fish and a vibrant and diverse coral reef community. A photo of a trawled location **(b)** nearby shows a flattened and lifeless expanse of broken coral.

FIGURE 12.29
In the North Atlantic off the coast of Newfoundland and Labrador, commercial catches of Atlantic cod increased with intensified fishing by industrial trawlers in the 1960s and 1970s. The fishery subsequently crashed, and moratoria imposed in 1992 and 2003 have not yet brought it back. The first peak and decline (before 1977) resulted from foreign fishing fleets; the second peak and decline (after 1977) resulted from Canadian fleets, after Canada laid claim to the 200-nautical-mile Exclusive Economic Zone. *Source:* From Millennium Ecosystem Assessment, 2005. Retrieved from http://www .millenniumassessment.org/documents/ document.356.aspx.pdf. Copyright © 2005 by World Resources Institute. Reprinted by permission.

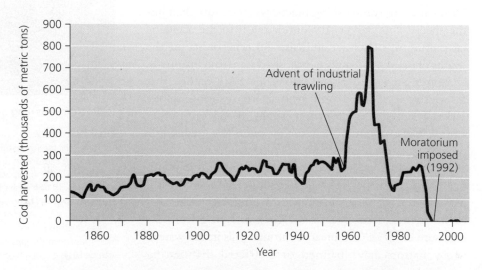

Overfishing, habitat destruction, and other factors that deplete biodiversity can threaten the ecosystem services we derive from the ocean. In the 2006 study in *Science* that predicted global fisheries' collapse by 2048,[10] the authors analyzed the scientific literature to summarize the effects of biodiversity loss on ecosystem function and ecosystem services. They found that across 32 different controlled experiments conducted by various researchers, systems with less species diversity or genetic diversity were less productive, and less able to withstand disturbance. The team also found that when biodiversity was reduced, so were habitats that serve as nurseries for fish and shellfish. Moreover, biodiversity loss was correlated with reduced filtering and detoxification (as from wetland vegetation and oyster beds), which can lead to harmful algal blooms, dead zones, fish kills, and beach closures.

Several factors can mask stock declines

Today's industrialized fishing fleets have the capacity to deplete marine populations very quickly. In a widely quoted 2003 study, Canadian fisheries biologists Ransom Myers and Boris Worm analyzed fisheries data from FAO archives, looking for changes in the catch rates of fish in various regions of ocean since they were first exploited by industrialized fishing. For one region after another, they found the same pattern: Catch rates dropped precipitously after the initiation of large-scale commercial fishing, with 90% of large-bodied fish and sharks eliminated within only a decade. This suggests that the composition of some marine communities may have been very different prior to the advent of industrialized fishing.

Although industrialized fishing has depleted fish stocks in region after region, the overall global catch has remained roughly stable for two decades (see **FIGURE 12.25**). The seeming stability of the total global catch can be explained by several factors that may be masking population declines. One is that fishing fleets have been travelling longer distances to reach less-fished portions of the ocean. They also have been fishing in deeper waters; average depth of catches was 150 m in 1970 and 250 m in 2000. Moreover, fishing fleets have been spending more time fishing and have been setting out more nets and lines—expending increasing effort just to catch the same number of fish.

Improved technology also helps explain large catches despite declining stocks. Modern fleets can reach almost any spot on the globe with high-speed vessels. They have access to an array of modern fish-finding technologies, including advanced sonar mapping equipment, satellite navigation, and thermal sensing systems. Some fleets rely on aerial spotters to find schools of commercially valuable fish, such as bluefin tuna. One final cause of misleading stability

in global catch numbers is that not all data supplied to international monitoring agencies may be accurate, for a variety of reasons, including economic pressure to falsify data.

A further problem is that the species, age, and size of fish harvested are often not reported. Careful analyses of fisheries data have revealed that as fishing increases, the size and age of fish caught tend to decline. Cod caught in the Northwest Atlantic today are on average *much* smaller than they were decades ago, and it is now rare to find a cod more than 10 years of age, although cod of this age formerly were common. The reproductive potential of today's smaller cod is much less—an order of magnitude less—than that of the giant cod of previous decades. However, this loss is easily masked in the way the harvest is reported.

As particular species become too rare to fish profitably, fleets begin targeting other species that are in greater abundance. Generally this means shifting from large, desirable species to smaller, less desirable ones. Because this often entails catching species at lower trophic levels, this phenomenon has been termed "fishing down the food chain," a concept proposed by University of British Columbia (UBC) fisheries scientist Daniel Pauly and colleagues in 1998. Some undesirable species that fishers formerly threw back when fishing for more marketable species have undergone "image makeovers" to aid their sale to consumers. For example, the species of fish now called "orange roughy" was called "slimehead" by fishers because of its mucus canals. Similarly, the toad-coloured "toothfish" that fishers once threw overboard has found new life as Chilean sea bass, even though the species is not biologically classified as a sea bass.

Fisheries managers, government and industry representatives, and marine scientists are now working together to try to ensure that these problems and practices do not lead to more stock declines. Scientists conduct surveys, study fish population biology, and monitor catches. Managers then use that knowledge to regulate the timing of harvests, the techniques used to catch fish, and the scale of the allowable harvest.

The goal is to allow for maximal harvests of particular populations, while keeping enough fish available to breed new stock for the future—the concept of **maximum sustainable yield (MSY)** (**FIGURE 12.30**). MSY is a resource management principle that can be applied to any renewable resource, including fish and trees. However, its successful application in real situations is complicated. If the yield is set too high, the fish population is overestimated, or environmental factors limit the recovery of the population after harvesting, there is a real danger that stocks may be overharvested. If data indicate that current yields are unsustainable, managers can limit the number or total biomass of that species that can be harvested, or they might restrict the type of gear that fishers are allowed to use or the amount of time they can spend fishing. In extreme cases of

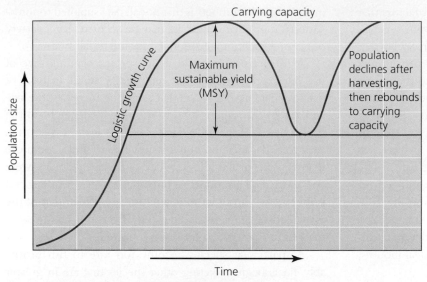

(a) Harvesting to the maximum sustainable yield

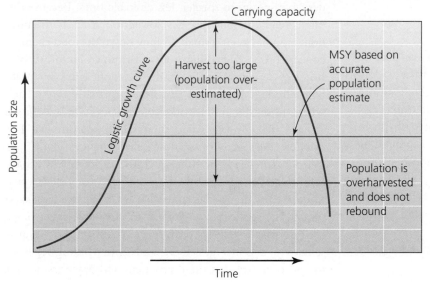

(b) Harvesting past maximum sustainable yield

FIGURE 12.30
Maximum sustainable yield (MSY) is a management principle that can be applied to any renewable resource. Before harvesting, the fish population grows exponentially until it is limited by carrying capacity, yielding a typical logistic growth curve (as discussed in Chapter 4, *Evolution, Biodiversity, and Population Ecology*). In **(a)**, scientists and fisheries managers have accurately estimated the population, which rebounds to carrying capacity after harvesting. If MSY is set accurately, the harvest can be repeated sustainably. In **(b)**, the fish population has been overestimated and the yield is too high; the stock is overharvested, and the population does not rebound.

stock collapse, such as in the Atlantic cod fishery, it may be necessary to close the stock to commercial fishing altogether.

One key change in fishery management where successes have been achieved has been to shift the focus away from individual species and toward viewing marine resources as elements of larger ecological systems. This ecosystem-based approach means keeping in mind the biological limitations and environmental pressures on ecosystems. It requires adaptive management and consideration of the impacts of fishing practices on habitat quality, on species interactions, and on other factors that may have indirect or long-term effects on populations.

Aquaculture has benefits and drawbacks

There has been a dramatic increase in farm fisheries, or **aquaculture**, over the past few decades, leading to speculation that aquaculture might one day replace wild fisheries altogether. From a small percentage of the total fish harvest only 20 years ago, aquaculture now accounts for about 50% of the world production of fish and seafood for human consumption. This dramatic growth is expected to continue. Freshwater fish species—including rainbow trout, brook trout, arctic char, and tilapia, as well as Atlantic coho and chinook salmon—are currently farmed in Canada. Canada is the fourth-largest producer of farmed salmon in the world.[13] The Canadian aquaculture industry now generates $2 billion in total economic activity ($1 billion in GDP), with British Columbia as the largest producer. Shellfish—including clams, oysters, scallops, and mussels—are also raised in controlled marine environments, as are seaweed and other aquatic algae and plants, and oysters for pearl production.

Aquaculture can help improve a region's or nation's food security by increasing overall amounts of fish available.

It also reduces pressure on overharvested and declining wild stocks, as well as reducing by-catch and providing employment for fishers who can no longer fish from depleted natural stocks. Aquaculture also relies far less on fossil fuels than do fishing vessels and provides a safer work environment. Fish farming can be remarkably energy-efficient, producing 10 times as much fish per unit area as is harvested from oceanic waters on the continental shelf, and up to 1000 times as much as is harvested from the open ocean.

Along with its benefits, though, aquaculture can have some serious disadvantages, as discussed in Chapter 8, *Agriculture, Food, and Biotechnology*. Dense concentrations of farmed animals can increase the incidence of disease. This necessitates antibiotic treatments, which can have negative impacts on water quality and human health, and results in additional expense for owners. If farmed aquatic organisms escape into ecosystems where they are not native, they may spread disease to native stocks or may outcompete native organisms for food or habitat. The opposite has also occurred—recent research suggests that wild Pacific salmon swimming near aquaculture pens may pass parasites, which then spread rapidly as a result of the high population densities in the pens.

The high-density fish populations involved in aquaculture also produce a significant amount of waste, both from the farmed organisms and from the feed that goes uneaten and decomposes in the water column. Farmed fish often are fed grain, and growing grain to feed animals that we then eat can reduce the energy efficiency of food production and consumption. In other cases, farmed fish are fed fishmeal made from wild ocean fish, such as herring and anchovies, whose harvest may place additional stress on wild fish populations.

Aquaculture has also led to damaging landscape changes in some coastal ecosystems, including the removal of protective mangrove forests. Coastlines evolve geologically as natural barriers to the battering of storm waves and winds. When these barriers are removed or disrupted, shorelines are left defenceless. The world saw the effects of shoreline modification graphically illustrated during the devastating Sumatra–Andaman tsunami of December 24, 2004, a seismic sea wave generated by a major submarine earthquake off the coast of Indonesia. Many of the coastlines in this part of South Asia and the Pacific had been significantly altered for aquaculture pen construction. The tsunami met with little natural resistance as it rushed onshore.

Modern aquaculture facilities have worked hard in recent years to overcome these problems and address the negative impacts of the industry. New approaches have been implemented to prevent escapes, and to report on them when they do occur. Standards have been established for the siting of new facilities, and fish population densities are regulated to reduce stress, disease, and the production

of waste. Vaccines and disease treatments used on farmed fish must now be approved by Health Canada, with the overall goal of minimizing their use. You can read more about these and other steps taken by the Canadian aquaculture sector in the DFO's 2012 report on aquaculture in Canada.[13]

Marine Conservation

Because we bear responsibility and stand to lose a great deal if valuable marine ecological systems collapse, many marine scientists, government regulators, and fisheries managers have been working to develop solutions to the problems that threaten the ocean. Part of the solution may lie with consumers and the food choices we make.

Consumer choice can influence marine harvest practices

To most of us, marine fishing practices may seem a distant phenomenon over which we have no control. Yet by exercising careful choice when we buy seafood, we, as consumers, can influence the ways in which fisheries function. Purchasing eco-labelled seafood, such as dolphin-safe tuna, is one way to exercise choice, but in most cases consumers have no readily available information about how their seafood was caught.

Several nonprofit organizations have recently devised concise guides to help consumers make informed choices. These guides differentiate fish and shellfish that are over-fished or whose capture is ecologically damaging from those that are harvested more sustainably; **TABLE 12.1** provides some examples.

Modern whaling provides an example of consumer influences on marine harvesting practices. Whaling has been carried out for centuries by traditional societies from Scandinavia to northern Canada to Japan. Modern whaling began in the nineteenth century off the eastern coast of North America in response to the demand for whale oil and meat. By the middle of the twentieth century, the catch exceeded sustainable limits, and some species were close to extinction. In 1986 the International Whaling Commission banned whaling in an effort to allow stocks to replenish; this moratorium probably averted the extinction of several species, at least temporarily. There is a large market for whale meat in Japan, with the result that Japan officially objected to the moratorium. Japan continues whaling today, ostensibly for scientific research purposes, although it is widely believed that it is actually illegal commercial whaling. Whaling is carried out legally by Inuit communities in Canada's North, where whale meat is an important part of the traditional diet.

TABLE 12.1 Seafood Choices for Consumers

On the left are fish or shellfish from healthy, well-managed populations that are certified as caught or farmed in environmentally sustainable ways. On the right are fish or shellfish from wild sources that are overfished, have high by-catch, cause extensive habitat damage, or are farmed in ways that harm other marine life or the environment.

Sustainable Choices	Seafood to Avoid
Albacore tuna (U.S., Canada)	Bluefin tuna*
Atlantic mackerel (Canada)	Caviar (sturgeon, imported, wild)
Dungeness crab	Chilean seabass/toothfish*
Longfin squid (U.S.)	Atlantic halibut, flounders, soles
Mediterranean mussel	Mahi mahi/dolphinfish (imported, longline caught)
Oysters (wild, farmed)	Orange roughy
Pacific sardines (U.S., Canada)	Salmon (farmed or Atlantic)
Rainbow trout (farmed)	Blacktip shark*
Sablefish/Black cod (U.S., Canada)	Shrimp (imported)
Salmon (canned)	Tilapia (farmed in Asia)
Wild Alaskan salmon	Swordfish (imported)*

*Limit consumption because of concerns about mercury or other contaminants.

Source: Based on Environmental Defense Fund, March 2013. Safe Seafood and Responsible Fisheries. http://seafood.edf.org/guide/best.

More recently, illegal shark finning has raised consumer concerns. Shark fins are in great demand, especially in Asia, for use in shark fin soup and traditional cures. They are obtained by capturing the shark, cutting off its fins while it is still alive, and then releasing it to die, typically by suffocation, starvation, or predation. Some countries have banned shark finning, and some have even banned the use of products that use shark fins. In Canada, shark finning is illegal but it is not illegal to import the fins from elsewhere. However, some municipalities in Canada (Brantford, Ontario, was the first) have taken it upon themselves to ban the sale of shark fin products through by-laws. Other countries have instituted full or partial protections against shark finning, largely in response to pressure from consumers and environmentalists.

We can protect vulnerable areas in the ocean

Hundreds of **marine protected areas (MPAs)** have been established, most of them along the coastlines of developed countries[14] (**FIGURE 12.31**). However, despite their name, marine protected areas do not necessarily protect all their natural resources, because nearly all MPAs around the world allow fishing and other extractive activities, even oil drilling.

Because of the lack of true refuges from fishing pressure, many scientists—and some fishers—want to establish areas where fishing is prohibited. Such "no-take" areas have come to be called **marine reserves**. Designed to preserve entire ecosystems intact without human interference, marine reserves are also intended to improve fisheries. Scientists argue that marine reserves can act as production factories for fish for surrounding areas, because fish larvae produced inside reserves will disperse outside and stock other parts of the ocean. Proponents maintain that by serving both purposes, marine reserves are a win–win proposition for environmentalists and fishers alike.

Many fishers dislike the idea of no-take reserves, however. Nearly every marine reserve that has been established or proposed has met with pockets of intense opposition from people and businesses that use the area for fishing or recreation. Opposition comes from commercial fishing fleets as well as from individuals who fish recreationally. Both types of fishers are concerned that marine reserves will simply put more areas off-limits to fishing. In some parts of the world, protests have become violent. For instance, to protest fishing restrictions, fishers in the Galápagos Islands destroyed offices at Galápagos National Park and threatened researchers and park managers with death.

Reserves can work for both fish and fishers

In the past decade, data synthesized from marine reserves around the world have been indicating that reserves *can* work as win–win solutions that benefit ecosystems, fish populations, and fishing economies. A comprehensive review of data from existing marine reserves as of 2001 revealed that just one to two years after their establishment, marine reserves had increased densities of organisms, on average, by 91%; increased biomass of organisms by 192%; increased average size of organisms by 31%; and increased species diversity by 23%.

That same year, 161 prominent marine scientists signed a "consensus statement" summarizing the effects of marine reserves. Besides boosting fish biomass, total catch, and record-sized fish, the report stated, marine reserves yield several benefits. Within reserve boundaries, they

- produce rapid and long-term increases in abundance, diversity, and productivity of marine organisms

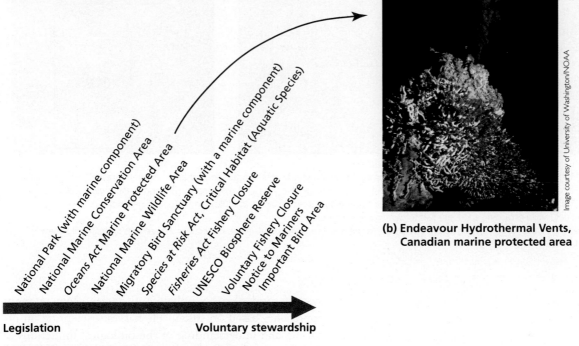

(b) **Endeavour Hydrothermal Vents, Canadian marine protected area**

(a) **Tools for protecting marine areas in Canada**

FIGURE 12.31 Tools that can be used to protect marine areas in Canada range from legislated to voluntary measures **(a)**. Canada's first official marine protected area, designated in 2003, was the Endeavour Hydrothermal Vents area on the Juan da Fuca Ridge off the west coast of Vancouver Island **(b)**.

■ decrease mortality and habitat destruction

■ lessen the likelihood of extirpation of species

Outside the reserve boundaries, marine reserves

■ can create a "spillover effect" when individuals of protected species spread outside reserves

■ allow larvae of species protected within reserves to "seed the seas" outside reserves

The consensus statement was backed up by research into reserves worldwide. At Apo Island in the Philippines, biomass of large predators increased eightfold inside a marine reserve, and fishing improved outside the reserve. At two coral reef sites in Kenya, commercially fished and keystone species were up to 10 times as abundant in the protected area as in the fished area. At Leigh Marine Reserve in New Zealand, snapper increased fortyfold, and spiny lobsters were increasing by 5–11% yearly. Spillover from this reserve improved fishing and ecotourism, and local residents who once opposed the reserve now support it. Since that time, further research has shown that reserves create a fourfold increase in catch per unit of effort in fished areas surrounding reserves, and that they can greatly increase ecotourism by divers and snorkellers.

If marine reserves work in principle, the question becomes how best to design the reserves and arrange them into networks. Scientists today are asking how large reserves need to be, how many there need to be, and where they need to be placed. Involving fishers directly in the planning process is crucial for coming up with answers to such questions. In Canada, marine reserves are managed by Parks Canada through the National Marine Conservation Areas Program, and their management plans are designed as partnerships among coastal communities; Aboriginal people; provincial, territorial, and federal government agencies; and other stakeholders.

Of several dozen studies that have estimated how much area of the ocean should be protected in no-take reserves, estimates range from 10% to 65%, with most falling between 20% and 50%. Other studies are modelling how to optimize the size and spacing of individual reserves so that ecosystems are protected, fisheries are sustained, and people are not overly excluded from marine areas (**FIGURE 12.32**). If marine reserves are designed strategically to take advantage of ocean currents, many scientists say, then they may well seed the seas and help lead us toward solutions to one of our most pressing environmental problems.

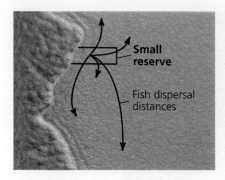 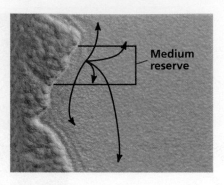 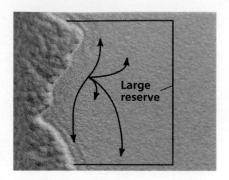

FIGURE 12.32   Marine reserves of different sizes have varying effects on ecological communities and fisheries. A small reserve (left) may fail to protect animals because too many disperse out of the reserve (red arrows). A large reserve (right) may protect fish and shellfish very well but will provide relatively less "spillover" into areas where people can legally fish. Medium-sized reserves (middle) may offer the best hope of preserving species and ecological communities, while providing adequate fish to human communities.
Source: Based on Halpern, B.S., and R.R. Warner (2003). Matching Marine Reserve Design to Reserve Objectives. *Proceedings of the Royal Society of London,* B 270, pp.1871–1878.

Conclusion

The ocean covers most of our planet and contains diverse topography and ecosystems, some of which we are only now beginning to explore and understand. We are learning more about the ocean and coastal environments while intensifying our use of their resources and causing these areas more severe impacts. In so doing, we are coming to understand better how to use these resources without depleting them or causing undue ecological harm to the marine and coastal systems on which we depend.

Today, scientists are demonstrating that setting aside protected areas of the ocean can serve to maintain natural systems and also to enhance fisheries. Management approaches based on sustainable harvests, scientific monitoring, and understanding of the biological limitations of ecosystems are demonstrating that marine resources can be managed successfully, and even rescued from steep declines. This is vital at a time when we are depleting many of the world's marine fish stocks. As historical studies reveal more information on how much biodiversity the ocean formerly contained and has now lost, we may increasingly look beyond simply making fisheries stable and instead consider restoring the ecological systems that once flourished in our waters.

REVIEWING OBJECTIVES

You should now be able to:

Identify physical, geographical, chemical, and biological aspects of the ocean and various marine environments

- The ocean covers 71% of Earth's surface and contains more than 97% of its surface water.
- Much of seafloor topography is rugged and complex.
- Ocean water contains 96.5% H_2O by mass and various dissolved salts.
- Cold, salty water is dense and sinks. The warmest surface water and the greatest difference between surface and deep water are in the tropics.
- Horizontal and vertical currents in the ocean are driven by variations in temperature, density, and wind. Surface and deep currents are connected by the global thermohaline circulation system.

- El Niño, La Niña, and the Southern Oscillation are part of a complex, cyclically varying regional weather pattern that demonstrates the important connections between ocean and atmosphere.

Describe major types of marine ecosystems

- Open-ocean pelagic and benthic systems vary widely in the type and abundance of life that they support. The photic zone is most productive, but even the deep ocean hosts uniquely adapted life forms.
- Shallow-water systems, including kelp forests and coral reefs, are particularly productive habitats.
- Intertidal zones are diverse but environmentally challenging. They undergo dramatic daily variations in temperature, salinity, and moisture.

- Coastal environments like salt marshes and mangroves provide unique habitats and stabilize shorelines, protecting them against storm surges.
- In estuaries, where rivers flow into the ocean, salt water and fresh water mix. Estuaries are particularly susceptible to the impacts of coastal development and human activities.

Outline human uses of marine resources and their impacts on marine environments

- For millennia, people have used ocean waters for transportation, withdrawn resources, and created settlements in coastal areas.
- Today we extract energy and some mineral resources from the ocean.
- People pollute ocean waters with trash, including plastic and nets that harm marine life and accumulate in oceanic gyres.
- Marine oil pollution results from nonpoint sources on land, as well as from tanker spills at sea.
- Heavy metal contaminants in seafood affect human health, and nutrient pollution can lead to harmful algal blooms.

Review the current state of ocean fisheries and reasons for their decline

- Half the world's marine fish populations are fully exploited, and 25% are overexploited. People began harvesting marine resources long ago but impacts have intensified in recent decades, and the global catch levelled off as of the late 1980s.

- Modern fishing practices and fish-finding technologies are extremely efficient. Commercial vessels range from enormous factory vessels to small-scale and artisanal boats. Techniques include driftnetting, longline fishing, and trawling.
- Many methods of industrialized fishing have negative environmental impacts, including by-catch and destruction of benthic ecosystems. Marine biodiversity loss negatively affects ecosystem services.
- Problems with data collection, sequential exploitation of stocks, and fishing down the food chain can mask stock declines. The newest approaches to fisheries management are integrated, adaptive, and ecosystem-based.
- Aquaculture has increased dramatically, now reaching 50% of world fish and seafood production. The industry is working to overcome the negative environmental impacts associated with intensive farming of marine organisms.

Evaluate marine protected areas and reserves as innovative solutions for marine conservation

- Consumers can encourage good fishery practices by shopping for sustainable seafood.
- We have established far fewer protected areas in the ocean than we have on land, and most marine protected areas still allow many extractive activities.
- Marine reserves, if thoughtfully designed and implemented, can protect ecosystems while also boosting fish populations and making fisheries sustainable.

TESTING YOUR COMPREHENSION

1. What is the average salinity of ocean water? How are density, salinity, and temperature related in each layer of ocean water?
2. What factors drive the system of ocean currents? In what ways do these movements affect conditions for life in the ocean?
3. Where in the ocean are the most productive areas of biological activity located?
4. Describe three kinds of ecosystems found near coastal areas and the kinds of life they support.
5. Why are coral reefs biologically valuable? How are they being degraded by human impact?

6. What is causing the disappearance of mangrove forests and salt marshes?
7. Discuss three ways in which people are combating pollution in the ocean and in coastal zones.
8. Describe an example of how overfishing can lead to ecological damage and fishery collapse.
9. Name three industrial fishing practices, and explain how they create by-catch and harm marine life.
10. How does a marine reserve differ from a marine protected area? Why do many fishers oppose marine reserves? Explain why many scientists say no-take reserves can still be good for fishers.

THINKING IT THROUGH

1. What benefits do you derive from the ocean? How does your behaviour affect the ocean, both directly and indirectly? Give specific examples.

2. We have been able to reduce the amount of oil we spill into the ocean, but petroleum-based products, such as plastic, continue to litter the ocean and shore-lines. Discuss some ways that we can reduce this threat to the marine environment.

3. Describe the trends in global fish capture over the past 50 years, and explain several factors that account for these trends.

4. Let's say that you are mayor of a coastal town where some residents are employed as commercial fishers and others make a living serving ecotourists who come to snorkel and scuba dive at the nearby coral reef. In recent years, several fish stocks have crashed, and ecotourism is dropping off as fish disappear from the increasingly degraded reef. Scientists are urging you to help establish a marine reserve around portions of the reef, but most commercial and recreational fishers are opposed to this idea. What steps would you take to restore your community's economy and environment?

INTERPRETING GRAPHS AND DATA

This graph illustrates how community-based management practices can protect marine organisms and benefit fishers, too. In Kenya, a network of protected areas was implemented in the 1990s, in cooperation with local communities. In some cases, the use of seine nets with high rates of by-catch was also prohibited. These measures led to a recovery of the biomass and size of available fish.

1. Describe the trends in fisher incomes after the various restrictions were implemented, and compare them to the trend and magnitude of fisher earnings in the area where no restrictions were implemented.

2. Which set of restrictions (or no restrictions) resulted in the highest earnings for fishers?

3. Explain how the establishment of no-take reserves and banning of highly unselective fishing gear could contribute to an increase in fishers' earnings.

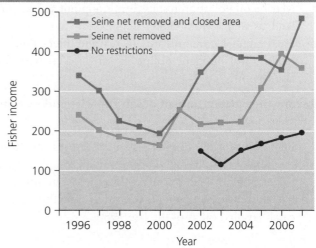

The graph shows changes in income of fishers in Kenya, subsequent to the implementation of some restrictions on fishing location and gear.
Source: Republished with permission of American Association for the Advancement of Science (AAAS), from Rebuilding Global Fisheries, *Science 325, 578.* Copyright 1980; permission conveyed through Copyright Clearance Center, Inc.

MasteringEnvironmentalScience®

STUDENTS
Go to **MasteringEnvironmentalScience** for assignments, the eText, and the Study Area with practice tests, videos, current events, and activities.

INSTRUCTORS
Go to **MasteringEnvironmentalScience** for automatically graded activities, current events, videos, and reading questions that you can assign to your students, plus Instructor Resources.

13 Atmospheric Science and Air Pollution

Atmospheric gases scatter blue wavelengths of light, giving Earth a blue halo. This astronaut photograph shows the outer edge of the atmosphere fading into the blackness of space.

NASA/JSC

Upon successfully completing this chapter, you will be able to

- Describe the composition, structure, and major processes of Earth's atmosphere
- Outline the scope of outdoor air pollution and assess potential solutions
- Explain stratospheric ozone depletion and identify steps taken to address it
- Characterize the main types of indoor air pollution and assess potential solutions

These photos, taken in 2012 (left) and 2013 (right) at the Palace Museum in Beijing, China, show the heavy pollution during the "airpocalypse" of 2013. Poor air quality conditions in some regions have continued or even worsened, and the government of China is now aggressively attacking this chronic problem.

Kyodo/Newscom

RUSSIA

MONGOLIA

Beijing

CHINA

INDIA

CENTRAL CASE
"AIRPOCALYPSE" IN BEIJING

"In a move to improve air quality, protect the climate, and reduce the health burden of air pollution, the Chinese Government is taking tough measures to prevent and control air pollution."

—ZHU CHEN, PRESIDENT OF THE CHINESE MEDICAL ASSOCIATION
(FORMER CHINESE MINISTER OF HEALTH)[1]

Although earlier-industrialized nations have been improving their air quality steadily over the past few decades, outdoor air pollution is growing worse in many currently industrializing countries. With its growing population and rush to economic and industrial development, China is now experiencing some of the world's worst air pollution.

Poor air quality became a serious concern for the health and performance of athletes during the Beijing Olympics of 2008. In January–February 2013, the situation became so severe that plane flights were cancelled, expressways closed, and people wore face masks in the streets to facilitate breathing. Levels of airborne particulate matter were literally off the charts. A pollution monitor atop the U.S. Embassy detected record-breaking readings of 755 on the Air Quality Index (the scale normally tops out at 500, "Hazardous"). The government of China denounced these readings as inaccurate, attributed the low visibility to fog, and asked the U.S. Embassy to stop publishing air pollution data.

During this so-called "airpocalypse," a fire at a factory went unnoticed for three hours because the smog was so thick that no one saw the smoke! Thousands of people suffered ill health, as pollution levels soared 30 times past the World Health Organization (WHO) safe limits. A resident of one northern city in China became the first person to sue the government for failing to perform its duty to control air pollution. Conditions became so bad that the government and official state media were finally forced to admit the problem and begin a public discussion about solutions.

China has fuelled its rapid industrial development with its abundant reserves of coal, the most polluting of the fossil fuels; 67% of China's current energy mix comes from coal. Power plants and factories have sprung up across the nation, often using outdated, inefficient, heavily polluting technology because it is quicker and cheaper to use. Car ownership is skyrocketing, too; in the capital of Beijing alone, 1500 new cars hit the streets each day. Industrial and economic development combined with the pressure of population growth have resulted in an intractable pollution problem that affects the entire nation; in many Chinese cities, the haze is often too thick for people to see the sun.

The health impacts of daily air pollution across China are enormous. A 2013 international research report blamed outdoor air pollution for 1.2 million premature deaths in China each year; the global burden of premature deaths from outdoor air pollution is estimated at more than 3 million annually.[2] Most large cities in China routinely exceed the "safe" standards for particulate air pollution, established in consultation with each nation by WHO. A report by Gonghuan Yang of the Chinese Academy of Medical Sciences, with colleagues,[3] cites outdoor and indoor air pollution as (respectively) the fourth and fifth most important contributors to premature deaths in China, based on *Disability-Adjusted Life Years*, an indicator used by WHO to express the overall burden of disease in terms of the number of years of life lost as a result of poor health, disability, or early death.

The air quality situation, especially in northern cities like Beijing, is exacerbated by a significant natural source of pollution: wind-blown dust. Desertification has been a chronic problem for decades in northern China and Mongolia, where land clearing and agricultural activities such as deep tillage and overgrazing expose fine glacial soils to the drying effects of the atmosphere. Dust that blows from dry lakebeds and cleared land on the borders of deserts is carried southward, where it blows into cities like Beijing in the form of massive, choking dust storms. These dust storms cover enormous areas and cause negative impacts on agriculture, industry, and human health.

Minhong Tan and colleagues[4] from the Chinese Academy of Sciences, Institute of Geographic Sciences and Natural Resources Research, studied the frequency, visibility, and intensity of dust storms in China over several decades. They found that dust storm activity was stable for many years but has been highly variable since 2000. Comparing these results to other environmental and agricultural data, including rainfall, temperature, and wind speed, they concluded that climatic conditions and land use changes are the most important factors in recent dust storm activity.

Some of the dust and particulate air pollution that causes problems in northern China is transported across the Pacific Ocean by major wind systems; Asian dust has even been found in ice cores from Greenland. Particulate matter from cars, industrial and agricultural sources, and wood-burning stoves in China, India, and other industrializing nations of Asia has formed a persistent 3-km-thick layer of pollution that hangs over southern Asia from December through April each year. Dubbed the *Asian Brown Cloud*, this massive layer of brownish haze has been estimated to reduce the sunlight reaching the surface in southern Asia by 10–20%, alter the monsoon rains, decrease rice productivity by 5–10%, speed the melting of Himalayan glaciers by depositing dark soot that absorbs sunlight, and contribute to many thousands of premature deaths each year.

China's government has now acknowledged the severity of the problem and is working hard to reduce air pollution. It has closed down some heavily polluting factories and mines, phased out some subsidies for polluting industries, installed pollution controls in power plants, and encouraged the development of wind, solar, and nuclear power to substitute for power produced by burning coal. The government now subsidizes efficient electric space heaters to replace dirty, inefficient coal stoves. It has mandated cleaner formulations for gasoline and diesel and has raised standards for fuel efficiency and emissions for cars. In Beijing, mass transit is being expanded, many buses run on natural gas, and heavily polluting vehicles are restricted from operating in the central city. The government has also undertaken many projects to try to combat desertification in the north, the major natural source of wind-blown dust. The nation likely will need to accelerate such steps, as its 1.35 billion citizens are becoming increasingly fed up with pollution.

The Atmosphere and Weather

Every breath we take reaffirms our connection to the **atmosphere**, the thin layer of gases that surrounds Earth. We live at the bottom of this layer, which provides us with oxygen, absorbs hazardous solar radiation, burns up incoming meteors, transports and recycles water and nutrients, and moderates climate.

Earth's atmosphere consists of roughly 78% nitrogen gas (N_2) and 21% oxygen gas (O_2). The remaining 1% is composed of argon gas (Ar) and minute concentrations of several other gases (**FIGURE 13.1**). These include *permanent gases* that remain at stable concentrations and *variable gases* that vary in concentration from time to time or place to place as a result of natural processes or human activities.

Over Earth's long history, the atmosphere's chemical composition has changed. Oxygen gas began to build up in an atmosphere dominated by carbon dioxide (CO_2), nitrogen, carbon monoxide (CO), and hydrogen (H_2) about 2.7 billion years ago, as a result of the emergence of autotrophic microbes that emitted oxygen as a by-product of photosynthesis. Today, human activity is altering the quantities of some atmospheric gases, such as carbon dioxide, methane (CH_4), and ozone (O_3). In this chapter we will explore some of the atmospheric changes brought about by pollutants, but we must first begin with an overview of Earth's atmosphere.

The atmosphere is layered

The atmosphere that stretches so high above us and seems so vast is actually just a thin coating about 1/100 of Earth's diameter, like the fuzzy skin of a peach. This coating

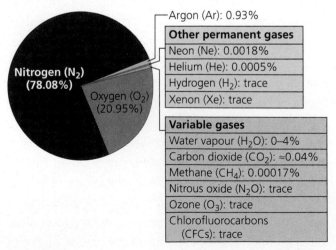

FIGURE 13.1
Earth's atmosphere consists mostly of nitrogen and oxygen. Permanent gases are fixed in concentration. Variable gases vary in concentration as a result of either natural processes or human activities.
Source: Data from Ahrens, C.D. (2007). *Meteorology Today*, 8th ed. Belmont, CA: Brooks/Cole.

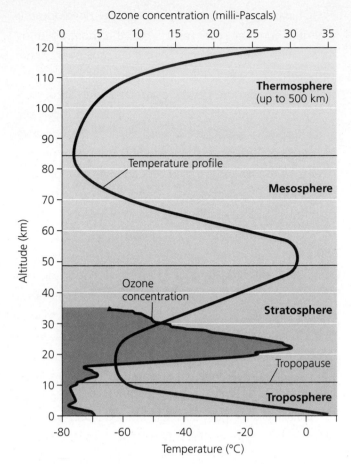

FIGURE 13.2
Temperature drops with altitude in the troposphere, rises with altitude in the stratosphere, drops in the mesosphere, and rises again in the thermosphere. The tropopause separates the troposphere from the stratosphere. Ozone reaches a peak in the stratospheric ozone layer.
Source: Adapted from Jacobson, M.Z. (2002). *Atmospheric Pollution: History, Science, and Regulation.* Cambridge, UK: Cambridge University Press; and Parson, E.A. (2003). *Protecting the Ozone Layer: Science and Strategy.* Oxford, UK: Oxford University Press.

DATA Q Above an altitude of 11 km, the temperature of the troposphere levels out at about −57°C. The average temperature of air at the surface is about +15°C. Based on these figures, what is the approximate change in temperature per km of altitude?

consists of four layers that atmospheric scientists recognize by measuring differences in temperature, density, and composition (**FIGURE 13.2**).

The bottommost layer, the **troposphere**, blankets Earth's surface and provides us with the air we need to live. The movement of air within the troposphere is also largely responsible for the planet's weather. Although it is thin (averaging 11 km high) relative to the atmosphere's other layers, the troposphere contains three-quarters of the mass of the atmosphere, because air is denser near Earth's surface.

The troposphere is warmest near the surface, because it is heated from below by surface materials that absorb and reradiate solar energy. On average, tropospheric air temperature declines by about 6.5°C for each kilometre in altitude, dropping to roughly −57°C at its highest point. The rate at which temperature decreases with height in the troposphere is referred to as the *environmental lapse rate*. At the top of the troposphere, temperatures cease to decline with altitude, marking a boundary called the *tropopause*. The tropopause acts like a cap, limiting mixing between the troposphere and the atmospheric layer above it, the stratosphere.

The **stratosphere** extends from approximately 11 km to 50 km above sea level. Although similar in composition to the troposphere, the stratosphere is 1000 times as dry and much less dense. Its gases experience little vertical mixing, so once substances (including pollutants) enter it, they tend to remain for a long time. The stratosphere attains a maximum temperature of −3°C at its highest altitude but is colder in its lower reaches. The reason is that **ozone (O_3)** and oxygen (O_2) absorb and scatter the Sun's ultraviolet (UV) radiation, so that much of the UV radiation penetrating the upper stratosphere fails to reach the lower stratosphere.

Most of the atmosphere's minute amount of ozone is concentrated in a portion of the stratosphere roughly from 17 km to 30 km above sea level, a region that has come to be called Earth's **ozone layer**. The ozone layer greatly reduces the amount of UV radiation that reaches Earth's surface. Because UV light can damage living tissue and induce mutations in DNA, the ozone layer's protective effects are vital for life on Earth.

Above the stratosphere is the **mesosphere**, which extends from 50 km to 85 km above sea level. Air pressure is extremely low in the mesosphere, and temperatures decrease with altitude, reaching their lowest point at the top of the mesosphere. From here, the **thermosphere**, our atmosphere's top layer, extends upward to an altitude of 500 km. The thermosphere is heated from above by direct exposure to sunlight. As you can see in our chapter-opening photo, the thermosphere does not end abruptly, but fades gradually away. The somewhat arbitrary altitude of 100 km—just above the bottom of the thermosphere—is defined as the boundary between the atmosphere and outer space (it's called the *Karman Line*).

Atmospheric properties include temperature, pressure, and humidity

Although the lower atmosphere is stable in its chemical composition, the air within it is dynamically mixed and constantly moving. Air movement in the troposphere is caused by differences in the physical properties of air masses. Among these properties are pressure and density, relative humidity, and temperature, which we have seen is responsible for defining the major layers of the atmosphere with altitude.

The temperature of air varies with time and location on the surface. At a global scale, temperature varies over Earth's surface because the Sun's rays strike some areas more directly than others. At more local scales, temperature varies because of topography, plant cover, proximity of land to water, and many other factors. Sometimes these local variations are striking—the side of a hill that is sheltered from wind or direct sunlight can have a totally different weather pattern, or **microclimate**, from the side facing into the wind or sunlight. Temperature is one of the fundamental characteristics that distinguish one air mass from another.

Like temperature, **atmospheric pressure**, the force per unit of area produced by a column of air, also decreases with altitude in the troposphere. Gravity pulls gas molecules toward Earth's surface, causing air to be most dense near the surface, and less dense at higher altitudes (**FIGURE 13.3**). At sea level, atmospheric pressure is 1013 millibars (mb). Mountain climbers trekking to Mount Everest, the world's highest mountain at 8.85 km, can look up and view their destination from Kala Patthar, a nearby peak, at roughly 5.5 km in altitude. At this altitude, pressure is 500 mb—half the atmosphere's air molecules are above the climber, and half are below. A climber who reaches Everest's peak, where the "thin air" pressure is just

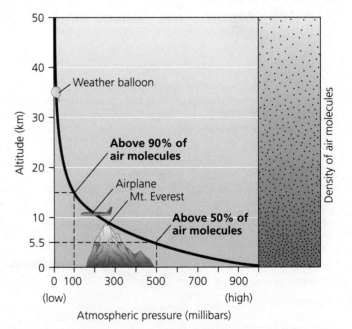

FIGURE 13.3
With height in the atmosphere, gas molecules become less densely packed. As density decreases, so does atmospheric pressure. A person standing at an altitude of 5.5 km would be above half the planet's air molecules.
Source: From Ahrens, C.D. (2007). *Meteorology Today* (with printed Access Card ThomsonNOW(T)), 8th ed. © 2007 Wadsworth, a part of Cengage Learning, Inc. Reproduced by permission. www.cengage.com/permissions.

 Use this graph to determine the approximate atmospheric pressure, in mb, at the average altitude of the tropopause.

over 300 mb, stands above two-thirds of the molecules in the atmosphere. When we fly on a commercial jet airliner at a cruising altitude of 11 km, we are above roughly 80% of the atmosphere's molecules.

Another property of air is **relative humidity**, the ratio of water vapour in a volume of air to the maximum amount it could potentially contain at a given temperature. Average daytime relative humidity in a desert might be only 25–30%, which means that the air contains only one-quarter to one-third of the water vapour it possibly could at that temperature. In tropical rain forests, in contrast, relative humidity rarely drops below 80%. People are sensitive to changes in relative humidity because we perspire to cool our bodies. When humidity is high, the air is holding nearly as much water vapour as it can, so sweat evaporates slowly and the body cannot cool itself efficiently. This is why high humidity makes it feel hotter than it is. Low humidity speeds evaporation and makes it feel cooler.

A large volume of air that is fairly uniform internally, in temperature, relative humidity, and air pressure, is called an **air mass**. The movement and interactions of air masses in global and regional circulation systems bring us the variations in weather that we experience on a daily basis.

Solar energy heats the atmosphere, creates seasons, and causes air to circulate

An enormous amount of solar energy continuously bombards the upper atmosphere—more than 1000 watts/m^2 where it hits directly—many thousands of times greater than the total output of electricity generated by human society. Of that solar energy, about 70% is absorbed by the atmosphere and planetary surface, while the rest is reflected back into space. The sunlight that comes into the Earth system heats air in the atmosphere, drives air movement, helps create seasons, and influences weather and climate, in addition to heating surface materials and driving photosynthesis.

The spatial relationship between Earth and the Sun determines how much solar radiation strikes each point on Earth's surface. Sunlight is most intense when it shines directly overhead and meets the planet's surface at a perpendicular angle. At this angle, sunlight passes through a minimum of energy-absorbing atmosphere, and Earth's surface receives a maximum of solar energy per unit of surface area. Conversely, solar energy that approaches Earth's surface at an oblique angle loses intensity as it traverses a longer distance through the atmosphere, and it is less intense when it reaches the surface, where the oblique angle of the incoming solar energy also is spread over a larger surface area. This is why, on average, solar radiation intensity is highest near the equator and weakest near the poles (**FIGURE 13.4**).

Because Earth is tilted on its *axis* (an imaginary line connecting the poles) by about 23.5°, the Northern and Southern Hemispheres each tilt toward the Sun for half the year, resulting in the change in seasons (**FIGURE 13.5**). Regions near the equator are largely unaffected by this tilt; they experience about 12 hours each of sunlight and darkness every day throughout the year. Near the poles, however, the effect is strong, and seasonality is pronounced.

Land, water, and all materials on Earth's surface absorb solar energy, causing reradiation of heat and evaporation

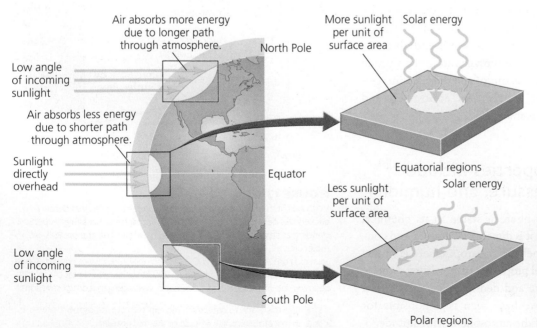

FIGURE 13.4

Because of Earth's curvature, polar regions receive less solar energy than equatorial regions. Sunlight gets spread over a larger area when striking the surface at an angle. Also, sunlight approaching at a lower angle near the poles must traverse a longer distance through the atmosphere, during which more energy is absorbed or reflected.

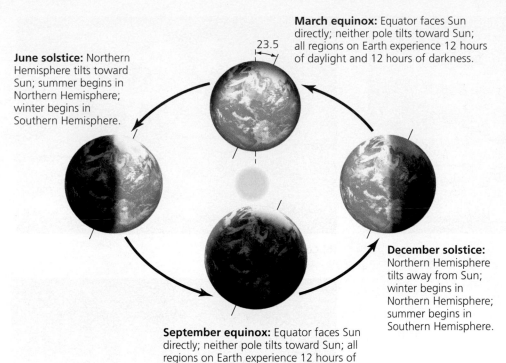

June solstice: Northern Hemisphere tilts toward Sun; summer begins in Northern Hemisphere; winter begins in Southern Hemisphere.

23.5

March equinox: Equator faces Sun directly; neither pole tilts toward Sun; all regions on Earth experience 12 hours of daylight and 12 hours of darkness.

December solstice: Northern Hemisphere tilts away from Sun; winter begins in Northern Hemisphere; summer begins in Southern Hemisphere.

September equinox: Equator faces Sun directly; neither pole tilts toward Sun; all regions on Earth experience 12 hours of daylight and 12 hours of darkness.

FIGURE 13.5
Seasons occur because Earth is tilted on its axis. As Earth revolves around the Sun, the Northern Hemisphere tilts toward the Sun for one half of the year, and the Southern Hemisphere tilts toward the Sun for the other half of the year. In each hemisphere, summer occurs during the period in which the hemisphere is tilted toward the Sun.

of water. Air near Earth's surface therefore tends to be warmer and moister than air at higher altitudes. These differences set into motion a process of convective circulation (**FIGURE 13.6**). Warm air is less dense than cool air, so it rises; this creates vertical air currents. As air rises into regions of lower atmospheric pressure, it expands and

cools. Once the air cools, it becomes denser and descends, replacing warm air that is rising. The air picks up heat and moisture near ground level and prepares to rise again, continuing the process. Similar convective circulation patterns occur in ocean waters and in magma beneath Earth's surface.

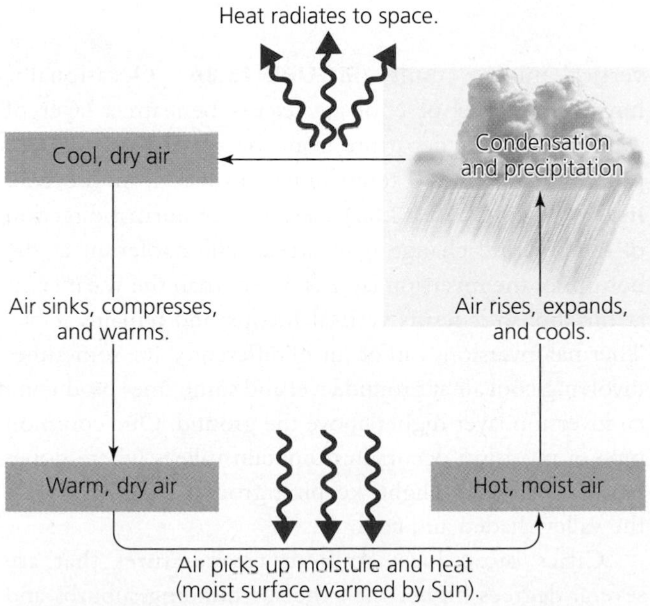

The physical properties of the troposphere, including temperature, pressure, humidity, cloudiness, and wind, vary spatially and temporally to produce the atmospheric conditions that we experience on a daily basis as *weather*. **Weather** specifically refers to atmospheric conditions over short time periods, typically hours or days, and in relatively small, local geographical areas. **Climate**, on the other hand, describes the pattern of atmospheric conditions found across large geographical regions over long periods—seasons, years, decades, or millennia. Writer Mark Twain noted the distinction between climate and weather by saying, "Climate is what we expect; weather is what we get."

Heat radiates to space.

Cool, dry air

Condensation and precipitation

Air sinks, compresses, and warms.

Air rises, expands, and cools.

Warm, dry air

Hot, moist air

Air picks up moisture and heat (moist surface warmed by Sun).

FIGURE 13.6
Weather is driven by the convective circulation of air in the atmosphere. Air being heated near Earth's surface picks up moisture and rises. Once aloft, this air cools, and moisture condenses, forming clouds and precipitation. The cool, dry air descends, compressing and warming in the process. Warm, dry air near the surface begins the cycle anew.

Air masses interact to produce weather

Weather can change quickly when air masses with different physical properties meet. The boundary between air masses that differ in moisture content and temperature (and therefore density) is called a **front**. Fronts are where we typically experience the most active weather. The boundary along which a mass of warmer, moister air replaces a mass of colder, drier air is termed a **warm front**

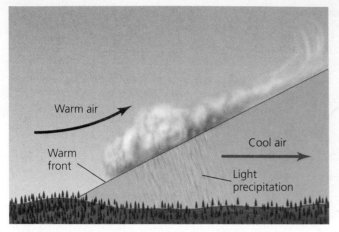

(a) Warm front

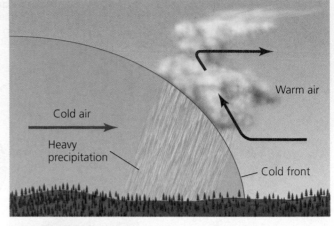

(b) Cold front

FIGURE 13.7
At a warm front **(a)**, warmer air rises over cooler air, causing light or moderate precipitation as moisture in the warmer air condenses. At a cold front **(b)**, colder air pushes beneath warmer air, and the warmer air rises, resulting in condensation and heavy precipitation. **(c)** Cold fronts often spawn thunderstorms and even tornadoes, like this one in 2011.

(c) Active weather

(**FIGURE 13.7A**). Some of the warm, moist air behind a warm front rises over the cold air mass and then cools and condenses to form clouds that may produce light rain. A **cold front** (**FIGURE 13.7B**) is the boundary along which a colder, drier air mass displaces a warmer, moister air mass. The colder air, being denser, tends to wedge beneath the warmer air. The warmer air rises, expands, and then cools to form clouds that can produce thunderstorms and even tornadoes (**FIGURE 13.7C**). Once a cold front passes through, the sky usually clears, and the temperature and humidity drop.

Adjacent air masses may also differ in atmospheric pressure. A **high-pressure system** contains air that moves outward away from a centre of high pressure as it descends. High-pressure systems typically bring fair weather. In a **low-pressure system**, air moves toward the low atmospheric pressure at the centre of the system and spirals upward. The air expands and cools, and clouds and precipitation often result.

Under most conditions, air in the troposphere decreases in temperature as altitude increases. Because warm air rises,

vertical mixing results (**FIGURE 13.8A**). Occasionally, however, a layer of cool air occurs beneath a layer of warmer air. This departure from the normal temperature profile is known as a temperature inversion, or **thermal inversion** (**FIGURE 13.8B**), because the normal direction of temperature change is inverted. The cooler air at the bottom of the inversion layer is denser than the warmer air at the top, so it resists vertical mixing and remains stable. Thermal inversions can occur in different ways, sometimes involving cool air at ground level and sometimes producing an inversion layer higher above the ground. One common type of inversion occurs in mountain valleys where slopes block morning sunlight, keeping ground-level air within the valley shaded and cool.

Cities often have ambient temperatures that are several degrees higher than the surrounding suburbs and rural areas. This is called the **urban heat island effect**, and it results from the concentration of heat-generating buildings, cars, factories, and people in the city centre. Tall buildings and paved surfaces are important contributors to the urban heat island effect by absorbing heat and

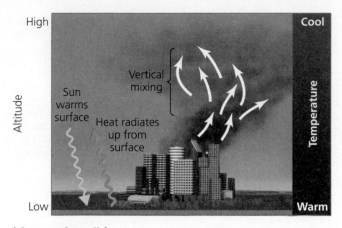

(a) Normal conditions

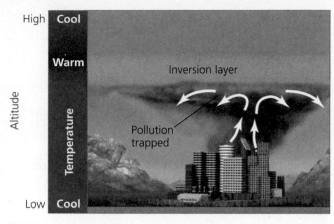

(b) Thermal inversion

FIGURE 13.8 Under normal conditions **(a)**, temperature decreases with altitude. Vertical mixing disperses pollutants upward and outward from their sources. During a thermal inversion **(b)**, cool air remains near the ground underneath a layer of warm air. Little mixing occurs, and pollutants can be trapped near the surface.

releasing it slowly and by interfering with the convective circulation of air that would otherwise cool the city.

Vertical mixing normally allows air pollution to be diluted upward, but thermal inversions concentrate both heat and pollutants near the ground. When heated air becomes trapped over cities, the smog and particulate air pollution it carries become trapped as well; this can lead to a phenomenon called a **dust dome**. If you drive toward a large city on a hot, hazy summer day—especially if the city is located in a basin or a valley—you will likely see particulate air pollution and smog hovering like a brown or blue-grey cloud over the city skyline. A thermal inversion sparked a "killer smog" crisis in London, England, in 1952. A high-pressure system settled over the city, acting like a cap on the pollution; at least 4000 people—possibly as many as 12 000—died as a result of this event. Inversions commonly cause smog buildups in large metropolitan areas in valleys ringed by mountains, such as Los Angeles, Mexico City, Seoul, and Rio de Janeiro.

Large-scale circulation systems produce global climate patterns

At larger geographical scales, convective air currents contribute to broad climatic patterns (**FIGURE 13.9A**). Near the equator, solar radiation sets in motion a pair of convective cells known as **Hadley cells**. Here, where sunlight is most intense, surface air warms, rises, and expands. As it does so, it releases moisture, producing the heavy rainfall that gives rise to tropical rain forests near the equator.

After releasing much of its moisture, this air diverges and moves in currents heading northward and southward.

The air in these currents cools and descends back to Earth at about 30° latitude north and south. Because the descending air has low relative humidity, the regions around 30° latitude are quite arid, giving rise to deserts. Two pairs of similar but less intense convective cells, called **Ferrel cells** and **polar cells**, lift air and create precipitation around 60° latitude north and south and cause air to descend at around 30° latitude and in the polar regions.

These three pairs of cells account for most of the latitudinal distribution of moisture across Earth's surface: warm, wet climates near the equator; arid climates and major deserts near 30° latitude; moist, temperate regions near 60° latitude; and dry, cold conditions near the poles. These patterns, combined with temperature variation, also help explain why biomes tend to be arrayed in latitudinal bands.

The Hadley, Ferrel, and polar cells interact with Earth's rotation to produce the global wind patterns shown in **FIGURE 13.9B**. As Earth rotates on its axis, locations on the equator spin faster than locations near the poles. As a result, the north–south air currents of the convective cells are deflected from a straight path as some portions of the globe move beneath them more quickly than others. This deflection is called the **Coriolis effect**, and it results in the curving global wind patterns evident in **FIGURE 13.9B**. The Coriolis effect influences the circulation of any freely moving fluid on Earth's surface, including ocean water, but its influence is not noticeable unless the scale of the circulation is quite large.

Just north and south of the equator, the **trade winds** blow from east to west. Where the trade winds meet and are deflected toward the west, just north and south of the equator, lies a region with little wind known as the

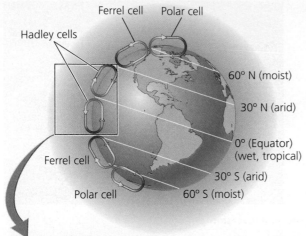

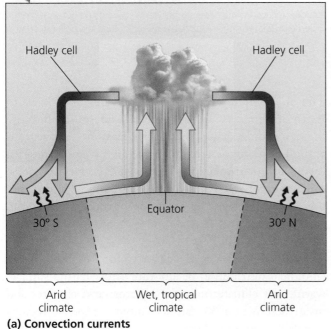

(a) Convection currents

FIGURE 13.9 Large-scale convective cells **(a)** influence global patterns of humidity and aridity. Warm air near the equator rises, expands, and cools; moisture condenses, giving rise to wet climates in tropical regions. The dry air travels toward the poles and descends around 30° latitude, causing dry climates. Hadley cells occur on both sides of the equator between 0° and 30°. From 30° to 60° latitude north and south are the Ferrel cells, and from 60° to 90° latitude the polar cells occur. Global wind currents **(b)** show latitudinal patterns as well. Trade winds between the equator and 30° latitude blow westward; westerlies between 30° and 60° latitude blow eastward. Global air circulation patterns are further modified by the influence of continental land masses and by the Coriolis force.

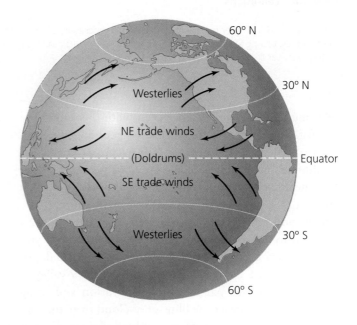

(b) Global wind patterns

doldrums. This zone where winds from the north and south come together is called the *Intertopical Convergence Zone*, or *ITCZ*; it is typically cloud covered, because the warm, moist air rises, clouds form, and rain falls. Farther from the equator, between 30° and 60° north and south latitude, the *westerlies* originate from the west and blow east. People used these global circulation patterns for centuries to facilitate ocean travel by wind-powered sailing ships.

The atmosphere interacts with the oceans to affect weather, climate, and the distribution of biomes. For instance, winds and convective circulation in ocean water together maintain ocean currents. Trade winds weaken periodically, leading to El Niño conditions (see Chapter 12). The atmosphere's interactions with other systems of the planet are complex, but even a basic understanding of how the atmosphere functions can help us comprehend how our

pollution of the atmosphere can affect ecological systems, economies, and human health.

Outdoor Air Pollution

Throughout human history, we have used the atmosphere as a dumping ground for airborne wastes. Whether from primitive wood fires or modern coal-burning power plants, people have generated pollutants, gases, and particulate material that can affect climate or harm people and other organisms. **Air pollution** refers to the release of harmful substances into the air. In recent decades, government policies, improved technologies, and international agreements have helped us substantially diminish outdoor air pollution, also called *ambient air pollution*. However, outdoor air pollution remains a

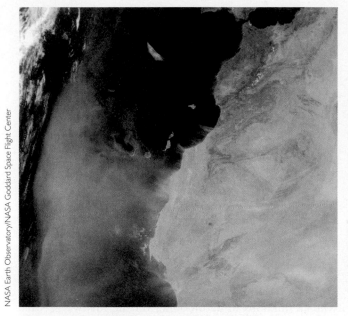

(a) Dust storm off west coast of Africa

(b) Grimsvötn volcano eruption

(c) Forest fire in Alberta

FIGURE 13.10
Massive dust storms, such as this one shown in a satellite image, blowing across the Atlantic Ocean from Africa **(a)**, are one source of natural air pollution. Volcanic eruptions are another source, as shown by Grimsvötn volcano in Iceland **(b)**, which erupted in 2011. A third natural source is forest fires, like this one in Alberta **(c)**.

problem in some urban areas, particularly in developing nations, as discussed in our Central Case.

Natural sources can pollute

When we think of outdoor air pollution, we tend to envision smokestacks belching black smoke from industrial plants. However, natural processes produce a great deal of air pollution. Some of these natural impacts can be exacerbated by human activity and poor land use policies and practices.

Winds sweeping over bare, arid terrain can send huge amounts of dust aloft. As discussed in our Central Case, massive dust storms contribute to the problem of air pollution in northern China. Strong winds lift loose, dry soil from deserts in Mongolia and China, carrying it southward to Beijing and eastward across the Pacific and Atlantic Oceans, settling as far away as Greenland and the Alps. Millions of tonnes of dust are also blown to the west by trade winds across the Atlantic Ocean from northern Africa to the Americas (**FIGURE 13.10A**). These dust storms bring nutrients to the Amazon Basin, cause algal blooms off the west coast of Africa, and bring fungal and bacterial spores that have been linked to die-offs in Caribbean coral reef systems. Although dust storms are natural, the immense scale of these events is exacerbated by unsustainable farming and grazing practices that strip vegetation from the soil, promote wind erosion, and lead to desertification.

Volcanic eruptions release large quantities of particulate matter, as well as sulphur dioxide and other gases, into the troposphere. In 2010 and again in 2011, Icelandic volcanoes erupted voluminous clouds of ash, which shut down airports and delayed air travel in many parts of Britain and Europe (**FIGURE 13.10B**). Major eruptions may blow matter into the stratosphere, where it can circle the globe and remain aloft for months or years. Sulphur dioxide reacts with water and oxygen, and then condenses into fine droplets called **aerosols**, which reflect sunlight back into space and thereby cool the atmosphere and surface. The 1991 eruption of Mount Pinatubo in the Philippines ejected nearly 20 million tonnes of ash and aerosols and cooled global temperatures by roughly 0.5°C over a period of two years.

Burning vegetation also pollutes the atmosphere with soot and gases. More than 60 million hectares of forest and grassland burn in a typical year (**FIGURE 13.10C**). Fires occur naturally, but many are made more severe by human action. In North America, fuel buildup from decades of fire suppression has caused damaging forest fires in recent years. In the tropics, many fires result from the clearing of forests for farming and grazing by "slash-and-burn." Australia experienced its hottest year ever in 2013, which contributed to a record-breaking wildfire season. In 1997, a severe drought brought on by the twentieth century's strongest El Niño event caused forest fires in Indonesia to rage out of control. Their smoke sickened 20 million Indonesians and caused a plane to crash and ships to collide. Combined with tens of thousands of fires in drought-plagued Mexico, Central America, and Africa, these fires released more carbon monoxide into the atmosphere during 1997–1998 than did our worldwide combustion of fossil fuels.

There are various sources and types of outdoor air pollution

Since the onset of industrialization, human activity has introduced a variety of sources of air pollution. As with water pollution, air pollution can emanate from mobile or stationary sources, and from *point sources* or *nonpoint sources*. A point source describes a specific spot where large quantities of pollutants are discharged. Nonpoint sources are more diffuse, often consisting of many small sources. Power plants and factories act as stationary point sources, whereas millions of automobiles on the roadways—each one a tiny point source—together create a massive, mobile nonpoint source of pollutants (**FIGURE 13.11**).

We mainly associate poor air quality with industrial and urbanized regions, but air pollution can also be a rural issue. In rural areas, people may suffer from drift of airborne pesticides from farms, as well as from industrial pollutants transported from cities, factories, and power plants. A great deal of rural air pollution emanates from feedlots, where cattle, hogs, or chickens are raised in dense concentrations. The huge numbers of animals at feedlots produce dust as well as methane, hydrogen sulphide, and ammonia. These gases create objectionable odours and can lead to respiratory health problems. Ammonia also contributes to nitrogen deposition across wide areas.

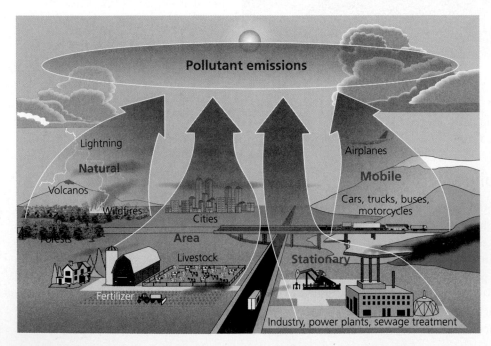

FIGURE 13.11
Mobile, stationary, natural, and broad-area sources like cities and agricultural regions all contribute to air pollution. Mobile sources like cars are small, individual point sources of pollution, but they combine to form a large, diffuse nonpoint source.
Source: From U.S. Department of the Interior, http://www.nature.nps.gov/air/aqbasics/sources.cfm.

Once pollutants are in the atmosphere, they may do harm directly or they may induce chemical reactions that produce harmful compounds. **Primary pollutants**, such as soot and carbon monoxide, are pollutants emitted into the troposphere in a form that can be directly harmful or that can react to form harmful substances. Harmful substances produced when primary pollutants interact or react with constituents of the atmosphere are called **secondary pollutants**. Secondary pollutants include tropospheric ozone, sulphuric acid, and others.

Arguably the greatest human-induced air pollution problem today is our emission of greenhouse gases that contribute to global climate change. Addressing our release of excess carbon dioxide, methane, and other gases that warm the atmosphere stands as one of our civilization's primary challenges. We will discuss this issue separately and in depth in the next chapter, *Global Climate Change*.

CEPA identifies harmful airborne substances

The *Canadian Environmental Protection Act (CEPA)* (1999) provides a list of air pollutants that are subject to legislative control and management. These pollutants differ widely in their chemical composition, chemical reactivity, emission sources, residence time (how long they remain in various environmental reservoirs, including organisms), persistence (how long they last before breaking down), transportability (ability to be moved long or short distances), and impacts on the natural and built environments, and on human and ecosystem health.

Environment Canada groups the pollutants of greatest concern into four categories:[5]

1. criteria air contaminants
2. persistent organic pollutants
3. heavy metals
4. toxic air pollutants

Criteria air contaminants   In Canada, the **criteria air contaminants (CACs)** include sulphur oxides, nitrogen oxides, particulate matter, volatile organic compounds, carbon monoxide, ammonia, and tropospheric ozone. These are referred to as "criteria" contaminants because they were the first to come under government regulations, because of concerns about their potential impacts on human health.

Sulphur dioxide (SO$_2$) is a colourless gas with a strong odour. The vast majority of SO$_2$ and other emissions of **sulphur oxides (SO$_x$ or SOX)** result from the combustion of coal for electricity generation and industry. During combustion, elemental sulphur (S)

in coal reacts with oxygen gas (O$_2$) to form SO$_2$. Once in the atmosphere, SO$_2$ may react to form sulphuric acid (H$_2$SO$_4$), which may then fall back to Earth as acid precipitation.

Nitrogen dioxide (NO$_2$) is a highly reactive, foul-smelling reddish brown gas that contributes to smog and acid precipitation. Along with nitric oxide (NO), NO$_2$ belongs to a family of compounds called **nitrous oxides (NO$_x$ or NOX)**. Nitrogen oxides result when atmospheric nitrogen and oxygen react at the high temperatures created by combustion engines. More than half of NO$_x$ emissions result from combustion in motor vehicle engines; electrical utility and industrial combustion account for most of the rest.

Particulate matter (PM) is composed of solid or liquid particles small enough to be suspended in the atmosphere. References to PM often include a number, specifying the size of the particles. PM$_{10}$, for example, refers to particles less than 10 μm (microns) in diameter; PM$_{2.5}$ refers to extremely fine particles, with diameters less than 2.5 μm; and so on. Particulate matter (which can also be called *suspended particulates* or *SP*) includes primary pollutants, such as dust and soot, as well as secondary pollutants, such as sulphates and nitrates, which form as a result of the alteration of primary sulphur and nitrous oxides. Particulates can damage respiratory tissues when inhaled.

Most particulate matter (60%) in the atmosphere is wind-blown dust; human activity and industrial emissions account for much of the rest. Along with SO$_2$, it was largely particulate matter from coal combustion that produced London's 1952 killer smog. Coal is also the main culprit in China's air quality problems. The high pollution readings obtained by the U.S. Embassy in Beijing in 2013, discussed in our Central Case, were based on measurements of PM$_{2.5}$. China's target for PM$_{2.5}$, established in conjunction with WHO, is 75 μg/m^3 of air by 2016, equivalent to about 160 on the U.S. Environmental Protection Agency's Air Quality Index. For comparison, in early 2013 the Embassy in Beijing had AQI readings up to 755.

Volatile organic compounds (VOCs or VOX) are carbon-containing chemicals used in and emitted by vehicle engines and a wide variety of solvents and industrial processes, as well as many household chemicals and consumer items. One group of VOCs consists of hydrocarbons, such as methane (CH$_4$, the primary component of natural gas), propane (C$_3$H$_8$, used as a portable fuel), butane (C$_4$H$_{10}$, found in cigarette lighters), and octane (C$_8$H$_{18}$, a component of gasoline). Human activities account for about half of VOC emissions, and the remainder comes from natural sources. For example, plants produce isoprene (C$_5$H$_8$) and terpene (C$_{10}$H$_{15}$).

Carbon monoxide (CO) is a colourless, odourless gas produced by the incomplete combustion of fuel. Vehicles and engines are the main source, but others include industrial processes, combustion of waste, and residential wood burning. Carbon monoxide poses risk to humans and other animals, even in low concentrations. It can bind to hemoglobin in red blood cells, preventing it from becoming oxygenated.

Ammonia (NH₃) is a colourless gas with a pungent odour—it is the smell associated with urine. Most NH_3 is generated from livestock waste and fertilizer production. Ammonia is poisonous if inhaled in great quantities and is irritating to the eyes, nose, and throat in lesser concentrations. In the atmosphere it combines with sulphates and nitrates to form secondary fine particulate matter ($PM_{2.5}$). NH_3 can also contribute to the nitrification and eutrophication of aquatic systems.

Tropospheric ozone is also called *ground-level ozone* to distinguish it from the ozone in the stratosphere, which shields us from the dangers of UV radiation. In contrast to stratospheric ozone, O_3 that accumulates at ground level is a pollutant. In the troposphere, this colourless gas results from the interaction of sunlight, heat, nitrogen oxides, and carbon-containing chemicals; it is therefore a secondary pollutant. A major component of smog, ozone can pose health risks because of its instability as a molecule; this triplet of oxygen atoms will readily release one, leaving a molecule of oxygen gas (O_2) and a free oxygen atom. The free oxygen atom may then participate in reactions that can injure living tissues and cause respiratory problems. Tropospheric ozone is the pollutant that most frequently exceeds its air quality standard.

Persistent organic pollutants

Persistent organic pollutants (POPs) can last in the environment for long periods of time. They are capable of travelling great distances by air because they are *volatile*, which means that they evaporate readily. The term **persistent** refers to substances that have long residence times because they remain in environmental reservoirs for a long time, or because they take a long time to degrade or break down, or both.

POPs are of particular concern because they can enter the food supply, bioaccumulate in body tissues, and have significant impacts on human health and the environment, even in low concentrations. They have few natural sources and come primarily from human activity. Examples include industrial chemicals, such as PCBs (polychlorinated biphenyls); pesticides, such as DDT (dichloro-diphenyl-trichloroethane, **FIGURE 13.12**), chlordane, and toxaphene; and contaminants and by-products, such as dioxins and furans, which come from incomplete combustion processes.

The indiscriminate spraying and persistent buildup of DDT caused widespread deaths of birds and other fauna

FIGURE 13.12
Learning about the persistence and negative ecological impacts of some chemical pesticides catalyzed the public and helped launch the modern environmental movement in the late 1960s and early 1970s. Here, a farmer sprays pesticide on his field without adequate personal protection against chemical exposure.

in the 1950s and 1960s, leading Rachel Carson to write her famous book *Silent Spring*, one of the pivotal events in the modern environmental movement. DDT was banned for agricultural use in most industrialized nations in the 1970s and 1980s, but its use continues today in some parts of the developing world. As recently as 2002, DDT and its by-products were still widely detectable in human blood, tissue, and breast milk samples from subjects in North America.[6] Blood levels of POPs, although generally decreasing, remain above WHO safe levels in some agricultural regions where these pesticides are still widely used, notably in India.[7]

Heavy metals

Heavy metals can be transported by the air, enter our water and food supply, and reside for long periods in sediment. Metals tend to be associated with particulate matter, either occurring in particulate form or attaching to small particles that can then be transported atmospherically. Heavy metals are poisonous, even in low concentrations, and can bioaccumulate in body tissues. These pollutants occur even in Canada's far north—they are carried from the industrial south by continent-scale atmospheric currents—where they are deposited on land and water surfaces. This is called *long-range transport of atmospheric pollutants (LRTAP)*, and it has been a concern in Canada, especially in the Arctic, since the early 1970s.

Mercury is a heavy metal of considerable concern in Canada. Mercury is *volatile* (evaporates readily) and occurs in a number of different chemical forms, some more toxic than others. It has natural as well as human sources. It has been used for a variety of industrial purposes, partly because of its unusual property of remaining liquid at surface temperatures and pressures. Mercury is *lipophilic*, which means "fat-loving"; it is chemically capable of binding to fatty tissues in organisms. Like other heavy metals, mercury can enter the food chain, accumulate in body tissues, and cause central nervous system malfunction and other ailments.

Lead, also a heavy metal, enters the atmosphere as a particulate pollutant. The lead-containing compounds tetraethyl lead and tetramethyl lead, when added to gasoline, improve engine performance. However, exhaust from the combustion of leaded gasoline emits lead into the atmosphere, from which it can be inhaled or deposited on land or water. Like mercury, lead is bioaccumulative and can cause damage to the central nervous system. Once people recognized the dangers of lead, leaded gasoline began to be phased out in most industrialized nations in the 1970s. The use of leaded gas ended in Canada in 1993.[8] Today the greatest source of atmospheric lead pollution in industrialized nations is metal smelting. However, many developing nations still add lead to gasoline and experience significant lead pollution.

Toxic air pollutants   Toxic air pollutants constitute a broad category of "other" pollutants identified as being harmful or toxic, and therefore are subject to regulation, control, and monitoring. They include substances known to cause cancer, reproductive defects, or neurological, developmental, immune system, or respiratory problems. This category overlaps with the other types of air pollutants (for example, lead, mercury, dioxins, furans, and ozone all appear on the list of toxic pollutants, as well as in other categories), but it includes additional substances such as asbestos and chlorofluorocarbons (CFCs). Most toxic air pollutants are produced by human activities.

Government agencies share responsibility for air pollution

In Canada the management of air-related issues is the responsibility of the federal government—primarily Environment Canada—and the provincial and territorial governments, through their environment ministries. For the most part, municipal governments do not have direct regulatory control over activities that affect air quality. However, municipal governments manage so many activities that influence air quality that their role is central to the collaborative effort.

Federal   The principal federal legislation under which air quality is regulated is the *Canadian Environmental Protection Act (CEPA)* (1999), described officially as "an Act respecting pollution prevention and the protection of the environment and human health in order to contribute to sustainable development."[9] Although CEPA gives the lead responsibility for air quality to Environment Canada, it also defines an important role for Health Canada, which we will investigate in greater detail in Chapter 19, *Environmental Health and Hazards*. Other federal agencies, such as Transport Canada and Natural Resources Canada, also have programs and activities with important linkages to air quality issues.

The federal government is also responsible for entering Canada into international agreements concerning air quality. Canada has a long history of international agreements with the United States to control **transboundary pollution**, pollution that crosses political borders. For air quality, these date back to agreements made in the early 1900s, although the modern era began in 1979 with agreements concerning the long-range transport of pollutants related to acid deposition. The present bilateral international agreement on transboundary air pollution, coordinated primarily through the International Joint Commission, is the *Canada–United States Air Quality Agreement*, signed in 1991. This agreement has three annexes, the first dealing with acid rain precursors, the second with coordination of international scientific research, and the third with ground-level ozone.

The *Montreal Protocol* and the *Kyoto Protocol* are examples of multilateral international agreements signed by Canada that are intended to address air pollution issues of global concern (stratospheric ozone depletion and global warming, respectively). In 2011 Canada became the first signatory nation from the developed world to announce that it would withdraw from the Kyoto Protocol (the *United Nations Framework Convention on Climate Change*). Aside from formal agreements, international organizations like the World Health Organization also influence air quality management in Canada. For example, WHO sets minimum standards for safe air quality in consultation with governments around the world, and participates in monitoring and evaluating air quality. Canada's urban areas routinely perform well on WHO air quality evaluations, although there are still some problem areas.

Provincial/territorial   Each provincial and territorial government approaches air quality issues with its own agenda and set of rules, through its environment ministry. This makes sense—not just politically but scientifically, too—since issues vary dramatically from one region to another. For example, acid deposition and smog caused by both local pollution and pollutants transported from the industrial midwestern states are major air pollution issues in Ontario. In Saskatchewan, in contrast, air quality issues associated with the handling of grain, feed, and livestock are of greater concern.

Regional differences in the handling of air quality issues can lead to problems. In the past, standards for acceptable levels and the protocols of measurement for pollutants varied significantly from one jurisdiction to another. This makes it difficult to compare the results of monitoring across borders; the number that represents "good" air quality in one province might be "poor" in another, not because the air quality actually differs but because a different scale is being used. The government has been trying to bring these standards and protocols into conformity across the nation, by working through the Canadian Council of Ministers of the Environment (CCME).

Municipal   Only two municipalities in Canada—Montreal and Greater Vancouver—have been given direct regulatory authority over sources of air pollution by their respective provincial governments. However, all municipalities manage programs and activities that directly influence air quality on a daily basis—public transportation and land use zoning come to mind, for example. Therefore, most municipalities have programs aimed at improving air quality and raising public awareness of air quality issues.

The top pollution concerns differ from one location to another, and air quality may or may not be the main issue in a given municipality. For example, in Mississauga, Ontario—a geographically spread-out, largely suburban city—transportation and related air quality concerns are the central environmental issue, both for government decision-makers and for the general public. In Halifax, with a smaller population, shorter travel distances, and fewer cars, wastewater management takes priority over air quality concerns. Air pollution sources differ, too, from one locality to another. In Sudbury, smelters have been a central concern; in Mississauga, the main issue is car exhaust. Elsewhere, pulp-and-paper mills are of primary concern, or grain-handling operations, or large feedlots, or dust from cement manufacturers.

Monitoring and reporting are standardized through the AQI and AQHI

Most nations and organizations that monitor and report on air quality, including the World Health Organization, simplify the results of scientific monitoring for communication purposes. Raw data from monitoring can be confusing, especially to the general public. Many different units of measurement are used, most of which are not familiar to us in daily life. For example, particulate pollution, such as $PM_{2.5}$, is reported in units of micro-

grams per cubic metre of air ($\mu g/m^3$); sulphur dioxide is measured in parts per million (ppm); radon (an indoor air pollutant) is measured in picocuries per litre of air (pCi/L); and so on. The public and personal health implications of such measurements are difficult to interpret, even if there is a standard or safe limit with which to compare them. Add in the other contaminants, each with its own units of measurement and monitoring protocols, and you have a very complicated mix that doesn't translate well to the general public.

Consequently, most countries have implemented reporting processes in which the scientific results of monitoring are combined and translated into a scale that is simpler and more straightforward to interpret—an *Air Quality Index*. Unfortunately, these scales vary from one place to another so they are not very comparable, but typically they represent "good" or "safe" air quality as a low number, with "poor" air quality increasing as the scale goes up. For example, the Air Quality Index in Ontario takes measurements of a number of pollutants from many monitoring stations, and translates them into a simple representation of air quality on a scale of 0 to 100+ (**FIGURE 13.13**).

As of May 2013, the various standards and protocols for monitoring and reporting of air pollutants were brought into alignment across the country, through the new Canadian **Air Quality Health Index (AQHI)**, established by the federal government. This index focuses principally on pollutants that are known to cause problems for human health. The standards set out in the AQHI are more stringent than previous national air quality standards; they also contain long-term targets, as well as providing safe limits for long-term exposure. Most importantly, the targets, referred to as the *Canada-Wide Standards (CWS)*, are consistent across the provinces and territories. Quebec is not a signatory to these harmonization agreements but is following similar protocols. Other provinces and territories are currently submitting plans for implementation of AQHI reporting, and in the meantime continue to issue reports using their own AQI protocols.

Monitoring shows that many forms of air pollution have decreased

CEPA not only lists the pollutants of interest but also requires that any releases of these pollutants be reported to the *National Pollutant Release Inventory (NPRI)*, which is maintained by Environment Canada. The NPRI is thus one important vehicle that can be used to keep track of air quality in Canada. The data on NPRI are submitted, under law, directly by those who emit harmful substances into the atmosphere.

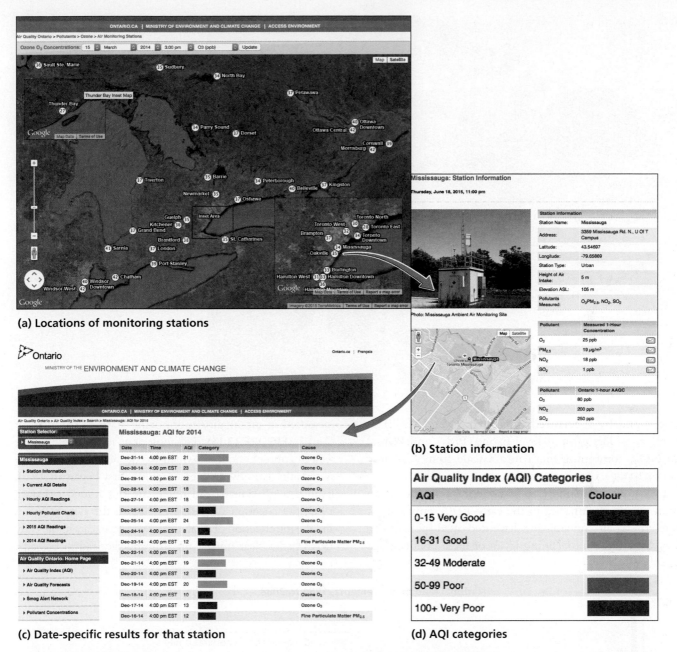

(a) Locations of monitoring stations

(b) Station information

(c) Date-specific results for that station

(d) AQI categories

FIGURE 13.13 If you visit the Ontario Ministry of Environment website (http://airqualityontario.com), you will find a map of locations where monitoring occurs **(a)**. Clicking on "Mississauga," for example, you will find that the air quality monitoring station for Mississauga is located on the campus of the University of Toronto **(b)**. You can request date-specific results of monitoring for that station **(c)**, which are expressed in terms of the Air Quality Index **(d)**.

Source: Images courtesy of Air Quality Ontario. © Queen's Printer for Ontario, 2014. Reproduced with permission.

DATA Q What pollutants are monitored at the Mississauga station? What are the best and worst results for this station, among the dates shown in the figure, and what do they mean in terms of AQI categories? Note that the AQI readings mention the specific pollutant that had the highest reading for that date ("Cause," in **FIGURE 13.13C**). On the basis of the readings shown in this figure, which are fairly typical for Mississauga, what is the most common air pollutant in this part of Ontario? Which one is second?

Pollutants also are measured and monitored through a nationwide set of monitoring stations that compose the *National Air Pollution Surveillance (NAPS)* network, coordinated by Environment Canada (**FIGURE 13.14**). Many of the stations in the network are NAPS-designated sites. Stations managed by other federal agencies (such as the Meteorological Service of Canada), as well as provincial,

territorial, municipal, or other types of agencies (such as universities) also contribute data to the network; the Mississauga station shown in **FIGURE 13.13** is an example of the latter.

The main focus of air quality monitoring in Canada is on the criteria air contaminants, but other categories of pollutants (such as mercury) are monitored at certain

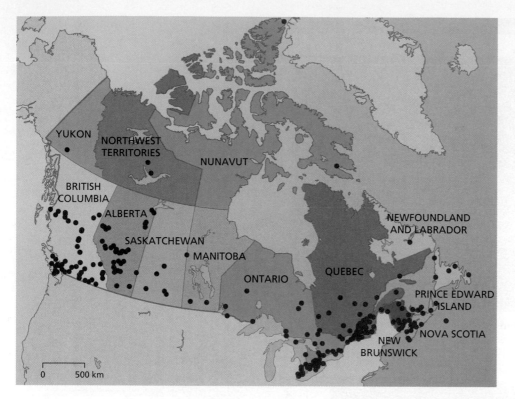

FIGURE 13.14
This map shows the National Air Pollution Surveillance (NAPS) network of real-time air quality monitoring stations across Canada.

stations and by specialized monitoring networks. The AQHI considers ground-level ozone, PM_{10} and $PM_{2.5}$, and NO_2, combining them into a scale from 1 (= low risk) to 10+ (= very high risk).

In the decades since the first modern anti-pollution actions in North America in the early 1970s, emissions of some criteria air contaminants have decreased substantially. This has resulted in declining levels of these pollutants, as measured at air quality monitoring stations throughout Canada (**FIGURE 13.15**). The most dramatic decrease can be seen in atmospheric lead, which has decreased by 97% since systematic nationwide measurements first began in 1970. Even though Canada did not phase out leaded gas until 1993, the cessation of leaded gas use in the United States and the introduction of catalytic converters in cars in the early 1970s had a clear impact on atmospheric lead in Canada.

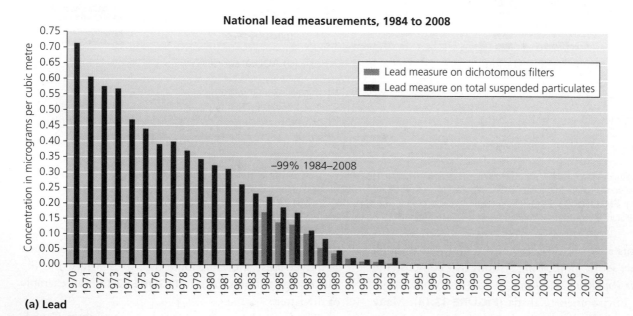

(a) Lead

FIGURE 13.15 Mean annual levels of many criteria air contaminants have declined since the early 1970s, some quite dramatically (such as lead). The graphs show annual averages across Canada, as measured at air quality monitoring stations in the National Air Pollution Surveillance network.
Source: © Her Majesty the Queen in Right of Canada, as represented by the Minister of the Environment, 2013.

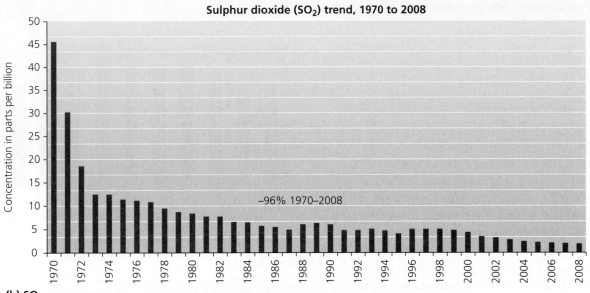

(b) SO₂

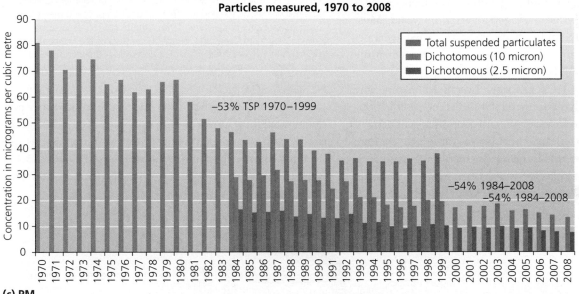

(c) PM

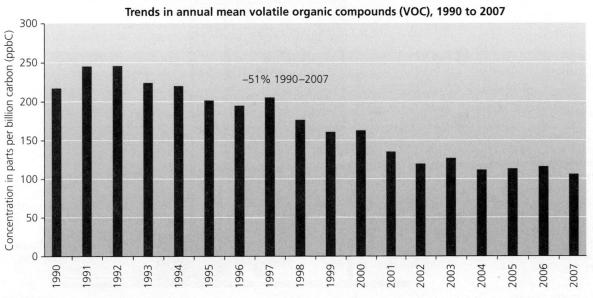

(d) VOCs

FIGURE 13.15 *(Continued)*

Over the same period, SO$_2$ decreased by 96%, and particulate matter by more than 50%. Volatile organic compounds, which have only been measured systematically since 1990, also show a downward trend, although not as dramatic.

The graphs in **FIGURE 13.15** tell a positive story of improvements in air quality over time across Canada; however, they only show a small part of the story. These are average values—averaged over an entire year and a very large geographical area. For these materials, and for other pollutants not represented here, there are hotspots where reported values locally or temporarily exceed targets or safe levels. For example, ground-level ozone, O$_3$ (a secondary pollutant, derived from primary pollutants by chemical reactions in the atmosphere) falls well below the Canada-Wide Standards for most reporting stations when averaged over three years (**FIGURE 13.16**). But in some localities (including Toronto), ground-level ozone exceeds the CWS, even when averaged over three years. Most other reporting locations have short-term exceedances in ground-level ozone from time to time, even when the long-term average is below the target level.

There are several reasons for the overall declines in some pollutants, which have occurred in spite of increases in population, energy use, vehicle use, and economic productivity in North America. Cleaner-burning motor vehicle engines, unleaded gasoline, and automotive technologies such as catalytic converters have played a large part in decreasing

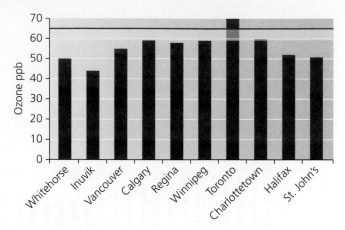

FIGURE 13.16
Levels of ground-level ozone are below the Canada-Wide Standard for most Canadian cities, when averaged over three years. The data shown here are averages from 2009 to 2011, for selected reporting sites in the National Air Pollution Surveillance Network of Environment Canada. The red line shows the ozone CWS target of 65 ppb, and the yellow band indicates the range within 10% of the target (65 to 59 ppb). *Source:* Data from Canadian Council of Ministers of the Environment (2013). *2011 Progress Report on the Canada-Wide Standards for Particulate Matter and Ozone.*

emissions of lead, carbon monoxide, and several other pollutants. Sulphur dioxide permit–trading and clean coal technologies have reduced SO$_2$ emissions. Technologies like electrostatic precipitators and **scrubbers** (**FIGURE 13.17**), which chemically convert or physically remove airborne

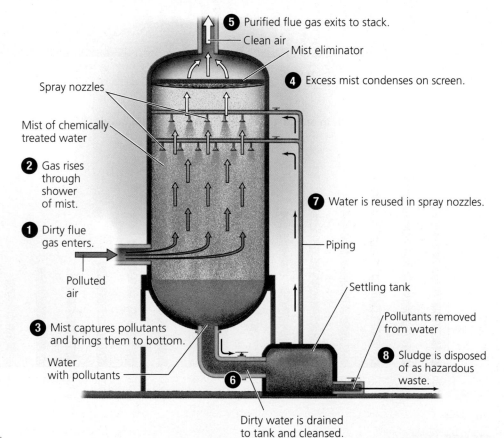

FIGURE 13.17
In a spray-tower wet scrubber, polluted air (**1**) rises through a chamber while nozzles spray a mist of water mixed with active chemicals (**2**). The falling mist captures pollutants and carries them to the bottom of the chamber (**3**). Excess mist is captured on a screen (**4**), and clean air is emitted from the scrubber (**5**). The dirty water is drained from the chamber (**6**), cleansed in a settling tank, and recirculated (**7**). The resulting sludge must be disposed of (**8**) as hazardous waste. This type of scrubber typically removes at least 90% of particulate matter and gases, such as sulphur dioxide.

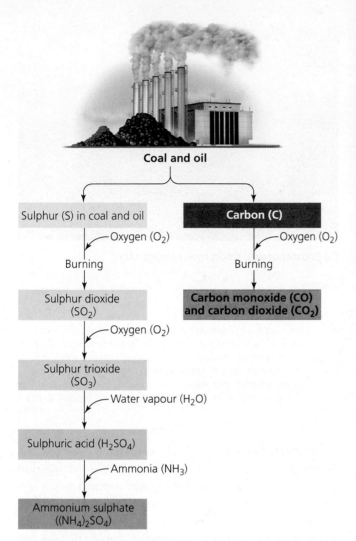

Coal and oil

Sulphur (S) in coal and oil → Oxygen (O_2) → Burning → Sulphur dioxide (SO_2) → Oxygen (O_2) → Sulphur trioxide (SO_3) → Water vapour (H_2O) → Sulphuric acid (H_2SO_4) → Ammonia (NH_3) → Ammonium sulphate ((NH_4)$_2SO_4$)

Carbon (C) → Oxygen (O_2) → Burning → Carbon monoxide (CO) and carbon dioxide (CO_2)

(a) Formation of industrial smog

(b) Smog event, midday in Donora, Pennsylvania, 1948

FIGURE 13.18

Emissions from the combustion of coal and oil in plants without pollution control technologies can create industrial smog. When fossil fuels are combusted, sulphur contaminants give rise to sulphur dioxide, which in the presence of other chemicals in the atmosphere can produce several other sulphur compounds **(a)**. Under certain weather conditions, industrial smog can blanket whole towns or regions, as it did in Donora, Pennsylvania, during a deadly 1948 smog episode **(b)**.

pollutants before they are emitted from smokestacks, have allowed factories, power plants, and refineries to decrease emissions of several pollutants.

Smog is the most common and widespread air quality problem

In response to the increasing incidence of fogs polluted by the smoke of Britain's Industrial Revolution, an early British scientist coined the term **smog**. Today the term is used worldwide to describe unhealthy mixtures of air pollutants that often form over urban areas.

The deadly smog that enveloped London in 1952 was what we today call **industrial smog**, or grey-air smog. When coal or oil is burned, some portion is completely combusted, forming CO_2; some is partially combusted, producing CO; and some remains unburned and is released as soot, or particles of carbon. Coal also contains contaminants, including mercury and sulphur. Sulphur reacts with oxygen to form sulphur dioxide, which can

undergo a series of reactions to form sulphuric acid and ammonium sulphate (**FIGURE 13.18A**). These chemicals and others produced by reactions in the atmosphere, along with soot, are the main components of industrial smog.

Although coal combustion and other industrial sources supply the chemical constituents for industrial smog, weather also plays a role, as it did in London in 1952, and in a similar event in Donora, Pennsylvania, in 1948. Here, air near the ground cooled during the night, and because Donora is located in hilly terrain, too little morning sun reached the valley floor to warm and disperse the cold air. The resulting thermal inversion trapped smog containing particulate matter emissions from a steel and wire factory; 21 people died, and more than 6000 people—nearly half the town—became ill (**FIGURE 13.18B**).

A somewhat different type of smog is formed by photochemical processes, which are activated by sunlight. **Photochemical smog**, or brown-air smog, is formed through light-driven chemical reactions between primary pollutants and natural atmospheric compounds. Such reactions produce a mix of more than 100 different chemicals, tropospheric ozone often being the most abundant (**FIGURE 13.19A**). High levels of NO_2 cause photochemical smog to form a brownish haze over cities (**FIGURE 13.19B**). Hot, sunny, windless days provide perfect conditions for the formation of photochemical smog. Exhaust from morning traffic releases NOX and

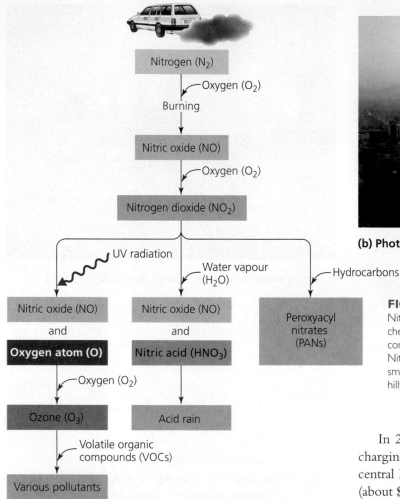

(b) Photochemical smog over Mexico City

FIGURE 13.19

Nitric oxide, a key element of photochemical smog, can start a chemical chain reaction **(a)** that results in the production of other compounds that can damage living tissues in animals and plants. Nitric acid also contributes to acidic deposition. Photochemical smog is common over many urban areas, especially those with hilly topography or frequent inversions, such as Mexico City **(b)**.

(a) Formation of photochemical smog

VOCs; sunlight then promotes the production of ozone and other constituents of photochemical smog, which typically peak in midafternoon. Photochemical pollutants can irritate people's eyes, noses, and throats.

The cities most afflicted by photochemical smog are those with weather and topography that promote it. The geographical area associated with a particular air mass is called an **airshed**. People who live within the same airshed tend to experience similar weather and "bad air" days. Airsheds that are topographically constrained, such as those that occupy basins or valleys, allow less natural circulation and renewal of the air, and are more prone to inversions and prolonged smog events.

Some provinces have controlled photochemical smog by cutting emissions through vehicle inspection programs, such as AirCare in British Columbia or Drive Clean in Ontario, where drivers are required to have their vehicle exhaust inspected regularly to maintain their registrations. Although a failed "smog check" means inconvenience for the car owner, these programs help maintain vehicle condition and make the air measurably cleaner for all of us.

In 2003, London, England, instituted a "congestion-charging" program to control smog. People driving into central London during weekdays were required to pay £8 (about $15) per day. The money was to be used to enhance bus service and encourage transport by rail, taxi, bicycle, and foot. Many citizens were outraged, arguing that the fees were too high; others complained that it would not work or that the promised improvements in public transport were too slow in coming. However, in the first year traffic congestion in the zone decreased by nearly 30%, particulate matter declined by 15.5%, nitrogen oxide emissions decreased by 13.4%, and carbon dioxide emissions fell by 16.4%. The charges are still in place. The fee is now higher, although the congestion charge zone is smaller. Similar schemes have been successfully implemented in other cities, such as Singapore and (temporarily, in 2014) Paris. London's most recent Air Quality Strategy (2011–2015)[10] focuses on getting older, more polluting vehicles off the roads, utilizing cleaner hybrid vehicles for transit routes, and encouraging businesses and individuals to adopt less polluting habits.

Acidic deposition is a transboundary problem

Acidic deposition is the settling of acidic or acid-forming pollutants from the atmosphere onto the surface. This can take place by **acidic precipitation** (commonly referred to as **acid rain**, but also including acid snow, sleet, hail,

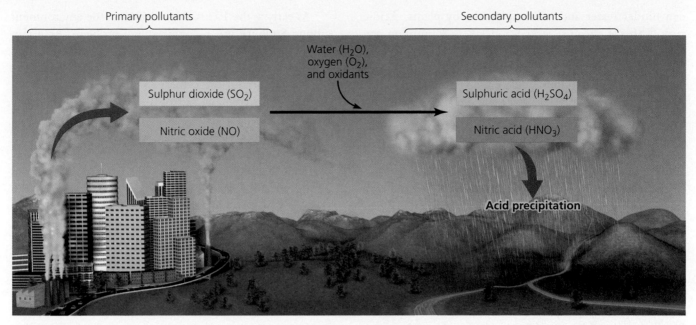

FIGURE 13.20 Sulphur dioxide and nitric oxide emitted by industries, utilities, and motor vehicles can be carried by the atmosphere and transformed by chemical reactions into sulphuric acid and nitric acid. The acids are deposited by rain, snow, fog, and dry deposition.

and fog) or by the deposition of dry particles. Acidic deposition is one type of **atmospheric deposition**, which refers more broadly to the wet or dry deposition on land of a wide variety of pollutants.

Acidic deposition originates with the emission of primary or *precursor* pollutants, especially sulphur dioxide and nitrogen oxides, which come from fossil fuel combustion by cars, power plants, and industrial facilities. Once airborne, these primary pollutants can react with water, oxygen, and oxidants to produce secondary compounds of low pH, primarily sulphuric and nitric acid. Suspended in the troposphere, acid droplets may travel for days or weeks, covering hundreds or even thousands of kilometres in a particular airshed before settling (**FIGURE 13.20**). Because the precursors may originate many kilometres from where deposition eventually occurs, acid rain is a common type of transboundary pollution.

Natural rain water is not neutral; it is slightly acidic, with a typical pH of 5.6. This is mainly because rain water reacts with naturally occurring carbon dioxide in the air, forming carbonic acid. Rain and other forms of precipitation with pH less than about 5.1 are considered to be acidified. In some regions of Britain and the United States, acid fogs with pH of 2.3 (equivalent to vinegar) have occurred. Acidification can occur as a result of natural processes, such as sulphur-rich volcanic eruptions, but the main cause is human-generated air pollution.

Acidic deposition can have wide-ranging, cumulative detrimental effects on ecosystems and the built environment (**TABLE 13.1**). Acids leach nutrients, such as calcium, magnesium, and potassium, from topsoil, altering soil

chemistry and harming plants and soil organisms. Hydrogen ions from acidic precipitation take the place of calcium, magnesium, and potassium ions in soil compounds, and these valuable nutrients are leached away into the subsoil, where they become inaccessible to plant roots.

Acidic precipitation also "mobilizes" toxic metal ions, such as aluminum, zinc, mercury, and copper, by chemically converting them from insoluble forms to soluble forms that are *bioavailable* (that is, more easily absorbed by living organisms). Elevated soil concentrations of metals

TABLE 13.1 **Effects of Acidic Deposition on Ecosystems in Northeastern North America**

Acidic deposition in northeastern forests has
■ accelerated leaching of base cations (ions that counteract acidic deposition) from soil
■ allowed sulphur and nitrogen to accumulate in soil
■ increased dissolved inorganic aluminum in soil, hindering plant uptake of water and nutrients
■ caused calcium to leach from needles of red spruce, leading to tree mortality from wintertime freezing
■ increased mortality of sugar maples because of leaching of base cations from soil and leaves
■ acidified many lakes, especially those situated on non-calcareous soils and bedrock of granitic composition
■ lowered lakes' capacity to neutralize acids
■ elevated aluminum levels in surface waters
■ reduced species diversity and abundance of aquatic life, and negatively affected entire food webs

Source: Adapted from Driscoll, C.T., et al. (2001). *Acid Rain Revisited.* Hubbard Brook Research Foundation.

can hinder water and nutrient uptake by plants. Acidic runoff affects streams, rivers, and lakes. Aluminum can kill fish by damaging their gills and disrupting their salt balance, water balance, breathing, and circulation. Besides altering aquatic ecosystems, acid precipitation can damage agricultural crops. It erodes stone buildings, corrodes cars, and erases the writing from tombstones. Ancient cathedrals, monuments, and stone statues in many parts of the world are experiencing irreversible damage as their features gradually wear away.

Because the pollutants leading to acid deposition can travel long distances, their effects may be felt far from their sources. For instance, much of the pollution from power plants and factories in Pennsylvania, Ohio, and Illinois falls out in southeastern Canada. The bedrock geology and soil chemistry of the area that is receiving the acidic deposition also play an important role in the acid tolerance and ecological response to acidification. Areas underlain by silica-rich rock types like granite are less tolerant to acidic deposition than areas underlain by limestone, which can have a buffering effect on the acid. As **FIGURE 13.21** shows, many regions in southeastern Canada have experienced acid deposition in excess of their critical loads.

One place in Canada where acid rain has been problematic is Sudbury, Ontario. Mining started in the Sudbury area in the late 1800s and continues today. Mining, refining, and smelting of ores generate emissions high in sulphur dioxide (SO_2), which can result in acidic deposition. The rocks and soils of the Sudbury area are naturally susceptible to acidification because of their silica-rich composition.

As early as the 1920s, degradation of the environment around Sudbury was recognized as a problem. Acid rain had devastated forests and water bodies, and the treeless soil was stained black. In 1969, the government of Ontario informed Inco, the principal mining company in the area, that it would be required to substantially decrease emissions from its facilities. Inco's response, cutting-edge in its day, was to build a "Superstack," 380 m tall, to carry emissions from the smelter far away from the immediate Sudbury area. The Superstack was completed in 1972, and remains the tallest smokestack in the Western Hemisphere.

The Superstack was only a partial solution to the acid rain problem for Sudbury. It did disperse the sulphur emissions, but rather than ending the acidic deposition it spread the problem farther afield. Environment Canada has estimated that 7000 lakes in northern Ontario and Quebec were damaged by acid-causing emissions from smelters in Sudbury.[11] The sensitivity of the underlying granitic rocks contributed to the acidification problem, which affected not only forests and soils, but also fish and the sport fishing industry in an area of approximately 17 000 km².[12]

In the early 1980s, Inco and Falconbridge (the other major producer of smelter emissions in the area) undertook vigorous efforts to clean their emissions prior to releasing them to the atmosphere, and SO_2 emissions today have been reduced by as much as 90%. Lakes, forests, and soils in the region have shown significant biological and chemical improvements in the more than

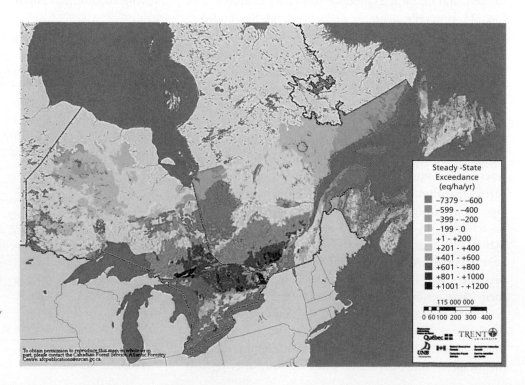

FIGURE 13.21 The maximum amount of acid deposition that a region can receive without damage to its ecosystems—its critical load—depends on the acid neutralizing capacity of water, rocks, and soils. This map shows areas of eastern Canada where the levels of acid deposition have exceeded the capacity of the soils to neutralize the acid. *Source:* Environment Canada, *Acid Rain and Forests,* www.ec.gc.ca/acidrain/images/Exceedance_E.jpg.

30 years since the Superstack was constructed.[13] The most important legacy of the Superstack is that it ushered in an era of ecological awareness, recovery, and restoration in Ontario, and of pride in the natural environment in the Sudbury area.

Although many improvements have been made, acidic deposition overall has not been reduced as much as many scientists had hoped. New technologies, such as scrubbers, have helped. As a result of declining emissions of SO_2, average sulphate precipitation decreased in northeastern North America from the early 1980s to the late 1990s (**FIGURE 13.22A**). However, because of increasing NO_x emissions, average nitrate precipitation changed little in the same period (**FIGURE 13.22B**). Long-term work at Hubbard Brook Experimental Forest supports the conclusion that there is still work to be done to limit the impacts of acid deposition (see "The Science behind the Story: Acid Rain at Hubbard Brook Experimental Forest").

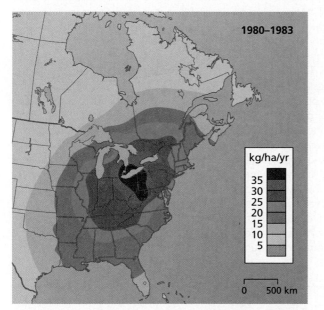
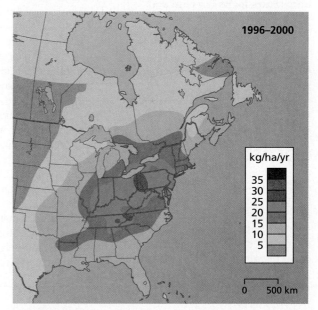

(a) Depositions of wet sulphate decreased from the early 1980s to the late 1990s

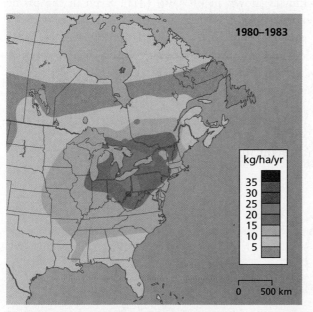
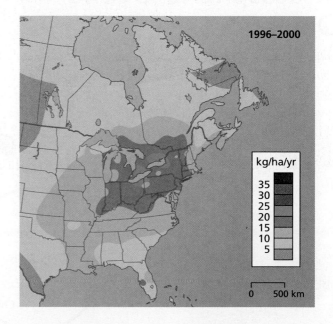

(b) Wet nitrate deposition: little change since early 1980s

FIGURE 13.22 Acidic deposition of wet sulphates **(a)** has declined in eastern North America since Canada and the United States signed a bilateral air quality agreement, aimed at reducing primary pollutants, in 1991, but deposition of wet nitrates **(b)** hasn't changed as much.
Source: Canadian National Atmospheric Chemistry Database, Meteorological Service of Canada, © Environment Canada, Science and Technology Branch, 4905 Dufferin Street, Toronto, Ontario, Canada, M3H 5T4.

Dr. Gene E. Likens is with the Cary Institute of Ecosystem Studies. He received the National Medal of Science, the highest scientific honour in the United States, for his long-term work at Hubbard Brook Experimental Forest.

Photo by HO Photography/Courtesy of Gene Likens

Acid Rain at Hubbard Brook Experimental Forest

Acidic deposition involves subtle and incremental changes in pH levels that take place over long periods, so short-term studies give an incomplete picture of the effects. One long-term research project conducted in the Hubbard Brook

Experimental Forest in New Hampshire's White Mountains has been critically important to our understanding of acidic deposition.

Established by the U.S. Forest Service in 1955, Hubbard Brook was initially devoted to research on hydrology, the study of water flow through forests and streams. In 1963, Hubbard Brook researchers broadened their focus to include a long-term study of nutrient cycling in forest ecosystems. Since then, they have collected and analyzed weekly samples of precipitation. These measurements make up the longest-running North American record of acid precipitation.

Throughout Hubbard Brook's 3160 ha, small plastic collecting funnels channel precipitation into clean bottles, which researchers retrieve and replace each week. Hubbard Brook's laboratory measures acidity and conductivity, which indicate the amounts of salts and other electrolytic contaminants dissolved in the water. Concentrations of sulphuric acid, nitrates, ammonia, and other compounds are measured elsewhere.

By the late 1960s, ecologists Gene Likens, F. Herbert Bormann, and others had found that precipitation at Hubbard Brook was several hundred times as acidic as natural rain water. By the early 1970s, a number of other studies had corroborated their findings. Together, these studies indicated that the precipitation had pH values averaging around 4 and that individual rainstorms showed values as low as 2.1—almost 10 000 times as acidic as ordinary rain water. Since then, clean air legislation and international agreements between the United States and Canada have helped reduce the emission of acid precursors, therefore also reducing the acidity of precipitation (see **FIGURE 1**). Nonetheless, acidic deposition continues to be a serious problem in the northeastern United States and southeastern Canada.

Some long-term consequences of acidic deposition are now becoming clear. Studies of soil nutrients showed that up to 50% of the calcium and magnesium in Hubbard Brook's soils had leached out. Meanwhile, acidic

Stratospheric Ozone Depletion

In the troposphere, ozone (O_3) is a respiratory irritant and a component of smog, but in the stratosphere it performs a crucial role for life on this planet. Ozone is concentrated in the stratospheric ozone layer at an altitude of about 25 km (as shown in **FIGURE 13.2**). Even in the ozone layer, the concentration of ozone is very low—only about 12 parts per million. However, ozone molecules are so effective at absorbing incoming sunlight that even this low concentration helps protect Earth's surface from the damaging effects of ultraviolet (UV) radiation.

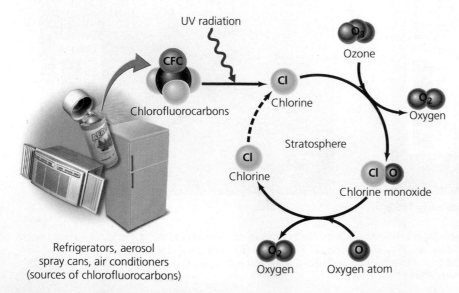

FIGURE 13.23

A chlorine atom released from a CFC molecule in the presence of UV radiation reacts with an ozone molecule, forming one molecule of oxygen gas and one chlorine monoxide (ClO) molecule. The oxygen atom in the ClO molecule binds with a stray oxygen atom to form oxygen gas (O_2), leaving the chlorine atom to begin the destructive cycle anew. A given chlorine atom may destroy up to 100 000 ozone molecules.

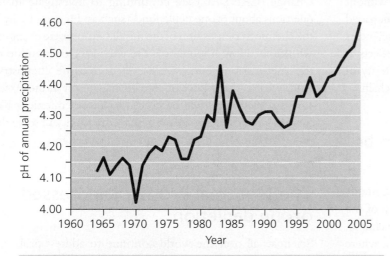

FIGURE I Over the past 40 years, precipitation at the Hubbard Brook Experimental Forest has become slightly less acidic. However, it is still more acidic than natural precipitation.
Source: Data from Likens, G.E. (2004). *Ecology, 85*, pp. 2355–2362.

kill trees, but it affects the availability of nutrients and the ability of trees to access them. It also reduces the ability of soil and water to neutralize acidity, making the ecosystem increasingly vulnerable to further inputs of acid.

In 1999, researchers used a helicopter to distribute 50 tons of a calcium-containing mineral called wollastonite over one of Hubbard Brook's watersheds. Their objective was to raise the concentration of base cations in the soil to estimated historical levels. Over the next 50 years, scientists plan to evaluate the impact of calcium addition on the watershed's soil, water, and life. By providing a comparison with watersheds in which calcium remains depleted, the results should provide new insights into the consequences of acid rain and the possibilities for reversing its negative effects. In honour of their long-term work at Hubbard Brooks, researchers Bormann and Likens received the 2003 Blue Planet Prize for outstanding scientific research that helps solve global environmental problems.

deposition increased the concentration of aluminum in the soil, which can prevent tree roots from absorbing nutrients. The resulting nutrient deficiency slows forest growth and weakens trees, making them more vulnerable to drought and insects. In other words, acidification of soil does not directly

Synthetic chemicals deplete stratospheric ozone

In the 1960s, atmospheric scientists began wondering why their measurements of stratospheric ozone were lower than theoretical models predicted. Researchers hypothesizing that natural or artificial chemicals were depleting the ozone finally pinpointed a group of human-made compounds derived from simple hydrocarbons, such as ethane and methane, in which hydrogen atoms are replaced by chlorine, bromine, or fluorine. One class of such compounds, **chlorofluorocarbons (CFCs)**, was being mass-produced by industry at a rate of 1 million metric tons per year in the early 1970s, and this rate was growing by 20% a year.

Soon researchers showed that CFCs could deplete stratospheric ozone by releasing chlorine atoms that split ozone molecules, creating from each of them an O_2 molecule and a ClO molecule (**FIGURE 13.23**). In 1985, scientists announced that stratospheric ozone levels over Antarctica had declined by 40–60% in the previous decade, leaving a thinned ozone concentration that was soon dubbed the **ozone hole** (**FIGURE 13.24**).

Research over the next few years confirmed the link between CFCs and ozone loss in the Antarctic, and

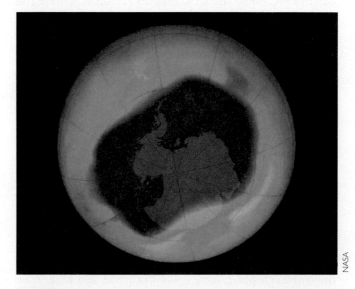

FIGURE 13.24
The "ozone hole" is a region of ozone depletion in the stratosphere over Antarctica. It reappears seasonally each September. Satellite imagery from September 24, 2006, shows the largest-ever recorded extent and level of depletion. The purple colour at the centre represents a level of about 85 DU; the global average is about 300 DU.

NASA

indicated that depletion was also occurring in the Arctic and perhaps globally. The depletion was shown to be growing both in severity and in areal extent, almost without exception, year to year. Already concerned that increased UV radiation would lead to more skin cancer, scientists were becoming anxious over possible ecological effects as well, including harm to crops and to the productivity of ocean phytoplankton, the base of the marine food chain.

There are still many questions to be resolved about ozone depletion

Significant progress has been made in understanding stratospheric ozone depletion since the initial discovery of the role of CFCs, but many questions remain. For example, will ozone depletion spread from the polar regions, where it is most severe, to encompass mid-latitude and even low-latitude regions (and if so, how quickly)? What is the actual relationship between ozone depletion and human health impacts, such as skin cancer? What are the other potential impacts of ozone depletion, for example, on marine and terrestrial ecosystems and on other types of materials? And—importantly, for policy decisions—are the substitute chemicals that are being proposed in international agreements definitely less damaging to the stratospheric ozone layer, or do they raise concerns of their own?

The question of polar ozone depletion is of particular interest, especially to Canadian scientists working in the Arctic. Scientists now believe that the ozone-depleting chemical reactions may find ideal sites on tiny ice crystals that are found only where the air is extremely cold. These conditions are optimal over Antarctica, where a circular wind pattern called the *polar vortex* traps extremely cold air over the pole.

The polar vortex in the northern polar region is not quite as strong as the system around the southern polar region; its influence on the ozone layer is often characterized as causing a "dent" rather than a "hole." In 2011, the northern polar vortex was particularly strong, cold, and undisturbed; early that year, Canadian scientists at the Polar Environment Atmospheric Research Laboratory (PEARL) on Ellesmere Island recorded the first actual ozone "hole" in the Northern Hemisphere.[14] In 2014, the northern polar vortex was disturbed and broken up by a phenomenon called sudden stratospheric warming. This brought jet stream winds and intensely cold temperatures much farther south than usual, leading to the long, cold winter experienced across much of Canada that year. Winters with sudden stratospheric warming, like 2014, tend to show lesser ozone depletion, because the lack of a strong vortex prevents the air from getting cold enough for ozone-depleting reactions to occur.

PEARL scientists and other researchers from the Canadian Network for the Detection of Atmospheric Change (CANDAC) are continuing to investigate many questions about ozone depletion,[15] such as: How and why is the chemical composition of the middle atmosphere changing with time? How important is the chemistry of bromine to the Arctic ozone budget, relative to chlorine chemistry? And, what is the impact of climate change on future Arctic ozone depletion? It will be crucial to answer questions like these as we move forward as a global community to solve the problems of ozone depletion and climate change.

The Montreal Protocol addressed ozone depletion

Scientists all over the world continue to address problems like these in an attempt to refine our understanding of the complex chemical process of ozone depletion. In the meantime, the international community has been unwilling to wait for catastrophe and has moved forward to strike an agreement to reduce and eventually eliminate the production of known ozone-depleting substances, also known as *ODS*.

In response to the scientific concerns, international policy efforts to restrict CFC production finally bore fruit in 1987 with the **Montreal Protocol**. In this treaty, signatory nations (eventually numbering 197, including Canada) agreed to cut CFC production in half (see "The Science behind the Story: Discovering Ozone Depletion and Its Causes"). Follow-up agreements strengthened the pact by deepening the cuts, advancing timetables for compliance, and addressing related ozone-depleting chemicals.

Today the production and use of ozone-depleting compounds have fallen by 95% since the late 1980s, and scientists can discern the beginnings of long-term recovery of the stratospheric ozone layer. Industry has been able to shift to alternative chemicals that have largely turned out to be cheaper and more efficient. Whether any of the replacement chemicals will have a negative impact on the ozone layer remains a subject of much scientific interest.

There are still additional challenges to overcome. Much of the 5 billion kilograms of CFCs emitted into the troposphere has yet to diffuse up into the stratosphere, and CFCs are slow to dissipate or break down. Thus, we can expect a considerable lag time between the implementation of policy and the desired environmental effect. (This is one reason scientists often argue for proactive policy guided by the precautionary principle, rather than reactive policy that may respond too late.) Moreover, nations can plead for some ozone-depleting chemicals to be exempted from the ban; for example, the United States recently was allowed to continue using methyl bromide, a fumigant used to control pests on strawberries.

Despite the remaining challenges, the Montreal Protocol and its follow-up amendments are widely considered the biggest success story so far in addressing any global environmental problem. Environmental scientists have attributed this success primarily to two factors:

1. Policymakers engaged industry in helping to solve the problem, and government and industry worked together on developing replacement chemicals. This cooperation reduced the battles that typically erupt between environmentalists and industry.
2. Implementation of the Montreal Protocol followed an adaptive management approach, altering strategies midstream in response to new scientific data, technological advances, or economic figures.

Because of its success in addressing ozone depletion, the Montreal Protocol is widely seen as a model for international cooperation in addressing other pressing global problems, such as persistent organic pollutants, climate change, and biodiversity loss.

Indoor Air Pollution

Indoor air generally contains higher concentrations of pollutants than does outdoor air. As a result, the health effects from indoor air pollution in workplaces, schools, and homes outweigh those from outdoor air pollution. The World Health Organization (WHO) attributes almost 4.3 million premature deaths worldwide each year to indoor air pollution (compared to 3.7 million from outdoor air pollution).[16] Indoor air pollution alone, then, takes roughly 11 000 lives each day.

If this seems surprising, consider that the average person in North America is indoors at least 90% of the time. Then consider that in the past half century a dizzying array of consumer products have been manufactured and sold, many of which we keep in our homes and offices and use extensively in our daily lives. Many of these products are made of synthetic materials that are not comprehensively tested for health effects before being brought to market. Products as diverse as insecticides, cleaning fluids, plastics, and chemically treated wood can all exude volatile chemicals into the air.

In an ironic twist, some attempts to be environmentally prudent during the "energy crisis" of 1973–1974 worsened indoor air pollution in industrialized countries. To reduce heat loss and improve energy efficiency, building managers sealed off most ventilation in existing buildings, and building designers constructed new buildings with limited ventilation and with windows that did not open.

David Turnley/Corbis

FIGURE 13.25
In the developing world, many people build fires inside their homes for cooking and heating, as seen here in a South African kitchen. Indoor fires expose family members to pollution from particulate matter and carbon monoxide.

These steps may have saved energy, but they also worsened indoor air pollution by trapping stable, unmixed air—and its pollutants—inside.

Indoor air pollution in the developing world arises from fuelwood burning

Indoor air pollution has the greatest impact in the developing world. Millions of people in developing nations burn wood, charcoal, animal dung, or crop waste inside their homes for cooking and heating with little or no ventilation (**FIGURE 13.25**). In the process, they inhale dangerous amounts of soot and carbon monoxide. In the air of such homes, concentrations of particulate matter are commonly 20 times as high as WHO standards. Many people are not aware of the health risks, and of those who are, many are too poor to have viable alternatives.

Tobacco smoke and radon are the most dangerous indoor pollutants in the industrialized world

In industrialized nations, the top risks associated with indoor air pollution are cigarette smoke and radon. The health effects of smoking cigarettes are well known, but only recently have scientists quantified the risks of inhaling secondhand smoke. *Secondhand smoke*, or environmental tobacco smoke, is smoke inhaled by a nonsmoker who is nearby or shares an enclosed airspace with a smoker. Secondhand smoke has been found to cause many of the same problems as directly inhaled cigarette smoke, ranging

THE SCIENCE BEHIND THE STORY

Drs. F. Sherwood Rowland (left), Mario Molina (centre), and Paul Crutzen (right) jointly received the 1995 Nobel Prize in chemistry for their work on the depletion of stratospheric ozone.

File/AP Images

Discovering Ozone Depletion and Its Causes

In discovering the depletion of stratospheric ozone and coming to understand the roles of halocarbons and other substances, scientists from around the world have relied on historical records, field observations, laboratory experiments, computer models, and satellite technology. The story starts back in 1924, when British scientist G.M.B. Dobson built an instrument that measured atmospheric ozone concentrations by sampling sunlight at ground level and comparing the intensities of wavelengths that ozone does and does not absorb. By the 1970s, the Dobson ozone spectrophotometer was being used by a global network of observation stations. One of the main units of measurement for ozone, the Dobson Unit (DU), is named in his honour.

Meanwhile, atmospheric chemists were learning how stratospheric ozone is created and destroyed. Ozone and oxygen exist in a natural balance, with one occasionally reacting to form the other, and oxygen being far more abundant. Researchers found that certain chemicals naturally present in the atmosphere, such as hydroxyl (OH) and nitric oxide (NO), destroy ozone, keeping the ozone layer thinner than it would otherwise be. Nitrous oxide (N_2O) produced by soil bacteria can make its way to the stratosphere and produce NO, Dutch meteorologist Paul Crutzen reported in 1970. This observation was important, because human activities, such as fertilizer application, were increasing emissions of N_2O.

Following Crutzen's report, scientists Richard Stolarski and Ralph Cicerone showed in 1973 that chlorine atoms can catalyze the destruction of ozone even more effectively than N_2O can. British scientist James Lovelock developed an instrument to measure trace amounts of atmospheric gases, and found that virtually all the chlorofluorocarbons (CFCs) humanity had produced in the past four decades were still aloft, accumulating in the stratosphere.

This set the stage for the key insight. In 1974, American chemist F. Sherwood Rowland and his Mexican postdoctoral associate Mario Molina took note of all the preceding research and realized that CFCs were rising into the stratosphere, being broken down by UV radiation, and releasing chlorine atoms that ravaged the ozone layer (see **FIGURE 13.23**). Molina and Rowland's analysis, published in the journal *Nature*,[17] earned them the 1995 Nobel Prize in chemistry, jointly with Crutzen.

The paper sparked discussion about setting limits on CFC emissions. Industry leaders vigorously attacked the research, but measurements in the lab and in the stratosphere by numerous researchers soon confirmed that CFCs and other halocarbons were indeed depleting ozone. In response, several nations (including Canada) banned the use of CFCs in aerosol spray cans in 1978. Other uses continued, however, and by the early 1980s global production of CFCs was again on the rise.

Then, a shocking new finding spurred the international community to take action. In 1985, Joseph Farman and colleagues analyzed data from a British research station in Antarctica that had been recording ozone concentrations since the 1950s. Farman's team reported in *Nature*[18] that springtime Antarctic ozone concentrations had plummeted by 40–60% just since the 1970s (see **FIGURE 1**). NASA scientists had been sitting on years' worth of data that would have led them to the same conclusions, and were forced to admit that they had been beaten out by Farman and his team.

By 1987, the mass of scientific evidence helped convince the world's nations to agree on the Montreal Protocol, which aimed to cut CFC production in half by 1998. Within two years, further scientific evidence and computer modelling

---

from irritation of the eyes, nose, and throat, to exacerbation of asthma and other respiratory ailments, to lung cancer. This hardly seems surprising when one considers that environmental tobacco smoke consists of a brew of more than 4000 chemical compounds, many of which are known or suspected to be toxic or carcinogenic.

Although smoking remains common in many parts of the developing world, its popularity has declined greatly in industrialized nations in recent years. Many public and private venues now ban smoking. In Canada, smoking is now prohibited in all indoor workplaces and public places, such as airports. The last of the provinces or territories to implement the ban was Yukon, in 2008.

After cigarette smoke, radon is the second-leading cause of lung cancer in the industrialized world, responsible for 15% of lung cancer cases worldwide. Radon is a radioactive gas that occurs naturally and results from the decay of uranium and thorium (by way of radium, an intermediate decay product) in soil, rock, or water. Once released, the gas seeps up from the ground and can infiltrate the basements of buildings through pipes and cracks in the foundation. Radon is colourless and odourless, and it can be impossible to predict where it will occur without knowing details of an area's underlying geology (**FIGURE 13.26**). As a result, the only way to determine whether radon is entering a building is to measure radon with a test kit.

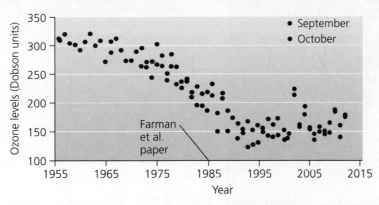

(a) Monthly mean ozone levels at Halley, Antarctica

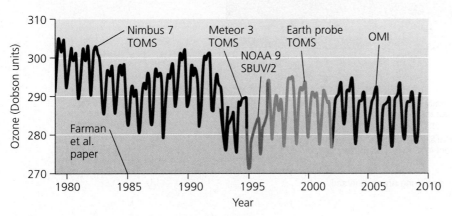

(b) Global ozone readings from five satellites

showed that more drastic measures were needed. In 1990, the Montreal Protocol was strengthened to include a complete phaseout of CFCs by 2000, in the first of several follow-up agreements.

Today, amounts of ozone-depleting substances in the stratosphere are beginning to level off. Nitrous oxide may now be the main ozone-depleting substance we emit, and it has less impact than CFCs and

other halocarbons did in 1987 (although more than any other known substance today). Scientists are hopeful that the stratospheric ozone layer has started on the long road to recovery.

Indoor radon is taken very seriously in the United States, where it is difficult to purchase a home without tests showing that radon is not a problem. The U.S. Environmental Protection Agency (EPA) has recommended a maximum acceptable level of radon that is essentially the same as the ambient outdoor level. This has brought thousands of houses into the unacceptable range; the EPA has estimated that 6% of U.S. homes exceed the maximum recommended level.

Historically, Health Canada has taken the approach that some indoor concentration of naturally occurring soil gases, such as radon, is unavoidable. The maximum acceptable, or "action," level for indoor radon in homes and other

non-occupational settings recommended by Health Canada is 0.02 WL (a Working Level is a unit of measurement that is specific to radon and its daughter by–products), which is roughly equivalent to 200 Bq/m^3 air (a Becquerel is a unit that measures the number of radioactive disintegrations per second), or 5.4 pCi/L air (picoCurie also measures the number of radioactive disintegrations per second). This is the same as the maximum permissible annual average concentration of radon daughters from the operation of a nuclear facility. The actionable level defined by the U.S. EPA is equivalent to 0.016 WL (about 4 pCi/L, or 150 Bq/m^3 of air), considerably lower than the permissible level defined by Health Canada.

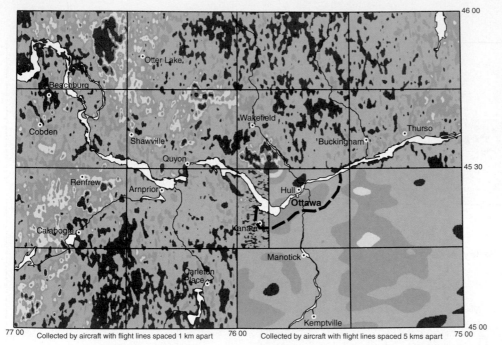

FIGURE 13.26
One's risk from radon depends largely on geology. There is much fine-scale geographical variation, and construction materials and techniques also have an impact on the processes that concentrate radon indoors. The potential for elevated levels of radon can be estimated on the basis of the amount of naturally occurring uranium in rocks and soils, as shown on this map. *Source:* Reproduced with the permission of Natural Resources Canada, 2015.

Collected by aircraft with flight lines spaced 1 km apart   Collected by aircraft with flight lines spaced 5 kms apart

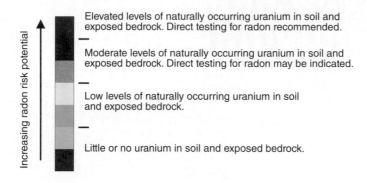

Increasing radon risk potential

Elevated levels of naturally occurring uranium in soil and exposed bedrock. Direct testing for radon recommended.

Moderate levels of naturally occurring uranium in soil and exposed bedrock. Direct testing for radon may be indicated.

Low levels of naturally occurring uranium in soil and exposed bedrock.

Little or no uranium in soil and exposed bedrock.

Many VOCs pollute indoor air

In our daily lives, we are exposed to indoor air pollutants from many sources (**FIGURE 13.27**). The most diverse indoor pollutants are volatile organic compounds (VOCs), airborne carbon-containing compounds that are released by everything from plastics to oils to perfumes to paints to cleaning fluids to adhesives to pesticides. VOCs evaporate from furnishings, building materials, colour film, carpets, laser printers, fax machines, and sheets of paper. Some products, such as chemically treated furniture, release large amounts of VOCs when new and progressively less as they age. Other items, such as photocopying machines, emit VOCs each time they are used. Seemingly innocent indoor activities, like cooking, can even contribute pollutants; dioxins and furans, the toxic by-products of incomplete combustion that are problematic outdoor pollutants, can come from burning foods during meal preparation.

Although we are surrounded by products that emit VOCs, they are released in very small amounts. Studies have found overall levels of VOCs in buildings nearly always to be less than 0.1 parts per million—very low, but a substantially greater concentration than is generally found outdoors. The implications for human health of chronic exposure to VOCs are not well known. Because they exist in such low concentrations and because individuals are regularly exposed to mixtures of many different types, it is extremely difficult to study the effects of any one pollutant.

An exception is formaldehyde, which does have clear and known health impacts. This VOC, one of the most common synthetically produced chemicals, irritates mucous membranes, induces skin allergies, and causes other ailments. Formaldehyde is used in numerous products, but health complaints have mainly resulted from its leakage from pressed wood and insulation. The use of plywood has decreased in the last decade because of health concerns over formaldehyde.

Pesticides are also VOCs. Many homes use at least one pesticide indoors during an average year, but most are

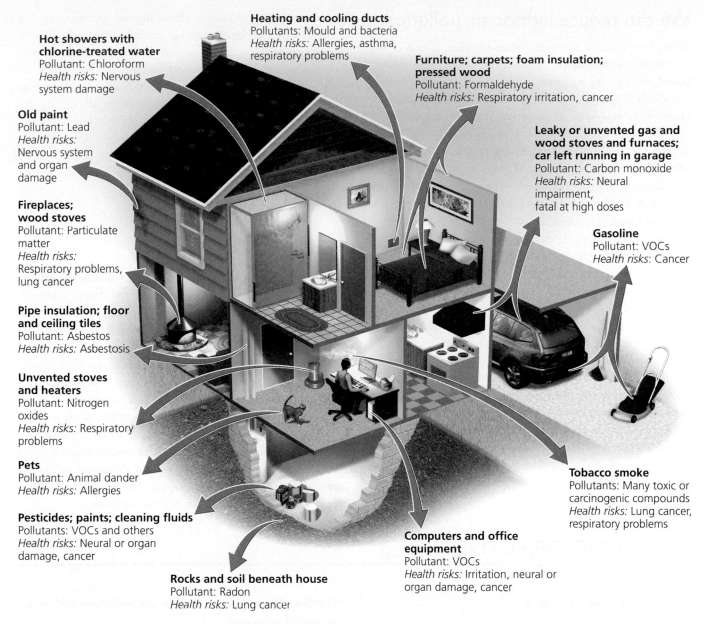

Heating and cooling ducts
Pollutants: Mould and bacteria
Health risks: Allergies, asthma,
respiratory problems

Hot showers with chlorine-treated water
Pollutant: Chloroform
Health risks: Nervous
system damage

Old paint
Pollutant: Lead
Health risks:
Nervous system
and organ
damage

Fireplaces; wood stoves
Pollutant: Particulate
matter
Health risks:
Respiratory problems,
lung cancer

Pipe insulation; floor and ceiling tiles
Pollutant: Asbestos
Health risks: Asbestosis

Unvented stoves and heaters
Pollutant: Nitrogen
oxides
Health risks: Respiratory
problems

Pets
Pollutant: Animal dander
Health risks: Allergies

Pesticides; paints; cleaning fluids
Pollutants: VOCs and others
Health risks: Neural or organ
damage, cancer

Rocks and soil beneath house
Pollutant: Radon
Health risks: Lung cancer

Furniture; carpets; foam insulation; pressed wood
Pollutant: Formaldehyde
Health risks: Respiratory irritation, cancer

Leaky or unvented gas and wood stoves and furnaces; car left running in garage
Pollutant: Carbon monoxide
Health risks: Neural
impairment,
fatal at high doses

Gasoline
Pollutant: VOCs
Health risks: Cancer

Tobacco smoke
Pollutants: Many toxic or
carcinogenic compounds
Health risks: Lung cancer,
respiratory problems

Computers and office equipment
Pollutant: VOCs
Health risks: Irritation, neural or
organ damage, cancer

FIGURE 13.27 The typical North American home contains a variety of potential sources of indoor air pollution. Shown are some of the most common sources, the major pollutants they emit, and some of the health risks they pose.

used outdoors. Thus it may seem surprising that the U.S. Environmental Protection Agency has reported that 90% of people's pesticide exposure comes from indoor sources. Pesticides that are used against termites may then seep into houses years later through floors and walls, or be carried in from outdoors on the soles of occupants' shoes.

Living organisms can pollute indoor spaces

Although it may seem unlikely, tiny living organisms constitute one of the most widespread sources of indoor air pollution. Dust mites and animal dander can worsen asthma in children. Some fungi, mould, and mildew (in particular, their airborne spores) can cause severe health problems, including allergies, asthma, and other respiratory ailments.

Some airborne bacteria can cause infectious disease. One example is the bacterium that causes Legionnaires' disease. Heating and cooling systems in buildings make ideal breeding grounds for microbes, providing moisture, dust, and foam insulation as substrates, as well as air currents to carry the organisms aloft. Microbes that induce allergic responses are thought to be a major cause of sicknesses produced by indoor pollution. When the cause of such an illness is a mystery, and when symptoms are general and nonspecific, it is often called **sick-building syndrome** or *building-related illness*.

We can reduce indoor air pollution

Using low-toxicity materials, when possible, is one of the keys to alleviating indoor air pollution. In the industrialized world, we can try to limit our use of plastics and treated wood where possible, and limit our exposure to pesticides, cleaning fluids, and other known toxicants by keeping them in a garage or an outdoor shed, rather than in the house. Keeping rooms and air ducts clean and free of mildew and other biological pollutants will reduce potential irritants and allergens. Testing is advised where mould or radon are suspected. Finally, it is important to keep indoor spaces as clean and well ventilated as possible, to minimize concentrations of the pollutants among which we live.

Health Canada and other organizations concerned with public safety recommend that we monitor homes continuously for carbon monoxide. Carbon monoxide is colourless and odourless, although it can be associated with leaks of natural gas, which does have a characteristic smell. Because carbon monoxide is so deadly and so hard to detect, many homes are equipped with detectors that sound an alarm if incomplete combustion produces dangerous levels of CO. However, at this time CO detectors are required in homes in only two jurisdictions in Canada: Yukon and Ontario.

Remedies for fuelwood pollution in the developing world include drying wood before burning (which reduces the amount of smoke produced), cooking outside, shifting to less-polluting fuels (such as natural gas), and replacing inefficient fires with cleaner stoves that burn fuel more efficiently. For example, the Chinese government has invested in a program that has placed more-fuel-efficient stoves in millions of homes in China. Installing hoods, chimneys, or cooking windows can increase ventilation for little cost, alleviating the majority of indoor smoke pollution.

Progress is being made worldwide in alleviating the health toll of indoor air pollution. Researchers calculate that rates of premature death from indoor air pollution dropped nearly 40 percent from 1990 to 2010. Taking steps like those described here should bring us further progress in safeguarding people's health.

Conclusion

Indoor air pollution is a potentially serious health threat; however, by keeping informed of the latest scientific findings and taking appropriate precautions, we as individuals can significantly minimize the risks to our families and ourselves. Outdoor air pollution has been addressed more effectively by government legislation and regulation. In fact, reductions in outdoor air pollution in Canada, the United States, Great Britain, and other industrialized nations represent some of the greatest strides made in environmental protection to date. Much room for improvement remains, particularly in reducing the photochemical smog that results from urban congestion. Avoiding unhealthy pollutant levels in the developing world will continue to pose a challenge as those nations industrialize.

REVIEWING OBJECTIVES

You should now be able to

Describe the composition, structure, and major processes of Earth's atmosphere

- The atmosphere consists of 78% nitrogen gas, 21% oxygen gas, and small amounts of other gases. There are four principal layers, which vary in temperature and other properties; they are the troposphere, stratosphere, mesosphere, and thermosphere. Ozone is concentrated in the stratosphere.

- The Sun's energy heats the atmosphere, drives air circulation, and helps determine weather, climate, and the seasons.

- Weather is a short-term phenomenon, whereas climate is a long-term phenomenon. Fronts, pressure systems, and the interactions among air masses influence weather.

- Global convective cells called Hadley, Ferrel, and polar cells create major regional wind systems and latitudinal climate zones.

Outline the scope of outdoor air pollution and assess potential solutions

- Natural sources such as windblown dust, volcanoes, and fires account for much atmospheric pollution, but human activity can worsen some of these phenomena.

- Human-emitted pollutants include primary and secondary pollutants from point and nonpoint sources.

- The principal legislation under which air quality and emissions are regulated in Canada is the *Canadian Environmental Protection Act* (1999).

- To safeguard public health, Environment Canada and various other federal, provincial, and territorial agencies monitor a number of air contaminants, including sulphur oxides, nitrogen oxides, particulate matter, volatile organic compounds, carbon monoxide, ammonia, and tropospheric ozone.

- Monitoring shows that many outdoor air pollutants have diminished over the past few decades, as a result of legislation and new, less-polluting technologies.

- Industrial smog is produced by fossil fuel combustion and is still a problem in urban and industrial areas of many developing nations. Photochemical smog is created by chemical reactions of pollutants in the presence of sunlight. It impairs visibility and negatively affects human health in urban areas.

- Acidic deposition results when pollutants, such as SO_2 and NO_x, react with water in the atmosphere to produce strong acids that are deposited on Earth's surface, often a long distance from the source of pollution. Water bodies, soils, trees, and ecosystems can experience negative impacts from acidic deposition.

Explain stratospheric ozone depletion and identify steps taken to address it

- CFCs destroy stratospheric ozone, and thinning ozone concentrations pose dangers to life because they allow more ultraviolet radiation to reach Earth's surface.

- The long residence time of CFCs in the atmosphere means a time lag between the protocol and the actual restoration of stratospheric ozone. Many scientific questions remain, both about the mechanisms of ozone depletion and the impacts of replacement chemicals for CFCs.

- The Montreal Protocol and its follow-up agreements have proven remarkably successful in reducing emissions of ozone-depleting compounds.

Characterize the main types of indoor air pollution and assess potential solutions

- Indoor air pollution causes far more deaths and health problems worldwide than outdoor air pollution.

- Indoor burning of fuelwood is the developing world's primary indoor air pollution risk.

- Tobacco smoke and radon are the deadliest indoor pollutants in the industrialized world.

- Volatile organic compounds and living organisms can pollute indoor air.

- Using low-toxicity building materials, keeping spaces clean, monitoring air quality, and maximizing ventilation are some of the steps we can take to reduce indoor air pollution.

TESTING YOUR COMPREHENSION

1. How thick is Earth's atmosphere? Name one characteristic of each of the four atmospheric layers.
2. How and why is stratospheric ozone beneficial for people and tropospheric ozone harmful?
3. How does solar energy influence weather and climate? How do Hadley, Ferrel, and polar cells help determine long-term climatic patterns and the location of biomes?
4. Describe a thermal inversion. How can thermal inversions contribute to the severity of pollution episodes?
5. Name three natural sources of outdoor air pollution and three sources caused by human activity.
6. What is the difference between a primary and a secondary pollutant? Give an example of each.
7. What is smog? How is smog formation influenced by the weather? By topography? How does photochemical smog differ from industrial smog?
8. Why are the effects of acidic deposition often felt in areas far from where the primary pollutants are produced? List three impacts of acidic deposition.
9. How do chlorofluorocarbons (CFCs) deplete stratospheric ozone? Why is this considered a long-term international problem? What has been done to address the problem, and why is it considered (at least tentatively) to be an environmental "success story"?
10. Name five common sources of indoor pollution. For each, describe one way to reduce one's exposure to this source.

THINKING IT THROUGH

1. Describe several factors that make it difficult to determine the specific causes of air pollution and to develop solutions. Are there any cities near you that have tried different approaches to mitigating air pollution?
2. Describe how and why emissions of some major pollutants have been reduced in North America since the 1970s, despite increases in population and economic activity.
3. Let's say that you have just been elected mayor of the largest city in your province. Your city's residents are complaining about photochemical smog and traffic congestion. Traffic engineers and city planners project that population and traffic will grow by 20% in the next decade. A citizens' group is urging you to implement a congestion-charging program like London's, but businesses are fearful of losing money if shoppers are discouraged from visiting. Consider the particulars of your city, and then decide whether you will pursue a congestion-charging program. If so, how would you do it? If not, why not, and what other steps would you take to address your city's problems?
4. You have just been hired to lead your region's office of public health, and the environment ministry has

informed you that your jurisdiction failed to meet the recommended air quality standards for ozone, sulphur dioxide, and nitrogen dioxide. The area is partly rural but is home to a city of 200 000 people, with sprawling suburbs. There are several large and ageing coal-fired power plants, a number of factories with advanced pollution control technology, and no public transportation system. What steps would you urge the municipal or provincial government to take, in order to meet the air quality standards? Explain how you would prioritize these steps.

INTERPRETING GRAPHS AND DATA

The U.S. National Oceanographic and Atmospheric Administration maintains an active website with ozone monitoring data at **www.ozonelayer.noaa.gov/**. Go to the website and choose "Data" from the menu.

Under "Current Data," click on "Antarctic (South Pole)," and then choose "Size of Ozone Hole (CPC)."

You should get a graph that looks something like the one below. Use your downloaded graph to answer the following questions (or you can use the graph shown below, if you don't have internet access right now).

1. On the basis of the graph, what was the area (in millions km^2) of the Southern Hemisphere ozone hole at the beginning of October 2015 (or the most recent year displayed on the graph that you downloaded)?

2. Is the most recent ozone hole larger or smaller than it was at the beginning of October in 2013? What about 2014? (Compare the area of the most recent ozone hole to the earlier ones shown on your downloaded graph, or on the graph below.)

3. How does the size of the most recent ozone hole in October compare to the mean (average) size of the ozone hole for the years 2005-2014 (or the range of years for which the mean size is shown on your downloaded graph)?

4. Which months constitute spring in the Southern Hemisphere (i.e., "austral spring")?

5. Try clicking on other parts of the website to see what else you can find. Under "Current" you will find summaries of recent data for the Arctic, Antarctica, and other locations, and under "Historical" you will find archives of data from various organizations involved in both stratospheric and tropospheric ozone monitoring.

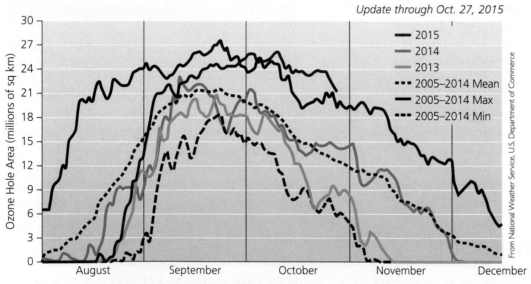

Update through Oct. 27, 2015

This graph shows the areal extent of the Southern Hemisphere ozone hole during the austral spring of 2015 (red line), compared to average, minimum, and maximum values for previous years.

MasteringEnvironmentalScience®

STUDENTS

Go to **MasteringEnvironmentalScience** for assignments, the eText, and the Study Area with practice tests, videos, current events, and activities.

INSTRUCTORS

Go to **MasteringEnvironmentalScience** for automatically graded activities, current events, videos, and reading questions that you can assign to your students, plus Instructor Resources.

14 Global Climate Change

This is the end, also called the *toe* or *terminus*, of the Athabasca Glacier of the Columbia Icefield in the Canadian Rockies.

Scottbeard/Dreamstime

Upon successfully completing this chapter, you will be able to

- Describe Earth's climate system and the many interacting factors that influence the atmosphere and climate, including human factors
- Summarize modern approaches to climate research

- Outline current and future trends and impacts of global climate change
- Suggest ways we can respond to climate change

The Athabasca Glacier has retreated about 1.5 km since 1844. Markers show the annual retreat of the glacier.

The glacier was here in
Le glacier était ici en
1844

CANADA
Greenland (DENMARK)
Ocean
Hudson Bay
Athabasca Glacier

CENTRAL CASE
THE RETREAT OF THE ATHABASCA

"Human influence on the climate system is clear. Warming of the climate is unequivocal."

—INTERGOVERNMENTAL PANEL ON CLIMATE CHANGE, *FIFTH ASSESSMENT REPORT* (2013)

"We're in a giant car heading towards a brick wall and everyone's arguing about where they're going to sit."

—DAVID SUZUKI

The Athabasca Glacier is one of six valley glaciers that flow like huge rivers of ice from the Columbia Icefield, which straddles the border between Jasper and Banff National Parks. The Columbia Icefield, which sits among some of the highest peaks in the Canadian Rockies, is a remnant of the massive ice sheet that covered much of Canada during the Pleistocene glaciation. The Athabasca Glacier itself is about 6 km long and between 90 and 300 m in thickness.

Glaciers, although they consist of solid ice, flow (very slowly) under the influence of gravity. There are two main mechanisms for glacial flow. The first involves *internal deformation* of the crystals of ice. This is something like what would happen if you were to press down and slightly sideways on a deck of cards; the cards would slide past one another. Similarly, the ice crystals in the glacier slide translationally along internal planes of weakness. The second main flow mechanism is called *basal sliding*. This happens when a layer of melt water lubricates the contact between the glacial ice and the underlying rock. Both mechanisms are probably at work in the Athabasca Glacier, which is moving downslope at a rate of a few centimetres per day.

Glaciers also grow and shrink in both volume and area, driven by changes in temperature and precipitation. When more snow falls than can melt in a year, the snow becomes compressed by additional snowfall and eventually recrystallizes to form glacial ice. This process, called

This archival photo shows the Athabasca Glacier in 1919. By this time, the toe of the glacier had already been retreating from its maximum historical extent (in 1844) for 75 years. The blue box shows the approximate location of the chapter-opening photo.

accumulation, is the mechanism by which glaciers grow in both volume and extent. However, if the annual snowfall melts entirely, the glacier will either remain constant or will slowly recede in volume and extent (through both melting and *sublimation*, in which the ice evaporates without actually melting); this loss of ice volume is called *ablation*.

The Athabasca Glacier has been retreating since 1844, when it was at its maximum historical extent (see photos). Since that time, the glacier has lost half of its volume and has retreated more than 1.5 km.[1] In 1844, the end (the toe, or terminus) of the glacier extended well past the area where the Icefields Parkway (Hwy. 93) is now located. The glacier is currently retreating at a rate of about 2 to 3 metres per year.

This area of the Rockies was of significant interest during the 1800s for scientific research (mainly geology) and for fur trading, carried out chiefly by the Hudson's Bay Company. Wilcox Pass was discovered in 1896 by an exploration party that included Walter Wilcox, for whom the pass and Wilcox Peak were eventually named; Robert Lemoyne Barrett, one of the founders of the American Association of Geographers; and William Peyto, for whom Peyto Glacier was named.[2] At that time, the toe of the Athabasca butted right up against the steep slopes of Wilcox Peak. Although they attempted to find easier routes through the valley (where Hwy. 93 now lies), explorers and fur traders were blocked by the glacier and were forced to follow steep mountain trails such as the Wilcox Pass and the Athabasca Pass.

Is the retreat of the Athabasca Glacier a recent side effect of global warming caused by human activity? Or is it just the natural continuation of the retreat of glacial ice that began 12 000 years ago with the end of the Pleistocene glaciation? Like most climate-related questions, this is difficult to answer. By definition, *climate* is a long-term phenomenon. Records that span a few decades are insufficient to reveal true climatic trends. However, evidence is very clear now that mountain glaciers throughout the world have been in retreat since the end of the Little Ice Age, a geologically brief period (300 years) of unusually cold weather that ended in the mid-1800s. This has led some scientists to predict that the Rockies will be ice-free by the end of this century.

Wilcox Peak

Athabasca Glacier

Icefields Parkway (Hwy. 93)

Columbia Icefield

Mount Athabasca

This satellite image shows part of the Columbia Icefield, the Athabasca Glacier, Icefields Parkway, and Wilcox Peak. The rest of the Columbia Icefield is off the left side of the photo.

2011 Google/Image Parks Canada/2012 Cnes/Spot Image

Our Dynamic Climate

Climate influences virtually everything around us, from the day's weather to major storms, from crop success to human health, from national security to the ecosystems that support our economies. If you are a student in your twenties, climate change may well be the major event of your lifetime and the phenomenon that most shapes your future.

By the time of the publication of the 2007 *Fourth Assessment Report* of the Intergovernmental Panel on Climate Change (IPCC),[3] there was a strong scientific consensus that climate is changing, that we are the main cause, and that climate change is already exerting impacts that will become increasingly severe if we do not take action. This consensus and the language used to communicate the information to the global community were strengthened even further in the IPCC's *Fifth Assessment Report*, released in 2013–2014.[4]

Climate change is the fastest-moving area of environmental science today. New scientific papers that refine our understanding of climate are published every week, and policymakers and businesspeople make headlines with decisions and announcements just as quickly. By the time you read this chapter, new issues will have developed, and new problems and solutions may have been found. We urge you to explore further, with your instructor and on your own, the most recent information on climate change and the impacts it will have on your future.

What is climate change?

Climate describes an area's long-term atmospheric conditions, including temperature, moisture content, wind, precipitation, barometric pressure, solar radiation, and other characteristics. As you have learned, climate differs from weather in that weather specifies conditions at localized sites over hours or days, whereas climate describes conditions across broader regions over seasons, years, or millennia. **Global climate change** describes trends and variations in Earth's climate, involving such aspects as temperature, precipitation, and storm frequency and intensity. People often use the term **global warming** synonymously with *climate change* in casual conversation, but global warming refers specifically to an increase in Earth's average surface temperature; it is thus only one aspect of global climate change.

Our planet's climate varies naturally through time, but the climatic changes taking place today are unfolding at an exceedingly rapid rate. Moreover, scientists agree that human activities, notably fossil fuel combustion and deforestation, are largely responsible. Understanding how and why climate is changing requires understanding how our planet's

climate functions. Thus, we first will survey the fundamentals of Earth's climate system—a complex and finely tuned system that has nurtured life for billions of years.

The Sun and atmosphere keep Earth warm

Three factors exert more influence on Earth's climate than all others combined. The first is the Sun; without it, Earth would be dark and frozen. The second is the atmosphere; without it, Earth would be as much as 33°C colder on average, and temperature differences between night and day would be far greater than they are. The third is the ocean, which shapes climate by storing and transporting heat and moisture.

The Sun supplies most of our planet's energy. Earth's atmosphere, clouds, land, ice, and water together absorb about 70% of incoming solar radiation and reflect the remaining 30% back into space (**FIGURE 14.1**); the reflectivity of a surface is called **albedo**. The 70% that is absorbed into the system powers a wide variety of Earth's processes, from photosynthesis to winds, waves, and evaporation.

Greenhouse gases warm the lower atmosphere

As Earth's surface absorbs the incoming short-wavelength solar radiation, surface materials increase in temperature and emit **infrared radiation**, radiation with longer wavelengths than visible light. In this longer-wavelength form, the radiation emitted by Earth's surface begins to make its way back to outer space.

However, some gases that are naturally present in the lower part of the atmosphere (the troposphere) absorb this infrared radiation very effectively. These include water vapour, ozone (O_3), carbon dioxide (CO_2), nitrous oxide (N_2O), and methane (CH_4), as well as halocarbons, a diverse group that includes chlorofluorocarbons (CFCs). Such gases are known as **greenhouse gases (GHGs)** or, technically, **radiatively active gases**. After absorbing radiation emitted from the surface, greenhouse gases subsequently re-emit infrared energy of slightly longer (lower-energy) wavelengths. Some of this re-emitted energy is lost to space, but some travels back downward, warming the troposphere and the planet's surface in a phenomenon known as the **greenhouse effect**.

The greenhouse effect is a natural phenomenon, and greenhouse gases have been present in our atmosphere for all of Earth's history. It's a good thing, too; without the natural greenhouse effect, our planet would have a much colder surface temperature—probably an average of around −18°C—and life on Earth would be impossible,

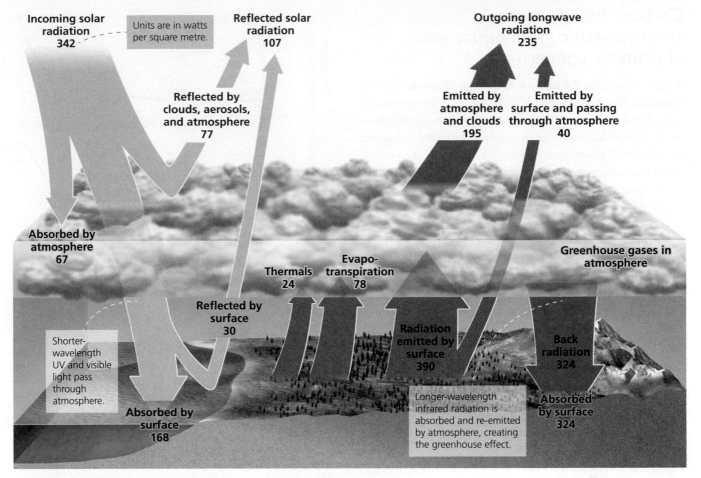

FIGURE 14.1 Our planet absorbs nearly 70% of the solar radiation it receives from the Sun and reflects the rest back into space (yellow arrows). About 30% is reflected off of the atmosphere, clouds, and surfaces of icecaps, ocean, and land as a function of the planet's albedo. Absorbed radiation is re-emitted (orange arrows) as heat. Greenhouse gases in the atmosphere absorb some of this radiation and re-emit it again, sending some downward to warm the atmosphere and the surface through the greenhouse effect. This illustration, showing major pathways of energy flow in watts per square metre, indicates that our planet naturally emits and reflects 342 watts/m², the same amount it receives from the Sun. Arrow thicknesses in the diagram are proportional to flows of energy in each pathway.
Source: Data from Kiehl, J.T., and K.E. Trenberth (1997). Earth's Annual Global Mean Energy Budget. *Bulletin of the American Meteorological Society, 78,* pp. 197–208.

or at least very, very different. Thus, it is not the natural greenhouse effect that is the cause of current concerns, but the **anthropogenic**, or human-generated, contributions to the greenhouse effect.

There are both natural and anthropogenic sources for almost all greenhouse gases—with the exception of chlorofluorocarbons (CFCs) and other halocarbons, such as HFC-23, which are wholly anthropogenic. However, human activities have increased the concentrations of many greenhouse gases in the past 250–300 years, thereby enhancing the greenhouse effect.

Greenhouse gases differ not only in their concentrations in the atmosphere but also in their ability to warm the troposphere and the surface. **Global warming potential** refers to the relative ability of one molecule of a given greenhouse gas to contribute to warming. **TABLE 14.1** shows the global warming potentials for several greenhouse gases. Values are expressed in relation to carbon

dioxide, which is assigned a global warming potential of 1. Thus, a molecule of methane is 28 times as potent as a molecule of carbon dioxide, and a molecule of nitrous oxide is 265 times as potent as a CO_2 molecule.

TABLE 14.1 Global Warming Potentials of Four Greenhouse Gases

Greenhouse gas	Relative heat-trapping ability (in CO_2 equivalents)
Carbon dioxide	1
Methane	28
Nitrous oxide	265
CFC-1	4660

Source: Data are for a 100-year time horizon, from IPCC (2013). Climate Change 2013: The Physical Science Basis. *Contribution of Working Group I to the Fifth Assessment Report of the Intergovernmental Panel on Climate Change,* p. 712.

Carbon dioxide is the anthropogenic greenhouse gas of primary concern

As you can see in **TABLE 14.1**, **carbon dioxide (CO$_2$)** is not the most potent greenhouse gas on a per-molecule basis, but it is far more abundant in the atmosphere than other GHGs. Emissions of greenhouse gases from human activity in Canada and other economically developed countries consist mostly of carbon dioxide. Even after accounting for the greater global warming potential of molecules of the other gases, carbon dioxide's abundance in our emissions makes it the major anthropogenic contributor to global warming. It is estimated that anthropogenic carbon dioxide is causing nearly six times more warming than methane, nitrous oxide, and wholly anthropogenic chemicals such as CFCs, combined.

The main natural source of carbon dioxide moving into the atmosphere is the decay of organic material; volcanoes also emit a significant amount of CO$_2$. Natural sources greatly outweigh the human contribution; however, human activities have boosted the atmospheric concentration of carbon dioxide from around 280 parts per million (ppm), as recently as the late 1700s, to just over 400 ppm in 2015[5] (**FIGURE 14.2**). The atmospheric CO$_2$ concentration is now higher than at any time in the entire ice-core record of data going back 800 000 years, and likely the highest in the last 20 million years. More importantly, researchers have established a very strong link between atmospheric carbon dioxide and temperature, as you will see. When atmospheric GHGs are high, so is temperature.

Human activity has released carbon from sequestration in long-term reservoirs

Why has atmospheric CO$_2$ increased by so much, if the natural sources still outweigh the human sources? The simple answer is that the natural fluxes (inputs and outputs) cancel each other out. Human sources, which do not cancel out, have thus shifted the overall balance of fluxes in the carbon cycle.

Recall our discussion of the carbon cycle and other biogeochemical cycles (Chapter 3, *Earth Systems and Ecosystems*), where we examined the idea that human activities often accelerate the movements or **fluxes** of material from one **reservoir** to another in biogeochemical cycles. Through the biogeochemical cycling of carbon, some carbon is stored or **sequestered** for long periods in the lithosphere. The deposition, partial decay, and compression of organic matter (mostly plants) that grew in

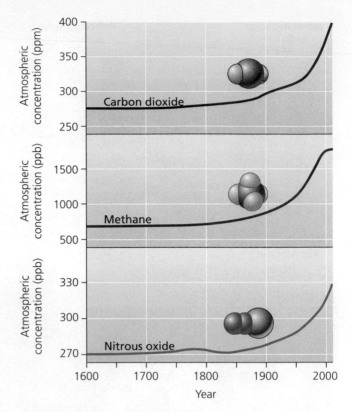

FIGURE 14.2

Since the start of the Industrial Revolution (late 1700s to early 1800s), global atmospheric concentrations of carbon dioxide, methane, and nitrous oxide have increased markedly.

Source: Based on data from Intergovernmental Panel on Climate Change (2007) *Fourth Assessment Report*, and (2013) *Fifth Assessment Report: The Physical Science Basis, Technical Summary.*

 By what percentage has atmospheric carbon dioxide increased since the year 1750?

wetland or marine areas led to the formation of coal, oil, and natural gas in sediments (as you will learn in greater detail in Chapter 15, *Fossil Fuels: Energy Use and Impacts*). It takes millions of years for the organic precursors of fossil fuels to be buried, chemically altered, and trapped deep underground in their rock hosts; the main chapter of fossil fuel formation occurred during the Carboniferous Period, 290–354 million years ago.

In the absence of human activity, these lithospheric carbon reservoirs would be practically permanent. However, over the past two centuries we have extracted fossil fuels and burned them in our homes, factories, and automobiles, transferring large amounts of carbon from one reservoir (the long-term underground deposits) to another (the atmosphere).

The human-modified flux of carbon from lithospheric reservoirs into the atmosphere is much faster than the natural flux (**FIGURE 14.3**). It is also faster than the combined total of all the fluxes of carbon out of the atmosphere and into carbon **sinks** (reservoirs that accept more of the material than they release), such as, notably,

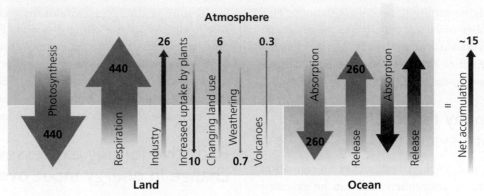

FIGURE 14.3 Human activities since the Industrial Revolution have sent more carbon dioxide from Earth to its atmosphere than is moving back from the atmosphere to Earth. Shown here are all the current fluxes of carbon dioxide, with arrows sized according to their mass of CO_2. Green arrows indicate natural fluxes, and red arrows indicate anthropogenic fluxes.
Source: Adapted from Intergovernmental Panel on Climate Change (2007). *Fourth Assessment Report.* Retrieved from https://www.ipcc.ch/publications_and_data/ar4/wg1/en/faq-2-1.html. Copyright © 2007 by IPCC Secretariat.

DATA Q For every metric ton of carbon dioxide we emit due to changing land use (e.g., deforestation), how much do we emit from industry?

the ocean. The release of carbon from long-term reservoirs and the acceleration of the carbon flux from the lithospheric reservoir to the atmospheric reservoir are the main reasons atmospheric carbon dioxide concentrations have increased so dramatically since the Industrial Revolution.

At the same time, people have cleared and burned forests to make room for crops, pastures, villages, and cities. Forests, soils, and crops also serve as storage reservoirs and sinks for carbon—in this case, for recently active carbon in the short-term carbon cycle. Their removal reduces the biosphere's ability to absorb carbon dioxide from the atmosphere. In this way, deforestation also has modified the flux of carbon from terrestrial reservoirs to the atmospheric reservoir and has contributed to rising atmospheric CO_2 concentrations.

Other greenhouse gases contribute to warming

Carbon dioxide is not the only greenhouse gas increasing in concentration in the atmosphere as a result of our activities. We release **methane (CH_4)** by tapping into fossil fuel deposits, by raising livestock that emit methane as a metabolic waste product, by disposing of organic matter in landfills, and by growing certain crops, such as rice. Since 1750, the atmospheric methane concentration has increased 2.5-fold, from 700 to 1800 ppm (see **FIGURE 14.2**), and today's concentration is the highest by far in more than 800 000 years. Buried organic matter and natural gas deposits are the main natural sources of methane.

Human activities have also augmented atmospheric concentrations of *nitrous oxides*. These greenhouse gases, by-products of feedlots, chemical manufacturing plants, auto emissions, and synthetic nitrogen fertilizers, have risen by more than 20% since 1750. The atmosphere is the largest nitrogen reservoir, but soils and soil-forming processes are the principal drivers of the nitrogen biogeochemical cycle. This is one reason that environmental scientists are taking a great interest in the impacts of climatic warming on soils.

Ozone—so important for life on Earth because of its function in the stratosphere as a UV filter—is also a radiatively active gas, contributing to warming both near the surface and up in the stratosphere. The concentration of ozone in the troposphere, which also contributes to photochemical smog, is extremely variable both spatially and temporally, but it has likely risen by about 36% since 1750. In a confusing twist of chemistry, CFCs and HFCs—the anthropogenic group of chemicals known as *halocarbons*, which have been implicated in stratospheric ozone depletion—are also radiatively active. The contribution of halocarbons to global warming has begun to slow as a result of the Montreal Protocol and subsequent controls on their production and release.

Water vapour is by far the most abundant naturally occurring greenhouse gas in our atmosphere, and contributes most to the natural greenhouse effect. Its concentrations vary both spatially and temporally, but its global concentration has not changed much over recent centuries, so it is not viewed as having driven industrial-age climate change. However, the concentration of water in the atmosphere is connected to temperature; as tropospheric temperatures continue to increase, Earth's water bodies should transfer more water vapour into the atmosphere, contributing to higher atmospheric concentrations. This could have a number of effects on climate, weather, and ecosystems. It also means that the effects of other greenhouse gases are amplified by the influence of added water vapour, contributing to further greenhouse warming.

There are many feedback cycles in the climate system

If global warming leads to an increase in the concentration of water vapour in the atmosphere, this can be expected to cause further warming because water is a radiatively active gas. This additional warming, in turn, could cause still more evaporation, leading to further increases in water vapour in the atmosphere, and so on. This is called a **positive feedback loop** or positive feedback cycle (discussed in Chapter 3, *Earth Systems and Ecosystems*), and the climate system is loaded with them.

The climate system also has lots of **negative feedback loops**, which are self-regulating or self-limiting feedbacks. For example, if global warming leads to increased evaporation and more water vapour in the atmosphere, it could give rise to increased cloudiness. Increased cloudiness could slow global warming by reflecting more solar radiation back into space. However, the complexity of cloud cover doesn't end there; depending on whether low- or high-elevation clouds resulted, they might shade and cool Earth's surface (a negative feedback), or contribute to warming, thus accelerating evaporation and further cloud formation (a positive feedback).

Water vapour is not the only natural constituent involved in feedbacks in the climate system. For example, warming of soils could cause an accelerated flux of soil gases to the atmosphere; some of these are GHGs, and their release could lead to further warming (a positive feedback). On the other hand, soil formation is accelerated by warmer, wetter weather, and soils function as a major sink for organic matter, removing carbon from the atmospheric reservoir (a negative feedback).

Aerosols—microscopic droplets and particles suspended in the air—also can have either a warming or cooling effect. Soot, or black carbon aerosols, can cause warming by absorbing solar energy. Most tropospheric aerosols are whiter, and cool the atmosphere by reflecting the Sun's rays. When sulphur dioxide enters the atmosphere, it undergoes various reactions, some of which can contribute to the formation of a sulphur-rich aerosol haze in the upper atmosphere. Sulphate aerosols reduce the amount of sunlight that reaches Earth's surface. Sulphur-rich aerosols released by major volcanic eruptions, such as the eruption of Mt. Pinatubo in 1991, also can exert short-term cooling effects on Earth's climate over periods of up to several years.

Because of the complexity of these competing effects and feedbacks in the climate system, and the interactions among the various processes, minor modifications of components of the atmosphere can potentially lead to major effects on climate. Sorting out feedbacks and their relative importance and influence on one another is one of the greatest challenges in modern climate science.

Radiative forcing expresses change in energy input over time

Scientists have made quantitative estimates of the degree of influence that aerosols, greenhouse gases, and other factors exert over Earth's energy balance (**FIGURE 14.4**). The amount of change in energy that a given factor causes is called its **radiative forcing**. Positive forcing warms the surface, whereas negative forcing cools it. Scientists'

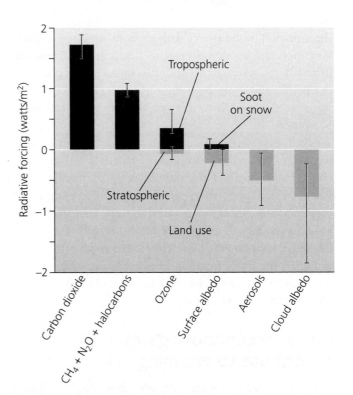

FIGURE 14.4

For each emitted gas or other human impact on the atmosphere since the Industrial Revolution, we can estimate the warming or cooling effect on Earth's climate. We express this as radiative forcing, which in this graph is shown as the amount of influence on climate today relative to 1750, in watts per square metre. Red bars indicate positive forcing (warming), and blue bars indicate negative forcing (cooling). A number of influences are not shown. IPCC scientists estimate (with a high degree of certainty) that human impacts on the atmosphere currently exert a net positive radiative forcing of 2.29 watts/m² compared to 1750.

Source: Based on data from Intergovernmental Panel on Climate Change (2007) *Fourth Assessment Report,* and (2013) *Fifth Assessment Report: The Physical Science Basis.*

best estimate is that Earth today compared with the pre-industrial Earth of the year 1750 is experiencing overall net positive (that is, warming) radiative forcing of about 2.29 watts/m². For context, look back at **FIGURE 14.1** and note that Earth is estimated to receive and give off 342 watts/m² of energy. Although 2.29 may seem like a small proportion of 342, over time it is enough to alter climate significantly.

The atmosphere is not the only factor that influences climate

Our climate is influenced by factors other than atmospheric composition. Among these are cyclic changes in Earth's rotation and orbit, variation in energy released by the Sun, absorption of carbon dioxide by the oceans, and oceanic circulation patterns. We will look at each of these factors in turn.

Orbital factors During the 1920s, Serbian mathematician Milutin Milankovitch described the influence of periodic changes in Earth's rotation and orbit around the Sun on **insolation**, the amount of solar energy that reaches Earth's surface per unit area in a given period. These variations, which include wobbling (or *precession*) of Earth's rotational axis, tilt of the axis (or *obliquity*), and change in the shape of Earth's orbit around the Sun (or *eccentricity*), alter the way solar radiation is distributed over Earth's surface (**FIGURE 14.5**).

The collective impact of these orbital variations causes cyclical changes in insolation and, therefore, in atmospheric heating. These so-called **Milankovitch cycles** (**FIGURE 14.6**) lead to variations that are sufficient to trigger climatic changes such as periodic episodes of **glaciation**, during which global surface temperatures drop and ice sheets advance from the poles toward the mid-latitudes, and the intervening warm **interglaciations**.

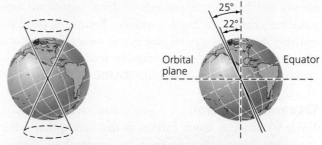

(a) Axial wobble (or *precession*)   **(b) Variation of tilt (or *obliquity*)**

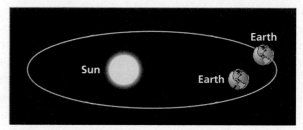

(c) Variation of orbit (or *eccentricity*)

FIGURE 14.5
There are three orbital factors whose variations have a significant influence on climate. The first is an axial wobble (or *precession*) **(a)**, which occurs on a 19 000- to 23 000-year cycle. The second is a 3° shift in the tilt of Earth's axis, called *obliquity* **(b)**, which occurs on a 41 000-year cycle. The third is the departure from circularity of Earth's orbit, called *eccentricity*, which varies from almost circular to slightly elliptical **(c)** and repeats every 100 000 years. These variations individually and collectively affect the intensity of solar radiation that reaches portions of Earth at different times, contributing to long-term changes in global climate.

Solar output The Sun varies in the amount of radiation it emits (its luminosity), over both short and long timescales. For example, at each peak of its 11-year sunspot cycle, the Sun may emit solar flares, bursts of energy strong enough to disrupt satellite communications. However, scientists are concluding that the variation in solar energy reaching our planet in recent centuries has

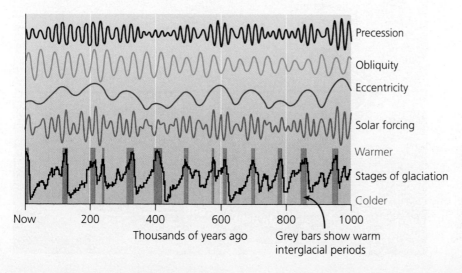

FIGURE 14.6
The collective influence of variations in axial tilt, wobble, and orbital shape causes cyclical variations in solar insolation and heating of Earth's atmosphere and surface, which force cooling and warming trends in global climate on a variety of timescales and initiate major climatic shifts, such as glaciations.

simply not been great enough to drive significant temperature change on Earth's surface. Estimates place the radiative forcing of natural changes in solar output at only about 0.12 watts/m^2—considerably less than any of the anthropogenic causes shown in **FIGURE 14.4**.

Ocean absorption

The ocean, acting as a sink, holds 50 times as much carbon as the atmosphere holds. It absorbs carbon dioxide from the atmosphere through direct solubility of gas in water, uptake by marine phytoplankton for photosynthesis, and incorporation into calcareous shells and skeletons by various marine organisms. As the concentration of carbon dioxide in the atmosphere increases, ocean water absorbs more carbon dioxide, but the rate of absorption increases more slowly than does the concentration in the atmosphere. Thus, carbon absorption by the ocean is slowing global warming but not preventing it. Furthermore, as ocean water warms, it absorbs less CO_2 because gases are less soluble in warmer water—a positive feedback effect that accelerates warming. The absorption of excess CO_2 by the ocean is also problematic because it causes the acidification of ocean water, which, in turn, causes calcareous shells and skeletons to dissolve, releasing more CO_2—another positive feedback.

Ocean circulation

Ocean water exchanges tremendous amounts of heat with the atmosphere, and ocean currents move energy from place to place. In equatorial regions, the oceans receive more heat from the Sun and atmosphere than they emit. Near the poles, ocean water emits more than it receives. Because cooler water is denser than warmer water, the cool water at the poles tends to sink, and the warmer surface water from the equator moves poleward to take its place. This is one of the principles underlying global ocean circulation patterns.

One ocean–atmosphere interaction that influences climate is the El Niño–Southern Oscillation (ENSO), discussed in Chapter 12, *Marine and Coastal Systems and Fisheries*. This system arises from the interactions of atmospheric pressure, sea surface temperature, and ocean circulation in the tropical Pacific Ocean. El Niño conditions are triggered when air pressure increases in the western Pacific and decreases in the eastern Pacific, causing the equatorial winds to weaken. La Niña events are the opposite; under these conditions, cold surface waters extend far westward in the equatorial Pacific. Both El Niño and La Niña have dramatic influences on global weather patterns. ENSO cycles are periodic but irregular, occurring every two to eight years. Scientists are getting better at deciphering the triggers for these events and predicting their impacts on weather. They are also investigating whether globally warming air and sea temperatures may be increasing the frequency and strength of these cycles.

Ocean currents and climate also interact through the **thermohaline circulation**, also discussed in Chapter 12. The thermohaline circulation is a worldwide current system in which warmer, fresher water moves along the surface and colder, saltier water (which is more dense) moves deep beneath the surface (**FIGURE 14.7**). In the Atlantic Ocean, warm surface water flows northward from the equator in the Gulf Stream, carrying heat to high latitudes and keeping Europe warmer than it would otherwise be. As the surface water of this conveyor belt system releases heat energy and cools, it becomes denser and sinks, creating *North Atlantic Deep Water* (*NADW*).

Scientists hypothesize that interruptions in the thermohaline circulation could trigger rapid climate change. If global warming causes much of Greenland's ice sheet to melt, freshwater runoff into the North Atlantic would dilute surface waters, making them less dense (because

FIGURE 14.7

Warm surface currents carry heat from equatorial waters of the ocean north toward Europe and Greenland, where they release heat into the atmosphere and then cool and sink, forming the North Atlantic Deep Water. Scientists debate whether rapid melting of Greenland's ice sheet could interrupt the flow of warm water from equatorial regions and cause Europe to cool dramatically.

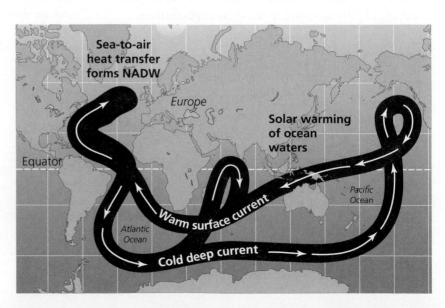

fresh water is less dense than salt water). This could potentially stop NADW formation and shut down the thermohaline circulation and the associated northward flow of warm equatorial water (that is, the Gulf Stream), causing Europe to cool rapidly. Such a change may have had a role in the "Little Ice Age," a period of unusually cold climate that began in the fifteenth century. This scenario inspired the 2004 film *The Day After Tomorrow*, although the filmmakers chose entertainment value over science and grossly exaggerated the potential impacts and the speed with which the effect might occur.

Some data suggest that the thermohaline circulation and formation of NADW in the North Atlantic is already slowing; other researchers argue that Greenland's runoff will not be enough to cause a shutdown this century, and that the thermohaline circulation system in the North Atlantic is quite robust. However, there is some evidence that interruptions of the thermohaline circulation have historically been responsible for extremely rapid shifts in climate in the areas bordering the North Atlantic. These climatic "flip-flops" are now thought to have occurred in the geological past in periods as short as a decade or so. (See "The Science behind the Story: Tip Jets and NADW off the Coast of Greenland," in Chapter 12.)

The Science of Climate Change

To comprehend any phenomenon that is changing, we must study its past and present, and try to project its future. Climate scientists monitor present-day climatic conditions, but they also have devised clever means of inferring past change and have developed sophisticated methods to predict future change.

Proxy indicators tell us about the past

To understand how climate is changing today, and to predict future change, scientists must learn what climatic conditions were like thousands or millions of years ago. Evidence about climate in the geological past—**paleoclimate**—is extremely important, because it gives us a baseline against which to measure the changes that we see happening in the climate system on a shorter timescale. The record of actual measurements of temperature, precipitation, and other indicators of climate is not very long—only a few hundred years' worth of data. Environmental scientists have developed a number of methods to decipher clues from the past. **Proxy indicators** are types of indirect evidence that serve as proxies, or substitutes, for direct measurement and that shed light on past climate.

Some human records of historical events can contribute to a longer database of weather information. For example, fishers have recorded the timing of sea ice formation for hundreds of years, and wine makers have kept meticulous records of precipitation and the length of the growing season.

To go back farther in time, we must begin to rely on the record-keeping ability of the natural world. For instance, growth rings in trees can give information about conditions of temperature and precipitation—they act as a proxy, or a stand-in, for actual measurements. The width of each ring of a tree trunk cut in cross-section reveals how much the tree grew in a particular growing season; a wide ring means more growth, generally indicating a wetter year. Long-lived trees, such as bristlecone pines, can provide records of precipitation and drought going back hundreds or thousands of years. Tree rings are also used to study fire history, since a charred ring indicates that a fire took place in that year.

The significance of annual growth rings is simple: Without growth rings, or annual or seasonal layering of some type, it would be difficult for scientists to know when a particularly cold or warm or dry or wet season occurred. If growth rings or layers are present, scientists can count back to determine the age of the tree (or any other organism that accumulates material in annual growth cycles). Other dating techniques such as carbon-14 dating can sometimes be used, but not all materials are amenable to those approaches.

Scientists continue to search for ways to extend the climate record back even farther. Researchers can gather data on past ocean conditions from fossils of marine microorganisms in layered sediments, and from corals, which also accumulate material in annual growth rings. As they build their reefs, living corals take in trace elements from ocean water, incorporating these chemical clues into their rings; the process is sensitive to changes in ocean temperature and composition. In arid regions, packrat middens are a valuable source of climate data. Packrats are rodents that carry seeds and plant parts back to their middens, or dens, in caves and rock crevices sheltered from rain. In an arid location, plant parts may last for centuries, allowing researchers to study the past flora of the region.

Researchers also remove core samples from sediments that lie beneath bodies of water. Sediments often preserve pollen grains and other remnants from plants that grew in the past, and analyzing these materials can illuminate the history of past vegetation. Because climate influences the types of plants that grow in an area, knowing what plants lived in a location at a given time can tell us much about the climate at that place and time.

Earth's icecaps, ice sheets, glaciers, and sediments hold clues to the much longer-term climate history. Over the ages, these huge expanses of snow and ice

THE SCIENCE BEHIND THE STORY

An EPICA researcher prepares a Dome C ice core sample for analysis.
Pasquale Sorrentino/Photo Researchers, Inc.

Reading History in the World's Longest Ice Core

Snow falling year after year compresses into ice and stacks up into immense sheets that scientists can mine for clues to Earth's climate history. The ice sheets of Antarctica and Greenland trap tiny air bubbles, dust particles, and other proxy indicators of past conditions. By drilling boreholes and extracting ice cores, researchers can tap into these valuable archives.

Recently, researchers drilled and analyzed the deepest core ever. At a remote, pristine site in Antarctica named Dome C, they drilled down 3270 m to bedrock and pulled out more than 800 000 years' worth of ice. The longest previous ice core (from Antarctica's Vostok station) had gone back 420 000 years. Ice near the top of these cores was laid down recently; ice at the bottom is oldest. By analyzing ice at intervals along the core's length, researchers can generate a timeline of environmental change.

The Dome C core was drilled by the European Project for Ice Coring in Antarctica (EPICA), a consortium of researchers from 10 European nations. Antarctic operations are expensive and logistically complicated: The environment

is harsh, ice drilling requires powerful technology, and the analysis requires a diverse assemblage of experts. When the team published its results in the journal *Nature* in 2004, the landmark paper had 56 authors.

The researchers obtained data on surface air temperature going back hundreds of thousands of years, by measuring the ratio of deuterium isotopes to normal hydrogen in the ice. This ratio is temperature dependent and it is preserved in the chemistry of the ice (bottom panel of graph in **FIGURE 1**).

By examining dust particles, they could tell when arid and/or windy climates sent more dust aloft. By analyzing air bubbles trapped in the ice, the researchers were later able to quantify atmospheric concentrations of carbon dioxide, methane, and nitrous oxide across 800 000 years.

One finding was expected—yet crucial. The researchers documented that temperature swings in the past were tightly correlated with concentrations of carbon dioxide (top panel of graph in **FIGURE 1**), as well as methane (middle panel of graph in **FIGURE 1**) and nitrous oxide. Also clear and expected from the data was that temperature varied with swings in solar radiation caused by Milankovitch cycles; the Dome C core spanned eight glacial cycles.

A tight correlation between greenhouse gas concentrations and temperature does not prove causality: It doesn't prove that high atmospheric GHG concentrations caused warming, nor does it prove that warming caused GHG concentrations to increase. What the correlation does demonstrate is that there is a close relationship between atmospheric GHGs and climatic warming: When atmospheric GHG concentrations are high, so is temperature, and vice versa. This is crucially important.

The data also demonstrate that by increasing greenhouse gas concentrations since the Industrial Revolution, we have brought them well above the highest levels they reached naturally in 800 000 years. Today's carbon dioxide concentration (400 ppm in 2015) is far above previous maximum values. Present-day concentrations of methane and nitrous oxide are likewise the highest in 800 000 years. These data show that we as a society have brought ourselves deep into uncharted territory.

Other findings from the ice core are not as easily explained. Intriguingly, earlier glacial cycles differ from recent cycles. For recent cycles, the Dome C core showed that glacial periods were long and interglacial periods were brief, with rapid rise and fall of temperature. Interglacials thus appear on a graph of temperature through time as tall, thin spikes. However, in older cycles the glacial and interglacial periods were of more equal duration, and the warm extremes of interglacials were not as great. This change in the nature of glacial cycles through time had been noted by researchers working with oxygen isotope data from the fossils of marine organisms. But why glacial cycles should be different before and after the 450 000-year mark, no one knows.

Today scientists are searching for a site that might provide an ice core stretching back more than 1 million years. Data from marine isotopes tell us that glacial cycles at that time switched from a periodicity of roughly 41 000 years (conforming to the influence of planetary tilt) to about 100 000 years (more similar to orbital changes). An ice core that captures cycles on both sides of the 1 million-year divide might help clarify the influence of Milankovitch cycles.

---

have accumulated to great depths, preserving within their layers tiny bubbles of the ancient atmosphere (**FIGURE 14.8**). Scientists can examine these trapped air bubbles by drilling into the ice and extracting long columns, or cores. The layered ice, accumulating season after season over thousands of years, provides a

timescale, something like the growth rings of the tree. From these ice cores, scientists can determine atmospheric composition, greenhouse gas concentrations, temperature trends, snowfall, solar activity, and even (from trapped soot particles) frequency of forest fires and volcanic eruptions.

The Dome C ice core research shows that we still have plenty to learn about our complex climate history. The close correlation between greenhouse gases and temperature evident in the EPICA data provides a strong indication that we would do well to bring greenhouse emissions under control before it is too late.

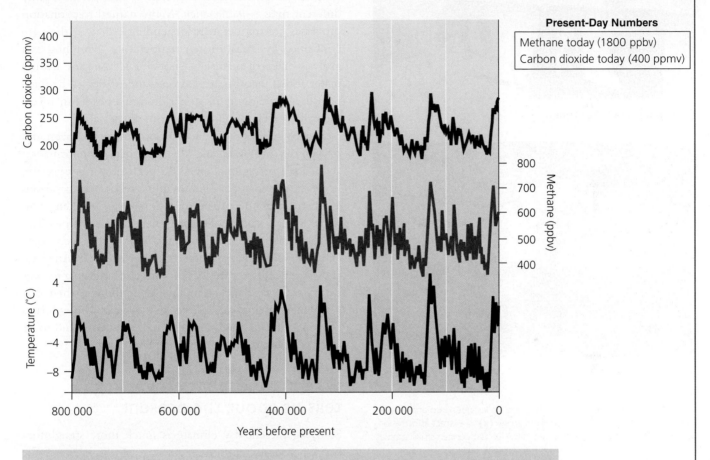

FIGURE 1

Data from the EPICA ice core reveal changes in surface temperature (black line), atmospheric methane concentration (green line), and atmospheric carbon dioxide concentration (red line). High peaks in temperature indicate warm interglacial periods, and low troughs indicate cold glacial periods. Atmospheric concentrations of carbon dioxide and methane rise and fall in tight correlation with temperature. Today's current values are included at the far top right of the graph, for comparison.

Source: Republished with permission of Nature Publishing Group, from Paleoclimate: Windows on the Greenhouse, Ed Brook, *Nature, 453,* pp. 291–291, 2008; permission conveyed through Copyright Clearance Center, Inc.

By extracting ice cores from Greenland and Antarctica, scientists have now been able to go back in time 800 000 years, reading Earth's global climatic history across eight glacial cycles (see "The Science behind the Story: Reading History in the World's Longest Ice Core"). During 2007–2009, such research was funded and promoted as part of the International Polar Year, a large international scientific program coordinating research in the Arctic and Antarctic. The main approach to these studies makes use of the fact that the chemistry of the H_2O in the ice itself holds clues about climatic conditions, specifically the temperature, at the time when it first formed as precipitation.

(a) Scientists with ice core

(b) Micrograph of ice core

FIGURE 14.8
In Greenland and Antarctica, scientists have drilled deep into ancient ice sheets and removed cores of ice like this one **(a)** to extract information about past climates. Bubbles trapped in the ice **(b)** contain small samples of the ancient atmosphere.

Stable isotope geochemistry is a powerful tool for studying paleoclimate

Clues about paleoclimates can be partly uncovered through the study of the **stable isotope geochemistry** of the ice. (You learned about isotopes in Chapter 2, *Matter, Energy, and the Physical Environment.*) Stable isotopes are naturally occurring variations of elements, which are not radioactive, and which vary just slightly from one another in mass but not in other chemical characteristics. This means, for example, that deuterium—a naturally occurring stable isotope of hydrogen that is heavier than simple hydrogen—behaves chemically like hydrogen. However, because of the difference in its mass, deuterium is separated or segregated from hydrogen and becomes differently concentrated in Earth materials (like ice) when acted upon by any process that is influenced by mass.

Consider, for example, the process of precipitation. Water that falls as precipitation is naturally enriched in the heavier isotopes of its components, hydrogen and oxygen. Water that remains behind in cloud form, on the other hand, will tend to be enriched in the lighter isotopes of its chemical constituents. The scientific term for the separation and differential concentration of isotopes of slightly different mass is *fractionation*. Many natural fractionation processes are temperature dependent—that is, they are controlled by variations in temperature. Sampling and analysis of Earth materials—not just ice but any natural material that incorporates isotopes and is affected by temperature-dependent fractionation processes—can reveal the past temperature history of those materials.

Proxy indicators like stable isotopes of hydrogen and oxygen can give useful and often very detailed information about local or regional areas. To get a global perspective, however, scientists combine multiple records from various areas. In fact, one of the most interesting challenges in stable isotope studies of paleoclimate has been to explain the slight mismatch in timing (*asynchronicity*) between major climatic events recorded in ice core records from Antarctic and Greenland ice cores. The number of available records decreases the farther back in time we go (because there isn't very much million-year-old ice left on Earth), so estimates of global climate conditions for the recent past tend to be more reliable than those for the distant past.

Direct atmospheric sampling tells us about the present

Studying present-day climate is much more straightforward, because scientists can measure atmospheric conditions directly. As mentioned above, our records of direct measurements of temperature and precipitation date back several hundred years, although the more recent records are obviously the most reliable. Atmospheric carbon dioxide concentrations have been measured continuously at the Mauna Loa Observatory in Hawaii, starting in 1958 (**FIGURE 14.9**). These data show that atmospheric CO_2 concentrations have increased from 315 ppm in 1958 to 392 ppm in 2011. Today scientists at Mauna Loa continue to make these measurements, building upon the best long-term data set we have of direct atmospheric sampling of any greenhouse gas.

Models help us understand climate

To understand how climate systems function and to predict future climate change, scientists simulate climate processes with sophisticated computer programs. *Climate models* are programs that combine what is known about atmospheric circulation, ocean circulation, atmosphere–ocean–land interactions, and feedback mechanisms to

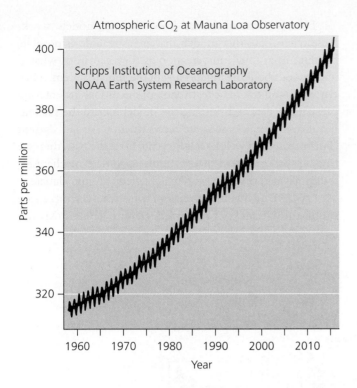

Atmospheric CO₂ at Mauna Loa Observatory

Scripps Institution of Oceanography
NOAA Earth System Research Laboratory

FIGURE 14.9
Atmospheric concentrations of carbon dioxide have risen steeply since these measurements began to be taken at the Mauna Loa Observatory in 1958. The jaggedness is caused by seasonal variations in photosynthetic uptake. The Northern Hemisphere has more land area and thus more vegetation than the Southern Hemisphere. Thus, more carbon dioxide is absorbed during the northern summer, when Northern Hemisphere plants are more photosynthetically active. *Source:* Data from National Oceanic and Atmospheric Administration, Earth System Research Laboratory, Global Monitoring Division (2015). Mauna Loa: Full Record. U.S. Department of Commerce.

with both horizontal and vertical variations—it was simply not possible until the advent of modern computers.

Canadian scientists are internationally recognized as leaders in the development of such models, also called *general circulation models* or **global climate models (GCMs)**. Research utilizing climate models developed at the Canadian Centre for Climate Modelling and Analysis (CCCma), part of Environment Canada, has contributed to coordinated experiments run by the World Climate Research Programme, and to reports by the IPCC, including the recent *Fifth Assessment Report*. Scenarios based on Canadian models have also been used by the government of the United States in its *National Assessment of the Potential Consequences of Climate Variability and Change*, providing a basis for strategic planning for managing the consequences of climate change.

Researchers develop different climate models for different purposes. You can think of these models as being

simulate climate processes (**FIGURE 14.10**). They couple, or combine, the climate influences of the atmosphere and oceans into a single simulation. This requires manipulating vast amounts of data and complex mathematical equations repeated in each element of a detailed geographical grid,

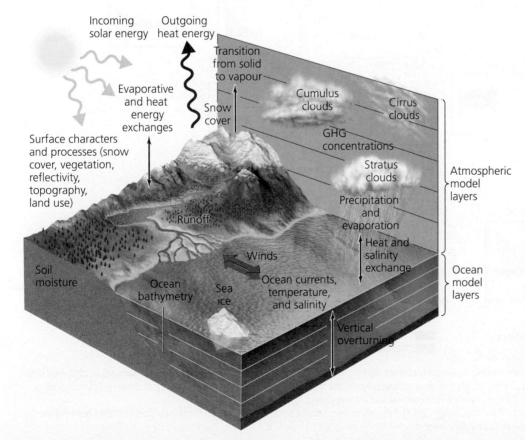

FIGURE 14.10
Modern climate models incorporate many factors, including processes involving the atmosphere, land, ocean, ice, and biosphere. Such factors are shown graphically here, but actual models deal with them as mathematical equations in computer simulations, carried in each individual element of a detailed geographical grid.

sort of like very sophisticated versions of computer simulation games, like *SimEarth*, *SimCity*, *SimAnt*, and so on. If you have played any of these games (especially the earlier versions, which required more hands-on intervention by the gamer), then you know something about how they work. You provide some starting information to the game (or computer model), setting up the preconditions for the simulation, and then you let it run. You can tweak it as the game goes on—constructing more buildings, taking away food sources, or adding more predators to the ant colony—but eventually the model will run to its natural completion.

That is something like how climate models work. Researchers construct models that are as realistic as possible, building in as much information as they can from what is understood about the functioning of the climate system. They can verify and test the effectiveness of the models by entering past climate data and running the model toward the present. If a model produces accurate reconstructions of our current climate, based on well-established data from the past, then we have reason to believe that it simulates climate mechanisms realistically and that it may accurately predict future climate.

FIGURE 14.11 shows temperature results from three such simulations, as reported by the IPCC. Model

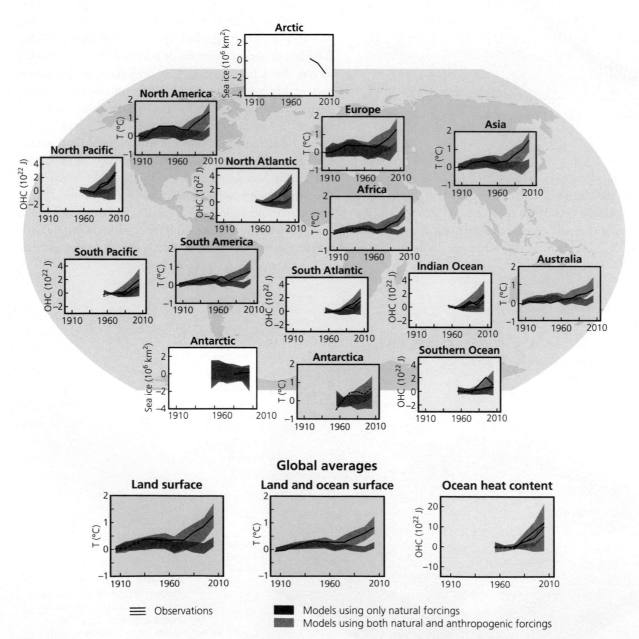

FIGURE 14.11 Scientists test climate models by entering climate data from past years and comparing model predictions with actual observed data (black lines in each diagram). The results from models that incorporate only natural factors (shaded blue) do not predict real, observed climate trends (the black lines) as accurately as models that incorporate both natural and anthropogenic factors (shaded pink).
Source: From the Intergovernmental Panel on Climate Change (2013). *Fifth Assessment Report. Climate Change 2013: The Physical Science Basis*, Figure SPM.6 Copyright © 2013 by IPCC Secretariat. Reprinted by permission.

results shaded in blue are based only on natural climate-changing factors alone (such as volcanic activity and variation in solar energy). Model results shaded in pink are based on both natural and anthropogenic factors (such as human emissions of greenhouse gases and sulphate aerosols). Models that incorporate both natural and anthropogenic factors produce the closest match between model predictions and actual climate observations, as shown by the black lines on the graphs.

Such results as those in **FIGURE 14.11** clearly support the finding that both natural and human factors contribute to climate dynamics, and they indicate that global climate models can produce reliable predictions. As computing power increases and we glean more and better data from proxies, these models become increasingly reliable. They also are improving in resolution and are beginning to predict climate change region by region for various areas of the world.

Current and Future Trends and Impacts

Evidence that climate conditions have changed worldwide since industrialization is now overwhelming and indisputable. Climate change in recent years has already had numerous effects on the physical properties of our planet, on organisms and ecosystems, and on human well-being. If we continue to emit greenhouse gases into the atmosphere, the impacts of climate change will only grow more severe.

Future impacts of climate change are subject to regional variation, so the way each of us experiences these impacts over the coming decades will vary tremendously, depending on where we live. The impacts on Canada could be particularly severe. According to the CCCma,

Over the next century, the Canadian climate model indicates that the most northerly regions of the Earth will experience the greatest warming, with potentially serious impacts on Arctic communities and ecosystems. Major changes in climate are also expected in the Prairies, the east and west coasts and the Great Lakes basin. Many communities and climate-sensitive industries will be profoundly affected, including forestry, agriculture, marine transportation, fishing and oil and gas development ... The magnitude and rate of change projected for the twenty-first century will put enormous pressure on our environment, our infrastructure and our social fabric.[6]

The IPCC summarizes evidence of climate change and predicts future impacts

In recent years, it seems that virtually everyone is becoming more aware of climatic changes around us. A fisher in the Indian Ocean island of the Maldives notes the seas encroaching on his home island. A rancher in Alberta suffers a multiyear drought. A homeowner in Florida finds it impossible to obtain insurance against the hurricanes and storm surges that increasingly threaten. Residents of Montreal worry about freakish winter weather events. But can we conclude that all of these impressions are part of a real pattern, or are they just ephemeral fluctuations in the data?

A true trend is a pattern that persists within a data set, even after short-term fluctuations, background noise, and local anomalies have been removed or accounted for. Is there solid scientific evidence to confirm that we are seeing a trend, and that the global climate is indeed already changing? A wide variety of data sets do appear to show significant trends in climate conditions over the past century, and particularly in recent years. The most thoroughly reviewed and widely accepted synthesis of scientific information concerning climate change is a series of reports issued by the **Intergovernmental Panel on Climate Change (IPCC)**. This international panel of scientists and government officials was established in 1988 by the United Nations Environment Programme (UNEP) and the World Meteorological Organization.

In 2013–2014 the IPCC released its *Fifth Assessment Report*, which represents the consensus of scientific climate research from around the world.[7] The first part of the report covers just the physical science basis of climate change; Part 1 alone is over 1500 pages long, had 209 lead authors and 50 review editors, thousands of expert reviewers, and cites over 9200 scientific publications. Parts 2 and 3, equally comprehensive and authoritative, address impacts, adaptation, and vulnerability; and mitigation of climate change, respectively.

This multipart report documents the observed trends in surface temperature, precipitation patterns, snow and ice cover, sea levels, storm intensity, and other factors, all based on thousands of scientific studies. The report also predicts future changes in these phenomena after considering a range of potential scenarios for future greenhouse gas emissions. The report addresses impacts of current and future climate change on wildlife, ecosystems, and human societies. Finally, it discusses possible strategies we might pursue in response to climate change. **FIGURE 14.12** summarizes a selection of the IPCC report's major observed impacts of climate change.

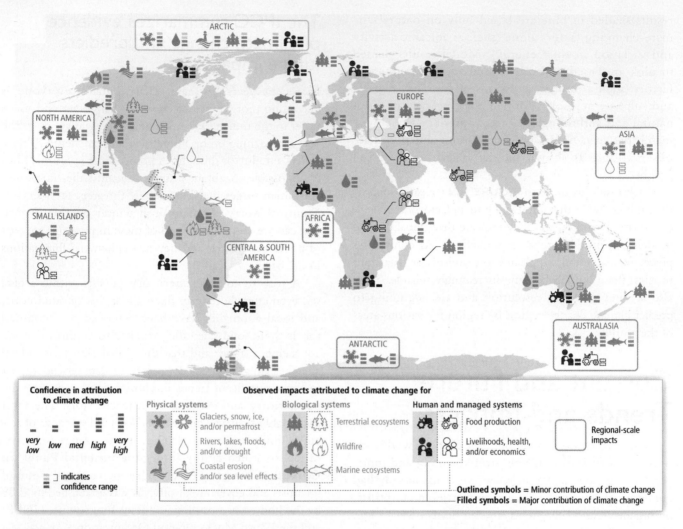

FIGURE 14.12 Climate change has had consequences already and is predicted to have many more. This figure shows some impacts around the world that have been attributed to climate change, based on studies published since the IPCC's *Fourth Assessment Report* in 2007. Symbols indicate the nature of the observed impacts (including impacts on physical, biological, and human or managed systems); the relative contribution of climate change to the observed impacts (major or minor, shown by the outlined and filled symbols); and the degree of confidence of scientists in attributing the impact to climate change (shown by the horizontal bar symbols).
Source: From Impacts, Adaptation, and Vulnerability. Summary for Policymakers. Part A: Global and Sectoral Aspects. Contribution of Working Group II to the Fifth Assessment Report of the Intergovernmental Panel on Climate Change, Figure SPM.2. Copyright © IPCC Secretariat. Reprinted by permission.

The IPCC reports are comprehensive and authoritative, but—like all science—they deal in uncertainties. Its authors have therefore taken great care to assign statistical probabilities to all conclusions and predictions. In the report (which is freely available online from **www.ipcc.ch/index.htm**), you will see words like "very likely" or "with highest confidence." Such phrases are not used lightly; they all have specific statistical probabilities and levels of confidence associated with them. For example, "very likely" carries the specific connotation of a 90–100% probability. In addition, the IPCC's estimates regarding the potential impacts of change on human societies are conservative, because its scientific conclusions had to be approved by representatives of the world's national governments, some of which are reluctant to move away from a fossil-fuel-based economy.

Surface temperature increases will continue

The IPCC report concludes that average surface temperatures on Earth increased by an estimated 0.85°C in the period from 1880 to 2012, based on multiple datasets. Most of this increase occurred in the last few decades, as seen in **FIGURE 14.13A**. The period from 1983 to 2012 was likely the warmest 30-year period of the last 1400 years in the Northern Hemisphere, with each of the past three decades successively warmer than the last. As **FIGURE 14.13B** shows, the changes have been greater over the land than over the ocean, which is consistent with what we know about the climate-moderating capacity of the ocean.

In the future, we can expect average surface temperatures on Earth to rise by roughly 0.3°C to 0.7°C more over

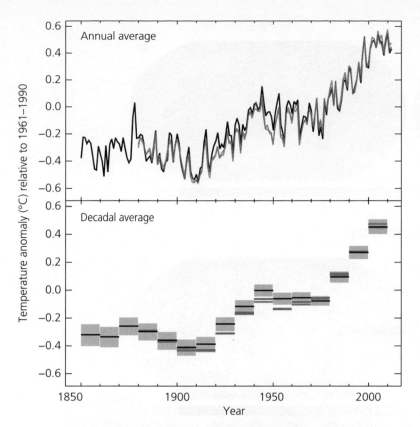

(a) Observed global land and ocean surface temperature anomalies since 1850

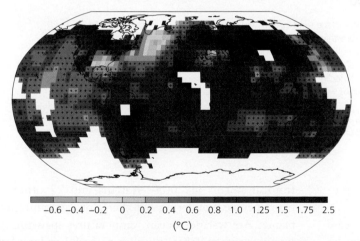

(b) Observed change in surface temperature from 1901 to 2012

FIGURE 14.13
Data from actual measurements **(a)** show changes in Earth's average surface temperature since 1850. The top part of this graph shows annually and globally averaged combined land and ocean surface temperature anomalies, that is, how much the average temperature each year differed from the average temperature for 1961 to 1990. Colours indicate different data sets, which show a high level of consistency The bottom graph is the same, but shows decadal (instead of annual) averages. The map in **(b)** shows the observed changes in temperature from 1902 to 2012. Grid boxes where the temperature trend is statistically significant to within 10% are marked with "+". Observed temperature changes up to 2.5°C are indicated on the map.
Source: From *Fifth Assessment Report. Climate Change 2013: The Physical Science Basis. Contribution of Working Group I to the Fifth Assessment Report of the Intergovernmental Panel on Climate Change*, Fig. SPM.1. Copyright © 2013 IPCC Secretariat. Reprinted by permission.

the period 2016 to 2035. If we were to cease greenhouse gas emissions today, temperatures would still rise 0.1°C per decade because of the time lag from gases already in the atmosphere that have yet to exert their full influence. At the end of the twenty-first century, the IPCC predicts that average global surface temperatures will likely be 1.5°C–3.0°C higher than today's, depending mainly upon future anthropogenic greenhouse gas emissions. Unusually hot days and heat waves, and extreme weather events in general, will become more frequent. Temperature changes will vary from region to region in ways that parallel existing regional differences (**FIGURE 14.14A**).

Changes in precipitation and storm activity will vary by region

A warmer atmosphere holds more water vapour; however, changes in precipitation patterns have been complex, with some regions of the world receiving more precipitation than usual and others receiving less. Some precipitation changes are already occurring. In regions from the African Sahel to the Western Prairies, droughts have become more frequent and severe, harming agriculture, promoting soil erosion, reducing drinking water supplies, and encouraging forest fires. Meanwhile, in both dry and humid

RCP 2.6

RCP 8.5

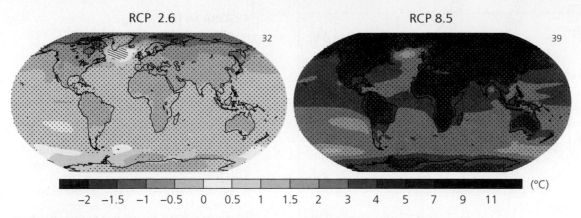

(a) Projected increases in surface temperature, 2081–2100

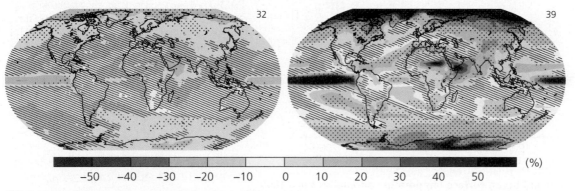

(b) Projected increases in precipitation, 2081–2100

FIGURE 14.14   The maps in **(a)** shows projected increases in surface temperature for the period 2081–2100, relative to the average temperature in 1986–2005. Land masses will warm more than the ocean, and the Arctic will warm the most. The maps in **(b)** show changes in average precipitation. The IPCC uses multiple climate models when predicting regional variations in how temperature and precipitation will change. The maps on the left are based on model scenario RCP2.6, which assumes that global annual GHG emissions will peak between 2010–2020 and will decline substantially after that. The maps on the right are based on model scenario RCP8.5, which assumes that GHG emissions will continue to rise, as they have done in the past few decades, throughout the rest of the twenty-first century.
Source: From IPCC (2013). *Fifth Assessment Report. Climate Change 2013: The Physical Science Basis. Contribution of Working Group I to the Fifth Assessment Report of the Intergovernmental Panel on Climate Change*, Fig. SPM.8. Copyright © 2013 by IPCC Secretariat. Reprinted by permission.

regions, heavy rain events have increased, contributing to damaging floods.

In the future, very broadly, precipitation will increase at high latitudes and decrease at low and middle latitudes (**FIGURE 14.14B**). This will magnify differences in rainfall that already exist, and worsen water shortages in many developing countries of the arid subtropics. In many areas, heavy precipitation events will become more frequent, increasing the risk of flooding. The increase in precipitation at very high latitudes will translate into greater snowfall in the High Arctic, Greenland, and Antarctica. This has implications for the behaviour of the polar ice caps, their response to warming, and even their role in controlling the overall reflectivity of the planet.

Sea surface temperatures also are increasing as the oceans absorb heat; this further contributes to evaporation and adds both heat energy and water vapour to the atmosphere. A record number of hurricanes and

tropical storms in 2005—Katrina and 27 others—left many people wondering whether global warming was to blame. Are warmer ocean temperatures spawning more storms, or storms that are more powerful or long lasting? But the 2006 hurricane season, in contrast, was quieter than normal. Recent analyses of storm data suggest that warmer seas may not be increasing the number of storms, but could be increasing the power of storms and possibly their duration.

Then, in 2012, Superstorm Sandy devastated the eastern coast of the United States, and the question arose again: Is climatic warming causing more frequent or stronger storms? Climate scientist Gerald Meehl is credited with having commented that warming is like "steroids in a baseball player." Steroids do not, on their own, produce home runs, but they make the conditions for home runs more likely. The pumped-up greenhouse gases in our atmosphere, similarly, do not on their own produce superstorms, but they create conditions that are conducive to the occurrence of more frequent and stronger storms.

Extreme weather is becoming "the new normal"

A study by climate scientist James Hansen and others revealed that summer temperatures since the 1950s have become not only warmer, but also more variable, with extreme summers (some of them unusually cool, more of them unusually warm) occurring more and more frequently. The sheer number of extreme weather events—droughts, floods, tornadoes, hurricanes, snowstorms, heat waves—in recent years has caught the public's attention, and weather records are being broken left and right.[8]

Scientists and hard-hit residents are not the only ones to notice the increase in extreme weather events. The insurance industry is finely attuned to such patterns, since insurers are the ones paying out money each time a major storm, drought, or flood hits. The major German insurer MunichRe calculates that from 1980 to 2013, the number of extreme weather events causing losses increased fully five times in North America, and four times worldwide, with losses totalling almost $1 trillion over that period.[9]

Researchers have long conservatively stated that although climate trends influence the probability of what the weather may be like on any given day, no single particular weather event—even a Hurricane Katrina or Superstorm Sandy—can reliably be attributed to climate change. However, as we gain a better understanding of what causes extreme weather events, it is starting to become more possible to link specific extreme weather events to climate change and changes in atmospheric circulation patterns, such as changes in the jet stream and the polar vortex.

The *jet stream* is a high-altitude air current that blows west to east and meanders north and south, influencing much of the weather from day to day across North America and Eurasia. There are two jet streams each in the Northern and Southern Hemispheres, between the poles and the equator (**FIGURE 14.15A**). The one that most influences the weather in Canada and the United States is the polar jet stream in the Northern Hemisphere (**FIGURE 14.15B**).

Because warming has been greater in the Arctic than at lower latitudes, this has weakened the intensity of the Northern Hemisphere's polar jet stream. As the jet stream slows down, its meandering loops become longer. These

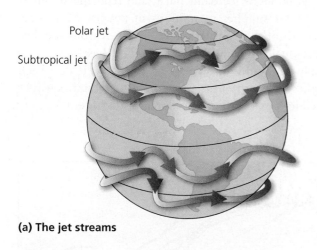

(a) The jet streams

FIGURE 14.15
There are four major jet streams **(a)**, which are high-altitude, high-speed winds that blow from west to east. Weather in Canada and the United States is normally influenced by the polar jet stream in the Northern Hemisphere **(b)**. In the past few years, warming in the High Arctic has caused the jet stream to be pushed out of its normal path and into blocking patterns **(c)**. Depending on its exact position and configuration, this can bring prolonged periods of extreme weather—sometimes hot, sometimes cold—to different parts of North America. The pattern shown here occurred in 2012, and resulted in extreme hot weather events in the eastern United States.

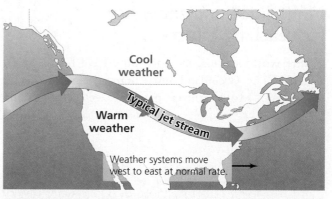

(b) Normal Northern Hemisphere polar jet stream

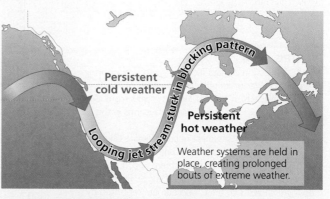

(c) Jet stream modified by blocking pattern

long lazy loops sometimes get stuck in one position for long periods of time. This is what meteorologists call an *atmospheric blocking pattern*, because it can block the movement of weather systems and hold them in place (**FIGURE 14.15C**). When this happens, a rainy system that would normally move past a city in a day or two might instead be held in place for several days, resulting in flooding. Or dry conditions over a farming region might last two weeks instead of two days, causing a drought. Cold spells and hot spells last longer, too.

In conjunction with these changes in the jet stream, episodes of sudden stratospheric warming in the Arctic have caused changes in the polar vortex. The *polar vortex* is a tight atmospheric circulation system that normally sits over the poles; it creates the extremely cold conditions conducive to stratospheric ozone depletion, as discussed in Chapter 13, *Atmospheric Science and Air Pollution*. When warming occurs in the stratosphere at high latitudes, it can cause the polar vortex (and with it, the jet stream) to break up and migrate southward, bringing pockets of extremely cold air along with it. The icy winters of 2013–2014 and 2014–2015 in eastern Canada and some of the unusual snowfall events farther south in the United States have been attributed to this factor.[10] Whether events like this will increase in frequency as a result of global climate change remains to be seen.

Melting ice and snow have far-reaching effects

As the world warms, ice and snow have been greatly affected. The Greenland and Antarctic ice sheets have lost mass. Mountaintop glaciers are disappearing

persistently, worldwide. Arctic sea ice has lost areal extent much faster than anticipated. Indeed, the mass of the entire world ice inventory has decreased. These changes, in turn, have an influence on sea level and on other parts of the climate system.

Glaciers—land-based deposits of perennially frozen ice—can be quite variable in length and overall mass from season to season and year to year. However, the World Glacier Monitoring Service[11] estimates that, since 1945, major glaciers have each lost an average of 20 m in vertical thickness of water mass equivalent; this conclusion is based on many thousands of length change observations and mass balance measurements on hundreds of glaciers (**FIGURE 14.16**). As discussed at the beginning of this chapter, the Athabasca Glacier has retreated 1.5 km since the late nineteenth century, with the rate of retreat accelerating dramatically after 1980. Peyto Glacier in Banff National Park, Alberta, retreated rapidly during the first half of the twentieth century, stabilized briefly in 1966, and then resumed its retreat in 1979. Illecillewaet Glacier in Glacier National Park, British Columbia, has retreated 2 km in the past century.

Mountains accumulate snow in the winter and release melt water gradually during the summer. Throughout high-elevation areas of the world, warming temperatures will continue to melt mountain glaciers, posing risks of sudden floods as ice dams burst, and reducing summertime water supplies to millions of people. More than one-sixth of the world's people live in regions supplied by mountain melt water, and some of these people are already beginning to face water shortages. If this water vanishes during drier months, whole communities will be forced to look elsewhere for water, or to move.

FIGURE 14.16
This is a collection of data from long-term monitoring of changes in glacier lengths for a number of North American glaciers, in metres. Similar graphs from all regions of the world show similar reductions in the lengths of glaciers.
Source: From *Fifth Assessment Report. Climate Change 2013: The Physical Science Basis.* Contribution of Working Group I to the Fifth Assessment Report of the Intergovernmental Panel on Climate Change, Fig. 4.9. Copyright © 2013 by IPCC Secretariat. Reprinted by permission.

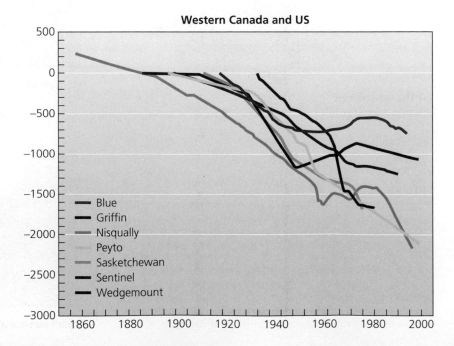

Change in Winter Snow

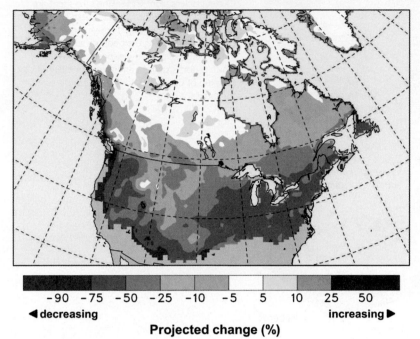

◄ decreasing   increasing ►

Projected change (%)

FIGURE 14.17

Change in the average winter snow amount (December-January-February) between the end of the twentieth century (1986-2005) and the middle of the twenty-first century (2046-2065), as simulated by climate model scenarios in which future greenhouse gas emissions are expected to peak around the year 2040 and decline after that (that is, the "RCP4.5" scenario used by the IPCC). *Source:* © Her Majesty the Queen in Right of Canada, as represented by the Minister of the Environment, 2005.

Warming temperatures are also affecting seasonal snow cover. Models by the Canadian Centre for Climate Modelling and Analysis (**FIGURE 14.17**) predict that seasonal snow cover will decrease throughout much of North America by the 2050s, while increasing in the High Arctic and the centre of the Greenland ice sheet (as well as Antarctica, not covered in this particular model). The increased snowfall at high latitudes is a result of increased precipitation in regions with temperatures that are warming, but still below freezing.

Warming will have an impact not only on surface snow and ice, but also on subsurface ice and **permafrost** (perennially frozen ground). Permafrost is the most characteristic feature of Arctic soils, and its expected behaviour in a changing climate regime is not thoroughly understood from a scientific perspective. The foundations of roads and buildings in the north would be at risk if permafrost undergoes major changes as a result of warming. Permafrost also plays a major role in slope stability, with greatly increased chances for landslides as permafrost begins to melt. There are also concerns that the warming of permanently frozen soils would lead to the accelerated release of soil gases, such as methane, which could contribute to a positive feedback cycle in the climate system, leading to further warming.

The biggest uncertainties related to snow and ice in the climate system concern the behaviour of the huge polar ice caps in a warming world. In Antarctica, ice shelves almost the size of Prince Edward Island have disintegrated as a result of contact with warmer ocean water. Ice sheets, like all perennial ice masses, gain mass by accumulating snow

during cold weather, which becomes packed into ice over time. They lose mass as surface ice melts in warm weather, generally at the periphery, where ice is thinnest or contacts sea water. If melting and runoff outpace accumulation, the ice sheet shrinks. In Antarctica, increased precipitation so far seems to have supplied the continent with enough extra snow to compensate for the loss of ice around its edges.

Greenland hosts the world's other massive polar ice sheet, nearly 3 km at its thickest point, and covering approximately 1.7 million km^2—an area larger than the entire province of Quebec. Scientists have known for years that the Arctic is bearing the brunt of global warming, and that Greenland's ice sheet is melting around its edges. But scientists studying how ice moves are now learning that ice sheets can collapse more quickly than expected, and that Greenland's ice loss is accelerating. As a result, the likely speed and extent of melting and its impact on future sea-level rise may have been underestimated. If the entire Greenland ice sheet were to melt (an unlikely scenario in the near future, but much more likely if warming of the climate system is sustained over the next millennium), global sea level would rise by a whopping 7 m.

The internal physical dynamics of glacial ice are important in determining ice sheet behaviour. Immense volumes of ice flow in outlet glaciers from land downhill toward the coast, where eroding ice sloughs off and melts into the sea or floats away as icebergs. In warm months, ice in outlet glaciers flows more quickly than in cold months. Pools of melt water form on the surface and leak down through crevasses and vertical tunnels (called *moulins*) to the bottom of the glacier. There, the water runs downhill in a layer between bedrock

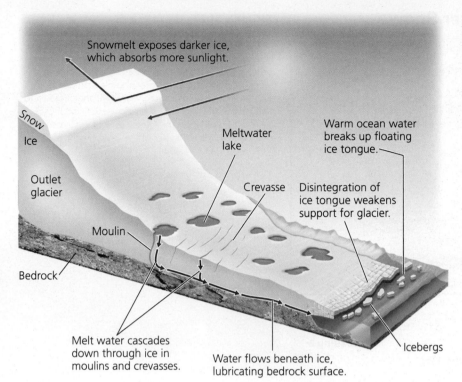

Snowmelt exposes darker ice, which absorbs more sunlight.

Snow

Ice

Outlet glacier

Moulin

Bedrock

Meltwater lake

Crevasse

Warm ocean water breaks up floating ice tongue.

Disintegration of ice tongue weakens support for glacier.

Melt water cascades down through ice in moulins and crevasses.

Water flows beneath ice, lubricating bedrock surface.

Icebergs

FIGURE 14.18
Outlet glaciers from Greenland's ice sheet are accelerating their slide into the ocean because melt water is descending through moulins and crevasses, lubricating the bedrock surface. Moreover, melting snow exposes darker ice, which absorbs sunlight and speeds melting. Finally, breakup of floating ice tongues weakens support for the glacier.

and ice, lubricating the bedrock surface and enabling the ice to slide downhill like a car hydroplaning on a wet road. The resulting rapid downslope movement of the glacier is called a *surge*. In addition, the melt water weakens ice on its way down and warms the base of the glacier, melting some of it to create more water (**FIGURE 14.18**).

Warming ocean water also melts ice shelves along the coast, depriving outlet glaciers of the supports that hold them in place. Without a floating ice tongue at its terminus, a glacier slides into the ocean more readily. These physical dynamics represent positive feedbacks; once global warming initiates these processes, they encourage further melting. As such, it is likely that Greenland's melting will continue to accelerate. Both Greenland and Antarctica have experienced a net loss of ice mass in the past two decades, and their contributions to sea-level rise are increasing.[12] These trends are expected to continue or accelerate.

Canada's Arctic is changing dramatically

Temperature changes, both current and those projected by climate models, are greatest in the High Arctic. Climate scientists anticipate that this will likely continue to be the case throughout the rest of this century. In Canada's Arctic regions, sea ice is thinning and decreasing in areal extent, storms are increasing, and altered conditions are posing challenges for people and wildlife. As sea ice melts earlier, freezes later, and recedes from shore, it becomes harder for

both Inuit and polar bears to hunt the seals they each rely on for food (**FIGURE 14.19**). Thin sea ice is dangerous for people to travel and hunt upon. In recent years, polar bears have been dying of exhaustion and starvation as they try to swim long distances between ice floes. Thawing of permafrost is destabilizing countless buildings. The strong Arctic warming is contributing to sea-level rise by the melting of icecaps and ice sheets.

According to Dr. Luke Copland of the University of Ottawa, the area of Canada's ice shelves has shrunk by approximately 90% over the past 100 years. One major event that contributed to this loss was the collapse of the Ayles Ice Shelf in 2005 (**FIGURE 14.20**). The Ayles Ice Shelf was one of Canada's six major ice shelves, located off the northern coast of Ellesmere Island, Nunavut, about 800 km south of the North Pole. On August 13, 2005, it broke off, creating an ice island 14 km by 5 km, and 37 m in thickness. The oldest ice in the ice island is more than 3000 years old. This ice breakup was so large that it was detected by seismometers. The breakup happened very quickly, lasting less than an hour according to reconstructions based on satellite images. Since the breakup, the ice island has drifted, becoming temporarily locked in place during winter freeze-ups and then breaking free and drifting again, and it has broken into two large parts. The large ice island fragments could pose a risk to oil rigs and drilling in the Beaufort Sea.

One reason warming is accelerating in the Arctic is that as snow and ice cover are melted, darker, less-reflective

Jan Martin Will/Shutterstock

Gordon Wiltsie/National Geographic Stock

Andrew Bain/Getty Images

FIGURE 14.19
The Arctic has borne the brunt of climate change's impacts so far. As Arctic sea ice melts it recedes from large areas, as shown by the map indicating the mean minimum summertime extent of sea ice for the recent past, present, and future. Inuit find it difficult to hunt and travel in their traditional ways, and polar bears starve because they are less able to hunt seals. As permafrost thaws beneath them, human-made structures are damaged: Phone poles topple and buildings can lean, buckle, crack, and fall. *Source:* Map data from National Center for Atmospheric Research and National Snow and Ice Data Center.

surfaces are exposed. As a result, Earth's albedo, or capacity to reflect light, decreases. As a result, more of the Sun's rays are absorbed at the surface, fewer reflect back into space, and the surface warms. In a positive feedback, this warming causes more ice and snow to melt, which in turn causes more absorption of radiation and more warming.

Near the poles, snow cover, permafrost, and ice sheets are projected to decrease, and sea ice will continue to shrink in the Arctic (**FIGURE 14.21**). Shipping vessels have already started to move through the Northwest Passage, creating the potential for sovereignty conflicts over Arctic waters, much of which is claimed by Canada. The opening of the Arctic Ocean will likely initiate a rush to exploit underwater oil and mineral reserves that may exist there.

Rising sea levels will affect millions in coastal zones and on islands

As glaciers and ice sheets melt, increased runoff into the oceans causes sea levels to rise. Sea levels also are rising because ocean water is warming; liquid water expands in volume as its temperature increases. In fact, recent sea-level rise has resulted primarily from the thermal expansion of sea water. Worldwide, average sea levels rose at least 19 cm

from 1901 to 2010 (**FIGURE 14.22**). The rate of sea-level rise has increased in the past decade, as well. This trend is expected to continue or accelerate. Note that these numbers represent vertical rise in water level; on most coastlines, especially those with shallowly dipping shores, a vertical rise of centimetres means metres of horizontal incursion inland.

As shown in **FIGURE 14.22**, there are several causes of global sea-level rise, both present and future, associated with climate warming. According to the IPCC,[13] in all model scenarios thermal expansion of sea water is the largest contributor, accounting for 30 to 55% of projected sea-level change to 2100. Melting of glaciers is next, although this contribution will wane as land-based glaciers disappear. Last, but most complicated and uncertain, is the contribution from melting of the Greenland and Antarctic ice sheets.

Note that, although the extent and mass of floating sea ice are strongly affected by the warming of ocean water, when sea ice melts it makes almost no contribution to sea-level rise. Sea ice is like an ice cube that is already floating in a glass of water; if it melts, it won't cause the glass to overflow. Land ice is like water that you pour into the glass from a pitcher, which adds mass and can make the water level rise and cause the glass to overflow.

NASA

(a) Ayles Ice Shelf before breakup

FIGURE 14.20

As seen in these satellite images, the Ayles ice shelf **(a)**, one of Canada's major ice sheets north of Ellesmere Island, collapsed in 2005 **(b)**, forming a huge ice island 14 km by 5 km in size.

NASA

(b) Ice island formed after breakup

FIGURE 14.21

The curves on this graph show various data sets, all of which have demonstrated a decrease in the average annual extent of sea ice in the Arctic Ocean, especially since about 1960. Climate models predict this trend to continue or accelerate in the next few decades, depending on the emissions scenario used.

Source: From *Fifth Assessment Report. Climate Change 2013: The Physical Science Basis. Contribution of Working Group I to the Fifth Assessment Report of the Intergovernmental Panel on Climate Change,* Fig. 4.3a. Copyright © 2013 by IPCC Secretariat. Reprinted by permission.

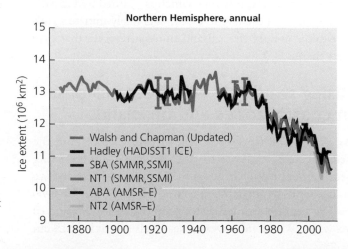

Northern Hemisphere, annual

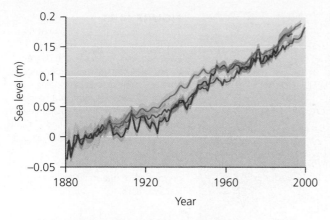

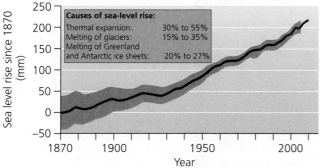

FIGURE 14.22

This graph represents a compilation of sea-level data from tidal gauges, from 1880 to 2010. The three colours represent three different studies of the data. Estimates of the causes of sea-level rise to 2100 are shown. *Source:* Data shown in orange are from Church and White (2011); data shown in blue are from Jevrejeva, et al. (2008); data shown in green are from Ray and Douglas (2011). All reported in IPCC (2013), *Fifth Assessment Report. Climate Change 2013: The Physical Science Basis. Contribution of Working Group I to the Fifth Assessment Report of the Intergovernmental Panel on Climate Change*, Figure 13.3C.

Higher sea levels lead to beach erosion, coastal flooding, intrusion of salt water into aquifers, and other impacts. In 1987, unusually high waves struck the island nation of the Maldives and triggered a campaign to build a large seawall around Male, the nation's capital. Known as "The Great Wall of Male," the seawall is intended to protect buildings and roads by dissipating the energy of incoming waves during storm surges. A storm surge is a temporary and localized rise in sea level brought on by the low atmospheric pressure and winds associated with storms. The higher sea level is to begin with, the farther inland a destructive storm surge can reach. "With a mere one-metre rise [in sea level]," Maldivian president Maumoon Abdul Gayoom warned in 2001, "a storm surge would be catastrophic, and possibly fatal to the nation."

The Maldives is not the only nation with such concerns. In fact, among island nations, the Maldives has fared better than many others. It saw a sea-level rise of about 2.5 mm per year throughout the 1990s, but most Pacific islands are experiencing greater changes. Islands and coastal regions experience differing amounts of

sea-level change because land elevations may be rising or subsiding naturally, depending on local geological conditions. In New Orleans, where the subsidence of the Mississippi Delta has caused the loss of protective coastal wetlands, most of the damage from Hurricane Katrina in 2005 was from the storm surge.

Although there are many variations among model results, the IPCC predicts with medium confidence that by the end of the twenty-first century, global mean sea level will be at least 0.40 m and as much as 0.63 m higher than it is today, depending upon the emission scenario. If melting continues to accelerate in the Greenland or Antarctic ice sheets, increasing their contribution of melt water, then sea levels could rise more quickly.

Rising sea levels will force hundreds of millions of people to choose between moving upland or investing in costly protections against high tides and storm surges. Canada's coastal regions will be variably affected, as shown in **FIGURE 14.23**. Coastal locations, such as parts of the St. Lawrence estuary, mainland Nova Scotia, Cape Breton Island, and Prince Edward Island, are relatively low-lying and particularly vulnerable to flooding. Some areas of the far north, including the coasts of the Beaufort Sea, and low-lying areas of the BC coast are also at risk of both flooding and accelerated coastal erosion.

Worldwide, densely populated regions on low-lying river deltas, such as the Ganges–Brahmaputra Delta in Bangladesh and the Irrawaddy Delta of Burma, will be most affected. So will storm-prone regions, coastal cities, and areas around the world where land is subsiding. Many Pacific islands will need to be evacuated, and some nations, such as Tuvalu and the Maldives, fear for their very existence. In the meantime, they may suffer from shortages of fresh water because rising seas threaten to bring salt water into the nation's wells, as the 2004 tsunami did. The contamination of ground water and soils by sea water threatens not only island nations but also coastal areas that depend on small lenses of fresh water floating atop saline ground water in coastal aquifers.

Climate change affects organisms and ecosystems

To summarize the physical changes associated with a changing climate, we can look at **FIGURE 14.24**, from the IPCC *Fifth Assessment Report*. This figure summarizes seven major indicators that scientists would expect to be increasing, if the global climate is warming. They are air temperature in the lower part of the troposphere, water vapour in the atmosphere, temperature over land, marine air temperature, sea-surface temperature, ocean heat content, and sea level. These seven indicators are, in fact, increasing. The figure also summarizes three major

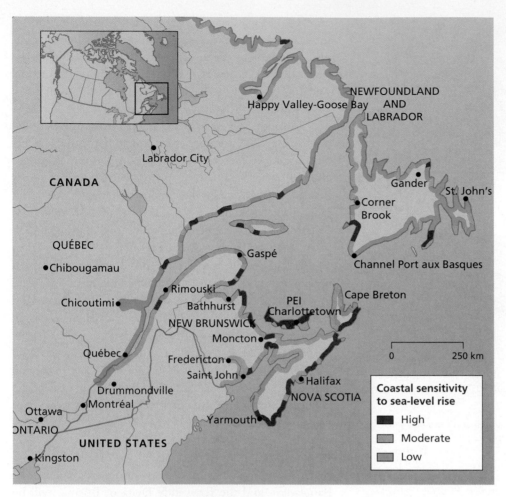

FIGURE 14.23
This map shows the sensitivity of the coastlines of Canada to rising sea level. Sensitivity refers to the degree to which a coastline may experience physical changes, such as flooding, erosion, beach migration, and coastal dune destabilization. Two major regions of high sensitivity are identified: Atlantic Canada, and parts of the Beaufort Sea coast.
Source: Reproduced with the permission of Natural Resources Canada, 2015.

indicators that scientists would expect to be decreasing, if the global climate is warming. They are snow cover; glacier volume; and sea ice area. These three indicators are, in fact, decreasing. What this means is that the world's physical systems are behaving in a manner that is entirely consistent with a warming climate.

The many changes in Earth's physical systems will have, and are already having, direct consequences for life on our planet. Organisms are adapted to their environments, so they are affected when those environments are altered. As global warming proceeds, it modifies temperature-dependent biological phenomena. For instance, in the spring, birds are migrating earlier, insects are hatching earlier, and animals are breeding earlier. Plants are leafing out earlier, too— an effect confirmed by satellite photography that records whole landscapes "greening up" each year.

These changes in seasonal timing are expected to continue, and they are having complex effects. For instance, European birds known as great tits time their breeding cycle so that their young hatch and grow at the time of peak caterpillar abundance. However, as plants leaf out earlier and insects emerge earlier, research shows that great tits are not breeding earlier. As a result of the mismatch in timing, fewer caterpillars are available when young birds need them, and fewer birds survive. Although some organisms will no doubt adapt to such changes in seasonal timing, research so far shows that in most cases mismatches occur.

A similar situation provided an unexpected benefit and is helping to prevent outbreaks of the invasive parasitic spruce budworm on Vancouver Island. Budworm development is triggered by temperature, while the development of the Douglas fir, the preferred food of the budworm larva, is triggered more by light and day length. As a result of a 90-year increase in winter temperatures on southern Vancouver Island, the budworm larvae are emerging from their shelters earlier, at a time when the Douglas fir, their main food source, is not yet available.

Biologists have recorded spatial shifts in the ranges of organisms, with plants and animals moving toward the poles or upward in elevation (toward cooler regions) as temperatures warm. As these trends continue, some organisms will not be able to cope, and as many as 20–30% of all plant and animal species could be threatened with extinction, the IPCC estimates. Trees may not be able to shift their distributions fast enough. Animals adapted to montane environments may be forced uphill until there is nowhere left to go. Rare species finding refuge in

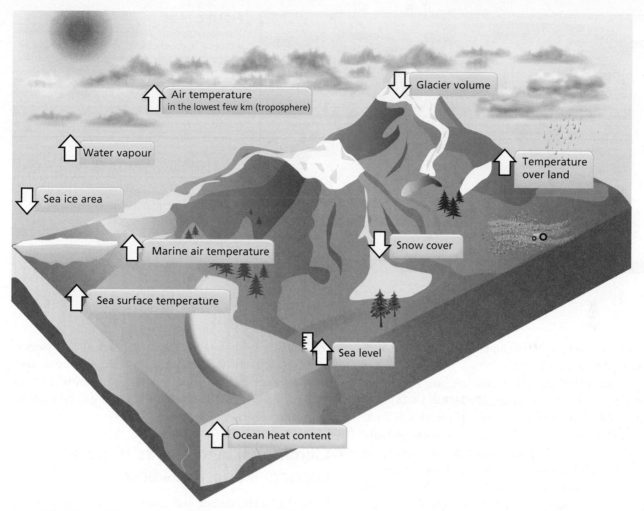

FIGURE 14.24 This figure shows seven major indicators that scientists would expect to increase if the climate is warming; they are all increasing. The figure also shows three major indicators that scientists would expect to decrease if the climate is warming; they are all decreasing. This means that Earth's physical systems are behaving in a manner that is consistent with a warming climate.
Source: From *Fifth Assessment Report. Climate Change 2013: The Physical Science Basis.* Chapter 2, FAQ Figure 2.1. Copyright © 2013 by IPCC Secretariat. Reprinted by permission.

protected preserves may be forced out of preserves into developed areas, making such refuges far less effective tools for conservation.

The American pika (*Ochotona princeps*) is a small mammal that lives on rocky slopes near mountain glaciers in southern Canada. Pikas (**FIGURE 14.25**) are clearly not the only species likely to suffer from warming (polar bears come to mind). However, pikas are extremely sensitive to environmental conditions, especially temperature, and cannot survive if temperatures are too high. Their favoured habitat, in the vicinity of glaciers, is threatened as a result of rapidly changing conditions in glacial environments.

Climatic changes will greatly affect species interactions, and scientists foresee major modifications in the structure and function of communities and ecosystems. In regions where precipitation and stream flow increase, erosion and flooding will pollute and alter aquatic systems. In regions where precipitation decreases, lakes, ponds, wetlands, and

FIGURE 14.25
Pikas live near glaciers in southwestern Canada. They are extremely sensitive to temperature, and may become one of the first animals whose habitat is completely eliminated by climatic warming.

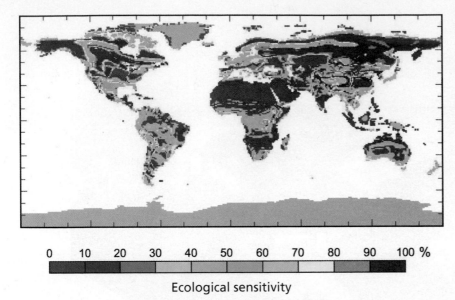

FIGURE 14.26
This map shows the predicted percentage of the ecological landscape that is being driven toward changes in plant species as a result of projected human-induced climate change, by 2100, assuming a climate system in which atmospheric CO_2 has doubled. The extreme ecological sensitivity in Alaska, Canada, and Russia results from a combination of the intense impacts of climate change on the Arctic and the fragility of the ecosystems.
Source: NASA–JPL/Caltech. *21st Century Ecological Sensitivity—Changes in Plant Species.* http://www.nasa .gov/topics/earth/features/climate20111214i.html. Accessed 13 March 2015.

0   10   20   30   40   50   60   70   80   90   100 %

Ecological sensitivity

streams will diminish, affecting aquatic organisms, as well as human health and well-being.

Effects on plant communities are an important component of climate change (**FIGURE 14.26**). By drawing in CO_2 for photosynthesis, plants act as sinks for carbon. If climate change increases vegetative growth, this could help mitigate carbon emissions, in a process of negative feedback. However, if climate change decreases plant growth (through drought or fire, for instance), then positive feedback could increase the carbon flux to the atmosphere. The many impacts on ecological systems likely will reduce the ecosystem goods and services we receive from nature and that our societies depend on, from food to clean air to drinking water.

Ecosystem impacts are not limited to terrestrial environments. Coastal and island residents worry about damage to the marine ecosystems that are critical for their economy, including coral reefs. Coral reefs provide habitat for important food fish that are consumed locally and exported; offer snorkelling and scuba diving sites for tourism; and reduce wave intensity, protecting coastlines from erosion. Around the world, rising seas will eat away at the coral reefs, mangrove forests, and salt marshes that serve as barriers protecting our coasts.

Climate change poses two additional threats to coral reefs: Warmer waters are causing coral bleaching, and enhanced CO_2 concentrations in the atmosphere are changing ocean chemistry. As ocean water absorbs atmospheric CO_2, it becomes more acidic, which impairs the growth of coral and other organisms whose exoskeletons consist of calcium carbonate. Ocean water has already decreased by 0.1 pH unit overall (a 26% decrease from pre-industrial levels), and is predicted to decline in pH by 0.06–0.32 more units over the next 100 years.[14] This

could easily be enough to destroy most of our planet's living coral reefs. Any significant decline of coral reefs would reduce marine biodiversity significantly, because so many other organisms depend on living coral reefs for food and shelter.

Climate change exerts societal impacts—and vice versa

Both the environment and human society have begun to feel the impacts of climate change, as shown in **FIGURE 14.12**. Damages from drought, flooding, hurricanes, storm surges, and sea-level rise have already taken a toll on the lives and livelihoods of millions of people. However, climate change will have additional consequences for humans, including impacts on agriculture, forestry, economics, and health.

FIGURE 14.27 gives an overview of the complex interactions between the natural world and human systems in determining and reacting to climatic change. In this section we will look in a bit more detail at some of the most important human drivers and impacts of climate change.

Agriculture   For farmers, earlier springs require earlier crop planting. For some crops in the temperate zones, production may increase slightly with moderate warming, because growing seasons become longer and because more carbon dioxide is available to plants for photosynthesis. However, rainfall will shift in space and time, and in areas where droughts and floods become more severe, these will cut into agricultural productivity. Overall, global crop yields are predicted to increase somewhat, but beyond a rise of 3°C, the IPCC expects crop yields to

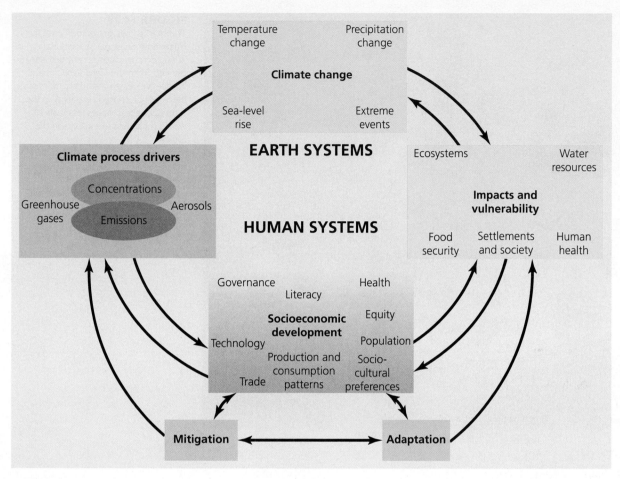

FIGURE 14.27 Human and natural systems interact in complex ways to both influence and react to climatic change, as shown in this diagram.
Source: From *Fourth Assessment Report: Climate Change 2007: Synthesis Report,* Figure 1. Copyright © 2007 by IPCC Secretariat. Reprinted by permission.

decline. In seasonally dry tropical and subtropical regions, growing seasons may be shortened, and harvests may be more susceptible to drought and crop failure. Thus, scientists predict that crop production will fall in these regions even with minor warming. This would worsen hunger in many of the world's developing nations.

It is tempting to think that a warmer North would mean that Canada's agricultural region would expand, or, at worst, shift toward the north. This is unlikely to happen. It takes a very long time for weathering and the accumulation of organic material to lead to the formation of soil profiles. Much of Canada's North is bare rock, covered by a thin, rocky, immature soil. The rock of the Canadian Shield was scraped clean by advancing glaciers during the last major glaciation in the Pleistocene epoch. The Pleistocene glaciers retreated from Canada's North around 7000 years ago, but in all that time there has been no significant development of a tillable soil horizon over much of the land area. In other words, our breadbasket, the Prairies, will still have good soils but will become too hot and dry for successful farming, while

Canada's North, although warmer and wetter, won't have the soil to take advantage of the better climatic conditions.

Forestry   Forest managers increasingly find themselves having to battle insect and disease outbreaks, invasive species, and catastrophic fires, which are promoted by longer, warmer, drier fire seasons (**FIGURE 14.28**). Natural Resources Canada attributes several recent forest impacts to climate change, including the mountain pine beetle infestation in British Columbia, increased fire activity in boreal forests of Western Canada, and increased dieback of aspen trees in the Prairies.[15]

For timber and forest products, enriched atmospheric CO_2 may spur greater growth in the near term, but this will vary substantially from region to region. Other climatic effects, such as drought, may eliminate these gains. For instance, droughts brought about by a strong El Niño in 1997–1998 allowed immense forest fires to destroy millions of hectares of rain forest in Indonesia, Brazil, Mexico, and elsewhere.

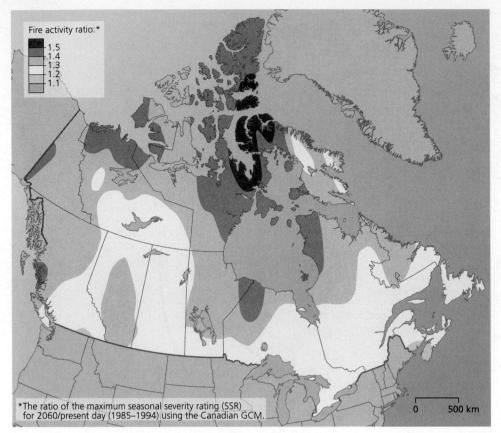

Fire activity ratio:*

- 1.5
- 1.4
- 1.3
- 1.2
- 1.1

*The ratio of the maximum seasonal severity rating (SSR) for 2060/present day (1985–1994) using the Canadian GCM.

0   500 km

FIGURE 14.28
This map shows projected fire activity (Maximum Seasonal Severity Rating) in 2060 compared to current fire activity, derived from the Canadian Climate Model. Values greater than 1.0 mean there will be more fire activity in the future, whereas values less than 1.0 would mean less fire activity in the future.
Source: Based on Flannigan, M.D., B.J. Stocks, and B.M. Wotton (2000). Climate Change and Forest Fires, *The Science of the Total Environment*, 262, pp. 221–229.

Health   As a result of climate change, we will face more heat waves—and heat stress can cause death, especially among older adults. A 2003 heat wave killed at least 35 000 people in Europe, for example, and this type of extreme event is expected to increase in frequency and intensity. In addition, a warmer climate exposes us to other health problems:

- respiratory ailments that result from air pollution, as hotter temperatures promote formation of photo-chemical smog
- expansion of tropical diseases, such as dengue fever, into temperate regions as vectors of infectious disease (such as mosquitoes) move toward the poles; the advent of mosquito-borne West Nile Virus in southern Canada may be a preview of this scenario
- disease and sanitation problems that occur when floods overcome sewage treatment systems
- injuries and drownings that will likely increase if storms become more frequent or intense
- hunger-related ailments that will worsen as the human population grows and climate-related stresses on agricultural systems increase
- health hazards from cold weather will likely decrease, but researchers feel that the increase from warm-weather hazards will more than offset these gains

Economics   People will experience a variety of economic costs—and some benefits, too—from the many impacts of climate change. On the whole, researchers predict that costs will outweigh benefits and that this gap will widen as climate change grows more severe. Climate change will also widen the gap between rich and poor, both within and among nations. Poorer people have less wealth and technology with which to adapt to climate change, and poorer people rely more on resources (such as local food and water) that are particularly sensitive to climatic conditions. As well, the poor often have the fewest livelihood options; if a source of livelihood, such as fishing or a forest-based occupation, is endangered by the impacts of climate change, there may be no other income-generating options available.

From a wide variety of economic studies, the IPCC has estimated that climate change will cost 1–5% of GDP on average globally, although poor nations would lose proportionally more than rich nations. Economists trying to quantify damages from climate change by measuring the "social cost of carbon" (i.e., external costs) have proposed costs of anywhere from $10 to $350 per tonne of carbon. The highest-profile economic study to date has been the Stern Review commissioned by the British government.[16] This study maintained that climate change could cost roughly 5–20% of world GDP by the year 2200, but that investing just 1% of GDP starting now could enable us to avoid these future costs. Regardless of the precise numbers, many economists and policymakers are concluding that spending money now to mitigate climate change will save us a great deal more in the future.

All these physical, biological, and social impacts of climate change are consequences of the warming effect of our greenhouse gas emissions. We are bound to experience further consequences, but by addressing the root causes of anthropogenic climate change we can still prevent the most severe future impacts.

Are we responsible for climate change?

The IPCC's 2007 *Fourth Assessment Report* concluded that it is more than 90% likely that most of the global warming recorded over the past half-century is due to the well-documented increase in greenhouse gas concentrations in our atmosphere. The language around warming and the

human causes and impacts of warming was strengthened even further in the *Fifth Assessment Report* (2013–2014):

> *Warming of the climate system is unequivocal, and since the 1950s, many of the observed changes are unprecedented over decades to millennia. The atmosphere and ocean have warmed, the amounts of snow and ice have diminished, and sea level has risen ... Human influence on the climate system is clear, and recent anthropogenic emissions of greenhouse gases are the highest in history. Recent climate changes have had widespread impacts on human and natural systems.[17]*

Scientists agree that the recent increase in greenhouse gases in the atmosphere is responsible for the observed warming. The increase has resulted primarily from our combustion of fossil fuels for energy and transportation, and secondarily from land use changes, including deforestation and agriculture (**FIGURE 14.29**).

According to the National Academies of Science, 97% of climate scientists agree that climate-warming trends

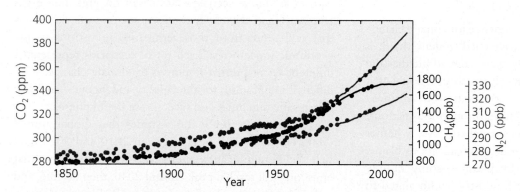

(a) Globally averaged greenhouse gas concentrations

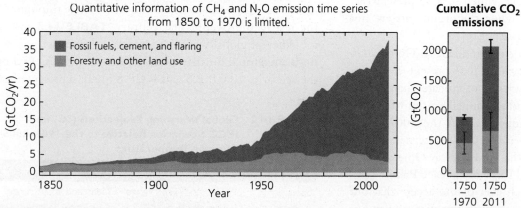

(b) Global anthropogenic CO₂ emissions

FIGURE 14.29 The top graph **(a)** shows atmospheric concentrations of the greenhouse gases carbon dioxide (CO₂, green), methane (CH₄, orange), and nitrous oxide (N₂O, red) determined from ice core data (dots) and from direct atmospheric measurements (lines). The bottom graph **(b)** shows global anthropogenic CO₂ emissions from forestry and other land use, as well as from burning of fossil fuel, cement production, and flaring of natural gas. Cumulative emissions of CO₂ from these sources and their uncertainties are shown as bars and whiskers, respectively, on the right hand side.
Source: From *Fifth Assessment Synthesis Report: Summary for Policymakers*, Fig. SPM.1 Rev1-01. Copyright © 2014 by IPCC Secretariat. Reprinted by permission.

over the past century are very likely due to human activities. Even by the time of the IPCC's *Fourth Assessment Report*, many scientists had become concerned enough to put themselves on record urging governments to address climate change. In June 2005, the National Academies of Science from 11 nations (Canada, Brazil, China, France, Germany, India, Italy, Japan, Russia, the United Kingdom, and the United States) issued a joint statement urging international political leaders to take action. Such a broad consensus statement from the world's scientists is virtually unprecedented, on any issue. The statement read, in part,

> *The scientific understanding of climate change is now sufficiently clear to justify nations taking prompt action. It is vital that all nations identify cost-effective steps that they can take now, to contribute to substantial and long-term reduction in net global greenhouse gas emissions … A lack of full scientific certainty about some aspects of climate change is not a reason for delaying an immediate response that will, at a reasonable cost, prevent dangerous anthropogenic interference with the climate system.[18]*

Despite the overwhelming evidence for climate change and its impacts, many people have tried to deny that it is happening. Many of these naysayers now admit that the climate is changing but a few still try to cast doubt that we are the cause. Most of the world's nations have moved forward to confront climate change through international dialogue, leaving behind outdated debates over whether climate change is real and whether humans are to blame. These debates were fanned by people—some quietly representing interests in the oil industry—who aimed to cast doubt on the scientific consensus. Their views were amplified by the news media, which seek to present two sides to every issue, even when the arguments are not equally supported by evidence.

For many, former U.S. Vice President Al Gore's 2006 movie and book *An Inconvenient Truth* presented an eye-opening summary of the science of climate change, and became a turning point and a compelling call to action. Awareness of climate change grew as the 2007 IPCC report was made publicly available on the internet and was widely covered in the media, and later as Gore and the IPCC were jointly awarded the Nobel Peace Prize.

Today, more people than ever before accept that our fossil fuel consumption is altering the planet that our children will inherit. As a result of this shift in public opinion, and in response to demand from their shareholders, many corporations and industries began offering support for reductions in greenhouse gas emissions. Students and others have urged universities to divest their financial holdings in oil companies. Corporate leaders joined ranks with the insurance industry, which many

years earlier had grown concerned with climate change as it foresaw increased payouts for damage caused by coastal storms, drought, and floods.

Responding to Climate Change

Today we possess a broadened consensus that climate change is a clear and present challenge to our society. Precisely how we should respond to climate change is a difficult question, however, and one we will likely be wrestling with for decades. The strategy that we choose, as a society, will determine how successful we are at curbing climate change, and how quickly (or *if*) we can make it happen.

As discussed previously, the IPCC bases its climate predictions on model results, which are built around a number of potential scenarios of differing human responses to climate change. The scenarios in current use are denoted "RCP," which stands for "representative concentration pathways." These scenarios are based on greenhouse gas emissions, and the concentrations of atmospheric GHGs that will result from those emissions in various socioeconomic trajectories. Each set of scenarios represents a different set of human responses to climate change, with different implications for the rapidity and vigour—and thus the specific outcomes and success—of the response.

For example, RCP 2.6 assumes that global annual GHG emissions will peak between 2010 and 2020, and will decline significantly after that. RCP 4.5 assumes that emissions will reach a peak around 2040, then decline, and in RCP 6, emissions will peak around 2080, then decline. In RCP 8.5, the "business-as-usual" or worst-case scenario, greenhouse gas emissions will continue to rise throughout the rest of the twenty-first century. **TABLE 14.2** shows the different global warming temperature increase projections anticipated for these four different scenarios. Two of the scenarios (RCP 2.6 and RCP 8.5) are represented in

TABLE 14.2 Global Warming Projections (°C) for IPCC Scenarios Relative to the 1986–2005 Average Temperature

Scenario	2046–2065 Mean and *likely* range	2081–2100 Mean and *likely* range
RCP 2.6	+1.0 (0.4 to 1.6)	+1.0 (0.3 to 1.7)
RCP 4.5	+1.4 (0.9 to 2.0)	+1.8 (1.1 to 2.6)
RCP 6.0	+1.3 (0.8 to 1.8)	+2.2 (1.4 to 3.1)
RCP 8.5	+2.0 (1.4 to 2.6)	+3.7 (2.6 to 4.8)

*The term *likely* carries a defined probability of >66%.
Source: From IPCC (2013). *Fifth Assessment Report, Working Group I: Summary for Policymakers*, Table SPM-2. Copyright © 2013 by IPCC Secretariat. Reprinted by permission.

FIGURE 14.14 showing the potential for changes in temperature and precipitation.

Shall we pursue mitigation, adaptation, or intervention?

We can respond to climate change in two fundamental ways: mitigation or adaptation. A third, which many climate scientists are reluctant to consider, is direct intervention in the global climate system.

For **mitigation**, the aim is to mitigate, or alleviate, the problem. In this case we would choose to pursue actions that reduce greenhouse gas emissions in order to lessen the severity of future climate change. For example, mitigation strategies could focus on reducing greenhouse gas emissions by improving energy efficiency, switching to clean and renewable energy sources, encouraging farm practices that protect soil quality, recovering landfill gas, and preventing deforestation. Unfortunately, as a global community we have failed to meet the challenge of mitigation. Many climate experts believe that, although we shouldn't abandon our efforts, it is likely too late for mitigation to be our only approach to dealing with climate change.

The second type of response is to accept that climate change is happening and to pursue strategies to minimize its impacts on us. This strategy is called **adaptation** because the goal is to adapt to change by finding ways to cushion oneself from its blows. Erecting a seawall, like Maldives residents did with the Great Wall of Male, is an example of adaptation using technology and engineering. The people of Tuvalu also adapted, but with a behavioural choice—some chose to leave their island and make a new life in New Zealand. Other examples of adaptation include restricting coastal development; adjusting farming practices to cope with drought; and modifying water management practices to deal with reduced river flows, glacial outburst floods, or salt contamination of ground water.

James Ford from McGill University, along with other Canadian researchers, has studied Inuit adaptation to climate change in Canada's North. Ford is a geographer, with an interest in global environmental change and natural hazards. He focuses his research specifically on the Arctic, where his doctoral thesis work investigated the vulnerability of a particular community in Nunavut to the effects of climate change. This included, for example, working with hunters to determine how they have been affected by changes in the extent and thickness of sea ice and in the migratory behaviour of wildlife.[19] This research has contributed to adaptation plans aimed at reducing the vulnerability of communities in Nunavut and elsewhere in the Arctic to the effects of climate change.

The third category of possible responses, **intervention**, is one that you don't hear as much about. This refers to a set of possible large-scale technological modifications to the global climate system (**FIGURE 14.30**). This option makes climate scientists extremely nervous. For one thing, interfering with global-scale Earth systems—especially a system as complex and riddled with feedbacks as the climate system—could lead to all sorts of unanticipated and unwanted side effects. And scientists fear that even the suggestion that the climate system could be engineered to "fix" human-induced climate change would cause people to abandon efforts at mitigation.

However, intervention strategies have been quietly discussed for years, and some scientists feel that we need to begin discussing them more seriously and openly, so that we can be ready if the day ever comes when these strategies are needed. Grand-scale climate intervention strategies, often referred to as **geoengineering**, fall into two main categories: (1) strategies that aim to increase fluxes of greenhouse gases out of the atmosphere, and (2) strategies that aim to reduce the incoming solar radiation.

An example of the first set of strategies is *ocean fertilization*. We know that the ocean serves as a sink for carbon dioxide. It has been hypothesized that adding iron to ocean water would boost the primary productivity of oceanic phytoplankton, increasing their ability to intake carbon dioxide through photosynthesis and thus enhancing the ability of the ocean to act as a sink for atmospheric carbon dioxide. Experimentation around this concept remains inconclusive. Similar approaches that could be applied on land include the planting of billions of trees to act as a carbon sink and genetic modification of plants to become more active photosynthesizers.

The second set of strategies revolves around technologies that aim to increase the albedo of Earth's atmosphere, so that a greater proportion of incoming solar radiation could be reflected. Huge rotatable mirrors orbiting in fixed position above Earth's surface have been proposed, for example. Another proposition is to alter the composition of Earth's atmosphere by injecting a mist of water or sulphur aerosols to increase atmospheric reflectivity. Sulphur aerosols would have the additional unwanted side effect of causing acidic precipitation, and would certainly do nothing to help ease the problem of ocean acidification.

Intervention through geoengineering remains hypothetical. Many environmental advocates have criticized intervention and even adaptation, because they view them as escapist—a way of sidestepping the hard work of mitigation that must be done to protect future generations from climate change. However, adaptation and mitigation are not mutually exclusive approaches; a person or a nation can pursue both. Indeed, both approaches are necessary.

Adaptation strategies will be needed. Even if we were to halt all our emissions now, global warming would continue until the planet's systems reached a new equilibrium, with

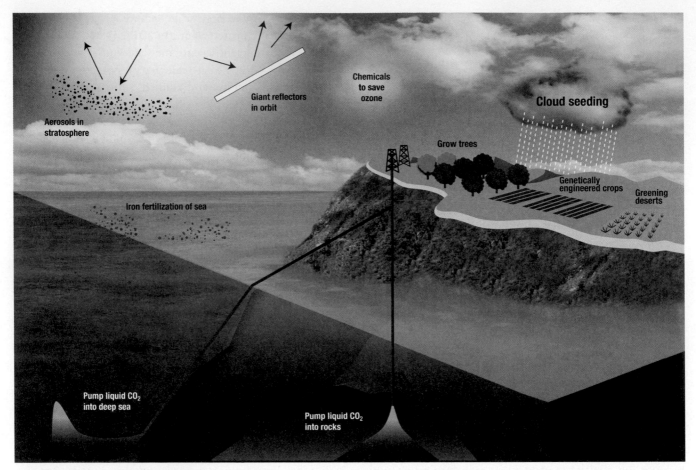

FIGURE 14.30 Intervention strategies involve grand-scale geoengineering of the climate system. Proposals have included putting huge reflecting mirrors in orbit, or fertilizing the ocean surface with iron to enhance photosynthesis in oceanic phytoplankton.
Source: From Geoengineering Could Slow Down the Global Water Cycle, News Release NR-08-05-04, May 27, 2008. https://www.llnl.gov/news/ geoengineering-could-slow-down-global-water-cycle. Accessed 13 March 2015. Copyright © by Lawrence Livermore National Laboratory. Reprinted by permission.

temperature rising an additional degree, at least, and considerably more in such areas as Canada's North, by the end of this century. Because we will face this change no matter what we do, it is wise to develop ways to minimize its impacts.

Mitigation is also necessary; we have to keep trying to mitigate the impacts of climate change, in spite of past failures. If we do nothing to slow climate change, it will eventually overwhelm any efforts at adaptation we could make. To leave a sustainable future for our civilization and to safeguard the living planet that we know, we will need to pursue mitigation. Moreover, the faster we begin reducing our emissions, the lower the level at which they will peak, and the less we will alter climate. We will spend the remainder of this chapter examining approaches to the mitigation of climate change.

We should look more closely at our lifestyle

In Canada, major sources of GHG emissions include power generation facilities that produce electricity, heat,

or steam by using fossil fuels; oil and gas extraction; mining, smelting, and refining of metals, and steel production; pulp, paper, and saw mills; petroleum refineries; and chemical producers. Other activities that contribute significantly to GHG emissions include transportation, waste disposal, and agriculture-related activities. Some of these come from large-scale industrial sources, but many can be influenced by the simple choices we all make in our everyday lives.

The generation of electricity is one significant source of carbon dioxide emissions. From cooking and heating to the clothes we wear, much of what we own and do depends on electricity. Although Canada's electricity comes mainly from hydroelectric sources (58%), 23% still comes from fossil-fuel-related processes. Reducing the volume of fossil fuels we burn to generate electricity would lessen greenhouse gas emissions, as would decreasing electricity consumption.

Conservation and efficiency in energy use can arise from new technologies, such as high-efficiency light bulbs and appliances, or from individual ethical choices to reduce electricity consumption. For instance, replacing an old washing machine with an energy-efficient washer can cut your CO_2 emissions by 200 kg annually. Replacing standard light bulbs with compact fluorescent lights reduces energy use for lighting by 40%. Such technological solutions are popular, and they can be profitable for manufacturers while also saving consumers money.

Consumers can also opt for lifestyle choices. For nearly all of human history, people managed without the electrical appliances that most of us take for granted today. It is possible for each of us to choose to use fewer greenhouse gas–producing appliances and technologies and to take practical steps to use electricity more efficiently.

We can reduce greenhouse gas emissions by altering the types of energy we use. Among fossil fuels, natural gas burns more cleanly than oil, and oil is cleaner-burning than coal. Using natural gas instead of coal produces the same amount of energy with roughly one-half the emissions. Moreover, approaches to boosting the efficiency of fossil fuel use, such as cogeneration, produce fewer emissions per unit of energy generated. We will examine fossil fuels and their origin, occurrence, use, and impacts in Chapter 15, *Fossil Fuels: Energy Use and Impacts*.

Currently, interest in carbon capture, sequestration, and storage is intensifying. **Carbon capture and storage (CCS)** refers to technologies or approaches that remove carbon dioxide from power plant emissions. Successful carbon capture technology would allow power plants to continue using fossil fuels while cutting greenhouse gas pollution. The carbon would then be stored somewhere— perhaps underground, under pressure, in locations where it would not seep out. However, we are still a long way from developing adequate technology and secure storage space to accomplish this, and some experts doubt that we will ever be able to sequester enough carbon to make a dent in our emissions.

Technologies and energy sources that generate electricity without using fossil fuels represent another means of reducing greenhouse gas emissions. These include nuclear power, hydroelectric power, geothermal energy, photovoltaic cells, wind power, and ocean energy sources. These energy sources give off no emissions during their use (but some in the production of their infrastructure). We will examine these clean and renewable energy sources in much greater detail in Chapter 16, *Energy Alternatives*.

Transportation is a significant source of greenhouse gases

The typical automobile is highly inefficient. Close to 85% of the fuel you pump into your gas tank does something other than move your car down the road (**FIGURE 14.31**). Although more aerodynamic designs, increased engine efficiency, and improved tire design could help reduce these losses, gasoline-fuelled automobiles may always remain somewhat inefficient.

The technology exists to make our vehicles more fuel-efficient than they currently are. Raising fuel efficiency will require government mandate and/or consumer demand. As gasoline prices rise, demand for more fuel-efficient automobiles will intensify. Advances in technology are also bringing us alternatives to the traditional combustion-engine automobile. These include hybrid vehicles that combine electric motors and gasoline-powered engines for greater efficiency. They also include fully electric vehicles, alternative fuels, such as compressed natural gas and biodiesel, and hydrogen fuel cells.

People can also opt to make lifestyle choices that reduce their reliance on cars. For example, some people are choosing to live nearer to their workplaces. Others use mass transit, such as buses, subway trains, and light rail. Still others bike or walk to work or for their errands (**FIGURE 14.32**). Canadians use public transportation for approximately 7% of daily transportation needs.

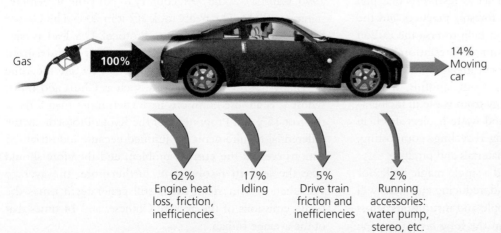

Gas 100% 14% Moving car

62% Engine heat loss, friction, inefficiencies

17% Idling

5% Drive train friction and inefficiencies

2% Running accessories: water pump, stereo, etc.

FIGURE 14.31
Conventional automobiles are extremely inefficient. Almost 85% of useful energy is lost, and only 14% actually moves the car down the road.

FIGURE 14.32
By choosing human-powered transportation methods, such as bicycles, we can greatly reduce our transportation-related greenhouse gas emissions. More people are choosing to live closer to their workplaces and to enjoy the dual benefits of exercise and reduced emissions by walking or cycling to work or school.

Increasing this level of use would be an extremely efficient way to reduce energy use and pollution. Unfortunately, reliable and convenient public transit is not available in many communities. Making automobile-based cities and suburbs more friendly to pedestrian and bicycle traffic and improving people's access to public transportation are central challenges for city and regional planners.

We will need to follow multiple strategies to reduce emissions

Other pathways toward mitigating climate change include advances in agriculture, forestry, and waste management. In agriculture, sustainable land management that protects the integrity of soil on cropland and rangeland enables soil to store more carbon. Techniques have also been developed to reduce the emission of methane from rice cultivation and from cattle and their manure, and to reduce nitrous oxide emissions from fertilizers.

In forest management, the rapid reforestation of cleared areas helps restore forests, which act as reservoirs that pull carbon from the air. Sustainable forestry practices and the preservation of existing forests can help reverse the carbon dioxide emissions resulting from deforestation. Waste managers are doing their part to cut greenhouse emissions by recovering methane seeping from landfills, treating waste water, and generating energy from waste in incinerators. Individuals, communities, and waste haulers also help reduce emissions by encouraging recycling, composting, and the reduction and reuse of materials and products.

We should not expect to find a single magic bullet for mitigating climate change. Instead, reducing emissions will require many steps by many people and institutions across many sectors of our economy. In the long term, any one approach will not be enough. To mitigate climate change, we will need to reduce emissions (as opposed to stabilizing them). How quickly and successfully we translate science and technology into practical solutions for reducing emissions depends largely on the policies we urge our leaders to pursue and on how government and the market economy interact. Governmental command-and-control policy has been vital in safeguarding environmental quality and promoting human well-being. However, government mandates are often resisted by industry, and market incentives can sometimes be more effective in driving change.

With climate change policy, we are in the midst of a dynamic period of debate and experimentation. At all levels—international, national, provincial/territorial, regional, and local—policymakers, industry, commerce, and citizens are searching for ways to employ government and the market to reduce emissions in ways that are fair, economically palatable, effective, and enforceable.

We have tried to tackle climate change by international treaty

In 1992, the United Nations convened the UN Conference on Environment and Development Earth Summit in Rio de Janeiro, Brazil. Nations represented at the Earth Summit signed the UN Framework Convention on Climate Change (FCCC), which outlined a plan for reducing greenhouse gas emissions to 1990 levels by the year 2000 through a voluntary, nation-by-nation approach.

By the late 1990s, it was already clear that the voluntary approach was not likely to succeed. After watching the seas rise and observing the failure of most industrialized nations to cut their emissions, nations of the developing world—the Maldives among them—helped initiate an effort to create a binding international treaty that would require all signatory nations to reduce their emissions. This effort led to the **Kyoto Protocol**, an outgrowth of the FCCC that required signatory nations, by the period 2012, to reduce emissions of six greenhouse gases to levels below those of 1990. Canada was the 99th country to ratify the agreement, signing in 2002. The treaty took effect in 2005. The United States refused to ratify the Kyoto Protocol. U.S. leaders have called the treaty unfair because it requires industrialized nations to reduce emissions but does not require the same of rapidly industrializing nations, such as China and India, whose greenhouse emissions have risen more than 50% in the past 15 years. Proponents of the Kyoto Protocol say the differential requirements are justified because industrialized nations created the current problem and therefore should take the lead in resolving it; furthermore, the average Canadian or American today still generates 3 times the GHG emissions of the average Chinese, and 14 times that of the average Indian.

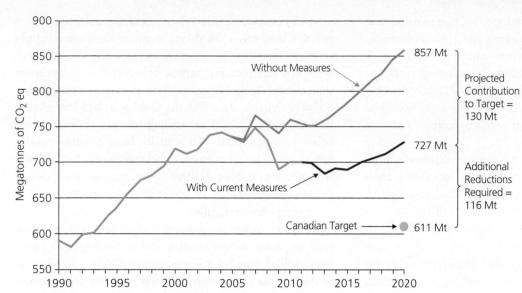

FIGURE 14.33

According to this projection made by Environment Canada, if no action from governments, consumers and businesses had been are taken to reduce emissions since 2005, GHG emissions would be 857 Mt by 2020.
Source: © Her Majesty the Queen in Right of Canada, as represented by the Minister of the Environment, 2014. The Environment Canada data is available online, at no cost, by visiting http://www.ec.gc.ca.

Because resource use and per capita emissions are far greater in the industrialized world, governments and industries there often feel they have more to lose economically from restrictions on emissions. Ironically, industrialized nations are also the ones most likely to gain economically, because they are best positioned to invent, develop, and market new technologies to power the world in a post–fossil fuel era. The Kyoto Protocol relied upon the *principle of common but differentiated responsibility*, stating that all nations of the world have a responsibility to address issues related to climate change, but not all nations have the same capacity to do this, whether financially, politically, or technologically.

As a signatory to the Kyoto Protocol, Canada was obliged to submit an inventory report of its GHG emissions on an annual basis, using an internationally agreed-to format. By 2009, however, it had become clear that Canada was not going to meet its mandated targets. Canada is not just off-target; it would take radical technological and economic changes to fulfill the commitment we made, as a nation, in signing the Kyoto Protocol. At the 15th Conference of the Parties (COP) of the Kyoto Protocol in Copenhagen in 2009, Canada received the so-called "Dinosaur Award" and was accused of obstructing global progress toward a solution for climate change, opting instead to protect the interests of the fossil fuel industry.

Then, in 2011 at the 17th COP in Durban, South Africa, Canada became the first nation to withdraw from the Kyoto agreement. Environment Minister Peter Kent cited Canada's liability to incur large financial penalties for failure to reach emission targets as the reason for the withdrawal. Canada then became a signatory to the Copenhagen Accord, under which we committed to reducing our GHG emissions to 17% below the 2005 level by the year 2020 (**FIGURE 14.33**).

For its part, the United States emits fully one-fifth of the world's greenhouse gases, so its refusal to join international efforts to curb greenhouse emissions generated widespread resentment and has undercut the effectiveness of global efforts. At the 2007 COP in Bali, Indonesia, the delegate from Papua New Guinea drew thunderous applause and cheers when he requested of the U.S. delegation, "If for some reason you are not willing to lead, please get out of the way."

The Kyoto Protocol was never a perfect instrument for dealing with climate change, but climate change activists did not lose hope that a replacement agreement could be implemented. In December of 2015, after a long period during which global public opinion seemed to undergo a noticeable shift in support of international action on climate change, a historic accord was reached among delegates from 195 nations and the European Union at COP-21. This accord quickly became known as the Paris Agreement. Both Canada and the United States were active participants in the negotiations.

The Paris Agreement cites the nations' collaborative intention to maintain an increase in global average surface temperature of less than 2° C above pre-industrial levels, to reach a peak in GHG emissions "as soon as possible," to foster climate resilience and prioritize food security, and to work toward effective financing mechanisms for all of these goals.[20]

The agreement is considered to be very ambitious, but it is not legally binding on the participating nations, which has drawn some criticism.

Market-based tools are being used to address climate change

Market mechanisms such as permit trading programs represent a means of harnessing the economic efficiency of the free market to achieve policy goals while allowing business, industry, or utilities flexibility in how they meet those goals. Supporters of permit trading programs argue that they provide the fairest, least expensive, and most effective method of achieving emissions reductions. Polluters get to choose how to reduce their emissions and are given financial incentives for reducing emissions below the legally required amount. We will likely discover how successful these ventures are over the next decade as various carbon trading programs are implemented around the world.

The world's first legally binding emissions trading program for greenhouse gas reduction (operating from 2003 to 2010) was the Chicago Climate Exchange, which had hundreds of member corporations, institutions, and municipalities, mainly in the United States, Canada, and Brazil. The operation was shut down in 2010, after the trading price for carbon credits dropped too low for effective trading. The organization still facilitates carbon exchanges between companies.

The world's largest *cap-and-trade* program is the European Union Emission Trading Scheme, which began on January 1, 2005. All EU member states participate, and each submits for approval a national allocation plan that conforms to the nation's obligations under the Kyoto Protocol. This market got off to a successful start but has seen some bumps along the way as the prices for carbon have fluctuated wildly to keep up with regulatory changes. The organization has now entered its third trading period, which is expected to last until 2020. The caps set for 2020 represent more than a 20% decrease in carbon emissions, compared to 2005 when the program was started.

Emissions trading programs have allowed participants who cannot or will not adequately reduce their own emissions to use carbon offsets instead. A *carbon offset* is a voluntary payment to another entity intended to enable that entity to reduce the greenhouse emissions that one is unable or unwilling to reduce oneself. The payment thus offsets one's own emissions. For instance, a coal-burning power plant could pay a reforestation project to plant trees that will soak up as much carbon as the coal plant emits. Or a university could fund the development of clean and renewable energy projects to make up for fossil fuel energy the university uses.

Carbon offsets are becoming popular among utilities, businesses, universities, governments, and individuals trying to achieve **carbon-neutrality**, a state in which no net carbon is emitted. For time-stressed people with enough wealth, offsets represent a simple and convenient way to reduce emissions without investing in efforts to change one's habits. For example, you can calculate your emissions for travel by car or plane or for your residence using an online tool, and purchase offsets for those emissions. Your money funds renewable energy and efficiency projects, and you can advertise your donation with bumper stickers and decals.

In principle, carbon offsets seem a great idea, but in practice they often fall short. Without rigorous oversight to make sure that the offset money actually accomplishes what it is intended for, carbon offsets risk being no more than a way for wealthy consumers to assuage a guilty conscience. Organizations like the Gold Standard Foundation have published guidelines to help people choose carbon offset programs that will actually have positive results. Efforts to create a transparent and enforceable offset infrastructure are ongoing. If these efforts succeed, then carbon offsets could become an effective and important key to mitigating climate change.

Carbon offsets, emissions trading schemes, national policies, international treaties, and technological innovations will all play roles in mitigating climate change. But in the end the most influential factor may be the collective decisions of millions of regular people. In our everyday lives, each one of us can take steps to approach a carbon-neutral lifestyle by reducing greenhouse emissions that result from our decisions and activities. Just as we each have an ecological footprint, we each have a **carbon footprint** that expresses the amount of carbon we are responsible for emitting.

You can apply many of the strategies discussed in this chapter in your everyday life—from deciding where to live and how to get to work to choosing appliances. You will encounter still more solutions in our discussions of energy sources, conservation, and renewable energy in subsequent chapters.

Conclusion

Many factors influence Earth's climate, and human activities have come to play a major role. Climate change is well underway, and further greenhouse gas emissions will increase global warming and cause increasingly severe and diverse impacts. Sea-level rise and other consequences of global climate change will affect locations worldwide from the Maldives to Bangladesh to Ellesmere Island to the Canadian Prairies. As scientists and policymakers come to better understand anthropogenic climate change and its environmental, economic, and social consequences, more and more of them are urging immediate action.

Global climate change is the biggest environmental challenge facing us and our children. It is still early enough that we may be able to avert the most severe impacts. Taking immediate, resolute action to reduce greenhouse gas emissions and mitigate climate change is one of the most important things that we, personally and as a society, can do. Hopefully, the Paris Agreement will propel us along that path.

REVIEWING OBJECTIVES

You should now be able to

Describe Earth's climate system and the many interacting factors that influence the atmosphere and climate, including human factors

- Earth's climate changes naturally over time, but it is now changing rapidly because of human influence.
- The Sun provides most of Earth's energy and interacts with the atmosphere, land, and oceans to drive climate processes. Earth absorbs about 70% of incoming solar radiation and reflects about 30% back into space.
- Greenhouse gases, such as carbon dioxide, methane, water vapour, nitrous oxide, ozone, and halocarbons, warm the atmosphere by absorbing infrared radiation and re-emitting infrared radiation of different wavelengths.
- Carbon dioxide (CO_2) is not the most potent greenhouse gas, but it is the most problematic for several reasons.
- By burning fossil fuels, clearing forests, and manufacturing cement, among other activities, humans are increasing atmospheric concentrations of many greenhouse gases. Carbon is released from long-term sequestration as a result of the burning of fossil fuels.
- Other greenhouse gases also affect the climate system, including methane (CH_4), nitrous oxide (N_2O), and various wholly anthropogenic chemicals, such as CFCs.
- The climate system is characterized by complexity, multiple interacting parts, and both positive and negative feedbacks.
- Positive radiative forcing warms the surface; negative forcing cools the surface. Both natural and human processes, such as the release of aerosols, can contribute to climate forcing.
- Atmospheric chemistry is not the only influence on our climate system. Other factors are orbital characteristics, resulting in Milankovitch cycles; variations in solar output; ocean absorption of greenhouse gases; and oceanic thermohaline circulation.

Summarize modern approaches to climate research

- Proxy indicators from sediments, ice cores, shells, and other materials reveal information about past climatic conditions.
- Stable isotope analysis of ice cores has proven to be a very powerful tool for paleoclimate studies.
- Direct atmospheric sampling tells us about the current composition of the atmosphere.
- Coupled general circulation models serve to predict future changes in climate.

Outline current and future trends and impacts of global climate change

- The IPCC has comprehensively synthesized current climate research, and its periodic reports represent the strong consensus of the scientific community.
- Temperatures on Earth have warmed by an average of 0.85°C over the past century and are predicted to rise even more over the next century.
- Changes in precipitation and storm activity will also result from climatic warming, and will vary by region. Increase in the frequency and intensity of storms is probable.
- Extreme weather events are becoming more frequent, likely in part because of alterations in the polar vortex and the jet stream.
- Warming has resulted in significant impacts on snow and ice cover, leading to the loss of mass from glaciers and the entire ice inventory. The response of the Greenland and Antarctic ice sheets to warming is complicated, but they have lost mass overall and will likely continue to do so.
- The effects of climatic change have been, and will likely continue to be, the most pronounced in the Arctic, including the melting of permafrost and thinning of sea ice.
- Global sea level has risen by approximately 19 cm over the past century. The main causes of sea-level rise are thermal expansion of sea water; melting of glaciers; and melting of the polar ice sheets.
- Climate change is affecting plants, animals, and ecosystems, both terrestrial and marine. Oceanic warming and acidification will likely have a negative effect on corals and other marine organisms.
- The socioeconomic effects of climate change will include impacts on agriculture and food security, forestry, human health, and economic well-being.
- Despite some remaining uncertainties, the scientific community feels that evidence for humans' role in influencing climate is strong enough to justify governments' taking action to reduce greenhouse emissions.

Suggest ways we can respond to climate change

- Both adaptation and mitigation are necessary for responding to climate change. A much less desirable option, but one that we may need to consider one day, is direct intervention in the global climate system.
- Conserving electricity, improving efficiency of energy use, and switching to clean and renewable energy

sources will help reduce fossil fuel consumption and greenhouse emissions.

■ Encouraging new automotive technologies and investment in public transportation will help reduce greenhouse emissions.

■ Solving the climate problem will require the deployment of multiple strategies.

■ The Kyoto Protocol provided a first step for nations to begin addressing climate change, but has proven largely ineffective. In 2011, just prior to the deadline for meeting greenhouse gas reduction targets, Canada became the first nation to withdraw from the international agreement.

■ Emissions trading programs are providing a way to harness the free market and engage industry in reducing emissions. Individuals are increasingly exploring carbon offsets and other means of reducing personal carbon footprints.

TESTING YOUR COMPREHENSION

1. What happens to solar radiation after it reaches Earth? How do greenhouse gases warm the lower atmosphere?

2. Why is carbon dioxide considered the main greenhouse gas? How could an increase in water vapour create either a positive or negative feedback effect?

3. How do scientists study the ancient atmosphere?

4. Has simulating climate change with computer programs been effective in helping us predict climate? How do these programs work?

5. List five major trends in climate that scientists have documented so far. Now list five future trends or impacts that they are predicting.

6. Describe how rising sea levels, caused by global warming, can create problems for people. How may climate change affect marine ecosystems?

7. How might a warmer climate affect agriculture? How is it affecting distributions of plants and animals? How might it affect human health?

8. What are the main sources of greenhouse gas emissions in Canada? In what ways can we reduce these emissions?

9. What roles have international treaties played in addressing climate change? Give two specific examples.

10. Describe one market-based approach for reducing greenhouse emissions. Explain one reason it may work well and one reason it may not work well.

THINKING IT THROUGH

1. Some people argue that we need "more proof" or "better science" before we commit to substantial changes in our energy economy. How much science, or certainty, do you think we need before we should take action regarding climate change? How much certainty do you need in your own life before you make a major decision? Should nations and elected officials follow a different standard? Do you believe that the precautionary principle is an appropriate standard in the case of global climate change? Why or why not?

2. Transportation is a big contributor to greenhouse gas emissions in industrialized nations, and especially in Canada, with our enormous geographical extent. Describe several ways in which we can reduce greenhouse gas emissions from transportation. Which approach do you think is most realistic, which approach do you think is least realistic, and why?

3. Imagine that you would like to make your own lifestyle carbon-neutral and that you aim to begin by reducing the emissions you are responsible for by 25%. What actions would you take first to achieve this reduction?

4. You have just been elected premier of a province. Polls show that the public wants you to take bold action to reduce greenhouse gas emissions. However, polls also show that the public does not want prices of gasoline or electricity to rise more than they already have. Carbon-emitting industries in your province are wary of emissions reductions being required of them but are willing to explore ideas with you. Your parliament will support you in your efforts as long as you remain popular with voters. The province to the west has just passed ambitious legislation mandating steep greenhouse gas emissions reductions. The province to the east has joined a new regional emissions-trading consortium. What actions will you take in your first year as premier?

INTERPRETING GRAPHS AND DATA

Have another look at **FIGURE 14.33** on page 457. The graph illustrates Canada's GHG emissions (in mega-tonnes (Mt) of CO_2 equivalent) between 1990 and 2012 (the latest information available as of mid-2015 from Environment Canada[21]), with projections to 2020. From the information provided on the graph, answer the following questions.

1. Calculate the approximate percentage change in GHG emissions between 1990 and 2012.

2. Between 1990 and 2012, the population of Canada increased from 27.7 million to 34.8 million and the inflation-adjusted gross domestic product (GDP) went from $989 billion to $1654 billion (inflation-adjusted, reported in 2007 dollars). Calculate the per capita GDP for 1990 and 2012, and then calculate the percentage increases in both population and GDP per capita over this time period.

3. What quantitative conclusions can you draw from these data about GHG emissions per capita? About GHG emissions per unit of total economic activity? Create a graph and sketch a trend line of GHG emissions per capita from 1990 to 2012. Now sketch a trend line of GHG emissions per unit of total economic activity from 1990 to 2012. Describe your graph, and speculate on its implications for Canada's economic and environmental future.

4. The Copenhagen target for CO_2 reduction, shown on the graph, is 17% below the 2005 level. The Kyoto target for Canada was 6% below the 1990 emissions level; what would Canada's Kyoto target have been? How far off was Canada from meeting this goal in 2012, before pulling out of the Kyoto agreement?

MasteringEnvironmentalScience®

STUDENTS
Go to **MasteringEnvironmentalScience** for assignments, the eText, and the Study Area with practice tests, videos, current events, and activities.

INSTRUCTORS
Go to **MasteringEnvironmentalScience** for automatically graded activities, current events, videos, and reading questions that you can assign to your students, plus Instructor Resources.

15 Fossil Fuels: Energy Use and Impacts

Caribou migrate across the Mackenzie River delta in the Northwest Territories.

Terry A. Parker/All Canada Photos/Corbis

Upon successfully completing this chapter, you will be able to

- Summarize the principal energy sources that we use and the role of fossil fuels in our energy mix
- Describe the nature and origin of coal and evaluate its extraction and use
- Describe the nature and origin of natural gas and evaluate its extraction and use
- Describe the nature and origin of oil and evaluate its extraction, use, and future availability

- Describe the nature, origin, and potential of unconventional fossil fuel types and technologies
- Outline and assess the environmental impacts of continued fossil fuel dependency
- Evaluate political, social, and economic impacts of fossil fuel use, and strategies for conserving energy and enhancing efficiency

The unspoiled natural environment and wildlife habitats of the Mackenzie Delta would be affected if a natural gas pipeline were constructed through the Mackenzie Valley.

Arctic Ocean

Greenland (DENMARK)

Mackenzie Delta

Northwest Territories

Hudson Bay

CANADA

CENTRAL CASE
ON, OFF, ON AGAIN? THE MACKENZIE VALLEY NATURAL GAS PIPELINE

"We've embarked on the beginning of the last days of the age of oil."

—DR. MICHAEL R. BOWLIN, FORMER CEO, ATLANTIC RICHFIELD COMPANY (ARCO)

"I listened to a brief by northern businessmen in Yellowknife who favour a pipeline through the North. Later, in a native village far away, I heard virtually the whole community express vehement opposition to such a pipeline. Both were talking about the same pipeline; both were talking about the same region—but for one group it is a frontier, for the other a homeland."

—JUSTICE THOMAS BERGER, *NORTHERN FRONTIER, NORTHERN HOMELAND*, 1977

The Mackenzie Valley Gas Project is a proposal to develop three major natural gas fields located in the Mackenzie Delta, Northwest Territories. The natural gas would be delivered to southern markets via a 1196-km pipeline system, to be constructed along the Mackenzie Valley from Inuvik, NWT, to the Alberta border, where it would connect with the TransCanada Pipeline system (see **FIGURE 1**).[1] The idea of a major pipeline running from the Beaufort Sea down the Mackenzie Valley is not new. A pipeline was proposed and examined seriously in the early 1970s; it was called, at the time, the "biggest private construction project in history."[2]

In 1974 the federal government appointed Justice Thomas Berger to carry out an inquiry into the potential impacts of the project on Canada's North, since the proposed route crosses four Aboriginal regions. Justice Berger travelled throughout the North, meeting with Dene, Inuit, Métis, and white residents and leaders. The report of the inquiry (released in two volumes in 1977) was called *Northern Frontier, Northern Homeland*.

FIGURE 1
These maps show the location of the proposed Mackenzie River valley natural gas pipeline. The Mackenzie Basin is outlined in red on the lower map.

It recommended that pipeline development in the North be delayed by at least 10 years because of deep opposition by native leaders and the potential for negative impacts. The report further recommended that such development not proceed until native land claims in the area had been successfully resolved.

Among the main concerns were the potential impacts on people and animals that would result from the infrastructure (roads, airports, towns) likely to accompany the construction of the pipeline. As reported by CBC News at the time, "Some dismissed the impact of a pipeline, saying it would be like a thread stretched across a football field. Those close to the land said the impact would be more like a razor slash across the Mona Lisa."[3]

In the end, the pipeline project has been delayed for much longer than 10 years. The idea was on-again, off-again for many years, but was revitalized at the beginning of the twenty-first century. The partners in the current project (as of 2015) are four major Canadian oil and gas companies—Imperial Oil Resources Ventures, ConocoPhillips Canada, Shell Canada, and ExxonMobil Canada, known collectively as the "Producer Group," and the Aboriginal Pipeline Group (APG), representing the interests of Aboriginal people in the project.[4]

Interestingly, many of the native leaders who vehemently opposed the pipeline in the 1970s eventually became supporters, realizing that it would bring much-needed jobs and revenue to the North. Chief Frank T'Seleie of Fort Good Hope, Northwest Territories, had been a vocal opponent of the project as a young man in the 1970s. More recently he has supported the pipeline project, since many of the land claims in the area have been settled. He supports the right of native people to benefit from development on their own land, and has even participated in project negotiations on behalf of native communities.

The project has been no less controversial in its more recent go-around than it was in the 1970s. Some native groups now support the project; others, notably some Dene from the Deh Cho region, still vow to stop the project, citing the potential for negative social impacts. Environmentalists have stressed the potential for the pipeline to fragment habitat for bears, caribou, and wolves; harm fish and fish habitat by increasing sediment deposition into rivers; damage breeding areas for geese, swans, and other migratory birds in the tundra; require forests to be cut and heavy machinery deployed for infrastructure; trigger a rush of oil and gas development in the Mackenzie Valley (this was also predicted by the Berger report); and increase greenhouse gas emissions. It remains to be proven whether these impacts would occur, and to what degree.

An interesting side issue concerns the ultimate fate of the gas itself. Much of the natural gas produced in the Mackenzie Delta gas fields may never make it to the southern Canadian and American consumers for whom it was originally intended. This is because of the enormous acceleration in development of oil sand (or tar sand) deposits in Alberta. Production of oil from these deposits requires significant inputs of energy, and complete, successful development is dependent on the production and delivery of natural gas from the North. The first 20 years' worth of natural gas from the Mackenzie Delta would potentially go straight into the production of oil from Alberta's oil sands.

Environmental and social justice groups have argued that Imperial Oil and its partners would earn billions of dollars from the project and should not receive federal subsidization. However, principal partner Imperial Oil claims that the project is only marginally economic and will not proceed without assurance of at least $1.2 billion from the federal government. The cost of the project has risen dramatically, market prices

are increasingly unfavourable, and the scheduled start date for production was delayed until at least the end of 2015. As of mid-2015, the proponents are "hopeful" and "continuing to monitor" the situation, but it looks increasingly unlikely that the project will go ahead in the near future.[5]

The Mackenzie pipeline would transport natural gas, but pipelines are also used to transport oil. Oil pipelines are subject to as much controversy as gas pipelines, primarily because of the risk of spills. Leaks and spills of oil being transported by pipeline from Alberta have contributed to heated controversies, both in Canada and in the United States, surrounding the Keystone XL and Northern Gateway pipelines. We will examine these controversies more closely in this chapter.

Sources of Energy

Humanity has devised many ways to harness the renewable and nonrenewable forms of energy available on our planet. We use a wide variety of energy sources to heat and light our homes, power our machinery, fuel our vehicles, and provide the comforts and conveniences to which we've grown accustomed in the modern industrial age.

We use a variety of energy sources

A great deal of energy emanates from Earth's core, making geothermal power available for our use. A transfer of energy also results from the gravitational pull of the Moon and Sun, and we are just beginning to harness the power from the ocean tides that these forces generate. An immense amount of energy resides within the bonds among protons and neutrons in atoms, and this energy, when released in controlled circumstances, provides us with nuclear power.

Most of our energy, however, comes ultimately from the Sun. We can harness solar energy from the Sun's radiation directly in a number of ways. Solar radiation also helps drive atmospheric winds and the hydrologic cycle, making possible such forms of energy as wind power and hydroelectric power. And of course, sunlight drives photosynthesis and the growth of plants, from which we take fuelwood and other biomass as fuel sources. Ultimately, when organisms die and are preserved in sediment under particular conditions, the chemical energy stored in their tissues may be transferred into the form of **fossil fuels**, highly combustible substances formed from the remains of

organisms from past geological ages. The three fossil fuels we use most widely today are oil, coal, and natural gas.

Since the Industrial Revolution, fossil fuels have replaced traditional biomass (like fuelwood) as our society's dominant source of energy. Global consumption of the three main fossil fuels has risen steadily for years and is now at its highest level ever (**FIGURE 15.1A**). The high energy content of fossil fuels makes them efficient to burn, ship, and store. Besides providing for transportation, heating, and cooking, these fuels are used to generate electricity, a secondary form of energy that is easier to transfer over long distances and apply to a variety of uses.

Canada's energy stream is complex (**FIGURE 15.1B**). We collect the primary materials, whether by mining, drilling, harvesting (e.g., fuelwood), or otherwise capturing or storing (e.g., solar, wind, and hydro). We extract energy from these sources through a very wide variety of chemical and physical transformations and capture technologies. We utilize energy in different forms for different purposes in the residential, commercial, industrial, and transportation sectors. We also both import and export energy—in some cases, importing and exporting different grades of the same form of energy (such as coal). But overall, Canada is a net exporter of energy.

Energy sources such as sunlight, geothermal energy, and tidal energy are considered perpetually **renewable** or even inexhaustible, because their supplies are not depleted by our use. Other sources, such as timber for fuelwood, are renewable only if we do not harvest them at too great a rate. In contrast, such energy sources as oil, coal, and natural gas are considered **nonrenewable** because they cannot be regenerated quickly enough to offset depletion. Nuclear power as currently

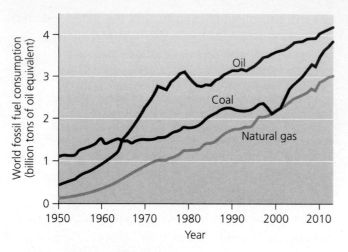

(a) Global consumption of fossil fuels

FIGURE 15.1

Global consumption of fossil fuels **(a)** has risen greatly over the past half century. Oil use rose steeply during the 1960s to overtake coal, and today it remains our leading energy source. The units are billion tonnes (for oil) or billion tonnes of oil equivalent (for coal and natural gas); the latter is abbreviated Gtoe ("G" or "giga" means "billion"). Tonnes of oil equivalent (toe) is defined as the amount of energy released by burning one tonne of crude oil. This unit allows comparisons among different energy sources. Canada's energy mix **(b)** illustrates our fundamental dependence on fossil fuels, as well as the importance of uranium (for nuclear energy) and hydropower.

Source: Data for (a) from U.S. Energy Information Administration and International Energy Agency and British Petroleum (2014). *Statistical Review of World Energy 2014,* report and statistical tables. Data in (b) from Natural Resources Canada (2006). Schematic (b) is from Natural Resources Canada (2006). *Report of the National Advisory Panel on Sustainable Energy Science and Technology,* Office of Energy Research and Development.

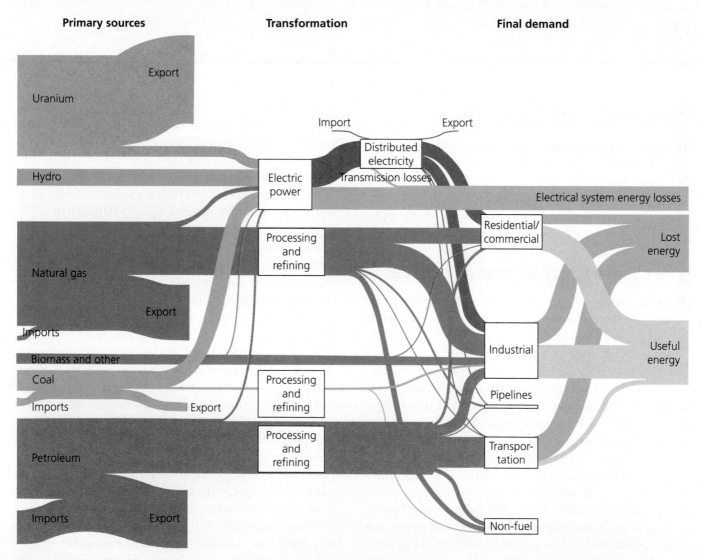

(b) Canada's energy stream

DATA Q By what percentage has the annual use of oil risen since the year your mother or father was born? Since the year when you were born? (As shown on the graph, oil use worldwide as of 2013 was about 4.2 billion tonnes; that is your starting point.)

harnessed through the *fission*, or splitting, of uranium is also nonrenewable, because uranium is a mineral resource, and because fission is an irreversible process.

Although nonrenewable fuels are constantly being generated by ongoing natural geological processes, the timescales on which they are created are so long that, once depleted, they cannot be replaced within any time span useful to our civilization. It takes about a thousand years for the biosphere to generate the amount of organic matter that must be buried to produce a single day's worth of fossil fuels for our society; to turn that material into fossil fuel by natural processes takes at least a million years. For this reason, and because fossil fuel use exerts severe environmental impacts, renewable energy sources increasingly are being developed as alternatives to fossil fuels, as we will see in Chapter 16, *Energy Alternatives*.

Industrialized nations consume more energy than developing nations

Citizens of developed regions generally consume far more energy than do those of developing regions (**FIGURE 15.2**). Per person, the most-industrialized nations use up to 100 times as much energy as do the least-industrialized nations. The United States, with only 4.4% of the world's population, accounts for 18% of the world's energy use.[6] Even so, Canada's per capita energy use is higher; with just 0.5% of the world's population, Canada accounts for 2.6%

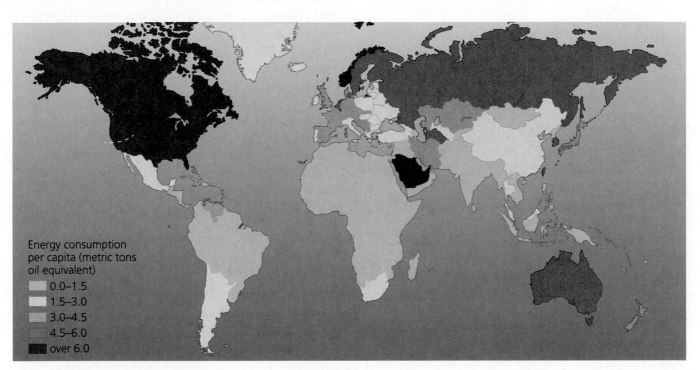

Energy consumption per capita (metric tons oil equivalent)

- 0.0–1.5
- 1.5–3.0
- 3.0–4.5
- 4.5–6.0
- over 6.0

(a) Map of energy consumption per person

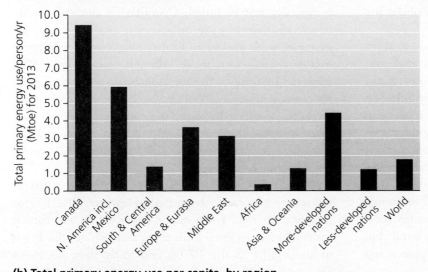

(b) Total primary energy use per capita, by region

FIGURE 15.2

Regions differ greatly in their consumption of energy per person. The map **(a)** and bar graph **(b)** show all types of energy, in Mtoe (million metric tons of oil equivalent), a unit that allows comparisons among different fuel types, as described in **FIGURE 15.1**. The bar graph compares energy use per capita between Canada and other regions of the world. Note that the map and graph include <u>all</u> energy use, not just energy use in the home or workplace.

Source: (a) Data from British Petroleum (2014). *Statistical Review of World Energy 2014,* report and statistical tables. (b) Data from World Population Data Sheet 2013. Population Reference Bureau (2014).

DATA Q How many times more energy (in oil equivalents) is used, per capita per year, in Canada, compared to the average for the world as a whole?

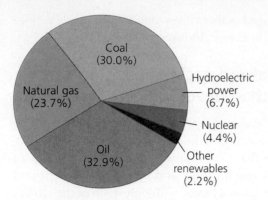

(a) World total energy consumption by fuel type

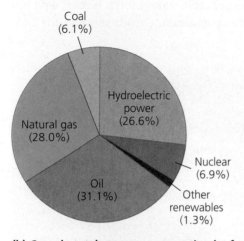

(b) Canada total energy consumption by fuel type

FIGURE 15.3
Fossil fuels dominate energy consumption worldwide **(a)** and in Canada **(b)**, as in other industrialized nations, in spite of the important role of hydroelectric power in Canada. (Contrast this chart with **FIGURE 16.2** in Chapter 16, which looks at energy sources for electricity generation in Canada.)
Source: Data from British Petroleum (2014). *Statistical Review of World Energy 2014,* report and statistical tables.

of the world's primary energy use. This is partly a result of the cold climate, long distances, and energy-intensive industries that characterize our nation, but careless use patterns are part of the story, too.

Developed and developing nations also tend to apportion their energy use differently. More industrialized nations use roughly one-third of their energy on transportation, one-third on industry, and one-third on all other uses (the industrial portion is a bit higher than this in Canada). Developing nations devote a greater proportion of their energy to subsistence activities, such as agriculture, food preparation, and home heating, and substantially less on transportation. In addition, people in developing countries often rely on manual or animal energy sources instead of automated ones. For instance, rice farmers in Bali plant and harvest rice by hand, but industrial rice growers in California use airplanes and mechanized harvesters that require fuel.

In Canada, where hydroelectric resources are particularly abundant, oil, coal, and natural gas still supply 65% of energy needs (**FIGURE 15.3**). Coal is much more important globally than in Canada, however, providing almost one-third of the world's total energy consumption.

It takes energy to make energy

We don't simply get energy for free. To harness, extract, process, and deliver the energy that we use requires that we invest substantial inputs of energy. Extracting coal, oil, and natural gas requires the construction of an immense infrastructure of roads, mines, wells, vehicles, storage tanks, pipelines, housing for workers, and more—all of which necessitate the use of energy. Piping and shipping the raw products for processing and subsequent delivery to the market for use require further energy inputs. Thus, when evaluating an energy source, it is important to subtract costs in energy invested from benefits in energy received.

Net energy expresses the difference between energy returned or acquired (that is, the resulting consumable energy) and energy invested to acquire it. When comparing energy sources, the ratio is often denoted as *EROEI* (*energy returned on energy invested*) or **EROI (energy return on investment)**, which can be defined as follows:

$$\text{EROI} = \text{Usable Energy Returned/Energy Invested}$$
$$= E_R/E_I$$

When EROI $= 1$, it means that the amount of energy invested is the same as the amount of energy extracted. Higher ratios mean that we acquire more usable energy from each unit of energy that we invest.

EROIs are notoriously difficult to calculate, because many assumptions are involved. For example, when calculating the EROI of biomass fuels, the energy involved in the photosynthesis that originally created the biomass is typically not included. Considering the efficiencies of energy-generating technologies is also important; for example, older nuclear and solar technologies yield lower EROIs than newer, more efficient technologies. The expected lifetime of the technology or generating plant also comes into play. And finally, EROI calculations are sometimes criticized because they don't take into account other benefits and drawbacks of different energy sources. Even so, it is useful and revealing to compare EROIs for different energy sources.

Fossil fuels are widely used because their EROI ratios have historically been high. However, EROI ratios change over time. For instance, those for oil and natural gas combined declined from more than 50:1 in the 1950s to about 30:1 in the 1970s, and today they are about 15:1 or less, depending on the specifics of the deposit and extraction technologies. This means that we are only getting back about half as much useful energy now for our energy investment in oil and natural gas, compared to 45 or so years ago.

The EROI ratios for oil and natural gas have declined because we harvested the easiest deposits first, and now must work harder to extract the remaining oil and gas. Oil extracted from bitumen in Canadian oil sands has an EROI ratio of about 3:1. Hydroelectric power has an EROI of 40:1 or even higher, depending on how the calculations are done. Solar panels are approximately 8:1 or 9:1. Extending the calculation over the lifetime of the technology greatly increases the calculated EROI for technologies like solar, which require the biggest energy (and financial) inputs up front.[7]

Fossil fuels are indeed fuels created from fossils

The fossil fuels we burn today in our vehicles, homes, industries, and power plants were formed from the tissues of organisms that lived 100 million to 500 million years ago. The energy these fuels contain came originally from the Sun and was converted to chemical-bond energy as a result of photosynthesis. The chemical energy in the organisms' tissues became concentrated as they decomposed, and the constituent **hydrocarbon** compounds (organic molecules dominated by hydrogen and carbon) were chemically altered. The carbon that was incorporated into plant tissue by photosynthesis was stored, first in the biosphere and later in the geosphere.

Fossil fuels thus are part of the global biogeochemical cycle of carbon. Most organisms, after death, do not end up as part of a coal, gas, or oil deposit. A tree that falls and decays as a rotting log undergoes mostly **aerobic** decomposition; in the presence of air, bacteria and other organisms that use oxygen break down plant and animal remains into simpler carbon-based molecules that are recycled through the ecosystem as part of the short-term carbon cycle.

If the hydrocarbons in the starting organic matter are partially broken down, buried, chemically altered, and stored deep underground, the result is fossil fuels, which represent a long-term reservoir for carbon in the global biogeochemical cycle. Fossil fuels result from decomposition only when the starting organic material is broken down in an **anaerobic** environment, one that has little or no oxygen. Such environments include the bottoms of shallow seas, deep lakes, and swamps (**FIGURE 15.4**).

Over millions of years, organic matter that accumulates at the bottoms of such water bodies undergoes decomposition and deep burial by sediment, eventually resulting

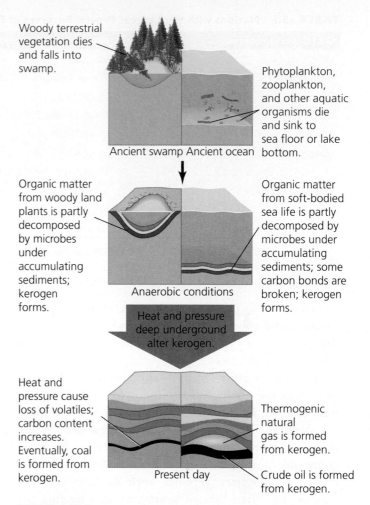

FIGURE 15.4
Fossil fuels begin to form when organisms die and end up in oxygen-poor conditions. This can occur when trees fall into lakes or swamps and are covered by mud, or when phytoplankton and zooplankton drift to the floor of a shallow sea or lake and become buried (**top diagram**). Organic matter that is deeply buried and compressed under layers of sediment undergoes very slow anaerobic decomposition to form kerogen (**middle diagram**). If the starting organic matter was mainly derived from plants, the result may be coal (**bottom diagram, left**). If the starting organic matter was mainly derived from marine phytoplankton and zooplankton, the result may be oil or natural gas (**bottom diagram, right**).

in an oil precursor called **kerogen**. Geothermal heating underground acts on the kerogen, altering it both chemically and physically to form the various fossil fuels. (Natural gas also can be produced nearer the surface by anaerobic bacterial decomposition of organic matter.) The specific type of fuel that forms in any given place is dependent on the chemical composition of the starting organic materials, the temperatures and pressures to which the material is subjected, the presence or absence of anaerobic decomposers, and the passage of time. Coal is formed primarily from the remains of vascular plant matter, whereas oil and

TABLE 15.1   **Nations with the Largest Proven Reserves of Fossil Fuels, End of 2013**

Oil (% world reserves)	Natural gas (% world reserves)	Coal (% world reserves)
Venezuela, 17.7	Iran, 18.2	United States, 26.6
Saudi Arabia, 15.8	Russian Federation, 16.8	Russian Federation, 17.6
Canada, 10.3	Qatar, 13.3	China, 12.8
Iran, 9.3	Turkmenistan, 9.4	Australia, 8.6
Iraq, 8.9	United States, 5.0	India, 6.8
Kuwait, 6.0	Saudi Arabia, 4.4	Germany, 4.5
United Arab Emirates, 5.8	United Arab Emirates, 3.3	Ukraine, 3.8
Russian Federation, 5.5	Venezuela, 3.0	Kazakhstan, 3.8
Libya, 2.9	Nigeria, 2.7	South Africa, 3.4
United States, 2.6	Algeria, 2.4	Indonesia, 3.1

Source: Data from British Petroleum (2014). *Statistical Review of World Energy 2014,* statistical tables.

natural gas are formed from the remains of marine phyto-plankton and zooplankton, including algae and bacteria. The compositions and properties of the various fossil fuels reflect these differences.

Fossil fuels are associated with specific rock types, related to the geological environments in which they were formed. Coal is typically associated with sedimentary deposits that accumulated in terrestrial wetland and shallow lake environments. Oil and natural gas are more commonly found in rock that formed from shallow marine sedimentary deposits, such as limestone; salt deposits are often located nearby, again reflecting the marine origin.

Consequently, deposits of fossil fuels are localized and unevenly distributed over Earth's surface. Some countries (such as Saudi Arabia, China, and Canada) have substantial recoverable reserves, whereas others (such as Japan) have almost none. Almost half of the world's proven reserves of crude oil lie in the Middle East, but Venezuela currently has the greatest proven reserves of any single country. The Middle East is also rich in natural gas, but Russia has significant proven reserves. The Russian Federation is also rich in coal, as is China, but the United States possesses more known coal reserves than any other nation (**TABLE 15.1**). How long each nation's fossil fuel reserves will last depends on how much the nation extracts, how much it consumes, and how much it imports from and exports to other nations.

Coal: The Most Abundant Fossil Fuel

Coal is organic matter (generally woody plant material) that was compressed under very high pressure to form a dense, combustible, carbon-rich solid material. During an

unusually warm global climate about 300 million years ago, swamps and marshes proliferated. In these environments, abundant plants lived, died, and became buried under sediment in anaerobic environments, eventually resulting in the formation of substantial coal deposits throughout the world.

Coal use has a long history

People have used coal longer than any other fossil fuel. The Romans used coal for heating in the second and third centuries in Britain, as have people in parts of China for 2000–3000 years. Native Americans of the Hopi Nation still follow ancestral traditions by using coal to fire pottery, cook food, and heat their homes.

Once commercial mining began in Europe in the 1700s, people began using coal more widely as a heating source. The market expanded after the invention of the steam engine, because coal was used to boil water to produce steam. Coal-fired steam engines helped drive the Industrial Revolution, powering factories, agriculture, trains, and ships. The birth of the steel industry in the late 1850s increased demand still further because coal fuelled the furnaces used to produce steel.

In the 1880s, people began to use coal to generate electricity. In coal-fired electric power plants, still widely used today around the world, coal combustion converts water to steam, which turns a turbine to create electricity. Today hydroelectricity is very important in electricity generation in Canada, but coal still provides more than 20% of the electrical-generating capacity. Canada is both an importer and an exporter of coal, importing some grades and types of coal for various purposes, and exporting others. Globally, China and the United States are the primary producers and consumers of coal (**TABLE 15.2**).

TABLE 15.2 Top Five Producers and Consumers of Coal

Production (% world production)	Consumption (% world consumption)
China, 47.4	China, 50.3
United States, 12.9	United States, 11.9
Australia, 6.9	India, 8.5
Indonesia, 6.7	Japan, 3.4
India, 5.9	Russian Federation, 2.4

Source: Data from British Petroleum (2014). *Statistical Review of World Energy 2014,* statistical tables.

Peat is the first step

The precursor to coal is *peat*, technically a moist soil composed of compressed organic matter (as discussed in Chapter 7, *Soils and Soil Resources*). Peat has been widely used as a fuel and is still used in some parts of the world, including Britain, Ireland, Indonesia, and the Russian Federation. Canada is home to some of the most extensive peat deposits in the world, and 90% of our exported peat goes to the United States. Peat is also of interest as a northern soil, and scientists are closely tracking its response to global climate change.

As peat is subjected to increasing pressure and temperature over time, the water and other volatile constituents are progressively driven off (**FIGURE 15.5**). With compression and heating, the thickness and moisture content of the deposit decrease, the hardness increases, and the colour changes from brown to glossy black. The carbon

content and energy concentration (that is, the caloric content) also increase, and the result is coal. Because of its origin as horizontal layers of organic matter and sediment, coal occurs in the form of extensive flat layers, called *coal seams*.

Peat is a soil, but coal is a rock—a sedimentary rock in the first stages, but technically a metamorphic rock at the highest temperatures and pressures of formation. Scientists categorize coal into several main classes, called *grades*. These include *lignite*, or brown coal, the lowest and softest grade of coal; *bituminous coal*, often used to produce steam for electricity generation; and *anthracite*, a hard, shiny, black metamorphic rock that is the highest grade of coal. Lignite is the least compressed, and anthracite is the most compressed.

Coal is mined at the surface and underground

We extract coal by using two major methods: subsurface mining for deep deposits, and surface mining for shallower deposits. There are currently 25 active coal mines in Canada, located in Nova Scotia, Saskatchewan, Alberta, and British Columbia.

We reach underground deposits with **subsurface mining**. Shafts are dug deep into the ground, and networks of tunnels are dug or blasted out to follow the coal seams. The coal is removed and shipped to the surface. Underground coal mining is hazardous, not least because the material itself is highly combustible. The long history of underground mine disasters

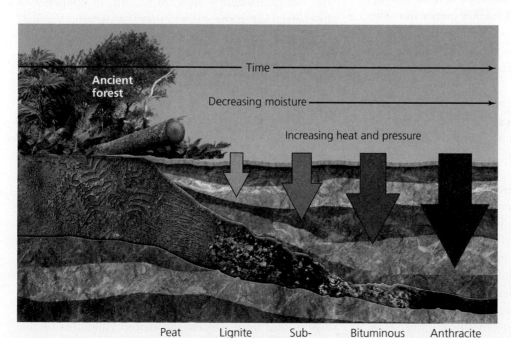

FIGURE 15.5
Coal forms as a result of the compaction of ancient plant matter underground. Peat is part of this continuum, representing plant matter that is minimally compacted. Lignite coal is formed under conditions of low pressure and heat, with much of the moisture retained. Anthracite coal is formed under the greatest pressure, where temperatures are high and most of the moisture is driven away.

in Nova Scotia (Springhill, 424 killed in mining-related incidents between 1881 and 1969), Alberta (Hillcrest, 189 killed in 1914), and British Columbia (Nanaimo, 150 killed in 1887) demonstrate the hazards posed to underground coal miners. Besides risking injury or death from collapsing shafts and tunnels and from dynamite blasts and coal dust or methane explosions, miners may inhale coal dust, which can lead to respiratory problems like black lung disease. Underground mine safety has improved dramatically in the past 50 years; however, today less than 2% of Canada's coal is extracted by underground mining.

When coal deposits are at or near the surface, open-pit or strip-mining methods are used. Open-pit mining involves large excavations, which are deepened and widened as mining proceeds. In **strip mining**, heavy machinery removes earth in long, horizontal strips to expose the seams and extract the coal. The pits are subsequently refilled with the soil that had been removed. Strip-mining operations can occur on immense scales; in some cases, particularly in the Appalachian region of the United States, entire mountaintops are lopped off to expose the coal. We will look more closely at both underground and strip-mining methods in Chapter 17, *Mineral Resources and Mining*.

In addition to posing health hazards for miners, coal mining can have substantial negative impacts on terrestrial and aquatic ecosystems. Surface strip mining can destroy large swaths of habitat and cause extensive soil erosion. It also can cause chemical runoff into waterways through the process of **acid drainage**. This occurs when sulphide minerals in newly exposed rock surfaces react with oxygen and rain water to produce sulphuric acid. As the sulphuric acid runs off, it leaches metals from the rocks, many of which are toxic to organisms in high concentrations. Acid drainage can be a natural phenomenon, but it is greatly accelerated when mining exposes many new rock surfaces at once.

One impact of coal that has recently raised concerns is selenium pollution. Selenium is a naturally occurring element, but it can be toxic to animals (including humans) even in small amounts; it is also bioaccumulative. Selenium occurs as a trace element in coal, and it is concentrated during nearly every phase of coal use, from mining and processing to combustion. Selenium in waste rock piles at coal mines is leached out and carried away by rain water. Coal is washed before it is transported to power plants; selenium can become highly concentrated in water used in this cleaning process.[8] In Canada, selenium contamination has been identified in Elk and Peace rivers, British Columbia, and in the Athabasca River in Alberta. Dennis Lemly of the United States

Forest Service has described several acute episodes of selenium contamination elsewhere in the world, in which thousands of fish died and many others were found to have abnormalities such as twisted spines, attributable to selenium exposure.[9]

The costs of alleviating all these health and environmental impacts are high, and the public eventually pays for them, one way or another, in an inefficient manner. Government regulations require mining companies to restore strip-mined land following mining, but impacts can be severe and long-lasting just the same.

Coal is used to generate electricity

In our modern society we use coal—lots of it—to generate electricity, in a process that dates back more than a century. Once mined, coal is cleaned and then hauled to power plants, where it is pulverized (**FIGURE 15.6**). The crushed coal is blown into a boiler furnace on a superheated stream of air and burned in a blaze of intense heat—typical furnace's flares may reach 815°C.

Water circulating around the boiler absorbs the heat and is converted into high-pressure steam. This steam is injected into a **turbine**, a rotary device that converts the kinetic energy of a moving substance, such as steam, into mechanical energy. As steam from the boiler exerts pressure on the blades of the turbine, they spin, turning the turbine's drive shaft.

The drive shaft is connected to a generator, which features a *rotor* that rotates and a *stator* that remains stationary. Generators make use of a phenomenon that you may have experimented with in your high school physics class: Moving magnets adjacent to coils of copper wire causes electrons in the wires to move, generating alternating electric current. The current flows into transmission lines that travel from the power plant out to the customers who use the plant's electricity.

"Clean coal" technologies can mitigate some environmental impacts

Coal contains impurities, which can include sulphur, mercury, arsenic, and other trace metals, which vary from deposit to deposit. Sulphur content depends in part on whether the coal was formed in freshwater or saltwater sediment. Coal from eastern provinces of Canada tends to be relatively high in sulphur (e.g., 2%) because it was formed from marine environments, where sulphur was incorporated from the surrounding sea water. In comparison, coal from Alberta and British Columbia (where

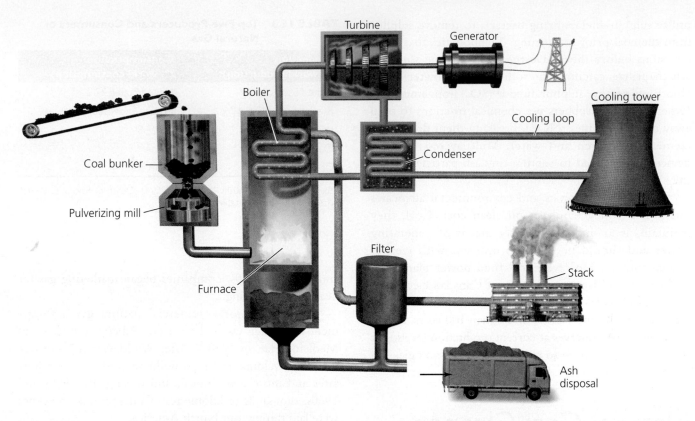

FIGURE 15.6 Coal is used as a fuel source to generate electricity. Coal is pulverized and blown into a high-temperature furnace. Heat from combustion boils water, and the resulting steam turns a turbine, generating electricity by passing magnets past copper coils. The steam is then cooled, condensed, and returned to the furnace.

the vast majority of Canada's coal reserves is located) is typically lower in sulphur content, ranging down to about 0.5%. Coal from China can be very sulphur-rich, ranging up to about 3%.

When coal is burned, pollutants are released. Burning high-sulphur coal produces sulphates, which contribute to industrial smog and acidic deposition. China's use of high-sulphur coal has contributed to severe air-quality problems in that country. Combustion of coal also can emit mercury, which bioaccumulates in organisms and bioconcentrates at high trophic levels in food chains. Such pollution problems commonly occur downwind of coal-fired power plants.

Scientists and engineers are seeking ways to cleanse coal of its impurities so that it can continue to be used as an energy source while minimizing impact on health and the environment. **Clean coal** technologies focus on approaches to rid toxic chemicals before, during, or after the burning of the coal. Reducing pollution from coal is important because society's demand for this relatively abundant fossil fuel may soon rise as supplies of oil and natural gas decline. Two principal clean-coal pathways are being investigated by researchers: cleaner combustion; and

gasification, or the production of cleaner synthetic fuels from coal.[10]

Combustion-focused technologies start with the pulverized-coal process and apply improvements aimed at making combustion more efficient and more complete, and thus cleaner. One example is *fluidized bed* technologies, which bathe the finely pulverized coal in jets of air during combustion. This leads to a turbulent, fluid-like environment, which allows the temperature to be increased and raises the efficiency of the chemical reactions that occur during combustion.

Gasification and liquefication technologies involve creating clean synthetic fuels, called *syngas*, from mixtures of pulverized coal, methane, water vapour, and other components. Gasifying or liquefying the coal not only makes it burn more cleanly, but also renders it more useful as fuel in a wider variety of applications.

Either approach can be combined with technologies that clean the coal prior to combustion; cleaning emissions after burning, and before they leave the smokestack; cogeneration technologies to maximize efficiency; and carbon capture and storage technologies. For example, some precombustion technologies utilize sulphur-metabolizing bacteria to remove sulphur from the coal prior to burning. Technologies that clean emissions before they leave the stack include *scrubbers*, which utilize calcium- or sodium-based materials to absorb and remove sulphur dioxide (SO_2) from emissions. Other types of scrubbers use chemical reactions to strip away nitrogen oxides (NO_X), breaking them down into elemental nitrogen and water. Multilayered filtering devices can be used to capture tiny ash particles before they leave the stack.

Some energy analysts and environmental advocates question a policy emphasis on clean coal. Coal, they maintain, is an inherently dirty means of generating power and should be replaced outright with cleaner energy sources. However, coal-fired power plants still generate a significant proportion of Canada's electricity, as well as Canada's greenhouse gas emissions. There is a lot of coal left in the world—more than 100 years' worth, based on proven reserves at current production levels. For these reasons, the push to clean up coal technologies still makes sense.

Natural Gas: Cleaner-Burning Fossil Fuel

Natural gas consists primarily of methane (CH_4), with varying amounts of other volatile hydrocarbons. (Natural "gas" is actually a misnomer, as it can be liquid at the ambient pressures and temperatures in subsurface reservoirs.) Natural gas provides almost one-quarter of total global energy consumption. It is a much cleaner-burning fuel than coal or oil, so it produces less pollution. World supplies of natural gas are projected to last for at least 55 more years (proven global reserves, at current levels of production), and much longer if more unconventional sources, such as shale gas, are brought into production.

Natural gas has only recently been widely used

Throughout history, naturally occurring seeps of gas would occasionally be ignited by lightning and could be seen burning in parts of what is now Iraq, inspiring the Greek essayist Plutarch around 100 CE to describe their "eternal fires." The first commercial extraction of natural gas took place in 1821, but until recently its use was localized because the technology did not exist to pipe gas safely over long distances. Natural gas was used to fuel streetlamps, but when electric lights replaced most gas

TABLE 15.3   Top Five Producers and Consumers of Natural Gas

Production (% world production)	Consumption (% world consumption)
United States, 20.6	United States, 22.2
Russian Federation, 17.9	Russian Federation, 12.0
Iran, 4.9	Iran, 4.8
Qatar, 4.7	China, 4.8
Canada, 4.6	Japan, 3.5

Source: Data from British Petroleum (2014). *Statistical Review of World Energy 2014,* statistical tables.

lamps in the 1890s, companies began marketing gas for heating and cooking.

The first major commercial natural gas development in Canada was at Bow Lake, Alberta, southwest of Medicine Hat, in 1908.[11] After World War II, improvements in welding and pipe building made gas transport safer and more economical, and during the 1950s and 1960s, thousands of kilometres of underground pipelines were laid throughout North America.

Today natural gas is increasingly favoured as an energy source. It is easy to transport, technologically versatile, and relatively clean-burning, emitting about half as much carbon dioxide per unit of energy produced as coal, and two-thirds as much as oil. Converted to a liquid at low temperatures as *liquefied natural gas* (or *LNG*), it can be shipped long distances in refrigerated tankers, although this poses risks of catastrophic explosions. The United States and the Russian Federation lead the world in natural gas production and consumption (**TABLE 15.3**); Canada is among the top five producers.

Natural gas is formed in two main ways

Natural gas can arise from either of two processes. *Biogenic* gas is created at shallow depths by the anaerobic decomposition of organic matter by bacteria. An example is the "swamp gas" you can sometimes smell when stepping into the muck of a swamp. Another source of biogenic natural gas is the decay process in landfills. Many landfill operators are now capturing this gas to sell as fuel (discussed in Chapter 18, *Managing Our Waste*). This practice decreases energy waste, can be profitable for the operator, and helps reduce the atmospheric release of methane.

In contrast, *thermogenic* gas results from compression of organic material, accompanied by heating deep underground. The organic precursor materials come most commonly from animal and plant matter, such as zooplankton and phytoplankton in shallow marine waters. As the organic matter is buried more and more deeply under sediments, the pressure exerted by the overlying sediments grows, and temperatures increase. This process, with its accompanying chemical changes, is called *maturation*, and it can result in the formation of natural gas as well as oil and other types of hydrocarbon deposits.

During maturation, carbon bonds in the precursor organic matter break, and the organic matter turns into kerogen, the source material for both natural gas and crude oil. Further heat and pressure act on the kerogen to degrade complex organic molecules into simpler hydrocarbon molecules. At very deep levels—below about 3 km—the high temperatures and pressures lead to the formation of natural gas. Whereas biogenic gas may be nearly pure methane, thermogenic gas contains small amounts of other gases. Natural gas is odourless and colourless; the distinctive smell that we associate with the gas is actually added for safety purposes.

Thermogenic gas may be formed directly, along with coal or crude oil, or from coal or oil that is altered by heating. The natural gas found in the Mackenzie River delta, discussed in the Central Case, originated from the thermogenic decomposition of organic matter in shallow marine sediments. Most natural gas that is extracted commercially is thermogenic and is found above deposits of crude oil or seams of coal, so it often accompanies the extraction of those fossil fuels.

Often, natural gas goes to waste as it escapes from coal mines or oil wells. In remote oil-drilling areas, where the transport of natural gas remains prohibitively expensive, the gas is flared—that is, simply burned off. In some cases, gas captured during drilling can be reinjected into the ground for potential future extraction or to maintain the subsurface pressure needed to bring the oil to the surface. Natural gas from coal seams, called *coalbed methane*, commonly leaks to the atmosphere during coal mining. To avoid this waste, and because methane is a potent greenhouse gas that contributes to climate change, mining engineers try to capture and use the gas. At extraction sites for coalbed methane, ground water is pumped out to free the gas to rise, but salty ground water dumped on the surface can contaminate soil and kill vegetation over large areas.

Natural gas extraction becomes more challenging with time

To access some natural gas deposits, prospectors need only drill an opening because pressure and low molecular

FIGURE 15.7
Horsehead pumps, like this one in Drayton Valley, Alberta, are used to extract natural gas as well as oil. The pumping motion of the machinery draws gas and oil upward from below ground.

Courtesy of Nathan Schneider/Wikipedia

weight drive the gas upward naturally. The first gas fields to be tapped were of this type. Most fields remaining today, however, require the gas to be pumped to the surface. In Alberta and parts of the United States it is common to see a device called a *horsehead pump* (**FIGURE 15.7**). The pump moves a rod in and out of a shaft, creating pressure to pull both natural gas and crude oil to the surface.

Many of the most accessible natural gas reserves have already been exhausted, causing their production to decline. Thus, deposits located in more remote areas, such as the Mackenzie River delta, are becoming more attractive economically. Much extraction today also makes use of sophisticated techniques to break into rock formations and pump gas to the surface.

One such fracturing technique involves pumping a fluid (usually water) under high pressure into the rocks to crack them. Sand or small glass beads are injected to hold the cracks open once the water is withdrawn. This type of extraction, called **hydraulic fracturing** or **fracking**, has extensive environmental impacts (**FIGURE 15.8**). Fracking is used to recover *shale gas*, natural gas produced from shale, which is the fastest growing part of the natural gas industry. It is also used for oil contained in rock of low permeability, and for bitumen from oil sands.

Noise has been one complaint associated with fracking, but a bigger concern is the water-intensity of the process, especially in water-stressed regions in places like Alberta, Texas, California, and Colorado. Another problem is earthquakes caused by rock cracking. The generation of (usually) small earthquakes by the injection of pressurized fluids is a process that has been known

(a) Protest against fracking

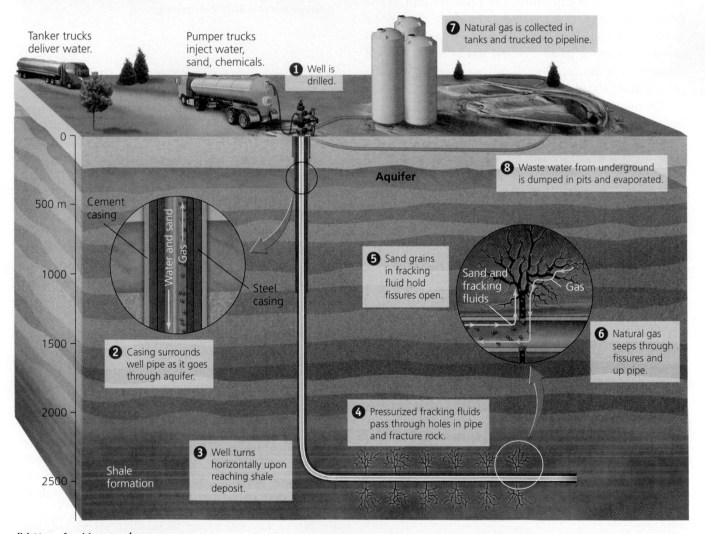

(b) How fracking works

FIGURE 15.8 Hydraulic fracturing, or fracking, is used to extract natural gas, some types of oil, and bitumen from oil sands (also called tar sands). Fracking often sparks controversy, as seen in this photo of a protest on the Kainai Blood Tribe Reserve in southern Alberta, 2011 **(a)**. Fracking **(b)** involves injecting a fluid (mainly water) into the ground under pressure to induce the rock to fracture and release the petroleum that it holds. Fluids called proppants are injected to keep the fractures open and allow the gas or oil to flow out.

for many decades, mainly in association with the underground injection of hazardous fluid wastes. It is referred to as *induced seismicity,* and it may result from the lubrication of long-inactive faults by pressurized fluids. Confirmed instances of induced seismicity caused by fracking are still somewhat rare. However, a swarm of earthquakes in Fox Creek, Alberta, in 2015, the largest of which had a magnitude of 4.4, along with earlier swarms in the same area in 2013 and 2014, have been linked to nearby fracking operations.

The fluids that carry out the work of fracking are 90% water, with a small proportion of sand and chemical additives. Fractures are initiated and then extended in the rock by injecting water under pressure. The purpose of the sand is to prop open the fractures once they have formed; the sand-containing fluids are called *proppants.* The other chemicals are added to change the viscosity and other characteristics of the gas, bitumen, or oil for ease of extraction, and to lubricate and limit corrosion in the pipes.

The nature of the additives and the possibility that they may escape and contaminate water supplies have generated the most controversy. A wide range of chemicals is used for various purposes in fracking. Many of them also have common household uses; for example, guar gum (used as a thickener in ice cream and toothpaste) thickens the fluid in which the proppant sand is suspended, and ethylene glycol (used in antifreeze and deicing fluid) prevents the buildup of deposits in the pipes. However, their use in large quantities in the natural environment, and the possibility of spills or leakage, are questioned by opponents of fracking.

Oil: World's Most-Used Fuel

Oil has dominated world energy use since the 1960s, when it eclipsed coal. It now accounts for one-third of the world's total energy consumption. Its use worldwide in just the past decade has risen more than 12%.[12]

People have used solid forms of oil (tar and asphalt) from easily accessible, near-surface deposits for at least 6000 years. The modern extraction and use of petroleum for energy began in the 1850s, when miners drilling for ground water or salt occasionally encountered oily rocks instead. At first, entrepreneurs bottled the crude oil from these deposits and sold it as a healing aid, unaware that crude oil is carcinogenic when applied to the skin and poisonous when ingested. Soon, however, it was realized that this "rock oil" could be used to light lamps and

lubricate machinery. Edwin Drake is generally credited with drilling the world's first oil well, in Titusville, Pennsylvania, in 1859. In fact, however, the first oil well was drilled a full year earlier, in 1858, at Oil Springs, Ontario, by James Miller Williams, who struck free liquid oil only 20 m below the surface while attempting to drill a water well.[13]

Today our global society produces and consumes nearly 675 L (more than half a tonne) of oil each year for each living person. The United States, with 4.4% of the global population, consumes nearly 20% of the world's oil. Canadians—less than 0.5% of the world's population—consume 2.5% of the oil.[14] **TABLE 15.4** shows the top oil-producing and oil-consuming nations.

Heat and pressure underground lead to petroleum formation

The sludgelike liquid we know as **oil, crude oil**, or **petroleum** (a term that includes both oil and natural gas) forms within temperature and pressure conditions most commonly often found 1.5–3 km below the surface. These conditions are sometimes referred to as the *oil window* or *petroleum window*. At lower temperatures and pressures, organic matter would remain as kerogen, and would fail to mature into oil. At higher temperatures, natural gas might be formed instead of oil.

Like natural gas, most of the crude oil we now extract was formed when dead marine organisms (mostly algae and plant matter, with small amounts of animal matter) drifted down through shallow coastal waters millions of years ago and became buried in sediment on the ocean

TABLE 15.4 Top Five Producers and Consumers of Oil

Production (% world production)	Consumption (% world consumption)
Saudi Arabia, 13.1	United States, 19.9
Russian Federation, 12.9	China, 12.1
United States, 10.8	Japan, 5.0
China, 5.0	India, 4.2
Canada, 4.7	Russian Federation, 3.7

Source: Data from British Petroleum (2014). *Statistical Review of World Energy 2014,* statistical tables.

DATA Q Examine the three tables in this chapter that show the top producers and consumers of coal, gas, and oil (**TABLES 15.2, 15.3,** and **15.4**). In each case, which countries are among the top consumers, but not among the top producers? Are nations necessarily net importers of energy if they are on the "top five" list as consumers, but not as producers?

floor. Over millions of years in the pressure and temperature conditions of the oil window, through the complex chemical changes involved in maturation, crude oil is formed.

Crude oil is an unrefined mixture of hundreds of different types of hydrocarbon molecules. The specific properties of the oil depend on the chemistry of the organic starting materials, the characteristics of the geological environment of formation, and the details of the maturation process. In spite of the chemical complexity of oil, there are only a few main elemental components: carbon, hydrogen, nitrogen, oxygen, and sulphur, with traces of other elements, mainly metals. Crude oils that contain appreciable amounts of sulphur are called "sour crude," and those with less sulphur are called "sweet crude."

These constituents combine to form many different types of hydrocarbon molecules, which fall into four main groups:

- *paraffins*, which are basically waxes
- *naphthenes*, which are chainlike hydrocarbon molecules
- *aromatics*, which are based on benzene rings
- *asphaltics*, which are solid and semi-solid hydrocarbons, such as bitumen

The resulting crude oil mixtures can range from black and nearly solid, to a thin, clear liquid that requires very little processing.

Petroleum geologists infer the location and size of deposits

Because petroleum forms only under certain conditions, it occurs in isolated deposits. Once the crude oil has been formed, it migrates from the *source rock* upward through pore spaces in rocks, facilitated by faults and fractures. Most of the oil that forms underground seeps out at the surface and volatilizes naturally, escaping into the atmosphere. If capped by an impermeable layer of rock, called a *cap rock*, the oil can collect in porous units called *reservoir rocks*. The combination of a source rock, a reservoir rock, and a cap rock is referred to as a *hydrocarbon* or *petroleum trap*, and this combination is what petroleum geologists seek.

Geologists searching for oil drill rock cores and conduct ground, air, and seismic surveys to map underground rock formations, understand geological history, and predict where fossil fuel deposits might lie (**FIGURE 15.9A**). One commonly used method is to create powerful vibrations at the surface (by exploding dynamite, thumping the ground with a large weight, or using an electric vibrating machine) and measure how long it takes the seismic waves to reach receivers at other surface locations (**FIGURE 15.9B**). Density differences in the substrate cause waves to reflect, refract, or bend as they pass from one layer to another. Scientists and engineers interpret the patterns of wave reception to infer the density, thickness, and location of underlying geological layers—which in turn provide clues about the location and size of oil and natural gas deposits.

(a) Petroleum geologists

(b) Seismic survey

Vibration source

Receivers

Less dense layer (sound travels more slowly)

More dense layer (sound travels more quickly)

Reflection paths (red)

Refraction paths (blue)

FIGURE 15.9 Petroleum geologists study geological maps **(a)** to determine where oil and gas might be found. Seismic surveys **(b)** provide clues to the location and size of fossil fuel deposits. Powerful vibrations are created, and receivers measure how long it takes the seismic waves to reach other locations. Waves travel more quickly through denser layers, and density differences cause waves to reflect or refract. Scientists interpret the patterns of wave reception to infer the densities, thicknesses, and locations of underlying rock layers.

Over the past few decades, geologists have greatly improved their methods for locating new deposits. However, with their scientific understanding of Earth processes, petroleum geologists are generally quick to acknowledge that oil is ultimately a finite and nonrenewable resource.

A portion of the oil that is located by geologists is impossible to extract using current technology, and will have to wait for future advances in extraction equipment or methods. Thus, estimates of reserves are based on the *technically recoverable* oil. Oil companies are not willing to extract these entire amounts because the expense would exceed the income the company would receive from the oil's sale. The amount a company chooses to produce will be determined by the costs of extraction and transportation, together with the current price of oil on the world market. Because the price of oil fluctuates, the portion of oil from a given deposit that is *economically recoverable* also fluctuates.

Thus, technology sets a limit on the amount that *can* be extracted, and economics determines how much *will* be extracted. The amount of oil, or any other fossil fuel, in a deposit that is technologically and economically feasible to remove under current conditions is the proven recoverable **reserve** of that fuel.

We drill and pump to extract oil

Once geologists have identified an oil deposit, an oil company will typically conduct exploratory drilling. Holes drilled during this phase are usually small in circumference and descend to great depths. If enough oil is encountered, extraction begins. Oil that has not yet been drilled for extraction is typically under pressure—from above by rock or trapped gas, from below by ground water, or internally from natural gas dissolved in the oil. All these forces are held in place by surrounding rock until drilling reaches the deposit, whereupon the oil will often rise to the surface of its own accord (a so-called *gusher*).

The initial extraction of oil from the well is called **primary recovery** or *primary extraction* (**FIGURE 15.10A**). Once the initial pressure is relieved, both oil and natural gas become more difficult to extract and typically must be pumped out with increasing pressure. Even after pumping, as much as two-thirds of a deposit may remain in the ground, stuck between rock particles. Companies may then begin **secondary recovery** or *secondary extraction*, in which fluids are injected into the ground to remove additional oil (**FIGURE 15.10B**). The fluids most commonly used for this purpose are water, sea water, steam, carbon dioxide, and natural gas.

Even after secondary recovery, a lot of oil remains underground; we lack the technology to remove every last

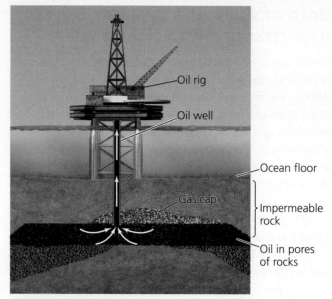

(a) Primary recovery of oil

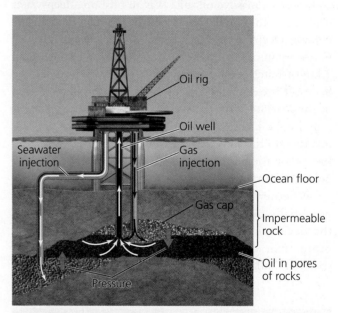

(b) Secondary recovery of oil

FIGURE 15.10
In primary recovery **(a)**, oil is drawn up through the well by keeping pressure at the top lower than pressure at the level of the oil deposit. Once the pressure in the deposit drops, fluids must be injected into the deposit to increase the pressure. Secondary recovery **(b)** involves injecting water, carbon dioxide, or another fluid to force more oil up and out of the deposit.

drop. Secondary recovery is more expensive than primary recovery; many oil deposits did not undergo secondary recovery when they were first drilled because the price of oil was too low to make the procedure economical. When oil prices rose in the 1970s, many drilling sites were reopened for secondary extraction. Still more are being reopened today, subject to the fluctuations of market price. In addition to being more expensive, secondary recovery is harder on the environment than primary extraction.

Both offshore and onshore drilling have risks and impacts

Extraction of oil and natural gas takes place not just on land, but also on the continental shelves in the ocean. Offshore drilling has required the development of technologies that can withstand the forces of wind, waves, icebergs, cold temperatures, and ocean currents. Some drilling platforms are fixed standing platforms built with unusual strength. Others are resilient floating platforms, anchored in place above the drilling site. Most of the offshore gas and oil development in Canada is located in the Beaufort Sea and in the North Atlantic Ocean off the coasts of Newfoundland and Labrador, and Nova Scotia (**FIGURE 15.11**).

Offshore drilling is risky because of stormy weather and icy waters, as well as the extreme depths of drilling (more than 2.5 km, in the case of Hibernia). British Petroleum's Deepwater Horizon, which exploded in 2010 and caused a massive oil spill, was an offshore, deep-water drill rig in the Gulf of Mexico. Prior to its catastrophic demise, Deepwater Horizon drilled the deepest oil well in history (more than 10 km deep), working in more than 1 km of water depth off the southern coast of the United States. It is extremely difficult to clean up spills and leaks in the marine environment, as demonstrated so dramatically by the Deepwater Horizon disaster; it took months just to stop the flow of oil from the destroyed well head located on the bottom of the Gulf, more than a kilometre below the surface.

Whether onshore or offshore, drilling has many potentially negative environmental impacts. However, the development of an oil or gas field involves much more than drilling. Road networks must be constructed, and many sites may be explored in the course of prospecting. Pipelines are needed to transport the oil or gas. The extensive infrastructure needed to support a full-scale drilling operation typically includes housing for workers, access roads, airstrips, and landing pads, as well as waste piles for removed soil. Ponds and holding dams may be constructed for collecting the toxic sludge that remains after the useful components of oil have been removed.

Many onshore oil and gas reserves in Canada, the United States, and Russia are located in the Arctic. Plants grow very slowly in tundra ecosystems, so even minor impacts can have long-lasting repercussions. For example, tundra vegetation at Prudhoe Bay Oilfield in Alaska has not yet recovered from roads installed for exploratory purposes more than 40 years ago. Numerous studies of wildlife have been carried out at Prudhoe Bay; the results are inconclusive and contradictory. Some studies have shown that female caribou and their calves avoid the oil complex, detouring many kilometres to do so, and that their reproductive success is affected. Other studies suggest that oilfield development affects individual animals but does not influence herds as a whole, and that fluctuations in herd size are influenced by many factors beyond just the oil development. The same researchers determined that the influence of oilfield development on the genetic diversity of grizzly bears was not a cause for concern (**FIGURE 15.12**).[15]

Martha Raynolds of the University of Alaska and her colleagues have carried out a comprehensive, long-term study of the cumulative effects of 62 years of oil and gas development at the Prudhoe Bay. They describe direct effects of oil-related infrastructure, such as roads, airstrips,

Ron Watts/Firstlight

FIGURE 15.11
The Hibernia Offshore Drilling Platform, shown here, is located in the North Atlantic about 300 km off the coast of Newfoundland in the Grand Banks. It is the world's largest offshore platform and began production in 1997.

Steven J. Kazlowski/Alamy

FIGURE 15.12
Like the Mackenzie River valley, Alaska's North Slope is home to a variety of large mammals, including grizzly bears, polar bears, wolves, Arctic foxes, and large herds of caribou. How oil development may affect these animals is a controversial issue, and scientific studies are ongoing. Grizzly bears, such as those shown here, have been found near, or even walking atop, the Trans-Alaska Pipeline.

pads, and gravel quarries, as well as indirect or unintentional effects, such as from offroad vehicle traffic, roadside flooding, and roadside dust. Of particular concern is the progressive development of *thermokarst*, a terrain associated with melting permafrost, in areas modified by oil infrastructure. Thermokarst, also influenced by climatic change in the Arctic, significantly changes the character of the tundra ecosystem, affecting wildlife as well as human interests. The conclusion of the researchers is that a combination of oil development and climatic change has led to a "new geoecological regime," which will continue to affect wildlife and human inhabitants in the Prudhoe Bay area for a long time.[16]

To address the wildlife management questions and challenges associated with far-north oil development, Brian Person of the University of Alaska and colleagues studied a large herd of caribou on Alaska's North Slope. The Teshekpuk caribou herd had little to no exposure to oil and gas development prior to their study. The researchers tracked the herd for 15 years to provide baseline data in advance of any oil or gas development in the area, so that the impacts of industrial activity will be distinguishable from natural influences on the herd.[17]

There is no way of knowing how ecosystems would have fared in the absence of oil development at Prudhoe Bay, the oldest and most extensive industrial development in the Arctic. It is difficult, therefore, to draw conclusions about the impacts of oil or gas development on caribou and other wild animals at remote places like the Mackenzie River valley. Long-term and baseline studies will help. It can be anticipated that activities like road building, oil pad construction, worker presence, oil spills, accidental fires, trash buildup, offroad vehicle trails, and dust from roads, combined with terrain changes due to permafrost melting, will have significant impacts on both vegetation and wildlife in these areas.

Petroleum is refined and made into many products

Once crude oil has been extracted, it must be processed. Crude oil is a complex mixture of many different kinds of hydrocarbon molecules characterized by carbon chains of different lengths. A hydrocarbon chain's length affects its chemical properties, which has consequences for human use, such as whether a given fuel burns cleanly in a car engine. Oil refineries sort the various hydrocarbons of crude oil, separating those intended for use in gasoline engines from those, such as tar and asphalt, used for other purposes.

Through **refining**, the hydrocarbons are separated into classes of different sizes and chemically transformed to create specialized fuels for heating, cooking, and transportation and to create lubricating oils, asphalts, and the precursors of plastics and other petrochemical products. To maximize the production of marketable products while minimizing negative environmental impacts, petroleum engineers have developed a variety of refining techniques.

The first step in processing crude oil is *distillation*, or *fractionation*. This process is based on the fact that different components of crude oil boil at different temperatures. In refineries, the distillation process takes place in tall columns filled with perforated horizontal trays (**FIGURE 15.13**). The columns are cooler at the top than at the bottom. When heated crude oil is introduced into the column, lighter components rise as vapour to the upper trays, condensing into liquid as they cool, while heavier components sink to the lower trays. Light gases, such as butane, boil at less than 32°C, and heavier oils, such as industrial fuel oil, boil only at temperatures above 343°C.

Since the early twentieth century, light gasoline, used in automobiles, has been in much higher demand than most other derivatives of crude oil. The demand for high-performance, clean-burning gasoline has also risen. To meet these demands, refiners have developed several techniques to convert heavy hydrocarbons into gasoline, which are referred to as *cracking*. One of the simplest methods is thermal cracking, in which long-chained molecules are broken into smaller chains by heating in the absence of oxygen. (The oil would ignite if oxygen were present.) Catalytic cracking, a related method, uses *catalysts*—substances that promote chemical reactions without being consumed by them—to control the cracking process. In either case, the result is an increase in the amount of a desired lighter product from a given amount of heavy oil. The products of cracking are then fed into a distillation column.

Besides distilling crude oil and altering the chemical structure of some of its components, refineries also remove harmful contaminants. Government regulations have forced refineries to implement scrubbers and other methods of removing contaminants, particularly sulphur. Some methods successfully remove up to 98% of sulphur.

Because crude oil is such a complex chemical mixture, many types of petroleum products can be created by separating its various components. Since the 1920s, refining techniques and chemical manufacturing have greatly expanded our uses of petroleum to include a wide array of products and applications, from lubricants to plastics to fabrics to pharmaceuticals. Today, petroleum-based products are all around us in our everyday lives (**FIGURE 15.14**). Our fundamental dependence on petroleum products, in every aspect of life, adds to concerns about the eventual depletion of oil resources.

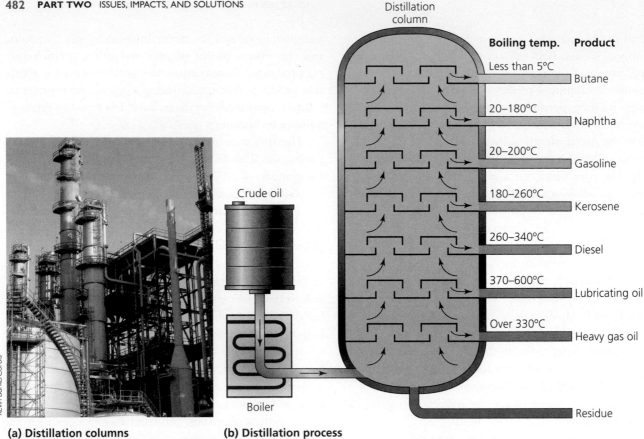

Kevin Burke/Corbis

(a) Distillation columns

(b) Distillation process

FIGURE 15.13 At a refinery, crude oil is boiled, causing its hydrocarbon constituents to volatilize and proceed upward through a distillation column **(a)**. Constituents that boil at the highest temperatures will condense at low levels in the column **(b)**. Constituents that volatilize at lower temperatures will continue rising through the column, condensing at higher levels where temperatures are lower. In this way, heavy oils (consisting of long hydrocarbon molecules) are separated from lighter oils (generally those with shorter hydrocarbon molecules).

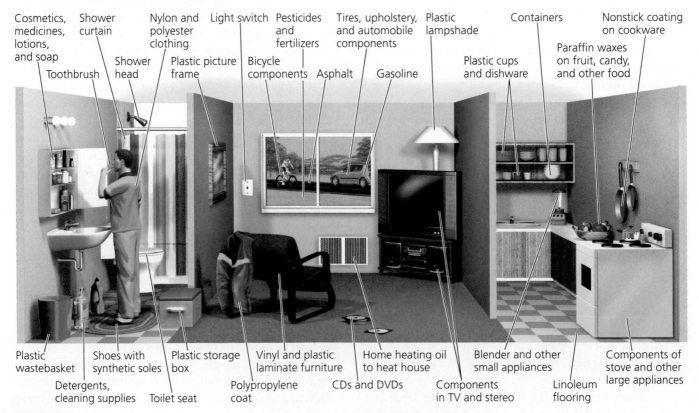

FIGURE 15.14 Petroleum products are everywhere in our daily lives. Besides the gasoline and other fuels we use for transportation and heating, petroleum products include many of the fabrics that we wear and most of the plastics that help make up countless items we use every day.

We may have already used almost half of our oil reserves

Some scientists and oil industry analysts calculate that we have already extracted nearly half of the world's oil reserves. So far we have used up about 1.1 trillion barrels* of oil. Estimates of how much recoverable oil is left vary widely, depending on the level of certainty about the size and quality of the remaining deposits, and whether they are defined as recoverable (that is, the technological challenges involved in extracting the oil), among other complexities. Nick Owen and colleagues from the Low Carbon Mobility Centre at the University of Oxford provides a well-balanced examination of these complexities in the journal *Energy Policy*.[18]

British Petroleum, one of the industry standards for current and historical data about oil production and consumption, estimates proven reserves of about 1.7 trillion barrels. To estimate how long this remaining oil will last, analysts calculate the *reserves-to-production ratio*, or *R/P ratio*, by dividing the amount of total remaining reserves by the annual rate of production (that is, extraction and processing). At current levels of production (about 30 billion barrels globally per year), 1.7 trillion barrels would last about 53 more years. The R/P ratio for natural gas is about 55 years, and for coal it is about 113 years.

This does not mean that we will absolutely run out of oil in 50 years or so. Geologists will find new deposits; petroleum engineers will develop new technologies for extracting oil from more challenging deposits, and for refining oil of lower quality. The market price will go up, and companies will be willing to spend more to extract the remaining oil.

On the other hand, it also does not mean that we have a full 53 years in which to figure out what to do once the oil runs out. A growing number of scientists and analysts insist that we will face a crisis not when the last drop of oil is pumped, but when the rate of production begins to decline. They point out that when production declines as demand continues to increase (because of rising global population and consumption), we will experience an oil shortage immediately. Because production tends to decline once reserves are depleted halfway, most of these experts calculate that this crisis will likely begin within the next several years.

To understand the basis of these concerns, we need to turn back the clock to 1956. In that year, Shell Oil geologist M. King Hubbert calculated that U.S. oil production would peak around 1970. His prediction was ridiculed at the time, but it proved to be accurate; U.S. production peaked in that very year and has continued to fall since then (**FIGURE 15.15A**). The peak in production came to be known as **Hubbert's peak**.

In 1974, Hubbert analyzed data on technology, economics, and geology, predicting that global oil production would peak in 1995. Production, in fact, continued to grow past 1995, but many scientists using newer, better data today predict that at some point in the coming decade, production will begin to decline (**FIGURE 15.15B**). Some industry experts even contend that we have already passed the peak. Indeed, because of year-to-year variability in production, we will only be able to recognize that we have passed the peak of oil production several years after it has happened.

Predicting an exact date for **"peak oil"** and the coming decline in production is difficult. Many companies and governments do not reveal their true data on oil reserves, and estimates differ greatly as to how much oil we can extract secondarily from existing deposits. A report by the U.S. General Accounting Office reviewed 21 studies and found that estimates for the timing of the oil production peak ranged from 2005 through 2040. Regardless of the exact timing, however, a peak in global oil production will occur. Meanwhile, global demand continues to rise, particularly as China and India industrialize rapidly. Because of the long time scales, uncertainties, and deep impacts involved in this transition, some peak oil analysts have referred to it as the "long emergency."

The coming divergence of demand and supply will likely have momentous economic, social, and political consequences that will profoundly affect the lives of each and every one of us. Pessimists predict the collapse of modern agriculture and industrial society as fossil fuel supplies become increasingly insufficient. More optimistic observers argue that as oil supplies dwindle, rising prices will create powerful incentives for businesses, governments, and individuals to conserve energy and to develop alternative energy sources, and that these developments will save us from major disruptions caused by the coming oil peak.

To achieve a sustainable society, we will need to switch to renewable energy sources; we will examine the options more closely in Chapter 16, *Energy Alternatives*. Energy conservation can extend the time we have in which to make this transition. However, the research and development needed to construct the infrastructure for a new energy economy will take investment, and the time we will have to make this enormous transition will be limited.

---

*A barrel is not a metric unit of measurement, but it is commonly used in the oil industry. It is equivalent to 0.158987 cubic metres, or 159 litres, or 42 U.S. gallons, or 0.1364 tonnes of oil.

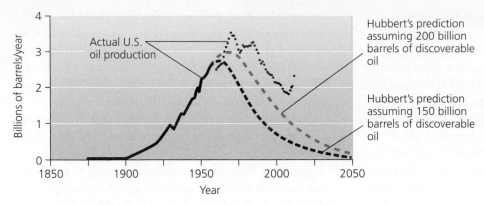

(a) Hubbert's prediction of peak in U.S. oil production, with actual data

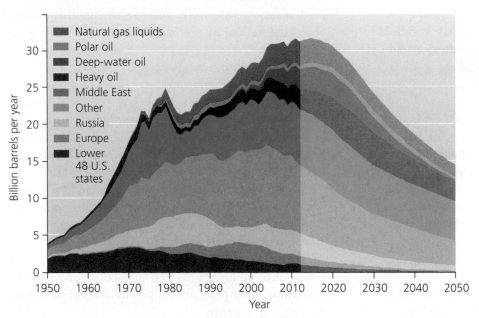

(b) Modern prediction of peak in global oil production

FIGURE 15.15 Oil production peaked in the United States in 1970 **(a)**, just as geologist M. King Hubbert had predicted. Since then, success in Alaska and the Gulf of Mexico, and with secondary extraction has enhanced production during the decline. (Note that curves shift to the right and peaks increase in height as more oil is discovered.) Today U.S. oil production is spiking upward as deep offshore drilling and hydraulic fracturing make new deposits accessible. Still, some analysts believe global oil production will soon peak. Shown in **(b)** is one recent projection, from a 2011 analysis by scientists at the Association for the Study of Peak Oil.
Source: Data from (a) Hubbert, M.K. (1956). Nuclear Energy and the Fossil Fuels. Shell Development Co. Publ. No. 95, Houston, TX; and U.S. Energy Information Administration; and (b) Reprinted by permission from Dr. Colin Campbell.

"Unconventional" Fossil Fuels

As oil production declines, we will rely more on natural gas and coal—yet these in turn will peak and decline in future years. We are turning increasingly to newer alternatives, which includes at least three types of unconventional fossil fuels that exist in large amounts: oil sands (or tar sands), shale oil and gas, and methane hydrates.

Canada owns massive deposits of oil sands

Oil sands (also called **tar sands**) are deposits of moist sand and clay that contain concentrations of a heavy, viscous hydrocarbon. Oil sands contain neither oil nor tar, but rather 1–20% **bitumen**, a thick, black, naturally occurring form of petroleum that is rich in carbon and relatively poor in hydrogen; technically, they should be referred to as *bituminous sands*. On average, two metric tons of material from oil sands are required to produce one barrel of synthetic crude oil.

Depending on how oil sands or tar sands are defined, almost three-quarters (more than 70%) of the world's proven oil reserves from such deposits are located in northeastern Alberta. Equally large deposits in eastern Venezuela's Orinoco Belt are often included, but the hydrocarbons they contain are technically closer to "heavy oil" than to bitumen. Oil sands in the two regions hold at least 168 and 221 billion barrels, respectively. In Alberta, strip-mining began in 1967. As rising crude oil prices made the deposits more profitable to extract, dozens of companies angled to begin mining projects in the region. In 2013, tar sands produced 1.8 million barrels of oil per day, contributing 56% of Canada's petroleum production.[19] The oil sands move Canada into a strong position for oil production, with the third-largest proven reserves in the world (**FIGURE 15.16**).

Because bitumen is too thick to extract by conventional oil drilling, oil sands closer to the surface (about 20% of the deposits in Alberta) are generally removed by strip mining (**FIGURE 15.17A, B**), using methods similar to coal strip mining. For deposits 75 m or more below ground, a variety of *in situ* extraction techniques are being used (**FIGURE 15.17C**). Most of these involve steam stimulation, in which steam is injected into a well, causing the bitumen to loosen to the point where it can be extracted through another well. After extraction, bitumen is sent to specialized refineries, where several types of chemical reactions that add hydrogen or remove carbon can upgrade it into more valuable synthetic crude oil.

In addition to being difficult to extract, bitumen also is too thick (that is, too viscous) to transport by pipelines that would be used to carry regular crude oil. For this reason, bitumen is often diluted with lighter oil or natural gas by-products; the resulting diluted bitumen is called *dilbit*. Concerns about the toxicity of dilbit and the potential for spills and environmental damage generated fierce opposition to the Keystone XL and Northern Gateway pipelines, which would take dilbit from the Alberta oil sands either south into the United States, or west to the Pacific Ocean through British Columbia. We look at this controversy more closely in "The Science behind the Story: Keystone XL, Northern Gateway, and the Dilbit Controversy."

Oil shale is abundant in the United States

Another important unconventional hydrocarbon is **oil shale**, a sedimentary rock (not always shale) that contains abundant kerogen, the organic precursor to oil and natural gas. Oil shale is formed by the same maturation processes that form crude oil and oil sands, but occurs when the kerogen was not buried deeply enough or subjected to enough heat and pressure to fully mature into oil.

It is worth noting that oil shale is not the same as shale oil, although the two are often confused (and certainly their names are confusing). *Shale oil* is simply an occurrence of crude oil within the sedimentary rock shale. Because shale is highly impermeable and fine-grained, shale oil is commonly what is referred to as *tight oil*, meaning that the oil is tightly bound to the mineral grains, and therefore difficult to

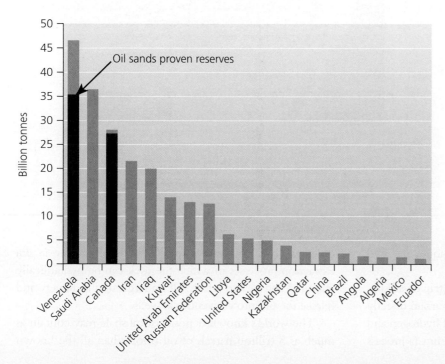

FIGURE 15.16
Canada's oil sands are a significant proven oil reserve, even in an international context. The diagram shows proven oil reserves, highlighting the significant heavy oil and bitumen deposits in Venezuela and Canada.
Source: Based on British Petroleum (2014). *Statistical Review of World Energy, 2014,* report and statistical tables.

THE SCIENCE BEHIND THE STORY

The Keystone XL Pipeline proposal has generated enormous controversy, both in the United States (as seen here in front of the White House) and in Canada.
J. Scott Applewhite/AP Images

Keystone XL, Northern Gateway, and the Dilbit Controversy

Alberta's oil sand deposits are huge. The open pit mines dug to extract the fuel are many kilometres across. The vehicles moving inside them have 4-metre-high tires, and shovels that are five stories high, and carry more than 300 tonnes of oil sand per load. The reservoirs that contain the left-over sludge from their extraction, and the dams that hold the reservoirs, are some of the largest in the world, of any type. The economic value of the extracted oil fuels the economy of Alberta, not to mention that of Canada as a whole. To some people the oil sands represent wealth and security, a key to maintaining our fossil-fuel-based lifestyle far into the future. To others they are a source of appalling pollution and threaten to radically alter Earth's climate.

Mining for oil sands (originally called tar sands in the petroleum industry) began in Alberta in 1967, but for many years it was hard to make money extracting these low-quality deposits. Rising oil prices turned it into a profitable venture, and today dozens of companies are mining here. According to the Canadian Association of Petroleum Producers, Canadian oil sands produce 1.8 million barrels of oil per day, more than

half of Canada's petroleum production.[20] Thanks to the oil sands, Canada boasts the world's third largest proven reserves of oil, after Saudi Arabia and Venezuela.

The obvious place to look for buyers for the oil was south of the border. In 2010 the TransCanada Corporation began operating the Keystone Pipeline, shipping diluted bitumen, or dilbit, 3500 km from Alberta to Illinois and Oklahoma. At the Oklahoma terminus in the town of Cushing, a bottleneck created a glut of oil that was unable to reach refineries on the Texas coast fast enough to meet demand. TransCanada then proposed the Keystone XL extension (see **FIGURE 1**), a two-part project consisting of a southern leg to connect Cushing to the Texas refineries, and a northern leg that would cut across the Great Plains to shave off distance and add capacity to the existing line.

The Keystone XL proposal soon met opposition from people living along the proposed route who were concerned about health, environmental protection, and property rights. Opponents expressed dismay at the destruction of boreal forest associated with strip mining of the oil sands and anxiety about transporting oil over the continent's largest aquifer, the Ogallala. The pipeline extension also faced opposition from climate change activists, who want to avoid extracting a vast new source of fossil fuels whose combustion would release greenhouse gases that will intensify climate change.

Pipeline supporters, on the other hand, feel the Keystone XL project will create jobs for workers both in Canada and in the United States, and guarantee a dependable oil supply for decades to come. They stress that if the United

FIGURE 1
The map shows the TransCanada Keystone XL Pipeline extensions and the western branch of the Enbridge Northern Gateway pipeline, proposed to transport diluted bitumen from Alberta to markets in the United States and Asia, respectively.

extract; typically secondary extraction techniques are used with drilling.

Oil shale, on the other hand, is mined by strip or subsurface mining. Once mined, oil shale can be burned directly like coal, or it can be baked in the presence of hydrogen and in the absence of air to extract liquid petroleum (a process

called *pyrolysis*). Industry researchers are developing *in situ* extraction processes in which the rock can be hydraulically fractured and heated while still underground to liquefy and release the oil into conventional wells.

The world's known deposits of oil shale may contain as much as 5 trillion barrels of oil (more than all the known

States buys oil from Canada it could end reliance on oil-producing nations such as Saudi Arabia and Venezuela that have had authoritarian governments and poor human rights records. They also argue that Canada will extract and sell its oil in any case, so the United States might as well be the purchaser.

Under pressure from all sides, the administration of U.S. President Barack Obama walked a fine line. Because the northern leg of the Keystone XL extension would have crossed an international border, it required a presidential permit from the U.S. Department of State, which TransCanada applied for in 2008. After three years of review, the State Department declined to approve the project because of concerns about damage to the ecologically sensitive Sandhills area of Nebraska, and potential contamination of the Ogallala Aquifer. Facing street protests at the White House (see photo), Obama postponed the permit decision indefinitely. The debate intensified, and tens of thousands of Americans protested against the pipeline in front of the White House. Obama announced that he would approve the pipeline only if it "does not significantly exacerbate the problem of carbon pollution."

In February 2015, after many legal challenges and amendments, the proposal was approved by the U.S. House of Representatives but subsequently vetoed by President Obama. A vote in the Senate in March of 2015 failed to achieve the two-thirds majority required to overturn the presidential veto. Republicans promised to find a way to revive the project. Even President Obama signalled the possibility that his veto could be lifted if his concerns were answered. In November, 2015, the application was finally rejected, and the project is now at a standstill for the foreseeable future.

Meanwhile, once the Keystone XL project appeared to have stalled, the Government of Canada actively pursued plans to move ahead with the construction of the Enbridge Northern Gateway Pipelines, a set of pipelines that would carry natural gas to the east and oil to the west from Alberta to British Columbia (see **FIGURE 1**; the western route is shown). The Northern Gateway is seen by petroleum producers as opening up new possibilities for the sale of oil from the oil sands to China and other Asian markets.

The western route of Northern Gateway passes for 520 km through Alberta and 657 km through British Columbia, terminating at Kitimat. The proposal has ignited a firestorm of controversy and opposition among environmentalists and Aboriginal groups. Opponents are particularly concerned that the pipeline route crosses hundreds of salmon-bearing streams. The terminus at Kitimat is in the Great Bear Rainforest, one of the largest areas of unspoiled temperate rain forest left in the world. From Kitimat, enormous oil tankers would transport the oil through the winding passageways of the Douglas Channel and the Inland Passage to the Pacific Ocean and on to markets in Asia.

A major part of the controversy for both pipelines surrounds the diluted bitumen, or dilbit, that would be transported through both the Keystone XL and Northern Gateway pipelines. Dilbit is a very controversial substance. Naphtha, the most common agent used to dilute the thick, heavy bitumen, is both carcinogenic and flammable. Some studies have suggested that dilbit is more corrosive than regular oil, and thus is more likely to damage pipelines, causing leaks; the petroleum industry argues that the temperatures in pipelines are not high enough to activate the corrosiveness of the dilbit.

Thomas King of the Centre for Offshore Oil, Gas, and Energy Research of Fisheries and Oceans Canada and colleagues investigated the possible fate of dilbit if it were to be spilled in the marine environment.[21] In their 2014 study, the researchers acknowledged large gaps in understanding of the behaviour of dilbit during oceanic spills. In particular, they investigated conditions that might lead the product to sink. Oil that sinks during a spill is difficult to manage; most approaches to the cleanup of marine oil spills are easier to implement with oil that floats to the surface. If the product sinks, it is more difficult to locate and collect, and thus has greater potential to cause environmental damage. Laboratory investigations and modelling by these researchers confirmed that some types of dilbit will indeed sink, even in low-energy marine environments (such as inland channels without heavy wave action).

On June 17, 2014, the Government of Canada approved the proposal for the Northern Gateway Pipeline. As part of the approval, more than 200 issues of concern identified in 2013 by a Joint Review Panel must be resolved. Among these is a requirement for Enbridge to carry out consultations with First Nations. Arguments continue as to whether the pipelines will be safe or will lead inevitably to spills that will affect sensitive environments; whether the projects will decrease or increase the price of oil; whether jobs will be created or lost, and whether any jobs would be permanent or temporary; and whether this enormous new supply of oil will have a permanent impact on the climate system.

The divergent views on Canada's oil sands reflect our confounding relationship with fossil fuels. These energy sources power our civilization and have enabled our modern standard of living—yet as climate change worsens, we face the need to wean ourselves from them and shift to clean renewable energy sources. The way in which we handle this complex transition will determine a great deal about the quality of our lives and the future of our society and our planet.

---

conventional crude oil in the world), but most of this will not easily be extracted. About 40% of global oil shale reserves are in the United States, mostly on federally owned land in Colorado, Wyoming, and Utah. Oil shale is costly to extract, and its EROI is very low, with the best estimates ranging from just 1.1:1 to 4:1. Historically, low prices for crude oil have kept investors away, but when crude oil prices rise, oil shale again attracts attention.

An increasingly important unconventional source of fossil fuel is **shale gas**. Natural gas can be produced from shale by fracking. When fracking is used for the extraction of shale gas, however, it is typical for a very large horizontal

Blackfox Images/Alamy

(a) Alberta strip mining

FIGURE 15.17

In Alberta, companies strip-mine oil sands with the world's largest dump trucks and power shovels **(a)**. Near-surface deposits **(b)** are dug out (1) with gigantic shovels and poured into a crushing machine (2). The material is mixed with hot water (3), and the slurry is piped to a facility (4) where bitumen is separated from sand and clay (5). It is then sent by pipeline (6) to a refinery. Deeper deposits **(c)** are extracted through well shafts. Pressurized steam is injected into the well (1) in the oil sand formation, liquefying the bitumen and allowing it to be pumped to the surface (2).

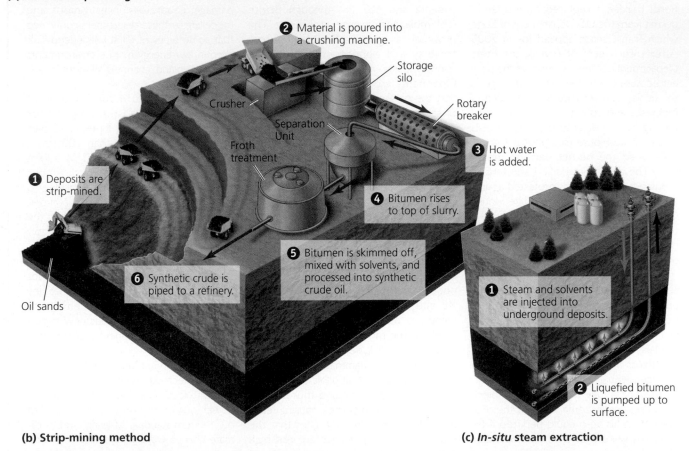

(b) Strip-mining method

(c) *In-situ* steam extraction

area to be fractured, leading to significant negative environmental impacts. It is likely that the amount of natural gas available to be extracted from shale and other unconventional sources will turn out to be overwhelmingly larger than the amount of oil that may be derived from these sources.

Methane hydrate is another form of natural gas

Methane hydrate (also called *methane clathrate* or *methane ice*) is a solid substance that consists of molecules of methane within a crystal lattice of water ice molecules (**FIGURE 15.18**). It occurs underground in some Arctic locations and more widely under the seafloor on the continental shelves. Most methane in these gas hydrates was formed by bacterial decomposition in anaerobic environments, but some results from thermogenic formation deeper below the surface. Scientists believe there to be immense amounts of methane hydrate on Earth—from perhaps as much as 20 times the amount of natural gas from all other sources.

However, there is currently no technology to extract methane hydrate safely. Japan recently showed that it could

FIGURE 15.18
Methane hydrate or methane ice, shown here, is methane gas contained within a lattice of water ice molecules. Methane gas bubbles up naturally from areas where partially decomposed organic matter is submerged on the continental shelves. Warming of ocean water raises concerns that methane, a powerful greenhouse gas, may be released from seafloor sediments that are currently at their freezing temperatures.

extract methane hydrate from the seafloor by sending down a pipe and lowering pressure within it so that the methane turned to gas and rose to the surface. However, we do not yet know whether such extraction is safe and reliable. Destabilizing a methane hydrate deposit on the seafloor during extraction could lead to a catastrophic release of gas. This could cause a massive landslide and tsunami, and would release huge amounts of methane, a potent greenhouse gas, into the atmosphere, worsening global climate change.

Alternative fossil fuels have significant environmental impacts

Alternative and unconventional fossil fuels are abundant, but they are no panacea for our energy challenges. For one thing, their net energy values are typically low because they are expensive to extract and process. Thus the energy returned on energy invested (EROI) ratio is low. For instance, at least 40% of the energy content of oil shale is consumed in its production, and oil shale's EROI is only about 2:1 or 3:1, compared with 15:1 for conventional crude oil. Natural gas extracted from the gas fields of the Mackenzie River delta might never make it all the way to southern consumers if it must be sidetracked to support the extraction of oil from the Athabasca oil sand deposits.

Second, extraction of these fuels exerts severe environmental impacts, as you have seen. Oil sands and oil shale require strip mining and/or fracking, which utterly devastate landscapes over large areas and pollute

waterways that run into other areas. Although most governments require mining companies to restore mined areas to their original condition, regions denuded by the very first oil sand mine in Alberta 30 years ago have still not recovered.

Canadian environmentalists also are worried about the intensive water use that typically accompanies the extraction of unconventional fossil fuels through methods such as fracking, as well as the impacts on water quality in surrounding regions. The impacts on wildlife are also of concern; this concern was brought into stark focus in 2008 when hundreds of migratory birds died when their feathers became fouled with oil after landing on Syncrude's massive tailings pond near Fort McMurray, Alberta (**FIGURE 15.19**). A similar event happened in 2010, just days after Syncrude agreed to pay a penalty of $3 million for the previous event.

(a) An oil sand tailings pond in Alberta

(b) Rescuing oiled sea ducks

FIGURE 15.19
Concerns about the impacts of oil sand exploitation on wildlife came to a head in 2008 when hundreds of migratory birds died after landing in an oily, toxic tailings pond at Fort McMurray, Alberta, similar to the one shown here **(a)**. Only a few of the oiled sea ducks managed to survive **(b)**.

To give you an idea of the magnitude of these tailings ponds, which can hold up to 540 000 000 m³ of oily sludge, consider that the Syncrude tailings dam is the largest dam in Canada, and second in the world only to the Three Gorges hydroelectric dam in China. Canada's largest hydroelectric dam, the Gardiner Dam in Saskatchewan, has just over one-tenth the reservoir capacity of the Syncrude tailings dam.[22]

Besides impacts from their extraction, the combustion of alternative fossil fuels emits at least as much carbon dioxide, methane, and other air pollutants as does our use of coal, oil, and gas. Thus, they contribute to the effects that fossil fuels are already causing, including air pollution, global climate change, and ocean acidification. These are not alternative energy sources: They are a continuation of the fossil fuel scenario that we already have.

Environmental Aspects of Fossil Fuel Dependency

Our society's love affair with fossil fuels and the many petrochemical products we have developed from them has boosted our material standard of living beyond what our ancestors could have dreamed, has eased constraints on travel, and has helped lengthen our lifespans. It has, however, also caused harm to the environment and human health. Concern over these impacts is a prime reason many scientists, environmental advocates, businesspeople, and policymakers are increasingly looking toward renewable sources of energy that exert less impact on natural systems.

Fossil fuel emissions cause air pollution and drive climate change

When we burn fossil fuels, we alter flux rates in Earth's carbon cycle. We essentially take carbon that has been retired into a long-term reservoir underground and release it into the air. This occurs as carbon from within the hydrocarbon molecules of fossil fuels unites with oxygen from the atmosphere during combustion, producing carbon dioxide (CO_2). Carbon dioxide is a greenhouse gas, and CO_2 released from fossil fuel combustion warms our planet and drives changes in global climate. Because global climate change may have diverse, severe, and widespread ecological and socioeconomic impacts, carbon dioxide pollution (**FIGURE 15.20**) is becoming recognized as the greatest environmental impact of fossil fuel use.

Fossil fuels release more than carbon dioxide when they burn. Methane is a potent greenhouse gas, and other air pollutants resulting from fossil fuel combustion can have serious consequences for human health and the environment. Deposition of mercury and other pollutants from coal-fired power plants is increasingly recognized as a substantial health risk. The burning of fossil fuels in power plants and vehicles releases sulphur dioxide and nitrogen oxides, which contribute to industrial and photochemical smog and to acidic deposition.

We have already employed technologies, such as catalytic converters, to cut down on vehicle exhaust pollution. Gasoline combustion in automobiles releases pollutants that irritate the nose, throat, and lungs. Some hydrocarbons, such as benzene and toluene, are carcinogenic to laboratory animals and likely also to people. In addition, gases, such as hydrogen sulphide, can evaporate from crude oil, irritate the eyes and throat, and cause asphyxiation. Crude oil also often contains trace amounts of known poisons, such as

FIGURE 15.20
As industrialization has proceeded, and as population and energy consumption have grown, emissions from fossil fuels have risen dramatically. Here, worldwide emissions of carbon are subdivided by their source (liquid oil, solid coal, or natural gas). Also included are cement manufacture and flaring of natural gas.
Source: Global, Regional, and National Fossil-Fuel CO_2 Emissions. Carbon Dioxide Information Analysis Center, Oak Ridge National Laboratory, U.S. Department of Energy, Oak Ridge, Tenn., U.S.A.

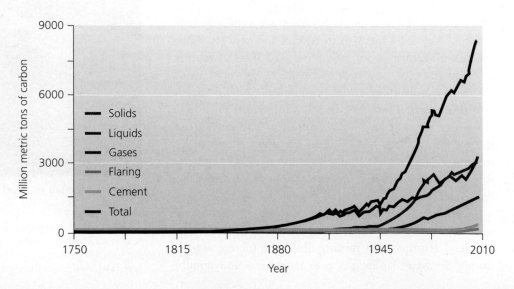

lead and arsenic. As a result, workers at drilling operations, refineries, and other jobs that entail frequent exposure to oil or its products can develop serious health problems, including cancer.

Some emissions from fossil fuel burning can be "captured"

One relatively new technology for "cleaning up" carbon-based fuel sources is **carbon capture and storage (CCS)** or *carbon capture and sequestration*. Recall from our discussion of biogeochemical cycles that *sequestration* refers to the storage of materials in geological reservoirs on a long timescale. In this case, the material of interest is carbon—primarily in the form of CO_2—and the goal is to prevent some of the carbon generated by the burning of fossil fuels from entering the atmosphere and contributing to global warming.

In a nutshell, what would happen is that the CO_2 emitted by, for example, a traditional coal-burning power plant would be captured before it reached the atmosphere and then diverted to a storage reservoir. The most likely reservoirs are saline aquifers deep underground, or spent oil and gas deposits (**FIGURE 15.21**). Deep-ocean disposal has been proposed, but this might exacerbate the problem of ocean acidification, already a concern associated with increased levels of carbon dioxide (because carbon dioxide mixed with water yields carbonic acid).

Some industry analysts have predicted that 80–90% of the carbon dioxide emissions from large emitters like coal-fired generators could be captured and diverted, as compared with a power plant without CCS technology. But many environmentalists are skeptical about CCS, arguing that the technology is unproven and that the true environmental impacts of reinjecting carbon dioxide into the ground are not known.

With regard to storage underground, some experts have expressed concern that extraction processes (especially fracking) have so extensively fractured rock units near Alberta's oil sands that they would no longer be suitable for the purpose of CCS. Still others argue that the approach is fundamentally flawed because it takes the burden off large emitters and serves only to prolong our dependence on fossil fuels rather than facilitating a shift to renewables.

Impacts reach terrestrial and aquatic environments, too

Fossil fuels can pollute water and land, as well as air. Oil that comes from nonpoint sources—such as industries, homes, automobiles, gas stations, and businesses—runs off roadways and enters rivers and sewage treatment facilities, to be discharged eventually into the ocean. The Ocean Review reports that oil from spectacular spills like that of the BP Deepwater Horizon only

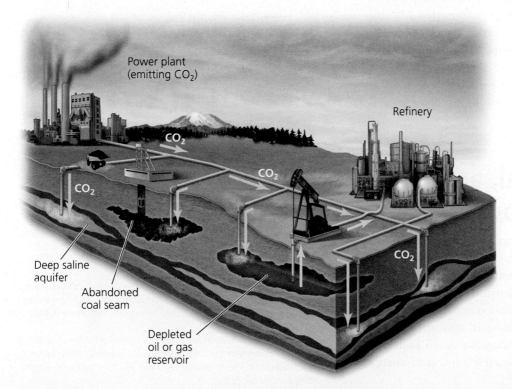

Power plant (emitting CO_2)

Refinery

CO_2

CO_2

CO_2

CO_2

CO_2

Deep saline aquifer

Abandoned coal seam

Depleted oil or gas reservoir

FIGURE 15.21
Carbon capture and storage schemes propose to inject liquefied carbon dioxide underground into depleted fossil fuels deposits, deep saline aquifers, or oil and gas deposits undergoing secondary extraction. If these technologie can be developed, it could provide a possible mechanism for reducing some of the harmful environmental impacts of fossil fuel use.

accounts for about 10% of the oil that enters the ocean. Around 5% comes from natural sources (seeps), 35% comes from oil tankers and other shipping operations. Atmospheric fallout of oil from burning, municipal and industrial runoff, and noncatastrophic releases from oil rigs account for 45%, and the remaining portion comes from undefined sources.[23]

Although most oil input into the ocean comes from nonpoint sources, catastrophic oil spills have significant impacts on the marine environment. Crude oil's toxicity to most organisms can lead to high mortality. This was the case with the massive BP Deepwater Horizon oil spill and leak of 2010, which caused the immediate deaths and injuries of thousands of birds, turtles, fish, and shellfish, as well as harming marine mammals and corals. Longer-term concerns centre on the ultimate fate of the oil as well as the *dispersants*—chemicals (also toxic) that were used to induce the oil slicks to break up and disperse through the water column.

Oil can also contaminate soil and groundwater supplies, such as when leaks from oil operations penetrate deeply into the subsurface. Of great concern are the millions of underground storage tanks at gas stations containing petroleum products that have leaked, contaminating soil and threatening drinking water supplies in some cases.

Pipeline spills and leaks are more common than you might think, or might hear about in the mainstream media. Canada's National Energy Board (NEB) tracks the reporting of leaks, spills, and other incidents associated with thousands of kilometres of oil and gas pipelines under its regulation; 76 incidents were reported to NEB in 2014, 121 in 2013, and 153 in 2012, either in Canada or in offshore areas, or in the United States on Canadian-owned pipelines.[24] Most of these are not disastrous or voluminous; even so, they can cause significant local environmental damage and raise the possibility of more catastrophic leaks in the future.

Political, Social, and Economic Aspects

The political, social, and economic consequences of fossil fuel use are numerous, varied, and far-reaching. Our discussion focuses on several negative consequences of fossil fuel use and dependence, but it is important to bear in mind that their use has enabled much of the world's population to achieve a higher material standard of living than ever before. It is also important to ask in each case whether switching to more renewable sources of energy would solve existing problems.

Oil supply and prices affect the economies of nations

Virtually *all* of our modern technologies and services depend in some way on fossil fuels. Putting all one's eggs in one basket is always a risky strategy; Canadians got a sense of this in early 2015 when the market price of oil plummeted, causing negative economic impacts. The fact that economies around the world are so closely tied to fossil fuels means that we are vulnerable to market fluctuations like these. Nations that lack adequate fossil fuel reserves of their own are especially vulnerable; for instance, Germany, France, South Korea, and Japan consume far more energy than they produce and thus rely almost entirely on imports for their continued economic well-being.

Reliance on foreign oil means that seller nations can control energy prices, forcing buyer nations to pay more and more as supplies dwindle. This became clear in 1973, when the *Organization of the Petroleum Exporting Countries (OPEC)* resolved to stop selling oil to the United States as a consequence of U.S. support of Israel. The embargo created panic in the West and caused oil prices to skyrocket (**FIGURE 15.22**), spurring inflation.

At the end of 2014, some financial observers were hoping that OPEC would cut production, thereby limiting supply and forcing oil prices to increase. That didn't happen, and Saudi Arabia, in particular, refused to cut production, resulting in an oversupply of oil. This sent the price of oil into a free fall, from over $100 US per barrel in

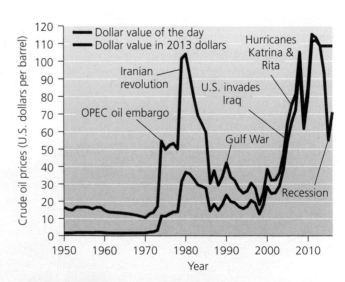

FIGURE 15.22

World oil prices have fluctuated greatly over the decades, often because of political and economic events in oil-producing countries. The greatest price hikes in recent times have resulted from wars and unrest in the oil-rich Middle East.

Source: Data from British Petroleum (2014). *Statistical Review of World Energy 2014,* report and statistical tables; and U.S. Energy Information Administration Short-Term Energy Outlook (2015). http://www.eia.gov/forecasts/steo/report/prices.cfm.

mid-2014 to $45 US per barrel in early 2015. Meanwhile, as of 2013 the United States had greatly increased its production of oil from oil shales through extensive fracking. This also contributed to oversupply. If you think the price of oil doesn't affect you directly except at the gas station, think again—the price of oil affects everything from real estate values to food prices to government spending on social services, and the job market as well. It also will affect the positions of companies with regard to investment in pipelines, and so may be the deciding factor in the fate of projects like Keystone XL, Northern Gateway, and Mackenzie Valley.

With so much of world oil reserves located in the politically volatile Middle East, crises in this region are of particular concern. This is one reason why turning to Canada's oil sands as a stable, friendly source of oil seems like an attractive alternative for U.S. policymakers. As Canadians, we will have to decide how much of our economy we want to be linked to fossil fuel production in the future; perhaps it is time to move Canada into the position of a world leader in the transition away from fossil fuels.

Residents may or may not benefit from fossil fuel resources

The extraction of fossil fuels can be extremely lucrative; many of the world's wealthiest corporations deal in fossil fuel energy or related industries. These industries provide jobs to millions of employees and supply dividends to millions of investors. Development can potentially yield economic benefits for people who live in petroleum-bearing areas, as well.

However, in many parts of the world where fossil fuels have been extracted, local residents have not seen great benefits but instead have suffered. When multinational corporations extract oil or gas in developing countries, paying those countries' governments for access, the money often does not trickle down to residents of the regions where the extraction takes place. Moreover, oil-rich developing countries, such as Ecuador, Venezuela, and Nigeria, tend to have few environmental regulations, and existing regulations may go unenforced if a government does not want to risk losing the large sums of money associated with oil development.

In Nigeria, oil was discovered in 1958 in the territory of the Ogoni, one of Nigeria's native peoples, and the Shell Oil Company moved in to develop oil fields. Although Shell extracted $30 billion of oil from Ogoni land over the years, the Ogoni still live in poverty, with no running water or electricity. The profits from oil extraction on Ogoni land went to Shell and to the military dictatorships of Nigeria. The development resulted in oil spills, noise, and constantly burning gas flares, all of which caused illness among people living nearby. From 1962 until his death in 1995, Ogoni activist and leader Ken Saro-Wiwa worked for fair compensation to the Ogoni for oil extraction and environmental degradation on their land. After years of persecution by the Nigerian government, Saro-Wiwa was arrested in 1994, given a trial universally regarded as a sham, and put to death by a military tribunal.

In 2007, President Rafael Correa of Ecuador made international news by proposing the Yasuní-ITT project, a plan to suspend oil extraction in the Ishpingo-Tambococha-Tiputini (ITT) area of Yasuní National Park in the Ecuadorian Amazon. The area holds more than 800 million barrels of oil, about 20% of the country's proven reserves. It is also one of the most biodiverse places in the world, and the homeland of the Waorani Aboriginal people, some of whom are as yet uncontacted by modern civilization. What was unusual about Correa's proposal was the offer to the global community to offset greenhouse gas emissions by leaving the oil unexploited, in exchange for commitments by world governments of half of the value of the oil. In 2013, having only raised a small portion of the required pledges, Correa announced that his government would abandon the Yasuní-ITT Project. Consequently, oil exploitation commenced in Yasuní National Park in mid-2014; it is as yet unclear what the impacts will be for Aboriginal people, biodiversity, and the nearly pristine tropical rain forest of the region.

We need to conserve energy and find renewable sources

Fossil fuel supplies are limited, and their use and our continued dependence on them have health, environmental, political, and socioeconomic consequences. Until our society makes the transition to renewable energy sources, we will need to find ways to minimize the expenditure of energy from our dwindling fossil fuel resources. **Energy conservation** is the practice of reducing energy use to extend the lifetimes of our nonrenewable energy supplies, to be less wasteful, and to reduce our environmental impact.

Many people first saw the value of conserving energy following the OPEC embargo of 1973–1974. In the subsequent three decades, however, many of the conservation initiatives that followed the oil crisis were abandoned. Without high market prices and an immediate threat of shortages, people lacked economic motivation to conserve. Government funding for research into alternative energy sources decreased, speed limits increased, and countless proposals to raise the mandated average fuel efficiency of vehicles failed. The average fuel efficiency of new vehicles worsened significantly, primarily as a result of increased sales of light trucks and sport utility vehicles.

All of this is changing, though, particularly in light of major fluctuations in oil prices. In 2008, oil traded at a record high price of $147 US per barrel, then declined, soaring again in 2011, and plummeting to less than $45 per barrel in early 2015. A survey of Canadian spending habits conducted by Investors Group revealed that 83% of Canadians planned to buy a more fuel-efficient car next time around, 51% had been cutting down on driving, and 44% changed their holiday plans in response to fluctuations in fuel prices.[25] These types of consumer changes are happening with equal fury in the United States, and as a result the light truck and SUV industry in North America has undergone a sudden, severe contraction, with a number of production facilities closing.

Transportation accounts for two-thirds of oil use and more than a quarter of energy use in Canada, and passenger vehicles consume more than half this energy. The vast distances of the Canadian landscape add to the problem. Transportation also accounts for about a third of Canada's greenhouse gas emissions, an increasing concern as we strive to meet our goals regarding climate change.[26]

Personal choice and increased efficiency are two routes to conservation

Energy conservation can be accomplished in two primary ways. As individuals, we can make conscious choices to reduce our own energy consumption. Examples include driving less, turning off lights when rooms are not being used, turning down thermostats, and investing in more-efficient machines and appliances. For any given individual or business, reducing energy consumption can save money while also helping to conserve resources.

As a society, we can conserve energy by making our energy-consuming devices and processes more efficient. Currently, more than two-thirds of the fossil fuel energy we use is simply lost, as waste heat, in automobiles and power plants. In the case of automobiles, we already possess the technology to increase fuel efficiency far above the current North American average. We could accomplish this with more efficient gasoline engines, lightweight materials, continuously variable transmissions, and alternative technology vehicles, such as electric/gasoline hybrids or vehicles that use hydrogen fuel cells.

We can also vastly improve the efficiency of our power plants. One way is to use *cogeneration*, in which excess heat produced during the generation of electricity is captured

FIGURE 15.23
One way to determine how much heat a building is losing is to take a photograph that records energy in the infrared portion of the electromagnetic spectrum. In such a photograph, or *thermogram* (shown here), the warmest areas are white, and yellow and red also signify warm temperatures at the surface of the house, where heat is escaping. Blue and green shades signify cold and cool temperatures, where the building is not allowing heat to escape.

and used to heat workplaces and homes and to produce other kinds of power. Cogeneration can almost double the efficiency of a power plant. The same is true of *coal gasification* and *combined cycle* generation. In this process, coal is treated to create hot gases that turn a gas turbine, while the hot exhaust of this turbine heats water to drive a conventional steam turbine.

In homes and public buildings, a significant amount of heat is lost in winter and gained in summer because of inadequate insulation (**FIGURE 15.23**). Improvements in the design of homes and offices can reduce the energy required to heat and cool them. Such design changes can involve the building's location, the colour of its roof (light colours keep buildings cooler by reflecting the Sun's rays), and its insulation.

Among consumer products, scores of appliances, from refrigerators to lightbulbs, have been reengineered through the years to increase energy efficiency. Energy-efficient lighting, for example, can reduce energy use by 80%, and new energy-efficient appliances have already reduced per-person home electricity use below what it was in the 1970s. Even so, there remains room for further improvement.

While manufacturers can improve the energy efficiency of appliances, consumers need to "vote with their wallets" by purchasing these energy-efficient appliances. Decisions by consumers to purchase energy-efficient products are crucial in keeping those products commercially available. For the individual consumer, studies show that the slightly higher cost of buying energy-efficient washing machines is rapidly offset by savings on water and electricity bills. On the national level, France, Denmark, and many other developed countries have standards of living equal to that of Canada, but they use much less energy per capita. This disparity indicates that Canadian citizens could significantly reduce their energy consumption without decreasing their quality of life.

Conclusion

It is often said that reducing our energy use is equivalent to finding a new oil reserve. Indeed, conserving energy is better than finding a new reserve because it lessens impacts on the environment while extending our access to fossil fuels.

However, energy conservation does not add to our supply of available fuel. Regardless of how much we conserve, we will still need energy, and it will need to come from somewhere. The only sustainable way of guaranteeing ourselves a reliable long-term supply of energy is to ensure sufficiently rapid development of renewable energy sources, which we will consider in greater detail in the next chapter.

Over the past 200 years, fossil fuels have helped us build the complex industrialized societies we enjoy today. However, we are now approaching a turning point in history: Our production of fossil fuels will begin to decline. We can respond to this new challenge in creative ways, encouraging conservation and developing alternative energy sources. Or we can continue our current dependence on fossil fuels and wait until the near depletion before we try to develop new technologies and ways of life. The path we choose will have far-reaching consequences for human health and well-being, for Earth's climate, and for our environment.

The ongoing debates over projects like the Keystone XL, Northern Gateway, and Mackenzie River valley pipelines are microcosms of this debate over our energy future. Fortunately, there is not simply a trade-off between benefits of energy for us and harm to the environment, climate, and health. Instead, as evidence builds that renewable energy sources are becoming increasingly feasible and economical, it becomes easier to envision giving up our reliance on fossil fuels and charting a win–win future for humanity and the environment.

REVIEWING OBJECTIVES

You should now be able to

Summarize the principal energy sources that we use and the role of fossil fuels in our energy mix

- We use a wide variety of energy, different types for different purposes. Some are renewable; others are nonrenewable on a humanly accessible timescale. Since the Industrial Revolution our energy mix has been dominated by fossil fuels: coal, natural gas, and oil.

- Industrialized nations typically use far more energy, overall and per capita, than developing nations. Industrialized nations also depend more heavily on fossil fuels, especially oil, in their energy mix.

- It takes energy to make energy. When evaluating an energy source, it is important to consider the EROI, that is, how much energy must be invested to extract usable energy from that source.

- Fossil fuels are a long-term reservoir in the global carbon biogeochemical cycle. They are formed when organic matter is partially decomposed in anaerobic conditions and then subjected to elevated temperatures and pressures underground over time, with resulting undergoing chemical transformations. The organic precursors for coal come mainly

from terrestrial plant matter; for natural gas and oil, they come mainly from marine phytoplankton and zooplankton.

Describe the nature and origin of coal and evaluate its extraction and use

- People have used coal as a fuel longer than any other fossil fuel. Since the late 1800s it has been widely used to generate electricity.

- The first step in formation of coal is peat, which is a wet, compressed organic soil. Coal, a rock, develops with further heat and pressure deep underground, as volatiles are expelled and the hardness and energy context of the material increase.

- Coal is mined in the subsurface for deeper deposits, and at the surface by open-pit or strip mines for shallower deposits. Coal mining presents many hazards, both to the health of miners and in terms of environmental impacts, including acid drainage and selenium contamination of waterways.

- In coal-fired power plants, coal is burned to create steam, which turns a turbine to generate electricity.

- "Clean coal" technologies can be implemented to mitigate some of the negative environmental impacts of coal, the dirtiest of the fossil fuels. Approaches can include cleaning prior to combustion, more efficient combustion, and capturing emissions before they leave the stack.

Describe the nature and origin of natural gas and evaluate its extraction and use

- Natural gas is the most recent of the fossil fuels to be widely used. It is increasingly favoured because it is cleaner-burning than coal or natural gas.

- Biogenic natural gas is formed at shallow depths, by anaerobic decomposition of organic matter. Thermogenic natural gas is formed at greater depths in a manner and environment similar to those of the formation of oil.

- Natural gas is extracted by drilling and pumping. When extraction becomes more challenging, hydraulic fracturing, or fracking, is used. Fracking, also used for bitumen and some types of oil, involves injecting fluids into the ground to facilitate extraction.

Describe the nature and origin of oil and evaluate its extraction, use, and future availability

- Oil is the main fuel in the world's energy mix today. Kerogen, the precursor to oil, undergoes chemical transformations during the process of maturation underground, resulting in the complex mixture of hydrocarbon compounds that is crude oil.

- Geologists use mapping, seismic surveys, and exploratory drilling to find and assess new deposits of oil. New finds may or may not be technologically or economically recoverable.

- Oil is extracted by drilling and pumping in primary recovery. If this becomes too difficult, the company may turn to secondary recovery, which involves the injection of fluids to facilitate the release and extraction of the oil.

- Both onshore and offshore drilling entail risks and negative environmental impacts, which can include failures due to challenging weather conditions, leaks and spills, impacts on wildlife, and impacts on sensitive environments such as Arctic tundra.

- In refineries, distillation columns are used to separate heavier (long-chain) oil components from lighter (short-chain) components. Favoured lighter oil products can be developed by cracking, which turns long-chain hydrocarbons into short-chain molecules.

- Oil and other fossil fuels are nonrenewable resources. The "peak oil" scenario suggests that we will inevitably see a decline in world oil production as recoverable resources are depleted.

Describe the nature, origin, and potential of unconventional fossil fuel types and technologies

- Oil sands (also called tar sands) are deposits of sand and clay that contain bitumen, a tarry substance. Canada's oil sand deposits are massive, and account for our position as third in the world for proven oil reserves.

- Oil shale, tight oil, and shale gas are three different forms of unconventional hydrocarbons. In the United States, shale gas is now being recovered with the use of fracking.

- Methane hydrate is natural gas trapped in frozen water molecules. It represents an enormous reservoir of natural gas, but is not technologically recoverable at this time. The possibility that large quantities of methane might be released as a result of the warming of seafloor sediments is worrisome because of the potential climate impacts.

Outline and assess the environmental impacts of continued fossil fuel dependency

- The combustion of fossil fuels emits pollutants that can lead to acid deposition, but more importantly, contributes greenhouse gases that lead to climate change.

- Carbon capture and storage technologies have been proposed as a way of mitigating some of the impacts of carbon dioxide releases from the combustion of fossil fuels. So far these technologies have not been widely applied.

- Fossil fuel extraction and use also have impacts on terrestrial and aquatic environments. Spills and leaks are common. Most of the oil that enters the ocean actually comes from nonpoint sources on land, although catastrophic spills can have devastating local impacts.

Evaluate political, social, and economic impacts of fossil fuel use, and strategies for conserving energy and enhancing efficiency

- Supply and price of fossil fuels, especially oil, affect economies around the world.

- When oil extraction occurs, especially where democratic governance and environmental regulations are not strong, residents do not necessarily benefit.

- Conservation can help us extend fossil fuel availability while we transition to more renewable energy sources. Price and availability affect consumer behaviour, especially with regard to transportation choices.

- We can make individual choices to conserve energy, such as choosing more energy-efficient vehicles or checking the energy efficiency of our homes and offices. Governments and companies can make choices to improve the energy efficiency of industry.

TESTING YOUR COMPREHENSION

1. Why are fossil fuels our most prevalent source of energy today? Why are they considered nonrenewable sources of energy?
2. How are fossil fuels formed? How do environmental conditions determine what type of fossil fuel is formed in a given location? Why are fossil fuels often concentrated in localized deposits?
3. Describe how net energy differs from energy returned on investment (EROI). Why are these concepts important when evaluating energy sources?
4. Describe how coal is used to generate electricity.
5. Why is natural gas often extracted simultaneously with other fossil fuels? What constraints on its extraction does it share with oil?
6. How do geologists estimate the total amount of oil reserves that remain underground? How is the

"technically recoverable" oil different from the "economically recoverable" oil?
7. How do we create petroleum products? Provide examples of several of these products.
8. What is Hubbert's peak? Why do many experts think we are about to pass the global production peak for oil? What consequences could there be for our society if we do not transition soon to renewable energy sources?
9. List three environmental impacts of fossil fuel production and consumption. Compare some of the contrasting views of scientists regarding the environmental impacts of the Mackenzie River valley natural gas pipeline.
10. Describe two main approaches to energy conservation; give specific examples of each.

THINKING IT THROUGH

1. Compare the effects of coal and oil consumption on the environment. Which process do you think has ultimately been more detrimental to the environment, oil extraction or coal mining, and why? What steps could governments, industries, and individuals take to reduce environmental impacts?
2. You have been elected to be a negotiator on behalf of Aboriginal interests in the ongoing discussions about the development of the Mackenzie River valley natural gas pipeline. What will your position be? Explain it to another negotiator who disagrees with you.
3. Would it be more difficult to contain and remediate a major oil spill, like the BP Deepwater Horizon spill of 2010, if it were to happen in the Beaufort Sea or in

deep water off the coast of Newfoundland, compared to the Gulf of Mexico? What would be some of the important differences?
4. Throughout this book we have asked you to imagine yourself in various roles. This time we ask you simply to be yourself. Given the information in this chapter on petroleum supplies, consumption, and depletion, what actions, if any, do you plan to take to prepare yourself for changes in our society that may come about as oil production declines? Describe how you think your life may change, and suggest one thing you could do to help reduce negative impacts of both oil dependence and oil depletion on our society.

INTERPRETING GRAPHS AND DATA

The fossil fuels that we burn today were formed long ago from buried organic matter. However, only a small fraction of the original organic carbon remains in the coal, oil, or natural gas that is formed. Thus, it requires approximately 90 metric tons of ancient organic matter—so-called paleoproduction—to result in just 3.8 L of gasoline. The

graph presents estimates of the amount of paleoproduction required to produce the fossil fuels humans have used each year over the past 250 years.

1. Estimate in what year the annual consumption of paleoproduction, represented by our combustion

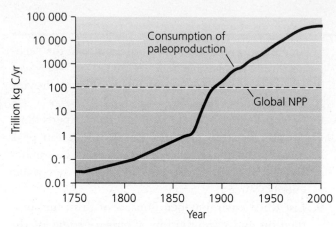

Annual human consumption of paleoproduction by fossil fuel combustion (red line), 1750–2000. The dashed line indicates current annual net primary production (NPP) for the entire planet.
Source: Data from Dukes, J. (2003). Burning Buried Sunshine: Human Consumption of Ancient Solar Energy. *Climatic Change, 61*, pp. 31–44.

of fossil fuels, surpassed Earth's current annual net primary production (NPP).

2. In 2000, approximately how many times greater than global NPP was our consumption of paleoproduction?

3. If on average it takes 7000 units of paleoproduction to produce 1 unit of fossil fuel, estimate the total carbon content of the fossil fuel consumed in 2000. How does this amount compare to global NPP?

MasteringEnvironmentalScience®

STUDENTS
Go to **MasteringEnvironmentalScience** for assignments, the eText, and the Study Area with practice tests, videos, current events, and activities.

INSTRUCTORS
Go to **MasteringEnvironmentalScience** for automatically graded activities, current events, videos, and reading questions that you can assign to your students, plus Instructor Resources.

16 Energy Alternatives

(a) Low tide

(b) High tide

The Bay of Fundy is shown here at low tide **(a)** and high tide **(b)**.

Laszlo Podor/Alamy

Upon successfully completing this chapter, you will be able to

- Outline the reasons for seeking alternatives to fossil fuels, and the conventional and new alternatives that are being developed
- Describe the technologies, environmental impacts, and energy contributions of the conventional alternatives to fossil fuels: hydroelectric power, nuclear power, and traditional biomass energy

- Summarize the major "new renewable" alternative sources of energy, and assess their potential for growth
- Describe a variety of new biomass, solar, wind, geothermal, ocean energy, and hydrogen fuel cell technologies, and outline their advantages and disadvantages

Water rushes from the sluice gates at the Annapolis Tidal Power Station in Nova Scotia.

CENTRAL CASE
HARNESSING TIDAL ENERGY AT THE BAY OF FUNDY

"I believe that water will one day be employed as fuel, that hydrogen and oxygen which constitute it, used singly or together, will furnish an inexhaustible source of heat and light. . . . Water will be the coal of the future."

—JULES VERNE, IN *THE MYSTERIOUS ISLAND*, 1874

"Not only will atomic power be released, but someday we will harness the rise and fall of the tides and imprison the rays of the Sun."

—THOMAS A. EDISON, 1921

"EXTREME HAZARD," reads a posted sign. "Incoming tide rises 5 feet [1.5 m] per hour and may leave you stranded for 8 hours." Not only do ocean tide waters rise quickly, but nowhere else in the world is the difference between low tide and high tide as great as here, in the Bay of Fundy on Canada's Atlantic coast (see chapter-opening photo).

The Bay of Fundy is a long, narrow bay that separates the provinces of Nova Scotia and New Brunswick, just touching the U.S. state of Maine. The name "fundy" is thought to have come from early Portuguese explorers, who called it the "deep river," or *rio fundo*.

The long, narrow, deep configuration of the bay is responsible for its extreme tidal variation. When the tide comes in, a large volume of ocean water (about 100 billion tonnes) rushes in, against the outgoing freshwater flow, up the length of the narrowly constricted bay (see satellite photos). This phenomenon is called a *tidal bore*. The result is a very large vertical difference between high and low tide (17 m and more near the extreme head of the bay, but more typically 14 m in the main part of the bay). The large tidal range makes this one of the most suitable locations in the world for generating power from ocean tides.

The Bay of Fundy is the site of one of only seven operating tidal power plants in the world (though more

(a) Low tide   **(b) High tide**

Satellite images show low tide **(a)** and high tide **(b)** in the Bay of Fundy.

are proposed or under construction). The Annapolis Royal Tidal Power Generating Station, owned by Nova Scotia Power, is located on a small side-basin on the Nova Scotia side of the main bay, by the town of Annapolis Royal. The power station, which opened in 1984, works on the same principle as other hydroelectric power generation—the movement of rushing water turns the blades of a turbine, which, in turn, runs a generator to produce electricity.

In the case of tidal power, a *sluice*—like a dam with movable openings—is constructed across a narrow part of the water body. When the tide comes in, the sluice gates are opened and the water flows through. At high tide, the gates are closed. When the tide starts to go out, the water above the sluice is held behind the closed gates. Once there is a sufficient difference in water level (1 m, in the case of the Annapolis power station), the gates are opened. The water rushes out from behind the sluice to join the rest of the outgoing tide (see photo), turning the turbine blades in the process. At this station, Nova Scotia Power generates enough power to run about 4000 homes.[1]

Many of the negative impacts of traditional hydro-electric power generation are associated with the creation of large reservoirs of standing water behind dams; tidal power does not entail these negative impacts, because the water has to be retained behind the sluice gates only for brief periods. The few negative environmental impacts of tidal power generation are mainly associated with interference in the normal currents of the water body. In one case, for example, a whale is thought to have died after following some fish through the sluice gates and becoming trapped.

Nova Scotia Power and its partners are among those working on new *in-stream* tidal power technologies. The Fundy Ocean Research Center for Energy (FORCE), located at the Bay of Fundy, is a research centre that focuses mainly on the development of in-stream tidal energy technologies. These approaches involve under-water turbines—like underwater windmills—that would eliminate the need to construct a visible dam or sluice across the waterway. As in-stream technologies are developed, the environmental impacts of tidal power should become even less significant than they are currently.

Alternatives to Fossil Fuels

Fossil fuels helped drive the Industrial Revolution, feed the explosively growing world population, and create the unprecedented material prosperity we enjoy today. Our global economy is largely powered by fossil fuels: Almost 80% of the world's energy comes from oil, coal, and natural gas, and these three fuels generate two-thirds of the world's electricity (**FIGURE 16.1**). However, they are nonrenewable energy sources that will not last forever. Oil production may already be peaking, and easily extractable supplies of oil and natural gas may not last more than half a century longer. Moreover, the use of coal, oil, and natural gas entails substantial negative environmental impacts. The market prices for fossil fuels do not incorporate the costs of these impacts, referred to in economic terms as *externalities*; this makes it more challenging for alternative energy sources to be competitive financially.

For these reasons, most scientists and energy experts, many economists and policymakers, and certainly most environmentalists accept that we will need to shift from fossil fuels to energy sources that are less readily depleted and gentler on our environment. Developing alternatives to fossil fuels has the added benefit of helping to diversify an economy's mix of energy, thereby lessening the hazards of price volatility and dependence on foreign fuel imports.

There is a diverse range of possible alternatives to fossil fuels; the most widely used are mentioned in **FIGURE 16.1**. Most of these alternative energy sources are renewable, and most have less impact on the environment than fossil fuels, although no energy source is completely without environmental impacts. In this chapter we will explore the "conventional" alternatives to fossil fuels; then we will look more closely at some "new renewable" alternatives.

Hydro, nuclear, and biomass are "conventional" alternatives

Three alternative energy sources are currently the most widely used worldwide; they are hydroelectric power, nuclear power, and traditional biomass fuels (see **FIGURE 16.1**). Each of these well-established energy sources plays a substantial role in the energy and electricity budgets of nations today. We therefore will refer to them as "conventional" alternatives to fossil fuels.

Fuelwood and other traditional biomass sources provide about 9% of the world's energy; nuclear power about 2.6% (but about 11% of the energy used for electricity generation); and hydropower about 3.8% (about 16.4% of the energy used for electricity generation).[2] In some respects, this trio of conventional energy alternatives is an odd collection. They are generally considered to exert less environmental impact than fossil fuels, but more than the "new renewable" alternatives we will discuss later in this chapter. Yet, as you will see, they each involve a complex mix of benefits and drawbacks for human well-being and the environment.

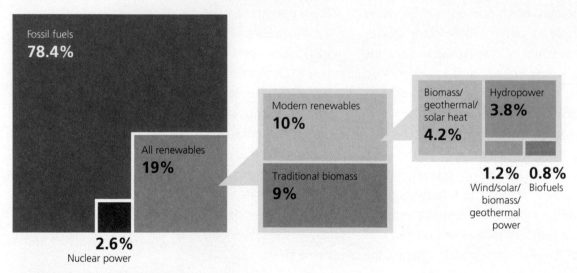

(a) Global final energy consumption, 2012

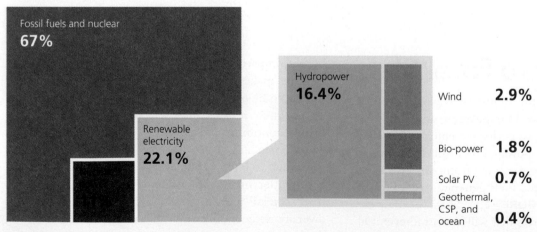

(b) Global electricity production, 2013

FIGURE 16.1 **(a)** Alternatives to fossil fuels accounted for almost 22% of final energy consumption globally, and this market share is growing. The scenario is slightly different for electricity generation **(b)**, where nuclear and hydro are considerably more important; however, fossil fuels still account for the lion's share of energy.
Source: From *Renewables 2014 Global Status Report,* Renewable Energy Policy Network for the 21st Century, http://www.ren21.net/ren21activities/globalstatusreport.aspx. Estimated Renewable Energy Share of Global Final Energy Consumption, 2012 (Figure 1, p. 21).

The "new renewables" are still being developed

Some of the energy sources that are potential alternatives to fossil fuels include energy from sunlight, wind, Earth's geothermal heat, ocean water, and hydrogen, as well as new technologies for producing biomass energy. These "new renewables" are increasingly being used for power generation, heating and cooling, and transport, and their installed capacity is growing quickly. At this time, however, most of the alternatives are still more costly than fossil fuels, and many still depend on technologies that are not yet fully developed.

These energy sources are not truly new; they are as old as our planet, and people have used them for millennia for heating their homes, turning windmills and waterwheels, moving vessels, and countless other applications. We call them "new" because they are not yet used on a wide scale in our modern industrial society. This is partly because many of the new technologies cannot yet rival fossil fuels from a financial perspective, and partly because so many modern technologies are specifically designed to function with fossil fuels.

The new renewable energy sources are still in a rapid phase of development. As of 2012–2013, "new renewables" (that is, not including hydro and traditional biomass) accounted for more than 16% of total energy consumption and almost 6% of the energy used for electricity generation. They will likely come to play a much larger role in the future as they become more technologically adaptable, and more financially competitive.

Hydroelectric Power

People worldwide draw more energy from the motion of water than from any other renewable source except biomass. In **hydroelectric power** (or **hydropower**), the kinetic energy of moving water is used to turn turbines and generate electricity. Canada's energy mix is somewhat different from that of the rest of the world, because almost 60% of our electricity is generated from hydroelectric power (**FIGURE 16.2A**). In the United States, for comparison, it is much more common for electricity to be generated by coal-fired power plants or nuclear energy than by running water. Hydropower accounts for more than 12% of Canada's total primary energy (which includes fuels for transportation, heating, cooling, and industrial applications, as well as electricity generation).[3]

Hydropower now accounts for 3.8% of the world's total energy supply but 16.4% of the energy used for electricity generation[4] (and, again, a considerably larger proportion in Canada). For nations with large amounts of river water and the economic resources to build dams, hydroelectric power has been a keystone of their development. The largest

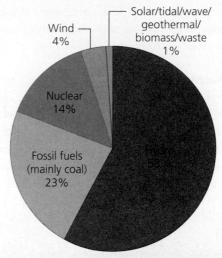

(a) Electricity generation in Canada by fuel type, 2013

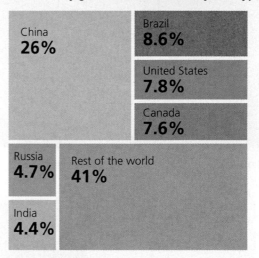

(b) Installed hydropower capacity, global market share, 2013

FIGURE 16.2

In Canada we depend heavily on abundant hydropower sources for electricity generation **(a)**; nuclear energy and coal-fired power generation are also important in some provinces. The chart **(b)** shows hydropower capacity in the top six countries for 2013, in terms of the global market share of hydro as an energy source for power generation. China leads the world in installed capacity, and in growth per year. *Source:* For (a), adapted from The Shift Project Data Portal: Breakdown of Electricity Generation by Energy Source, http://www.tsp-data-portal. org/Breakdown-of-Electricity-Generation-by-Energy-Source#tspQvChart. Accessed 7 February 2015; additional data from IAEA PRIS (Power Reactor Information System): Nuclear Share of Electricity Generation in 2013, http://www.iaea.org/PRIS/home.aspx. Accessed 7 February 2015. For (b), from REN21 (2014). *Renewables 2014 Global Status Report*, Renewable Energy Policy Network for the 21st Century, Paris: REN21 Secretariat, Figure 10, p. 44, Hydropower Global Capacity, Shares of Top Six Countries, 2013.

producing region is China, followed fairly distantly by Brazil, the United States, Russia, and Canada (**FIGURE 16.2B**).

Modern hydropower uses two approaches

People have long harnessed the power of moving water. Waterwheels spun by swiftly flowing river water powered millwheels in past centuries. Today we utilize the kinetic energy of water in two major ways: with large dams, and with "run-of-river" technologies.

Impoundment Most hydroelectric power today comes from holding water in reservoirs behind large concrete dams that block the flow of river water, and then controlling the flow of the water through the dam. Because the water is held in reservoirs behind dams, this approach is variously referred to as a *reservoir*, *storage*, or *impoundment* approach. If you have ever seen the Gardiner Dam on the South Saskatchewan River (Canada's largest hydroelectric dam), the Sir Adam Beck Generating Stations on the Niagara River (**FIGURE 16.3A**), the Hoover Dam on the Colorado River, or any other large dam, you have seen an impoundment approach being used for hydroelectric power generation.

When reservoir water passes through a dam, it turns the blades of turbines, causing a generator to generate electricity (**FIGURE 16.3B, C**). Electricity generated in the powerhouse of a dam is transmitted to the electric grid by transmission lines, and the water is allowed to flow into the riverbed below the dam to continue downriver. The

(a) Beck II Generating Station

(b) Turbine generator

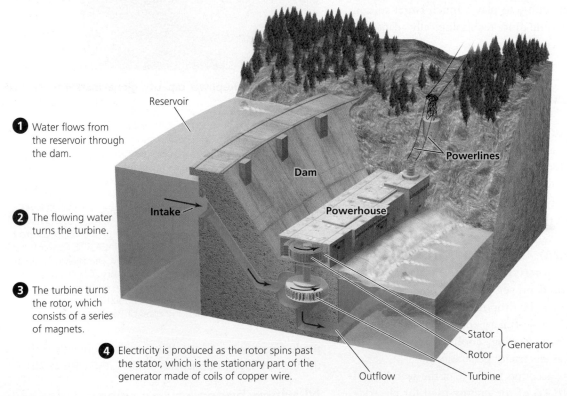

❶ Water flows from the reservoir through the dam.

❷ The flowing water turns the turbine.

❸ The turbine turns the rotor, which consists of a series of magnets.

❹ Electricity is produced as the rotor spins past the stator, which is the stationary part of the generator made of coils of copper wire.

(c) Hydroelectric power

FIGURE 16.3 Large dams, such as the Sir Adam Beck Stations on the Niagara River **(a)**, generate substantial amounts of hydroelectric power. Flowing water is used to turn turbines similar to those shown in **(b)**. As shown in **(c)**, the water is funnelled through a portion of the dam (1) to rotate the turbine (2), which turns rotors containing magnets (3). The spinning rotors generate electricity (4) as their magnets pass coils of copper wire. Electrical current is transmitted through power lines, and the river's water flows out through the base of the dam.

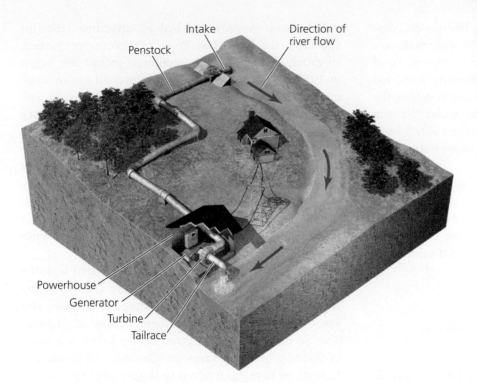

Penstock
Intake
Direction of river flow

Powerhouse
Generator
Turbine
Tailrace

FIGURE 16.4
Run-of-river systems divert a portion of a river's water for electricity generation. Some designs involve piping water downhill through a powerhouse and releasing it downriver, and some involve using water as it flows over shallow dams.

amount of power generated depends on the distance the water falls and the volume of water released. By storing water in reservoirs, dam operators can ensure a steady and predictable supply of electricity at all times, even during seasons of naturally low river flow.

The great age of dam building for hydroelectric power (as well as for flood control and irrigation) began in the 1930s, when the U.S. federal government began constructing many large dams as public-works projects. Dam-building technologies subsequently were exported to the developing world. Indian Prime Minister Pandit Nehru commented in 1963 that "dams are the temples of modern India," referring to their central importance in the development of energy capacity in that nation.

China's Three Gorges Dam on the Yangtze River is the world's largest hydroelectric installation (Itaipu Dam in Brazil is second). The final turbine out of a total of 32 was brought into production in 2012, and Three Gorges alone is now producing as much electricity as several coal-fired or nuclear plants put together (22 500 MW generating capacity; an average coal-fired plant would have a generating capacity of 1000–5000 MW; a multireactor nuclear plant might be capable of generating 6000–8000 MW). The reservoir, more than 600 m long, required the displacement of over a million people, and there are some lingering concerns about the stability and ecological impacts of the dam.

Run-of-river   An alternative to large dams is the **run-of-river** (or *diversion*) approach, which generates electricity without water impoundment and without greatly disrupting the flow of river water. On the one

hand, this approach sacrifices the reliability of water flow that the storage approach guarantees, but on the other hand it minimizes many of the negative impacts of large dams. Run-of-river can use various methods, one of which diverts a portion of a river's flow through a pipe or channel, passing it through a powerhouse and then returning it to the river (**FIGURE 16.4**). The pipe or channel can be run along the surface or underground. Another method involves the river water flowing over a dam small enough not to impede fish passage, using the water to turn turbines and then returning the water to the river.

Run-of-river systems are particularly useful in areas remote from established electrical grids and in regions without the economic resources to build and maintain large dams. Run-of-river is often called *small hydro* or even *micro-hydro* (referring to installations that produce less than 1000 kW of power), in contrast to the *large hydro* of traditional hydroelectric dams. Some environmentalists worry that the impacts of run-of-river systems on water flow and other aspects of the aquatic system are not sufficiently understood, but the impacts are surely less than those of large dams.

Hydropower generates relatively little air pollution

Hydropower has two clear advantages over fossil fuels for the production of electricity. First, it is renewable. Technically, we could refer to it as a *perpetual* energy source: As long as precipitation falls from the sky and fills rivers and reservoirs, we can use water to turn turbines.

The second advantage is cleanliness. Because no combustion is involved, no carbon dioxide or other pollutants are emitted during the production of hydropower. Fossil fuels *are* used in constructing and maintaining dams, and recent evidence indicates that large reservoirs release the greenhouse gas methane as a result of anaerobic decay in deep water. But overall, hydropower creates only a small fraction of the greenhouse gas emissions typical of fossil fuel combustion.

Hydropower is also efficient, with an EROI (*energy returned on investment*) of 10:1 or more, at least as high as any other modern-day energy source. Fossil fuels had higher EROI in the past, but as the cost of reaching remaining deposits has increased, their EROI values have dipped below that of hydropower.

Hydropower has many negative impacts, too

Although it is renewable, efficient, and produces little air pollution, hydropower does have other negative impacts. Damming rivers destroys habitat for wildlife as riparian areas above dam sites are submerged and those below are starved of water. Because water discharge is regulated to optimize electricity generation, the natural flooding cycles of rivers are disrupted. Suppressing floods prevents floodplains from receiving fresh nutrient-laden sediments. Instead, sediments become trapped behind dams, where they gradually fill the reservoir.

Dams also cause **thermal pollution**, because water downstream may become warmer if water levels are kept unnaturally shallow. Moreover, periodic flushes of cold water occur from the release of reservoir water; such thermal shocks, together with habitat alteration, have damaged many native fish populations in dammed waterways. Dams also block the passage of fish and other aquatic creatures, effectively fragmenting the river and reducing biodiversity.

The weight of water in a large reservoir can cause geological impacts, such as earthquakes, particularly where water has seeped into fractures in the bedrock underlying the reservoir. Dam collapses—whether as a result of earthquakes or landslides or from degradation of the construction materials—are regrettably not uncommon, and have resulted in many deaths in the decades since the great surge in large dam construction. These ecological and physical impacts also translate into negative social and economic impacts on local communities. There also has been concern about the generation of greenhouse gases in large hydroelectric reservoirs. As mentioned above, large reservoirs release methane, as well as carbon dioxide, as a result of anaerobic decay in deep water, particularly in warm climates.

The most controversial hydroelectric installation in Canada is the James Bay Project in northern Quebec, an extremely complicated, multiphase project that began in 1970.[5] James Bay has been controversial for environmental, political, social, and economic reasons, as well as being intimately connected with the modern history of Aboriginal land claims in Canada. The project has involved significant river diversions, resulting in the merging of several major watersheds on the eastern shore of the Hudson Bay, from the southern tip of James Bay to Ungava Bay in the north. Just two of these diversions, of the Caniapiscau and Eastmain rivers, inundated about 11 000 km^2 of boreal forest, with major impacts on water resources in nearby Cree villages.

Additional environmental impacts associated with the James Bay Project have included massive changes to the landscape, seismic tremors, fluctuating water levels, and loss of wetlands, with consequent ecological impacts. The decomposition of organic matter killed as a result of flooding has also proven problematic, and flooding is thought to have contributed to higher levels of methylmercury in local waterways. (*Methylmercury* is a particularly toxic and bioaccumulative form of mercury, which forms in aquatic systems through bacterially mediated processes. These occur most readily in oxygen-poor conditions, which are created by reservoir flooding and the subsequent decay of flooded organic matter.)

The river diversions associated with the James Bay Project also have interfered with animal migration patterns (e.g., caribou) and fish spawning habitats (e.g., salmon). This has been of concern, not only to conservationists and wildlife biologists but also to both sport and subsistence hunters.

Partly because of such concerns, hydropower is expanding much more slowly than before in North America. One reason is that most of the large rivers that offer excellent opportunities for hydropower are already dammed. Another reason is that people have grown more aware of the negative environmental impacts of dams. In some regions residents are resisting dam construction, or even (primarily in the United States) dismantling some dams to restore river habitats. Hydropower will likely continue to increase substantially in Asia, however, as China develops its capacity.

Nuclear Power

Nuclear power—usable energy extracted from the force that binds atomic nuclei together—occupies a conflicted position in our modern debate over energy. It is free of the air pollution produced by fossil fuel combustion, so it has been put forth as an environmentally friendly alternative to fossil fuels. Yet nuclear power's great promise has been clouded by nuclear weaponry, by the challenge of radioactive waste disposal, and by the long shadows of Chernobyl, Fukushima Daiichi, and other power plant accidents. Public safety concerns and the costs of addressing them have constrained the development and spread of nuclear power in Canada and many other nations.

First developed commercially in the 1950s, nuclear power experienced most of its growth in the 1970s and 1980s. The United States generates by far the most electricity from nuclear power—nearly a third of the world's production—followed distantly by France and Russia (**TABLE 16.1**). Although the United States is the leader in quantity of electricity generated by nuclear energy (and in the number of operating plants), less than 20% of its total electricity comes from nuclear sources. A number of other nations rely much more heavily on nuclear power; France leads the list, obtaining almost 75% of its electricity from nuclear power. Canada generates approximately 14–15% of its electricity with nuclear power.[6] The province of Ontario leads the nuclear industry in Canada, with over half of the electricity used in the province coming from its 18 operating nuclear generators; New Brunswick also has one operating reactor.

Japan, which used to be among the top 10 producers, dropped precipitously in nuclear electricity generation after the major meltdowns at the Fukushima Daiichi nuclear plants in 2011. Public support for nuclear energy in Japan declined significantly after that event, and at this point about four-fifths of Japan's nuclear plants are still not operating.

Globally, nuclear power generation dropped by 7% in 2012, following a similarly large decline of 4% in 2011; some analysts attribute this continuing slowdown to the after-effects of Fukushima Daiichi, affecting public attitudes toward nuclear energy far beyond the borders of Japan.[7]

Fission releases nuclear energy

Strictly defined, *nuclear energy* is the energy that holds together protons and neutrons in the nucleus of an atom. We harness this energy by releasing it and converting it to thermal energy, which can then be used to generate electricity.

The reaction that drives the release of nuclear energy within the **nuclear reactors** of power plants is **fission**, the splitting apart of atomic nuclei (**FIGURE 16.5**). To induce fission, the nuclei of large, heavy atoms, such as uranium or plutonium, are bombarded with neutrons, causing them to break apart. Each split nucleus produces heat, radiation, and multiple neutrons, as well as the leftover fragments of the original nucleus. The emitted neutrons then bombard other nearby *fissionable* (or *fissile*) atomic nuclei, resulting in a self-sustaining chain reaction.

If not controlled, this chain reaction becomes a runaway process that releases enormous amounts of energy; it is responsible for the explosive power of a nuclear bomb. Inside a nuclear power plant, however, fission is controlled so that, on average, only one of the two or three neutrons emitted with each fission event goes on to induce another fission event. In this way, the chain reaction maintains a constant output of energy at a controlled rate.

TABLE 16.1   Top Ten Producers of Nuclear Power, 2013

Nation	Nuclear power produced*	Plants operating + under construction[†]	Share of electricity from nuclear power*
United States	789.0	99 + 1	19%
France	403.7	58 + 1	73%
Russia	169.1	34 + 9	19%
South Korea	132.5	23 + 5	28%
China	110.7	24 + 25	2%
Canada	97.0	19	15%
Germany	92.1	9	16%
Ukraine	78.0	15 + 2	44%
United Kingdom	64.1	16	18%
Sweden	63.7	10	43%
World total	2358	439	10%

*In gigawatt-hours (GWh)

[†]As of 2013. Data from the IAEA-Power Reactor Information System. http://www.iaea.org/PRIS/WorldStatistics/OperationalReactorsByCountry.aspx.

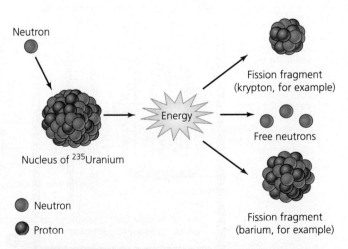

FIGURE 16.5
In nuclear fission, an atom of fissionable material (in this case, ^{235}U) is bombarded with a neutron. The collision splits the uranium atom into smaller atoms and releases two or three neutrons, radiation, and energy. The neutrons can continue to split other uranium atoms in a chain reaction. Engineers at nuclear plants use control rods to absorb excess neutrons and regulate the rate of the reaction.

Enriched uranium is used as fuel in nuclear reactors

The generation of electricity by controlled fission is just one step in a longer process called the *nuclear fuel cycle*, which begins when the naturally occurring element uranium is mined from underground deposits. Uranium minerals are uncommon and must be mined, which is why nuclear power is generally considered to be a nonrenewable energy source. However, Canada is particularly rich in uranium resources, and currently produces about one-third of the world's uranium. Canada also has developed reactor technologies that are used in many countries.

Uranium is useful for nuclear power because it is radioactive. Radioactive isotopes, or *radioisotopes*, emit subatomic particles and high-energy radiation as they decay into lighter radioisotopes, until they ultimately become stable isotopes. The isotope uranium-235 decays into a series of daughter isotopes, eventually forming stable lead-207. Each radioisotope decays at a rate determined by that isotope's *half-life*, the time it takes for half of the atoms to give off radiation and decay.

Over 99% of the uranium in nature occurs as the isotope uranium-238; uranium-235 (with three fewer neutrons) makes up less than 1% of the total. Because ^{238}U does not emit enough neutrons to maintain a chain reaction when it splits, ^{235}U is more commonly used for commercial nuclear

Cameco Corporation

FIGURE 16.6
Enriched uranium fuel is formed into pellets and then packaged into fuel rods, shown here. The fuel rods are encased in metal and inserted into the cores of nuclear reactors.

power. Mined uranium ore therefore must be processed or *enriched*, so that the concentration of ^{235}U is at least 3%. The enriched uranium is formed into pellets of uranium dioxide (UO_2), which are then incorporated into *fuel rods* (**FIGURE 16.6**) that are used in nuclear reactors.

The neutrons that bombard the uranium fuel rods in a reactor are slowed down with a substance called a *moderator*, most often water (**FIGURE 16.7**). As fission

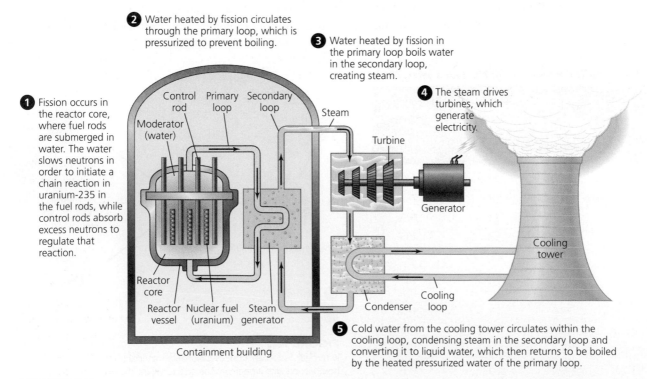

❷ Water heated by fission circulates through the primary loop, which is pressurized to prevent boiling.

❸ Water heated by fission in the primary loop boils water in the secondary loop, creating steam.

❶ Fission occurs in the reactor core, where fuel rods are submerged in water. The water slows neutrons in order to initiate a chain reaction in uranium-235 in the fuel rods, while control rods absorb excess neutrons to regulate that reaction.

❹ The steam drives turbines, which generate electricity.

❺ Cold water from the cooling tower circulates within the cooling loop, condensing steam in the secondary loop and converting it to liquid water, which then returns to be boiled by the heated pressurized water of the primary loop.

Control rod · Primary loop · Secondary loop · Moderator (water) · Steam · Turbine · Generator · Cooling tower · Reactor core · Reactor vessel · Nuclear fuel (uranium) · Steam generator · Condenser · Cooling loop · Containment building

FIGURE 16.7 In a pressurized light water reactor, uranium fuel rods are placed in water, a moderator that slows neutrons so fission can occur (1). Control rods are moved into and out of the reactor core to absorb excess neutrons and regulate the chain reaction. Water heated by fission circulates through the primary loop (2) and warms water in the secondary loop, which turns to steam (3). Steam drives turbines to generate electricity (4). The steam is then cooled in the cooling tower by water from an adjacent water body and returns to the containment building (5) to be heated again in the primary loop.

proceeds, it becomes necessary to soak up the excess neutrons produced when uranium nuclei divide, so that on average only a single neutron from each nucleus goes on to split another nucleus. For this purpose, *control rods*, made of a material that absorbs neutrons, are placed into the reactor among the fuel rods. Engineers move these control rods into and out of the water to maintain the fission reaction at the desired rate.

Instead of using ^{235}U-enriched fuel, some reactors (including the Canadian CANDU reactor) are designed to make use of ^{238}U and other fissile materials present in natural (unenriched) ores. Some of these are so-called **breeder reactors**, which generate new fissile material at a faster rate than they use it up. Because 99% of all uranium is ^{238}U, breeder reactors make better use of fuel, produce less waste for a given amount of energy generation, and eliminate the costly necessity to enrich nuclear fuels. However, breeder reactors are more expensive to construct than conventional reactors, and some technologies may be more susceptible to explosive accidents.

All of this takes place within the reactor core as part of the electricity-generating processes of a nuclear power plant. Water warmed by fission circulates, and warms water in a secondary loop, creating steam. This steam is used to turn a turbine, which generates electricity. The reactor core is housed within a reactor vessel, and the vessel, steam generator, and associated plumbing are all protected within a containment building. Most nuclear power plants are located near natural surface water bodies, to ensure a ready source of water for all processes. Containment buildings, with their metre-thick concrete and steel walls, are constructed to prevent leaks of radioactivity due to accidents or natural catastrophes such as earthquakes.

After several years in a reactor, the uranium fuel will have decayed to the point where it can no longer generate adequate energy, and must be replaced with new fuel rods. In some countries (not Canada or the United States), the spent fuel is reprocessed to recover what usable energy may be left. Most spent fuel, however, must be disposed of as radioactive waste.

Nuclear power delivers energy more cleanly than fossil fuels

Nuclear power plants generate electricity without creating air pollution from stack emissions. In contrast, combusting coal, oil, or natural gas emits sulphur dioxide that contributes to acidic deposition and particulate air pollution, as well as greenhouse gases.

Scientists from the International Atomic Energy Agency (IAEA) calculated that nuclear power produces emissions up to 150 times lower than those from fossil fuel combustion. They conducted a "cradle-to-grave" analysis that included emissions not just from power generation, but also from the mining, processing, and transport of fuel; manufacturing of equipment; construction of power plants; disposal of wastes; and decommissioning of plants. The results showed that, per unit of energy produced, fossil fuels produce much higher emissions than either renewable energy sources or nuclear energy. Worldwide, nuclear power generation avoids emissions of about 2.5 billion metric tons of carbon dioxide per year, comparable to the emissions avoided through the use of hydropower.[8]

Because the IAEA is responsible for promoting nuclear energy, critics point out that the agency is motivated to show nuclear power in a favourable light. However, few experts would argue with the conclusion that nuclear and renewable energy sources are cleaner than fossil fuels. In a series of "roadmaps" on various energy sources, the International Energy Agency (IEA) and Organisation for Economic Co-operation and Development (OECD) pointed to strong uptake of nuclear power generation as one way to forge a lower-emission pathway to 2050.[9]

Nuclear power has additional environmental advantages over fossil fuels—coal in particular. Because uranium generates far more power than coal by weight or volume, less of it needs to be mined, so uranium mining (which is commonly done in underground mines) causes less damage to landscapes and generates less solid waste than coal mining (which is often, though not always, done at the surface by strip mining). Moreover, in the course of normal operation, nuclear power plants are safer for workers than coal-fired plants.

Nuclear power also has serious drawbacks, though. One is that the waste it produces is radioactive; it must be handled with great care, and disposed of in a way that minimizes danger to present and future generations. A second major concern is that if an accident occurs at a power plant, or if a plant is sabotaged, or if a government were to use its nuclear capabilities for aggressive purposes, the consequences can be catastrophic. Given this mix of advantages and disadvantages (**TABLE 16.2**), many governments (although not necessarily most citizens) have judged the good to outweigh the bad, and today the world has 439 operating nuclear plants in 30 nations.[10]

Nuclear power poses small risk of large accidents

Although nuclear power poses fewer chronic health risks than does fossil fuel combustion, the possibility of catastrophic accidents has spawned a great deal of public anxiety over nuclear power. As plants around the world age, they require more maintenance. New concerns have also surfaced that nuclear plants could become targets for terrorism, or radioactive material could be stolen and used in terrorist attacks. This possibility is especially worrisome

TABLE 16.2   Environmental Impacts* of Coal-Fired and Nuclear Power

Type of Impact	Coal	Nuclear
Land and ecosystem disturbance from mining	Extensive, on surface or underground	Less extensive
Impacts on water bodies of adjacent water bodies	Can lead to acid mine drainage	Thermal pollution
Greenhouse gas emissions	Considerable emissions	None from plant operation; much less than coal over the entire life cycle
Other air pollutants	Sulphur dioxide, nitrogen oxides, particulate matter, and other pollutants	No pollutant emissions
Radioactive emissions	No appreciable emissions	No appreciable emissions during normal operation; possibility of emissions during severe accident
Occupational health among workers	More known health problems and fatalities	Fewer known health problems and fatalities
Health impacts on nearby residents	Air pollution impairs health	No appreciable known health impacts under normal operation
Effects of accident or sabotage	No widespread effects	Potentially catastrophic widespread effects
Solid waste	More generated	Less generated
Radioactive waste	None	Radioactive waste generated
Fuel supplies remaining	Should last several hundred more years	Uncertain; supplies could last for a longer or shorter time than coal supplies

*The more severe impacts are shown in red.

in countries of the former Soviet Union, where former nuclear sites have gone without adequate security for years.

Three significant events have been most influential in shaping public opinion about nuclear energy. They are Three Mile Island, Chernobyl, and Fukushima Daiichi.

Three Mile Island   The first of these events took place at the Three Mile Island plant in Pennsylvania, where in 1979 the United States experienced its most serious nuclear power plant accident. Through a combination of mechanical failure and human error, coolant water drained from the reactor vessel, temperatures rose inside the reactor core, and metal surrounding the uranium fuel rods began to melt, releasing radiation. This process, called a **meltdown**, proceeded through half of one reactor core at Three Mile Island. Area residents stood ready to be evacuated, but fortunately most radiation remained trapped inside the containment building.

The accident was brought under control within days, the damaged reactor was shut down, and a multibillion-dollar cleanup lasted for years. Three Mile Island is regarded as a near-miss; the emergency could have been far worse had the meltdown proceeded through the entire stock of uranium fuel or had the containment building not contained the radiation. Although residents have shown no significant health impacts in the years since, the event put safety concerns squarely on the map for both citizens and policymakers.

Chernobyl   In 1986 an explosion at the Chernobyl plant in Ukraine (then part of the Soviet Union) caused the most severe nuclear power plant accident the world has yet seen. Engineers had turned off safety systems to conduct tests. Human error, combined with technological failures and unsafe reactor design, led to explosions that destroyed the reactor and sent clouds of radioactive debris into the atmosphere. For 10 days radiation escaped from the plant while emergency crews risked their lives (some later died from radiation exposure) to put out fires. Most residents of the surrounding countryside remained at home for these 10 days, exposed to radiation, before the Soviet government belatedly began evacuating more than 100 000 people.

In the months and years afterward, workers erected a gigantic concrete sarcophagus around the demolished reactor, scrubbed buildings and roads, and removed irradiated materials (**FIGURE 16.8**). However, the landscape for at least 30 km around the plant remains contaminated today. An international team plans to build a larger sarcophagus around the original one, which is seriously deteriorating.

Atmospheric currents carried radioactive fallout from Chernobyl across much of the Northern Hemisphere, particularly Ukraine, Belarus, and parts of Russia and Europe (**FIGURE 16.9**). Fallout was greatest where rainstorms

(a) The Chernobyl sarcophagus

(b) Technicians measuring radiation

FIGURE 16.8 The world's worst nuclear power plant accident unfolded in 1986 at Chernobyl, in present-day Ukraine. As part of the extensive cleanup operation, the destroyed reactor was encased in a massive concrete sarcophagus **(a)** to contain further radiation leakage. Technicians scoured the landscape surrounding the plant **(b)**, measuring radiation levels, removing soil, and scrubbing roads and buildings.

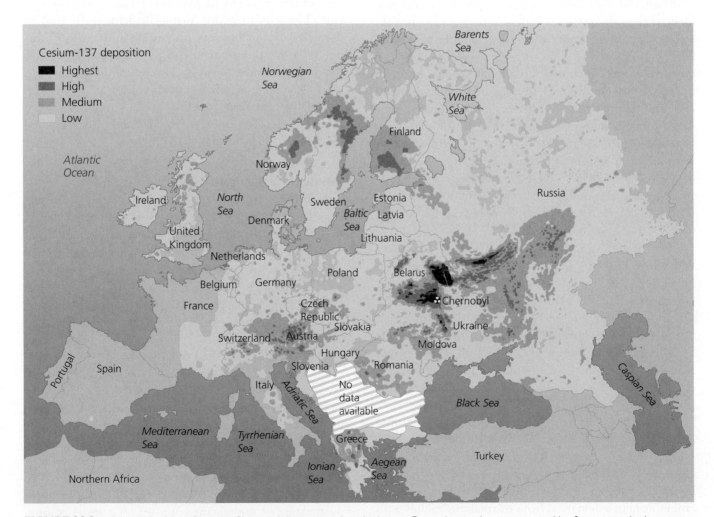

FIGURE 16.9 Radioactive fallout from the Chernobyl disaster was deposited across Europe in complex patterns resulting from atmospheric currents and rainstorms in the days following the accident. Darker colours in this map of ^{137}Cs deposition indicate higher levels of radioactivity. Although Chernobyl produced 100 times more fallout than the U.S. bombs dropped on Hiroshima and Nagasaki in World War II, it was distributed over a much wider area. *Source:* Data from Swiss Agency for Development and Cooperation, Bern, 2005.

brought radioisotopes down from the radioactive cloud. Parts of Sweden and Finland received high amounts of fallout, and the accident reinforced the public's fears about nuclear power. A survey taken in Sweden after the event showed that the proportion of respondents opposed to nuclear energy jumped from 21% before Chernobyl to 41% afterward.[11]

The accident killed 31 people directly and sickened or caused cancer in thousands more. Exact numbers are uncertain because of inadequate data and the difficulty of determining long-term radiation effects. The major long-term health consequence of Chernobyl's radiation has been thyroid cancer in children and those who were children at the time of the accident. The thyroid gland is where our bodies concentrate iodine, and one of the most common radioactive isotopes released early in the disaster was iodine-131 (^{131}I). Children have large and active thyroid glands, so they are especially vulnerable to thyroid cancer induced by radioisotopes.

Predicting that thyroid cancer might be a problem, medical workers measured iodine activity in the thyroid glands of several hundred thousand people in Russia, Ukraine, and Belarus following the accident. They also measured food contamination and surveyed people on their food consumption. These data showed that drinking milk from cows that had grazed on contaminated grass was the main route of exposure to ^{131}I, although fresh vegetables also contributed. Rates of thyroid cancer were found to have increased among children in regions of highest exposure.[12] By 2006, medical professionals estimated the number of cases at 6000 and rising. Fortunately, treatment of thyroid cancer has a high success rate.

Studies addressing other health aspects of the accident have found limited impact. Some research has shown an increase in cataracts due to radiation, especially among emergency workers. But neither the World Health Organization nor United Nations found evidence that rates of any other cancers (aside from thyroid cancer) had risen among people exposed to Chernobyl's radiation. Still, some cancers may appear decades after exposure, so it is possible that some illnesses have yet to arise.

After the Chernobyl event, the International Atomic Energy Association (IAEA) introduced the *International Nuclear and Radiological Event Scale*, in an effort to standardize communications and reporting of nuclear incidents worldwide (**FIGURE 16.10**). The scale ranges from 0, a deviation from normal procedures with no safety significance; through levels 1–3, incidents of increasing severity; through levels 5–7, accidents

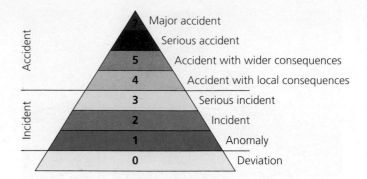

FIGURE 16.10 The International Nuclear and Radiological Event Scale was introduced by IAEA in 1990. The Chernobyl and Fukushima Daiichi meltdowns are both categorized as level-7 events; Three Mile Island and a 1952 accident at Chalk River, Ontario, in which the reactor core was damaged, were both level-5 events.

of increasing severity and impacts. There are many specific characteristics that define each level, including how much radiation (if any) was released, the toxicity and radioactivity of the material, damage (if any) to the reactor, and impacts on people and the environment. So far there have been just two level-7 events: the meltdowns at Chernobyl and Fukushima Daiichi. A level-7 accident involves "major release of radioactive material with widespread health and environment effects, requiring implementation of planned and extended counter-measures."

Fukushima Daiichi   On Friday, March 11, 2011, a megathrust earthquake of magnitude 9.0—one of the largest earthquakes ever measured—struck in a subduction zone just off the northeast coast of Tōhoku, Japan. The earthquake, devastating in its own right, generated a massive tsunami that swept ashore, killing more than 15 000 people, injuring many more, and wiping whole villages off the map. (See also the Central Case in Chapter 2, "The Tōhoku Earthquake: Shaking Japan's Trust in Nuclear Power.")

The tsunami inundated the Fukushima Daiichi power plant, located on the eastern coast of Japan not far from the earthquake's epicentre. The nuclear reactors shut down automatically as a result of the earthquake, as they were intended to do; however, fission continued in the reactor cores, requiring constant cooling to maintain a manageable temperature. The backup generators that should have kept the coolant fluids circulating were flooded and incapacitated by the tsunami (**FIGURE 16.11**). The subsequent overheating of the reactor cores set off a series of fires, explosions, and core meltdowns, accompanied by several releases of radioactive materials.

FIGURE 16.11
This photo from March 2011 shows some of the nuclear reactor buildings at Fukushima Daiichi that were damaged by the massive Tōhoku earthquake and tsunami.

The radioactivity that was released during and after these events were at levels about one-tenth of those from Chernobyl, but minor releases of radioactivity continued up to two years after the event. Japanese authorities applied lessons from Chernobyl. They provided iodine tablets to young people evacuated from the region, and restricted agriculture near the plant, stopping contaminated food and milk from entering the market. These measures to prevent uptake of iodine-131 apparently were effective. In late March 2011, 1000 evacuated children were tested for ^{131}I exposure, and none showed evidence of a high dose affecting the thyroid.

As the crisis was gradually brought under control, some researchers began to wonder what long-term health impacts might arise. Using data on cancer rates from studies on survivors of the atomic bombs dropped on Hiroshima and Nagasaki, researchers extrapolated to predict cancer rates at the lower radiation doses from Fukushima. In a 2011 study, for example, Frank Von Hippel from Princeton University calculated that about 1 million people were exposed to more than 1 curie per km^2 of radiation from cesium-137, which should result in a 0.1% increase in cancer risk among them, or 1000 extra cancer deaths.[13]

After the meltdown, much of the radioactive material was spread by air or water into the Pacific Ocean (**FIGURE 16.12**), and trace amounts were eventually detected around the world. Scientists from Stanford University used a global atmospheric model together with data on radiation,

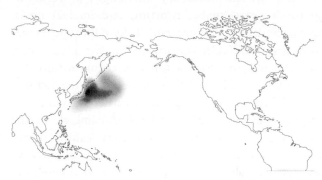

(a) After 36 hours

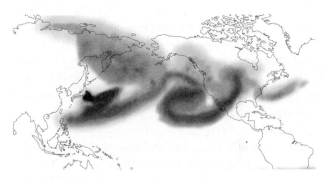

(b) After 180 hours

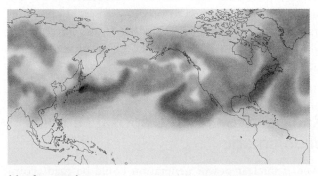

(c) After 468 hours

$$10^0 \quad 10^1 \quad 10^2 \quad 10^3 \quad 10^4 \quad 10^5 \quad 10^6$$
Cesium-137 concentration (micro-Becquerels per m^3)

FIGURE 16.12
These maps show atmospheric modelling results of the spread of radiation from ^{137}Cs around the world in the three weeks following the Fukushima Daiichi nuclear accident. (Contrast these maps with those in **FIGURE 16.9**, which show actual measurements of fallout from the Chernobyl accident.)
Source: Data from Ten Hoeve, J., and M. Jacobson (2012). Worldwide Health Effects of the Fukushima Daiichi Nuclear Accident. *Energy and Environmental Science, 5*(9), p. 8743.

population dispersion, and weather in the days following the event to estimate radiation levels people received.[14] The researchers predicted that Fukushima's radiation would eventually produce 125 cancer-related deaths and 178 nonfatal cancer cases. Large uncertainties attended both estimates; the projected number of total cases ranged from 39 to 2900.

Amid the extraordinary challenges of responding to the earthquake, tsunami, and nuclear crisis, the Japanese government and Tokyo Electric Power Company (TEPCO), the owners of the plant, were criticized for not sharing full and accurate information with the public. This, combined with the fact that TEPCO had failed to respond to earlier warnings that the plant was vulnerable to a large tsunami, shook the Japanese people's confidence in their leaders and in nuclear power. In the aftermath of the disaster, the government idled all 50 of the nation's nuclear reactors and embarked on safety inspections. In 2012, the government restarted a number of plants, arguing that electricity from them was necessary to avoid summertime blackouts. In response, large crowds of people demonstrated in protest, and some urged a national referendum on phasing out nuclear power. Across the world, many nations have since reassessed their nuclear programs, and Germany, Belgium, and Spain proposed to phase out nuclear power.

Time will tell what health impacts the Fukushima crisis causes. Long-term health effects on the region's people remain uncertain and debated. It will prove difficult to measure statistically rare instances of cancer in a large population exposed to very low doses, but researchers will try. They will be assisted by Japan's government, which budgeted $1.2 billion for coordinated research into long-term health effects of radiation.

Radioactive waste disposal remains problematic

Even if nuclear power generation could be made completely safe, and accidents like those at Chernobyl and Fukushima Daiichi were completely avoidable, we would be left with the conundrum of what to do with spent fuel rods and other radioactive waste. Recall that fission utilizes ^{235}U as fuel, leaving as waste the 97% of uranium that is ^{238}U. This ^{238}U, as well as all irradiated material and equipment that is no longer being used, is radioactive and must be disposed of in a location where radiation will not escape. Because the half-lives of uranium, plutonium, and many other radioisotopes are far longer than human lifetimes, this *high-level waste* will continue emitting

radiation for many thousands of years. Thus, radioactive waste must be placed in unusually stable and secure locations and monitored for many, many years to protect future generations.

Currently, nuclear waste from power generation is held in temporary storage at nuclear power plants in Canada and other places around the world. Spent fuel rods are sunken in pools of cooling water to minimize radiation leakage (**FIGURE 16.13A**). However, plants have only a limited capacity for this type of on-site storage. Many plants are now expanding their storage capacity by storing waste in thick casks of steel, lead, and concrete (**FIGURE 16.13B**).

Kyodo/Newscom

(a) Wet storage

Patrick Landmann/Photo Researchers, Inc./Science Source

(b) Dry storage

FIGURE 16.13
Spent uranium fuel rods are stored on-site at nuclear power plants and will likely remain at these scattered sites until a central repository for commercial radioactive waste is fully developed. Spent fuel rods are most often kept in "wet storage" in pools of water **(a)**, which keep them cool and reduce radiation release. Lower-level radioactive waste may be kept in "dry storage" in thick-walled casks layered with lead, concrete, and steel **(b)**.

In addition to keeping the waste isolated from ground water and the biosphere, and safe from geological disruptions such as earthquakes, another concern is that nuclear waste eventually must be transported to a permanent repository where it can be heavily protected. Because this would involve shipment by rail or truck, many people worry that the risk of an accident or of sabotage is unacceptably high. A single, central repository would minimize the need for multiple highly engineered facilities, but would increase the required transport distance and associated risks. It would also require effectively permanent cooperation and collaboration among nations, which might be an unreasonable expectation.

Sweden has established a single repository for low-level waste near one power plant, and is searching for a single disposal site deep within bedrock for spent fuel rods and other high-level waste. In the United States, a multiyear search for a disposal site homed in on Yucca Mountain, in the desert of southern Nevada; however, the project is now all but dead as a result of political delays and funding cutbacks. No site has yet been chosen in Canada, but at an underground research facility in Manitoba (closed and slated for decommissioning in 2010), scientists tested proposals for long-term storage deep in the stable, ancient crystalline rocks of the Canadian Shield. This approach is called *geological isolation* (**FIGURE 16.14**), and it is the disposal method of choice among nations that are seeking permanent repositories for high-level waste.

Multiple dilemmas have slowed nuclear power's growth

Dogged by concerns over waste disposal, safety, and cost overruns, nuclear power's growth has slowed. Many plants have been much more expensive than anticipated, and public anxiety in the wake of the Chernobyl and Fukushima Daiichi accidents has made utilities less willing to invest in new plants. In addition, some plants have aged more quickly than expected because of problems that were underestimated, such as corrosion in coolant pipes. Decommissioning a plant can sometimes be more expensive than the original construction.

Some experts predict that nuclear power will decrease because three-quarters of Western Europe's capacity is scheduled to be retired by 2030. Germany and Belgium have declared an intention to phase out nuclear power altogether, and appear to be continuing in that direction. Others (e.g., Spain, Italy) have recently reinforced bans on the construction of new plants, or have recommitted to remaining non-nuclear (e.g., Austria). Some Asian

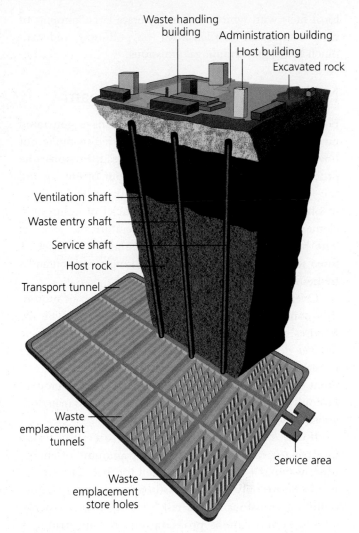

FIGURE 16.14
There is no permanent repository for high-level radioactive waste in Canada. Scientific testing of various proposals is ongoing. At a site similar to the former Underground Research Laboratory in Manitoba, waste could be buried in a network of tunnels deep underground in the stable, crystalline rocks of the Canadian Shield, as shown here.

nations, in contrast, are adding nuclear capacity. China, India, and South Korea are expanding their nuclear programs to help power their rapidly growing economies. Japan is so reliant on imported oil that it is eager to diversify its energy options, but since Fukushimna Daiichi, public opinion has turned away from nuclear as the top option. Asia hosts two-thirds of the most recent nuclear plants to go into operation, and by far the majority of plants now under construction are in China.

Electricity from nuclear power today remains more expensive than electricity from coal and other sources as a result of these challenges. Nonetheless, nuclear power remains one of the few currently viable alternatives to fossil fuels with which we can generate large amounts of electricity in short order, with less pollution, and with much lower greenhouse gas emissions.

Nuclear fusion remains a dream

For as long as scientists and engineers have generated power from nuclear fission, they have tried to figure out how they might use nuclear fusion instead. **Fusion**—the process that drives our Sun's vast output of energy and the force behind hydrogen or thermonuclear bombs—involves forcing together the small nuclei of lightweight elements under extremely high temperature and pressure. The hydrogen isotopes deuterium and tritium can be fused together to create helium, releasing a neutron and a tremendous amount of energy.

Overcoming the mutually repulsive forces of protons in a controlled manner is difficult, and fusion typically requires extremely high temperatures. Researchers have not yet developed "cold" fusion for commercial power generation; fusion experiments still require scientists to input more energy than they produce from the process. Fusion's potentially huge payoffs, though, make many scientists eager to keep trying.

If one day we were to find a way to control fusion in a reactor, we could produce vast amounts of energy using water as a source of hydrogen for fuel. The process would create only low-level radioactive wastes, without polluting emissions or the risk of dangerous accidents, sabotage, or weapons proliferation. A consortium of industrialized nations, including the United States, the European Union, India, China, Japan, Russia, and South Korea, is collaborating to build a prototype fusion reactor called the International Thermonuclear Experimental Reactor (ITER) in southern France. Even if this multibillion-dollar effort succeeds, however, power from fusion seems likely to remain a dream until many years in the future.

Traditional Biomass Energy

Although technologies for many renewable energy sources are still early in their stages of development, biomass energy is widely used in traditional formats. Worldwide, it is a "conventional" alternative to fossil fuels. Indeed, biomass was the very first source of energy used by our human ancestors, whose mastery of fire represents one of the first great steps toward the control of the environment. Today, traditional biomass accounts for about 9% of energy use globally.[15]

When people use the term **biomass energy**, they can mean very different things. To a subsistence farmer in Africa, biomass energy means cutting wood from trees or collecting livestock manure and burning it to heat and cook for her family. To an industrialized farmer in Saskatchewan, biomass energy might mean shipping his grain to a high-tech refinery that converts it into liquid fuel to run automobiles.

People harness biomass energy from many types of plant matter, including wood from trees, charcoal (which is actually charred wood, not coal), peat, and agricultural crops, as well as combustible animal waste products such as cattle manure, and even decomposing garbage in landfills. Fossil fuels are not considered biomass energy sources because their organic matter has not been part of living organisms for millions of years and has undergone considerable chemical alteration.

Traditional biomass sources are widely used in the developing world

Over 2 billion people still use and depend on wood from trees as their principal energy source.[16] In developing nations, especially in rural areas, families gather fuelwood to burn in their homes for heating and cooking (**FIGURE 16.15**). In these nations, the direct combustion of fuelwood, charcoal, and manure accounts for fully 35% of energy use—in the poorest nations, up to 90%.

Fuelwood and other traditional biomass sources constitute almost half of all renewable energy used worldwide (two-thirds, if hydropower is not included). Many new approaches and technologies are being developed to make biomass energy more efficient and sustainable, but these still constitute a very small proportion of overall energy use. As developing nations industrialize, fossil fuels are typically replacing traditional energy sources (**FIGURE 16.16**); as a result, traditional biomass use has grown more slowly worldwide than overall energy use. However, the International Energy Agency still estimates that by the year 2030, 2.6 billion people will be using traditional fuels for heating and cooking in ways that are both unsustainable and potentially damaging to human health.

Traditional biomass energy has environmental pros and cons

Biomass energy has one overarching environmental benefit: It is essentially *carbon-neutral*, releasing no net carbon into the atmosphere. Although burning biomass emits plenty of carbon, it is simply the carbon that

FIGURE 16.15
About 2 billion people in developing countries still rely on fuelwood and charcoal for heating and cooking, according to the United Nations. Wood cut from trees remains the major source of biomass energy used in the world today. In theory, biomass is renewable; in practice, it may not be renewable if forests are overharvested.

photosynthesis had pulled from the atmosphere to create the biomass in the first place. That is, the carbon that biomass combustion emits is balanced by the carbon that photosynthesis had sequestered within the biomass just years, months, or weeks before. Therefore, when we replace fossil fuels with bioenergy, we reduce net carbon flux to the atmosphere, helping to mitigate global climate change.

However, this holds only if biomass sources are not overharvested. Harvesting fuelwood at an unsustainably rapid rate can lead to local deforestation, although it has now been demonstrated that this is probably not a major driving force in deforestation globally, compared to logging and clearing land for ranching and agriculture. Additional problems that can be associated with over-harvesting of fuelwood include damaged landscapes, and diminished habitat and biodiversity. In arid regions that are heavily populated and support meagre woodlands, fuelwood harvesting can have enormous impacts; such is

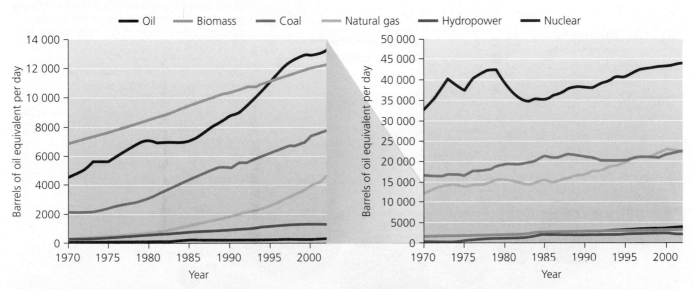

(a) Energy consumption in developing nations

(b) Energy consumption in industrialized nations

FIGURE 16.16 Energy consumption patterns vary greatly between developing nations **(a)** and wealthier industrialized nations, here represented by nations of the Organisation for Economic Co-operation and Development (OECD) **(b)**. Note the large role that traditional biomass (primarily fuelwood) has played in supplying energy to developing countries. Note also that the scales of the y axes differ; people in developing nations consume far less energy than those in industrialized nations.
Source: From Energy Information Administration, U.S. Department of Energy.

the case with many regions of Africa and Asia. Burning fuelwood and other biomass in traditional ways for cooking and heating also leads to health hazards from indoor air pollution.

"New" Renewable Energy Sources

Next we will explore a group of alternative energy sources that are often called the "new renewables," including solar, wind, tidal, geothermal, and some other energy technologies that are being developed. In addition to growing rural and off-grid use, there are three major applications for new renewables in the world's energy market today:

■ power generation (using wind, solar, and other renewable energy sources to generate electricity)

■ space heating (using solar or terrestrial energy sources to heat buildings) and *district heating/cooling* (using one distributed system to heat or cool many buildings in a community or urban area)

■ fuels for transportation (using hydrogen fuels, crops, agricultural and forestry wastes, and other materials to manufacture ethanol and biodiesel)

All of these applications are very important in Canada, and promising alternative energy resources are available for all three. Electricity generation powers our modern lifestyle and industry; space heating is important in our northern climate; and fuel for transportation is vital, given the need for mobility in our large country. Energy solutions and new technologies for these applications differ from one region of the country to another, depending on available resources.

"New" renewable contributions are small but growing quickly

As a global community, we obtain less than 6% of our total energy from new renewable energy sources. Only about 22% of our electricity worldwide comes from renewable energy sources, and traditional "large hydro" accounts for three-quarters of this.[17] In Canada, less than 6% of electricity generation comes from renewable sources *other* than traditional large hydro. Most non–large hydro renewable electricity generation in Canada now comes from wind, followed by small hydro (run-of-river) installations and biomass (**FIGURE 16.17**).

Although they comprise only a minuscule proportion of our energy budget, the new renewable energy sources are growing at much faster rates than conventional

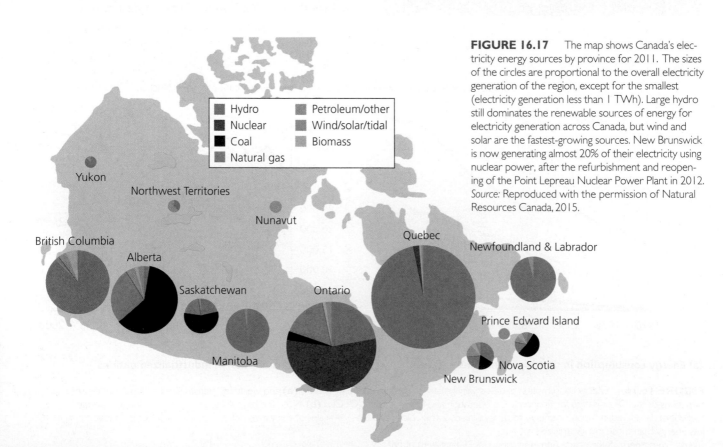

FIGURE 16.17 The map shows Canada's electricity energy sources by province for 2011. The sizes of the circles are proportional to the overall electricity generation of the region, except for the smallest (electricity generation less than 1 TWh). Large hydro still dominates the renewable sources of energy for electricity generation across Canada, but wind and solar are the fastest-growing sources. New Brunswick is now generating almost 20% of their electricity using nuclear power, after the refurbishment and reopening of the Point Lepreau Nuclear Power Plant in 2012. *Source:* Reproduced with the permission of Natural Resources Canada, 2015.

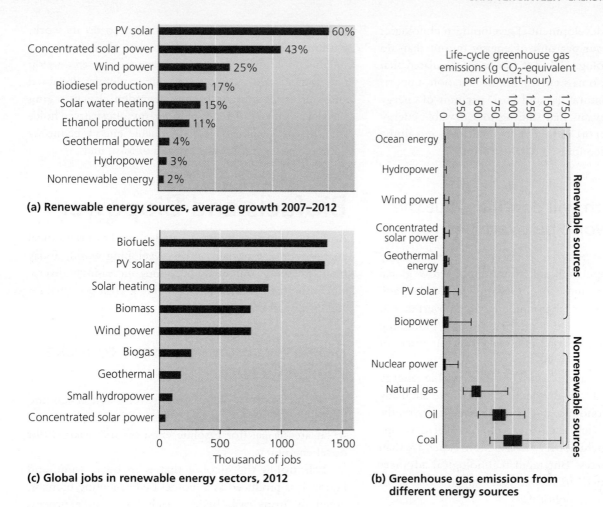

(a) Renewable energy sources, average growth 2007–2012

(c) Global jobs in renewable energy sectors, 2012

(b) Greenhouse gas emissions from different energy sources

FIGURE 16.18 The "new renewable" energy sources are growing far faster than conventional energy sources **(a)**. Shown are the average rates of growth each year between 2007 and 2012. Renewable energy sources release far fewer greenhouse gas emissions than fossil fuels **(b)**. Shown are ranges of estimates from scientific studies of each source when used to generate electricity. Renewable energy also creates new green-collar jobs **(c)**. Nearly 6 million people worldwide were employed in renewable energy jobs as of 2012, with more than half of these in solar, wind, and geothermal energy. *Source:* Data for (a) and (c) from REN21 (2013). *Renewables 2013: Global Status Report.* REN21, UNEP, Paris. (b) Data from Intergovernmental Panel on Climate Change (2012). *Renewable Energy Sources and Climate Change Mitigation. Special report.* Cambridge University Press, New York.

DATA Q At the yearly growth rate shown in **(a)**, if you began with 10 units of PV solar capacity, how much would there be after 5 years?

energy sources. Over the past three decades, solar, wind, and geothermal energy sources have grown far faster than the overall energy supply (**FIGURE 16.18A**). The leader in growth is wind power, which has expanded by nearly 50% each year since the 1970s; in recent years solar technologies have exceeded this, with annual growth rates of 60% and more (in Canada, about 40% per year).[18] Because these sources started from such low levels of use, however, it will take them some time to catch up to conventional sources. For instance, the absolute amount of energy added by a 50% increase in wind power is still far less than the amount added by just a 1% increase in oil, coal, or natural gas.

Use of new renewables has been expanding quickly for a number of reasons, including growing concerns

over diminishing fossil fuel supplies and national energy security, the substantial environmental impacts of fossil fuel combustion (**FIGURE 16.18B**), and public anxiety about nuclear energy following the Fukushima Daiichi event. Unlike fossil fuels, renewable sources are inexhaustible on timescales relevant to human societies. Advances in technology are also making it easier and less expensive to harness renewable energy sources. Developing renewables can diversify an economy's mix of energy, lowering price volatility and protecting against supply restrictions.

New energy sources also can create employment opportunities and sources of income and property tax for communities, especially in rural areas passed over by other economic development. Developing technologies require more labour per unit of energy output than do established technologies, and it has been calculated that solar, wind, and biomass energy implementation supports more jobs than natural gas and coal, per unit of energy generated. This means that shifting to renewable energy could actually support more employment than remaining with a fossil fuel economy (**FIGURE 16.18C**).[19]

Rapid growth will continue, but the transition won't be overnight

Rapid growth in renewable energy sectors seems likely to continue as population and consumption grow, global energy demand expands, fossil fuel supplies decline, and people demand cleaner environments. More governments, utilities, corporations, and consumers are now promoting and using renewable energy, and, as a result, the prices of renewables are falling.

However, we cannot switch to renewable energy sources from one day to the next, because there are many technological, economic, and social barriers. Currently, most renewables lack adequate technological development and infrastructure; they are also more expensive than conventional sources. But rapid technological advances and declining prices in recent years suggest that most remaining barriers are political.

For decades, research and development of renewable sources have received far less in subsidies, tax breaks, and other incentives from governments than have conventional sources. One analysis showed that in the United States, from the early days of oil and gas development, oil and gas have received in total 75 times more subsidies and nuclear energy 31 times more subsidies than new renewable energy sources. In only one single year over that entire century have solar, wind, and geothermal combined *ever* received as much support as the *lowest* amount ever offered to oil, gas, or nuclear power.[20]

The funding situation has not been much different in Canada. As recently as 2006, the Report of the National Advisory Panel on Sustainable Energy Science and Technology[21] included the somewhat surprising statement that "renewable and nuclear technologies were not considered by the Panel to be key priorities for a national energy science and technology effort." In that same report, however, the Panel stated that, "A major effort to develop new energy technologies and new methods for fostering their development and application in Canada will be critical to . . . determining our energy future."

If there is to be a transition to renewable energy sources, it will likely be a gradual shift driven largely by economic supply and demand. If the transition proceeds too slowly—if we wait for the market to do its work, without government encouragement—then fossil fuel supplies could dwindle faster than we are able to develop new sources, and we could find our economies disrupted and our environment highly degraded. Encouraging the development of renewable energy alternatives holds promise for a vigorous and sustainable energy economy without the environmental impacts of fossil fuels.

Biofuels and Biopower

As discussed earlier, biomass is widely used as a traditional energy source, especially in the developing world. Today biomass energy sources are becoming increasingly diverse and innovative (**TABLE 16.3**), providing great potential for addressing our energy challenges.

Biomass can be processed to make vehicle fuels

An important use of biomass energy is for conversion into **biofuels**. The two principal types of biofuels developed so far are *ethanol* (for gasoline engines) and *biodiesel* (for diesel engines).

Ethanol is the alcohol that is in beer, wine, and liquor. It is produced as a biofuel by fermenting biomass, generally from carbohydrate-rich crops, in a process similar to brewing beer. The carbohydrates contained in the plants are converted to sugars and then to ethanol, which then must be completely dehydrated (that is, all water removed) before it can be used as a vehicle fuel. Ethanol is now widely added to gasoline in the United

TABLE 16.3   New Sources of Biomass Energy

Biofuels for powering vehicles
- corn grown for ethanol
- *bagasse* (sugarcane residue) grown for ethanol
- soybeans, rapeseed (canola), and other crops grown for biodiesel
- used cooking oil for biodiesel
- plant matter treated with enzymes to produce cellulosic ethanol
- algae fuels

Biopower for generating electricity
- crop residues (such as cornstalks) burned at power plants
- forestry residues (such as wood waste from logging) burned at power plants
- processing wastes (such as solid or liquid waste from sawmills, pulp mills, and paper mills) burned at power plants
- "landfill gas" burned at power plants
- livestock waste from feedlots for gas from anaerobic digesters
- organic components of municipal solid waste from landfills

FIGURE 16.19

An increasing proportion of the corn crop in Canada, as in the United States, is used to produce ethanol. Brazil produces most of the rest of the world's ethanol, from *bagasse* (sugarcane residue). Production of ethanol in Canada and worldwide has grown rapidly in the last decade.

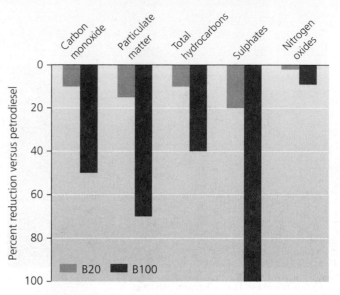

FIGURE 16.20

Burning biodiesel in a diesel engine emits less pollution than burning conventional petroleum-based diesel. Shown are the percentage reductions in several major automotive pollutants that one can attain by using B20 (a mix of 20% biodiesel and 80% petrodiesel) and B100 (pure biodiesel). *Source:* From U.S. Environmental Protection Agency.

DATA Q Which type of pollutant is reduced most effectively by the use of biodiesel, relative to petrodiesel?

States to reduce automotive emissions, spurred by the 1990 U.S. *Clean Air Act*. The total production capacity for ethanol in Canada, as of 2014, was about 1 billion litres; most of this is produced from corn (**FIGURE 16.19**).

Any vehicle with a gasoline engine runs well on gasoline blended with up to 10% ethanol. Many automakers are now producing *flexible-fuel vehicles* that run on E-85, a mix of 85% ethanol and 15% gasoline. Fewer gas stations offer E-85, so drivers are sometimes forced to fill these cars with conventional gasoline; however, increasing infrastructure for ethanol will change this. In Brazil, for many years, sugarcane residue has been crushed to make *bagasse*, a material that is then used to make ethanol. Sugarcane ethanol accounts for about 40% of all automotive fuel in Brazil.

Vehicles with diesel engines can run on **biodiesel**, produced from vegetable oil mixed with small amounts of ethanol or methanol (wood alcohol). In Canada and in Europe, where most biodiesel is used, rapeseed (canola) oil is the oil of choice, whereas U.S. biodiesel producers use mostly soybean oil. Biodiesel producers can even utilize animal fats and used grease and cooking oil from restaurants.

Most frequently, biodiesel is mixed with conventional *petrodiesel* (petroleum-based diesel fuel); a 20% biodiesel mix (called B20) is common today. To run on straight vegetable oil, a diesel engine needs to be modified. Although the parts needed for these modifications can be bought for as little as $800, it remains to be seen whether using straight vegetable oil might entail further costs, such as reduced longevity or greater engine maintenance. Biodiesel cuts down on emissions compared with petro-diesel (**FIGURE 16.20**). Its fuel economy is almost as good, and it costs just slightly more, at today's oil prices. It is also nontoxic and biodegradable.

Growing crops specifically to produce ethanol or biodiesel may not be sustainable, so researchers are refining some new techniques for biofuel production. So-called second-generation biofuel technologies use enzymes to produce ethanol from the cellulose that gives structure to all plant material. If we can produce this *cellulosic ethanol* in commercially feasible ways, ethanol could be made from low-value crop waste (indigestible residues, such as corn stalks and husks), rather than from high-value crops (**FIGURE 16.21A**).

Other research projects are underway that aim to produce biofuels from algae, called third-generation biofuels. Lipids and carbohydrates from several species of algae can be converted into a variety of biofuels. Algae can be grown for the purpose in open ponds or in closed transparent tubes called *photobioreactors* (**FIGURE 16.21B**). Algae grow much faster than terrestrial crops, can be harvested every other day, and produce many times more biofuel than other crops. Algae farms can be set up just about anywhere, and can use waste water or salt water. Because algae need nutrients, waste water from sewage treatment plants can be a good source of water. And because carbon dioxide is required for algal growth, algae farms can even be used to capture smokestack emissions. Production of algal fuels is still expensive, but the cost will likely come down as technologies improve.

(a) Switchgrass for cellulosic ethanol

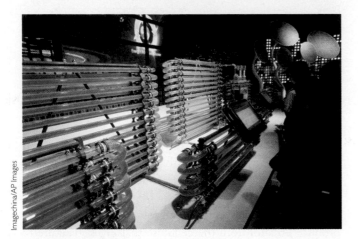

(b) Algae for third-generation biofuels

FIGURE 16.21

(a) Switchgrass, a fast-growing plant native to the North American prairies, provides fuel for biopower and is being studied as a crop to provide cellulosic ethanol. (b) Algae are a leading candidate for a next-generation biofuel. Here algae are growing in networks of lighted tubes at a demonstration facility. Algae grow quickly, can be farmed in many places, and can be used to produce biodiesel, ethanol, and other fuels.

(a) Forestry residues

(b) Fast-growing poplars

FIGURE 16.22

(a) Forestry residues (here from a Swedish logging operation on Vancouver Island) are a major source material for the production of biopower. (b) Hybrid poplars are specially bred for fast growth, and they are harvested for use in biopower production. The monocultural plantations, which may replace natural systems over large areas, do not function ecologically as forests.

Electricity can be generated from biomass

We can harness biomass energy by combusting biomass to generate electricity. Many of the sources used for this **biopower** are the waste products of existing industries. For instance, the forest products industry generates large amounts of woody debris in logging operations and at sawmills, pulp mills, and paper mills (**FIGURE 16.22A**), which can be used for biopower. Other waste sources include organic waste from municipal landfills, animal waste from agricultural feedlots, and crop residues (such as cornstalks and husks).

Some crops are grown specifically to produce biopower. These include fast-growing trees, such as willows and poplars (**FIGURE 16.22B**), and fast-growing grasses, such as bamboo, fescue, and switchgrass. Many of these plants are also being grown to produce liquid biofuels.

Power plants built to combust biomass operate similarly to those fired by fossil fuels; the combustion heats water, creating steam to turn turbines and generators, thereby generating electricity. Much of the biopower produced so far comes from power plants that generate both electricity and heat. The simultaneous production of different forms of energy through two mechanisms operating at the same plant is called **cogeneration**. A common example of cogeneration is the use of excess steam heat from electricity generation for space heating of homes and workplaces. Biopower cogeneration or *co-gen* plants are often located where they can take advantage of organic waste from forestry, agriculture, or landfills to generate both electricity and heat.

Biomass is also increasingly being combined with coal in coal-fired power plants in a process called *co-firing*. Wood chips, wood pellets, or other biomass is introduced with coal into a high-efficiency boiler that uses one of several technologies. Biomass can substitute for up to 15% of the coal with only minor equipment modification and no appreciable loss of efficiency. Co-firing can be a relatively easy and inexpensive way for fossil-fuel-based utilities to expand their use of renewable energy.

The anaerobic bacterial breakdown of waste by microbes in landfills produces methane and other gases (to be discussed in the chapter, *Managing Our Waste*). The capture, recovery, and use of such **landfill gas** for electricity generation (**FIGURE 16.23**) can yield from 50% to 70% of the energy of natural gas, in addition to cutting down on greenhouse gas emissions that would otherwise result from its release into the atmosphere.[22] The main component of landfill gas is methane (CH_4), and carbon dioxide (CO_2) is typically the next most abundant component. Capture is particularly important in the case of methane, which is approximately 21 times more effective than carbon dioxide as a greenhouse gas. Methane and other gases also can be produced in a more controlled way in anaerobic digestion facilities. In either case, the resulting **biogas** can be burned in a boiler to generate electricity.

Capturing and using methane can help power and even finance the operations of a landfill facility. The Beare Road plant produces enough electricity to service 4000 homes. The total cost of this *LFGTE (landfill gas-to-electricity)* facility was about $8.5 million. It has a 10-year anticipated lifespan, based on the amount of garbage in the landfill, and generates approximately $2 million in revenues each year for its owner, E.S. Fox, who has an agreement with Ontario Power Generation to purchase the electricity.[23] Landfill gas utilization also solves local environmental problems by

FIGURE 16.23
This is the landfill gas cleanup room at the Beare Road Landfill Power Plant in Scarborough, Ontario, where particulates and moisture are removed from the landfill gas prior to its use for electricity generation.

cutting down on odours and limiting the possibility of fires and explosions caused by escaping methane gas.

We can also harness biopower through *gasification*, in which biomass is vaporized at extremely high temperatures in the absence of oxygen, creating a gaseous mixture that includes hydrogen, carbon monoxide, carbon dioxide, and methane. This mixture can generate electricity when used in power plants to turn a gas turbine to propel a generator. Gas from gasification can be treated in various ways to produce methanol, synthesize a type of diesel fuel, or isolate hydrogen for use in hydrogen fuel cells. An alternative method of heating biomass in the absence of oxygen can produce a liquid fuel called *pyrolysis oil*, which can be burned to generate electricity.

Biomass energy has environmental and economic benefits and drawbacks

Adding biofuels such as ethanol and biodiesel to gasoline or petrodiesel helps those fuels combust more completely, reducing pollution. Replacing gasoline or petrodiesel with biofuels reduces emissions of nitrogen oxides, greenhouse gases, and other pollutants; however, these reductions and the climate change mitigation potential of biofuels in general depend heavily on underlying factors such as land management on the farms where the raw materials are produced. When used instead of coal in co-firing and direct combustion, biopower significantly reduces emissions of sulphur dioxide.

Shifting from fossil fuels to biomass energy also can have economic benefits. The forest products industry in North America now obtains much of its energy by combusting the waste it recycles, including woody waste from pulp mill processing. Biomass tends to be the least expensive type of fuel for burning in power plants, and improved energy efficiency brings lower prices for consumers. As a resource, biomass tends to be well spread geographically, so using it should help support rural economies and reduce many nations' dependence on imported fuels.

However, there are also some significant drawbacks to some biomass energy technologies. Growing crops to produce biofuels and biopower exerts tremendous impacts on ecosystems. Although crops grown for energy typically receive lower inputs of pesticides and fertilizers than those grown for food, cultivating biofuel crops is land-intensive and brings with it all the impacts of monocultural agriculture. Biofuel crops take up land that might otherwise be left in its natural condition or developed for other purposes.

If we were to try to produce all the automotive fuel currently used in North America with ethanol from corn, we would need to expand the already immense corn acreage by more than 60%, with no decrease in productivity and without producing any additional corn for food. Even at current levels of production, biofuel is competing with food production. As farmers shift more corn crops to ethanol, corn supplies for food have dropped, and international corn prices have skyrocketed. In Mexico, where corn tortillas are a staple food, average citizens found themselves struggling and protests erupted across the country over the inflated price of corn.

Growing bioenergy crops also requires substantial inputs of energy. We currently operate farm equipment using fossil fuels, and farmers apply petroleum-based pesticides and fertilizers to increase yields. Moreover, fossil fuels are used in refineries to heat water so that we can distill pure ethanol. Thus, shifting from gasoline to ethanol for our transportation needs would not eliminate our reliance on fossil fuels.

Furthermore, growing corn for ethanol yields only a modest amount of energy relative to the energy that needs to be input. Recall our discussion of *energy returned on investment* (EROI) in the previous chapter on fossil fuels. The EROI, or ratio of energy returned to energy invested, for corn-based ethanol is controversial, but the best recent estimates place it around 1.5:1. This means that to gain 1.5 units of energy from ethanol, we need to expend 1 unit of energy. The EROI of Brazilian *bagasse* ethanol is much higher, but the low ratio for corn-based ethanol makes this fuel relatively quite inefficient. For this reason, many critics do not view ethanol as an effective path to sustainable energy use.

Future advances in cellulosic ethanol may ease the environmental impacts of biofuel crops considerably, and third-generation biofuels such as algal fuels are promising, too. In the meantime, until these technologies develop and costs decline, biomass energy use involves a complex mix of advantages and disadvantages.

Solar Energy

The Sun provides energy for almost all biological activity on Earth by converting hydrogen to helium through nuclear fusion. On average, each square metre of Earth's surface receives about 1 kilowatt of solar energy—17 times the energy of a light bulb. As a result, a typical house has enough roof area to generate all its power needs with rooftop panels that harness **solar energy**. The amount of energy Earth receives from the Sun *each day*, if it could be collected in full for our use, would be enough to power human consumption for a quarter of a century.

The potential for using sunlight to meet our energy needs is tremendous. However, we are still in the process of developing solar technologies and learning the most effective and cost-efficient ways to put the Sun's energy to use. Most solar technologies rely on the collection and, in some cases, concentration, of the Sun's rays. The principal uses for solar energy today are heating and cooling air; heating water for home and industrial uses; drying crops, such as tea, coffee, fruit, and others; generating electricity for off-grid and distributed energy applications; detoxifying water and air; cooking food; and, of course, daytime lighting.

Passive solar heating is simple and effective

The most commonly used way to harness solar energy is through **passive solar** energy collection. In this approach, buildings are designed and building materials are chosen to maximize absorption of sunlight in winter, even as they keep the interior cool in the summer.

Passive solar design usually involves installing south-facing windows to maximize sunlight capture in the winter (in the Northern Hemisphere; north-facing windows are used in the Southern Hemisphere). Overhangs block light from above, shading windows in the summer when the Sun is high in the sky and when cooling is needed. Passive solar techniques also include the use of heat-absorbing construction materials (called *thermal mass*) that absorb heat, store it, and release it later. Thermal mass made of straw, brick, concrete, or other materials most often is used in floors, ceilings, and walls.

Thermal mass is strategically located to capture sunlight in cold weather and radiate heat to the interior of the building. In warm weather the mass absorbs warm air in the interior to cool the building. Passive solar design can also involve planting vegetation in particular locations around a building. By heating buildings in cold weather and cooling them in warm weather, passive solar methods conserve energy and reduce energy costs.

Active solar technologies can heat, cool, and produce electricity

Active solar approaches make use of technological devices to focus, move, and store solar energy. One active technology involves using **solar panels** or *flat-plate solar collectors*, most often installed on rooftops. The panels generally consist of dark-coloured, heat-absorbing metal plates mounted in flat boxes covered with glass panes. Water, air, or antifreeze solutions are run through tubes that pass through the collectors, transferring heat throughout a building. Heated water can be pumped to tanks to store the heat for later use. Active solar systems are especially effective for heating water, and can be used even in remote areas (**FIGURE 16.24**).

We can magnify the strength of solar energy by gathering sunlight from a wide area and focusing it on a single point. This is the principle behind *solar cookers*, simple portable ovens that use reflectors to focus sunlight

FIGURE 16.24
This inexpensive solar installation in a remote village in West Bengal, India, provides power for domestic uses such as cooking and lighting.

onto food and cook it. Such cookers are proving extremely useful in parts of the developing world.

Utilities have put the solar cooker principle to work in large-scale, high-tech approaches to generating electricity. In one approach, called *concentrated solar power* or *CSP*, mirrors concentrate sunlight onto a receiver atop a tall "power-tower" (**FIGURE 16.25**). From the receiver, heat is transported by fluids that are piped to a steam-driven generator to create electricity. These solar power plants can harness light from large mirrors spread across many hectares of land. The largest such plant—a collaboration among government, industry, and utility companies in the California desert—produces power for 10 000 households.

FIGURE 16.25
At the Solar Two facility in the desert of southern California, the largest such facility in the world, mirrors are spread across wide expanses of land to concentrate sunlight onto a receiver atop a "power-tower." Heat is then transported through fluid-filled pipes to a steam-driven generator that produces electricity.

An interesting combination of active and passive solar technologies is the "Solarwall" that provides heating and cooling at the Bombardier's Canadair plant in Montreal. This is a south-facing cladding wall with millions of tiny holes, each about 1 mm in diameter, which allow outside air to pass through. Behind the cladding is a space for air flow. The air passing along the back of the wall absorbs solar-generated heat and rises to the roof, where it is collected and circulated through the building. This installation, the largest in the world, eliminated the need for fossil fuels at this facility, thereby cutting atmospheric emissions, and improved interior air quality in the plant as well.

Active solar technology dates from the eighteenth century, but largely because of lack of investment it contributes only a minuscule portion of energy production in Canada and worldwide. However, the growth in solar energy use worldwide has been second only to that of wind power. The federal government and some provincial and territorial governments have provided financial incentives for homeowners and business owners to switch to solar technologies, through *feed-in-tariff* (*FIT* or *MicroFIT*) programs. In feed-in-tariff programs, electricity generators—even individual farmers and homeowners with solar panels on their roofs—are paid for supplying renewable electricity to the power grid. This type of program acts as a financial incentive for the adoption of solar and other renewable electricity-generating technologies.

PV cells generate electricity directly

A direct approach to producing electricity from sunlight involves **photovoltaic (PV) cells**, which collect sunlight and convert it to electrical energy by making use of the *photovoltaic* or *photoelectric effect*. This effect occurs when light strikes one of a pair of metal plates in a PV cell, causing the release of electrons, which are attracted by electrostatic forces to the opposing plate. The flow of electrons from one plate to the other creates an electrical current (direct current, DC), which can be converted into alternating current (AC) and used for residential and commercial electrical power (**FIGURE 16.26**).

The plates of a typical PV cell are made of silicon, which conducts electricity. One silicon plate (the *n-type layer*) is rich in electrons; the other (the *p-type layer*) is electron-poor. When sunlight strikes the PV cell, it knocks electrons loose from some of the silicon atoms. Connecting the two plates with wires generates electricity as electrons flow from the n-type layer back to the p-type layer. Photovoltaic cells can be connected to batteries that store the accumulated charge until it is needed.

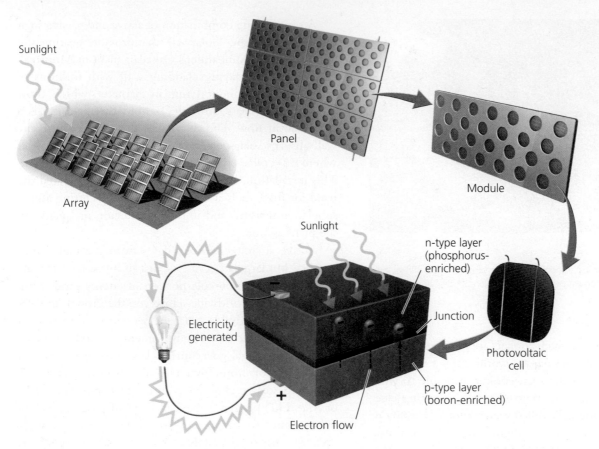

FIGURE 16.26 A photovoltaic (PV) cell converts sunlight to electrical energy. When sunlight hits the silicon layers of the cell, electrons are knocked loose and move from layer to layer. Connecting the layers with wiring allows electrical current to flow between them. This direct current (DC) is converted to alternating current (AC) to produce usable electricity. PV cells are grouped in modules, which comprise panels that can be erected in arrays.

Small PV cells probably power your watch or solar calculator. Atop the roofs of homes and other buildings, PV cells can be arranged in arrays; increasingly, PV roofing tiles are being used in place of these arrays. In some remote areas, PV systems are used in combination with wind turbines and a diesel generator to power entire villages. The use of PV cells is growing fast, and should continue to increase as prices fall, technologies improve, and governments enact economic incentives to spur investment.

Solar power offers many benefits, but location and cost can be challenges

The Sun is effectively inexhaustible as an energy source for human civilization. Moreover, the amount of solar energy reaching Earth's surface should be enough to power our civilization once the technology is adequately developed. These primary benefits of solar energy are clear, but the technologies themselves also provide benefits. PV cells and other solar technologies use no fuel, are quiet and safe, contain no moving parts, require little maintenance, and do not even require a turbine or generator to create

electricity. An average unit can produce energy for 20 to 30 years.

Solar systems allow for local, decentralized power. Homes, businesses, and isolated communities can use solar power to produce their own electricity without being near a power plant or connected to the power grid. In developing nations, solar cookers enable families to cook food without gathering fuelwood, lessening the daily workload and reducing deforestation. The low cost of solar cookers has made them available to many impoverished areas.

In the developed world, most PV systems are connected to the regional electric grid. This may enable owners of houses with PV systems to sell their excess solar energy to their local power utility. In this process, called **net metering**, the value of the power the consumer sells to the utility is subtracted from the consumer's monthly utility bill. As of this writing, net metering is available in most Canadian provinces.

Another advantage of solar power is that its development is producing new jobs. Currently, among major energy sources, PV technology employs the most people per unit of energy output. Finally, a major advantage of solar power over fossil fuels is that once a PV system is up

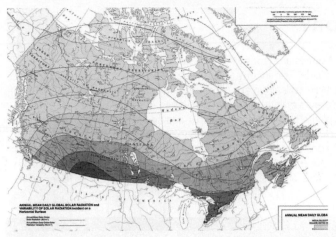

(a) Mean daily solar radiation

(b) Solar panels in remote Nunavut

FIGURE 16.27 Because some locations receive more sunlight than others, harnessing solar energy is more profitable in some areas than in others. The yearly solar average of Canada's populated areas **(a)** exceeds that in both Germany and Japan, the world's solar leaders. Solar energy can be used to power remote applications, even in locations with cold climates, as in this solar installation along the Alaska Highway in northern British Columbia **(b)**. *Source:* Courtesy of Canadian Geographic, canadiangeographic.ca/atlas.

and running, it produces no greenhouse gases or other polluting emissions.

Solar energy currently has two major disadvantages. One is that not all regions are sunny enough to provide adequate power, given current technology. Daily or seasonal variation in sunlight can also pose problems for stand-alone solar systems if storage capacity in batteries or fuel cells is not adequate or if backup power is not available from a municipal power grid.

In far northern locations, for example, although the total number of bright sunshine hours is very high over the course of a year, seasonal variations in the number of hours of daylight can make solar power more difficult to utilize (**FIGURE 16.27**). Intermittent supply is an argument that is also commonly levied against wind and other alternative energy sources. This can be mitigated by having backup systems available, and by coupling different energy sources together. The use of alternative energy sources doesn't have to be an all-or-nothing venture; even if solar is only supplying 50% of the energy to a building, for example, GHG emissions will still be cut in half.

The primary disadvantage of current solar technology is the upfront cost of equipment. Proponents of solar power argue that decades of government promotion of fossil fuels and nuclear power have made solar power unable to compete. However, decreases in price and improvements in energy efficiency of solar technologies so far are encouraging, even in the absence of significant financial commitment from government and industry.

At their advent in the 1950s, solar technologies had efficiencies of around 6% while costing $600 per installed watt (the cost is now closer to $6 per watt). Recent single-crystal silicon PV cells are showing 15% efficiency com-

mercially and 24% efficiency in lab research, suggesting that future solar technologies may be more efficient than any energy technologies we have today. Solar systems have become much less expensive over the years and now can pay for themselves within 10 to 20 years. After that time, they provide energy virtually for free as long as the equipment lasts. With future technological advances, some experts believe that the time to recoup investment could fall to one to three years.

Major solar installations also take up a lot of space, and modify the landscape significantly. This is even more of a concern in scenic or fragile landscapes. In "The Science behind the Story: Weighing the Impacts of Solar and Wind Development," we weigh some of the pros and cons of these up-and-coming technologies.

Wind Energy

Wind energy—energy derived from moving air masses—is an indirect form of solar energy, because it is the Sun's differential heating of air masses that causes wind to blow. We can harness power from wind using **wind turbines**, mechanical assemblies that convert wind's kinetic energy, or energy of motion, into electrical energy.

Today's wind turbines have their historical roots in Europe, where windmills have been used for 800 years to drain wetlands, irrigate crops, and grind grain into flour. Wind causes the windmill's blades to turn, driving a shaft connected to several cogs that turn wheels, which perform the required work. In North America, countless ranches in the Prairies and the Great Plains have long used windmills to draw ground water up for thirsty cattle.

THE SCIENCE ⟩ BEHIND THE STORY

Weighing the Impacts of Solar and Wind Development

Renewable energy sources and technologies alleviate many of the negative impacts of fossil fuel combustion, and may one day sustainably fulfill our energy needs. However, this does not mean that renewable energy is a panacea free of costs. As our society decides how to pursue energy sources such as solar and wind power, we will need to consider their impacts as well as their benefits. Scientific study of these impacts is just getting underway and will be important as energy development proceeds.

Major solar power installations can cover many hundreds of hectares of land (see **FIGURE 1**). Desert environments are particularly sensitive, so researchers say we should expect substantial impacts. Besides altering the pristine appearance of an undeveloped landscape, arrays of thousands of mirrors or panels affect communities of plants and animals by casting shade and altering microclimate. Altered conditions may hurt native desert-adapted species while helping invasive weeds. At

existing solar facilities, the sites are graded and sprayed with herbicide, eliminating plants and damaging fragile soils. Human presence increases as workers maintain the facilities. Solar power plants also require water for cooling and cleaning, and water is scarce in the arid regions hosting most of these facilities. All these impacts will have consequences for plants, animals, and ecosystems.

Given the impacts of large-scale solar facilities, researchers have determined that installing photovoltaic panels on rooftops of buildings is a low-impact alternative. Simply adding PV panels or roofing tiles to a rooftop has no effect on the landscape. When researchers assessed impacts over the systems' entire life cycles (from production to installation to operation), they found that besides avoiding land use

FIGURE 1 Solar arrays require large areas of land and exert substantial environmental impacts. Still, researchers estimate that the overall impacts are less than those demanded by fossil fuels; burning coal for energy uses at least as much land, once one includes the strip mining needed to obtain the coal. This solar plant in California is one of nine that spread across more than 650 ha in the Mojave Desert, providing power for over 230 000 homes.

Hank Morgan/RGB Ventures/SuperStock/Alamy

After the 1973 oil crisis, governments in North America and Europe began funding research and development for wind power. This moderate infusion of funding boosted technological progress, and the cost of wind power was cut in half within 10 years. Today wind power at favourable locations generates electricity for nearly as little cost per kilowatt-hour as conventional sources. The first wind power producer in Canada was Cowley Ridge Wind Plant in Alberta, the first phase of which opened in 1993. The largest is the Le Nordais project in the Gaspé Peninsula.

Modern wind turbines convert kinetic energy to electrical energy

In modern wind turbines, the wind turns blades that rotate machinery inside a compartment called a *nacelle*, which sits atop a tall tower (**FIGURE 16.28**). Inside the nacelle are a gearbox, a generator, and equipment to monitor and

control the turbine's activity. Most towers range from 40 to 100 m tall. Higher is generally better, to minimize turbulence and maximize wind speed. Most rotors consist of three blades and measure 42 to 80 m across. Turbines are designed to yaw, or rotate back and forth, in response to changes in wind direction, ensuring that the motor faces into the wind at all times. Turbines can be erected singly, but they are often erected in groups called **wind farms**. The world's largest wind farms contain hundreds of turbines spread across the landscape.

Slight differences in wind speed yield substantial differences in power output. The energy content of a given amount of wind increases as the square of its velocity; for this reason, if wind velocity doubles, energy quadruples. Some turbines are designed to generate low levels of electricity by turning in light breezes; others are programmed to rotate only in stronger winds, operating less frequently but generating large amounts of electricity in short time periods.

impacts, the rooftop systems also emitted significantly fewer greenhouse gases.[24]

Scientists have also studied health impacts of PV panels. PV cells can pose risks to workers manufacturing them, because they contain toxic chemicals such as cadmium and arsenic, and release fine silicon dust that can damage lung tissue. Researchers have concluded that proper attention to worker safety can greatly reduce these risks and that the exposure risks are not much different from other exposure risks we all face in our modern industrialized society.

The overall messages from studies so far are that (1) solar power, even with its impacts, is still cleaner and more sustainable than fossil fuel power; and (2) we can minimize the impacts of solar power by using rooftop panels and developing better technologies.

Similar messages are emerging from research on wind power. One major concern is that birds and bats are killed when they fly into the spinning blades of turbines. Studies suggest that bird deaths may be a less severe problem than was initially feared, but uncertainty remains. For instance, one European study indicated that migrating seabirds fly past offshore

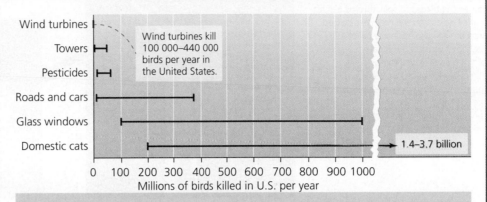

FIGURE 2 Wind turbines kill birds that fly into them. Yet far more birds are killed by other human causes. Shown are ranges of recent estimates of yearly bird mortality in the United States from several main causes. Habitat alteration is responsible for still more than any of the causes shown. *Source:* Data from American Bird Conservancy, from sources dating through 2010; and Loss, S.R., et al., (2012). The Impact of Free-Ranging Domestic Cats on Wildlife of the United States. *Nature Communications*, 4, Article 1396.

turbines without problem, but other data show that resident seabird densities have declined near turbines.

On land, the wind industry estimates that about two birds are killed per 1-megawatt-turbine per year. This is far fewer than the hundreds of millions of birds being killed each year by television, radio, and cell phone towers; pesticides; automobiles; glass windows; and domestic cats (**FIGURE 2**). If you own a cat and let it outside, you may be killing more wildlife than most wind farms.

At this point, bat mortality appears to be a more severe problem at wind turbines, but further research is needed. We can reduce wildlife impacts by avoiding development at sites on known migratory flyways or amid prime habitat for bat and bird species that are likely to fly into the blades.

Continued studies on the impacts of wind and solar development should help us find ways to harness renewable energy and attain a sustainable energy future while minimizing the environmental and social impacts of this development.

Wind power is the fastest-growing energy sector

Like solar energy, wind provides only a small proportion of the world's power needs, but wind power capacity has been growing fast—by 26% per year globally between 2000 and 2005, and then quadrupling between 2006 and 2009. Germany has been the world leader in installed wind capacity; however, in Denmark a series of wind farms supplies over 20% of the nation's electricity needs, the highest proportion in the world. Experts agree that wind power's rapid growth will continue. Meteorological studies suggest that wind power could be expanded in Canada to meet 15% of the nation's electrical needs.

Offshore and high-altitude sites are particularly promising. Wind speeds on average are roughly 20% greater over water than over land, and there is less air turbulence over water. It is clear from **FIGURE 16.29** that wind conditions over Hudson Bay are stronger than almost anywhere over land in Canada, for example. Costs to erect and maintain offshore wind turbines are higher, but the stronger, less turbulent winds produce more power and make them potentially more profitable. Currently, offshore wind farms are limited to shallow water, where towers are sunk into sediments to stabilize them. In the future, towers may be placed on floating pads anchored to the seafloor in deep water. There are as yet no operational offshore wind installations in Canada.

Winter weather also promotes stronger wind conditions, as do high elevations. Studies in Yukon, for example, have found many elevated sites suitable for wind power development, and the territory has recently invested in new installations (**FIGURE 16.29C**). Wind turbines in cold climates face specific challenges, however, such as icing of the turbine blades.

FIGURE 16.28

A wind turbine converts wind's energy of motion into electrical energy. Wind causes the blades to spin, turning a shaft that extends into the nacelle, perched atop the tower. Inside the nacelle, a gearbox converts the rotational speed of the blades, which can be up to 20 revolutions per minute (rpm) or more, into much higher rotational speeds (over 1500 rpm). These high speeds provide motion for a generator inside the nacelle to produce electricity.

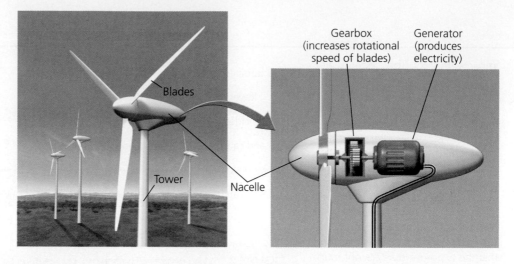

Wind power has many benefits

Like solar power, wind produces no emissions once the necessary equipment is manufactured and installed. The graph in **FIGURE 16.30** shows emissions per kWh of electricity produced over the entire installed lifetimes of various technologies. Emissions of CO_2 (the main greenhouse gas associated with global warming), SO_2 (the main precursor of acid rain), and NO_x (the main precursor of photochemical smog) are significantly lower for wind and other renewables, as well as nuclear and hydro, compared to fossil-fuel-based electricity generation. Other types of harmful emissions also can be avoided; for example, running a 1-megawatt wind turbine for 1 year prevents the release of approximately 30 kg of mercury, in comparison to the generation of electricity by a typical coal-fired power plant.

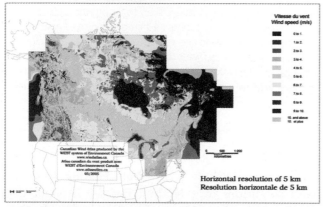

(a) Distribution of wind in Canada

(b) Cowley Ridge Wind Plant, Alberta

(c) Remote wind installation, Yukon

FIGURE 16.29

Southern Alberta and Saskatchewan enjoy some of Canada's best conditions for wind power on land, but offshore wind speeds are typically highest **(a)**. Cowley Ridge Wind Plant in southwestern Alberta **(b)** was Canada's first commercial wind power plant, opened in 1993. Winter winds and high elevations combine to make conditions appropriate for wind power generation in Yukon **(c)**, although icing of turbine blades can be a challenge.

Source: (a) Courtesy of Canadian Geographic, canadiangeographic.ca/atlas.

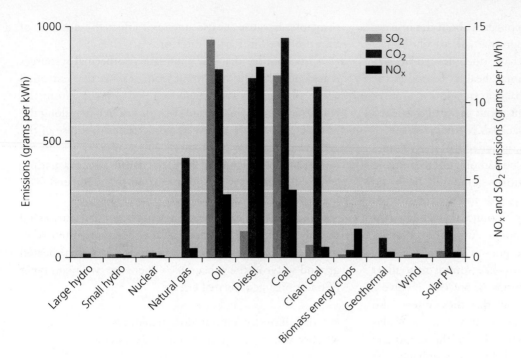

FIGURE 16.30 The lifetime emissions of CO_2, SO_2, and NO_x from wind, solar PV, hydro, nuclear, and even biomass-sourced electricity generation are dramatically lower per kWh than for all of the fossil-fuel-based sources.
Source: Data from International Energy, 1998. Image courtesy of Canadian Electricity Association.

Wind power, under optimal conditions, is considerably more efficient than conventional power sources in its energy returned on investment (EROI). Calculations of EROI typically find that wind turbines produce about 18 to 20 times more energy than they consume. For nuclear energy, the ratio is typically 15:1 or 16:1; for coal it can be anywhere from 8:1 to 30:1, depending on the type of coal technology and how the calculation is carried out. These calculations are very challenging, with many variables and factors that must be considered (such as economic fluctuations and the expected lifetimes of the technologies), so these are not conclusive results.

Wind turbine technology can be used on many scales, from a single tower for local use to fields of hundreds that supply large regions. Small-scale turbine development can help make local areas more energy self-sufficient. Farmers and ranchers can lease their land for wind development, which provides extra revenue while increasing property tax income for rural communities. Because each turbine takes up only a small area, most of the land between can still be used for farming, ranching, or other uses.

The Wind Energy Institute of Canada carries out research on these and other topics and challenges relevant to wind energy, including long-term studies at the North Cape Wind Farm in PEI. This is a research facility as well as an active, electricity-generating wind farm that is owned by the Prince Edward Island Energy Corporation, and was established in 1980.[25] Its location in a cold, harsh, coastal environment allows testing of wind equipment under very challenging conditions.

Wind power has some downsides— but not many

Wind is an intermittent resource; we have no control over when wind will occur. Wind speeds vary a lot from place to place, both geographically and with altitude, and over time, both seasonally and over the course of a day and night. Global and regional wind systems combine with local topography to create local wind patterns, and the best wind resources are not always near population centres that need the energy. Transmission networks would need to be greatly expanded to get wind power to where people live.

Unpredictability and the mismatch between location and need are major limitations in relying on wind as an electricity source; it poses less of a problem if wind is only one of several sources contributing to a utility's power generation. The issue of storage and transport of energy is not unique to wind energy; it is also an issue for solar and for hydrogen cell technologies, in particular. Several new technologies are being developed to store energy generated by wind and release it when and where it is most needed.

When wind farms *are* proposed near population centres, local residents often oppose them. Some people object to wind farms for aesthetic reasons, feeling that the structures clutter the landscape. Although polls show wide public approval of existing wind projects and of the concept of wind power in general, newly proposed wind projects often elicit the **"not-in-my-backyard" (NIMBY)** syndrome among people living nearby.

Neighbours have expressed concerns about the impacts of noise pollution on their cognitive and psychological well-being. Numerous studies have documented no significant impacts of noise on human health, as long as the noise source is adequately set back from populated areas; this was reinforced in a statement about the health impacts of wind turbines in 2012 by Canada's Medical Officer of Health. Other researchers have concluded that the volumes involved are not loud enough to be pathological to humans.

However, the *volume* of noise generated by wind turbines may not be the only cause for concern; wind turbines also generate some types of sounds that are different in their *characteristics* from other noises. An interesting study by Francesco Ruotolo and colleagues in 2012 in *Cognitive Processing* found the wind turbine-like sounds may affect certain types of cognitive functions (those that involve a high degree of concentration), but that these effects also may be intensified by the presence of a visual stimulation.[26] In other words, the combination of the sound and the visual presence of the wind turbine may enhance the negative effects for some people. This suggests that there is still research to be done, and impacts to be taken into consideration when examining proposed sites for wind energy.

Wind turbines may also pose a threat to birds and bats, as discussed in "The Science behind the Story: Weighing the Impacts of Solar and Wind Energy." Additional research is needed on this topic, but it may be possible to protect birds and bats by selecting turbine sites that are not on known migratory pathways or in the midst of prime habitat.

Geothermal Energy

Geothermal energy is one form of renewable energy that does not originate from the Sun; it is generated deep within Earth. The radioactive decay of elements deep in the interior of our planet generates heat that rises to the surface, heating rock and ground water, and generating steam. Geysers, volcanic eruptions, and submarine hydrothermal vents are the surface manifestations of these processes. Classic geothermal installations depend upon this localized heating as their energy source, but new technologies are taking advantage of much smaller temperature differences between surface and subsurface air and water as sources of energy.

We can harness geothermal energy for heating and electricity

Traditional geothermal power plants use the energy of naturally heated underground water and steam for direct heating and to turn turbines and generate electricity (**FIGURE 16.31**). Geothermal energy is renewable, in principle, because its use does not affect the amount of heat produced in Earth's interior.

One country that has abundant geothermal resources is Iceland. It is an island built from magma that extruded above the ocean's surface and cooled—magma from the Mid-Atlantic Ridge, the area of volcanic activity along the spreading boundary of two tectonic plates. Because of the geothermal heat in this region, volcanoes and geysers are numerous in Iceland. In fact, the word *geyser* (a natural spring that discharges hot water or steam) originated from the Icelandic *Geysir*, the island's largest geyser.

Geothermal energy sometimes can be harnessed directly from geysers at the surface, but most often wells must be drilled down hundreds of metres toward heated ground water in the subsurface. There are two main types of traditional geothermal energy:

- dry steam (or vapour-dominated) systems
- hot water (or liquid-dominated) systems

In liquid-dominated systems, which are the most common type, hot water is brought to the surface and converted to steam very quickly by lowering the pressure in specialized compartments. The steam is then used to turn a turbine to generate electricity. In dry steam systems, which are much less common, naturally occurring steam is piped directly from the geothermal well and used to turn the turbine to generate electricity. In both types, the steam is then condensed into water, which may be pumped back underground.

Hot ground water also can be used directly for district heating of homes, offices, and greenhouses; for industrial processes; and for drying crops. Iceland heats most of its homes through direct heating with piped hot water. Iceland began using geothermal energy in the 1940s, and today geothermal sources account for 66% of the nation's primary energy supply, and 25% of electricity generation.[27]

In principle, geothermal energy is inexhaustible. However, geothermal power plants may not be capable of operating indefinitely. If a geothermal plant uses heated water more quickly than the ground water is recharged, the plant will eventually run out of water. This is occurring at The Geysers, in Napa Valley, California, a very early geothermal application where the first generator was built in 1960. In response, operators have begun injecting municipal waste water into the ground to replenish the supply. More geothermal plants are now reinjecting water after it is used to help maintain pressure in the aquifer and thereby sustain the resource.

Geothermal **ground-source heat pumps (GSHPs)** are different from these classic geothermal applications. GSHPs use thermal energy from near-surface land and water (**FIGURE 16.32**); more precisely, they utilize *differences* in

FIGURE 16.31

At the Nesjavellir geothermal power station in Iceland **(a)**, steam is piped from four wells to a condenser, where cold water from lakeshore wells 6 km away is heated to 83°C. The heated water is sent through an insulated 270-km pipeline to Reykjavik, where residents use it for washing and space heating. At a typical geothermal site **(b)**, magma heats ground water deep underground (1), some of which escapes naturally through surface vents such as geysers (2). Geothermal facilities tap into heated water below ground and channel steam through turbines to generate electricity (3). After being used, the steam is condensed and the water pumped back into the aquifer to maintain pressure (4).

(a) Geothermal plant, Nesjavellir, Iceland

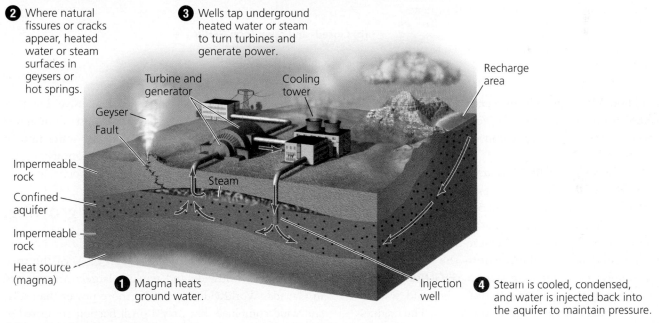

2 Where natural fissures or cracks appear, heated water or steam surfaces in geysers or hot springs.

3 Wells tap underground heated water or steam to turn turbines and generate power.

Turbine and generator

Cooling tower

Recharge area

Geyser

Fault

Impermeable rock

Confined aquifer

Impermeable rock

Heat source (magma)

Steam

1 Magma heats ground water.

Injection well

4 Steam is cooled, condensed, and water is injected back into the aquifer to maintain pressure.

(b) Geothermal power plant schematic

temperature between the surface and the subsurface. A few metres below the surface, ground materials tend to be relatively stable in temperature, varying much less from season to season than the air does. GSHPs can be used to provide heat in the winter by transferring heat from the ground into buildings. Heat pumps are reversible, which means that they also can be used in the summer to provide cooling by transferring heat from buildings into the ground. These installations are sometimes referred to as *Earth-energy* systems, to distinguish them from traditional geothermal systems.

The ground-to-air heat transfer in Earth-energy systems is accomplished by using a network of underground plastic pipes that circulate water or another liquid heat-transferring medium, commonly ethylene glycol (antifreeze) because it has a lower freezing temperature

than water. Because heat is simply moved from place to place rather than being produced using outside energy inputs, heat pumps are highly energy-efficient. The final component is the air duct system for delivering the heated or cooled air to the building. Ground-source heat pumps can significantly reduce the costs of heating and cooling a building, while also eliminating greenhouse gas emissions.

Water-temperature differences also can be used with heat pumps in a similar manner. In Toronto, the cold, deep waters of Lake Ontario provide air conditioning to more than 2.5 million m² of downtown office and living spaces in Toronto. The Enwave Deep Lake Water Cooling project is a district cooling technology that takes advantage of low-temperature lake water immediately adjacent to the downtown area. The project saves approximately

Enwave Energy Corporation

(a) Enwave Deep Lake Water Cooling Plant, Toronto

FIGURE 16.32
The Enwave Deep Lake Water Cooling Plant **(a)** takes advantage of the cold, deep waters of Lake Ontario to provide district cooling. Geothermal heat pump systems **(b)** operate on either closed-loop or open-loop systems, in which fluids are used to carry, supply, and store thermal energy. Heat pumps extract energy from the fluids, which are then returned to the underground reservoir.

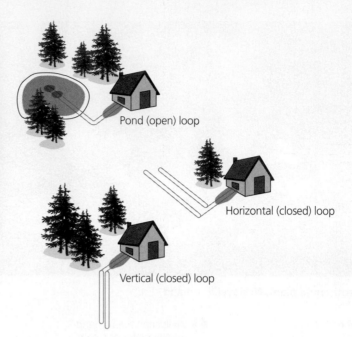

Pond (open) loop

Horizontal (closed) loop

Vertical (closed) loop

(b) Closed- and open-loop GSHPs

128 million kWh annually in electricity, reducing CO_2 emissions by 79 000 tonnes and reducing electricity consumption by 90% compared to conventional air-conditioning technologies.[28]

The Enwave system in Toronto works by taking cold water in through three large intake pipes that lie 83 m below the surface of the lake and extend 5.1 km from the shore. Each day, millions of litres of water at 4°C are pumped to the Toronto Island filtration plant, where the water is treated and circulated for distribution into the city drinking water supply via a pumping station (**FIGURE 16.32A**). Before leaving the pumping station, the cold water is diverted through a series of heat exchangers. The coldness of the water is used to remove the thermal energy from warm water passing through pipes on the other side of the heat exchangers. The newly cooled water moves through a system of many kilometres of underground pipes, providing cooling to customers in more than 140 buildings. The water is then returned to the pumping station, and the circuit is repeated.

The chilled water and the city's drinking water always remain on opposite sides of the heat exchanger and thus never come into contact with each other. The chilled water is cooled and recirculated through a *closed-loop system*, as shown in **FIGURE 16.32B**. In closed-loop systems, the heat-exchanging fluid is continuously recirculated through a pipeline circuit without discharging it back into the aquifer or water body from which it was obtained. In an *open-loop* system, in contrast, heat energy is acquired from the water via heat exchangers, and the water is then discharged back into the aquifer or water

body. Open-loop systems can be less expensive, but they require a sustainable water supply and must be designed to mitigate environmental impact. Injecting water into an aquatic system at a temperature that differs from its original temperature may damage organisms in the ecosystem.

Use of geothermal power is growing, though there are limitations

Geothermal energy provides less than 0.5% of the total energy used worldwide and remains largely unexploited in Canada. Worldwide it provides more power than solar and wind combined, but only a small fraction compared to hydropower and biomass. In Canada the geological settings are such that true geothermal energy is commercially viable only in British Columbia because of its proximity to the boundary of the Juan da Fuca tectonic plate.

In the right setting, geothermal power can be among the cheapest sources of electricity. Currently Japan, China, and the United States lead the world in use of geothermal power. However, at the world's largest geothermal power plants, The Geysers in northern California, generating capacity has declined by more than 50% since 1989 as steam pressure has declined.

Like other renewable sources, geothermal power greatly reduces polluting emissions, relative to fossil fuel combustion. Geothermal sources can release variable amounts of gases dissolved in their water, including carbon dioxide, methane, ammonia, and hydrogen sulphide. However, these gases are generally in very small quantities, and geothermal facilities using the latest filtering technologies produce even fewer emissions. By one estimate, each megawatt of geo-thermal power prevents the emission of 7.0 million kg of carbon dioxide emissions each year.

On the negative side, traditional geothermal sources, as we have seen, may not always be truly sustainable. In addition, the water of many hot springs is laced with salts and minerals that corrode equipment. These factors may shorten the lifetime of plants, increase maintenance costs, and add to pollution. Moreover, use of geothermal energy is limited to areas where the energy can be tapped. Unless technology is developed to penetrate far more deeply into the ground, geothermal energy use will remain extremely localized. Nonetheless, many geothermal resources remain unexploited, awaiting improved technology and govern-mental support for their development.

Ground-source and other types of heat pump systems have fewer drawbacks, especially where they are used as supplementary or backup sources for heating and cooling. The installations are typically unobtrusive, and people are often not even aware that the building they occupy is heated by a GSHP system. They are useful even in cold climates; there are many thousands of heat pump instal-lations throughout Canada, used both for individual buildings and for district heating and cooling. Compared to conventional electric heating and cooling systems, GSHPs heat spaces 50–70% more efficiently, cool them 20–40% more efficiently, can reduce electricity use by 25–60%, and can reduce emissions by up to 70%.[29]

Ocean Energy

The ocean hosts several underexploited energy sources. Each involves continuous natural processes that could potentially provide sustainable, predictable energy. Of the four approaches being developed, three involve motion, and one involves temperature.

We can harness energy from tides, waves, and currents

Just as dams on rivers use the kinetic energy of flowing water to generate hydroelectric power, the natural motion of ocean water can be used to generate electrical power. This motion can take the form of tides, waves, or ocean currents.

The rising and falling of ocean tides twice each day at coastal sites throughout the world moves large amounts of water past any given point on the world's coastlines (see "Central Case: Harnessing Tidal Energy at the Bay of Fundy"). Differences in height between low and high tides are especially great in long, narrow bays like the Bay of Fundy. Such locations are best for harnessing **tidal energy**, which is accomplished by erecting sluices across the outlets of tidal basins. The incoming tide flows through the sluice

gates and is trapped behind. Then, as the outgoing tide passes through the gates, it turns turbines to generate elec-tricity (**FIGURE 16.33**). Some designs allow for generating electricity from water moving in both directions.

The world's largest tidal generating facility is the La Rance facility in France, which has operated for over 40 years. Smaller facilities operate in China and Russia and in Canada at the Bay of Fundy. Tidal stations release few or no pollutant emissions, but they can have impacts on the ecology of estuaries and tidal basins. As discussed in the Central Case, a major focus of scientific research now is on the development of in-stream applications, which would be mostly under water and have the potential to minimize disruptions to marine and estuarine ecosystems.

Wave energy, harnessed from wind-driven waves at the ocean's surface, could be developed at a variety of sites to produce electricity. Many designs for machinery to harness wave energy exist, but few have been ade-quately tested and no commercial facilities are operating yet. Some designs are for offshore facilities and involve floating devices that move up and down with the waves. Wave energy is greatest at deep-ocean sites, but transmit-ting the electricity to shore would be expensive. Other wave-energy technologies are designed for coastal installa-tion. Some designs funnel waves into narrow channels and elevated reservoirs from which the water is then allowed to flow out, generating electricity in much the same manner as hydroelectric dams do. Other designs use rising and falling waves to push air into and out of chambers, turning turbines to generate electricity.

In addition to tides and waves, a third potential source of marine kinetic energy is the flow of ocean currents, such as the Gulf Stream. Devices that look essentially like underwater wind turbines have been erected in European waters to test the intriguing idea of harnessing the energy from the flow of surface currents.

The ocean also stores thermal energy

Each day the tropical oceans absorb an amount of solar radiation equivalent to the heat content of 250 billion barrels of oil—enough to provide about 100 000 times the electricity used daily in Canada. The ocean's sun-warmed surface is higher in temperature than its deep water, and **ocean thermal energy conversion (OTEC)** is based on this gradient in temperature.

In a closed-cycle approach (similar to the systems discussed in the context of ground-source heat pump systems), warm surface water is piped into a facility and used to evaporate chemicals that boil at low temperatures, such as ammonia. The evaporated gases spin turbines to generate electricity. Cold water piped in from ocean depths

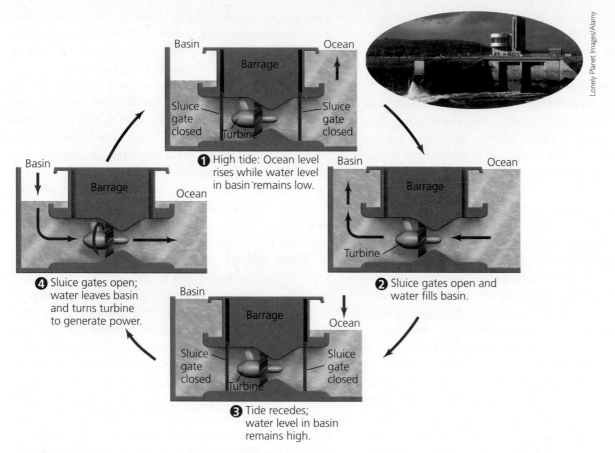

FIGURE 16.33 Energy can be extracted from the movement of the tides at coastal sites where tidal flux is great enough. One way of doing so involves using bulb turbines in concert with the outgoing tide. At high tide (1), ocean water is let through the sluice gates, filling an interior basin (2). At low tide (3), the basin water is let out into the ocean, spinning turbines to generate electricity (4). This technology is similar to what is used at the Annapolis Tidal Power Station (photo).

then causes the gases to condense so they can be reused. In the open-cycle approach, warm surface water is evaporated in a vacuum, and its steam turns turbines and then is condensed by cold water. Because ocean water loses its salts as it evaporates, the water can be recovered, condensed, and sold as desalinized fresh water for drinking or agriculture.

OTEC systems require not only a large temperature difference between the surface and deeper waters, but also a rapid dropoff of underwater topography near the coast, so that sufficiently cold temperatures can be accessed within a reasonable distance of the shore. Research on OTEC systems has been conducted in Hawaii and Japan, where conditions are optimal, but costs remain high, and as yet no facility is commercially operational.

Hydrogen Fuel and Power Storage

At the beginning of the chapter we mentioned that there are three main categories of applications for renewables: electricity generation, space heating (and cooling), and

fuels. All the renewable energy sources we have discussed so far can be used to generate electricity more cleanly than can fossil fuels. Many of them can be applied locally to provide space heating or cooling for buildings and districts.

As useful as these applications are, however, a major drawback common to most of them is that the energy and electricity cannot be stored and transported easily in large quantities for use when and where they are needed. This is one of the main reasons why most vehicles still rely on fossil fuels for power. The development of hydrogen fuels, fuel cells, and related energy storage technologies shows promise for storing and transporting energy conveniently.

Some energy experts believe that a "hydrogen economy" will be the next big energy wave worldwide. In this vision, hydrogen fuel—together with electricity generated from renewable sources—would serve as the basis for a clean, safe, and efficient energy system. A hydrogen-based system would use as fuel the universe's simplest and most abundant element. Electricity generated from renewable sources that are intermittent, such as wind or solar energy, could be used to produce the

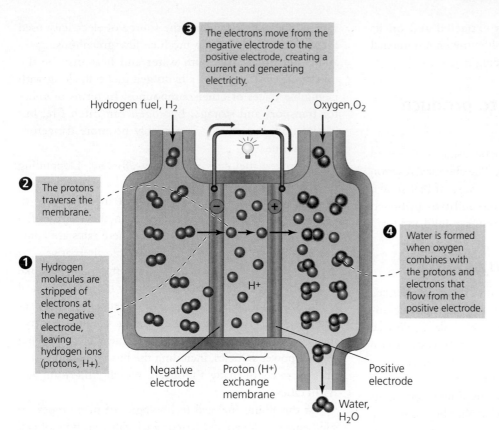

❸ The electrons move from the negative electrode to the positive electrode, creating a current and generating electricity.

Hydrogen fuel, H_2

Oxygen, O_2

❷ The protons traverse the membrane.

❶ Hydrogen molecules are stripped of electrons at the negative electrode, leaving hydrogen ions (protons, H+).

H^+

❹ Water is formed when oxygen combines with the protons and electrons that flow from the positive electrode.

Negative electrode

Proton (H+) exchange membrane

Positive electrode

Water, H_2O

FIGURE 16.34
Hydrogen fuel drives electricity generation in a fuel cell, creating water as a waste product. There are many different types of fuel cells. In this example, atoms of hydrogen first are stripped of their electrons (1). The electrons move from a negative electrode to a positive one, creating a current and generating electricity (2). Meanwhile, the hydrogen ions pass through a proton exchange membrane (3) and combine with oxygen to form water molecules (4).

hydrogen. **Fuel cells**—which utilize chemical energy, like batteries—could then employ the hydrogen to produce electrical energy as needed to power vehicles, computers, cell phones, home heating, and countless other applications (**FIGURE 16.34**).

Basing an energy system on hydrogen could alleviate dependence on foreign fuels and help fight climate change. For these reasons, many governments, including the federal and provincial governments in Canada, are funding research into hydrogen and fuel cell technology. Automobile companies and other private corporations also are investing significant amounts in research and development to produce vehicles that run on hydrogen and to develop fuel cell technologies and infrastructure.

Hydrogen may be produced from water or from other matter

Hydrogen gas (H_2) does not tend to exist freely on Earth; rather, hydrogen atoms bind to other molecules, becoming incorporated in everything from water to organic molecules. To obtain hydrogen, we must force these substances to release their hydrogen atoms, and this requires an input of energy. Several ways of producing hydrogen are being studied. In **electrolysis**, electricity is

input to split hydrogen atoms from the oxygen atoms of water molecules:

$$2H_2O \rightarrow 2H_2 + O_2$$

Whether this strategy for producing hydrogen would cause pollution over its entire life cycle depends primarily on the source of the electricity used for the electrolysis. If coal is burned to create the electricity, then the process may not reduce emissions compared with fossil fuels. If, however, the electricity is produced by a less-polluting renewable source, then hydrogen production by electrolysis would create much less pollution than fossil fuels. The "cleanliness" of a future hydrogen economy, therefore, depends largely on the source of electricity used in electrolysis.

The environmental impact of hydrogen production also depends on the source material for the hydrogen. Besides water, hydrogen can be obtained from biomass and fossil fuels. Obtaining hydrogen from these sources generally requires less energy input, but results in emissions of carbon-based pollutants. For instance, extracting hydrogen from the methane (CH_4) in natural gas entails producing one molecule of the greenhouse gas carbon dioxide for every four molecules of hydrogen gas:

$$CH_4 + 2H_2O \rightarrow 4H_2 + CO_2$$

Thus, whether a hydrogen-based energy system would be environmentally cleaner than a fossil fuel system will depend on how the hydrogen is extracted and on its source. Other questions about the lifetime environmental impacts of hydrogen fuel are still being investigated.

Fuel cells can be used to produce electricity

Once isolated, hydrogen gas can be used as a fuel to produce electricity within fuel cells. The chemical reaction involved in a fuel cell is simply the reverse of that shown for electrolysis; an oxygen molecule and two hydrogen molecules each split so that their atoms can bind and form two water molecules:

$$2H_2 + O_2 \rightarrow 2H_2O$$

The way this occurs in one common type of fuel cell is shown in **FIGURE 16.34**. Hydrogen gas (usually compressed and stored in an attached fuel tank) is allowed into one side of the cell, whose middle consists of two electrodes that sandwich a membrane that only protons (hydrogen ions) can move across. One electrode, helped by a chemical catalyst, strips the hydrogen gas of its electrons, creating two hydrogen ions that begin moving across the membrane.

Meanwhile, on the other side of the cell, oxygen molecules are split into their component atoms along the other electrode. These oxygen ions soon bind to pairs of hydrogen ions travelling across the membrane, forming molecules of water that are expelled as waste, along with heat. While this is occurring, the electrons from the hydrogen atoms have travelled to a device that completes an electric current between the two electrodes. The movement of the hydrogen's electrons from one electrode to the other creates the output of electricity.

The main differences among the various types of hydrogen fuel cell technologies that are being developed today are in the material used as the catalyst (which speeds up the reactions at the electrodes) and the electrolyte, which carries the electrons from one electrode to the other. Other differences include the source material for the hydrogen (typically either water or natural gas) and the technology used to produce the hydrogen fuel.

Hydrogen fuel cells have many benefits but require further development

As a fuel, hydrogen offers a number of potential benefits. We will never run out of hydrogen; it is the most abundant element in the universe and occurs everywhere on Earth as a fundamental component of water. It can be clean and nontoxic to use, and—depending on the source of the hydrogen and the source of electricity used for its extraction—it may produce few greenhouse gases and other pollutants. Pure water and heat may be the only waste products from a hydrogen fuel cell, along with negligible traces of other compounds. In terms of safety for transport and storage, hydrogen can catch fire, but if kept under pressure it is probably no more dangerous than gasoline.

Hydrogen fuel cells are energy-efficient. Depending on the type of fuel cell, 35% to 70% of the energy released in the reaction can be used. If the system is designed to capture heat as well as electricity, then the energy efficiency of fuel cells can rise to 90%. These rates are comparable or superior to most nonrenewable alternatives.

Fuel cells are also silent and nonpolluting. Unlike batteries (which also produce electricity through chemical reactions), fuel cells will generate electricity whenever hydrogen fuel is supplied, without ever needing recharging. For all these reasons, hydrogen fuel cells are being used to power vehicles, including the buses now operating on the streets of many European, North American, and Asian cities.

At this point, fuel cell technologies of many types are still being developed and tested. Each different type of fuel cell has benefits and drawbacks in terms of efficiency, adaptability, cost, and environmental impacts, and no one technology has emerged as the "fuel cell of the future." One obvious limitation to widespread adoption is the small number of hydrogen refuelling stations, worldwide. But the main limitation, still, is that fuel cell technologies are not yet economically competitive with other energy technologies.

Conclusion

The potential for decline of fossil fuel supplies and the increasing concern over air pollution and global climate change have convinced many people that we will need to shift to renewable energy sources that will not run out and will pollute far less. Hydropower is a renewable, low-emission alternative, but it is not without its own negative ecological impacts. Nuclear power showed promise, but high costs, the continuing challenge of radioactive waste management, and public fears over safety have stalled its growth. Biomass energy sources include traditional fuelwood, as well as newer biofuels and biopower; these sources can be carbon-neutral but are not all strictly renewable, especially if biomass sources are overharvested.

One way that industrial developers and decision-makers can evaluate the efficiency and overall impacts of clean-energy technologies and initiatives is to use a software package called *RETScreen*, which was developed by the Government of Canada but is now used internationally.[30]

RETScreen is a clean-energy project analysis tool that is managed by the CanmetENERGY research centre of Natural Resources Canada. It is a free download that runs from an Excel platform; you can download it and try it out yourself. RETScreen has been used to evaluate a wide range of projects internationally, including energy retrofitting of the Empire State Building in New York City, analysis of Manitoba Hydro's bioenergy program applications, assessment of solar photovoltaic performance in Toronto, and a variety of academic and developmental applications, such as assessing the viability of wind farming in Argentina.

Most renewable energy sources have been held back by inadequate funding for research and development, and by artificially cheap market prices for nonrenewable resources that do not include external costs. Despite these obstacles, renewable technologies have progressed far enough to offer hope that we can shift from fossil fuels to renewable energy with a minimum of economic and social disruption. Whether we can also limit environmental impact will depend on how soon and how quickly we make the transition and to what extent we put efficiency and conservation measures into place.

REVIEWING OBJECTIVES

You should now be able to

Outline the reasons for seeking alternatives to fossil fuels, and the conventional and new alternatives that are being developed

- Fossil fuels are nonrenewable resources, and we are gradually depleting them. Fossil fuel combustion causes air pollution that results in many environmental and health impacts and contributes to global climate change.
- "Conventional" energy sources are the alternatives to fossil fuels that are most widely used today. They include hydroelectric power, nuclear power, and biomass energy.
- "New renewable" energy technologies are still being developed, and for the most part are not yet financially competitive with fossil fuels and conventional energy sources.

Describe the technologies, environmental impacts, and energy contributions of the conventional alternatives to fossil fuels: hydroelectric power, nuclear power, and traditional biomass energy

- Hydroelectric power is generated when water from a river runs through a powerhouse and turns turbines. Hydropower produces little air pollution, but dams and reservoirs can greatly alter riverine ecology. Run-of-river is an alternative approach. Hydropower accounts for about 16% of electricity generation globally, but almost 60% in Canada.
- Nuclear power comes from converting the energy of subatomic bonds into thermal energy. Uranium is mined, enriched, processed, and used as fuel in controlled fission reactions. This process, carried out in nuclear reactors, produces heat that powers electricity generation. Many advocates of "clean" energy support nuclear power because it lacks the polluting emissions of fossil fuels. For others the risks, costs, and challenges, especially the management of radioactive waste, outweigh the benefits.
- Traditional biomass (mainly fuelwood, charcoal, peat, and animal manure) provides 9% of global primary energy use, but much more, proportionately, in the poorest developing countries. Biomass energy is theoretically renewable, unless overharvesting occurs.

Summarize the major "new" renewable alternative sources of energy, and assess their potential for growth

- "New" renewable energy sources include solar, wind, geothermal, ocean, and new biomass technologies, as well as hydrogen fuels and fuel cells. Most of these are not truly "new," but they are currently in a stage of rapid development of modern technologies.
- The new renewables currently provide far less energy and electricity than fossil fuels and other conventional energy sources. However, their use is growing quickly, and this growth is expected to continue as people move away from fossil fuels.

Describe a variety of new biomass, solar, wind, geothermal, ocean energy, and hydrogen fuel cell technologies, and outline their advantages and disadvantages

- Biofuels, including ethanol and biodiesel, are used to power automobiles. Some crops are grown specifically for this purpose, and waste oils are also used. There are concerns about the impacts of using food crops to produce biofuels. Biomass from special crops, agricultural waste, forestry waste, and landfill gas can be used to generate electricity (biopower).

- Energy from the Sun's radiation can be harnessed using passive methods, or by active methods involving powered technology. Solar energy is perpetually renewable, creates no emissions, and enables decentralized power.

- Wind energy is harnessed using turbines mounted on towers. They are often erected in arrays at wind farms on land or offshore, where wind conditions are optimal. Wind energy is renewable and creates no emissions. The cost is competitive with that of electricity from fossil fuels, but wind energy is intermittent and often faces local opposition.

- Thermal energy from inside the planet rises toward the surface and heats ground water. Earth-energy systems use heat pumps to exchange heat between the air and water, or materials in the subsurface. The use of geothermal energy and Earth energy for direct heating of water, for electricity generation, for space heating and cooling can be efficient and clean.

- Major ocean energy sources include the motion of tides, waves, and currents, and the thermal heat of ocean water. Ocean energy is perpetually renewable, but so far technologies have seen only limited development.

- Hydrogen fuel cells create electricity by controlling an interaction between hydrogen and oxygen, and they produce only water as a waste product. Fuel cells are silent, nonpolluting, and do not need recharging. Hydrogen can be produced through electrolysis, or by using fossil fuels; in the latter case, the environmental benefits are reduced.

TESTING YOUR COMPREHENSION

1. How much of our global energy supply do nuclear power, biomass energy, and hydroelectric power contribute? How much of our global electricity do these three conventional energy alternatives generate?

2. Describe how nuclear fission works. How do nuclear plant engineers control fission and prevent a runaway chain reaction? What has been done so far about disposing of radioactive waste?

3. In terms of greenhouse gas emissions, how does nuclear power compare to coal, oil, and natural gas? How do hydropower and biomass energy compare?

4. Contrast the two major approaches to generating hydroelectric power, and compare their environmental impacts.

5. List five sources of biomass energy. What is the world's most-used source of biomass energy? How does biomass energy use differ between developed and developing nations?

6. About how much of our energy now comes from renewable sources? What is the most prevalent form of renewable energy we use? What form of renewable energy is most used to generate electricity? Which "new" renewable source is experiencing the most rapid growth?

7. Contrast passive and active solar heating. Describe how each works, and give examples of each. What are the environmental and economic advantages and disadvantages of solar power?

8. How do modern wind turbines generate electricity? How does wind speed affect the process? What factors affect where wind turbines are placed? What are the environmental and economic benefits and drawbacks of wind power?

9. Define geothermal energy, and explain how it is used. In what ways is it renewable, or not renewable? How does it differ from what Natural Resources Canada calls "Earth energy"?

10. List and describe four approaches to obtaining energy from ocean water.

THINKING IT THROUGH

1. Given what you have learned about some of the energy alternatives discussed in this chapter, do you think it is important for Canadians to move ahead now with minimizing our use of fossil fuels and maximizing our use of renewable alternatives? What challenges or obstacles do we need to overcome in order to transition smoothly to alternative sources of energy?

2. Nuclear power has by now been widely used for over three decades, and the world has experienced only two major accidents (Chernobyl and Fukushima Daiichi). Would you call this a good safety record? Does the safety record and low atmospheric pollution associated with nuclear power outweigh the financial costs and the problem of disposing safely of radioactive wastes?

3. There are many different sources of biomass and many ways of harnessing energy from biomass. Discuss one that seems particularly beneficial to you, and one with which you see problems. What biomass energy sources and strategies do you think our society should focus on investing in?

4. Let's say that you are the CEO of a company that develops wind farms. Your staff is presenting you with three options, listed next, for sites for your next development. Describe at least one likely advantage and at least one likely disadvantage you would expect to encounter with each option. What further information would you like to know before deciding which to pursue? What information will you share with neighbours of your preferred site?

- Option A: A remote rural site in Yukon
- Option B: A ridge-top site among the suburbs of Saskatoon
- Option C: An offshore site off the Nova Scotia coast

INTERPRETING GRAPHS AND DATA

It is not clear that growing crops such as corn for the purpose of producing ethanol is the most efficient use of crop lands. In 2005, David Pimentel and Tad Patzek estimated that replacing just one-third of the total gasoline used in North America with ethanol would require more cropland than is needed to feed the population![31] They calculated that 0.6 hectares of corn yield enough ethanol to displace one-third of the gasoline needed to run one average North American car for one year; for comparison, it would require 0.5 hectares of corn to feed one person for one year.

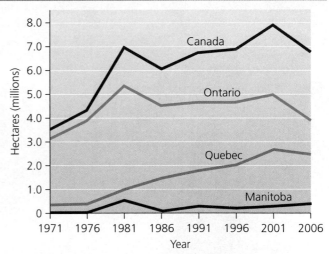

This graph shows the corn area planted in Canada, over time.

Source: Based on Pimentel, D., and T.W. Patzek (2005). Ethanol Production Using Corn, Switchgrass, and Wood; Biodiesel Production Using Soybean and Sunflower. *Natural Resources Research, 14*(1).

1. Using the estimates made by Pimentel and Patzek, how many hectares of corn grown for ethanol would it take to completely displace the need for gasoline to run one car for a year?

2. As shown on the graph, the area of corn planted in Canada peaked at about 1 315 000 hectares in 2001. Given that there are approximately 12 650 000 cars in Canada, by how much would the hectares planted in corn need to be increased in order to run all of the cars in Canada on ethanol? One-third of the cars? One-fourth of the cars?

MasteringEnvironmentalScience®

STUDENTS
Go to **MasteringEnvironmentalScience** for assignments, the eText, and the Study Area with practice tests, videos, current events, and activities.

INSTRUCTORS
Go to **MasteringEnvironmentalScience** for automatically graded activities, current events, videos, and reading questions that you can assign to your students, plus Instructor Resources.

17 Mineral Resources and Mining

These workers are processing coltan ore mined in central Africa.

Hereward Holland/Reuters/Landov

Upon successfully completing this chapter, you will be able to

- Summarize the main types of mineral resources and the ways in which mineral resources contribute to our economy and society
- Describe the major methods of mining

- Characterize the impacts of mining, both environmental and social, and approaches to management, restoration, and reclamation
- Evaluate ways to encourage sustainable use of mineral resources

A young artisanal miner works to extract coltan ore in North Kivu, Democratic Republic of Congo.

AFRICA

Region of coltan mining

Democratic Republic of the Congo

CENTRAL CASE
MINING FOR ... CELL PHONES?

> "The conflict in the Democratic Republic of the Congo has become mainly about access, control, and trade of five key mineral resources: coltan, diamonds, copper, cobalt, and gold."
>
> —REPORT TO THE UNITED NATIONS SECURITY COUNCIL, APRIL 2001

> "Coltan ... is not helping the local people. In fact, it is the curse of the Congo."
>
> —AFRICAN JOURNALIST KOFI AKOSAH-SARPONG

A student on a university campus in Canada uses her cell phone to text a friend. Inside the phone is a little-known metal called tantalum—just a tiny amount, but no cell phone could operate without it. Half a world away, a dirt-poor miner in the heart of Africa toils in a jungle streambed (see photo), sifting sediment for nuggets of ore that contain tantalum. At nightfall, rebel soldiers take most of his ore, leaving him to sell what little remains to buy food for his family at the squalid mining camp where they live.

Tantalum links our glossy, high-tech economy with one of the most badly wrecked regions on Earth. The Democratic Republic of the Congo (often referred to as DRC, to distinguish it from the Republic of Congo) has been embroiled in a sprawling conflict that has involved six nations and various rebel militias. Over 5 million people have lost their lives in DRC since 1998—a devastating loss for a country with a population just twice the size of Canada's. It is the latest chapter in the sad history of a nation rich in natural resources—copper, cobalt, gold, diamonds, uranium, and timber—whose impoverished people keep losing control of those resources to others.

Central to this conflict is tantalum (Ta), element number 73 on the Periodic Table of the Elements. We rely on this metal for our cell phones, computer chips, DVD players, game consoles, and digital cameras. Tantalum powder is ideal for capacitors (the components that store energy and regulate current in miniature circuit boards) because it is highly heat resistant and readily conducts electricity. Tantalum comes from a dull, blackish mineral called tantalite. Tantalite often occurs with a mineral called columbite, so the ore is referred to as columbite-tantalite, or *coltan* for short. In eastern Congo, men and boys dig craters in rain forest stream-beds, panning for coltan much as early miners in the Yukon panned for gold.

As information technology boomed in the late 1990s, global demand for tantalum rose. Market prices for the metal shot up, reaching as high as $770/kg in 2000.[1] High prices led some to mine coltan by choice, but many more were forced into it. Local militias, supported by forces from neighbouring Rwanda and Uganda, overran eastern Congo. Farmers were chased off their land, villages were burned, and civilians were raped, tortured, and killed. Soldiers seized control of mining operations, forcing farmers, refugees, prisoners, and children to work, and skimming off the profits. AIDS and other sexually transmitted diseases spread through the mining camps. The turmoil caused ecological havoc as miners and soldiers streamed into national parks, clearing rain forests and killing wildlife for food, including elephants, hippos, endangered gorillas, and the okapi, a rare relative of the giraffe.

Most miners ended up with little, while rebels, soldiers, and bandits enriched themselves selling coltan to traders, who sold it to processing companies in Europe and North America. These companies refine and sell tantalum powder to capacitor manufacturers, which in turn sell capacitors to Nokia, Motorola, Sony, Intel, Compaq, Dell, and other high-tech corporations. In 2001, an expert panel commissioned by the United Nations Security Council concluded that coltan was fuelling, financing, and prolonging the war.[2] The panel urged a UN embargo on coltan and other minerals smuggled from DRC. A grassroots activist movement advanced the slogan, "No blood on my cell phone!"

Sony, Nokia, Ericsson, and other corporations rushed to assure consumers that they were not using tantalum from eastern Congo—and the region was in fact producing less than 10% of the world's supply. Meanwhile, some observers felt an embargo on minerals could hurt the long-suffering Congolese people, rather than help them. The mining life may be miserable, but it pays better than most alternatives in a land where the average income is less than $2 a day.

Soon, however, the high-tech boom went bust, and global demand for tantalum diminished. This occurred just as Australia and other countries were ramping up industrial-scale tantalite mining. As supply outpaced demand, the market price of tantalum plummeted to below $100/kg, and several major producers quit mining tantalum. But nations began to work through their stockpiles, and by 2010 demand had grown, driving prices up once again; it is currently (mid-2015) around $360/kg.[3]

As of 2014, rebel groups had lost control of most of the coltan mining operations in DRC, but internal factions continue to fight. The United Nations Security Council voted in 2015 to continue an arms embargo on the region. Western electronics companies avoid knowingly purchasing tantalum from DRC, but the coltan trade has so many middlemen and so little transparency that many companies find it difficult to determine where their tantalum actually comes from. Much ends up being sold to China. Black markets for coltan also have sprung up in Colombia, Venezuela, and Brazil, in the northern part of the Amazon.

Similar stories are playing out with other "conflict minerals," such as gold and diamonds. The final report of the UN Expert Group cited significant progress in improving traceability and due diligence by those who purchase tantalum and other conflict minerals.[4] However, the report also lamented the continued lack of progress in stemming the illegal trade and smuggling of wildlife, including ivory, from DRC.

Earth's Mineral Resources

Coltan provides just one example of how we extract raw materials and turn them into products we use in our everyday lives. In previous chapters we saw how the rock cycle creates new rock and alters existing rock. We saw how plate tectonics builds mountains; shapes the geography of oceans, islands, and continents; and gives rise to earthquakes and volcanoes. The coltan mining areas of eastern Congo are situated along the western edge of Africa's Great Rift Valley system, a region where the African tectonic plate is slowly pulling itself apart. Some of the world's largest lakes have formed in the immense valley floors, far below towering volcanoes such as Mount Kilimanjaro.

These geological processes are fundamental to shaping the world around us; they also control the distribution of rocks and minerals in the lithosphere and their availability to us. In this chapter we take a closer look at how resources from the lithosphere contribute directly to our economies and our lives. We will first examine mineral resources and the products they provide us, and the various ways we extract them. We will then examine the many social and environmental impacts

that our mining efforts exert, and see what we can do to mitigate these impacts. Finally, because mineral resources are nonrenewable on human timescales, we need to be attentive to finite and decreasing supplies of economically important minerals, so we will examine solutions we can pursue to make our mineral use more sustainable.

Rocks provide the minerals we use

A **rock** is a solid aggregation of minerals, and a **mineral** is a naturally occurring solid chemical element or inorganic compound with a crystal structure, a specific chemical composition, and distinct physical properties. For instance, the mineral tantalite consists of the elements tantalum, oxygen, iron, and manganese. Tantalite and its common partner, columbite, occur most commonly in *pegmatite*, a type of coarse-grained igneous rock that is similar in composition to granite. In addition to tantalite, pegmatite generally contains the minerals feldspar, quartz, and mica, and occasionally even includes gemstones and other rare minerals.

We depend on a wide array of mineral resources as raw materials for our products, so we mine and process these resources. Consider a typical scene from a student lounge at a college or university (**FIGURE 17.1**), and note how

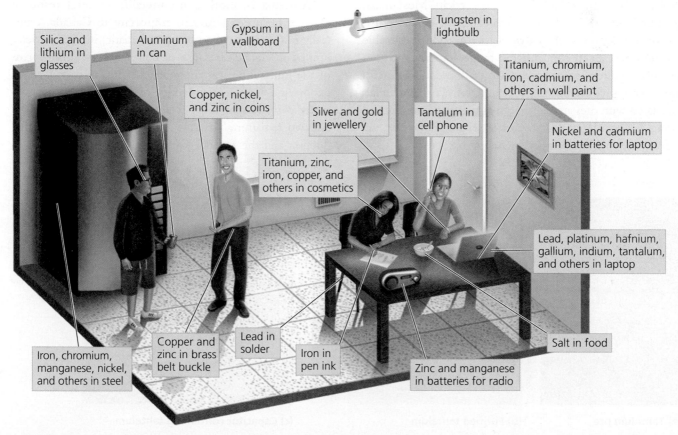

Silica and lithium in glasses

Aluminum in can

Gypsum in wallboard

Tungsten in lightbulb

Titanium, chromium, iron, cadmium, and others in wall paint

Copper, nickel, and zinc in coins

Silver and gold in jewellery

Tantalum in cell phone

Nickel and cadmium in batteries for laptop

Titanium, zinc, iron, copper, and others in cosmetics

Lead, platinum, hafnium, gallium, indium, tantalum, and others in laptop

Iron, chromium, manganese, nickel, and others in steel

Copper and zinc in brass belt buckle

Lead in solder

Iron in pen ink

Zinc and manganese in batteries for radio

Salt in food

FIGURE 17.1 Elements from minerals that we mine are everywhere in the products we use in our everyday lives. This scene from a typical student lounge points out just a few of the many elements from minerals that surround us.

many items are made with elements from the minerals we take from Earth. Without the resources from beneath the ground that we use to make building materials, wiring, clothing, appliances, fertilizers for crops, and so much more, civilization as we know it could not exist.

We obtain minerals and metals by mining

We obtain the minerals we use and need through the process of mining, and Canada is one of the world's leading mining nations. The term *mining* in the broad sense describes the extraction of any resource that is nonrenewable on the timescale of our society. In this sense, we mine fossil fuels and ground water, as well as minerals. It can even be said that we mine fish, soils, and trees, if we withdraw from the stocks of those renewable resources at rates that are faster than the rate of recharge. When used specifically in relation to minerals, **mining** refers to the systematic removal of rock, soil, or other material for the purpose of extracting minerals of economic interest. Because most minerals of interest are widely spread but in low concentrations, miners and mining geologists first try to locate concentrated sources of minerals before mining begins.

Some minerals can be mined for metals. A **metal** is a type of chemical element that typically is shiny, opaque, and malleable, and can conduct heat and electricity. Most metals are not found in a pure (or *native*) state in Earth's crust, but instead are present in **ore**, a rock in which valuable minerals have been concentrated by geological processes, and from which we can extract metals and other useful mineral resources. The waste rock and nonvaluable minerals associated with ores are referred to as **gangue** (pronounced "gang"). The higher the proportion of gangue to ore, the more waste rock is generated by the mining process.

Copper, iron, lead, gold, and nickel are among the many economically valuable **metallic resources** we extract from mined ores; Canada has been a leading producer of all these metallic resources, and many others, including tantalum. These metals and others serve so many purposes that our modern lives would be impossible without them. For example, tantalum, used in the electronic components of computers, cell phones, DVD players, and other devices, is a metal that comes from the mineral tantalite (**FIGURE 17.2**), which occurs, along with the mineral columbite, in the form of coltan ore. Columbite contains the metal niobium, formerly known as columbium, which is utilized in ways similar to tantalum.

We also mine and use many minerals that do not contain metals. **Nonmetallic resources** such as limestone, salt, and many others are mined for a number of diverse purposes. **FIGURE 17.3** illustrates a selection of economically useful mineral resources, both metallic and nonmetallic. For each one, its major nation of origin and several main uses are shown.

Sand and gravel together are the most commonly mined nonmetallic mineral resources. They are used as **aggregate** for road-building fill and construction materials and for the manufacture of concrete. Each year over 225 million tonnes of sand and gravel are mined in Canada overall, from active mines in almost every province and territory.

Asbestos is another nonmetallic mineral resource that has been economically important in Canada, especially in Quebec. Asbestos is problematic; its use is strictly limited in Canada because of concerns that some forms of the mineral are carcinogenic. Yet in the past, Canada exported a significant quantity of asbestos every year to many countries in the developing world. In 2011, the last two remaining asbestos mines in Canada halted production and the federal government committed funds to

(a) Tantalum ore   **(b) Purified tantalum**   **(c) Capacitor containing tantalum**

FIGURE 17.2   Tantalum ore **(a)** is mined from the ground and then processed to extract the pure metal tantalum **(b)**. This metal is used in capacitors **(c)** and other electronic components in computer chips, cell phones, and many other devices.

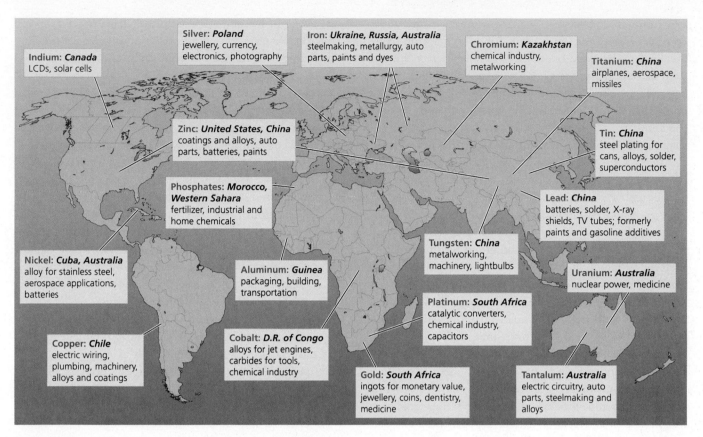

Indium: *Canada*
LCDs, solar cells

Silver: *Poland*
jewellery, currency, electronics, photography

Iron: *Ukraine, Russia, Australia*
steelmaking, metallurgy, auto parts, paints and dyes

Chromium: *Kazakhstan*
chemical industry, metalworking

Titanium: *China*
airplanes, aerospace, missiles

Zinc: *United States, China*
coatings and alloys, auto parts, batteries, paints

Tin: *China*
steel plating for cans, alloys, solder, superconductors

Phosphates: *Morocco, Western Sahara*
fertilizer, industrial and home chemicals

Lead: *China*
batteries, solder, X-ray shields, TV tubes; formerly paints and gasoline additives

Nickel: *Cuba, Australia*
alloy for stainless steel, aerospace applications, batteries

Aluminum: *Guinea*
packaging, building, transportation

Tungsten: *China*
metalworking, machinery, lightbulbs

Uranium: *Australia*
nuclear power, medicine

Platinum: *South Africa*
catalytic converters, chemical industry, capacitors

Copper: *Chile*
electric wiring, plumbing, machinery, alloys and coatings

Cobalt: *D.R. of Congo*
alloys for jet engines, carbides for tools, chemical industry

Gold: *South Africa*
ingots for monetary value, jewellery, coins, dentistry, medicine

Tantalum: *Australia*
electric circuitry, auto parts, steelmaking and alloys

FIGURE 17.3 The minerals we use come from all over the world. Shown is a very small selection of economically important elements (mostly metals), together with their major uses and one or more important producer nations.

restore the economic well-being of the communities that depended on them for employment.

Gemstones also are important nonmetallic mineral resources, treasured for their rarity and beauty. For instance, diamonds have long been prized—and, like coltan, they have fuelled resource wars. The diamond trade has acted to fund, prolong, and intensify wars in Congo, Angola, Sierra Leone, Liberia, and elsewhere, as armies exploit local people for mine labour, and sell the diamonds for profit. This is why you may hear the phrase "blood diamonds," just as coltan has been called a "conflict mineral."

We also mine nonmetallic substances that we use for fuel. One of the most common fuels that we mine is coal. Coal is technically a rock, not a mineral, because it is an aggregate that consists of organic matter mixed with up to 50% mineral matter. We consider coal mining in this chapter because it has been economically important in Canada and has relevance for many issues related to mining and the environment. Other fossil fuels—petroleum, natural gas, and alternative fossil fuels such as oil sands, oil shale, and methane hydrates—are also extracted from the ground, as discussed in Chapter 15, *Fossil Fuels*. Uranium is mined and used as fuel in nuclear power generation. Technically, uranium is a metal, but it is mined as a fuel, not as a metal, so it is considered a

nonmetallic resource. Canada has traditionally been the world's largest producer of uranium, although the top spot was taken over by Kazakhstan in 2009.

We process ores after mining

Extracting minerals from the ground is the first step in putting them to use. Most of these minerals then need to be processed in some way to become useful for our products. Typically, after ores are mined, the rock is crushed and pulverized, and the desired metals or other materials are isolated by chemical or physical means. The material is then processed to purify the metals we desire through a set of processes called **refining**. With coltan, for example, processing facilities use acid solvents to separate tantalite from columbite. Other chemicals are then used to produce metallic tantalum powder. This powder can be consolidated by various melting techniques and then shaped into wire, sheets, or other forms.

Sometimes metals are melted and mixed with another metal or a nonmetal substance to form an **alloy**. For example, steel is an alloy of the metal iron that has been fused with a small quantity of carbon. The strength and malleability of this particular alloy make steel ideal for its many applications in buildings, vehicles, appliances, and more. To make steel, we first mine iron ore, which consists of iron-containing compounds such as iron oxide. Steelmakers heat the ore and chemically extract the iron with carbon in a process known as **smelting** (heating ore beyond its melting point for the purpose of extracting base metal or ore from gangue). They then melt and reprocess the mixture, removing precise amounts of carbon and shaping the product into rods, sheets, or wires. During this melting process, certain other metals may be added to modify the strength, malleability, or other characteristics of the steel, as desired.

Processing minerals exerts environmental impacts. Most methods are water-intensive and energy-intensive. Moreover, many chemical reactions and heating processes used for extracting metals from ore emit air pollution, and smelting plants in particular have long been hotspots of toxic air pollution. In addition, soil and water may become polluted by **tailings**, the remnants that are left over after metals have been extracted from ores. Tailings consist of finely milled rock waste that may contain toxic metals, such as arsenic and selenium, as well as chemicals used in the extraction process. For instance, miners use cyanide and (in a different process) mercury to extract gold from ore, and sulphuric acid to extract copper; these chemical reagents can end up in tailings and must be carefully managed.

Canada is a world leader in mining

Just as we are fundamentally dependent on mineral resources in our daily lives, the economy of Canada has historically been centrally linked to the mining sector. Some 60 different minerals are mined in Canada. The Mining Association of Canada reports that the mining and mineral processing industries today employ 380 000 people, and mining is the largest private-sector employer of Aboriginal people in Canada. Mining (including coal, but excluding oil and gas extraction) contributed $54 billion to Canada's GDP in 2013, accounting for almost 20% of the value of exports.[5]

There are hundreds of producing mines and over a thousand mining-related establishments in Canada, in every province and territory. Canadian mining companies also carry out mineral exploration and mining in many other countries around the world. According to Natural Resources Canada, there are more mining companies based in Canada than anywhere else.[6] Most Canadian mining companies use ethical mining practices and follow Canadian standards for environmental protection and worker health and safety, even when operating in countries that have lower standards. The few that do not have the potential to cause harm to workers and to the environment, and damage to the reputation of Canadian companies worldwide.

As of 2013, and traditionally, Canada ranked among the top five producers of a number of minerals worldwide, including both metals and nonmetals:[7]

- first in potash (a nonmetallic resource, used to make fertilizers)
- second in uranium (used as fuel for nuclear energy) and cobalt (used in many alloys, catalysts, and dyes)
- third in aluminum (a low-density, low-corrosion metal that has many applications) and tungsten (a hard metal used to strengthen alloys and in many electrical applications)
- fourth in platinum (a precious metal that has other uses in alloys and electronics) and elemental sulphur (a bright yellow, nonmetallic resource used to make sulphuric acid and as a component of fertilizers, pesticides, and other chemical compounds)
- fifth in titanium (another hard metal, used in many alloys, including steel), diamonds (a nonmetallic gemstone that is also used industrially as an abrasive), nickel (a magnetic mineral used in stainless steel, coins, batteries, and many other familiar products), and salt (which occurs in three immense deposits in Western Canada, Ontario, and the Atlantic Provinces)

Canada is also among the top 10 producers of gold, silver, zinc, copper, molybdenum, cadmium, and several other important mineral resources.

Diamond mining in Canada's North is particularly interesting. For many years the conventional wisdom among geologists held that no diamond deposits would be found in Canada. That turned out to be wrong. In 1991, two exploration geologists found trace minerals from what would eventually become the Ekati Diamond Mine, the first in Canada, which started production in 1998. The Diavik Mine (featured in the Central Case in Chapter 20, *Environmental Ethics and Economics*) and several others quickly followed. Canada is now among the top producers of gem-quality diamonds in the world.

Canada is a party to the *Kimberley Process Certification Scheme*, an international agreement set up in 2003 to certify the country of origin of rough diamonds in an effort to stop the trade in "blood" diamonds. Consequently, diamonds produced in Canada are laser-inscribed with a tiny logo (such as the maple leaf or a polar bear) or the words "ice on fire," and their information is entered into a database certifying that they are nonconflict minerals.

There are five producing diamond mines in Canada, and all are in the North: three in the Northwest Territories, one in Nunavut, and one in Northern Ontario. Diamond mining companies enjoy unique and, for the most part, positive relationships and partnerships with communities and groups in the far North, including Aboriginal organizations that promote social and economic development in remote communities. Some mining operations have partnered with Aboriginal groups in managing mining operations on lands used for subsistence hunting and fishing, by combining traditional and modern approaches.

Mining Methods

Depending on the nature of the mineral deposit, any of several mining methods may be employed to extract the resource from the ground. Mining companies select which method to use based largely on economic efficiency. We will examine the major mining approaches commonly used throughout the world.

Subsurface mining takes place in underground tunnels

When a resource occurs in concentrated pockets or seams deep underground, and the rock allows for safe tunnelling, mining companies pursue **subsurface mining**. In this approach, shafts are excavated deep into the ground, and networks of tunnels are dug or blasted out to follow deposits of the mineral (**FIGURE 17.4**). Miners remove the resource systematically and ship it to the surface.

We use subsurface mining for metals such as zinc, lead, nickel, tin, gold, copper, and uranium, as well as for diamonds, phosphate, salt, and potash. In addition, a

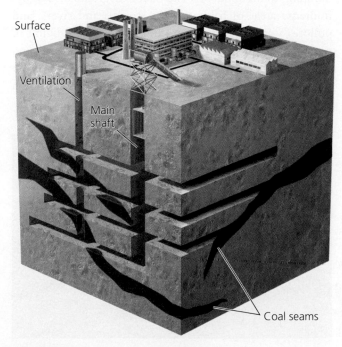

FIGURE 17.4
In subsurface mining, miners work below ground in shafts and tunnels blasted through the rock. These passageways provide access to underground seams of coal or minerals.

great deal of coal is mined using the subsurface technique. The scale of subsurface mining can be mind-boggling; the world's deepest mines (certain gold mines in South Africa) extend nearly 4 km underground. A well-publicized incident occurred in August, 2010, when 33 miners were trapped in a subsurface copper–gold mine in Copiapó, Chile. The miners were trapped 700 m under the surface by a cave-in. The world watched intently until they were finally rescued more than two months later.

Subsurface mining is the most dangerous form of mining and indeed one of society's most dangerous occupations. Fatal accidents are not unusual, even today; for instance, 29 coal miners died underground after an explosion in the Upper Big Branch mine in Montcoal, West Virginia, in 2010. In China, coal-mining conditions are so dangerous that in 2009 over 2600 miners lost their lives. Besides risking injury or death from dynamite blasts, natural gas explosions, and collapsing shafts and tunnels, miners inhale toxic fumes and coal dust, which can lead to respiratory diseases, including fatal black lung disease.

In Canada, some of the most famous historical mine disasters have occurred in subsurface coal mines. An example is the Springhill Mine Disaster, memorialized in popular culture, which actually refers to three different incidents in the Springhill coalfield in Nova Scotia. First was an underground coal-dust fire in 1891, which killed 125 miners. Next was an explosion in 1956, set off when a spark ignited coal dust in an underground mine shaft; the explosion killed 39 in total. The third Springhill disaster (in 1958) was a mining-induced earthquake, called a *mine tremor*, a *rock burst*, or, colloquially, a "bump." The mine tremor was caused by the buildup of stresses in the rock, related to the opening of underground tunnels and shafts more than 4 km in length. Seventy-four miners died in the resulting mine collapse.

The Canadian mining industry has a major focus on worker health and safety. Government regulations are strict, and the industry has done a good job of sharing "best practices" around mine safety. Mine-related accidents involving loss of life are now a rare occurrence in Canada.

Solution mining dissolves and extracts resources in place

When a deposit is especially deep and the resource can be dissolved in a liquid, miners may use a different subsurface technique called **solution mining** or *in situ recovery*. In this approach, a narrow borehole is drilled deep into the ground to reach the deposit, and water, carbon dioxide, acid, or another liquid is injected down the borehole to leach the resource from the surrounding rock and dissolve it in the liquid. The resulting solution is sucked out, and the desired resource is then isolated from the solution.

Salt (sodium chloride, the mineral *halite*) can be mined in this way; water is pumped into deep salt caverns, the salt dissolves in the water, and the salty solution is extracted. Other salts mined in this manner include salts of lithium, boron, bromine, magnesium, and potash. This type of approach is referred to as in-situ recovery (*in situ* means "in place"). In-situ approaches are sometimes used for copper (dissolved with acids) and uranium (dissolved with acids, or alkalis such as hydrogen peroxide, if carbonate minerals are present). In-situ recovery, in combination with hydraulic fracturing or *fracking*, are also used for natural gas, some forms of oil, and bitumen from oil sands; as discussed in Chapter 15, *Fossil Fuels*.

Strip mining removes surface layers of soil and rock

When a resource occurs in shallow horizontal deposits closer to the surface, the most effective mining method is often **strip mining**, whereby layers of surface soil and rock are removed from large areas to expose the resource. Heavy machinery removes the overlying soil and rock (termed *overburden*) from a strip of land, and the resource is extracted. This strip is then refilled with the overburden that had been removed, and miners proceed to an adjacent strip of land and repeat the process. Strip mining is commonly used for coal (**FIGURE 17.5**) and oil sands, and sometimes for sand and gravel.

If the resource occurs in seams near the tops of ridges or mountains, companies may practise an extension of strip mining called **mountaintop removal**, in which a hundred or more vertical metres of mountaintop may be removed to allow recovery of seams of the resource (**FIGURE 17.6A**). This method of mining is used primarily for coal in the Appalachian Mountains of the eastern

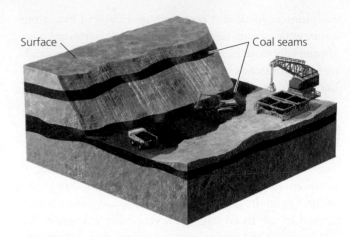

FIGURE 17.5
In strip mining, often used for mining coal and oil sands, the soil is removed from the surface in strips, exposing seams from which the coal or bitumen is mined.

United States; probably the closest equivalent in Canada is the removal of forests, wetlands, and overburden for oil sand mining in the Athabasca region of Alberta.

In mountaintop removal mining, forests are clear-cut and the timber is sold, topsoil is removed, and then rock is repeatedly blasted away to expose the coal for extraction. Overburden is placed back onto the mountaintop, but this waste rock is unstable and typically takes up more volume than the original rock, so generally a great deal of waste rock is dumped into adjacent valleys (a practice called "valley filling").

Mountaintop removal mining is an economically efficient way for companies to extract coal. However, dumping tonnes of debris into valleys degrades or destroys immense areas of habitat, clogs streams and rivers, and pollutes waterways with acid drainage. People living in communities near the sites also experience social and health impacts (**FIGURE 17.6B**). Blasts from mines crack house foundations and wells, loose rock tumbles down

(a) Mountaintop removal mining in West Virginia

(b) Train hauls coal past homes in West Virginia

FIGURE 17.6 In the Appalachians, mountaintop mining for coal **(a)** takes place on massive scales. Communities near these sites **(b)** experience negative environmental, social, and health impacts.

into yards and homes, overloaded coal trucks speed down once-peaceful rural roads, and floods tear through properties. Negative health impacts come from inhaling gaseous emissions or airborne dust, drinking contaminated ground water, or eating fish contaminated with toxic substances.

Open-pit mining creates immense holes at the surface

When a mineral is spread widely and evenly throughout a rock formation, or when the rock is unsuitable for tunnelling, the method of choice is **open-pit mining**. This essentially involves digging a large hole and removing the desired ore, along with waste rock that surrounds the ore. Some open-pit mines are enormous. The world's largest, the Bingham Canyon Mine near Salt Lake City, Utah, is about 4 km across and 1.2 km deep (**FIGURE 17.7**). Conveyor systems and immense trucks with tires taller than a person carry out nearly half a million tonnes of ore and waste rock each day.

Open-pit mines are terraced so that miners and machinery can move about, and waste rock is left in massive heaps outside the pit. The pit is expanded until the resource runs out or becomes unprofitable to mine. Open-pit mining is used for copper, iron, gold, diamonds, and coal, among other resources. We also use this technique to extract clay, gravel, sand, and stone such as limestone, granite, marble, and slate; a pit used for this purpose is called a **quarry**.

Open-pit mines are so large because huge volumes of waste rock need to be removed in order to extract relatively small amounts of ore, which, in turn, contain still smaller traces of valuable minerals. For example, a 3:1 waste-to-ore ratio, or *stripping ratio*, indicates that miners have to extract 3 tonnes of waste rock for every 1 tonne of ore obtained. For materials like gravel and sand, the stripping ratio is typically very low or they would not be economically extractable. For materials like gold, diamonds, or copper, the ratio can be much higher; a stripping ratio of 5:1 or 6:1 would not be unusual for a gold mine, which means that such mines can generate very large waste-rock piles. Note, too, that this ratio only looks at waste rock compared to ore—the actual amount of gold extracted from that ore during processing would be much smaller.

Placer mining uses running water to isolate minerals

Some relatively heavy metals and gems accumulate in riverbed deposits, having been displaced from a main ore body upstream and carried along by flowing water. To search for these metals and gems, miners sift through material in modern or ancient riverbed deposits, generally using running water to separate lightweight mud and gravel from heavier minerals of value (**FIGURE 17.8**), commonly gold and gemstones, as well as coltan. This technique is called **placer mining** (pronounced "plasser").

Placer mining is the method used by Congo's coltan miners, who wade through streambeds, sifting through large amounts of debris by hand with a pan or simple tools, searching for high-density tantalite that settles to the bottom while low-density material washes away. Today's African miners practise small-scale placer mining similar to the method used long ago by miners who ventured to California in the Gold Rush of 1849, Fraser Canyon in British Columbia in 1858, and the Yukon in the Klondike Gold Rush of 1896–1899. Placer mining for gold is still practised in areas of Alaska and Canada, although today it

David R. Frazier/The Image Works

FIGURE 17.7
The Bingham Canyon open-pit mine outside Salt Lake City, Utah, is one of the world's largest human-made holes. This immense mine produces mostly copper. For scale, each of the terraces visible in the photo is a full-scale road. The trucks that are just barely discernible in the foreground portion of the pit are enormous mining haulers, with tires almost twice the height of an average person.

FIGURE 17.8
Miners in eastern Congo find coltan by placer mining. Sediment is placed in plastic tubs and water is run through them. A mixing motion allows the sediment to be poured off while the heavy coltan settles to the bottom.

uses large dredges, sluices, and heavy machinery instead of individual gold pans.

Besides the many social impacts of placer mining in places like Congo, placer mining is environmentally destructive. Most methods wash large amounts of debris into streams, making them uninhabitable for fish and other life for many miles downstream. Gold mining in northern California's rivers in the decades following the Gold Rush washed so much debris all the way to San Francisco Bay that a U.S. district court ruling in 1884 finally halted the practice. Placer mining also disturbs stream banks, causing erosion and harming ecologically important riparian plant communities.

Some mining occurs in the ocean

The ocean holds many minerals useful to our society. We extract some minerals from sea water, such as magnesium from salts held in solution. We extract other minerals from the ocean floor. Using large vacuum-cleaner-like hydraulic dredges, miners collect sand and gravel from beneath the sea. Other valuable minerals found on or beneath the seafloor include calcium carbonate (used in making cement) and silica (used as fire-resistant insulation and in manufacturing glass), as well as copper, sulphur, zinc, silver, and gold.

One potentially important seafloor mineral resource is manganese nodules, small ball-shaped accretions that are scattered across parts of the ocean floor. Over 1.5 trillion tonnes of manganese nodules may exist in the Pacific Ocean alone, and their reserves of metal may exceed all terrestrial reserves. A significant new discovery of manganese nodules, ranging in from golf ball to bowling ball sizes, was made in 2015 in the Atlantic Ocean by German researchers exploring for deep-sea organisms. The logistical difficulty of mining them, however, has kept their extraction uneconomical so far; most manganese nodules lie 3.5 to 6.5 km below the surface of the ocean.

As land resources become scarcer and as undersea mining technology develops, mining companies may turn increasingly to the seas. Some companies already are exploring hydrothermal vents as potentially concentrated sources of metals such as gold, silver, and zinc, because these vents emit dissolved metals resulting from underground volcanic activity. Jurisdiction over seafloor mining is complicated, at best. If the mine occurs within the Exclusive Economic Zone (EEZ) of a nation (as discussed in the context of fishing in Chapter 12, *Marine and Coastal Systems and Fisheries*), then that nation would have mining rights according to the United Nations Law of the Sea. But seafloor deposits in the open ocean are technically outside of national jurisdiction and mining activity would be very difficult to regulate.

Impacts of undersea mining are largely unknown, but such mining would undoubtedly destroy marine habitats and organisms that have not yet been studied. It would also likely cause some metals to diffuse into the water column at toxic concentrations and enter the food chain.

Mining Impacts and Reclamation

Mining for minerals is an important industry that provides jobs for people and revenue for communities in many regions. Mining supplies raw materials for countless products we use daily, so it is necessary for the lives we lead. At the same time, mining also exacts a price in environmental and social impacts. Because minerals of interest often make up only a small portion of the rock in a given area, typically very large amounts of material must be removed in order to obtain the desired minerals. This frequently means that mining disturbs large areas of land, thereby exerting severe impacts on the environment and on people living nearby.

Environmental impacts vary with the stage of mining

Mining occurs in stages, and the environmental impacts associated with each stage vary greatly. The four main stages of mining are (1) exploration; (2) mining and milling; (3) smelting and refining; and (4) post-operational management, reclamation, and restoration. We can also think about the impacts of mining in terms of the environmental media that are potentially affected: air, water, land, and wildlife, as well as human health.

Exploration The exploration phase of mining is typically the least impactful from an environmental perspective. Exploration geologists or *prospectors* explore an area, taking samples and making seismic surveys. This might involve airborne surveying and helicopter landings, which can disturb wildlife. If the area looks promising, the company might decide to put in some exploratory drill holes. Roads might be constructed, which can have significant impacts, especially in remote areas or environmentally sensitive areas, such as the Arctic. Drilling also involves noise, and there can be some minor discharges of oil, slurries, and fine sediment that can enter waterways.

Mining and milling   The next stage of mining is the extraction of the ore, usually accompanied by some degree of on-site *milling* (that is, crushing) and processing of the ore. Mining and milling probably entail the most negative impacts of the four stages of mining. However, the impacts depend greatly on the mining techniques used, the type of ore involved, the geological setting, the sensitivity of the environment, and the care taken by the mining company. Environmental regulations on mining in Canada today are strict, and for the past couple of decades Canadian mining companies have prioritized both environmental safety and the health of workers. Even so, accidents and impacts can still occur.

Solution mining generally exerts less environmental impact than other mining techniques, because less area at the surface is disturbed. The main potential impacts involve accidental leakage of acids into ground water surrounding the borehole, contamination of aquifers with acids or heavy metals leached from the rock, and surface collapse into large cavities opened up by solution underground. This happened at the Retsof Salt Mine in New York, once the largest salt mine in North America. In 1994, the roofs of the underground mine cavities began to collapse, and the open spaces filled with ground water. This caused water tables to be lowered, and wells to run dry. Large subsidence pits opened up at the surface, and the U.S. Geological Survey expects further subsidence and collapse to occur over the next century.

Strip mining and open-pit mining can be economically efficient, but they cause severe environmental impacts on land, water, and wildlife habitat. By completely removing vegetative cover and nutrient-rich topsoil, strip mining obliterates natural communities over large areas, and soil from refilled areas easily erodes away. The sheer size of most open-pit mines means that the degree of land disruption, habitat loss, and aesthetic degradation is considerable. Mountaintop removal mining is particularly damaging to land and water; with slopes deforested and valleys filled with debris, erosion intensifies, mudslides become frequent, and flash floods ravage the lower valleys. Even placer mining, though smaller in scale, can lead to siltation of waterways and erosion of stream banks.

Subsurface, open-pit, and strip mines also can pollute surface and ground water by generating **acid mine drainage (or AMD)** through the chemical interaction of water with exposed tailings and waste rock piles. Acid drainage is one example of a contaminated liquid, or **effluent**, that can be generated at mine sites, with the potential for causing problems in both terrestrial and aquatic environments, on-site and offsite. Acid drainage arises from a natural process, but mining greatly accelerates and exacerbates the process by exposing many new rock surfaces at once.

AMD occurs when sulphide minerals in newly exposed rock surfaces react with oxygen to form sulphates, and then with rain water to produce sulphuric acid. As the acidic effluent runs off, it leaches metals from the rocks, many of which can be toxic to organisms (**FIGURE 17.9**), such as arsenic, zinc, aluminum, and cadmium. This toxic

Ronald Karpilo/Alamy

FIGURE 17.9
The discoloured water in this stream, downstream from a mine site, is a sign of acid drainage. Acid mine drainage can make stream water toxic and reduce aquatic biodiversity. The red and yellow colours typical of acidic effluents result from the precipitation of various iron oxides and hydroxides.

THE SCIENCE BEHIND THE STORY

This photo shows some of the material that flowed out of the tailings pond at Mount Polley Mine in British Columbia after a dam breach in August, 2014.

Chris Harris/All Canada Photos/Glow Images

Mount Polley Tailings Dam Failure

Mount Polley Mine is an open-pit copper–gold mine located in south central British Columbia, owned by Imperial Metals Corporation. In 2013, it produced almost 85 000 tonnes of copper, 1.3 tonnes (46 000 oz) of gold, and 3.5 tonnes (124 000 oz) of silver. The company was well on its way to matching that in 2014, and had plans to expand into an underground mining operation on the site, when disaster struck.

On August 4, 2014, a 4-km^2 dam located in south central British Columbia, that was holding back a large tailings pond, failed dramatically. The pond sludge contained at least 326 tonnes of nickel, 400 tonnes of arsenic, 177 tonnes of lead, and 18 400 tonnes of copper and copper compounds. When the dam failed, approximately 25 million m^3 of toxic sludge rushed from the tailings pond and into nearby Polley Lake (see **FIGURE 1**). The torrent rushed down Hazeltine Creek, turning the 2-m-wide stream into a 50-m-wide raging wasteland. Reaching the end of Hazeltine Creek, the spilled materials entered larger Quesnel Lake, the source of drinking water for the nearby town of Likely, BC, and a tributary of the Fraser River.[8]

The *Australasian Mine Safety Journal* was not alone in referring to the event as the "largest environmental disaster in modern Canadian history."[9] The company immediately placed the mine under "care and maintenance" for an "indeterminate" amount of time while they struggled to deal with the consequences of the spill.[10]

With a spill of this type, there are both short-term and long-term environmental and public health concerns that must be taken into consideration. The spill caused immediate damage in the adjacent waterways, including

- erosion and scour of the embankment separating the tailings pond from Polley Lake and along Hazeltine Creek and Cariboo Creek
- deposition of trees and debris in Polley Lake, along the sides of Hazeltine Creek, and in Quesnel Lake at the mouth of Hazeltine Creek, including two large, floating "islands" of debris
- deposition of tailings and eroded earth materials in Polley Lake, Hazeltine Creek, and Quesnel Lake.[11]

The material that spilled from the tailings pond consisted of slurry (a mixture of crushed rock and chemical reagents); interstitial water (that is, the water that was mixed with crushed rock in the slurry); supernatant water (that is, water that had separated from the slurry and was floating on top of it); and construction materials and other woody debris that broke loose when the dam failed.

Water quality was an immediate concern both for drinking water and for aquatic wildlife, such as salmon and trout. Cariboo Regional District issued a local state of emergency and "do not use" order for water, indicating that boiling the water would not solve any contamination issues. This was scaled back over the following week and was mostly lifted by August 12. However, the Region continued to caution residents who obtain their drinking water from Quesnel Lake to avoid cloudy or odd-tasting water.[12]

A consulting company gathered and analyzed 108 samples of the spilled tailings on behalf of Polley Mines. They did find enrichment of some elements beyond "background" levels (taken to be the typical amount in basalt rock), notably arsenic, copper, molybdenum, selenium, and sulphur. Of these, selenium and copper were the most enriched, at about 10 times background levels. Selenium and copper are both trace elements that are essential to human health in small quantities, but toxic in higher doses. They also can be problematic for fish, birds, and other wildlife. An additional concern regarding trace metals is the possibility of bioaccumulation in wildlife over time, also resulting in potential health impacts for people who consume them.

Another follow-up concern was that the spilled materials would cause acidifi-

liquid is called *leachate*, a term that is also used to describe the toxic liquids that form in landfills. Leachates from landfills are caused by anaerobic reactions in organic substances, but leachates from mining sites result from aerobic reactions in inorganic substances.

Traditionally, mining engineers have treated acid mine drainage by adding a strongly alkaline substance, such as lime, bicarbonate, or sodium hydroxide, to raise the water's pH and precipitate the dissolved heavy metals. In recent years, researchers have begun using sulphate-reducing bacteria to convert the metal sulphates to more harmless sulphide compounds that allow the metals to be removed from the effluent. Sulphate-reducing bacteria are microbes that thrive in the absence of oxygen by chemically reducing sulphates to support their metabolism, creating sulphides that they expel as waste. In nature, such bacteria degrade organic materials in the mud of swamps and produce hydrogen sulphide, the gas that gives swamps and mudflats their rotten-egg odour. Hydrogen sulphide in turn reacts with metal sulphates to create metal sulphides, which give swamp mud its blackish colour. The approach has been used, in conjunction with bicarbonate, cation of the waterways. As it turned out, the tailings were quite alkaline, with a pH of about 8.5. Of the 108 samples tested shortly after the spill, only one was identified as being "potentially acid-generating." This allayed early fears that the material would cause widespread acidification of the waterways (although highly alkaline substances also can cause environmental damage in some circumstances).

Prior to the dam failure, Quesnel Lake had been considered the most pristine deep-water lake in the world. Local residents, especially in the town of Likely, depend on the lake for their drinking water, but also for tourism-related activities, especially fishing. However, the residents also depend on the mine as a major source of employment; the dam failure was a disaster for them on many fronts.

John Marsden, 2014 President of the Society for Mining, Metallurgy, and Exploration, described the mining industry as "stunned" and "frustrated" by the dam failure. In a paper for the journal *Mining Engineering*, published shortly after the spill, he stated the position of many in the mining industry, which has made enormous efforts toward better environmental and health protection: "Apart from the direct impacts caused at Mount Polley, this type of event poses other potential risks to people, wildlife and the environment, but there are also serious implications for many others involved within and around the mining industry … The credibility of the mining industry as a whole is done great harm by this event. As an industry, we must learn from this event."[13]

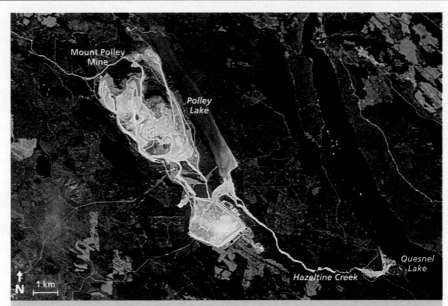

FIGURE 1 This satellite image shows the location of the tailings pond and the dam breach (arrow), relative to Polley Lake, Hazeltine Creek, and Quesnel Lake.
NASA Earth Observatory

After the spill, accusations flew. Former employees and others, including the engineers responsible for the design of the tailings pond and dam, accused the company of operating the pond above its capacity prior to the spill, and stated that they knew that the dam was aging and posed a hazard.[14] It will take a long time to sort out the environmental impacts of the spill, which will require monitoring for the foreseeable future. It will take even longer to sort out questions of responsibility, and the social and eco- nomic impacts of the spill—both for the company, and for the residents of Likely and the broader area. In December of 2015, BC's Chief Inspector of Mines found no evidence that the company failed to comply with BC mining legislation; nevertheless it is likely that litigation will continue for many years. The mine reopened, with some restrictions, in August, 2015. Imperial Metals has produced a phased cleanup and rehabilitation plan, and is providing regular updates on its company website.[15]

to clean heavy metals from mine effluents and make it less acidic.

AMD (also called *acid rock drainage*, or ARD) is considered the most significant environmental liability facing the mining industry in North America. In Canada, a collaborative program called *Mine Environment Neutral Drainage (MEND)* was established in 1989, and continues its work today. The initiative was originally funded by the Mining Association of Canada and Natural Resources Canada to facilitate the development of new technologies and approaches to prevent, control, and remediate

the effects of acid mine drainage. It is now part of the International Network for Acid Prevention, an industry-led group.[16]

Tailings are another significant source of environmental impacts associated with mining and milling. As mentioned previously, tailings are the fine-grained remains of milled ore and waste rock left over after the material of interest has been removed. Tailings usually occur in the form of a thick slurry, in which the crushed rock is mixed with water and any chemical reagents used in on-site processing. Mining operations often pump the toxic slurry of tailings into a holding reservoir called a *tailings pond*, from which the water is allowed to evaporate. It is difficult to isolate such ponds well enough to prevent leakage into the environment. In August of 2014, a dam that contained a large tailings pond at Mount Polley Mine in south central British Columbia failed, releasing as much as 25 million m^3 of tailings and associated water into adjacent waterways all at once. We explore this environmental disaster in "The Science behind the Story: Mount Polley Tailings Dam Failure."

Smelting and refining
After an ore has been mined and has undergone milling and initial processing at the mine site, it will typically be shipped (often by train) to a processing facility. At the processing facility, the material will be concentrated and refined into a product of sufficiently high grade or quality that is marketable.

Metals typically undergo smelting, in which the ore is melted and mixed with various chemical agents to cause the metal to separate from the ore. The solid remnant of smelting is a clinkery-looking substance called *slag*. Slag is a waste material, but it is not typically hazardous and can sometimes be used in roadbed construction or as a base for train rails.

Smelting is a high-temperature process, and it leads to gaseous emissions that can be very harmful in the environment. If sulphide minerals are present—and they often are, in metal ores—sulphur-bearing gases can be a major component of the emissions. If emitted into the atmosphere, the sulphur reacts with water vapour in the air to form acids, which can then be deposited as acidic precipitation. Smelter emissions also can contain lead, cadmium, and other substances that are hazardous for people and other organisms.

As discussed in Chapter 13, *Atmospheric Science and Air Pollution*, decades of nickel mining and smelting in Sudbury, Ontario, generated emissions high in sulphur dioxide (SO_2), which resulted in significant acidification of soils and lakes, devastating fish and forests in the area. In the early 1980s, Inco and Falconbridge (the two major producers of smelter emissions in the area) undertook vigorous efforts to clean their emissions prior to releasing them to the atmosphere. Today, sulphur emissions from smelters in Sudbury have been reduced by as much as 90%, and lakes, forests, and soils have shown significant biological and chemical improvements. Scrubbers on smelter stacks are ubiquitous now in Canada.

Post-operational
Mineral resources are nonrenewable, so mines eventually become depleted and must be shut down. The post-operational phase of mining, that is, the management of the mine site following its closure, is one of the most important phases of mining for limiting negative environmental impacts.

Mines, both subsurface and open-pit or strip mines, can affect people years after they are closed. In Centralia, Pennsylvania, a fire that started in 1962 continues to burn in underground coal mine workings, even today. Toxic gas and smoke from the fire rise through cracks in the surface, and all efforts to extinguish the fire have been unsuccessful. The town has been abandoned and the few remaining buildings are scheduled for demolition.

Once mining has been completed, abandoned pits and subsurface tunnels generally fill up with ground water, which can react with sulphides to produce acidic effluents. Acidic water not only harms wildlife and aquatic systems at the surface, it can also percolate into aquifers and spread regionally in the subsurface. The Berkeley Pit, a former open-pit copper mine near Butte, Montana, filled with ground water after its closure in 1982. The water became so acidic (pH = 2.2) and concentrated with toxic metals that microbiologists discovered new species of microbes in the water that were specifically adapted to the harsh conditions.

Canada's *Metal Mining Effluent Regulation,* or *MMER* (2002), governs the composition of any contaminated liquid issuing from a mine site. The MMER established limits for cyanide, arsenic, and other hazardous constituents of mine runoff, as well as limiting pH and suspended solids. These regulations help protect both natural ecosystems and human health.

In spite of tighter regulations, however, many old mine sites will likely continue to leach acid for many decades. One site in Canada that has suffered from acid mine drainage is Britannia Beach, British Columbia. The Britannia Copper Mine was opened in the early 1900s, and there followed a long history of pollution and mine-related accidents. Mining ceased in 1974, but acid mine drainage continued to be a severe problem until remediation efforts were undertaken in the early 2000s by scientists from the University of British Columbia in conjunction with the provincial government. These efforts finally appear to have had an impact on water quality in the area.

AMD is not the only problem associated with old mine sites. Waste rock piles and tailings heaps also can continue to be problematic following the closure of the mine. One major problem is that dust containing toxic metals can blow off of exposed rock and tailings piles. Regulations in developed nations now require that waste heaps and dry tailings piles be capped with a plastic or geotechnical coating while still active, then covered with clay and soil, and then planted with vegetation once the mine has been closed.

When mines are shut down properly, according to modern regulations and using best practices, we say that the site is *decommissioned*. The same term is used for other types of industrial sites after they are shut down. However, many mine sites have been abandoned over the hundreds of years of mining in Canada before the advent—only very recently—of modern legislation for decommissioning. Some abandoned sites are very old, but mines can still be abandoned if, for example, the company goes bankrupt, or the mine is exhausted and the company ceases to exist. If no one has clear responsibility for an abandoned site, it is referred to as an *orphaned* mine site. In Canada the *National Orphaned/Abandoned Mines Initiative* (or *NAOMI*) has worked to establish a national inventory of orphaned and abandoned mines. About 10 000 abandoned mine sites are on record in Canada, but it is likely that many more remain undocumented.[17] Many of these have the potential to cause hazards (such as cave-ins) or exert negative impacts on the natural environment.

Restoring mined sites can be very challenging

Because of the environmental impacts of mining, governments of Canada and other developed nations now require that mining companies rehabilitate surface-mined sites following the cessation of mining and the decommissioning of the mine site. The aim of such **reclamation** may be to restore the site to a condition similar to its pre-mining condition, but more commonly the goal is simply to remove hazards (both human and environmental) so the site may be safely used for other purposes. To reclaim a mine site, companies must remove buildings and other structures used for mining, replace overburden, fill in shafts, and replant the area with vegetation (**FIGURE 17.10**). In Canada, mining companies are required to post bonds to cover reclamation costs before the development of a mine can even be approved. This ensures that if the company fails to restore the land for any reason, the government will have the money to do so. Most other nations exercise less oversight, and in nations such as Congo, there is no regulation at all.

The mining industry has made great strides in reclaiming mined land and employs many ecologists and engineers to conduct these efforts. However, even on reclaimed sites, the impacts from mining (such as soil and water damage from acid drainage) can be severe and long-lasting. With strip mining, open-pit mining, and mountaintop removal, even replacing the removed

Greenshoots Communications/Alamy

(a) Mine site reclamation in Ghana

FIGURE 17.10
More mine sites are being reclaimed today, but mine site reclamation rarely is able to recreate the natural community present before mining. In **(a)**, reclamation workers in Ghana, West Africa, plant trees on the benches and floor of an abandoned gold-mining pit.

materials will not restore the landscape to anything like its former condition.

Moreover, reclaimed sites do not generally regain the same biotic communities that were naturally present before mining. One reason is that fast-growing grasses are generally used to initiate and anchor restoration efforts. This helps control erosion quickly from the outset, but it can hinder the longer-term establishment of forests, wetlands, or other complex natural communities. Instead, grasses may outcompete slower-growing native plants in the acidic, compacted, nutrient-poor soils that usually result from mining. Inconspicuous but vital symbiotic relationships that maintain ecosystems—such as specialized relationships between plants and fungi, and plants and insects—can be eliminated by mining, and they are difficult or impossible to restore.

Water polluted by mining and acid drainage also can be reclaimed if pH can be moderated and if toxic heavy metals can be removed. Like the reclamation of land, this is a challenging and imperfect process, but researchers and the mining industry are making progress in improving techniques.

Mining also has social and economic impacts

In the Central Case, we touched on some of the extensive social and economic impacts that conflict minerals, including coltan, gold, and diamonds, have had in the Democratic Republic of the Congo. Unfortunately such cases are not rare, although DRC presents a particularly challenging case. The desire to avoid purchasing conflict minerals has led many companies to participate in international partnerships to enhance transparency and document supply chains for minerals that are susceptible. An example is the *Kimberley Process* set up to certify diamond sources and stop the global trade in blood diamonds.

Artisanal mining, or small-scale mining, is an occupation of "last resort" for many in the poorest nations of the world. Small-scale miners often set up their operations illegally, sometimes in national parks or at the sites of abandoned mines. The operations are typically very dangerous and environmentally damaging.

Placer mining is the method of choice among many artisanal miners, because it can be carried out with relatively accessible technologies. This is unfortunate, because placer mining is often carried out with the help of extremely hazardous materials. Gold, for example, is unreactive with most chemicals. One easy way to remove gold from ore in placer operations is to use mercury, which reacts with the gold to form a mixture called an *amalgam*. Another commonly used mixture is caustic soda and cyanide. Accidental releases of these extremely toxic combinations into waterways have occurred at small mining operations in Central Europe, South America, and Africa. Miners typically have little or no training in the handling of these toxic materials, and many fall ill.

In North America, mining has exerted social and economic impacts—both positive and negative—on communities almost since mining began on this continent. Small rural and remote communities located near a mine site can benefit greatly from the employment and economic spinoffs of the operation. In 25 or 30 years, however, when the mine is played out and the town's only large employer closes up shop, the impacts can be devastating for the community.

Today in Canada, most mining companies set up funds at the beginning stage of mining, which grow throughout the lifetime of the mine. When the mine closes, the fund is used to retrain workers and to help the community find new sources of income to recover from the loss of its key employer.

Toward Sustainable Mineral Use

Mining exerts plenty of environmental impacts, but we also have another concern to keep in mind: Minerals are nonrenewable resources in finite supply. Like fossil fuels, they form far more slowly than we use them. If we continue to mine them, they will eventually be depleted. As a result, it will benefit us to find ways to conserve the supplies we have left and to make them last. Reducing waste and developing means of recovering and recycling used mineral resources are ways we can pursue to use the mineral resources more sustainably. We will never achieve 100% recovery, but we can do much better than we are doing today.

Minerals are nonrenewable resources in limited supply

Unlike sunlight or water or forests, minerals do not regenerate fast enough to provide us a new supply once we have mined all known reserves. They are therefore considered nonrenewable resources. Some minerals we use are abundant in their supply and will likely never run out; an example is uranium, which supplies fuel for nuclear reactors. It is likely that there are sufficient uranium reserves to provide fuel for nuclear power generation for millions of years.

Other mineral resources are in scarcer supply. For example, analysts have calculated that the world's known reserves of tantalum will last about 129 more years at today's rate of consumption. If demand for tantalum increases, however, it could run out much faster. If everyone in the world began consuming tantalum at the rate of North Americans, the known reserves would last for only 18 years! The supply of indium, an obscure metal used for LCD screens, might last only another 32 years. A lack of indium would threaten the production of high-efficiency cells for solar power. Because of these supply concerns and price volatility, industries now are working hard to develop ways of substituting other materials for indium. Platinum is dwindling, too, and if it became unavailable, this would make it harder to

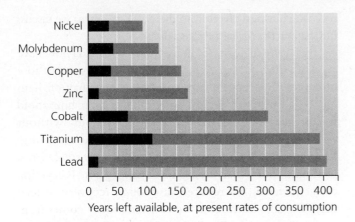

FIGURE 17.11

Shown in dark red are the numbers of remaining years that certain metals are estimated to be economically recoverable at current prices, given known global reserves and assuming current rates of consumption. The entire lengths of the bars (dark red plus orange) show the numbers of remaining years, using current technology on all known deposits, whether or not they are currently economically recoverable. All these time periods could increase if more reserves are found or new extraction technologies are developed, or decrease if consumption rates rise. There are possible substitutes for some of these commodities; others have properties that would be more difficult to replace.

Source: Data are for 2014, from U.S. Geological Survey (2015). *Mineral Commodity Summaries 2015.* USGS, Reston, Virginia.

DATA Q Which metal has the highest proportion of its technically recoverable reserves that are currently economically recoverable? Approximately what percentage is economically recoverable? Which metal has the smallest proportion of its technically recoverable reserves that are currently economically recoverable and what is this value?

manufacture fuel cells for vehicles. However, platinum's high market price encourages recycling, which may keep it available longer.

FIGURE 17.11 shows estimated years remaining for several selected minerals at today's consumption rates. Each bar in the figure consists of two parts. The left (dark red) portion of the bar shows the number of years researchers calculate that we will have the mineral available under today's economic conditions, that is, the portion of the resource that is *economically recoverable* at today's prices, using current extraction and processing technologies. As minerals become scarcer, demand for them increases and prices rise. Higher market prices make it more profitable for companies to mine the resource, so they become willing to spend more to reach further deposits that were not economically worthwhile originally. The entire length of each bar (red + orange) shows the number of years researchers calculate that we will have the mineral available in total. The orange bars thus represent the *technically recoverable* portion of the resource, which would come into play if prices rise, or if

new, more efficient extraction and processing technologies are developed.

Several factors affect how long mineral deposits may last

Mineral resources are widespread, but the entirety of a mineral resource is not necessarily mineable. In the context of mineral resources, we define a **reserve** as that portion of the resource that is economically (and legally) mineable using current technologies. Calculating how long a given mineral resource will be available to us (as in **FIGURE 17.11**) involves estimating the sizes of known reserves, and comparing these to current (and estimated future) levels of consumption. There are causes of uncertainty in such calculations, and several major reasons why estimates may increase or decrease over time.

Discovery of new reserves As we discover new deposits of a mineral, the known reserves—and thus how long the mineral should remain available to us—increase. For this reason, some previously predicted shortages have not come to pass, and we may have access to these minerals for longer than currently estimated. As one recent example, in 2010 geologists associated with the U.S. military in Afghanistan discovered that Afghanistan holds immense mineral riches that were previously unknown. The newly discovered reserves of iron, copper, niobium, lithium, and many other metals are estimated to be worth over $900 billion—enough to realign the entire Afghan economy around mining. (Note, however, that this find is not guaranteed to make Afghanistan a wealthy nation; history teaches us that regions rich in nonrenewable resources, such as Congo, have often been unable to prosper from them.)

New extraction technologies Just as rising prices of scarce minerals encourage companies to expend more effort to reach difficult deposits, rising prices also may favour the development of enhanced mining technologies that can reach more minerals at less expense. If more powerful technologies are developed, making it easier (and also, therefore, less expensive) to extract or process the ore, it may increase the amounts of minerals that are technically feasible for us to mine.

Changing social dynamics New societal developments, political and legal factors, and new technologies in the marketplace can modify demand for minerals in unpredictable ways. Just as cell phones and computer chips boosted demand for tantalum, fibre optic cables decreased demand for copper as they replaced copper wiring in communications applications. The advent of legal requirements for catalytic converters, which limit polluting automobile emissions and contain platinum as a crucial component, greatly increased demand for platinum. Lithium-ion batteries are replacing nickel-cadmium batteries in many devices; this will have an impact on demand for lithium (and nickel-cadmium). Synthetically made diamonds are driving down prices of natural diamonds and extending their availability. Additionally, health concerns sometimes motivate change. For example, we have replaced toxic substances such as lead and mercury with safer materials in many applications.

Changing consumption patterns   Changes in the rates and patterns of consumption also alter the speed with which we exploit mineral resources. For instance, economic recession depressed demand and caused the production and consumption of most minerals to decrease in 2008 and 2009. However, over the long term, demand has been rising. This is especially true today as China, India, and other major industrializing nations rapidly increase their consumption.

Recycling   Advances in recycling technologies and increased prices for the recovered substances have helped make recycling more technically and economically feasible. Recycling rates will continue to increase and, in turn, will allow us to extend the lifetimes of many mineral resources.

Even so, we would be wise to be concerned about Earth's finite supplies of mineral resources, and to try to use them more sustainably. Sustainable use will benefit future generations by conserving resources for them to use. It will also benefit us today, because conserving mineral resources through reuse and recycling can prevent price hikes that result from reduced supply. Also, domestic conservation of resources helps make our national economy less vulnerable in instances when other nations decide to withhold resources.

We can make our mineral use more sustainable

We can begin to address both major challenges facing us regarding mineral resources—finite supply and environmental damage—by encouraging recycling of these resources. Metal-processing industries regularly save resources and money by reusing some of the waste products produced during their refining processes. In addition, municipal recycling programs help provide metals by handling used items that we as consumers place in

recycling bins and divert from the waste stream, as we discuss in Chapter 18, *Managing Our Waste*.

By far the majority of households in Canada now have access to curbside recycling programs, and most of those households do recycle, although only about half of them recycle all of the recyclable materials in their household waste each week.[18] Even among households without access to curbside pickup, most do some recycling. Programs have been established to recycle steel, iron, platinum, and other metals from auto parts. Recycling of car batteries has resulted in enough recovery of lead that fully 80% of the lead we consume today comes from recycled materials. Similarly, 35% of our copper comes from recycled copper sources such as pipes and wires. We have also found ways to recycle much of our gold, lead, iron and steel scrap, chromium, zinc, aluminum, and nickel.

In many cases, recycling can decrease energy use substantially. For instance, making steel by re-melting recycled iron and steel scrap requires much less energy than producing steel from virgin iron ore. Because this practice saves money, the steel industry today is designed to make efficient use of iron and steel scrap. Over half its scrap comes from discarded consumer items like cars, cans, and appliances; scrap produced within their plants and scrap produced by other types of plants in the industry each account for nearly one-fourth of the total. Similarly, recycling of aluminum is a good thing because it takes over 20 times more energy to extract virgin aluminum from ore (called *bauxite*) than it does to obtain it from recycled sources. Every tonne of aluminum cans your community recycles saves the energy equivalent of more than 6000 litres of gasoline (**FIGURE 17.12**).

Consider the complicated case of platinum. It is very difficult and expensive to locate new platinum deposits; they are rare. There also are some large sociopolitical and economic forces at play in the platinum market. For one thing, much of the platinum in the world comes from South Africa, where strikes, workplace violence, and mine shutdowns have recently occurred. Russia was sitting on a very large stockpile of platinum for many years, but most of that has slowly made its way into the marketplace. Technologies also come into play: The catalytic converter in your car utilizes platinum, and this technology is required by law in many countries. Catalytic converters have not been required in China, but now that China is addressing its air pollution issue, such regulations may be coming. These factors will all limit production and drive up demand (and, therefore, price). Industry is extremely efficient at recycling platinum—many platinum refineries now produce more platinum from recycled sources than from newly

FIGURE 17.12
When you recycle aluminum cans, you contribute to valuable efforts to save mineral resources, money, and energy.

mined sources. But recycling of platinum at the personal level—mainly through proper dismantling and recycling of automobiles—still has a long way to go.

Tantalum, which we have followed through our story about cell phone components and conflict minerals, is recycled from scrap by-products generated during the manufacture of electronic components and also from scrap from tantalum-containing alloys and manufactured materials. Recycling accounts for 20–25% of the tantalum available for use in products. This percentage has been growing quickly, but its future growth will depend on how quickly we expand recycling efforts for used cell phones and other electronic waste and on how well we enable recycling facilities to recover metals from these products. One way to achieve this would be to make changes at the design stage, so that our electronic accessories are designed to make the disassembly and recycling process more straightforward.

We can recycle metals from e-waste

Electronic waste, or **e-waste**, from discarded computers, printers, cell phones, handheld devices, and other electronic products is rising fast—and e-waste contains hazardous substances. Recycling old electronic devices helps keep them out of landfills and also helps us conserve valuable minerals such as tantalum. About 96% of the material in old cell phones is recyclable by one process or another (**FIGURE 17.13**).

Today in Canada more than 75% of all people own cell phones, but only about 12% of old cell phones are recycled. That leaves a long way to go! Most people (93%) purchase a new phone instead of a refurbished one when they replace their phone. About 42% of people hang onto

Amplifier and receiver:
Arsenic and gallium

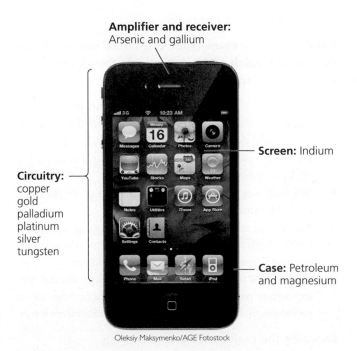

Screen: Indium

Circuitry:
copper
gold
palladium
platinum
silver
tungsten

Case: Petroleum and magnesium

Oleksiy Maksymenko/AGE Fotostock

FIGURE 17.13
Your cell phone contains a diversity of mined materials from around the world. Many old cell phones can be refurbished and sold for use in developing countries.

their old phone, once the new one has been purchased; another 41% make use of the phone in some way, either passing it along to someone, selling it, or recycling it.[19] It appears that the main impediment to recycling is simply lack of information about the reasons and options for recycling cell phones.

In fact, each of the almost 2 billion cell phones sold globally each year contains about 200 chemical compounds and close to a dollar's worth of precious metals. When you turn in your old phone to a recycling and reuse centre rather than discarding it, the phone may be refurbished and resold in a developing country. People in African nations, in particular, readily buy used cell phones because they are inexpensive, and because land-line phone service does not always exist in rural areas. Alternatively, the phone may be dismantled in a developing country and the various parts refurbished and reused or recycled for their metals. Either way, you are helping to extend the availability of resources through reuse and recycling and to decrease waste of valuable minerals.

As more of us recycle our phones, computers, and other electronic items, more tantalum and other metals may be recovered and reused. By recycling more, we can reduce demand for virgin ore and decrease pressure on the health of people and ecosystems in places where coltan and other components of cell phones are mined. Throughout the world, using recycling to make better use of the mineral resources we have already mined will help minimize the impacts of mining and assure us access to resources farther into the future.

Conclusion

We depend on a diversity of minerals and metals to help manufacture products widely used in our society. We mine these nonrenewable resources by various methods, according to how the minerals are distributed. Economically efficient mining methods have greatly contributed to our material wealth, but they have also resulted in extensive environmental impacts, ranging from habitat loss to acid drainage. Restoration efforts and enhanced regulation help minimize the environmental and social impacts of mining, although to some extent these impacts will always exist. We can lengthen our access to mineral resources and make our mineral use more sustainable by maximizing the recovery and recycling of key minerals.

REVIEWING OBJECTIVES

You should now be able to

Summarize the main types of mineral resources and the ways in which mineral resources contribute to our economy and society

- Rocks are composed of minerals. The minerals we mine provide raw materials for most of the products we use and depend on every day. The valuable parts of a rock are called ore; the nonvaluable parts are waste rock, or gangue.
- We mine ore to extract metals; we also mine for non-metallic minerals and fuels.
- Processing and refining mineral resources through methods such as smelting is an important step between mining the ore and manufacturing products.
- Canada has traditionally been a mineral producing and exporting nation, and is currently among the top 5 producers in the world for 10 important mineral resources.

Describe the major methods of mining

- Subsurface mining involves tunnelling underground.
- Solution mining uses water to dissolve minerals in place and extract them.

- Strip mining removes surface layers of soil and rock to expose resources that occur in seams, close to the surface. Mountaintop removal mining is an extreme form of strip mining.
- Open-pit mining involves creating immense holes; it is efficient for the extraction of ores that are close to the surface but diffuse.
- Placer mining uses running water to isolate dense minerals, including gold and gemstones.
- A great deal of mineral wealth exists in the oceans, including metals contained in manganese nodules, but it is mostly uneconomical (so far) to extract.

Characterize the impacts of mining, both environmental and social, and approaches to management, restoration, and reclamation

- The four stages of mining are exploration, mining and milling, refining and smelting, and post-operational management. All stages can have negative impacts on land, air, or water, but the exploration phase is the least impactful.
- Many methods of mining completely remove vegetation, soil, and habitat, leading to land disturbance and

erosion. Acid mine drainage occurs when water leaches compounds from freshly exposed waste rock. It is often toxic to aquatic organisms. Tailings piles and ponds can be problematic in a number of ways. Smelting contributes to air pollution; scrubbers are used to prevent gaseous contaminants from leaving the stack.

■ Although mining companies have made great advances environmentally, reclamation is challenging, and efforts generally fall short of effective ecological restoration.

■ Mining may have health impacts on miners and diverse health, economic, and social impacts on people living nearby. In Canada, mining companies are required to post a bond to ensure that funds will be available for reclamation purposes and to help the community adjust to the eventual closure of the mine.

Evaluate ways to encourage sustainable use of mineral resources

■ Minerals are nonrenewable resources, and some are limited in supply.

■ Several factors affect supply and demand, and thus influence how long a given mineral resource will last. These factors include new discoveries of the resource; new extraction technologies; changing social, political, and legal dynamics; consumption patterns; and recycling behaviours.

■ Reuse and recycling by industry and consumers are key to more sustainable practices of mineral use.

■ E-waste, including computers and cell phones, can be a source of recovered metals.

TESTING YOUR COMPREHENSION

1. Define each of the following, and contrast them with one another: (1) mineral, (2) metal, (3) ore, (4) alloy, (5) gangue.
2. A mining geologist locates a horizontal seam of coal very near the surface of the land. What type of mining method will the mining company use to extract it? What is one common environmental impact of this type of mining?
3. How does strip mining differ from subsurface mining? How does each of these approaches differ from open-pit mining?
4. What type of mining is used for both coltan and gold? What does a miner do to conduct this type of mining?
5. Describe and contrast how water is used in placer mining and solution mining.
6. What is acid drainage, and what are some ways of managing or preventing it?
7. Describe three major environmental or social impacts of mountaintop removal mining.
8. Explain why reclamation efforts after mining frequently fail to effectively restore natural communities. Include reference to both soil and vegetation in your answer.
9. List five factors that can influence how long global supplies of a given mineral will last, and explain how each might increase or decrease the time span the mineral will be available to us.
10. Name three types of metal that we currently recycle, and identify the products or materials that are recycled to recover these metals.

THINKING IT THROUGH

1. List three impacts of mining on the natural environment, and describe how particular mining practices can lead to each of these impacts. How are these impacts being addressed? Can you think of additional solutions to prevent, reduce, or mitigate these impacts?
2. You have been hired for a summer job by Environment Canada to work with a mining company to develop a more effective way of restoring a mine site that has been abandoned. Describe a few preliminary ideas for preventing problems from occurring and for carrying out effective site restoration. Now describe a field experiment you would run to test one of your ideas.
3. If you wanted to recycle your old cell phone, where would you take it? Do you know what would happen to the materials in the phone? You can get more information about cell phone recycling at **www .recyclemycell.ca**.
4. The story of coltan in the Congo is just one example of how an abundance of exploitable resources can often worsen or prolong military conflicts in nations that are too poor or ineffectively governed to protect these resources. In such "resource wars," civilians often suffer the most as civil society breaks down. Suppose you are the head of an international aid agency that has earmarked $10 million to help address conflicts

related to mining in the Democratic Republic of the Congo. You have access to government and rebel leaders in the Congo and neighbouring countries, to ambassadors of the world's nations in the United Nations, and to representatives of international mining corporations. Based on what you know from this chapter, what steps would you consider taking to help improve the situation in the Congo?

INTERPRETING GRAPHS AND DATA

"Coltan" is the industrial name for two minerals—columbite (also known as niobium) and tantalite, which usually occur together in a coarse-grained igneous rock type called *pegmatite*. Tantalum from coltan is an important component in a variety of electronic devices, including cell phones, DVD players, and video game systems.

The Democratic Republic of the Congo is a relatively small producer of coltan (see table). However, as discussed in the Central Case, coltan mining in DRC (and other African nations) has been cited as a factor that contributes to civil conflicts, because it provides funds for military groups and encourages smuggling and other illegal activities. Because coltan mining in Africa is largely an unregulated, artisanal (small-scale) activity, it also contributes to environmental degradation and unsafe working conditions.

In the past, Canada has been a significant producer of both tantalum (mainly from Manitoba) and the closely associated element niobium (mainly from Quebec). Canadian production has been promoted as a source of "conflict-free coltan" to electronics manufacturers concerned about the ethics of coltan mining in Africa.

Based on the data provided in the table, answer the questions below.

Metric Tons of Tantalum Mined													
	1990	1991	1992	1993	1994	1995	1996	1997	1998	1999	2000	2001	2002
Australia	165	218	224	170	238	274	276	302	330	350	485	660	940
Brazil	90	84	60	50	50	50	55	55	310	165	190	210	200
Canada	86	93	48	25	36	33	55	49	57	54	57	77	58
D.R. Congo	10	16	8	6	1	1	–	–	NA	NA	130	60	30
Rwanda	–	–	–	–	–	–	–	–	–	–	–	–	–
Africa, other	45	66	59	59	8	3	3	3	82	76	208	173	242
WORLD	**396**	**477**	**399**	**310**	**333**	**361**	**389**	**409**	**779**	**645**	**1070**	**1180**	**1470**
	2003	2004	2005	2006	2007	2008	2009	2010	2011	2012	2013	2014	
Australia	765	807	854	478	441	557	81	0	80	0	0	0	
Brazil	200	213	216	176	180	180	180	180	180	180	140	98	
Canada	55	57	63	56	45	40	25	0	25	50	50	0	
D.R. Congo	15	20	33	14	71	100	87	NA	NA	100	110	180	
Rwanda	–	–	–	–	–	–	–	110	120	150	250	250	
Africa, other	245	333	214	146	135	313	297	391	390	230	276	199	
WORLD	**1280**	**1430**	**1380**	**870**	**872**	**1190**	**670**	**681**	**790**	**670**	**786**	**787**	

Source: From 1990–2009: U.S. Minerals Yearbooks; 2010–2014 USGS Mineral Commodities Summaries.

NA = data not available

– = zero

1. In which of the years represented in the table was Canada's production of tantalum the highest? How many metric tons were produced, and what proportion of total world production did that represent?

2. For the years represented on this chart, in which year was Canada's relative contribution to world production highest? In which year was it lowest?

3. What about the Democratic Republic of the Congo—what was its proportion of world production in 1990, compared to the proportion in 2011, for example?

4. Australia has traditionally been the world's largest producer of tantalum, reaching 75.9% of world production in 1995. How many metric tons of tantalum did Australia produce in that year? In what year did

Australia produce the highest number of metric tons? What was Australia's production in 2009 and later years, in terms of metric tons and proportion of world production? Do you think the decline in Australian production (which resulted from the bankruptcy of the major Australian producer) had a major impact on world production? Refer to the graph to check your answer, and describe what seems to have happened to world production in 2009.

5. Do you think you might be able to predict where production will go after 2014, the last year represented on the graph? What are some of the questions you would need to answer to feel comfortable with your prediction?

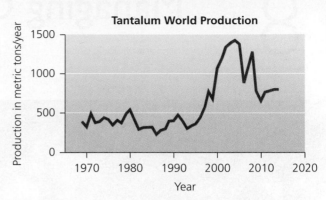

Tantalum World Production

World tantalum production, 1990–2014
Source: Based on estimates from the U.S. Geological Survey's Mineral Yearbooks and Mineral Commodities Summaries for those years.

MasteringEnvironmentalScience®

STUDENTS
Go to **MasteringEnvironmentalScience** for assignments, the eText, and the Study Area with practice tests, videos, current events, and activities.

INSTRUCTORS
Go to **MasteringEnvironmentalScience** for automatically graded activities, current events, videos, and reading questions that you can assign to your students, plus Instructor Resources.

18 Managing Our Waste

These containers are on their way to a recycling facility.

Goodshoot/Thinkstock

Upon successfully completing this chapter, you will be able to

- Summarize and compare the types of waste we generate, the scale of the waste dilemma, and the major approaches to managing waste
- Describe conventional municipal waste disposal methods: landfills and incineration
- Evaluate approaches for reducing waste: source reduction, reuse, composting, and recycling

- Discuss industrial solid waste management and principles of industrial ecology
- Assess issues in managing hazardous waste and e-waste

After only a couple of years, the newly established North Garden at Beare Road, planted by community volunteers, was flourishing.

Jim Robb/Friends of the Rouge Watershed

CANADA
Hudson Bay
Beare Road Landfill
UNITED STATES

CENTRAL CASE
THE BEARE ROAD LANDFILL: MAKING GOOD USE OF OLD GARBAGE

"An extraterrestrial observer might conclude that conversion of raw materials to wastes is the real purpose of human economic activity."

—GARY GARDNER AND PAYAL SAMPAT, WORLDWATCH INSTITUTE

"We can't have an economy that uses our air, water, and soil as a garbage can."

—DAVID SUZUKI

In the eastern part of Toronto, not far from the University of Toronto Scarborough campus, there is a park with a grassy hill. On the hill there are trees, bike trails, and fields of wildflowers. This is the highest spot in the neighbourhood, overlooking the Rouge River. Visitors stroll, chat, admire the view, and walk their dogs. There are few clues that beneath this grassy hill lie 9.6 million tonnes of garbage.[1] This is the old Beare Road Landfill.

The Beare Road site is located on gravel that was deposited around 12 000 years ago. When the last glaciation was drawing to a close, melt water from the Laurentide Ice Sheet flowed into the Lake Ontario basin, creating a glacial lake much larger than the present-day Lake Ontario. Geologists refer to this ancient lake as Lake Iroquois (see map).

The coarse sandy and gravelly deposits of the former shoreline of Lake Iroquois are now bluffs stranded high above the current lake level. These porous and permeable units host significant aquifers, including the aquifer of the Oak Ridges Moraine, as well as the headwaters for hundreds of streams and rivers. The deposits have been profitable for producers of aggregate—gravel, sand, and crushed stone used for construction purposes, including the building of roads and production of concrete. Many small towns scattered along the ancient shoreline in Ontario owe their economic beginnings to the exploitation of gravel and sand from these deposits.

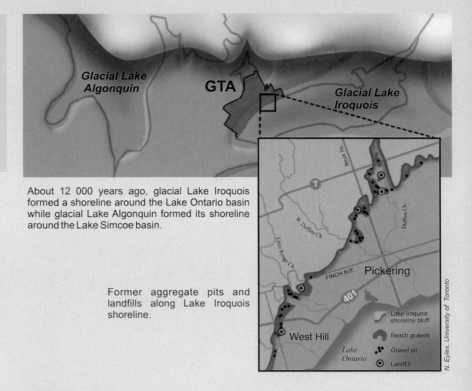

The location of the ancient shoreline of glacial Lake Iroquois is marked by a series of gravel deposits and bluffs. Old gravel pits are shown where aggregate extraction has occurred in these deposits over the past 100+ years. Many worked-out gravel pits were later utilized as dumpsites for municipal solid waste.

Source: Reprinted by permission from Nicholas Eyles, University of Toronto.

About 12 000 years ago, glacial Lake Iroquois formed a shoreline around the Lake Ontario basin while glacial Lake Algonquin formed its shoreline around the Lake Simcoe basin.

Former aggregate pits and landfills along Lake Iroquois shoreline.

Fast-forward to the latter half of the 1900s, when worked-out gravel pits scattered along the ancient shoreline sat empty. Some of the pits filled with water, serving as recreational lakes for fishing and swimming. In other cases, they beckoned to local residents—what better place to dispose of municipal solid waste? In that era, the negative impacts of tossing waste into such a porous and permeable medium were little known, and a number of pits were used for this purpose (see maps), including the Beare Road gravel pit.

The Beare Road pit officially began receiving municipal garbage in 1968. It began with a capacity of 3 million tonnes, but this was increased several times over the years as the urgency grew for places to put Toronto's ever-increasing garbage. The final increase in capacity was accompanied by a promise of funding from the government to be used toward the rehabilitation of the landscape, ultimately making it available for recreational use by the community.[2]

The landfill was eventually closed in 1983. At that time, a system for passive flaring of landfill gases was installed. Some landscape restoration was undertaken by the government and by local residents (see photo), and the site was opened as a park.

In 1996, E.S. Fox Ltd., in agreement with the City of Toronto (owner of the site) and Ontario Power Generation, began to collect the methane-rich gas being generated by decomposing garbage at Beare Road. This is called *LFGTE (landfill gas-to-electricity)*, and it makes use of what would otherwise be a harmful by-product of the garbage. Methane gas smells bad, damages vegetation, and is explosive and flammable. The Beare Landfill Power Plant currently produces enough electricity to service 4000 homes. Methane is also a highly effective greenhouse gas; collection and active management of the gas can help Canada minimize GHG emissions to control global warming in the future.

Some environmental problems persist at the Beare Road site; they are typical of old dumpsites and will require active management for years. For example, early engineering installations designed to control the collection and movement of leachate failed years ago. The impermeable liner that had been installed to prevent leakage filled with leachate, which began to seep from the side of the hill at the level where the liner topped out—the "bathtub effect." The possibility persists that leachate may threaten community developments immediately downstream from the site. The exact content and composition of the waste also are unknown.

In spite of these problems, the Beare Road project provides a hope-inspiring model for the management of old landfill sites. Gas collection and utilization has helped resolve a number of local environmental problems (such as odour and damage to vegetation), including reduction of GHG emissions. Although LFGTE sounds like a win-win idea, some environmentalists are fundamentally opposed to its development. They fear that generating something positive from a pile of garbage will derail attempts to get people to cut down on the amount of waste that they generate. This is a reasonable concern; however, in the case of old sites like Beare Road, which hasn't received any new garbage since 1983, the pollution-reducing benefits seem to make it a winning proposition.

Waste Generation and Management

As the world's population increases, and as we produce and consume more material goods, we generate more waste. **Waste** refers to any unwanted material or substance that results from a human activity or process. The federal government of Canada has adopted a definition that states, in part, that waste is "any substance for which the owner/generator has no further use."[3] Another popular definition suggests that waste is "resources out of place," emphasizing the fact that most waste still contains a significant proportion of useful materials. These definitions represent a changing perception of waste—that there is much of value that can be recovered from our waste.

Consumption leads to increased waste generation

In North America since 1960, total waste generation has increased by almost 300%, and per capita waste generation has risen by about 70%.[4] Plastics, which came into wide consumer use only after 1970, have accounted for the greatest relative increase in the waste stream during last several decades. (Note that plastics are more voluminous but weigh less than most other types of solid waste materials, so calculations of waste composition and generation rates differ considerably, depending on whether weight or volume is considered.)

In the past decade or so, waste generation in Canada has kept pace with the growth rate of the population but has lagged slightly behind the growth in gross domestic product (GDP). Residential diversion rates (for recycling or composting) also are beginning to outpace population growth. This suggests a promising trend of waste diversion and producing more for less, perhaps because of recycling and a shift to more efficient waste management processes.

The intensive consumption that has long characterized wealthy nations is now increasing rapidly in developing nations. To some extent, this reflects rising material standards of living, but an increase in packaging is also to blame. Items made for temporary use and poor-quality goods designed to be inexpensive wear out and pile up quickly as trash, littering the landscapes of countries from Mexico to Kenya to Indonesia. Over the past three decades, per capita waste generation rates have more than doubled in Latin American nations and have increased more than fivefold in the Middle East. Like consumers in "the throwaway society," wealthy consumers in developing nations often discard items that can still be used. At many dumps and landfills throughout the developing world, poor people still support themselves by selling items they scavenge. This contributes to greater recovery of resources, but subjects the workers to extremely hazardous working and living conditions (**FIGURE 18.1**).

In many industrialized nations, per capita waste generation rates have levelled off or decreased in recent years. This is due largely to the increased popularity of recycling, composting, reduction, and reuse. We will examine these aspects of waste management shortly, but let us first assess the basic aims and approaches of waste management and disposal.

We have several aims in managing waste

Waste can degrade water quality, soil quality, and air quality, thereby degrading human health and the environment. Waste is also a measure of inefficiency, so reducing waste can potentially save industry, municipalities, and consumers both money and resources. In addition, waste is unpleasant aesthetically. For these and other reasons, waste management has become a vital pursuit.

For management purposes, waste is divided into several main categories. *Municipal solid waste* is nonliquid waste that comes from homes, institutions, and small businesses. *Industrial waste* includes waste from production of consumer goods, mining, agriculture, and petroleum extraction and refining. *Hazardous waste* refers to solid or liquid waste that is toxic, chemically reactive, flammable, corrosive, or radioactive. It can include everything from

FIGURE 18.1
Tens of thousands of people used to scavenge each day from this dump outside Manila in the Philippines, selling recyclable material to junk dealers for 100–200 pesos ($2–$4) per day. That so many people could support themselves this way testifies to the immense amount of usable material discarded by wealthier portions of the population. The dump was closed in 2000 after an avalanche of trash killed hundreds of people.

Romeo GACAD Agence/Newscom

paint and household cleaners to medical waste to industrial solvents. Another type of waste is *waste water*, water we use in our households, businesses, industries, or public facilities and drain or flush down our pipes, as well as the polluted runoff from our streets and storm drains.

There are three main steps in **waste management**:

1. minimizing the amount of waste we generate
2. recovering waste materials and finding ways to recycle them
3. disposing of waste safely and effectively

Minimizing waste at its source—called *source reduction*— is the preferred approach (**FIGURE 18.2A**). There are several ways to reduce the amount of waste that enters the **waste stream**, the flow of waste as it moves from its sources toward disposal destinations (**FIGURE 18.2B**). Manufacturers can use materials more efficiently. Consumers can buy fewer goods, buy goods with less packaging, and use those goods longer. Reusing goods you already own, purchasing used items, and donating your used items for others also help reduce the amount of material entering the waste stream.

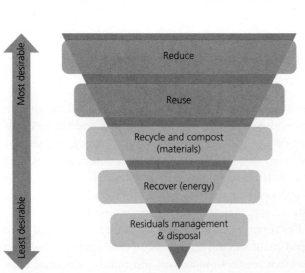

(a) The waste management hierarchy

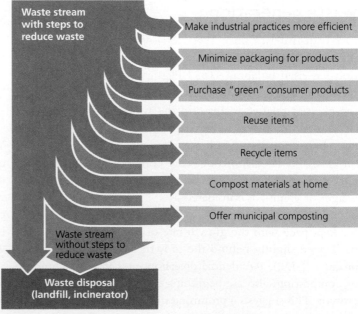

(b) The waste stream

FIGURE 18.2 The waste hierarchy **(a)** describes the most and least desirable approaches to waste management. The most effective way to manage waste is to reduce the amount of material that enters the waste stream **(b)**. Manufacturers can increase efficiency, and consumers can buy products that have minimal packaging. We can recycle, compost organics at home, and reuse items rather than buying new ones. For all remaining waste, we must find disposal methods that minimize impacts on human health and the environment.

Recovery (which includes recycling and composting) is widely viewed as the next best strategy in waste management. *Recycling* involves sending used goods to facilities that extract and reprocess raw materials to manufacture new goods. Newspapers, white paper, cardboard, glass, metal cans, appliances, and some plastic containers have all become increasingly recyclable as new technologies have been developed and as markets for recycled materials have grown. Organic waste can be recovered through *composting*, or biological decomposition. Recycling is not a concept that humans invented; recall that all materials are recycled in ecosystems. Recycling is a fundamental feature of the way natural systems function.

Regardless of how effectively we reduce our waste stream, there will always be some waste left to dispose of. Disposal methods include burying waste in landfills and burning waste in incinerators. In this chapter we first examine how these approaches are used to manage municipal solid waste, and then we address industrial solid waste and hazardous waste.

Municipal Solid Waste

Municipal solid waste (or **MSW**) is waste produced by consumers, public facilities, and small businesses. It is what we commonly refer to as "trash" or "garbage." Everything from paper to food scraps to roadside litter to old appliances and furniture is considered municipal solid waste.

Patterns in municipal solid waste vary from place to place

In Canada, paper, organics (mainly yard debris and food scraps), and plastics are the principal components of municipal solid waste, together accounting for about two-thirds of the waste stream (**FIGURE 18.3A**). Even after recycling, paper is the largest component of municipal solid waste. Patterns differ in developing countries (**FIGURE 18.3A**); there, food scraps are often the primary contributor to solid waste, and paper makes up a smaller proportion.

Most municipal solid waste comes from packaging and nondurable goods (products meant to be discarded

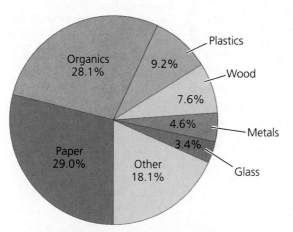

(a) Composition of MSW in Canada

FIGURE 18.3
Organics (yard trimmings and food scraps) and paper products are the largest components of the municipal solid waste stream in Canada, followed by plastics **(a)**. Much of the waste generated by people in economically developed nations, including Canadians, such as organics, paper, plastics, metal, and glass, is recyclable or compostable **(b)**. *Source:* (a) Based on Environment Canada, *Waste Management: Municipal Solid Waste*; Statistics Canada (2013). *Waste Management Industry Survey*, 2010; and UNEP (2011) *Waste: Investing in Energy and Resource Efficiency.* (b) From UNEP (2011). *Towards a Green Economy: Waste: Investing in Energy and Resource Efficiency.* Copyright 2011 by United Nations Environment Programme (UNEP). Reprinted by permission.

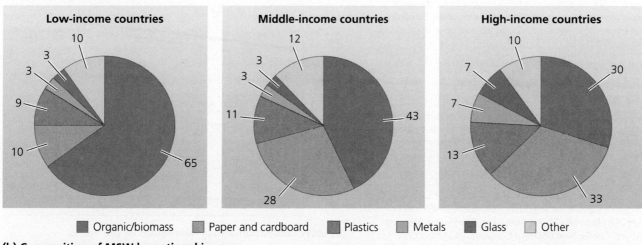

(b) Composition of MSW by national income

after a short period of use). In addition, consumers throw away old durable goods (such as large appliances, also called *white goods* in the waste management industry) and outdated equipment as they purchase new products. As we acquire more goods, we generate more waste. According to Statistics Canada, which tracks a variety of social, economic, and environmental indicators, Canadian citizens produced about 25 million tonnes of municipal solid waste in 2010 (down about 3% from 2008), for a population of 34 million. This is the first decline in total solid waste generation in Canada for many years (**FIGURE 18.4A**). However, it still means that Canadians are generating over 700 kg of trash per person per year, or more than 2 kg per person per day.[5]

Matching Canada in per capita solid waste generation is the United States, with about 2 kg per person per

day.[6] Among developed nations, Germany and Sweden produce the least waste per capita, generating under 0.9 kg per person per day. Differences among nations result in part from differences in the cost of waste disposal; where disposal is expensive, people have incentive to waste less. The wastefulness of the North American lifestyle, with its excess packaging and reliance on nondurable goods, has caused critics to label this as "the throwaway society."

In developing nations, people consume less and generate considerably less waste; the composition of MSW is also different, typically with much more organics and less paper (see **FIGURE 18.3B**). One study found that people of high-income nations waste more than twice as much as people of low-income nations; certainly the per capita rates of waste generation in these

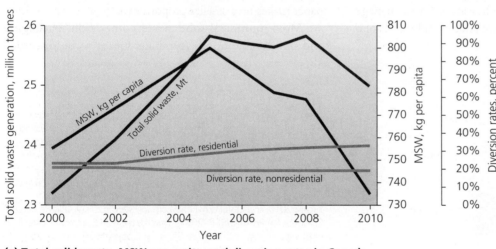

(a) Total solid waste, MSW per capita, and diversion rates in Canada

FIGURE 18.4

Total solid waste generation may be levelling off in Canada **(a)**: After increasing for decades, it stalled for a few years and then declined slightly between 2008 and 2010. Each Canadian generates more than 700 kg (including both residential and industrial components) of solid waste each year. The residential diversion rate for solid waste (to recycling and composting) increased from 23% to 33% over that period, but the nonresidential (i.e., industrial and agricultural) diversion rate was steady at about 19–20%. Elsewhere in the world, the long-term trend of increasing total municipal solid waste generation continues **(b)**. EU on the graph includes 27 European Union nations.

DATA Q In graph **(a)**, compare the curve for total solid waste generation to the curve for MSW per capita. What can you conclude about the population of Canada between 2005 and 2010? How do you know this?

DATA Q In graph **(b)**, the plot reflects MSW generation for the entire populations of these regions. The population of the European Union was around 711 million in 2010, and the population of the United States was about 309 million. That same year, the population of Asia was about 4.1 billion. What influence would these numbers have if you were to change the graph to reflect per capita MSW generation instead of total solid waste generation?

Source: Data from Statistics Canada (2010). CANSIM Table 153-0041; Statistics Canada (2012). *Human Activity and the Environment: Waste Management in Canada.* Government of Canada, catalogue no. 16-201-X; Giroux, L. (2014). *State of Waste Management in Canada.* Prepared for Canadian Council of Ministers of Environment; Statistics Canada Environment Accounts and Statistics Division, Environmental Protection Accounts and Surveys (2013). *Waste Management Industry Survey: Business and Government Sectors, 2010;* Government of Canada, catalogue no. 16F0023X; UNEP (2011). *Towards a Green Economy: Waste: Investing in Energy and Resource Efficiency;* and Conference Board of Canada (2013). *How Canada Performs: Municipal Waste Generation.*

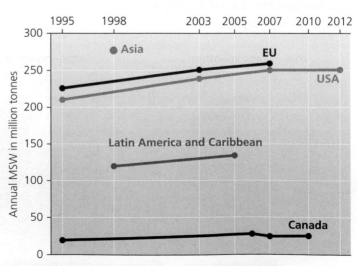

(b) MSW generation across the regions of the world

countries are much higher (**FIGURE 18.4B**). However, wealthier nations also invest more in waste collection and disposal technologies, so they are often better able to manage their waste proliferation and minimize impacts on human health and the environment.

Waste disposal is regulated by three levels of government

In Canada, municipal governments (including regional municipalities) are responsible for the collection, diversion, and disposal of solid waste from residential and small commercial and industrial sources. If you put waste by the side of the road on garbage collection day and it is picked up by city garbage trucks, then it is municipal solid waste. Many municipalities now also provide dropoff facilities for special categories of waste, including household hazardous wastes, such as leftover paint and batteries.

Provincial and territorial governments have control over the movement of waste materials within the jurisdiction, and over the licensing of treatment facilities and waste-to-energy installations. Thus, each province or territory has its own legislation and guidelines regulating the design, siting, licensing, operations, and expansion of landfill sites.

The federal government, meanwhile, is responsible for looking after international agreements about waste and regulating transboundary movements of waste materials. Federal involvement in waste management within Canada occurs mainly through the *Canadian Environmental Protection Act*. In the international context, Canada is a signatory to the *Basel Convention on Control of Transboundary Movements of Hazardous Wastes and Their Disposal*, adopted in 1989.

Open dumping has given way to improved disposal methods

Historically, people dumped their garbage wherever it suited them. Until the mid-nineteenth century, New York City's official method of garbage disposal was to dump it off piers into the East River. As population densities increased, municipalities took on the task of consolidating trash into open dumps at specified locations to keep other areas clean. This is how the worked-out gravel pits of the ancient Lake Iroquois shoreline in southern Ontario came to be used as dumpsites. To decrease the volume of trash, the dumps would be burned from time to time. Open dumping and burning still occur throughout much of the world.

As population and consumption rose in developed nations, more packaging and the use of nondegrad-

able materials increased, waste production increased, and dumps accordingly grew larger. At the same time, expanding cities and suburbs forced more people into the vicinity of dumps and exposed them to the noxious smoke of dump burning. Reacting to opposition from residents living near dumps and to a rising awareness of health and environmental threats posed by unregulated dumping and burning, many nations improved their methods of waste disposal. Most industrialized nations now bury waste in lined and covered disposal sites called **landfills**, or burn the waste in specially engineered incineration facilities.

Since the late 1980s, the recovery of materials for recycling has expanded, slightly decreasing the pressure on landfills. The total rate of diversion of municipal solid waste in Canada (that is, diversion away from disposal or incineration, to recycling or composting) increased from 21% in 2000 to 24.5% in 2010.[7] Change has been slow, however, and recyclable materials worth millions of dollars are still thrown into landfills in Canada each year.

Sanitary landfills are engineered to minimize leakage of contaminants

In a modern **sanitary landfill**, waste is buried in the ground or piled up in large, carefully engineered mounds. In contrast to open dumps, sanitary landfills are designed to prevent waste from contaminating the environment and threatening public health (**FIGURE 18.5**). In a sanitary landfill, waste is partially decomposed by bacteria and compresses under its own weight to take up less space. Waste is typically interlayered with soil, a method that speeds decomposition, reduces odours, and reduces infestation by pests. Limited infiltration of rain water allows for biodegradation by aerobic and anaerobic bacteria.

To protect against environmental contamination, landfills must be located away from wetlands and situated well above the local water table. The bottoms and sides of modern sanitary landfills are lined with heavy-duty plastic or other high-tech, specially engineered fabrics, called *geotextiles*. The liner is typically underlain by a thick layer (a metre or more) of impermeable clay to help prevent contaminants from seeping into aquifers. Sanitary landfills also have systems of pipes, collection ponds, and treatment facilities to collect and treat **leachate**, the liquid that results when substances from the trash dissolve in water as rain water percolates downward. Leachate collection systems must be maintained throughout the lifetime of the landfill and for many years after closure. Regulations also require that

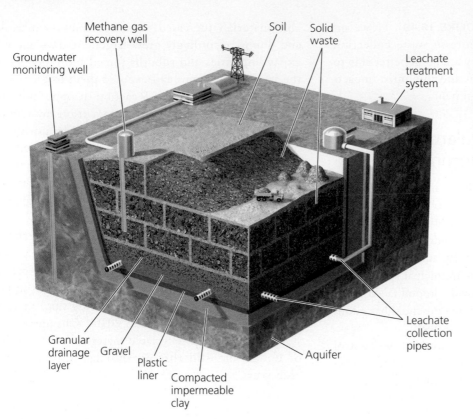

Groundwater monitoring well

Methane gas recovery well

Soil

Solid waste

Leachate treatment system

Granular drainage layer

Gravel

Plastic liner

Compacted impermeable clay

Aquifer

Leachate collection pipes

FIGURE 18.5
In sanitary landfills, waste is laid in a large depression underlain by an impervious clay layer and lined with a specially engineered plastic designed to prevent liquids from leaching out. Pipes in a leachate collection system draw liquids from the bottom of the landfill. Waste is interlayered with soil until the depression is filled, and the landfill is eventually capped. Landfill gas produced by anaerobic bacteria may be recovered, and waste managers monitor ground water for contamination.

area ground water and soils be monitored regularly for contamination.

At the Beare Road Landfill, some of the underlying rock and sediment units are clay-rich shales, creating a barrier that slows the subsurface movement of leachate from the site. However, the Iroquois sand and gravel deposits—mostly removed at an earlier stage for aggregate production, forming the pit itself—are much more permeable, providing a pathway for leachate migration.[8] Although it met provincial standards for landfill technology at the time of its construction, the Beare Road Landfill predated most current regulations and guidelines. As a result, it has caused some environmental contamination, and residents of adjacent areas continue to express some concerns about the possibility of gas and leachate migration. There is no functional engineering to control leachate migration at the site aside from a collection ditch around the perimeter and the gas collection technologies. The refuse mound is surrounded by a number of wells installed for the purpose of monitoring ground water for any signs of leachate migration into the surrounding areas.[9]

After a landfill is closed, it is capped with an engineered cover that must be maintained. The cap usually consists of a hydraulic barrier that prevents water from seeping down and gas from seeping up; a gravel layer above the hydraulic barrier that drains water, lessening pressure on the hydraulic barrier; a soil barrier that stores water and protects the hydraulic layer from weather extremes; and a topsoil layer that encourages plant growth, helping to prevent erosion. At Beare Road, the mound was covered after the closure of the site to inhibit the infiltration of rain water and the production and migration of leachate.

Landfills can be transformed after closure

Today many landfills lie abandoned. One reason is that waste managers have closed many smaller landfills and consolidated the waste stream into fewer, much larger, and more modern facilities. Meanwhile, growing numbers of cities have converted closed landfills into public parks (**FIGURE 18.6**) like the Rouge Park in Toronto, which includes the Beare Road Landfill site. Shutting down an industrial site and getting it ready for cleanup and post-industrial repurposing is called **decommissioning**. Such efforts date back at least to 1938, when an ash landfill at Flushing Meadows, in Queens, New York, was redeveloped to host the 1939 World's Fair. Designated a park in 1967, today the site hosts Shea Stadium, the Queens Museum of Art, the New York Hall of Science, and the Queens Botanical Garden.

FIGURE 18.6
Residents and members of Friends of the Rouge plant wildflowers at the Beare Road North Garden.

The Fresh Kills landfill in New York City's Staten Island was the largest landfill in the world and the primary repository for New York's garbage for half a century. The 890-ha landfill featured six gigantic mounds of trash and soil. The highest, at 69 m, was higher than the nearby Statue of Liberty. In March of 2001, a barge arrived to dump the final load of trash at Fresh Kills. The landfill's closure was a welcome event for Staten Island's 450 000 residents, who had long viewed it as a bad-smelling eyesore, health threat, and civic blemish.

The redevelopment endeavour at Fresh Kills will be the world's largest landfill conversion project. New York City planned to transform the old landfill into a public park three times the size of Central Park, with rolling hills, wildlife, and wetlands. Later in 2001, however, the landfill had to be reopened temporarily. After the 9/11 terrorist attacks, the 1.8 million tonnes of rubble from the collapsed World Trade Center towers, including unrecoverable human remains, were taken to Fresh Kills, where the material was sorted and buried. A monument will be erected at the site as part of the new park.

Landfills have drawbacks

Despite improvements in liner and cover technologies and landfill siting, it is possible that leachate could eventually escape, even from well-engineered landfills. Liners can be punctured, and leachate collection systems eventually cease to be maintained. Moreover, landfills are kept dry to reduce leachate, but the bacteria that break down material thrive in wet conditions. Dryness,

therefore, slows waste decomposition. In fact, it is surprising how slowly some materials biodegrade when they are tightly compressed in a landfill. Innovative research by archaeological William Rathje and colleagues has revealed that landfills often contain food that has not decomposed, and 40-year-old newspapers that are still legible.

Another problem is finding suitable areas to locate landfills, because most communities do not want them nearby. This *not-in-my-backyard* or **NIMBY syndrome** is one reason why Toronto, New York, and many other cities export their waste and why residents of areas that are receiving the waste are increasingly protesting. The past practice in Toronto—exporting garbage to Michigan—became increasingly unpopular among residents there. In 2007 an average of 74 truckloads per day of solid waste (approximately 441 363 tonnes) went to a Michigan landfill from Toronto (down from 142 daily truckloads in 2003). This amount continued to decline as the city surpassed its waste export reduction targets. Toronto's agreement with the receiving landfill in Michigan (by which the city of Toronto was contractually obligated to continue to deliver garbage to the privately maintained landfill) expired at the end of 2010. In 2007, the City of Toronto acquired the Green Lane Landfill site located southwest of London, Ontario. This landfill is about 200 km from downtown Toronto, but it provides an alternative to the Michigan exports. The site features the latest landfill engineering technology, including onsite leachate treatment and methane gas collection and flaring systems.[10]

As a result of the NIMBY syndrome, landfills are rarely sited in neighbourhoods that are home to wealthy and educated people who have the political clout to keep them out. Instead, they are disproportionately sited in poor and minority communities, as environmental justice advocates have frequently made clear.

The unwillingness of most communities to accept waste became apparent with the famed case of the "garbage barge," the *Mobro 4000*. In Islip, New York, in 1987, the town's landfills were full, prompting town administrators to ship waste by barge to a methane production plant in North Carolina. Prior to the barge's arrival, it became known that the shipment was contaminated with medical waste, including syringes, hospital gowns, and diapers. Because of the medical waste, the plant rejected the entire load. The barge sat in a North Carolina harbour for 11 days before heading for Louisiana. However, Louisiana would not permit the barge to dock. The barge travelled toward Mexico, but the Mexican navy prevented it from entering that nation's waters. In the end, the barge travelled 9700 km

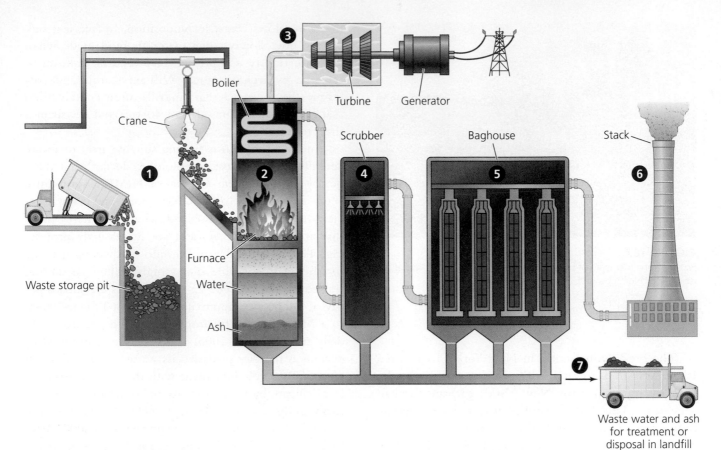

FIGURE 18.7 Incinerators reduce the volume of solid waste by burning it, but may emit toxic compounds into the air. In a waste-to-energy (WTE) facility, solid waste (1) is burned at extremely high temperatures (2), heating water, which turns to steam. The steam turns a turbine (3), which powers a generator to create electricity. In an incinerator outfitted with pollution-control technology, toxic gases produced by combustion are mitigated chemically by a scrubber (4), and airborne particulate matter is filtered physically in a baghouse (5) before air is emitted from the stack (6). Ash remaining from the combustion process is toxic and must be disposed of (7) in a hazardous-waste landfill.

before eventually returning to New York, where, after several court battles, the waste was finally incinerated at a facility in Queens.

Incinerating trash can reduce pressure on landfills

Incineration, or combustion, is a controlled process in which mixed garbage is burned at very high temperatures (**FIGURE 18.7**). At incineration facilities, waste is generally sorted and metals removed. Metal-free waste is chopped into small pieces to aid combustion and then is burned in a furnace. Incinerating waste reduces its weight by up to 75% and its volume by up to 90%.

However, simply reducing the volume and weight does not rid trash of components that are toxic. The ash remaining after trash is incinerated is toxic, and therefore must be disposed of in special landfills engineered to

receive hazardous waste. Moreover, when trash is burned, hazardous chemicals—including dioxins, heavy metals, and PCBs—can be released into the atmosphere. Such releases have caused a backlash against incineration from citizens concerned about health hazards. Opponents also feel that incineration is incompatible with the more sustainable path of reducing consumption, producing less waste, and diverting more of the waste we produce into recycling and composting.

Most developed nations now regulate incinerator emissions, and some have banned incineration outright. In Canada the ability to ban incineration rests with the provinces and territories. In some provinces where incineration is allowed, such as Ontario (where a previous ban was lifted in 1995), some municipalities continue to ban incineration within their own jurisdictions, or do not include incineration as part of their waste reduction and diversion plans. For example, the City of Toronto's aggressive plan for reducing the amount of waste produced and shipped to Michigan does not include incineration.

As a result of real and perceived health threats from incinerator emissions, and to respond to community concerns, engineers have developed several technologies to try to mitigate hazardous airborne emissions. *Scrubbers* chemically treat the gases produced in combustion to remove hazardous components and neutralize acidic gases, such as sulphur dioxide and hydrochloric acid, turning them into water and salt. Scrubbers generally do this either by spraying liquids formulated to neutralize the gases or by passing the gases through dry lime. Particulate matter is physically removed from incinerator emissions in a system of huge filters known as a *baghouse*. These tiny particles, called fly ash, often contain some of the worst dioxin and heavy metal pollutants. Burning the garbage at especially high temperatures can destroy some pollutants, such as PCBs. Even these measures, however, do not fully eliminate toxic emissions.

Landfills and incinerators can be used to generate power

Incineration was initially practised simply to reduce the volume of waste, but today it often serves to generate electricity as well. When burned, waste generates about 35% of the energy generated by burning coal, per unit weight. Most North American incinerators today are **waste-to-energy (WTE)** (also called *EFW, energy-from-waste*) facilities that use the heat produced by waste combustion to boil water, creating steam that can drive electricity generation or fuel systems for space heating or water heating. *Cogeneration* or *co-gen* is another approach, in which landfill facilities generate both electricity and thermal energy.

Combustion in WTE plants is not the only means of gaining energy from waste. Deep inside landfills, bacteria decompose waste in an oxygen-deficient environment. This anaerobic decomposition produces **landfill gas** (also called *biogas* or *LFG*). Methane (CH_4) is the main component of landfill gas. (Carbon dioxide, CO_2, is typically a close second.) Landfill gas can be collected, processed, and used in the same way as natural gas. The capture, recovery, and use of landfill gas for electricity generation also cut down greenhouse gas emissions that would otherwise result from its release into the atmosphere. Capture is particularly important in the case of methane, which is approximately 21 times more effective than carbon dioxide as a greenhouse gas.

Municipalities around the world have long captured and burned off the gases that are naturally generated in landfills. In the past, this has been done mainly to avoid odours, fires, and the potential for explosions. For years at Beare Road, gas leaked from the landfill via pipes called "candlesticks"—vertical pipes that allow the gas to flow passively out of the ground, with flames occasionally flaring from the tops of the pipes. The passive flaring system has now been replaced by over 80 vertical wells that extract landfill gas from the ground. The gas is brought to the power plant on site, where particulate matter and moisture are removed, and then to a series of reciprocating gas furnaces where it is burned to generate electricity.

Today there are hundreds of WTE and LFGTE facilities in Canada that are collecting landfill gas and generating electricity, or steam, or both via co-gen.[11] Other countries are doing the same; there are about 800 large facilities and thousands of smaller facilities worldwide.[12] In the United States alone, more than 80 incinerators and 590 landfill sites are generating thermal or electrical energy, or both.[13] This includes Fresh Kills, which supplies electricity to 22 000 Staten Island homes.

Revenues from power generation are usually not enough to offset the considerable financial cost of building and running incinerators. Because it can take many years for a WTE facility to become profitable, many companies that build and operate these facilities require communities contracting with them to guarantee the facility a minimum amount of garbage. In a number of cases, such long-term commitments have interfered with communities' later efforts to reduce their waste through recycling and other waste-reduction strategies. This is a continuing source of concern about waste incineration in some communities.

Waste Reduction and Recycling

Reducing the amount of material entering the waste stream avoids costs of disposal and recycling, helps conserve resources, minimizes pollution, and can often save consumers and businesses money. Preventing waste generation in this way is known as **source reduction**.

Reducing waste at its source is a better option

Much of our waste stream consists of materials used to package goods. Packaging serves worthwhile purposes—

TABLE 18.1   Provincial–Territorial Initiatives for Source Reduction

British Columbia	**Analysis:** A report commissioned on the business case for zero waste is currently under review and will help drive the Government's next steps and approach to waste prevention and management.
Alberta	**Formal aggreement and goal:** The Government has a memorandum of understanding with three major retail associations to reduce the distribution of one-way plastic bags by 50% by December 31, 2013. **Disposal target:** The Government also sets annual per capita waste disposal targets.
Manitoba	**Waste reduction and pollution prevention fund:** Funding is available to support waste reduction projects. **Goal:** Government sets goal to reduce the use of one-way plastic bags by 50% by 2015 and a beverage container recycling target of 75% by 2016.
Quebec	**Formal agreement:** The Government has an agreement with the industrial, commercial, and institutional (ICI) sectors to reduce waste upstream at the manufacturing level. **Disposal upper limit max:** The Government also sets annual disposal upper limit targets of waste disposal per capita, which could help drive reduction activity.
Nova Scotia	**Policy:** The Government has a sustainable procurement policy for government purchasing. **Disposal upper limit max:** The Government also sets annual disposal upper limit targets of waste disposal per capita, which could help drive reduction activity.
Yukon	**Waste reduction and recycling fund:** Funding is available to support waste reduction and recycling projects.
Northwest Territories	**Goal:** Government sets goal to reduce the use of one-way single-use bags (includes all single-use retail bags such as plastic, paper, biodegradable) by 75% (no year identified).

Remaining jurisdictions in Canada do not have jurisdiction-specific initiatives targeting waste reduction upstream.

Source: Giroux, L., Environmental Consulting (2014). *State of Waste Management in Canada.* Exhibit 13: Waste-reduction initiatives currently in place. www.ccme.ca (accessed July 2015).

preserving freshness, preventing breakage, protecting against tampering, and providing information—but much packaging is extraneous. Consumers can give manufacturers incentive to reduce packaging by choosing minimally packaged goods, buying unwrapped fruit and vegetables, and buying food in bulk. In addition, manufacturers can use packaging that is more recyclable. They can also reduce the size or weight of goods and materials, as they already have with many items, such as aluminum cans, plastic soft drink bottles, and personal computers.

In Canada, at the provincial–territorial level, only seven jurisdictions have formal policies or agreements in place to promote source reduction of wastes. These are summarized in **TABLE 18.1**.

Some governments have recently taken aim at single-use plastic grocery bags. As many as 5 trillion bags are produced worldwide each year.[14] These lightweight polyethylene bags can persist for centuries in the environment, choking and entangling wildlife and littering the landscape. Several nations have now banned their use. When Ireland began taxing these bags, their use dropped 90%; revenues from the tax are put into an environmental fund. The IKEA Company began charging for them and saw similar drops in usage. In 2007 the small Manitoba town of Leaf Rapids became the first municipality in Canada to ban plastic bags; only a handful of municipalities have followed their lead, but some cities and even

provinces have instituted voluntary bans or reduction targets (see **TABLE 18.1**). The City of Toronto approved a per-bag charge of five cents for new plastic bags, but Council then overturned it in 2012.

Increasing the longevity of goods also helps reduce waste. Consumers generally choose goods that last longer, all else being equal. To maximize sales, however, companies often produce short-lived goods that need to be replaced frequently, particularly in electronics and fashion. Thus, increasing the longevity of goods is largely up to the consumer. If demand is great enough, manufacturers will respond.

Reuse is one main strategy for waste reduction

To reduce waste, you can save items to use again or substitute disposable goods with durable ones. Habits as simple as bringing your own coffee cup to coffee shops or bringing sturdy reusable cloth bags to the grocery store can, over time, have substantial impact. You can also donate unwanted items and shop for used items yourself at yard sales and resale centres. Besides doing good for the environment, reusing items is often economically advantageous. Used items are quite often every bit as functional as new ones, and much cheaper. **TABLE 18.2** presents a sampling of actions that we all can take to reduce the waste we generate.

TABLE 18.2 Some Everyday Things You Can Do to Reduce and Reuse

- Donate used items to charity.
- Reuse boxes, paper, plastic wrap, plastic containers, aluminum foil, bags, wrapping paper, fabric, packing material, and so on.
- Rent or borrow items instead of buying them when possible, and lend your items to friends.
- Buy groceries in bulk.
- Decline bags at stores when you don't need them.
- Bring reusable cloth bags for shopping.
- Make double-sided photocopies.
- Bring your own coffee cup to coffee shops.
- Pay a bit extra for durable, long-lasting, reusable goods rather than disposable ones.
- Buy rechargeable batteries.
- Select goods with less packaging.
- Compost kitchen and yard wastes in a compost bin or worm bin (often available from your community or waste hauler).
- Buy clothing and other items at resale stores and garage sales.
- Use cloth napkins and rags rather than paper napkins and towels.
- Write to companies to tell them what you think about their packaging and products.
- When solid waste policy is being debated, let your government representatives know your thoughts.
- Support organizations that promote waste reduction.

Source: From United States Environmental Protection Agency.

Charles Stephenson/Alamy

FIGURE 18.8
We can mimic the efficiency of nature, return organic matter to the soil, and reduce solid waste through curbside or centralized composting, or by using a backyard composter, as shown here.

Composting recovers organic waste

Composting is the conversion of organic waste into mulch or humus through natural biological decomposition. The resulting compost can be used to enrich soil. Householders can place food waste (**FIGURE 18.8**) and yard waste in compost piles or specially constructed containers. As wastes are added, heat from microbial action builds in the interior, and decomposition proceeds. Banana peels, coffee grounds, grass clippings, autumn leaves, and countless other organic items, including paper, can be converted into rich, high-quality compost through the actions of earthworms, bacteria, soil mites, sow bugs, and other detritivores and decomposers. Home composting is a prime example of how we can live more sustainably by mimicking natural cycles and incorporating them into our daily lives.

Centralized composting programs—there are now more than 160 of them in Canada[15]—divert food and yard waste from the waste stream to composting facilities, where they are turned into mulch that community residents can use for gardens and landscaping. Some municipalities now ban yard waste from the municipal waste stream, helping accelerate the drive toward composting. Approximately 28–29% of the Canadian solid waste stream is made up of organics, but paper products also can be composted; in all, about 40% of Canadian household waste consists of compostable materials (see **FIGURE 18.3A**).

Composting reduces landfill waste, enriches soil and helps it resist erosion, encourages soil biodiversity, makes for healthier plants and more pleasing gardens, and reduces the need for chemical fertilizers and water. According to Statistics Canada and the Canadian Council of Environment Ministers, composting of food and yard waste has resulted in a 125% increase in diversion of compostable materials from landfills between 2005 and 2010. About 57% of Canadians do some form of composting, although only about 40% of households have access to curbside food waste collection for composting at centralized facilities[16] (**TABLE 18.3**).

TABLE 18.3 **Households That Compost in Canada**

Composted either kitchen or yard waste	Percent
Canada	**57**
Newfoundland and Labrador	46
Prince Edward Island	98
Nova Scotia	92
New Brunswick	56
Quebec	35
Ontario	71
Manitoba	50
Saskatchewan	52
Alberta	54
British Columbia	59

Source: Based on Statistics Canada, *Households and the Environment Survey,* 2007.

Recycling has grown rapidly and can expand further

Recycling, too, offers many benefits. **Recycling** consists of collecting materials that can be broken down and reprocessed to manufacture new items. In 2010,

8.0 million tonnes of materials were prepared for recycling by waste management organizations and companies. Of this, paper was the main component.[17]

The recycling loop consists of three basic steps (**FIGURE 18.9A**). The first step is collecting and processing used recyclable goods and materials. Communities may designate locations where residents can drop off recyclables or receive money for them. Many of these have now been replaced by the more convenient option of curbside recycling, in which trucks pick up recyclable items in front of houses, usually in conjunction with municipal trash pickup. Curbside recycling has grown rapidly, and its convenience has helped boost household recycling rates across Canada.

Items collected are taken to *materials recovery facilities (MRFs),* where workers and machines sort items using automated processes including magnetic pulleys, optical sensors, water currents, and air classifiers that separate items by weight and size. The code in the recycling logo helps identify the materials from which the products were made. Just having a recycling code on a product doesn't necessarily mean that it is recyclable within that jurisdiction. For example, code "4 LDPE" represents low-density polyethylene plastic, commonly used to make squeezable bottles, dry cleaning and shopping

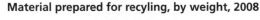

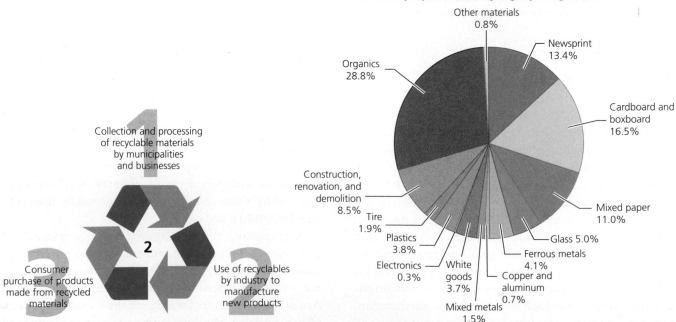

(a) The familiar recycling logo

(b) Composition of materials recycled in Canada

FIGURE 18.9 The three arrows of the familiar recycling symbol **(a)** represent the three steps of recycling: collection and processing of recyclable materials, use of the materials to make new products, and consumer purchase of the products. The number in the middle designates the nature of the material; for example, "2" refers to high-density polyethylene, used in plastic bottles, plastic bags, trash cans, and imitation wood products. A wide variety of materials are prepared for recycling each year in Canada **(b)**.
Source: Data for (b) from Statistics Canada (2012). *Human Activity and the Environment: Waste Management in Canada.* Government of Canada, catalogue no. 16-201-X.

bags, and some other products; plastics with this code are usually not accepted by curbside recycling operations.

After sorting, the MRFs clean the materials, shred them, and prepare them for reprocessing or resale (see **FIGURE 18.9B**) to be used in manufacturing new goods. Newspapers and many other paper products use recycled paper, many glass and metal containers are now made from recycled materials, and some plastic containers are of recycled origin. Some large objects, such as benches and bridges in city parks, are now made from recycled plastics, and glass is sometimes mixed with asphalt (creating "glassphalt") for paving roads and paths. The pages in the print version of this textbook are made from recycled paper that is up to 10% post-consumer waste.

The thousands of curbside recycling programs and MRFs in operation today have sprung up only in the last 20 years. According to Statistics Canada, 95% of Canadian households had access to recycling programs by 2007, and most programs covered all four recyclables: paper, plastic, glass, and metal.[18]

Recycling rates vary greatly from one product or material type to another and from one location to another. Rates for different types of materials and products range from nearly zero to almost 100%. The increase in recycling has been propelled in part by economic forces as established businesses see opportunities to save money and as entrepreneurs see opportunities to start new businesses. It has also been driven by the desire of municipalities to reduce waste and by the satisfaction people take in recycling. These two forces have driven recycling's rise, even though it has often not been financially profitable. In fact, many of the increasingly popular municipal recycling programs are run at an economic loss. The expense required to collect, sort, and process recycled goods is often more than recyclables are worth in the market. Furthermore, the more people recycle, the more glass, paper, and plastic are available to manufacturers for purchase, driving down prices.

Recycling advocates, however, point out that market prices do not take into account external costs—in particular, the environmental and health impacts of *not* recycling. For instance, it has been estimated that globally, recycling saves enough energy to power 6 million households per year. And recycling aluminum cans saves 95% of the energy required to make the same amount of aluminum from mined virgin bauxite, its source material.

If the recycling loop is to function, consumers and businesses must complete the third step in the cycle by purchasing products made from recycled materials. Buying post-consumer recycled goods provides economic incentive for industries to recycle materials, and for new recycling facilities to open or existing ones to expand. In this arena, individual consumers have the power to encourage environmentally friendly options through the free market. Many businesses now advertise their use of recycled materials, a widespread instance of *ecolabelling*. As markets for products made with recycled materials expand, prices will continue to fall.

As more manufacturers use recycled products and as more technologies and methods are developed to use recycled materials in new ways, markets should continue to expand, and new business opportunities may arise. We are still at an early stage in the shift from an economy that moves linearly from raw materials to products to waste to an economy that moves circularly, using waste products as raw materials for new manufacturing processes. The steps we have taken in recycling so far are central to this transition, which many analysts view as key to building a sustainable economy (see "The Science behind the Story: Edmonton Showcases Reduction and Recycling").

Financial incentives can help reduce waste

Waste managers have employed economic incentives to reduce the waste stream. The "pay-as-you-throw" approach to garbage collection uses a financial incentive to influence consumer behaviour. In these programs, municipalities charge residents for home trash pickup according to the amount of trash they put out. The less waste the household generates, the less the resident has to pay.

Return-for-refund schemes ("bottle bills" in the United States) represent another approach that hinges on financial incentive. To date, all provinces and territories except Nunavut have such programs. Consumers pay a deposit, return bottles and cans to stores after use, and receive a refund—generally $0.05 to $0.20 per bottle or can. The first bottle bills were passed in the 1970s to cut down on litter, but they have also served to decrease the waste stream. Where they have been enacted, these laws have proven effective and popular; they are recognized as among the most successful recycling programs of recent decades. Jurisdictions with bottle and can refund programs have reported that their beverage container litter has decreased by 69–84%, their total litter has decreased by 30–64%, and their per capita container recycling rate has risen 260%.[19]

Sometimes financial incentives and penalties can backfire. Years ago, the city of Seattle tried to limit curbside pickup to one bag per household, levying a fee of $10 per bag for any extra bags. The result was an

THE SCIENCE **BEHIND THE STORY**

Edmonton's gigantic composting facility, seen here, is the size of eight football fields.

City of Edmonton

Edmonton Showcases Reduction and Recycling

Edmonton, Alberta, has one of the world's most advanced waste management programs. As recently as 1998, fully 85% of the city's waste was being landfilled, and space was running out. Today, just 40% goes to the new sanitary landfill, whereas 15% is recycled, and an impressive 40% is composted. The rest (large items and household hazardous waste) can be dropped at one of the city's Eco-Station dropoff waste processing centres. Edmonton's citizens are proud of the program, and 88% of them participate in its curbside recycling program. Where blue recycling bins are available at apartments

and condominiums, the participation rate of residents is 91%.[20]

When Edmonton's residents put out their trash, city trucks take it to their new *co-composting* plant, the largest in North America (see photo at left). The waste is dumped on the floor of the facility, and large items, such as furniture, are removed and landfilled. The bulk of the waste is mixed with dried sewage sludge for one to two days in five large rotating drums, each the length of six buses.

The resulting mix travels on a conveyor to a screen that removes nonbiodegradable items. It is aerated for several weeks in the largest stainless steel building in North America (**FIGURE 1**). The mix is then passed through a finer screen and finally is left outside for four to six months. The resulting compost—80 000 tonnes annually—is made available to area farmers and residents. The facility even filters the air it emits with a 1-m layer of compost, bark, and wood chips, which eliminates the release of unpleasant odours into the community. Christmas tree composting and "grasscycling" programs are now included as well.

Edmonton's municipal program also includes a state-of-the-art MRF, a leachate treatment plant, a research centre, public education programs, and a wet-

land and landfill revegetation program. The program even includes an intensive training course, through which citizens can become "Master Composter Recyclers," who teach and mentor their neighbours and newcomers in recycling and other waste management practices.

In addition, pipes at the composting facility collect enough landfill gas to power 4000 homes, bringing thousands of dollars to the city and helping power the new waste management centre. Five area businesses reprocess the city's recycled items, including e-wastes. Newsprint and magazines are turned into new newsprint and cellulose insulation, and cardboard and paper are converted into building paper and shingles. Household metal is made into rebar and blades for tractors and graders, and recycled glass is used for reflective paint and signs.

The city's waste management programs are backed by research carried out at the Edmonton Waste Management Centre of Excellence (EWMCE). This organization is a public–private partnership, established in 2003, where scientists and engineers carry out applied research and technology development projects, and run training programs and technical workshops for waste managers.

---

upsurge in garbage dropped off illegally in ravines and by the roadside, and the limit and fee were repealed. But attitudes and times have changed: As of 2014, the city allows just one garbage can per household, with the lid completely closed. The fee for any extra garbage, which is also weight- and size-limited, is $10.20 per bag, bundle, or can.

For small businesses like local restaurants and grocery stores, financial incentives can drive waste management practices. When institutions (and, commonly, residents of rural areas with no curbside pickup) have to dispose of garbage at landfills, they pay what is known as a *tipping fee*. The tipping fee is a charge, usually based on weight, for the amount of garbage that is being dumped at the landfill. When tipping fees go up, even a little bit, it can make it more profitable for restaurants to try to reduce their food waste by composting, or by engaging with a partner like

a food bank or Second Harvest, which can reroute still-edible products to the hungry.

Industrial Solid Waste

In Canada, disposal of wastes from nonresidential sources (the industrial, commercial, and institutional sectors, sometimes referred to as *ICI*) increased from 14.6 million to about 17 million tonnes between 2002 and 2008, falling again to 15.6 million tonnes in 2010.[22] Much of the waste from ICI sources is handled as part of the municipal waste stream, which complicates monitoring and accounting of waste generation from this source. In principle, **industrial solid waste** includes not only ICI waste, but also waste from factories, mining activities, agriculture, petroleum extraction, construction, and more.

Some of the current research projects at EWMCE include[21]

- biosolids management and application
- solid waste collection, sorting, and recycling
- construction and demolition waste recovery
- odour and gas reduction from operations
- energy recovery from solid waste (including gasification)
- thermal, chemical, and bio-conversion of waste to value-added products
- alternative uses for shredded tires

One study, for example, is examining the temperature conditions experienced over time by actual particles of compost in full-scale composting systems. Using specially designed temperature probes, the researchers are measuring the effects of temperature on human pathogens and organisms within the organic waste. Another study is looking at how Edmonton handles drywall and other construction materials in the waste stream—a big concern, given the major fluctuations in Alberta's housing market. Still another project is investigating how various pre-treatment approaches might limit the transfer of metal contaminants such as batteries into compost products from municipal garbage sources.

Edmonton's waste management program is not perfect, but it is backed by science. Through commitment, education, and investment, the city has made steady progress toward its goals, and has become a model for success and sustainability in waste management.

City of Edmonton

Aeration building, Edmonton composting facility

FIGURE 1
Inside the composting building, which is the size of 14 professional hockey rinks, mixtures of solid waste and sewage sludge are exposed to oxygen and composted for 14–21 days.

Waste is generated at various points along the process from raw materials extraction to manufacturing to sale and distribution (**FIGURE 18.10**). When heavy industry sources are included in the totals, the amounts generated increase dramatically.

Regulation and economics influence industrial waste generation

Most methods and strategies of waste disposal, reduction, and recycling by the ICI sectors are similar to those for municipal solid waste. For instance, businesses that manage their own waste on-site most often dispose of it in landfills, and companies must design and manage their landfills in ways that meet provincial/territorial, local, or tribal guidelines. Other businesses pay to have their waste disposed of at municipal disposal sites. Regulation and enforcement vary across provinces and territories and from municipality to municipality.

The amount of waste generated by a manufacturing process is one measure of its efficiency; the less waste produced per unit or volume of product, the more efficient that process is, from a physical standpoint. However, physical efficiency is not always equivalent to economic efficiency. Often it is cheaper for industry to manufacture its products or perform its services quickly but messily. That is, it can be cheaper to generate waste than to avoid generating waste. In such cases, economic efficiency is maximized, but physical efficiency is not. The frequent mismatch between these two types of efficiency is a major reason that the output of industrial waste is so great.

Rising costs of waste disposal and tipping fees, however, enhance the financial incentives to decrease waste and increase physical efficiency. If government regulations or the market make the physically efficient use of raw materials economically efficient as well, businesses have greater financial incentives to reduce their waste.

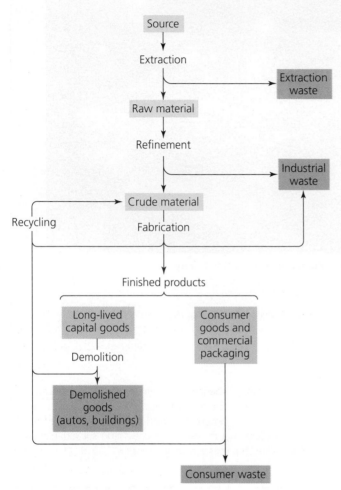

(a) The industrial waste stream

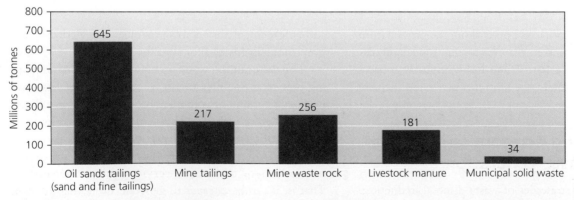

(b) Additional sources of waste in Canada

FIGURE 18.10 Both industrial and municipal wastes are generated throughout the life-cycles of products, from the extraction of raw materials needed for production, to processing, and manufacturing of the products **(a)**. Waste also results from the demolition or disposal of products by businesses and individuals. At each stage, there are opportunities for waste reduction or recycling. Waste from the industrial, commercial, and institutional sectors typically is included in municipal waste, as it is here **(b)**. Massive amounts of waste are also generated by mining, fossil fuel production, and agriculture, shown here by weight for 2008. *Source:* Data for (b) from Statistics Canada (2012). *Human Activity and the Environment: Waste Management in Canada.* Government of Canada, catalogue no. 16-201-X.

Industrial ecology seeks to make industry more sustainable

To reduce waste, growing numbers of industries today are experimenting with industrial ecology. A holistic approach that integrates principles from engineering, chemistry, ecology, and economics, **industrial ecology** seeks to redesign industrial systems to reduce resource inputs and to minimize physical inefficiency while maximizing economic efficiency. Industrial ecologists would reshape industry so that nearly everything produced in a manufacturing process is used either within that process or in a different one.

The larger idea behind industrial ecology is that industrial systems should function more like ecological systems, in which almost everything produced is used by some organism, with very little being wasted. This principle brings industry closer to the ideal of ecological economists, in which human economies attain sustainability by functioning in a circular fashion rather than a linear one.

Industrial ecologists pursue their goals in several ways. For one, they examine the entire life-cycle of a given product—from its origins in raw materials, through its manufacturing, to its use, and finally its disposal—and look for ways to make the process more ecologically efficient. This strategy is called **life-cycle analysis** (**FIGURE 18.11**). In addition, industrial ecologists examine industrial processes with an eye toward eliminating environmentally harmful products and materials.

Related to life-cycle analysis are the concepts of *embodied energy* and *embodied water*. In this case, the analysis

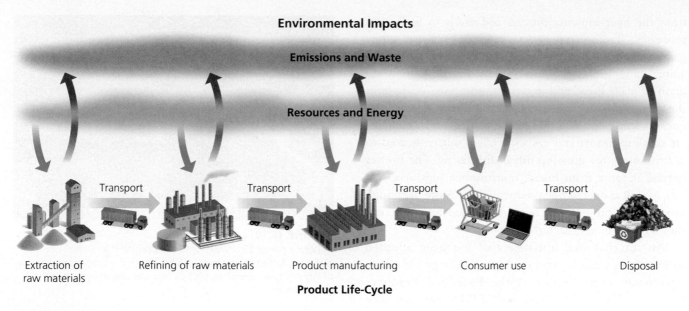

Environmental Impacts

Emissions and Waste

Resources and Energy

Transport　　Transport　　Transport　　Transport

Extraction of raw materials　　Refining of raw materials　　Product manufacturing　　Consumer use　　Disposal

Product Life-Cycle

FIGURE 18.11

In life-cycle analysis, the environmental impacts of all stages of product manufacture are accounted for, from the extraction, transport, and processing of raw materials, to the manufacture and use of the product, to its final disposal.

is undertaken from the specific perspective of the energy or water involved in production. This approach is often used in analyzing construction materials, where, for example, it can help to make decisions more environmentally responsible by revealing the differences in "virtual energy" content of different building materials. It is also used fairly extensively to analyze the water required to produce various foods, both in the field and in manufacturing. For example, beef and (consequently) leather products are extremely water-intensive to produce; they have much higher embodied water per kg than tomatoes, potatoes, or even chicken, for example.

Industrial ecologists also study the flow of materials through industrial systems to look for both processes and materials that can help manufacturers create products that are more durable, recyclable, or reusable. Goods that are currently thrown away when they become obsolete, such as computers, automobiles, and some appliances, could be designed to be more easily disassembled and their component parts reused or recycled. In this way, industrial ecology also helps close the loop by minimizing wastes at both the industry end and the consumer end of the process.

By applying strategies aimed at reducing waste and preventing pollution at its source—commonly referred to as **pollution prevention (P2) strategies**—companies can significantly reduce their waste output.

Businesses are adopting industrial ecology perspectives

Businesses are taking advantage of the insights of industrial ecology to reduce waste and lessen their impact on

health and the environment while saving money. A good example is the carpet tile company Interface, which asks customers to return used tiles for recycling and reuse as backing for new carpet. Interface also modified its tile design and its production methods to reduce waste. It adapted its boilers to use landfill gas for its energy needs and has the goal of sourcing all energy from renewable sources by 2020. Through such steps, the company has cut its waste generation by 80%, its fossil fuel use by 45%, and its water use by 70%—all while saving $30 million per year, holding prices steady for its customers, and raising profits by 49%.

Among many other initiatives are such programs as Canadian Tire's auto parts return initiatives; the "Car Heaven" program from Summerhill Impact, for recycling and management of older vehicles; and Xerox's take-back/lease programs for its used photocopiers. Programs like these are founded in good business principles, and they have the added benefit of building customer loyalty. An interesting variation of the take-back concept is the ENVIRx program run by the Pharmacists Association of Alberta, in which consumers are able to return unused medications to participating pharmacies for proper disposal. This helps prevent pharmaceuticals from entering the municipal waste stream.

The Swiss Zero Emissions Research and Initiatives (ZERI) Foundation sponsors dozens of innovative projects worldwide that attempt to create goods and services without generating waste. One example involving breweries is currently being pursued in Canada, Sweden, Japan, and Namibia. Brewers in these projects take waste from the beer-brewing process and use it to fuel other processes. Traditional breweries produce only beer while generating much waste, some of which goes toward animal feed. ZERI-sponsored breweries use their waste grain to make bread and to farm mushrooms. Waste from the mushroom farming, along with brewery waste water, goes to feed pigs. The pigs' waste is digested in containers that capture natural gas and collect nutrients used to nourish algae for growing fish in fish farms. The brewer derives income from bread, mushrooms, pigs, gas, and fish, as well as beer, while producing little waste. Although most ZERI projects are not fully closed-loop systems, they attempt to approach this ideal.

An international initiative that has been adopted nationally in Canada as an overriding strategy for industrial waste minimization is the Extended Producer Responsibility and Stewardship (or EPRS) program. The central goal of the program is to transfer a large part of the responsibility for waste minimization, both physical and financial, to producers.[23] This gives producers the economic incentives to design more environmentally efficient products and processes and to take greater responsibility for the product at the end of its life-cycle. It also encourages producers to build the environmental costs of a product into its market price—one of the basic tenets of ecological economics.

The concept of industrial ecology is based on a "closed loop" in which wastes are recycled back through the system. Following the definition of wastes as "resources out of place," industrial ecologists strive to find practical, economical uses for waste materials. To achieve this goal, they try to identify how waste products from one manufacturing process can be used as raw materials for a different process. For instance, used plastic beverage containers cannot be refilled because of the potential for contamination, but they can be shredded and reprocessed to make other plastic items, such as benches, tables, and decks.

Many network services have emerged with the goal of linking producers of waste with industries or individuals that can make use of the waste as raw materials. Such a network is called a **waste exchange**. You can check out an example of a nationwide waste exchange by visiting the website of The Waste Exchange of Canada at **www.recyclexchange.com**. Other waste exchanges operate locally or internationally.

Hazardous Waste

Hazardous wastes are diverse in their chemical composition and may be liquid, solid, or gaseous. In Canada, according to the *Canadian Environmental Protection Act*

FIGURE 18.12
In 1990, a pile of 15 million discarded tires in Hagersville, Ontario, caught fire and burned for 17 days, releasing thousands of litres of toxic chemicals, such as benzene and toluene, into the air and water.

(CEPA) (1999), **hazardous waste** is waste that has one or more of the following properties:

■ **Flammable:** substances that easily catch fire (for example, natural gas or alcohol)
■ **Corrosive:** substances that corrode metals in storage tanks or equipment
■ **Reactive:** substances that are chemically unstable and readily react with other compounds, often explosively or by producing noxious fumes
■ **Toxic:** substances that harm the health of humans or other organisms when they are inhaled, are ingested, or come into contact with skin

Materials with these characteristics can harm human health and degrade environmental quality. Flammable and explosive materials can cause ecological damage and atmospheric pollution. For instance, fires at tire dumps, such as the one in Hagersville, Ontario, in 1990 (**FIGURE 18.12**), have caused air pollution and highway closures. Toxic wastes in lakes and rivers have caused fish die-offs, endangered aquatic mammals (see "Central Case: The Plight of the St. Lawrence Belugas" in the chapter on Earth systems and ecosystems), and closed important domestic fisheries.

Certain categories of materials that are clearly "dangerous" are nevertheless not included in the official definition of hazardous waste. An example is biomedical waste, which includes things like human tissues and fluids, and discarded medical sharps. These materials are excluded from the definition of hazardous waste not because they are without risk but because they require specialized handling, treatment, and disposal methods and are therefore controlled under different legislation. Similarly, radioactive waste requires a special set of management approaches.

Hazardous wastes have diverse sources

Industry, mining, households, small businesses, agriculture, utilities, and building demolition all create hazardous waste. Industry produces the largest amounts of hazardous waste, but in most developed nations industrial waste generation and disposal are highly regulated. This regulation has reduced the amount of hazardous waste entering the environment from industrial activities. As a result, households currently are the largest source of unregulated hazardous waste.

Household hazardous waste (HHW) includes a wide range of items, such as paints, batteries, oils, solvents, cleaning agents, lubricants, and pesticides. (**TABLE 18.4**). Canadians improperly dispose of approximately 27 000 tonnes of HHW each year,[24] and the average home contains close to 45 kg of it in sheds, basements, closets, and garages (**TABLE 18.5**). Although many hazardous substances become less hazardous over time as they degrade chemically, two classes of chemicals are particularly hazardous because their toxicity persists over time: organic compounds and heavy metals.

Organic compounds and heavy metals can be hazardous

In our day-to-day lives, we rely on the capacity of synthetic organic compounds and petroleum-derived compounds to resist bacterial, fungal, and insect activity.

TABLE 18.4 Common Types of Household Hazardous Waste

- Antifreeze
- Fertilizers
- Lubricating oils, engine and gear oils, transmission fluids
- Paints and coatings, including nail polish
- Pesticides, fungicides, herbicides, insecticides
- Pressurized containers (such as propane tanks and cylinders, and oxygen tanks)
- Single-use dry cell batteries (regular and button batteries)
- Solvents (turpentine, mineral spirits, linseed oils, paint and lacquer thinners, nail polish remover, automotive body resin solvents, contact cement thinners, paint strippers, and degreasers)
- Used oil filters
- Medical waste (such as syringes and leftover medications)
- Household cleaning products (such as abrasives, bleach, ammonia, drain cleaners, window cleaners, oven cleaners, and silver polishes)
- Compact fluorescent bulbs and tubes

Such items as plastic containers, rubber tires, pesticides, solvents, and wood preservatives are useful to us precisely because they resist decomposition. We use these substances to protect our buildings from decay, kill pests that attack crops, and keep stored goods intact. However, the resistance of these compounds to decay is a double-edged sword, for it also makes them persistent pollutants. Many synthetic organic compounds are toxic because

TABLE 18.5 Household Hazardous Waste Disposal in Canada, 2009

| | Households that had HHWW items for disposal | How the identified hazardous waste item was disposed of | | | | | | |
		Put them in the garbage	Took or sent them to a depot or drop-off centre	Returned them to a supplier or retailer	Poured them down the drain, toilet, or sink	Donated them or gave them away	Still had them	Other
		Percent						
Medication	39	22	6	57	8	...	15	1
Paint or solvents	39	4	62	8	...	...	31	2
Unwanted engine oil or antifreeze	15	1	61	19	...	...	18	4
Dead or unwanted car batteries	12	F	46	31	...	...	20	5
Other dead or unwanted batteries	58	42	35	7	...	...	18	4
Unwanted electronic devices	36	11	45	5	...	22	28	2
Dead or unwanted compact fluorescent lights (CFLs)	22	56	24	...	...	4	13	3

Source: Statistics Canada (2011). *Households and the Environment (2009)*. Government of Canada, catalogue no. 11-526-X.

 Which category of HHW shown on this table was most often disposed of improperly, compared to the others? Which category was second?

they can be absorbed readily through the skin of humans and other animals and can act as mutagens, carcinogens, teratogens, and endocrine disruptors.

Heavy metals, such as lead, chromium, mercury, arsenic, cadmium, tin, and copper, are used widely in industry for wiring, electronics, metal plating, metal fabrication, pigments, and dyes. Heavy metals enter the environment when paints, electronic devices, batteries, and other materials are disposed of improperly. Lead from fishing weights and from hunters' lead shot has accumulated in many rivers, lakes, and forests. In older homes, lead from pipes contaminates drinking water, and lead paint remains a problem, especially for infants. Heavy metals, which are fat-soluble and break down slowly, bioaccumulate and bioconcentrate. Such materials in streams, for example, can render fish unsafe to eat.

"E-waste" is a new and growing problem

When we first began to conduct much of our business, learning, and communication with computers and other electronic devices, many people predicted that our paper waste would decrease. Instead, the proliferation of computers, printers, VCRs, fax machines, cell phones, GPS devices, MP3 players, and other gadgets has created a substantial new source of waste. These products have short lifetimes before people judge them obsolete, and most are discarded after only a few years.

The amount of **electronic waste**—often called **e-waste**—is growing rapidly. Statistics Canada reports that in 2009, 59% of Canadian households had unwanted electronic devices (and 42% of them disposed of these items

(a) Electronic waste

FIGURE 18.13

Discarded electronic waste can leach heavy metals and should be considered hazardous waste **(a)**. Some proportion of all electronic devices tested exceeded 5 mg/L, the U.S. EPA's regulatory threshold for lead leachate **(b)**. (Canada'a threshold for drinking water is 0.010 mg/L.) Devices with higher ferrous metal content tended to leach less lead. Where both standard and modified TCLPs were used, results are averaged.

Source: Based on the data from Townsend, T. G., et al. (2004). RCRA Toxicity Characterization of Computer CPUs and Other Discarded Electronic Devices. July 15, 2004, report to the U.S. EPA.

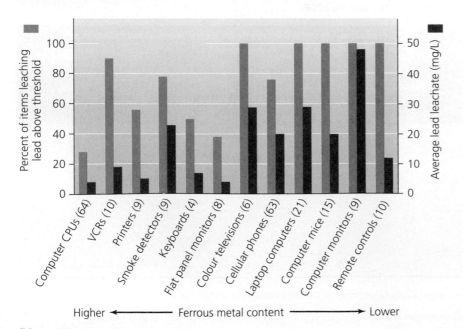

(b) Lead leaching from e-waste items

in the regular garbage can).[25] Canadians discarded 74 000 tonnes of computer waste in 2002, including 1.7 million desktop computers, 1.9 million cell phones, 2 million television sets, and 1.1 million VCRs (see also **TABLE 18.5**). The U.S. Environmental Protection Agency (EPA) reports that 70% of the heavy metals found in U.S. landfills came from discarded electronic products, two-thirds of which were still functional.[26] Most e-waste is disposed of in landfills as conventional solid waste. However, most electronic products contain heavy metals and toxic flame retardants, and recent research suggests that e-waste should instead be treated as hazardous waste (**FIGURE 18.13**). For instance, more than 6% of a typical computer is composed of lead. The cathode ray tubes in televisions and computer screens can hold up to 5 kg of heavy metals, such as lead and cadmium. These represent the second-largest source of lead in landfills today, behind auto batteries.

In Canada there are no federal programs or legislation aimed specifically at dealing with e-waste, although initiatives have been started in most provinces. More and more electronics are now being recycled. The devices are taken apart, and parts are either reused or disposed of more safely. Roughly one-fifth of the nearly 2 million tonnes of electronics discarded in 2005 in the United States was recycled, the EPA estimates. The Electronics Products Recycling Association reports that in 2012 over 122 000 tonnes of end-of-life electronics and electronic products were diverts from landfills in 2012.[27]

There are serious concerns about the health risks that recycling e-waste may pose to workers doing the disassembly. Wealthy nations consequently ship much of their e-waste to developing countries, where the disassembly is done by poor workers with no training and minimal safety regulation and protection. These environmental justice concerns need to be resolved, but if electronics recycling can be done responsibly, it seems likely to be the way of the future.

In many North American cities, businesses, nonprofit organizations, or municipal services now collect used electronics for reuse or recycling. The next time you upgrade to a new computer, TV, DVD player, VCR, or cell phone, find out about opportunities to recycle your old ones. Electronics Product Stewardship Canada (EPSC) is an example of a national-level, not-for-profit organization that was established to promote better recycling and sustainable solutions for end-of-life electronics.

Several steps precede the disposal of hazardous waste

For many years we discarded hazardous waste without special treatment. In many cases, people did not know that certain substances were harmful to human health. In other cases, the danger posed by these substances was known or suspected, but it was assumed that the substances would disappear or be sufficiently diluted in the environment. The resurfacing of toxic chemicals in a residential area years after their burial at Love Canal in upstate New York provided a dramatic demonstration to the North American public that hazardous waste deserves special attention and treatment.

Since the 1980s, many communities have designated sites or special collection days to gather household hazardous waste or have designated facilities for the exchange and reuse of substances (**FIGURE 18.14**). In Ontario, for example, the nonprofit, industry-sponsored organization Stewardship Ontario runs dropoff sites earmarked for household hazardous and special wastes, called the Orange Drop program. Once consolidated

Joe Sohm/Alamy

(a) Community dropoffs for household hazardous waste

Jonathan Hayward/AP Images

(b) Recycling hazardous e-wastes

FIGURE 18.14
Many communities designate dropoff sites or collection days for household hazardous waste. Here **(a)**, workers handle waste from an Earth Day collection event. The medals awarded to athletes at the 2010 Vancouver Winter Olympic Games were manufactured partly from precious metals recycled from discarded e-waste **(b)**.

Robert Brook/Photo Researchers, Inc.

FIGURE 18.15
Unscrupulous individuals or businesses sometimes dump hazardous waste illegally to avoid disposal costs.

in such sites, the waste materials are sorted and then transported for treatment and ultimate disposal.

The management and control of hazardous waste and hazardous recyclable materials are a shared responsibility in Canada. The federal government regulates international agreements and transport. Provincial and territorial governments regulate intraprovincial transport and are responsible for licensing hazardous waste generators, carriers, and treatment facilities.[28] As hazardous waste is generated, transported, and disposed of, the producer, carrier, and disposal facility must each report the type and amount of material generated; its location, origin, and destination; and its handling. This is intended to prevent illegal dumping and encourage the use of reputable waste carriers and disposal facilities.

It can be quite costly to dispose of hazardous waste. For this reason, irresponsible companies sometimes illegally and anonymously dump waste. This creates health risks for residents and long-lasting technical and financial headaches for local governments forced to deal with the mess (**FIGURE 18.15**).

Hazardous waste from industrialized nations is also sometimes dumped illegally in developing nations—a major environmental justice issue. This practice occurs despite the *Basel Convention*, the international treaty that was specifically designed to prevent such acts. One of the central tenets of the *Basel Convention* is the principle of *prior informed consent*, which in this case says that a recipient may not be asked to receive or be offered money to receive hazardous materials unless that person is fully informed, trained to handle the materials, and aware of the risks involved.

In 2006, a ship secretly dumped toxic wastes in Abidjan, the capital of Ivory Coast, after being told by Dutch authorities that the Netherlands would charge money to dispose of the waste in Amsterdam. The waste caused several deaths and thousands of illnesses in Abidjan, and street protests forced the government to resign over the scandal. Some jail sentences and fines were eventually handed down in this case, although thousands of victims still have received no compensation for their suffering.

Fortunately, high costs of disposal have also encouraged conscientious businesses to invest in reducing their hazardous waste. Many biologically hazardous materials can be broken down by incineration at high temperatures in cement kilns. Some hazardous materials can be treated by exposure to bacteria that break down harmful components and synthesize them into new compounds. Besides bacterial bioremediation, phytoremediation is also used. Various plants have now been bred or engineered to take up specific contaminants from soil; they then break down organic contaminants into safer compounds or concentrate heavy metals in their tissues. The plants are eventually harvested and disposed of.

There are three disposal methods for hazardous waste

There are three primary means of hazardous waste disposal: secure landfills, surface impoundments, and injection wells. These do nothing to lessen the hazards of the substances, but they do help keep the waste isolated from people, wildlife, and ecosystems.

Secure landfills Design and construction standards for landfills that receive hazardous waste are similar to, but much stricter than, those for ordinary sanitary landfills (see **FIGURE 18.5**). Hazardous waste landfills, also called **secure landfills**, must have several engineered geotextile liners, leachate removal systems, and extensive monitoring wells, and they must be located on a thick, clay-rich substrate, far from aquifers. Dumping of hazardous waste in ordinary landfills is particularly problematic in closed-down landfills that received wastes prior to the advent of more-secure disposal options for hazardous materials.

Surface impoundments Liquid hazardous waste, or waste in dissolved form, may be stored in ponds or **surface impoundments**, shallow depressions lined with plastic and an impervious material, such as clay. Water containing dilute hazardous waste is placed in the pond and allowed to evaporate, leaving a residue of solid hazardous waste on the bottom (**FIGURE 18.16**). This process is repeated until the dry material is removed and transported elsewhere for permanent disposal. Impoundments are not ideal. The underlying layer can crack and leak waste. Some material may evaporate or blow into surrounding areas. Rainstorms may cause waste to overflow and contaminate nearby areas. For these reasons, surface impoundments are used only for temporary storage.

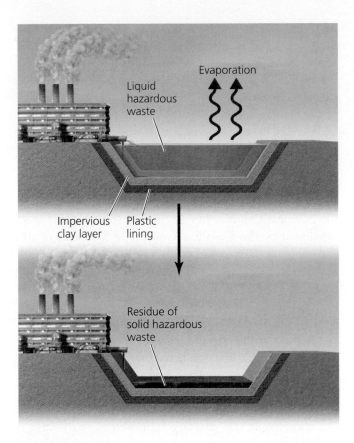

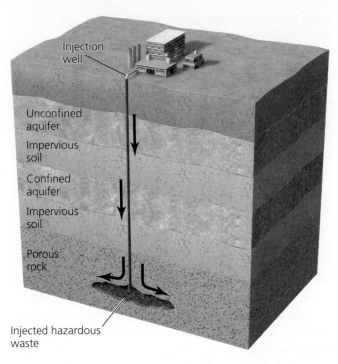

FIGURE 18.17
Liquid hazardous waste may be pumped deep underground by deep-well injection. The well must be drilled below any aquifers, into porous rock separated by impervious clay. The technique is expensive, and waste may leak from the well shaft into ground water.

FIGURE 18.16
Surface impoundments are a strategy for temporarily disposing of liquid hazardous waste. The waste, mixed with water, is poured into a shallow depression lined with plastic and clay to prevent leakage. When the water evaporates, leaving a crust of the hazardous substance, new liquid is poured in and the process repeated. This method alone is not satisfactory because waste can potentially leak, overflow, evaporate, or blow away.

The potential for problems with surface impoundment of hazardous wastes became abundantly clear in the small rural town of Elmira, Ontario, in the late 1980s. UniRoyal (now the Crompton Company) had operated a chemical production facility at Elmira since 1942. One of the substances produced at the plant was the so-called Agent Orange, an extremely powerful herbicide used by the United States during the Vietnam War to defoliate large areas of forest. (U.S. and Vietnamese soldiers and civilians later suffered serious health impacts as a result of exposure to this chemical.) Rubber and agrichemicals were also produced at the plant. The waste by-products of chemical production were disposed of—legally, and within Ontario Ministry of Environment regulations for the day—in a clay-lined surface impoundment pit on site, starting in the 1960s.

Then, in 1989, traces of chemical markers, notably the carcinogen N-nitrosodimethylamine (NDMA), began to appear in municipal drinking water wells downslope of the impoundment site. Apparently the clay liner of the impoundment pit had failed, perhaps because it had become saturated, and contaminants were leaking out of the pit and joining the ground water. The company undertook remediation of both the impoundment pit and the surrounding aquifers in the early 1990s. However, the town of Elmira no longer withdraws its drinking water from these aquifers; instead, it pipes water in from another municipality.

Deep-well injection The third method is intended for long-term disposal. In **deep-well injection**, a well is drilled deep beneath the water table into porous rock, and wastes are injected under pressure (**FIGURE 18.17**). The waste is meant to remain deep underground, isolated from ground water and human contact. This idea seems attractive in principle, but in practice wells become corroded and can leak wastes into soil, allowing them to enter aquifers. Deep-well injection also has been linked to earthquakes, resulting from the reactivation of deep-seated faults by pressurized fluids. Alberta accounts for approximately 90% of all deep-well injection of hazardous wastes in Canada.

Radioactive waste is especially hazardous

Radioactive waste is particularly dangerous to human health. It is, by definition, **radioactive** (that is, giving off energetic particles and radiation in the process of spontaneous radioactive decay) and thus potentially hazardous to human and animal health. In addition, some of it is highly toxic, and some of it is extremely persistent. The dilemma of disposal has dogged the nuclear energy industry for decades.

In Canada, nuclear waste is regulated by the Canadian Nuclear Safety Commission (CNSC). The government has signed a contract with the Nuclear Waste Management Organization (NWMO) to assume the phased-in responsibility for Canada's nuclear fuel waste. There is as yet no identified permanent repository for high-level radioactive waste in Canada; the Ontario Power Generation has proposed a permanent repository for low- and medium-level waste at the site of the Bruce Nuclear Power Plant in Tiverton, Ontario. Yucca Mountain in Nevada was approved as the single-site repository for all nuclear waste in the United States, but the licence application was withdrawn in 2010, partly because of environmental concerns. Other countries, including Germany, have designated permanent disposal sites for radioactive waste.

Most proposals for permanent disposal of high-level radioactive waste involve some form of **geological isolation**; that is, using the absorptive capacity and impermeability of some naturally occurring rock units to block the movement of contaminants away from the disposal site. Canadian proposals focus mainly on disposal in automated facilities located deep underground in the stable, ancient, plutonic igneous rocks of the Canadian Shield. Other geological settings that are amenable to geological isolation include salt domes and thick shale units.

Geological isolation would be combined with the chemical immobilization of the waste, sophisticated engineering design of the facility itself, and multiple-layered impervious containers for storage of the waste. In Canada, this is referred to as a **multiple-barrier approach**—that is, engineering the entire facility to place as many barriers as possible, both physical and chemical, in the pathway of any escaping contaminants.

Contaminated sites are being cleaned up, slowly

Many thousands of former military and industrial sites remain contaminated with hazardous waste in Canada and virtually every other nation on Earth. For most nations, dealing with these messes is simply too difficult, time-consuming, and expensive. Some contaminated sites, especially in the United States, have reached iconic status for the roles they played in raising public awareness, spurring local residents to action, and kick-starting the modern environmental movement. In Love Canal, a residential neighbourhood in Niagara Falls, New York, families were evacuated after toxic chemicals buried by a company and the city in past decades rose to the surface, contaminating homes and an elementary school. In Missouri, the entire town

of Times Beach—another name that is practically synonymous with poorly managed toxic waste—was evacuated and its buildings demolished after being contaminated by dioxin from waste oil sprayed on its roads. The beneficial outcome is that these horrific cases led to the establishment of the first legislation to deal with liability, compensation, and cleanup costs associated with contaminated sites.

In Canada, many contaminated sites have been abandoned and thus fall under federal jurisdiction. Approximately 18 000 sites are currently listed in Canada's Federal Contaminated Sites Inventory and have been assessed or classified under the National Classification System for contaminated sites, developed by the Canadian Council of Ministers of the Environment.[29] Examples of the nearly 1197 priority sites identified for cleanup include the following (costs may have changed as projects proceed and are reassessed):

- *Faro Mine, Yukon:* $14.6 million for the largest and highest priority of the federal contaminated sites, an old lead, zinc, silver, and gold mine that was shut down when the company went into receivership. There are now approximately 70 million tonnes of tailings and 320 million tonnes of waste rock at the site, with consequent acid drainage, wind-blown particulates, and other environmental hazards. The site has been assessed, and remediation plans are proceeding.[30]
- *Canadian Forces Base Esquimalt, British Columbia:* $4.56 million for remediation of hydrocarbon and heavy metal–contaminated soil, assessment, and risk management. In 2012 it was determined that an environmental assessment is required for the project.[31]
- *Port Radium Mine, Northwest Territories:* $7.1 million funding for sealing of mine openings, covering of areas of elevated radiation levels, stabilization of tailings areas, demolition, and hazardous waste disposal. The Government of Canada considers the remediation project to have been completed.[32]
- *Belleville Small Craft Harbour, Ontario:* $6.8 million to treat contaminated soil and prevent contaminants in ground water from discharging into the adjacent Bay of Quinte. Belleville Harbour was used for more than 50 years for the storage of coal and fuel products, which led to petroleum hydrocarbon and heavy metal contamination.[33]

You can examine the complete Federal Contaminated Sites Inventory for yourself. It is maintained (interestingly, perhaps because of the enormous costs associated with remediation of these sites) by the Treasury Board of Canada Secretariat, and the website is **www.tbs-sct.gc.ca/fcsi-rscf**.

Sites that have been contaminated but have the potential to be cleaned up and remediated for other purposes are called **brownfields**. However, many sites are contaminated with hazardous chemicals we have no effective way to deal with. In such cases, cleanups simply involve trying to isolate waste from human contact, either by building trenches and clay or concrete barriers around a site or by excavating contaminated material, placing it in industrial-strength containers, and shipping it to a hazardous waste disposal facility. For all these reasons, the current emphasis is on preventing hazardous waste contamination in the first place.

All three North American countries monitor industrial pollutants by using Pollutant Release and Transfer Registers (PRTRs), which combine reports from industrial facilities with information about transfers, off-site treatments, and disposal or recycling of pollutants. In Canada the PRTR is the National Pollutant Release Inventory (NPRI, **www.ec.gc.ca/inrp-npri/**), established in 1992, which covers more than 300 chemicals plus the criteria air contaminants. In Mexico, it is the *Registro de Emisiones y Transferencia de Contaminantes (RETC)*, which covers 100 chemicals, and in the United States it is the Toxics Release Inventory (TRI), started in 1987, which tracks data for more than 600 chemicals.[34] By learning where these pollutants come from, where they end up, and how they are transferred, we may ultimately be in a better position to control them.

Conclusion

Our societies have made great strides in addressing our waste problems. Modern methods of waste management are far safer for people and gentler on the environment than past practices of open dumping and open burning. In many countries, recycling and composting efforts are making rapid progress. Canada has changed in a few decades from a country that did virtually no recycling to a nation in which nearly one-quarter of all solid waste is diverted from disposal. The continuing growth of recycling, composting, and pollution-prevention initiatives, driven by market forces, government policy, and consumer behaviour, shows potential to further alleviate our waste problems.

Despite these advances, our prodigious consumption habits have created more waste than ever before. Our waste management efforts are marked by a number of difficult dilemmas, including the cleanup of highly contaminated sites, safe disposal of hazardous and radioactive waste, and frequent local opposition to disposal sites. These dilemmas make clear that the best solution to our waste problem is to reduce our generation of waste. Finding ways to reduce, reuse, and efficiently recycle the materials and goods that we use stands as a key challenge for this century.

REVIEWING OBJECTIVES

You should now be able to

Summarize and compare the types of waste we generate, the scale of the waste dilemma, and the major approaches to managing waste

- Waste is increasing everywhere as a result of growth in population and consumption, but developed nations generate far more waste than developing nations do.
- Municipal and industrial solid waste, hazardous waste, and waste water are major types of waste. Source reduction, recovery, and disposal are the three main components of waste management.

Describe conventional municipal waste disposal methods: landfills and incineration

- Municipal solid waste varies from place to place, both in amount and in composition.
- In Canada, waste is regulated by all three levels of government, each of which has its own regulatory tasks. Municipal solid waste is managed by cities, towns, and regional levels of government.

- In the past, people dumped their garbage wherever it was convenient. Those attitudes have changed, and most garbage is now managed in modern landfill facilities.
- Sanitary landfills are engineered to guard against contamination of ground water, air, and soil. Nonetheless, such contamination can occur.
- Landfills can be decommissioned and, in some cases, the land can be reclaimed for other uses after closure and remediation of the site.
- Incinerators reduce waste volume by burning it. Pollution control technology removes most pollutants from emissions, but some escape, and highly toxic ash needs to be disposed of in landfills.
- More and more facilities are harnessing energy from landfill gas and generating electricity from incineration.

Evaluate approaches for reducing waste: source reduction, reuse, composting, and recycling

- Source reduction, that is, reducing waste before it is generated, is the best waste management approach. Recovery is the next-best option.

- Consumers can take simple steps to reduce their waste output, including reusing items that are still functional instead of buying new ones.
- Composting reduces waste while creating organic matter for gardening and agriculture.
- Recycling has grown slowly in recent years and now removes about 33% of household waste from the Canadian waste stream.
- Regulations, including waste limits, and financial incentives, such as tipping fees and charges for excess waste, can help modify behaviours at the household and commercial contexts.

Discuss industrial solid waste management and principles of industrial ecology

- Regulations and financial incentive structures differ, but light industrial waste management is similar to that for municipal solid waste. Agriculture, mining, and oil extraction produce an enormous quantity of industrial solid waste each year.
- Industrial ecology urges industrial systems to mimic ecological systems and provides ways for industry to increase its efficiency.
- Some applications of industrial ecology principles intended to reduce waste and harmful environmental impacts include life-cycle analysis, pollution prevention (P2) strategies, embodied energy and water analysis, and the Extended Producer Responsibility and Stewardship (EPRS) program.

Assess issues in managing hazardous waste and e-waste

- Hazardous wastes are flammable, corrosive, reactive, or toxic. They come from diverse sources, including industry and households.
- Organic compounds and heavy metals are particularly hazardous to human health; they also typically bioaccumulate and biomagnify.
- Electronic waste has several hazardous components, and is just now beginning to be adequately controlled.
- Hazardous waste is strictly regulated, yet illegal dumping remains a problem.
- There are three main approaches to the disposal of hazardous wastes: secure landfills, surface impoundment, and deep-well injection.
- Radioactive waste is particularly hazardous and requires specific approaches; there is as yet no permanent facility for radioactive waste disposal in Canada or the United States.
- Cleanup of hazardous waste sites is a long and expensive process. Old, abandoned sites are slowly undergoing remediation.

TESTING YOUR COMPREHENSION

1. Describe five major methods of managing waste. Why do we practise waste management?
2. Why have some people labelled modern North America as "the throwaway society"? How much solid waste do Canadians generate, and how does this amount compare with that of people from other countries?
3. Name several technologies designed to make sanitary landfills safe places for the disposal of waste. Describe three problems with landfills.
4. Describe the process of incineration or combustion. What happens to the resulting ash? What is one major drawback of incineration?
5. What is composting, and how does it help reduce input to the waste stream?

6. What are the three elements of a sustainable process of recycling?
7. What are the goals of industrial ecology?
8. What four criteria are used to define hazardous waste? Why are heavy metals and synthetic organic compounds particularly hazardous?
9. What are the largest sources of hazardous waste? Describe three ways to dispose of hazardous waste.
10. How is waste regulated in Canada? What are some of the similarities and differences between the regulation of nonhazardous wastes and that of hazardous, biomedical, and radioactive wastes?

THINKING IT THROUGH

1. How much waste do you generate? Look into your waste bin at the end of the day, and categorize and measure the waste there. List all other waste you may have generated in other places throughout the day. How much of this waste could you have avoided generating? How much could have been reused or recycled?

2. Of the various waste management approaches covered in this chapter, which ones are your community and campus pursuing, and which are they not pursuing? Would you suggest that your community or campus start pursuing any new approaches? If so, which ones, and why?

3. You are the CEO of a corporation that produces containers for soft drinks and a wide variety of other consumer products. Your company's shareholders are asking that you improve the company's image—while not cutting into profits—by taking steps to reduce waste. What steps would you consider taking?

4. Now let's say that you are the president of your college or university. Your trustees want you to engage with local businesses and industries in ways that benefit both the school and the community. Your faculty and students want you to make the school a leader in waste reduction and industrial ecology. Consider the industries and businesses in your community and the ways they interact with facilities on your campus. Bearing in mind the principles of industrial ecology, can you think of any novel ways in which your school and local businesses might mutually benefit from one another's services, products, or waste materials? Are there waste products from one business, industry, or campus facility that another might put to good use? Can you design an eco-industrial park that might work on your campus? What steps would you propose to take as president?

INTERPRETING GRAPHS AND DATA

As of 2010, more than 90% of all households in Canada reported using some sort of recycling program.[35] This sounds like fantastic news—the "reduce, reuse, recycle" message appears to have reached the population, and the work of protecting the environment is mostly done. Is this true? Find out by taking a closer look at Canadian statistics.

The following table shows how many kilograms of residential waste were produced per person and how much waste was diverted from landfill sites through recycling in Canada in 2006. It also shows the percentage of households that used any type of recycling program in 2006.

Definitions

- **Waste disposed:** waste that is landfilled, incinerated, or treated for final disposal (does not include materials destined for recycling and composting)

- **Recyclable materials diverted:** materials diverted from the waste stream and remanufactured into a new product or used as a raw material substitute

- **Total materials discarded:** the combined total amount of waste disposed and recyclable materials diverted

Disposal and Diversion of Residential Waste and Household Use of Recycling Programs, by Province and Territory, 2006

	1	2	3	4	5
	Total residential waste disposed,[36] per capita	Residential recyclable materials diverted,[37] per capita	Total materials discarded, per capita (Column 1 + Column 2)	Diversion rate	Households that used any recycling program in 2006[38]
	Kilograms per capita				Percent %
Newfoundland and Labrador	446	×*			82
Prince Edward Island	×	×			98
Nova Scotia	181	149			95
New Brunswick	289	44			83
Quebec	285	122			86
Ontario	292	119			93
Manitoba	386	60			79
Saskatchewan	300	39			87
Alberta	289	98			85
British Columbia	222	145			93
Yukon	214	×			—
Northwest Territories	347	×			—
Nunavut	×	×			—
Canada	283	115			90

*"x" means data is suppressed to meet the confidentiality requirements of the *Statistics Act*.
Source: From Statistics Canada.

Use the table to answer these questions:

1. For each province and territory, calculate the total materials discarded per capita, including waste disposed and recyclable materials diverted. Record your answers in column 3 of the table.

2. For each region, calculate the diversion rate (the amount of residential recyclable materials diverted, per capita, compared to total materials discarded per capita), using the following formula:

$$\text{Diversion rate} = \frac{\text{Column 2}}{\text{Column 1}} \times 100$$

Record your answers in column 4 of the table.

3. Which province had the highest diversion rate?

4. Do you think there is a connection between the diversion rate and the percentage of households that use a recycling program in the provinces and territories? Explain your answer.

5. Create a scatter plot to compare the percentages of households that used a recycling program (column 5 from the table above) with the diversion rates (column 4). Identify each data point by labelling it with the provincial or territorial abbreviation (e.g., MB for Manitoba, NB for New Brunswick).

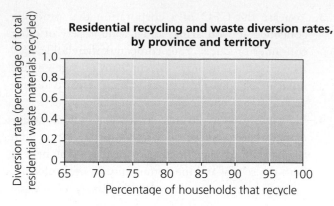

Source: From Statistics Canada.

6. Draw the line of best fit (the trend line that fits the majority of data points) for the above graph.

7. Describe the trend in the graph. Does this match your prediction in question 4?

8. How can you explain the connection between the percentage of households that recycle and the diversion rate?

MasteringEnvironmentalScience®

STUDENTS
Go to **MasteringEnvironmentalScience** for assignments, the eText, and the Study Area with practice tests, videos, current events, and activities.

INSTRUCTORS
Go to **MasteringEnvironmentalScience** for automatically graded activities, current events, videos, and reading questions that you can assign to your students, plus Instructor Resources.

19 Environmental Health and Hazards

Plastic pollution is now ubiquitous. Of particular concern lately is microplastics—plastic "microbeads" and extremely tiny eroded particles of plastic.

NOAA Marine Debris Program

Upon successfully completing this chapter, you will be able to

- Identify the major types of environmental health hazards and the goals of environmental health
- Describe the types, abundance, distribution, and movement of toxicants in the environment
- Summarize how hazards and contaminants are studied scientifically, including case histories,

epidemiology, animal testing, and dose–response analysis
- Summarize approaches to risk assessment and risk management
- Compare philosophical approaches to risk and their role in environmental health policy

Plastic microbeads like the ones shown here are added to shower gels, toothpaste, and facial scrubs, as seen in the product label (inset photo). Concern is rising about the accumulation of microplastics in the environment, and the possible health impacts. Some cosmetics companies have begun voluntary initiatives to eliminate microbeads from consumer products.

CENTRAL CASE
MICROPLASTICS: BIG CONCERNS ABOUT TINY PARTICLES

"Over the past 50 years we have all been unwitting participants in a vast, uncontrolled, worldwide chemistry experiment involving the oceans, air, soils, plants, animals, and human beings."

—UNITED NATIONS ENVIRONMENT PROGRAMME, IN *A GUIDE TO THE STOCKHOLM CONVENTION ON PERSISTENT ORGANIC POLLUTANTS*

[Moved that,] "in the opinion of the House, microbeads in consumer products entering the environment could have serious harmful effects, and therefore the government should take immediate measures to add microbeads to the list of toxic substances managed by the government under the *Canadian Environmental Protection Act, 1999*."

—MOTION TO THE HOUSE OF COMMONS OF THE GOVERNMENT OF CANADA ON 24 MARCH 2015, BY MEGAN LESLIE (HALIFAX, NDP)

P lastic microbeads are tiny spheres made of polyethylene (PE), polypropylene (PP), or other types of polymer plastic; most are less than 100 μm, or 0.1 mm in size. Since the late 1990s they have been added to personal care products such as shower gels, toothpaste, and facial scrubs—either to create a silky feel, or for their exfoliant properties (see photos above). They are now found in literally thousands of products, and can make up ~10% of the product by volume. Health Canada has deemed them to be safe for use in cosmetics, toothpaste, and even foods. But problems arise when they leave our houses as part of our waste water. Of particular concern recently are the environmental impacts of plastic microbeads entering and accumulating in sediment at the bottom of water bodies and in aquatic and marine food chains.

Due to their tiny size, microbeads in the water waste stream are not captured by filters in sewage treatment plants, and thus directly pass into lakes or oceans. After a heavy rain, waste water with microbeads can also overflow directly into surface waters. As a result, microbeads are making up an increasing amount of the plastic pollution—dubbed "plastic soup"—in both freshwater and marine systems.

University of Auckland researchers Lisa Fendall and Mary Sewell did a study in which they observed that tiny planktonic organisms at the base of ocean food chains readily ingest microbeads.[1] While this is damaging in itself, microbeads also attract and absorb persistent organic pollutants (POPs)[2], such as polychlorinated biphenyls (PCBs) and dichlorodiphenyltrichloroethane (DDT), concentrating and transporting these toxic substances into the food chain. This leads to additional concerns about the health impacts for people who consume fish and other aquatic organisms that may contain concentrations of microbeads.

The problem is not restricted to oceans, however. In 2014, Anthony Ricciardi and his research group at McGill University found over 1000 microbeads per litre of sediment in some locations in the St. Lawrence River—amounts similar to what has been found in the most contaminated ocean sediments.[3] One research group sampled surface waters in the Great Lakes and found high concentrations of "multicoloured spheres" from consumer cosmetic products; one sampling area, downstream from an urban area, contained almost half a million microplastic particles per square kilometre.[4] Another research group has found high concentrations of microplastics (in this case, airborne eroded fragments of plastic rather than microbeads) in a remote mountain lake in Mongolia, testifying to the ubiquity of microplastic pollution.[5]

Several nongovernmental organizations and environmental lobby groups have come together to put pressure on industry and government policymakers to ban plastic microbeads from consumer products. In Netherlands, The North Sea Foundation and Plastic Soup Foundation launched the "Beat the Microbead" campaign to bring awareness to consumers, even creating a smartphone app that allowed consumers to check products for microbeads.[6] A Twitter campaign targeted at Unilever resulted in a promise to phase microbeads out of their products worldwide. Germany soon followed the Netherlands, with many enterprises also promising to phase out microbeads. By mid-2013, Johnson & Johnson, Procter & Gamble, The Body Shop,

and IKEA all voluntarily committed to stop having microbeads in their products.

These strong, voluntary responses from industry leaders helped mobilize policymakers. In March 2013, the European Commission published the *Green Paper on a European Strategy on Plastic Waste in the Environment*, specifically addressing the issue of microbeads.[7] (Although the name "Green Paper" seems to suggest a document about the environment, it's not; it is the name given to a document intended to stimulate discussion, begin a consultation process, and make proposals. These proceedings may lead to a policy document called a "White Paper," which is typically consulted when drafting legislation.)

In early 2014, the nongovernmental organization 5Gyres collaborated with Tulane Law School to create model legislation prohibiting the sale of personal care products containing plastic microbeads. Soon after, the *Microbead-Free Waters Act* passed the New York State Assembly by a vote of 108 to 0. By March 2015, Illinois, New Jersey, Colorado, and Maine passed bills into law banning the sale of products with microbeads by 2019 at the latest.

Canada, however, still lags behind in policy and legislation. The process is somewhat complicated by the fact that waste management in Canada is mostly a provincial and territorial issue, not a federal issue. On March 24, 2015, the House of Commons debated a motion to ban plastic microbeads from consumer products, brought forward by the opposition (quoted, see page 598). The opposition asked the government to apply the "precautionary principle," and to classify microbeads as a "toxic substance" under the *Environmental Protection Act*. This would give the federal government the authority to control or prohibit their use, including a complete ban on their use in consumer products.

In response to the motion, the government committed to discuss the issue at an upcoming Environment Ministers' meeting. Environment Canada was tasked with undertaking a review of scientific evidence, and the government indicated that they are watching closely what actions have been taken by various states in the United States toward banning microbeads.[8]

Developing sound environmental policy on harmful substances in the environment is a moving target, and it is important for policies to be founded in science. But new technologies, products, and processes are creating a constantly changing stream of new chemicals and substances entering our soil, water, and air; it is difficult for both scientists and policymakers to keep up with all of the potential impacts.

The concern among scientists and the general public is now quite strong regarding the hazards associated with microbeads, and policymakers are beginning to listen to those concerns. Time will tell what action, if any, Canada will take on limiting the use of these substances.

Environmental Health

The study and practice of **environmental health** assesses factors in both the natural and built environments that affect human health and quality of life. Examining the impacts of human-made chemicals and materials such as microplastics on both people and wildlife is an important aspect of the broad field of environmental health. Environmental factors that influence health include wholly natural aspects of the environment over which we have little or no control, as well as *anthropogenic* (human-caused) factors for which we ourselves are responsible. Practitioners of environmental health seek to prevent adverse effects on human health and on the ecological systems that are essential to environmental quality and human well-being.

Environmental hazards can be chemical, physical, biological, or cultural

There are innumerable environmental health threats, or hazards, in the world around us; we are exposed to them every day. Some—like earthquakes, heat waves, poison ivy, and venomous snakes—occur naturally. Others arise from our exposure to activities and materials that are wholly anthropogenic, such as smoking, or airplane flight, or synthetic chemicals like DDT. And some hazards are halfway between—a wide array of hazardous processes that are mostly natural but are triggered or exacerbated by human activity. Examples would be a landslide triggered by an oversteep road cut, or a naturally occurring substance, such as the mineral asbestos, that can become hazardous when it is used indoors in the built environment.

For ease of understanding, we can categorize hazards into four main types: chemical, physical, biological, and cultural (**FIGURE 19.1**). For each of these four types of hazards, there is some amount of risk that we cannot avoid—but there is also some amount of risk that we

can avoid by taking precautions. Much of environmental health consists of taking steps to minimize the impacts of hazards and the risks of encountering them.

Chemical hazards   **Chemical hazards** include spills and other exposures to the many synthetic chemicals that our society produces, such as disinfectants, plastics, pharmaceuticals, and pesticides (**FIGURE 19.1A**). Chemicals produced naturally by organisms, such as snake venom or poisonous alkaloid compounds produced by plants as defence mechanisms, also can be hazardous. Toxic metals and other potentially harmful chemical substances occur naturally in rocks and soils, too, such as arsenic, chromium, mercury, and many others. Following our overview of environmental health and the various types of hazards, much of the remainder of this chapter will focus on chemical health hazards and the ways people study and regulate them.

Physical hazards   **Physical hazards** can arise from processes that occur naturally and pose risks to human life, health, and property (and to the natural environment as well). These include earthquakes, volcanic eruptions, fires, floods, blizzards, landslides, hurricanes, and droughts (**FIGURE 19.2**).

Most earthquakes are barely perceptible, but occasionally they are powerful enough to do tremendous damage to human life and property (**TABLE 19.1**). Damage is generally greatest where soils are loose or saturated with water; areas built on landfill are particularly susceptible, as witnessed during the 1995 earthquake in Kobe, Japan. To minimize damage, engineers have developed ways to protect buildings from shaking. They do this by strengthening structural components while designing points at which a structure can move and sway harmlessly with ground motion. Just as a flexible tree trunk bends in a storm whereas a brittle one breaks, buildings with built-in flexibility are more likely to withstand an earthquake's violent shaking. Such designs are now an important part of building codes in quake-prone regions.

(a) Chemical hazard

(b) Physical hazard

(c) Biological hazard

(d) Cultural hazard

FIGURE 19.1

Environmental health hazards can be divided into four types. Chemical hazards **(a)** include both artificial and natural chemicals. Much of our exposure comes from household chemical products, such as pesticides. The sun's ultraviolet radiation is an example of a physical hazard **(b)**; excessive exposure increases the risk of skin cancer. Biological hazards **(c)** include exposure to organisms that transmit disease, such as mosquitoes that transmit malaria. Cultural or lifestyle hazards **(d)** include decisions we make about how to behave, as well as constraints forced on us by socioeconomic factors. Smoking, for example, is a lifestyle choice that raises one's risk of lung cancer and other diseases.

Erupting volcanoes affect both people and the natural environment. At some volcanoes, lava flows slowly downhill, such as Mount Kilauea in Hawaii, which has been erupting continuously since 1983. Other volcanoes may let loose large amounts of ash and cinder in a sudden explosion, such as during the 1991 eruption of Mount Pinatubo in the Philippines. Besides affecting people, volcanic eruptions exert environmental impacts (**TABLE 19.2**). Ash blocks sunlight, and sulphur emissions lead to a sulphuric acid haze that blocks radiation and cools the atmosphere. Large eruptions can depress temperatures throughout the world. When Indonesia's Mount Tambora erupted in 1815, it cooled average global temperatures by 0.4°C to 0.7°C, enough to cause crop failures worldwide and make 1816 "the year without a summer."

Earthquakes, volcanic eruptions, and large coastal landslides can all displace huge volumes of ocean water instantaneously, triggering a massive tsunami, such as the one that devastated coastlines all around the Indian Ocean in 2004, killing about 230 000 people. In addition to the enormous human toll, coral reefs, coastal forests, and wetlands were damaged, and salt water contaminated soil and aquifers.

Those of us who live in Canada should not consider tsunamis to be something that occurs only in faraway places; a large tsunami struck North America's Pacific coast in 1700 following a huge earthquake in the Pacific

TABLE 19.1 Examples of Large or Recent Earthquakes

Location	Year	Fatalities	Magnitude[1]
Shaanxi Province, China	1556	830 000	~8
Lisbon, Portugal	1755	70 000[2]	8.7
San Francisco, California	1906	3 000	7.8
Kwanto, Japan	1923	143 000	7.9
Anchorage, Alaska	1964	128[2]	9.2
Tangshan, China	1976	255 000+	7.5
Michoacan, Mexico	1985	9 500	8.0
Kobe, Japan	1995	5 502	6.9
Northern Sumatra	2004	~230 000[2]	9.1
Kashmir, Pakistan	2005	86 000	7.6
Sichuan Province, China	2008	50 000+	7.9
Port-au-Prince, Haiti	2010	236 000	7.0
Maule, Chile	2010	521	8.8
Tōhoku, Japan	2011	15 890[2]	9.0

[1]Measured by moment magnitude; each full unit is roughly 32 times as powerful as the preceding full unit.
[2]Includes deaths from the resulting tsunami.

FIGURE 19.2 This map shows the geographical distribution of significant natural disasters in Canada.
Source: Natural Hazards, Courtesy of Canadian Geographic.

TABLE 19.2 Examples of Notable Volcanic Eruptions

Location	Year	Impacts	Magnitude*
Yellowstone Caldera, Wyoming, United States	640 000 BP**	Most recent "mega-eruption" at site of Yellowstone National Park	8
Mount Mazama, Oregon, United States	6870 BP	Created Crater Lake	7
Mount Vesuvius, Italy	79 CE***	Buried Pompeii and Herculaneum	5
Mount Tambora, Indonesia	1815	Created "the year without a summer"; killed at least 70 000 people	7
Krakatau, Indonesia	1883	Killed over 36 000 people; heard 5000 km away; affected weather for 5 years	6
Mount Saint Helens, Washington, United States	1980	Blew top off mountain; sent ash 19 km into the sky; 57 people killed	5
Kilauea, Hawaii, United States	1983–present	Continuous lava flow; threatens towns	1
Mount Pinatubo, Philippines	1991	Sulphuric aerosols lowered world temperature 0.5 °C for two years	6
Eyjafjallajökull, Iceland	2010	Ash cloud disrupted air travel throughout Europe	1

*Measured by the Volcanic Explosivity Index, which ranges from 0 (least powerful) to 8 (most powerful)

**BP = years before the present

***CE = common era

Northwest. This could happen again if a large earthquake were to occur in the subduction zone of the Juan da Fuca Plate, offshore from Vancouver Island. Since the 2004 tsunami, nations and international agencies have stepped up efforts to develop systems to give coastal residents advance warning of approaching tsunamis. We can lessen the impacts of tsunamis if we preserve natural shorelines, which help protect coastal areas by absorbing wave energy.

Aside from these geological hazards, people face other types of natural hazards, many of which are linked to weather and climate. Heavy rains can lead to flooding that ravages low-lying areas near rivers and streams; drought can devastate agricultural fields; coastal erosion can eat away at beaches; wildfires can threaten life and property in fire-prone areas; tornadoes and hurricanes can cause extensive property damage, ecological damage, and loss of life.

We can do little to predict the timing of a natural disaster, such as an earthquake, and nothing to prevent one. However, scientists can map geological faults to determine areas at risk of earthquakes; engineers can design buildings in ways that help them resist damage; and citizens and governments can take steps to prepare for the aftermath of a severe quake. Other types of natural disasters are more amenable to prediction, forecasting, and early warning. Meteorologists have combined their skills and knowledge with sophisticated GIS-based terrain models to improve the forecasting of major storms and floods, for example.

Other physical hazards arise from our exposure to ongoing natural processes, such as ultraviolet (UV) radiation from sunlight (**FIGURE 19.1B**). Excessive exposure to UV radiation damages DNA, and has been tied to skin cancer, cataracts, and immune suppression in humans. We can reduce UV exposure and risk by using clothing and sunscreen to shield our skin from intense sunlight.

Some regions are naturally prone to certain types of hazards, by virtue of their geology, topography, climate, and other physical characteristics. For example, you can see from **FIGURE 19.2** that the West Coast, southern Ontario, and south-central Quebec are the most vulnerable to earthquakes, whereas the Rockies and West Coast are susceptible to landslides, and the Prairies and southern Ontario to tornadoes. However, some common processes and practices increase our vulnerability to physical hazards in Canada, and around the world:

■ As our human population grows, more people live in areas susceptible to natural hazards.

■ Many of us choose to live in areas that are prone to specific hazards. For instance, coastlines are vulnerable to tsunamis and erosion by storms, and mountainous areas are prone to landslides.

■ We engineer landscapes around us in ways that can increase the frequency or severity of natural hazards. Damming and diking rivers to control floods can sometimes lead to catastrophic flooding, and suppressing natural wildfires puts forests at risk of larger, truly damaging fires. Mining, clearing forests for agriculture, and clear-cutting on slopes can increase susceptibility to landslides, increase runoff, compact soil, and change drainage patterns.

■ As we change Earth's climate by emitting greenhouse gases, we are altering patterns of precipitation, which, in turn, increases risks of drought, fire, flooding, and mudslides locally and regionally. Rising sea levels induced by global warming increase susceptibility to coastal erosion. Warming ocean temperatures may increase the power and duration of hurricanes.

We can reduce or mitigate the impacts of hazards through the thoughtful use of technology, engineering, and policy, informed by a solid understanding of geology and ecology. Examples include building earthquake-resistant structures; designing early warning systems for tsunamis, floods, and volcanic eruptions; and conserving reefs and shoreline vegetation to protect against tsunamis and coastal erosion. In addition, better forestry, agriculture, and mining practices can reduce exposure of soil to the atmosphere, which can help prevent landslides of steep slopes. Zoning regulations, building codes, and insurance incentives that discourage development in areas prone to landslides, floods, fires, and storm surges can help keep us out of harm's way, and can decrease taxpayer expense in cleaning up after natural disasters. Finally, mitigating global climate change may help reduce the frequency of natural hazards in many regions.

Biological hazards **Biological hazards** result from ecological interactions among organisms (**FIGURE 19.1C**). When we become sick from a virus, bacterial infection, or other **pathogen**, we are suffering parasitism by other species that are simply fulfilling their ecological roles. This is *infectious disease*, also called *communicable* or *transmissible disease*. Infectious diseases, such as malaria, cholera, tuberculosis, and influenza (flu), all are considered environmental health hazards. As with physical and chemical hazards, it is impossible for us to avoid risk from biological agents completely, but we can take steps to reduce the likelihood of infection.

Cultural hazards Hazards that result from the place we live or our socioeconomic status, occupation, or behavioural choices can be thought of as **cultural hazards** or **lifestyle hazards**. For instance, engaging in extreme sports increases the risk of injury. Choosing to smoke cigarettes, or living or working with people who smoke, greatly increases the risk of lung cancer (**FIGURE 19.1D**). Choosing to smoke is a personal behavioural decision, but exposure to secondhand smoke in the home or workplace may not be under one's control.

Much the same might be said for diet and nutrition, crime, high-risk occupations, transportation choices, and even drug use, all of which have elements of choice but are strongly influenced by socioeconomic circumstances. Simply living or working in a large urban centre can expose people to unexpected hazards, such as **noise pollution**, which results from undesired ambient sound. Excess noise degrades one's surroundings aesthetically, can induce stress, and at intense levels can even harm hearing. **Light pollution** in urban areas is also of increasing concern. The glow of city lights obscures the night sky and disorients birds, which die by the thousands each year as a result of crashing into illuminated buildings. Excessively bright or inappropriate lighting can cause stress and can disrupt circadian rhythms in humans and other animals. *Dark-sky reserves* have been established in some areas to combat light pollution.

The various forms of exposure to cultural hazards and the health and environmental threats they pose are not evenly shared. Those who bear the brunt of urban and cultural hazards are often (though not always) those who are too poor to live in cleaner or safer areas. As advocates of environmental justice argue, such health-determining factors as living in proximity to a toxic waste dump, making poor nutritional choices, or having a risky job are often correlated with lack of education and awareness, and with socioeconomic deprivation.

Disease is a major focus of environmental health

Among the hazards people face, disease stands preeminent. Despite all our technological advances, we still find ourselves battling disease, which causes the vast majority of human deaths worldwide (**FIGURE 19.3A**). Many major killers, such as cancer, heart disease, stroke, and respiratory disorders, have genetic bases but are also influenced by environmental factors. For instance, whether a person develops asthma is influenced not only by genes but also by environmental conditions. Pollutants from fossil fuel combustion worsen asthma, and fewer children raised on farms suffer asthma than children raised in cities, studies have shown. Malnutrition can foster a wide variety of illnesses, as can poverty and poor hygiene. Moreover, lifestyle choices can affect risks of acquiring some noncommunicable diseases: Smoking can lead to lung cancer, and lack of exercise can increase the risk of heart disease, for example.

Infectious diseases today account for more than 18% of deaths that occur worldwide each year—nearly 10 million people[9] (**FIGURE 19.3B**). Some pathogenic microbes attack us directly, and sometimes infection occurs through a **vector**, an organism that transfers the pathogen to the host. Mosquitoes are common vectors; they can carry malaria, West Nile virus, and many other diseases. So are rats; they can carry Hanta virus, and *Yersinia pestis*, which causes bubonic plague. Many infectious agents are transported by water, or spend part or all of their life-cycle in or

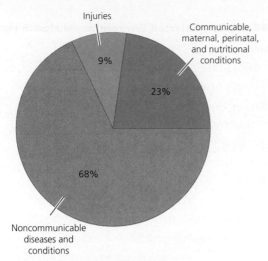

(a) Causes of death, worldwide

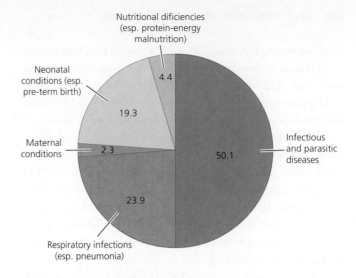

(c) Top causes of death: communicable, maternal, neonatal, and nutritional diseases and conditions

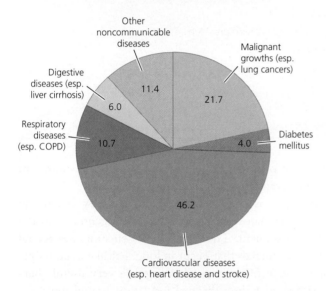

(b) Top causes of death: noncommunicable diseases and conditions

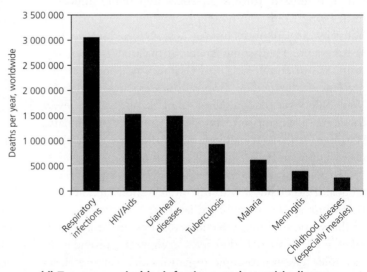

(d) Top communicable, infectious, and parasitic diseases

FIGURE 19.3 The main cause of deaths worldwide, overall **(a)**, is noncommunicable diseases and conditions; the main ones **(b)** are cancers (especially lung cancer) and cardiovascular diseases (especially heart disease and stroke). Cardiovascular diseases, as a group, account for more than 31% of all deaths per year. Communicable, maternal, neonatal, and nutritional diseases **(c)** cause 23% of deaths worldwide. Of these, seven types of infectious and parasitic diseases—respiratory infections, HIV/AIDS, diarrhea, tuberculosis (TB), malaria, meningitis, and childhood diseases, especially measles, are the most important, together causing more than 8 million deaths each year **(d)**.
Source: Data are from World Health Organization (2014). *Global Health Estimates (GHE) 2014: Deaths by Age, Sex, and Cause. Global Summary Estimates.* Retrieved from http://www.who.int/healthinfo/global_burden_disease/estimates/en/index1.html.

DATA
Q
Here are some additional statistics about some of the diseases represented in **FIGURE 19.3**. Complete the parts of the table that are missing. Then answer the following questions:

2000			2012		
Cause	**Deaths (000s)**	**% Deaths**	**Cause**	**Deaths (000s)**	**% Deaths**
Coronary artery disease and stroke		22.0	Coronary artery disease and stroke	14 027	25.1
HIV/AIDS	1 678		HIV/AIDS	1 534	
Diarrheal diseases	2 171		Diarrheal diseases		2.7
All causes	**52 806**		**All causes**	**55 859**	

1. Has the number of deaths overall increased between 2000 and 2012? Let's say that the global population in 2000 was about 6.12 billion, and in 2012 it was 7.08 billion. Did the death *rate* (that is, number of deaths in the population as a whole for that year) increase or decrease between 2000 and 2012?

2. Has the percentage of deaths caused by coronary heart disease and stroke increased or decreased? What about HIV/AIDS and diarrheal diseases? To what might you attribute the change in the rate of deaths from diarrhea between 2000 and 2012?

near the water. Infectious diseases account for close to half of all deaths in developing countries but very few deaths in developed nations. This discrepancy is due to differences in hygiene conditions and access to nutrition, health care, and medicine, which are tightly correlated with wealth.

Public health efforts have lessened the impact of infectious disease in developed nations and even have eradicated some diseases, notably smallpox. Nevertheless, other diseases—among them tuberculosis, cholera, dengue fever, plague, and malaria—are increasing. Others, such as the ebola virus that broke out in 2013 in West Africa and the avian flu that began spreading worldwide in 2005, remain as threats for a possible global epidemic.

Many diseases are spreading because of the mobility we have achieved in our era of globalization. A virus for influenza or SARS (Severe Acute Respiratory Syndrome), which paralyzed Toronto in 2003 after being imported from Hong Kong, can now hop continents in a matter of hours by airplane in its human host. Other diseases, such as tuberculosis and some strains of malaria, are evolving resistance to antibiotics, in the same way as pests evolve resistance to pesticides. Tropical diseases, such as malaria, West Nile virus, dengue fever, cholera, and yellow fever, threaten to expand into temperate zones with climatic warming. And habitat alteration can affect the abundance, distribution, and movement of certain disease vectors.

To predict and prevent infectious diseases, environmental health experts deal with the often complicated interrelationships among technology, land use, and ecology. One of the world's leading infectious diseases, malaria (which takes more than 600 000 lives each year), provides an example. The microscopic organisms that cause malaria (four species of *Plasmodium*) depend on mosquitoes as a vector. These microorganisms can sexually reproduce only within a mosquito, and it is the mosquito that injects them into a human or other host. Thus the primary mode of malaria control has been to use insecticides, such as DDT, to kill mosquitoes. Large-scale eradication projects involving insecticide use and draining of wetlands have removed malaria from areas of the temperate world where it used to occur, such as throughout the southern United States. However, land uses that create pools of standing water in formerly well-drained areas can boost mosquito populations, and warmer winters allow wintering-over of mosquito populations, potentially allowing malaria and other vector-borne diseases to reinvade.

Environmental health hazards exist indoors as well as outdoors

Outdoor hazards are generally more familiar to us, but we spend most of our lives indoors. Therefore, we must consider the spaces inside our homes and workplaces to

TABLE 19.3 Selected Environmental Health Hazards

Air
Smoking and secondhand smoke
Chemicals from automotive exhaust
Chemicals from industrial pollution
Tropospheric ozone (smog)
Pesticide drift
Dust and particulate matter
Water
Pesticide and herbicide runoff
Nitrates and fertilizer runoff
Mercury, arsenic, and other heavy metals in ground water and surface water
Food
Natural toxins
Pesticide and herbicide residues
Indoors
Asbestos
Radon
Lead in paint and pipes
Toxicants (such as VOCs) in paints, plastics, and consumer products

be part of our environment and, as such, the sources of potential environmental hazards (**TABLE 19.3**).

Asbestos is one example of a naturally occurring mineral that can be hazardous when concentrated in the built environment. The term *asbestos* encompasses several groups of minerals that occur in long, hairlike microscopic fibres. These fibres give asbestos some very useful characteristics, including thermal insulation, sound muffling, tensile strength, and fire resistance, so it was used widely in buildings and products, beginning in the 1870s. Two main groups of minerals produce asbestos fibres—the amphibole group and the serpentine group (specifically, the mineral chrysotile). Until recently, Canada was the world's largest producer and exporter of chrysotile asbestos, mainly from open-pit mines clustered in the Eastern Townships of southern Quebec.

Unfortunately, the fibrous structure of asbestos that gives it its tensile strength and insulating properties also makes it dangerous when inhaled. When lodged in lung tissue, asbestos induces the body to produce acid to combat it. The acid scars the lung tissue but doesn't dislodge or dissolve the asbestos. Within a few decades the scarred lungs may cease to function, a disorder called *asbestosis*. The rare but highly aggressive cancer *mesothelioma* is caused primarily by exposure to asbestos; Canada has one of the highest mesothelioma rates in the world. There is widespread agreement that exposure to amphibole asbestos is extremely hazardous, but for many decades there was both political and scientific controversy over the safety of chrysotile asbestos.[10] For some, this was mainly a scientific debate; for others, it seemed like government agencies were ignoring the scientific evidence for political and financial purposes, as the federal government stepped in repeatedly to support the asbestos mining industry.

Because of the health risks, asbestos has been removed from most schools, offices, and homes in North America, often at great cost. In some cases, however, removing asbestos can be more hazardous than leaving it in place, if dry asbestos fibres are exposed, disturbed, or displaced during removal. Once asbestos enters the air circulation system of a building it can be almost impossible to eliminate. To prevent this, the decision is sometimes taken to leave the material and encapsulate it.

Risks from asbestos exposure are greatest for miners and those involved in the production of asbestos products, and their families (because the fibres can be carried home on the miners' clothing). There are also concerns for residents of towns like Thetford Mines and Asbestos, Quebec, where everyday, lifelong exposures have occurred. Since asbestos-related health risks arise from inhalation, the risk depends on the concentration of the fibres in the air, how long the exposure lasts, how often exposure occurs, the size of the fibres inhaled, and the amount of time since the initial exposure (because some asbestos-related diseases are dormant for many years after the initial exposure). After years of pressure to stop producing and exporting asbestos, the last Canadian asbestos mines were shut down in 2011–2012. Mining of asbestos is not banned in Canada, but the market for the product no longer supports mine operations.

Radon is another example of an indoor hazard that is of natural, outdoor origin. *Radon* is a radioactive gas that is colourless, odourless, and undetectable without specialized kits. It is most common in areas with bedrock that contains a lot of uranium. Uranium is naturally radioactive, and one of its decay products is radon. Since it comes from minerals, radon commonly occurs as a soil gas. From the soil, it seeps out into the air, where it is dispersed into the atmosphere; for this reason, radon is generally not hazardous outdoors. However, it can become problematic if it enters a building, such as through cracks in the foundation. Radon is a heavy gas—heavier than air—so it tends to settle in basements, where it continues to decay radioactively and can be inhaled by people occupying the space. Exposure to radon over long periods is thought to be the second most important cause of lung cancer, after smoking. It can be detected by home testing and remediated by improving the air circulation in the building.

Smoking—a lifestyle choice—exacerbates the effects of exposure to radon and other harmful substances, especially in the indoor environment. This type of combined factor makes it very complicated to estimate the number of deaths attributable to a single cause, such as radon exposure. For example, according to Health Canada, the risk of getting lung cancer is 1 in 10 for a lifelong smoker. If the smoker is also exposed to a high level of radon over much of his or her lifetime, the risk becomes 1 in 3. For nonsmokers, in contrast, the lifetime lung cancer risk at the same high level of radon exposure would be about 1 in 20.[11] This is further complicated by the fact that a person might be exposed to radon or secondhand smoke (or both) in the workplace or in the home (or both).

Lead poisoning represents another indoor health hazard; it, too, originates in the natural environment but has been incorporated in products that are used in everyday life. *Lead* is a heavy metal. When ingested, especially over time, it can cause damage to the brain, liver, kidney, and stomach; learning problems and behavioural abnormalities; anemia; hearing loss; and even death. It has been suggested that the downfall of ancient Rome was caused in part by chronic lead poisoning. Today, lead poisoning can result from drinking water that has passed through the lead pipes common in older houses, especially those built before about 1950. Even in newer pipes, lead solder was widely used into the 1980s and is still sold in stores.

Lead is perhaps most dangerous to children through its presence in paint. Until 1976, most paints contained lead, and interiors in most houses were painted with lead-based paint. If you strip layers of paint from woodwork in an older home, you are likely to be exposed to airborne lead and should take appropriate precautions. Babies and young children often take an interest in peeling paint from walls, and may ingest or inhale fragments of it. There is also a lot of lead in the environment, mostly left over from previous use of leaded gasoline, which settled from the atmosphere onto land surfaces. This lead is now in soils and may be ingested by children playing in yards, or brought indoors in the form of airborne dust, or clinging to shoes and clothing. About 5% of children in North America are thought to have been affected by lead poisoning; the number exposed may be much higher, as the health impacts from casual exposure may not be immediately apparent or easily diagnosed.

Lead in paint has been strictly limited in Canada since 1976. The Government of Canada has responded to lead paint in imported toys and other items intended for children through legislation, such as the *Children's Jewellery Regulations*, which came into force in 2005. This law, among the toughest regulations in the world on lead, made it illegal to import, advertise, or sell children's jewellery or accessories that contain more than 600 mg/kg of total lead, or more than 90 mg/kg of migratable lead.[12] Lead that is *migratable* will leach from the item under certain circumstances; for children's toys, jewellery, and other accessories, the circumstance of greatest concern would occur when the child puts the item into his or her mouth.

In 2007, concerns were heightened when it was discovered that millions of children's toys, play jewellery, baby bibs, and other items manufactured in China contained lead-based paint, including many toys sold under the popular Fisher-Price and Mattel brand names. China agreed to limit the lead content of paint and to closely monitor the use of lead-based paints in manufacturing. Health Canada is working on further reducing the risk of children's exposure by limiting the lead contents of consumer products through the *Hazardous Products Act*. These limits will focus mainly on categories of products with which children are most likely to come into close contact, such as cribs and other children's furniture; toys; kitchen utensils; candles; and other items.

A recently recognized indoor hazard is a group of chemicals known as *polybrominated diphenyl ethers* (*PBDEs*). These compounds provide fire-retardant properties and are used in a diverse array of consumer products, including computers, televisions, plastics, and furniture. They appear to be released by evaporation at very slow rates throughout the lifetime of products. These chemicals persist and accumulate in living tissue, and their abundance in the environment and in people is doubling every few years. PBDEs appear to be *endocrine disruptors*, which means that they interfere with hormone levels; lab testing with animals shows them to affect thyroid hormones.

Animal testing also shows limited evidence that PBDEs may affect brain and nervous system development and might possibly cause cancer. Concern about PBDEs rose after a study showed that concentrations in the breast milk of Swedish mothers had increased exponentially from 1972 to 1997. The European Union decided in 2003 to ban PBDEs, and industries in Europe had already begun to phase them out. As a result, concentrations in breast milk of European mothers have fallen substantially. So far, Health Canada does not consider that the amounts of PBDEs consumed by Canadians pose a health risk.

Another common indoor air pollutant is *formaldehyde* (CH_2O), a colourless gas with a sharp, irritating odour. Formaldehyde is a *volatile organic compound (VOC)*, which we discussed in Chapter 13, *Atmospheric Science and Air Pollution*. Health Canada states that some formaldehyde is present in the air in virtually all Canadian homes, in varying concentrations.[13] It is widely used as a disinfectant and preservative. In the home, it can come from a wide variety of plastics and adhesives, textiles (such as permanent press clothing), and wood products—that "new furniture" or "new carpet" smell is often related to the presence of formaldehyde. Cigarette smoke also contributes formaldehyde to the indoor environment.

Just as outdoor hazards sometimes find their way into buildings, indoor contaminants can end up outside in the natural environment. One example of this (among many)

is *triclosan*, a common constituent of antibacterial hand soaps and cleansers. When we wash our hands with these products, triclosan is washed straight down the drain. Because of its antibacterial and antifungal properties, triclosan is damaging to aquatic microorganisms, especially algae.[14] A large proportion of triclosan is removable during sewage treatment, if appropriate procedures are used, but a significant amount still enters water bodies along with the sewage effluent and sludge.[15]

Triclosan has been linked to allergies and skin conditions, but much more troublesome is its possible role in the development of antibiotic resistance. This is of concern not just in hospitals, where it is used in surgical scrubs and antibacterial cleansers, but also in the wider natural environment. Exposure to triclosan has been linked to the development of multidrug resistance in natural microbial communities.[16] This could ultimately be extremely problematic in the context of decreasing effectiveness of antibiotic drug treatments for a wide variety of diseases in the human population.

Toxicology is the study of chemical hazards

Studying the health effects of chemical agents suspected to be harmful is the focus of **toxicology**, the science that examines the effects of poisonous substances on humans and other organisms. Toxicologists assess and compare substances to determine their **toxicity**, the degree of harm a chemical substance can inflict. The concept of toxicity in chemical hazards is analogous to that of *pathogenicity* or *virulence* of biological hazards that spread infectious disease. Just as types of microbes differ in their ability to cause disease, chemical hazards differ in their capacity to endanger us.

However, *any* chemical substance—even water!—may exert negative effects if it is ingested in great enough quantities, or if exposure is extensive enough. Conversely, a **toxic** agent (also known as a **toxin** or **toxicant**) present or ingested in a minute enough quantity may pose no health risk at all. These facts are often summarized in the catchphrase, "The dose makes the poison." In other words, a substance's toxicity depends not only on its chemical identity but also on its quantity or concentration and on the length of exposure to the toxin.

During the past century, our ability to produce new chemicals has expanded, concentrations of chemical contaminants in the environment have increased, and public concern for health and the environment has grown. These trends have driven the rise of *environmental toxicology*, which deals specifically with toxic substances that come from or are discharged into the environment. Environmental toxicology includes the study of health effects on humans, other animals, and ecosystems, and it represents one approach within the broader scope of environmental health.

Toxicologists generally focus on human health, using other organisms as models and test subjects. In environmental toxicology, animals are also studied out of concern for their welfare and because—like canaries in a coal mine—animals can serve as indicators of health threats that could soon affect humans.

As we review the effects of human-made chemicals throughout this chapter, it is important to keep in mind that many artificially produced chemicals have played a crucial role in giving us the standard of living we enjoy today. These chemicals have helped create the industrial agriculture that produces our food, the medical advances that protect our health and prolong our lives, and many of the modern materials and conveniences we use every day. It is important to remember these benefits as we examine some of the unfortunate side effects of these advances, and look for better alternatives.

Toxic Agents in the Environment

The environment contains countless natural chemical substances that may pose health risks. These substances include oil oozing naturally from the ground, radon gas seeping up from bedrock, and toxic chemicals stored or manufactured in the tissues of living organisms—for example, toxins that plants use to ward off herbivores and toxins that insects use to defend themselves from predators. We also are exposed to many synthetic (artificial, or human-made) chemicals, both outdoors and indoors. In this section we will look at the sources and characteristics of some of these materials.

Synthetic chemicals are ubiquitous in our environment

Synthetic chemicals are all around us in our daily lives (**FIGURE 19.4**). It's difficult to get a handle on how many synthetic chemicals are in current use in industry and commerce; the total number probably exceeds 100 000. Health Canada lists 23 000 so-called existing substances on the *Domestic Substances List*.[17] The U.S. EPA has 84 000 synthetic chemicals on the *Chemical Substance Inventory*, maintained and updated regularly as a provision of the *Toxic Substances Control Act*; around 1700 new chemicals are assessed and added yearly.[18] By any reckoning, tens of thousands of synthetic chemicals are in common use, and many have found their way (intentionally or accidentally) into soil, air, and water.

Synthetic chemicals that have been identified in Canadian lakes and streams include antibiotics and antibac-

FIGURE 19.4
Synthetic chemicals, such as those in many common household products, are everywhere around us in our everyday lives. Some of these compounds may pose environmental or human health risks.

terial agents, detergents, steroids, plasticizers, disinfectants, solvents, perfumes, and ingredients from many other substances. Cosmetics, personal hygiene products, and pharmaceuticals contain chemicals that are washed down the drain every day, and these are of increasing concern, as discussed in the context of triclosan in antibacterial soaps. The pesticides we use to kill insects and weeds on farms, lawns, and golf courses are some of the most widespread synthetic chemicals. As a result of all this exposure, every one of us carries traces of numerous industrial chemicals in our body.

This should not *necessarily* be a cause for generalized alarm. Not all synthetic chemicals pose health risks, and relatively few are known with certainty to be toxicants. Even for those that are known toxicants, we may not be routinely exposed to them. Of the 23 000 chemicals on Health Canada's Domestic Substances List, only about 4000 have been identified as warranting further action or concern. These are chemicals that have been determined to be inherently toxic to humans or other organisms; that are **persistent** (that is, they take a very long time to break down); and/or that build up in organisms and the food chain through **bioaccumulation** and therefore pose the greatest potential for human exposure and negative impacts.[19]

However, there are still many gaps in scientific knowledge, both about the effects of synthetic chemicals and their behaviour in the environment. Of the many thousands of synthetic chemicals on the market today, very few (perhaps as low as 10%) have been thoroughly tested for harmful effects. For the vast majority, we simply do not know what effects, if any, they may have.

Why are there so many synthetic chemicals around us? Consider pesticides and herbicides, made widespread by advances in chemistry and production capacity during and following World War II. As material prosperity grew in Westernized nations in the decades following the war, people began using pesticides not only for agriculture but also to improve the look of their lawns and golf courses and to fight termites, ants, and other insects inside their homes and offices. Pesticides were viewed as means toward a better quality of life. In fact, "Better Living through Chemistry" was a marketing slogan for the DuPont chemical company, starting in 1935. It was not until the 1960s that people began to learn about the risks of exposure to pesticides. A key event was the publication of Rachel Carson's 1962 book, *Silent Spring*, which brought the pesticide DDT to the public's attention.

Silent Spring began the public debate over synthetic chemicals

Rachel Carson was a naturalist, author, and government scientist. In *Silent Spring*, she brought together a diverse collection of scientific studies, medical case histories, and other data that no one had previously synthesized and presented to the general public. Her message was that DDT in particular, and artificial pesticides in general, are hazardous to people's health, the health of wildlife, and the well-being of ecosystems.

Carson began her study at the request of a friend who owned a bird sanctuary that had been sprayed with pesticides, against her wishes, causing the deaths of many songbirds (hence the title "Silent Spring"). The book was written at a time when large amounts of pesticides, virtually untested for health effects, were sprayed over residential neighbourhoods and public areas, on the assumption that the chemicals would do no harm to people (**FIGURE 19.5**). Most consumers had no idea that the store-bought chemicals they used in their houses, gardens, and fields might be toxic. An important part of Rachel Carson's own views on the subject was that being exposed to pesticides against one's wishes is not just a health or environmental hazard, but an issue of personal freedom and choice.

Although challenged vigorously by spokespeople for the chemical industry, who attempted to discredit both the author's science and her personal reputation, Carson's book was a best-seller. Carson suffered from cancer as she finished *Silent Spring*, and she lived only briefly after its publication. However, the book helped generate significant social change in views and actions toward the environment, and is widely credited as one of a handful of seminal events that launched the modern environmental movement.

The use of DDT is now illegal in a number of nations and was banned in Canada in 1985. The United States banned DDT for almost all uses (except public emergencies) in 1972 but still manufactures and exports DDT to countries that do use it. (Some of the agricultural products

FIGURE 19.5
Before the 1960s, the environmental and health effects of potent pesticides, such as DDT, were not widely studied or publicly known. Public areas, such as parks, neighbourhoods, and beaches, were regularly sprayed for insect control without safeguards against excessive human exposure. Here children on a beach are fogged with DDT from a pesticide spray machine in 1945.

Bettmann/Corbis

grown in other countries with the use of DDT make their way back to the United States as food imports, in what has been dubbed a "circle of poison.") Many developing countries with tropical climates use DDT to control human disease vectors, such as mosquitoes that transmit malaria. In these countries, malaria represents a far greater health threat than do the toxic effects of the pesticide.

Toxicants can have different effects

Toxic substances, whether natural or synthetic, can be classified into different types based on their particular effects on health. Some chemicals can have multiple, overlapping effects. Some of the effects occur mainly in humans, and some mainly in other organisms.

Probably the group of toxic substances that is the best known to the general public is the **carcinogens**, which includes chemicals and types of radiation that cause cancer. In cancer, malignant cells grow uncontrollably, creating tumours, damaging the body's functioning, and sometimes leading to death. In our society today, the greatest number of cancer cases is thought to result from carcinogenic chemicals contained in cigarette smoke.

Carcinogen exposure can be difficult to identify because there may be a long lag time between exposure to the agent and the detectable onset of cancer. Cancer as a health problem is not limited to humans; for example, in Chapter 3, *Earth Systems and Ecosystems*, you learned about the problem of cancers in the beluga whales of the St. Lawrence Estuary.

Mutagens are chemicals that cause mutations in the DNA of organisms. Most genetic mutations actually have little or no effect, but some can lead to severe problems, including cancer and other disorders. If mutations occur in an individual's sperm or egg cells, then the individual's offspring may suffer from the associated effects.

Chemicals that cause harm to the unborn are called **teratogens**. Teratogens that affect the development of embryos in the womb can cause birth defects. One well-known human example involves the drug thalidomide, developed in the 1950s as a sleeping pill and to prevent nausea during pregnancy. Tragically, the drug turned out to be a powerful teratogen, and its use caused birth defects in thousands of babies (**FIGURE 19.6A**). Even a single exposure during pregnancy could result in limb deformities and organ defects. Thalidomide was banned in the 1960s once scientists recognized its connection with

(a) Effects of thalidomide poisoning in utero

(b) Effects of mercury poisoning

FIGURE 19.6 Two episodes of almost iconic status in the history of synthetic toxicants are the use of thalidomide by pregnant women in the late 1950s to early 1960s and the mercury poisoning of the residents of Minamata, Japan. The drug thalidomide turned out to be a potent teratogen. It was banned in the 1960s, but not before causing thousands of birth defects. Butch Lumpkin was a "thalidomide baby." He overcame his disability, eventually becoming a professional tennis instructor **(a)**. The name of Minamata, Japan, is synonymous with the mercury poisoning that occurred there between the 1930s and 1960s **(b)**. Mercury poisoning, also called Minamata disease, causes severe neurological problems and can be fatal, as well as teratogenic. A victim of Minamata disease is shown in this famous photograph by Eugene Smith.

birth defects. Ironically, today the drug shows promise in treating other diseases, including Alzheimer's disease, AIDS, and various types of cancer.

The human immune system protects our bodies from disease. Some toxicants weaken the immune system, reducing the body's ability to defend itself against bacteria, viruses, allergy-causing agents, and other attackers. Others, called **allergens**, overactivate the immune system, causing an immune response when one is not necessary. Allergic reactions can be serious, and even fatal; a familiar case is the increasingly common peanut allergy. The prevalence of allergenic synthetic chemicals in our environment is one proposed cause for the increase in asthma in recent years. Continuous low-concentration chemical exposure and the overwhelming of peoples' immune systems may also be the cause of *multiple chemical hypersensitivity*. In this condition, people may develop severe allergic responses to many common synthetic chemicals, especially in indoor environments.

Neurotoxins are chemicals that assault the nervous system. These include various heavy metals, such as lead, mercury, and cadmium, as well as pesticides and some chemical weapons developed for use in war. A famous case of neurotoxin poisoning occurred in Japan, where a chemical factory dumped waste laden with mercury, a heavy metal, into Minamata Bay between the 1930s and 1960s. Thousands of people in and around the town on the bay were poisoned by eating fish contaminated with mercury (**FIGURE 19.6B**). People began to show symptoms that included slurred speech, loss of muscle control, disfiguring birth defects, and even death. The company and the government eventually paid about $5000 in compensation to each affected resident.

Endocrine-disrupting chemicals are of increasing concern

Scientists have recently begun to recognize the importance of **endocrine disruptors**, toxicants that interfere with the endocrine system. The endocrine system sends chemical messengers (*hormones*) through the body at extremely low concentrations. They perform many vital functions, including stimulating growth, development, and sexual maturity, as well as regulating brain function, appetite, sex drive, and many other aspects of our physiology and behaviour.

Hormone-disrupting toxicants can affect an animal's endocrine system by blocking the action of hormones, or by accelerating their breakdown. Many endocrine disruptors are similar to hormones in their molecular structure and chemistry, with the result that they "mimic" the hormone by interacting with receptor molecules, just as the actual hormone would (**FIGURE 19.7**). Such

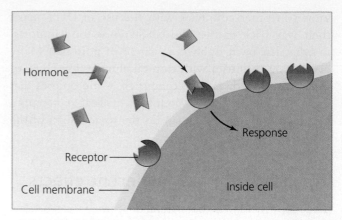

(a) Normal hormone binding

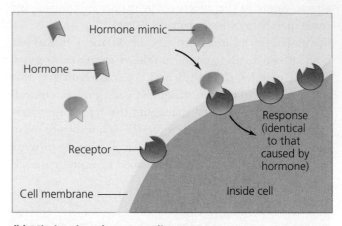

(b) Mimicry by a hormone disruptor

FIGURE 19.7
Many endocrine-disrupting substances mimic the structure of hormone molecules. Like a key similar enough to fit into another key's lock, the hormone mimic binds to a cellular receptor for the hormone, causing the cell to react as though it had encountered the hormone.

effects were first noted as far back as the 1960s, with the pesticide DDT.

One common type of endocrine disruption involves the feminization of male animals, as shown by studies on alligators, fish, frogs, and other organisms. Feminization may be widespread because a number of chemicals appear to mimic the female sex hormone estrogen and bind to estrogen receptors. For example, when biologist Louis Guillette began studying the reproductive biology of alligators in Florida lakes (**FIGURE 19.8**), he discovered that alligators from one lake in particular showed a number of bizarre reproductive problems.[20] Females were having trouble producing viable eggs. Male hatchling alligators had severely depressed levels of the male sex hormone testosterone, and female hatchlings showed greatly elevated levels of the female sex hormone estrogen. Young male animals had abnormal gonads and smaller penises. Males produced sperm at a premature age,

(a) Louis Guillette taking blood samples from an alligator

FIGURE 19.8

Both wildlife studies in the field and experimental studies in the laboratory help scientists examine the effects of toxic substances on organisms. Researchers Louis Guillette **(a)** and Tyrone Hayes **(b)** found that alligators and frogs, respectively, show reproductive anomalies that they attribute to endocrine disruption by pesticides.

(b) Tyrone Hayes with a laboratory frog

and ovaries of the females contained multiple eggs per follicle instead of the expected one egg.[21]

Guillette and his co-workers grew to suspect that environmental contaminants might be responsible for the reproductive abnormalities. The lake had suffered a major spill of pesticides in 1980, and yearly surveys thereafter showed a precipitous decline in the number of juvenile alligators in the lake. In addition, the lake received high levels of chemical runoff from agriculture and was experiencing eutrophication from nutrient input from fertilizers. Comparing alligators from the heavily polluted lake with those from cleaner lakes nearby, the team found that alligators in the contaminated lake had abnormally low hatching rates in the years after the pesticide spill. Even as hatching rates recovered in the 1990s, the alligators continued to show aberrant hormone levels and bizarre gonad abnormalities.

Similar problems began cropping up in other lakes that experienced runoff of chemical pesticides. In the lab, researchers found that several contaminants detected in alligator eggs and young could bind to receptors for estrogen and exist in concentrations great enough to cause sex reversal of male embryos. One chemical in particular, atrazine—a widely used herbicide—appeared to disrupt hormones by inducing production of aromatase, an enzyme that converts testosterone to estrogen.

Following up on Guillette's work, researcher Tyrone Hayes (**FIGURE 19.8B**) found similar reproductive problems in frogs and was able to attribute them to atrazine.[22] In lab experiments, male frogs raised in water containing very low doses of the herbicide became feminized and hermaphroditic, developing both testes and ovaries. Hayes then moved to the field to look for correlations between herbicide use and reproductive impacts in the wild. His field surveys showed that leopard frogs across North America experienced hormonal problems in areas of heavy atrazine usage. His work indicated that atrazine, which kills plants by blocking biochemical pathways in photosynthesis, can also act as an endocrine disruptor.

To date, endocrine effects have been found most widely in non-human animals, but the research has raised concerns about hormone-disrupting impacts on human health. For example, some scientists attribute a striking drop in sperm counts among men worldwide to endocrine disruptors. Danish researchers reported in 1992 that the number and motility of sperm in men's semen had declined by 50% since 1938. Subsequent studies by other researchers—including some who set out to disprove the findings—have largely confirmed the results by using other methods and other populations (although there is tremendous geographical variation that remains unexplained).

Some researchers have also voiced concerns about rising rates of testicular cancer, undescended testicles, and genital birth defects in men. Other scientists have proposed that the rise in breast cancer rates (one in nine Canadian

THE SCIENCE BEHIND THE STORY

Banning Bisphenol A

Can a compound in everyday products damage the most basic processes necessary for healthy pregnancies and births? Canada has been a leader in the international effort to determine whether bisphenol A—found in plastic baby bottles, reusable water bottles, pitchers, tableware, storage containers, and other products— might be damaging the health of Canadians, especially children and infants.

Canada's *Chemicals Management Plan* was introduced in 2006[25] to review the safety of hundreds of chemicals that have been on the market for many years without adequate scientific knowledge of their impacts. Bisphenol A, or BPA, is one of these chemicals, and its potential for reproductive impacts made it a high priority for investigation.

Business analysts estimate that the plastics industry produces more than 2.2 million tonnes of BPA every year— more than a third of a kilogram for every person on the planet.[26] BPA has been known to be an estrogen mimic since the 1930s and more recently has been linked to reproductive abnormalities in mice. Research has shown that BPA can leach out of plastic into water and food when

the plastic is treated with extreme heat, acidity, or harsh soap. One study conducted by researchers for the magazine *Consumer Reports* found that BPA seeped out of the plastic walls of heated baby bottles into infant formula.[27]

One early investigation took place almost by accident at a laboratory at Case Western Reserve University in 1998. The study was initiated when geneticist Patricia Hunt was making a routine check of female lab mice, which included extracting and examining developing eggs from the ovaries. The results on this occasion showed chromosome problems in about 40% of the eggs.

A bit of sleuthing revealed that a lab assistant had mistakenly washed the plastic mouse cages and water bottles with an especially harsh soap. The soap damaged the cages so badly that parts of them melted. The cages were made from plastic containing BPA. To recreate the accidental BPA exposure from the cage-washing incident, Hunt and other researchers washed the polycarbonate cages and water bottles by using varying levels of the harsh soap. They then compared mice kept in damaged cages with plastic water bottles with mice kept in undamaged cages with glass water bottles. The developing eggs of mice exposed to BPA through the damaged plastic showed significant problems dur-

(a)   (b)

Elsevier Science Ltd.

FIGURE 1
During normal cell division **(a)**, chromosomes align properly. Exposure to bisphenol A can cause abnormal cell division **(b)**, whereby chromosomes scatter and are distributed improperly and unevenly between daughter cells.

ing meiosis, the division of chromosomes during egg formation—just as they had in the original incident (see **FIGURE 1**). In contrast, the eggs of mice in the control cages were normal.

In additional tests, three sets of female mice were given oral doses of BPA over three, five, and seven days, and the same abnormalities were observed, although at lower levels (see **FIGURE 2**). The mice given BPA for seven days were most severely affected.

Published in 2003 in the journal *Current Biology*, Hunt's findings set off

women alive today will develop breast cancer) may also be due to hormone disruption, because an excess of estrogen appears to feed tumour development in older women.

Endocrine disruptors can affect more than just the reproductive system; some impair the brain and nervous system. North American studies have shown neurological problems associated with PCB contamination. In one study, mothers who ate Great Lakes fish contaminated with PCBs had babies with lower birth weights and smaller heads, compared with mothers who did not eat fish. Research on endocrine-disrupting chemicals has even postulated connections with obesity.

Research into hormone disruption has generated strident debate.[23] This is partly because a great deal of scientific uncertainty is inherent in any young and developing field. Another reason is that negative findings about chemicals pose an economic threat. Our society has invested heavily in some chemicals that now are suspect. One example is bisphenol A (see "The Science behind

the Story: Banning Bisphenol A"). A building block of polycarbonate plastic, bisphenol A (BPA) occurs in a wide variety of plastic products we use daily, from drink containers to soft plastic toys, eating utensils, auto parts, CDs, and DVDs (**FIGURE 19.9**). It also is used in epoxy resins, including those that coat the insides of metal food and drink cans, and even in dental sealants.

BPA often occurs in conjunction with a group of chemicals called *phthalates*, which also are added to plastics and other products to increase flexibility and softness. Phthalates, which also have been shown to have endocrine-disrupting effects, are so ubiquitous in everyday products that they have been dubbed the "everywhere chemical." Like the research on other hormone-disrupting chemicals, research on the health effects of both BPA and phthalates has been concerning, but controversial and sometimes inconclusive.[24]

A major concern about these chemicals is that they may leach out of products into water and food. Experimental

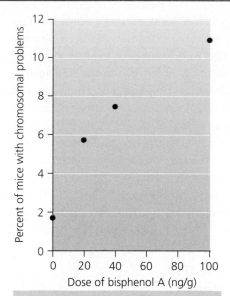

FIGURE 2

In this dose-response experiment, the percentage of mice showing chromosomal problems during cell division rose with increasing doses of bisphenol A.
Source: Data from Hunt, P. A., et al. (2003). Bisphenol A Exposure Causes Meiotic Aneuploidy in the Female Mouse. *Current Biology, 13,* pp. 546–553.

a wave of concern over the safety of bisphenol A.[28] Dozens of other studies have since come out; most have shown harmful effects in lab animals. A diversity of reproductive and other effects has been shown, and findings with obese mice have led some to suggest that bisphenol A may be contributing to the current obesity epidemic. Health Canada's recent updated assessment of BPA does not support a link to obesity but does conclude that there are still uncertainties, and that even low levels of exposure can affect neural development and behaviour if exposure occurs early in life. The recommendation is that the dose for vulnerable people, especially infants and young children, be kept at a level termed *ALARA* ("as low as reasonably achievable").[29]

The result was a recommendation by Health Canada that BPA be declared a "toxic substance" by Environment Canada, as defined by the *Canadian Environmental Protection Act* (1999). Health Canada still maintains that typical BPA exposures for Canadians are far below harmful levels but urges caution for children and infants under 18 months. The assessment also concluded that BPA can harm fish and other organisms, and confirmed that the chemical has been identified in municipal waste water.

In 2015, Cassandra Kinch, from the Alberta Children's Hospital Research Institute at the University of Calgary, and her colleagues published a study of low-dose exposure of fish to BPA and the substitute, BPS. They found that exposure to these chemicals at larval stages, even at very low doses, caused neurodevelopmental and later behavioural abnormalities in zebrafish. The critical finding here is the potential for a relationship between BPA exposure and observed hyperactivity in children, as well as the potential for chemical exposure to cause problems for fetuses *in utero*.[30] The researchers suggest a precautionary approach and the removal of all bisphenols from consumer products.

In the United States and Europe, regulators have set safe intake levels for people at doses of 50 ng/g (nanograms per gram) of body weight, per day. The acceptable level in Canada has been set at half that amount, 25 ng/g of body weight per day. For comparison, lab tests show concentrations of 5–8 ng/mL of liquid in heated plastic baby bottles containing BPA, suggesting significant potential for exposure at harmful levels.[31] Studies have found BPA in the urine of 91% of Canadians tested.[32] With a chemical that is so widely present in our lives, the scientific and social debate over bisphenol A seems set to continue building.

evidence has tied BPA to birth defects and other abnormalities in lab animals. The plastics industry has vehemently protested that the chemical is safe; however, there are now many published studies on laboratory animals that report harmful effects, even from small amounts of bisphenol A.

In 2008, Canada became the first country to declare bisphenol A to be a dangerous substance. Many retailers responded by pulling BPA-containing products off their store shelves. A related chemical, bisphenol S (BPS), is now being used in some products as a substitute for BPA; however, researchers are beginning to raise questions about the safety of this substitute.[33] Some retailers and municipalities have even moved to change or ban thermal receipt paper. This is the soft white paper that you commonly receive as a receipt from stores and restaurants. It turns out that some thermal papers are coated with BPA, which can be transferred to the skin and from there absorbed into the body in appreciable doses, just by holding the paper.[34] Some researchers have suggested that pregnant women should avoid extensive exposure to thermal receipt papers.

Research results on bisphenol A in mice and fish, and those on atrazine in alligators and frogs, have shown detrimental effects at extremely low levels of chemical exposure. This is also the case with research on other known or purported endocrine disruptors. The apparent reason is that the endocrine system is specifically geared to respond to minute concentrations of substances (normally, hormones in the bloodstream). For this reason, it is especially vulnerable to environmental contaminants that are dispersed and diluted in the environment, even though they reach our bodies in very low concentrations.

Some toxicants are concentrated in water

Toxicants are not evenly distributed in the environment, and they move about in specific ways (**FIGURE 19.10**).

Olga Sweet/Getstock

(a) Exposure through toys

CandyBox Images/Fotolia

(b) Exposure through cosmetics

FIGURE 19.9
Soft plastic children's toys **(a)** and cosmetics **(b)** may contain bisphenol A (BPA) and phthalates, which have been shown to be hormone disruptors. In 2010, both Environment Canada and Health Canada placed limitations on the use of BPA.

For instance, water, in the form of runoff, often carries toxicants from large areas of land and concentrates them in small volumes of surface water. If chemicals persist in soil, they can leach into ground water and contaminate drinking water supplies.

Many chemicals are soluble in water and enter organisms' tissues through drinking or absorption. For this reason, aquatic animals, such as fish, frogs, and stream invertebrates, are effective indicators of pollution. The process by which chemicals carried in water become concentrated or magnified in organisms or in the food chain in aquatic environments is called **bioconcentration**; it is equivalent to biomagnification and bioaccumulation in terrestrial environments. When aquatic organisms become sick, we can take it as an early warning that something is amiss. This is why many scientists see findings that show impacts of low concentrations of pesticides on frogs, fish, and invertebrates as a warning that humans could be next. The contaminants that wash into streams and rivers also flow and seep into the water we drink and drift through the air we breathe.

Airborne toxicants can travel widely

Because many chemical substances can be transported by air (see Chapter 13), toxicological effects can occur far from the site of direct chemical use. We saw an example of this in the Central Case, when researchers found high concentrations of wind-transported microplastic fragments in a high-altitude lake in remote Mongolia. When air currents transport contaminants very far from their source, it is called *long-range atmospheric transport*.

Airborne transport of pesticides is termed *pesticide drift*. For example, roughly 143 million kilograms of pesticide active ingredients are used in California each year, mostly in the Central Valley, which is one of the most productive agricultural regions in the world. Because it is a naturally arid area, food production in the Central Valley depends on intensive use of irrigation, fertilizers, and pesticides. The region's frequent winds often blow the airborne spray and dust particles containing pesticide residue for long distances. In the Sierra Nevada Mountains, research has associated pesticide drift from the Central Valley with population declines in frogs. Families living in the Central Valley have suffered health impacts, and activists for farm workers maintain that thousands of residents are at risk.

Despite being manufactured and applied mainly in temperate and tropical zones, synthetic chemical contaminants appear in substantial quantities in the tissues of Arctic polar bears, Antarctic penguins, and people living in Greenland. Scientists can travel to the most remote and seemingly pristine alpine lakes in British Columbia and find them contaminated with foreign toxicants, such as PCBs. The surprisingly high concentrations in polar regions result from patterns of global atmospheric circulation that

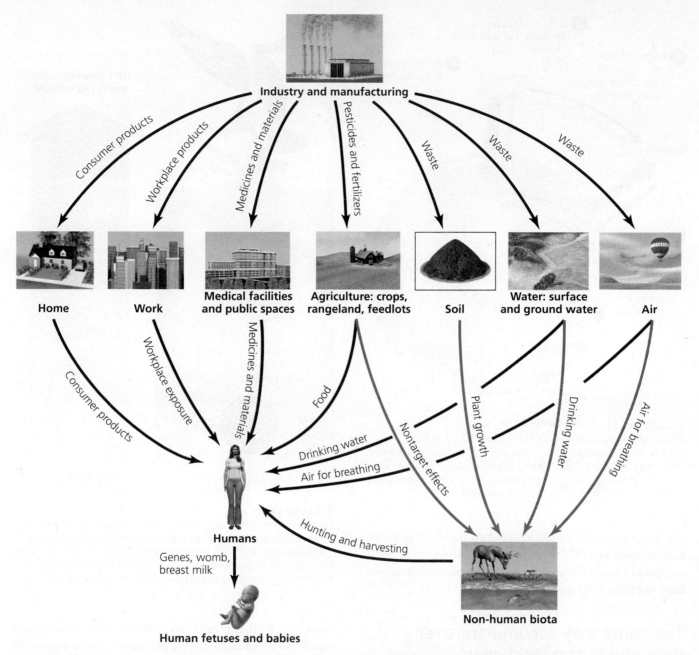

FIGURE 19.10   Synthetic chemicals take many routes in travelling through the environment. Although humans take in only a tiny proportion of these compounds, and although many compounds are harmless, we humans receive small amounts of toxicants from many sources throughout our lifetimes. Toxins present at low levels of concentration are often more harmful for infants because of their small body size.

move airborne chemicals systematically toward the poles (**FIGURE 19.11**).

Some toxicants are persistent

Once toxic agents arrive somewhere, they may degrade quickly and become harmless, or they may remain unaltered and persist for many months, years, or decades. Many such persistent synthetic chemicals exist in our environment today because we have designed them to persist. The synthetic chemicals used in plastics, for instance, are used precisely because they resist breakdown.

The rate at which chemicals degrade depends on such factors as temperature, moisture, and Sun exposure and on how these factors interact with the chemistry of the toxicant. Toxicants that persist in the environment have the greatest potential to harm many organisms over long periods of time. A major reason people have been so concerned about some toxic chemicals, such as DDT and PCBs, is that they have long persistence times.

Most toxicants eventually degrade into simpler compounds called **breakdown products**. Often these are less harmful than the original substance, but sometimes they are just as toxic as the original chemical, or even more so.

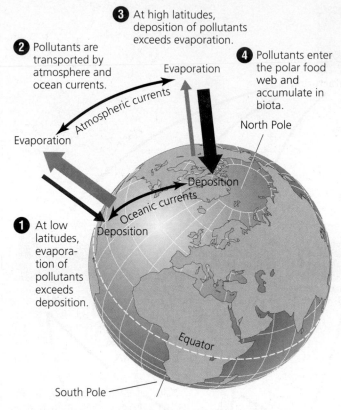

② Pollutants are transported by atmosphere and ocean currents.

③ At high latitudes, deposition of pollutants exceeds evaporation.

④ Pollutants enter the polar food web and accumulate in biota.

Evaporation

Atmospheric currents

Evaporation

Oceanic currents

Deposition

North Pole

Deposition

① At low latitudes, evaporation of pollutants exceeds deposition.

Deposition

Equator

South Pole

FIGURE 19.11
Pollutants that evaporate and rise into the atmosphere at lower latitudes, or are deposited in the ocean, are carried toward the poles by atmospheric and oceanic currents. For this reason, polar organisms take in more than their share of toxicants, despite the fact that relatively few synthetic chemicals are manufactured or used near the poles.

For instance, DDT breaks down into dichlorodiphenyl-dichloroethylene (DDE), a highly persistent and toxic compound in its own right. Many breakdown products have not been fully studied.

Toxicants may accumulate over time and in the food chain

Of the toxicants that organisms breathe, consume, or absorb through the skin, some are quickly excreted; others are degraded into harmless breakdown products. However, some remain intact in the body. Toxicants that are fat-soluble or oil-soluble (often organic compounds, such as DDT and DDE) can be absorbed and stored in fatty tissues. Others, such as methylmercury, may be stored in muscle tissue. If the rate of ingestion of such toxicants is greater than the rate of excretion, they may build up in the animal over time, through the process of bioaccumulation.

Toxicants that bioaccumulate over time in the tissues of one organism may be transferred to other organisms when predators consume prey. When one organism consumes another, it takes in any stored toxicants and stores them itself, along with the toxicants it has received

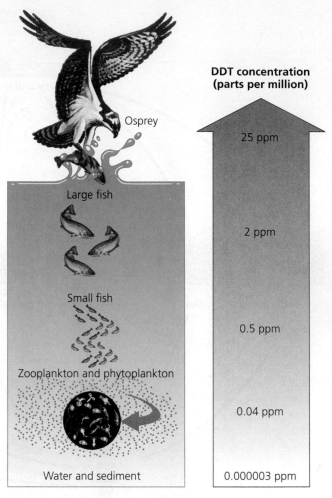

Osprey

Large fish

Small fish

Zooplankton and phytoplankton

Water and sediment

DDT concentration (parts per million)

25 ppm

2 ppm

0.5 ppm

0.04 ppm

0.000003 ppm

FIGURE 19.12
Fat-soluble compounds, such as DDT, bioaccumulate in the tissues of organisms. As animals at higher trophic levels eat organisms lower on the food chain, the latter's load of toxicants passes up to each consumer. DDT moves from zooplankton through various types of fish, finally becoming highly concentrated in fish-eating birds, such as ospreys.

from eating other prey. Thus with each step up the food chain, from producer to primary consumer to secondary consumer and so on, concentrations of toxicants can be greatly magnified. This process, called **biomagnification** or *food chain concentration*, has occurred perhaps most famously with DDT. Top-level predators ended up with high concentrations of the pesticide when concentrations were magnified as DDT moved from water to algae to plankton, and then to small fish, bigger fish, and finally fish-eating birds (**FIGURE 19.12**).

Biomagnification caused populations of many North American birds of prey, including the peregrine falcon, bald eagle, osprey, and brown pelican, to decline precipitously from the 1950s to the 1970s. Eventually scientists determined that DDT was causing the birds' eggshells to grow thinner, so that eggs were breaking while in the nest.

All these birds' populations have rebounded since DDT was banned in North America, but such scenarios are by no means a thing of the past. The polar bears of Svalbard Island in Arctic Norway are at the top of the food chain and feed on seals that have biomagnified toxicants. Despite their remote Arctic location, Svalbard's polar bears show some of the highest levels of PCB contamination of any wild animals tested, as a result of biomagnification and the processes of global transport shown in **FIGURE 19.11**. The contaminants are likely responsible for the immune suppression, hormone disruption, and high cub mortality from which the bears seem to be suffering. Cubs that survive receive PCBs through their mothers' milk, so the contamination persists and accumulates over generations.

Not all toxicants are synthetic

Although we have focused mainly on synthetic chemicals, chemical toxicants also exist naturally in the environment around us and in the foods we eat. We have good reason as citizens and consumers to insist on being informed about risks synthetic chemicals may pose, but it is a mistake to assume that all artificial chemicals are unhealthful and that all natural chemicals are healthful.

In fact, the plants and animals we eat contain many chemicals that can cause us harm. Recall that plants naturally produce toxins to ward off animals that might eat them. In domesticating crop plants, we have selected for strains with reduced toxin content, but we have not eliminated these dangers. Furthermore, when we consume meat, we take in toxins the animals have ingested from plants or animals they have eaten.

Scientists are actively debating just how much risk natural toxicants pose. Some even maintain that the amounts of synthetic chemicals in our food from pesticide residues are dwarfed by the quantities of natural toxicants. Fish and shellfish can contain algae, some of which are toxic to people (as seen in "red tides"). Potatoes can produce toxic moulds and glycoalkaloids that cause stomach aches and other symptoms, and some mushrooms (not those grown commercially) are highly poisonous. Rhubarb leaves contain high levels of oxalic acid, which is poisonous to people. And the list goes on and on. Wheat is even an allergen for some people with special sensitivity to the naturally occurring gluten that it contains.

Others argue that natural toxicants are more readily metabolized and excreted by the body than synthetic ones, that synthetic toxicants persist and accumulate in the environment, and that synthetic chemicals expose people (such as farm workers and factory workers) to risks in ways other than the ingestion of food. In addition, our species may have been exposed to natural toxicants for millions of years, so we may have evolved partial resistance to some of them, which is not the case for the newer synthetic toxicants. What is clear is that more research is required in this area.

Studying the Effects of Hazards

Determining health effects of particular environmental hazards is a challenging job, especially because any given person or organism has a complex history of exposure to many hazards throughout life. Scientists rely on several different methods, including correlative surveys and manipulative experiments.

Wildlife studies use careful observations in the field and lab

When scientists were zeroing in on the health and environmental impacts of DDT, one key piece of evidence came from museum collections of wild birds' eggs from the decades before synthetic pesticides were manufactured. Eggs from museum collections had measurably thicker shells than the eggs scientists were studying in the field from present-day birds. Scientists have pieced together the puzzle of toxicant effects on alligators by taking measurements from animals in the wild and then doing controlled experiments in the lab to test hypotheses (as shown in **FIGURE 19.8**). With frogs and atrazine, scientists first measured toxicological effects in lab experiments and then sought to demonstrate correlations with herbicide use in the wild.

Often the study of wildlife advances in the wake of a conspicuous die-off of many individuals at once, called an *unusual mortality event*. For example, off the California coast in 1998–2001, populations of sea otters fell noticeably and many dead otters washed ashore. Field biologists documented the population decline, and specialists went to work in the lab performing autopsies to determine causes of death. The most common cause of death was found to be infection with the parasite *Toxoplasma*, which killed otters directly and also made them vulnerable to shark attack. *Toxoplasma* occurs in the feces of cats, so scientists hypothesized that sewage runoff containing waste from litter boxes was entering the ocean from urban areas and infecting the otters.

Recent deaths of bees and the collapse of entire colonies have spurred research into the health of honeybees, so crucial for humans in the pollination of food crops. Their recent poor health and die-offs of colonies have now been linked tentatively to negative effects from a commonly used class of chemical fungicide called *neonicotinoids*. There are many other examples of unusual mortality events that have ultimately been linked to chemical toxins in the environment. However, animals—like people—are exposed to an extremely wide range of chemicals and other environmental factors, so it can be very difficult to try to untangle the true cause of a negative effect on an organism.

Human studies rely on case histories, epidemiology, and animal testing

In studies of human health, we gain much knowledge by studying sickened individuals directly. Medical professionals have long treated victims of poisonings, so the effects of common poisons are well known. Autopsies help us understand what constitutes a lethal dose. This process of observation and analysis of individual patients is known as a *case history* approach. Case histories have advanced our understanding of human illness, but they do not always help us infer the effects of rare hazards, newly manufactured compounds, or chemicals that exist at low environmental concentrations and exert minor long-term effects. Case histories also tell us little about probability and risk, such as how many extra deaths we might expect in a population because of a particular cause.

For such situations, which are common in environmental toxicology, epidemiological studies are necessary. Studies in the field of **epidemiology** involve large-scale comparisons among groups of people, usually contrasting a group known to have been exposed to some hazard and a control group that has not. Epidemiologists track the fate of all people in the study, generally for a long period of time (often years or decades), and measure the rate at which deaths, cancers, or other health problems occur in each group. The epidemiologist then analyzes the data, looking for observable differences between the groups, and statistically tests hypotheses accounting for differences. When a group exposed to a hazard shows a significantly greater degree of harm, it suggests that the hazard may be responsible. For example, asbestos miners have been tracked for asbestosis, lung cancer, and mesothelioma rates. Survivors of the Chernobyl disaster in Ukraine have been monitored for thyroid and other cancers. Currently, levels of BPA in humans are being tracked as part of the Canadian Health Measures Survey.

This type of human tracking is part of an ongoing approach taken by Health Canada to the management and tracking of toxic chemicals, called *human biomonitoring*. The process is akin to a natural experiment, in which the experimenter takes advantage of the presence of groups of subjects made possible by some event or long-term exposure that has already occurred. Health Canada has carried out several studies of this type, and continues to collect data in partnership with Statistics Canada and the Public Health Agency of Canada.[35]

Another common approach to epidemiology is to examine outbreaks or *clusters* of diseases in local populations, to determine whether their incidence is higher than or similar to the incidence in the population at large. For example, much attention has been paid recently to cancer rates in Fort Chipewyan, Alberta, after speculation by residents and some researchers that living downstream from oil sand developments might expose them to hazardous chemicals. After much controversy, a report by the Province of Alberta, which examined cases from 1992 to 2011, has now established that rates for three types of cancer are, indeed, higher than expected for residents of Fort Chipewyan, and higher than in the rest of the province; however, Alberta Health Services declined to link the cancers to environmental exposure to toxins.[36]

The advantages of epidemiological studies are their realism and ability to yield relatively accurate predictions about risk. The drawbacks can include the need to wait a long time for results. In addition, participants in epidemiological studies encounter many factors that affect their health besides the one under study. Epidemiological studies measure a statistical association between a health hazard and an effect, but usually they cannot confirm that that particular hazard *caused* the effect.

It can also be difficult in epidemiological studies to disentangle the contributions of various factors to any observed negative health impacts. In cases where a number of factors are present, they may interact, affecting results. An example is smoking, which complicates the interpretation of epidemiological studies on the linkages between cancer, radon exposure, and exposure to other carcinogens. Smoking acts not only as an additional factor that may cause cancer in subjects but also as a reinforcing or *synergistic* factor.

Establishing direct causation for an effect requires the use of manipulative experiments. However, subjecting people to massive amounts of toxic substances in a lab experiment would clearly be unethical, so researchers have traditionally used other animals as subjects to test toxicity. Foremost among these animal models have been laboratory strains of rats, mice, and other mammals. Because of our shared evolutionary history, the bodies of other mammals function similarly to ours. Fish and even some insects also have aspects of their developmental biology and physiology that are very similar to those of humans. The extent to which results from animal lab tests apply to humans varies from one study to the next.

Some people feel the use of rats and mice for testing is unethical; certainly there are frivolous uses that should be avoided. However, animal testing enables scientific and medical advances that would be impossible or far more difficult otherwise. New techniques that are now being devised, using human cell cultures, bacteria, or tissue from chicken eggs, among other possibilities, may one day replace some live-animal testing.

Dose–response analysis is a mainstay of toxicology

The standard method of testing with lab animals in toxicology is dose–response analysis. Scientists quantify the toxicity of a given substance by measuring how much effect a toxicant produces at different doses or how many animals are affected by different doses of the toxic agent. The **dose** is the amount of toxicant the test animal receives or is exposed to and absorbs; the **response** is the type or magnitude of negative (or positive) effects the animal exhibits as a result. The response is generally quantified by measuring the proportion of animals exhibiting negative effects. The data are plotted on a graph, with dose on the x-axis and response on the y-axis (**FIGURE 19.13A**). The resulting curve is called a **dose–response curve**.

Once they have plotted a dose–response curve, toxicologists can calculate a convenient shorthand gauge of a substance's toxicity: the amount of toxicant it takes to kill half the population of study animals used. The dose that is lethal for 50% of individuals in the population is termed the **lethal-dose-50%** or **LD_{50}**. A high LD_{50} indicates low toxicity, and a low LD_{50} indicates high toxicity.

If the experimenter is instead interested in nonlethal health effects, he or she may want to document the level of toxicant at which 50% of a population of test animals is affected in some other way (for instance, at what level of the material does 50% of the population of lab mice lose their fur, or at what level is the beneficial effect of a substance observed in 50% of the population). This is called the **effective-dose-50%, or ED_{50}**.

Sometimes no response is observed until a certain dose has been exceeded. Such a **threshold dose** (**FIGURE 19.13B**) might be expected if the body's organs can fully metabolize or excrete a toxicant at low doses, but become overwhelmed at higher concentrations. It might also occur if cells can repair damage to their DNA, but only up to a certain point.

Sometimes responses *decrease* with increased dose. For example, the effectiveness of some vitamins increases with increasing dose until a threshold is reached, beyond which the benefits cease to accrue. With still higher doses, a turning point is eventually reached at which negative health effects begin to occur as dose is increased (**FIGURE 19.13C**). This type of U- or J-shaped dose–response curve is typical of materials that are harmful to health if they are deficient or missing altogether, but toxic if the dose is too high.

Some dose–response curves are even shaped like an upside-down U. Such counterintuitive curves often occur with endocrine disruptors, likely because the hormone system is geared to function with extremely low concentrations of

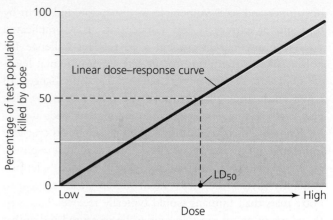

(a) Linear dose–response curve

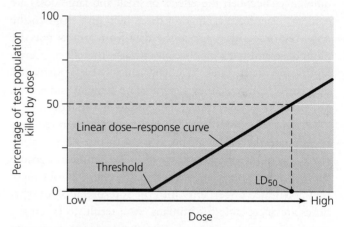

(b) Dose–response curve with threshold

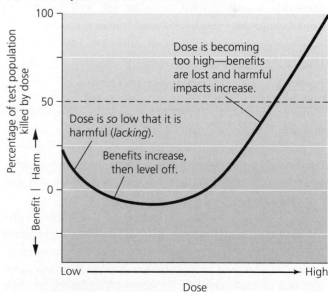

(c) J- or U-shaped dose–response curve

FIGURE 19.13

In a classic linear dose–response curve **(a)**, the percentage of animals killed or otherwise affected by a substance rises with the dose. The point at which 50% of the animals are killed is labelled the lethal-dose-50, or LD_{50}. For some toxic agents, a threshold dose **(b)** exists, below which doses have no measurable effect. Some substances have dose–response curves shaped like a J or a U. In example **(c)**, at extremely low doses the organism is experiencing harm because the substance is lacking (a *deficiency*). At slightly higher doses, beneficial impacts are seen. At some point, the dose becomes so high that negative impacts again begin to occur.

Very low levels of vitamin A (that is, vitamin A deficiency) can cause blindness in people. On the other hand, very large doses of vitamin A (such as might be ingested accidentally by a child) can cause vitamin A toxicity, with symptoms such as nausea, vomiting, liver problems, and additional serious effects, especially if the exposure to high doses lasts a long time. Which of the curves in **FIGURE 19.13** do you think best describes the dose–response characteristics of vitamin A? Curve **(a)**, **(b)**, or **(c)**?

hormones and so is vulnerable to disruption and harm by toxicants at extremely low concentrations. Complicated dose–response curves present a particular challenge for policymakers attempting to set safe environmental levels for toxicants. We may have underestimated the dangers of compounds that behave in these ways, because many such chemicals exist in very low concentrations over wide areas.

Knowing the shape of dose–response curves is crucial if one is planning to extrapolate from them to predict responses at doses below those that have been tested. Scientists generally give lab animals much higher doses relative to body mass than humans would typically receive. This is so that the response is great enough to be measured and so that differences between the effects of small and large doses are evident. Data from a range of doses help give shape to the dose–response curve. Once the data from animal tests are plotted, scientists extrapolate to estimate the effect of even higher and lower doses on a hypothetically large population of animals. This way, they can come up with an estimate of, say, what dose causes cancer in 1 mouse in 1 million.

A second extrapolation from the data is then required to estimate the effect on humans, with our greater body mass and different physiology from lab animals. Because these extrapolations go beyond the actual data obtained, they introduce uncertainty into the interpretation of what doses are acceptable for humans. As a result, to be on the safe side, regulatory agencies set standards for maximum allowable levels of toxicants that are well below the minimum toxicity levels estimated from lab studies.

Individuals vary in their responses to contaminants

Different individuals may respond quite differently to identical exposures to contaminants. These differences can be genetically based or can be due to a person's current condition. People in poorer health are often more sensitive to biological and chemical hazards. Sensitivity also can vary with sex, age, and weight.

Because of their smaller size (and thus higher concentration of toxin per kg of body weight for the same level of exposure) and rapidly developing organ systems, fetuses, infants, and young children tend to be much more sensitive to toxicants than are adults. Regulatory agencies like Health Canada set standards for adults and then extrapolate downward for infants and children. However, in many cases the linear extrapolations still do not protect babies adequately. Many critics today contend that despite improvements, regulatory agencies still do not account explicitly enough for risks to fetuses, infants, children, older adults, and the immunocompromised.

Anthropologist Elizabeth Guillette and colleagues used some unusual techniques to investigate the effects of pesticides on children, in studies focusing on the farming region of the Yaqui Valley in northwestern Mexico.[37] Synthetic pesticides were introduced in the Yaqui Valley area in the 1940s. Some farmers in the valley embraced the agricultural innovations, spraying their farms in the valley to increase their yields. Farmers in the surrounding foothills, however, generally chose to bypass the chemicals and to continue more traditional farming practices. Although differing in farming techniques, Yaqui in the valley and foothills continued to share the same culture, diet, education system, income levels, and family structure.

Guillette and fellow researchers studied 50 children aged four to five—33 from the valley, where pesticide use was common, and 17 from the foothills. Each child underwent a half-hour exam with a series of physical and mental tests. The two groups of children were not significantly different in height and weight, but they differed markedly in other areas of development. Valley children were far behind the foothill children developmentally in coordination, physical endurance, long-term memory, and fine motor skills. The children's drawings of people exhibited the most dramatic differences (**FIGURE 19.14**). By developmental standards, the four- and five-year-old valley children drew human figures at the level of a two-year-old.

Some scientists greeted Guillette's study with skepticism, pointing out that its sample size was too small to be meaningful. Others said that such factors as different parenting styles or unknown health problems could be to blame. Toxicologists argued that without blood or tissue tests on the children, the study results couldn't be tied to agrichemicals. Regardless of these criticisms, the study shows that nontraditional study methods may be useful in determining the effects of environmental toxins like pesticides on human growth and health.

The type of exposure can affect the response

The risk posed by a hazard varies according to whether a person experiences high exposure for short periods of time, known as **acute exposure**, or lower exposure over longer periods of time, known as **chronic exposure**. Incidences of acute exposure are easier to recognize, because they often stem from discrete events, such as accidental ingestion, an oil spill, a chemical spill, or a nuclear accident. Lab tests and LD_{50} values generally reflect acute toxicity effects. The symptoms of acute, high-level exposure may be rapid and noticeable as well.

However, chronic, low-dose exposure is more common than acute, accidental doses—and often much more difficult to detect and diagnose. Chronic exposure often affects organs gradually, as when smoking causes lung cancer, or when alcohol abuse induces liver or kidney

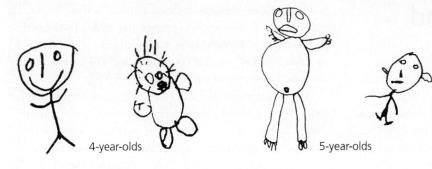

4-year-olds   5-year-olds

(a) Drawings by children from the foothills

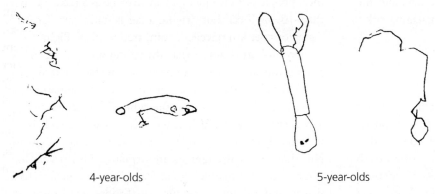

4-year-olds   5-year-olds

(b) Drawings by children from the valley

FIGURE 19.14
Both sets of drawings are intended to depict people. The main difference between the two groups of young artists is exposure to pesticides. The study offers a startling example of apparent neurological effects of pesticide poisoning. Young children from areas where pesticides were not commonly used **(a)** drew recognizable figures of people. Children the same age from the valley areas **(b)**, where pesticides were used heavily in industrialized agriculture, drew only scribbles. *Source:* Adapted from Guillette, E. A., M. M. Meza, M. G. Aquilar, A. D. Soto, and I. E. Garcia (1998). Anthropological Approach to the Evaluation of Preschool Children Exposed to Pesticides in Mexico. *Environmental Health Perspectives, 106(6),* pp. 347–353.

damage. Pesticide residues on food, low levels of heavy metals in drinking water or soil, or exposure to toxins like BPA and POPs at low levels in the home or workplace and in consumer goods also can pose chronic risks. Because of the long time periods involved, relationships between cause and effect may not be readily apparent.

Mixes may be more than the sum of their parts

It is difficult enough to determine the impact of a single hazard on an organism, but the task becomes astronomically more difficult when multiple hazards interact. For instance, chemical substances, when mixed, may act in concert in ways that cannot be predicted from the effects of each in isolation. Mixed toxicants may sum each other's effects, cancel out each other's effects, or multiply each other's effects. The effects of small amounts of toxicants can add up over time, yielding **cumulative effects**. And whole new types of impacts may arise when toxicants are mixed together.

Such interactive impacts—those that are more than or different from the simple sum of their constituent effects—are called **synergistic effects**. For example, lab experiments with alligators have indicated that DDE can either cause or inhibit sex reversal, depending on the presence of other chemicals. Mice exposed to a mixture of nitrate, atrazine, and aldicarb (an insecticide) have been found to

show immune, hormone, and nervous system effects that were not evident from exposure to each of these chemicals alone. Being near an agricultural field with pesticide runoff increases the rate of parasitic infection in frogs, because pesticides suppress the frog's immune response, making it more vulnerable to parasites.

The existence of synergistic effects highlights one of the main problems associated with epidemiological studies that follow groups of people with potential exposure to hazards over a period of time. Most people are routinely exposed to a complex mixture of hazards from the home, the workplace, the environment, and the lifestyle choices they make, including such activities as smoking. In addition to its own negative health impacts, for example, smoking can synergistically reinforce the impacts of other factors. Disentangling the health effects of these various hazards is challenging and always brings a certain degree of uncertainty to the conclusions of epidemiological studies.

Traditionally, environmental health has tackled the effects of single hazards, one at a time. In toxicology, the complex experimental designs required to test interactions and synergies, and the sheer number of possible chemical combinations, have meant that single-substance tests have received priority. Scientists in environmental health and toxicology will never be able to test all possible combinations; there are simply too many chemical hazards and combinations in the environment.

Risk Assessment and Risk Management

Policy decisions on whether to ban chemicals or restrict their use generally follow years of rigorous testing for toxicity. Likewise, strategies for combating disease and other health threats are often based on extensive research. Policy and management decisions reach beyond the scientific results on health to incorporate considerations about economics and ethics. All too often, they are influenced by political pressure from powerful interests. The steps between the collection and interpretation of scientific data and the formulation of policy involve assessing and managing risk.

Risk is expressed in terms of probability

Exposure to an environmental health threat does not invariably produce some given effect. Rather, it causes some probability of harm, some statistical chance that damage will result. To understand the impact of a health threat, a scientist must know more than just its identity and strength. He or she must also know the chance that an organism will be exposed to the hazard, the frequency at which the organism may encounter it, the amount of substance or degree of threat to which the organism is exposed, and the organism's sensitivity to the threat.

The probability that some harmful outcome (for instance, injury, death, environmental damage, or economic loss) will result from a given action, event, or exposure to a substance expresses the **risk** posed by that phenomenon. *Probability* is a quantitative description of the likelihood of a certain outcome, and risks are generally expressed in terms of probabilities.

Every action we take and every decision we make involve some element of risk, some (generally small) probability that things will go wrong. We try in everyday life to behave in ways that minimize risk, but our perceptions of risk do not always match statistical reality (**FIGURE 19.15**). People often worry unduly about negligibly small risks but happily engage in other activities that pose high risks. For instance, most people perceive flying in an airplane as a riskier activity than driving a car, but driving a car is statistically far more dangerous per km travelled, compared to plane flight.

Psychologists agree that this difference between risk perception and reality stems from the fact that we feel more at risk when we are not controlling a situation and more safe when we are "at the wheel"—regardless of the actual risk involved. When we drive a car, we feel we are in control, even though statistics show we are at greater risk than as a passenger in an airplane. This psychology can account for people's great fear of nuclear power, toxic waste, and pesticide residues on foods—environmental hazards that are invisible or little understood, and whose presence in their lives is largely outside their personal control. In contrast, people are more ready to accept and ignore the risks of smoking cigarettes, overeating, and not exercising, all voluntary activities statistically shown to pose greater risks to health.

FIGURE 19.15
Our perceptions of risk do not always match the reality of risk. Shown here are several risks leading to death (the figure is based on data for the United Kingdom; it would differ slightly for Canada). The larger the area of the circle in the figure, the greater the risk of dying from that cause. Note that the risk of death from transport accidents is many times less than that from smoking, or even physical inactivity. *Source:* From United Kingdom National Health Services (2013). *Atlas of Risk.* http://www.nhs.uk/tools/pages/nhsatlasofrisk.aspx.

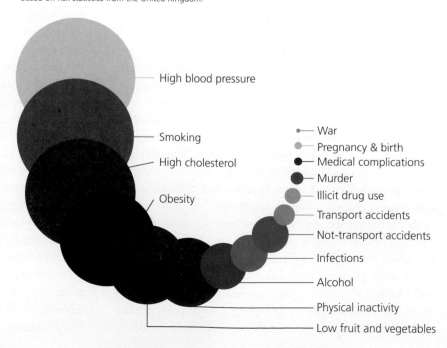

Risks leading to death in perspective*
*Based on risk statistics from the United Kingdom.

High blood pressure

Smoking

High cholesterol

Obesity

War
Pregnancy & birth
Medical complications
Murder
Illicit drug use
Transport accidents
Not-transport accidents
Infections
Alcohol
Physical inactivity
Low fruit and vegetables

Risk assessment analyzes risk quantitatively

The quantitative measurement of risk and the probability and severity of impacts, and the comparison of risks involved in different activities, hazards, or substances, are termed **risk assessment**. Risk assessment is a way of identifying problems and gauging the potential extent of their impacts. In environmental health assessment, it helps decision makers ascertain which substances and activities pose health threats to people or wildlife, and which are largely safe.

In a risk matrix or probability-impact matrix like the one shown in **FIGURE 19.16A**, both the likelihood of occurrence and the severity of the expected impact are taken into account. If both the probability and severity of impacts are low, it might indicate that no preventive or preparatory action is needed, or that it is safe to proceed. If both the probability and severity of impacts are high, preventive or preparatory action is clearly required. If the probability of a negative impact is very low but the impacts would be catastrophic or unacceptably severe (bottom right quadrant of the matrix), then there is a moral imperative to take actions to mitigate or prevent that outcome. This is referred to as the **precautionary principle**, and it is a fundamental concept in risk management.

Assessing the health and environmental risks posed by a chemical substance involves several steps. The first steps involve the scientific studies of toxicity already outlined—determining whether a given substance has toxic effects and, through dose–response analysis, measuring how effects on an organism vary with the degree of toxicant exposure. Subsequent steps involve assessing the individual's or population's likely extent of exposure to the substance, including the frequency of contact, the concentrations likely encountered, and the length of time the substance is expected to be encountered. Characteristics of the individual or sub-population that might increase vulnerability are also considered.

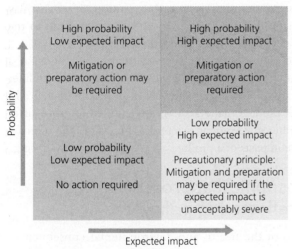

(a) Risk matrix

FIGURE 19.16 The first step in addressing an environmental hazard is risk assessment. A risk matrix or probability-impact matrix like the one shown here **(a)** helps risk managers assess the likelihood that the impact will happen, and how severe it will be. Once science identifies and measures risks, then risk management can proceed **(b)**. In this process, economic, political, social, and ethical issues are considered in light of the scientific data from risk assessment. The consideration of all these types of information is intended to result in policy decisions that minimize the risk of the environmental hazard.

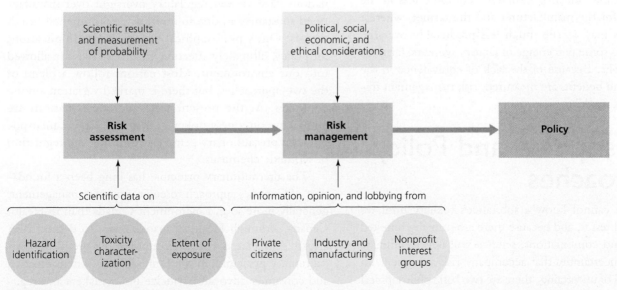

(b) Risk management

Risk assessment studies may be performed by scientists associated with the same industries that manufacture the potentially harmful substances. This, in many people's minds, undermines the objectivity of the process. Ideally, risk assessment should be carried out or at least monitored by government.

Risk management combines science and other social factors

Accurate risk assessment is a vital step toward effective **risk management**, which consists of decisions and strategies to minimize risk (**FIGURE 19.16B**). In most developed nations, risk management is handled largely by federal agencies, such as Environment Canada, Health Canada, Public Safety Canada, and Public Health Canada, as well as the Canadian Council of Ministers of the Environment (CCME), which comprises the federal, territorial, and provincial environment ministers. In risk management, scientific assessments of risk are considered in light of economic, social, and political needs and values. The costs and benefits of addressing risk in various ways are assessed with regard to both scientific and nonscientific concerns. Decisions whether to reduce or eliminate risk are then made.

In environmental health and toxicology, comparing costs and benefits can be difficult because the benefits are often economic, whereas the costs often pertain to health. Moreover, economic benefits are generally known, easily quantified, and of a discrete and stable amount, whereas health risks are hard-to-measure probabilities, often involving a very small percentage of people likely to suffer greatly and a large majority likely to experience little effect.

For example, when a government agency bans a pesticide, it may mean measurable economic loss in the short term for the manufacturer and the farmer, whereas the benefits may accrue much less predictably over the long term to some percentage of factory workers, farmers, and the public. Because of the lack of equivalence in the way costs and benefits are measured, risk management frequently stirs up debate.

Philosophical and Policy Approaches

Because we cannot know a substance's toxicity until we measure and test it, and because there are so many untested chemicals and combinations, science will never eliminate the many uncertainties that accompany risk assessment. In such a world of uncertainty, there are two basic philosophical approaches to categorizing substances as safe or dangerous.

Two approaches exist for determining safety

One approach is to assume that substances are harmless until shown to be harmful. We might nickname this the *innocent-until-proven-guilty approach* (**FIGURE 19.17**). Because thoroughly testing every existing substance (and combination of substances) for its effects is a hopelessly long, complicated, and expensive pursuit, the innocent-until-proven-guilty approach has the benefit of not slowing down technological innovation and economic advancement. However, it has the disadvantage of putting into wide use some substances that may later turn out to be dangerous.

The other approach is to assume that substances are harmful until they are shown to be harmless. This approach, shown on the right in **FIGURE 19.17**, follows the precautionary principle. This more cautious approach should enable us to identify troublesome toxicants, and those which might pose improbable but unacceptably severe risks, before they are released into the environment. However, there is a trade-off: Taking such a cautious approach may impede the pace of technological advance and add to the cost of development of new marketable products.

These two approaches are actually two ends of a continuum of possible approaches. The two endpoints differ mainly in where they lay the burden of proof—specifically, whether product manufacturers are required to prove safety or whether government, scientists, citizens, or the ultimate negative impacts of a product are required to prove danger.

Philosophical approaches are reflected in policy

Because of the health and environmental consequences of hazardous substances, governments of developed nations have viewed regulatory oversight over manufactured substances as one solution to environmental health threats. One's philosophical approach has implications for policy, ultimately affecting what materials are allowed into our environment. Most nations follow a blend of the two approaches, but there is marked variation among countries. At the present time, European nations are embarking on a new policy course that largely incorporates the precautionary principle regarding the regulation of synthetic chemicals.

The precautionary principle has long been a foundation of Canada's approach to environmental management, generally more so in Environment Canada than in Health Canada, although Health Canada's move to declare BPA a toxic substance is a clear example of the use of the precautionary principle. In the United States, environmental and consumer advocates criticize policymakers and regulatory agencies for largely following the innocent-until-proven-guilty approach. For instance, in the United States, compounds involved in cosmetics require no FDA review or approval before being sold to the public.

In Canada, several federal agencies are jointly responsible for tracking and regulating synthetic chemicals, emissions, and effluents. The centrepiece of the laws that govern the environment and environmental health issues in Canada, the *Fisheries Act*, used to prohibit the introduction of deleterious substances into any water body that provides habitat for fish. This was one of the most powerful tools available for the protection of water quality in Canada, until it was largely dismantled in 2012 by the federal government's Bill C-38, the *Jobs, Growth and Long-Term Prosperity Act*. We will turn our attention to Bill C-38 and its impacts on environmental policy in Canada in Chapter 21, *Environmental Policy: Decision-Making and Problem-Solving*.

Toxicants are also regulated internationally

Nations have sought to address chemical pollution with international treaties. One example is the *Basel Convention*

[of the United Nations] on the Control of Transboundary Movements of Hazardous Wastes and Their Disposal. The Basel Convention (mentioned in the context of waste management) is probably the most comprehensive international agreement on hazardous materials. The convention came into effect in 1992 and has 181 parties (signatory nations that have ratified the agreement) as of 2015. Its central goal is to protect human health and the environment against the adverse effects of hazardous and other wastes.[38]

Another example of a UN-moderated vehicle is the *International Code of Conduct on the Distribution and Use of Pesticides*. This is not an international law but a set of voluntary guidelines and best practices for both private and public entities. The World Health Organization also plays a role in establishing these guidelines and monitoring adherence to them. The *Rotterdam Convention on the Prior Informed Consent Procedure for Certain Hazardous Chemicals and Pesticides in International Trade* is an example of an international legal instrument that operates according to rules established by consensus of the parties.

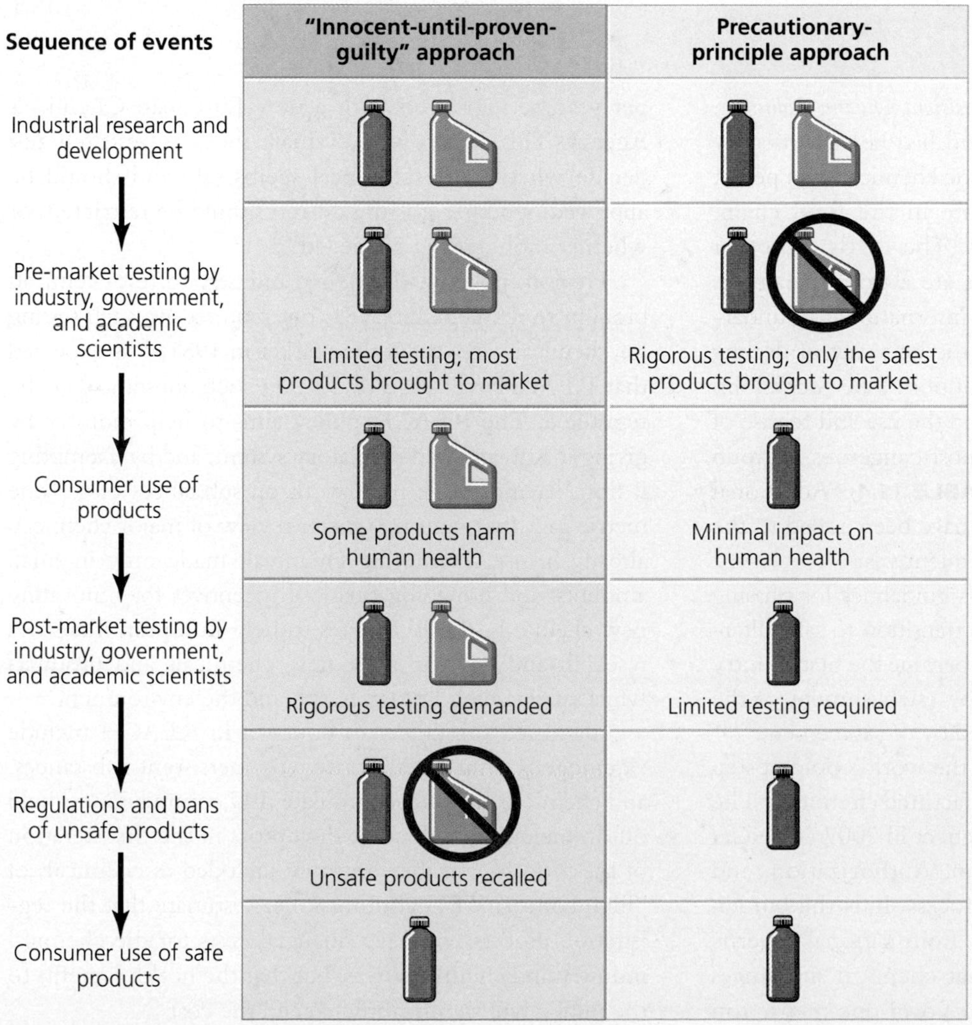

FIGURE 19.17

Two main approaches can be taken to introduce new substances to the market. In one approach, substances are "innocent until proven guilty" and brought to market relatively quickly after limited testing. Products reach consumers more quickly, but some fraction of them (blue bottle in diagram) may cause harm to some fraction of people. The other approach adopts the precautionary principle, bringing substances to market cautiously, and only after extensive testing. Products that reach the market should be safe, but many perfectly safe products (purple bottle in diagram) will be delayed in reaching consumers.

Sequence of events	"Innocent-until-proven-guilty" approach	Precautionary-principle approach
Industrial research and development		
Pre-market testing by industry, government, and academic scientists	Limited testing; most products brought to market	Rigorous testing; only the safest products brought to market
Consumer use of products	Some products harm human health	Minimal impact on human health
Post-market testing by industry, government, and academic scientists	Rigorous testing demanded	Limited testing required
Regulations and bans of unsafe products	Unsafe products recalled	
Consumer use of safe products		

TABLE 19.4 The "Dirty Dozen" Persistent Organic Pollutants (POPs) Targeted by the Stockholm Convention

Toxicant	Type	Description
Aldrin	Pesticide	Kills termites, grasshoppers, corn rootworm, and other soil insects
Chlordane	Pesticide	Kills termites and is a broad-spectrum insecticide on various crops
DDT	Pesticide	Widely used in the past to protect against insect-spread diseases; continues to be applied in several countries to control malaria
Dieldrin	Pesticide	Controls termites and textile pests; also used against insect-borne diseases and insects in agricultural soil
Dioxins	Unintentional by-product	Produced by incomplete combustion and in chemical manufacturing; released in some kinds of metal recycling, pulp and paper bleaching, automobile exhaust, tobacco smoke, and wood and coal smoke
Endrin	Pesticide	Kills insects on cotton and grains; also used against rodents
Furans	Unintentional by-product	Result from the same processes that release dioxins; also found in commercial mixtures of PCBs
Heptachlor	Pesticide	Kills soil insects, termites, cotton insects, grasshoppers, and mosquitoes
Hexachlorobenzene	Fungicide; unintentional by-product	Kills fungi that affect crops; released during chemical manufacture and from processes that give rise to dioxins and furans
Mirex	Pesticide	Combats ants and termites; also is a fire retardant in plastics, rubber, and electronics
PCBs	Industrial chemical	Used in industry as heat-exchange fluids, in electrical transformers and capacitors, and as additives in paint, sealants, and plastics
Toxaphene	Pesticide	Kills insects on crops; kills ticks and mites on livestock

Source: Data from United Nations Environment Programme (UNEP), Secretariat of the Stockholm Convention (2001). *Measures to Reduce or Eliminate POPs.*

The *Stockholm Convention on Persistent Organic Pollutants* (POPs) came into force in 2004 and had been ratified by 179 nations as of 2015. POPs are toxic chemicals that persist in the environment, bioaccumulate in the food chain, and often can travel long distances. The PCBs and other contaminants found in polar bears are a prime example. Because contaminants often cross international boundaries, an international treaty seemed the best way of dealing fairly with such transboundary pollution. The Stockholm Convention was designed first to end the use and release of 12 of the POPs shown to be the most dangerous, a group nicknamed the "dirty dozen" (**TABLE 19.4**). Additional chemicals and groups of chemicals have been added to the agreement by amendments in subsequent years.

The Stockholm Convention sets guidelines for phasing out these chemicals and encourages transition to safer alternatives. In September 2006, Canada became the first country to prioritize domestic chemicals, by establishment of the Domestic Substances List, as required by the agreement.[39]

The European Union is taking the world's boldest step toward testing and regulating manufactured chemicals. The EU's REACH program went into effect in 2007 (*REACH* stands for Registration, Evaluation, Authorization, and Restriction of Chemicals). This process shifts the burden of proof for testing chemical safety from national governments to industry and requires that chemical substances produced or imported in amounts of over one metric ton

per year be registered with a new European Chemicals Agency. This agency will evaluate industry research and decide whether the chemical seems safe and should be approved, whether it is unsafe and should be restricted, or whether more testing is needed.[40]

Previous policy had required industry to test chemicals brought to market after 1981, but required no such testing for chemicals already on the market in 1981. It is expected that REACH will require 30 000 such substances to be registered. The REACH policy aims to help industry by giving it a streamlined regulatory system, and by exempting it from having to file paperwork on substances under one metric ton. By requiring stricter review of major chemicals already in use, exempting chemicals made only in small amounts, and providing financial incentives for innovating new chemicals, the EU hopes to help European industries research and develop safer new chemicals and products while safeguarding human health and the environment.

Identified substances of concern in REACH include carcinogens, mutagens, teratogens, persistent substances, and chemicals that bioaccumulate. EU commissioners held off on including endocrine disruptors in the first iteration of the process, but they are now included as chemicals of "high concern." EU commissioners estimate that the registration process will have financial costs for the chemical industry and chemical users, but that the health benefits to the public will significantly exceed the costs.

Conclusion

International agreements, such as the Rotterdam, Basel, and Stockholm Conventions, inspire hope that governments will act to protect the world's people, wildlife, and ecosystems from toxic chemicals and other environmental hazards. At the same time, solutions often come more easily when they do not arise from government regulation alone. To many minds, consumer choice, exercised through the market, may be the best way to influence industry's decision-making. Consumers of products can make decisions that influence industry when they have full information from scientific research regarding the risks involved. Once scientific results are in, a society's philosophical approach to risk management will determine policy decisions.

Whether the burden of proof is laid at the door of industry or of government, it is important to realize that we will never attain complete scientific knowledge of any risk. Rather, we must make choices based on the information available. Synthetic chemicals have brought us innumerable modern conveniences, a larger food supply, and medical advances that save and extend human lives. Human society would be very different without these chemicals. Yet a safer and happier future, one that safeguards the well-being of both humans and the environment, depends on knowing the risks that some hazards pose and on having means in place to phase out harmful substances and replace them with safer ones.

REVIEWING OBJECTIVES

You should now be able to

Identify the major types of environmental health hazards and the goals of environmental health

- Environmental health seeks to assess and mitigate environmental factors that adversely affect human health and ecological systems. Environmental health threats include physical, chemical, biological, and cultural hazards.
- Disease is a major focus of environmental health. We have successfully fought some infectious diseases, but others are spreading.
- Some materials that originate outside become hazardous when concentrated indoors. Other hazards come directly from products that we use in everyday life. Contaminants that originate indoors can cause problems when they are released into the natural environment.
- Toxicology is the study of poisonous substances, their occurrence, and their impacts.

Describe the types, abundance, distribution, and movement of toxicants in the environment

- Thousands of potentially toxic substances, both synthetic and natural, exist in the environment all around us.
- *Silent Spring*, published in 1962, began a public conversation about the hazards of some of the chemicals we use and are exposed to in daily life.
- Toxicants can act as carcinogens, mutagens, teratogens, allergens, neurotoxins, or endocrine disruptors.
- Endocrine-disrupting chemicals mimic hormones, with the potential for damaging the endocrine system, which regulates many important functions in people and animals.

- Toxicants may enter and move through surface- and groundwater reservoirs.
- Some toxicants are transported and travel long distances through the atmosphere, reaching areas far from their original source.
- Some chemicals break down very slowly and thus persist in the environment.
- Some toxic materials bioaccumulate and move up the food chain, poisoning consumers at high trophic levels through the process of biomagnification.
- Natural toxins are produced as defence mechanisms by plants and animals; we are exposed to some of these through the food we eat.

Summarize how hazards and contaminants are studied scientifically, including case histories, epidemiology, animal testing, and dose–response analysis

- Animal studies can be carried out in the field or in the laboratory. These can inform efforts to improve human health.
- Through the use of case histories, researchers study health problems in individual people. Epidemiology involves gathering data from large groups of people over long periods of time, and comparing groups with and without exposure to the threat being assessed.
- In dose–response analysis, scientists measure the response of test animals to various doses of the suspected toxicant.
- Individuals differ in their responses to contaminant exposure. Children, infants, and fetuses are often at greatest risk because of their small body mass and rapidly developing physiological systems.

- Toxicity or strength of response may be influenced by the dose or amount of exposure, the nature of exposure (acute or chronic), individual variation, and synergistic interactions with other hazards.
- Combinations of chemicals may have synergistic effects, that is, greater impacts than the individual chemicals on their own. Small effects also can build into cumulative effects over time.

Summarize approaches to risk assessment and risk management

- Risk, or the likelihood of a negative impact, is usually expressed in terms of probability.
- Risk assessment involves quantifying and comparing the risks involved in different activities or substances.
- Risk management integrates science with political, social, and economic concerns in order to design strategies to minimize risk.

Compare philosophical approaches to risk and their role in environmental health policy

- An innocent-until-proven-guilty approach assumes that a substance is not harmful unless it is shown to be so. The precautionary approach entails assuming that a substance may be harmful unless proven otherwise.
- Health Canada, Environment Canada, and other federal and provincial/territorial agencies are jointly responsible for regulating environmental health threats under Canadian policy.
- Canada and European nations tend to take a precautionary approach to environmental hazards and the testing of chemical products, compared with the United States. Canada is a party to many important international agreements that regulate hazardous substances.

TESTING YOUR COMPREHENSION

1. What four major types of health hazards does research in the field of environmental health encompass?
2. In what way is disease the greatest hazard that people face? What kinds of interrelationships must environmental health experts study to learn about how diseases affect human health?
3. Where does most exposure to lead, asbestos, radon, and PBDEs occur? How has each been addressed?
4. When did concern over the effects of pesticides start to emerge? Describe the argument presented by Rachel Carson in *Silent Spring*. What impact did it have on public perception of risk from synthetic chemicals? Is DDT still used?
5. List and describe six important types or general categories of toxicants described in this chapter, including their occurrence, characteristics, and health or environmental impacts.

6. How do toxicants travel through the environment, and where are they most likely to be found? What are the lifespans of toxic agents? Describe the processes of bioaccumulation and biomagnification.
7. What are epidemiological studies, and how are they most often conducted?
8. Why are animals used in laboratory experiments in toxicology? Explain the dose–response curve. Why is a substance with a high LD_{50} considered safer than one with a low LD_{50}?
9. What factors may affect an individual's response to a toxic substance? Why is chronic exposure to toxic agents often more difficult to measure and diagnose than acute exposure? What are synergistic effects, and why are they difficult to measure and diagnose?
10. How do scientists identify and assess risks from substances or activities that may pose health threats?

THINKING IT THROUGH

1. Describe some environmental hazards that you think you may be living with indoors. How do you think you may have been affected by indoor or outdoor environmental hazards in the past? What approach do you take in dealing with potential toxic hazards in your own life?
2. Why is it that research on endocrine disruption has spurred so much debate? What steps do you think could be taken to help establish more consensus

among scientists, industry, regulators, policymakers, and the public?
3. Industry's critics say chemical manufacturers should bear the burden of proof for the safety of their products before they hit the market. Industry's supporters say that mandating more safety research will hamper the introduction of products that consumers want, increase the price of products as research costs are passed on to consumers, and cause companies to

move to nations where standards are more lax. What do you think? Should government follow the precautionary principle and require proof of safety prior to a chemical's introduction to the market?

4. You have just been hired as the office manager for a high-tech startup company that employs bright and motivated young people but is located in an old, dilapidated building. Despite their youth, the company's employees seem perpetually sick with other unexplained illnesses. Looking into the building's history, you discover that the water pipes and ventilation system are many decades old, there have been repeated termite infestations, and part of the building was remodelled just before your company moved in but there are no records of what was done in the remodelling. Your company has all the latest furniture, computers, and other electronics. Most windows are sealed shut.

You want to figure out what is making the employees sick, and you want to convince your boss to give you a budget to hire professionals to examine the building for hazards. What hazards might you expect from this scenario? What arguments will you use to convince your employer to fund tests and inspections? What questions will you ask employees to help focus and prioritize any funds you are granted for testing and inspections?

INTERPRETING GRAPHS AND DATA

To minimize exposure to ultraviolet (UV) radiation and thus the risk of skin cancer, people have increased their use of sunscreen lotions in recent decades. Recently, however, research has shown that chemicals in sunscreens may themselves pose some risk to human health.[41] The compounds commonly used as UV protectants are fat soluble, environmentally persistent, and prone to bioaccumulation. Moreover, they exhibit estrogen-disrupting effects in laboratory rats. Although the benefits of sunscreen use are substantial, the risks are not yet well understood. A hypothetical trade-off between the risk factors of UV exposure and those of sunscreen use illustrates the balancing act of risk management on a very personal level.

1. What dosage of applied sunscreen on the graph corresponds to the greatest risk caused by UV exposure? What dosage corresponds to the greatest risk caused by chemicals in the sunscreen? Which of these two points on the graph is associated with the greater risk?

2. What dosage of applied sunscreen on the graph corresponds to the least risk caused by UV exposure? What dosage corresponds to the least risk caused by chemicals in the sunscreen? Which of these two points is associated with the greater risk?

3. The total risk to the individual is the sum of the two individual risks. What point on the graph corresponds

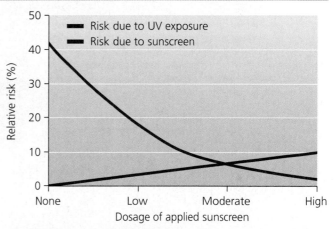

Hypothetical risk distributions for individuals using an estrogenic sunscreen to prevent skin cancer.
Source: Schlumpf, M., et al. (2001). *In vitro* and *in vivo* Estrogenicity of UV Screens. *Environmental Health Perspectives, 109*, pp. 239–244.

to the greatest total risk? What sunscreen dosage corresponds to the least total risk, according to this graph? (Is the answer unambiguous?)

4. What other types of information would you like to know before you might consider changing the way you use sunscreen? Can you think of any other cases that illustrate this sort of trade-off between dose-dependent risk factors?

MasteringEnvironmentalScience®

STUDENTS
Go to **MasteringEnvironmentalScience** for assignments, the eText, and the Study Area with practice tests, videos, current events, and activities.

INSTRUCTORS
Go to **MasteringEnvironmentalScience** for automatically graded activities, current events, videos, and reading questions that you can assign to your students, plus Instructor Resources.

20 Environmental Ethics and Economics: Values and Choices

This is an aerial view of Diavik Diamond Mine, Northwest Territories.

The Diavik Diamond Mine

Upon successfully completing this chapter, you will be able to

- Characterize the influences of culture and worldview on the choices people make
- Outline the nature, evolution, and expansion of environmental ethics in Western cultures
- Describe some basic principles and approaches of economics and how they relate to resource use and environmental impacts of production

- Explain the fundamentals of environmental economics, ecological economics, and natural resource accounting, and their implications for economic growth and sustainability

This is the mining settlement at Port Radium, Northwest Territories, in 1930.

Library and Archives Canada

Arctic
Ocean

Greenland
(DENMARK)

Denendeh

*Northwest
Territories*

Hudson
Bay

CANADA

CENTRAL CASE
MINING DENENDEH[1]

"This forest is where our people have fished and trapped for generations—so we have this responsibility to take care of it, and that responsibility came from the elders, our ancestors, who told us to do whatever we have to do to protect the land."

—SOPHIA RABLIAUSKAS, GOLDMAN ENVIRONMENT PRIZE WINNER, POPLAR RIVER FIRST NATION COMMUNITY

"Let us not forget that, in the end, all economies, whether new or old, are built on foundations of access to land, natural wealth and resources."

—MATTHEW COON COME, FORMER GRAND CHIEF OF THE CREE NATION AND NATIONAL CHIEF OF THE ASSEMBLY OF FIRST NATIONS

The traditional lands of the Dene and Métis form the Northwest Territories of Canada, known as *Denendeh*. The people of Denendeh have mined, manufactured, and traded metals—especially copper—since long before the arrival of Europeans.

Industrial-scale mining has played a significant role in the history and development of the Northwest Territories since 1930, when the first modern mining operation, Eldorado uranium mine, was established at Port Radium on Great Bear Lake (*Sahtu*, in the Dene language; see **FIGURE 1** and photo). The Con and Giant gold mines, which began production in Yellowknife in 1938 and 1948, respectively, were important to the economy of the area for more than 50 years. Mining continues today in the North but the emphasis has shifted to diamonds, with the development of mines like Ekati and Diavik, the first diamond mines in Canada.

Recent staking rushes in the Northwest Territories, particularly for diamonds, have generated concerns over unsettled land claims. For example, the public hearings of the Mackenzie Valley Environmental Impact Review Board for Wool Bay and Drybones Bay in 2003 centred on conflicts between unresolved Aboriginal land claims and mineral claims made by exploration companies. The environmental impact assessments halted exploration.

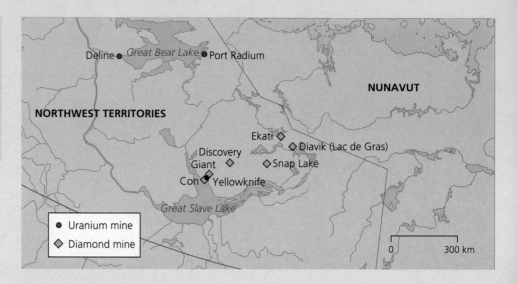

FIGURE 1
Great Bear Lake in the Northwest Territories was the site of uranium mining at the Eldorado mine in Port Radium from 1930 until the 1960s. More recent diamond mining centres on an area to the northeast of Yellowknife.

The Dene claimed that their spiritual and cultural use of the land was in jeopardy. This included many practical issues, such as the locations of gravesites, as well as improperly managed waste materials from exploration. In this case, the Aboriginal rights and title to the land were given precedence over mining exploration.

The environmental, social, and economic impacts of past mining development in the North have been considerable, including contamination of land and water, devastating economic boom-and-bust cycles, influxes of large numbers of non-native temporary workers, and persistent health issues for workers and residents. An early case in the Aboriginal rights movement involved Eldorado's alleged discrimination against Dene workers, discussed later in this chapter, which came to light when former mine workers developed health problems that may have resulted from inadequate protection against radiation.

Mines can leave a legacy of contamination, and Canada's North is strewn with closed and abandoned mines. Discovery gold mine, 80 km northeast of Yellowknife (see **FIGURE 1**), began production in 1944 and closed in 1969. Left behind were uncontrolled tailings and waste rock piles, mercury and cyanide contamination of the nearby lake, and an abandoned and crumbling town site that was finally demolished in 2005.

Mining also can cause profound social and economic changes. The inability to maintain traditional food systems

and cultural practices can undermine community health and wellness; impacts can include substance abuse and family breakdown. When Giant closed, the economic impacts on Yellowknife were devastating, causing housing prices to plummet. The recent diamond rush has created a new economic boom in Yellowknife and a construction push to house incoming workers, with the result that rents and housing prices are now some of the highest in Canada.

The life-cycle of a mine, and the involvement of a mining company with a site and its people, no longer end when the supply/demand for metal and minerals is depleted. Today, greater care is legally required when new mines are developed, to ensure that they will have minimal environmental, social, and health impacts. Mitigation, closure, and restoration plans, and financing for those plans, must be in place before a mine even begins production. The decision to transform an area into a mine comes with the responsibility of post-operational management and remediation, and a commitment to ongoing quality of life for workers and the community.

The new diamond mines at Ekati, Diavik, and Snap Lake have been designed so that, once mining is finished, the land will be reclaimed and restored. There will be no town, no permanent roads, no waste rock or tailings left behind. The new mines also distinguish themselves in the consideration they give to traditional knowledge, including learning from elders about environmental conditions, and applying this knowledge to construction, aquatic monitoring, and other aspects of mine management.

There is no single perspective or consensus on mining among Dene people today. Some call for greater involvement of Dene leaders in mine development. Some wish for a greater share of economic benefits and reduced risks. Others are concerned about the preservation of their land and culture. Decisions about mining in Canada's North must be based not only on scientific assessments but on social and cultural concerns as well. Mining in Denendeh provides an example of how values, beliefs, lifestyles, and traditions interact with economic interests to influence decisions about how to live within our environment.

Culture, Worldviews, and the Environment

The Dene have an opportunity to gain substantial economic benefits by allowing mining to proceed on their lands, but they are cognizant of the long history of environmental and cultural damage associated with mining in Canada's North. Trade-offs in which economic benefits and social or ethical concerns appear to be in conflict crop up frequently in environmental issues. In this chapter we will examine some of the underlying causes of such conflicts, how they may be resolved, and how they influence the environmental choices we all make every day.

Culture, worldviews, and values influence our understanding of the environment

Environmental science examines Earth's natural systems, how they affect humans, and how humans affect them. To address environmental problems, however, requires input from disciplines beyond the natural sciences. It is necessary to understand how people perceive their environment, how they relate to it philosophically and pragmatically, and how they value its elements. Ethics and economics are quite different disciplines, but each deals with questions of what we value and how those values influence our decisions and actions.

Anyone trying to address an environmental problem must try to understand not only how natural systems work but also how values shape human behaviour. Every action we take affects our environment. Growing food requires soil, cultivation, and irrigation. Building homes requires land, lumber, and metal. Manufacturing and fuelling vehicles require metal, plastic, glass, and petroleum. From nutrition to housing to transportation, we meet our needs by withdrawing resources and altering our surroundings.

Decisions about how we manipulate and exploit our environment to meet our needs depend in part on rational assessments of costs and benefits.

Our decisions are also heavily influenced by the culture of which we are a part, and by our particular worldview. Culture—the ensemble of knowledge, beliefs, values, and learned ways of life shared by a group of people—together with personal experience, influences each person's perception of the world and his or her place within it, something described as the person's worldview. A worldview reflects a person's (or group's) beliefs about the meaning, operation, and essence of the world.

People with different worldviews can study the same situation and review identical data yet draw dramatically different conclusions. For example, many well-meaning people support mining in Canada's North while many other well-meaning people oppose it. The officers, employees, and shareholders of the mining companies, and the government officials who support mining, view it as a source of jobs, income, energy, and economic growth. They believe mining will benefit the North in general and the Dene in particular. Opponents, in contrast, foresee environmental problems and negative social consequences. They recognize that mining disturbs the landscape and can pollute air and water, while community disruption, substance abuse, and crime can accompany mining booms.

The same is true of wildlife management in Canada's North. For a number of years, scientists have been sounding an alarm about polar bear populations, pressing for their inclusion on the Endangered Species lists. In the same period, many native hunters have disputed this, claiming that they see no change or even improvements in the status of polar bears in traditional hunting areas. The scientists feel that their modern tracking equipment tells a more complete story; some traditional hunters are scornful of these approaches, which may not be backed up by long cultural experience and expertise in the northern wilderness.

In Australia, too, there have been conflicts between corporations seeking to develop mining operations and

The Gundjehmi Aboriginal Corporation

(a) Ranger Uranium Mine, Kakadu region, Australia

Vale Archive

(b) Voisey's Bay Nickel Mine, northern Labrador, Canada

FIGURE 20.1

The Ranger mine **(a)**, located on Australian Aboriginal lands amid sacred sites, caused enough environmental impacts to spark fierce opposition to the proposed Jabiluka mine nearby. **(b)** The Voisey's Bay mine and concentrator site in northern Labrador faced similar challenges from both Innu Nation and Nunatsiavut Government. The mine's owner negotiated special agreements with the two groups prior to the opening of the mine, dealing with land rights, social and economic considerations, and environmental concerns.

Aboriginal people trying to maintain their traditional culture. One such group is the Mirarr clan, native or **indigenous** to Australia's Northern Territory. Several uranium deposits, key to the national economy, occur on Mirarr lands in the remote region of Kakadu. The region also features Kakadu National Park, a United Nations World Heritage Site recognized for its irreplaceable natural and cultural resources. The Ranger Uranium Mine (**FIGURE 20.1A**) was established on Mirarr land in 1978. When the owners proposed to open another uranium mine, Jabiluka, the Mirarr launched into an environmental battle that would rage over a number of years and several continents.

The Mirarr saw the Jabiluka mine as a threat not only to their health and the integrity of the environment but also to their culture and religion, which are deeply tied to the landscape. The proposed mine site was near traditional hunting and gathering sites in the floodplain of a river that provides the clan with food and water. Like many other Aborigines, the Mirarr hold the landscape to be sacred, and they depend on its resources for their daily needs. After having experienced repeated radioactive spills at the Ranger mine, many feared that contaminated water would be released into area creeks and that radioactive radon gas would emanate from stored waste materials, or that that dams holding mine waste could fail catastrophically in an earthquake. "We are talking about a uranium mine inside our largest national park," said Peter Robertson, coordinator of the Environment Centre of the Northern Territory at the time. "This is not a place to cut corners."[2]

Environmental activists worldwide joined the Mirarr's struggle. In late 2002 their efforts finally succeeded when chief executive officer of Rio Tinto, the corporation holding rights to the ore body, announced the cancellation of mining plans at Jabiluka. In 2007 the CEO of Rio Tinto reiterated the company's intention not to pressure the government for approval to develop Jabiluka without the prior informed consent of the traditional landowners.[3] **Prior informed consent**, one of the hallmark principles of modern environmental ethics, means that consent or acceptance of an activity (such as land development or waste disposal) is not legally valid unless the consenting person or group has been properly and adequately informed *and* can be shown to have a reasonable understanding of all potential impacts before giving consent.

In Aboriginal lands in Canada's North, the Voisey's Bay mine and concentrator (**FIGURE 20.1B**) in northern Labrador faced similar challenges from both Innu Nation and Nunatsiavut Government. Among others were concerns about the potential release of toxic materials into the environment; destruction of wildlife habitat and disturbance of animal migration routes; and contamination of food, water, and other resources used by nearby communities. The mine's owner, Vale, negotiated special agreements with the two Aboriginal groups prior to the opening of the mine. The agreements dealt with land rights, social and economic considerations, and numerous environmental concerns.

The Mirarr and the Innu Nation opposed mine development in these cases, despite the economic benefits promised to them in the form of jobs, income, development, and a higher material standard of living. The decision to fight was not easy; indeed, other Aboriginal groups, both in Australia and in Canada's North, have supported mine development in other circumstances.

In evaluating mine proposals, they have had to weigh economic, social, cultural, and philosophical questions as well as scientific ones. After agreement was reached, open-pit mining started at Voisey's Bay in 2005 and an underground mine is scheduled to begin construction in 2016. However, Mirarr leaders have recently reconfirmed their opposition to mining at Jabiluka.

Many factors shape our perceptions of the environment

The traditional culture and worldview of the Mirarr clan helped shape its response to the proposed Jabiluka mine. Australian Aborigines view the landscape around them as the physical embodiment of stories that express the beliefs and values central to their culture. The landscape to them is a sacred text. They believe that spirit ancestors travelled routes called "dreaming tracks," leaving signs and lessons in the landscape. By explaining the origins of specific landscape features, dreaming-track stories assign meaning to notable landmarks and help Aborigines construct detailed mental maps of their surroundings. The stories also teach lessons concerning family relations, hunting, food gathering, and conflict resolution. The Mirarr who opposed the Jabiluka uranium mine believed that it would desecrate sacred sites and compromise their culture.

Similarly, many Aboriginal people of Canada's North, including the Dene and Innu, believe that the landscape is inhabited by spirits, both benevolent and malevolent, and thus must be honoured and protected.[4] They worry that mining may have negative impacts on their lands, traditional hunting routes, water sources, or resource-gathering sites. There is also concern that noise, disruption, and emissions from such operations might cause harm to sacred animals or to sites that have spiritual or cultural significance.

Religion and spiritual beliefs are among many factors that can shape people's worldviews and perception of the environment. A community may also share a particular view of the environment if its members have lived similar experiences. For example, European Christian settlers in both Australia and North America viewed their environment as a hostile force because inclement weather, wild animals, and other natural forces frequently destroyed crops, killed livestock, and took settlers' lives. Such experiences were shared in stories and in songs and helped shape social attitudes in frontier communities. The view of nature as a hostile force and an adversary to be overcome passed from one generation to the next and still influences the way many North Americans and Australians view their surroundings.

Political ideology also shapes a person's worldview and attitude toward the environment. For instance, one's views on the role of government will influence whether or not one wants government to intervene in a market economy to protect environmental quality. Economic factors also sway how people perceive their environment and make decisions. An individual with a strong interest in the outcome of a decision that may result in his or her private gain or loss is said to have a *vested interest*. Mining company executives and shareholders have a vested interest in a decision to open an area to mining because a new mine can increase profits. Vested interests may lead people to view a proposed mine as a source of economic gain, while minimizing (whether consciously or subconsciously) the potential for negative environmental impacts.

There are many ways to understand the environment

An interesting aspect of the relationship of Indigenous peoples with their local environment is **traditional ecological knowledge (TEK)**, or *Indigenous ecological knowledge*, intimate knowledge of a particular environment possessed and passed along by those who have inhabited an area for many generations.[5] Examples include knowledge of the medicinal properties of local plants, wintering-over or migration habits of local animals, local geographical and microclimatic variations, or the sequence of tasks required to carry out a traditional task, such as trapping and butchering a large animal. Such a deep understanding is gained through generations of hunters, fishers, gatherers, and harvesters sharing their knowledge of the natural world, usually by way of oral teachings, songs, and storytelling.

In some circumstances, TEK can be assigned a market value. For example, Indigenous knowledge of local plants might be extremely valuable to a pharmaceutical company searching for plants with modern medicinal applications. In recent years the value of TEK has become more widely recognized, acknowledged, and remunerated by governments and industry. The Nunavut Wildlife Management Board, which meets annually to set limits on the annual polar bear hunt for Inuit traditional hunters, receives and weighs information both from government scientists and from the hunters themselves as part of the decision-making process.

Throughout this book you have encountered scientific data regarding the environmental impacts of our choices (where to make our homes, how to make a living, what to wear, what to eat, how to travel, how to spend our leisure time, and so on). Culture, worldviews, and values play critical roles in such choices and even can influence the interpretation of scientific data. Thus, acquiring a foundation of scientific understanding is only one part of the search for solutions to environmental problems. Attention to ethics and economics helps us understand why and how we value those things we value.

Environmental Ethics

The field of **ethics** is a branch of philosophy that involves the study of good and bad, right and wrong. The term *ethics* can also refer to the set of moral principles or values held by a person or a society. Ethicists help clarify how people judge right from wrong by elucidating the criteria, standards, or rules that people use in making these judgments. Such criteria are grounded in values—for instance, promoting human welfare, maximizing individual freedom, or minimizing pain and suffering.

People of different cultures or with different worldviews may differ in their fundamental values and thus may differ in the specific actions they consider to be right or wrong. Ethical standards help differentiate right from wrong across different cultures and situations. One classic ethical standard is the *categorical imperative* proposed by philosopher Immanuel Kant, which roughly approximates the "golden rule" common to many of the world's great religions. For example, Hindus learn that they should "not do to others what would cause pain if done to you." A central tenet of Buddhism is to "hurt not others in ways that you yourself would find hurtful." Christians are encouraged to "do unto others as you would have others do unto you." The universality of the golden rule or categorical imperative makes it a fundamental ethical standard.

Another ethical standard is the principle of *utility*, elaborated by British philosopher John Stuart Mill, among others. The **utilitarian principle** holds that something is right when it produces the greatest practical benefits for the most people. For example, a utilitarian might argue that forest biodiversity should be conserved because the possibility exists that a cure for cancer might be found there among the naturally occurring biological compounds. The argument that forest species should be preserved because they have an **intrinsic value**, or an inherent right to exist, would be much less convincing to a utilitarian thinker. This is particularly true in the context of marine protected areas, which are often justified on the basis of utilitarian ends.

Environmental ethics pertains to humans and the environment

The application of ethical standards to relationships between humans and non-human entities is known as **environmental ethics**. This branch of ethics arose once people began to perceive environmental changes brought about by industrialization. Human interactions with the environment frequently give rise to ethical questions that can be difficult to resolve. Consider some examples:

1. Does the present generation have an obligation to conserve resources for future generations? If so, how should this influence our decision-making, and how much are we obligated to sacrifice?
2. Are there situations that justify exposing some communities to a disproportionate share of pollution? If not, what actions are warranted in preventing this problem? By extension, if a certain community stands to gain the most from a particular activity, should that community be expected to take on most of the risk associated with the activity?
3. Are humans justified in driving species to extinction? Are we justified in causing other permanent changes in ecological systems? If destroying a forest would drive extinct an insect species few people have heard of, but would create jobs for 10 000 people, would that action be ethically admissible? What if it were an owl species, or an ape, or a whale? What if only 100 jobs would be created? What if it were a species that is harmful to humans, such as mosquitoes? What about a bacterium or a virus?

The intergenerational question—whether we owe consideration to those who will live on this planet and make use of its resources years from now—is of particular interest. The most common definition of **sustainable development** says that we must meet our current needs without compromising the availability of natural resources or the quality of life for future generations. But how can we tell what future generations may need or want, or what they will value or hold sacred?

For example, in 2006 construction crews began digging trenches for the laying of a 7-metre-wide extension of the Trans Mountain oil pipeline through pristine wilderness areas of Jasper National Park in the Canadian Rockies. The extension, which has allowed for the movement of an extra 40 000 barrels of oil each day from Alberta to markets in the United States, was approved in 1952. If the pipeline extension had been requested today, it is unlikely that it would have been approved because of the environmental disruption involved. Would it have been possible to know, in 1952, how this decision would be viewed more than 50 years later? And how can we determine how the environmental decisions that we make today will be viewed 50 years into the future?

We extend ethical consideration to non-human entities

Answers to questions like those above depend partly on what ethical standard(s) a person chooses to use. They also depend on the breadth and inclusiveness of the person's domain of ethical concern. A person who feels responsibility for the welfare of insects would answer the third question very differently from a person whose domain of ethical concern ends with humans. Most of us feel moral obligations to some entities in the world but by no means to all.

Throughout the history of Western cultures, people have gradually enlarged the array of entities they feel deserve ethical consideration. The enslavement of human beings by other human beings was common in many societies until recently, for instance. Women were not allowed to vote in Canada until 1916 (even then, only in Manitoba), and many still receive lower pay for equal work.[6] Consider, too, how little ethical consideration citizens of one nation generally extend to those of another on which their government has declared war. Human societies are only now beginning to embrace the principle that all people should be granted equal ethical consideration.

Our expanding domain of ethical concern has begun to include non-human entities as well. Mahatma Gandhi reportedly said, "The greatness of a nation and its moral progress can be judged by the way its animals are treated." Concern for the welfare of domesticated animals is evident today in humane societies and in the way many people provide for their pets. Animal-rights activists voice concern for animals that are hunted or used in laboratory testing. Most people now accept that wild animals (at least obviously sentient animals, such as primates and other large vertebrates, with which we share similarities) merit ethical consideration.

Today many environmentalists are concerned not only with certain animals but also with the well-being of whole natural communities. Some have gone still further, suggesting that all of nature—living and nonliving things, even rocks—should be ethically represented (**FIGURE 20.2**). If you think this is a silly idea, consider how you might react if someone put a fast-food restaurant on the top of Mt. Everest, or if a multinational corporation decided to paint a gigantic corporate logo on the Moon. Do the unique landmarks of the natural environment deserve ethical consideration? Do they have any inherent value, or the "right" to exist unaltered? Is their value distinct from the services they may render to humans, or inseparable from human interests? These questions are all worth considering.

What is behind this ongoing expansion of ethical consideration? Rising economic prosperity in Western cultures, as people became less anxious about their day-to-day survival, has helped enlarge our ethical domain. Science has also played a role in demonstrating that humans do not stand apart from nature but rather are part of it. Ecology, as it has developed over the past 75 years, has made clear that all organisms are intercon-

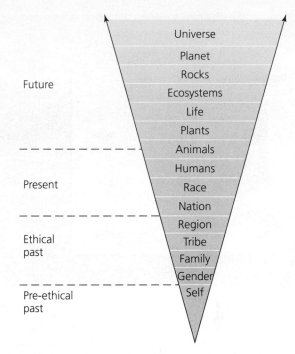

FIGURE 20.2
Through time, people in Western cultures have broadened the scope of their ethical consideration for others. We can view ethics progressing through time in a generalized way outward from the self.

nected, and what affects plants, animals, and ecosystems can in turn affect humans. Evolutionary biology has shown that humans are merely one species out of millions and have evolved subject to the same pressures as other organisms.

For many non-Western cultures, expansive ethical domains are nothing new. Many traditional cultures have long granted ethical standing to non-human entities. Aboriginal groups, like the Mirarr, Innu, and Dene, many of whom view their landscape as sacred and alive, are a case in point. However, it is worthwhile to examine Western ethical expansion because it underlies so many of modern secular society's beliefs and actions regarding the environment. We can simplify the continuum of attitudes toward the natural world by dividing it into three ethical perspectives: anthropocentrism, biocentrism, and ecocentrism.

Anthropocentrism **Anthropocentrism** takes a human-centred view of our relationship with the environment. An anthropocentrist denies or ignores the notion that non-human entities can have rights and measures the costs and benefits of actions solely according to their impact on people (**FIGURE 20.3**). To evaluate an action that affects the environment, an anthropocentrist might use such criteria as impacts on human health, economic costs and benefits, and aesthetic concerns.

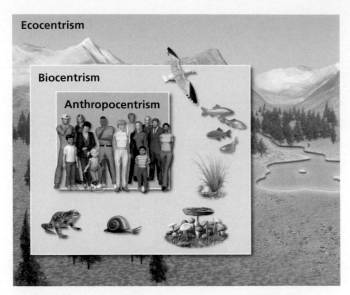

FIGURE 20.3
An anthropocentrist extends ethical standing only to humans and judges actions in terms of their effects on humans. A biocentrist values and considers all living things, human and otherwise. An ecocentrist extends ethical consideration to living and nonliving components of the environment and takes a holistic view of their interconnections, valuing the larger functional systems of which they are a part.

For example, if a mine provides a net economic benefit while doing no harm to human health and having little aesthetic impact, the anthropocentrist would conclude it was a worthwhile venture, even if it might drive some native species extinct. If protecting the area would provide spiritual, economic, or other benefits to humans now or in the future, an anthropocentrist might favour its protection. In the anthropocentric perspective, anything not providing benefit to people is considered to be of negligible value.

Biocentrism   In contrast to anthropocentrism, **biocentrism** ascribes values to actions, entities, or properties on the basis of their effects on all living things or on the integrity of the biotic realm in general (see **FIGURE 20.3**). In this perspective, all life has ethical standing. A biocentrist evaluates actions in terms of their overall impact on living things, including—but not exclusively focusing on—human beings. In the case of a mine proposal, a biocentrist might oppose the mine if it posed a serious threat to the abundance and variety of living things in the area, even if it would create jobs, generate economic growth, and pose no threat to human health. Some biocentrists advocate equal consideration of all living things, whereas others advocate that some types of organisms should receive more than others.

Ecocentrism   Ecocentrism judges actions in terms of their benefit or harm to the integrity of whole ecological systems, which consist of biotic and abiotic elements and the relationships among them (see **FIGURE 20.3**). An ecocentrist would value the well-being of entire species, communities, or ecosystems over the welfare of a given individual. Implicit in this view is that the preservation of larger systems generally protects their components, whereas selective protection of the components may not always safeguard the entire system. Ecocentrism is a *holistic* perspective. Not only does it encompass a wide variety of entities, but it also stresses preserving the connections that tie the entities together into functional systems.

Environmental ethics has ancient roots

Environmental ethics arose as an academic discipline in the early 1970s, but people have contemplated our relationship with nature for thousands of years. Ancient Aboriginal oral traditions and dreaming-track stories treat the environment as a source of sacred teachings, worthy of contemplation and protection. In the Jain Dharma, one of the oldest religious traditions in the world, compassion for all life—human and non-human—is a core belief. Jains are typically vegetarian or vegan and try to avoid foods obtained with unnecessary cruelty; for some, this even involves refusing to eat root vegetables (such as potatoes and onions) to avoid killing the plant from which they were obtained. In the Western tradition, the ancient Greek philosopher Plato expressed what he considered humans' moral obligation to the environment, writing, "The land is our ancestral home and we must cherish it even more than children cherish their mother."

Some ethicists and theologians have pointed to the religious traditions of Christianity, Judaism, and Islam as sources of anthropocentric hostility toward the environment. They point out biblical passages such as, "Be fruitful and multiply, and fill the earth and subdue it; and have dominion over the fish of the sea and over the birds of the air and over every living thing that moves upon the earth." Such wording has justified and encouraged separation from and animosity toward nature over the centuries, some scholars say.

Others emphasize sacred texts that encourage benevolent human stewardship over nature. For example, consider the biblical directive, "You shall not defile the land in which you live . . ." Although people have held differing views of their ethical relationship with the environment for millennia, environmental impacts that became apparent during the Industrial Revolution intensified debate about our responsibility toward the environment.

The Industrial Revolution inspired environmental philosophers

As the Industrial Revolution spread from Great Britain throughout Europe and elsewhere, its technological advances and resultant population growth amplified human impacts on the environment. In this period of social and economic transformation, agricultural economies became industrialized, machines enhanced or replaced human and animal labour, and much of the rural population moved into cities. Consumption of natural resources accelerated, and pollution increased dramatically as coal combustion fuelled railroads, steamships, ironworks, and factories.

Many writers and philosophers of the time criticized industrialization. British critic John Ruskin (1819–1900) called cities "little more than laboratories for the distillation into heaven of venomous smokes and smells." He complained that people prized the material benefits that nature could provide, but no longer appreciated its spiritual and aesthetic benefits. A number of citizens' groups sprang up in nineteenth-century England that could be considered some of the first environmental organizations. These included the Commons Preservation Society, the Coal Smoke Abatement Society, and the Selborne League, dedicated to the protection of rare plants, birds, and landscapes.

During the 1840s, a philosophical movement called *transcendentalism* flourished, espoused by American philosophers Ralph Waldo Emerson and Henry David Thoreau and poet Walt Whitman. The transcendentalists viewed nature as a direct manifestation of the divine, emphasizing the soul's oneness with nature and God. They objected to what they saw as their fellow citizens' obsession with material things. Through their writings the transcendentalists promoted a holistic view of nature. They identified a need to experience wild nature and portrayed natural entities as symbols or messengers of a deeper truth. Thoreau viewed nature as divine, but he also observed the natural world closely, in the manner of a scientist; he was in many ways one of the first ecologists. His book *Walden*, in which he recorded his observations and thoughts while he lived at Walden Pond away from the bustle of urban Massachusetts, is a classic of philosophical and environmental literature.

Conservation and preservation arose at the start of the twentieth century

One admirer of Emerson and Thoreau was John Muir (1838–1914), a Scottish immigrant who eventually settled in California and made the Yosemite Valley

FIGURE 20.4

A pioneering advocate of the preservation ethic, John Muir is remembered for his efforts to protect the Sierra Nevada from development and for his role in founding the Sierra Club. Muir (right) is shown with President Theodore Roosevelt in Yosemite National Park. After his 1903 wilderness camping trip with Muir, the president instructed his interior secretary to increase protected areas in the Sierra Nevada.

Library of Congress Prints and Photographs Division [LC-DIG-ppmsca-36413]

his wilderness home. Although Muir chose to live in isolation in his beloved Sierra Nevada for long stretches of time, he nonetheless became politically active and won fame as a tireless advocate for the preservation of wilderness (**FIGURE 20.4**).

Muir was motivated by the rapid deforestation and environmental degradation he witnessed throughout North America and by his belief that the natural world should be treated with the same respect that cathedrals receive. Today he is associated with the **preservation ethic**, which holds that we should protect the natural environment in a pristine, unaltered state. Muir argued that nature deserved protection for its own inherent value (an ecocentrist argument), but he also maintained that nature played a large role in human happiness and fulfillment (an anthropocentrist argument). "Everybody needs beauty as well as bread," he wrote in his 1912 work *The Yosemite*, "places to play in and pray in, where nature may heal and give strength to body and soul alike."

Canadian James Bernard Harkin (1875–1955) was strongly influenced by the writings of Muir, his contemporary. Harkin was the first commissioner of Dominion Parks (which eventually became Parks Canada) and is credited with saving vast areas of Canadian wilderness from development (**FIGURE 20.5**). He believed in the preservation of nature and in its spiritual, healing, and restorative powers. He was drawn to mountains, which, he said, "elevate the mind and purify the spirit." Harkin

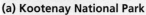

(a) Kootenay National Park

(b) James Harkin

FIGURE 20.5
Kootenay National Park **(a)** was designated as a national park in 1920. It was one of 13 national parks designated during the tenure of James Harkin **(b)** as Canada's commissioner of Dominion Parks. Mount Harkin in the Mitchell Range is named in his honour.

believed that setting aside land as national parks was only the beginning of preservation; the real challenge would be to maintain them as wilderness. Resource extraction and even some vehicle use were limited in national parks during his service. He eventually founded 13 national parks, drafted legislation for the protection of parks, established the Historic Sites and Monuments Board, and promoted the construction of scenic roadways through the Rockies. Harkin eventually came to be known as the "Father of National Parks" of Canada.[7]

Some of the factors that motivated Muir and Harkin also inspired American forester Gifford Pinchot (1865–1946), who opposed rapid deforestation and unregulated economic development of land. However, Pinchot took a more anthropocentric view of how and why nature should be valued. He is today the person most closely associated with the **conservation ethic**, which holds that humans should put natural resources to use but also that we have a responsibility to manage them wisely. Whereas preservation aims to preserve nature for its own sake, conservation promotes the prudent, efficient, and sustainable extraction and use of natural resources for the benefit of present and future generations. The conservation ethic uses a utilitarian standard, stating that in using resources, humans should attempt to provide the greatest good to the greatest number of people for the longest time.

Pinchot's counterpart in Canada was Clifford Sifton (1861–1929), a controversial politician and conservationist. As minister of the interior, appointed by Prime Minister Wilfrid Laurier, Sifton aggressively lured immigrants to settle and farm in the West, even going so far as to advertise Canadian lands in Europe. He was deeply committed to the agricultural development of land as Canada's principal natural resource—not an obvious conservationist position. However, Sifton was also a "champion" of Canada's natural resources, and was

particularly devoted to forest conservation and reforestation. He was the first chairman of the Commission for the Conservation of Natural Resources, which undertook detailed inventories of Canadian natural resources.[8]

Conservation and preservation are rooted in fundamentally different ethical approaches, which often meant that advocates were pitted against one another on policy issues of the day. Nonetheless, both branches represented reactions against the prevailing "frontier development ethic," which held that humans should be masters of nature, and which promoted economic development without regard to its negative consequences. Those who led the conservation and preservation movements in the nineteenth and early twentieth centuries left legacies that reverberate today in our ethical approaches to the environment.

The land ethic and deep ecology enlarged ethical boundaries

As a young forester and wildlife manager, Aldo Leopold (1887–1949) (**FIGURE 20.6**) began his career as a conservationist. At first, he embraced the government policy of shooting predators, such as wolves, to increase populations of deer and other game animals. At the same time, however, Leopold was following the development of ecological science. He eventually ceased to view certain species as "good" or "bad" and instead came to see that healthy ecological systems depend on the protection of all their interacting parts, including predators as well as prey. Drawing an analogy to mechanical maintenance, he wrote, "to keep every cog and wheel is the first precaution of intelligent tinkering."

Leopold argued that humans should view themselves and "the land" as members of the same community, and that people are obliged to treat the land in an ethical

FIGURE 20.6
Aldo Leopold, wildlife manager and pioneering environmental philosopher, articulated a new relationship between people and the environment. In his essay "The Land Ethic," he called on people to include the environment in their ethical framework.

manner based on mutual respect. In his 1949 essay "The Land Ethic," he wrote

> All ethics so far evolved rest upon a single premise: that the individual is a member of a community of interdependent parts . . . The land ethic simply enlarges the boundaries of the community to include soils, waters, plants, and animals, or collectively: the land . . . A land ethic changes the role of Homo sapiens from conqueror of the land-community to plain member and citizen of it.[9]

Leopold intended that the land ethic would help guide decision-making. "A thing is right," he wrote in his 1949 book *A Sand County Almanac*, "when it tends to preserve the integrity, stability, and beauty of the biotic community. It is wrong when it tends otherwise." Many today view Aldo Leopold as the most eloquent and important philosopher of environmental ethics.

One philosophical perspective that goes beyond even ecocentrism is **deep ecology**, which was first discussed in the 1970s by Norwegian philosopher Arne Naess, who coined the term. Proponents describe deep ecology as a holistic movement resting on principles of "self-realization" and biocentric equality. They define self-realization as the awareness that humans are inseparable from nature and that the air we breathe, the water we drink, and the foods we consume are both products of the environment and integral parts of us. Biocentric equality is the concept

that all living beings have equal value and that because we are truly inseparable from our environment, we should protect all living things as we would protect ourselves.

Ecofeminists see parallels between the oppression of nature and of women

As deep ecology and mainstream environmentalism were extending people's ethical domains during the 1960s and 1970s, major social movements, such as the civil rights movement and the feminist movement, were gaining prominence. A number of feminist scholars saw parallels in human behaviour toward nature and men's behaviour toward women. The degradation of nature and the social oppression of women shared common roots, these scholars asserted.

Ecofeminism argues that the patriarchal (male-dominated) structure of society—which tradition-ally grants more power and prestige to men than to women—is a root cause of both social and environmen-tal problems. Ecofeminists hold that a worldview tra-ditionally associated with women, which interprets the world in terms of interrelationships and cooperation, is more compatible with nature than a worldview tradi-tionally associated with men, which interprets the world in terms of hierarchies and competition. Ecofeminists maintain that a tendency to try to dominate and conquer has historically been exercised against both women and the natural environment.

One of the most interesting environmental movements of our time is Chipko Andolan, which had its philosophi-cal grounding in the principles of Gandhian nonviolent resistance, grassroots social activism, and ecofeminism.[10] The name *Chipko Andolan* ("Chipko Movement") comes from the Hindi root word *chipka*, "embrace," and *andolan* is Hindi for a "movement" or "campaign." Its use by the women of northern Uttarakhand in their nonviolent resis-tance against logging may have been the origin of the term *tree-hugger*, commonly used (often condescendingly) in reference to environmentalists. Chipko emerged in the early 1970s in the northern Uttarakhand region of India as an effort to stop clear-cutting from decimating the vast forests of northern India.

The Chipko movement came to a dramatic climax in 1973, when government workers turned up unan-nounced to cut trees but were met by a group of village women who refused to allow the work to proceed (**FIGURE 20.7A**). Leader Gaura Devi reportedly stated, "The forest is like our mother. You will have to shoot us before you can cut it down." The women stood watch over the forest, wrapping their arms around the trees.

Right Livelihood Award Foundation. www.rightlivelihood.org

Patrick Robert/Corbis

(a) The Chipko Movement   **(b) Wangari Maathai**

FIGURE 20.7   In the early 1970s, grassroots resistance efforts by village women led to the establishment of the Chipko Movement, dedicated to preventing deforestation in northern India's Himalayan foothills **(a)**. The founder of Kenya's Green Belt Movement, Professor Wangari Maathai **(b)**, was awarded the Nobel Peace Prize in 2004 for her work to empower women and fight deforestation; she died in 2011.

Eventually, after considerable negotiation, the government declared the region to be an environmentally "sensitive" area. Today, Chipko is an international icon of grassroots environmentalism.

Another movement rooted in ecofeminism and the empowerment of the poor is the Green Belt Movement of Kenya. This organization, which began by paying impoverished village women to plant tree seedlings, was founded in 1977. The Nobel Peace Prize was awarded to Green Belt's founder, Wangari Maathai, in 2004 (**FIGURE 20.7B**).[11]

Environmental justice seeks equitable distribution of costs and benefits

Our society's domain of ethical concern has been expanding from rich to poor and from majority races and ethnic groups to minority ones. This ethical expansion involves applying a standard of fairness and equality and has given rise to the environmental justice

movement. **Environmental justice** is based on the principle that all people have the right to live and work in a clean, healthy environment; to receive protection from the risks and impacts of environmental degradation, and to be compensated for having suffered such impacts; and to have equitable access to environmental resources of high quality.

The environmental justice movement is fuelled by the fact that the poor and minorities tend to be exposed to a greater share of pollution, hazards, and environmental degradation than are richer people and whites. This has been supported by scientific research (**FIGURE 20.8**). For example, studies have found the percentage of minorities in areas with toxic waste sites to be twice that in areas without toxic waste sites. Researchers in many parts of the world who study air pollution, lead poisoning, pesticide exposure, and workplace hazards have found similar patterns.

A protest in the early 1980s by African Americans in Warren County, North Carolina, against a toxic waste dump in their community is widely seen as the beginning of the movement in North America. The state had chosen

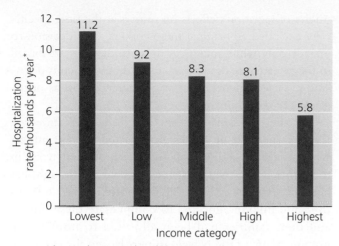

*due to asthma, croup, bronchitis, or pneumonia

FIGURE 20.8
Many studies have demonstrated links between exposure to pollution and socioeconomic status. This is shown by this graph comparing incidences of hospitalization for respiratory illnesses—a common health impact of exposure to environmental pollutants—among children of different income levels in Toronto.

to site the dump in the county with the highest percentage of blacks, prompting residents to suspect "environmental racism." Environmental justice grew in prominence as more people began fighting environmental hazards in their communities.

The early environmental justice movement—like the Chipko and Green Belt movements internationally but in contrast to early environmental movements in North America—was made up largely of low-income people and minorities. Today the movement has broadened to encompass worker health and safety; education and financial services for preparation for, response to, and recovery from natural disasters; and access to land, water, and other environmental resources worldwide.

There are two basic ways that environmental injustice is manifested. First, a community or group can be denied equitable access to environmental resources. This often occurs where poverty is common, or where land of high quality is scarce. For example, in many developing countries rich landowners control the best agricultural land, while poor subsistence farmers are left to scratch out a living on small patches of unsuitable land. In richer countries we may see this inequity expressed differently: as expensive beachfront or mountain-view properties that are available only to the wealthiest few, for example, or as limited access to an expensive diet rich in vitamins and nutrients.

Second, a community or group can be subjected to environmental injustice by having disproportionate risks or costs of pollution or degradation transferred to them, as in the example (above) of toxic waste dumps sited in communities of racial or ethnic minorities. In less

economically developed parts of the world, the poor and other marginalized groups are directly dependent on the environment for survival and therefore suffer the harshest and most immediate impacts of environmental degradation. **Environmental refugeeism** (or *environmental migrantism*) may result in cases where the degradation is so intense that residents are unable to survive and are forced to leave their land. Cultural differences can lead to disproportionate exposure of specific groups to the effects of environmental degradation. For example, the traditional diet of many of Canada's Aboriginal people is heavily reliant on fish and marine mammals, which tend to have high concentrations of harmful pollutants, such as mercury.

Aboriginal groups struggling to maintain a traditional lifestyle have been linked with the environmental justice movement. Many such struggles originate not as environmental causes but as social justice causes. The rubbertappers or *seringueiros* of the Brazilian Amazon are representative of causes that originate at this intersection of environmental justice and social justice. The *seringueiros* and their leader, Chico Mendes (**FIGURE 20.9**), caught the world's attention when it became apparent that the battle to save their traditional economy—based on the sustainable extraction of latex and other forest resources—was intimately connected with efforts to stop deforestation of the Amazonian rain forest for ranching.

Antonio Scorza/Getty Images

FIGURE 20.9
Chico Mendes, seen here shortly before his assassination in 1988, was a social activist on behalf of the traditional lifestyle of his people, but he caught the eye of the world because of his struggle to preserve the Amazon rain forest.

The *seringueiros* forged a partnership with Indigenous Amazonians, who face many of the same challenges to their traditional lands, resources, and lifestyle. Chico Mendes thought of himself first as a social activist, fighting on behalf of his people and their way of life. Would the international environmental movement have embraced his cause if it had *not* been intimately connected with the rain forest? Sadly, Chico Mendes's iconic status and effectiveness as an activist on behalf of his people and their beloved forests and traditions were dramatically heightened by his assassination, in 1988, by Brazilian ranchers.

Critics have characterized the attempts of the predominantly white Australian government and uranium mining companies to open mines on traditional lands of the Mirarr as environmental injustice. In North America, as well, uranium mining has been a focus of environmental justice concerns. As discussed in the Central Case, Dene mineworkers from the Northwest Territories and Saskatchewan, as well as Navajo mineworkers from the United States, suffered delayed health effects that may have been caused by working in uranium mines with minimal safeguards.

Dene workers were hired as early as 1930 to haul radioactive uranium ore wrapped in cloth bags at Eldorado. The governments of Canada (then the world's largest supplier of uranium) and the United States (the main purchaser of the ore) were aware of the health hazards of dealing with radioactive ores, as documented in government publications from the time.[12] However, these concerns were not communicated to Dene workers, many of whom did not speak English, and protective equipment was not provided. For decades, until the mine was shut down in the 1960s, Dene miners "slept on the ore, ate fish from water contaminated by radioactive tailings and breathed radioactive dust while on the barges, docks and portages," according to the *Calgary Herald*.[13]

The later deaths, from lung cancer, of many Port Radium miners led to the nickname "Village of Widows" for the settlement of Deline on the western edge of Great Bear Lake, and the Dene Nation called for an official response from the federal government. In 2005 the government of Canada issued the results of its study, which concluded that there was insufficient evidence to link the deaths of the miners with exposure to radiation during mining work at Port Radium. The report acknowledged that the mine had had an impact on soil and water quality in the immediate vicinity of Great Bear Lake and called for the proper decommissioning of the site, with provisions for continued environmental monitoring. Compensation for residents and the remaining miners was not addressed in the report.[14]

Although cases of lung cancer began to appear among uranium mineworkers in the early 1960s, scientific studies of radiation's effects on miners at the time specifically excluded Aboriginal mineworkers. The decision to include only white miners in those studies was attributed to the researchers' desire to study a "homogeneous population." A later generation would perceive this as negligence and discrimination.

In the United States, the *Radiation Exposure Compensation Act* of 1990 compensated Navajo miners who suffered health effects from unprotected work in the mines. Even the compensation process, though, has been controversial; as of 2006, approximately 80% of the $300 million allocated to compensate these miners and their families had been allocated to non-Native miners. Many of the Navajo miners lacked the extensive documentation required to apply for compensation, including medical documentation and records of the exact dates and conditions in which they worked. Some Navajo miners also have been denied compensation because they fail to qualify as "nonsmokers" as a result of having participated in traditional Native ceremonies involving smoke.[15] Cases like these illustrate the interplay between changing ethical values and resultant policymaking.

Economics and Environmental Goods and Services

Economics, like ethics, addresses people's values, influences behaviour, and widely informs policy. People who oppose mining and other developments in wilderness areas or on traditional lands typically do so on the basis of worries over environmental impacts, or perhaps ethical concerns. Few challenge such activities on economic grounds; even opponents generally recognize them as lucrative activities that generate jobs and income. On the other hand, support for mining and other development and resource extraction activities is primarily founded in economic reasoning.

Conflict between ethical and economic motivations is thus a recurrent theme in environmental issues. We often hear it said that environmental protection is in opposition to economic progress. Arguments are made that environmental protection costs too much money, interferes with progress, and leads to job loss. But is this necessarily the case? Growing numbers of economists assert that there need be no such trade-off—that, in fact, environmental protection can be *good* for the economy. The position one takes often depends on whether one thinks in the short term or the long term, and whether one holds to traditional economic schools of thought or to newer ones that view human economies as coupled to the natural environment.

Economics studies the allocation of scarce resources

Like ethics, economics examines factors that guide human behaviour. **Economics** is the study of how people decide to use scarce resources to provide **goods**, material commodities manufactured for and bought by individuals and businesses, and **services**, work done for a financial return. By this definition, environmental problems are economic problems that can intensify as population and resource consumption increase. For example, pollution may be viewed as depleting the scarce resources of clean air, water, or soil. The word *economics* and the word *ecology* come from the same Greek root, *oikos,* meaning "household." Economists traditionally have studied the household of human society, and ecologists the broader household of all life.

The economic judgments we make often pertain to short time scales, and in the short term many activities that cause environmental harm, such as natural resource extraction, may be economically profitable. In the longer term, however, environmental degradation generally feeds back and harms economies. The view one takes also depends on whether one stands to benefit directly. Often when resource extraction or development causes environmental degradation, a few private parties benefit economically, but the broader public is harmed.

Traditional economic schools of thought have underestimated or overlooked the contributions of the environment to our economies, and as a result, many people have equated environmental protection with economic sacrifice. Newer schools of thought recognize that economies are coupled to the environment and reliant on its goods and services. For people holding this worldview, our economic health depends on protecting the environment and conserving natural resources.

Several types of economies exist today

An **economy** is a social system that converts resources into goods and services. The resources that are turned into goods and services can be materials that are derived directly from the environment, such as lumber or minerals, or energy resources such as oil. There are other types of resources, too, such as human resources.

The oldest type of economy is the *subsistence economy*. People in subsistence economies—who still compose much of the human population—meet most or all of their daily needs directly from nature and do not purchase or trade for most of life's necessities. You have already learned about subsistence agriculture and subsistence fishing in earlier chapters of this book.

A second type of economy is the *capitalist market economy*. In this system, buyers and sellers interact to determine which goods and services to produce, how much to produce, and how these should be produced and distributed. Capitalist economies are often contrasted with a third type, state socialist economies, or *centrally planned economies*, in which government determines in a top-down manner how to allocate resources.

A utopian *pure market economy* would operate without any government intervention. In reality, all capitalist market economies today, including that of Canada, are hybrid systems (often called *mixed economies*) that involve at least some government intervention in the functioning of the market. In modern market economies, governments intervene for several reasons: (1) to eliminate unfair advantages held by monopoly buyers or sellers; (2) to provide social services, such as national defence, medical care, and education; (3) to provide "safety nets" (for older adults, the unemployed, those with chronic illnesses or disabilities, victims of natural disasters, and so on); (4) to manage commonly owned resources (the "commons"); and (5) to mitigate pollution and other types of environmental damage.

Environment and economy are intricately linked

All human economies exist within the larger environment and depend on it in important ways. Economies receive inputs from the environment, process them in complex ways that enable human society to function, and then discharge outputs of waste from this process into the environment. Economies are thus *open systems* integrated with the larger environmental system of which they are a part. Earth, in turn, is a *closed system*. This means that the material inputs Earth can provide to economies are ultimately finite and so is the waste-absorbing capacity of the planet.

Although the interactions between human economies and the non-human environment are readily apparent, traditional economic schools of thought have long overlooked the importance of these connections. Indeed, most conventional economists today still adhere to a worldview that largely ignores the environment (**FIGURE 20.10A**), and this worldview continues to drive most policy decisions. A conventional economic worldview essentially holds that environmental resources (the inputs into the economy) are limitless and free and that wastes (outputs) can be endlessly exported and absorbed by the environment, at no cost.

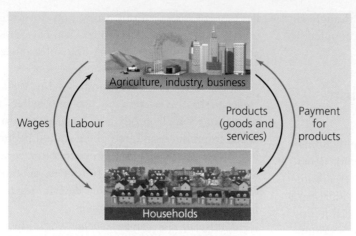

(a) Conventional view of economic activity

FIGURE 20.10
Conventional economics focuses on the processes of production and consumption between households and businesses **(a)**, viewing the environment as a "factor of production" that helps enable the production of goods. Environmental and ecological economists view the human economy as existing within the natural environment **(b)**, receiving resources from it, discharging waste into it, and interacting with it through ecosystem services.

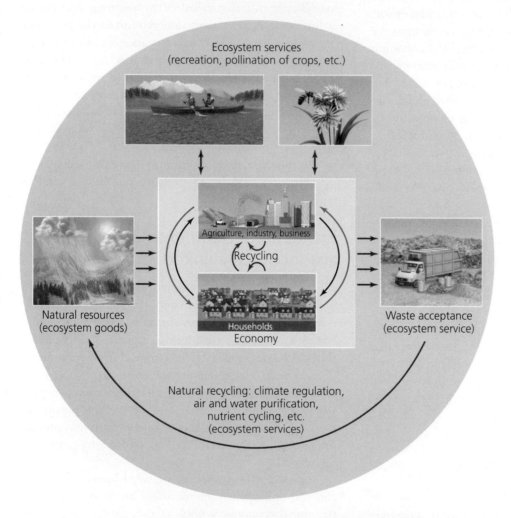

(b) Economic activity as viewed by environmental and ecological economists

However, modern economists belonging to the fast-growing fields of environmental economics, ecological economics, and natural resource accounting explicitly accept that human economies are subsets of the environment and depend crucially on the environment (**FIGURE 20.10B**). **Ecological economics** takes a holistic view of the linkages between environment and economy, applying the principles of ecology to the study of economics. **Environmental economics** and natural resource accounting are both seated within traditional economics, but pay particular attention to valuing environmental goods and services and giving credit for actions taken on behalf of the environment.

Economic activity uses resources from the environment. Natural resources are the substances and forces we need to survive: the Sun's energy, fresh water, trees that provide lumber, rocks that provide metals, and fossil fuels and other energy sources that power our machines. We can think of natural resources as "goods" produced by nature. Without Earth's natural resources, there would be no human economies and no human beings.

Environmental systems also function in a manner that naturally supports economies. Earth's ecological systems purify air and water, cycle nutrients, provide for the pollination of plants by animals, and serve as receptacles and recycling systems for the wastes generated by our economic activity. Such essential services, often called **ecosystem services** (**TABLE 20.1**), sustain the life that makes our economic activity possible. Some ecosystem services represent the very nuts-and-bolts of our survival; others enhance our quality of life.

Although the environment allows economic activity to occur by providing ecosystem goods and services, that economic activity can affect the environment in return. When we deplete natural resources or produce pollution, we degrade the ability of ecological systems to function. The *Millennium Ecosystem Assessment* concluded in 2005 that 15 of 24 ecosystem services surveyed globally were being degraded or used unsustainably. The degradation of ecosystem services can in turn negatively affect economies. Ecological degradation harms poor people more than wealthy people, the Millennium Ecosystem Assessment found. As a result, restoring ecosystem services is a prime objective for alleviating poverty in much of the world.

These interrelationships have only recently become widely recognized. Let us briefly examine how economic thought has changed over the years, tracing the path that is now beginning to lead economies to become more compatible with natural systems.

TABLE 20.1 Ecosystem Services

Type of ecosystem service*	Example(s)
Regulation of atmospheric gases	Maintaining the ozone layer; balancing oxygen, carbon dioxide, and other gases
Regulation of climate	Controlling global temperature and precipitation through oceanic and atmospheric currents, greenhouse gases, cloud formation, and so on
Protection and buffering	Providing storm protection, flood control, and drought recovery, mainly through vegetation and shoreline structure
Regulation of water flow	Providing water for agriculture, industry, transportation
Storage of water	Providing water through watersheds, reservoirs, aquifers
Control of erosion	Preventing soil loss from wind or runoff; storing silt in lakes and wetlands
Formation of soil	Weathering rock; accumulating organic material
Cycling of nutrients	Cycling carbon, nitrogen, phosphorus, sulphur, and other nutrients through ecosystems
Waste treatment	Removing toxins, recovering nutrients, controlling pollution
Pollination of plants	Transporting floral gametes by wind or pollinating animals, enabling crops and wild plants to reproduce
Population control	Controlling prey with predators; controlling hosts with parasites; controlling herbivory on crops with predators and parasites
Provision of habitat	Providing ecological settings in which creatures can breed, feed, rest, migrate, winter
Provision of food	Producing fish, game, crops, nuts, and fruits that humans obtain by hunting, gathering, fishing, subsistence farming
Supply of raw materials	Producing lumber, fuel, metals, fodder
Genetic resources	Providing unique biological sources for medicine, materials science, genes for resistance to plant pathogens and crop pests, ornamental species (pets and horticultural plant varieties)
Recreational opportunities	Ecotourism, sport fishing, hiking, birding, kayaking, other outdoor recreation
Noncommercial services	Aesthetic, artistic, educational, spiritual, and/or scientific values of ecosystems

*Ecosystem "goods" are here included in ecosystem services.

Source: Based on Costanza, R., et al. (1997). The Value of the World's Ecosystem Services and Natural Capital. *Nature, 387*, pp. 253–260.

Classical economics promoted the free market

Economics shares a common intellectual heritage with ethics, and practitioners of both have long been interested in the concept of human progress, and the relationship between individual action and societal well-being. Some philosophers argued that individuals acting in their own self-interest would harm society (as in the tragedy of the commons). Others believed that such behaviour could benefit society, as long as the behaviour was constrained by the rule of law and private property rights and operated within fairly competitive markets.

The latter view was articulated by Scottish philosopher Adam Smith (1723–1790). Known today as the father of classical economics, Smith believed that when people are free to pursue their own economic self-interest in a competitive marketplace, the marketplace will behave as will benefit society as a whole. In his 1776 book *Inquiry into the Nature and Causes of the Wealth of Nations*, Smith wrote

> *[Each individual] intends only his own security, only his own gain. And he is led in this by an invisible hand to promote an end which was no part of intention. By pursuing his own interests he frequently promotes that of society more effectually than when he really intends to.[16]*

Smith's philosophy remains a pillar of free market thought today, and many credit it for the tremendous gains in material prosperity that industrialized nations have experienced in the past few centuries. Others argue that the policies spawned by free-market thought worsen inequalities between rich and poor and contribute to environmental degradation. Market capitalism, these critics assert, should be constrained and regulated by democratic government.

Neoclassical economics considers price, supply, and demand

Economists subsequently took more quantitative approaches to issues related to price, supply, and demand. Modern *neoclassical economics* examines the psychological factors underlying consumer choices, explaining market prices in terms of consumer preferences for units of particular commodities. In neoclassical economic theory, buyers desire the lowest possible price, whereas sellers desire the highest possible price. This conflict between buyers and sellers results in a compromise price being reached and the "right" quantity of commodities being bought and sold.

This is often phrased in terms of *supply*, the amount of a product that is available and offered for sale at a given price, and *demand*, the amount of a product people need and will buy at a given price if free to do so. Theoretically, when supply goes up, prices fall; when demand is high

and supplies are limited, prices increase. In theory, the market automatically moves toward an equilibrium point, a price at which supply equals demand (**FIGURE 20.11A**).

Similar reasoning can be applied to environmental issues, whereby economists can determine "optimal" levels of resource use or pollution control, either for private companies or for society as a whole. In **FIGURE 20.11B**, the cost per unit of resource use or environmental cleanup (blue line) rises as the resource extraction or remediation proceeds, and it becomes expensive to extract or clean up

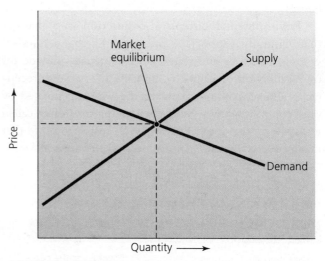

(a) Classic supply–demand curve

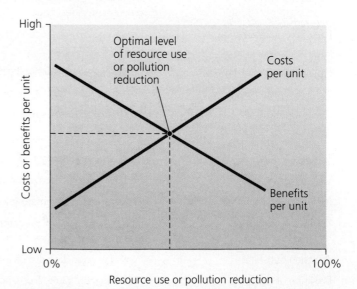

(b) Marginal benefit and cost curves

FIGURE 20.11

This basic supply-and-demand curve **(a)** illustrates the relationship among supply, demand, and market equilibrium, the "balance point" at which demand is equal to supply. We can use a similar graph **(b)** to determine an "optimal" level of resource use or pollution mitigation for society.

In graph **(b)**, what would happen if the costs to society per unit of production of this item were to increase? Sketch the curve. What kinds of costs might they be? What happens to the optimal levels of pollution and resource use in this case?

the remaining amounts. Meanwhile, the benefits per unit of resource use or pollution cleanup (red line) decrease. The point where the lines intersect gives the optimal levels of environmental benefit versus environmental harm. In other words, this diagram is showing the optimal balance—from the perspective of society as a whole—between the costs and benefits of resource use and environmental protection and degradation. Note that the balance point will never be at zero because we will always need to extract resources, and human activities will always have an impact on the environment. But economics can help us determine the best possible balance between these costs and benefits.

Cost–benefit analysis is a useful tool

Neoclassical economists commonly use a method referred to as **cost–benefit analysis**. In this approach, estimated costs for a proposed action are totalled and compared with the sum of benefits anticipated to result from the action. If total benefits exceed costs, the action should be pursued; if costs exceed benefits, it should not. When choosing among multiple alternative actions, the one with the greatest excess of benefits over costs should be chosen.

This reasoning seems eminently logical; however, problems often arise in applying it to environmental situations because not all costs and benefits are easily quantified, or even easily identified. It may be simple to quantify wages paid to uranium miners or the market value of uranium extracted from a mine or even the cost of measures to minimize health risks for miners. But it is much more difficult to assess the cost of a landscape scarred by mine development, or the cost of radioactive contamination of a stream, or the costs in health and emotional well-being to the families of miners who die from mining-related cancers.

Because some costs and benefits cannot easily be assigned monetary values, cost–benefit analysis is often controversial. Moreover, because economic benefits usually are more easily quantified than environmental costs, economic benefits tend to be overrepresented in traditional cost–benefit analyses. As a result, environmental advocates often feel these analyses are biased in favour of economic development and against environmental protection.

A corollary to cost–benefit analysis has been the development of a variety of mechanisms for understanding, calculating, and defining the value, to society and to individuals, of environmental costs and benefits. In general, **valuation** refers to the attempt to quantify the value of a particular environmental good or service—even if it cannot easily be expressed in monetary terms. We will discuss next how different valuation approaches can be used to more accurately represent the environment in cost–benefit analysis and asset accounting.

Environmental Implications of Economics

Today's capitalist market systems have generated unprecedented material wealth, employment, and many other desirable outcomes, but they have also contributed to environmental problems. Four of the fundamental assumptions of neoclassical economics have implications for the environment:

1. Resources are infinite or substitutable.
2. Long-term effects should be discounted.
3. Costs and benefits are internal.
4. Growth is good.

Let us critically examine each of these assumptions in turn.

Are resources infinite or substitutable?

Neoclassical economic models generally treat the supply of workers and other resources as being either infinite or largely "substitutable and interchangeable." This implies that once we have depleted a resource—natural, human, or otherwise—we should be able to find a replacement for it. Human resources can substitute for financial resources, for instance, or manufactured resources can substitute for natural resources. Theory allows that the substituted resource may be less efficient or more costly, but some degree of substitutability is generally assumed. In other words, traditional economists have considered environmental goods and services to be so-called free gifts of nature—infinitely abundant and resilient, and ultimately substitutable by human technological ingenuity.

The question of the substitutability of environmental resources—let's call them *natural capital*—is central to our concepts of sustainability (the goal) and sustainable development (the process for getting to the goal). Recall that we defined sustainable development as meeting our present needs without compromising the ability of future generations to meet their needs. If we use up a critical, life-supporting resource, can we substitute something else—perhaps some kind of manufactured capital asset—in its place, for the use of future generations? Would that be sufficient to ensure that future generations will have the resources they need to meet their needs?

In fact, it is true that many resources can be replaced; societies have transitioned from manual labour to animal labour to steam-driven power to fossil fuel power and may yet transition to renewable power sources,

Sir Nicholas Stern led a team to assess the economic impacts of climate change.

Alastair Grant/AP Images

THE SCIENCE BEHIND THE STORY

Ethics and Inter-generationality in Economics: Discounting, Climate Change, and the Stern Review

How much will it cost our society if we do nothing about global climate change? To address this question and to help decide how to respond to climate change, the British government commissioned economist Nicholas Stern to lead a team to assess the economic costs that a changing climate may impose on society. But once Stern's report was published in 2006, a debate ensued that had as much to do

with ethics as with economics. The dispute centred on discounting, an economic practice heavy with ethical implications.

To produce the *Stern Review on the Economics of Climate Change*, Stern's research team surveyed the burgeoning scientific literature on the impacts of rising temperatures, changing rainfall patterns, rising sea level, and increasing storminess (discussed in the chapter on climate change). The team then estimated the economic consequences of these climatic changes and tried to put a price tag on the global cost. The report concluded that without action to forestall it, climate change would cause losses in annual global gross domestic product (GDP) of 5–20% by the year 2200 (see **FIGURE 1**).

Stern's team also calculated that by paying just 1% of GDP annually starting now, our society could stabilize atmospheric greenhouse gas concentrations and prevent most of these future economic losses. The bottom line: Spending a relatively small amount of money now will save us much larger expenses in the future. This conclusion caught the attention of governments worldwide; for the first time, leading economists were advancing a strong economic argument for tackling climate change immediately.

The *Stern Review's* conclusions depend partly on how one chooses to weigh future impacts versus current impacts. Stern used two discount factors.

One accounted for the likelihood that people in the future will be richer than we are today and thus better able to handle economic costs. The other considered whether the future should be discounted simply because it is the future. This latter discount factor (called a *pure time discount factor*) is essentially an ethical issue because it assigns explicit values to the welfare of future versus current generations, valuing the present more highly than the future.

The *Stern Review* used a pure time discount rate of 0.1%. This means that an impact occurring next year is judged 99.9% as important as one occurring this year. It means that the welfare of a person born 100 years from now is valued at 90% of one born today. This discount rate treats current and future generations *nearly* equally. Future generations are down-weighted only because of the (very small) possibility that our species could go extinct (in which case, there would be no future generations to be concerned about).

Some economists viewed this discount rate as too low. Yale University economist William Nordhaus proposed a discount rate starting at 3% and falling to 1% in 300 years. He maintained that such numbers are more objective because they reflect how people value things, as revealed by prices we pay in the marketplace. Indeed, when economists assess capital investments or construction projects (say, development of a railroad, dam, or highway)

---

like solar energy. However, Earth's material resources are ultimately limited. Nonrenewable resources, such as fossil fuels, can be depleted, and even renewable resources can be used up if we exploit them more quickly than they can be replenished. This has been experienced by many ancient civilizations, including the inhabitants of Rapa Nui, who harvested wood faster than their forests could regrow.

However, some resources are simply not replaceable. Once a species has disappeared from this planet by extinction, we can't bring it back. We don't know for sure that future generations will need or value that species, but—by definition—we are not meeting the requirements of our definition of sustainable development if we do not leave them the choice and the possibility.

Should long-term effects be discounted?

Although few people would dispute that resources are *ultimately* limited, many assume that their depletion will take place so far in the future that there is no need for current generations to worry. For economists in the neoclassical tradition, an event far in the future counts much less than one in the present; in economic terminology, we say that these future effects are "discounted."

Through discounting, short-term and present-day costs and benefits are granted more importance than long-term costs and benefits. This encourages policymakers to play down long-term consequences of decisions we make today. For a discussion of how

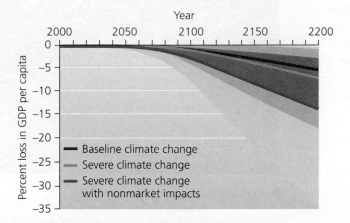

FIGURE 1

Baseline climate change forecast by the IPCC could decrease global per capita GDP by 5.3% annually by the year 2200. Severe climate change could bring annual losses of 7.3%—and adding nonmarket values raises this figure to 13.8%. Grey-shaded areas show ranges of future values judged statistically to be 95% likely; darkest grey indicates where ranges overlap for all three lines, and medium grey shows where ranges overlap for two lines. *Source:* Data from HM Treasury (2007). *Stern Review on the Economics of Climate Change.* London, U.K.

they typically choose discount rates close to those Nordhaus suggests. Nordhaus argued that the *Stern Review*'s near-zero discount rate overweights the future, forcing people today to pay too much to address hypothetical future impacts.

In their response to Nordhaus and other critics, Stern and his team argued that discount rates of 3% or 1% may be useful for assessing development projects but are too high for long-term environmental problems that directly affect human well-being. A 3% rate means that a person born in 2015 is valued only half as much as a person born in 1990. It means a grandchild is judged to be worth far less than a grandparent simply because of the dates they were born. For various reasons, Stern argued, the market should not be used to guide ethical decisions.

Stern's group also published sensitivity analyses that examined how their conclusions would vary under different discount rates. These showed that the report's main message—that it's cheaper to deal with climate change now than later—was robust across all discount rates up to at least 1.5%.

At the end of the day, the choice of a discount rate is an ethical decision on which well-intentioned people may differ. As governments, businesses, and individuals begin to invest in addressing climate change, the debate over the *Stern Review* reveals how ethics and economics remain intertwined.

Addressing global climate change is becoming urgent, as the climate is changing faster than scientists had predicted just a few years ago. Stern announced that newer scientific forecasts of accelerated change mean that it is becoming even more expensive to mitigate global warming than the *Stern Review* had calculated. Instead of 1% of GDP annually, the world would need to spend 2% of GDP each year in order to control climate change, Stern said—and that was in 2008. Since then, Hurricane Sandy and other climate-related disasters have made clear that the costs of inaction could be immense. These developments are all helping drive home the same message: The more quickly we get going, the better off we'll be.

discounting affects approaches to the mitigation of climate change, see "The Science behind the Story: Ethics and Intergenerationality in Economics: Discounting, Climate Change, and the Stern Review."

For example, a stock of uranium ore in the ground or living trees standing in the forest or fish in a stream represent potential commodities that would have monetary or **market values** if they were extracted or harvested and sold. In cost–benefit analysis, however, such commodities are typically discounted for future use to the extent that they appear to be of economic value only if they can be used up as quickly as possible. Some governments and businesses use a 10% annual discount rate for decisions on resource use. This means that the long-term value of a stand of ancient trees worth $500 000

for the timber it contains would drop by 10% each year; after 10 years of discounting, it would be worth only $174 339.22. By this logic, the more quickly the trees are cut down, the more they are worth.

A related problem is that accounting procedures at the national level typically do not assign an *asset value* to intact natural resources as they would, for example, for a physical asset, such as a factory or a system of roads. This means that an unexploited natural resource has *no discernible value* to the nation that possesses it, providing a powerful incentive for nations to exploit their natural resources to the greatest extent and as quickly as possible.

This was stated eloquently by Robert Repetto, one of the pioneers of **natural resource accounting**, which seeks mechanisms by which to incorporate the economic asset values of natural resources into national accounting systems. According to Repetto

> *A country can cut down its forests, erode its soils, pollute its aquifers and hunt its wildlife and fisheries to extinction, but its measured income is not affected as its assets disappear . . . By failing to recognize the asset value of natural resources, the accounting framework that underlies the principal tools of economic analysis misrepresents the policy choices nations face.*[17]

To further complicate matters, national accounting systems also fail to incorporate certain categories of activities and expenses into the calculations. For example, an individual or a nation might spend money to prevent future environmental degradation or to restore a degraded natural environment. This could include, for example, launching a program of pollution abatement, constructing windbreaks to prevent soil erosion, or installing heavy-walled containers in oil tankers to prevent spills. Such costs, which economists call *defensive expenditures*, are calculated only as current expenses; the future value of the resource these activities may be helping to preserve or restore is of no economic value to the system.

Are costs and benefits strictly internal?

A third assumption of neoclassical economics is that all costs and benefits associated with a particular exchange of goods or services are borne by individuals engaging directly in the transaction. In other words, it is assumed that the costs and benefits of a transaction are "internal" to the transaction, experienced by the buyer and seller alone, and do not affect other members of society.

However, in many situations this is simply not the case. Pollution from a factory can harm people living nearby. In such cases, someone—often taxpayers not involved in producing the pollution—ends up paying the costs of alleviating it. Market prices do not take the social, environmental, or economic costs of this pollution into account. Costs or benefits of a transaction that involve people other than the buyer or seller are known as **externalities**. A positive externality is a benefit enjoyed by someone not involved in a transaction, and a negative externality, or *external cost*, is a cost borne by someone not involved in a transaction (**FIGURE 20.12**). Negative externalities often harm groups of people or society as a whole, while allowing certain individuals private gain.

Some examples of negative environmental externalities are

- property damage
- declines in desirable elements of the environment, such as poorer air quality, or fewer fish in a stream

FIGURE 20.12

An Indonesian boy wading through a polluted river suffers external costs, costs that are not borne by the buyer or seller of the product that caused the pollution. External costs may include water pollution, human health problems, property damage, harm to aquatic life, aesthetic degradation, declining real estate values, and other impacts.

- aesthetic damage, such as that resulting from air pollution or clear-cutting
- stress and anxiety experienced by people downstream or downwind from a pollution source
- declining real estate values resulting from these problems

The Mirarr clan experienced external costs in the form of pollution from the Ranger mine. In March 2002, radioactive material from the Ranger mine contaminated a stream on Mirarr land with uranium concentrations 4000 times as high as allowed by law. According to an Aboriginal representative, this was the fourth such violation in less than three months. In 2004 the government temporarily shut down the mine and brought the owners to court for these violations.[18] For the Dene of Great Bear Lake the external costs of uranium mining have come in the form of health impacts on workers and, by extension, economic and psychological impacts on their families.

By ignoring external costs, economies create a false idea of the true and complete costs of particular choices, and unjustly subject people to the consequences of transactions in which they did not participate. External costs are one reason governments develop environmental legislation and regulations. Unfortunately, external costs are difficult to account for and eliminate. It is tough to assign a monetary value to illness, premature death, or degradation of an aesthetically or spiritually significant site.

Is growth always good?

A fourth assumption of the neoclassical economic approach is that economic growth is required to keep employment high and maintain social order. The argument goes something like this: By making the overall economic pie larger, everyone's slice becomes larger, even if some people still have much smaller slices than others.

The idea that economic growth is good has been encouraged over the centuries by the concept of material progress, espoused by Western cultures since the Enlightenment. Progress started out as a moral and religious precept; after the Industrial Revolution, the concept of "human progress" slowly began to change, becoming more linked to consumption and economic gain. Everywhere, every day in the modern industrialized world we see expressions of the view that "more and bigger" is always better. We hear constantly in business news of increases in an industry's output or percentage growth in a country's economy, with increases touted as good news and decreases, stability, or even a minor drop in the *rate* of growth presented as bad news. Economic growth has become the quantitative ruler by which progress is measured.

The rate of economic growth in recent decades is unprecedented in human history. As a result, the world economy is seven times the size it was half a century ago. All measures of economic activity—trade, rates of production, amount and value of goods manufactured—are higher than they have ever been and are still increasing. This growth has brought many people much greater material wealth (although gaps between rich and poor are immense, and growing).

To the extent that economic growth is a means to an end—a tool with which we can achieve greater human happiness—it can be a good thing. However, many observers today worry that growth has become an end in itself and is no longer necessarily the best tool with which to pursue happiness. Critics of the growth paradigm often note that runaway growth resembles the multiplication of cancer cells, which eventually overwhelm and destroy the organism in which they grow. These critics fear that runaway economic growth will likewise destroy the economic system on which we all depend. Resources for growth are ultimately limited, they argue, so nonstop growth is not sustainable and will fail as a long-term strategy.

Defenders of traditional economic approaches reply that critics have been saying for decades that limited resources would doom growth-oriented economies, yet most of these economies are still expanding dramatically. If resources are dwindling, why are we witnessing the most rapid growth of material wealth in human history?

One prime reason is technological innovation. In case after case, improved technology has enabled us to push back the limits on growth, effectively expanding the carrying capacity of the environment. More powerful technology for extracting minerals, fossil fuels, and ground water has expanded the amounts of these natural resources

available to us. Technological developments, such as automated farm machinery, fertilizers, and chemical pesticides, have allowed us to grow more food per unit area of land, boosting agricultural output. Faster, more powerful machines in our factories have enabled us to translate our enhanced resource extraction and agricultural production into faster rates of manufacturing.

Economists disagree on whether economic growth is sustainable

Can we conclude, then, that endless improvements in technology are possible and that we will never run into shortages of resources? At one end of the spectrum are those who believe that technology can solve everything— a philosophy that has greatly influenced economic policy in market economies over the past century.

At the other end of the spectrum, *ecological economists* argue that a couple of centuries is not a very long period of time and that history suggests that civilizations do not, in the long run, overcome their environmental limitations. Ecological economics, which has emerged as a discipline only in the past decade or two, applies the principles of ecology and systems science to the analysis of economic systems. Earth's natural systems generally operate in self-renewing cycles, not in a linear or progressive manner. Ecological economists advocate sustainability in economies and see natural systems as good models.

To evaluate an economy's sustainability, ecological economists take a long-term perspective and ask, "Could we continue this activity forever and be happy with the outcome?" Most ecological economists argue that the growth paradigm will eventually fail and that if nothing is done to rein in population growth and resource consumption, depleted natural systems could plunge our economies into ruin. Many advocate economies that do not grow and do not shrink but rather are stable. Such steady-state economies are intended to mirror natural ecological systems.

Environmental economists tend to agree that economies are unsustainable if population growth is not reduced and resource use is not made more efficient. However, they maintain that we can accomplish these changes and attain sustainability within our current economic systems. By retaining the principles of neoclassical economics but modifying them to address environmental challenges, environmental economists argue that we can keep our economies growing and that technology can continue to improve efficiency. Environmental economists were the first to develop ways to tackle the problems of external costs and discounting and to weigh the true costs and benefits associated with resource use. They then went farther, proposing that sustainability requires far-reaching

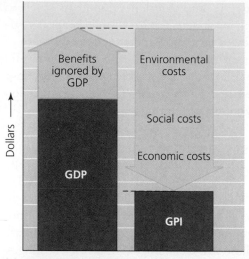

(a) Components of GDP vs. GPI

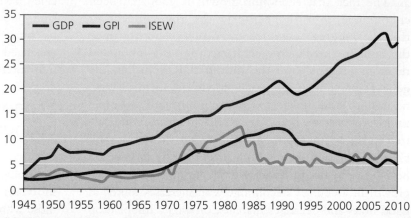

(b) Change in GDP vs. alternative indexes

FIGURE 20.13   Gross domestic product (GDP) sums all economic activity, whether good or bad **(a)**. As a result, many researchers believe that GDP is a poor indicator of overall well-being. The Genuine Progress Indicator (GPI) is an alternative indicator of progress and well-being. GPI adds to GDP benefits by accounting for the value of activities such as volunteering and parenting. GPI also subtracts external environmental costs, such as pollution and resource depletion; social costs, such as crime; and economic costs, such as borrowing and the gap between rich and poor. The result, many believe, is a more accurate reflection of overall well-being. The graph in **(b)** compares GDP, GPI, and the Index of Sustainable Economic Well-Being (ISEW), another alternative index that accounts more fully for costs and benefits than GDP. This graph shows the three indexes over time for the country of Finland. *Source: From Future Trends of Genuine Progress in Finland : Trends and Future of Sustainable Development.* Tampere. 9–10 June 2011.

In **(b)**, in what year (approximately) do the curves for Finland's GPI and ISEW begin to show a downturn? What happens to GDP at the same time? By 2010, how many times higher is GDP, compared to ISEW and GPI?

changes leading ultimately to a steady-state economy. The idea of a steady-state economy did not originate with the rise of ecological economics. Back in the nineteenth century, British economist John Stuart Mill (1806–1873) hypothesized that as resources became harder to find and extract, economic growth would slow and eventually stabilize. Economies would carry on in a state in which individuals and society subsist on steady flows of natural resources and on savings accrued during occasional productive but finite periods of growth.

Modern proponents of a steady-state global economy, such as the pioneering economist Herman Daly, believe we need to rethink our assumptions and fundamentally change the way we conduct economic transactions. They argue that quality of life can continue to rise under a steady-state economy and, in fact, may be more likely to do so. Technological advances will not cease just because growth stabilizes, they argue, and neither will behavioural changes (such as greater use of recycling) that enhance sustainability. Instead, wealth and human happiness can continue to rise after economic growth has levelled off.

Attaining sustainability will certainly require reforms and may well require fundamental shifts in thinking, values, and behaviour. How can these goals be attained in a world whose economic policies are still largely swayed by a cornucopian worldview that barely takes the environment into account? While keeping in mind that ecological and environmental economic approaches are still actively being developed, we will now survey a few strategies for sustainability that have been offered so far.

We can measure economic progress differently

For decades, economists have assessed the economic robustness of a nation by calculating its gross domestic product (GDP), the total monetary value of final goods and services produced in a country each year (**FIGURE 20.13A**). GDP is an extremely powerful indicator, used to make financial policy decisions by federal governments worldwide, with fundamental impacts on quality of life and well-being for billions of people.

However, there are problems with using this measure of economic activity to represent a nation's economic well-being. For one, GDP does not account for the nonmarket values of ecosystem goods and services, nor is GDP necessarily an expression of *desirable* economic activity. In fact, GDP can increase, even if the economic activities driving it hurt the environment or society.

For example, a large oil spill would increase GDP because oil spills require cleanups, which cost money and, as a result, increase the production of goods and services. Such activities generate income and are therefore reflected by a positive change in GDP, but the negative health and ecological costs of disasters like these are typically *not* reflected in the GDP—unless they generate jobs or cash transactions, in which case they may show up as positive changes. A radiation leak at a uranium mine on Mirarr homelands would likely add to the Australian GDP because of the many monetary transactions required for cleanup and medical care. Similarly the $6.7 million in contracts offered by the Government of Canada for the cleanup of abandoned uranium mine sites in the Northwest Territories will add to the GDP of Canada. Even Hurricane Katrina—which caused the deaths of almost 2000 people, the loss of tens of thousands of homes and jobs, and the contamination or permanent flooding of millions of square kilometres of coastal marshlands and forests—generated positive growth in GDP of the United States.

Some economists have attempted to develop economic indicators that differentiate between desirable and undesirable economic activity. Such indicators can function as more accurate guides to nations' welfare. One alternative to the GDP is the genuine progress indicator (GPI), introduced in 1995 by Redefining Progress, a nonprofit organization that develops economic and policy tools to promote accurate market prices and sustainability (see **FIGURE 20.13**). The GPI has not gained widespread acceptance, partly because it is tremendously data-intensive, but it has generated a great deal of discussion that has drawn attention to the weaknesses of the GDP.

To calculate GPI, economists begin with conventional economic activity and then add to it all *positive* contributions to the economy that do not have to be paid for with money, such as volunteer work and parenting. They then subtract the *negative* impacts, such as crime, pollution, gaps between rich and poor, and other detrimental social, environmental, and economic factors. The GPI thereby summarizes many more forms of economic activity than does GDP and differentiates between economic activity that increases societal well-being and economic activity that decreases it.

Thus, whereas GDP increases when fossil fuel use increases, GPI declines because of the adverse environmental and social impacts of such consumption, including air and water pollution, increased road congestion and traffic accidents, and global climate change. **FIGURE 20.13B** compares changes in GDP and GPI over time for the country of Finland. The country's GDP increased greatly

over the decades on the graph, as a result of increased economic activity; however, GPI has stagnated and then declined in recent years.

GPI is not the only alternative to GDP. The *Index of Sustainable Economic Welfare (ISEW)* is based on income, wealth distribution, the value of volunteerism, and natural resource depletion; ISEW is compared to GPI and GDP in **FIGURE 20.13B**. The *Net Economic Welfare (NEW)* index adjusts GDP by deducting the costs of environmental degradation. The United Nations uses a tool called *Human Development Index*, calculated on the basis of a nation's standard of living, life expectancy, and education. Any of these indexes, ecological economists maintain, should give a more accurate portrait of a nation's welfare than GDP, which policymakers currently use so widely.

We can give ecosystem goods and services monetary values

Economies receive from the environment vital resources and ecosystem services. However, any survey of environmental problems today—deforestation, biodiversity loss, pollution, collapsed fisheries, climate change, and so on—makes it immediately apparent that our society often mistreats the very systems that keep it alive and healthy. Furthermore, the values of environmental goods and services are routinely underrepresented in cost–benefit analyses, one of the most powerful tools of economic decision-making. Why is this? From the economist's perspective, humans overexploit natural resources and systems because the market assigns them no quantitative monetary value or, at best, assigns values that underestimate their true worth.

Think for a minute about the nature of some of these services. The aesthetic and recreational pleasure we obtain from natural landscapes, whether wildernesses or city parks, is something of real value. Yet this value is hard to quantify and appears in no traditional measures of economic worth. Or consider Earth's water cycle, by which rain fills our reservoirs with drinking water, rivers give us hydropower and flush away our waste, and water evaporates, purifying itself of contaminants and readying itself to fall again as rain. This natural cycle is absolutely vital to our existence, yet because its value is not quantified, markets impose no financial penalties when we interfere with it.

Ecosystem services are said to have nonmarket values, values not usually included in the price of a good or service (**FIGURE 20.14**). Because the market does not assign value to ecosystem services, debates, such as that over the Jabiluka mine, often involve comparing apples

FIGURE 20.14 Values That Modern Market Economies Generally Do Not Address
Accounting for nonmarket values, such as those shown here, may help us make better environmental and economic decisions.

(a) Existence value includes things that have value just because they exist, even though we may never experience them directly (e.g., remote wilderness or endangered species in a far-off place).

(d) Scientific value includes things that may be important as subjects for scientific research and environmental decision-making.

(b) Option value includes things that we do not use now but might use or find a use for at a later time.

(e) Educational value includes things that may teach us about ourselves and about the world.

(c) Aesthetic value includes things that we appreciate for their beauty or emotional appeal.

(f) Cultural value includes things that sustain or help define our cultures..

FIGURE 20.14   *(Continued)*

Tyler Olson/Thinkstock/Getty Images

(g) Use value includes things that we use directly but that may not have a market-based value.

Nikolai Lebedev/123RF

(h) Spiritual value includes things that are sacred to certain groups, or evoke spiritual, religious, or philosophical responses in us.

and oranges—in this case, the intangible cultural, ecological, and spiritual arguments of the Mirarr versus the hard numbers of mine proponents.

To partially resolve this dilemma, environmental and ecological economists have sought ways to assign values to ecosystem goods and services. One technique, *contingent valuation*, uses surveys to determine how much people are willing to pay to protect a resource or to restore it after damage has been done. Contingent valuation also considers how much people would want to be compensated before agreeing to lose or forego access to an environmental resource or service that they value.

Such an exercise was conducted with a mining proposal in the Kakadu region in the early 1990s that preceded the Jabiluka proposal. The Kakadu Conservation Zone, a government-owned 50 km^2 plot of land surrounded by Kakadu National Park, was either to be developed for mining or to be preserved and added to the park. To determine the degree of public support for environmental protection versus mining, a government commission sponsored a contingent valuation study to determine how much Australian citizens valued keeping the Kakadu Conservation Zone

preserved and undeveloped. Researchers interviewed 2034 citizens, asking them how much money they would be willing to pay to stop mine development.

The interviewers presented two scenarios: (1) a "major-impact" scenario based on predictions of environmentalists who held that mining would cause great harm, and (2) a "minor-impact" scenario based on predictions of mining executives who held that development would have few downsides. After presenting both scenarios in detail, complete with photographs, the interviewers asked the respondents how much their households would pay if each scenario, in turn, were to occur. Multiplying the results by the number of households in Australia (5.4 million at the time), the researchers found that preservation was "worth" $435 million annually to the Australian population under the minor-impact scenario, and $777 million under the major-impact scenario. Because both of these numbers significantly exceeded the $102 million in annual economic benefits expected from mine development, the researchers concluded that preserving the land in its undeveloped state was worth more than mining it.[19]

Because contingent valuation relies on survey questions, critics complain that in such cases people will volunteer idealistic (inflated) values rather than realistic ones, knowing that they will not actually have to pay the price they name. In part because of such concerns, the Australian government commission decided not to use the Kakadu contingent valuation study's results. (The mine was stopped, ultimately, but mainly as a result of Aboriginal opposition.)

Whereas contingent valuation measures people's *expressed* preferences, other methods aim to measure people's *revealed* preferences—preferences as revealed by data on actual behaviour. For example, the amount of money, time, or effort people expend to travel to parks for recreation has been used to measure the value people place on parks. Economists have also analyzed housing prices, comparing homes with similar characteristics but different environmental settings to infer the dollar value of landscapes, views, greenspace, and peace and quiet. Another approach assigns environmental amenities value by measuring the cost required to restore natural systems that have been damaged or to mitigate harm from pollution.

In 1997 a research team led by Robert Costanza reviewed ecosystem valuation studies, with the goal of calculating the global economic value of all the services that ecosystems provide (**FIGURE 20.15**). The team identified more than 100 studies that estimated the worth of such ecosystem services as water purification, greenhouse gas regulation, plant pollination, and pollution cleanup. The studies used such methods as contingent valuation to estimate the values of such aspects of natural systems as biodiversity and aesthetics.

To estimate the worth of ecosystem services more accurately, the team reevaluated the data using alternative valuation techniques. One method was to calculate the cost of replacing ecosystem services with technology. For example, marshes protect people from floods and filter out water pollutants. If a marsh were destroyed, the researchers would calculate the value of the services it had provided by measuring the cost of the levees and water-purification technology that would be needed to assume those tasks. The researchers then calculated the global monetary value of such wetlands by multiplying those totals by the global area occupied by the ecosystem. By calculating similar totals from other ecosystems, they arrived at a global value for ecosystem services.[20] The total figure calculated in the original study was $33 trillion per year (1997 dollars)—greater than the combined gross domestic products of all nations in the world.

A follow-up study in 2014 by Costanza and colleagues[21] resulted in an estimate of $145 trillion per year for the value of ecosystems services, with an estimated value loss of about $4.3–20.2 trillion/yr since their earlier study, as a result of environmental change and degradation. The researchers concluded that the economic benefits of preserving the world's remaining natural areas outweigh the benefits of exploiting them by a factor of 100 to 1.

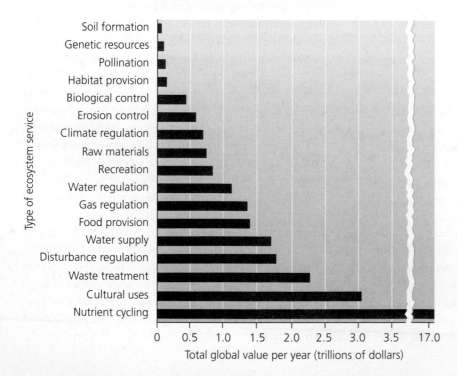

FIGURE 20.15

Costanza and colleagues estimated the total value of the world's ecosystem services at approximately $33 trillion. Shown are subtotals for each major class of ecosystem service. The $33 trillion figure does not include values from some ecosystems, such as deserts and tundra, for which adequate data were unavailable.

Source: Data from Costanza, R., et al. (2007). The Value of the World's Ecosystem Services and Natural Capital. *Nature, 387,* pp. 253–260.

When markets do not reflect the full costs and benefits of actions, they are said to fail. Market failure occurs when markets do not take into account the environment's positive effects on economies (such as ecosystem services) or when they do not reflect the negative effects of economic activity on the environment or on people (external costs). Traditionally, market failure has been countered by government intervention. Governments can dictate limits on corporate behaviour through laws and regulations. They can institute *green taxes,* which penalize environmentally harmful activities. Or they can design economic incentives that put market mechanisms to work to promote fairness, resource conservation, and economic sustainability. These are just some of the economic and regulatory actions that governments can take to try to counteract market failure in the context of environmental goods and services.

Corporations are responding to sustainability concerns

As more consumers and investors express preferences for sustainable products and services, more industries, businesses, and corporations are finding that they can make money by "greening" their operations (**FIGURE 20.16**).

Some companies, such as The Body Shop, cultivate eco-conscious images; others donate a portion of their proceeds to environmental and other progressive nonprofit groups. Today, some newer businesses are trying to go even further than these pioneers. The outdoor apparel company Nau manufactures items from materials made of corn biomass and recycled bottles and helps fund environmental nonprofits. Entrepreneurs are starting thousands of local sustainability-oriented businesses across the world.

In the past few years, corporate sustainability has gone mainstream, and some of the world's largest corporations have joined in, including McDonald's, Starbucks, IKEA, Dow, and British Petroleum. Nike collects millions of used sneakers each year and recycles the materials to create synthetic surfaces for basketball courts, tennis courts, and running tracks. Nike also uses more organic cotton and has developed less-toxic rubber and adhesives. In response to media attention and consumer concern, such corporations as Nike and the Gap are also working to improve labour conditions in their factories overseas.

Another example of a corporation that seems to take its environmental and social impacts seriously is Patagonia, which sells outdoor adventure gear and clothing. The corporation donates 1 percent of its sales to grassroots environmental organizations; to date, this has added up to $31 million in grants and in-kind donations. They helped start the Sustainable Apparel Coalition, focusing on improving the environmental and social performance of clothing and footwear manufacturers. They recently launched a new initiative called *$20 Million and Change,* a fund to help start-up companies bring about positive benefit to the environment. Patagonia describes its efforts at sustainability as a work in progress, and their ultimate goal is to become a "responsible company."

Of course, corporations exist to make money for their shareholders, so they cannot be expected to pursue goals that are not profitable. Moreover, some corporate greening efforts are more rhetoric than reality, and corporate *greenwashing* may mislead some consumers into thinking that companies are acting more sustainably than they are. However, as consumer preferences turn increasingly to sustainable products and practices, many corporations are seeing the economic wisdom of moving toward a more sustainable model of operation.

Perhaps the most celebrated corporate greening is that of Walmart. Environmentalists and labour activists have long criticized the world's largest retailer for its environmental and social impacts. In 2006 the company began a quest to sell organic and sustainable products, reduce packaging and use recycled materials, enhance fuel efficiency in its truck fleets, reduce energy use in its stores, cut carbon dioxide emissions, and preserve an equivalent area of natural land for every parcel of land developed. Many observers remain highly skeptical of Walmart's commitment, calling it superficial greenwashing; yet, if the company achieves only a fraction of its stated goals, the environmental benefits could be substantial because of the corporation's vast global reach.

The bottom line is that corporate actions are responsive to consumer behaviour and pressures. It is up to all of us as consumers to encourage trends in sustainability by rewarding corporations that truly promote sustainable solutions.

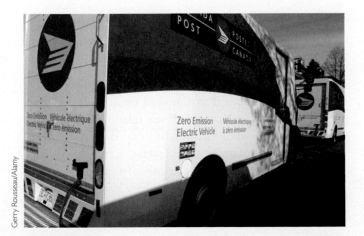

FIGURE 20.16
Canada Post boasts a fleet of "green" fuel-efficient hybrid and all-electric trucks.

Gerry Rousseau/Alamy

Conclusion

Corporate sustainability, alternative ways of measuring growth, and the valuation of ecosystem goods and services are a few of the recent developments that have brought economic approaches to bear on environmental protection and resource conservation. As economics becomes more environmentally friendly, it renews some of its historic ties to ethics.

Environmental ethics has expanded people's sphere of ethical consideration outward to encompass other societies and cultures, other creatures, and even nonliving entities that were formerly outside the realm of ethical concern. This ethical expansion involves the concept of *distributional equity*, or equal treatment for all, which is the aim of environmental justice. One type of distributional equity is equity among generations. Such concern by current generations for the welfare of future generations is the basis for the notion of sustainability.

Many of us think of sustainability as the intersection or maximization of three sets of goals: environmental goals, social goals, and economic goals, all with equal value (as shown in **FIGURE 1.8A** on page 16). However, ecological economists, among others, posit that neither society nor the economy could continue for very long without the underlying support of environmental resources and services. They refer to the "three equal spheres" concept as a *weak sustainability* model. To emphasize the fundamental importance of the environmental foundations for human society and economy, they conceived the latter two as being subsets of the environment, in what has come to be called a *strong sustainability* model (as shown in **FIGURE 1.8B** on page 16). The central tenet of strong sustainability is the acknowledgment of the fundamental dependence of human society and economy on the environment and natural resources, known as the principle of *environmental primacy*. These concepts are discussed in greater detail in Chapter 22, *Strategies for Sustainability*.

Although we tend to think of sustainability as a "modern" idea, it is actually inherent to some of the most basic concepts of neoclassical economics. In his 1939 work *Value and Capital*, Sir John Hicks (1904–1989), one of the most influential economists of the twentieth century, defined income as "the maximum value [a person] can consume during a week, and still expect to be as well off at the end of the week as he was at the beginning." The implication is that *true* income is *sustainable* income—if your spending compromises your resource base and reduces your future ability to produce, then you are depleting your capital. As former World Bank economist Herman E. Daly explains,

> *Why all the fuss about sustainability? Because, contrary to the theoretical definition of income, we are in fact consuming productive capacity and counting it as income in our national accounts. . . . Depletion of natural capital and consequent reduction of its life-sustaining services is the meaning of unsustainability.*[22]

Is sustainable development a pragmatic pursuit for us, both in Canada and as a global community, and is sustainability a practical goal? The answer largely depends on whether we believe that economic well-being and environmental well-being are opposed to each other, or whether we accept that they can work in tandem. Equating economic well-being with economic growth, as most economists traditionally have, suggests that economic welfare entails a trade-off with environmental quality. However, if economic welfare can be enhanced in the absence of growth, we can envision economies and environmental quality benefiting from one another.

REVIEWING OBJECTIVES

You should now be able to

Characterize the influences of culture and worldview on the choices people make

- An individual's culture and personal experiences strongly influence his or her worldview.
- Such factors as religion, traditional beliefs, and political ideology are especially influential in shaping our perspectives on the environment.
- Traditional ecological knowledge, passed from generation to generation among Indigenous inhabitants, is valuable in the management of complex ecosystems.

Outline the nature, evolution, and expansion of environmental ethics in Western cultures

- Questions about distributional equity, intergenerational considerations, and our relationship to other species are central to environmental ethics.
- Our society's domain of ethical concern has expanded to grant ethical consideration to more non-human entities. Anthropocentrism values humans above all else, whereas biocentrism values all life, and ecocentrism values whole ecological systems.
- Environmental ethics is relatively new as an academic pursuit, but it has ancient religious and philosophical roots.

- The preservation ethic and the conservation ethic have guided resource managers and the environmental movement during the past century.

- The land ethic and deep ecology both speak to the fundamental right to exist and the value of other organisms and the natural environment, and our deep connection to them as humans.

- Ecofeminism sees a feminine force in the natural world and draws parallels between the oppression of the environment and the oppression of women.

- The environmental justice movement, seeking equal distribution of environmental costs and benefits for all people, is a recent offshoot of environmental ethics.

Describe some basic principles and approaches of economics and how they relate to resource use and environmental impacts of production

- Economics is concerned with human values and behaviour in the context of production, consumption, and the allocation of resources.

- There are several types of economies in the world today. Among them are subsistence economies, capitalist economies, and centrally planned economies.

- Classical economic theory proposes that individuals acting for their own economic good can benefit society as a whole. This view has provided a philosophical basis for free-market capitalism.

- Neoclassical economics focuses on consumer behaviour and supply and demand as forces that drive economic activity.

- Cost–benefit analysis helps neoclassical economists weigh costs, such as resource depletion or environmental damage, against potential benefits.

Explain the fundamentals of environmental economics, ecological economics, and natural resource accounting, and their implications for economic growth and sustainability

- Several assumptions of neoclassical economic theory contribute to environmental impacts. First, it has been assumed that environmental goods and services are either free and infinite, or (once lost) can be replaced by other types of capital; in some circumstances, such as the loss of a species, this assumption is faulty.

- It is conventional to discount benefits that will not come to pass until far in the future. This applies a value judgment in which the future is valued less than the present. Discounting tends to disfavour long-term projects, such as many environmental restoration or protection plans, that would have present-day costs but future benefits.

- It is common for production to be associated with environmental and social costs that are not internalized by the private corporation, including pollution. Taxes and government regulations are among the mechanisms for managing these externalities.

- Conventional economic theory has promoted never-ending economic growth, with little regard to possible environmental impacts or the sustainability of growth.

- Environmental economists advocate reforming standard economic practices to promote sustainability, and to make market prices reflect real costs and benefits. Ecological economists take a more holistic view, applying the fundamental principles of ecology to human and economic systems.

- One area of interest for environmental economists is to find better ways of measuring economic progress and overall human well-being. The alternatives typically incorporate more environmental and social costs and benefits, compared to the standard measure, GDP.

- Another contribution of environmental economics has been in the area of valuation, that is, determining a quantitative value for environmental goods and services that may be not be measurable in standard market terms.

- Economic growth is not necessarily required for overall economic well-being. In the long run, some economists believe that a steady-state economy will be necessary to achieve sustainability.

- Consumer choice in the marketplace can help drive businesses and corporations to pursue sustainability goals.

TESTING YOUR COMPREHENSION

1. What does the study of ethics encompass? Describe the three classic ethical standards. What is environmental ethics?

2. Why in Western cultures have ethical considerations expanded to include non-human entities?

3. Describe the philosophical perspectives of anthropocentrism, biocentrism, and ecocentrism. How would you characterize the perspective of the Mirarr clan?

4. Differentiate between the preservation ethic and the conservation ethic. Explain the contributions of John

Muir, James Harkin, Gifford Pinchot, and Clifford Sifton in the history of environmental ethics.

5. Describe Aldo Leopold's "land ethic." How did Leopold define the "community" to which ethical standards should be applied?

6. Name four key contributions the environment makes to the economy.

7. For each of these basic tenets of neoclassical economics, explain the potential impacts on the environment and provide a hypothetical example:
   - Resources are infinite or substitutable.
   - Long-term effects should be discounted.
   - Costs and benefits are internal.
   - Growth is good.

8. Neoclassical economists have moved away from Adam Smith's original definition of *income* as "economic gains made with no negative impacts on the resource base." What are the environmental implications of straying from this definition of *income*?

9. Compare and contrast the views of neoclassical economists, environmental economists, and ecological economists.

10. What is contingent valuation, and what is one of its weaknesses? Describe an alternative method that addresses this weakness.

THINKING IT THROUGH

1. How would you analyze the Keystone XL Pipeline (discussed in the chapter on fossil fuels) from each of the following perspectives? In your description, list two questions that a person of each perspective would likely ask when attempting to decide whether the pipeline should have been built or should have been halted. Be as specific as possible, and be sure to identify similarities and differences in approaches:
   - preservationist
   - conservationist
   - deep ecologist
   - environmental justice advocate
   - Indigenous land rights activist
   - ecofeminist
   - neoclassical economist
   - ecological economist

2. Do you think we should attempt to quantify and assign market values to ecosystem services and other entities that have only nonmarket values? Why or why not? What is a steady-state economy? Do you think this model is a practical alternative to the growth paradigm? Why or why not?

3. A manufacturing facility on a river near your home provides jobs for 200 people in your community and pays $2 million in taxes to the local government each year. Sales taxes from purchases made by plant employees and their families contribute an additional $1 million to local government coffers. However, a recent peer-reviewed study in a well-respected scientific journal revealed that the plant has been disposing large amounts of waste into the river, causing a 25% increase in cancer rates, a 30% reduction in riverfront property values, and a 75% decrease in native fish populations.

   The plant owner says the facility can stay in business only because there are no regulations mandating expensive treatment of waste from the plant. If such regulations were imposed, he says he would close the plant, lay off its employees, and relocate to a more business-friendly community. How would you recommend resolving this situation? What further information would you want to know before making a recommendation? In arriving at your recommendation, how did you weigh the costs and benefits associated with each of the plant's impacts?

4. You are a researcher working for a large pharmaceutical company. You are doing botanical fieldwork, searching for a plant that may offer a new cure for cancer. Some of the Indigenous people in the area have a deep understanding of the medicinal properties of local plants, and you would like to ask them some questions. Under what circumstances should you do this? What if you were to make a major discovery on the basis of something you learned from them, with potential earnings of billions of dollars for your company—would the people who passed along the crucial information have a legitimate claim to part of those earnings?

INTERPRETING GRAPHS AND DATA

Economists use various indicators of economic well-being. One that has been used for decades is the gross domestic product (GDP), the total monetary value of final goods and services produced each year. An alternative measure called the genuine progress indicator (GPI) is calculated as follows:

$$GPI = GDP + (\text{Benefits Ignored by GDP}) - (\text{Environmental Costs}) - (\text{Social and Economic Costs})$$

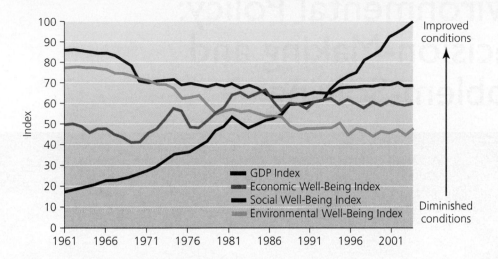

This graph compares the components of the genuine progress indicator (GPI), comprising indexes for economic, social, environmental well-being, and to the gross domestic product (GDP) index in Alberta, 1961–2003.
Source: Data from Pembina Institute, Alberta GPI Accounts, 1961–2003.

Benefits include such things as the value of parenting and volunteer work. Environmental costs include the costs of water, air, and noise pollution; loss of wetlands; depletion of nonrenewable resources; and other environmental damage. Social and economic costs include investment, lending, and borrowing costs; costs of crime, family breakdown, underemployment, commuting, pollution abatement, and automobile accidents; and loss of leisure time.

1. Describe economic growth as measured by GDP for the province of Alberta from 1961 to 2003. Now describe economic growth as measured by GPI over the same time period. To what factors would you attribute the growing difference between these measures?

2. For GPI to grow, one or more things must happen: Either GDP must grow faster; benefits must grow faster; or social, economic, and environmental costs must shrink relative to the other terms. How would you explain the changes in GDP and GPI over the past few decades for the province of Alberta? How do the data in the graph support your answer?

3. Even with regulations for air and water pollution control, hazardous waste disposal, solid waste management, forestry practices, and species protection, environmental costs continue to increase. Why do you suppose the trend is still in that direction?

4. Alberta in the 2000s was in the midst of a huge economic boom, related largely to the development of the Athabasca oil sands and other fossil fuel deposits. How do you think this ongoing resource development will affect Alberta's GDP and GPI for the province over the next few decades? Think about all of the factors—economic, environmental, and social—that will play a role, and whether the impacts will likely be positive or negative.

5. In 2014 the price of oil plummeted as a result of high levels of production of oil by Saudi Arabia. Speculate on how you think this drop in oil revenues would affect the graphs of GDP and the other three indexes of well-being.

MasteringEnvironmentalScience®

STUDENTS
Go to **MasteringEnvironmentalScience** for assignments, the eText, and the Study Area with practice tests, videos, current events, and activities.

INSTRUCTORS
Go to **MasteringEnvironmentalScience** for automatically graded activities, current events, videos, and reading questions that you can assign to your students, plus Instructor Resources.

21 Environmental Policy: Decision-Making and Problem-Solving

Sagebrush prairie, as seen here in Saskatchewan Landing Provincial Park, is typical habitat for the endangered greater sage-grouse.

Bill Brooks/Alamy

Upon successfully completing this chapter, you will be able to

- Describe environmental policy and its main goals and challenges
- Identify institutions, instruments, laws, and processes that are central to Canadian environmental policy
- Discuss some of the institutions and guiding principles that are important for international

environmental policy, and for the resolution of transboundary environmental problems
- Outline various approaches to environmental policy and the role of science in formulating policy

The greater sage-grouse has become extirpated from British Columbia and is endangered in Alberta and Saskatchewan, mostly because of habitat loss and degradation. The Government of Canada issued an emergency order regulating activities in sage-grouse habitat in Alberta and Saskatchewan, in an effort to save the species.

CANADA

Alberta   Saskatchewan

CENTRAL CASE
SARA AND THE SAGE-GROUSE

"Creating evidence-based policy has become an increasingly complex process. The ongoing generation of new knowledge continues to increase both the number and variety of potential policy issues and challenges."

—SHAUN P. YOUNG, SENIOR FELLOW OF THE YORK
CENTRE FOR PUBLIC POLICY AND LAW[1]

"When you get ready to vote, make sure you know what you're doing."

—ROBERT L. HUNTER, JOURNALIST, ACTIVIST,
AND CO-FOUNDER OF GREENPEACE

The greater sage-grouse (*Centrocercus urophasianus*) is one of the iconic bird species of the Canadian prairies. The bird is native to sagebrush prairies in southern Alberta and Saskatchewan; there are also some populations in the grasslands of the United States. Another sub-species was native to the southern Okanagan Valley in British Columbia, but that population is now extirpated. Much of the prairie grassland areas where sagebrush grows has been converted for agriculture or ranching, or degraded as a result of oil exploitation.

Canada's public registry of threatened species, maintained as part of the *Species at Risk Act* (*SARA*), estimates that there has been a 90% reduction in range and a substantial decrease in sage-grouse breeding locations as a result of habitat loss. Between 1988 and 2015, the total Canadian population decreased by 98%, to the point where perhaps only 100 adults are now left in the wild.[2]

The sage-grouse became the focus of particular attention when, in November 2013, it became the very first endangered species for which Environment Canada issued an emergency order under SARA. The order, still the only one of its kind in Canada (as of 2015), was issued because of "imminent threats to the survival and

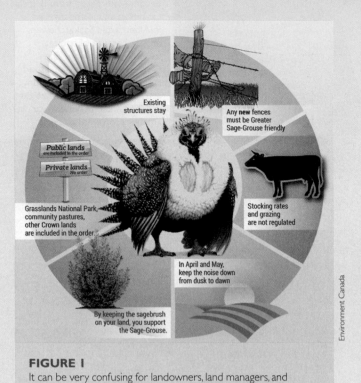

Environment Canada

FIGURE 1

It can be very confusing for landowners, land managers, and others to keep track of the various levels of legislation regarding threatened species. Environment Canada produced this infographic to help stakeholders understand which activities are regulated under the emergency order to protect the sage-grouse.

recovery"[3] of the species. It lists several types of activities to be regulated or no longer allowed on public lands because of negative impacts on the sage-grouse and its habitat (see **FIGURE 1**).

The main focus of the emergency order is on protecting the breeding areas of the birds, known as *leks* or "dancing grounds." Sage-grouse have an elaborate mating ritual in which the males gather in the lek and undertake a competitive strutting display to show off for the females, who then select the most attractive mates. (The term "lek" can refer to the breeding location, or to the mating ritual itself.) Usually only one or two of the males mate with the majority of females in a given year. Subsequent nesting takes place in sagebrush grasslands. The birds tend to return to the same lek and nesting grounds year after year, which is one reason why habitat loss and degradation are so damaging to their populations.

Most of the activities that are limited by the emergency order are directly relevant to farming, including installing new fences (must be "sage-grouse friendly"), altering the hydrology of a sagebrush area, running of noisy machinery, and construction of new buildings or roads in sage-grouse habitat. The order also

prohibits the killing of sagebrush and encourages farmers to protect and promote the growth of sagebrush on their land. The order was amended in March of 2014 to include privately owned land. This amendment came after the federal government discovered that public land had been sold to private landowners in Saskatchewan during the final approval stage of the original order.[4]

As part of Canada's *National Strategy for the Protection of Species at Risk,*[5] the federal government runs the *Habitat Stewardship Program,* which allocates money each year for projects and partnerships aimed at protecting threatened species and their habitats on private and Crown lands, Aboriginal lands, and aquatic or marine areas in Canada. Additional funds for restoration and recovery of endangered species and habitats are offered by the federal government through the Interdepartmental Recovery Fund and the Aboriginal Fund for Species at Risk.[6] The province of Alberta is sharing with the federal government the cost of running a captive breeding program for the sage-grouse at the Calgary Zoo.

The *Species at Risk Act* is the principal piece of legislation in Canada that directly facilitates the classification and protection of threated and endangered species. SARA is generally perceived to be a more consultative process than the comparable, more "top-down" legislation in the United States, the *Endangered Species Act.* At its core, SARA's decision-making process is based on the scientific understanding of species. However, there is a political aspect, too, through which the federal government has the authority to reject the recommendations made by COSEWIC, the *Committee on the Status of Endangered Wildlife in Canada.* The Act also specifies a central role for Aboriginal knowledge as part of the process of identifying a species as endangered; you learned about an example of this in the "Central Case: Saving the Polar Bear: What Will It Take?" in Chapter 9.[7] The relationship between scientific understanding and Aboriginal traditional knowledge is not always simple.

One specific weakness of SARA is that the legislation typically only applies mandatory legal protections for species on federal lands, which accounts for only about 5% of the total area of Canada.[8] An unusual exception to this was the case of the sage-grouse, in which the federal government stepped in to apply protections on public and private land at the provincial level. Another perceived shortcoming is in the timing of the process: SARA contains no mechanism for protecting habitat early in the species identification process. Even if a species is eventually listed as threatened or endangered, its habitat is not legally protected until the recovery strategy stage (which for most species takes many years to reach, if ever). In the lag time, only the species and its immediate residence are protected. We saw this in the case of the sage-grouse, too, as sagebrush habitat continued to pass from public into private hands, where it was unprotected, even as the federal emergency order was in process of being finalized.

Stakeholder consultation and science-based decision-making have been hallmarks of Canadian environmental policy for decades. As informed citizens, it is important that we all work to maintain a consultative, evidence-based, transparent, and accountable decision-making process, so that we can continue to protect and restore the natural environment on which we and our economy and industries depend.

Environmental Policy

When a society reaches broad agreement that a problem exists, it may persuade its leaders to try to resolve the problem through the making of policy. *Policy* is a formal set of general plans and principles intended to address problems and guide decision-making in specific instances. *Public policy* is policy made by governments, consisting of laws, regulations, orders, incentives, and practices intended to advance societal welfare. **Environmental policy** is policy that pertains to human interactions with the environment. It generally aims to regulate resource use or reduce pollution to promote human welfare and/or protect natural systems.

Environmental policy requires inputs from multiple sources

Forging effective policy requires input from science, ethics, economics, and culture, taking into account many different perspectives. Science provides the information and analysis needed to identify and understand environmental problems and devise potential solutions to them. Ethics, economics, and cultural perspectives offer criteria to assess the extent and nature of problems and to help clarify how society might like to address them. We explored a number of these perspectives in detail in Chapter 20, *Environmental Ethics and Economics: Values and Choices.*

Different levels and types of organizations also are involved in the development of environmental policy. Government at all levels interacts with individual citizens, nongovernmental organizations, and the private sector in a variety of ways to formulate and then implement policy (**FIGURE 21.1**).

In real life, we operate within the constraints of our modern economies, which are largely driven by incentives for short-term market gains rather than long-term social and environmental sustainability. The market provides little incentive for businesses or individuals to behave in ways

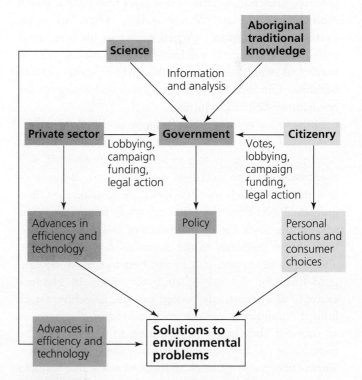

FIGURE 21.1
Policy plays a central role in how we as a society address environmental problems. Voters, the private sector, and groups representing various interests influence government representatives. Scientific research and Aboriginal traditional and ecological knowledge also inform government decisions in Canada. Government representatives and agencies formulate policy that aims to address social problems, including environmental problems. Public policy—along with improvements in technology and efficiency and personal actions exercised by citizens—can produce lasting solutions to environmental problems.

that minimize environmental impacts or equalize costs and benefits among parties. This has traditionally been viewed as justification for government intervention in some areas of society, such as in social welfare. Environmental policy is one such government intervention, which aims to protect environmental quality and natural resources and to promote equity in people's use of resources.

Three of the particular challenges of environmental policy are (1) to protect resources and environmental quality, especially in the context of resources that are commonly owned; (2) to deal with the equitable distribution of resources and access to resources; and (3) to manage downstream environmental costs and costs that are not internalized by the market, and to ensure that those costs are distributed equitably. Let's look briefly at each of these three challenges, which are interrelated.

The tragedy of the commons

Resources that are owned or accessed by groups of people in common can be particularly challenging to manage and regulate. Policies to protect such resources are intended to safeguard them from depletion or degradation. Examples of commonly owned resources on a global scale include open-ocean fisheries, the stratospheric ozone layer, Antarctica, global biodiversity, and the climate system. There are many examples of commonly owned resources on more local scales, too, from the Great Lakes to our national parks to municipal swimming pools to the party room in a condo building. Commonly owned and accessed resources are notoriously difficult to protect.

In an essay entitled "The Tragedy of the Commons,"[9] ecologist Garrett Hardin wrote that a resource held in common, that is accessible to all and is unregulated, will eventually become overused and degraded. Because of human nature, each user has incentive to maximize their individual gain from the resource, while minimizing their part of the costs for maintenance of the resource. This leads almost inevitably to the degradation of the resource. Therefore, he argued, it is in our best interest to develop guidelines for the use of such resources. In Hardin's example of a commonly owned pasture, guidelines might limit the number of animals each individual can graze; or an overseer body might require users to take turns at the resource or pay to restore and manage the shared resource. These concepts—regulation of use, and active management and regulation—are central to environmental policy today.

The tragedy of the commons does not always play itself out as Hardin predicted. Some societies have devised safeguards against overexploitation, and resource users sometimes cooperate successfully to prevent overexploitation. Sometimes this can be achieved through voluntary cooperation; sometimes through regulation and enforcement. However, in many cases Hardin's starting assumptions are not even met;

resources on public lands may not be equally accessible to everyone but may instead be more accessible to wealthier or more established resource extraction industries. In any case, the threat of overexploiting public resources is real and has been a driving force behind much environmental policy.

Free riders

Another reason to develop policies to govern publicly held and publicly accessible resources is the so-called *free rider* problem. This is an economic concept that arises from an actual case of a "free rider," that is, someone who rides the bus without paying. The free rider is reaping the benefits of public transit without paying the costs that others have paid for the privilege. Analogously, if someone uses more than their fair share of environmental resources, or uses resources without paying for them, the costs devolve to the rest of society and it becomes a "free rider" problem.

This concept has applications in resource extraction, such as someone who is fishing and not reporting their full catch to the regulators. But it also applies to other uses of environmental goods and services. For example, let's say that a community on a river suffers from water pollution that emanates from 10 different factories. The problem could in theory be solved if every factory voluntarily agreed to reduce its own pollution. However, once they all begin reducing their pollution, it becomes tempting for any one of them to stop doing so, to save money. That factory, by avoiding the costs that the others are paying to control pollution, would in essence get a "free ride" on the efforts of others. If enough factories take a free ride, the whole effort will collapse.

Because of the free rider problem, private voluntary initiatives are often less effective than efforts mandated by public policy. Public policy can prevent free riders and ensure that all parties sacrifice equitably (or use only their fair share of a resource) by enforcing compliance with laws and regulations or by taxing parties to attain funds with which to pursue societal goals.

Externalities

Environmental policy is also developed to ensure that some parties do not use resources in ways that harm others. *Externalities* are harmful impacts that result from market transactions, but are borne by people not involved in the transactions. For example, a factory may reap greater profits by discharging waste freely into a river and avoiding paying for proper waste disposal. Those actions, however, would impose external costs (water pollution, decreased fish populations, aesthetic degradation, or other problems) on downstream users of the river (**FIGURE 21.2**). These costs are not captured by the market; they are borne by others downstream, or by society as a whole (through the costs of negative environmental and health impacts). As such, externalities constitute what is called a *market failure*.

FIGURE 21.2
Downstream pollution raises many issues that have been viewed as justification for environmental policy. This woman washing clothes in the river may suffer pollution from factories located upstream, and her own use of detergents may cause pollution for people living downstream.

Downstream pollution is a problem in the estuary of the St. Lawrence River, as we saw in "Central Case: The Plight of the St. Lawrence Belugas" (Chapter 3). The pollution that ends up in the estuary originates far upstream; it comes from the industries, towns, and farms that line the river, all the way from the shores of the Great Lakes and the Greater Toronto Area and down into the estuary, where it can cause problems for wildlife like the beluga whales. The costs—both economic and ecological—of downstream pollution are passed along to someone else, or to society as a whole, and are not directly internalized to the activities that generated the pollution in the first place. Protecting resources and promoting equitable distribution of environmental costs and benefits by addressing externalities and downstream costs are some of the core objectives of environmental policy.

Environmental policy has changed along with society

Environmental policy in North America has evolved, along with social and economic conditions, in tandem with and influenced by the evolution of environmental ethics. From the 1780s to the late 1800s, environmental law dealt primarily with the management of public lands. It grew from early explorations and accompanied the westward expansion of settlers across the continent. This period is associated with a "frontier ethic," characterized by efforts to tame and conquer the wilderness. Environmental laws of this period were intended to promote settlement and the extraction and use of the continent's abundant natural resources.

In the late 1800s, as the continent became more populated and its resources were increasingly exploited, public perception and government policy toward natural resources began to shift. Laws of this period aimed to regulate resource use and mitigate some of the environmental problems associated with westward expansion. Policies were influenced by the emergence of the conservation and preservation ethics. This period saw the opening of the first national parks, including Banff National Park in 1885.

As discussed in Chapter 20, *Environmental Ethics and Economics*, probably no person is more emblematic of this period in Canada than Clifford Sifton, minister of the interior in the late 1800s and early 1900s. Under Sifton's policies, wave upon wave of immigrants were encouraged to move into Canada's West to settle the Prairie grasslands and convert them into farms. Sifton recognized the value of Canada's natural resources, but unlike some others he also realized that those resources were not unlimited. He saw conservation and reforestation as an economic necessity and created a forestry branch of the department of the interior with the goal of regulating logging (**FIGURE 21.3**) and conserving federal forests. Later he commissioned detailed studies of all Canadian natural resources and came to be known as the "father of conservation" in Canada.

Policies focusing on land management and resource extraction continued to develop through the twentieth century. The next wave of environmental policy came about in the 1960s and 1970s. It responded mainly to pollution and environmental crises and built upon public awareness of the impacts of environmental degradation. During those decades, a series of major events triggered increased awareness of environmental problems and brought about a shift in public priorities and important changes in public policy.

One landmark event was the 1962 publication of *Silent Spring* by American scientist and writer Rachel Carson

Library and Archives Canada

FIGURE 21.3
Early natural resource laws in North America were intended to encourage westward expansion, settlement, land clearing, and resource development. Here, loggers in the late 1800s fell large cedars in British Columbia.

Erich Hartmann/Magnum

FIGURE 21.4
Scientist, writer, and citizen activist Rachel Carson illuminated the problem of pollution from DDT and other pesticides in her 1962 book *Silent Spring*. The book eventually led to the banning of DDT in North America and elsewhere, and is widely credited with having launched the modern environmental movement.

(**FIGURE 21.4**). *Silent Spring* awakened the public to the negative ecological and health effects of pesticides and industrial chemicals. (The book's title refers to Carson's warning that pesticides might kill so many birds that few would be left to sing in springtime.)

Several other books brought to the public consciousness environmental issues such as the limitations of resources, the impacts of human activities, and the health implications of environmental degradation. Among these were *The Limits to Growth* (1972),[10] one of the very first efforts to use computers to develop quantitative models to explore the interplay among population growth, resource use, and resource depletion; *The Population Bomb* (1968),[11] an extreme neo-Malthusian call for action against uncontrolled human population growth; and *Diet for a Small Planet* (1971),[12] probably the first widely read book linking food and vegetarianism with the responsible use of the planet's resources.

The impacts of pollution on surface water bodies also became starkly evident to the average citizen through some major incidents that happened around the same time. The "death" of Lake Erie in the early 1970s is one example of a highly publicized environmental disaster. The Cuyahoga River (**FIGURE 21.5**), which flows into Lake Erie, also did

its part to bring attention to the hazards of pollution. The Cuyahoga near Cleveland, Ohio, was so polluted with oil and industrial waste that the river actually caught fire more than half a dozen times during the 1950s and 1960s. This spectacle moved the public throughout North America to do more to protect the environment.

Other iconic eco-disasters of the 1960s and 1970s included the leakage of hazardous wastes from the old Hooker Chemical waste dump at Love Canal, New York, which led to the first-ever declaration of a federal state of emergency in the United States from an environmental cause. The Amoco *Cadiz* oil spill off the coast of France in 1978—still one of the largest oil spills ever—also served to raise public awareness.

In response to these events, and armed with unprecedented access to knowledge and information about the environmental and health impacts of pollution, citizens were moved to action. Several young Canadian activists, including Bob Hunter (quoted at the beginning of

FIGURE 21.5
The Cuyahoga River (located in northern Ohio, near Lake Erie) caught fire several times in the 1950s and 1960s. The Cuyahoga, which flows into the Great Lakes, was so polluted with oil and industrial waste that the river would burn for days at a time.

this chapter), founded a tiny environmental organization called Greenpeace (**FIGURE 21.6**). Their original intent was to stage daring, high-profile protests against submarine testing of nuclear devices by the United States. They soon branched out into protests against whaling, bottom trawling, and other industrial activities they saw as exploitive and unsustainable. Greenpeace has since grown into one of the most powerful environmental organizations in the world, still known for the daring nature of its protests.

Today, largely because of grassroots activism and environmental policies enacted since the 1960s, pollution is more strictly regulated, and air and water are considerably cleaner. The public enthusiasm for environmental protection that spurred such advances remains strong today. Polls repeatedly show that an overwhelming majority of Canadians favour environmental protection—even if it means paying more. Such support is evident each year in April, when millions of people worldwide celebrate Earth Day in thousands of locally based events featuring speeches, demonstrations, hikes, bird walks, cleanup parties, and more. Since the first Earth Day, on April 22, 1970, participation in this event has grown and spread to nearly every country in the world (**FIGURE 21.7**).

The economic context influences environmental policy

Historians have suggested that major advances in environmental policy occurred in the 1960s and 1970s because three factors converged. First, evidence of environmental problems became widely apparent. Second, people could

FIGURE 21.6
Jim Bohlen, Greenpeace skipper John Cormack, Erving Stowe, and Paul Cote of the British Columbia branch of the Sierra Club (l-r) at a Vancouver harbour in 1971, prior to setting sail for Amchitka Island where the United States was scheduled to set off an underground nuclear blast.

visualize policies to deal with the problems. Third, the political climate in North America was ripe, with a supportive public and leaders who were willing to act.

There was a fourth reason for the advancement in environmental policy: economic confidence. By the 1960s and 1970s, people in North America had reached a point in their economic development where life was reasonably comfortable for more people than ever before. The basic necessities for survival were ensured, and people found themselves willing to make sacrifices—notably financial sacrifices but also behavioural changes—to obtain a cleaner, healthier environment for themselves and their children.

(a) The first Earth Day, Washington, D.C., 1970

(b) Schoolchildren celebrating Earth Day

(c) Earth Day Canada's familiar logo

The word mark "Earth Day" and the Earth Day logo are registered trademarks of Earth Day Canada (1991) Inc. Use of these trademarks for mercantile, promotional and communication purposes are strictly forbidden without the written approval of Earth Day Canada.

FIGURE 21.7

April 22, 1970, saw the first Earth Day celebration **(a)**. This public outpouring of support for environmental protection sparked a wave of environmental policy to address pollution. Earth Day is still celebrated by millions of people across the world **(b)**, as shown here in 2015 in Vancouver, at an Earth Day parade organized by Youth for Climate Justice Now. The familiar Earth Day Canada© logo **(c)** symbolizes leadership in environmental education and action in addition to traditional awareness-raising Earth Day activities.

Economists have a name for this change in attitude: It is called a *willingness-to-pay transition,* and it comes along with economic development in a society. Most basic environmental problems, such as lack of sanitation or clean water, improve with increasing economic status of societies (**FIGURE 21.8A**). However, for some types of environmental problems there is an initial stage of worsening of environmental quality while industrial development accelerates (**FIGURE 21.8B**). Eventually a social turning point is reached, and the idea of regulating industry or paying for cleaner technologies becomes more attractive.

During the Industrial Revolution in England, for example, and later in North America, air pollution from factories and refineries was a huge problem. Financial progress eventually led to environmental policies and investment in technologies to limit air pollution, with the result that air quality has improved dramatically. A modern example can be seen in China, where rapid industrial development has led to serious problems with urban air pollution. Both the government and the public in China now seem committed to solving this problem.

For the past century, waste generation and carbon emissions have been two environmental problems for which there seemed to be no turning point and no willingness-to-pay threshold—they just continued to increase in severity as incomes rose (**FIGURE 21.8C**). In recent years, however, many people appear to be more willing to make financial sacrifices and modify their behaviour to address these problems.

Different environmental media require different regulatory approaches

The laws governing different environmental media—water, land, forests, mineral resources, and even plants, animals, and air—have built upon different legal precedents. Let us briefly consider how some of these historical precedents influence the legal landscape today.

Water law in Canada (and other countries) has developed from two different historical/legal concepts. The first is called **riparian** law, in reference to the *riparian zone,* or water's edge. In riparian law, anyone who has legal access to the water's edge (such as by owning property on a river bank) has the right to withdraw water from the resource. The second legal concept in water law is **prior appropriation**. This refers to the "first come, first right" principle, in which one's right to withdraw water is established by historical precedent—if you have always withdrawn water from this river, then your right to do so has been established historically.

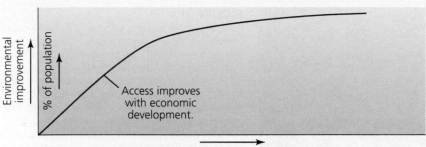

(a) Access to sanitation and clean water

FIGURE 21.8
Access to sanitation and clean water generally improve with increasing income in a society undergoing development **(a)**. Air pollution tends to worsen with increasing income early in industrial development **(b)**, which typically involves increased use of fossil fuels. As incomes increase, a willingness-to-pay threshold is reached; pollution control technologies are employed, and air quality improves. Waste generation and carbon emissions have traditionally worsened with increasing income; until recently, a willingness-to-pay transition had not been evident **(c)**. We may now be willing to alter our behaviour to see some improvements in these two areas.

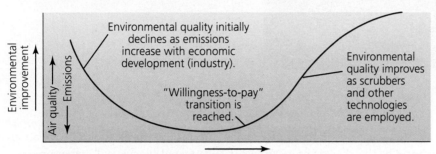

(b) Air quality

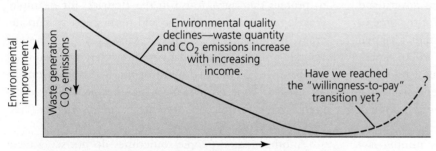

(c) Waste generation and carbon emissions

Water is particularly complicated, because it varies in time and space. Everyone needs water, and the potential uses for a body of water are often competing. The specific approach taken to water law will depend on whether the resource is publicly owned, privately owned, or a common property. In Canada, the Crown owns all water, and water rights are granted by licence; this stems partly from the fact that the federal government regulates transport. The provinces gain control over water rights through various ecosystem and land management responsibilities. In Canada, Aboriginal water rights are also an important part of the water management landscape.

Many of the environmental problems in water management today were not evident when these historical precedents were set. For example, groundwater depletion wasn't a problem a hundred years ago, and neither was pollution. These have necessitated some interesting reconsiderations of water law. For example, if you spill a toxic chemical on land that you own, you might not be held liable for any damages; however, if the toxic chemical infiltrates, joins the ground water, and then migrates in the subsurface into the ground water in your neighbour's property, contaminating the aquifer and wells there, you would be liable for damages.

Environmental policy to protect the atmosphere is particularly tricky because the atmosphere is a commonly owned and commonly accessed resource. Some atmospheric impacts are local and can be managed and regulated at a local level. But more often than not, what we put into the atmosphere is transported by wind and weather systems and may eventually affect important functions such as the climate system and the stratospheric ozone layer. As a global community, we have had to come together and hammer out our differences to reach international agreements for effective governance of these and other global commons, such as the open oceans and Antarctica.

FIGURE 21.9
Aboriginal representatives gathered at Nelson House, Manitoba, in 1910 for a treaty signing, where the representatives asked many questions about the continuation of their resource rights. The Treaty Commissioner is reported to have assured them that "not for many years to come, probably not in the lifetime of any of them, would their hunting rights be interfered with."
Source: Photo: Library and Archives Canada. Quote: Frank Tough, Economic Aspects of Aboriginal Title in Northern Manitoba: Treaty 5 Adhesions and Métis Scrip. *Manitoba History, 15,* Spring 1988.

In Canada, the right to govern and allocate water and other natural resources was granted to the provinces by the federal government, which continues to play an important role (in the management of fisheries, for example). For the eastern provinces, the transfer of rights to govern and allocate access to natural resources happened at the time of Confederation; for the western provinces, it didn't happen until the *Natural Resources Transfer Act* of 1930. These transfers of jurisdiction over land and natural resources from the federal to the provincial governments happened with little or no regard for the rights of prior Aboriginal inhabitants of the land.

The history of land law and especially mining law in Canada is intricately connected to Aboriginal land claims. There are two categories of Aboriginal land claims in Canada—those for which treaties exist and those for which no treaty exists. Resolving these claims often rests on establishing traditional occupancy and continuous use of lands and resources (**FIGURE 21.9**).

Many factors hinder implementation of environmental policy

If the goals of environmental policy are seemingly so noble—protecting resources and promoting equitable distribution of environmental costs and benefits—and reflective of the values of our society, why is it that environmental laws are so often challenged, and the ideas of environmental activists ignored or rejected by citizens and policymakers?

In Canada, most environmental policy has come in the form of laws instituted by government regulators. Some businesses and individuals view these regulations as overly restrictive, bureaucratic, or unresponsive to human needs.

For instance, many landowners fear that zoning regulations or protections for endangered species will impose restrictions on their activities or on the use of their land; this response was seen in the case of the emergency order to protect the sage-grouse in the Prairies, for example. Developers complain of time and money lost to bureaucracy in obtaining permits; reviews by government agencies; surveys for endangered species; and required environmental controls, monitoring, and mitigation. In the eyes of property owners and businesspeople, environmental regulation all too often means inconvenience ("red tape") and economic loss.

Another reason people sometimes do not see a need for environmental policy stems from the nature of most environmental problems, which often develop gradually. The degradation of ecosystems caused by human impacts on the environment is a long-term process. Human behaviour is sometimes geared toward addressing short-term needs, and this is reflected in many of our social institutions. Businesses may opt for short-term financial gains over long-term considerations. The news media have a short attention span based on the daily news cycle, whereby new events are given more coverage than slowly developing long-term trends. Politicians often act out of short-term interest because they depend on getting re-elected every few years. Governments may opt to spend money to solve immediate, pressing social problems, such as hunger, rather than investing in a long-term future that may seem vague.

Sometimes people simply don't want to change their lifestyle or their behaviour. Sometimes they are unaware, unconvinced, or simply confused by scientific evidence pointing to urgent environmental problems. For all of these reasons, many environmental policy goals that seem admirable, and that attract wide support in theory, may be obstructed in their practical implementation in our everyday lives. Making and successfully implementing new environmental policies involve the consideration of environmental goals within a much broader context of the legal and regulatory framework; economic, social, and political realities; technological possibilities; and the weight of scientific evidence and consensus.

Canadian Environmental Law and Policy

Canadians have been leaders in the development of environmental management approaches and policies. However, our approach to environmental management, law, and policy has been inevitably influenced by the United States. This is partly because we are next-door neighbours; it is also because we have economies that are closely linked, as well as a shared history of life on this continent.

Canada's environmental policies are influenced by our neighbour

The United States was a leader in the early development of national-level environmental laws, passing the first comprehensive environmental protection law (*National Environmental Protection Act, NEPA*) in 1970. This led, in turn, to the creation of the Environmental Protection Agency (EPA). Canada followed shortly thereafter, with the establishment of Environment Canada, mandated by the *Department of the Environment Act* in 1971. Most countries in the world have now followed the lead of Canada and the United States with regard to establishing federal-level environmental legislation.

Canada also is heavily influenced by the United States in its environmental management approach because of our trading relationship, now governed largely by the North American Free Trade Agreement (NAFTA), which regulates how marketable commodities are handled. This includes things like wood, agricultural crops, and animal products. NAFTA even identifies water resources as a marketable commodity that should not be restricted from export. We will look at NAFTA and its environmental implications in greater detail when we turn to international agreements in environmental policy.

Canadian environmental management also is strongly influenced by the sheer extent of the environmental resources we share with our southern neighbour—across the 8891 km of the world's longest undefended border we share countless rivers and watersheds, rock and soil masses, airsheds and weather systems, animal migration paths, and ecosystems. In the past, and still today, human activities on both sides of the border have negatively affected people and ecosystems on the other side. For example, industrial emissions from the central and northeastern states are transported into southern Canada, leading to smog and acid precipitation.

Sometimes Americans and Canadians don't agree on how our shared natural systems and cross-border impacts should be managed. For the most part, however, Canada–U.S. binational management of transboundary pollution and shared environmental resources has been characterized by cooperation and dialogue, and serves as a successful model for international environmental management.

Legal instruments are used to ensure environmental goals are achieved

Government agencies rely on several types of instruments to achieve environmental goals; chief among these are legal instruments. *Laws* (acts or statutes) are proposed and voted upon by the Parliament of Canada. Laws that influence the environment in various ways also can be passed by provincial and territorial parliaments and by municipalities, but they cannot contradict or bypass federal-level laws.

Regulations are legal instruments, too, but they are more specific; they are detailed sets of requirements (such as numerical limits, licensing requirements, performance specifications, and exemption criteria) established by governments to allow them to implement, enforce, and achieve the objectives of environmental acts. For example, the *Passenger Automobile and Light Truck Greenhouse Gas Emission Regulations* establish emission standards for greenhouse gases that must be met by all cars and light trucks in model years from 2011 onward. Typically, regulations operate under the authority of a particular piece of legislation; in this case, the Passenger Automobile Emission Regulations operate under the authority of the *Canadian Environmental Protection Act* (CEPA). Some provinces have emissions standards that are stricter than those of the federal government, and these regulations are enforced at the provincial level; an example is the *Drive Clean* program, a provincial regulation under Ontario's *Environmental Protection Act*.

In Canada there are several different ways in which individuals and organizations can be required or compelled to obey environmental laws, or penalized if they do not obey the law. They are criminal enforcement (usually based on failure to meet due diligence requirements); penalties or fees; administrative orders to investigate, clean up, or otherwise address an environmental situation of concern; and finally, civil actions, through which an individual or corporation that causes environmental damage to another's property or person may be held responsible for the damage via a lawsuit.[13]

Agreements are another type of instrument for achieving the goals of environmental policy. Agreements can be either enforceable or voluntary; they are entered into by agencies of government, often with industry representatives. Agreements are usually time-limited and are assessed for their performance in achieving their goals once the assigned time frame has expired. An example of a current environmental agreement is the one between Environment Canada and the Refractory Ceramic Fibre Industry (2013, for five years). Refractory ceramic fibre is essentially an insulating "wool" made of silica. It is used for many of the same purposes as asbestos, but has fewer known health risks. There is some concern, based on animal studies, that inhalation of the tiny silica fibres could cause mesothelioma (as discussed for asbestos in Chapter 19, *Environmental Health and Hazards*). Therefore, the industry has entered into an agreement with Environment Canada to limit the release of microscopic fibres into the air. The success of the agreement will be assessed at the end of the five-year period, and it may be renewed at that time.

Permits are documents that grant a group or an individual legal permission to carry out an activity that will have environmental impacts, usually within certain limitations and for a specified period of time. Examples of activities that require permits at the federal level include the disposal of substances at sea, import and export of hazardous wastes, controlled release of toxic substances, and hunting of migratory birds.[14] Scientists who study birds and other wildlife also need permits for activities such as banding and tagging, which might interfere with the animal's condition, even if temporarily. Additional permits are often required, too, at the level of the province or the specific location where the bird-banding or other wildlife study is to take place.

Environmental goals also can be promoted by voluntary initiatives

One alternative to environmental legislation that appeals to many in the private sector is the adoption, by consensus, of *voluntary guidelines* that are sector-based and self-enforced. A familiar example is self-policing by

television and music producers for content deemed too violent for children. Some sectors have had success with this approach; others have not. The Canadian mining industry, for example, which is heavily regulated by both federal and provincial environmental legislation, has undertaken some voluntary initiatives and self-imposed guidelines on behalf of the environment. Examples include guidelines on the use of traditional ecological knowledge in mining, and sharing of best practices for the management of acid drainage at mine sites. While many of these programs are considered successful, they could not replace environmental legislation and regulatory requirements in the mining industry.

The Accelerated Reduction and Elimination of Toxics (ARET) Program is an example of a major voluntary initiative in Canada that involved industries from a number of sectors, including mining, as well as government agencies and nongovernmental organizations. The program, which was started in 1994, issued a proactive challenge to industry to reduce emissions of toxic substances even before such reductions were legislated. By the time the program ended in 2000, it claimed many successes in reaching its emission reduction targets. However, UBC researchers Werner Antweiler and Kathryn Harrison carried out an in-depth analysis of the program and found many discrepancies between the voluntary reporting of emission reductions through ARET and subsequent legally mandated reporting through Canada's National Pollutant Release Inventory (NPRI) after 2001. Their conclusion was that ARET was successful in some cases but not in others, and suffered from lack of verifiability.[15]

Another example of voluntary guidelines for environmental practice that have been widely adopted, both in Canada and internationally, is the ISO 14001 standards for environmental management. *ISO* is the *International Organization for Standardization*, a nongovernmental organization with 163 member nations that is headquartered in Geneva, Switzerland.[16] Probably the best-known set of international standards is the ISO 9000 series aimed at quality assurance in industrial processes. The ISO 14000 series was similarly designed to promote consistency and best practices, but in the specific context of environmental management. If a factory, office, government, or other type of organization chooses to become certified under ISO 14001, it voluntarily agrees to meet the standards and follow internationally accepted procedures for environmental management. In turn, it becomes eligible to promote itself as an ISO 14001-certified organization.

It is open to debate whether voluntary initiatives and self-policing are truly effective in achieving environmental goals. Some critics argue that by asking for self-regulation, industry is simply seeking to avoid legislation or—more cynically—trying to divert attention from its environmental shortcomings. Even those in industry who favour voluntary initiatives acknowledge that these generally represent the "best practices" of the most proactive, environmentally responsible companies, and that legislation is needed in order to avoid free riders that do not follow the voluntary guidelines.

Canadian environmental policy arises from all levels of government

Even with the influence of such a populous nation and heavyweight economy right next door, we Canadians have developed our own approach to environmental management and regulation. This approach has tended to make environmental management collaborative, cooperative, and consultative. It is codified in the *Canadian Environmental Protection Act* (1999) that the administration of the Act falls to Environment Canada, but all other aspects of its implementation are collaborative. Public consultation also is part of the wording of the Act.

In Canada, the federal government shares responsibility for environmental protection with provincial/territorial, Aboriginal, and municipal/local governments. The principal responsibility for the environment falls to the provincial/territorial governments in most situations. This multilevel approach necessitates close collaboration among the levels and agencies of government; sometimes it results in overlapping and redundancies, confusion of jurisdictions, and even contradictions between provincial and federal law.

One way the government attempts to overcome the overlapping of environmental jurisdictions is through the Canadian Council of Ministers of the Environment (CCME), which includes the federal, provincial, and territorial environment ministers. They meet regularly to work on projects of mutual and overlapping interest. For example, the CCME is developing a set of nationwide standards, the Canadian Environmental Quality Guidelines. The first harmonized standards, for water quality, were signed into accord by the ministers in 1987. Since then, several sets of standards and refinements for water quality (for various uses) and for sediment and soil quality have been added. Similarly, the Natural Ambient Air Quality Objectives (NAAQOs) have established goals, benchmarks, and testing protocols for common air pollutants. Even if standards are in place, however, the provinces and territories may interpret and implement the standards differently.

Federal government There are many federal-level laws and regulations that affect the environment, some directly and others indirectly. The federal government also influences environmental policy by choosing to fund certain programs, such as the Habitat Stewardship Program mentioned in the case of the greater sage-grouse, or the Accelerated Reduction and Elimination of Toxics program. Another example is the Northern Contaminants Program, established in 1991 to monitor contaminants in wildlife species that are important to the traditional diets of Aboriginal peoples in Canada's North.[17]

Some of the environmental laws that form the basis for environmental policy at the federal level in Canada include the following:

- *Canadian Environmental Protection Act* (CEPA) (1999), the foundation of Canada's federal-level environment legislation. CEPA focuses on preventing pollution and protecting the environment and human health. The Act gives the government broad enforcement powers and mandates public participation and consultation, giving citizens input into environmental policy decisions.[18]
- *Canadian Environmental Assessment Act* (CEAA) (2012). CEAA outlines the requirements for **environmental impact assessment (EIA)**, that is, a study of the potential environmental impacts of a proposed project. The depth or magnitude of the assessment, the specific requirements, and the mitigation efforts that may be required depend on the extent and type of project. CEAA was originally enacted in 1992; it was superseded by CEAA 2012 (see **TABLE 21.1**).
- *Fisheries Act* (1985). The *Fisheries Act* originally prohibited the "disruption" or the "release of deleterious substances" into any body of water where fish may be present at any time. It was one of the most powerful laws for the protection of freshwater resources and aquatic environments in Canada; its provisions have been modified by Bill C-38 (see **TABLE 21.1**).
- *Transportation of Dangerous Goods Act* (1992). This federal law regulates the transport of hazardous materials in Canada by air, road, rail, or ship, no matter what the purpose or point of origin of the transfer. Hazardous waste materials are also covered under this act.
- *Canadian Wildlife Act* (1985), *Migratory Birds Act* (1994), and *Species at Risk Act* (SARA) (2002). All of these laws aim, in various ways, to respect the needs of and identify risk factors for organisms in Canada, and to protect them from harm or extinction.

Several longstanding environmental laws and procedures, notably the *Fisheries Act* and the *Canadian Environmental Assessment Act*, were affected or modified by

TABLE 21.1 The Environmental Implications of Bill C-38[19]

Bill C-38 was passed into law by the federal government in 2012. Although its title is *Jobs, Growth, and Long-Term Prosperity Act*, much of the content replaces or modifies previous environmental legislation. The Act is multifaceted, but part of the purpose was to streamline the often-cumbersome environmental assessment process, with the goal of facilitating the economic development of Canada's natural resources.

Here are some examples, in regular, nonlegal language, of changes in Bill C-38 that are relevant to environmental policy. Don't take this analysis at face value; it is obviously and unavoidably biased. You should read the legislation and decide for yourself whether you think the changes in Bill C-38 are beneficial and what impacts they will have on Canada's environment and economy.

Bill C-38

- Establishes a new *Canadian Environmental Assessment Act*, superseding *CEAA (1992)*
  - The new *Canadian Environmental Assessment Act (2012)* exempts projects proposed or regulated by the federal government from undergoing an environmental assessment, unless an assessment is specifically required by the environment minister.
  - New time limits have been imposed for completion of environmental assessments. The minister has the power to disband a panel if the assessment is not likely to be completed within the time limits.
  - Public participation in environmental assessments is now limited to those who will be directly affected by a project or who have specifically applicable technical expertise.
  - Pipelines and nuclear project proposals are specifically expedited. They will no longer be referred to an independent review panel, even if they are designated to undergo an environmental assessment. Instead, they will be assessed by the National Energy Board or Canadian Nuclear Safety Commission.

- Modifies the definition of "environmental effects"
  - Environmental assessments are now only required to consider the potential effects of a project on fish, fish habitat, other aquatic species, and migratory birds directly linked to the government's role in the project. Other effects (such as on other endangered species or habitats) might not be considered unless required by the minister.
  - In the case of a project that is deemed to have potential adverse effects, the Act gives decision-making power to Cabinet, instead of to the department or agency directly involved in regulating the project and its impacts.

- Modifies the *Fisheries Act*
  - The *Fisheries Act* prohibited the "disruption" of fish habitat and the release of "deleterious substances" into environments where fish live. In the new Act, protection only applies to fish that contribute to commercial, recreational, or Aboriginal fisheries.
  - The focus of regulation is limited to activities that could cause permanent alteration or destruction of habitat; "disruption" is no longer prohibited.

- Repeals the *Kyoto Implementation Act*
  - Canada withdrew from the Kyoto Protocol in 2011, but the government was still required by this legislation to develop a plan for meeting greenhouse gas reduction targets, and to report annually on progress toward those goals. The legal requirement for regular reporting on progress toward GHG reductions has now been eliminated.

- Modifies the *Species at Risk Act (SARA)*
  - The Act removes time limitations on permits for activities that affect species at risk or their habitat, formerly limited to 3–5 years under SARA.
  - For pipeline applications, the National Energy Board is now exempt from the SARA requirement to consider and seek to minimize impacts on the habitat of species at risk.

- Eliminates the *National Round Table on Environment and Economy (NRTEE)*
  - Funding for the NRTEE, established in 1998, ended and the organization ceased to exist in 2013 as a result of Bill C-38.

- Changes the environmental reporting requirements for federal government agencies
  - Parks Canada is now required to report less frequently to the government regarding the status of national parks and marine reserves.

Based on information from Government of Canada (2012). *Bill C-38: Jobs, Growth, and Long-Term Prosperity Act*. First Session, Forty-first Parliament, 60-61 Elizabeth II, 2011–2012. An Act to implement certain provisions of the budget tabled in Parliament on March 29, 2012 and other measures. Assented to 29th June 2012. http://www.parl.gc.ca/HousePublications/Publication.aspx?DocId=5697420. Accessed 6 April 2015. EcoJustice & West Coast Environmental Law (2012). What Bill C-38 Means for the Environment and Top Ten Environmental Concerns of Budget Bill C-38. http://wcel.org/resources/publication/what-bill-c-38-means-environment. Accessed 6 April 2015. David Suzuki Foundation (2012). Bill C-38: What You Need to Know. http://www.davidsuzuki.org/publications/resources/2012/bill-c38-what-you-need-to-know/. Accessed 6 April 2015.

Bill C-38, the *Jobs, Growth, and Long-Term Prosperity Act*, passed in 2012. Some of the environmental impacts of this bill are summarized in **TABLE 21.1**.

Environment Canada has the duty to administer, implement, and enforce CEPA and several other acts, collaborating in this effort with a wide variety of other federal-level departments. For example, Environment Canada and Health Canada together decide on the allowable levels for various contaminants in drinking water; the regulations are then published by Health Canada and enforced by Environment Canada. Other federal-level agencies that commonly work with Environment Canada to set environmental policies and implement environmental legislation are listed in **TABLE 21.2**.

Aboriginal governments

Aboriginal governments in Canada participate in environmental governance through a wide variety of mechanisms. In some cases, this involvement stems directly from land claims, both treaty and non-treaty. Aboriginal governments are often centrally involved in decision-making about where, when, and under what conditions to allow resource extraction activities, such as mining, forestry, or oil drilling, to proceed on Aboriginal lands. The settlement of land claims and legal establishment of resource rights are central to such activities.

In other cases, involvement stems from the desire (and, sometimes, the legal mandate) to ensure adequate representation for Aboriginal ecological knowledge in environmental decision-making. For example, decisions made under SARA are required to take into account input from Aboriginal environmental experts about the species in question. Aboriginal communities have been represented in decision-making regarding the Mackenzie River valley natural gas pipeline. Local Aboriginal groups also participate in stakeholder consultations and scientific efforts to document traditional resource use patterns and traditional ecological knowledge of the environment.

Even corporate decision-making has begun to integrate the Aboriginal perspective; for example, modern diamond mining companies in Canada's North depend heavily on collaborations with Aboriginal stakeholders. The effectiveness with which Aboriginal perspectives and traditional knowledge have been incorporated into environmental management processes has not always been optimal, but the foundations for the process are there.

Much of Aboriginal involvement in environmental management in Canada is based on the struggle to ensure equitable access to natural resources, and to prevent excessive exposure of Aboriginal communities to environmental degradation. These are the two central concerns of the environmental justice movement. For

TABLE 21.2 Federal Agencies That Influence and Implement Environmental Policy and Natural Resource Management

- **Environment Canada**, which encompasses a number of agencies, including the Canadian Environmental Assessment Agency, Parks Canada, Canadian Wildlife Service, Meteorological Service of Canada, and other agencies and regional offices
- **Health Canada**, which participates in scientific research concerning disease and the health risks of exposure to substances and in the setting of regulatory guidelines for potentially harmful substances in water, air, soil, and food
- **Fisheries and Oceans Canada**, which has the central responsibility for the implementation and enforcement of the *Fisheries Act*
- **Natural Resources Canada** (including the Office of Energy Efficiency), which promotes the responsible use of Canada's natural resources
- **Agriculture and Agri-Foods Canada**, which provides information, research, technology, policies, and programs for security of the food system, soils, and agricultural lands
- **Natural Sciences and Engineering Research Council (NSERC) of Canada**, which funds scientific research on the environment and related technologies
- **Aboriginal Affairs and Northern Development Canada**, which (among other roles) promotes the sustainable development of Aboriginal communities and their natural resources
- **Transport Canada**, which regulates the transport of hazardous materials by air, rail, road, and sea in Canada, as well as plays a major role in transportation system design and in Canada's efforts to reduce greenhouse gas emissions
- **Statistics Canada**, which maintains extensive databases on population, patterns of travel, consumption, trade, lifestyle, and other human activities that affect the environment
- Many other agencies, some of which come into play primarily or exclusively when it becomes necessary to respond to an emergency, such as a major flood

example, there is an ongoing struggle in Canada to ensure adequate drinking-water quality in Aboriginal communities—a resource that most of the rest of Canada takes for granted.

Provincial/territorial governments

The provincial and territorial governments are very active in environmental protection and regulation. The specific details of content and procedure vary from one provincial jurisdiction to the next, but there is some consistency. For example, all of the provincial and territorial governments have some form of legislation that sets limits on the amounts and concentrations of specific harmful or potentially harmful substances that can be released (or *discharged*) into the environment. (The exact limits and

Frank Gunn/The Canadian Press/AP Images

FIGURE 21.10
There are many ways that municipalities can use bylaws to promote environmental goals. Toronto was the first city in North America to have a bylaw requiring green roofs on new commercial, institutional, residential, and industrial developments that exceed a minimum floor space. This is a green roof at the Pan Am Games Athletes' Village in Toronto.

the specific lists of controlled substances vary, but the approaches are similar.)

All of the provinces and territories also require organizations and individuals to obtain approvals or permits before undertaking activities that might prove damaging to the environment. For example, the taking of water from aquifers or surface water bodies in amounts over a certain threshold requires a special permit. Construction, dredging, mining, and logging are other examples of activities that are regulated provincially and territorially, and require approval before being undertaken.

In Ontario, citizens are protected by an *Environmental Bill of Rights* (EBR, 1994), one of the few such laws in the world. The EBR, administered by the appointed Environmental Commissioner of Ontario, establishes that the people of Ontario have as a common goal the protection of the natural environment. It guarantees to all Ontarians the right to a clean, healthy environment and the right to participate in and be made aware of government decisions regarding the environment. The EBR mandates the establishment and maintenance of an Environmental Registry whereby any citizen may enter a query against an activity that may be damaging the environment, to which the government and/or industry agents responsible for that activity are required to respond.[20] The Environmental Commissioner also reports annually to the people of Ontario on the progress (or lack of progress) by the provincial government in meeting its environmental goals.

Municipal/local governments   The legal and policy landscape is further complicated by the role taken by municipal governments in environmental regulation and enforcement, a role that is growing in importance.

Municipalities in Canada have traditionally taken responsibility for the management of water and sewage systems, noise issues, waste disposal, land-use zoning to regulate development, and local air quality concerns, such as requiring permits for backyard burning of yard wastes. All of these traditional municipal-level roles have some environmental influence.

Expanding their traditional environmental role, many municipalities have passed bylaws that restrict or prohibit the use of pesticides and herbicides on lawns in urban and suburban areas, even while their use was still approved by the provincial government. A number of provinces and territories have now followed the lead of these municipalities in banning the cosmetic use of pesticides. Other municipalities have adopted initiatives to deal with sprawl and transportation issues across municipal boundaries. It is also becoming more common for municipalities to have "green roof" policies; Toronto was the first city in North America to enact such a law, in 2009[21] (**FIGURE 21.10**). Many urban municipalities are becoming more involved in the documentation, monitoring, and even regulation of *brownfields*, contaminated sites that may become available for redevelopment once they have been rehabilitated to provincial standards. Some municipalities have even entered into agreements to limit CO_2 emissions—the type of action that traditionally has fallen within the jurisdiction of the federal government.

Government and ENGOs can work together on environmental issues

The Canadian approach to environmental management has traditionally been consultative; in fact, as mentioned, public participation and consultation are mandated by

CEPA. This means that government normally should not undertake any major revisions to policy without extensive consultation with stakeholders. A **stakeholder** is any person or group that has an interest in, or might be affected by, the outcome of a particular undertaking. This includes both environmental projects and policies.

Individual citizen stakeholders in environmental policy development are often represented by **environmental nongovernmental organizations (ENGOs)**. These can be activist groups, such as Greenpeace; political advocacy groups, like the Sierra Club, Riverkeepers, or the Environmental Defence Fund; or groups with specific mandates, such as the Nature Conservancy, which aims to preserve land, habitat, and ecosystems by acquiring and setting aside land in trust, or the Mountain Gorilla Conservation Fund, which raises money to preserve habitat for endangered primates.

Stakeholders in the industrial sector are often represented by sector-specific nongovernmental agencies. For example, the Protected Areas Initiative in Manitoba employs an extensive process of public consultation on the environmental impacts of mining, in which the mining industry is represented by the Mining Association of Manitoba and the Mineral Exploration Liaison Committee; the government of Manitoba is represented by staff from the Department of Conservation and the Manitoba Geological Survey and Mines Branch; and other stakeholders are represented by World Wildlife Fund Canada.

Another consultative vehicle that has been used in Canada is the **round table**, a multistakeholder working group established to consult on a particular issue, generally within a particular sector or area of concern. Round tables involve representatives from a variety of governmental, nongovernmental, and private-sector organizations. The *National Round Table on the Environment and the Economy (NRTEE)* was established in 1988 as an advisory council to the prime minister. With appointed members that included business, labour, environmental organizations, academic institutions, and First Nations, the NRTEE was a model for best practices in environmental decision-making. NRTEE funding from the federal government ended with the passage of Bill C-38, and the organization was disbanded in 2013.

Scientific monitoring and reporting help with environmental policy decisions

All levels of government report to the Canadian public about their activities and any changes, positive or negative, in the condition of the environment within their jurisdiction.

State-of-the-environment reporting (SOER) refers to the collection, organization, and reporting of information that can be used to measure and monitor changes in the environment and in processes or factors that have impacts on the environment over time. The information is reported by using **indicators**, values that can be measured and in comparison to which changes can be assessed. Some indicators are simple numbers (e.g., number of species on the endangered species list, or hectares of forested area). Others are complex, composite, or derived numbers (e.g., the Air Quality Health Index, which combines several indicators into one number).

The intention of environmental reporting is typically to answer five key questions:[22]

1. What is happening in the environment?
2. Why is it happening?
3. Why is it significant?
4. What is being done about it?
5. Is this response sustainable?

The beginning of formal state-of-the-environment reporting in Canada dates back to *Our Common Future*, the 1987 Report of the United Nations Commission on Environment and Development, when a response to the report was assembled by the Canadian Council of Ministers of the Environment (CCME). Around the same time, several provinces and territories were beginning to develop their own sets of indicators and plans for sustainability (Yukon, Alberta, British Columbia, Nova Scotia, and Saskatchewan were the earliest). Not all provinces and territories have regular programs for environmental or sustainability reporting, though all of them collect and report on environmental information in one way or another. It is time-consuming and expensive to collect information on a comprehensive set of indicators and keep the information up to date.

Environment Canada now reports to the public through the Canadian Environmental Sustainability Indicators Program (CESI), as part of the government's *Federal Sustainable Development Strategy*.[23] The CESI reporting program has three main themes, with sets of indicators within each theme. Theme I, Climate Change and Air Quality, includes indicators such as greenhouse gas emissions, particulates in the air, and mercury and lead emissions. Theme II, Maintaining Water Quality and Availability, includes indicators such as water quality in Canadian rivers, drinking-water advisories, and phosphorus in the Great Lakes. Theme III, Protecting Nature and Canadians, includes indicators such as species-at-risk population trends, the status of major fish stocks, and the ecological integrity of national parks.

State-of-the-environment reporting can take place on many levels, not always corresponding to a particular jurisdiction. For example, Fisheries and Oceans Canada periodically produces a State of the Ocean Report for Canada.[24] A number of provinces, regions, and municipalities in Canada produce SOE or sustainability reports on a fairly regular basis. Municipal-level reports tend to focus on community-level indicators of well-being for both ecosystems and people. Many corporations, too, have adopted environmental or sustainability reporting as sections or addenda to their annual business-reporting framework.

There is so much scientific information about the environment that it can be very challenging to organize and analyze it, and to decide which pieces of information are the most important. The Province of Saskatchewan organizes its state-of-the-environment reporting under five themes, centred on environmental media and processes; they are air, climate, land, forest, and water (**FIGURE 21.11**).[25] Underlying these five themes are a

number of diverse indicators of environmental quality. It is challenging (and expensive) to keep on top of the data collection required for environmental reporting on an entire suite of indicators; therefore, it is not typical for SOE reports to be published every year. Saskatchewan has reported every two years since 1991.

Saskatchewan also utilizes an organizational framework for SOER that is widely used internationally, called the **pressure–state–response (PSR) model**, sometimes called the *DPSIR model*, which stands for driving force–pressure–state–impact–response. (In the Saskatchewan SOER it is called "condition–stressor–response.")[27] This framework is based on establishing linkages and causalities (**FIGURE 21.12**) for environmental impacts and tracking our responses to them.

As shown in **FIGURE 21.12**, driving forces such as population growth and climate change, and human activities

air	climate	land	forest	water
Air pollutant concentration	Greenhouse gas emissions	Agricultural land cover	Forest type and age class	Surfacewater quality
		Mineral disposition activity	Forest wildfire disturbance	
Air pollutant volume		Area under zero-tillage	Forest insect and disease disturbance	Surfacewater quantity
		Private land stewardship	Proportion of sustainable harvest level utilized	
Air zone management		Waste recycling	Forest regeneration	Water consumption and conservation

FIGURE 21.11   The Province of Saskatchewan has produced state-of-the-environment reports every two years since 1991. The reporting is organized under a framework of the five themes shown here, which correspond to environmental media and processes rather than political boundaries. The five themes are underlain by a suite of indicators.
Source: Image courtesy of Environment—Government of Saskatchewan.

DATA Q   In Chapter 22, *Strategies for Sustainability*, we will examine the *Environmental Performance Index (EPI)*, a framework of 9 issues and 20 indicators that can be combined into a single weighted index to assess and track the environmental performances of nations.[26] Visit the EPI website (http://epi.yale.edu/our-methods) and compare the EPI indicators to the Saskatchewan SOER indicators. Are any major areas of environmental quality missing from either set of indicators? Is the missing indicator perhaps captured indirectly, through one of the other indicators or groups of indicators? Which set of indicators do you think does the best job, overall, at capturing environmental quality?

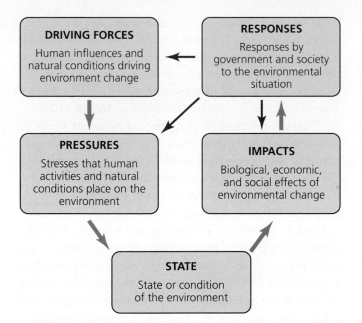

(a) DPSIR model

FIGURE 21.12
The driving force–pressure–state–impact–response (DPSIR) model **(a)** allows researchers and decision-makers to assess linkages and causalities and to assess the effectiveness of responses to environmental change. The table **(b)** shows how the model is used as an organizational framework for environmental indicators in the State of the Scotian Shelf Report. The Scotian Shelf is a part of the continental shelf of the Atlantic Ocean, off the east coast of Canada, that supports many ecological processes and human activities.
Source: MacLean M., H. Breeze, J. Walmsley, and J. Corkum (eds.) (2013). State of the Scotian Shelf Report. *Can. Tech. Rep. Fish. Aquat. Sci.* 3074; and Coastal and Ocean Information Network (2010). *State of the Scotian Shelf Report.* http://coinatlantic.ca/index.php/state-of-coast-and-ocean/state-of-the-scotian-shelf/217-driving-forces-pressures-state-impacts-response-framework. Copyright © 2015 by Atlantic Coastal Zone Information Steering Committee (ACZISC). Reprinted by permission.

INDICATOR	POLICY ISSUE	DPSIR	ASSESSMENT	TREND
Invasive Species: e.g.: Number of established invasive species	Growth in global trade and other human activities	Driving force, pressure	Poor	–
Waste and Debris: e.g.: Shoreline clean-up results	Human behaviour and waste management	State	Poor	+
Climate Change and Its Effects on Ecosystems, Habitats, and Biota: e.g.: Shifts in species distribution	Change in availability of fisheries resources	Impact	Fair	/
Water and Sediment Quality: e.g.: Number of regulated chemicals and substances	Public health, environmental protection, and regulation	Actions and responses	Good	+

Description	Symbol
Negative trend	–
Positive trend	+
Unclear or neutral trend	/

(b) Using the DPSIR framework for environmental reporting

such as agriculture, mining, water extraction, or logging, exert stresses on the environment. This affects the state or condition of the environment, with impacts on the quality of the environment, natural resources, and human well-being. Society responds to the changes by adopting environmental and economic policies aimed at reducing some of the stressors. Scientists and environmental managers at Environment Canada were instrumental in the early development of this model, which is now used as a reporting framework by a number of organizations internationally.

Another organizational challenge in environmental reporting is how to organize the information spatially. One option is to base the report on geographical or geopolitical boundaries, but these often don't correspond to natural ecosystem boundaries. For example, it would be more meaningful to produce a report for the Great Lakes system as a whole than for any of the provinces or states that border it to produce a report based only on its own jurisdictional boundary. Therefore, some reports focus on a biogeographical feature, such as a specific basin or a location with unique characteristics, such as the Arctic, or the Scotian Shelf, as shown in **FIGURE 21.12**. Others focus on a particular sector, such as the forestry sector; a specific current issue, such as climate change; or environmental media, such as air, water, or land.

International Environmental Law and Policy

Canadians (and anyone operating a business within Canada) are also governed by the various international agreements that Canada has entered into. For example, it is illegal to import elephant ivory into Canada because Canada is a party to the *United Nations Convention on International Trade in Endangered Species*. International environmental agreements are not easy to create, and can be even harder to enforce. Nevertheless, they are important because they represent a position taken by the entire global community on the importance of maintaining environmental quality.

Transboundary problems are a big focus of international policy

Natural systems pay no heed to political boundaries, so environmental problems often are not restricted to the confines of particular countries. For instance, most of the world's major rivers straddle or cross international borders, including the St. Lawrence, the Rhine, the Jordan, the Red River, and many others. Because Canada's laws have no authority in the United States or any other nation outside of Canada, international law is vital to solving transboundary problems.

Often nations make progress on international issues not through legislation, but through creative bilateral or multilateral agreements hammered out after a lot of hard work and diplomacy. Such was the case with the effort to develop a long-term plan to manage water quality in the Great Lakes through the International Joint Commission (see "The Science behind the Story: The Great Lakes and the International Joint Commission.")

Other examples of important North American environmental agreements, some of which we have mentioned elsewhere in this book, include the *Agreement on Air Quality*, *Migratory Birds Convention*, *North American Agreement on Environmental Cooperation*, *Canada–U.S. Agreement on the Transboundary Movement of Hazardous Waste*, and *Canada–U.S. Joint Marine Pollution Contingency Plan*. Canada is a signatory to 81 international

environmental agreements (there are some 500 such agreements worldwide).[28] One example is the *Montreal Protocol*, a 1987 accord among more than 160 nations to reduce the emission of airborne chemicals that deplete the ozone layer. Some examples of important international environmental laws and agreements are listed in **TABLE 21.3**.

The environment is also affected by other types of international agreements—not just those that focus specifically on environmental problems. In 1992, for example, the United States, Mexico, and Canada signed the *North American Free Trade Agreement*, better known as *NAFTA*, a comprehensive trade and investment agreement with aggressive measures to reduce barriers to trade. After the signing of NAFTA, it was a major concern of ENGOs and other organizations to determine the impacts—positive or negative—of NAFTA on the environment. The critical section of the agreement is the infamous "Chapter 11," which aims to prevent the mistreatment of investors by foreign governments. This means, for example, that companies may challenge a government's handling of a particular industry, including its environmental regulations, if those are seen as a threat to investment.

Canada has participated in a number of Chapter 11 disputes, including some with environmental implications. Probably the most famous is the softwood lumber dispute, in which the United States has claimed that Canada gives unfair advantage to domestic logging companies by subsidizing the forestry industry. The softwood lumber dispute dates back many decades and has been brought to NAFTA and WTO tribunals on numerous occasions. The United States has maintained that stumpage fees for logging on public lands were set too low by the Canadian government, giving unfair advantage to Canadian loggers. The dispute was settled temporarily in 2006 to the satisfaction of the two federal and three provincial governments involved (British Columbia, Ontario, and Quebec). That agreement has been extended twice and, at this writing (end of 2015), is set to expire again with no guarantee of renewal.

The greatest environmental concern arising from Chapter 11 has been the possibility that domestic environmental legislation, such as laws protecting water, could be interpreted as barriers to international free trade, and thus be overruled by NAFTA. An example was the lawsuit brought by Vancouver-based Methanex, the manufacturers of the gasoline additive MTBE, against the state of California. Methanex sought almost $1 billion in damages, claiming that California's ban on the additive amounted to trade protectionism. The Methanex claim was dismissed by a NAFTA tribunal in 2005, but other cases remain to be settled.

To address some of the environmental concerns raised by NAFTA, the three nations reached a corollary

TABLE 21.3 Some Important International Environmental Laws and Agreements in Which Canada Is a Party

Air

- Stockholm Convention on Persistent Organic Pollutants (POPs)
- Vienna Convention for the Protection of the Ozone Layer–Protocol on Substances that Deplete the Ozone Layer (Montreal Protocol)
- Convention of the World Meteorological Organization (WMO)
- United Nations Framework Convention on Climate Change (UNFCCC)–Kyoto Protocol (Canada formally withdrew in 2011)

Biodiversity

- Agreement on the Conservation of Polar Bears
- Convention on Biological Diversity
- Convention on International Trade in Endangered Species of Wild Fauna and Flora (CITES)
- Convention on Wetlands of International Importance (Ramsar)

Ecosystems

- Antarctic Treaty
- Arctic Council

Environmental Cooperation

- UNECE Convention on Environmental Impact Assessment in a Transboundary Context (Espoo Convention)

Hazardous Materials and Wastes

- Basel Convention on the Control of Transboundary Movements of Hazardous Wastes and Their Disposal
- Rotterdam Convention on the Prior Informed Consent (PIC) Procedure for Certain Hazardous Chemicals and Pesticides in International Trade

Lakes and Rivers

- Canada–U.S. Agreement on Great Lakes Water Quality
- Treaty Relating to the Boundary Waters and Questions Arising along the Border between the United States and Canada

Oceans

- Convention on the Prevention of Marine Pollution by Dumping of Waste and Other Matter (LC72)
- International Convention for the Prevention of Pollution from Ships (MARPOL 73/78)
- International Convention on Oil Pollution Preparedness, Response, and Cooperation
- United Nations Fish Stocks Agreement
- Northwest Atlantic Fisheries Organization (NAFO)

Source: Based on Environment Canada > International Environmental Agreements (Archived Content). https://www.ec.gc.ca/international/default.asp?lang= En&n= 0E5CED79-1.

agreement called the *North American Agreement on Environmental Cooperation (NAAEC*, 1993). The objectives are to promote sustainable development, encourage pollution prevention policies and practices, and enhance compliance with environmental laws and regulations. The agreement also promotes transparency and public participation in the development of environmental policies.[29] The agreement mandated the establishment of the Commission for Environmental Cooperation, provided a mechanism for the resolution of disputes, and established a cooperative work plan for the environment among the three nations.[30] The ultimate goal is to ensure that the parties do not lower their environmental standards to attract investment.

Now, 25 years later, it is difficult to assess whether NAFTA has been good for the environment or bad. In a paper focusing on this topic, produced for the Commission for Environmental Cooperation, Soleded Aguilar and her co-authors concluded that, "At a general level, the answer to what effects the NAFTA has had on the environment is: *It depends*. That is not as shallow an answer as it sounds. Various studies *[have]* found differing sorts of impact, both positive and negative, depending on the sector, the firm characteristics, and the associated institutions of governance and regulatory oversight . . . At a general level, the worst pre-NAFTA fears did not materialize."[31]

Several organizations shape international environmental policy

Most international environmental laws involve written legal contracts (treaties) into which nations agree to enter; an example is the *United Nations Framework Convention on Climate Change*. There is no real, binding mechanism for enforcing international environmental law. However, many important international organizations regularly act to influence the behaviour of nations by providing funding, applying peer pressure among nations, setting guidelines, monitoring progress toward agreed-upon goals, and/or directing media attention.

The United Nations   In 1945, representatives of 50 countries founded the United Nations. Headquartered in New York City, this organization's purpose is to maintain international peace and security; to develop friendly relations among nations; to cooperate in solving international economic, social, cultural, and humanitarian problems and in promoting respect for human rights and fundamental freedoms; and to be a centre for harmonizing the actions of nations in attaining these ends.[32]

THE SCIENCE BEHIND THE STORY

In this satellite image, nutrient and sediment plumes can be seen swirling through the water of Lake Erie (centre of photo). Rows of crops are visible in the surrounding areas, and the Canada–U.S. international border runs lengthwise through the middle of the lake.

Jeff Schmaltz/Modis Land Rapid Response Team/NASA GSFC

The Great Lakes and the International Joint Commission

During the 1970s, Lake Erie became infamous when it effectively "died" as a result of pollution. Sewage, fertilizers, phosphate-containing detergents, and other chemicals in runoff combined to result in overfertilization, algal blooms, eutrophication, and oxygen depletion. The water turned a sickly green colour, and the lake was widely declared to be dead. The water subsequently became so depleted in oxygen that portions of it became crystal clear—not because it was "clean," but because it was incapable of supporting aquatic life.

Lake Erie is the smallest of the Great Lakes in volume and depth, though not in surface area. Erie therefore is lacking the very cold, deep waters that characterize the larger lakes (especially Superior). It is slightly warmer, circulates nutrients more effectively, and is the most biologically productive of the Great Lakes. Erie is surrounded by agricultural land, with significant industrial development, and a population that has surged. These factors combined to

expose this most sensitive of the Great Lakes to an onslaught of chemicals that overloaded its capacity.

A concentrated international effort brought Lake Erie back from the brink of disaster. After tough legal restrictions on nutrient runoff had been implemented on both sides of the border, improvements began to be seen. By the 1980s, nutrient levels in the lake had decreased sufficiently that aquatic populations began to rebound. Achieving this required significant cross-border communication and collaboration on policy and decision-making.

The principal binational agreement governing the Great Lakes is the *Canada–U.S. Great Lakes Water Quality Agreement (GLWQA)*. Its precursor was the *Boundary Waters Treaty* of 1909, which set up a mechanism for resolving international disputes over transboundary water resources and established the *International Joint Commission (IJC)* to ensure the respective and common interests of Canada and the United States. Given the significance of the Great Lakes, which contain about 22% of the world's fresh water, with a total surface area of 244 100 km^2, it is not surprising that this is where much of the work of the IJC has been focused.

In 1972, in response to the "death" of Lake Erie, other pollution crises of the early 1970s, and the increasing environmental awareness of citizens, Canada and the United States signed the *GLWQA*. Its main objective is to restore and maintain the chemical, physical, and biological integrity of the waters of the Great Lakes basin ecosystem, including the St. Lawrence River system.

Under the *GLWQA*, Canada and the United States designated 43 Areas of Concern (AOCs) in the lakes and connecting channels. Sixteen are either in Canada or binational; eight still-active AOCs are associated with Lake Erie (see **FIGURE 1**). By definition, AOCs are areas in which environmental degradation is pronounced, such that restrictions must be placed on swimming, fishing, and drinking-water consumption. Some AOCs, especially those associated with connecting channels, are of particular

concern because they contribute to the overall degradation of the Great Lakes. For each AOC, an individually tailored *Remedial Action Plan*, or RAP, has been developed.

The RAPs focus on specific locations, but more holistic *Lakewide Management Plans (LaMPs)* have been created for Erie, Ontario, Michigan, and Superior; Lake Huron has a binational agreement modelled on a LaMP.[33] The LaMPs typically involve all relevant stakeholders; for example, the Lake Erie LaMP involves the province of Ontario, Great Lakes states, federal and local governments, and other stakeholders. The goal of a LaMP is to describe the lake's problems, identify the sources of the problems, and work toward solving the problems and improving the resource.

The Canadian and U.S. federal governments recommitted to the *GLWQA* in 2012. However, efforts to cooperate on behalf of the Great Lakes have not been restricted to the federal level. For example, the Government of Canada reached an agreement with the province of Ontario called the *Canada–Ontario Agreement Respecting the Great Lakes Basin Ecosystem (COA)*. Its intention is to promote cooperative action between Canada and Ontario to restore and sustain the environmental quality of the Great Lakes. The Great Lakes states and a number of Tribal Nations in the United States have participated in similar agreements.

Lake Erie recovered from the worst of its pollution after the *GLWQA* came into effect. However, the environmental problems of the lake are far from over. For example, phosphate levels in the lake—largely under control since the 1980s—began to increase in the mid-1990s. The recurring algal blooms (see photo) have resulted in beach closures and loss of tourism revenue. Invasive species, such as the zebra mussel, are having complex impacts on native populations. Invasive species have contributed to the degradation of habitat for native species. Bioaccumulative toxins in the lake are also of concern. The Lake Erie *Biodiversity Conservation Strategy (BCS)*, part of the

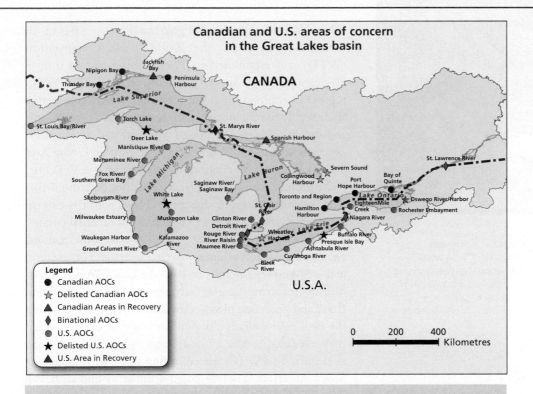

Canadian and U.S. areas of concern in the Great Lakes basin

CANADA

U.S.A.

Legend
- ● Canadian AOCs
- ☆ Delisted Canadian AOCs
- ▲ Canadian Areas in Recovery
- ◆ Binational AOCs
- ● U.S. AOCs
- ★ Delisted U.S. AOCs
- ▲ U.S. Area in Recovery

0   200   400
Kilometres

FIGURE I

Through the *Canada–U.S. Great Lakes Water Quality Agreement (GLWQA)*, Canada and the United States have identified 43 Areas of Concern (AOCs) around the Great Lakes. Each of these has associated with it a Remedial Action Plan, or RAP. There are currently eight active AOCs associated with Lake Erie.
Source: Environment Canada > Great Lakes Areas of Concern. http://www.ec.gc.ca/raps-pas/.

LaMP, identifies the following current goals and challenges:[34]

- reducing the impact of agricultural pollutants
- preventing and reducing the impact of invasive species
- coastal conservation
- reducing the impacts of urban pollutants
- improving habitat connectivity by reducing the impact of dams and other barriers

These challenges are multifaceted; they involve many different stakeholders, and the goals cannot be achieved just by applying scientific principles to the recovery strategy. For example, the goal of coastal conservation involves building a business case and an educational outreach plan on the importance of maintaining healthy coastlines. The goal of controlling invasive species necessitates the development of a common framework; that is, the two nations involved in the agreement must use the same guidelines and protocols, otherwise they may be working at cross-purposes.

Environment Canada reports that cleanup projects have been completed in five Canadian AOCs, and they anticipate completing remedial actions in five more. Three Canadian and four U.S. AOCs have now been delisted, indicating site recovery; so progress is being made. But who knew that it would take so many decades, so many organizations (and acronyms), and so much effort to get two nations to work together for the good of this

A wave, green from a harmful algal bloom consisting mainly of cyanobacteria, breaks on the shore of Lake Erie.

crucial resource? The rescue of Lake Erie is one example of how people and organizations work together to achieve environmental goals and how laws and policies directly influence the environment.

FIGURE 21.13
Many nations are shifting their policies to support sustainable development efforts, trying to increase standards of living while safeguarding the environment. The United Nations has been carrying out consultations in nations around the world to set the post-2015 sustainable development agenda. Here, participants gather at a consultation in Kigali, Rwanda, to discuss their own national sustainable development challenges and priorities.

The United Nations has taken an active role in shaping international environmental policy (**FIGURE 21.13**), and particularly in spearheading the post-2015 agenda for sustainable development goals. Of several agencies within it that influence environmental policy, most notable is the **United Nations Environment Programme (UNEP)**, created in 1972, which helps nations understand and solve environmental problems. Based in Nairobi, Kenya, its mission is sustainability, enabling countries and their citizens "to improve their quality of life without compromising that of future generations." UNEP's extensive research and outreach activities provide a wealth of information useful to policymakers and scientists throughout the world.

The European Union   The European Union (EU) seeks to promote Europe's economic and social progress (including environmental protection) and to "assert Europe's role in the world." The EU can sign binding treaties on behalf of its 28 member nations, and can enact regulations that have the same authority as national laws in each member nation. It can also issue *directives*, which are more advisory in nature. The EU's European Environment Agency works to address waste management, noise pollution, water pollution, air pollution, habitat degradation, and natural hazards. The EU also seeks to remove trade barriers among member nations. It has classified some nations' environmental regulations as barriers to trade because some northern European nations have traditionally had more stringent environmental laws that

prevent the import and sale of environmentally harmful products from other member nations.

The World Trade Organization   Based in Geneva, Switzerland, the World Trade Organization (WTO) was established in 1995, having grown from a 50-year-old international trade agreement. The WTO represents multinational corporations and promotes free trade by reducing obstacles to international commerce and enforcing fairness among nations in trading practices. Whereas the United Nations and the European Union have limited influence over nations' internal affairs, the WTO has real authority to impose financial penalties on nations that do not comply with its directives. These penalties can on occasion play major roles in shaping environmental policy.

Like the EU, the WTO has interpreted some national environmental laws as unfair barriers to trade. For instance, in a well-known example, in 1995 the U.S. EPA issued regulations requiring cleaner-burning gasoline in U.S. cities. Brazil and Venezuela filed a complaint with the WTO, saying the new rules unfairly discriminated against the petroleum they exported to the United States, which did not burn as cleanly. The WTO agreed, ruling that even though the South American gasoline posed a threat to human health in the United States, the EPA rules represented an illegal trade barrier. Not surprisingly, critics of globalization have frequently charged that the WTO and trade agreements like NAFTA aggravate environmental problems.

The World Bank   Established in 1944 and based in Washington, D.C., the World Bank is one of the world's largest sources of funding for economic development. This institution has shaped environmental policy through its funding of dams, irrigation infrastructure, and other major development projects. In fiscal year 2014, the World Bank provided $65.6 billion in loans and grants for projects in middle- and low-income countries around the world.[35]

The World Bank has frequently been criticized for funding unsustainable projects that cause more environmental problems than they solve. Providing for the needs of growing human populations in poor nations while minimizing damage to the environmental systems on which people depend can be a tough balancing act. National debt has an enormous impact on the environment; if a country has loan payments to be met, sometimes the only option is to sell off environmental resources. The Bank also does *policy-based lending*, which requires countries to make fiscal changes in order to qualify for loans. At times in the past, these required policy changes have had severe negative environmental impacts.

In recent years the Bank has recommitted itself to its founding mandate of reducing poverty and to the sustainable management of the environment and natural resources, which it sees as critical for building the economies of poor nations. Starting in the 1990s, the Bank initiated programs to address the burden of debt in the world's poorest countries. The Bank also has recently started a climate finance initiative that is designed to help poor countries deal with climate change mitigation and adaptation challenges.[36]

Organization of the Petroleum Exporting Countries (OPEC)

OPEC is an intergovernmental organization of oil-producing and -exporting nations, currently with 12 members.[37] It was originally founded to protect the interests of its members in the international marketplace, which it has accomplished by controlling the world price of oil. The position of OPEC with regard to recent developments in our understanding of global climate change and its relationship to fossil fuel use is understandably wary. At the United Nations Conference of the Parties (COP-18) for the UN Framework Convention on Climate Change in 2012, the OPEC representative stated, "Climate change is a threat to sustainable development and concerns all of us."[38] According to the OPEC website, the three guiding themes for the organization in the twenty-first century are "stable energy markets, sustainable development, and the environment."[39]

International ENGOs

A number of ENGOs have become international in scope and exert influence over international environmental policy. The nature of these advocacy groups is diverse. Groups such as Greenpeace, Conservation International, the World Wildlife Fund, and many others, attempt to shape policy directly or indirectly through research, education, lobbying, or protest. ENGOs apply more funding and expertise to environmental problems—and conduct more scientific research on them—than do many national governments. In some Communist and post-Communist countries, nongovernmental organizations that have been illegal or severely restricted are now becoming much more visible and are influencing both awareness and public policy.

This is just a small sampling of the very many organizations that play central roles in determining international environmental policy that affects us all. As globalization proceeds, our world is becoming ever more interconnected. As a result, human societies and Earth's ecological systems are being altered at unprecedented rates. Trade and technology have expanded the global reach of all societies, especially those that consume resources from across the world, including Canada. Highly consumptive nations that import goods and resources from far and wide exert extensive impacts on the planet's environmental systems. Multinational corporations operate outside the reach of national laws and rarely have incentive to conserve resources or conduct their business sustainably in the nations where they operate. For all these reasons, in today's globalizing world the organizations and institutions that influence international policy are becoming increasingly vital.

Sustainable development now guides international environmental policy

The concept of sustainable development gained popularity as a result of the 1987 report of the United Nations Commission on Environment and Development, led by the (then) prime minister of Norway, Gro Harlem Brundtland. The commission's report, entitled *Our Common Future*, defined **sustainable development** as "development that meets the needs of the present without compromising the ability of future generations to meet their own needs."[40] The concept of sustainable development got a further boost at the 1992 Earth Summit at Río de Janeiro, Brazil. This was the largest international diplomatic conference ever held, drawing representatives from 179 nations and unifying these leaders around the idea of sustainable development.

The idea of sustainable development has not been without controversy. Some people find it too vague. Others find it prone to misuse and misinterpretation, or a contradiction in terms; is it "sustainable" development, or "sustained" development? But if nothing else, the concept of sustainable development has led to enthusiastic discussion and debate among the people of the world. An alternative definition that has emerged more recently defines sustainable development as maximizing the co-achievement of economic, environmental, and social goals.

On an international level, sustainable development as a policy approach tries to find ways to safeguard the functionality of natural systems while raising living standards for the world's poorer people (**FIGURE 21.14**). As the

FIGURE 21.14
At the Barefoot College in Tilonia, India, women construct solar parabolas that track and focus the sun's rays onto cooking pots, to rapidly boil water and other ingredients without the need for firewood. All parts are locally available. Solar cookers made at the college are being set up in remote rural villages and maintained by local women. Solutions like these help both the environment and the economy in the developing world.

Robert Wallis/Panos

world's nations continue to feel the social, economic, and ecological effects of environmental degradation, environmental policy will without doubt become a more central part of governance and everyday life in all nations in the years ahead.

Other Approaches to Environmental Policy

We have discussed environmental policy in this chapter mainly from the regulatory or legislative perspective, and from the perspective of public awareness, pressure from nongovernmental organizations, and grassroots environmental activism. There are other important approaches that can inform environmental policy development and help modify both corporate and consumer behaviour. Let us explore some of these.

Command-and-control policy has improved our lives, but it is not perfect

A great deal of environmental policy has functioned by setting rules or limits and threatening punishment for violating these rules or limits. This is a "top-down" approach to environmental policy, which is often called **command-and-control**. Without doubt, our environment would be in far worse shape today were it not for this type of government regulatory intervention.

Most of the major environmental laws of recent decades, and most regulations enforced by agencies today, use the command-and-control approach. This simple and direct approach to policymaking has brought citizens of Canada and many other nations cleaner air, cleaner water, safer workplaces, healthier neighbourhoods, and many other improvements in quality of life. The relatively safe, healthy, comfortable lives most of us enjoy today owe much to the environmental policy of the past few decades.

Despite the successes of command-and-control policy, it is not without its drawbacks. Policy can fail if a government does not live up to its responsibilities to protect its citizens or treat them equitably. In other cases, government actions may be well intentioned but not well-enough informed, so they can lead to negative consequences. One example of this might be seen in the collapse of the Canadian cod fishery. (See "Lessons Learned: The Collapse of the Cod Fisheries" in the Central Case that opens Chapter 12.) In the late 1980s, catches were dwindling and scientists were sounding alarms about overfishing of the Grand Banks cod stocks. Government, meanwhile, delayed or refused to implement the drastic restrictions

that were called for; this was at least partly because of scientific uncertainties about the situation, and partly because politicians were concerned about the economic and social hardships such measures would bring to fishing communities. The delays were damaging, though, and by the early 1990s the cod populations had crashed, probably never to recover to their former status.

Although Canada of course has its share of environmental laws and regulations, Canada's approach to environmental policy has generally been a little bit more consultative and "bottom-up," compared to the more "top-down" regulatory approach often taken in the United States. For example, when SARA was passed in 2002, there was a specific effort to avoid a command-and-control approach in the wording of the bill. The idea was to avoid the kind of conflict that had been created by the *Endangered Species Act* in the United States, which was perceived as taking a more top-down approach.

Economic tools can be used to achieve environmental goals

The most common critique of command-and-control policy is that it achieves its goals in a more costly and less efficient manner than the free market can. By mandating particular solutions to problems, command-and-control policy fails to take advantage of the fact that private entities competing in the free market can sometimes produce better solutions at lower cost. Many minds that are economically motivated to compete in the market are more likely to innovate and find optimal solutions than a smaller number of policymakers with no such economic incentive.

The most widely developed alternatives to command-and-control policies therefore involve the creative use of economic incentives to encourage desired outcomes, discourage undesired outcomes, and set market dynamics in motion to achieve goals in an economically efficient manner. Policymakers now often try to combine the perspective of government and the private sector through partnerships. The challenge of crafting economic policy tools is to channel the innovation and economic efficiency of the free market in directions that benefit the environment and the public.

Subsidies Subsidies are one type of economic policy tool that aims to encourage industries or activities that are deemed desirable. Governments may give tax breaks to certain types of businesses or individuals, for instance, and national governments commonly provide subsidies to industries they judge to benefit the nation in some way. Subsidies can be used to promote environmentally sustainable activities, although all too often they have been used to prop up unsustainable ones. It is very difficult to figure out the extent of environmental and human health impacts of subsidies in agriculture, mining, fishing, oil and gas, and other sectors, because many of the negative impacts are indirect or unintended. Some studies have suggested that subsidies that are harmful to the environment total in the *trillions* of dollars yearly, across the globe.[41]

Although there are many examples of government subsidies in Canada that result—either directly or indirectly—in harm to the environment (subsidies that promote logging in old-growth forests come to mind), perhaps the most controversial are those that support the oil and gas industry (**FIGURE 21.15**). According to the International Monetary Fund, Canada's subsidies to the oil industry total about $34 billion annually.[42] Are these subsidies harmful? It depends on your perspective. The amount of financial support provided to the oil industry in Canada certainly *far* outweighs the amount spent on

FIGURE 21.15
Subsidies designed to support industry can have a negative impact on the environment, either directly or indirectly. The billions of dollars of subsidies to the oil and gas industry in Canada have so far greatly outweighed subsidies for alternative energy sources or investment in climate change solutions.

limiting or even tracking greenhouse gas emissions, or subsidies to the renewable and emerging energy technologies sector.

Green taxes and "polluter pays" Another economic policy tool—taxation—can be used to discourage undesirable activities. Taxing undesirable activities helps "internalize" external costs by making them part of the overall cost of doing business. Taxes on environmentally harmful activities and products are generally called **green taxes**. By taxing activities and products that cause undesirable environmental impacts, a tax becomes a tool for policy as well as simply a way to fund government.

In a green taxation scenario, a factory that pollutes would pay taxes based on the amount of pollution it discharges. The idea is to give companies a financial incentive to reduce pollution while allowing the polluter the freedom to decide how best to minimize its expenses. One polluter might choose to invest in technologies to reduce its pollution if doing so is less costly than paying the taxes. Another polluter might find abating its pollution more costly and could choose to pay the taxes instead—funds the government might then apply toward mitigating pollution in some other way. Bottle-return programs, which charge a small fee for glass bottles that are not returned to the distributor, are an example of this idea applied at the consumer end.

Taxes on pollution have been widely instituted in Europe, where many nations have adopted this **polluter-pays principle**. This principle specifies that the price of a good or service should include all its costs, including costs of environmental degradation that would otherwise be passed on as external costs. Green taxes have yet to gain widespread support in Canada, although similar "sin taxes" on cigarettes and alcohol have been common tools of Canadian social policy for a long time.

The Province of British Columbia became the first in Canada to institute a carbon tax, which was passed in 2008. The tax was phased in slowly, to allow businesses and individuals time to figure out how to reduce their emissions; it came into final enforcement in 2012. According to the BC Ministry of Finance, the goals of taxing carbon are to "encourage individuals, businesses, industry, and others to use less fossil fuel and reduce their greenhouse gas emissions; send a consistent price signal; ensure those who produce emissions pay for them; and make clean energy alternatives more attractive."[43] BC's carbon tax is "revenue neutral," meaning any savings generated by the tax are returned to British Columbians through other tax reductions.

Green taxation provides incentives for industry to lower emissions not merely to a level specified in a regulation but to still-lower levels. However, green taxes do have disadvantages. One is that businesses may pass on their tax expenses to consumers, and these increased costs may affect low-income consumers disproportionately more than high-income ones.

Permit trading   A different, market-based approach to the management of environmental impacts is permit trading. In a **permit trading** system, the government creates a market in permits for an environmentally harmful activity, and companies, utilities, or industries are allowed to buy, sell, or trade rights to conduct the activity. For instance, to decrease emissions of air pollutants, a government might grant emissions permits and set up an **emissions trading system**. The government first determines the overall amount of pollution it will accept and then issues marketable permits to polluters that allow them each to emit a certain fraction of that amount. Polluters may buy, sell, and trade these permits with other polluters. Each year, the government may reduce the amount of overall emissions allowed.

In such a **cap-and-trade system**, a polluting party that is able to reduce its pollution receives credit for the amount it did not emit and can sell this credit to other parties. Suppose, for example, you are a plant owner with permits to release 10 units of pollution, but you find that you can become more efficient and release only 5 units of pollution instead. You then have a surplus of permits, which might be very valuable to some other plant owner who is having trouble reducing pollution or who wants to expand production. In such a case, you can sell your extra permits. Doing so generates income for you and meets the needs of the other plant, while preventing any increase in the total amount of pollution. Moreover, environmental organizations can buy up surplus permits and "retire" them, thus reducing the overall amount of pollution.

Marketable permits provide companies with an economic incentive to find ways to reduce emissions. If successful, permit trading can end up costing both industry and government much less than a conventional regulatory system. Ontario has had a capped emissions trading system in place since 2001, covering NO_x and SO_2 emissions (both of which contribute to poor air quality and to the generation of acid rain). The Ontario Emissions Trading Registry serves as a mechanism for tracking emissions, as well as facilitating the transfer and use of allowances and emission-reduction credits, which are issued under the Emissions Trading Regulation. It also provides a forum for public notification and commentary on emissions trading transactions.[44]

Cap-and-trade programs are no panacea. They can reduce pollution overall, but they do allow hotspots of pollution to occur around plants that buy permits to pollute more. Moreover, large firms can hoard permits, deterring smaller new firms from entering the market and thereby suppressing competition. Nevertheless, permit trading has shown promise for safeguarding environmental quality while granting industries the flexibility to lessen their impacts in ways that are economically palatable.

Presently, a market in carbon emissions is operating among European nations as a result of the Kyoto Protocol to address climate change. Under the Kyoto Protocol, nations have targets for reducing their carbon emissions from power plants, automobiles, and other sources that are driving climate change. Each nation participating in the European Union Emission Trading Scheme takes the emissions permits it is allowed and allocates them to its industries according to their emissions at the start of the program. The industries then can trade permits freely, establishing a market whereby the price of a carbon emissions permit fluctuates according to supply and demand.

The EU Emissions Trading System got off to a rocky start when it allocated too many permits in the program's first phase, destroying industries' financial incentive to cut emissions and causing the permits to become nearly worthless. The over-allocation was corrected. According to the EU, the carbon trading system is now by far the largest in the world, covering more than 11 000 power stations and industrial plants in 31 countries, and impacting 45% of EU countries' greenhouse gas emissions.[45]

Market incentives are being tried widely on the local level

You probably have already taken part in transactions involving financial incentives as policy tools. For example, many municipalities charge residents for waste disposal according to the amount of waste they generate. Some cities place taxes or disposal fees on items that require costly safe disposal, such as tires and motor oil. Still others give rebates to residents who buy water-efficient toilets and appliances because the rebates can cost the city less than upgrading its sewage treatment system. Many grocery stores are now charging 5 cents for plastic bags. Power companies sometimes offer discounts to customers who buy high-efficiency light-bulbs and appliances, because doing so is cheaper for the utilities than expanding the generating capacity of their plants. All of these are incentive programs aimed at modifying behaviour at the individual level, to benefit the environment.

ENGOs and even private companies are involved in this process at the local level, too. Many new programs have emerged in the past five years or so, aimed at providing rewards for behavioural changes that will benefit the environment. These include rebate programs for purchasing energy-efficient appliances; vouchers for turning in old, gasoline-guzzling lawnmowers; mechanisms for properly retiring and recycling an older vehicle; and financial incentives for businesses that turn down their lighting or air conditioning, among countless other examples.

More and more, not just in response to legislation but voluntarily, manufacturers are designating on their labels how their products were grown, harvested, or manufactured. This method, called **ecolabelling**, serves to tell consumers which brands use environmentally benign processes. It can also raise consumer awareness of environmental issues, as well as encouraging companies to choose more environmentally friendly production methods. By preferentially buying ecolabelled products, consumers can provide businesses with a powerful incentive to switch to more sustainable processes. One early example was labelling cans of tuna as "dolphin-safe," indicating that the methods used to catch the tuna avoid the accidental capture of dolphins. Other examples include labelling recycled paper, organically grown foods, genetically modified foods (widely done in Europe but not in North America), fair-trade and shade-grown coffee, lumber harvested through sustainable forestry, clothing made from organically grown cotton, and products made from recycled materials (**FIGURE 21.16**).

The big challenge with ecolabelling is not convincing companies to do it, but finding ways to regulate the use of ecolabelling and ensure that the content is truthful and that the labelled product does have a meaningful, positive impact on the environment. The Ecolabel Index is the world's largest inventory of ecolabels, tracking

more than 450 ecolabels in 197 countries.[46] We will return to the idea of ecolabelling, and other incentives aimed at individual consumers, in Chapter 22, *Strategies for Sustainability*.

Science plays a role in policy, but it can be politicized

Ethical values, economic interests, and political ideology influence most policy decisions. However, environmental policy decisions that are most effective are generally those informed by scientific research. In general, Canada has been committed to evidence-based decision-making in setting environmental policy.

For instance, when deciding whether and how to regulate a substance that may pose a public health risk, regulatory agencies like Health Canada deeply investigate the scientific literature. They may even commission new studies, or have their own scientists carry out studies, seeking to gain as full an understanding of the health risks as science can reasonably provide. When trying to win support for a bill to reduce pollution, a government representative will often call upon data from scientific studies to quantify the human and economic costs of the pollution, or the predicted benefits of its reduction. We saw this in the Central Case for Chapter 19, "Microplastics: Big Concerns about Tiny Particles." The more information a policymaker can glean from science, the better the policy he or she will be able to create. When designating a species as threatened or endangered, the process under SARA begins with input from both scientists and Aboriginal environmental experts.

Unfortunately, sometimes policymakers choose to ignore science and instead allow political ideology to determine policy. In past years, some scientists—particularly those working on politically sensitive issues such as climate change or endangered species protection—have occasionally found their work suppressed or discredited, or their jobs threatened. Examples of cases in which the work or opinions of Canadian scientists have been ignored or overlooked by some politicians, who may have been under social and economic pressure, include the collapse of the cod fishery in Atlantic Canada; the connections between asbestos and disease; and in recent years, scientific evidence about global warming and the potential impacts of climate change.

When scientists are gagged, and when taxpayer-funded science is suppressed or distorted for political ends, we all lose. We cannot simply take for granted that science will play a role in policy. As scientifically literate citizens of a democracy, we all need to make sure our government representatives are making proper use of the tremendous scientific assets we have at our disposal.

FIGURE 21.16
Ecolabelling allows businesses to promote products that minimize environmental impacts and gives consumers the opportunity to choose healthier, low-impact products.

Conclusion

Environmental policy is a problem-solving tool that makes use of science, ethics, and economics and that requires an astute understanding of the political process. Conventional command-and-control approaches of legislation and regulation are the most common approaches to policymaking, but various innovative economic policy tools are also being developed. As we have seen in the case of the Great Lakes and other examples discussed in this chapter, environmental issues often overlap political boundaries and require international cooperation. Through the hard work of concerned citizens interacting with their government representatives, the political process has produced some promising solutions.

The central focus of this book has been the science behind the pressing environmental issues of our day. In this chapter, we have departed somewhat from this central focus to consider the fundamentals of environmental law and policy, and some of the approaches to environmental management that have been used in the past and are currently emerging in Canada. By understanding these fundamentals and combining them with your understanding of science, you will be well equipped to develop your own creative solutions to many of the challenging environmental problems you will encounter, both in your life and in your career.

REVIEWING OBJECTIVES

You should now be able to

Describe environmental policy and its main goals and challenges

- The goals of environmental policy are to protect natural resources and environmental quality from depletion or degradation, and to promote equitable treatment of people. Three particular challenges for environmental policy are the management of commonly owned resources, the free rider problem, and dealing with externalities or downstream costs.

- Early environmental policy encouraged frontier expansion and resource extraction. Policies in the late 1800s and early 1900s aimed to regulate resource use. In the 1960s and subsequent decades, high-profile publications and pollution events raised public awareness and gave us many of today's major environmental laws.

- The willingness-to-pay transition represents a shift in societal attitudes, in which people at a particular level of economic development are willing to pay for the technologies and processes that will ensure better environmental quality.

- Laws and policies governing land, water, air, and species are built upon historical precedents.

- Environmental policy fulfills many lofty societal goals; however, some people see environmental laws and regulations as impediments to business or to their lifestyle choices.

Identify institutions, instruments, laws, and processes that are central to Canadian environmental policy

- Canada's environmental policy is influenced by proximity to the United States, both because of our trade and political relationship and because of our shared history on this continent.

- Laws, regulations, agreements, and permits are the legal and semi-legal instruments that can be used to ensure compliance with environmental policies.

- Voluntary initiatives, often sector-based, can be used to promote environmental goals, but they are often not as successful as legislation and regulation.

- Federal, Aboriginal, provincial/territorial, and municipal/local governments, together with administrative agencies, all play important roles in Canadian environmental policy. The *Canadian Environmental Protection Act* is the foundation of Canada's environmental legislation.

- Environmental nongovernmental organizations represent people and ecosystems at a grassroots level, through political activism, education, and awareness-raising, or through efforts to protect a particular species or environment.

- In Canada we have traditionally relied on scientific evidence-based decision-making in setting and implementing environmental policy.

Discuss some of the institutions and guiding principles that are important for international environmental policy, and for the resolution of transboundary environmental problems

- Many activities have environmental impacts that cut across jurisdictional and political boundaries. Transboundary issues, though often complicated, can be managed through international law and the implementation of bilateral agreements.

- Institutions such as the United Nations, European Union, World Bank, World Trade Organization, and nongovernmental organizations all play important roles in shaping and enforcing international environmental policy.

■ Sustainable development is now the driving force behind environmental goals and policies, both internationally and in Canada.

Outline various approaches to environmental policy and the role of science in formulating policy

■ Legislation that comes from a central government agency is referred to as a top-down or command-and-control approach.

■ Shortcomings of the command-and-control approach have led many to advocate economic policy tools, including green taxation and permit trading. Government subsidies can have both positive and negative effects on the environment.

■ Market-based approaches at the local and consumer level include charging for waste disposal or plastic bags, and bottle deposit programs.

■ Science plays a role in policymaking, although some policymakers may ignore or distort it for political ends.

TESTING YOUR COMPREHENSION

1. Describe two common justifications for environmental policy, and discuss three problems that environmental policy commonly seeks to address.

2. What are some factors that may hinder the implementation or enforcement of environmental policy?

3. Environmental policy and management in Canada is shared among all levels of government. What does this mean, in practice, and what are some of the most important agencies that work on behalf of the environment at all levels of government?

4. Summarize the evolution of environmental law from the early settling of the West to the present day. What is the approach that appears to characterize the present wave of environmental policy?

5. What are the central goals of the *Canadian Environmental Protection Act*?

6. What is the difference between laws, regulations, agreements, and permits? What special difficulties do transboundary environmental problems present for lawmakers and enforcement agencies?

7. What are some of the important international environmental agreements to which Canada is a party?

8. Why are environmental regulations sometimes considered to be unfair barriers to trade?

9. Differentiate among a green tax, a subsidy, a tax break, and a marketable emissions permit.

10. Many recent environmental initiatives have focused on giving incentives to individuals and corporations to change their environmental behaviour. Explain what this means, and give some examples of programs that do this.

THINKING IT THROUGH

1. Reflect on the causes for the historical transitions from one type of environmental policy to another. Now peer into the future, and think about how life might be different in 25, 50, or 100 years. What would you speculate about the environmental policy of the future? What issues might it address?

2. Compare the roles of the United Nations, the European Union, the World Bank, the World Trade Organization, and nongovernmental organizations. If you could gain the support of just one of these institutions for an environmental policy that you favour, which would you choose? Why?

3. Think of one environmental problem that you would like to see solved. From what you have learned about the policymaking process, describe how you think

you could best shepherd your ideas through the process to address this problem.

4. Compare the main approaches to environmental policy—command-and-control legislation and regulation, and economic or market-based approaches. Can you describe an advantage and a disadvantage of each? Do you think any one approach is most effective? Could we do with just one approach, or does it help to have more than one?

5. You are the Prime Minister of Canada. You must represent Canada at an international meeting of leaders, and they want to know what Canada's position will be on the future of international agreements to limit climate change. What will you say?

INTERPRETING GRAPHS AND DATA

Here is a threat assessment from Environment Canada's Recovery Strategy for the greater sage-grouse, which we looked at in the "Central Case: SARA and the Sage-Grouse."

Look at the table and then answer these questions:

1. Taking into account only the level of concern, extent, and severity of impacts in the threat assessment, which would you say are the four most problematic threats to the sage-grouse?

2. Why is it important to identify the causal certainty for the various threats?

3. Based on the degree of severity, what was the most significant threat to the sage-grouse that happened *historically* and is not anticipated to happen again?

4. Which of the threats identified in this table do you think are wholly natural, that is, not caused directly or indirectly by human activities?

5. Do you think the regulation of activities in the Emergency Order (restricting noise, prohibiting fences and new structures that are not sage-grouse friendly, and promoting the preservation of sage grasslands) go far enough toward addressing the identified threats?

Threat Assessment Table for the Greater Sage-Grouse

Threat	Level of concern[1]	Extent	Occurrence	Frequency	Severity[2]	Causal certainty[3]
Climate and natural disasters						
Drought	High[4]	Widespread	Anticipated (historical)	Seasonal	High	Medium
Severe or inclement weather conditions	High	Widespread	Current/ anticipated	Seasonal	High	High
Exotic, invasive, or introduced species						
Disease (West Nile virus)	High	Widespread	Current/ anticipated	Seasonal	High	High
Disturbances						
Facilities associated with noise	High[4]	Localized	Current/ anticipated	Continuous	Moderate	High
Vehicle noise	High[4]	Localized	Current/ anticipated	Recurrent	Moderate	High
Vertical structures	Medium[4]	Localized	Current/ anticipated	Continuous	Moderate	Medium
Human presence at or near a lek	Low	Localized	Current/ anticipated	Seasonal	Moderate	Medium
Natural processes or activities						
Small population size	High	Widespread	Current	Continuous	High	Low
Reduced genetic diversity	Low	Unknown	Anticipated	Unknown	Low	Unknown
Habitat loss or degradation						
Habitat conversion to crop and forage production	Medium[4]	Localized (widespread)	Anticipated (historical)	Recurrent	High	High
Habitat conversion to energy development infrastructure	Medium[4]	Localized	Current/ anticipated	Recurrent	High	High
Habitat loss or degradation from conversion to roads	Medium[4]	Localized	Current/ anticipated (historical)	Recurrent	High	High
Degradation of vegetative cover from grazing levels inappropriate for sage-grouse	Medium[4]	Localized (widespread)	Current (historical)	Recurrent	High	High
Removal of sage-brush or other shrubs	Low	(Localized)	(Historical)	One-time	High	Medium

Threat	Level of concern[1]	Extent	Occurrence	Frequency	Severity[2]	Causal certainty[3]
Changes in ecological dynamics or natural processes						
Increased predator pressure	High[4]	Widespread	Current (historical)	Continuous	Moderate	Medium
Alteration of natural hydrology	Medium[4]	Localized (widespread)	Current (historical)	Recurrent	Moderate	Medium
Alteration to natural fire and grazing regimes	Low	(Widespread)	(historical)	Recurrent	Unknown	Low
Accidental mortality						
Collisions with traffic	Low[4]	Localized	Current (historical)	Continuous	Moderate	Medium
Collisions with infrastructure	Low	Localized	Current (historical)	Continuous	Moderate	Medium

[1] Level of concern signifies that managing the threat is of (high, medium, low) concern for species recovery, consistent with population and distribution objectives. This overall criterion takes into account all the other individual criteria in the table.

[2] Severity reflects the population-level effect (High: very large population-level effect, Moderate, Low, Unknown)

[3] Causal certainty reflects the degree of evidence that is known for the threat (High: available evidence strongly links the threat to stresses on population viability; Medium: there is a correlation between the threat and population viability; e.g. expert opinion; Low: the threat is assumed or plausible)

[4] Though threats are assessed individually in this table, multiple threats may co-occur in some locations or in association with particular activities. Such combinations of threats can lead to very high levels of concern.

Source: Environment Canada (2014). Amended Recovery Strategy for the Greater Sage-Grouse (Centrocercus urophasianus urophasianus) in Canada. Species at Risk Act Recovery Strategy Series. Environment Canada, Ottawa. vi + 53 pp., Table 2. © Environment Canada, 2014.

MasteringEnvironmentalScience®

STUDENTS
Go to **MasteringEnvironmentalScience** for assignments, the eText, and the Study Area with practice tests, videos, current events, and activities.

INSTRUCTORS
Go to **MasteringEnvironmentalScience** for automatically graded activities, current events, videos, and reading questions that you can assign to your students, plus Instructor Resources.

22 Strategies for Sustainability

For its Post-2015 Development Agenda, the United Nations has adopted the 17 Sustainable Development Goals shown here. These goals are intended to provide a follow-up framework to the 2015 Millennium Development Goals.

Upon successfully completing this chapter, you will be able to

- Explain the concept of sustainability and some of the key approaches to sustainable development
- Describe and assess some sustainable development initiatives on college and university campuses
- Discuss the need and urgency for action on behalf of the environment, and the tremendous human potential to solve problems

Car Heaven is a program of Summerhill Impact, which offers incentives for people to retire and properly recycle their older, polluting vehicles.

Car Heaven/Summerhill Impact

CENTRAL CASE
A DIFFERENT WAY OF DOING BUSINESS

"We are now poised to take a major step towards ushering in a new era of sustainable development for all—an era of transformation."

—BAN KI-MOON, SECRETARY GENERAL OF THE UNITED NATIONS[1]

"We have to move decisively to protect our children's future."

—ELIZABETH MAY, LEADER OF THE GREEN PARTY OF CANADA

The Summerhill Group runs a different kind of business. For one thing, Summerhill is a social enterprise. Definitions of *social enterprise* vary, but they generally focus on the idea that such organizations—whether not-for-profit or for-profit—have, as their core mission, achieving social goals or solving social problems using an entrepreneurial approach.

Summerhill was started as a social enterprise in 2001 by Ian Morton, with the vision of "transforming markets to sustainability." The organization now consists of Summerhill, an environmental consulting company that works with large corporations and other organizations, helping them design and implement initiatives for energy efficiency and sustainability; and Summerhill Impact (formerly the Clean Air Foundation), the not-for-profit arm of the group. Their mission is to develop innovative market-based solutions, with the goal of generating "sustained environmental and social improvements."[2] The company is fundamentally entrepreneurial and business-oriented, but committed to achieving social and environmental goals through creative partnerships with industry, using events and programs that engage consumers.

At the heart of Summerhill's activities is a fundamentally different way of doing business. It has been traditional to view environmental, social, and economic goals as mutually exclusive. Summerhill makes it their mission to show their large corporate clients (which include The

Home Depot Canada, Walmart Canada, and SaskPower) that adopting environmental programs and strategies will, in fact, help them promote their business, meet their targets, strengthen their brand, engage their customers, and boost the bottom line—all while promoting more environmentally sustainable products and choices. It is an approach that creates economic value in a way that also creates value for society.

The Home Depot Canada discovered the usefulness of this approach through Mow Down Pollution, Go Low Flow, Bright Ideas, and other programs developed by Summerhill (some in conjunction with partners). The programs create a customer experience that builds loyalty, enhances the company's reputation as an environmental leader, and drives sales, while helping consumers reduce their environmental impacts. For example, the Mow Down Pollution program, which ran from 2001 until 2012, offered rebates to customers who turned in their gas-powered equipment for new, energy-efficient electric models. The program retired 63 789 old mowers and trimmers, reducing greenhouse gas and smog-forming emissions by more than 2149 tonnes.[3] This is important because, according to Statistics Canada, 70% of Canadian homes own gas-powered lawn equipment, releasing about 80 000 tonnes of polluting emissions every year.[4]

Another successful and long-standing program, started by Summerhill but now an independent registered charity, is called Car Heaven. The program encourages people to turn in their old cars so that they can be safely and properly recycled, while raising funds for charity through the donation of retired vehicles. Car Heaven is an end-of-life vehicle charitable program that relies only on auto recyclers who comply with the environmental process guidelines outlined by the *Canadian Automotive Recyclers' Environmental Code*.[5] So far more than 123 000 vehicles have been recycled through this program, which, in addition to reducing greenhouse gases and smog-causing emissions, has raised over $3 million for affiliated charities.

Programs like these show how small actions by individual people can add up to big changes for the environment. Influencing individual behaviour is at the core of many of Summerhill's programs, which strive to make people aware of easy and effective actions they can take to reduce their impacts on the environment. The opportunity to become engaged doesn't apply just to consumers; for example, the EcoExecutives program gets corporate execs out of the boardroom and into the Toronto Zoo, where they learn about alternative energy, waste management, and climate change and about how these affect the bottom line of their businesses.

Summerhill employees also act as "ambassadors of the improvements we want to see in the world." Summerhill's success in the corporate sector provides hope that we may be moving toward a business model in which social, environmental, and economic goals can be viewed as mutually compatible, rather than mutually exclusive.

Sustainability and Sustainable Development

As more people come to appreciate Earth's limited capacity to accommodate our rising population and consumption, they are voicing concerns that we will need to modify our behaviours, institutions, and technologies if we wish to sustain our civilization and the natural environment on which it depends, for generations to come.

When people speak of **sustainability**, what precisely do they mean, and what are they hoping to sustain? Generally they mean to sustain human well-being, at the individual, community, and global levels, which requires maintaining our institutions in a healthy and functional state. Typically they also recognize that sustaining human economic and social activities will require sustaining the planet's ecological systems and physical resources. The contributions of biodiversity and ecosystem goods and services to human welfare are tremendous. Indeed, they are so fundamental (some would say infinitely valuable, thus literally priceless) that we have long taken them for granted.

Sustainable development starts with the triple bottom line

The United Nations has defined **sustainable development** as "development that meets the needs of the present without compromising the ability of future generations to meet their own needs."[6] Today, it is widely recognized that sustainability does not just mean protecting the environment against the ravages of human development. Instead, it means finding ways to promote social justice, economic well-being, and environmental quality at the same time (as discussed in Chapter 20, *Environmental Ethics and Economics*). Meeting this "triple bottom line" has been the starting point for modern discussions of sustainable development (see **FIGURE 22.1A**). Achieving this goal is most pressing in nations of the developing world, but it is a vital need everywhere. It is our primary challenge for this century and likely for the rest of our species' time on Earth.

However, ecological economists, among others, have questioned the "triple bottom line" model of sustainability. If we place the same importance on social and economic goals as on environmental goals, what are our chances of succeeding in the long term? Without functioning ecosystems and a healthy environmental resource base, we have little hope of achieving our social goals or sustaining our economic activities for very long. The concept that human society and economy function within and depend fundamentally upon the environment is referred to as the principle of *environmental primacy*. This idea suggests that we must protect our resource base and the integrity of the environment first; then we can move on to social and economic goals. This is often illustrated as "nested" systems, in which society and economy are portrayed as functioning within and depending upon the environmental system (**FIGURE 22.1B**).

Yet, as a global community and even at local levels, we seem to struggle to meet environmental goals. "Environmental sustainability" was one of the United Nations' *Millennium Development Goals (MDGs)* set by the international community at the turn of this century (2000). The Millennium Development Goals were laid out through a global process of consensus-building coordinated by the United Nations Development Program. They were

1. Eradicate extreme poverty and hunger.
2. Achieve universal primary education.
3. Promote gender equality and empower women.
4. Reduce child mortality.
5. Improve maternal health.
6. Combat HIV/AIDS, malaria, and other diseases.
7. Ensure environmental sustainability.
8. Develop a global partnership for development.[7]

Much progress was made on these goals prior to the 2015 target date—some goals saw more progress than others, and some regions were more successful than others at achieving the goals. Probably the goal on which the *least* progress was made overall was #7, environmental sustainability. In assessing progress toward the 2015 MDG targets, the Millennium Project and the Millennium Ecosystem Assessment determined the following:[8]

- Environmental degradation is a major barrier to achieving the Millennium Development Goals.
- Investing in environmental assets and management is vital to relieving poverty, hunger, and disease.

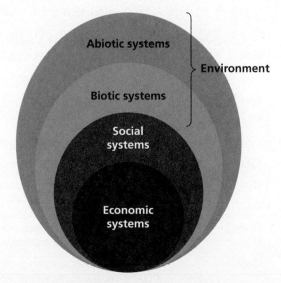

(a) "Triple bottom line": Sustainability as intersecting goals

(b) "Strong sustainability": A system of nested dependencies

FIGURE 22.1 A standard definition describes sustainability as the intersection or maximization of three sets of goals: environmental, economic, and social goals. We first introduced this definition of sustainability in Chapter 1, *An Introduction to Environmental Science* (see **FIGURE 1.8** on page 16). When the three sets of goals are depicted as equal, carrying the same relative value **(a)**, it is called a "triple bottom line" or "weak sustainability" model. Ecological economists hold that society and economy are, in fact, subsets of the environment, and that neither social nor economic activities could go on for very long without the underlying support of environmental goods and services **(b)**. This is referred to as a "strong sustainability" model.

- Forests are disappearing at an alarming rate.
- Global emissions of carbon dioxide have increased.
- The targeted reduction in the rate of biodiversity and ecosystem loss has not been achieved.
- Progress has been made in reducing ozone-depleting substances and in making improved water sources more available.

In preparing for the MDG target year of 2015, the global community again convened to set goals for 2030. In this process, much learning was incorporated from the Millennium Development Goals—both with regard to targets that were met on schedule, or ahead of time, and for targets that were missed. The new goals for the Post-2015 Development Agenda are being called **Sustainable Development Goals (SDGs)**. The adopted goals—each of which has many underlying targets and indicators—are portrayed in our chapter-opening photo, and shown in **TABLE 22.1**.

There are 17 Sustainable Development Goals. Many of them have components that are specifically environmental, recognizing the importance of maintaining our

planetary resources if we hope to achieve our social and economic goals. Can we achieve these goals by the target date of 2030? To have the future we want, for everyone in the world, we must work toward that end.

Environmental protection can enhance economic opportunity

Achieving sustainable development goals as a global community requires adopting integrated approaches, finding new synergies, developing new technologies, and sometimes just thinking differently. For example, reducing resource consumption and waste isn't just a way to meet an environmental goal; it often saves money, too, as many businesses, municipalities, colleges, and universities discover when they embark on sustainability initiatives. Sometimes savings accrue immediately, and other times an upfront investment brings long-term savings.

For society as a whole, attention to environmental quality can enhance economic opportunity by providing new types of employment. This reality contrasts with the common perception that environmental protection hurts

TABLE 22.1 Sustainable Development Goals for 2030

Goal 1	End poverty in all its forms everywhere.
Goal 2	End hunger, achieve food security and improved nutrition, and promote sustainable agriculture.
Goal 3	Ensure healthy lives and promote well-being for all at all ages.
Goal 4	Ensure inclusive and equitable quality education and promote lifelong learning opportunities for all.
Goal 5	Achieve gender equality and empower all women and girls.
Goal 6	Ensure availability and sustainable management of water and sanitation for all.
Goal 7	Ensure access to affordable, reliable, sustainable, and modern energy for all.
Goal 8	Promote sustained, inclusive, and sustainable economic growth; full and productive employment and decent work for all.
Goal 9	Build resilient infrastructure, promote inclusive and sustainable industrialization, and foster innovation.
Goal 10	Reduce inequality within and among countries.
Goal 11	Make cities and human settlements inclusive, safe, resilient, and sustainable.
Goal 12	Ensure sustainable consumption and production patterns.
Goal 13	Take urgent action to combat climate change and its impacts.*
Goal 14	Conserve and sustainably use the oceans, seas, and marine resources for sustainable development.
Goal 15	Protect, restore, and promote sustainable use of terrestrial ecosystems; sustainably manage forests; combat desertification; halt and reverse land degradation; and halt biodiversity loss.
Goal 16	Promote peaceful and inclusive societies for sustainable development; provide access to justice for all; and build effective, accountable, and inclusive institutions at all levels.
Goal 17	Strengthen the means of implementation and revitalize the global partnership for sustainable development.

Acknowledging that the United Nations Framework Convention on Climate Change is the primary international, intergovernmental forum for negotiating the global response to climate change

Source: United Nations Sustainable Development Knowledge Platform, Open Working Group Proposal for Sustainable Development Goals, https://sustainabledevelopment.un.org/sdgsproposal (proposal accepted by the United Nations General Assembly in 2015). © 2015 United Nations. Reprinted with the permission of the United Nations.

FIGURE 22.2

The northern spotted owl (*Strix occidentalis occidentalis*) has become an iconic symbol of the "jobs-versus-environment" debate. This bird of the Pacific rain forest is considered endangered because of the logging of mature forests. Proponents of logging argue that laws protecting endangered species cause economic harm and job loss. Advocates of endangered species protection argue that unsustainable logging practices pose a larger risk of job loss.

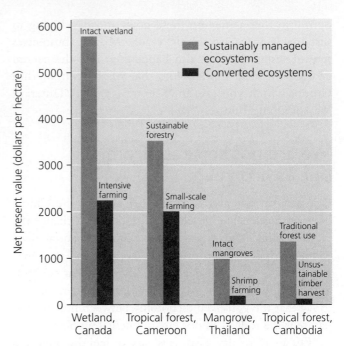

FIGURE 22.3

Once external costs and benefits are factored in, the economic value of sustainably managed ecosystems generally exceeds the economic value of ecosystems that have been converted for intensive private resource harvesting. Shown are land values calculated by researchers in four such comparisons from sites around the world.

Source: From Economic Benefits under Alternate Management Practices in Millennium Ecosystem Assessment (MA). Copyright © 2007 by World Resources Institute. Reprinted by permission.

DATA Q How does the value of 1 hectare of intact wetland in Canada compare to the value of 1 hectare of intensively farmed land in Canada?

the economy by costing people jobs. For example, in the controversy over logging of old-growth forest in the Pacific Northwest, protection for the endangered northern spotted owl (**FIGURE 22.2**), whose natural habitat ranges from Northern California to British Columbia, set limits on timber extraction. Proponents of logging claim that such restrictions cost local loggers their jobs. However, loggers' jobs are far more at risk when timber companies cut trees at unsustainable rates and then leave a region, seeking mature forests elsewhere, as has happened in region after region throughout history.

The jobs-versus-environment debate frequently overlooks the fact that as some industries decline, others spring up to take their place. As jobs in logging, mining, and manufacturing have disappeared in developed nations over the past few decades (some, such as logging jobs, as a direct result of mechanization of the industry), jobs have proliferated in service occupations and high-technology sectors. Similarly, as we decrease our dependence on fossil fuels, jobs and investment opportunities are opening up in renewable energy sectors, such as wind power and fuel-cell technology. Some analyses predict greater employment in alternative energy sectors than in conventional energy sectors, due to the labour-intensive nature of some alternative energy installations.

Will the alternative energy technology industry provide jobs for all displaced workers? Not single-handedly, and not right away. However, environmental protection need not lead to economic stagnation, but instead can enhance economic opportunity. This connection is also suggested by the fact that global economies have expanded rapidly in the past 30 years, the very period during which

environmental protection measures have proliferated. Pitting the environment against the economy is a false dichotomy.

Moreover, if we look beyond conventional economic accounting (which measures only private economic gain and loss) and instead include external costs and benefits that affect people at large, then environmental protection becomes still more valuable. Consider several studies reviewed by the Millennium Ecosystem Assessment: They each show how overall economic value is maximized by conserving natural resources rather than exploiting them for short-term private gain (**FIGURE 22.3**).

Although people desire private monetary gain, they also desire to live in areas that have clean air, clean water, intact forests, and parks and open space. Environmental protection increases a region's attractiveness, drawing more residents and increasing property values and the tax revenues that help fund social services. As a result, regions that act to protect their environments are generally the ones that retain and increase their wealth and quality of life.

The many actions being taken today by organizations and individuals across the globe are giving people optimism that achieving these goals and developing in sustainable ways are within reach. Moving businesses toward sustainability is also happening; it is the central goal of business models like that of the Summerhill Group, which you saw in "Central Case: A Different Way of Doing Business."

We are not separate from our environment

Part of what is required in order to achieve the United Nations' Sustainable Development Goals is to change the way we think of our relationship with the natural environment. It is common to hear "humans and the environment" or "people and nature" being set in contrast, as though they were separate. Some philosophers venture to say that the perceived dichotomy between humans and nature is the root of all our environmental problems.

On a day-to-day basis, it is easy to feel disconnected from the natural environment, particularly in industrialized nations and large cities. We live inside houses, work in shuttered buildings, travel in enclosed vehicles, and generally know little about the plants and animals around us. Millions of urban citizens have never set foot in an undeveloped area. Just a few centuries or even decades ago, most of the world's people were able to name and describe the habits of the plants and animals that lived nearby. They knew exactly where their food, water, and clothing came from. Today it seems that water comes from the faucet, clothing from the mall, and food from the grocery store. It is not surprising that we have lost track of the connections that tie us to our natural environment.

However, this doesn't make our connections to the environment any less real. Consider a thoroughly un-"natural" invention of the human species: the banana split (**FIGURE 22.4**), in which each and every element has ties to the resources of the natural environment, and each exerts environmental impacts.

Once we learn to consider where the things we use and value each day actually come from, it becomes easier to see how people are an integral part of the environment. And once we reestablish this connection, it becomes readily apparent that our own interests are best served by preservation or responsible stewardship of the natural systems around us. Because what is good for the environment can also be good for people, win–win solutions are very much within reach, if we learn from what science can teach us, think creatively, and act on our ideas.

We can follow a number of strategies toward sustainable solutions

Sustainable solutions to environmental problems are numerous, and we have seen many specific examples throughout this book. The challenges lie in being imaginative enough to think of solutions, and being shrewd

Maraschino cherry
Royal Ann cherry from Washington, treated with food colouring, lemon juice, brine, alum, and almond extract from many other locales

Strawberries
Grown in coastal California

Banana
Grown in Ecuador or Panama

Ice cream
Milk from dairy cows in Vermont or Wisconsin; sugar from sugarcane in Florida or Hawaii; eggs from hens in Indiana or Georgia; vanilla extract from Mexico or Tahiti

Walnuts
Grown in California's Central Valley

Chocolate sauce
Cocoa grown in Latin America or West Africa, with butter, sugars, salt, milk, and additional ingredients from many other locales

Spoon
Stainless steel, a complex alloy of metals mined in the U.S., Canada, South Africa, and Asia (or plastic from petroleum from Saudi Arabia or Venezuela)

Bowl
Glass, from silica sand, soda ash, limestone, etc., from U.S. Midwest

FIGURE 22.4
Though it seems to be an un-"natural" construction, a banana split consists of ingredients from around the world, whose production has impacts on the environments of many distant locations. Transport of products to the marketplace, their packaging, storage (such as refrigeration), use, and disposal further contribute greenhouse gas emissions and other environmental impacts.

TABLE 22.2 Some Strategies for Sustainability

- Engage politically and exercise our power as consumers
- Rethink economic growth and quality of life
- Stabilize population
- Encourage the development of green technologies
- Mimic natural systems by promoting closed-loop industrial processes
- Think in the long term
- Enhance local self-sufficiency, yet embrace some aspects of globalization
- Use systems thinking to find holistic solutions
- Promote research, education, and awareness

and dogged enough to overcome political or economic obstacles that may lie in the path of their implementation. Let's examine some of the strategies, tools, and approaches that we can use to help us move toward a sustainable future (**TABLE 22.2**).

Political engagement and purchasing power

Sustainable solutions often require leaders and policymakers to usher them through, and politicians respond to whoever exerts influence. Individually and collectively, we can guide our political leaders to enact policies for sustainability. Corporations and interest groups employ lobbyists to influence politicians all the time. Citizens in a democratic country have the same power, if they choose to exercise it. You can exercise your power at the ballot box, by attending public hearings, by donating to advocacy groups that promote positions you favour, and by writing letters and making phone calls to officeholders. You might be surprised how little input policymakers receive from the public; sometimes a single letter, email, or phone call can make a big difference.

Today's environmental and consumer protection laws came about because citizens pressured their representatives to act. The raft of legislation enacted in the 1960s and 1970s in North America might never have come about had ordinary citizens not stepped up and demanded action. We owe it to future generations to be engaged and to act responsibly now so that they will have a better world in which to live. The words of anthropologist Margaret Mead are worth repeating: "Never doubt that a small group of thoughtful, committed people can change the world."[9]

In addition to expressing our preferences through the political system, we also wield influence through the choices we make as consumers. When products produced sustainably are ecolabelled, consumers can "vote with their wallets" by purchasing these products. Consumer choice has helped drive sales of everything from recycled paper to organic produce to "dolphin-safe" tuna. Individuals can multiply their own influence by promoting "green" purchasing habits at their school or workplace.

Rethinking growth and quality of life   It is conventional among economists and policymakers to speak of economic growth as an ultimate goal. Many politicians view nurturing an expanding economy as their prime responsibility while in office. Yet economic growth is merely a tool with which we try to attain the real goal of maximizing human happiness and well-being. If economic growth depends on an ever-increasing consumption and depletion of nonrenewable resources, then we will not be able to attain long-term happiness by endlessly expanding the size of our economy.

Currently, goods and services are priced as though externalities such as pollution and resource depletion (discussed in Chapter 20, *Environmental Ethics and Economics*) involve no costs to society. If we can make our accounting practices reflect indirect consequences to the public, then we can provide a clearer view of the full costs and benefits of any given action or product. In that case, the free market could become the optimal tool for improving environmental quality, our economy, and quality of life, rather than just a vehicle for endless growth. Implementing green taxes and phasing out harmful subsidies could hasten our attainment of prosperous and sustainable economies. The political obstacles to this are considerable, and such changes will require educated citizens to push for them and courageous policymakers to implement them.

Economic growth is driven by consumption: the purchase of material goods and services (and thus the resources and waste involved in their manufacture) by consumers (**FIGURE 22.5**). Our tendency to believe that more, bigger, and faster are always better is reinforced by advertisers seeking to sell more goods more quickly. Consumption has grown tremendously, with the wealthiest nations leading the way. Our houses are larger than ever, gas-guzzling sport-utility vehicles are among the most popular cars, and many citizens have more material belongings than they know what to do with. We think nothing today of having multiple computers with high-speed internet access, let alone the televisions, telephones, refrigerators, and dishwashers that were marvels just decades ago.

Because many of Earth's natural resources are limited and nonrenewable, consumption cannot continue growing forever. Eventually, if we do not shift to more sustainable resource use, per capita consumption will drop for rich and poor alike as resources dwindle. Critics may scoff at the notion that resources are limited, but we must remember that our perspective in time is limited, and that our consumption is taking place within an extraordinarily brief

FIGURE 22.5
Citizens of Canada consume more material goods than the people of any other nation, with the exception (but only marginally) of the United States. We also love to shop in luxury surroundings. Unless we find ways to decrease waste and increase the sustainability of our manufacturing processes, this rate of consumption cannot be maintained in the long run.

slice of time in the long course of history. Our lavishly consumptive lifestyles are a brand-new phenomenon on Earth. We are enjoying the greatest material prosperity in all of history, but if we do not find ways to make our wealth sustainable, the party may not last much longer.

Fortunately, material consumption alone does not reflect a person's quality of life. For many people in industrialized nations, the accumulation of possessions has not brought contentment. Observing how affluent people often fail to find happiness in their material wealth, social critics have given this phenomenon a name: "affluenza." Scientific research backs up the contention that money cannot buy nearly as much happiness as people typically believe (**FIGURE 22.6**). Although economic growth is generally equated with "progress," true progress consists of an increase in human happiness and overall well-being. In the end we are, one would hope, more than just the sum of what we buy.

We can reduce our consumption while enhancing our quality of life—squeezing more from less—in at least three ways. One way is to improve the technology of materials and the efficiency of manufacturing processes, so that industry produces goods using fewer natural resources. Another way is to develop a sustainable manufacturing system—one that is circular and based on recycling, in which the waste from a process becomes raw material for input into that process or others. A third way is to modify our behaviour, attitudes, and lifestyles to minimize consumption.

At the outset, such choices may seem like sacrifices, but people who have slowed down the pace of their busy lives and freed themselves of an attachment to material possessions say it can feel tremendously liberating. Fans

of local foods, homegrown foods, and the "slow-food" movement (**FIGURE 22.7**) feel that they are helping themselves as much as they are helping the environment.

Population stability   Just as continued growth in consumption is not sustainable, neither is growth in the human population. We have seen that populations may grow exponentially for a time but eventually encounter limiting factors and decline or level off. We have used technology to increase Earth's carrying capacity for our species, but our population cannot continue growing forever; sooner or later, human population growth will end. The question is whether it will happen involuntarily, when we exceed environmental limitations, or through voluntary means, as a result of wealth and education.

The demographic transition is already far along in many industrialized nations thanks to urbanization, wealth, education, and the empowerment of women. If today's developing nations also pass through a demographic transition, then there is hope that humanity may halt its population growth while creating a more prosperous and equitable society.

Green technologies   It is largely technology—developed with the Agricultural Revolution, the Industrial Revolution, and advances in medicine and health—that has spurred our population increase. Technology has magnified our impact on Earth's environmental systems, yet it can also give us ways to reduce our impact. Recall the IPAT equation (Chapter 6, *Human Population*), which summarizes human environmental

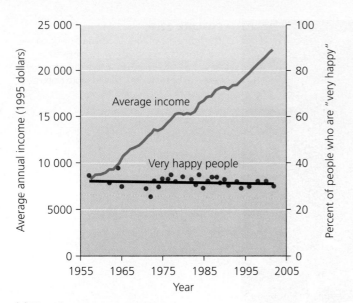

(a) Happiness versus income

FIGURE 22.6
Although average income of North Americans has risen steadily in the past half-century, the percentage of people reporting themselves as being "very happy" has remained stable or declined slightly **(a)**. Respondents in "happiness" studies guessed on average that a fivefold increase in income would improve happiness by 32%, but the actual improvement is only 12%. Canada's GDP (black line) has improved over time **(b)**. So has overall well-being, as measured by the Canadian Index of Wellbeing (green line); however, well-being has increased more slowly than income. The eight domains of the Index, shown here, are education, living standards, community vitality, democratic engagement, healthy population, time use, leisure and culture, and environment. *Source:* Data for (a) from Myers, D.G. (2000). *The American Paradox: Spiritual Hunger in an Age of Plenty.* New Haven, CT: Yale University Press; Gardner, G., and E. Assadourian, (2004). Rethinking the Good Life, pp. 164–179 in *State of the World 2004,* Worldwatch Institute; and Kahneman, D., et al. (2006). Would You Be Happier if You Were Richer? A Focusing Illusion. *Science, 312,* pp. 1908–1910. Graph in (b) from Canadian Index of Wellbeing. (2012). *How Are Canadians Really Doing? The 2012 CIW Report.* Waterloo, ON: Canadian Index of Wellbeing and University of Waterloo.

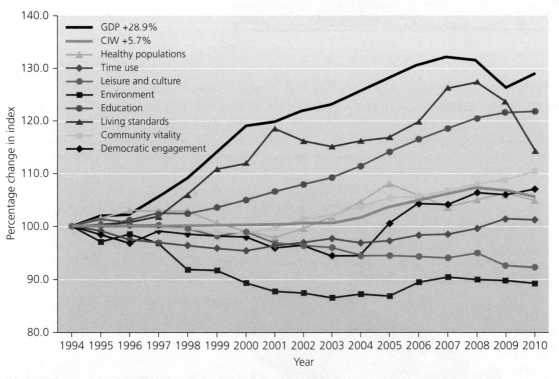

(b) Canadian Index of Wellbeing versus gross domestic product

DATA Q According to this graph, Canada's income (GDP) increased by about 29% between 1994 and 2010. Over the same period, well-being, as measured by the weighted aggregate of eight sets of indicators in the Canadian Index of Wellbeing, increased by about 5.7%. Environment is one of the eight subdomains that is weighted with the others to yield the Index. By how much did the environment subdomain change over the same time period?

impact (I) as the interaction of population (P), consumption or affluence (A), and technology (T). Technology can exert either a positive or negative value in this equation. The short-sighted use of technology may have gotten us into an environmentally damaging situation, but wiser use of environmentally friendly, or "green," technologies can help get us out.

In recent years, technology has intensified environmental impacts in developing countries, as industrial technologies from the developed world have been exported to poorer nations eager to industrialize. The lack of a legislative and regulatory framework to protect the environment from these impacts has exacerbated the situation. In developed nations, meanwhile, green

FIGURE 22.7
The homegrown, local food, and slow-food movements are growing all over North America. "Kitchen gardens," like the one shown here, are becoming more and more popular, especially in urban areas.

John Howard/Getty Images

technologies have begun mitigating our environmental impact. Catalytic converters on cars have reduced emissions (**FIGURE 22.8**), as have scrubbers on industrial smokestacks. Recycling technology and advances in wastewater treatment are helping reduce the impacts of our waste. Solar, wind, and geothermal energy technologies are producing cleaner renewable energy. Countless technological advances such as these are one reason that people of North America and Western Europe today enjoy cleaner environments—although they consume far

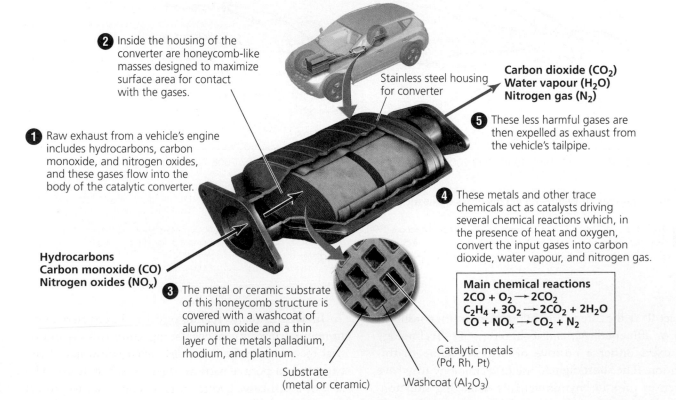

2 Inside the housing of the converter are honeycomb-like masses designed to maximize surface area for contact with the gases.

Stainless steel housing for converter

Carbon dioxide (CO₂)
Water vapour (H₂O)
Nitrogen gas (N₂)

5 These less harmful gases are then expelled as exhaust from the vehicle's tailpipe.

1 Raw exhaust from a vehicle's engine includes hydrocarbons, carbon monoxide, and nitrogen oxides, and these gases flow into the body of the catalytic converter.

4 These metals and other trace chemicals act as catalysts driving several chemical reactions which, in the presence of heat and oxygen, convert the input gases into carbon dioxide, water vapour, and nitrogen gas.

Hydrocarbons
Carbon monoxide (CO)
Nitrogen oxides (NO$_x$)

3 The metal or ceramic substrate of this honeycomb structure is covered with a washcoat of aluminum oxide and a thin layer of the metals palladium, rhodium, and platinum.

Main chemical reactions
$2CO + O_2 \longrightarrow 2CO_2$
$C_2H_4 + 3O_2 \longrightarrow 2CO_2 + 2H_2O$
$CO + NO_x \longrightarrow CO_2 + N_2$

Catalytic metals
(Pd, Rh, Pt)

Substrate
(metal or ceramic)

Washcoat (Al$_2$O$_3$)

FIGURE 22.8 The catalytic converter is a classic example of green technology. This device filters air pollutants from vehicle exhaust and has helped improve air quality in Canada and many other nations.

more—than people of Eastern Europe or rapidly industrializing nations, such as China.

In addition to new technologies, we are developing new approaches and new ways of thinking about our activities and the way we construct the built environment. For example, taking into consideration the embodied or "virtual" energy and materials that go into the construction of buildings and other infrastructure (Chapter 18, *Managing Our Waste*) is one way to work toward reducing the consumption of resources.

Mimicking natural systems

As industries seek to develop green technologies and sustainable practices, they have an excellent model: nature itself. As you have learned, environmental systems tend to operate in cycles consisting of feedback loops and the circular flow of materials. In natural systems, output is recycled into input. In contrast, human manufacturing processes traditionally run on a linear model in which raw materials are input and processed to create a product, while by-products and waste are generated and discarded.

Some forward-thinking entrepreneurs, inventers, and industrialists are making their processes more sustainable by transforming linear pathways into closed cycles in which waste is recovered, recycled, and reused. For instance, several companies now produce carpets that can be retrieved from the consumer when they wear out, and these materials are then recycled to create new carpeting. Some automobile manufacturers are planning cars that can be disassembled and recycled into new cars. Proponents of this industrial model see little reason why virtually all appliances and other products cannot be recycled, given the right technology. Their ultimate vision is to create truly closed loop industrial processes, generating almost no waste.

We also saw the "closed-loop" approach in the context of new geothermal energy technologies for district heating and cooling (Chapter 16, *Energy Alternatives*). Ground-source heat pumps circulate fluids to extract and transfer heat, without ever allowing the fluids to come into contact with any Earth materials. Systems like these will help us reduce negative impacts and resource depletion, and maintain standards of living in the future.

Long-term perspectives

To be sustainable, a solution must work in the long term. Often the best long-term solution is not the best short-term solution, which explains why much of what we currently do is not sustainable. Policymakers in democracies often act for short-term good because they aim to produce immediate, positive results so that they will be reelected.

This poses a major hurdle for addressing environmental dilemmas, many of which are cumulative, worsen gradually, and can be resolved only over long periods. Often the costs of addressing an environmental problem are short term, whereas the benefits are long term, giving politicians little incentive to tackle the problem. In such a situation, citizen pressure on policymakers is especially vital, because policy is an essential tool for pursuing sustainability.

Businesses may act according to either long-term or short-term interests. A business committed to operating in a particular community for a long time has incentive to sustain environmental quality. However, a business merely attempting to make a profit and move on has little incentive to invest in environmental protection measures that involve short-term costs.

The local and the global

As our societies become more globally interconnected, we experience a diversity of impacts, positive and negative. To many people, encouraging local self-sufficiency is an important element of building sustainable societies. When people feel closely tied to the area in which they live, they tend to value the area and seek to sustain its environment and human communities. This line of reasoning is frequently made in relation to locally based organic or sustainable agriculture.

Many advocates of local self-sufficiency criticize globalization. However, as ecological economist Herman Daly has explained, globalization means different things to different people. Those who view it as a positive phenomenon generally focus on how people of the world's diverse cultures are increasingly communicating and learning about one another. Books, airplanes, television, and the internet have made us more aware of one another's cultures and more likely to respect and celebrate—rather than fear—differences among cultures.

Those who view globalization in a negative light generally cite the homogenization of the world's cultures, by which a few cultures and worldviews displace many others. For instance, the world's many languages are going extinct with astonishing speed. Traditional ways of life in many areas are being abandoned as more people take up the material and cultural trappings of a few dominant Western cultures.

In recent years, many people have reacted against this homogenization and the growing power of large multinational corporations. In France, farm activist Jose Bove became a popular hero when he wrecked a McDonald's restaurant with his tractor to protest what many French farmers view as a threat to local French cuisine. In Quebec in 2003, hundreds of protesters picketed a meeting of the

FIGURE 22.9
Protesters picketed this World Trade Organization meeting in Quebec in 2003, criticizing the homogenizing effects of globalization, as well as relaxations in labour and environmental protections brought about by free trade.

World Trade Organization, which they viewed as a symbol of Western market capitalism. Since then, protesters have picketed every WTO meeting (**FIGURE 22.9**).

Daly and others argue that globalization entails a process in which multinational corporations attain greater and greater power over global trade while governments retain less and less. Most critics of globalization consider corporations less likely than governments to support environmental protection, so they feel that globalization will hinder progress toward sustainability. Moreover, market capitalism, almost by definition, promotes a high-consumption lifestyle, which does indeed threaten efforts to attain sustainable solutions.

On the positive side, however, globalization may foster sustainability because Western democracy, as imperfect as it is, serves as a model for people living under repressive governments. Open societies allow for entrepreneurship and the flowering of creativity in business, research, and academia. Millions of minds thinking about issues are more likely to come up with sustainable solutions than the minds of a few holding authoritarian power. Politically open democracies offer a compelling route for pursuing sustainability: the power of the vote.

One way to enhance transparency is by comparing the effectiveness of environmental policy in nations around the world. This is what the Environmental Performance Index (EPI) was designed to do, by researchers from Yale University. The EPI is a selectively weighted combination of 20 indicators in two main categories: environmental health and ecosystem vitality. It has been tracked now for more than 10 years, and is becoming a useful tool for comparative analysis of approaches to sustainability around the world (see "The Science behind the Story: Rating the Environmental Performance of Nations").

Systems thinking   There are two general ways to respond to a problem. One is to address the symptoms of the problem, and the other is to address the root cause of the problem. All too often our society takes a symptomatic approach (addressing the symptoms as we view things one part at a time) rather than a systemic approach (viewing the sum of the parts as a whole system and addressing the root cause). Addressing symptoms one by one as they appear is often easier in the short term, but is generally not effective in resolving the problem.

For instance, when pests evolve resistance to a pesticide, we generally call on chemists to develop more potent pesticides. This addresses a symptom but does not resolve the overall problem, because pests will likely proceed to evolve resistance to the new chemical as well. A systemic solution would be to develop agricultural approaches that rely less heavily on chemical pesticides. Similarly, as we deplete easily accessible fossil fuel deposits, we are choosing to reach farther and deeper for new fossil fuel sources, even as environmental and health repercussions mount. A systemic solution to our energy demands would involve developing clean renewable energy sources instead. For a great many issues that face our society, it will prove worthwhile to pursue systemic solutions.

This type of "systems thinking" has grown out of a field of study called *general systems theory*, which seeks to apply the concepts and principles of systems in all fields of research. We considered systemic approach and its many applications to environmental science in Chapter 3, *Earth Systems and Ecosystems*. One example that we considered was the plight of the beluga whales in the St. Lawrence Estuary, which have been dying of cancers. To understand why, we cannot study or treat one whale, or even one population of whales, in isolation. We cannot consider just one possible cause, or one possible source for a chemical that might be causing the problem. To solve the problem we have to take a step back and consider the whole system, with all of its complexity and interconnections, if we hope to find holistic answers and potential solutions for the problem.

The idea that the Earth system is an integrated whole, and that all parts are interacting and interconnected, is sometimes called the *principle of environmental unity*. In an interconnected system, the only way to find effective, long-lasting solutions for problems is to approach them systemically.

Research and education   None of these approaches will succeed fully if they are not founded in a solid understanding about environmental processes, and if the public is not aware of their importance. An individual's decision to reduce consumption and waste, purchase ecolabelled products, or vote for candidates who support sustainable approaches will have limited impact unless many others do the same. Individuals can influence large numbers of people by educating others and by serving as role models through their actions.

The campus sustainability efforts at many colleges and universities accomplish both of these approaches. Moreover, the discipline of environmental science plays a key role in providing information that people can use to make wise decisions about environmental issues. By promoting scientific and social scientific research, by raising the awareness of decision-makers, by exerting pressure on corporations to create social and environmental value along with economic value, and by educating the public about environmental science, we can all assist in the pursuit of sustainable solutions.

Sustainability on Campus

If we are to attain a sustainable civilization, we will need to make efforts at every level, from the individual to the household to the community to the nation to the world. Governments, corporations, and organizations must all encourage and pursue sustainable practices. Among the institutions that can contribute the most to sustainability efforts are colleges and universities.

Why strive for campus sustainability?

You enrolled at your college or university to gain an education, not to transform the institution. Why, then, are increasing numbers of students promoting sustainable practices on their campuses?

First, reducing the ecological footprint of a campus really can make a difference. The consumptive impact of educating, feeding, and housing hundreds or thousands of students is immense. Second, campus sustainability efforts make students aware of the need to address environmental problems, and students who act to promote campus sustainability serve as models for their peers.

Support from faculty, staff, and administrators is crucial for success, but students are often the ones who initiate change. Students often feel freer than faculty or staff to express themselves. Students also arrive on campus with new ideas and perspectives, and they generally are less attached to traditional ways of doing things. And finally, students who engage in sustainability efforts learn and grow as a result. The challenges, successes, and failures that they encounter can serve as

valuable preparation for similar efforts in transforming inertia-bound institutions in the broader society. Becoming involved in the push for more sustainable technologies and approaches can help students acquire many practical skills that will be useful in the job market and later in their career paths.

Campuses are great places to implement sustainable approaches

We tend to think of colleges and universities as enlightened and progressive institutions that generate benefits for society. They are places of learning and knowledge-sharing, where experimentation, exploration, and entrepreneurship are encouraged.

However, colleges and universities are also centres of lavish resource consumption. Institutions of higher education feature extensive infrastructure, including classrooms, offices, research labs, and residential housing. Most have dining establishments, sports arenas, vehicle fleets, and road networks. The ecological footprint of a typical college or university is substantial. Reducing the size of this footprint is challenging. Colleges and universities tend to be bastions of tradition, where institutional habits are deeply ingrained and where bureaucratic inertia and financial constraints can often block the best intentions for positive change. Nonetheless, faculty, staff, administrators, and especially students are progressing on a variety of fronts to make the operations of educational institutions more sustainable.

At UBC Vancouver, the Alma Mater Society (AMS), which today represents 48 000 students, created the AMS Lighter Footprint Strategy, based on the ecological footprint concept of William Rees and Mathis Wackernagel.[10] The intent of the strategy is to coordinate and build upon the successes of student-run sustainability initiatives. The strategy defines internal goals that can be achieved by a student organization working on its own and interactive goals that require communication or collaboration with other agencies, both within the university and outside of it. The goals are divided into such categories as food and beverage, communications, transportation, and campus policies. Each category has suggested actions and timeframes, as well as indicators with which progress can be measured.

Student-run organizations at other campuses are beginning to adopt some of the actions and principles of the Lighter Footprint Strategy and similar approaches. Even university and college presidents have become involved. The *Talloires Declaration* is a commitment on the part of college and university presidents around the world to pursue and foster sustainability. The declaration provides

THE SCIENCE ⟩ BEHIND THE STORY

Rating the Environmental Performance of Nations

To measure the progress of nations toward environmental sustainability, researchers have devised the Environmental Performance Index (EPI), which rates countries using data from 20 indicators of environmental conditions for which governments can be held accountable. A report detailing the results is published online and updated periodically by Daniel Esty and colleagues at the Yale Center for Environmental Law and Policy and Columbia University's Center for International Earth Science Information Network, in collaboration with the World Economic Forum and the Joint Research Centre of the European Commission. At **http://epi.yale.edu** you will find the details of the 2014 EPI for all of the 178 nations included, representing 99% of the world's population. (A few countries are excluded because of lack of information.)

The UN's Millennium Development Goals did not define how to quantify progress on environmental measures. The lack of quantitative measures, these researchers believe, has stymied progress on how to implement effective environmental policy. To address this problem, the EPI researchers have aimed to track the performance of environmental policy with the same quantitative rigour as statistics are tracked for health, poverty reduction, and other development goals. Giving nations scores and ranking them should reveal "leaders and laggards," showing which nations are on the right track and which are not.

The researchers gather internationally available data from the UN and other sources on 20 indicators. These indicators are grouped into nine "issues" within two overarching objectives, environmental health and ecosystem vitality, as shown in **FIGURE 1**. Nations are scored on their overall performance on each objective. Scores range from 0 to 100, with 100 representing a target value for that objective, established by international

ENVIRONMENTAL HEALTH

- Health impacts
- Air quality
- Water and sanitation

ECOSYSTEM VITALITY

- Water resources
- Agriculture
- Forests
- Fisheries
- Biodiversity and habitat
- Climate and energy

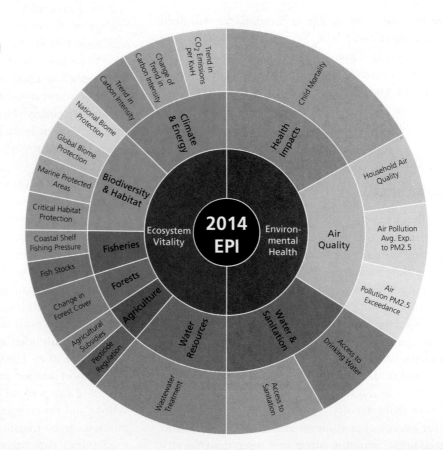

FIGURE 1   For the Environmental Performance Index, researchers score nations on their performance in approaching environmental sustainability goals for two broad objectives, environmental health and ecosystem vitality, which are subdivided into nine "issues," as shown here. The issues, in turn, are underlain by 20 indicators, which are selectively weighted to produce the overall score.
Source: From Environmental Performance Index, Our Methods, The 2014 EPI Framework—What Does the EPI Measure? Retrieved from http://epi.yale.edu/our-methods/. Copyright © 2014 Yale Center for Environmental Law & Policy.

consensus. Twenty underlying indicators are selectively weighted to yield separate scores for each of the nine issues. You can find out more about how this weighting is done by visiting **http://epi.yale.edu/our-methods**.

The main pattern in the results, for both 2014 and previous years, is conspicuous: Nations with the highest scores (see **TABLE I**) are mostly wealthy, industrialized nations with the capacity to commit substantial resources toward environmental protection. Nations with the lowest scores are largely developing nations with few resources to invest. They include countries with dense populations and stressed ecosystems that are trying to industrialize, those with arid environments and limited natural resources, and those facing extreme poverty.

The correlation between economic vitality and environmental protection was expected, but the researchers were more interested in the variation around this trend. For any given level of income or development, some nations vastly outperformed others. The researchers say this demonstrates that other factors are at work—namely, the political choices national leaders make. Indeed, their data show a strong correlation between EPI scores and "good governance," which involves aspects such as rule of law, open political debate, and lack of corruption.

These interpretations touched sensitive nerves in nations that did not rank as highly as they would have liked. For example, in the 2014 rankings Canada placed 24th, with a score of 73.14, and the United States placed 33rd with a score of 67.52. Critics say the rankings have little meaning because only certain data sets were measured, and these were then weighted subjectively. Supporters counter that although no one will ever agree on exact numbers, the EPI provides a useful tool for evaluating actions and investments, highlighting policies that work, and identifying future priorities.

Conversely, poorer nations tended to score low on environmental health and higher on indicators that reflected the availability of remaining unexploited natural resources. Overall across all nations, performance was worst for the indicators of renewable energy and wilderness protection.

TABLE I Ranking of Nations in the 2014 Environmental Performance Index

EPI ranking	Country	EPI score	Ten-year change
Top-ranked nations			
1.	Switzerland	87.67	0.80%
2.	Luxembourg	83.29	3.02%
3.	Australia	82.40	2.32%
4.	Singapore	81.78	0.94%
5.	Czech Republic	81.47	3.47%
6.	Germany	80.47	1.89%
7.	Spain	79.79	1.82%
8.	Austria	78.32	1.82%
9.	Sweden	78.09	1.30%
10.	Norway	78.04	2.79%
24.	Canada	73.14	2.58%
Bottom-ranked nations			
169.	Bangladesh	25.61	3.98%
170.	Democratic Republic of Congo	25.01	3.56%
171.	Sudan	24.64	0.49%
172.	Liberia	23.95	11.03%
173.	Sierra Leone	21.74	21.79%
174.	Afghanistan	21.57	12.17%
175.	Lesotho	20.81	4.36%
176.	Haiti	19.01	6.08%
177.	Mali	18.43	8.67%
178.	Somalia	15.47	6.62%

Source: From Environmental Performance Index 2014, Country Rankings. http://epi.yale.edu/epi/country-rankings/. Copyright © 2014 by Yale Center for Environmental Law & Policy. Reprinted by permission.

DATA Q Countries like Sierra Leone and Afghanistan rank very low in the EPI, but at least their scores have improved dramatically over the past 10 years. Given that Sierra Leone's 2014 EPI score is 21.74 and the 10-year improvement has been 21.79%, what was Sierra Leone's EPI score 10 years ago? What kinds of changes do you think might account for the improvement in the score? Visit the EPI Data Explorer at **http://epi.yale.edu/epi/data-explorer** to dig deeper into the data underlying the country rankings.

Whereas controversy generated by the overall rankings drew media attention, other trends were apparent for those who examined the data more deeply. For instance, wealthy industrialized nations tended to score highest in environmental health and lowest in ecosystem vitality indicators, especially biodiversity and habitat. This is presumably because they possess money to invest in protecting the health of their citizens but achieved economic development by exploiting their natural resources and degrading their natural environment.

The EPI scores are only as solid as the data that go into them, and not every nation provides reliably accurate data. Moreover, the research group is constantly exploring ways to improve both the data and the methodology for the calculations. However, the Environmental Performance Index provides a unique means of assessing current environmental status and what is working in environmental policy worldwide. As the methodology is refined year by year, the EPI promises to become a more accurate performance indicator and a more helpful policy tool.

TABLE 22.3　The 10-Step Plan of the Talloires Declaration

1. Raise public, government, industry, foundation, and university awareness by publicly addressing the urgent need to move toward an environmentally sustainable future.
2. Engage in education, research, policy formation, and information exchange on population, environment, and development to move toward a sustainable future.
3. Establish programs to produce expertise in environmental management, sustainable economic development, population, and related fields to ensure that all university graduates are environmentally literate and responsible citizens.
4. Develop the capability of university faculty to teach environmental literacy.
5. Set an example of environmental responsibility by establishing programs of resource conservation, recycling, and waste reduction at the universities.
6. Encourage the involvement of government (at all levels), foundations, and industry in supporting university research, education, policy formation, and information exchange in environmentally sustainable development. Expand work with nongovernmental organizations to assist in finding solutions to environmental problems.
7. Convene school deans and environmental practitioners to develop research, policy, information exchange programs, and curricula for an environmentally sustainable future.
8. Establish partnerships with primary and secondary schools to help develop the capability of their faculty to teach about population, environment, and sustainable development issues.
9. Work with the UN Conference on Environment and Development, the UN Environment Programme, and other national and international organizations.
10. Establish a steering committee and a secretariat to continue this momentum and inform and support each other's efforts in carrying out this declaration.

Source: University Leaders for a Sustainable Future, Programs: Talloires Declaration, www.ulsf.org/programs_talloires.html. Retrieved from link http://www.ulsf.org/pdf/TD.pdf. Copyright by University Leaders for a Sustainable Future. Reprinted by permission.

a 10-point action plan for incorporating sustainability and environmental literacy in teaching, research, operations, and outreach at colleges and universities (**TABLE 22.3**). As of 2014, the Talloires Declaration has been signed by 472 presidents and chancellors from 54 nations, including 39 Canadian universities and colleges.[11]

Campus efforts often begin with an audit and grow from there

Campus sustainability efforts often begin with a quantitative assessment of the institution's operations. Such audits provide baseline information on what an institution is doing or how much it is consuming and help set priorities and goals. It is useful in such an **environmental audit** to target items that can lead directly to specific recommendations to reduce impacts and enhance sustainability. Once changes are implemented, the institution can monitor progress by comparing future measurements to the audit's baseline data.

For example, student auditors at Mount Allison University undertake a comprehensive environmental audit every two years. The results of the audits have contributed to the development of a green action plan, which provides ideas for campus groups looking for possible activities and serves as a "tool kit" for the environmental issues committee. The 2014 audit showed that the University has successfully reduced its waste production over the past three years.[12]

Where can campus sustainability efforts go after students, faculty, and staff take stock through environmental and waste audits? Sustainability programs on campuses are often an interesting mix, ranging from student-run programs; to official, university-wide policies on energy standards or materials used in the construction of new buildings; to educational initiatives implemented through the curriculum. Let's have a look at some examples from campuses around Canada and in other countries.

Recycling and waste reduction　Campus sustainability often begins with a waste audit, so follow-up efforts frequently involve waste reduction, recycling, or composting. Waste management initiatives are relatively easy to start and maintain because they offer many opportunities for small-scale improvements and because people generally enjoy recycling and reducing waste. Waste and recycling audits, battery dropoffs, and residence composting programs are popular on Canadian campuses.

For example, students, faculty, and staff at Concordia University divert waste from landfills through recycling and composting initiatives. This includes dropoffs for old computers and other electronic wastes, which are becoming an increasingly problematic component of municipal waste. Concordia was the first of (now) five universities certified under the *Ici On Recycle* program, sponsored by Recyc-Québec and the government of Quebec, which recognizes institutions that divert at least 65% of their waste from landfills.[13] McGill University students—inspired by what they saw happening at Concordia and other universities—started Gorilla Composting, a student-run initiative that promotes composting on campus.[14] (**FIGURE 22.10**).

Students at some campuses run events that promote the reuse of items. At the University of Calgary, the U-Bike program was started when 10 discarded bicycles were salvaged, repaired, and pressed into service. The bikes, painted with bright red and yellow stripes, are

Gorilla Composting

FIGURE 22.10
The mascot of McGill University's student-run Gorilla Composting Team (inset shows the team's logo) introduces new students to on-campus composting.

available for use free of charge for any university student. Programs similar to this, which apparently originated in the Netherlands in 1998, are springing up at campuses and in cities across North America.

One very popular initiative on campuses in Canada lately has been Bottled Water Free Day, a partnership among the Canadian Federation of Students, the Polaris Institute, the Sierra Youth Coalition, the Canadian Union of Public Employees, and Development and Peace. The program urges students and staff to pledge not to drink bottled water if tap water is available, with the goal of reducing waste from plastic water bottles. As of 2014, 14 campuses in Canada have taken the pledge to ban the sale of plastic water bottles altogether; the University of Winnipeg was first, in early 2009. More than 80 municipalities in Canada have also taken steps to reduce waste from plastic water bottles.[15]

The Renew Project at UBC looks at renovate-versus-replace decisions on campus, and is aimed at incorporating the full long-term costs—environmental and social, as well as economic—into those decisions. As a result of this project, the university has avoided $89 million in facilities replacement costs and diverted 1458 tonnes of construction waste from landfills. It also saved 97 million megajoules of primary energy, 27 million litres of water, and more than 6000 tonnes of greenhouse gas emissions, among other impacts.[16]

Green building design   Dozens of campuses now boast "green" buildings that are constructed from sustainable building materials and whose design and technologies encourage energy efficiency, water efficiency, renewable

energy, and the reduction of pollution. As with any type of ecolabelling, agreed-upon standards are needed, and for sustainable buildings, these are the *Leadership in Energy and Environmental Design (LEED)* standards. Developed and maintained by the nonprofit Green Building Council, LEED standards guide the design and certification of new construction and the renovation of existing structures on campuses and elsewhere.

Buildings earn LEED credits in various categories if they are constructed using materials that were recycled or reused, took little energy to produce, or were locally harvested, produced, or distributed. Other common features are energy-efficient lighting, heating, and appliances; state-of-the-art ventilation systems; and paints, adhesives, and carpeting that emit few volatile organic compounds.

Vancouver Island University was an early leader in green buildings and environmental retrofits in Canada. The university adopted a holistic, environmentally sustainable approach to building design in 1990, when it began planning for the Nanaimo campus. Since then, the university has realized millions of dollars in energy savings. The movement for "green buildings" continues to grow. Pavillons Lassonde at the École Polytechnique de Montréal was the first LEED-certified building in Canada, achieving its gold status in October 2005, followed closely by the Life Sciences Centre at UBC in December 2005 and the University of Victoria Medical Sciences Centre in August 2006.

Green building design approaches can also be applied to existing buildings, not just to new buildings. In fact, LEED Platinum certification requires the re-use of an

Child Development Centre/University of Calgary

FIGURE 22.11
The Centre for Child Development at the University of Calgary is Canada's largest platinum LEED-certified building.

existing structure. Only a handful of buildings in Canada have earned platinum LEED certification; these include the Centre for Child Development at the University of Calgary, certified in 2007 (**FIGURE 22.11**).

Administrators are more easily convinced to enact institutional changes if they save money. The EcoTrek Project accomplished this at UBC. This building retrofit project began in the fall of 2002 and was completed in 2008. The objectives were to generate savings, reduce energy use in university buildings by 20%, and reduce water use by 30%; these goals were met and exceeded. The project saved UBC an estimated $2.6 million annually, making the payback period approximately 17 years for the $35 million project. The work involved retrofits to nearly 300 campus buildings, including installing more efficient lighting, weather-stripping nearly 2000 doors, and replacing old steam boilers with energy-efficient burners.[17]

Environmentally sensitive landscaping

Sustainable architecture doesn't stop at the walls of a building. Careful design of campus landscaping can create livable spaces that promote social interaction, and where plantings supply shade, prevent soil erosion, create attractive settings, and provide wildlife habitat. No campus sustainability program would be complete without some attempt to enhance the campus's natural environment. Such efforts remove invasive species, restore native plants and communities, improve habitat for wildlife, enhance soil and water quality, reduce pesticide use, and create healthier, more attractive surroundings.

The University of Victoria, known for the beauty of its 160-ha campus grounds, protects certain areas from development and has made a commitment to restoration and monitoring of naturalized areas. The university has adopted the concept of "naturescaping," a landscaping approach that emphasizes restoring, preserving, and enhancing wildlife habitat in urban and rural areas. The intent is to create new habitats, utilize native plant species, reduce the need for watering, and eliminate chemical pesticides and herbicides. There are several natural landscaping projects on campus, including the Native Plant Study Garden and the Lorene Kennedy Memorial Native Plant Garden (**FIGURE 22.12A**).[18]

The University of Saskatchewan College of Education has constructed a Prairie Habitat Garden with native grasses, shrubs, and wildflowers. The garden helps preserve the natural heritage and habitat of prairie grasslands while acting as an educational resource (**FIGURE 22.12B**). The University of Toronto Mississauga joined with Evergreen's Learning Grounds program for a major campus greening initiative that has won local environmental awards. Students, faculty, staff, and neighbours turn out for regular volunteer plantings of native shrubs and flowers. Besides providing wildlife habitat, the restored areas reduce runoff, erosion, and maintenance costs, and provide opportunities for research and education. For example, student researchers have assessed threatened Jefferson salamander populations and habitat conditions at the campus. The university also undertook a controlled burn to restore native vegetation on a small area of the campus, planting experimental plots of native plants where students monitored ecological changes in the wake of the restoration.

Efficient water use   Managing water efficiently is a key element of sustainable campuses. Simon Fraser University was recognized with an Earth Award from the Building Owners and Managers Association (BOMA) of British Columbia for their Arts and Social Sciences Complex. The building has a cistern under the inner courtyard that can store more than 227 000 L of rain water to be used for landscape irrigation purposes, in addition to many energy-saving features.[19] The building has won engineering and architectural design awards; it is certified under the BOMA-sponsored Go Green program, a green building certification program similar to LEED.

The Living with the Lakes Centre at Laurentian University demonstrates another approach to water management integrated with green building design. This project, which will house the university's aquatic ecology programs, will be completely integrated into its lakeside setting, featuring a flow-through aquatic laboratory with natural lake water. A green roof will store and filter rain water, and two constructed wetlands, one outdoors and one indoors, will treat and purify storm water. The building and site are designed to improve the quality of water entering Ramsay Lake, a drinking water reservoir.[20]

Water conservation is just as important indoors as outdoors. Water-saving technologies, such as waterless

(a) Native plant garden

Don Pierce/UVic Photo Services

(b) Prairie habitat garden

FIGURE 22.12
The Lorene Kennedy Memorial Native Plant Garden **(a)** is one of a number of "naturescaping" projects at the University of Victoria. Many schools have embarked on habitat restoration projects to beautify their campuses, provide wildlife habitat, restore native plants, and filter water runoff. These flowers **(b)** are in the University of Saskatchewan's Prairie Habitat Garden.

Jay Wilson

urinals and "living machines" to treat waste water, are being installed at a number of campuses. UBC's EcoTrek program reduced water use on campus by 30%.[21] A study by the sustainability office of the University of Toronto found that urinals in some buildings on the downtown campus were set to flush automatically at timed intervals, 24 hours a day, 7 days a week, resulting in what the study called an "astronomical" waste of water. The problem was resolved by installing motion-sensor flush systems in 15 washrooms, resulting in estimated savings of 7 million litres of water and $10 000 per year.[22]

Energy conservation and alternatives Students are finding many ways to conserve energy on campus. Campuses can realize large energy savings simply by not powering unused buildings. How many times have you walked through a classroom building at night or in the summer when it was totally empty, yet found that all the lights were on and the heating or air conditioning was running full blast? Large buildings are expensive to heat, cool, and light, so powering them down when not in use saves a great deal of energy, money, and greenhouse gas emissions.

One successful on-campus energy conservation program is the University of Toronto's ReWire Project, which aims to help students, staff, and faculty reduce their energy consumption through behavioural changes. Components of the project include sustainability pledges, environmental education initiatives, and "tool kits" that provide information, resources, and strategies to users.

The tool kits contain an assortment of "action tools" that allow users to pass along information to promote sustainable behaviour. The tool kits are designed for use in specific campus settings such as residence rooms, common areas, and offices. The result has been an energy savings of about 10–13% in residences where the tool kit has been introduced. The project received financial support from a number of sources outside the university, including the Toronto Atmospheric Fund, the Better Buildings Partnership, and the Ontario Power Authority.[23]

Campuses also can reduce energy consumption and greenhouse emissions by altering the types of energy they use. This goes hand in hand with the green building initiatives described in this chapter. The University of Toronto Mississauga's award-winning Instructional Centre, completed in 2011 and certified LEED Silver, was built on the site of an existing parking lot to avoid loss of green space. The building is heated and cooled almost entirely by a ground-source heat pump system, boosted by a solar electric system, and assisted by energy-efficient lighting, computers, and AV equipment.[24] Carleton University has made use of a similar system for space heating since the 1980s.

Student initiatives on campus also can influence the types of energy we use in our society. Each year university- and college-based teams compete at the Formula Sun Grand Prix, a solar car competition, with solar-powered vehicles of their own design and construction. In 2014, Canada was represented by teams from Western, McMaster, and Montréal Polytechnique; the latter came in fourth overall in the competition.[25]

Every two years, teams of students compete in the Solar Decathlon hosted by the U.S. Department of Energy. Teams from all over the world, including Canada, are involved. In this remarkable event, teams of students travel to Washington, D.C., bringing material for solar-powered homes they have spent months designing. They erect their homes on the National Mall in Washington, where the buildings stand for three weeks. The homes are judged on 10 criteria, and prizes are awarded to winners in each category. In 2013, Canada was represented by Team Alberta, from the University of Calgary, and Team Ontario, a partnership between Queens University, Carleton University, and Algonquin College. Team Ontario's house was called ECHO, a solar-powered starter home that produced more energy than it used. Team Alberta produced a solar-powered house they called *Borealis*. The building was designed to address living conditions in remote, far northern locations, and featured an easily transportable, modular design, passive and active solar technologies, and a "living wall" (**FIGURE 22.13**).[26] Team Ontario finished in sixth place and Team Alberta in ninth place in the 2013 competition.

FIGURE 22.13

Teams from colleges and universities around the world compete in the biennial Solar Decathlon in Washington, D.C. Each team erects an entire house, of the students' own design, fully powered by solar energy. In the 2013 event, Canada was represented by the University of Calgary. Their solar-powered entry, called *Borealis*, was designed for remote living conditions in the far north.

Trent University, located on the fast-flowing Otanabe River, has its own hydroelectric power plant, the Stan Adamson Powerhouse, which houses three turbine generators. The plant supplies most of the electricity used on campus in a given year, and excess power from low-use times is exported to the power grid. The power plant technology dates from the original construction in the late 1800s, but recently underwent a major update involving the installation of three new turbines; the plant now produces 3.9MW of power. The Trent campus is essentially self-sufficient on a source of energy that is associated with virtually no pollutants or emissions.[27]

Carbon-neutrality　　Now that global climate change has vaulted to the forefront of society's concerns, reducing greenhouse gas emissions from fossil fuel combustion has become a top priority for campus sustainability proponents. Today many campuses are aiming to become carbon-neutral.

As of 2015, 695 university and college presidents in the United States and other countries had signed on to the American College and University Presidents Climate Commitment (including Confederation College, the only Canadian signatory). This pledge to reduce campus greenhouse emissions commits presidents to undertake inventory of emissions, set target dates for becoming carbon-neutral, and take immediate steps to lower emissions with short-term actions, while also integrating sustainability into the curriculum.[28] In March 2008, a similar Climate Change Statement of Action was initiated and signed by university and college presidents in British Columbia. The statement commits each institution to initiate a comprehensive plan to reduce greenhouse gases. As of 2015, 23 postsecondary institutions from across Canada had signed the pledge.[29]

Student pressure and petitions at many campuses have nudged administrators and trustees to set targets for reducing greenhouse emissions and to strive to do more to counteract climate change than they do to contribute to it. Common Energy is a network of students, faculty, staff, and community stakeholders working to encourage their universities to move beyond climate-neutrality by providing resources and undertaking actions to promote community involvement in a wide variety of climate-related activities and other environmental initiatives.[30]

Campus gardens and local food　　Campus food service operations can promote sustainable practices by buying organic produce, composting food scraps, and purchasing food in bulk or with less packaging. Buying locally grown or produced food supports local economies and

Christopher DeWolf

FIGURE 22.14
A number of campuses now include gardens where students can grow organic vegetables that are used for meals in dining halls. This is a view from a garden run by McGill University's Campus Crops Initiative.

cuts down on fuel use from long-distance transportation. At the University of Waterloo, students and the University Food Services teamed up to start an on-campus Farm Market to promote local foods. The market features 100% local produce, preserves and honeys made in Waterloo County, and fresh-baked goods made right on campus. McGill University's Organic Campus is a student-run nonprofit organization that delivers local organic fruits and vegetables on campus.

Some college campuses even have community gardens where students can grow food. McGill's Campus Crops is a student-run urban gardening cooperative (**FIGURE 22.14**). Faced with a shortage of community garden plots, the students have identified unused spaces on the McGill campus that could support urban agriculture. They also run an on-campus vegan collective called the Midnight Kitchen. The students see their efforts not just as environmental or food-related, but also political, because decision-making around access to land and food security in urban areas affects the poor in urban areas most intensively.[31]

Institutional green purchasing The kinds of purchasing decisions made in dining halls favouring local food, organic food, and biodegradable products can be applied across the entire spectrum of a campus's needs. When campus purchasing departments buy recycled paper, certified sustainable wood, energy-efficient appliances, goods with less packaging, and other ecolabelled products, they send signals to manufacturers and increase the demand for such items. Students are working with campus bookstores to carry, promote, and sell more environmentally and socially sustainable books, paper products, and school supplies.

The University of Guelph in Ontario is one of many universities that have undertaken green purchasing policies for their computing services departments, with a pledge to take into consideration the entire life-cycle of electronic products. "Green computing" considers everything from "cradle" (the materials and processes used in the manufacturing and shipping of the product), through operational use (the energy efficiency of the product), to "grave" (the effective recycling and disposal of the product).

At Chatham College in Pennsylvania, students chose to honour their school's best-known alumnus, Rachel Carson, by seeking to eliminate toxic chemicals on campus. Administrators agreed to this effort, provided that alternative products to replace the toxic ones worked just as well and were not more expensive. Students brought in the CEO of a company that produces nontoxic cleaning products, who demonstrated to the janitorial staff that his company's products were superior. The university switched to the nontoxic products, which were also cheaper, and proceeded to save $10 000 per year. Chatham students then found a company offering paint without volatile organic compounds and negotiated with it for a free paint job and discounted prices on later purchases. Students also worked with grounds staff to eliminate herbicides and fertilizers used on campus lawns and to find alternative treatments.

Cindy Lopez/The Concordian

FIGURE 22.15
The Sustainable Concordia *Allégo* program holds free bike tune-up events for students, staff, and faculty.

Transportation alternatives   Many campuses struggle with traffic congestion, parking shortages, commuting delays, and pollution from vehicle exhaust. Some are addressing these issues by establishing or expanding bus and shuttle systems; encouraging bicycling, walking, and carpooling; and introducing alternative vehicles to university fleets (**FIGURE 22.15**). The University of Calgary has installed covered, secure bike storage on campus for cyclists, in support of their sustainable transportation policy. The *Allégo* program at Concordia University, established in 2002, makes sustainable transportation choices available to those who are commuting to and from the campus, offering a free rideshare board, bicycle safety and maintenance workshops, and a Guaranteed Ride Home program for carpoolers.[32]

The University of British Columbia is also a leader in transportation efforts. For less than $40 a month, UBC's U-Pass program provides students with biking programs and facilities, expanded campus bus and shuttle services, unlimited use of some city transit systems, rides home in emergencies, merchant discounts, and priority parking spaces and ride-matching services for carpoolers. The program has boosted transit ridership to campus and decreased single-person car use.[33]

Campus sustainability also belongs in the curriculum

Campus sustainability activities provide students with active, hands-on ways to influence how their campuses function. Sustainability concerns are also transforming academic curricula and course offerings. The course for which you are using this book right now likely did not exist a generation ago. As our society comes to appreciate the looming challenge of sustainability, colleges and universities are attempting to train students to confront this challenge more effectively.

Almost all universities and colleges in Canada now offer courses and programs on environment and sustainability issues; the earliest date back to the early 1970s. Trent University, which has offered courses in environmental and resource management since 1975, now partners with Fleming College to offer a practical, field-based program in ecological restoration. Faculty at Queen's University have received awards for excellence in education for the promotion of sustainable practices, in recognition of their contributions to sustainability education. There are many other examples of sustainability-focused programs and courses at campuses across Canada.

Some universities specialize on specific topics within the general fields of environment and sustainability. For example, McMaster University is widely recognized for excellence in the field of environmental health. The Institute of Island Studies at Prince Edward Island University emphasizes the study of the culture, environment, and economy of small islands. The institute's associates undertake studies on environmental policy, water quality, land use, and other topics from the particular perspective of island economies, and partner with other island researchers from around the world.

Besides offering new courses and programs, schools are incorporating sustainability issues into established courses across many disciplines. UBC, for example, now offers more than 350 courses with sustainability-related themes. Many of these are also *interdisciplinary* courses and programs, meaning that they integrate knowledge from a variety of different disciplines. Sustainability courses and programs often have a practical slant, too. For example, the University of Waterloo's ERS310 "Environmental Analysis and Solutions III: Greening Communities" course and the University of Toronto's ENV421 "Environmental Research" course both offer students opportunities to carry out practical research on campus for course credits. Often the results of student research are used in developing campus policies; for example, past ENV421 projects have included "Recommendations for Reducing Energy Consumption" and "The Natural Step and the University of Toronto: A Value Added Approach to Energy Efficiency Planning."[34]

The UBC SEEDS (Social Ecological Economic Development Studies) program links students, staff, and faculty to research projects that enhance sustainability on campus, and allows students to get course credits for their projects. SEEDS projects have benefited the campus through a number of research projects that have led to policy changes and/or initiatives, including

■ a biodiesel project that led UBC's Vancouver Plant Operations to use 20% biodiesel fuel in its diesel fleet vehicles

■ research that led to UBC's becoming a pesticide-free campus

■ reassessment of landscape techniques to reduce heavy metal contaminants in storm water

■ research that led to new seafood purchasing policies[35]

In the best tradition of colleges and universities, classroom learning and real-world learning go hand in hand in SEEDS and similar programs on campuses across Canada.

Organizations are available to assist campus efforts

Many campus sustainability initiatives are supported by outside, coordinating bodies, such as University Leaders for a Sustainable Future and the National Wildlife Federation's Campus Ecology program. These organizations act as information clearing houses for campus sustainability efforts, making resources and advice readily available. With the assistance of these organizations, it is easier than ever to start sustainability efforts on your own campus and obtain the support to carry them through to completion.

The Sierra Youth Coalition is the youth arm of the Sierra Club and is youth-led.[36] The organization offers a framework for sustainability assessment that has provided a starting point for many initiatives on campuses across Canada. The framework incorporates indicators of both human and ecosystem well-being in a post-secondary context.

One organization that has emerged as a leader in the campus sustainability movement in North America is the Association for the Advancement of Sustainability in Higher Education (AASHE), founded in 2005 to help coordinate sustainability efforts across North America.[37] Its members are university and college campuses, although businesses and other organizations can also join as associates. The mission of AASHE is to "inspire and catalyze higher education to lead the global sustainability transformation," and their vision is for AASHE to "lead higher education to be a foundation for a thriving, equitable, and ecologically healthy world."[38]

Sustainability Tracking, Assessment, and Rating System, or STARS, is one program coordinated by AASHE.[39] STARS is an online reporting framework that allows universities and colleges to self-assess and report on their environmental performance, highlighting both their successes and areas in need of improvement. As of 2014, almost 700 institutions had signed up to use the STARS reporting tool.

Precious Time

The pace of our lives is getting faster, and life's commotion can make it hard to give attention to problems we don't need to deal with on a daily basis. The world's sheer load of environmental dilemmas can feel overwhelming, and even the best intentioned among us may feel we have little time to devote to saving the planet.

The systems we depend on are changing

The natural systems we depend on are changing quickly. Many human impacts continue to intensify, including deforestation, overfishing, land clearing, wetland draining, atmospheric emissions, and resource extraction. Our window of opportunity for turning some of these trends around is shrinking. Even if we can visualize sustainable solutions to our many problems, how can we possibly find the time to implement them before we do irreparable damage to our environment and our own future?

Today, humanity faces a challenge more important than any previous one—the challenge to achieve sustainability. Attaining sustainability will be a massive and complex process, but it is one to which every single person on Earth can contribute; in which government, industry, and citizens can all cooperate; and toward which all nations can work together. Human ingenuity is capable of it; we merely need to rally public resolve and engage our governments, institutions, and entrepreneurs in the race.

However, we must be realistic about the challenges that lie ahead. As we deplete the natural capital we can draw on, we give ourselves, and the rest of the world's creatures, less room to manoeuvre. Until we implement sustainable solutions, we will be squeezing ourselves through a progressively tighter space, like being forced through the neck of a bottle. The key question for the future of our species and our planet is whether we can make it safely through this bottleneck. Biologist Edward O. Wilson has written eloquently of this view:

> At best, an environmental bottleneck is coming in the twenty-first century. It will cause the unfolding of a new kind of history driven by environmental change. Or perhaps an unfolding on a global scale of more of the old kind of history, which saw the collapse of regional civilizations [in] Mesopotamia, and subsequently Egypt, then the Mayan and many others . . . Somehow humanity must find a way to squeeze through the bottleneck without destroying the environments on which the rest of life depends.[40]

Earth is an island

We began this book with the vision of Earth as an island, and indeed that is what it is (**FIGURE 22.16**). Islands can be paradise, as Rapa Nui (Easter Island) likely was when the Polynesians first reached it. But when Europeans arrived at Easter Island, they either witnessed or contributed to the downfall of a civilization that collapsed once its island's resources had been depleted.

As Rapa Nui's trees disappeared, some individuals must have spoken out for conservation and for finding ways to live sustainably amid dwindling resources. Perhaps some people ignored those calls and went on extracting more than the land could bear, assuming that somehow things would turn out all right. Whoever cut the last tree atop the most remote mountaintop, whether it was an islander or a colonizer, could have looked out across the island and seen that it was the last tree. And yet that person cut it down.

It would be tragic folly to let such a fate befall our planet as a whole. By recognizing this, by deciding to shift our individual behaviour and our cultural institu-

FIGURE 22.16
We end the book as we began the first chapter, with a photo of Earth from space. This photograph of Earth and Earth's Moon, taken by the spacecraft *Voyager* as it sped away from home, shows our planet as it truly is—an island in space. Everything we know, need, love, and value comes from and resides on this small sphere, so we had best treat it well.

tions in ways that encourage sustainable practices, and by employing science to help us achieve these ends, we may yet be able to live happily and sustainably on our wondrous island, Earth.

Conclusion

In any society facing dwindling resources and environmental degradation, there will be those who raise alarms and those who ignore them. Fortunately, in our global society today we have many thousands of scientists who study Earth's processes and resources.

For this reason, we are amassing a detailed knowledge and an ever-developing understanding of our dynamic planet, what it offers us, and what impacts it can bear. The challenge for our global society today, our one-world island of humanity, is to support that science so that we may judge false alarms from real problems and distinguish legitimate concerns from thoughtless denial. This science, this study of Earth and of ourselves, offers us hope for our future.

> *Unless someone like you cares a whole awful lot, nothing is going to get better. It's not.*
>
> —from *The Lorax,* by Dr. Seuss (1971, Random House).

REVIEWING OBJECTIVES

You should now be able to

Explain the concept of sustainability and some of the key approaches to sustainable development

- Sustainable development entails environmental protection, economic development, and social justice, which are mutually supportive goals. The principle of environmental primacy suggests that we must first protect the environment, as it is what sustains our economic and social activities.

- Economic development and environmental quality can enhance one another. Environmental protection and green technologies and industries can create rich sources of new jobs. Protecting environmental quality enhances a community's desirability and economy.

- It is easy to forget that we, as humans, are not separate from our environment but an integral part of it, and dependent upon its goods and services.

- Approaches that can inspire sustainable solutions include refining our ideas about quality of life, reducing unnecessary consumption and waste, limiting population growth, encouraging green technologies, mimicking natural systems, thinking long-term, enhancing local self-sufficiency, being politically active, voting with our wallets, thinking systemically, and promoting research and education.

Describe and assess some sustainable development initiatives on college and university campuses

- Reducing the ecological footprints of post-secondary campuses has a positive impact. Being involved in sustainability initiatives raises the awareness of students, faculty, and staff, who then become ambassadors for sustainability beyond the campus.

- Campuses are a great place to strive for sustainability. They are places of learning and knowledge, and places where exploration and experimentation are encouraged.

- Audits produce baseline data, and campus sustainability efforts grow from there. Campus programs often focus on recycling and waste reduction, or water and energy efficiency. Green buildings are being constructed on a growing number of campuses, and natural habitats are being restored. Carbon-neutrality is an emerging goal. Dining services can provide local food and reduce waste. Institutional purchasing can favour sustainable products in institutional purchasing. Campuses can use alternative fuels and vehicles, and encourage bicycling, walking, and public transportation.

- Curricula are adding issues of sustainability and enabling students to earn course credits while working on campus and community sustainability projects.

- National and international organizations have emerged to help students and others on post-secondary campuses work toward sustainable solutions.

Discuss the need and urgency for action on behalf of the environment, and the tremendous human potential to solve problems

- Time for turning around our increasing environmental impacts is running short. The Earth systems on which we depend are changing very quickly.

- Canada and other nations have met tremendous challenges before, so we have reason to hope that we will be able to attain a sustainable society.

TESTING YOUR COMPREHENSION

1. What do environmental scientists and others mean by *sustainable development*?
2. Describe three ways in which environmental protection can enhance economic well-being.
3. Why are many people now living at the highest level of material prosperity in history? Is this level of consumption sustainable? How can it feel good to consume less?
4. In what ways can technology help us achieve sustainability? How do natural processes provide good models of sustainability for manufacturing? Provide examples.
5. Why do many people feel that local self-sufficiency is important? What consequences of globalization may threaten sustainability? How can open democratic societies help promote sustainability?
6. In what ways are campus sustainability efforts relevant to sustainability efforts in the broader society?
7. Name one way in which campus sustainability proponents have addressed each of the following areas:
   (a) recycling and waste reduction
   (b) "green" buildings
   (c) water conservation
   (d) energy efficiency
   (e) renewable energy
   (f) global climate change

8. Name one way in which campus sustainability proponents have addressed each of the following areas:
   (a) dining services
   (b) institutional purchasing
   (c) transportation
   (d) habitat restoration
   (e) curricula

9. Explain Edward O. Wilson's metaphor of the "environmental bottleneck."

10. How can thinking of Earth as an island help prevent us from repeating the mistakes of previous civilizations?

THINKING IT THROUGH

1. Choose one item or product that you enjoy, and consider how it came to be. Think of as many components of the item or product as you can, and determine how each of them was obtained or created. What steps were involved in creating your item's components, and where did the raw materials come from? How was your item manufactured? How was it delivered to you? Compare your findings to the examples in **FIGURE 22.4** and **FIGURE 18.11**. Did you forget any components or steps?

2. Reflect on the experiences of prior human civilizations and how they came to an end. What is your prognosis for our current human civilization? Do you see a vast world of independent cultures all individually responsible for themselves, or do you consider human civilization to be one great entity? If we accept that all people depend on the same environmental systems for sustenance, what resources and strategies do we have to ensure that the actions of a few do not determine the outcome for all and that sustainable solutions are a common global goal?

3. You have been elected president of your student government, and your school's administrators promise to be responsive to student concerns. Many of your fellow students are asking you to promote sustainability initiatives on your campus. Consider the many approaches and activities pursued by the colleges and universities mentioned in this chapter, and now think about your own school. Which of these approaches and activities are most needed at your school? Which might be most effective? What ideas would you prioritize and promote during your term as president?

4. In our final "Think It Through" question, you are . . . *you!* In this chapter and throughout this book, you have encountered a diversity of ideas for sustainable solutions to environmental problems. Many of these are approaches you can pursue in your own life. Name at least five ways in which you think you can make a difference—and would most like to make a difference—in helping to attain a more sustainable society. For each approach, describe one specific thing you could do today or tomorrow or next week to begin.

INTERPRETING GRAPHS AND DATA

As we have seen throughout this book, individuals can contribute to sustainable solutions for our society and our planet in many ways. Some of these involve advocating for change at high levels of government or business or academia. But plenty of others involve the countless small choices we make in how we live our lives day to day. Where we live, what we buy, how we travel—these types of choices we each make as citizens, consumers, and human beings determine how we affect the environment and the people around us. As you know, such personal choices are summarized (crudely, but usefully) in an ecological footprint.

Visit an online ecological footprint calculator and calculate your personal ecological footprint. Examples of sites with online footprint calculators include **www.myfootprint.org** (requires a subscription) and **www.footprintnetwork.org**.

1. Enter your current footprint, as determined by the online calculator, in the table. What changes have you made in your lifestyle since beginning this course that influence your environmental impact?

2. How does your personal footprint compare with the average footprint of a Canadian resident? How does it compare with that of the average person in the world? What do you think would be an admirable yet realistic goal for you to set as a target value for your own footprint?

3. Now think of three changes in your lifestyle that would decrease your footprint. These should be changes that you would like to make and that you believe you could reasonably make. Take the footprint quiz again, incorporating these three changes. Enter the resulting footprint in the table.

4. Now set a goal of reducing your footprint by 25%, and experiment by changing various answers in your footprint quiz. What changes would allow you to attain a 25% reduction in your footprint? What types of activities is your footprint the most sensitive to, and which changes make the biggest difference in reducing your footprint? What changes would be needed to reduce your footprint to the hypothetical target value you set in Question 2?

	Footprint value* (hectares per person)
World average	2.7
Canadian average	7.0
Your footprint	
Your footprint with three changes	

*These numbers vary widely depending on which calculator is used. Based on the data from Global Footprint Network, **www.footprintnetwork.org.**

MasteringEnvironmentalScience®

STUDENTS
Go to **MasteringEnvironmentalScience** for assignments, the eText, and the Study Area with practice tests, videos, current events, and activities.

INSTRUCTORS
Go to **MasteringEnvironmentalScience** for automatically graded activities, current events, videos, and reading questions that you can assign to your students, plus Instructor Resources.

Appendices

Chapter 1

Figure 1.7: In graph **(b)**, the components of the world ecological footprint are all shown relative to the biocapacity, which is presented as constant. In other words, the graph shows the amount by which EF differed from biocapacity in a given year.

Chapter 2

Table 2.1: The most abundant element in the crust by volume is oxygen, by far. Oxygen accounts for over 94% of the crust by volume. (That's why well-known geochemist Brian Mason, who contributed greatly to what is known about the overall composition of the crust, said that the crust could just as well be called the "oxysphere.")

This question is actually considerably trickier than it might seem—especially if you want to carry out an exact calculation—but there are several lines of reasoning that you can use to make a rough estimate. The density of silicon is 2.3296 g/cm³ and that of oxygen is 0.001429 g/cm³ (from the Periodic Table), which means that oxygen is *much* less dense than silicon—we can reason that there must be a lot more atoms of oxygen than silicon for oxygen to account for almost half of the mass of the crust! Note, however, that these densities are measured at standard temperature and pressure—conditions that are nothing like the conditions below Earth's surface.

Oxygen actually has a *smaller* atomic radius than silicon (60 pm compared to 111 pm, from the Periodic Table), which might suggest that it takes up less volume than silicon. However, the atomic radii are not relevant here because the atoms are tightly bonded to one another, and their "size" is a function of the lengths of the bonds that join them. Oxygen and silicon commonly occur in the form of the anionic oxide compound SiO_4^{4-}. (In mineral structures, two of the four oxygens are commonly shared with adjacent complexes; this results in the chemical formula SiO_2, which is how the silica component of minerals is most commonly denoted.) The respective ionic radii of the oxygen and silicon in these structures depend on the bond lengths within the compound. So the relevant measure of "size" is the one given in the hint—the ionic radius—and by that measure, oxygen is more than twice as large as silicon.

So to summarize, we know that oxygen is much more abundant than silicon in the crust by weight; that there are far more oxygen atoms present in the crust to account for this mass, based on the low density of oxygen; and that oxygen atoms take up more volume in compounds bonded with silicon, which are common in the crust. All of these facts point to one answer: Oxygen takes up much more volume in the crust than silicon does.

Figure 2.8: A substance with pH = 6 contains 10 times as many hydrogen ions as a substance with pH = 7, and a substance with pH = 5 contains 100 times as many hydrogen ions as one with pH = 7.

Chapter 3

Figure 3.6: The width of the arrows in each figure represents magnitude, so the wider the arrow, the larger the value. For chemical energy, the widest arrow goes from producers to detritus. This means that the largest directional flux of chemical energy in the ecosystem is from producers to detritus. For nutrients, the arrow widths show that the largest flux, by far, is from detritus to producers.

Figure 3.10 (i): The trend from lower left to upper right shows that processes that operate on the smaller end of the spatial scale (i.e., local processes) tend to be shorter in time scale, too. On the other hand, processes that are greater in spatial extent (i.e., regional and global processes) tend to have longer characteristic time scales.

Figure 3.10 (ii): Soil erosion can happen pretty quickly; if vegetation is removed or lost from a forested area, there can be significant loss of topsoil in just a few years. In contrast, it can take thousands or even tens of thousands of years for soil to form on bedrock, depending on the climate and other physical environmental conditions. This is problematic, and it raises management concerns. Because soil can be eroded so quickly but takes so long to reestablish itself, it is extremely important to protect soil resources and to prevent—or at least not promote—accelerated soil erosion.

Figure 3.15: No, precipitation to and evaporation from the ocean don't balance. There are two main reasons for this. (1) Not surprisingly, there is a lot more evaporation of water from the surface of the ocean than there is from the land surface. If you use Precipitation + Runoff to represent the inputs of water into the ocean, do they now balance with Evaporation? (2) The other reason that the numbers sometimes don't balance is that diagrams like this are models of enormous, highly complicated, global Earth processes. It is

very difficult to determine the exact magnitude of fluxes operating on this magnitude in temporal and spatial scales. Some of the numbers are necessarily estimates.

Chapter 4

Figure 4.14: A human being has the highest rate of survival at a young age. As shown by the type I survivorship curve, the vast majority of individuals survive during youth, and then mortality rates rise during older age.

Figure 4.17: The St. Paul population of *R. tarandus* crashed in the span of just one generation. The fluctuating population cycles illustrated for *E. sexmaculatus* lasted, on average, just over 15 days—less than one generation.

Chapter 5

Figure 5.10: In the generalized example shown here, there is a ratio of 1:10 between levels in the trophic pyramid. Thus, if there are 3000 grasshoppers we can expect to find approximately 300 rodents in the system (300 ÷ 3000 = 1:10).

The Science behind the Story, Figure 2: The mudflow substrate gained the highest species richness, as seen by the fact that its curve is highest on the graph in part **(a)**. The maximum number of species that it had in any one year was 22 or 23, about 17–18 years after the eruption. The pumice substrate showed the slowest increase in percent plant cover, as seen by the fact that its curve is the lowest on the graph in part **(b)**.

Chapter 6

Central Case, Figure 1: Populations have momentum. Even though the *rate* of growth decreased after 1980, the people of childbearing age and younger who were alive at that time still grew up and reproduced, contributing to population growth. Note, too, that the growth rate is positive (above zero) until about 2030, meaning that growth in the absolute population number will still be occurring. After 2070 the rate of growth is projected to begin to rise again, but it is still below zero so the absolute number in the population will continue to decline.

Figure 6.2: Answers for the first two questions will vary. Read the year and the number directly from the graph. Note that the population axis is in millions, rather than billions. According to this projection and graph, the growth rate will drop to 1%/yr in about 2018. Reading from the graph, the population that year will be about 7.5 billion (7500 million). 7500 × 0.01 = 75, so about 75 million people will be added to the world's population that year.

The Science behind the Story, Figure 1: In 1860 the average household size in Canada was about 4.2 people. By 2000, it had decreased to about 2.2 people. This is a decrease of almost 50%: [(4.2 − 2.2) ÷ 4.2] × 100 = 47.6%

Figure 6.13: The one-child policy was enforced in 1980. In 2010, children born at that time would be in the 30-year-old cohort. In 2050, they would be in the 70-year-old cohort.

Figure 6.17: It is a negative correlation. As female literacy increases, birth rate decreases. It is a statistical correlation; it does not, on its own, demonstrate or explain how an increase in female literacy would cause a decrease in birth rate.

Chapter 7

Figure 7.5: Following the white lines inward from these three values along the edges of the triangle, they come together at a point within the area labelled "silt loam." A soil with these relative proportions of particles would be classified as a silt loam.

Figure 7.18: In the diagram, arrows point from causes to consequences. Thus, according to the diagram, the immediate cause of exposure of bare topsoil is the removal of native grasses. The immediate consequence of exposing bare topsoil is wind and water erosion. Four arrows lead away from wind and water erosion to other items, so four immediate consequences of wind and water erosion are shown on the diagram.

Chapter 8

Figure 8.3: Follow the "2.0" line until it intersects with the item that you are interested in, and then straight down to find the year. The 1960 population doubled by the year 2000. (You can check this by looking up population statistics: The population in 1960 was over 3 billion; by 2000 it was about 6.1 billion.) The 1960 grain production doubled by 1984.

Figure 8.12: The value of rice production in 1986 was roughly 26.5 metric tons. By 1989, when pesticide subsidies were discontinued, production had increased to about 29.5 metric tons. The percent increase was $(29.5 - 26.5) \div 26.5 = 3 \div 26.5 = 11.3\%$.

Chapter 9

Figure 9.3: The right portion of the figure shows that there are 4680 species of mammals. The left portion of the figure shows that there are about 1 750 000 species in total that have been described. This means that mammals comprise just 0.27% of all organisms: $(4680 \div 1\ 750\ 000) \times 100$. In reality, the percentage is actually much lower than this, because virtually all mammal species have already been discovered, but most species of other types of organisms have not yet been discovered.

Chapter 10

Figure 10.16: Insects (19%) + fires (9%) = 28% and $0.28 \times 46\ 000$ kha = approximately 12 900 kha deforested by these causes. Urban settlement (54%) = $0.54 \times 46\ 000$ kha = approximately 25 000 kha. Areas deforested by natural causes such as fire or insect infestation are the most likely to regenerate over time. Areas deforested by human activities, especially urban settlement, agriculture, infrastructure, and oil and gas extraction, are the least likely to regenerate over time.

Chapter 11

Figure 11.1: $20\% \times 2.5\%$ (the proportion of total water that is fresh, rather than salt water) = 0.005%. When you consider that much of this ground water is stored too deep underground to be easily accessed, it becomes clear how rare accessible fresh water actually is.

Figure 11.7: As shown in the diagram, only indigo or violet light is able to penetrate clear water below a depth of about 46 m. By this definition, all of the Great Lakes should have an aphotic or profundal zone. The profundal zone in Lake Erie should be very limited in geographical extent, because the average depth of the lake is much shallower than 46 m.

All lakes have a benthic zone. "Benthic" refers to the bottom of the water body, and is not linked to depth in any specific way. A benthic zone can be very shallow (as it would be in the littoral zone) or very deep (as it would be in the profundal zone).

Chapter 12

Figure 12.1: TAC in 1973 was about 670 000 t. The reported catch was about 355 000 t. With the actual reported catch *so* much lower than the allowable catch, there were only two possible conclusions: (1) Fishers were putting less effort* into their fishing, so they were *catching* fewer fish; or (2) the fish were simply not there to be caught. The second conclusion turned out to be the correct one. (*Note: "Effort" is defined very broadly in the context of fishing, and it can include hours spent, distance travelled, equipment and technologies utilized, and other factors.)

Figure 12.5: Multiplying 1000 g by 3.5% (0.035) reveals that there are 35 g of salts in the 1000 g sample of sea water in the beaker. To determine the grams of negatively charged ions in the sample, sum the values for chloride (1.9%), sulphate (0.3%), and bicarbonate (0.01%) to find that 2.21% of the sample (or 22.1 g) consists of such ions.

Figure 12.7: Released off the southeast coast of Japan, the buoy would be carried northeast by the Kuroshiro Current and then eastward across the ocean on the North Pacific Current. Upon reaching North America, it could turn southward on the California Current, or it could turn northward toward Alaska and then return to Japan via the North Equatorial Current. So, although Japan is closer to Australia, the buoy would most likely reach North America first.

Chapter 13

Figure 13.2: From +15°C to −57°C is a drop of 72°C. To decrease by 72°C in 11 km, we divide 72 by 11 to get a decrease of 6.5°C per km. This simple change of temperature with height is called the environmental lapse rate.

Figure 13.3: The tropopause is the boundary between the troposphere and the stratosphere. Its average altitude is 11 km (although this varies, influenced by a number of factors, including topography). Reading from the graph, at a height of 11 km the atmospheric pressure should be about 200 mb.

Figure 13.13: This station monitors O_3, SO_2, NO_2, and $PM_{2.5}$ on an hourly basis. The best AQI result shown here is for December 24 (AQI = 8, or Very Good). The worst reading is for December 25 (AQI = 24, or Good). Ozone (ground-level, or tropospheric ozone) is typically the main air pollutant in southern Ontario, followed by fine particulate matter ($PM_{2.5}$).

Chapter 14

Figure 14.2: Atmospheric CO_2 was about 280 ppm in 1750; in 2015 it was 400 ppm. This is an increase of about 43%. This is how to solve the question: $(400 - 280) \div 280 \times 100$.

Figure 14.3: Changing land use accounts for about 6 billion metric tons of carbon dioxide emissions per year, and industry accounts for about 26 billion metric tons. Therefore, industry accounts for $26 \div 6 = 4.3$ times as much carbon dioxide. In other words, for every metric ton of CO_2 released as a result of land use change, 4.3 metric tons are released by industrial processes.

Chapter 15

Figure 15.1: Answers will vary. Start by figuring out the difference (that is, the increase) by subtracting oil use (red line) in the year of your parent's birth from the present-day use (4.2 billion tonnes, reading from the same curve, far right-hand side of the diagram). Divide the difference by the earlier number, and multiply the answer by 100 to get the percent increase. For example, if your parent were born in 1960, the answer would be 4.2 billion tonnes (current oil use) minus 1 billion tonnes (1960 oil use) = 3.2 billion tonnes difference, and $(3.2 \div 1) \times 100 = 320\%$ increase. For someone born in 1995, the answer would be $4.2 - 3.2$ = a difference of 1 billion tonnes, and $(1 \div 3.2) \times 100 = 31\%$ increase.

Figure 15.2: Reading from the graph in Figure 15.2B, total energy use per capita in Canada is about 9.4 Mtoe per person per year, and the world total per capita is about 1.8. So energy use in Canada is $9.4 \div 1.8 = 5.2$, more than 5 times higher than the world average.

Table 15.5: Countries that are among the top consumers but not among the top producers for coal are Japan and the Russian Federation. Japan and China are among the top consumers of natural gas. Japan and India are among the top consumers (but not the top producers) of oil. Nations that are on the "top five" list as consumers but not as producers are not necessarily net importers of energy. For example, they might be importing more of one type of energy and exporting more of other types.

Chapter 16

Figure 16.18: The yearly growth rate shown for PV solar is 60%, so you would multiply 10 starting units by 1.60, then multiply that number by 1.60, and so on for five years. The result is 104.9 units.

Figure 16.20: Biodiesel reduces sulphates most effectively, compared to petrodiesel, for both B20 and B100.

Chapter 17

Figure 17.11: At current rates of production and for known global reserves, as shown in the graph, nickel has the highest proportion that is economically extractable, compared to the reserve as a whole. Nickel has technically recoverable reserves (the whole bar) that would last about 54 years at current rates of production; the dark red bar shows that the proportion that is currently economically extractable would last about 34 years at current rates of production: $34 \div 54 = 0.63$ and multiply by 100 to express it as a percentage (63%). Molybdenum would be second, with relative proportion of economically to technically extractable reserves of $41 \div 73$, or 56%. Lead has the greatest technically recoverable reserves (about 366 years' worth) and the smallest proportion that is economically extractable (about 16 years): $16 \div 366 = $ about 4% of the reserve that is economically extractable today.

Chapter 18

Figure 18.4A: The population increased over that time. You can tell this because the curve for municipal solid waste (MSW) per person declined; so has the curve for total solid waste generation, but not as quickly. This means that the population must have increased.

Figure 18.4B: In 2010, the total MSW for the United States and the EU were about the same, but the population of the United States was half that of the EU, so the rate of waste generation per person would be about twice as high in the United States as in the EU. They would both dwarf the per capita waste generation rate for Asia, where the population was more than 13 times as large as that of the United States in 2010.

Table 18.5: The two improper disposal methods shown on the table are "put in the garbage" or "poured down the drain." (For medication, it could be argued that giving them away is also an improper practice, but no one took that route for disposal in this particular study. "Still had them" can even be considered an improper practice for hazardous materials such as paints and solvents.) Proportionately, the dead compact fluorescent lights (CFLs) were most often disposed of improperly (thrown in the garbage 56% of the time); this is problematic because CFLs contain mercury, which can enter leachate in landfills. The second category most commonly disposed of improperly was other dead batteries (aside from car batteries), which were put into the garbage 42% of the time; this is problematic because batteries contain acids and heavy metals such as cadmium and lithium.

Chapter 19

Figure 19.3: See completed table.

1. The death rate (deaths per total population) in 2000 was 52 806 000 ÷ 6 120 000 000 = 0.86%. The death rate in 2012 was 55 859 000 ÷ 7 080 000 000 = 0.79%. So the death rate has decreased, even though the absolute number of deaths has increased as a result of population increase.

2. The percentage of deaths caused by coronary heart disease and stroke has increased from 22% in 2000 to 25.1% in 2012. HIV/AIDS has decreased from 3.2% in 2000 to 2.8% in 2012 (and the absolute number of deaths decreased, as well). Diarrheal diseases decreased significantly, from 4.1% in 2000 to 2.7% in 2012, and the absolute number decreased, too. Diarrheal diseases are communicable, and they are often caused by contaminated water, dirty hands, and unwashed utensils. This suggests that perhaps there has been an improvement in the supply of clean water for drinking and hygiene in parts of the world where these diseases are still common.

	2000			2012	
Cause	**Deaths (000s)**	**% Deaths**	**Cause**	**Deaths (000s)**	**% Deaths**
All causes	52 806	100.0	**All causes**	55 859	100.0
Coronary artery disease and stroke	11 636	22.0	Coronary artery disease and stroke	14 027	25.1
HIV/AIDS	1 678	3.2	HIV/AIDS	1 534	2.8
Diarrheal diseases	2 171	4.1	Diarrheal diseases	1 498	2.7

Figure 19.13: Curve (c) is the best choice. It shows the negative impacts of deficiency (at left); increasing benefits with increasing dose, to a limit (centre); and increasing harm with much higher doses (at right).

Chapter 20

Figure 20.11: The curve would be a line that is located higher than the existing blue "cost" curve. The costs to society of producing this item could be things like excess pollution or resource depletion. The optimal level of resource use or pollution is read by drawing a vertical line from the point where the new blue line intersects the red line. The optimal level of resource use or pollution will be lower, in this case.

Figure 20.13: ISEW begins to decline in about 1983, and GPI a bit later in 1990. GDP continues to increase, with a small dip in the early 1990s. By 2010, GDP is about four times higher than ISEW, and about five times higher than GPI.

Chapter 21

Figure 21.12: The Province of Saskatchewan SOER uses 17 indicators, organized as follows:

■ Air
  • Air pollutant concentrations
  • Air pollutant volumes
  • Air zone management

■ Climate
  • Greenhouse gas emissions

■ Land
  • Agricultural land cover
  • Mineral disposition activities
  • Area under zero-tillage
  • Private land stewardship
  • Waste recycling

■ Forest
  • Forest age classes
  • Forest wildlife disturbance
  • Forest insect diseases and disturbance
  • Proportion of sustainable harvest level utilized
  • Forest regeneration

■ Water
  • Surfacewater quality
  • Surfacewater quantity
  • Water consumption and conservation

The five categories in the Saskatchewan framework compare to two in the EPI (environmental health and ecosystem vitality). The indicators from the Saskatchewan SOER are missing the following areas, which are covered by the EPI indicators: health impacts, sanitation (a bigger concern for developing nations than for Saskatchewan), fisheries (Saskatchewan is landlocked, so only freshwater fisheries and aquaculture are potentially involved), energy use, and biodiversity and habitat (partly captured in the forest and land categories in the Saskatchewan framework).

The EPI indicator set is missing the following areas, which are covered in the Saskatchewan framework: GHG emissions and some details of air pollution (which may be grouped together under the air quality category in EPI), mineral disposition activity, private land stewardship, and perhaps a few others.

Both sets of indicators have strengths and weaknesses. The indicator framework used by Saskatchewan is better suited to the province's situation in some respects, but probably wouldn't do as good a job at capturing the differences among nations, which is the intention of the EPI indicators.

Chapter 22

Figure 22.3: According to this analysis, a hectare of intact wetland is worth about $5800 in Canada (once external costs and benefits are considered; this is not the same as the real estate property value), whereas a hectare of intensive farmland is worth about $2250. Thus, a hectare of wetland is worth about $3550 more in absolute terms, or about 58% more in percentage terms.

Figure 22.6: Taking the 1994 value of the Index as a baseline, we see that the environment subdomain decreased from 100 to about 89% on the graph. So, it decreased by 11% over the period that saw an increase of 29% in GDP, and almost 6% in overall well-being as measured by the eight sets of indicators taken together. Do you think the increase in income and overall well-being are sustainable, if the environment subdomain continues to decline? If you are interested in learning more about how the Canadian Index of Wellbeing is researched and calculated, you should visit their website at **https://uwaterloo.ca/canadian-index-wellbeing/**.

The Science behind the Story, Table 1: Set up the calculation like this, denoting the old score as variable "x":

$$x + (0.2179x) = 21.74$$

Solve for x. You will find that Sierra Leone's score would have been 17.85 10 years ago. What accounts for the positive change? You will need to investigate the country more deeply if you want to find out.

Some Basics on Graphs

Presenting data in ways that help make trends and patterns visually apparent is a vital part of the scientific endeavour. For scientists, businesspeople, and others, the primary tool for expressing patterns in data is the graph. Thus, the ability to interpret graphs is a skill that you will want to cultivate. This appendix guides you in how to read graphs, introduces a few vital conceptual points, and surveys the most common types of graphs, giving rationales for their use.

Navigating a Graph

A graph is a diagram that shows relationships among *variables,* which are factors that can change in value. Many common graphs relate values of a *dependent variable* to those of an *independent variable.* A dependent variable is so named because its values "depend on" the values of an independent variable. In other words, as the values of an independent variable change, the values of the dependent variable change in response. In a manipulative experiment, changes that a researcher specifies in the value of the independent variable *cause* changes in the value of the dependent variable. In observational studies, there may be no causal relationship, and scientists may plot a correlation. In either case, the values of the independent variable are known or specified, and the values of the dependent variable are unknown and are what we are interested in observing or measuring.

By convention, independent variables are generally represented on the horizontal axis, or *x*-axis, of a graph, while dependent variables are represented on the vertical axis, or *y*-axis. Numerical values of variables generally become larger as one proceeds rightward on the *x*-axis or upward on the *y*-axis. In many cases, independent variables are not numerical at all, but categorical. For example, in a graph presenting population sizes of several nations, the nations comprise a categorical independent variable, whereas population size is a numerical dependent variable.

As a simple example, **FIGURE B.1** shows data from a scientist who ran an experiment to test the effects on laboratory mice of the chemical bisphenol A. The *x*-axis shows values of her independent variable, the dose of bisphenol A given to the mice. The values are expressed in units of nanograms per gram of water (ng/g). The researcher was interested in what proportion of mice

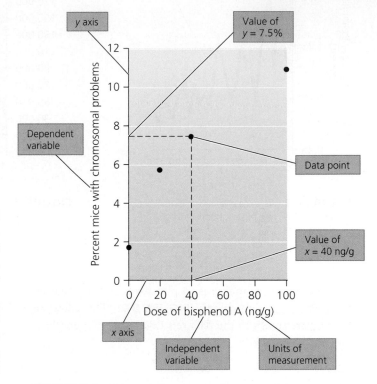

FIGURE B.1 This graph illustrates the frequency of chromosomal problems relative to dose of bisphenol A.

developed chromosomal problems as a result. Thus, her dependent variable, presented on the *y*-axis, is the percentage of mice found to develop chromosomal problems. For each of four doses of bisphenol A she supplied in her experiment, a data point on the graph is plotted to show the corresponding percentage of mice with chromosomal problems.

Now that you're familiar with the basic building blocks of a graph, let's survey the most common types of graphs you'll see, and examine a few vital concepts in graphing.

GRAPH TYPE: Line Graph

A line graph is drawn when a data set involves a sequence of some kind, such as a series of values that occur one by one and change through time or across distance (**FIGURE B.2**). Line graphs are most appropriate when the *y*-axis expresses a continuous numerical variable and the *x*-axis expresses either continuous numerical data or discrete sequential categories (such as years).

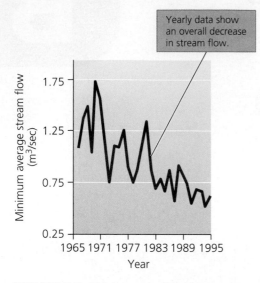

FIGURE B.2 This is a graph of minimum average stream flow per year for a particular stream.

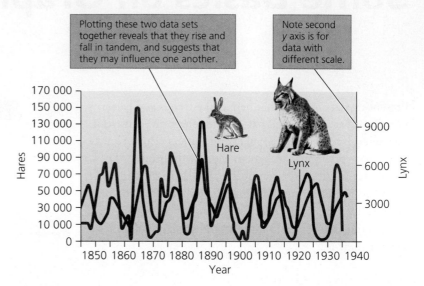

FIGURE B.3 Population fluctuations are shown here for both hare and lynx.

One useful technique is to plot two or more data sets together on the same graph (**FIGURE B.3**). This allows us to compare trends in the data sets to see whether and how they may be related.

KEY CONCEPT: Projections

Besides showing observed data, we can use graphs to show data that is predicted for the future, based on models, simulations, or extrapolations from past data. Often, projected future data on a line graph are shown with dashed lines, as in **FIGURE B.4**, to indicate that they are less certain than data that have already been observed.

GRAPH TYPE: Bar Chart

A bar chart is most often used when one variable is categorical and the other is numerical. In such a chart, the height (or length) of each bar represents the quantitative value of a given category of the categorical variable; longer bars mean larger values (**FIGURE B.5**). Bar charts allow us to visualize how a variable differs quantitatively among categories.

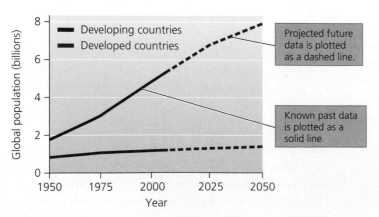

FIGURE B.4 The graph shows past and projected future population growth for developing and developed countries.

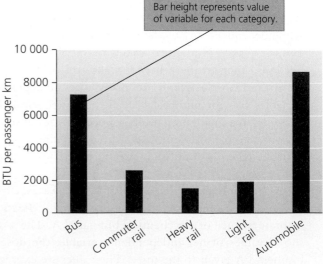

FIGURE B.5 The bars show energy consumption for different modes of transit, in units of BTU (British thermal units, a measure of power) per km travelled.

It is often instructive to graph two or more data sets together to reveal patterns and relationships. A bar chart, such as **FIGURE B.6**, allows us to compare two data sets (oil production and oil consumption) both within and among nations. A graph that does double duty in this way allows for higher-level analysis (in this case, suggesting which nations depend on others for petroleum imports). Most bar charts in this book illustrate multiple types of information at once in this manner.

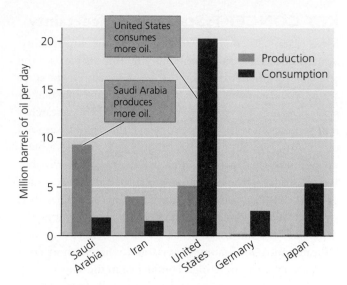

FIGURE B.6 The bars show oil production and consumption by selected nations.

FIGURE B.7 illustrates two ways in which a bar chart can be modified. First, note that the orientation of bars is horizontal instead of vertical. In this configuration, the categorical variable is along the *y*-axis and the numerical variable is along the *x*-axis. Second, the bars can extend in either direction from a central *x*-axis value of zero, representing either positive (right) or negative (left) values. Depending on the nature of one's data and the points one wants to make, sometimes such arrangements can make for a clearer presentation.

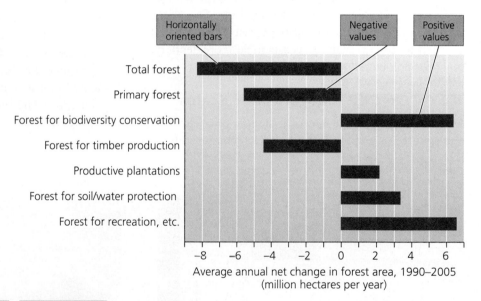

FIGURE B.7 The bars show loss (negative net change) or gain (positive annual net change) of forests in millions of hectares per year, categorized by forest type.

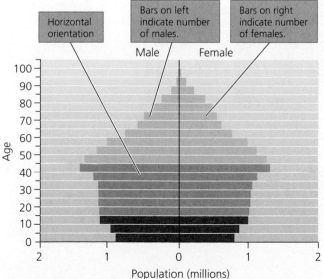

FIGURE B.8 The age structure of Canada's population in 2005 is shown in this population pyramid or age pyramid.

One special type of horizontally oriented bar chart is the age pyramid used by demographers (**FIGURE B.8**). Age categories are displayed on the *y*-axis, with bars representing the population size of each age group varying in their horizontal length.

KEY CONCEPT: Statistical Uncertainty

Most data sets involve some degree of uncertainty. Sometimes exact measurements are impossible, so the researcher estimates the likely range of measurement error around the data point. Other times, data points represent the *mean* (average) of many measurements, and the researcher may want to show the degree to which the data vary around this mean. Mathematical techniques are used to obtain precise statistical probabilities for degrees of variation. Results from such analyses are represented in bar charts, such as **FIGURE B.9**, as thin black lines called *error bars*. These bars extend past and within the end of each bar, representing the degree of variation around the bar's value. Longer error bars indicate more uncertainty or variation, whereas short error bars mean we can have high confidence in the value.

The statistical analysis of data is critically important in science. In this book we provide a broad and streamlined introduction to many topics, so we often omit error bars from our graphs and details of statistical significance from our discussions. This is for clarity of presentation only; the research we discuss analyzes its data in far more depth than any textbook could possibly cover.

KEY CONCEPT: Logarithmic Scales

When data span a large range of values, it can help change the scale on a graph from the standard linear scale to a logarithmic scale. In a logarithmic scale, each equal unit of distance on an axis corresponds to a ratio of values rather than an additive increase in values. Most often, logarithmic scales advance by factors of 10. **FIGURE B.10** uses a logarithmic scale on its *y*-axis for a graph in which we also show a linear scale using white horizontal lines crossing the graph space. The choice of scale does not affect data values, but does drastically affect the appearance of lines or bars on a graph.

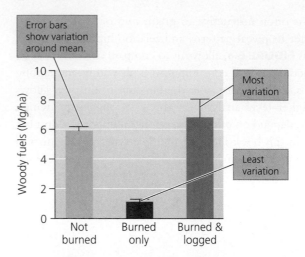

FIGURE B.9 The bars illustrate what happened to fine-scale woody debris left after treatments in salvage logging study.

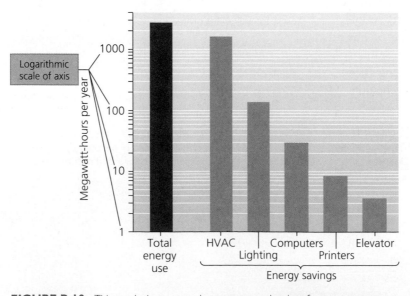

FIGURE B.10 This graph shows annual energy use and savings from a campus sustainability study, from various sources, in units of megawatt-hours per year on a logarithmic scale.

GRAPH TYPE: Scatter Plot

A scatter plot is often used when there is no sequential aspect to the data, when a given x-axis value could have multiple y-axis values, and when each data point is independent, having no particular connection to other data points (**FIGURE B.11**). Scatter plots allow us to visualize a broad positive or negative correlation between variables.

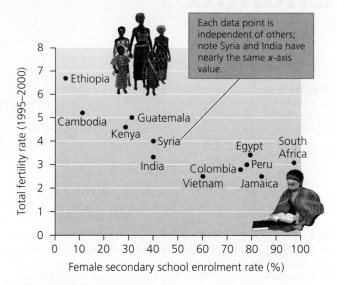

FIGURE B.11 This graph illustrates how fertility rate varies with female education in various countries. A correlation is shown, but the question of causality must be investigated and interpreted by the researcher.

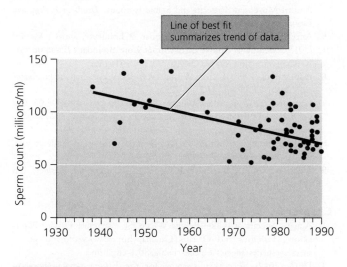

FIGURE B.12 The graph illustrates sperm count in men across the world, with a "best fit" line superimposed.

A "line of best fit" may be drawn through the data of a scatter plot in order to make a trend in the data clearer to the eye (**FIGURE B.12**). These lines are not drawn casually, however; their placement and slope are determined by precise mathematical analysis of the data through a statistical technique called linear regression.

GRAPH TYPE: Pie Chart

A pie chart is used when we wish to compare the proportions of some whole that are taken up by each of several categories (**FIGURE B.13**). A pie chart is appropriate when one variable is categorical and one is numerical. Each category is represented visually like a slice from a pie, with the size of the slice reflecting the percentage of the whole that is taken up by that category.

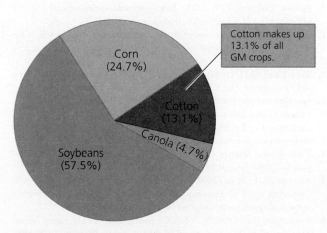

FIGURE B.13 This pie chart illustrates the relative proportions of genetically modified crops grown worldwide, by type.

Endnotes

Chapter 1

1. NASA, The *Apollo 8* Christmas Eve Broadcast, http://nssdc.gsfc.nasa
.gov/planetary/lunar/apollo8_xmas.html.
2. Information about the Blue Marble and other photographs of Earth
from space: http://earthobservatory.nasa.gov/Features/BlueMarble/
BlueMarble_history.php.
3. Meadows, D., D.L. Meadows, J. Randers, and W.W. Behrens, III,
(1972). *Limits to Growth*, New York: Universe Books.
4. Ehrlich, P.R. (1968). *The Population Bomb*, Sierra Club-Ballantine Books.
5. Schumacher, E.F. (1973). *Small Is Beautiful: Economics as if People Mat-
tered*, New York: Harper & Row.
6. Carson, R. (1962). *Silent Spring*, Houghton Mifflin.
7. McCormick, J. (1995). *The Global Environmental Movement*, 2nd edition,
Chichester: John Wiley & Sons.
8. Environment Canada > About Environment Canada > Our Mandate,
http://ec.gc.ca/default.asp?lang=En&n=BD3CE17D-1.
9. Sagan, C. (1995). *The Demon-Haunted World: Science as a Candle in the
Dark*, Random House/Ballantine Books.
10. Flenley, J., and P. Bahn (2003). *The Enigmas of Easter Island,* Oxford
University Press.
11. Orliac, C., and M. Orliac (1995). *The Silent Gods: Mysteries of Easter
Island*, Thames and Hudson.
12. See, for example, Steadman, D.W., P. Vargas, and C. Cristino (1994).
Stratigraphy, Chronology, and Cultural Context of an Early Faunal
Assemblage from Easter Island, *Asian Perspectives, 33*(1), pp. 79–96.
13. Brander, J.A., and M.S. Taylor (1998). The Simple Economics of Easter
Island: A Ricardo-Malthus Model of Renewable Resource Use, *The
American Economic Review, 88*(1), pp. 119–138.
14. Hunt, T.L., and C.P. Lipo (2011). *The Statues That Walked: Unraveling the
Mystery of Easter Island,* Simon & Schuster.
15. Diamond, J. (2005). *Collapse: How Societies Choose to Fail or Succeed*,
Penguin Books.
16. Hardin, G. (1968). The Tragedy of the Commons, *Science, 162*(3859),
pp. 1243–1248.
17. The United Nations officially marked the "Day of Seven Billion" on
October 31, 2011.
18. United Nations Department of Economic and Social Affairs, Population Division, http://esa.un.org/unpd/wpp/.
19. Eriksen, M., S. Mason, S. Wilson, C. Box, A. Zellers, W. Edwards,
H. Farley, and S. Amato (2013). Microplastic pollution in the surface
waters of the Laurentian Great Lakes, *Marine Pollution Bulletin*, 77(1–2),
pp. 177–182.
20. Global Footprint Network (2013). *National Footprint Accounts, 2012
Edition*, Oakland, CA, USA, http://www.footprintnetwork.org/en/
index.php/GFN/page/footprint_data_and_results.
21. Global hectare (gha) is a unit for the measurement of biocapacity. One
gha measures the average productivity of all biologically productive
areas in a given year.
22. Brundtland, G.H. (1987). *Report of the World Commission on Environment
and Development: Our Common Future*, Oxford University Press.
23. See, for example, Intergovernmental Panel on Climate Change (2014).
Climate Change 2013: The Physical Science Basis, Contribution of Working Group I to the Fifth Assessment Report of the IPCC, World Meteorological Organization and United Nations Environment Programme.
24. Earth System Science Committee (1986). *Earth System Science: A Program for Global Change*, NASA Advisory Council.
25. Rockström, J., et al. (2009). Planetary Boundaries: Exploring the Safe
Operating Space for Humanity, *Ecology and Society, 14*(2), p. 32, http://
www.ecologyandsociety.org/vol14/iss2/art32/.
26. Rockström, J., et al. (2009). A Safe Operating Space for Humanity, *Nature,
461*, pp. 472–475 (24 September 2009), doi:10.1038/461472a; published online.

Chapter 2

1. Stimpson, I. (2011). Japan's Tohoku Earthquake and Tsunami, *Geology
Today*, 27(3), pp. 96–98.
2. Hammer, J. (2011). Aftershocks: The Powerful Earthquake That Struck
Yokohama and Tokyo on September 1, 1923, Traumatized a Nation
and Unleashed Historic Consequences, *Smithsonian, 42*(2), p. 50.
3. McCurry, J. (2012). Japan's Tohoku Earthquake: 1 Year On, *The Lancet,
379*, pp. 880–881.
4. See, for example: Hobson, K.A., J.F. Piatt, and J. Pitocchelli (1994).
Using Stable Isotopes to Determine Seabird Trophic Relationships,
Journal of Animal Ecology, 63, pp. 786–798.
Hobson, K.A. (1999). Tracing Origins and Migration of Wildlife Using
Stable Isotopes: A Review, *Oecologia, 120*(3), pp. 314–326.
Hobson, K.A. (2005). Stable Isotopes and the Determination of Avian
Migratory Connectivity and Seasonal Interactions, *The Auk, 122*(4),
pp. 1037–1048.
Hobson, K.A., S. Wilgenburg, Y. Ferrand, F. Gossman, and C. Bastat
(2013). A Stable Isotope (δ^2H) Approach to Deriving Origins of Har-
vested Woodcock (*Scolopax rusticola*) Taken in France, *European Journal of
Wildlife Research, 59*(6), pp. 881–892.
Rubenstein, D.R., and K.A. Hobson (2004). From Birds to Butterflies:
Animal Movement Patterns and Stable Isotopes, *Trends in Ecology and
Evolution, 19*(5), pp. 256–263.
5. Anzellini, S., A. Dewaele, M. Mezouar, P. Loubeyre, and G. Morard
(2013). Melting of Iron at Earth's Inner Core Boundary Based on Fast
X-Ray Diffraction, *Science, 340*(6131), pp. 464–466.
6. Anonymous (1993). J. Tuzo Wilson 1908–1993. *Eos, Transactions American Geophysical Union, 74*(2), p. 225.
7. Bowring, S.A., and I.S. Williams (1999). Priscoan (4.00-4.03 Ga)
Orthogneisses from Northwestern Canada, *Contributions to Mineralogy
and Petrology, 134*, p. 3.
8. See, for example, Koonin, E. (2014). The Origins of Cellular Life,
Antonie van Leeuwenhoek, 106(1), pp. 27–41.
9. See, for example, Buesseler, K.O. (2012). Fishing for Answers Off
Fukushima, *Science, 338*(6106), pp. 480–482.

Chapter 3

1. Aquatic Species at Risk—Beluga Whale (St. Lawrence Estuary Population), Fisheries and Oceans Canada, http://www.dfo-mpo.gc.ca/
species-especes/species-especes/belugaStLa-eng.htm.
2. DFO (2012). Recovery strategy for the beluga whale (*Delphinapterus leucas*) St. Lawrence Estuary population in Canada, *Species at
Risk Act,* Recovery Strategy Series, Fisheries and Oceans Canada,
Ottawa. 88 pp.
3. Martineau, D., et al. (2002). St. Lawrence Beluga Whales, the River
Sweepers? *Environmental Health Perspectives, 110*(10), pp. A562–A564.
4. Martineau, D., et al. (2002). Cancer in Wildlife, a Case Study: Beluga
from the St. Lawrence Estuary, Québec, Canada. *Environmental Health
Perspectives, 110*(3), pp. 285–292.
5. COSEWIC Committee on the Status of Endangered Wildlife in
Canada, Wildlife Species Search > Whale, Beluga, *Delphinapterus leucas*,
St. Lawrence Estuary population, http://www.cosewic.gc.ca/eng/sct1/
searchform_e.cfm.
6. For example, Zalasiewicz, J., et al. (2008) Are We Now Living in the
Anthropocene? *GSA Today, 18*(2), pp. 4–8.
7. Ramanujan, K. (2004). NASA Scientists Get Fix on Food, Wood, and
Fiber Use, *NASA Goddard Space Flight Center, News & Features > Earth*,
http://www.nasa.gov/vision/earth/environment/0624_hanpp.html.
See also Imhoff, M.L., L. Bounoua, T.H. Ricketts, C.J. Loucks,
R. Harriss, and W.T. Lawrence (2004). Global Patterns in Human
Consumption of Net Primary Production, *Nature, 429*(6994),
pp. 870–873.

8. For example, Schindler, David W. (1977) Evolution of Phosphorus Limitation in Lakes: Natural Mechanisms Compensate for Deficiencies of Nitrogen and Carbon in Eutrophied Lakes, *Science, 195,* pp. 260–262.
See also Chapter 11, *Science Behind the Stories: Near-Death Experience at the Experimental Lakes Area.*

9. Granéli, E., K. Wallström, U. Larsson, W. Granéli, and R. Elmgren (1990). Nutrient Limitation of Primary Production in the Baltic Sea Area, *Ambio, 19*(3), Marine Eutrophication, pp. 142–151.

10. Parrott, L., et al. (2011). A Decision Support System to Assist the Sustainable Management of Navigation Activities in the St. Lawrence River Estuary, Canada, *Environmental Modelling and Software, 26*(12), pp. 1403–1418.

11. DFO (2012). Recovery strategy for the beluga whale (*Delphinapterus leucas*) St. Lawrence Estuary population in Canada, *Species at Risk Act,* Recovery Strategy Series, Fisheries and Oceans Canada, Ottawa. 88 pp.

12. Nature Conservancy of Canada, What We Do > Resource Centre > GIS in Conservation, http://www.natureconservancy.ca/en/what-we-do/resource-centre/101s/gis_in_conservation_101.html. Copyright © 2015 Nature Conservancy of Canada (NCC). Reprinted by permission.

13. See Rabalais, N.N. (2011). Troubled Waters of the Gulf of Mexico. *Oceanography, 24*(2), p. 200.

14. See, for example, Rabalais, N.N., R.J. Díaz, L.A. Levin, R.E. Turner, D. Gilbert, and J. Zhang (2010). Dynamics and Distribution of Natural and Human-Caused Hypoxia, *Biogeosciences, 7*, pp. 585–619.
See also Rabalais, N.N. (1994). Comparison of Continuous Records of Near-Bottom Dissolved Oxygen from the Hypoxia Zone along the Louisiana Coast, *Estuaries, 17*(4), p. 850.

15. Scavia, D., et al. (2014). Assessing and Addressing the Re-Eutrophication of Lake Erie: Central Basin Hypoxia, *Journal of Great Lakes Research, 40*(2) pp. 226–246.

Chapter 4

1. Savage, J.M. (1966). An Extraordinary New Toad (*Bufo*) from Costa Rica, *Revista de Biologia Tropical*, pp. 153–167.

2. Savage, J., J. Pounds, and F. Bolaños (2008). *Incilius periglenes*. The IUCN Red List of Threatened Species. Version 2014.2. http://www.iucnredlist.org/details/3172/0. Last known male of the species was found in 1989; officially declared extinct in 2007.

3. U.S. Fish and Wildlife Service (2010). Hakalau Forest National Wildlife Refuge Comprehensive Conservation Plan.

4. Bambach, R.K., A.H Knoll, and S.C. Wang (2004). Origination, Extinction, and Mass Depletions of Marine Diversity, *Paleobiology, 30*(4), pp. 522–542.

5. Odum, E. P. (1959). *Fundamentals of Ecology*, W. B. Saunders Co., Philadelphia and London.

6. Sillett, T. S. (1994). Foraging Ecology of Epiphyte-Searching Insectivorous Birds in Costa Rica, *The Condor, 96*(4), pp. 863–877.

7. Excerpt from Sullivan, J. (2004) *Hunting for Frogs on Elston, and Other Tales from Field & Street*, Ed. V.M. Cassidy, University of Chicago Press, 320 pages.

8. Pounds, J.A. and M.L. Crump (1994). Amphibian Declines and Climate Disturbance: The Case of the Golden Toad and the Harlequin Frog, *Conservation Biology, 8*(1), pp. 72–85.

9. Pounds, J., et al. (2010). *Atelopus varius*. The IUCN Red List of Threatened Species, Version 2014.2. http://www.iucnredlist.org/details/54560/0.

10. Pounds, J.A. and M.L. Crump (1994). Amphibian Declines and Climate Disturbance: The Case of the Golden toad and the Harlequin Frog, *Conservation Biology, 8*(1), pp. 72–85.

11. Pounds, J.A., M.P.L. Fogden, and J.H. Campbell (1999). Biological Response to Climate Change on a Tropical Mountain, *Nature, 398*(6728), pp. 611–615.

12. Pounds, J.A, et al. (2006). Widespread Amphibian Extinctions from Epidemic Disease Driven by Global Warming, *Nature, 439*(7073), pp. 161–167.

13. Lips, K.R., et al. (2008). Riding the Wave: Reconciling the Roles of Disease and Climate Change in Amphibian Declines, *PLoS Biology, 6,* pp. 441–454.

14. La Marca, E., et al. (2005). Catastrophic Population Declines and Extinctions in Neotropical Harlequin Frogs (Bufonidae: *Atelopus*), *Biotropica 37*(2), pp. 190–201.

15. Kiesecker, J.M., et al. (2001). Complex Causes of Amphibian Population Declines, *Nature, 410*, pp. 681–684.

Chapter 5

1. Bardwaj, M. (2003). 'Museling' in on an Ecosystem: Tracing the Natural History of the Zebra Mussel in Lake Erie, *Canadian Geographic*, September.

2. Benson, A.J., D. Raikow, J. Larson, A. Fusaro, and A.K. Bogdanoff (2014). *Dreissena polymorpha*. USGS Nonindigenous Aquatic Species Database, Gainesville, FL. http://nas.er.usgs.gov/queries/FactSheet.aspx?speciesID=5, Revision Date: 6/26/2014.

3. Transport Canada. *Alien Invasive Species: Economic Impact*. https://www.tc.gc.ca/eng/marinesafety/oep-environment-ballastwater-alienspecies-1055.htm.

4. Environment Canada. *Aquatic Invasive Alien Species,* http://ec.gc.ca/grandslacs-greatlakes/default.asp?lang=En&n=90661FCF-1.

5. See, for example: Holland, R.E. (1993). Changes in Planktonic Diatoms and Water Transparency in Hatchery Bay, Bass Island Area, Western Lake Erie Since the Establishment of the Zebra Mussel, *Journal of Great Lakes Research, 19*(3), pp. 617–624.
Nicholls, K.H., and G.J. Hopkins (1993). Recent Change in Lake Erie (North Shore) Phytoplankton: Cumulative Impacts of Phosphorus Loading Reductions and the Zebra Mussel Introduction, *Journal of Great Lakes Research, 19*(4), pp. 637–647.

6. Paine, R. (1969). The Pisaster-Tegula Interaction: Prey Patches, Predator Food Preference, and Intertidal Community Structure, *Ecology, 50*(6), pp. 950–961.

7. See, for example: Estes, J.A. and D.O. Duggins (1995). Sea Otters and Kelp Forests in Alaska: Generality and Variation in a Community Ecological Paradigm. *Ecological Monographs, 65*, pp. 75–100.
Watson J., and J.A. Estes (2011). Stability, Resilience, and Phase Shifts in Rocky Subtidal Communities Along the West Coast of Vancouver Island, Canada, *Ecological Monographs, 81*(2), pp. 215–239.

8. Parks Canada. Cape Breton Highlands National Park of Canada: *Moose – Our Largest Herbivore*. Date modified 07 November 2014, http://www.pc.gc.ca/eng/pn-np/ns/cbreton/natcul/natcul1/c/i/b.aspx.

9. See, for example: Frank, K.T., et al. (2007). The Ups and Downs of Trophic Control in Continental Shelf Ecosystems, *Trends in Ecology and Evolution, 22*, pp. 236–242.
Worm, B. and R.A. Myers (2003). Meta-Analysis of Cod-Shrimp Interactions Reveals Top-Down Control in Oceanic Food Webs, *Ecology, 84*, pp. 162–173.

10. See, for example: Dale, V.H. (1989). Wind Dispersed Seeds and Plant Recovery on the Mount St. Helens Debris Avalanche, *Canadian Journal of Botany, 67*(5), pp. 1434–1441.
Dale, V.H., and W.M. Adams (2014). Plant Reestablishment 15 Years after the Debris Avalanche at Mount St. Helens, Washington, *Science of the Total Environment, 313*(1), pp. 101–113.

11. See, for example: Wood, D.M. and R. del Moral (2012). Vegetation Development on Permanently Established Grids, Mount St. Helens (1986–2010), *Ecology, 93*(9), pp. 2125–2125.
Thomason, L.A., M.D. Abata, A.C. Wenke, N. Lozanoff, and R. del Moral (2012). Primary Succession Trajectories on Pumice at Mount St. Helens, Washington, *Journal of Vegetation Science, 23*(1), pp. 73–85.
del Moral, R., and B. Magnússon (2014). Surtsey and Mount St. Helens: A Comparison of Early Succession Rates, *Biogeosciences, 11*(7), pp. 2099–2111.

12. See, for example: Crawford, R.L., J.S. Edwards, and P.M. Sugg (1995). Spider Arrival and Primary Establishment on Terrain Depopulated by Volcanic Eruption at Mount St. Helens, Washington, *The American Midland Naturalist, 133*(1), p. 60.
Edwards, J.S., and P.M. Sugg (1993). Arthropod Fallout as a Resource in the Recolonization of Mount St. Helens, *Ecology, 74*(3), pp. 954–958.

13. Danoff-Burg, James, ed., *Introduced Species Summary Project*, Columbia University, Centre for Environmental Research and Conservation, and other sources.

14. Haber, Erich, *Guide to Monitoring Exotic and Invasive Plants*, Environment Canada Ecological Monitoring and Assessment Network (EMAN).

15. Carder, Judith E.W. *Restoration of Natural Systems Program*, Garry Oak Ecosystems Recovery Team, Research Colloquium 2006, University of Victoria, BC; and Garry Oak Restoration Project (GORP), www.gorpsaanich.com.

16. Roberts, K., S. Koole, and D. Clements, Garry Oak Ecosystems Recovery Team—Research Colloquium 2006, Trinity Western University, Langley, BC.

17. World Wildlife Fund, "Biomes: Ecoregions," http://www.worldwildlife.org/biomes.

18. Climatographs adapted from S.W. Breckle (1999). *Walter's Vegetation of the Earth: The ecological systems of the geo-biosphere*, 4th ed., Berlin: Springer-Verlag.

19. World Wildlife Fund, "Biomes: Temperate Broadleaf and Mixed Forests." http://www.worldwildlife.org/biomes/temperate-broadleaf-and-mixed-forests.

Chapter 6

1. There are many sources for reliable population data. Unless otherwise noted, population data in this chapter are current as of March 2015, and were retrieved from (or calculated from) the following sources:
The United Nations Department of Economic and Social Affairs, Population Division, Population Estimates and Projections Section, http://esa.un.org/unpd/wpp/unpp/panel_population.htm.
Population Reference Bureau, http://www.prb.org/. Especially the 2014 World Population Data Sheet, http://www.prb.org/Publications/Datasheets/2014/2014-world-population-data-sheet/data-sheet.aspx.
Geohive (www.geohive.com). Geohive is an open-access resource created in 1996 to tabulate global population statistics and geopolitical statistics, drawing from reporting by statistical agencies from all over the world.

2. The world population growth rate as of 2015 is about 1.04%/yr.

3. Richman, S. (1993). Population means progress, not poverty. *Washington Post*, op-ed, September 1.

4. United Nations Population Fund (UNFPA) (2012). *Sex Imbalances at Birth: Current Trends, Consequences, and Policy Implications*. Bangkok, Thailand.

5. World DataBank (2012). Gender Statistics Highlights from 2012 World Development Report. World DataBank, a compilation of databases by the World Bank. February 2012.

6. United Nations Department of Economic and Social Affairs, Population Division (2014). World Contraceptive Use 2014. POP/DB/CP/Rev2014, http://www.un.org/en/development/desa/population/publications/dataset/contraception/wcu2014.shtml.

7. Alkema, L., et al. (2013). National, Regional and Global Rates and Trends in Contraceptive Prevalence and Unmet Need for Family Planning between 1990 and 2015: A Systematic and Comprehensive Analysis, *The Lancet*, *381*(9878), pp. 1642–1652.

8. Women in National Parliaments, Inter-Parliamentary Union, 2015, http://www.ipu.org/wmn-e/classif.htm.

9. United Nations Department of Economic and Social Affairs, Population Division (2013). Probabilistic Projections Based on the World Population Prospects 2012 Revision: Highlights, http://esa.un.org/unpd/ppp/Documentation/highlights.htm.

10. Global Footprint Network, World Footprint: Do We Fit the Planet? http://www.footprintnetwork.org/en/index.php/GFN/page/world_footprint/.

11. World Bank Poverty Overview, http://www.worldbank.org/en/topic/poverty/overview.

12. WHO Global Health Observatory Data, Number of Deaths due to HIV/AIDS, http://www.who.int/gho/hiv/epidemic_status/deaths_text/en/.

13. United Nations, End Poverty 2015, Millennium Development Goals, http://www.un.org/millenniumgoals/.

Chapter 7

1. Convention on Wetlands (Ramsar Convention) Canada, www.ramsar.org/wetland/canada.

2. Based on information from www.trentu.ca/academic/bluelab/research_merbleue.html, and personal communications with Nigel Roulet, Tim Moore, Cameron Proctor, Varun Gupta, and Nathan Basiliko.
For work on peat soils and biogeochemical cycling at Mer Bleue, see for example:
Talbot, J., N.T. Roulet, O. Sonnentag, and T.R. Moore (2014). Increases in Aboveground Biomass and Leaf Area 85 Years after Drainage in a Bog, *Botany*, *92*(10), pp. 713–721.
Blodau, C., N. Basiliko, B. Mayer, and T.R. Moore (2006). The Fate of Experimentally Deposited Nitrogen in Mesocosms from Two Canadian Peatlands, *Science of the Total Environment*, *364*(1–3), pp. 215–228.
Blodau, C., and T.R. Moore (2003). Experimental Response of Peatland Carbon Dynamics to a Water Table Fluctuation, *Aquatic Sciences*, *65*(1), pp. 47–62.

3. Wall, G.J., D.R. Coote, E.A. Pringle, and I.J. Shelton (eds.) (2002). *RUSLEFAC—Revised Universal Soil Loss Equation for Application in Canada: A Handbook for Estimating Soil Loss from Water Erosion in Canada*. Research Branch, Agriculture and Agri-Food Canada, Ottawa, Contribution No. AAFC/AAC2244E. 117 pp.

4. Wilkinson, B. (2005). Humans as Geologic Agents: A Deep-Time Perspective, *Geology*, *33*(3), p. 161.

5. Boardman, J. (2006). Soil Erosion Science: Reflections on the Limitations of Current Approaches, *Catena*, *68*(2–3), pp. 73–86.

6. Eilers, W., R. MacKay, L. Graham, and A. Lefebvre (eds.) (2010). Environmental Sustainability of Canadian Agriculture, Agriculture and Agri-Food Canada, Agri-Environmental Indicator Report Series, Report #3.
Agriculture and Agri-Food Canada (1997). *Profile of Production Trends and Environmental Issues in Canada's Agriculture and Agri-Food Sector*, http://www4.agr.gc.ca/resources/prod/doc/policy/environment/pdfs/sds/profil_e.pdf.

7. van Vlietl, L.J.P., G.A. Padbury, and D.A. Lobb (2003). Soil Erosion Risk Indicators Used in Canada. Paper presented at the OECD Expert Meeting on Soil Erosion and Biodiversity.

8. UN Convention to Combat Desertification, as cited in the Millennium Ecosystem Assessment (2005). *Ecosystems and Human Well-Being: Desertification Synthesis*, www.millenniumassessment.org/documents/document.355.aspx.pdf.
UN FAO Soils Portal Key Definitions, www.fao.org/soils-portal/about/all-definitions/en/.

9. Global Environmental Facility (GEF) and United Nations Convention to Combat Desertification (UNCCD) secretariats (2011). Land for Life: Securing Our Common Future. https://www.thegef.org/gef/sites/thegef.org/files/publication/SLM-english-1.pdf.
Food and Agriculture Organization (FAO) (2011). The State of the World's Land and Water Resources for Food and Agriculture (SOLAW): Managing Systems at Risk. FAO, Rome and Earthscan, London. http://www.fao.org/docrep/015/i1688e/i1688e00.pdf.

10. United Nations Convention to Combat Desertification, http://www.unccd.int/.
Especially: UNCCD (2014). Land Degradation Neutrality: Resilience at Local, National, and Regional Levels, http://www.unccd.int/en/programmes/RioConventions/RioPlus20/Pages/Land-Degradation-NeutralWorld.aspx/.

11. United Nations World Day to Combat Desertification, http://www.un.org/en/events/desertificationday/background.shtml, and UNCCD Desertification Land Degradation and Drought: Some Global Facts and Figures.

12. CBC Radio Archives, Devastating Dry Spells: Drought on the Prairies. http://www.cbc.ca/archives/categories/environment/extreme-weather/devastating-dry-spells-drought-on-the-prairies/drought-on-the-prairies.html.

13. Government of Canada (1984). *Soil at Risk: Canada's Eroding Future*.

14. Soil Conservation Council of Canada, www.soilcc.ca.

15. Agriculture and Agri-Food Canada, Agri-Environmental Services Branch, publications.gc.ca/collections/collection_2011/agr/A125-7-2-2011-eng.pdf.

16. Statistics Canada (2012). *Farm and Farm Operator Data, 2011 Census of Agriculture,* Catalogue no. 95-640-X. http://www.statcan.gc.ca/pub/16-201-x/2014000/t009-eng.htm.

17. Soil Conservation Council of Canada, www.soilcc.ca.
    Toner, P., and W. Brown (2006). *Evaluating Tillage Options for Your Farm.* New Brunswick Department of Agriculture, Fisheries and Aquaculture, www.soilcc.ca/ggmp_fact_sheets/pdf/Tillage.pdf.

18. UN Food and Agriculture Organization (FAO). Aquastat: Water Uses, http://www.fao.org/nr/water/aquastat/water_use/index.stm.

19. UN Food and Agriculture Organization (FAO). Statistical Yearbook 2013: World Food and Agriculture. http://www.fao.org/economic/ess/ess-publications/ess-yearbook/en/#.VHd7SslZgxY.

Chapter 8

1. Cluis, C. (2011). Rounding Up the Schmeiser Case: Benefit and Liability Issues of Transgenic Crops, *The Science Creative Quarterly, 6.* Accessed online 28 November 2014.
   See also: Monsanto > Newsroom > Percy Schmeiser, http://www.monsanto.com/newsviews/pages/percy-schmeiser.aspx.

2. FAO Statistical Yearbook 2013 and World Bank, http://data.worldbank.org/indicator/AG.LND.AGRI.ZS/countries/1W?display=graph.

3. The species *Homo sapiens* is thought to have originated in Africa approximately 200 000 years BP (before present). The modern human subspecies or anatomically modern humans, *H. sapiens sapiens,* is the only subspecies of *H. sapiens* that survives today. Based on information from Smithsonian Institution Human Origins Project, www.mnh.si.edu/anthro/humanorigins/.

4. International Fund for Agricultural Development (IFAD). Food prices: Smallholder farmers can be part of the solution. UN FAO. http://www.ifad.org/operations/food/farmer.htm.

5. International Fund for Agricultural Development (IFAD) (2013). Smallholders, Food Security, and the Environment. UN FAO. www.ifad.org/climate/resources/smallholders_report.pdf.

6. UN FAO, IFAD, and WFP (2014). The State of Food Insecurity in the World 2014: Strengthening the Enabling Environment for Food Security and Nutrition. Rome, FAO.

7. Ibid.

8. Canadian Institute for Health Information (2011). Obesity in Canada. A Joint Report from the Public Health Agency of Canada and the Canadian Institute for Health Information.

9. World Bank Open Data, http://data.worldbank.org/.

10. Hunger Awareness Week: About Hunger, http://hungerawareness-week.ca/about_hunger.
    See also Food Banks Canada (2014). *Hunger Count 2014: A Comprehensive Report on Hunger and Food Bank Use in Canada, and Recommendations for Change,* http://www.foodbankscanada.ca/hungercount.

11. Tarasuk, V., A. Mitchell, and N. Dachner (2014). Household Food Insecurity in Canada, 2012. Toronto: Research to identify policy options to reduce food insecurity. Retrieved from http://nutritionalsciences.lamp.utoronto.ca/.

12. Ibid.

13. Health Canada Drinking Water Guidelines, http://www.hc-sc.gc.ca/ewh-semt/pubs/water-eau/nitrate_nitrite/index-eng.php#a211.
    See also U.S. EPA Drinking Water Standards > Contaminants, http://water.epa.gov/drink/contaminants/#List.

14. FAO Land and Plant Nutrition Management Service, Global Network on Integrated Soil Management for Sustainable Use of Salt-Affected Soils, http://www.fao.org/worldfoodsummit/english/newsroom/focus/focus1.htm.

15. UN FAO. The Sources of Food. http://www.fao.org/docrep/u8480e/u8480e07.htm.

16. Enserink, M., P.J. Hines, S.N. Vignieri, N.S. Wigginton, and J.S. Yeston (2013). The Pesticide Paradox: Introduction to Special Issue, *Science, 341*(6147), pp. 728–729.

17. Köhler, H.-R., and R. Triebskorn (2013). Wildlife Ecotoxicology of Pesticides: Can We Track Effects to the Population Level and Beyond? *Science, 341*(6147), pp. 759–765.

18. Agriculture and Agri-Foods Canada. An Overview of Selected Farm Input Prices in Ontario and Manitoba 2004-2008 (December 2009). http://www.agr.gc.ca/eng/industry-markets-and-trade/statistics-and-market-information/by-product-sector/crops/crops-market-information-canadian-industry/market-outlook-report/an-overview-of-selected-farm-input-prices-in-ontario-and-manitoba-2004-2008-december-2009/?id=1378841325191#img-1.
    World total is from FAOSTAT. http://faostat.fao.org/site/423/DesktopDefault.aspx?PageID=423#ancor.

19. Health Canada, Pest Management Regulatory Agency, www.pmra-arla.gc.ca/english/legis/pcpa-e.html, and Health Canada, The Regulation of Pesticides in Canada: Fact Sheet, http://www.hc-sc.gc.ca/cps-spc/pubs/pest/_fact-fiche/reg-pesticide/index-eng.php.
    See also Government of Canada (2000). *Pesticides: Making the Right Choice for Health and the Environment.* Report of the Standing Committee on Environment and Sustainable Development. http://cmte.parl.gc.ca/cmte/CommitteePublication.aspx?COM=173&Lang=1&SourceId=36396.

20. Beckie, H.J., and L.M. Hall (2014). Genetically-Modified Herbicide-Resistant (GMHR) Crops a Two-Edged Sword? An Americas Perspective on Development and Effect on Weed Management, *Crop Protection, 66,* pp. 4045. http://www.sciencedirect.com/science/article/pii/S0261219414002683.

21. Regulatory Directive DIR2013-04, Pesticide Resistance Management Labelling Based on Target Site/Mode of Action, Government of Canada, Pest Management Regulatory Agency, 17 December 2013. http://www.hc-sc.gc.ca/cps-spc/pubs/pest/_pol-guide/dir2013-04/index-eng.php.

22. Resosudarmo, B.P. (2012). Implementing a National Environmental Policy: Understanding the 'Success' of the 1989–1999 Integrated Pest Management Programme in Indonesia, *Singapore Journal of Tropical Geography, 33*(3), pp. 365–380.

23. Integrated Pest Management Council of Canada, http://www.ipm-councilcanada.org/epar/en-CA/Default.aspx.

24. NSERC-CANPOLIN, Ontario Ministry of Agriculture and Rural Affairs, and the University of Guelph with Seeds of Diversity. Best Management Practices for Pollination in Ontario Crops, http://www.pollinator.ca/bestpractices/alfalfa_lcb.html.

25. Pitts-Singer, T.L. (2013) Variation in Alfalfa Leafcutting Bee (*Hymenoptera Megachilidae*): Reproductive Success According to Location of Nests in United States Commercial Domiciles, *Journal of Economic Entomology, 106*(2), pp. 543–551.
    See also O'Neill, K.M., R.P O'Neill, W.P. Kemp, and C. M. Delphia (2011). Effect of Temperature on Post-Wintering Development and Total Lipid Content of Alfalfa Leafcutting Bees, *Environmental Entomology, 40*(4), pp. 917–930.

26. Based on information from Kevan, P.G., E.A. Clark, and V.G. Thomas (1990). Pollinators and Sustainable Agriculture, *American Journal of Alternative Agriculture, 5*(1), pp. 13–22.

27. Klein, A.M., et al. (2007). Importance of Pollinators in Changing Landscapes for World Crops, *Proc. R. Soc. B-Biol. Sci., 274*(1608), pp. 303–313.

28. Aizen, M.A., L.M. Garibaldi, S.A. Cunningham, and A.M. Klein (2009). How Much Does Agriculture Depend on Pollinators? Lessons from Long-Term Trends in Crop Production, *Ann. Bot., 103*(9), pp. 1579–1588.

29. Hanley, N., T.D. Breeze, C. Ellis, and D. Goulson (2014). Measuring the Economic Value of Pollination Services: Principles, Evidence and Knowledge Gaps, *Ecosystem Services,* Available online 22 October 2014 doi:10.1016/j.ecoser.2014.09.013.

30. Pettis J.S., et al. (2013). Crop Pollination Exposes Honey Bees to Pesticides which Alters Their Susceptibility to the Gut Pathogen *Nosema ceranae. PLoS ONE, 8*(7): e70182. doi:10.1371/journal.pone.0070182.

31. Garibaldi, L.A., et al. (2013). Wild Pollinators Enhance Fruit Set of Crops Regardless of Honey-Bee Abundance, *Science, 339,* pp. 1608–1611.

32. International Service for the Acquisition of Agri-Biotech Applications (ISAAA). Brief 46-2013: Executive Summary: Global Status of Commercialized Biotech/GM Crops: 2013. http://www.isaaa.org/resources/publications/briefs/46/executivesummary/.

33. Nicolia, A., A. Manzo, F. Veronesi, and D. Rosellini (2014). An Overview of the Last 10 Years of Genetically Engineered Crop Safety Research. *Critical Reviews in Biotechnology, 34*(1), pp. 77–88.

34. Spisák, S., et al. (2013). Complete Genes May Pass from Food to Human Blood, *PLoS ONE, 8*(7): e69805. doi:10.1371/journal.pone.0069805.

35. Monsanto, http://www.monsanto.com/newsviews/pages/saved-seed-farmer-lawsuits.aspx.

36. Convention on Biological Diversity, http://www.cbd.int/convention/. See also Agriculture and Agri-Foods Canada, http://www.agr.gc.ca/eng/industry-markets-and-trade/agri-food-trade-policy/trade-topics/the-biosafety-protocol/?id=1384287629516.

37. Eaton, D., J. Windig, S.J. Hiemstra, and M. Van Veller (2006). Indicators for Livestock and Crop Diversity, North-South Policy Brief 2001-6. Programme International Cooperation, Wageningen International.

38. Millennium Seed Bank, http://www.kew.org/science-conservation/millennium-seed-bank.

39. FAO (2014). Food Outlook: Biannual Report on Global Food Markets, October 2014. Accessed online 17 January 2015.
See also Global Agriculture: Agriculture at a Crossroads: Findings and Recommendations for Future Farming: Meat and Animal Feed, http://www.globalagriculture.org/report-topics/meat-and-animal-feed.html.

40. Commission on Genetic Resources for Food and Agriculture, Food and Agriculture Organization of the United Nations (2007). *The State of the World's Animal Genetic Resources for Food and Agriculture.*

41. UN FAO. Animal Production, http://www.fao.org/animal-production/en/.

42. MacKinnon, S. (2013). The National Organic Market Growth, Trends & Opportunities. Canada Organic Trade Association.

43. Ibid.

44. Sustainable Cotton Project. www.sustainablecotton.org/; Worldwide Fund for Nature (WWF). The Impact of a Cotton T-Shirt, http://www.worldwildlife.org/stories/the-impact-of-a-cotton-t-shirt. Accessed 3 August 2015.

45. Roots, Organic Cotton. www.roots.com.

46. Mäder, P., et al. (2002). Soil Fertility and Diversity in Organic Farming, *Science, 296*, pp. 1694–1697.

47. Rodale Institute (2011). The Farming Systems Trial: Celebrating 30 Years. Rodale Institute, Kutztown, Pennsylvania.

48. Smith, A., and J. MacKinnon (2007). *The 100-Mile Diet: Local Eating for Global Change.* Random House Canada.

49. Equiterre. www.equiterre.org/en/agriculture/paniersBios/index.php.

50. Fresh City Farms, About Our Farms. https://www.freshcityfarms.com/.

51. Arnold, T., K. Hammel, B. Hsueh, S. Robart, and L. Thomson (2013). Fresh City: Impacts of Local Food. Technical Report: Assessment of Greenhouse Gas Emissions by Food Box Distribution. Grey Bruce Centre for Agroecology, Park Head, Ontario, November 2013.

Chapter 9

1. Derocher, A.E., N.J. Lunn, and I. Stirling (2004). Polar Bears in a Warming Climate, *Integrative and Comparative Biology, 44*, pp. 163–176.

2. Consultations on the Proposed Listing of the Polar Bear as Special Concern under the *Species at Risk Act*, Conducted February–April 2009 by the Canadian Wildlife Service.

3. Alaska Department of Fish and Game > Species > Special Status > Polar Bear *Ursus maritimus*: Federally Threatened. http://www.adfg.alaska.gov/index.cfm?adfg=specialstatus.fedsummary&species=polarbear.

4. The quota for annual culls, called the Total Allowable Harvest, is set by the provinces and territories individually for each of the 13 subpopulations of polar bears that live in Canada.

5. IUCN (2014). *IUCN Red List of Threatened Species* Version 2014.3 *Ursus maritimus: Summary.* http://www.iucnredlist.org/details/22823/0.

6. Overland, J.E. and M. Wang (2013). When Will the Summer Arctic Be Nearly Sea Ice Free? *Geophysical Research Letters, 40*, pp. 2097–2101.

7. Alaska Department of Fish and Game > Species > Special Status > Polar Bear *Ursus maritimus*: Federally Threatened. http://www.adfg.alaska.gov/index.cfm?adfg=specialstatus.fedsummary&species=polarbear.

8. Derocher, A.E., N.J. Lunn, and I. Stirling (2004). Polar Bears in a Warming Climate, *Integrative and Comparative Biology, 44*, pp. 163–176.

9. IUCN (2014). *IUCN Red List of Threatened Species* Version 2014.3 *Ursus maritimus: Summary.* www.iucnredlist.org/apps/redlist/details/22823/0.

10. Stapleton et al. (2014). Polar Bears from Space: Assessing Satellite Imagery as a Tool to Track Arctic Wildlife, *PLoS One, 9.*7 (July 2014).

11. *CBC News* (January 5, 2011). Iqaluit Polar Bear Hunting Quota Unclear. www.cbc.ca/news/canada/north/story/2011/01/05/iqaluit-polar-bear-quota.html.

12. Peacock, E., et al. (2015) Implications of the Circumpolar Genetic Structure of Polar Bears for Their Conservation in a Rapidly Warming Arctic, *PLoS ONE, 10*(1): e112021

13. Rode, K.D., et al. (2014). Variation in the Response of an Arctic Top Predator Experiencing Habitat Loss: Feeding and Reproductive Ecology of Two Polar Bear Populations, *Global Change Biology, 20*, pp. 76–88.
Regehr, E.V., N.J. Lunn, S.C. Amstrup, and I. Stirling (2007). Effects of Earlier Sea Ice Breakup on Survival and Population Size of Polar Bears in Western Hudson Bay, *Journal of Wildlife Management, 71*(8), pp. 2673–2683.

14. Environment Canada (2011). *National Polar Bear Conservation Strategy for Canada.* http://ec.gc.ca/nature/default.asp?lang=En&n=60D0FDBD-1.

15. Campagna, C. (2008). *Mirounga angustirostris.* In IUCN (2008), *IUCN Red List of Threatened Species,* accessed 28 January 2009.
IUCN (2014). *IUCN Red List of Threatened Species,* Version 2014.3. www.iucnredlist.org.

16. IUCN (2014). *Summary of Statistics: Assessed Species: Table 1: Numbers of Threatened Species by Major Groups of Organisms (1996–2014).* http://www.iucnredlist.org/about/summary-statistics#Tables_1_2.

17. Caley, M.J., R. Fisher, and K. Mengersen (2014). Global Species Richness Estimates Have Not Converged, *Trends in Ecology and Evolution, 29*(4) pp. 187–188.

18. See, for example, Erwin, T.L. (1991). How Many Species Are There? Revisited, *Conservation Biology, 5,* pp. 1–4.

19. Tropical Ecology Assessment and Monitoring Network (TEAM). http://www.teamnetwork.org/.

20. AmphibiaWeb, http://www.amphibiaweb.org/.

21. Bass, M.S., et al. (2010). Global Conservation Significance of Ecuador's Yasuní National Park, *PLoS ONE, 5*: e8767.

22. Blake, J.G., D. Mosquera, J. Guerra, B.A. Loiselle, D. Romo, and J.K. Swing (2011). Mineral Licks as Diversity Hotspots in Lowland Forest of Eastern Ecuador, *Diversity, 3*(2), pp. 217–234.

23. Blake, J.G., D. Mosquera, J. Guerra, B.A. Loiselle, D. Romo, and J.K. Swing (2014). Yasuní – A Hotspot for Jaguars *Panthera onca* (Carnivora: Felidae)? Camera Traps and Jaguar Activity at Tiputini Biodiversity Station, Ecuador, *Revista de biología tropical, 62*(2), p. 689.

24. Galvis, N., A. Link, and A. Di Fiore (2014). A Novel Use of Camera Traps to Study Demography and Life History in Wild Animals: A Case Study of Spider Monkeys (*Ateles belzebuth*), *International Journal of Primatology, 35*, pp. 908–918.

25. According to IUCN Red List, there are 7302 known species of amphibians as of 2014.

26. Different sources give different total numbers of amphibian species in Canada, ranging from "more than 40" to 48.

27. Canada reference: Summary Table of Wildlife Species Assessed by COSEWIC as of November 2014. http://www.cosewic.gc.ca/eng/sct0/index_e.cfm.
Global Amphibian Assessment: Amphibian Assessment Forum. http://www.amphibians.org/redlist/.
IUCN Red List Initiatives > Amphibians. http://www.iucnredlist.org/initiatives/amphibians.
AmphibiaWeb: http://amphibiaweb.org/declines/declines.html.

28. Environment Canada SARA Public Registry: COSEWIC Status Report Vancouver Island Marmot (*Marmota vancouverensis*). http://www.cosewic.gc.ca/eng/sct1/searchdetail_e.cfm?id=136&StartRow=1&boxStatus=All&boxTaxonomic=All&location=1&change=All&board=All&commonName=&scienceName=&returnFlag=0&Page=1.

29. Millennium Ecosystem Assessment (2005). *Ecosystems and Human Well-Being: Synthesis,* Island Press, Washington, DC.

30. IUCN (2014). *IUCN Red List of Threatened Species,* Version 2014.3, *Red List Summary Statistics.* http://www.iucnredlist.org/about/summary-statistics#How_many_threatened.

31. On the IUCN Red List, "Threatened" species include species that are Critically Endangered, Endangered, or Vulnerable. For mammals and birds, close to 100% of the known species have been evaluated in the Red List process. The proportion of evaluated species is lower for amphibians, reptiles, and fishes, and much lower for invertebrates, plants, fungi, and protists. The percentages of threatened species in each category have increased since 1996; however, this statistic must be interpreted carefully, because the numbers of species described in each category also have increased, but not necessarily at the same rate. Note, too, that these statistics do not include extinct species, only extant threatened species. The numbers of extinct species and species that are extinct in the wild in the IUCN Red List only includes species that were at one time reported and verified alive in the wild.

32. Barnosky, A.D., et al. (2011). Has the Earth's Sixth Mass Extinction Already Arrived? *Nature, 471*(7336), pp. 51–57.

33. Ibid.

34. WWF, GFN, WFN, and ZSL (2014). *Living Planet Report 2014: Species and Spaces, People and Places.* http://wwf.panda.org/about_our_earth/all_publications/living_planet_report/.

35. Ibid.

36. Environment Canada, SARA Public Registry Greater Sage-Grouse *urophasianus* subspecies. http://www.registrelep-sararegistry.gc.ca/species/speciesDetails_e.cfm?sid=305.

37. Environment Canada (2011). *National Polar Bear Conservation Strategy for Canada.* http://ec.gc.ca/nature/default.asp?lang=En&n=60D0FDBD-1.

38. Ecosystem services and biodiversity services are not exactly the same, but they are linked closely, in many ways. These linkages have been explored by a number of researchers, including Hicks, C., Woroniecki S., Fancourt, M., Bieri, M., Garcia Robles, H., Trumper, K., Mant, R., (2014). The Relationship between Biodiversity, Carbon Storage and the Provision of Other Ecosystem Services: Critical Review for the Forestry Component of the International Climate Fund. Cambridge, UK.

39. deGroot, R., et al. (2012). Global Estimates of the Value of Ecosystems and Their Services in Monetary Units, *Ecosystem Services, 1*(2012), pp. 50–61.

40. Leopold, A. (1953) *Round River,* Oxford University Press, New York, NY, pp. 145–146.

41. Honey, M. (2008). *Ecotourism and Sustainable Development, Second Edition: Who Owns Paradise?* Island Press.

42. Wilson, E. O. (1984). *Biophilia,* Harvard University Press, Cambridge, MA.

43. Louv, R. *Last Child in the Woods: Saving Our Children from Nature-Deficit Disorder,* Algonquin Books of Chapel Hill, Chapel Hill, NC.

44. Wilson, E.O. (1994). *Naturalist,* Shearwater Books, Washington, DC.

45. MacArthur, R.H., and E.O. Wilson (1963). An Equilibrium Theory of Insular Zoogeography, *Evolution, 17,* pp. 373–387.
MacArthur, R.H., and E.O. Wilson (1967). *The Theory of Island Biogeography.* Princeton University Press, NJ.

46. Whooping Crane Eastern Partnership. www.bringbackthecranes.org.

47. Environment Canada, SARA Public Registry, Species Profile: Swift Fox. www.sararegistry.gc.ca/species/speciesDetails_e.cfm?sid=140.

48. Weagle, K., and C. Smeeton. Captive breeding of swift fox for reintroduction: Final report to funding bodies 1994 to 1997. Cochrane Ecological Institute. www.ceinst.org/Final%20Report%20to%20Funders%2097.pdf.

49. Cotterill, S.E. (1997). Status of the swift fox *(Vulpes velox)* in Alberta. Alberta Environmental Protection, Wildlife Management Division, Wildlife Status Report No. 7, Edmonton, AB.

50. Ibid.

51. Nature Conservancy Canada. Swift fox. http://www.natureconservancy.ca/en/what-we-do/resource-centre/featured-species/swift_fox.html.

52. Wildlife Preservation Canada: Swift Fox Recovery Team. http://wildlifepreservation.ca/swift-fox/.

53. Conservation International: Hotspots. http://www.conservation.org/How/Pages/Hotspots.aspx.

54. World Wide Fund for Nature. *Ecoregions.* http://wwf.panda.org/about_our_earth/ecoregions/about/.

55. Community Baboon Sanctuary. http://www.howlermonkeys.org/.

56. Runte, A. (1997). *National Parks: The American Experience,* 3rd ed. University of Nebraska Press, Lincoln, NE.

57. Parks Canada, Canada's National Parks and National Reserves. http://www.pc.gc.ca/eng/docs/v-g/nation/nation103.aspx.

58. Parks Canada Attendance 2013-2014. http://www.pc.gc.ca/eng/docs/pc/attend/table1.aspx.

59. Campbell, L., and C.D.A. Rubec (2006). *Land Trusts in Canada: Building Momentum for the Future.* Ottawa: Wildlife Habitat Canada and the Stewardship Section, Canadian Wildlife Service, Environment Canada.

60. UNEP World Conservation Monitoring Centre: Mapping the World's Special Places. http://www.unep-wcmc.org/featured-projects/mapping-the-worlds-special-places.

61. Global Conservation Fund. http://www.conservation.org/projects/Pages/global-conservation-fund.aspx.

62. Environment Canada > Canada's Protected Areas. https://www.ec.gc.ca/indicateurs-indicators/default.asp?lang=en&n=478A1D3D-1.

63. UNESCO World Heritage Sites. http://whc.unesco.org/en/list/.

64. UNESCO Ecological Sciences for Sustainable Development. *Biosphere Reserves: Learning Sites for Sustainable Development.* http://www.unesco.org/new/en/natural-sciences/environment/ecological-sciences/biosphere-reserves/.

65. UNESCO Ecological Sciences for Sustainable Development. *Biosphere Reserves: Learning Sites for Sustainable Development.* Bras d'Or Lake. http://www.unesco.org/new/en/natural-sciences/environment/ecological-sciences/biosphere-reserves/europe-north-america/canada/bras-dor-lake/.

66. Environment Canada, *Marine Protected Areas.* http://www.dfo-mpo.gc.ca/oceans/marineareas-zonesmarines/mpa-zpm/index-eng.htm.

67. Rails to Trails. http://www.railstotrails.org/about/.

Chapter 10

1. Food and Agriculture Organization of the United Nations, 2015, International Day of Forests 2015. Reproduced with permission.

2. UNESCO Ecological Sciences for Sustainable Development. http://www.unesco.org/new/en/natural-sciences/environment/ecological-sciences/biosphere-reserves/europe-north-america/canada/clayoquot-sound/.

3. *Friends of Clayoquot Sound Newsletter* (Fall 2007–Winter 2008).

4. Food and Agriculture Organization of the United Nations (2010). *Global Forest Resources Assessment 2010.*
NOTE: The next edition of the FAO's *Global Forest Resources Assessment* was scheduled for September 2015.

5. Food and Agriculture Organization of the United Nations (2010). *Global Forest Resources Assessment 2010: Canada Country Report (2009).* Natural Resources Canada (2014). *State of Canada's Forests: Annual Report 2014.*

6. Food and Agriculture Organization of the United Nations (2010). *Global Forest Resources Assessment 2010.*

7. Canadian Council of Forest Ministers (2003). National Forest Strategy 2003–2008, *A Sustainable Forest: The Canadian Commitment.* National Forest Strategy Steering Committee.
Canadian Council of Forest Ministers (2008). *A Vision for Canada's Forests: 2008 and Beyond.* www.ccfm.org/pdf/Vision_EN.pdf.

8. Food and Agriculture Organization of the United Nations (2010). *Global Forest Resources Assessment 2010: Canada Country Report.*
NOTE: Primary forests are defined by the FAO as forests of native species in which there are no clearly visible indications of human activity and ecological processes are not significantly disturbed.

9. Food and Agriculture Organization of the United Nations (2010). *Global Forest Resources Assessment 2010.*

10. Canadian Boreal Institute, Boreal Leadership Council. http://boreal-council.ca.

11. Ibid.

12. Armson, K. (1999). Knowledge of the Environment for Youth (KEY) Environmental Literacy Series, The KEY Foundation. *Canadian Forests: A Primer.*

13. Food and Agriculture Organization of the United Nations (2010). *Global Forest Resources Assessment 2010.*

14. Ibid.

15. Ibid.

16. United Nations Food and Agriculture Organization (2014). *State of the World's Forests: Enhancing the Socioeconomic Benefits from Forests.*

17. Natural Resources Canada (2014). *State of Canada's Forests: Annual Report 2014.*

18. United Nations Food and Agriculture Organization (2014). *World Agriculture: Towards 2015/2030: An FAO Perspective.*

19. Natural Resources Canada > Mountain Pine Beetle. https://www.nrcan.gc.ca/forests/insects-diseases/13381.

20. Turner, J. (2008). Changing Climate Stops Pest Outbreaks on Vancouver Island. Forest NewsTips.
    See also Régnière, J., R. St-Amant, and P. Duval (2012). Predicting Insect Distributions under Climate Change from Physiological Responses: Spruce Budworm as an Example. *Biological Invasions, 14*(8), pp. 1571–1586.

21. Natural Resources Canada > NRCAN Forest Topics > Asian Long-Horned Beetle. https://www.nrcan.gc.ca/forests/insects-diseases/13369.

22. Turgeon, J. (2011). *Asian Longhorned Beetle.* Natural Resources Canada. Canadian Forest Service. Great Lakes Forestry Centre. Sault Ste. Marie, Ontario. *Frontline Express,* 39, 2p.

23. Natural Resources Canada > NRCAN Forest Topics > Asian Long-Horned Beetle. https://www.nrcan.gc.ca/forests/insects-diseases/13369.

24. Food and Agriculture Organization of the United Nations (2010). *Global Forest Resources Assessment 2010.*
    United Nations Food and Agriculture Organization (2014). *State of the World's Forests: Enhancing the Socioeconomic Benefits from Forests.*

25. Natural Resources Canada (2014). *State of Canada's Forests: Annual Report 2014.*

26. Canadian Council of Forest Ministers (2003). National Forest Strategy 2003–2008, *A Sustainable Forest: The Canadian Commitment.* National Forest Strategy Steering Committee.

27. Canadian Council of Forest Ministers (2008). *A Vision for Canada's Forests: 2008 and Beyond.* www.ccfm.org/pdf/Vision_EN.pdf.

28. Ibid.

29. British Columbia, Ministry of Forestry and Range, Forest Practices Branch. *Adaptive Management Initiatives in the BC Forest Service.* https://www.for.gov.bc.ca/hfp/archives/amhome/amhome.htm.

30. Natural Resources Canada > Forest Topics > Fire. https://www.nrcan.gc.ca/forests/fire/13143.

31. See, for example, Westerling, A.L., and B.P. Bryant (2008). Climate Change and Wildfire in California, *Climatic Change, 87*(Suppl1), pp. 231–249.

32. International Institute for Sustainable Development, *What Is REDD+?* https://www.iisd.org/climate/land_use/redd/about.aspx.

33. Hewitt, N., et al. (2011). Taking Stock of the Assisted Migration Debate. *Biological Conservation, 144,* pp. 2560–2572.

34. Peters, R.L., and J.D.S. Darling (1985). The Greenhouse Effect and Nature Reserves: Global Warming Would Diminish Biological Diversity by Causing Extinctions among Reserve Species, *Bioscience, 35*(11), pp. 707–717.

35. Ibid.

36. Hewitt, N., and M. Kellman (2002). Tree Seed Dispersal among Forest Fragments: I. Conifer Plantations as Seed Traps, *Biogeography, 29*(3), pp. 337–349.
    Hewitt, N., and M. Kellman (2002). Tree Seed Dispersal among Forest Fragments: II. Dispersal Abilities and Biogeographical Controls, *Biogeography, 29*(3), pp. 351–363.

37. Hewitt, N., et al. (2011). Taking Stock of the Assisted Migration Debate, *Biological Conservation, 144,* pp. 2560–2572.

38. Smith, A.L., et al. (2012). Effects of Climate Change on the Distribution of Invasive Alien Species in Canada: A Knowledge Synthesis of Range Change Projections in a Warming World, *Environmental Reviews, 20,* pp. 1–16.

39. Hewitt, N., et al. (2011). Taking Stock of the Assisted Migration Debate, *Biological Conservation, 144,* pp. 2560–2572.

40. Ibid.

41. Torreya Guardians. www.torreyaguardians.org.

42. Natural Resources Canada > Forest Topics > Climate Change > Assisted Migration. http://www.nrcan.gc.ca/forests/climate-change/13121.

43. Hewitt, N., et al. (2011). Taking Stock of the Assisted Migration Debate, *Biological Conservation, 144,* pp. 2560–2572.

44. United Nations Food and Agriculture Organization (2014). *State of the World's Forests: Enhancing the Socioeconomic Benefits from Forests.*
    United Nations REDD. http://www.un-redd.org/aboutredd.

45. Natural Resources Canada > Forest Topics > Climate Change > REDD+. http://www.nrcan.gc.ca/forests/climate-change/13101.

46. Natural Resources Canada > Forest Topics > Climate Change. http://www.nrcan.gc.ca/forests/climate-change/13083. Reproduced with the permission of the Minister of Natural Resources Canada, 2015.

47. Natural Resources Canada (2014). *State of Canada's Forests: Annual Report 2014.*

48. Amiro, B.D., A. Cantin, M.D. Flanigan, and W.J. deGroot (2009). Future Emissions from Canadian Boreal Forest Fires, *Canadian Journal of Forest Research, 39,* pp. 383–395.

49. Natural Resources Canada > Forest Topics > Climate Change. http://www.nrcan.gc.ca/forests/climate-change/13085.

50. Westerling, A.L., and B.P. Bryant (2008). Climate Change and Wildfire in California, *Climatic Change, 87*(Suppl1), pp. 231–249.

Chapter 11

1. Agriculture and Agri-Foods Canada, *The Canadian Bottled Water Industry.* www4.agr.gc.ca/AAFC-AAC/display-afficher.do?id=1171644581795&lang=e#sig.

2. Quinn, F. (2007). *Water Diversion, Export, and Canada–U.S. Relations: A Brief History* (August). Program on Water Issues (POWI), Munk Centre for International Studies at the University of Toronto. www.powi.ca.

3. Ibid.

4. UNFAO (2012). *The State of World Fisheries and Aquaculture 2012, Part I: World Review of Fisheries and Aquaculture,* FAO Fisheries and Aquaculture Department, Rome. http://www.fao.org/docrep/016/i2727e/i2727e00.htm.

5. Environment Canada (2010). *Water and Climate Change.* https://www.ec.gc.ca/eau-water/default.asp?lang=En&n=3E75BC40-1#Section23.

6. Environment Canada (2004). *Threats to Water Availability in Canada.* National Water Research Institute, Burlington, Ontario. NWRI Scientific Assessment Report Series No. 3 and ACSD Science Assessment Series No. 1. 128 pp. https://www.ec.gc.ca/inre-nwri/default.asp?lang=En&n=0CD66675-1.

7. Ibid.

8. National Wetlands Working Group (1997). *The Canadian Wetlands Classification System,* 2nd Edition. Edited by B.G. Warner and C.D.A. Rubec, Wetlands Research Centre, Waterloo, ON.

9. Brink et al. (2013). *The Economics of Ecosystems and Biodiversity for Water and Wetlands. Executive Summary.*

10. M. Paterson (2014). Pers. Comm.

11. Experimental Lakes Area, Research Summaries and History. http://www.experimentallakesarea.ca/Summaries.html.
    http://www.ramsar.org/cda/en/ramsar-pubs-info-ecosystem-services/main/ramsar/1-30-103%5E24258_4000_0__.

12. Editorial, Death of Evidence: Changes to Canadian Science Raise Questions That the Government Must Answer, *Nature,* 487, p. 271, 19 July 2012.

13. Global Research Possibilities Expand as IISD Assumes Operation of Canada's Renowned Experimental Lakes Area, International Institute for Sustainable Development, 1 April 2014. http://www.iisd.org/media/press.aspx?id=274.

14. Orihel, D., and D. Schindler (2014). "Experimental Lakes Area Is Saved, but It's a Bittersweet Victory for Science." *Globe and Mail,* April. http://www.theglobeandmail.com/globe-debate/experimental-lakes-area-is-saved-but-its-a-bittersweet-victory-for-science/article17753956/. Reprinted by permission from Dr. Diane Orihel.

15. Natural Resources Canada, *Groundwater Mapping Program: Overview.* http://www.nrcan.gc.ca/science/story/1387?destination=node/6625.

16. Coté, F. (2006). *Freshwater Management in Canada IV: Groundwater.* Library of Parliament, Parliamentary Information and Research Service, Science and Technology Division, February 2006. http://www.parl.gc.ca/Content/LOP/ResearchPublications/prb0554-e.html.

17. World Bank Open Data, Renewable Internal Freshwater Resources per Capita (Cubic Meters). http://data.worldbank.org/indicator.

18. Environment Canada, *Water Management: Dams and Diversions.* http://ec.gc.ca/eau-water/default.asp?lang=En&n=9D404A01-1.

19. Prairie Pothole Joint Venture, *Introduction to the Prairie Pothole Region, 2005 Implementation Plan Section I.* www.ppjv.org.

20. From The Mission Statement of The Ramsar Convention on Wetlands, www.ramsar.org. Accessed April 18, 2014. Copyright © 2014 by The Ramsar Convention. Reprinted by permission.

21. McGuire, V.L. (2007). Changes in Water Level and Storage in the High Plains Aquifer, Predevelopment to 2005: U.S. Geological Survey Fact Sheet 2007-3029, 2 p.

22. Statistics Canada (2008). Against the Flow: Which Households Drink Bottled Water? *EnviroStats,* Summer, Vol. 2, No. 2. http://www.statcan.gc.ca/pub/16-002-x/2008002/article/10620-eng.htm.

23. Ibid.

24. Naidenko, O., et al., (2008). Bottled Water Quality Investigation: 10 Major Brands, 38 Pollutants, Environmental Working Group, Washington, D.C.

25. Wagner, M., and J. Oehlmann (2009). Endocrine Disruptors in Bottled Mineral Water: Total Estrogenic Burden and Migration from Plastic Bottles, *Environmental Science and Pollution Research, 16*(3), pp. 278–286.

26. Stamminger R. (May/June 2006). Is a Machine More Efficient than the Hand? Home Energy. http://www.homeenergy.org/show/article/nav/kitchen/id/180.

27. Environment Canada, *The Management of Water: Leaking Underground Storage Tanks and Pipelines.* https://www.ec.gc.ca/eau-water/default.asp?lang=En&n=6A7FB7B2-1.

28. Based partly on information from O'Connor, D.R. (2002). *Report of the Walkerton Inquiry: The Events of May 2000 and Related Issues.* Toronto: Government of Ontario.

29. Ibid.

30. Ibid.

31. For example, see Almadidy, A., J. Watterson, P.A.E. Piunno, I.V. Foulds, P.A. Horgen, and U.J. Krull (2003). A Fiber Optic Biosensor for Detection of Microbial Contamination, *Canadian Journal of Chemistry, 81*, pp. 339–349.

32. Nikiforuk, N. (2012). "Don't Gut Fisheries Act, Plead 625 Scientists." Letter in *The Tyee,* 24 March 2012. http://thetyee.ca/News/2012/03/24/Fisheries-Act-Gutting/.

33. WHO/UNICEF, Joint Monitoring Programme for Water Supply and Sanitation, Progress on Drinking Water and Sanitation: 2012 Update.

Chapter 12

1. Department of Fisheries and Oceans (2013). *Stock Assessment of Northern (2J3KL) Cod in 2013.* DFO Can. Sci. Advis. Sec. Sci. Advis. Rep. 2013/014.

2. Choi, J.S., Frank, K.T., Leggett, W.C., & Drinkwater, K. (2004). Transition to an Alternate State in a Continental Shelf Ecosystem, *Canadian Journal of Fisheries and Aquatic Sciences, 61*(4), pp. 505–510.

3. Fisheries and Oceans Canada (2012). *A Federal-Provincial Strategy for the Rebuilding of Atlantic Cod Stocks.* http://www.dfo-mpo.gc.ca/fm-gp/initiatives/cod-morue/strategy_overview-eng.htm.

4. Moore, G.W.K. (2012). A New Look at Greenland Flow Distortion and Its Impact on Barrier Flow, Tip Jets and Coastal Oceanography, *Geophys. Res. Lett., 39*, L22806. doi:10.1029/2012GL054017.

5. Moore, G.W.K., & R.S. Pickart (2012). Northern Bering Sea Tip Jets, *Geophys. Res. Lett., 39*, L08807. doi:10.1029/2012GL051537.

6. NOAA Coral Health and Monitoring Program, *Coral Bleaching.* www.coral.noaa.gov/cleo/coral_bleaching.shtml.

7. Schnoor, J.L. (2013). Ocean Acidification: Comment, *Environ. Sci. Technol., 47*(21), p. 11919. doi: 10.1021/es404263h.

8. Halifax Regional Municipality, *Halifax Harbour Project.* www.halifax.ca/harboursol/WhatistheHarbourSolutionsProject.html.

9. UNDP Issue Brief: Ocean Hypoxia—'Dead Zones,' 15 May 2013. http://www.undp.org/content/undp/en/home/librarypage/environment-energy/water_governance/ocean_and_coastalareagovernance/issue-brief—ocean-hypoxia—dead-zones-/.

10. Worm, B., et al. (2006). Impacts of Biodiversity Loss on Ocean Ecosystem Services, *Science New Series, 314*(5800), pp. 787–790.

11. Hilborn, R. (2007). Moving to Sustainability by Learning from Successful Fisheries, *Ambio, 36*(4), pp. 296–303.

12. Clucas, I. (1997). *A Study of the Options for Utilization of Bycatch and Discards from Marine and Capture Fisheries,* Fishery Industries Division, FAO Fisheries Department. http://www.fao.org/docrep/W6602E/w6602E10.htm#10.

13. Fisheries and Oceans Canada (2012). *Aquaculture in Canada 2012: A Report on Aquaculture Sustainability,* DFO/2012-1803, Cat. No. Fs 45-1/2012E, ISBN 978-1-100-10131-3.

14. Parks Canada, Fisheries and Oceans Canada, and Environment Canada (2005). *Canada's Marine Protected Areas Strategy.* www.dfo-mpo.gc.ca/oceans-habitat/oceans/mpa-zpm/fedmpa-zpmfed/pdf/mpa_e.pdf.

Chapter 13

1. Chen, Z., Wang, J.N., Ma, G.X., & Zhang Y.S. (2013). China Tackles the Health Effects of Air Pollution: Comment, *The Lancet, 382*, pp. 1959–1660.

2. Silva, R.A., et al. (2013). Global Premature Mortality Due to Anthropogenic Outdoor Air Pollution and the Contribution of Past Climate Change, *Environ. Res. Lett., 8*, 034005, doi:10.1088/1748-9326/8/3/034005.

3. Yang, G., et al. (2013). Rapid Health Transition in China, 1990–2010: Findings from the Global Burden of Disease Study 2010, *The Lancet, 381*, pp. 1987–2015.

4. Tan, M., Li, X., & Xin, L. (2014). Intensity of Dust Storms in China from 1980 to 2007: A New Definition, *Atmospheric Environment, 85*, pp. 215–222.

5. This section partially based on information from Environment Canada, *Air: Pollutants.* http://www.ec.gc.ca/air/default.asp?lang=En&n=BCC0B44A-1.

6. Centers for Disease Control, Agency for Toxic Substances and Disease Registry, *Division of Toxicology ToxFAQs,* September 2002. www.atsdr.cdc.gov/tfacts35.pdf.

7. Bedi, J.S., Gill, J.P.S., Aulakh, R.S., Kaur, P., Sharma, A., & Pooni, P.A. (2013). Pesticide Residues in Human Breast Milk: Risk Assessment for Infants from Punjab, India, *Science of the Total Environment, 463–464*, pp. 720–726.

8. UNEP, *Phasing Lead Out of Gasoline: An Examination of Policy Approaches in Different Countries,* 1999.

9. Environment Canada, *CEPA Environmental Registry.* www.ec.gc.ca/CEPARegistry/the_act/ © Environment Canada, 1999.

10. *City of London Air Quality Strategy 2011–2015* (2011).

11. Environment Canada Green Lane, *Acid Rain and . . . Case Studies.*

12. Ibid.

13. Environment Canada National Water Research Institute, 2001.

14. Kolonjari, F. (2014). The 2014 North American Cold Wave and the Polar Vortex, Create Arctic Science blog, March 14, 2014. http://createarctic-science.wordpress.com/2014/03/14/the-polar-vortex-of-2014/.

15. Canadian Network for the Detection of Atmospheric Change (CANDAC), *Research Themes: Arctic Middle Atmospheric Chemistry,* Leader: K. Strong. http://www.candac.ca/candac/Projects/project.php?type=AMAC.

16. World Health Organization (2014). *Burden of disease from the joint effects of Household and Ambient Air Pollution for 2012,* Public Health, Social and Environmental Determinants of Health Department, World Health Organization, Geneva, Switzerland

17. Molina, M.J., & F.S. Rowland (1974). Stratospheric Sink for Chlorofluoromethanes: Chlorine Atom-Catalysed Destruction of Ozone, *Nature, 249*, pp. 810–812. doi:10.1038/249810a0.

18. Farman, J. C., Gardiner, B.G., & Shanklin, J. D. (1985). Large Losses of Total Ozone in Antarctica Reveal Seasonal ClO_x/NO_x Interaction, *Nature, 315*, pp. 207–210.

Chapter 14

1. Parks Canada, Jasper National Park. http://www.pc.gc.ca/eng/pn-np/ab/jasper/activ/explore-interets/glacier-athabasca.aspx.

2. Dave Birrell, Peakfinder: People. www.peakfinder.com/people.asp?PersonsName=Wilcox%2C+Walter+D.

3. IPCC (2007). *Climate Change 2007: Contribution of Working Group I to the Fourth Assessment Report of the Intergovernmental Panel on Climate*

Change. (Solomon, S., D. Qin, M. Manning, Z. Chen, M. Marquis, K.B. Averyt, M. Tignor, and H.L. Miller, eds.). Cambridge University Press, Cambridge, United Kingdom and New York, NY, USA.

4. The four main components of the Intergovernmental Panel on Climate Change *Fifth Assessment Report* are

   i. IPCC (2013). *Climate Change 2013: The Physical Science Basis. Contribution of Working Group I to the Fifth Assessment Report of the Intergovernmental Panel on Climate Change* (Stocker, T.F., D. Qin, G.K. Plattner, M. Tignor, S.K. Allen, J. Boschung, A. Nauels, Y. Xia, V. Bex and P.M. Midgley, eds.). Cambridge University Press, Cambridge, United Kingdom and New York, NY, USA.

   ii. IPCC (2014). Climate Change 2014: Impacts, Adaptation, and Vulnerability. Summary for Policymakers. Part A: Global and Sectoral Aspects. *Contribution of Working Group II to the Fifth Assessment Report of the Intergovernmental Panel on Climate Change.* (Field, C.B., V.R. Barros, D.J. Dokken, K.J. Mach, M.D. Mastrandrea, T.E. Bilir, M. Chatterjee, K.L. Ebi, Y.O. Estrada, R.C. Genova, B. Girma, E.S. Kissel, A.N. Levy, S. MacCracken, P.R. Mastrandrea, and L.L. White, eds.). Cambridge University Press, Cambridge, United Kingdom and New York, NY, USA.

   iii. IPCC (2014). Climate Change 2014: Mitigation of Climate Change. *Contribution of Working Group III to the Fifth Assessment. Report of the Intergovernmental Panel on Climate Change* (Edenhofer, O., R. Pichs-Madruga, Y. Sokona, E. Farahani, S. Kadner, K. Seyboth, A. Adler, I. Baum, S. Brunner, P. Eickemeier, B. Kriemann, J. Savolainen, S. Schlömer, C. von Stechow, T. Zwickel, and J.C. Minx, eds.). Cambridge University Press, Cambridge, United Kingdom and New York, NY, USA.

   iv. IPCC (2014). *Climate Change 2014: Synthesis Report.* Summary for Policymakers. (The Core Writing Team, R.J. Pachauri, and L. Meyer, eds.). Cambridge University Press, Cambridge, United Kingdom and New York, NY, USA.

5. U.S. National Oceanic and Atmospheric Administration (NOAA). *Trends in Atmospheric Carbon Dioxide—Mauna Loa.* www.esrl.noaa.gov/gmd/ccgg/trends/.

6. Canadian Climate Change Modelling and Analysis (CCCma) (2005). As Canada's Climate Changes, and Weather Patterns Shift, Canadian Climate Models Provide Guidance in an Uncertain Future. Environment Canada. www.cccma.ec.gc.ca/20051116_brochure_e_pgs.pdf.

7. IPCC (2014). *Fifth Assessment Report.* http://www.ipcc.ch/report/ar5/index.shtml.

8. Hansen, J., M. Sato, and R. Ruedy (2012). Perception of Climate Change, *PNAS, 109*(37), pp. E2415–E2423.

9. MunichRe NatCatService, Loss Events Worldwide 1980–2014.

10. Kolonjari, F. (2014). The 2014 North American Cold Wave and the Polar Vortex. CREATE Arctic Science, Canada PEARL. https://createarcticscience.wordpress.com/2014/03/14/the-polar-vortex-of-2014/.

11. World Glasogutcier Monitoring Service (2008). *Global Glacier Changes: Facts and Figures.*

12. IPCC (2013). *Climate Change 2013: The Physical Science Basis. Contribution of Working Group I to the Fifth Assessment Report of the Intergovernmental Panel on Climate Change* (Stocker et al., eds.), Technical Summary, Technical Focus Summary, Thematic Focus Element 2, p. 47

13. Ibid., Chapter 13, p. 1180.

14. Ibid., Summary for Policymakers, p. 12.

15. NRCAN Forest Topics > Climate Change > Impacts. http://www.nrcan.gc.ca/forests/climate-change/13095.

16. Stern, N. (2006). *The Stern Review on the Economics of Climate Change.* HM Treasury, London, UK.

17. IPCC (2014). *Climate Change 2014: Synthesis Report.* Summary for Policymakers, Summary for Policymakers, p. 40.

18. You can read this and many other statements about climate change from scientists and scientific associations at NASA's Global Climate Change website: http://climate.nasa.gov/scientific-consensus/.

19. Dr. James Ford, *Geography—Global Environmental Change, Area of Specialization.* http://www.jamesford.ca/about.
    See also Ford, J. (2008). Emerging Trends in Climate Change Policy: The Role of Adaptation, *International Public Policy Review*, pp. 5–16.

20. United Nations Framework Convention on Climate Change (2015) Conference of the Parties, Twenty-first session, Paris, 30 November to 11 December 2015, GE.15-21930(E).

21. National GHG Inventory, Greenhouse Gas Emissions Reporting Program Online Data Search – Facility Reported Data. Environment Canada. www.ec.gc.ca/pdb/ghg/onlinedata/downloadDB_e.cfm#s5.

Chapter 15

1. Mackenzie Gas Project. www.mackenziegasproject.com/theProject/index.html#.

2. CBC News Archives, *The Mackenzie Valley Pipeline: 37 Years of Negotiation.* http://www.cbc.ca/news/business/mackenzie-valley-pipeline-37-years-of-negotiation-1.902366.

3. Ibid.

4. Mackenzie Gas Project. www.mackenziegasproject.com/whoWeAre/index.htm.

5. Ibid.

6. British Petroleum (2014). *Statistical Review of World Energy 2014,* report and data tables.

7. For a detailed discussion of the factors involved in these calculations see, for example, Weißbach, D., G. Ruprecht, A. Huke, K. Czerski, D. Gottlieb, and A. Hussein (2013). Energy Intensities, EROIs (Energy Returned on Invested), and Energy Payback Times of Electricity Generating Power Plants, *Energy, 52*, pp. 210–221.

8. Lemly, A.D. (2004). Aquatic Selenium Pollution Is a Global Environmental Safety Issue, *Ecotoxicology and Environmental Safety, 59*, pp. 44–56

9. Ibid.

10. CCTRM Canada's Clean Coal Technology Roadmap, Natural Resources Canada, 2005. http://www.nrcan.gc.ca/energy/coal/clean-coal/4283.

11. Petroleum History Society, *Six Historical Events in the First 100 Years of Canada's Petroleum Industry.* www.petroleumhistory.ca/history/wells.html#springs.

12. British Petroleum (2014). *Statistical Review of World Energy 2014,* report and data tables.

13. Petroleum History Society, *Six Historical Events in the First 100 Years of Canada's Petroleum Industry.* www.petroleumhistory.ca/history/wells.html#springs.

14. British Petroleum (2014). *Statistical Review of World Energy 2014,* report and data tables.

15. See, for example, Cronin, M.A., W.B. Ballard, J.D. Bryan, B.J. Pierson, and J.D. McKendrick (1998). Northern Alaska Oil Fields and Caribou: A Commentary, *Biological Conservation, 83*(2), pp. 195–208.
    and
    Cronin, M.A., R. Shideler, L. Waits, and R.J. Nelson (2005). Genetic Variation and Relatedness in Grizzly Bears in the Prudhoe Bay Region and Adjacent Areas in Northern Alaska, *Ursus, 16*(1), pp. 70–84.

16. Raynolds, M.K., et al. (2014). Cumulative Geoecological Effects of 62 Years of Infrastructure and Climate Change in Ice-Rich Permafrost Landscapes, Prudhoe Bay Oilfield, Alaska, *Global Change Biology, 20*, pp. 1211–1224.

17. Person, B. T., A.K. Prichard, G.M. Carroll, D.A. Yokel, R.S. Suydam, and J.C. George (2007). Distribution and Movements of the Teshekpuk Caribou Herd 1990–2005: Prior to Oil and Gas Development, *Arctic, 60*(3), pp. 238–250.

18. Owen, N.A., R.O.R. Inderwildi, and D.A. King (2010). The Status of Conventional World Oil Reserves—Hype or Cause for Concern? *Energy Policy, 38*, pp. 4743–4749.

19. Canadian Association of Petroleum Producers (CAPP) (2013). About Canada's Oil Sands. http://www.capp.ca/canadaIndustry/oilSands/Dialogue-Resources/US/Pages/default.aspx.

20. Ibid.

21. King, T.L., B. Robinson, M. Boufadel, and K. Lee (2014). Flume Tank Studies to Elucidate the Fate and Behavior of Diluted Bitumen Spilled at Sea, *Marine Pollution Bulletin, 83*, pp. 32–37.
    and
    Government of Canada (2013). Federal Government Technical Report: Properties, Composition and Marine Spill Behaviour, Fate and Transport of Two Diluted Bitumen Products from the Canadian Oil Sands, pp. 1–85, ISBN 978-1-100-23004-7. Cat. No. En84-96/2013E-pdf.

22. Piracha, T. (2008). Squeezing Water from Oil Sands—Resources Management in Petroleum Development. *Natural Elements: NRCan's Electronic Newsletter*, 22 February 2008, Issue 22.

23. World Ocean Review, Oil Pollution of Marine Habitats > How Oil Enters the Sea. http://worldoceanreview.com/en/wor-1/pollution/oil/.

24. National Energy Board of Canada (NEB). Safety and Environmental Performance Dashboard > Pipeline Incidents. http://www.neb-one.gc.ca/sftnvrnmnt/sft/dshbrd/index-eng.html.

25. CanWest News Service (2008). Energy and Food Inflation Spark Canadian Consumer Change, 17 June 2008. http://www.canada.com/topics/news/story.html?id=a2f55b78-a7ed-4f07-9afc-ff25a5f78f50.

26. Energy and Mines Ministers' Conference of Canada (2012). Moving Forward on Energy Efficiency in Canada: Achieving Results to 2020 and Beyond. Cat. No. M144-166/1-2012E (print).

Chapter 16

1. Nova Scotia Power > Annapolis Tidal Station. http://www.nspower.ca/en/home/about-us/how-we-make-electricity/renewable-electricity/annapolis-tidal-station.aspx.

2. REN21 (2014). *Renewables 2014 Global Status Report*, Renewable Energy Policy Network for the 21st Century, Paris, France: REN21 Secretariat.

3. NRCAN > About Renewable Energy. http://www.nrcan.gc.ca/energy/renewable-electricity/7295#hydro.

4. REN21 (2014). *Renewables 2014 Global Status Report*, Renewable Energy Policy Network for the 21st Century, Paris, France: REN21 Secretariat.

5. Grand Council of the Crees, Social Impact on the Crees of James Bay Project. http://www.gcc.ca/archive/article.php?id=38.

6. REN21 (2014). *Renewables 2014 Global Status Report*, Renewable Energy Policy Network for the 21st Century, Paris, France: REN21 Secretariat.
and
International Atomic Energy Agency (IAEA). Power Reactor Information System (PRIS). http://www.iaea.org/PRIS/WorldStatistics/OperationalReactorsByCountry.aspx.

7. Schneider, M., A. Froggatt, et al. (2014). World Nuclear Industry Status Report, 2014, with Fukushima Daiichi Update.

8. IAEA (2013). *Climate Change and Nuclear Power*, International Atomic Energy Agency, Vienna.

9. OECD/IEA and OECD/NEA (2010). Technology Roadmaps: Nuclear Energy.

10. International Atomic Energy Agency (IAEA). Power Reactor Information System (PRIS). http://www.iaea.org/PRIS/WorldStatistics/OperationalReactorsByCountry.aspx.

11. Walker, J. S. (2004). *Three Mile Island: A Nuclear Crisis in Historical Perspective*. University of California Press.

12. UN Scientific Committee on the Effects of Atomic Radiation (2011). *Sources and Effects of Ionizing Radiation*, 2, NY.

13. von Hippel, F.N. (2011). The Radiological and Psychological Consequences of the Fukushima Daiichi Accident, *Bulletin of the Atomic Scientists*, 67(5), pp. 27–36.

14. Ten Hoeve, J., and M. Jacobson (2012). Worldwide Health Effects of the Fukushima Daiichi Nuclear Accident, *Energy and Environmental Science*, 5(9), p. 8743.

15. REN21 (2014). *Renewables 2014 Global Status Report*, Renewable Energy Policy Network for the 21st Century, Paris, France: REN21 Secretariat.

16. UNFAO Wood Energy. http://www.fao.org/forestry/energy/en/.
and
Dobie, P., and N. Sharma (2014). *Trees as a Global Source of Energy: From Fuelwood and Charcoal to Pyrolysis-Driven Electricity Generation and Biofuels*, World Agroforestry Centre 2014.

17. REN21 (2014). *Renewables 2014 Global Status Report*, Renewable Energy Policy Network for the 21st Century, Paris, France: REN21 Secretariat.

18. Energy and Mines Ministers' Conference (2013). Canada – A Global Leader in Renewable Energy: Enhancing Collaboration on Renewable Energy Technologies. Yellowknife, Northwest Territories. Cat. No.M34-15/2013E-PDF.

19. Worldwatch Institute and Center for American Progress (2006). *American Energy: The Renewable Path to Energy Security*. Washington, D.C.

20. Pfund, N., and B. Healey (2011). What Would Jefferson Do? The Historical Role of Federal Subsidies in Shaping America's Energy Future. DBL Investors.

21. *Powerful Connections: Priorities and Directions in Energy Science and Technology in Canada* (2006). The Report of the National Advisory Committee on Sustainable Energy Science and Technology, Office of Energy Research and Development, Natural Resources Canada.

22. Canadian Biogas Association (2013). Canadian Biogas Study: Benefits to the Economy, Environment and Energy, Summary Document.

23. Enercogen, Beare Road Landfill Gas Power Plant. http://www.enercogen.com/information.htm.
and
City of Toronto Carbon Credits, Generating Clean Power from Waste at the Beare Road Landfill.

24. Fthenakis, V.M., and H.C. Kim (2011). Photovoltaics: Life-cycle analyses, *Solar Energy*, 85, pp. 1609–1628.

25. North Cape Wind Farm. http://www.gov.pe.ca/energy/index.php3?number=60458&lang=E.
and
Wind Energy Institute. http://www.weican.ca/ and http://www.weican.ca/about/facilities.php.

26. Ruotolo, F., V. Senese, G. Ruggiero, L. Maffei, M. Masullo, and T. Iachini (2012). Individual Reactions to a Multisensory Immersive Virtual Environment: The Impact of a Wind Farm on Individuals, *Cognitive Processing*, Special Issue ICSC 2012 5th International Conference on Spatial Cognition: Space and Embodied Cognition, pp. 319–323.

27. Iceland National Energy Authority. http://www.nea.is/geothermal/.

28. Enwave Energy Corporation, Deep Lake Water Cooling System in Toronto. http://www.enwave.com/district_cooling_system.html.

29. NRCAN Natural Resources Canada > Ground-Source Heat Pumps (Earth-Energy Systems). http://www.nrcan.gc.ca/energy/publications/efficiency/heating-heat-pump/6833.

30. RETScreen International, managed by CanmetENERGY Research Centre, Natural Resources Canada. http://www.retscreen.net/ang/home.php.

31. Pimentel, D., and T.W. Patzek (March 2005). Ethanol Production Using Corn, Switchgrass, and Wood; Biodiesel Production Using Soybean and Sunflower, *Natural Resources Research*, 14(1).

Chapter 17

1. Minor Metals Trade Association (2015). Tantalum Market Overview. http://www.mmta.co.uk/tantalum-market-overview.

2. United Nations Meeting Coverage and Press Releases (2001). Security Council Condemns Illegal Exploitation of Democratic Republic of Congo's Natural Resources. http://www.un.org/press/en/2001/sc7057.doc.htm.

3. Metal Pages (2015). http://www.metal-pages.com/metalprices/tantalum/.

4. United Nations (2015). Final Report of the Group of Experts on the Democratic Republic of the Congo. UN S/2015/19.

5. Marshall, B. (2014). *Facts and Figures of the Canadian Mining Industry*. The Mining Association of Canada. http://mining.ca/resources/mining-facts.

6. Natural Resources Canada, *Canadian Global Exploration Activity*. http://www.nrcan.gc.ca/mining-materials/exploration/8296.

7. Marshall, B. (2014). *Facts and Figures of the Canadian Mining Industry*. The Mining Association of Canada. http://mining.ca/resources/mining-facts.

8. Imperial Metals > Mount Polley Mine. http://www.imperialmetals.com/s/MountPolleyMine.asp.

9. *Australasian Mine Safety Journal* (2014). Breach of Tailings Pond Results in 'Largest Environmental Disaster in Modern Canadian History.' http://www.amsj.com.au/news/breach-tailings-pond-results-largest-environmental-disaster-modern-canadian-history/.

10. Imperial Metals > Mount Polley Mine. http://www.imperialmetals.com/s/MountPolleyMine.asp.

11. Imperial Metals > Mount Polley Updates > Tailings Breach Information. http://www.imperialmetals.com/s/Mt_Polley_Update.asp?ReportID=671041.

12. British Columbia Ministry of Environment, Ministry of Energy and Mines, Cariboo Regional District, Fact Sheet – Mount Polley Update 2014ENV0069-001196, August 15, 2014.

13. Marsden, J.O. (2014). Mount Polley Spill Impacts All of Us: Mining Industry's Credibility Is Challenged by Accidents, *Mining Engineering*, *66*(10), pp. 6, 18. Copyright by Society for Mining, Metallurgy & Exploration. Reprinted by permission.

14. Mackin, R. (2014). Government Under Investigation Over Alleged Mount Polley Secrecy, *The Tyee*, August 14, 2014 03:00 pm.

15. Imperial Metals > Mount Polley Updates > Tailings Breach Information: http://www.imperialmetals.com/s/Mt_Polley_Update. asp?ReportID=671041.

16. Mine Environment Neutral Drainage (MEND) Canada. http://mend-nedem.org/default/.

17. National Abandoned/Orphaned Mines Initiative. http://www.abandoned-mines.org/home-e.htm.
and
Mackasey, W.O. (2001). *Abandoned Mines in Canada*, Keynote address to the Orphaned/Abandoned Mines in Canada Workshop, Winnipeg, MB, June 26–27, 2001.

18. Statistics Canada. *Recycling by Canadian Households, 2007*. www.statcan .gc.ca/pub/16-001-m/2010013/aftertoc-aprestdm1-eng.htm.

19. Canadian Wireless Telecommunications Association (2014). 2013 National Cell Phone Recycling Study.

Chapter 18

1. Hogg, K. (2005). Canadian Biogas Industry Overview, *New Era Renewable Energy Solutions*, October 27.

2. City of Toronto Council and Committees, June 1, 1999. *Ontario Hydro Corridor Lands and Beare Road Ski Facility Trust Fund*. www.toronto.ca/ legdocs/1999/agendas/committees/sc/sc990622/it035b.htm.

3. Government of Canada Depository Services Program, *Hazardous Waste Management: Canadian Directions, 1992*. http://dsp-psd.tpsgc.gc.ca/ Collection-R/LoPBdP/BP/bp323-e.htm#A.%20Definitions%20and% 20Classification(txt).

4. United States Environmental Protection Agency: Municipal Solid Waste. http://www.epa.gov/epawaste/nonhaz/municipal/.

5. Statistics Canada (2012). *Human Activity and the Environment: Waste Management in Canada*. Government of Canada, catalogue no. 16-201-X.

6. UNEP (2011). *Towards a Green Economy: Waste: Investing in Energy and Resource Efficiency*. https://www.google.ca/search?q=population+of+ca nada+2010&ie=utf-8&oe=utf-8&gws_rd=cr&ei=0ZzOVP-oNce1yA SEzoBg#q=per+capita+waste+generation+by+country+2010.
See also Giroux, L. (2014). *State of Waste Management in Canada*. Prepared for Canadian Council of Ministers of Environment.

7. Giroux, L. (2014). *State of Waste Management in Canada*. Prepared for Canadian Council of Ministers of Environment.

8. Desrocher, S., and B. Sherwood-Lollar (1998). Isotopic Constraints on Off-Site Migration of Landfill CH_4, *Ground Water*, *36*(5), pp. 801–809. Research Library Core.

9. Ibid.

10. City of Toronto Solid Waste Management, *Facts About Toronto's Trash*, updated November 1, 2007. www.toronto.ca/garbage/facts.htm.

11. Green Energy Futures. http://www.greenenergyfutures.ca/episode/33-landfill-gas-how-old-garbage-can-generate-electricity, and Giroux, L. (2014). *State of Waste Management in Canada*. Prepared for Canadian Council of Ministers of Environment.

12. Giroux, L. (2014). *State of Waste Management in Canada*. Prepared for Canadian Council of Ministers of Environment.

13. US EPA Waste Management: LFG Energy Projects. http://www.epa .gov/outreach/lmop/faq/lfg.html.

14. Worldwatch Institute (2015). New Bans on Plastic Bags May Help Protect Marine Life. http://www.worldwatch.org/node/5565.

15. Composting Council of Canada FAQs. http://www.compost.org/English/qna.html.

16. Giroux, L. (2014). *State of Waste Management in Canada*. Prepared for Canadian Council of Ministers of Environment.

17. Statistics Canada, Summary Tables: Disposal and Diversion of Waste, by Province and Territory. http://www.statcan.gc.ca/tables-tableaux/sum-som/l01/cst01/envir32a-eng.htm.

18. Statistics Canada, Environment Accounts and Statistics Analytical and Technical Paper Series. *Recycling by Canadian Households*, 2007.

19. Felder, M., and C. Morawski (2003). Evaluating the Relationship Between Refund Values and Beverage Container Recovery. Prepared for Beverage Container Management Board.

20. City of Edmonton (2011). *Recycling Facts*. www.edmonton.ca/for_residents/Environment/City_recycle_factsheet(web).pdf.

21. *Solid Waste Research—Areas of Emerging Research*. Retrieved from http://www.ewmce.com/swresearch. Copyright © by Edmonton Waste Management Centre of Excellence. Reprinted by permission.

22. Statistics Canada, Environment Accounts and Statistics Division, CANSIM Table 153-0041, and Giroux, L. (2014). *State of Waste Management in Canada*. Prepared for Canadian Council of Ministers of Environment.

23. Environment Canada. *EPR: Extended Producer Responsibility*. https:// www.ec.gc.ca/gdd-mw/default.asp?lang=En&n=FB8E9973-1.

24. Canadian Institute for Environmental Law and Policy. *Understanding Hazardous Waste in Ontario, 2006*. www.cielap.org/pdf/HazWaste2007 .pdf.

25. Statistics Canada (2012). *Human Activity and the Environment: Waste management in Canada*. Government of Canada, catalogue no. 16-201-X.

26. Canadian Institute for Environmental Law and Policy. *Understanding Hazardous Waste in Ontario, 2006*. www.cielap.org/pdf/HazWaste2007 .pdf.

27. CM Consulting (2013). The Canadian WEEE Report: Waste Electrical and Electronic Equipment Reuse and Recycling in Canada 2013.

28. Environment Canada, Waste Management, *Hazardous Waste: Export and Import of Hazardous Waste and Hazardous Recyclable Material Regulations* (SOR/2005-149). http://www.ec.gc.ca/lcpe-cepa/eng/regulations/ detailReg.cfm?intReg=84.

29. Environment Canada, News Release, July 26, 2007, and Federal Contaminated Sites Inventory. www.tbs-sct.gc.ca/fcsi-rscf/home-accueil. aspx?Language=EN&sid=wu91133413870.

30. Faro Mine Remediation Project. http://www.faromine.ca/project/ steps.html.

31. Canadian Environmental Assessment Agency, Archived – COL-42 Remediation: CFB Esquimalt. http://www.ceaa.gc.ca/052/details-eng.cfm?pid=66173.

32. Aboriginal Affairs and Northern Development Canada, Port Radium Mine (Remediation Complete). https://www.aadnc-aandc.gc.ca/eng/ 1332423218253/1332441057035.

33. Office of the Auditor General of Canada (2002). October Report of the Commissioner of the Environment and Sustainable Development, Chapter 2, Exhibit 2.1, Examples of federal contaminated sites requiring action. http://www.oag-bvg.gc.ca/internet/English/att_ c20021002xe01_e_12327.html.

34. Commission for Environmental Cooperation (2008). *The North American Mosaic: An Overview of Key Environmental Issues: Industrial Pollution and Waste*, Communications Department of the CEC. http://www.cec .org/soe.

35. *EnviroStats* "Recycling in Canada" (July 1, 2007). As reported in *The Daily;* and Statistics Canada (2012). *Human Activity and the Environment: Waste Management in Canada*. Government of Canada, catalogue no. 16-201-X.

36. Statistics Canada (2006). *Waste Management Industry Survey: Business and Government Sectors*. Catalogue no. 16F0023X, Text Table 1: Disposal of waste, by source and by province and territory. Residential nonhazardous wastes disposed includes solid waste produced by all residences and includes waste that is picked up by the municipality (either using its own staff or through contracting firms) and waste from residential sources that is self-hauled to depots, transfer stations, and disposal facilities.

37. Statistics Canada. Table 153-0042, Materials prepared for recycling, by source, Canada, provinces and territories, every 2 years (tonnes),

CANSIM (database), Using E-STAT (distributor). This information covers only those companies and local waste management organizations that reported nonhazardous recyclable material preparation activities and refers only to that material entering the waste stream and does not cover any waste that may be managed on-site by a company or household. Additionally, these data do not include those materials transported by the generator directly to secondary processors such as pulp and paper mills while bypassing entirely any firm or local government involved in waste management activities.

38. Statistics Canada (2006). *Households and the Environment Survey*, Text table 3.7.

Chapter 19

1. Fendall, L.S., and M.A. Sewell (2009). Contributing to Marine Pollution by Washing Your Face: Microplastics in Facial Cleansers, *Marine Pollution Bulletin, 58*(8), pp. 1225–1228.

2. Note: We introduced POPs and many of the other chemicals discussed in this chapter in Chapter 13, *Atmospheric Science and Air Pollution*, and Chapter 11, *Freshwater Systems and Water Resources*.

3. Castañeda, R.A., S. Avlijas, M.A. Simard, and A. Ricciardi (2014). Microplastic Pollution in St. Lawrence River Sediments, *Canadian Journal of Fisheries and Aquatic Sciences, 71*, pp. 1767–1771.

4. Eriksen, M., et al. (2013). Microplastic Pollution in the Surface Waters of the Laurentian Great Lakes, *Marine Pollution Bulletin, 77*, pp. 177–182.

5. Free, C. M., O.P. Jensen, S.A. Mason, M. Eriksen, N.J. Williamson, and B. Boldgive (2014). High Levels of Microplastic Pollution in a Large, Remote, Mountain Lake, *Marine Pollution Bulletin, 85*(1) pp. 156–163.

6. International Campaign Against Microbeads in Cosmetics. http://www.beatthemicrobead.org/en/in-short.

7. European Union (2013). *Green Paper on a European Strategy on Plastic Waste in the Environment.* Brussels, 7.3.2013 COM(2013) 123 final.

8. Government of Canada, House of Commons (2015). *House of Commons Debates: Official Report (Hansard).* Volume 147, Number 188, 2nd Session, 41st Parliament, 24 March 2015.

9. World Health Organization (2014). The Top Ten Causes of Death in the World, Fact Sheet N°310. http://www.who.int/mediacentre/factsheets/fs310/en/.
Note: The calculation is heavily dependent on how the diseases are categorized. For example, neonatal sepsis and other newborn infections might be considered to be perinatal diseases or infectious diseases they are both.

10. Greenberg, M. (2008). The Defence of Chrysotile, 1912–2007, *International Journal of Occupational Environmental Health, 14*(1), pp. 57–66.
and
Health Canada > Health Risks of Asbestos. http://healthycanadians.gc.ca/healthy-living-vie-saine/environment-environnement/outdoor-air-exterieur/asbestos-amiante-eng.php.

11. Health Canada > Radon > What Are the Health Effects of Radon? http://www.hc-sc.gc.ca/ewh-semt/pubs/radiation/radon_brochure/index-eng.php.
Note: There are several units of measurement in common use for radon, and they are not directly convertible. WL (Working Level) is a measure of the concentration of the daughter products of radon; pCi (pico-Curie, usually measured per litre of air) and Bq (Becquerel, usually measured per cubic metre of air) are measures of the number of radioactive disintegrations or transformations per second. 1 Bq is equivalent to 27 pCi. 1 WL is approximately equivalent to 200 pCi/L. Health Canada's actionable limit is currently 200 Bq/m^3, or 5.4 pCi/L (as of 2015).

12. Health Canada > Lead > Frequently Asked Questions. http://www.hc-sc.gc.ca/ewh-semt/contaminants/lead-plomb/asked_questions-questions_posees-eng.php.

13. Health Canada > Environmental and Workplace Health > Formaldehyde. http://www.hc-sc.gc.ca/ewh-semt/pubs/air/formaldehyde-eng.php and http://www.hc-sc.gc.ca/ewh-semt/air/in/poll/construction/formaldehyde-eng.php.

14. Orvos, D.R., D.J. Versteeg, J. Inauen, M. Capdevielle, A. Rothenstein, and V. Cunningham (2002). Aquatic Toxicity of Triclosan, *Environmental Toxicology and Chemistry, 21*(7), pp. 1338–1349.

15. McAvoy, D.C., B. Schatowitz, M. Jacob, A. Hauk, and W.S. Eckhoff (2002). Measurement of Triclosan in Wastewater Treatment Systems, *Environmental Toxicology and Chemistry, 21*(7), pp. 1323–1329.

16. Carey, D.E., and P.J. McNamara (2015). The Impact of Triclosan on the Spread of Antibiotic Resistance in the Environment, *Frontiers in Microbiology*, 01/2015, Volume 5.

17. Health Canada > Canada's Approach on Chemicals > Categorization on the Domestic Substances List. http://www.chemicalsubstanceschimiques.gc.ca/approach-approche/categor-eng.php.

18. United States Environmental Protection Agency > Existing Chemicals > TSCA Existing Chemicals Inventory. http://www.epa.gov/oppt/existingchemicals/pubs/tscainventory/basic.html.

19. Health Canada > Canada's Approach on Chemicals > Categorization on the Domestic Substances List. http://www.chemicalsubstanceschimiques.gc.ca/approach-approche/categor-eng.php.

20. See, for example, Guillette, L.J. Jr., T.S. Gross, G.R. Masson, J.M. Matter, H.F. Percival, and A.R. Woodward (1994). Developmental Abnormalities of the Gonad and Abnormal Sex Hormone Concentrations in Juvenile Alligators from Contaminated and Control lakes in Florida, *Environmental Health Perspectives, 102*, pp. 680–688.
and
Guillette, L.J. Jr., D.A. Crain, and M.P. Gunderson (2000). Alligators and Endocrine-Disrupting Contaminants: A Current Perspective, *American Zoologist, 40*(3), p. 438.

21. For a recent update of this work, see Cruze, L., A.M. Roark, G. Rolland, M. Younas, N. Stacy, and L.J Guillette (2015). Endogenous and Exogenous Estrogens during Embryonic Development Affect Timing of Hatch and Growth in the American Alligator (*Alligator mississippiensis*), *Comparative Biochemistry and Physiology Part B: Biochemistry and Molecular Biology, 184*, pp. 10–18.

22. See, for example, Hayes, T.B. (2005). Welcome to the Revolution: Integrative Biology and Assessing the Impact of Endocrine Disruptors on Environmental and Public Health, *Integrative and Comparative Biology, 45*(2), pp. 321–329.
and
Hayes, T.B., P. Case, S. Chui, D. Chung, C. Haeffele, K. Haston, M. Lee, V.P. Mai, Y. Marjuoa, J. Parker, and M. Tsui (2006). Pesticide Mixtures, Endocrine Disruption, and Amphibian Declines: Are We Underestimating the Impact? *Environmental Health Perspectives, 114*(S-1), pp. 40–50.

23. Renner, R. (2002). Amphibian Declines: Conflict Brewing over Herbicide's Link to Frog Deformities, *Science, 298*(5595), pp. 938–939.

24. Brewer, P., and B. Ley (2014). Contested Evidence: Exposure to Competing Scientific Claims and Public Support for Banning Bisphenol A, *Public Understanding of Science, 23*(4) pp. 395–410.

25. Health Canada > Chemicals Management Plan. http://www.chemical-substanceschimiques.gc.ca/plan/index-eng.php.

26. Kelland, K. (2010). Experts Demand European Action on Plastics Chemical, *Reuters: Science, Health, Environment*, 22 June 2010. http://www.reuters.com/article/2010/06/22/us-chemical-bpa-health-idUSTRE65L6JN20100622?loomia_ow=t0:s0:a49:g43:r3:c0.084942:b35124310:z0.

27. Consumer Reports (2009). Concern over Canned Foods: Our Tests Find Wide Range of Bisphenol A in Soups, Juice, and More, *Consumer Reports*, December 2009.

28. Hunt, P.A., et al. (2003). Bisphenol A Exposure Causes Meiotic Aneuploidy in the Female Mouse, *Current Biology, 13*(7), pp. 546–553.

29. Health Canada > Food and Nutrition > Health Canada's Updated Assessment of Bisphenol A (BPA) Exposure from Food Sources. http://www.hc-sc.gc.ca/fn-an/securit/packag-emball/bpa/index-eng.php#fnb1 and http://www.hc-sc.gc.ca/fn-an/securit/packag-emball/bpa/index-eng.php#fnb1.

30. Kinch, C.D., K. Ibhazehiebo, J.-H. Jeong, H.R. Habibi, and D.M. Kurrasch (2015). Low-Dose Exposure to Bisphenol A and Replacement Bisphenol S Induces Precocious Hypothalamic Neurogenesis in Embryonic Zebrafish, *Proceedings of the National Academy of Science, 112*(5), pp. 1475–1480.

31. Environmental Defence, *Toxic Nation: Toxic Baby Bottles in Canada,* 2008. www.toxicnation.ca/files/toxicnation/report/ToxicBabyBottleReport.pdf.

32. Bushnik, T., et al. (2010). Lead and Bisphenol A Concentrations in the Canadian Population. Statistics Canada, Health Reports. 82-003-X, Vol. 21, No. 3.

33. Kinch, C.D., K. Ibhazehiebo, J.-H. Jeong, H.R. Habibi, and D.M. Kurrasch (2015). Low-Dose Exposure to Bisphenol A and Replacement Bisphenol S Induces Precocious Hypothalamic Neurogenesis in Embryonic Zebrafish, *Proceedings of the National Academy of Science,* *112*(5), pp. 1475–1480.

34. Biedermann, S., P. Tschudin, and K. Grob (2010). Transfer of Bisphenol A from Thermal Printer Paper to the Skin, *Analytical and Bioanalytical Chemistry,* *398*(1), pp. 571–576.

35. See, for example, Health Canada > Second Report on Human Bio-monitoring of Environmental Chemicals in Canada. http://www.hc-sc.gc.ca/ewh-semt/pubs/contaminants/chms-ecms-cycle2/index-eng.php.

36. Alberta Health Services (2014). Fort Chipewyan Update: Appendix I. Prepared by Surveillance & Reporting, Cancer Measurement Outcomes Research and Evaluation, Cancer Control, Alberta Health Services. February 7, 2014.

37. See, for example, Guillette, E.A. (2000). A Broad-Based Evaluation of Pesticide-Exposed Children, *Central European Journal of Public Health,* 07/2000, *8*, Suppl., pp. 58–59.
   and
   Guillette, E.A., M.M. Meza, M.G. Aquilar, A.D. Soto, and I. E. Garcia (1998). Anthropological Approach to the Evaluation of Pre-school Children Exposed to Pesticides in Mexico, *Environmental Health Perspectives,* *106*(6), pp. 347–353.

38. Basel Convention on the Transboundary Movement of Hazardous Wastes and Their Disposal. www.basel.int.

39. Chatterjee, Rhitu (2008). Hunting for Persistent Chemicals That Might Pollute the Arctic, *Environmental Science and Technology,* *42*(14), p. 5034.

40. EU Growth: Internal Market, Industry, Entrepreneurship, and SMEs. *How REACH Works.* http://ec.europa.eu/growth/sectors/chemicals/reach/about/index_en.htm.

41. Krause, M., et al. (2012). Sunscreens: Are They Beneficial for Health? An Overview of Endocrine Disrupting Properties of UV-Filters, *International Journal of Andrology,* *35*(3), pp. 424–436.
   and
   Schlumpf, M., et al. 2001. *In vitro* and *in vivo* Estrogenicity of UV Screens, *Environmental Health Perspectives,* *109*, pp. 239–244.

Chapter 20

1. This piece is based on a summary of *Mining Denendeh: A Dene Nation Perspective on Community Health Impacts of Mining,* by Chris Paci (Lands and Environment Department, Dene Nation) and Noeline Villebrun, Dene National Chief, prepared for the 2004 Mining Ministers Conference, Coppermine, Nunavut.

2. Regarding the incident of contamination that led to the mine's closure in 2004, as reported by Friends of the Environment, Australia: *A History of Duress—Uranium Mining on Mirarr Land.* http://www.foe.org.au/history-duress-uranium-mining-mirarr-land.

3. *The Age,* July 27, 2007, as cited by World Information Service on Energy and The WISE Uranium Project. www.wise-uranium.org/upjab.html.

4. Based partly on information from *Living with the Land: A Manual for Documenting Cultural Landscapes,* NWT Cultural Places Program, Government of NWT, 2007.

5. Parts of this paragraph are based on information from *Mining Denendeh: A Dene Nation Perspective on Community Health Impacts of Mining,* by Chris Paci (Lands and Environment Department, Dene Nation) and Noeline Villebrun, Dene National Chief, prepared for the 2004 Mining Ministers Conference, Coppermine, Nunavut.

6. The most recent data from Statistics Canada show the gender gap in Canada is 26%—this compares only full-time, full-year workers. This means that for every $1.00 earned by a man in Canada, a woman earns 74 cents. Statistics Canada. Pay Equity Commission. Gender Wage Gap, September 18, 2012. http://www.payequity.gov.on.ca/en/about/pubs/genderwage/wagegap.php.

7. Based on information from the Canadian Museum of History: *James Bernard Harkin.* http://www.historymuseum.ca/cmc/exhibitions/hist/biography/biographi204e.shtml.

8. Based on information from Canadian Museum of History: *Clifford Sifton.* http://www.historymuseum.ca/cmc/exhibitions/hist/advertis/ads2-06e.shtml#sifton.

9. Leopold, A. (1949). *A Sand County Almanac,* Oxford University Press, New York: NY.

10. Kakkar, Rahul (2014). *History of Chipko Movement.* http://www.importantindia.com/11686/history-of-chipko-movement/.

11. Green Belt Movement. http://www.greenbeltmovement.org/ and other sources.

12. Recall that radioactivity had only been discovered some 30 years earlier, in the late 1890s.

13. Nikiforuk, A. (1998). Echoes of the Atomic Age: Cancer Kills Fourteen Aboriginal Uranium Workers, *Calgary Herald,* Saturday, March 14.

14. World Information Service on Energy, *Decommissioning of Port Radium.* http://www.wise-uranium.org/udcdn.html#PORTRADIUM.

15. *Radiation Exposure Compensation Act.* http://www.justice.gov/civil/common/reca; and "Navajo President Joe Shirley, Jr., Updates Navajo Miners on Progress toward Getting Fair RECA Compensation," September 2006.

16. Smith, A., *An Inquiry into the Nature and Causes of the Wealth of Nations* ("The Wealth of Nations"), first published in 1776.

17. Repetto, Robert (1992). Accounting for Environment Assets, *Scientific American,* *266,* pp. 94–100.

18. Environment Centre Northern Territory (Australia) (2014). Reconsidering Ranger: A Brief on the Social, Environmental, and Economic Cost of Uranium Mining in Kakadu. http://www.ecnt.org/content/reconsidering-ranger-social-environmental-and-economic-cost-uranium-mining-kakadu.

19. Carson, R.T., L. Wilks, & D. Imber (1994). Valuing the Preservation of Australia's Kakadu Conservation Zone, *Oxford Economic Papers, 46* (Special Issue on Environmental Economics), pp. 727–749.

20. Costanza, R., et al. (1997). The Value of the World's Ecosystem Services and Natural Capital, *Nature, 387,* pp. 253–260.

21. Costanza, R., et al. (2014). Changes in the Global Value of Ecosystem Services, *Global Environmental Change, 26,* pp. 152–158.

22. Daly, Herman E. (2001). Sustainable Development and OPEC. Paper invited for the conference OPEC and the Global Energy Balance: Towards a Sustainable Energy Future. Vienna, Austria.

Chapter 21

1. Young, S.P. (2013). *Evidence-Based Policy-Making in Canada.* Oxford University Press.

2. Environment Canada, SARA Public Registry, Greater Sage-Grouse *urophasianus* Subspecies. http://www.registrelep-sararegistry.gc.ca/species/speciesDetails_e.cfm?sid=305.

3. Environment Canada, SARA Registry Emergency Order for the Protection of the Greater Sage-Grouse. http://www.registrelep-sararegistry.gc.ca/default.asp?lang=En&n=F25868B7-1.

4. Environment Canada, SARA Registry Emergency Order, Amending the Emergency Order for the Protection of the Greater Sage-Grouse. http://www.registrelep-sararegistry.gc.ca/default.asp?lang=En&n=DEBA2642-1.

5. Environment Canada, SARA Public Registry, *Canada's Strategy: Species at Risk Act.* http://www.sararegistry.gc.ca/default.asp?lang=En&n=6B319869-1.

6. Environment Canada, *Habitat Stewardship Program.* http://www.ec.gc.ca/hsp-pih/default.asp?lang=En&n=59BF488F-1.

7. Note: We discussed the specific meanings of terms like endangered, threatened, extirpated, and vulnerable in Chapter 9, *Conservation of Species and Habitats.*

8. Olive, A. (2012). Does Canada's Species at Risk Act Live Up to Article 8(J)? *The Canadian Journal of Native Studies, 32*(1), pp. 173–189.

9. Hardin, Garrett (1968). The Tragedy of the Commons, *Science, 162*(3859), pp. 1243–1248.

10. Meadows, D.H., D.L. Meadows, J. Randers, and W.W. Behrens III (1972). *Limits to Growth.* New York, NY: Universe Books.

11. Ehrlich, P.R. (1968). *The Population Bomb.* New York, NY: Ballantine Books.

12. Lappé, F.M. (1971). *Diet for a Small Planet.* New York, NY: Ballantine Books.

13. Based on information from Denstedt, S. and R.J. King (2014). *Doing Business in Canada.* Chapter 16: Environmental Law in Canada. Osler Canada.

14. Environment Canada, *Acts, Regulations, and Agreements.* www.ec.gc.ca/default.asp?lang=En&n=48D356C1-1.

15. Antweiler, W., and K. Harrison (2006). The ARET Challenge: Evaluating a Negotiated Voluntary Challenge Program. University of British Columbia.

16. International Organization for Standardization (ISO). http://www.iso.org/iso/home/about.htm and http://www.iso.org/iso/home/standards/management-standards/iso14000.htm.

17. Environment Canada > Northern Contaminants Program. http://www.science.gc.ca/default.asp?lang=En&n=7A463DBA-1.

18. CEPA Fact Sheet No. 2: *CEPA 1999 at a Glance: What It Is, What It Does, How It Works,* Government of Canada, October 2005.

19. Based on information from Government of Canada (2012). *Bill C-38: Jobs, Growth, and Long-Term Prosperity Act.* First Session, Forty-first Parliament, 60-61 Elizabeth II, 2011–2012. An Act to implement certain provisions of the budget tabled in Parliament on March 29, 2012, and other measures. Assented to 29th June 2012. http://www.parl.gc.ca/HousePublications/Publication.aspx?DocId=5697420.
EcoJustice & West Coast Environmental Law (2012). *What Bill C-38 Means for the Environment* and *Top Ten Environmental Concerns of Budget Bill C-38.* http://wcel.org/resources/publication/what-bill-c-38-means-environment.
David Suzuki Foundation (2012). *Bill C-38: What You Need to Know.* http://www.davidsuzuki.org/publications/resources/2012/bill-c38-what-you-need-to-know/.

20. Environmental Commissioner of Ontario. www.eco.on.ca.

21. City of Toronto > Green Roof Bylaw. http://www1.toronto.ca/wps/portal/contentonly?vgnextoid=83520621f3161410VgnVCM10000071d60f89RCRD&vgnextchannel=3a7a036318061410VgnVCM10000071d60f89RCRD.

22. Bond, W., D. O'Farrell, G. Ironside, B. Buckland, and R. Smith (2005). *Environmental Indicators and State of the Environment Reporting: An Overview for Canada—Background Paper to an Environmental Indicators and State of the Environment Reporting Strategy, 2004–2009, Environment Canada.* Environment Canada, Knowledge Integration Strategies Division, Environmental Reporting Branch, Strategic Information Integration Directorate, Ottawa.

23. Environment Canada Environmental Indicators > CESI. https://www.ec.gc.ca/indicateurs-indicators/default.asp?lang=En.

24. Fisheries and Oceans Canada > Canada's State of the Oceans Report 2012. http://dfo-mpo.gc.ca/science/coe-cde/soto/report-rapport-2012/index-eng.asp.

25. Province of Saskatchewan (2015). Saskatchewan's 2015 State of the Environment Report.

26. Environmental Performance Index. http://epi.yale.edu/.

27. Province of Saskatchewan (2015). Saskatchewan's 2015 State of the Environment Report.

28. Environment Canada > International Environmental Agreements (Archived Content). https://www.ec.gc.ca/international/default.asp?lang=En&n=0E5CED79-1.

29. Commission on Environmental Cooperation > North American Agreement on Environmental Cooperation. http://www.cec.org/Page.asp?PageID=1226&SiteNodeID=567.

30. Murray, William (1993). *NAFTA and the Environment.* Ottawa, ON: Environment Canada Science and Technology Division. http://dsp-psd.tpsgc.gc.ca/Collection-R/LoPBdP/MR/mr116-e.htm.

31. Aguilar, S., R. Constantino, A. Cosbey, and S. Shaw (2011). *Environmental Assessment of NAFTA by the Commission for Environmental Cooperation: An Assessment of the Practice and Results to Date.* International Institute for Sustainable Development.

32. Excerpt from the Charter of the United Nations (1945). http://www.un.org/en/documents/charter/index.shtml.

33. Environment Canada > What Is a Lakewide Management Plan? http://ec.gc.ca/grandslacs-greatlakes/default.asp?lang=En&n=0CB6DFA3-1.

34. Environment Canada > Lake Erie Lakewide Action and Management Plan Annual Report 2013. http://ec.gc.ca/grandslacs-greatlakes/default.asp?lang=En&n=5A2E69DC-1&offset=1&toc=show.

35. World Bank (2015). World Bank Lending Fiscal Year 2014 Annual Report.

36. World Bank > Closing the $70 Billion Climate Finance Gap. April 9, 2015. http://www.worldbank.org/en/news/feature/2015/04/09/closing-the-climate-finance-gap.

37. OPEC's 12 members (as of mid-2015) are Algeria, Angola, Ecuador, the Islamic Republic of Iran, Iraq, Kuwait, Libya, Nigeria, Qatar, Saudi Arabia, United Arab Emirates, and Venezuela. OPEC > Member Countries. http://www.opec.org/opec_web/en/about_us/25.htm.

38. OPEC Statement to the United Nations Climate Change Conference (COP-18), Doha, Qatar, 2012. http://www.opec.org/opec_web/en/2461.htm.

39. OPEC > A Brief History. http://www.opec.org/opec_web/en/about_us/24.htm.

40. *Our Common Future: The Report of the United Nations Commission on Environment and Development* (1987). Oxford, UK: Oxford University Press.

41. Withana, S., for Institute for European Environmental Policy (2014). Reforming Environmentally Harmful Subsidies. http://www.ieep.eu/work-areas/environmental-economics/market-based-instruments/2014/02/reforming-environmentally-harmful-subsidies–1276.
and
Withana, S., et al. (2012). Study Supporting the Phasing Out of Environmentally Harmful Subsidies. A report by the Institute for European Environmental Policy (IEEP), Institute for Environmental Studies, Vrije Universiteit (IVM), Ecologic Institute and VITO for the European Commission – DG Environment. Final Report. Brussels. 2012.

42. International Monetary Fund (IMF) (2013). Energy Subsidy Reform: Lessons and Implications. Paper available from http://www.imf.org/external/np/fad/subsidies/.

43. Province of British Columbia > Ministry of Finance > Carbon Tax. http://www.fin.gov.bc.ca/tbs/tp/climate/carbon_tax.htm.

44. Ontario Emissions Trading Registry. http://www.oetr.on.ca.

45. European Union Climate Action > The EU Emissions Trading System. http://ec.europa.eu/clima/policies/ets/index_en.htm.

46. Ecolabel Index. http://www.ecolabelindex.com/.

Chapter 22

1. New York, 4 December 2014—Secretary-General remarks to the General Assembly on the Synthesis Report on the Post-2015 Agenda.

2. Summerhill Group. www.summerhillgroup.ca/about-us.

3. Summerhill Impact > Projects > Mow Down Pollution. http://summerhillimpact.org/projects/mow-down-pollution/.

4. Statistics Canada, Lawnmower Use—Canada and Provinces. http://www.statcan.gc.ca/pub/16-002-x/2010001/article/lawnmowers-tondeuses/tbl/tbl001-eng.htm.

5. Car Heaven. http://carheaven.ca/.

6. *Our Common Future* (1987). Report of the World Commission on Environment and Development, Oxford University Press.

7. United Nations Development Program, Millennium Development Goals and Beyond. © 2014 United Nations. Reprinted with the permission of the United Nations.

8. United Nations Development Program, Millennium Development Goals and Beyond. http://www.un.org/millenniumgoals/environ.shtml.

9. Reprinted by permission from Sevanne Kassarjian.

10. AMS Lighter Footprint Strategy, updated 2014. www.ams.ubc.ca/.

11. University Leaders for a Sustainable Future, Talloires Declaration: http://www.ulsf.org/programs_talloires.html.

12. Comeau, J.-S. (2014). Environmental Audit Reveals Mt. A's Improvements, *The Argosy*, November 5, 2014. http://www.argosy.ca/article/environmental-audit-reveals-mt-a%E2%80%99s-improvement.

13. Peden, S. (2014). Bien Sûr, We Recycle Here! University Recognized by Recyc-Québec. *Concordia University News*, October 20, 2014. http://www.concordia.ca/cunews/main/stories/2014/10/20/bien-sur-we-recyclehere.html.

14. Gorilla Composting, McGill University. http://gorilla.mcgill.ca/.

15. Ban the Bottle: http://www.banthebottle.net/map-of-campaigns/ and Canadian Federation of Students > Back the Tap: http://cfs-fcee.ca/take-action/bottle-water-free-communities/.

16. University of British Columbia, UBC Sustainability, Campus Initiatives, UBC Renew. http://sustain.ubc.ca/campus-initiatives/green-buildings/ubc-renew.

17. UBC Campus Sustainability Office, EcoTrek Project Complete, *Sustainable Energy Management Program*, www.ecotrek.ubc.ca, and http://sustain.ubc.ca/campus-initiatives/climate-energy/ecotrek.

18. University of Victoria, Office of Campus Planning and Sustainability. http://www.uvic.ca/sustainability/.

19. Simon Fraser University > Community > ASSC1 Building Earns Earth Award. July 10, 2008. www.sfu.ca/sfunews/Stories/sfunews071008019.shtml.

20. Laurentian University, Cooperative Freshwater Ecology Unit, *Vale Living with Lakes Centre*. http://laurentian.ca/cooperative-freshwater-ecology-unit.

21. UBC EcoTrek. http://sustain.ubc.ca/campus-initiatives/climate-energy/ecotrek.

22. University of Toronto, Facilities and Services. http://sustain.fs.utoronto.ca/.

23. University of Toronto, Sustainability Office, *ReWire Project*. http://sustain.fs.utoronto.ca/get-involved/about-rewire/index.html.

24. University of Toronto Mississauga. http://www.utm.utoronto.ca/green/green-building.

25. Formula Sun Grand Prix > Teams 2014. http://americansolarchallenge.org/about/formula-sun-grand-prix/.

26. United States Department of Energy Solar Decathlon, 2013 Teams. http://www.solardecathlon.gov/past/2013/teams.html.

27. Trent University, *Physical Resources, Powerhouse*. www.trentu.ca/physicalresources/powerhouse.php.

28. American College and University Presidents' Climate Commitment. www.presidentsclimatecommitment.org.

29. University and College Presidents' Climate Change Statement of Action for Canada. http://www.climatechangeaction.ca/signatories.

30. Common Energy. http://commonenergyubc.org.

31. McGill University Campus Crops. http://campuscropsmcgill.blogspot.ca/.

32. Sustainable Concordia, *Allégo: Smarter Transport for a More Sustainable Future*. http://sustainableconcordia.ca/working-groups/allego-concordia/.

33. UBC U-Pass. http://planning.ubc.ca/vancouver/transportation-planning/u-pass-compass-card.

34. University of Toronto School of the Environment > Undergraduate Research. http://www.environment.utoronto.ca/Research/UndergraduateResearch.aspx.

35. *UBC Sustainability Report: SEEDS*. www.sustain.ubc.ca/ and http://sustain.ubc.ca/courses-teaching/seeds/project-highlights.

36. Sierra Youth Coalition. http://www.syc-cjs.org/.

37. Association for the Advancement of Sustainability in Higher Education (AASHE). www.aashe.org.

38. Mission and Vision Statements of AASHE. Copyright by Association for the Advancement of Sustainability in Higher Education. Reprinted by permission.

39. Association for the Advancement of Sustainability in Higher Education (AASHE), STARS Program. https://stars.aashe.org/pages/about/stars-annual-review-2014.html.

40. Wilson, E.O. (1999). *Consilience: The Unity of Knowledge*, Vintage Books, p. 319, Random House Inc., New York: NY.

Glossary

abiotic Refers to any nonliving component of the environment. Compare *biotic*.

abyssal plain The flat, mostly topographically featureless ocean floor.

acid drainage (or acid mine drainage) A process in which sulphide minerals in newly exposed rock surfaces react with oxygen and rain water to produce sulphuric acid, which causes chemical runoff as it leaches metals from the rocks. Acid drainage can be a natural phenomenon, but mining can greatly accelerate it by exposing many new rock surfaces.

acidic deposition The deposition of acidic or acid-forming pollutants from the atmosphere onto Earth's surface by the precipitation of unusually acidic rain, snow, or hail, fog, or settling of dry particles.

acidic precipitation See *acidic deposition*.

acid mine drainage (or **AMD**) See *acid drainage*.

acid rain Acidic precipitation deposited in the form of rain.

active solar (energy) An approach in which technological devices are used to focus, move, or store solar energy. Compare *passive solar energy collection*.

acute exposure Exposure to a toxicant occurring in high amounts for short periods of time. Compare *chronic exposure*.

adaptation A responsive strategy in which it is accepted that an event with potential impacts (such as climatic change) is going to occur, and plans are made to adjust to the impacts. Compare *intervention; mitigation*.

adaptive management The systematic, scientific testing of different management approaches to improve methods over time.

adaptive trait A trait that confers greater likelihood that an individual will reproduce. Compare *adaptation*.

aerobic Occurring in an environment where oxygen is present. For example, the decay of a rotting log proceeds by aerobic decomposition. Compare *anaerobic*.

aerosols Very fine liquid droplets or solid particles aloft in the atmosphere.

afforestation Planting of trees where the land has not been forested for a long time.

age distribution The relative numbers of organisms of each age within a population, often displayed on an age structure diagram or age pyramid.

age structure See *age distribution*.

aggregate Unconsolidated sediment (such as sand or gravel) or crushed stone used in construction.

Agricultural Revolution The shift around 10 000 years ago from a hunter-gatherer lifestyle to an agricultural way of life in which people began to grow their own crops and raise domesticated animals; this much more intensive, manipulative way of producing and extracting resources marked a permanent change in the relationship of people to the natural environment. Also called *Neolithic Revolution*. Compare *Industrial Revolution; Medical-Technological Revolution*.

agriculture The practice of cultivating soil, producing crops, and/or raising livestock for human use and consumption.

agroforestry Planting of trees in conjunction with crops. The trees benefit the crops by aiding in the biogeochemical cycling of nutrients and water, by contributing organic material to the soil, and by providing shade. The trees can also provide harvestable products, such as fruits and nuts.

A horizon A layer of soil found in a typical soil profile. It forms the top layer or lies below the O horizon (if one exists). It consists of mostly inorganic mineral components such as weathered substrate, with some organic matter and humus from above mixed in. Often referred to as *topsoil*. Compare *B horizon; C horizon; O horizon; R horizon*.

air The specific mixture of gaseous constituents that makes up Earth's atmosphere.

air mass A large volume of air that is internally uniform in temperature, relative humidity, and density.

air pollution The act of contaminating the air, or the condition of being contaminated by air pollutants.

Air Quality Health Index (AQHI) A standardized, national-level index that quantifies the level of air pollution (from 1 = very low to 10+ = very high) and associated health risks.

airshed The geographical area associated with a particular air mass.

albedo The reflectivity of a surface.

allergen A toxicant that over-activates the immune system, causing an immune response when one is not necessary.

alloy A mix of metals.

alpine Pertaining to high-altitude environments.

alpine tundra Tundra that occurs at high altitudes.

amensalism A relationship between members of different species in which one organism is harmed and the other is unaffected. Compare *commensalism, mutualism,* and *parasitism*.

ammonia (NH_3) A colourless gas with a pungent smell; an important precursor component of terrestrial plant nutrients.

anaerobic Occurring in an environment that is anoxic (i.e., has little or no oxygen). The conversion of organic matter to fossil fuels (crude oil, coal, natural gas) at the bottom of a deep lake, swamp, or shallow sea is an example of anaerobic decomposition. Compare *aerobic*.

Antarctic Bottom Water (AABW) A deep water mass originating in the Southern Ocean and linked to the global thermohaline circulation. Compare *North Atlantic Deep Water (NADW)*.

anthropocentrism A human-centred view of our relationship with the environment. Compare *biocentrism; ecocentrism*.

anthropogenic Human-generated.

anthroposphere The human sphere, and the built environment; the environment as modified by human actions.

aphotic zone A zone (usually in a water body) that is lacking in light, or where light does not penetrate. Compare *photic zone*.

aquaculture The raising of aquatic organisms for food in controlled environments.

aquifer An underground reservoir of water, typically a permeable rock or sediment unit.

arable Used in reference to land or soil that is suitable for growing crops.

artificial selection Trait selection conducted under human direction. Examples include selective breeding of crop plants, pets, and livestock.

artificial wetland See *constructed wetland*.

atmosphere The thin layer of gases surrounding planet Earth. Compare *biosphere; hydrosphere; geosphere*.

atmospheric deposition The wet or dry deposition on land of a wide variety of pollutants, including mercury, nitrates, organochlorides, and others. Acidic deposition is one type of atmospheric deposition.

atmospheric pressure The weight per unit area produced by a column of air.

atom The smallest component of an element that maintains the chemical properties of that element.

autotroph (or primary producer) An organism that can produce its own food from simple substances in the surrounding environment, most commonly utilizing energy from sunlight. Includes green plants, algae, and cyanobacteria.

bedrock The continuous mass of solid rock that makes up Earth's crust.

benthic Of, relating to, or living on the bottom of a water body. Compare *pelagic*.

benthic zone The bottom of a water body, including the lowermost layer of water and the bottom sediments. Compare *littoral zone; limnetic zone; profundal zone*.

B horizon The layer of soil that lies below the A horizon and above the C horizon. Minerals that leach out of the A horizon are carried down into the B horizon (or subsoil) and accumulate there. Sometimes called the subsoil, zone of accumulation, or zone of deposition. Compare *A horizon; C horizon; O horizon; R horizon*.

bioaccumulation The buildup of toxicants in the tissues of an organism.

bioconcentration The process by which chemicals carried in water can become magnified or concentrated in organisms or in the food chain in aquatic environments. Compare *bioaccumulation* and *biomagnification*.

biocapacity The capacity of a terrestrial or aquatic system to be biologically productive and to absorb waste.

biocentrism A philosophy that ascribes relative values to actions, entities, or properties on the basis of their effects on all living things or on the integrity of the biotic realm in general. The biocentrist evaluates an action in terms of its overall impact on living things, including—but not exclusively focusing on—human beings. Compare *anthropocentrism; ecocentrism*.

biodiesel Diesel fuel produced by mixing vegetable oil, used cooking grease, or animal fat with small amounts of ethanol or methanol (wood alcohol) in the presence of a chemical catalyst.

biodiversity (or **biological diversity**) The sum total of all organisms in an area, taking into account the diversity of species, their genes, their populations, and their communities.

biodiversity hotspot An area that supports an especially great diversity of species, particularly species that are endemic to the area.

biofuel Fuel produced from biomass energy sources and used primarily to power automobiles.

biogas Natural gas (methane) extracted from landfill gas.

biogeochemical cycle See *nutrient cycle*.

biological control (or **biocontrol**) The attempt to battle pests and weeds with organisms that prey on or parasitize them, rather than by using pesticides.

biological diversity See *biodiversity*.

biological hazard Human health hazards that result from ecological interactions among organisms. These include parasitism by viruses, bacteria, or other pathogens. Compare *chemical hazard; cultural hazard; physical hazard*.

biological weathering Weathering that occurs when living organisms break down parent material, either by physical or chemical means. Compare *chemical weathering; physical weathering*.

biomagnification The magnification of the concentration of toxicants in an organism caused by its consumption of other organisms in which toxicants have bioaccumulated. Also called *food chain concentration*.

biomass Biological material; consists of living and recently deceased organic matter.

biomass energy Energy harnessed from plant and animal matter, including peat, wood from trees, charcoal from burned wood, and combustible animal waste products, such as cattle manure. Fossil fuels are not considered biomass energy sources because their organic matter has not been part of living organisms for millions of years and has undergone considerable chemical alteration since that time.

biome A major regional complex of similar plant communities; a large ecological unit defined by its dominant plant type and vegetation structure. Compare *ecoregion*.

biophilia The concept that human beings subconsciously seek a connection with the rest of life.

biopower The burning of biomass energy sources to generate electricity.

biosphere The sum total of all the planet's living organisms and the abiotic portions of the environment with which they interact. Compare *atmosphere; hydrosphere; geosphere*.

biotechnology The material application of biological science to create products derived from organisms. The creation of transgenic organisms is one type of biotechnology.

biotic Refers to any living component of the environment. Compare *abiotic*.

biotic potential An organism's capacity to produce offspring.

bitumen A thick (that is, viscous) form of petroleum rich in carbon and poor in hydrogen.

bog A type of wetland in which a pond is thoroughly covered with a thick, floating mat of vegetation. Compare *marsh; swamp; fen; peatland*.

bond A chemical–physical connection between two or more atoms; an electrical force linking two atoms together.

boreal forest A biome of northern coniferous forest that stretches in a broad band across much of Canada, Alaska, Russia, and Scandinavia. Boreal forest consists of a limited number of tree species, dominated by evergreen conifers such as black and white spruce, in large regions of forest interspersed with occasional bogs and lakes. See also *taiga*.

breakdown product A compound that results from the degradation of a toxicant.

breeder reactor A nuclear reactor that generates fissile material at a rate that is faster than the rate at which it is consumed.

brownfield A site that has been contaminated by hazardous materials but that has the potential to be cleaned up and remediated for other purposes.

by-catch That portion of a commercial fishing catch consisting of animals caught unintentionally. By-catch kills many thousands of fish, marine mammals, and birds each year.

canopy (of a forest) The more or less continuous upper level of leaves and branches defined by the tree tops. Forests can have closed or open canopies.

cap-and-trade system A permit trading system in which a government determines an acceptable level of overall pollution and then issues individual companies permits to pollute. A company receives credit for amounts it does not emit and can then sell this credit to other companies. Essentially synonymous with *emissions trading system*.

captive breeding The practice of capturing members of threatened and endangered species so that their young can be bred and raised in controlled environments and subsequently reintroduced into the wild.

carbohydrate An organic compound consisting of atoms of carbon, hydrogen, and oxygen.

carbon (C) The chemical element with atomic number six. A key element in organic compounds.

carbon capture and storage (CCS) Technologies or approaches that remove carbon dioxide from power plant or other emissions and sequester it, in an effort to mitigate global climate change.

carbon cycle A major nutrient cycle consisting of the routes that carbon atoms take through the nested networks of environmental systems.

carbon dioxide (CO_2) A colourless gas used in photosynthesis, given off by respiration, and released by burning fossil fuels. A primary greenhouse gas whose buildup contributes to global climate change.

carbon footprint The cumulative amount of carbon, or carbon dioxide equivalent, that a person or institution emits, and is indirectly responsible for emitting, into the atmosphere, contributing to global climate change. Compare *ecological footprint*.

carbon monoxide (CO) A colourless, odourless gas produced primarily by the incomplete combustion of fuel. A criteria air pollutant.

carbon-neutrality The state in which an individual, business, or institution emits no net carbon to the atmosphere. This may be achieved by reducing carbon emissions and/or employing carbon offsets to compensate for emissions.

carcinogen A chemical or type of radiation that causes cancer.

carnivore An organism that consumes animals. Compare *detritivore; herbivore; omnivore*.

carrying capacity The maximum population size that a given environment can sustain.

cell The most basic organizational unit of organisms.

cellular respiration The process by which a cell uses the chemical reactivity of oxygen to split glucose into its constituent parts, water and carbon dioxide, and thereby release chemical energy that can be used to form chemical bonds or to perform other tasks within the cell. Compare *photosynthesis*.

channelization Modification of a river's channel or banks by straightening, widening, or concrete-lining, usually for the purposes of navigation, flood control, or diversion for irrigation or water supply.

chemical hazard Chemicals that pose human health hazards. These include toxins that occur or are produced naturally, such as some heavy metals and plant toxins, as well as many anthropogenic chemicals, including some disinfectants, pesticides, and other synthetic chemicals. Compare *biological hazard; cultural hazard; physical hazard*.

chemical weathering Weathering that results when water or other substances chemically interact with parent material, altering it and producing new minerals or dissolved ions. Compare *biological weathering; physical weathering*.

chemosynthesis The process by which bacteria in hydrothermal vents use the chemical energy of hydrogen sulphide (H_2S) to transform inorganic carbon into organic compounds. Compare *photosynthesis* and *autotroph*.

chlorofluorocarbon (CFC) One of a group of human-made organic compounds derived from simple hydrocarbons, such as ethane and methane, in which hydrogen atoms are replaced by halide elements such as chlorine, bromine, or fluorine. CFCs deplete the protective ozone layer in the stratosphere.

C horizon The layer of soil that lies below the B horizon and above the R horizon. It contains rock particles that are larger and less weathered than the layers above. It consists of parent material that has been altered only slightly or not at all by the process of soil formation. Compare *A horizon; B horizon; O horizon; R horizon*.

chronic exposure Exposure for long periods of time to a toxicant occurring in low amounts. Compare *acute exposure*.

clay Sediment particles less than 0.002 mm in diameter; also refers to clay minerals, a group of hydrated alumniosilicate minerals that are typically very fine-grained, with platy crystal structures. Compare *sand; silt*.

clean coal Term used to describe technologies and approaches that seek to reduce the generation and release of sulphur and other pollutants before, during, or after coal is burned for power.

clear-cutting The harvesting of timber by cutting all the trees in an area, leaving only stumps. Although it is the most cost-efficient method, clear-cutting is also the most damaging to the environment.

climate The pattern of atmospheric conditions found across large geographical regions over long periods of time. Compare *weather*.

climatograph (or climate diagram) A visual representation of a region's average monthly temperature and precipitation.

climax community In the traditional view of ecological succession, a community that remains in place with little modification until disturbance restarts the successional process. Today, ecologists recognize that community change is more variable and less predictable than originally thought, and that assemblages of species may instead form complex mosaics in space and time.

closed canopy (of a forest) A forest canopy that has very few openings where light can penetrate to the forest floor.

closed system A system that is self-contained with regard to exchanges of matter (but not energy) with its surroundings. Scientists may treat a system as closed to simplify some questions they are investigating, but no natural system is truly closed. Compare *open system*.

coal A fossil fuel composed of organic matter that was compressed under high pressure to form a dense, carbon-rich rock.

coevolution The circumstance in which two closely associated species influence each other's evolutionary process.

cogeneration A practice in which two forms of energy are produced simultaneously at the same plant; for example, the extra heat generated in the production of electricity can be captured and used to heat workplaces and homes, or to produce other kinds of power.

cold front The boundary where a mass of cold air displaces a mass of warmer air. Compare *warm front*.

command-and-control An approach to protecting the environment that sets strict legal limits and threatens punishment for violations of those limits.

commensalism A relationship between members of different species in which one organism benefits and the other is unaffected. Compare *amensalism, mutualism,* and *parasitism*.

community A group of populations of organisms that live in the same place at the same time.

community-supported agriculture (CSA) A practice in which consumers pay farmers in advance for a share of their yield, usually in the form of weekly deliveries of produce.

compaction (of soil) A decrease in the volume of soil and collapse of the soil structure, which can occur when fluids are withdrawn or when the soil is too heavily tilled, dries out as a result of the removal of vegetation, or bears too much weight.

competition A relationship in which multiple organisms seek the same limited resource.

competitive exclusion An outcome of interspecific competition in which one species excludes another species from resource use entirely.

composting The conversion of organic waste into mulch or humus by encouraging, in a controlled manner, the natural biological processes of decomposition.

compound A molecule composed of atoms of two or more elements.

confined aquifer A water-bearing, porous, and permeable layer of rock, sand, or gravel that is trapped between an upper and lower layer of less permeable substrate, such as clay or unfractured rock. The water in a confined aquifer (also called *artesian aquifer*) is under pressure because it is trapped between two impermeable layers. Compare *unconfined aquifer*.

coniferous Refers to trees that are "evergreen," that is, they do not lose their leaves in the fall. Coniferous trees produce cones to host their seeds, and typically have needles rather than broad, flat leaves. Compare *deciduous*.

conservation biology A scientific discipline devoted to understanding the factors, forces, and processes that influence the loss, protection, and restoration of biological diversity within and among ecosystems.

conservation ethic An ethical position, holding that humans should put natural resources to use but also have a responsibility to manage them wisely. Compare *preservation ethic*.

conservation of matter (law of) The basic principle that matter may be transformed from one type of substance into others, but it cannot be created or destroyed.

constructed wetland A wetland that is built, usually for the purpose of stormwater runoff management or wastewater management. Also called *artificial wetland* or *engineered wetland*.

consumer See *heterotroph*.

continental collision A convergent boundary between tectonic plates, both of which carry continental crust, leading to the formation of a continental collision zone, rock deformation and metamorphism, and uplift of high mountain chains.

continental shelf The very gently sloping underwater edge of a continent, varying in width from 100 m to 1300 km. At the shelf break, the continental shelf gives way to the steeper continental slope, which leads down to the ocean floor or abyssal plain.

continental slope The relatively steep offshore slope that extends downward from the continental shelf to the continental rise, which connects it to the abyssal plain of the ocean.

control The portion of an experiment in which a variable has been left unmanipulated or untreated.

controlled burn See *prescribed burn*.

convection A process of heat transfer in which materials that are heated from below rise (as a consequence of lower density), cool, and then sink again (as a consequence of higher density).

convergent evolution The process whereby two distinct and geographically separate species evolve similar traits, generally as a result of adaptation to selective pressures from similar environments or habitats.

convergent plate boundary A boundary along which tectonic plates are moving toward one another. Can result in subduction or collision, and mountain range formation.

coral bleaching The loss of the coloured symbiotic algae that normally inhabit corals; possibly caused by stress due to increased water temperature or turbidity.

coral reef A structure composed of the secreted calcium carbonate skeletons of tiny colonial marine organisms.

core (1) The innermost zone of Earth, made up mostly of iron, that lies beneath the crust and mantle. (2) The part of a forest or reserve that is isolated from the surrounding area by a transitional or buffer zone. Compare *edge*.

Coriolis effect The apparent deflection of north–south air currents to a partly east–west direction, caused by the Coriolis force, which results from the faster spin of regions near the equator than of regions near the poles as a result of Earth's rotation.

corridor A passageway of protected land established to allow animals to travel between islands of protected habitat.

corrosive Substances that corrode (gradually erode, or eat away) metals and other materials. One criterion for defining hazardous waste.

cost–benefit analysis A method commonly used by neoclassical economists, in which estimated costs for a proposed action are totalled and then compared to the sum of benefits estimated to result from the action.

criteria air contaminants (CACs) (or criteria pollutants) Six air pollutants—carbon monoxide, sulphur dioxide, nitrogen dioxide, tropospheric ozone, particulate matter, and lead—for which maximum allowable concentrations have been established for ambient outdoor air because of the threats they pose to human health.

cropland Land that humans use to raise plants for food and fibre.

crude birth rate The number of births per 1000 individuals for a given time period. Compare *crude death rate*.

crude death rate The number of deaths per 1000 individuals for a given time period. Compare *crude birth rate*.

crude oil (or petroleum) A fossil fuel produced by the conversion of organic compounds by heat and pressure. Crude oil is a mixture of hundreds of different types of hydrocarbon molecules characterized by carbon chains of different length.

crust The relatively low-density outer layer of Earth, consisting of rock that floats atop the malleable mantle, which in turn surrounds a mostly iron core.

cryosphere The temporarily and perennially frozen parts of the hydrosphere, including snow, sea ice, lake and river ice, glaciers, and ice caps and sheets. Compare *hydrosphere*.

cultivate Preparing and using soil for the purpose of raising crops (domesticated plants, or *cultivars*).

cultural hazard Human health hazards that result from the place we live, our socioeconomic status, our occupation, or our behavioural choices. These include choosing to smoke cigarettes, or living or working with people who do. Also called *lifestyle hazard*. Compare *biological hazard; chemical hazard; physical hazard*.

cumulative effects Repeated, local, short-timescale events, processes, or materials that add up to something bigger, more impactful, or more concentrated.

current The flow of a liquid (or gas, such as air) in a certain direction.

cycle Flows of elements, compounds, and energy from reservoir to reservoir through the Earth system.

dam Any obstruction placed in a river or stream to block the flow of water so that water can be stored in a reservoir. Dams are built to prevent floods, provide drinking water, facilitate irrigation, and generate electricity.

data Information that is generally quantitative in nature.

deciduous Refers to trees, usually broad-leafed, that lose their leaves each fall and remain dormant during the winter. Compare *coniferous*.

decommissioning The process by which an industrial facility is permanently removed from service and the land on which it resides is reclaimed, remediated, or stabilized.

decomposer Organisms, mainly fungi and bacteria, that break down leaf litter and other nonliving matter into simpler constituents that can then be taken up and used as nutrients by plants. Contrast with *detritivore*; decomposers are often detritivores, but the terms are not exactly synonymous.

deep ecology A philosophy established in the 1970s based on principles of self-realization (the awareness that humans are inseparable from nature) and biocentric equality (the precept that all living beings have equal value). Deep ecology holds that because we are truly inseparable from our environment, we must protect all other living things as we would protect ourselves.

deep-well injection A hazardous waste disposal method in which a well is drilled deep beneath an area's water table into porous rock below an impervious layer. Wastes are then injected into the well, so that they will be absorbed into the porous rock and remain deep underground, isolated from ground water and human contact. Compare *surface impoundment*.

deforestation The loss of forested land.

demographic transition A theoretical model of economic and cultural change that explains the declining death rates and birth rates that occurred in Western nations as they became industrialized. The model holds that industrialization caused these rates to fall naturally by decreasing mortality and by lessening the need for large families.

demography A social science that applies the principles of population ecology to the study of statistical change in human populations.

denitrification A multistep chemical process in which nitrates in soil are reduced by denitrifying bacteria, and ultimately released to the atmosphere.

deoxyribonucleic acid (DNA) A molecule that directs the production of proteins; self-replicating material that is the main constituent of chromosomes in almost all living organisms. Compare *ribonucleic acid (RNA)*.

deposition The arrival of transported material at a new location. For example, eroded sediment is deposited in streams, and wind-borne particulates are deposited on the surfaces of land and water bodies.

desalinization (or desalination) The removal of salt from sea water or from soil.

desert The driest biome on Earth, with annual precipitation of less than 25 cm. Because deserts have relatively little vegetation to insulate them from temperature extremes, sunlight readily heats them in the daytime, but daytime heat is quickly lost at night, so temperatures vary widely from day to night and in different seasons.

desertification A loss of more than 10% of a land's productivity due to erosion, soil compaction, forest removal, overgrazing, drought, salination, climate change, depletion of water sources, or other factors. Severe deserti-

fication can result in the expansion of desert areas or creation of new deserts in areas that once supported fertile land.

detritivore Organisms (heterotrophs) that feed on dead and decomposing plant or animal matter (i.e., detritus). Contrast with *decomposer*.

directional selection Mode of natural selection in which selection drives a feature in one direction rather than another—for example, toward larger or smaller body size, or longer or shorter tail length. Compare *disruptive selection; stabilizing selection*.

discharge zone An area where ground water emerges from the subsurface and flows out on the surface to become or join a surface water body.

disruptive selection Mode of natural selection in which a trait diverges from its starting condition in two or more directions. Compare *directional selection; stabilizing selection*.

divergent evolution The process whereby two populations of the same species, which are geographically separated or reproductively isolated, evolve different traits over time as a result of adaptation to selective pressures from different environments or habitats. Divergent evolution of two populations may be expected eventually to result in speciation—that is, the characteristics of the populations eventually become so different that they are no longer of the same species.

divergent plate boundary A boundary along which tectonic plates are moving apart from one another. If magma rising underneath the divergent boundary reaches the surface, new crust is formed as the emerging lava cools and solidifies in the rift. An example is the Mid-Atlantic ridge. Compare *transform plate boundary* and *convergent plate boundary*.

diversion (of water) Removing water from a river system or changing its flow for use in another location.

domesticate Breeding and raising of wild animals or plants in captivity, for human use.

dose The amount of toxicant a test animal receives in a dose–response test. Also, the amount of toxicant ingested, absorbed, or otherwise taken in by an organism. Compare *response*.

dose–response curve A curve that plots the response of test animals to different doses of a material, such as a toxicant. The response is generally quantified by measuring the proportion of animals exhibiting negative effects upon ingesting or being exposed to the material.

downwelling In the ocean, a flow of surface water toward the ocean floor. Downwelling can occur where surface currents converge, or where surface water cools sufficiently that it becomes dense enough to sink. Compare *upwelling*.

drainage basin Area of land drained by one river and its tributaries; sometimes called a *catchment*. Compare *watershed*.

dryland An area with generally low precipitation but not so arid as to be classified as a desert; grasslands, savannahs, and shrublands are typical dryland biomes.

dust dome A phenomenon in which smog and particulate air pollution become trapped in a layer overlying an urban centre. Can be exacerbated by the urban heat island effect.

dynamic equilibrium The state reached when processes within a system are moving in opposing directions at equivalent rates so that their effects balance out.

Earth system science The scientific study of the entire Earth system as an integrated whole, including how its component parts and their interactions have evolved, how they function, and how they may be expected to continue to evolve on all timescales.

earthquake Ground-shaking associated with the sudden release of strain energy stored in rock. The release of stored energy generates seismic waves that travel outward in all directions from the earthquake's focus.

ecocentrism A philosophy that considers actions in terms of their damage or benefit to the integrity of whole ecological systems, including both biotic and abiotic elements. For an ecocentrist, the well-being of an individual organism—human or otherwise—is less important than the long-term well-being of a larger integrated ecological system. Compare *biocentrism; anthropocentrism*.

ecofeminism A philosophy holding that the patriarchal structure of society is a root cause of both social and environmental problems. Ecofeminists hold that a worldview traditionally associated with women, which interprets the world in terms of interrelationships and cooperation, is more in tune with nature than a worldview traditionally associated with men, which interprets the world in terms of hierarchies and competition.

ecolabelling The practice of designating on a product's label how the product was grown, harvested, or manufactured, so that consumers buying it are aware of the processes involved and can differentiate between brands that claim to use environmentally beneficial processes and those that do not.

ecological economics A field of study that applies the principles of ecosystems and thermodynamics to economic and social processes. Ecological economists view society and the economy as being within and dependent upon the natural environment. Compare *environmental economics*.

ecological footprint The cumulative amount of land and water required to provide the raw materials a person or population consumes and to dispose of or recycle the waste that is produced.

ecological restoration Efforts to reverse the effects of human disruption of ecological systems and to restore communities to their "natural" state. Compare *restoration ecology*.

ecology The science that deals with the distribution and abundance of organisms, the interactions among them, and the interactions between organisms and their abiotic environments.

economics The study of how we decide to use scarce resources to satisfy the demand for goods and services.

economy A social system that converts resources into goods and services.

ecoregion A large area of land or water that contains a geographically distinct assemblage of natural communities that share a large majority of their species and ecological dynamics, share similar environmental conditions, and interact ecologically in ways that are critical for their long-term persistence. Compare *biome*.

ecosystem All of the organisms and nonliving entities that occur and interact with each other and with the abiotic environment in a particular area at the same time.

ecosystem-based management The attempt to manage the harvesting of resources in ways that minimize impact on the ecosystems and ecological processes that provide the resources.

ecosystem diversity The number and variety of ecosystems in a particular area. One way to express biodiversity. Related concepts consider the geographical arrangement of habitats, communities, or ecosystems at the landscape level, including the sizes, shapes, and interconnectedness of patches of these entities.

ecosystem ecology The study of how the living and nonliving components of ecosystems interact and transfer energy among themselves.

ecosystem service An essential service an ecosystem provides that supports life and makes economic activity possible. For example, ecosystems naturally purify air and water, cycle nutrients, provide for plants to be pollinated by animals, and serve as receptacles and recycling systems for the waste generated by our economic activity.

ecotone A transitional zone where ecosystems meet.

ecotourism Visitation of natural areas for tourism and recreation. Most often involves tourism by more affluent people, which may generate economic benefits for less affluent communities near natural areas and thus provide economic incentives for conservation of natural areas.

ED$_{50}$ (effective-dose-50%) The amount of a toxicant it takes to affect 50% of a population of test animals. Compare *threshold dose; LD$_{50}$.*

edge The part of a forest that is immediately adjacent to the surrounding area, and is separated from the forest core by a transitional or buffer zone.

effluent Water that flows out of a facility such as a wastewater treatment plant, mine, or power plant.

electrolysis A process in which electrical current is passed through a compound to release ions. Electrolysis offers one way to produce hydrogen for use as fuel: Electrical current is passed through water, splitting the water molecules into hydrogen and oxygen atoms.

electromagnetic spectrum The range of wavelengths of radiation from shortest (gamma rays) to longest (radio waves).

electronic waste (or **e-waste**) Discarded electronic products such as computers, monitors, printers, DVD players, cell phones, and other devices. Heavy metals in these products mean that this waste may require treatment as hazardous waste.

element A fundamental type of matter; a chemical substance with a given set of properties, which cannot be broken down into constituent substances with other properties. Chemists currently recognize 92 elements that occur in nature, as well as more than 20 others that have been artificially created. These are organized by atomic number and other properties into the Periodic Table of the Elements.

El Niño An exceptionally strong warming of the eastern Pacific Ocean associated with weakness or even reversal of the trade winds. El Niño patterns occur roughly every 2 to 7 years, affect regional weather systems by distorting the jet streams, and depress local fish and bird populations by altering the marine food web. Compare *La Niña*.

El Niño–Southern Oscillation (ENSO) A systematic shift in atmospheric pressure, sea-surface temperature, and ocean circulation in the tropical Pacific Ocean. ENSO cycles give rise to El Niño and La Niña conditions.

emergent property A characteristic that is not evident in a system's components individually.

emigration The departure of individuals from a population.

emissions trading system A permit trading system in which a government issues marketable emissions permits to conduct environmentally harmful activities. The government determines an acceptable level of overall emissions and then issues permits to individual companies. A company receives credit for amounts it does not emit and can then sell this credit to other companies. Essentially synonymous with *cap-and-trade*.

endangered Categorization of a species in danger of extirpation or extinction. Compare *threatened; vulnerable*.

endemic Native or restricted to a particular geographical region. An endemic species occurs in one area and nowhere else on Earth.

endocrine disruptor A toxicant that interferes with the endocrine (hormone) system.

"end-of-pipe" A response to pollution that deals only with effluents as they emerge from the "end of the pipe," rather than reducing or eliminating the pollution at its source.

energy An intangible phenomenon that can do work, that is, can change the position, physical state, or temperature of matter. Energy is a property of matter; it can be transferred or converted into a different form, but it cannot be created or destroyed.

energy conservation The practice of reducing energy use as a way of extending the lifetime of our fossil fuel supplies, of being less wasteful, and of reducing our impact on the environment.

engineered wetland See *constructed wetland*.

environment The sum total of our surroundings, including all of the living things and nonliving things with which we interact.

environmental audit An assessment undertaken to determine whether a facility's operations are in compliance with environmental regulations and/or whether there are changes that could be made to reduce the environmental impacts or enhance the sustainability of the operation.

environmental economics A branch of economics that acknowledges and attempts to account for the value of environmental goods and services through accounting procedures such as valuation. Compare *ecological economics*.

environmental ethics The application of ethical standards to environmental questions.

environmental health Environmental factors that influence human health and quality of life and the health of ecological systems essential to environmental quality and long-term human well-being.

environmental impact assessment (EIA) A multifaceted, multistakeholder assessment of the possible impacts, both positive and negative, that a proposed project might have on the environment; the outcome of an EIA is an environmental impact statement.

environmental justice The principle that all people (sometimes extended to all beings) have the right to a clean, healthy environment, and should therefore have equal access to environmental resources and protection from the impacts of environmental degradation.

environmental nongovernmental organization (ENGO) An environmental organization that is not a government or a private-sector agency. Well-known examples include the Sierra Club, Conservation International, and Greenpeace.

environmental policy Public policy that pertains to human interactions with the environment. It generally aims to regulate resource use or reduce pollution to promote human welfare and/or protect natural systems.

environmental refugeeism A situation in which a person has been driven from his or her homeland by natural disasters, resource shortages, or environmental change or degradation.

environmental science The study of how the natural world works and how humans and the environment interact.

environmentalism A social movement dedicated to protecting the natural world.

Environment Canada The department of the federal government that is mandated to preserve and enhance the quality of the natural environment; conserve Canada's renewable natural resources; conserve and protect Canada's water resources; forecast weather and environmental change; enforce rules relating to boundary waters; and coordinate environmental policies and programs; created in 1971 by the *Department of the Environment Act.*

epidemiology A field of study that involves large-scale comparisons among groups of people in order to determine patterns and determinants of health and disease and the effects of exposure to various health risks.

EROI (energy returned on investment) The ratio determined by dividing the quantity of energy returned from a process by the quantity of energy invested in the process. Higher EROI ratios mean that more energy is produced from each unit of energy invested.

erosion The removal of material from one place and its transport to another by the action of wind, water, or glacial ice.

estuary An area where a river flows into the ocean, mixing fresh water with salt water.

ethanol Alcohol produced as a biofuel by fermenting biomass, generally from carbohydrate-rich crops such as corn.

ethics The study of good and bad, right and wrong. The term can also refer to the set of moral principles or values of a group or an individual.

eutrophic Term describing a water body that has high-nutrient and (typically) low-oxygen conditions. Compare *oligotrophic.*

eutrophication The process of nutrient enrichment, increased production of organic matter, and subsequent ecosystem degradation.

evaporation The conversion of a substance from a liquid to a gaseous form.

even-aged A condition in timber plantations—generally monocultures of a single species—in which all trees in a stand are of the same age. Most ecologists view plantations of even-aged stands more as crop agriculture than as ecologically functional forests. Compare *uneven-aged.*

evenness See *relative abundance.*

evolution Genetically based change in the appearance, functioning, and/or behaviour of organisms across generations, often by the process of natural selection.

e-waste See *electronic waste.*

Exclusive Economic Zone (EEZ) The area of ocean under the jurisdiction of the bordering nation, which extends to 200 nautical miles from the shore, as per the United Nations' *Convention on the Law of the Sea.*

exotic (species) A species that is non-native to an area; alien. Compare *invasive species.*

experiment An activity designed to test the validity of a hypothesis by manipulating variables.

exponential growth The increase of a population (or of anything) by a fixed percentage each year; geometric growth. Contrast with *linear growth.*

extensification Increasing resource productivity by bringing more land into production. Compare *intensification.*

externality (or external cost) A cost or benefit of a transaction that affects people other than the buyer or seller. Examples of negative externalities or external costs include harm to citizens from water pollution or air pollution discharged by nearby factories.

extinction The disappearance of an entire species. Compare *extirpation.*

extirpation The disappearance of a particular population from a given area, but not the entire species globally. Compare *extinction.*

fault In rock, a fracture along which some movement has occurred.

feedback loop (or feedback cycle) A circular process in which a system's output serves as input to that same system. See *negative feedback loop; positive feedback loop.*

feedlot A huge barn or outdoor pen designed to deliver energy-rich food to animals living at extremely high densities. Also called a "factory farm" or *concentrated animal feeding operation (CAFO).*

fen A type of wetland covered with a thick mat of vegetation, and fed by ground water. Compare *bog, marsh, swamp, peatland.*

Ferrel cell One of a pair of cells of convective circulation between 30° and 60° north and south latitude that influence global climate patterns. Compare *Hadley cell; polar cell.*

fertilizer A substance that promotes plant growth by supplying essential nutrients such as nitrogen or phosphorus.

fission A reaction in which a nucleus splits into two smaller nuclei, with an accompanying release of energy. Compare *fusion.*

flammable Substances that easily catch fire. One criterion for defining hazardous waste.

floodplain The region of land over which a river has historically wandered, and which is periodically inundated.

floor (of a forest) The lowest level of a forest, consisting of the topsoil, organic litter, and humus.

flux The movement of materials or energy among pools or reservoirs in a cycle.

food chain A simple relationship in which primary producers are eaten by primary consumers, who are, in turn, eaten by secondary consumers, and so on. A significant amount of energy is typically lost in moving from the bottom of the food chain to higher levels, in turn.

food security An adequate, reliable, and available food supply to all people at all times.

food web A visual representation of feeding interactions within an ecological community that shows an array of relationships between organisms at different trophic levels.

forest A densely wooded area.

forestry The scientific study and professional management of forests; also called *silviculture.*

fossil The remains, impression, or trace of an ancient organism that has been preserved in rock or sediment.

fossil fuel Naturally occurring hydrocarbon fuels, including but not limited to oil (petroleum), natural gas, and coal.

fossil record The cumulative body of fossils worldwide that palaeontologists study to infer the history of past life on Earth.

fracking (or hydraulic fracturing) A hydrocarbon extraction technology that involves pumping a fluid under high pressure into the ground to crack the rock, then injecting sand or glass beads to hold the cracks open. Additional fluids are injected to facilitate the loosening and extraction of the hydrocarbons. Used for shale gas, oil contained in rock of low permeability, and bitumen from oil sands.

fresh water Water that has low salinity (generally, less than about 500 ppm dissolved salts).

front The boundary between air masses that differ in temperature and moisture (and therefore density). See *warm front; cold front.*

fuel cell A technology that produces energy from a chemical reaction similar to that in a battery, without combustion of the fuel.

fundamental niche The full niche of a species. Compare *realized niche.*

fusion A reaction in which two nuclei combine to form a single, more massive nucleus, with an accompanying release of energy. Compare *fission.*

gangue Waste rock and commercially nonvaluable minerals that occur with ore.

gene A stretch of DNA that represents a unit of hereditary information.

gene bank See *seed bank.*

generalist A species that can survive in a wide array of habitats or use a wide array of resources. Compare *specialist.*

genetic bottleneck An event in which a significant portion of the population of a species is killed or otherwise prevented from reproducing, leading to a decrease in genetic diversity in subsequent generations.

genetic diversity A measurement of the differences in DNA composition among individuals within a given species.

genetic engineering Any process scientists use to manipulate an organism's genetic material in the lab by adding, deleting, or changing segments of its DNA.

genetically modified (GM) organism An organism that has been genetically engineered using a technique called recombinant DNA technology.

geoengineering Any of a number of approaches to the artificial manipulation of the global climate system.

geographic information system (GIS) Computer software that takes multiple types of data (for instance, on geology, hydrology, vegetation, animal species, and human development) and overlays them on a common set of spatial coordinates; a common tool of geographers, landscape ecologists, resource managers, and conservation biologists.

geological isolation An approach to waste disposal that makes use of the natural buffering and containment properties of rocks and minerals to hold, absorb, and isolate the waste from contact with the hydrosphere and biosphere. Mainly used for hazardous or radioactive wastes.

geosphere The solid Earth; sometimes the word *lithosphere* is used with this connotation, but technically it has a distinct meaning. Compare *lithosphere*.

geothermal energy Energy derived from heat generated deep within Earth. The radioactive decay of elements at depth generates heat that rises to the surface in magma, through heat flow by conduction, and through fissures and cracks. Can be synonymous with *terrestrial energy*.

glaciation The process of extension of ice sheets from the polar regions far into Earth's temperate zones during cold periods of Earth's history; also, a period during which this process occurred in the past. Compare *interglaciation*.

glacier A perennially frozen body of ice that is formed, over many years, from compressed snow.

global climate change A change in planetary-scale aspects of Earth's climate, such as average temperature, precipitation, and storm activity. Generally refers to the current (post-Industrial Revolution) warming trend in global temperatures and associated climatic changes, which are at least partly of anthropogenic origin. Compare *global warming*.

global climate model (GCM) Quantitative, computer-based models designed to link processes in the atmosphere, hydrosphere, biosphere, and geosphere in three spatial dimensions on a detailed geographical grid. Such models allow for the study of interrelationships and feedbacks among the processes that control the global climate system, and changes in these processes over time, including projecting possible future changes.

global warming An increase in Earth's average surface temperature. The term is most frequently used in reference to the warming trend of recent years and decades. Global warming is one aspect of global climate change, and in turn drives other components of climate change. Compare *global climate change*.

global warming potential A quantity that specifies the ability of one molecule of a given greenhouse gas to contribute to atmospheric warming, relative to carbon dioxide.

goods Something material (such as a resource or a product) that has economic utility.

grassland An area of land in which grasses are the dominant plant species.

greenhouse effect The warming of Earth's surface and lower atmosphere, caused by infrared (thermal) energy that is absorbed and re-emitted by greenhouse gases in the troposphere.

greenhouse gas (GHG) A gas that selectively absorbs infrared (thermal) radiation released by Earth's surface and then warms the surface and troposphere by emitting energy, thus giving rise to the greenhouse effect. Important greenhouse gases include carbon dioxide (CO_2), water vapour, ozone (O_3), nitrous oxide (N_2O), halocarbon gases, and methane (CH_4). Compare *radiatively active gas*.

Green Revolution An intensification of the industrialization of agriculture in the latter half of the twentieth century, which has led to dramatically increased crop yields per unit area of farmland. Practices include devoting large areas to monocultures of crops specially bred for high yields and rapid growth; heavy use of fertilizers, pesticides, and irrigation water; and sowing and harvesting on the same piece of land more than once per year or per season.

green tax A levy on environmentally harmful activities and products aimed at providing a market-based incentive to correct for market failure.

gross primary production (GPP) The energy that results when autotrophs convert energy (most commonly from sunlight) to energy of chemical bonds in sugars. Autotrophs use a portion of this production to power their own metabolism, which entails oxidizing organic compounds by cellular respiration. Compare *net primary production; secondary production*.

ground-source heat pump (GSHP) A pump that harnesses geothermal energy from near-surface sources of earth and water, and that can help heat residences.

ground water Water held in aquifers underground.

gyre A rotating oceanic current, or vortex, that forms as a result of complex interactions involving wind, ocean water, currents, friction, and the Coriolis force.

Haber-Bosch process A chemical process for synthesizing ammonium, thereby fixing nitrogen, on an industrial scale.

habitat The specific environment in which an organism lives, including both biotic and abiotic factors.

habitat fragmentation The process by which large expanses of habitat are broken up into smaller, isolated pieces. The size of contiguous habitat is an issue for some large animals, but the character of the habitat can be an issue even for smaller animals; smaller habitat fragments have a greater proportion of edge, and the characteristics of core and edge habitat can differ substantially.

Hadley cell One of a pair of cells of convective circulation between the equator and 30 degrees north and south latitude that influence global climate patterns. Compare *Ferrel cell; polar cell*.

half-life The amount of time it takes for one-half the atoms of a radioisotope to emit radiation and decay. Different radioisotopes have different half-lives, ranging from fractions of a second to billions of years.

hardwood Wood derived from broad-leafed, mostly deciduous trees. Compare *softwood*.

harvesting Gathering, withdrawal, capture, or other method of removal of product from the stock of a resource.

hazardous waste Waste that is toxic, chemically reactive, flammable, or corrosive. Compare *industrial solid waste; municipal solid waste*.

heavy metals A metallic element with relatively high density or atomic mass. Some of these, such as lead, mercury, and arsenic, are neurotoxins and thus are of concern from a public health and environmental perspective.

herbivore An organism that consumes plants. Compare *carnivore; omnivore*.

herbivory The consumption of plants by animals.

heterotroph (or consumer) An organism that consumes other organisms. Includes all animals and fungi, as well as microbes (bacteria and protozoa) that decompose organic matter, and even some plants.

high-pressure system An air mass with elevated atmospheric pressure, containing air that descends, typically bringing fair weather. Compare *low-pressure system*.

homeostasis The characteristic whereby a system tends to maintain constant or stable internal conditions.

horizon A distinct layer of soil. See *A horizon; B horizon; C horizon; E horizon; O horizon; R horizon*.

host The organism in a parasitic relationship that suffers harm while providing the parasite with nourishment or some other benefit.

Hubbert's peak The peak in production of crude oil in the United States, which occurred in 1970, as Shell Oil geologist M. King Hubbert predicted in 1956.

humus A dark, spongy, crumbly material in the soil, composed of complex organic compounds, resulting from the partial decomposition of organic matter.

hydraulic fracturing See *fracking*.

hydrocarbon Organic molecules consisting of only the elements hydrogen and carbon; constituents of fossil fuels.

hydroelectric power (or hydropower) The generation of electricity using the kinetic energy of moving water.

hydrologic cycle The flow of water—in liquid, gaseous, and solid forms—through our biotic and abiotic environment.

hydropower See *hydroelectric power*.

hydrosphere Earth's water—salt or fresh, liquid, ice, or vapour—that resides in surface bodies, underground, and in the atmosphere. Compare *biosphere; geosphere*.

hypothesis An educated guess that explains a phenomenon or answers a scientific question. Compare *theory*.

hypoxia A state of oxygen deficiency.

igneous rock One of the three main categories of rock. Formed from the cooling and solidification of magma. Granite and basalt are examples of igneous rock. Compare *metamorphic rock; sedimentary rock*.

immigration The arrival of individuals from outside a population.

incineration A controlled process of burning solid waste for disposal in which mixed garbage is combusted at very high temperatures. Compare *sanitary landfill*.

indicator A fact or piece of data or information that reveals the status or level of something.

indigenous Originating in or native to a particular place.

industrial ecology A holistic approach to industry that integrates principles from engineering, chemistry, ecology, economics, and other disciplines and seeks to redesign industrial systems in order to reduce resource inputs and minimize inefficiency.

industrialized agriculture A form of agriculture that uses large-scale mechanization and fossil fuel combustion, enabling farmers to replace horses and oxen with faster and more powerful means of cultivating, harvesting, transporting, and processing crops. Other aspects include irrigation and the use of inorganic fertilizers. Use of chemical herbicides and pesticides reduces competition from weeds and herbivory by insects. Compare *traditional agriculture*.

Industrial Revolution The dramatic shift in the mid-1700s from rural life, animal-powered agriculture, and manufacturing by craftsmen to an urban society powered by fossil fuels such as coal and crude oil. The Industrial Revolution led to rapid industrialization and urbanization, with related economic and social changes in population, health, transportation, agricultural productivity, and environmental quality. Compare *Agricultural Revolution*; *Medical-Technological Revolution*.

industrial smog Grey-air smog caused by the incomplete combustion of coal or oil when burned. Compare *photochemical smog*.

industrial solid waste Nonliquid waste that is not especially hazardous and that comes from production of consumer goods, mining, petroleum extraction and refining, and agriculture. Compare *hazardous waste; municipal solid waste*.

infrared radiation Electromagnetic radiation with wavelengths longer than visible red, ranging from 700 nm to 1 mm.

inorganic compound Chemical compounds that are of mineral (rather than biological) origin. Inorganic compounds may contain carbon—the element that characterizes organic compounds—but lack the carbon–carbon bonds that are typical of organic compounds.

inorganic fertilizer A fertilizer that consists of mined or synthetically manufactured mineral supplements. Inorganic fertilizers are generally more susceptible than organic fertilizers to leaching and runoff and may be more likely to cause unintended off-site impacts.

insolation Solar radiation that reaches Earth's surface.

integrated pest management (IPM) The use of multiple techniques in combination to achieve long-term suppression of pests, including biocontrol, use of pesticides, close monitoring of populations, habitat alteration, crop rotation, transgenic crops, alternative tillage methods, and mechanical pest removal.

intensification Increasing the resource productivity of a given unit of land, usually by applying new technologies to enhance productivity. Compare *extensification*.

interdisciplinary (field) A field that borrows techniques from several more traditional fields of study and brings together research results from these fields into a broad synthesis.

interglaciation A period of global warming between glaciations. Compare *glaciation*.

Intergovernmental Panel on Climate Change (IPCC) An international panel of atmospheric scientists, climate experts, and government officials established in 1988 by the United Nations Environment Programme and the World Meteorological Organization, whose mission is to assess information relevant to questions of human-induced global climate change. The IPCC's reports summarize current and probable future global trends and represent the consensus of atmospheric scientists around the world.

intertidal Of, relating to, or living along shorelines between the highest reach of the highest tide and the lowest reach of the lowest tide.

intervention A set of possible actions that could be taken to modify the climate system on a global scale. Compare *adaptation, geoengineering, mitigation*.

intrinsic value The idea that organisms and even nonliving components of the natural world have their own worth and the right to exist, completely aside from any value they might provide to people.

invasive species A species that spreads widely and rapidly becomes dominant in a community, interfering with the community's normal functioning. Invasive species are generally *exotic*, or alien, with no natural predators in the new location.

ion An electrically charged atom or combination of atoms.

ion exchange Process in which soil exchanges positively charged ions (cations) and negatively charged ions (anions) with the soil solution. Negatively charged soil particles hold (absorb) positively charged cations, notably Al^{3+}, Ca^{2+}, Mg^{2+}, K^+, NH_4^+, and Na^+, releasing them to the soil solution where they become available as plant nutrients.

IPAT model A relationship that expresses human impact on the environment (I) as a function of population (P), affluence (A), and technology (T).

isotope One of several forms or variants of an element having differing numbers of neutrons in the nucleus of its atoms. Chemically, isotopes of an element behave almost identically, but they have different physical properties because they differ in mass. Compare *stable isotope geochemistry*.

kelp Large brown algae or seaweed that can form underwater "forests," providing habitat for marine organisms.

kerogen A substance derived from deeply buried organic matter that acts as a precursor or source material for both natural gas and crude oil.

keystone species A species that has an especially far-reaching effect on a community.

kinetic energy Energy of motion. Compare *potential energy*.

K-selected (or K-strategist) Term denoting a species with low biotic potential whose members produce a small number of offspring and take a long time to gestate and raise each of their young, but invest heavily in promoting the survival and growth of these few offspring. Populations of K-selected species are generally regulated by density-dependent factors. Compare *r-selected*.

Kyoto Protocol An agreement drafted in 1997 that calls for reducing, by 2012, emissions of six greenhouse gases to levels lower than their levels in 1990. Canada was the 99th country to ratify the agreement, which happened in 2002, and then become the first nation to withdraw from the agreement, in 2011.

land trust Local or regional organization that preserves lands valued by its members. In most cases, land trusts purchase land outright with the aim of preserving it in its natural condition. The Nature Conservancy may be considered the world's largest land trust.

landfill An engineered waste disposal site. Compare *sanitary landfill, secure landfill*.

landfill gas A mix of gases that consists of roughly half methane produced by anaerobic decomposition deep inside landfills, and which can be captured and used as a source of energy.

landscape ecology An approach to the study of organisms and their environments at the landscape scale, focusing on geographical areas that include multiple ecosystems.

landslide A type of mass wasting in which a mass of rock or sediment moves downslope, often as a relatively coherent block.

La Niña An exceptionally strong cooling of surface water in the eastern equatorial Pacific Ocean that occurs irregularly (roughly every 4 to 12 years), with climatic consequences that are widespread but generally not as damaging as those associated with El Niño. Compare *El Niño*.

lava Magma that flows or spatters across Earth's surface.

LD_{50} (lethal-dose-50%) The amount of a toxicant it takes to kill 50% of a population of test animals. Compare *ED_{50}; threshold dose*.

leachate Liquid that percolates through a sanitary landfill or other disposal installation or site, picking up soluble materials from the substance it passes through, and then escapes into the soil underneath.

leaching The process by which solid materials are dissolved in a liquid (usually water) and transported to another location.

lichen A mutualistic aggregate of fungi and algae in which the algal component provides food and energy via photosynthesis, while the fungal component takes a firm hold on rock and captures moisture.

life-cycle analysis In industrial ecology, the examination of the entire life-cycle of a given product—from its origins in raw materials, through its manufacturing, to its use, and finally its disposal—in an attempt to identify ways to make the process more ecologically efficient.

life expectancy The average number of years that individuals in particular age groups are likely to continue to live.

lifestyle hazard See *cultural hazard*.

light pollution Pollution from city lights that obscures the night sky.

limiting factor A physical, chemical, or biological characteristic of the environment that restrains population growth.

limnetic zone In a water body, the layer of open water through which sunlight penetrates. Compare *littoral zone; benthic zone; profundal zone.*

linear growth The increase of a population (or of anything) by a fixed amount each year; arithmetic growth. Contrast with *exponential growth.*

lipid One of a chemically diverse group of large, biologically important molecules that are classified together because they do not dissolve in water. Lipids include fats, phospholipids, waxes, pigments, and steroids.

lithosphere The solid part of Earth, including the rocks, sediment, and soil at the surface and extending down many kilometres underground; the crust and the outer portion of the mantle. Compare *geosphere.*

litter Leaves, twigs, and other organic refuse that accumulates on the land surface, especially in forested areas.

littoral Along the shoreline of a lake or the ocean. See *intertidal.*

littoral zone The region ringing the edge of a water body. Compare *benthic zone; limnetic zone; profundal zone.*

loam Soil with a relatively even mixture of clay-, silt-, and sand-sized particles.

logistic growth curve A plot that shows how the initial exponential growth of a population is slowed and finally brought to a standstill by limiting factors, yielding an S-shaped or sigmoidal growth curve.

low-pressure system An air mass in which the air moves toward the low atmospheric pressure at the centre of the system and spirals upward, typically bringing clouds and precipitation. Compare *high-pressure system.*

macronutrient A nutrient that organisms require in relatively large amounts. Compare *micronutrient.*

magma Molten (that is, liquid, melted) rock, typically containing dissolved gases, and sometimes mineral grains and rock fragments.

malnutrition The condition of lacking nutrients the body needs, including a complete complement of vitamins and minerals.

mangrove A tree with a unique type of roots that curve upward to obtain oxygen, which is lacking in the mud in which they grow, and that serve as stilts to support the tree in changing water levels. Mangrove forests grow on the coastlines of the tropics and subtropics.

mantle The layer of rock that lies beneath Earth's crust and surrounds a mostly iron core.

marine protected area (MPA) An area of the ocean set aside to protect marine life from fishing pressures. An MPA may be protected from some human activities but be open to others. Compare *marine reserve.*

marine reserve An area of the ocean designated as a "no-fishing" zone, allowing no extractive activities. Compare *marine protected area.*

market value (of natural resources) The monetary value that would be achievable by placing a natural resource for sale on the open market. It is relatively straightforward to quantify the potential market value of a resource (such as a standing forest or an ore deposit) that is already represented in the market as a traded commodity (timber or minerals); it is much more challenging to quantify the market value of an environmental good or service with no existing analogy in the marketplace, such as the value of a particular habitat, or a beautiful view, or the natural filtering function of ground water. Compare *valuation.*

marsh A type of wetland in which shallow water allows grasses and seasonal herbaceous plants to grow above the water's surface. Compare *bog, fen, swamp, peatland.*

mass extinction (event) The extinction of a large proportion of the world's species in a very short time period due to some extreme and rapid change or catastrophic event. Earth has seen five mass extinction events in the past half-billion years.

mass wasting The downslope movement of Earth material (such as rock or soil) under the influence of gravity.

matter Any material that has mass and occupies space.

maximum sustainable yield (MSY) The maximal harvest of a particular renewable natural resource that can be accomplished while still keeping the resource available for the future.

mechanical weathering See *physical weathering.*

Medical-Technological Revolution The modern era of medicine, public health innovations, and technological advances, including the shift to modern agriculture through what is known as the Green Revolution. Compare *Agricultural Revolution; Industrial Revolution; Green Revolution.*

Mediterranean A biome typical of the Mediterranean climate zone, which is characterized by warm, dry summers and mild, wet winters. *Chaparral* is one example of a Mediterranean-type biome.

mega-city A city with more than 10 million inhabitants.

meltdown The accidental melting of the uranium fuel rods inside the core of a nuclear reactor, causing the release of radiation.

mesosphere The atmospheric layer above the stratosphere, extending 50–80 km above sea level. (The term *mesosphere* also applies to part of Earth's mantle.)

metal An element or an alloy of elements that displays properties that typically include malleability, ductility, shiny (metallic) lustre, and good electrical conductivity.

metallic resources Ores that are mined and processed in order to extract the metals they contain. Compare *nonmetallic resources.*

metamorphic rock One of the three main categories of rock. Formed by great heat and/or pressure that reshapes crystals within the rock and changes its appearance, mineral assemblage, physical properties, and sometimes its overall composition. Common metamorphic rocks include marble and slate. Compare *igneous rock; sedimentary rock.*

methane (CH_4) The simplest hydrocarbon compound; the key component of natural gas, and a naturally occurring greenhouse gas.

methane hydrate An ice-like solid consisting of molecules of methane (CH_4) embedded in a crystal lattice of water molecules called a clathrate structure. Also called *methane ice.*

microclimate Variations in weather and climate that occur on an extremely local scale, such as from one side of a hill to the other.

micronutrient A nutrient that organisms require in relatively small amounts. Compare *macronutrient.*

Milankovitch cycle One of three types of variations in Earth's rotation and orbit around the sun that result in slight changes in the relative amount of solar radiation reaching Earth's surface at different latitudes. As the cycles proceed, they change the way solar radiation is distributed over Earth's surface and contribute to changes in atmospheric heating and circulation that have triggered the ice ages and other climate changes.

mineral A naturally occurring, solid, crystalline, inorganic compound; the building blocks of rocks.

mining The extraction of mineral resources and other valuable Earth materials. Compare *open-pit mining, strip mining, mountaintop removal, subsurface mining, solution mining,* and *placer mining.*

mitigation A responsive strategy in which efforts are made to forestall or minimize the anticipated impacts of environmental change. Compare *adaptation; intervention.*

model A simplified representation of a natural process. Models can be physical, graphical, or quantitative and computer-based. They allow scientists to study natural systems and processes that are highly complex or unwieldy in their spatial or temporal scale. They also can be used to predict future behaviour in complex systems.

molecule A combination of two or more atoms.

monoculture The uniform planting of a single crop over a large area. Characterizes modern industrialized agriculture.

Montreal Protocol International treaty ratified in 1987, in which the (now 197) signatory nations agreed to restrict production of chlorofluorocarbons (CFCs) in order to forestall stratospheric ozone depletion.

mortality Deaths within a population.

mountaintop removal A surface mining process in which the entire top part of a mountain is removed in order to extract the material of interest, commonly coal.

multiple-barrier approach (to waste containment) An approach to hazardous waste disposal and management that places as many impediments as possible in the way of any escaping waste, leachate, effluents, or other harmful emissions from waste (such as radioactivity), including both manufactured and natural barriers.

multiple use A principle that has nominally guided management policy for national forests and parks for the past half century or more. The multiple-use principle specifies that forests and other environmental resources be managed to allow for recreation, wildlife habitat, mineral extraction, and various other uses.

municipal solid waste (or MSW) Nonliquid waste that is not especially hazardous and that comes from homes, institutions, and small businesses. Compare *hazardous waste; industrial solid waste.*

mutagen A toxicant that causes mutations in the DNA of organisms.

mutation An accidental change in DNA that may range in magnitude from the deletion, substitution, or addition of a single nucleotide to a change affecting entire sets of chromosomes. Mutations provide the raw material for evolutionary change.

mutualism A relationship between members of different species in which all participating organisms benefit from their interaction. Compare *parasitism, amensalism,* and *commensalism.*

natality Births within a population.

natural gas A fossil fuel composed primarily of methane (CH_4), produced as a by-product when bacteria decompose organic material under anaerobic conditions.

natural rate of population growth The rate of change in a population's size resulting from birth and death rates alone, excluding migration. Compare *population growth rate.*

natural resource Any of the various substances and energy sources we need in order to survive and to lead our lives.

natural resource accounting A discipline that has the goal of adjusting national accounts and indicators, such as GDP, to reflect the true costs of economic production, including the costs of depleting natural resources and damaging the environment.

natural selection The process by which traits that enhance survival and reproduction are passed on more frequently to future generations of organisms than those that do not, thus altering the genetic makeup of populations through time. Natural selection acts on genetic variation and is a primary driver of evolution.

negative feedback loop (or cycle) A feedback loop in which output of one type acts as input that moves the system in the opposite direction. The input and output essentially neutralize each other's effects, stabilizing the system. Compare *positive feedback loop.*

nekton Aquatic animals that actively swim. Compare *plankton.*

net energy The quantitative difference between energy returned from a process and energy invested in the process. Positive net energy values mean that a process produces more energy than is invested. See also *EROI.*

net metering Process by which owners of houses with photovoltaic systems can sell their excess solar energy to their local power utility.

net primary production (NPP) The energy or biomass that remains in an ecosystem after autotrophs have metabolized enough for their own maintenance through cellular respiration. Net primary production is the energy or biomass available for consumption by heterotrophs. Compare *gross primary production; secondary production.*

neurotoxin A toxicant that assaults the nervous system. Examples of neurotoxins include heavy metals, pesticides, and some chemical weapons developed for use in war.

new forestry A set of ecosystem-based management approaches for harvesting timber that explicitly mimic natural disturbances. For instance, "sloppy clear-cuts" that leave a variety of trees standing mimics the changes a forest might experience if hit by a severe windstorm.

niche The functional role of a species in a community. See *fundamental niche; realized niche.*

NIMBY syndrome See *not-in-my-backyard.*

nitrification The conversion by bacteria of ammonium ions (NH_4^+) first into nitrite ions (NO_2^-) and then into nitrate ions (NO_3^-).

nitrogen (N) The chemical element with atomic number seven. The most abundant element in the atmosphere, a key element in macromolecules, and a crucial plant nutrient.

nitrogen cycle A major nutrient cycle consisting of the routes that nitrogen atoms take through the nested networks of environmental systems.

nitrogen dioxide (NO_2) A criteria air contaminant and a common by-product of internal combustion engines.

nitrogen fixation The process by which inert nitrogen gas combines with hydrogen to form ammonium ions (NH_4^+), which are chemically and biologically active and can be taken up by plants.

nitrous oxides (NO_x or NOX) Various nitrogen-based gaseous compounds, including NO_2 and NO_3, which commonly result from industrial processes involving combustion, especially in internal combustion engines.

noise pollution Ambient sound that is undesirable or unhealthy.

nonmetallic resources A diverse assortment of rocks and minerals that are mined for nonmetallic properties; examples include salt, gemstones, building stone, and aggregate. Compare *metallic resources.*

nonpoint source A diffuse source of pollutants, often consisting of many small sources. Compare *point source.*

nonrenewable natural resource A natural resource that is in limited supply and is formed much more slowly than we use it. Compare *renewable natural resource.*

North Atlantic Deep Water (NADW) A deep water mass originating in the northern Atlantic Ocean and linked to the global thermohaline circulation. Compare *Antarctic Bottom Water (AABW).*

not-in-my-backyard ("NIMBY") A common reaction by homeowners to siting proposals for waste disposal facilities, wind farms, or other installations.

nuclear power Power derived from the energy released during the splitting of an atom (fission) in a nuclear reactor.

nuclear reactor A facility within a nuclear power plant that initiates and controls the process of nuclear fission in order to generate electricity.

nucleic acid A molecule that directs the production of proteins; includes DNA and RNA.

nutrient An element or compound that organisms consume and require for survival.

nutrient cycle The comprehensive set of cyclical pathways by which a given nutrient moves through the environment.

ocean acidification The process whereby ocean waters becomes more acidic (lower pH), which can happen when the flux of atmospheric carbon dioxide into surface ocean water increases. Ocean acidification poses a threat to many marine organisms.

oceanography The scientific study of the physics, chemistry, biology, and geology of the ocean.

ocean thermal energy conversion (OTEC) A potential energy source that involves harnessing the solar radiation absorbed by tropical oceans in the tropics.

O horizon The top layer of soil in some soil profiles, made up of organic matter, such as decomposing branches, leaves, crop residue, and animal waste. Compare *A horizon; B horizon; C horizon; R horizon.*

oil See *crude oil.*

oil sands (or tar sands) Deposits consisting of moist sand and clay with 1–20% bitumen, which can be mined for the production of oil.

oil shale Sedimentary rock that contains kerogen, which can be processed to produce liquid petroleum. Oil shale is formed by the same processes that form crude oil but occurs when kerogen is not buried deeply enough or subjected to enough heat and pressure to form oil. Compare *shale gas.*

old-growth forest A complex, primary forest in which the trees are generally at least 150 years old.

oligotrophic Term describing a water body that has low-nutrient and (typically) high-oxygen conditions. Compare *eutrophic.*

omnivore An organism that consumes both plants and animals. Compare *carnivore; herbivore.*

open canopy (of a forest) A forest canopy with many openings that allow light to pass through to the forest floor.

open-pit mining Mining of rock or mineral resources through extraction at the surface from a large, open pit or hole.

open system A system that allows both energy and matter to cross its boundaries. Compare *closed system.*

ore Naturally occurring rock or mineral from which an economically valuable material can be profitably extracted, using current extraction and processing technologies.

organic agriculture Agriculture that uses no synthetic fertilizers or pesticides but instead relies on biological approaches such as composting and biocontrol.

organic compound A compound made up of carbon atoms (and, generally, hydrogen atoms) joined by covalent bonding and sometimes including other elements, such as nitrogen, oxygen, sulphur, or phosphorus. The unusual ability of carbon to build elaborate molecules has resulted in millions of different organic compounds showing various degrees of complexity. Compare *inorganic compound.*

organic fertilizer A fertilizer made up of natural materials (largely the remains or wastes of organisms), including animal manure, crop residues, fresh vegetation, and compost. Compare *inorganic fertilizer.*

ozone (O_3) A molecule consisting of three atoms of oxygen. Ozone in the stratosphere performs the important function of absorbing ultraviolet radiation, but occurring in the troposphere it is an air pollutant. Compare *ozone layer; tropospheric ozone.*

ozone hole Term popularly used to describe the annual depletion of the stratospheric ozone layer as a result of chemical interactions with anthropogenic pollutants such as chlorofluorocarbons.

ozone layer A portion of the stratosphere, roughly 17–30 km above sea level, that contains most of the ozone in the atmosphere.

paleoclimate A climate of Earth's past, prior to instrumental records of climate.

parasite The organism in a parasitic relationship that extracts nourishment or some other benefit from the host.

parasitism A relationship, usually between members of different species, in which one organism, the parasite, depends on another, the host, for nourishment or some other benefit while simultaneously doing the host harm. Compare *mutualism, amensalism,* and *commensalism.*

parent material The base geological material in a particular location.

particulate matter (PM) Solid and (sometimes) liquid and colloidal particles (aerosols) small enough to be suspended in the atmosphere and able to damage respiratory tissues when inhaled. Includes primary pollutants such as dust and soot, as well as secondary pollutants such as sulphates and nitrates. A criteria air contaminant; abbreviated PM and categorized by particle size (e.g., $PM_{2.5}$, particles less than 2.5 μm).

passive solar (energy) An approach in which buildings are designed and building materials are chosen to maximize their direct absorption of sunlight in winter, even as they keep the interior cool in the summer. Compare *active solar energy.*

pathogen A microbe that causes disease.

"peak oil" Term used to describe the point of maximum production of petroleum in the world (or for a given nation), after which oil production declines. This is also expected to be roughly the midway point of extraction of the world's oil supplies. The term is generally used in contexts suggesting that our society will face tremendous challenges once the peak has been passed. Compare *Hubbert's peak.*

peat An organic-rich soil; a precursor stage to coal, produced when wet, boggy organic material is compressed and partially broken down by anaerobic decomposition.

peatland A type of wetland characterized by a thick, waterlogged layer of organic soil (peat).

peer review The process by which a manuscript submitted for publication in an academic journal is examined by other specialists in the field, who provide comments and criticism (generally anonymously), and judge whether the work merits publication in the journal.

pelagic Of, relating to, or living between the surface and floor of the ocean. Compare *benthic.*

permafrost Permanently frozen ground; a layer of perennially frozen water under the surface, essentially the W horizon in an arctic soil.

permeability A measure of the interconnectedness of pore spaces in a rock, sediment, or soil, contributing to the ability of the material to transmit fluids. Compare *porosity.*

permit trading The practice of buying and selling government-issued marketable permits to conduct environmentally harmful activities. Under such a system, the government determines an acceptable level of pollution and then issues permits to pollute. A company receives credit for amounts it does not emit and can then sell this credit to other companies. Compare *emissions trading system* and *cap-and-trade.*

persistent Refers to a chemical that does not break down, degrade, or decompose easily, and that consequently may have a long residence time in a given environmental reservoir.

pest A species—typically, but not always, a non-native or alien invasive species—that has more harmful than beneficial impacts on an ecosystem or community (or on human interests, such as human health or crop health).

pesticide An artificial chemical used to kill insects (insecticide), plants (herbicide), or fungi (fungicide).

petroleum See *crude oil.*

pH (scale) A measure of the concentration of hydrogen ions in a solution. The pH scale ranges from 0 to 14: Solutions with pH of 7 are neutral; solutions with pH below 7 are acidic, and those with pH higher than 7 are basic. Because the pH scale is logarithmic, each step on the scale represents a tenfold difference in hydrogen ion concentration.

phosphorus (P) The chemical element with atomic number 15. An abundant element in the lithosphere, a key element in macromolecules, and a crucial plant nutrient.

phosphorus cycle A major nutrient cycle consisting of the routes that phosphorus atoms take through the nested networks of environmental systems.

photic zone In the ocean or a freshwater body, the well-lit top layer of water where photosynthesis occurs. Compare *aphotic zone.*

photochemical smog Brown-air smog caused by light-driven reactions of primary pollutants with normal atmospheric compounds that produce a mix of over 100 different chemicals, ground-level ozone often being the most abundant among them. Compare *industrial smog.*

photosynthesis The process by which autotrophs produce their own food. Sunlight powers a series of chemical reactions that convert carbon dioxide and water into sugar (glucose), thus transforming low-quality energy from the sun into high-quality energy the organism can use. Compare *cellular respiration.*

photovoltaic (PV) cell A device designed to collect sunlight and convert it to electrical energy directly by making use of the photoelectric effect.

phylogenetic tree A treelike diagram that represents the history of divergence of species or other taxonomic groups of organisms.

physical (or mechanical) weathering Weathering that breaks down rock without triggering a chemical change in the parent material. Wind, rain, and freeze-thaw cycles are important forces. Compare *biological weathering; chemical weathering.*

physical hazard Physical processes that occur naturally in our environment and pose human health hazards. These include discrete events such as earthquakes, volcanic eruptions, fires, floods, blizzards, landslides, hurricanes, and droughts, as well as ongoing natural phenomena such as ultraviolet radiation from sunlight. Compare *biological hazard; chemical hazard; cultural hazard.*

phytoplankton Microscopic floating plants and protists, mainly algae. Compare *plankton, zooplankton.*

pioneer species A species that arrives earliest, beginning the ecological process of succession in a terrestrial or aquatic community.

placer mining Mining of ore that has been concentrated by alluvial processes, that is, by the action of running water or waves.

plankton Microscopic aquatic organisms that float (rather than swim). Compare *phytoplankton, zooplankton, nekton.*

plastic A synthetic polymer material, mostly derived from the hydrocarbons in petroleum.

plate tectonics The process by which Earth's surface is shaped by the extremely slow movement of tectonic plates—large fragments of lithosphere. Earth's surface includes about 15 major tectonic plates. Their interactions give rise to processes that build mountains, cause earthquakes and volcanic eruptions, and otherwise influence the landscape.

point source A specific spot—such as a factory's smokestacks—where large quantities of pollutants are discharged. Compare *nonpoint source.*

polar cell One of a pair of cells of convective circulation between the poles and 60° north and south latitudes that influence global climate patterns. Compare *Ferrel cell; Hadley cell.*

pollination An interaction in which pollen (male plant reproductive cells) is transferred to the ova (female plant reproductive cells) of a flower, which is thereby fertilized. Pollination can occur when pollen is picked up and transferred by a vector organism, such as a bee, bird, or bat; or it can occur when pollen is distributed by abiotic processes, such as wind; or it can occur by self-pollination of a single flower or flowers of the same individual.

polluter-pays principle Principle in which the party that produces pollution pays the costs of cleaning up or mitigating the pollution.

pollution Any matter or energy released into the environment that causes undesirable impacts on the health and well-being of humans or other organisms. Pollution can be physical, chemical, biological, or thermal, and can affect water, air, or soil.

pollution prevention (P2) strategies Industrial strategies that aim to prevent pollution or mitigate pollution-generating processes beforehand, rather than having to clean up after the pollution has been generated.

population A group of organisms of the same species that live in the same area. Species are often composed of multiple populations.

population density The number of individuals within a population per unit area. Compare *population size.*

population dispersion See *population distribution*.

population distribution The spatial arrangement of organisms within a particular area. Also called *population dispersion*.

population growth rate An increase in the size of a population, or group of individuals of a particular species in a given location, in a given unit of time. Compare *natural rate of population growth*.

population pyramid A graph that represents the age cohorts and gender breakdown of a particular population at a given time.

population size The number of individual organisms present at a given time.

porosity The proportion of open pore space in a rock, sediment, or soil. Compare *permeability*.

positive feedback loop (or cycle) A feedback loop in which output of one type acts as input that moves the system in the same direction. The input and output drive the system further toward one extreme or another. Compare *negative feedback loop*.

potential energy Energy of position. Compare *kinetic energy*.

precautionary principle The idea that one should not undertake a new action until the ramifications of that action are well understood.

precipitation Water that condenses out of the atmosphere and falls to Earth in droplets or crystals.

predation The process in which one species (the predator) hunts, tracks, captures, and ultimately kills its prey.

predator An organism that hunts, captures, kills, and consumes individuals of another species, the prey.

prediction A specific statement that can be tested, generally arising from a hypothesis.

prescribed (or controlled) burns The practice of burning areas of forest or grassland under carefully controlled conditions to improve the health of ecosystems, return them to a more natural state, and help prevent uncontrolled catastrophic fires.

preservation ethic An ethical position, holding that we should protect the natural environment in a pristine, unaltered state. Compare *conservation ethic*.

pressure–state–response (PSR) model A framework for environmental monitoring and reporting, which considers activities and processes that cause pressure or stress on the environment, the resulting state or condition of the environment, and the effectiveness of human responses to the management of these stresses. Also called *DPSIR model*, for driver–pressure–state–impact–response.

prey An organism that is killed and consumed by a predator.

primary consumer An organism that consumes producers and feeds at the second trophic level. Compare *secondary consumer*.

primary forest Forest uncut by people. Compare *second-growth*.

primary pollutant A hazardous substance, such as soot or carbon monoxide, that is emitted into the troposphere in a form that is directly harmful. Compare *secondary pollutant*.

primary producer See *autotroph*.

primary recovery (or primary extraction) The initial drilling and pumping of available crude oil. Compare *secondary recovery*.

primary succession A series of changes that theoretically would be expected to occur as an ecological community develops over time, beginning with a lifeless substrate. In terrestrial systems, primary succession begins when a bare expanse of rock, sand, or sediment becomes newly exposed to the atmosphere and pioneer species arrive. Compare *secondary succession*.

prior appropriation A concept of water law in which a user who can demonstrate a history of earlier use is entitled to have access to the *resource*.

prior informed consent The ethical and legal principle that consent or acceptance of an activity (such as land development or waste disposal) is not legally valid unless the consenting person or group has been properly and adequately informed and can be shown to have a reasonable understanding of all potential impacts before giving consent.

producer See *autotroph*.

productivity The rate at which plants convert solar energy (sunlight) to biomass. Ecosystems whose plants convert solar energy to biomass rapidly are said to have high productivity. See *gross primary production; net primary production*.

profundal zone In a water body, the volume of open water that sunlight does not reach. Compare *littoral zone; benthic zone; limnetic zone*.

protein A large molecule made up of long chains of amino acids.

proxy indicator Indirect evidence, such as pollen from sediment cores, isotopic compositions of natural materials, and air bubbles from ice cores, of past environmental conditions, including climate.

pump-and-treat An approach to groundwater remediation that involves pumping out contaminated ground water, cleaning it in a surface facility, and then (typically) pumping it back into the aquifer.

quarry An open pit or hole from which rock, gravel, or other Earth materials are extracted.

radiative forcing The amount of change in energy that a given factor (such as aerosols, albedo, or greenhouse gases) exerts over Earth's energy balance. By convention, positive radiative forcing warms the surface, whereas negative radiative forcing cools it.

radiatively active gas Greenhouse gas; a gas in the atmosphere that is effective at absorbing infrared radiation.

radioactive The quality held by some isotopes in which they spontaneously decay, changing their chemical identity as they shed subatomic particles and emit high-energy radiation.

radioactive isotope (or radioisotope) An unstable isotope that emits subatomic particles and high-energy radiation as it decays into progressively lighter isotopes until becoming stable.

rain forest A forest biome that is characterized by high annual rainfall and, in general, aseasonality.

rangeland Land used for grazing livestock.

reactive Materials that are chemically unstable and readily able to react with other compounds, often explosively or by producing noxious fumes. One criterion for defining hazardous waste.

realized niche The portion of the fundamental niche that is fully realized (used) by a species. Compare *fundamental niche*.

recharge zone An area where precipitation falls and infiltrates the ground, percolating downward to eventually join and replenish ground water in an aquifer.

reclamation The process of remediating land degraded by mining, industrial pollution, desertification, or other damaging processes to make it suitable once again for agriculture, habitation, or recreational use.

recycling The collection of materials that can be broken down and reprocessed to manufacture new items.

Red List of Threatened Species A list of species facing unusually high risks of extinction. The list is maintained by the World Conservation Union (IUCN).

red tide A harmful algal bloom consisting of microorganisms called dinoflagellates that produce reddish pigments that discolour surface waters.

refining (of oil or mineral ores) Process of separating the molecules of the various hydrocarbons in crude oil into different-sized classes and transforming them into various fuels and other petrochemical products; or, the process of separating, under high heat, the metals from a metallic mineral ore.

reforestation Planting or replanting of trees in a previously forested area from which the trees had been removed by logging, fire, or some other cause.

relative abundance The extent to which the proportions of individuals in an area of different species are equal or skewed. One way to express species diversity. Essentially synonymous with *evenness*. Compare *species richness*.

relative humidity The ratio of the water vapour contained in a given volume of air to the maximum amount the air could contain, for a given temperature.

remediation Removing pollutants from contaminated ground water. More generally, cleaning or restoring a degraded habitat, environment, or ecosystem.

remote sensing The collection of information about a target, from a distance. Most commonly refers to the collection of information about the near-surface Earth environment by instruments carried on satellites.

renewable natural resource A natural resource that is virtually unlimited or that is replenished by the environment over relatively short periods (hours to weeks to years) so that it can be replenished at a rate that is faster than the rate of withdrawal by people. Compare *nonrenewable natural resource*.

replacement fertility The total fertility rate (TFR) that maintains a stable population size.

reserve That portion of a resource that is known with a high degree of certainty and is profitably extractable using present extraction and refining technologies.

reservoir A location where materials in a cycle remain for a period of time, before moving to another reservoir. Also called a *pool*. Compare *flux*.

residence time The amount of time a material in a cycle remains in a given pool or reservoir before moving to another reservoir. Compare *flux*, *turnover time*.

resilience The ability of an individual or a community (human or ecological) to change in response to disturbance, and then return to its original state or to a new state of equilibrium that is similar to the original state. Compare *resistance*.

resistance The ability of an ecological community to remain stable in the presence of a disturbance. Compare *resilience*.

resource management Strategic decision-making about who should extract resources and in what ways, so that resources are used wisely and not wasted.

resource partitioning The process by which species adapt to competition by evolving to use slightly different resources, or to use their shared resources in different ways, thus minimizing interference with one another.

response The type or magnitude of negative effects an animal exhibits in response to a dose of toxicant in a dose–response test. Compare *dose*.

restoration ecology The study of the historical conditions of ecological communities as they existed before humans altered them. Compare *ecological restoration*.

R horizon The bottommost layer of soil in a typical soil profile. Also called *bedrock*. Compare *A horizon; B horizon; C horizon; O horizon*.

ribonucleic acid (RNA) A molecule that is involved in protein synthesis and sometimes in the transmission of genetic information. Compare *deoxyribonucleic acid (DNA)*.

riparian Relating to a river or the area along a river.

risk The mathematical probability that some harmful outcome (for instance, injury, death, environmental damage, or economic loss) will result from a given action, event, or substance.

risk assessment The quantitative measurement of risk, together with the comparison of risks involved in different activities or substances.

risk management The process of considering information from scientific risk assessment in light of economic, social, and political needs and values, in order to make decisions and design strategies to minimize risk.

rock Durable Earth material made principally of minerals.

rock cycle The very slow process in which rocks and the minerals that make them up are heated, melted, cooled, broken, and reassembled, forming igneous, sedimentary, and metamorphic rocks.

round table An approach to decision-making in which stakeholders with differing views meet, exchange ideas, and seek consensus.

r-selected (or r-strategist) Term denoting a species with high biotic potential whose members produce a large number of offspring in a relatively short time but do not care for their young after birth. Populations of r-selected species are generally regulated by density-independent factors. Compare *K-selected*.

runoff The water from precipitation that flows over land into streams, rivers, lakes, and ponds, and eventually to the ocean.

run-of-river Any of several methods used to generate hydroelectric power without greatly disrupting the flow of river water. Run-of-river approaches eliminate much of the environmental impact of large dams.

salinity Saltiness; the concentration of dissolved salts in water.

salinization (or salination) The buildup of salts in surface soil layers.

salt marsh Flat land that is intermittently flooded by the ocean where the tide reaches inland. Salt marshes occur along temperate coastlines and are thickly vegetated with grasses, rushes, shrubs, and other herbaceous plants.

salvage logging The removal of dead trees following a natural disturbance. Although it may be economically beneficial, salvage logging can be ecologically destructive, because the dead trees provide food and shelter for a variety of insects and wildlife and because removing timber from recently burned land can cause severe erosion and damage to soil.

sand Sediment consisting of particles 0.005–2.0 mm in diameter. Compare *clay; silt*.

sanitary landfill A site at which solid waste is buried in the ground or piled up in large mounds for disposal, designed to prevent the waste from contaminating the environment. Compare *incineration; secure landfill*.

savannah A biome characterized by grassland interspersed with clusters of acacias and other trees. Savannah is found across parts of Africa (where it was the ancestral home of our species), South America, Australia, India, and other dry tropical regions.

science A systematic process for learning about the world and testing our understanding of it.

scientific method A formalized method for testing ideas with observations that involves several assumptions and a more or less consistent series of interrelated steps.

scrubber Technology used to chemically treat gases produced in combustion to remove hazardous components and neutralize acidic gases, such as sulphur dioxide and hydrochloric acid, turning them into water and salt, in order to reduce smokestack emissions.

secondary consumer An organism that consumes primary consumers and feeds at the third trophic level. Compare *primary consumer*.

secondary pollutant A hazardous substance produced through the reaction of substances added to the atmosphere with chemicals normally found in the atmosphere. Compare *primary pollutant*.

secondary production The total biomass that heterotrophs generate by consuming autotrophs. Compare *gross primary production; net primary production*.

secondary recovery (or secondary extraction) The extraction of crude oil remaining after primary extraction by using solvents or by flushing underground rocks with water or steam. Compare *primary recovery*.

secondary succession A series of changes that would theoretically be expected to occur as an ecological community develops over time, beginning when some event disrupts or dramatically alters an existing community. Compare *primary succession*.

second-growth Term describing trees that have sprouted and grown to partial maturity after virgin timber has been cut.

secure landfill A landfill that is especially engineered to receive and effectively isolate hazardous wastes. Compare *sanitary landfill*.

sediment Loose particles, mainly of weathered rock.

sedimentary rock One of the three main categories of rock. Formed when dissolved minerals seep through sediment layers and act as a kind of glue, crystallizing and binding sediment particles together. Sandstone and shale are examples of sedimentary rock. Compare *igneous rock; metamorphic rock*.

seed bank A storehouse for samples of the world's crop diversity.

selection systems (for forest harvesting) Timber harvesting approaches in which some trees are cut and others are selectively allowed to remain standing.

septic system A wastewater disposal method, common in rural areas, consisting of an underground tank and series of drainpipes. Waste water runs from the house to the tank, where solids precipitate out. The water proceeds downhill to a drain field of perforated pipes laid horizontally in gravel-filled trenches, where microbes decompose the remaining waste.

sequestration Isolation; very long-term storage in a *reservoir*.

services Something that is of economic value, but is not a tangible or material product; in economics, the intangible equivalent of goods.

sex ratio The ratio of males to females in a population.

shale gas Natural gas produced from shale. Compare *oil shale*.

shallow-water wetland A wetland with some open, flowing, or standing water; includes ponds, sloughs, oxbows, and vernal pools. Compare *wetland*.

shelterbelt A row of trees or other tall perennial plants that are planted along the edges of farm fields to break the wind and thereby minimize wind erosion.

shrubland An area of open woodland, characterized by low, bushy vegetation and occasional taller trees.

sick-building syndrome An illness produced by indoor air pollution of which the specific cause is not identifiable.

silt Sediment consisting of particles 0.002–0.005 mm in diameter. Compare *clay; sand*.

sink In a cycle, a reservoir that takes in more material than it releases. Compare *flux, source*.

sinkhole An area where the ground has given way with little warning as a result of subsidence, especially in areas underlain by soluble carbonate (limestone) bedrock, and often associated with depression of the water table or depletion of an underlying aquifer.

SLOSS ("single large or several small") dilemma The debate over whether it is better to make reserves large in size and few in number or many in number but small in size.

smelting The extraction of metals by heating and melting of ore.

smog See *industrial smog, photochemical smog.*

softwood Wood derived from coniferous or needle-bearing trees. Compare *hardwood.*

soil A complex plant-supporting system consisting of disintegrated rock, organic matter, air, water, nutrients, and microorganisms.

soil degradation Damage to soils (typically through loss of organic matter or moisture, or through chemical contamination), or loss of soils (typically through erosion).

soil profile The cross-section of a soil as a whole, from the surface to the bedrock.

soil structure A measure of the organization or "clumpiness" of soil.

soil texture A characteristic of soil that is determined by the relative proportions of clay, sand, and silt particles.

solar energy Energy from the sun. It is perpetually renewable and may be harnessed in several ways.

solar panels Panels generally consisting of dark-coloured, heat-absorbing metal plates mounted in flat boxes covered with glass panes, often installed on rooftops to harness solar energy. Also called *flat-plate solar collectors.*

solution A chemical mixture; most often used in reference to liquids, it can also be applied to solid and gaseous mixtures.

solution mining A subsurface mining technique in which a fluid is injected into the ground to dissolve the substance of interest, facilitating its extraction through a pipeline.

source In a cycle, a reservoir that releases more material than it takes in. Compare *sink; flux.*

source reduction The reduction of the amount of material that enters the waste stream to avoid the costs of disposal and recycling, help conserve resources, minimize pollution, and save money for consumers and businesses.

specialist A species that can survive only in a narrow range of habitats that contain very specific resources. Compare *generalist.*

speciation The process by which new species are generated.

species A population or group of populations of a particular type of organism, whose members share certain characteristics and can breed freely with one another and produce fertile offspring. Different biologists may have different approaches to diagnosing species boundaries.

***Species at Risk Act* (SARA)** Canada's endangered species protection law, enacted in 2002.

species diversity The number and variety of species in the world or in a particular region.

species richness The number of species in a particular region. One way to express species diversity. Compare *evenness; relative abundance.*

stabilizing selection Mode of natural selection in which selection favours intermediate traits, in essence preserving the status quo. Compare *directional selection; disruptive selection.*

stable isotope A nonradioactive form of an element.

stable isotope geochemistry The study of the behaviour of isotopes that are not radioactive. Stable isotopes can act as tracers to help scientists unravel the effects of past environmental processes.

stakeholder A party (an individual or organization) that has a specific interest or stake in a proceeding. Stakeholders have specific, legally defined rights and responsibilities in processes such as environmental impact assessments.

state-of-the-environment reporting (SOER) Monitoring and analysis of the conditions of the environment and periodic reporting on the results of the monitoring, usually undertaken by agencies at various levels of government.

steady state A state of dynamic equilibrium or balance in which there is no net change in the system. Compare *homeostasis.*

stock The harvestable portion of a resource.

storm surge A temporary and localized rise in sea level and associated large waves, brought on by the high tides, low atmospheric pressure, and winds associated with storms, especially hurricanes.

stratosphere The layer of the atmosphere above the troposphere and below the mesosphere; it extends from 11 km to 50 km above sea level.

strip mining The use of heavy machinery to remove huge amounts of rock and soil overburden at the surface to expose coal or ore minerals, which then can be mined out directly.

subduction The plate tectonic process by which a tectonic plate carrying dense oceanic crust slides (or is pushed or dragged) beneath another plate at a convergent plate boundary.

subsistence agriculture The oldest form of traditional agriculture, in which farming families produce only enough food for themselves.

subsurface mining Method of mining underground coal and mineral deposits, in which shafts are dug deeply into the ground and networks of tunnels are dug or blasted out to follow coal seams or ore bodies.

succession A stereotypical series of changes in the composition and structure of an ecological community through time. See *primary succession; secondary succession.*

sulphur dioxide (SO$_2$) A criteria air contaminant and a common by-product of the burning of fossil fuels, especially coal.

sulphur oxides (SO$_x$ or SOX) Various sulphur-based gaseous compounds, including SO$_2$ and SO$_3$, which commonly result from industrial processes involving combustion, especially of coal.

surface impoundment A hazardous waste disposal method in which a shallow depression is dug and lined with impervious material, such as clay. Water containing small amounts of hazardous waste is placed in the pond and allowed to evaporate, leaving a residue of solid hazardous waste on the bottom. Compare *deep-well injection.*

survivorship curve A graph that shows how the likelihood of death for members of a population varies with age.

sustainability A guiding principle of environmental science that requires us to live in such a way as to maintain Earth's systems and its natural resources for the foreseeable future.

sustainable agriculture Agriculture that does not deplete soils faster than they form.

sustainable development Development that satisfies our current needs without compromising the future availability of natural resources or our future quality of life.

Sustainable Development Goals (SDGs) A set of 17 goals established by the United Nations, in consultation with the global community, to guide world development toward sustainability in the post-2015 era.

sustainable forestry certification A form of ecolabelling that identifies timber products that have been produced using sustainable methods. Several organizations issue such certification.

swamp A type of wetland consisting of shallow water rich with vegetation, occurring in a forested area. Compare *bog; marsh; fen; peatland.*

symbiosis A parasitic or mutualistic relationship between different species of organisms that live in close physical proximity.

synergistic effect An interactive effect (for example, of toxicants) that is more than or different from the simple sum of their constituent effects.

system A network of relationships among a group of parts, elements, or components that interact with and influence one another through the exchange of energy, matter, and/or information.

taiga The boreal forest biome. In Canada and the United States, the term *taiga* more commonly refers to the northern part of the boreal forest biome.

tailings Residual materials that remain after processing of ore. Tailings, which are commonly stored at minesites, often consist of crushed waste rock contaminated with chemicals used in ore processing.

tar sands See *oil sands.*

temperate deciduous forest A biome consisting of midlatitude forests characterized by broad-leafed trees that lose their leaves each fall and remain dormant during winter. These forests occur in areas where precipitation is spread relatively evenly throughout the year: much of Europe, eastern China, and eastern North America.

temperate forest A forest biome in a temperate climate zone, that is, a seasonal climate.

temperate grassland A biome whose vegetation is dominated by grasses and features more extreme temperature differences between winter and summer, and less precipitation than temperate deciduous forests; also known as steppe or prairie.

temperate rain forest A biome consisting of tall coniferous trees, cooler and less species-rich than tropical rain forest and milder and wetter than temperate deciduous forest.

teratogen A toxicant that causes harm to the unborn, resulting in birth defects.

theory A widely accepted, well-tested explanation of one or more cause-and-effect relationships that has been extensively validated by a great amount of research. Compare *hypothesis*.

thermal inversion A departure from the normal temperature distribution in the atmosphere, in which a pocket of relatively cold air occurs near the ground, with warmer air above it. The cold air, denser than the air above it, traps pollutants near the ground and causes a buildup of smog.

thermal pollution Heat or heated water released into the environment, which can damage natural ecosystems.

thermocline A zone of water in the ocean (or large lake) in which temperature decreases rapidly with depth.

thermohaline circulation A worldwide system of ocean currents in which warmer, fresher water moves along the surface and colder, saltier (and therefore denser) water moves deep beneath the surface.

thermosphere The atmosphere's top layer, extending upward to an altitude of 500 km.

threatened Categorization of a species at risk; this category encompasses both vulnerable and endangered species.

threshold dose The dose at which a toxicant begins to affect a population of test animals. Compare ED_{50}; LD_{50}.

tidal energy Energy harnessed by erecting a dam (called a barrage) across the outlet of a tidal basin. Water flowing with the incoming or outgoing tide through sluices in the dam turns turbines to generate electricity.

tide The periodic rise and fall of the ocean's height at a given location, caused by the gravitational pull of the Moon and Sun.

topsoil That portion of the soil that is most nutritive for plants and is thus of the most direct importance to ecosystems and to agriculture. Also known as the *A horizon*.

total fertility rate (TFR) The average number of children born per female member of a population during her lifetime.

toxic A quality of materials that are poisonous; able to harm health of people or other organisms when a substance is inhaled, ingested, or touched. One criterion for defining hazardous waste. Compare *toxicant*.

toxic air pollutant Air pollutant that is known to cause cancer, reproductive defects, or neurological, developmental, immune system, or respiratory problems in humans, and/or to cause substantial ecological harm by affecting the health of nonhuman animals and plants.

toxicant (toxin or toxic agent) A substance that acts as a poison to humans or wildlife.

toxicity The degree of harm a chemical substance can inflict.

toxicology The scientific field that examines the effects of poisonous chemicals and other agents on humans and other organisms.

toxin See *toxicant*.

trade winds Prevailing winds between the equator and 30° latitude that blow from east to west.

traditional agriculture Biologically powered agriculture, in which human and animal muscle power, along with hand tools and simple machines, perform the work of cultivating, harvesting, storing, and distributing crops. Compare *industrialized agriculture*.

traditional ecological knowledge (TEK) Indigenous or Aboriginal ecological knowledge. The intimate knowledge of a particular environment, possessed and passed along by those who have inhabited an area for many generations, sometimes via an oral tradition of teachings, songs, and storytelling.

tragedy of the commons The scenario in which each individual withdraws whatever benefits are available from an unregulated or poorly regulated common property resource, as quickly as possible, until the resource becomes overused and depleted.

transboundary pollution Pollution that is transported from its source, by air, water, or some other mechanism, across a national boundary.

transform plate boundary A boundary along which two tectonic plates are moving past one another in a translational sense. For example, the Pacific Plate and the North American Plate are slipping and grinding past each other along California's San Andreas Fault.

transgenic Term describing an organism that contains DNA from another species.

transpiration The release of water vapour by plants through their leaves.

trawling Fishing method that entails dragging immense cone-shaped nets through the water, with weights at the bottom and floats at the top to keep the nets open.

tributary A smaller river that flows into a larger one.

trophic level Rank in the feeding hierarchy of a food chain. Organisms at higher trophic levels consume those at lower trophic levels.

trophic pyramid A diagram showing the trophic levels in a food chain from bottom to top, with the autotrophs at the bottom, moving up through the various levels of consumers. The diagram is typically pyramid-shaped because biomass, energy, and numbers of individuals decrease upward through the trophic levels.

tropical dry forest A biome that consists of deciduous trees and occurs at tropical and subtropical latitudes where wet and dry seasons each span about half the year. Widespread in India, Africa, South America, and northern Australia.

tropical forest A forest located in a low-latitude region with a tropical climate.

tropical rain forest A biome characterized by year-round rain and uniformly warm temperatures. Found in Central America, South America, Southeast Asia, West Africa, and other tropical regions. Tropical rain forests have dark, damp interiors; lush vegetation; and highly diverse biotic communities.

troposphere The lowest layer of the atmosphere; it extends to 11 km above sea level. See also *stratosphere*.

tropospheric ozone Ground-level ozone, an air pollutant and a component of smog.

tsunami A wave or series of waves produced by the vertical displacement of a large volume of water; also called seismic sea wave. Tsunamis differ from normal, wind-driven waves in their large amplitude, very long wavelength, high velocity of propagation, and potentially large onshore run-up heights.

tundra A biome that is nearly as dry as desert but is located at very high latitudes along the northern edges of Russia, Canada, and Scandinavia. Extremely cold winters with little daylight and moderately cool summers with lengthy days characterize this landscape of lichens and low, scrubby vegetation.

turbine A rotary device that converts the kinetic energy of a moving substance, such as steam, into mechanical energy. Used widely in commercial power generation from various types of energy sources.

turnover time The time it would take for a material to work its way through and out of a reservoir, if all of the sources or fluxes of that material into the reservoir were stopped. Compare *residence time*.

unconfined aquifer A water-bearing, porous, and permeable layer of rock, sand, or gravel that lies atop a less-permeable substrate. The water in an unconfined aquifer is not under pressure because there is no impermeable upper layer to confine it. Compare *confined aquifer*.

understorey (of a forest) The floor and the lowest levels of growth within a forest.

uneven-aged Term describing stands of trees in timber plantations that are of different ages. Uneven-aged stands more closely approximate a natural forest than do even-aged stands. Compare *even-aged*.

United Nations Environment Programme (UNEP) The department of the United Nations, headquartered in Nairobi, Kenya, established after the Conference on the Human Environment in 1972 to deal with international environmental concerns and encourage nations to adopt environmentally sound approaches to development. One of several agencies within the United Nations that influence international environmental policy and practice.

upwelling In the ocean, a flow of cold, deep water toward the surface. Upwelling can occur in areas where surface currents diverge, and in regions of prevailing offshore winds. Compare *downwelling*.

urban heat island effect A phenomenon in which cities are generally several degrees warmer than surrounding suburbs and rural areas, due to tall buildings that interfere with convective cooling, paved surfaces that absorb heat, and concentrated activities that generate waste heat. The urban heat island effect can contribute to the development of dust domes.

utilitarian principle The principle that something is right when it produces the greatest practical benefits for the most people.

valuation The attempt to quantify the value of a particular environmental good or service—even if it cannot easily be expressed in monetary terms.

variable In an experiment, a condition that can change.

vector An organism that transfers a pathogen to its host. An example is a mosquito that transfers the malaria pathogen to humans.

volatile organic compound (VOC) One of a large group of potentially harmful organic chemicals used in industrial processes.

volcano A vent through which lava, ash, and other volcanic materials are extruded to the surface.

vulnerable Categorization of a species that is likely to become endangered unless there is improvement in the circumstances threatening its survival or reproduction. Compare *endangered; threatened.*

warm front The boundary where a mass of warm air displaces a mass of colder air. Compare *cold front.*

waste Any unwanted product that results from a human activity or process.

waste exchange A system whereby producers of waste materials are matched with organizations in need of the same materials as inputs into other industrial processes.

waste management Strategic decision-making to minimize the amount of waste generated and to dispose of waste safely and effectively.

waste stream The flow of waste as it moves from its sources toward disposal destinations.

waste-to-energy (WTE) facility An incinerator that uses heat from its furnace to boil water to create steam that drives electricity generation or fuels heating systems.

waste water Water that is used in households, businesses, industries, or public facilities and is then drained or flushed down pipes, as well as polluted runoff from streets and storm drains; effluent.

water quality The suitability of water for various purposes (such as drinking or swimming), as determined by comparing the water's physical, chemical, and biological characteristics with a set of predetermined standards.

water table The upper limit of ground water held in an aquifer.

waterlogging The saturation of soil by water, in which the water table is raised to the point that water bathes plant roots. Waterlogging deprives roots of access to gases, essentially suffocating them and eventually damaging or killing the plants.

watershed The term *watershed* is often used as a synonym for *drainage basin*; sometimes it is used to refer to the boundaries that delineate a drainage basin; and sometimes it refers to the drainage basin and all of its component parts, including the physical (abiotic) resources, ecological communities, and human occupants and uses of the area.

wave energy Energy harnessed from the motion of wind-driven waves at the ocean's surface. Many designs for machinery to harness wave energy have been invented, but few have been adequately tested.

weather The local physical properties of the troposphere, such as temperature, pressure, humidity, cloudiness, and wind, over relatively short time periods. Compare *climate.*

weathering The physical, chemical, and biological processes that break down rocks and minerals.

weed A plant that competes with crops. The term is defined by economic interest and is not biologically meaningful. Compare *pest.*

wetland A system that combines elements of fresh water and dry land; usually there is standing water for at least part of the year. These biologically productive systems include freshwater marshes, swamps, peatlands, fens, bogs, and shallow-water wetlands such as vernal pools.

wildlife refuge An area set aside to serve as a haven for wildlife and also sometimes to encourage hunting, fishing, wildlife observation, photography, environmental education, and other public uses.

wind energy Energy (power) derived from wind.

wind farm A development involving a large group of wind turbines.

wind turbine A mechanical assembly that converts the wind's kinetic energy, or energy of motion, into electrical energy.

woodland Land that is covered with woody vegetation, including trees and shrubs. Compare *forest.*

zooplankton Very small or microscopic floating heterotrophic organisms. Compare *plankton; phytoplankton.*

Credits

Index